DISTANCE AND MIDPOINT FORMULAS

The **distance** between $P(x_1, y_1)$ and $Q(x_2, y_2)$ is
$$PQ = \sqrt{(x_2 - x_1)^2 + (y_2 - y_1)^2}$$
and the coordinates of the **midpoint** of line segment $\overline{PQ}$ are
$$\left(\frac{x_1 + x_2}{2}, \frac{y_1 + y_2}{2}\right)$$

EQUATION OF A CIRCLE

The equation of a circle with center (h, k) and radius r is
$$(x - h)^2 + (y - k)^2 = r^2$$

LINES

1. The **slope** of a line through $P(x_1, y_1)$ and $Q(x_2, y_2)$ is
$$m = \frac{y_2 - y_1}{x_2 - x_1}$$

2. The **slope-intercept form** of a line with slope m and y-intercept b is
$$y = mx + b$$

3. The **point-slope form** of a line through $P(x_1, y_1)$ with slope m is
$$y - y_1 = m(x - x_1)$$

VERTEX FORMULA

The graph of $f(x) = ax^2 + bx + c$ is a *parabola*. The coordinates of its **vertex** are
$$\left(-\frac{b}{2a}, f\left(-\frac{b}{2a}\right)\right)$$

LOGARITHMS

1. $y = \log_b x$ is equivalent to $b^y = x$
2. $\log_b 1 = 0$
3. $\log_b b = 1$
4. $\log_b b^x = x$
5. $b^{\log_b x} = x$
6. $\log_b xy = \log_b x + \log_b y$
7. $\log_b \frac{x}{y} = \log_b x - \log_b y$
8. $\log_b x^n = n \log_b x$
9. $\log_b x = \frac{\log_a x}{\log_a b}$

GRAPHS OF

1. **Constant Function**

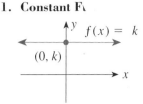

2. **Identity Function**

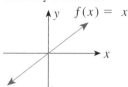

3. **Absolute Value Function** 4. **Squaring Function**

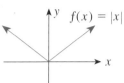

5. **Cubing Function** 6. **Reciprocal Function**

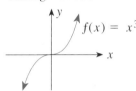
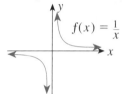

7. **Square Root Function** 8. **Cube Root Function**

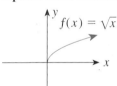
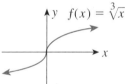

9. **Exponential Function** 10. **Logarithmic Function**

BINOMIAL THEOREM

$$(A + B)^n = \sum_{i=0}^{n} \binom{n}{i} A^{n-i} B^i$$

$\log a = \log_{10} a$

$\ln a = \log_e a$

$\log_b a = \frac{\ln a}{\ln b}$

$\log_b a = \frac{1}{\log_a b}$

$(\log_b a)(\log_a b) = 1$

Final   makeup 10th   open notebook

10 questions
1 from ea

5.3
5.4
7.1
7.2
7.3
7.4
7.5
7.7
8.1
9.3

Extra Credit

pg   #
482 —— 90
584 —— 66
584 —— 70

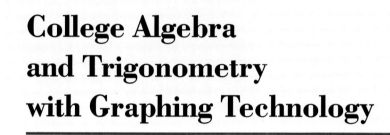

College Algebra
and Trigonometry
with Graphing Technology

College Algebra and Trigonometry with Graphing Technology

David E. Stevens

Wentworth Institute of Technology

West Publishing Company

Minneapolis/St. Paul ■ New York ■ Los Angeles ■ San Francisco

Production Credits

Text Design: Katherine Townes/TECHarts
Copyediting: Katherine Townes/TECHarts
Composition: G&S Typesetters, Inc.
Illustration: Scientific Illustrators
Cover Image: Graeme Outerbridge
Cover Design: Diane Beasley Design

West's Commitment to the Environment In 1906, West Publishing Company began recycling materials left over from the production of books. This began a tradition of efficient and responsible use of resources. Today, 100% percent of our legal bound volumes are printed on acid-free, recycled paper consisting of 50% new paper pulp and 50% paper that has undergone a de-inking process. We also use vegetable-based inks to print all of our books. West recycles nearly 22,650,000 pounds of scrap paper annually—the equivalent of 187,500 trees. Since the 1960s, West has devised ways to capture and recycle waste inks, solvents, oils, and vapors created in the printing process. We also recycle plastics of all kinds, wood, glass, corrugated cardboard, and batteries, and have eliminated the use of polystyrene book packaging. We at West are proud of the longevity and the scope of our commitment to the environment.

West pocket parts and advance sheets are printed on recyclable paper and can be collected and recycled with newspapers. Staples do not have to be removed. Bound volumes can be recycled after removing the cover.

Production, Prepress, Printing and Binding by West Publishing Company.

British Library Cataloguing-in-Publication Data. A catalogue record for this book is available from the British Library.

Copyright © 1996 by West Publishing Company
                610 Opperman Drive
                P. O. Box 64526
                St. Paul, MN 55164-0526

All rights reserved

Printed in the United States of America

03   02   01   00   99   98   97   96   8   7   6   5   4   3   2   1   0

Library of Congress Cataloging-in-Publication Data

    Stevens, David E.
      College algebra and trigonometry with graphing technology / David E. Stevens.
        p.   cm.
      Includes index.
      ISBN 0-314-06240-8 (hard : alk. paper)
      1. Algebra—Data processing.
      2. Trigonometry—Data processing.
      3. Graphic calculators.   I. Title.
    QA154.2.S75   1996
    512'.13—dc20   95-24067
                                                     CIP

CONTENTS

CHAPTER 1

◆◆ **Prerequisites of College Algebra**

1.1 Real Numbers 2
1.2 Integer Exponents and Scientific Notation 12
1.3 Rational Exponents and Radical Expressions 21
1.4 The Complex Numbers 29
1.5 Polynomials and Special Products 35
1.6 Factoring Techniques 44
1.7 Algebraic Fractions 50
1.8 Generating Equivalent Equations and Inequalities 60

Chapter 1 Review
    Questions for Writing or Group Discussion 70
    Review Exercises 71

CHAPTER 2

◆◆ **Functions and Graphs**

2.1 Working in the Cartesian Plane 76
2.2 Graphs of Equations and the Graphing Calculator 81
2.3 Functions 94
2.4 Graph of a Function 103
2.5 Shifting, Reflecting, and Stretching a Graph 112
2.6 Composite and Inverse Functions 124
2.7 Applied Functions and Variation 135

Chapter 2 Review
    Questions for Writing or Group Discussion 146
    Review Exercises 147

CHAPTER 3

◆◆ **Linear and Quadratic Functions**

3.1 Linear Functions 152
3.2 Determining the Equation of a Line 163
3.3 Quadratic Functions 172
3.4 Quadratic Equations and Inequalities 182
3.5 Intersection Points of Two Graphs 196

Chapter 3 Review
    Questions for Writing or Group Discussion 208
    Review Exercises 208

Cumulative Review Exercises for Chapters 1, 2, and 3 212

CHAPTER 4
Polynomial and Rational Functions

4.1 Polynomial Functions and Their Graphs   216
4.2 Polynomial Division   228
4.3 Factors and Zeros of Polynomial Functions   237
4.4 Polynomial Equations and Inequalities   249
4.5 Rational Functions and Their Graphs   262
4.6 Rational Equations and Inequalities   277

Chapter 4 Review
    Questions for Writing or Group Discussion   288
    Review Exercises   290

CHAPTER 5
Exponential and Logarithmic Functions

5.1 Exponential Functions and Their Graphs   294
5.2 Logarithmic Functions and Their Graphs   304
5.3 Properties of Logarithms   316
5.4 Exponential and Logarithmic Equations   325

Chapter 5 Review
    Questions for Writing or Group Discussion   338
    Review Exercises   339

CHAPTER 6
Conic Sections

6.1 The Circle   344
6.2 The Parabola   353
6.3 The Ellipse   365
6.4 The Hyperbola   378

Chapter 6 Review
    Questions for Writing or Group Discussion   389
    Review Exercises   390

Cumulative Review Exercises for Chapter 4, 5, and 6   393

CHAPTER 7
Introduction to Trigonometry

7.1 Angles and Their Measures   396
7.2 Defining the Trigonometric Functions   407
7.3 Evaluating the Trigonometric Functions   418
7.4 Properties and Graphs of the Sine and Cosine Functions   427
7.5 Properties and Graphs of the Other Trigonometric Functions   440
7.6 The Inverse Trigonometric Functions   453
7.7 Applications Involving Right Triangles and Harmonic Motion   467

Chapter 7 Review
    Questions for Writing or Group Discussion   479
    Review Exercises   480

CHAPTER 8
Analytic Trigonometry

8.1 Algebraic Manipulations of Trigonometric Expressions   486
8.2 Trigonometric Equations   496
8.3 Sum and Difference Formulas   511
8.4 Multiple-Angle Formulas   523
8.5 Product-to-Sum Formulas and Sum-to-Product Formulas   535

Chapter 8 Review
    Questions for Writing or Group Discussion   544
    Review Exercises   545

CHAPTER 9

◆◆ Additional Topics in Trigonometry

9.1 Law of Sines 550
9.2 Law of Cosines 561
9.3 Vectors 571
9.4 Trigonometric Form of Complex Numbers 585
9.5 Powers and Roots of Complex Numbers 594
9.6 Polar Coordinate System 604
9.7 Parametric Equations 615

Chapter 9 Review
    Questions for Writing or Group Discussion 627
    Review Exercises 628

Cumulative Review Exercises for Chapters 7, 8, and 9 633

CHAPTER 10

◆◆ Linear Systems and Matrices

10.1 Systems of Linear Equations 638
10.2 Matrices and Their Properties 649
10.3 Determinants and Inverses of Matrices 662
10.4 Systems of Linear Inequalities and Linear Programming 676

Chapter 10 Review
    Questions for Writing or Group Discussion 686
    Review Exercises 686

CHAPTER 11

◆◆ Sequences and Series

11.1 An Introduction to Sequences and Series 692
11.2 The Sum of a Series and Mathematical Induction 703
11.3 Arithmetic Sequences and Series 714
11.4 Geometric Sequences and Series 724
11.5 Infinite Geometric Series 735
11.6 Factorials and Their Applications 743

Chapter 11 Review
    Questions for Writing or Group Discussion 755
    Review Exercises 756

Cumulative Review Exercises for Chapters 10 and 11 759

Appendix: Significant Digits 763

Solutions to Problems and Answers to Odd-Numbered Exercises 765

Index 857

CONTENTS ◆ Graphing Technology Usage

Section	Graphing Technology Usage	Illustration
2.2	Generating the graph of an equation	Example 2, page 84
2.2	Verifying symmetry and intercepts of a graph	Example 3, page 86
2.3	Determining if an equation defines y as a function of x	Example 1, page 95
2.4	Finding the domain and range of a function	Example 1, page 103
2.4	Determining if a function is even, odd, or neither	Example 2, page 105
2.4	Solving an equation by the x-intercept method	Example 3, page 106
2.4	Solving an inequality by the x-intercept method	Example 4, page 107
2.5	Generating the graph of a piecewise-defined function	Problem 6, page 122
2.6	Determining if a function is a one-to-one function	Example 3, page 127
2.6	Verifying that a pair of functions are inverses of each other	Example 4, page 129
2.7	Determining the maximum functional value of a function	Example 1, page 135
3.1	Generating the graph of a nonvertical line $Ax + By = C$	Problem 4, page 159
3.2	Verifying a linear function with given characteristics	Example 3, page 166
3.3	Verifying a quadratic function with given characteristics	Example 2, page 174
3.3	Finding the maximum or minimum value of a quadratic function	Example 4, page 177
3.4	Finding the factors and roots of a quadratic equation	Example 1, page 183
3.4	Determining the nature of the roots of a quadratic equation	Example 2, page 186
3.4	Solving a quadratic inequality	Example 3, page 189
3.5	Approximately the intersection points of two curves	Example 5, page 203
3.5	Solving an equation by the intersection point method	Example 6, page 204
4.1	Generating a complete graph of a polynomial function	Problem 2, page 220
4.1	Locating relative extrema of a polynomial function	Problem 3, page 222
4.3	Determining if $x - a$ is a factor of a polynomial	Example 1, page 238
4.3	Verifying a polynomial function with given characteristics	Example 2, page 241

4.3	Finding the zeros of a polynomial function	Example 4, page 243
4.3	Factoring completely a polynomial over the set of integers	Example 5, page 244
4.4	Solving a polynomial equation	Example 2, page 250
4.4	Solving a polynomial inequality	Example 4, page 254
4.5	Locating relative extrema of a rational function	Problem 4, page 272
4.6	Solving a rational equation	Example 1, page 277
4.6	Solving a rational inequality	Example 3, page 281
5.1	Verifying an exponential identity	Example 3, page 298
5.2	Generating the graph of $f(x) = \log_b x$	Problem 3, page 310
5.2	Verifying a logarithmic identity	Example 5, page 311
5.3	Verifying the expanding of a logarithmic expression	Example 1, page 318
5.3	Verifying the condensing of a logarithmic expression	Example 3, page 320
5.4	Solving an exponential equation	Example 1, page 327
5.4	Solving a logarithmic equation	Example 3, page 329
6.1	Verifying the equation of a circle with given characteristics	Example 2, page 346
6.1	Finding the center and radius of a circle	Example 3, page 348
6.2	Verifying the equation of a parabola with given characteristics	Problem 1, page 356
6.2	Finding the vertex of a parabola	Example 2, page 357
6.2	Verifying the focus and directrix of a parabola	Example 3, page 360
6.3	Verifying the equation of an ellipse with given characteristics	Problem 1, page 368
6.3	Finding the center and vertices of an ellipse	Example 2, page 369
6.3	Verifying the foci of an ellipse	Example 3, page 372
6.4	Verifying the equation of a hyperbola with given characteristics	Problem 1, page 381
6.4	Finding the center and vertices of a hyperbola	Example 2, page 382
6.4	Verifying the foci of a hyperbola	Example 3, page 384
7.1	Converting degrees, minutes, seconds to decimal degrees	Example 2, page 398
7.1	Converting degrees to radians and vice versa	Example 3, page 401
7.3	Evaluating trigonometric functions	Example 3, page 424
7.4	Determining amplitude, period, and phase shift	Example 1, page 432
7.4	Verifying a trigonometric function with given characteristics	Example 2, page 433
7.6	Evaluating the inverse trigonometric functions	Problem 1, page 455
8.1	Simplifying a trigonometric expression	Problem 3, page 489
8.1	Verifying a trigonometric identity	Example 5, page 491
8.2	Solving a trigonometric equation	Example 1, page 498
8.3	Verifying sum and difference formulas	Example 1, page 512
8.4	Verifying multiple-angle formulas	Example 1, page 524
8.5	Verifying product-to-sum or sum-to-product formulas	Example 1, page 536
9.6	Generating the graph of a polar equation	Example 5, page 611

9.7	Generating the graph of a pair of parametric equations	Example 5, page 622
10.3	Finding the determinant of a matrix	Problem 1, page 666
10.3	Finding the inverse of a matrix	Problem 2, page 670
10.3	Solving a system of linear equations	Problem 3, page 672
10.4	Generating the graph of a system of linear inequalities	Problem 2, page 679
11.1	Generating a sequence from its general element	Example 2, page 695
11.1	Generating a series from its general term	Example 4, page 698
11.2	Finding the sum of a series	Example 1, page 704
11.3	Verifying the general element of an arithmetic sequence	Example 1, page 716
11.3	Finding the nth element of an arithmetic sequence	Example 2, page 717
11.3	Finding the sum of an arithmetic series	Example 4, page 719
11.4	Verifying the general element of a geometric sequence	Example 1, page 726
11.4	Finding the nth element of a geometric sequence	Example 2, page 727
11.4	Finding the sum of a geometric series	Example 4, page 729
11.5	Verifying the convergence of an infinite geometric series	Example 1, page 737
11.5	Converting a repeating decimal to a reduced fraction	Example 2, page 739
11.6	Evaluating a permutation	Example 1, page 746
11.6	Evaluating a combination	Example 3, page 748

CONTENTS ◆ Applied Problems and Models

Section	Applied Problem	Illustration
1.2	Light from Alpha Centauri to Earth	Example 2, page 18
1.2	Electrical power and conversion factors	Example 3, page 19
1.8	Radius of a sphere	Example 4, page 67
2.2	Temperature of a pizza as it cools	Example 5, page 91
2.4	Height of a flagpole	Example 5, page 109
2.7	Fencing a rectangular garden	Example 1, page 135
2.7	Water flowing into a conical funnel	Example 2, page 136
2.7	Real estate tax on a property (*direct variation model*)	Example 3, page 138
2.7	Time to travel a fixed distance (*inverse variation model*)	Example 4, page 139
3.1	Tax rate schedule from the IRS	Example 5, page 159
3.2	Linear depreciation of a car (*linear model*)	Example 5, page 169
3.3	Maximum area of a rectangular garden	Example 5, page 178
3.3	Law of motion for freely falling objects (*quadratic model*)	Example 6, page 179
3.4	Parking lot addition	Example 4, page 191
3.4	Height of a pop fly above the ground	Example 5, page 193
4.1	Bending moment of a beam (*polynomial model*)	Example 4, page 223
4.4	Dimensions of a silo for storing grain	Example 5, page 256
4.4	Design of an ash pan for a coal stove	Example 6, page 257
4.5	Minimum average cost of a log home (*rational model*)	Example 6, page 273
4.6	Speed of a crew team	Example 4, page 284
4.6	Spillway flooding a cranberry bog	Example 5, page 285
5.1	Compound interest, *n* compoundings per year	Example 4, page 300
5.1	Compound interest, continuous compounding	Example 5, page 301
5.2	Retention of learned material (*logarithmic model*)	Example 6, page 312
5.3	Earthquakes and the Richter scale	Example 5, page 323
5.4	Money deposited in a bank	Example 7, page 333
5.4	Growth of bacteria in a culture (*exponential growth model*)	Example 8, page 334
5.4	Waste product of a nuclear reactor (*exponential decay model*)	Example 9, page 335
6.1	Layout of a circular curve for a roadway	Example 4, page 350
6.2	Parabolic reflector in a searchlight	Example 5, page 362

7.1	Length of a circular curve for a roadway	Example 4, page 402
7.1	Angular speed of an automobile tire	Example 5, page 403
7.4	Populations in a predator-prey relationship (*trigonometric model*)	Example 3, page 435
7.6	Throwing out a runner at second base	Example 7, page 463
7.7	Sharpening the teeth of a saw blade	Example 3, page 469
7.7	Speed of a chair lift at a ski area	Example 4, page 470
7.7	Using a transit to find heights	Example 5, page 471
7.7	Weight in simple harmonic motion (*trigonometric model*)	Example 6. page 473
8.2	Daylight hours in the Northern Hemisphere (*trigonometric model*)	Example 6, page 507
8.3	Current in an electrical circuit	Example 8, page 520
8.4	Design of an A-frame structure	Example 9, page 531
9.1	Width of a river from a surveyor's notes	Example 4, page 555
9.2	Distance to the hole after driving a golf ball	Example 5, page 567
9.3	Forces acting on a dogsled	Example 7, page 579
9.3	Resultant force in a force system	Example 8, page 580
9.3	Aircraft displacement to a landing strip	Example 9, page 581
9.4	Alternating current (AC) circuits (*complex plane model*)	Example 4, page 592
9.6	Flight of a football (*projectile motion model*)	Example 6, page 623
10.1	Tickets to a rock concert	Example 5, page 645
10.2	Inventory and cost of brass beds	Example 5, page 659
10.4	Maximum profit of a ski company	Example 4, page 682
11.3	Installment plan for buying a stereo	Example 6, page 721
11.4	Simple annuity for college expenses (*geometric series model*)	Example 6, page 732
11.5	Hammering a nail	Example 4, page 740

PREFACE

◆ **Intent**

A course in college algebra and trigonometry is a prerequisite for more advanced courses in mathematics as well as a prerequisite for other college courses in astronomy, biology, physics, chemistry, computer science, business, and engineering. The goal of this text is to effectively prepare students for these courses and to illustrate through real-life applied problems that knowledge of college algebra and trigonometry is fundamental to these disciplines. Unlike a traditional text, however, this text integrates *graphing technology* (graphing calculator or graphing software) into the college algebra and trigonometry course. In each chapter after Chapter 1, graphing technology is used as a true teaching aid, without sacrificing mathematical thought or rigor. Whenever a graphical interpretation of the mathematics is appropriate, graphing technology is included. See *Table of Contents: Graphing Technology Usage* on page viii.

◆ **Approach**

For their success in a college algebra and trigonometry course, it is essential that students be active rather than passive readers of the text. Therefore, I have written this text using an *interactive approach* with graphing technology. Each key mathematical concept is supported by a step-by-step text example, with marginal annotations and explanatory notes, and each text example is followed by a practice problem for the student to work. The practice problem may ask the reader to check the preceding text example by using graphing technology, to work the preceding text example using an alternative algebraic approach, to extend the preceding text example by asking for additional information, or to try an entirely different problem that has similar mathematical steps. In effect, the practice problems require the student to become involved with the mathematics and thus constitute a built-in workbook for the student. *A complete, detailed solution of each practice problem is given in the back of the text.*

◆ **Features**

Written in a warm and user-friendly style, this text addresses the concerns of writing across the curriculum, group learning, critical thinking, and the use of modern technology in the math classroom. The following features distinguish this text from other texts in the market.

◆ Applied Problems

Student interest in mathematics is undoubtedly greatest when applied problems are integrated directly within the curriculum. For this reason, applied problems from the fields of science, engineering, and business are introduced at every reasonable opportunity.

Chapter opening applications To arouse student interest, each chapter opens with an applied problem and a related photograph. The solution to the problem is presented within the chapter, after the necessary mathematics has been developed.

Application subsections Separate material has been included at the end of most sections to apply the mathematics to real-life situations. The logo ◆ identifies these application subsections. See *Table of Contents: Applied Problems and Models* on page xi.

◆ Exercise Sets

The heart of any math textbook is its end-of-section exercise sets. It is here that students are given an opportunity to practice the mathematical concepts that have been developed. The exercise sets in this text are presented in two parts: *Basic Skills* and *Critical Thinking* exercises. Applied problems are included in both parts.

Basic Skills These exercises are routine in nature, and most of them mimic the text examples that are worked out in each section. The reader is encouraged to use graphing technology to confirm results and to solve problems to which standard algebraic methods do not apply.

Critical Thinking These exercises require the student to think critically and to transcend routine application of the basic skills to the next level of difficulty. Exercises in this group may require the student to draw upon skills developed in earlier chapters or to use graphing technology to explore new ideas.

Some of the exercise sets also contain problems that are *calculus related*. Designed for students who are taking a college algebra and trigonometry course as a prerequisite to calculus, these exercises illustrate the algebraic support that is needed in calculus. The logo $\frac{\Delta y}{\Delta x}$ identifies these problems.

◆ Chapter Reviews

To help students prepare for chapter exams, each chapter in this text concludes with an extensive chapter review, which is divided into two parts: *Questions for Writing and Group Discussion* and *Review Exercises*.

Questions for Writing and Group Discussion In keeping with the interactive approach, these questions allow students to state in their own words what they have learned in the chapter. Since many of these questions have open-ended answers, they are well suited for class or group discussions and are extremely valuable to those who believe in cooperative or collaborative learning.

Review Exercises These exercises reinforce the ideas discussed in the chapter and allow the instructor to indicate to the student the types of problems that may appear on a chapter test.

◆ Cumulative Reviews

To help students pull together ideas from several chapters, cumulative review exercises are placed strategically after Chapter 3, 6, 9, and 11. The problems in these exercises are ungraded by difficulty and are presented in a random order. Some problems are basic and similar to those already studied; others are more challenging and require creative thinking.

◆ Pedagogy

Every effort has been made to make this a text from which students can learn and succeed. The following pedagogical features attest to this fact.

Caution notes flagged by the symbol CAUTION, help eliminate misconceptions and bad mathematical habits by pointing out the errors most commonly made by students.

Introductory comments at the start of each section introduce vocabulary and inform the reader of the purpose of the section.

Boxed definitions, formulas, laws, and properties state key mathematical ideas and provide the reader with quick, easy access to this information.

Step-by-step procedural boxes indicate the sequence of steps that a student can follow for tasks such as simplifying algebraic expressions, solving certain types of equations and inequalities, or finding the inverse of a function or a matrix.

◆ Development

The prerequisite for a course in college algebra and trigonometry is two years of high school algebra or a semester of college intermediate algebra. Chapter 1, *Prerequisites of College Algebra,* provides a quick review of the topics in intermediate algebra, such as exponents, radicals, factoring, algebraic fractions, equations, inequalities, and formulas. Complex numbers are introduced in this chapter in preparation for their use with quadratic functions in Chapter 3 and polynomial functions in Chapter 4.

Chapter 2, *Functions and Graphs,* introduces the coordinate plane and the language of functions and graphs. Graphing technology is introduced in Section 2.2 and is used extensively throughout the remainder of the text to verify results, solve equations, and explore new ideas. Section 2.5 lists eight basic functions and their graphs (constant, identity, absolute value, squaring, cubing, reciprocal, square root, and cube root), and then applies the vertical and horizontal shift rules, the x- and y-axes reflection rules, and the vertical stretch and compress rules to sketch the graphs of several related functions. These eight basic functions and their graphs are then used to discuss composition of functions, inverse functions, applied functions, and variation.

Chapter 3, *Linear and Quadratic Functions,* uses the ideas of shifting, reflecting, stretching, and compressing to develop the linear and quadratic functions from the identity and squaring functions. The vertex formula for a parabola is developed and used to solve some max–min applied problems.

Section 3.4 discusses the algebraic and graphical methods of solving quadratic equations and inequalities, and Section 3.5 gives an introduction to solving a system of two equations in two unknowns by looking at the intersection points of their graphs.

Chapter 4, *Polynomial and Rational Functions*, focuses on the relationship between factors, roots, zeros, and x-intercepts. Graphing technology and the rational zero theorem are used to find the real zeros of a polynomial function. This chapter also discusses algebraic and graphical methods for solving polynomial equations and inequalities and rational equations and inequalities. With the aid of graphing technology, it is possible to solve max–min applied problems that deal with polynomial and rational functions.

Chapter 5, *Exponential and Logarithmic Functions*, discusses the properties of real exponents, defines the exponential function, and develops the logarithmic function as the inverse of the exponential function. By letting the number of compounding periods in the compound interest formula increase without bound, the reader is shown how the number e develops in a real-life situation. The properties of logarithms are used in graphing functions that contain logarithmic expressions (Section 5.3) and also in solving exponential and logarithmic equations (Section 5.4). Graphing technology is used to support the work and to help solve equations that are not solvable by ordinary algebraic methods.

Chapter 6, *Conic Sections*, discusses the conics and their related quadratic equations in two unknowns. This chapter states the geometric properties of the circle, parabola, ellipse, and hyperbola, as well as the distinguishing characteristics of their equations. The reflection properties of the conics are applied to various technical applications. Throughout the chapter, graphing technology is used to support algebraic manipulations.

Chapters 7, 8, and 9 represent the core of a college trigonometry course. Chapter 7, *Introduction to Trigonometry*, introduces the trigonometric functions using both angle domains and real number domains. After an introductory section on angles and their measure, Section 7.2 defines the six trigonometric ratios for any angle θ and develops the fundamental trigonometric identities from these definitions. This section also introduces the trigonometric ratios of right triangles and the cofunction relationships. Section 7.3 uses the unit circle to show that the trigonometric functions of a real number may be found by considering the real number as the radian measure of its corresponding central angle. The graphs of the sine and cosine functions (Section 7.4) are obtained by observing the changes in the x-coordinate and y-coordinate of a point $P(x, y)$ on the unit circle. These graphs are then used in listing the important properties of the sine and cosine functions and in developing the graphs of the other trigonometric functions. The restricted sine, restricted cosine, and restricted tangent functions are defined in Section 7.6, and the inverse of these functions are then discussed. The chapter concludes with a variety of applied problems involving angles of elevation, angles of depression, and simple harmonic motion.

Chapter 8, *Analytic Trigonometry*, discusses the branch of mathematics in which algebraic procedures are applied to trigonometry. Section 8.1 suggests a general scheme for verifying trigonometric identities, and Section 8.2 states a general procedure for solving trigonometric equations. Several important trigonometric formulas are developed in the later sections of this chapter.

These include the sum and difference formulas, multiple-angle formulas, product-to-sum formulas, and sum-to-product formulas. The chapter concludes with a summary of all the important trigonometric identities and formulas discussed in Chapters 7 and 8. Throughout this chapter, the graphing technology is used to support the work with trigonometric identities and trigonometric equations.

Chapter 9, *Additional Topics in Trigonometry*, discusses several applications of the trigonometric functions. The method of solving an oblique triangle by using the law of sines and law of cosines is developed in the first two sections of this chapter. Area formulas for oblique triangles, including Hero's formula, are developed and applied to several problems. Section 9.3 defines a vector, introduces the vector operations of addition, subtraction, and scalar multiplication, and applies vectors to problems involving forces and displacements. Sections 9.4 and 9.5 define the trigonometric form of a complex number and develop the rules for multiplying, dividing, and finding the powers and roots of complex numbers. The polar coordinate system is discussed in Section 9.6, and parametric equations of the conic sections are developed in Section 9.7. The polar and parametric modes on a graphing calculator are used to generate the graphs of polar equations.

Chapters 10 and 11 include other topics of interest in a college algebra course. Chapter 10, *Linear Systems and Matrices*, extends the discussion of 2×2 systems (from Section 3.5) to a procedure for solving $n \times n$ linear systems by the elimination method. Matrices are first introduced as an aid for solving a system of linear equations and are then applied to problems that require using the matrix operations of addition and multiplication. Determinants and inverses of matrices are evaluated algebraically and by using the matrix features of a graphing calculator. Section 10.4 introduces linear systems of inequalities and applies the results to linear programming problems in two unknowns.

Chapter 11, *Sequences and Series*, offers an introduction to these topics for the precalculus student. Section 11.2 introduces the idea of proof by mathematical induction. In this section, the reader is encouraged to use pattern recognition to guess a formula for the sum of a series and then to prove this guess by mathematical induction. Similarly, pattern recognition and mathematical induction are used to develop a formula for the general element of an arithmetic sequence, for the general element of a geometric sequence, and for the expansion of $(A + B)^n$. The sequence and series features of a graphing calculator are used to support the work.

◆ Supplements

The following supplements are available for users of this text.

Instructor's Solution Manual, by Eleanor Canter of Wentworth Institute of Technology, includes detailed solutions to all the even-numbered exercises.

Student's Solution Manual, by Eleanor Canter of Wentworth Institute of Technology, provides detailed solutions for the odd-numbered exercises from the text.

Instructor's Manual with Test Bank, by Cheryl Roberts of Northern Virginia Community College, includes sample syllabi, suggested course schedules, chapter outlines with references to videos, homework assignments, chapter tests, and a test bank of multiple-choice questions and open-ended problems.

Graphing Technology Laboratory Manual, by David Lawrence of Southwestern Oklahoma State University, includes keystroke instructions for various types of graphing calculators—Texas Instruments, Casio, Sharp, and Hewlett-Packard, as well as Derive software.

DERIVE©Laboratory Manual, by Lloyd R. Jaisingh of Morehead State University contains sixteen laboratory experiments using the computer algebra system DERIVE©.

WESTEST, computer-generated testing program, includes algorithmically generated questions and is available in both Macintosh and PC versions.

West Math Tutor Software by Mathens, an algorithmically based tutorial, is available for Macintosh and PC platforms. The instructor's version contains the complete bank of algorithms so that an instructor can create assignments, save them to disk for student use, and grade automatically when assignments are turned in. The student version is customized for each chapter of the appropriate text. The package is available to instructors, and a disk is available to students. It is also available as a site license for a math laboratory.

Videos, which are produced specifically for this text.

Fifty Transparency Masters illustrate the important figures, rules, and procedures in the text.

Please ask your West representative about qualifications for these supplements.

◆ Acknowledgments

Most of the material in this text has been class-tested with several hundred students at Wentworth Institute of Technology. I thank these students for their helpful comments and critiques. Special thanks go to my friend and colleague Eleanor Canter for her work in checking the answers and writing complete solutions to the more than 6000 exercises in this text. I also express my sincere thanks to the following reviewers—their ideas were extremely helpful in shaping this text into its present form.

Daniel D. Anderson,
University of Iowa

Lora L. Brewer,
Savannah State College

Frederick J. Carter,
St. Mary's University

John P. Edwards,
Anne Arundel Community College

Sarita Gupta,
University of Nevada–Reno

Harold Harutunian,
Salem State College

Carol Ann Janik,
Tompkins Cortland Community College

Ahmad Karami,
Essex Community College

William M. Mays,
Gloucester County College

Jack D. Murphy,
Pennsylvania College of Technology

James W. Newsom,
Tidewater Community College

Bonny J. Peters,
St. Petersburg Junior College

John M. Plachy,
Metropolitan State College of Denver

George W. Schultz,
St. Petersburg Jr. College

Ralph Selig,
Fairleigh Dickinson University

Thomas J. Sharp,
West Georgia College

James W. Lea,
Middle Tennessee State University

Jann W. MacInnes,
Florida Community College

Chris R. Siragusa,
Cypress College

Bruce Williamson,
University of Wisconsin–River Falls

The production of a textbook is a team effort between the editorial staff and the author. My editor, Nancy Hill-Whilton, offered the support and guidance that I needed to complete this project. Denise Bayko organized our reviewers' comments into a format that revealed where extra work was needed. Kathi Townes copyedited the manuscript and encouraged me to provide additional information that would benefit the reader, and Sandy Gangelhoff kept the project moving through the various stages of production. I thank each of you for the encouragement and enthusiasm that you provided in the preparation of this book.

D. E. Stevens
Boston, Massachusetts
1995

TO THE STUDENT

In order to use this text effectively, you should purchase a graphing calculator. As an alternative to a graphing calculator, you can use a computer equipped with graphing software. Your instructor has information on the type of calculator that you should purchase and the type of software that is available from West.

◆ **Get Involved**

The key to success in a college algebra and trigonometry course is to *get involved* with the mathematics. Do the homework exercises the day that your instructor assigns them. Before beginning the homework exercises, read the appropriate section in the text. Read carefully each text example in the section and try the practice problem that follows the text example. (A complete, detailed solution of each practice problem is given in the back of the text.) Form a study group with other members of your class and discuss the homework exercises with the group. You'll be surprised how much you can learn from each other.

◆ **Ask Questions**

You are in college to learn. You have an inquiring mind, and inquiring minds want to know. So, be sure to *ask questions*. Make note of the homework exercises that you did not understand, and come to class prepared to ask specific questions. Remember, there is no such thing as a "dumb" question. In fact, many other members of your class probably want to ask the same question that is causing you difficulty. Your instructor will welcome your participation in the discussion.

◆ **Have Confidence**

In order to do well in a mathematics test, you must *have confidence* in your ability to do the work. Building confidence requires study, intense work, and perseverance. If you have put in an honest effort, then you will never feel intimidated, defensive, or flustered during a test situation. You can enter the classroom knowing that you have given your very best.

◆ Never Give Up

Life is exciting when you're in college. You are full of optimism and enthusiasm for the opportunities that lie ahead. In the future, you will look back at your college experience as one of the best times of your life. So laugh, enjoy yourself, and *never give up!*

> Achievement consists of never giving up . . . If there is no dark and dogged will, there will be no shining accomplishment; if there is no dull and determined effort, there will be no brilliant achievement.
>
> HSUN TZU
> Chinese philosopher

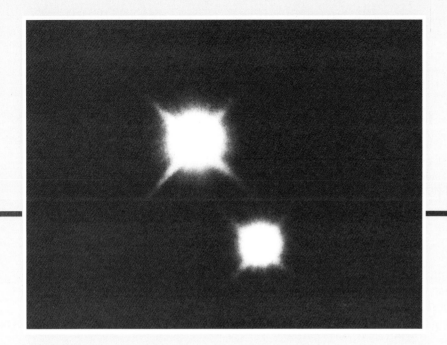

The distance from Earth to Alpha Centauri (the star closest to our Sun) is approximately 25,500,000,000,000 miles. How many years does it take light from this star to reach Earth? Assume that light travels at 186,000 miles per second.

For the solution, see Example 2 in Section 1.2.

Prerequisites of College Algebra

1.1 Real Numbers
1.2 Integer Exponents and Scientific Notation
1.3 Rational Exponents and Radical Expressions
1.4 The Complex Numbers
1.5 Polynomials and Special Products
1.6 Factoring Techniques
1.7 Algebraic Fractions
1.8 Generating Equivalent Equations and Inequalities

1.1 Real Numbers

◆ **Introductory Comments**

A chef requires 3 eggs and $\frac{2}{3}$ cup of sugar for a cake recipe. A meteorologist reports that the temperature is $-4°C$ and the barometric pressure is 29.35 inches. A student determines that the side of a right triangle is $\sqrt{5}$ units and the area of a circle is π square units. Each of the numbers

$$3 \quad \frac{2}{3} \quad -4 \quad 29.35 \quad \sqrt{5} \quad \text{and} \quad \pi$$

is an element of the **set of real numbers**.* These are the type of numbers that we work with every day. The set of real numbers has five important subsets:

1. Natural numbers $\{1, 2, 3, 4, \ldots\}$
2. Whole numbers $\{0, 1, 2, 3, \ldots\}$
3. Integers $\{\ldots, -3, -2, -1, 0, 1, 2, 3, \ldots\}$
4. Rational numbers $\{$All real numbers of the form $\frac{p}{q}$, where p and q are integers and $q \neq 0.\}$
    or
    $\{$All decimal numbers that either terminate or repeat the same block of digits.$\}$
    Examples:
    $\frac{3}{5}, \frac{-8}{3}, \frac{16}{1}, \frac{0}{4}, 0.75, -5.343434\ldots$
5. Irrational numbers $\{$All real numbers that are not rational.$\}$
    or
    $\{$All decimal numbers that neither terminate nor repeat the same block of digits.$\}$
    Examples:
    $\sqrt{2}, -\sqrt{3}, \pi, \sqrt[3]{6}, 3.050050005\ldots$

In algebra, real numbers such as 3, $\frac{2}{3}$, -4, 29.35, $\sqrt{5}$, and π are called **constants**, since each is a fixed number. Letters from the beginning of the alphabet, like a, b, c, and k, are often used as symbols for yet unspecified constants. A **variable** is a symbol that represents *any* member from a given set of numbers. Letters from the end of the alphabet, like x, y, z, and t, are often used as variables. By assuming that a variable represents any member from the set of real numbers, we can manipulate and simplify any expression that contains constants and variables by using the *fundamental properties of real numbers*.

---

*The concept of a set is often used in mathematics. A *set* is a collection of objects and these objects are called the *elements* of the set. A *subset* of a given set is formed by selecting particular elements of the set. Braces { } are used to enclose the elements of sets and subsets.

Fundamental Properties of Real Numbers

For any real numbers a, b, and c,

Property	Example
Commutative Property of Addition	
$a + b = b + a$	$2x + 4 = 4 + 2x$
Commutative Property of Multiplication	
$ab = ba$	$(x + y)2 = 2(x + y)$
Associative Property of Addition	
$(a + b) + c = a + (b + c)$	$(x + 2) + 3 = x + (2 + 3)$ $= x + 5$
Associative Property of Multiplication	
$(ab)c = a(bc)$	$3(2x) = (3 \cdot 2)x = 6x$
Distributive Property (left)	
$a(b + c) = ab + ac$	$7(x + y) = 7x + 7y$
Distributive Property (right)	
$(a + b)c = ac + bc$	$5x + 3x = (5 + 3)x = 8x$
Additive Identity Property	
$a + 0 = 0 + a = a$	$(4x + 3y) + 0 = 4x + 3y$
Multiplicative Identity Property	
$a(1) = (1)a = a$	$1(-3y) = -3y$
Additive Inverse Property	
$a + (-a) = (-a) + a = 0$	$5x + (-5x) = 0$
Multiplicative Inverse Property	
$a \cdot \dfrac{1}{a} = \dfrac{1}{a} \cdot a = 1, \quad a \neq 0$	$(x + 4) \cdot \dfrac{1}{x + 4} = 1, \quad x \neq -4$

The difference of two real numbers can be defined in terms of addition. To subtract b from a, we add a with the additive inverse, or opposite, of b.

◆ **Subtraction**

> For any real numbers a and b,
> $$a - b = a + (-b).$$

The quotient of two real numbers, denoted $a \div b$ or $\dfrac{a}{b}$, can be defined in terms of multiplication. To divide a by b, we multiply a by the multiplicative inverse (or *reciprocal*) of b. Since 0 does not have a multiplicative inverse, division by zero is *undefined*.

◆ **Division**

> For any real numbers a and b, except $b = 0$,
> $$a \div b = \frac{a}{b} = a \cdot \frac{1}{b} = \frac{1}{b} \cdot a.$$

Note that if a is a negative number, then its additive inverse, $-a$, is a positive number. For example, if $a = -4$, then $-a = -(-4) = 4$. Here are some other *properties of negation* that we use frequently throughout this text.

◆ **Properties of Negation**

> For any real numbers, a, b, and c,
>
> | Property | Example |
> |---|---|
> | $(-1)a = -a$ | $-1(3xy) = -3xy$ |
> | $-(-a) = a$ | $-(-2x) = 2x$ |
> | $(-a)b = a(-b) = -(ab)$ | $(-4x)2 = 4x(-2) = -(4x \cdot 2) = -8x$ |
> | $(-a)(-b) = ab$ | $(-3x)(-y) = 3xy$ |
> | $\dfrac{-a}{b} = \dfrac{a}{-b} = -\dfrac{a}{b},\ b \neq 0$ | $\dfrac{-2x}{3} = \dfrac{2x}{-3} = -\dfrac{2x}{3}$ |
> | $\dfrac{-a}{-b} = \dfrac{a}{b},\ b \neq 0$ | $\dfrac{-(x+3)}{-y} = \dfrac{x+3}{y}$ |
> | $-(a + b) = -a - b$ | $-(-2 + 7x) = 2 - 7x$ |
> | $-(a - b) = b - a$ | $-(2x - 3) = 3 - 2x$ |

◆ **The Real Number Line and Inequality Symbols**

A geometric interpretation of the real numbers can be shown on the *real number line*, as illustrated in Figure 1.1. For each point on this line there corresponds exactly one real number, and for each real number there corresponds exactly one point on this line. This type of relationship is called a **one-to-one correspondence**.

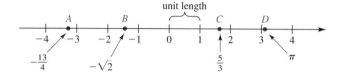

FIGURE 1.1
On the real number line there exists a *one-to-one correspondence* between the set of real numbers and the set of points on the line.

The real number associated with a point on the real number line is called the **coordinate** of the point. In Figure 1.1, $-\frac{13}{4}$ is the coordinate of point A, $-\sqrt{2}$ is the coordinate of point B, $\frac{5}{3}$ is the coordinate of point C, and π is the coordinate of point D.

The real number line gives us a convenient way to compare two distinct real numbers a and b. If a is to the *right* of b on the real number line, then ***a* is greater than *b***, and we write

$$a > b$$

If a is to the *left* of b on the real number line, then ***a* is less than *b***, and we write

$$a < b$$

Comparing the coordinates of points C and D in Figure 1.1, we may write either

$$\pi > \frac{5}{3} \quad \text{or} \quad \frac{5}{3} < \pi$$

In general, for real numbers a and b,

$$a < b \quad \text{if and only if} \quad b > a$$

Note: The phrase *if and only if* occurs frequently in mathematics. In the preceding statement, it implies two statements:

1. If $a < b$, then $b > a$ and   **2.** If $b > a$, then $a < b$.

The symbols $>$ and $<$ are called **inequality symbols**, and statements such as $a > b$ and $a < b$ are called **inequalities**. Two other inequality symbols are used frequently:

$\leq$ read "less than or equal to" and $\geq$ read "greater than or equal to."

Inequalities can be used to indicate whether a number is *positive, negative, nonnegative,* or *nonpositive.* Some inequalities and their meanings are shown in Table 1.1.

TABLE 1.1
Some inequalities and their meanings

Inequality	Meaning
$a > 0$ or $0 < a$	a is positive.
$a < 0$ or $0 > a$	a is negative.
$a \geq 0$ or $0 \leq a$	a is nonnegative.
$a \leq 0$ or $0 \geq a$	a is nonpositive.

FIGURE 1.2

Three distinct points on the real number line with $a < b < c$.

Figure 1.2 shows three distinct real numbers a, b, and c on a real number line. To indicate that b is between a and c on this line, we can write either

$$a < b < c \qquad \text{or} \qquad c > b > a$$

Each of these expressions is called a **double inequality**. When you use a statement such as $a < b < c$ or $c > b > a$, be sure that all the inequality symbols point in the same direction. For example, the expression $a < c > b$ is meaningless.

EXAMPLE 1 Write each statement as an inequality.

(a) a is at most 6. **(b)** b is at least -2.

(c) c is nonnegative and less than 10.

◆ **Solution**

(a) a is at most 6 is written as 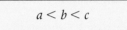 **(b)** b is at least -2 is written as

$$a \leq 6 \qquad\qquad b \geq -2$$

(c) c is nonnegative and less than 10 is written as

$$0 \leq c \quad \text{and} \quad c < 10$$

or, more compactly, $0 \leq c < 10$. ◆

PROBLEM 1 Write each statement as an inequality.

(a) a is not more than 8. **(b)** b is negative and at least -4. ◆

◆ **Absolute Value**

The distance between zero and a number a on the real number line, without regard to direction, is called the *absolute value* of a and is denoted $|a|$. Because distance is independent of direction and is always nonnegative, the absolute value of any real number is also nonnegative, that is, $|a| \geq 0$. A more formal definition of **absolute value** follows.

◆ **Absolute Value**

For any real number a,
$$|a| = \begin{cases} a & \text{if } a \geq 0 \\ -a & \text{if } a < 0. \end{cases}$$

To find $|4|$ from this definition, we use $|a| = a$, since $4 \geq 0$. Thus,
$$|4| = 4.$$

To find $|-4|$ from this definition, we use $|a| = -a$, since $-4 < 0$. Thus,
$$|-4| = -(-4) = 4.$$

In general, we have $|a| = |-a|$ for any real number a. We now give two *properties of absolute value,* which involve products and quotients.

◆ **Properties of Absolute Value**

For any real numbers a and b,

Property	Example
Product Property	
$|ab| = |a||b|$	$|2x| = |2||x| = 2|x|$
Quotient Property	
$\left|\dfrac{a}{b}\right| = \dfrac{|a|}{|b|}$, $b \neq 0$	$\left|\dfrac{x+2}{-4}\right| = \dfrac{|x+2|}{|-4|} = \dfrac{|x+2|}{4}$

If points A and B on the real number line have coordinates a and b, respectively, then the **distance** between points A and B, denoted AB, is defined as follows.

◆ **Distance Between Two Points on the Real Number Line**

For any points A and B with coordinates a and b, respectively,
$$AB = |a - b|.$$

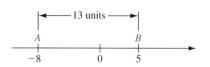

FIGURE 1.3
The distance between -8 and 5 is 13 units.

For example, if the coordinate of point A is -8 and the coordinate of point B is 5, then
$$AB = |-8 - 5| = |-13| = 13 \text{ units,}$$

as shown in Figure 1.3. Note that $BA = |5 - (-8)| = |13| = 13$ units. In general, for any points A and B, the distance is the same in both directions, that is, $AB = BA$.

EXAMPLE 2 Write each statement as an inequality with absolute value notation.

(a) The distance between a and 2 is at most 3.

(b) b is more than 3 units from -5.

◆ Solution

(a) The distance between a and 2 is written as $|a - 2|$, or as $|2 - a|$. Hence, the distance between a and 2 is at most 3 is written as

$$|a - 2| \leq 3.$$

(b) If b is more than 3 units from -5, then the distance from b to -5 is greater than 3, and we write

$$|b - (-5)| > 3 \quad \text{or} \quad |b + 5| > 3. \quad ◆$$

PROBLEM 2 Express "the distance between c and -1 is at least 5" as an inequality with absolute value notation. ◆

◆ Interval Notation

Any unbroken portion of the real number line is called an **interval**. We may describe the set of all real numbers in an interval by using either *set-builder notation* or *interval notation*. For example, consider the interval shown on the real number line in Figure 1.4. The *closed* dot indicates that -3 *is* contained in the interval, and the *open* circle indicates that 5 is *not* contained in the interval. Hence, this figure represents the set of all real numbers greater than or equal to -3 and less than 5. We may describe the set of real numbers x in this interval by writing

Set-builder notation: $\{x \mid -3 \leq x < 5\}$

or

Interval notation: $[-3, 5)$

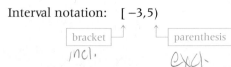

In interval notation, a bracket indicates than an endpoint is included in the interval and a parenthesis indicates that the endpoint is excluded from the interval.

An interval that includes both its endpoints is called a **closed interval** and an interval that excludes both its endpoints is called an **open interval**. An

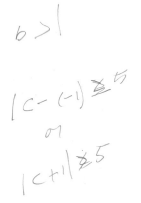

FIGURE 1.4

All real numbers x such that x is greater than or equal to -3 and less than 5.

interval that contains one of its endpoints but not the other, such as [−3, 5), is called a **half-open interval**.

CAUTION When using interval notation, be sure the lesser of the two numbers is written first. To express the interval shown in Figure 1.4 as

$$(5, -3] \quad \text{is WRONG!}$$

Remember, always record the numbers in the order they appear on the number line from left to right.

In interval notation, the symbol ∞ (*infinity*) indicates that an interval has no right-hand boundary and the symbol $-\infty$ (*negative infinity*) indicates that an interval has no left-hand boundary. We refer to an interval with no right-hand boundary, or no left-hand boundary, or neither, as an **unbounded interval**. The symbols ∞ and $-\infty$ are *not* real numbers and therefore cannot be included in an interval. Thus, in an unbounded interval, $-\infty$ is always preceded by a parenthesis and ∞ is always followed by a parenthesis.

In summary, we list nine types of intervals. In each case, a and b are constants and x is a variable. A closed dot indicates that the point is included as part of the interval, and an open circle indicates that the point is not part of the interval.

◀ **Types of Intervals**

Type of Interval	Set-Builder Notation	Interval Notation
Unbounded intervals		
	$\{x \mid x > a\}$	(a, ∞)
	$\{x \mid x \geq a\}$	$[a, \infty)$
	$\{x \mid x < a\}$	$(-\infty, a)$
	$\{x \mid x \leq a\}$	$(-\infty, a]$
	$\{x \mid x \text{ is real}\}$	$(-\infty, \infty)$
Open interval		
	$\{x \mid a < x < b\}$	(a, b)
Closed interval		
	$\{x \mid a \leq x \leq b\}$	$[a, b]$
Half-open interval		
	$\{x \mid a < x \leq b\}$	$(a, b]$
	$\{x \mid a \leq x < b\}$	$[a, b)$

Note: Throughout the remainder of this text, we shall use interval notation, instead of set-builder notation, to describe the set of all real numbers in an interval.

EXAMPLE 3 Use interval notation to describe the set of all real numbers designated on each real number line.

(a) ← —|—|—|—|—●—|—|—|—|—|—→ x
            −2

(b) ← —|—|—|—○—|—|—|—○—|—|—|—→ x
            −2   2

◆ **Solution**

(a) This real number line shows all real numbers x such that x is less than or equal to -2. We may describe this set of real numbers by writing

$$(-\infty, -2].$$

(b) This real number line shows all real numbers x that are greater than -2 and less than 2. We may describe this set of real numbers by writing

$$(-2, 2).$$

PROBLEM 3 Indicate each interval on a real number line.

(a) $(3, \infty)$ (b) $[4, 9)$

The **union** of two sets A and B is denoted $A \cup B$, which we read as "A union B." The union operation denotes the set of all elements that are members of set A, or members of set B, or members of both sets A and B. In the next example, we use the union operation in conjuction with interval notation to describe the set of all real numbers designated by two or more intervals on a real number line.

EXAMPLE 4 Use interval notation to describe the set of all real numbers designated by each real number line.

(a) ← —|—|—●—|—|—|—|—|—●—|—|—→ x
            −3   3

(b) ← —|—|—○—|—|—|—|—|—○—|—|—→ x
            −1   7

◆ **Solution**

(a) This real number line shows two intervals that represent all real numbers x such that x is less than or equal to -3 *or* greater than or equal to 3. We may describe the set of real numbers in these two intervals by writing

$$(-\infty, -3] \cup [3, \infty).$$

(b) This real number line shows three intervals that represent all real numbers x except -1 and 7. We may describe the set of real numbers in these three intervals by writing

$$(-\infty, -1) \cup (-1, 7) \cup (7, \infty).$$

PROBLEM 4 Indicate the given intervals on a real number line.

(a) $[-2, -1) \cup (-1, 2]$ **(b)** $(-\infty, -3] \cup [3, 5) \cup (5, \infty)$

Exercises 1.1

Basic Skills

In Exercises 1–10, write each statement as an inequality.

1. x is negative.
2. y is positive.
3. a is at most 7.
4. b is at least -9.
5. p is greater than 2 and less than or equal to 10.
6. q is less than or equal to 4 and greater than -1.
7. c is positive and less than 8.
8. d is negative and more than -5.
9. t is nonpositive and at least -2.
10. k is nonnegative and at most 4.

In Exercises 11–20, find AB, the distance from A to B on the real number line. Write each answer without absolute value notation.

11. The coordinate of point A is -9 and the coordinate of point B is 6.
12. The coordinate of point A is 13 and the coordinate of point B is -2.
13. The coordinate of point A is -1.9 and the coordinate of point B is -3.8.
14. The coordinate of point A is 0.16 and the coordinate of point B is 0.09.
15. The coordinate of point A is $\frac{7}{8}$ and the coordinate of point B is $-\frac{1}{3}$.
16. The coordinate of point A is $-\frac{3}{4}$ and the coordinate of point B is $-\frac{9}{10}$.
17. The coordinate of point A is $-3\frac{1}{2}$ and the coordinate of point B is $5\frac{2}{3}$.
18. The coordinate of point A is 19 and the coordinate of point B is $6\frac{5}{6}$.
19. The coordinate of point A is $\sqrt{2}$ and the coordinate of point B is π.
20. The coordinate of point A is $-\sqrt{2}$ and coordinate of point B is $\sqrt{3}$.

In Exercises 21–30, write each statement using an inequality with absolute value notation.

21. The distance between a and 7 is at least 3.
22. The distance between b and 2 is at most 7.
23. The distance between x and -4 is less than 2.
24. The distance between y and -6 is greater than 1.
25. a is more than 4 units from -1.
26. c is less than 6 units from 0.
27. y is closer to 1 than to 0.
28. y is closer to -3 than to -2.
29. b is farther from -2 than from 5.
30. a is farther from 5 than from -1.

In Exercises 31–50, use interval notation to describe the set of all real numbers designated on each real number line.

31. ───────●═══════●───────▶ x
          -4 -2

32. ───────●═══════●───────▶ x
       -1 4

33. ───────○═══════○───────▶ x
        2   6

34. ───────○═══════○───────▶ x
      -12 -8

35. ───────●═══════○───────▶ x
       -1 4

36. ───────○═══════●───────▶ x
      -5 0

12 CHAPTER 1 ♦ *Prerequisites of College Algebra*

37. [number line with open circle at −1, shaded left]
38. [number line with open circle at 8, shaded left]
39. [number line with open circle at 10, shaded right]
40. [number line with closed/shaded right from −5]
41. [number line with open circle at 0, shaded right]
42. [number line with open circle at 1, shaded right]
43. [number line segment between −7 and −4]
44. [number line with open circles at −2 and 2, shaded between]
45. [number line with open circle at 2, shaded between 0 and 2]

46. [number line between −3 and 2, open circles]
47. [number line between 1 and 6, open circles]
48. [number line between −1 and 0, open circles]
49. [number line with open circles at 0, 2, 3]
50. [number line with open circles at −2, −1, 1, 3]

In Exercises 51–60, indicate the given intervals on a real number line.

51. $(-3, 0]$
52. $[-6, -3]$
53. $(-\infty, \infty)$
54. $[2, \infty)$
55. $[0, 1) \cup (3, \infty)$
56. $(-\infty, 1) \cup [2, 5]$
57. $(-\infty, 7) \cup (7, \infty)$
58. $[-1, 0) \cup (0, 2]$
59. $(-\infty, 0] \cup [1, 2) \cup (2, \infty)$
60. $(0, 1) \cup (1, 3) \cup (3, \infty)$

◀ **Critical Thinking** ▶

In Exercises 61–66, rewrite each expression so that it does not contain absolute value bars.

61. $|\pi - x|$ if $x \geq \pi$
62. $|\pi - x|$ if $x < \pi$
63. $|x - 3| + |x - 4|$ if $3 < x < 4$
64. $|x - 3| - |x - 4|$ if $x < 3$
65. $|x| < 5$
66. $|y| > 3$

67. What meaning (if any) can be assigned to the interval notation $[a, a]$? to (a, a)?
68. Explain why each of the given interval notations is meaningless.
    (a) $(0, -6)$
    (b) $[-2, -\infty)$
    (c) $[-\infty, 5)$
    (d) $[1, \infty]$

69. Use interval notation to describe the set of all real numbers that are less than their reciprocals.
70. Use interval notation to describe the set of all real numbers that are less than or equal to their squares.
71. Use a calculator to help list the following real numbers in order from smallest to largest:

    $3.145 \quad \pi \quad \dfrac{22}{7} \quad 3.2 \quad \sqrt{10} \quad \dfrac{157}{50}$

72. Use a calculator to help list the following real numbers in order from largest to smallest:

    $\dfrac{7}{5} \quad \sqrt{2} \quad 1.414 \quad \dfrac{71}{50} \quad \dfrac{8\pi}{17} \quad 1.5$

1.2 Integer Exponents and Scientific Notation

◆ **Introductory Comments**

When we add or subtract two or more numbers, the numbers are called *terms*. To describe the repeated addition of a real number a, we use multiplication:

$$\underbrace{a + a + a + \cdots + a}_{n \text{ terms}} = na.$$

SECTION 1.2 ◆ Integer Exponents and Scientific Notation

When we multiply two or more numbers, the numbers are called *factors*. To describe the repeated multiplication of a real number *a*, we use a positive integer exponent:

◀ **Definition of a^n**

For any real number *a* and any positive integer *n*,

$$a^n = \underbrace{(a)(a)(a) \cdots (a)}_{n \text{ factors}}$$

The expression a^n is read "*a* to the *n*th power" and is referred to as the **exponential form** of the repeated multiplication. In the expression a^n, the number *a* is called the **base** and *n* is the **exponent**, or **power**.

Note the difference in the meanings of these exponential forms:

$$-2^4 = -(2)(2)(2)(2) = -16$$
whereas $\quad (-2)^4 = (-2)(-2)(-2)(-2) = 16$

Also,

$$2y^4 = 2(y)(y)(y)(y)$$
whereas $\quad (2y)^4 = (2y)(2y)(2y)(2y) = 16y^4$

Our definition of a^n gives us the following **properties of positive integer exponents**.

◀ **Properties of Positive Integer Exponents**

For any positive integers *m* and *n* and real numbers *a* and *b* that yield nonzero bases and denominators,

Property	*Example*
Product Property $a^m a^n = a^{m+n}$	$(-2x^2)(3x^5) = (-2 \cdot 3)x^{2+5}$ $= -6x^7$
Power Property $(a^m)^n = a^{mn}$	$[(x+y)^2]^3 = (x+y)^{2 \cdot 3}$ $= (x+y)^6$
Quotient Property $\dfrac{a^m}{a^n} = \begin{cases} a^{m-n} & \text{if } m > n \\ \dfrac{1}{a^{n-m}} & \text{if } m < n \\ 1 & \text{if } m = n \end{cases}$	$\dfrac{3x^5y^3z^2}{6x^2y^4z^2} = \dfrac{3}{6} \cdot x^{5-2} \cdot \dfrac{1}{y^{4-3}} \cdot 1$ $= \dfrac{x^3}{2y}$
Power of a Product Property $(ab)^n = a^n b^n$	$(-3xy)^2 = (-3)^2 x^2 y^2 = 9x^2y^2$
Power of a Quotient Property $\left(\dfrac{a}{b}\right)^n = \dfrac{a^n}{b^n}$	$\left(\dfrac{x^2+2}{y}\right)^3 = \dfrac{(x^2+2)^3}{y^3}$

CAUTION It is not possible to manipulate a power of a sum or difference in the same manner as a power of a product or quotient; that is,

$$(a + b)^n \neq a^n + b^n \quad \text{and} \quad (a - b)^n \neq a^n - b^n.$$

Note that

$$(3 \cdot 2)^2 = 3^2 \cdot 2^2 \quad \text{whereas} \quad (3 + 2)^2 \neq 3^2 + 2^2$$
$$36 = 36 \quad \quad \quad \quad \quad \quad 25 \neq 13$$

◆ Integer Exponents

Our definition of a^n applies only to *positive* integer exponents n. How should we define a^0 or a^{-n}? Certainly, we would like the properties of positive integer exponents to hold for zero and negative integer exponents as well. If the product property $a^m a^n = a^{m+n}$ is to hold for the zero exponent, then

$$a^0 \cdot a^n = a^{0+n} = a^n.$$

The only way $a^0 \cdot a^n$ can equal a^n is for a^0 to be the *identity element for multiplication*, that is, $a^0 = 1$. If the product property is to hold for negative integer exponents, then

$$a^{-n} \cdot a^n = a^{-n+n} = a^0 = 1.$$

The only way $a^{-n} \cdot a^n$ can equal 1 is for a^{-n} to be the *multiplicative inverse*, or *reciprocal*, of a^n, that is, $a^{-n} = \dfrac{1}{a^n}$.

◀ **Definitions of a^0 and a^{-n}**

> For any real number a, except 0, and any integer n,
>
> $$a^0 = 1 \quad \text{and} \quad a^{-n} = \frac{1}{a^n}.$$

Here are some illustrations of evaluating zero and negative integer exponents:

$$5^0 = 1 \quad 3^{-2} = \frac{1}{3^2} = \frac{1}{9} \quad \left(-\frac{1}{2}\right)^{-4} = \frac{1}{(-\frac{1}{2})^4} = \frac{1}{\frac{1}{16}} = 16$$

We can use the power key, $\boxed{\wedge}$ or $\boxed{y^x}$, on a calculator to raise a number to an integer power. Figure 1.5 shows a typical display on a calculator for evaluating 5^0, 3^{-2}, and $(-\frac{1}{2})^{-4}$ and confirms our work.

Now, consider the quotient a^{-m}/b^{-n}, where m and n are positive integers. To write a^{-m}/b^{-n} *without* negative exponents, we can proceed as follows:

$$\frac{a^{-m}}{b^{-n}} = \frac{1/a^m}{1/b^n} = \frac{1}{a^m} \cdot \frac{b^n}{1} = \frac{b^n}{a^m}.$$

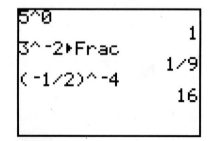

FIGURE 1.5

Typical display of evaluating 5^0, 3^{-2}, and $(-\frac{1}{2})^{-4}$

Notice that the factor a^{-m} in the numerator becomes a^m in the denominator, and the factor b^{-n} in the denominator becomes b^n in the numerator. In general, a factor may be moved from the numerator to the denominator—or from the denominator to the numerator—by changing the sign of the exponent.

CAUTION Only a *factor* of the numerator or denominator may be moved by changing the sign of its exponent. A *term* of the numerator or denominator cannot be moved in this fashion. For example, to write

$$\frac{3 \cdot 2^{-1}}{5^{-1}} \quad \text{as} \quad \frac{3 \cdot 5}{2} = \frac{17}{2} \quad \text{is CORRECT!}$$

(up: 2^{-1} moves up; down: 5^{-1} moves down)

However, to write

$$\frac{3 + 2^{-1}}{5^{-1}} \quad \text{as} \quad \frac{3 + 5}{2} = 4 \quad \text{is WRONG!}$$

(up: 2^{-1} moves up; down: 5^{-1} moves down)

To evaluate $(3 + 2^{-1})/5^{-1}$ correctly, we proceed as follows:

$$\frac{3 + 2^{-1}}{5^{-1}} = \frac{3 + \frac{1}{2}}{\frac{1}{5}} = \frac{\frac{7}{2}}{\frac{1}{5}} = \frac{35}{2}$$

The display in Figure 1.6 confirms our work.

It can be shown that each of the five properties of positive integer exponents are valid for *all* integer exponents. In fact, since we now know the meaning of zero and negative integer exponents, the quotient property for exponents can be written simply as

$$\frac{a^m}{a^n} = a^{m-n}, \quad a \neq 0.$$

The important definitions and properties of integer exponents are summarized next. We refer to them as the **laws of exponents**.

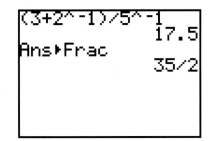

FIGURE 1.6
Typical display of evaluating $(3 + 2^{-1})/5^{-1}$.

◀ **Laws of Exponents**

For any integers m and n and real numbers a and b that yield nonzero bases and denominators,

1. $a^0 = 1$
2. $a^{-n} = \dfrac{1}{a^n}$
3. $a^m a^n = a^{m+n}$
4. $(a^m)^n = a^{mn}$
5. $\dfrac{a^m}{a^n} = a^{m-n}$
6. $(ab)^n = a^n b^n$
7. $\left(\dfrac{a}{b}\right)^n = \dfrac{a^n}{b^n}$
8. $\dfrac{a^{-m}}{b^{-n}} = \dfrac{b^n}{a^m}$

EXAMPLE 1 Simplify each expression. Express the answer with positive exponents.

(a) $(x + y)^{-4}(x + y)^3$

(b) $\left(\dfrac{3}{y}\right)^{-2}$

(c) $(-3x^{-3}y^4)^{-2}$

(d) $\dfrac{4^{-2}x^{-3}y^4}{4x^{-4}y^{-3}}$

◆ Solution

(a) $(x + y)^{-4}(x + y)^3 = (x + y)^{-1}$ Add exponents

$= \dfrac{1}{x + y}$ Change to a positive exponent

(b) $\left(\dfrac{3}{y}\right)^{-2} = \dfrac{3^{-2}}{y^{-2}}$ Apply $(a/b)^n = a^n/b^n$ (law 7)

$= \dfrac{y^2}{3^2}$ or $\dfrac{y^2}{9}$ Change to positive exponents

(c) $(-3x^{-3}y^4)^{-2} = (-3)^{-2}(x^{-3})^{-2}(y^4)^{-2}$ Apply $(ab)^n = a^n b^n$ (law 6)

$= (-3)^{-2}x^6 y^{-8}$ Multiply exponents

$= \dfrac{x^6}{(-3)^2 y^8}$ or $\dfrac{x^6}{9y^8}$ Change to positive exponents

(d) $\dfrac{4^{-2}x^{-3}y^4}{4x^{-4}y^{-3}} = 4^{-2-1}x^{-3-(-4)}y^{4-(-3)}$ Subtract exponents

$= 4^{-3}x^1 y^7$ Simplify

$= \dfrac{xy^7}{4^3}$ or $\dfrac{xy^7}{64}$ Change to a positive exponent ◆

Several alternate procedures may be used to simplfy the expressions in Example 1. For example, we may first change to positive exponents, and then apply the properties of positive integer exponents.

PROBLEM 1 Simplify the expression in Example 1(d) by first applying the property $a^{-m}/b^{-n} = b^n/a^m$ to eliminate the negative exponents. Then simplify by using the properties of positive integer exponents. You should obtain the same result as in Example 1. ◆

◆ **Application: Scientific Notation and Conversion Factors**

The laws of exponents may be used in conjunction with *scientific notation* to perform arithmetic calculations containing large and small numbers.

◆ Scientific Notation

A number is written in **scientific notation** if it is of the form

$k \times 10^n$ where $1 \leq |k| < 10$ and n is an integer.

SECTION 1.2 ◆ Integer Exponents and Scientific Notation

To change an ordinary decimal number to a number in scientific notation, or vice versa, remember the following three facts:

1. A decimal number greater than 10 has an exponent on base 10 that is a positive integer.
2. A decimal number between 0 and 1 has an exponent on base 10 that is a negative integer.
3. A decimal number between 1 and 10 has an exponent on base 10 that is 0.

Here are some examples of changing an ordinary decimal number to a number in scientific notation, and vice versa:

$$0.00000065 = 6.5 \times 10^{-7} \qquad 8{,}240{,}000{,}000 = 8.24 \times 10^9$$

7 places 9 places

$$3.2 \times 10^3 = 3200. \qquad 9.84 \times 10^{-4} = 0.000984$$

3 places 4 places

$$8.75 = 8.75 \times 10^0 \qquad 1.56 \times 10^0 = 1.56$$

To perform arithmetic calculations containing large and small numbers, we may first write the numbers in scientific notation and then apply the laws of exponents. For example, to evaluate

$$\frac{(0.0000650)(0.00000008)}{1{,}300{,}000}$$

we proceed as follows:

$$\frac{(0.0000650)(0.00000008)}{1{,}300{,}000} = \frac{(6.50 \times 10^{-5})(8 \times 10^{-8})}{1.3 \times 10^6}$$

$$= \frac{(6.50)(8)}{(1.3)} \times \frac{(10^{-5})(10^{-8})}{10^6} \qquad \textbf{Rearrange the factors}$$

$$= 40 \times 10^{-19} \qquad \textbf{Evaluate and simplify exponents}$$

$$= (4 \times 10^1) \times 10^{-19} = 4 \times 10^{-18} \qquad \textbf{Write in scientific notation}$$

$$= 0.000000000000000004 \qquad \textbf{Write as a decimal number}$$

We can use the scientific notation key, $\boxed{\text{EE}}$ or $\boxed{\text{EXP}}$, on a calculator to perform arithmetic calculations containing large and small numbers. Figure 1.7 shows a typical display on a calculator for evaluating the previous expression. This display confirms our work.

Scientist and engineers often use scientific notation along with *conversion factors* in their work. A **conversion factor** is formed by two equal values, one of the values placed in the numerator and the other value in the denominator of a fraction. Hence, a conversion factor is equal to 1. A conversion factor is considered an exact number and therefore does not affect the number of sig-

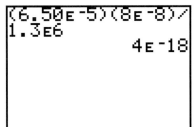

FIGURE 1.7
Typical display of evaluating
$\dfrac{(0.0000650)(0.00000008)}{1{,}300{,}000}$

nificant digits that we maintain in a calculation with measured data. (See Appendix for a discussion of significant digits).

EXAMPLE 2 *Chapter Opening Problem*
The distance from Earth to Alpha Centauri (the star closest to our Sun) is approximately 25,500,000,000,000 miles. How many years does it take light from this star to reach Earth? Assume that light travels at 186,000 miles per second.

◆ **Solution**   First, we express these large numbers in scientific notation and find the time in second(s) that it takes light from this star to reach Earth:

$$\frac{25{,}500{,}000{,}000{,}000 \text{ mi}}{186{,}000 \text{ mi/s}} = \frac{2.55 \times 10^{13}}{1.86 \times 10^{5}} \text{ s}$$

Now, we apply three conversion factors and change from seconds to years (yr):

$$\frac{2.55 \times 10^{13}}{1.86 \times 10^{5}} \cancel{s} \cdot \frac{1 \cancel{hr}}{3600 \cancel{s}} \cdot \frac{1 \cancel{day}}{24 \cancel{hr}} \cdot \frac{1 \text{ yr}}{365 \cancel{days}}$$

We can evaluate this expression by using the scientific notation key on a calculator. The calculator display in Figure 1.8 indicates that it takes approximately 4.35 years for light to reach Earth from Alpha Centauri. We round this answer to three significant digits since the measured data, 2.55×10^{13} mi and 1.86×10^{5} mi, are given to three significant digits.   ◆

PROBLEM 2   The distance from Earth to our Sun is approximately 93,000,000 mi. How many minutes does it take light from the Sun to reach Earth? Assume that light travels at 186,000 mi/s.   ◆

In engineering, prefixes indicating powers of ten, such as *mega-, kilo-, centi-, milli-,* and *micro-*, are attached to the basic units of measurement—meters, grams, volts, watts, and so on, in order to work with very large or small numbers. Listed in Table 1.2 are the symbols and meanings of these commonly used prefixes.

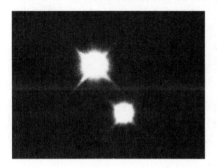

FIGURE 1.8
Typical display of the evaluation in Example 2.

TABLE 1.2
Symbols and meanings of the most commonly used prefixes

Prefixes	Symbol	Meaning
mega-	M	10^{6}
kilo-	k	10^{3}
centi-	c	10^{-2}
milli-	m	10^{-3}
micro-	μ	10^{-6}

Numerical quantities are usually substituted in their basic unit size into a formula. As illustrated in the next example, if a prefix is attached to a basic unit, then the prefix must be converted to a power of ten before substituting the unit into the formula. If necessary, the answer in basic units may then be converted to a specific prefix.

EXAMPLE 3 The formula $P = V^2/R$ gives the power P in watts (W) dissipated by a resistance of R ohms (Ω) when the voltage drop across the resistance is V volts (V). Find the power P in *milliwatts* (mW) if $V = 1.2$ kV and $R = 20$ MΩ.

Solution Converting to powers of ten, we have

$$V = 1.2 \text{ kV} = 1.2 \times 10^3 \text{ V} \quad \text{and} \quad R = 20 \text{ M}\Omega = 20 \times 10^6 \text{ }\Omega.$$

Thus,

$$P = \frac{V^2}{R} = \frac{(1.2 \times 10^3)^2}{20 \times 10^6} = \frac{(1.2)^2 \times (10^3)^2}{20 \times 10^6} \text{ W}$$
$$= \frac{1.44 \times 10^6}{20 \times 10^6} \text{ W}$$
$$= 0.072 \times 10^0 \text{ W}$$
$$= 0.072 \text{ W}$$
$$= 72 \times 10^{-3} \text{ W} = 72 \text{ mW}$$

The calculator display in Figure 1.9 supports our work.

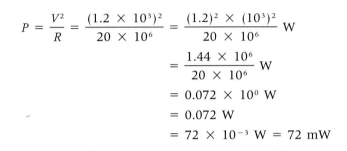

FIGURE 1.9

Typical display of the evaluation in Example 3.

PROBLEM 3 Use the power formula in Example 3 to calculate P in megawatts (MW) if $R = 30 \text{ }\Omega$ and $V = 600,000$ V.

Exercises 1.2

Basic Skills

In Exercises 1–20, use the definitions of a^n, a^{-n}, and a^0 to evaluate each expression. Use a calculator to verify each answer.

1. $(-6)^2$
2. 5^3
3. -4^2
4. -2^5
5. 4^{-3}
6. 3^{-4}
7. -8^{-2}
8. -2^{-3}
9. $\left(\dfrac{2}{3}\right)^{-4}$
10. $\left(\dfrac{2}{5}\right)^{-1}$
11. $(-9)^{-2}$
12. $\left(\dfrac{-3}{4}\right)^{-3}$
13. $\dfrac{(-\frac{1}{3})^2}{2^{-1}}$
14. $\dfrac{4^{-1}}{-(\frac{2}{3})^4}$
15. $\dfrac{2^{-1} - 3^0}{2^0 + 3^0}$
16. $\dfrac{3^0}{2^0 + 3^{-2}}$
17. $3^{-1} + 4^{-1}$
18. $\dfrac{8^{-2} - 4^{-1}}{2^{-1}}$
19. $\dfrac{3 \cdot 2^{-3}}{5 \cdot 4^{-2}}$
20. $\dfrac{3 + 2^{-3}}{5 + 4^{-2}}$

In Exercises 21–40, use the laws of exponents to simplify each expression. Express the answer with positive exponents only.

21. $-8x^{-1}y^{-3}$
22. $2^{-1}xy^{-4}$
23. $-5(1 - 2x)^{-6}(-2)$
24. $-3(x^2 + 1)^{-4}(2x)$
25. $(-2x^2y^3)(5xy^5)^2$
26. $(2x)^4(-x^2y)(-3xy^{-1})$
27. $(-2y^{-1})^4$
28. $(-3x^2y^{-2})^3$

29. $\dfrac{8x^4yz^3}{4xy^{-3}z^3}$

30. $\dfrac{9x^{-1}y^2z}{12x^3y^{-2}}$

31. $(x+3)^2(x+3)^{-3}$

32. $(x+1)(x+1)^4$

33. $[(x-2)^2]^4$

34. $[(t-6)^{-1}]^{-2}$

35. $\dfrac{(2y+3)^{-5}(2y+3)}{(2y+3)^{-6}}$

36. $\dfrac{(2a-5)(2a+5)^{-3}}{(2a+5)^3(2a-5)^2}$

37. $\dfrac{(3m^2n)^{-2}}{2m^{-3}}$

38. $\dfrac{(3x^{-1}y^2)^{-4}}{(-x^4y^3)^{-1}}$

39. $\dfrac{(2p)^{-3}(q-r)^3}{(q-r)^{-4}(4p^2)^{-3}}$

40. $\dfrac{[a^{-2}(b-c)^2]^{-1}}{a(b-c)^{-3}}$

In Exercises 41–46, express each number as an ordinary decimal number.

41. 6.9×10^{-9}

42. 2.33×10^6

43. -1750×10^8

44. 2200×10^{-5}

45. 0.00392×10^{-1}

46. -0.0698×10^0

In Exercises 47–52, express each number in scientific notation.

47. $54{,}300$

48. 0.00000294

49. 0.13×10^{-6}

50. 1730×10^5

51. 240×10^3

52. 0.005×10^{-8}

In Exercises 53–60, perform the indicated operations. Write each answer in scientific notation.

53. $(8{,}000{,}000{,}000)(0.00000025)$

54. $(170{,}000{,}000)(-20{,}000{,}000{,}000{,}000)$

55. $\dfrac{0.0000496}{16{,}000{,}000{,}000{,}000}$

56. $\dfrac{486{,}000{,}000}{0.000000006}$

57. $(0.0000000002)^{-5}$

58. $(25{,}000{,}000{,}000)^{-2}$

59. $(300{,}000{,}000)(0.00002)^3$

60. $\dfrac{(8{,}000{,}000)^2(2000)}{(0.0004)^4}$

In Exercises 61–64, use the power key, $\boxed{\wedge}$ or $\boxed{y^x}$, on a calculator to evaluate each expression. Record each answer as an ordinary decimal number rounded to four significant digits (see Appendix for a discussion of significant digits).

61. $(1.0025)^{45}$

62. $(-0.936)^{12}$

63. $(0.0287)^{-3}$

64. $(1.806)^{-10}$

In Exercises 65–68, use the scientific notation key, $\boxed{\text{EE}}$ or $\boxed{\text{EXP}}$, on a calculator to evaluate each expression. Record the answer in the scientific notation form $k \times 10^n$, rounding k to three significant digits (see Appendix for a discussion of significant digits).

65. $(2.731 \times 10^{-11})(3.924 \times 10^{-13})$

66. $\dfrac{(8.75 \times 10^{12})(167 \times 10^{14})}{0.000000000796}$

67. $\dfrac{(0.0000212)(0.000000567)^2}{3.42 \times 10^{12}}$

68. $(8.92 \times 10^{15})^2(0.0000000274)^3$

In Exercises 69–74, use scientific notation to help solve each problem.

69. The distance from Earth to the planet Pluto is approximately 3,700,000,000 mi. If a space ship leaves from Earth and averages 35,000 mi/hr, how many years will it take the ship to reach Pluto?

70. *Halley's comet* travels approximately 43,300,000,000 km in one orbit around the Sun and averages 65,100 km/hr. How many years does it take Halley's comet to orbit the Sun once?

71. The formula $I = V/R$ gives the current I in amps (A) through a resistance of R ohms (Ω) when the voltage drop across the resistance is V volts (V). Calculate the current in microamps (μA) if the resistance is 330 MΩ and the voltage drop is 2.2 kV.

72. When two resistances R_1 and R_2 are connected in parallel, their total resistance R_t is given by

$$R_t = \dfrac{R_1 R_2}{R_1 + R_2}.$$

If $R_1 = 4.2$ MΩ and $R_2 = 680$ kΩ, find the total resistance R_t in kΩ.

73. The land area of the earth is approximately 1.484×10^8 square kilometers (sq km). In 1980, the world population was approximately 4,483,000,000. At that time, what was the amount of land area in square meters (sq m) per person?

74. If our national debt is approximately one trillion dollars and the population of the United States is approximately 250,000,000, what is the amount of debt per person?

Critical Thinking

75. Given that x, y, and z are integers and $a \neq 0$, simplify $a^{x-y} \cdot a^{y-z} \cdot a^{z-x}$.

76. Given that n is an integer and $b \neq 0$, rewrite each expression as a power of b:

(a) $(b^{2-n} \cdot b^{n-4})^{-3}$

(b) $\dfrac{(b^2)^n}{b^n \cdot b^{n-2}}$

77. We have defined a^0 to be equal to 1, provided $a \neq 0$. If we erroneously define 0^0 to be equal to 1, then its

reciprocal $1/0^0$ must also be equal to 1 and, by the laws of exponents, we could write

$$1 = 0^0 = \frac{1}{0^0} = \frac{a^0}{0^0} = \left(\frac{a}{0}\right)^0.$$

Do you see the fallacy in this argument? Explain.

78. What is the value of $(-1)^n$ when n is an even integer? an odd integer? 0?

79. Suppose you have a home mortgage of $80,000 at an interest rate of 9% per year over 30 years. The monthly payment may be calculated by evaluating

$$\frac{(7.5 \times 10^{-3})(8 \times 10^4)(1.0075)^{360}}{(1.0075)^{360} - 1}$$

(a) Determine, to the nearest cent, the amount of the monthly payment.
(b) Determine the total amount you will pay to the lender.

80. You may exceed the computational range of your calculator when working with the power key or scientific notation key. Most calculators will not display positive values less than 1×10^{-99} or greater than $9.999999999 \times 10^{99}$. Determine the smallest and largest integer n for which you are able to obtain a value for 25^n on the display of your calculator.

1.3 Rational Exponents and Radical Expressions

◆ Introductory Comments

In Section 1.2 we defined the expression a^n for integer exponents. We now extend our definition to include exponents that are rational numbers; that is, we now assign meaning to expressions such as these:

$$49^{1/2} \qquad 32^{2/5} \qquad 16^{-3/4}$$

In order to make the transition from integer exponents to rational exponents, we must first discuss the meaning of the n*th root* of a real number.

◄ **Definition of nth root**

> If a and b are any real numbers and $n \geq 2$ is an integer such that $b^n = a$, then b is an **nth root** of a.

We now describe the nature of the nth root(s) of a for n even or odd and a positive or negative.

◄ **Nature of the nth root(s) of a**

n	a	nth root(s) of a	Example
even	positive	two real roots, one positve and one negative	The two real fourth roots of 16 are ± 2, since $(\pm 2)^4 = 16$.
even	negative	no real root	No real square (2nd) root of -9 exists, because no real number squared is -9.
odd	positive	one real root, a positive root	The one real cube (3rd) root of 8 is 2, since $2^3 = 8$.
odd	negative	one real root, a negative root	The one real cube root of -8 is -2, since $(-2)^3 = -8$.

Note: The nth root of zero is 0, since $0^n = 0$ for any integer $n \geq 2$.

To avoid ambiguity when working with even roots of positive numbers, we define the *principal nth root* of a number.

◂ **Principal nth Root of a Number**

> If a is any real number that has at least one real nth root, then the nth root that has the same sign as a is called the **principal nth root of a** and is denoted by $\sqrt[n]{a}$.

In the expression $\sqrt[n]{a}$, the symbol $\sqrt{}$ is called the **radical**, n is the **index**, and a is the **radicand**. We usually write the principal square root as $\sqrt{a}$ instead of $\sqrt[2]{a}$, but for $n \geq 3$ we must show the index. Here are some illustrations of evaluating nth roots:

$$\sqrt[4]{16} = 2 \qquad -\sqrt[4]{16} = -2 \qquad \sqrt{-9} \text{ is not a real number}$$
$$\sqrt[3]{8} = 2 \qquad \sqrt[3]{-8} = -2 \qquad \sqrt[5]{0} = 0$$

◆ **Rational Exponents**

We now define a rational exponent, $a^{1/n}$, where n is a positive integer. Certainly, we would like a definition of $a^{1/n}$ to be consistent with the laws of exponents listed in Section 1.2. If the power property for exponents is to hold true, then we must have

$$(a^{1/n})^n = a^{(1/n)(n)} = a^1 = a.$$

The only way $(a^{1/n})^n$ can equal a is for $a^{1/n}$ to be an nth root of a. We define the expression $a^{1/n}$ as the *exponential form of* $\sqrt[n]{a}$.

◂ **Definition of $a^{1/n}$**

> For any integer $n \geq 2$, where $a \geq 0$ when n is even,
>
> $$a^{1/n} = \sqrt[n]{a}.$$

Notice the difference between these examples of the definition of $a^{1/n}$:

$$49^{1/2} = \sqrt{49} = 7 \quad \text{and} \quad -49^{1/2} = -\sqrt{49} = -7$$
$$27^{1/3} = \sqrt[3]{27} = 3 \quad \text{and} \quad (-27)^{1/3} = \sqrt[3]{-27} = -3$$

How might we define $a^{m/n}$, where m and n are positive integers with m/n reduced to lowest terms and $a \geq 0$ when n is even? Again, if the power property for exponents is to hold true, we must define $a^{m/n}$ as follows:

Definition of $a^{m/n}$

If $a^{1/n}$ is any real number and m and n are positive integers such that m/n is reduced to lowest terms, then

$$a^{m/n} = (a^{1/n})^m = (\sqrt[n]{a})^m$$

or

$$a^{m/n} = (a^m)^{1/n} = \sqrt[n]{a^m}.$$

Thus, to evaluate an expression that contains a rational exponent, such as $9^{3/2}$, we can use either of two methods of reasoning:

1. $9^{3/2} = (9^{1/2})^3 = (\sqrt{9})^3 = (3)^3 = 27$ or
2. $9^{3/2} = (9^3)^{1/2} = \sqrt{9^3} = \sqrt{729} = 27$.

As you can see, to evaluate $a^{m/n}$ the first method is easier to apply: *first take the root, then raise to the power.* Here are two more examples of evaluating expressions with rational exponents:

$$32^{2/5} = (32^{1/5})^2 = (\sqrt[5]{32})^2 = (2)^2 = 4$$

$$16^{-3/4} = \frac{1}{16^{3/4}} = \frac{1}{(16^{1/4})^3} = \frac{1}{(\sqrt[4]{16})^3} = \frac{1}{2^3} = \frac{1}{8}$$

We can use the power key, $\boxed{\wedge}$ or $\boxed{y^x}$, on a calculator to raise a number to a rational power. Figure 1.10 shows a typical display on a calculator for evaluating $32^{2/5}$ and $16^{-3/4}$ and confirms our work.

FIGURE 1.10
Typical display of evaluating $32^{2/5}$ and $16^{-3/4}$.

◆ Exponential and Radical Expressions

To change a radical expression to an exponential expression, or vice versa, we use the definition of a rational exponent, $a^{m/n} = \sqrt[n]{a^m}$. The procedure is shown in the next example.

EXAMPLE 1

(a) Change to an exponential expression with a negative exponent:

$$\frac{1}{\sqrt{x^2 + 9}}$$

(b) Change to a radical expression: $(3x)^{2/3}$

◆ Solution

(a) The index of the radical becomes the denominator of the exponent. Hence,

$$\frac{1}{\sqrt{x^2 + 9}} = \frac{1}{(x^2 + 9)^{1/2}} = (x^2 + 9)^{-1/2}$$

(b) The denominator of the exponent becomes the index of the radical, and the numerator of the exponent is then the power of the radicand. Hence,

$$(3x)^{2/3} = \sqrt[3]{(3x)^2} = \sqrt[3]{9x^2}$$

PROBLEM 1

(a) Change to an exponential expression: $\sqrt[5]{(x^2 + 2)^2}$

(b) Change to a radical expression: $(2x^2)^{3/7}$

To simplify an expression containing rational exponents, we use the laws of exponents (Section 1.2). First, we look at problems in which each *variable base represents a positive real number*.

EXAMPLE 2 Simplify each exponential expression. Assume all variables represent *positive* real numbers.

(a) $(9x^2)^{1/2}$ **(b)** $(5x^{1/4})(-4x^{5/6})$ **(c)** $\left(\dfrac{x^{-3}}{-8y}\right)^{-2/3}$

Solution

(a) $(9x^2)^{1/2} = 9^{1/2}(x^2)^{1/2}$ Apply $(ab)^n = a^n b^n$

$\qquad\qquad\quad = 9^{1/2} x^{(2)(1/2)}$ Apply $(a^m)^n = a^{mn}$

$\qquad\qquad\quad = 3x$ Simplify

(b) $(5x^{1/4})(-4x^{5/6}) = (5)(-4)x^{1/4}x^{5/6}$ Regroup

$\qquad\qquad\qquad\quad = -20x^{1/4+5/6}$ Apply $a^m a^n = a^{m+n}$

$\qquad\qquad\qquad\quad = -20x^{3/12+10/12}$ Add the exponents and simplify

$\qquad\qquad\qquad\quad = -20x^{13/12}$

(c) $\left(\dfrac{x^{-3}}{-8y}\right)^{-2/3} = \dfrac{(x^{-3})^{-2/3}}{(-8)^{-2/3} y^{-2/3}}$ Apply $(a/b)^n = a^n/b^n$ and $(ab)^n = a^n b^n$

$\qquad\qquad\qquad = \dfrac{x^2}{(-8)^{-2/3} y^{-2/3}}$ Apply $(a^m)^n = a^{mn}$

$\qquad\qquad\qquad = (-8)^{2/3} x^2 y^{2/3}$ Change to positive exponents

$\qquad\qquad\qquad = 4x^2 y^{2/3}$ Simplify

When we simplify the expression $(9x^2)^{1/2}$ in Example 2(a), if we do *not* restrict x to *positive* real numbers, then we *cannot* state that $(9x^2)^{1/2}$ and $3x$ are equal. For example, suppose $x = -2$. Then

$(9x^2)^{1/2} = [9(-2)^2]^{1/2}$ whereas $3x = 3(-2)$
$\qquad\quad\; = [9(4)]^{1/2}$ $\qquad\qquad\qquad\qquad = -6$
$\qquad\quad\; = 36^{1/2} = 6$

If we allow x to be *any* real number, then we must write

$$(9x^2)^{1/2} = 3|x|.$$

In general, we have the following definition.

◀ **Definition of $(a^n)^{1/n}$**

If n is any positive integer and a is *any* real number, then

$$(a^n)^{1/n} = \sqrt[n]{a^n} = \begin{cases} a & \text{if } n \text{ is odd} \\ |a| & \text{if } n \text{ is even.} \end{cases}$$

PROBLEM 2 Simplify $[(-2)^4 x^4]^{1/4}$, assuming each of the following descriptions of x.

(a) x is a *positive* real number.   **(b)** x is *any* real number.   ◆

◆ **Simplifying Radical Expressions**

We can use the laws of exponents (Section 1.2) in conjunction with the definition of a rational exponent to derive the following *properties of radicals*.

◀ **Properties of Radicals**

For any real numbers a and b and positive integers $m \geq 2$ and $n \geq 2$ such that $\sqrt[n]{a}$ and $\sqrt[n]{b}$ are real numbers,

Property	Example
1. $(\sqrt[n]{a})^n = a$	$(\sqrt[3]{10})^3 = 10$
2. $\sqrt[m]{\sqrt[n]{a}} = \sqrt[mn]{a}$	$\sqrt[3]{\sqrt{7}} = \sqrt[3 \cdot 2]{7} = \sqrt[6]{7}$
3. $\sqrt[n]{ab} = \sqrt[n]{a} \cdot \sqrt[n]{b}$	$\sqrt{25 \cdot 2} = \sqrt{25} \cdot \sqrt{2} = 5\sqrt{2}$
4. $\sqrt[n]{\dfrac{a}{b}} = \dfrac{\sqrt[n]{a}}{\sqrt[n]{b}}$	$\sqrt[3]{\dfrac{-27}{64}} = \dfrac{\sqrt[3]{-27}}{\sqrt[3]{64}} = \dfrac{\sqrt[3]{(-3)^3}}{\sqrt[3]{4^3}} = \dfrac{-3}{4}$

CAUTION   It is not possible to take the nth root of a sum or difference in the same manner as taking the nth root of a product or quotient (as in property 3 or 4), that is,

$$\sqrt[n]{a+b} \neq \sqrt[n]{a} + \sqrt[n]{b} \quad \text{and} \quad \sqrt[n]{a-b} \neq \sqrt[n]{a} - \sqrt[n]{b}.$$

Notice that

$$\sqrt{9 \cdot 16} = \sqrt{9} \cdot \sqrt{16}, \quad \text{whereas} \quad \sqrt{9+16} \neq \sqrt{9} + \sqrt{16}$$
$$12 = 12 \qquad\qquad\qquad\qquad 5 \neq 7$$

We can use the properties of radicals to write a radical expression in *simplified form*.

Simplified Form of a Radical

A radical expression is in **simplified form** if both of the following conditions are satisfied.

Example

1. The radicand contains no factor to a power m greater than or equal to the index n of the radical; that is, $\sqrt[n]{a^m}$ has $m < n$.

   $\sqrt{75} = \sqrt{5^2 \cdot 3} = \sqrt{5^2} \cdot \sqrt{3} = 5\underbrace{\sqrt{3}}_{\text{Simplified form}}$

2. The power of the radicand m and the index n of the radical have no common factor other than 1; that is, for $\sqrt[n]{a^m}$, the exponent m/n is reduced to lowest terms.

   $\sqrt[4]{5^2} = 5^{2/4} = 5^{1/2} = \underbrace{\sqrt{5}}_{\text{Simplified form}}$

EXAMPLE 3 Write each radical expression in simplified form. Assume all variables represent *positive* real numbers.

(a) $\sqrt{18x^3}$

(b) $\sqrt{\sqrt[3]{16x^2}}$

(c) $\sqrt{\dfrac{8x^5}{y^6}}$

(d) $\sqrt[3]{3x^2y} \cdot \sqrt[3]{9xy^2}$

Solution

(a) $\sqrt{18x^3} = \sqrt{(3)^2 2 \cdot (x^2)x}$ Express each factor in terms of perfect squares

$= \sqrt{(3x)^2 \cdot 2x}$ Group the perfect square factors

$= \sqrt{(3x)^2} \cdot \sqrt{2x}$ Apply $\sqrt[n]{ab} = \sqrt[n]{a}\sqrt[n]{b}$

$= 3x\sqrt{2x}$ Simplify using $\sqrt[n]{a^n} = a$

(b) $\sqrt{\sqrt[3]{16x^2}} = \sqrt[6]{16x^2}$ Apply $\sqrt[m]{\sqrt[n]{a}} = \sqrt[mn]{a}$

$= \sqrt[6]{(4x)^2}$ Write the radicand as a perfect square

$= (4x)^{2/6}$ Change to exponential form

$= (4x)^{1/3} = \sqrt[3]{4x}$ Reduce the order of the radical

SECTION 1.3 ◆ *Rational Exponents and Radical Expressions*   27

(c) $\sqrt{\dfrac{8x^5}{y^6}} = \dfrac{\sqrt{8x^5}}{\sqrt{y^6}}$ Apply $\sqrt[n]{a/b} = \sqrt[n]{a}/\sqrt[n]{b}$

$\qquad = \dfrac{\sqrt{(2)(2)^2 \cdot (x)(x^2)^2}}{\sqrt{(y^3)^2}}$ Rewrite numerator and denominator in terms of perfect squares

$\qquad = \dfrac{\sqrt{(2x^2)^2} \cdot \sqrt{2x}}{\sqrt{(y^3)^2}}$ Regroup and apply $\sqrt[n]{ab} = \sqrt[n]{a}\,\sqrt[n]{b}$

$\qquad = \dfrac{2x^2\sqrt{2x}}{y^3}$ Simplify using $\sqrt[n]{a^n} = a$

(d) $\sqrt[3]{3x^2y} \cdot \sqrt[3]{9xy^2} = \sqrt[3]{(3x^2y)(9xy^2)}$ Apply $\sqrt[n]{a}\,\sqrt[n]{b} = \sqrt[n]{ab}$

$\qquad = \sqrt[3]{27x^3y^3}$ Multiply

$\qquad = \sqrt[3]{(3xy)^3}$ Rewrite as a perfect cube

$\qquad = 3xy$ Simplify using $\sqrt[n]{a^n} = a$ ◆

PROBLEM 3   Write each radical expression in simplified form. Assume all variables represent *positive* real numbers.

(a) $\sqrt[3]{32(x+y)^3}$ (b) $\sqrt[4]{\sqrt[3]{x^2y^6}}$ ◆

◆ Rationalizing Denominators and Numerators

In algebra we must in some instances remove the radical symbol from the denominator or numerator of a fraction in order to perform further simplifications. This process is called **rationalizing the denominator or numerator**. To rationalize a denominator (or a numerator) that contains a radical with an index of n, we multiply numerator and denominator of the expression by a **rationalizing factor**—a factor that produces a perfect nth power in the radicand.

EXAMPLE 4

(a) Rationalize the denominator: $\dfrac{2\sqrt{3}}{\sqrt{2}}$

(b) Rationalize the numerator: $\dfrac{\sqrt[3]{5x^2}}{\sqrt[3]{2y}}$

◀ **Solution**

(a) $\dfrac{2\sqrt{3}}{\sqrt{2}} = \dfrac{2\sqrt{3}}{\sqrt{2}} \cdot \boxed{\dfrac{\sqrt{2}}{\sqrt{2}}} = \dfrac{2\sqrt{3 \cdot 2}}{\sqrt{2^2}} = \dfrac{2\sqrt{6}}{2} = \sqrt{6}$

↑ This rationalizing factor produces a perfect square in the denominator.

(b) $\sqrt[3]{\dfrac{5x^2}{2y}} = \dfrac{\sqrt[3]{5x^2}}{\sqrt[3]{2y}} \cdot \sqrt[3]{\dfrac{5^2 x}{5^2 x}} = \dfrac{\sqrt[3]{(5x)^3}}{\sqrt[3]{50xy}} = \dfrac{5x}{\sqrt[3]{50xy}}$

This rationalizing factor produces a perfect cube in the numerator.

PROBLEM 4 Rationalize the denominator of the radical expression in Example 4(b), and simplify.

Exercises 1.3

Basic Skills

In Exercises 1–20, evaluate each exponential expression. Use a calculator to verify each answer.

1. $4^{1/2}$
2. $8^{1/3}$
3. $-64^{1/6}$
4. $-81^{1/2}$
5. $49^{3/2}$
6. $4^{5/2}$
7. $(-125)^{2/3}$
8. $(-64)^{2/3}$
9. $100^{-3/2}$
10. $27^{-2/3}$
11. $-81^{-3/4}$
12. $-36^{-3/2}$
13. $\left(-\dfrac{1}{8}\right)^{2/3}$
14. $\left(\dfrac{1}{16}\right)^{1/4}$
15. $(2.25)^{-1/2}$
16. $(-125)^{1/3} \cdot 25^{-3/2}$
17. $\dfrac{4^{-3/2} \cdot 32^{3/5}}{-27^{0/3}}$
18. $\dfrac{6^{-2}(0.04)^{1/2}}{(-3)^{-1}}$
19. $(2^{-2} - 8^{-2/3} + 4^{-3/2})^{-1}$
20. $4^{-3/2} - 4^{-1/2} + 4^0 - 4^{1/2} + 4^{3/2}$

In Exercises 21–28, use a calculator to evaluate each expression. Record each answer as a decimal number rounded to four significant digits. (See Appendix for a review of significant digits.)

21. $(27.5)^{1/4}$
22. $(-0.654)^{1/3}$
23. $(287)^{-6/5}$
24. $(2.345 \times 10^{-3})^{3/2}$
25. $\sqrt{275}$
26. $\sqrt[3]{-8.32}$
27. $\sqrt[4]{0.1844}$
28. $\sqrt{12.6 \times 10^8}$

In Exercises 29–36, rewrite each radical expression as an exponential expression. For Exercises 33 and 34, express the answer with a negative exponent.

29. $\sqrt{2a}$
30. $\sqrt[4]{(x^2+1)^3}$
31. $\sqrt[5]{(y^2-3)^4}$
32. $\sqrt[10]{(2y)^5}$
33. $\dfrac{x}{\sqrt{x^2+y^2}}$
34. $\dfrac{3}{\sqrt[3]{x^2}}$
35. $x\sqrt{x}$
36. $\sqrt[3]{y} \cdot \sqrt{y}$

In Exercises 37–44, rewrite each exponential expression as a radical expression.

37. $x^{1/2}$
38. $y^{3/4}$
39. $(3m^2)^{3/4}$
40. $(-2x)^{2/3}$
41. $(x+y)^{4/5}$
42. $[2(x-3)^2]^{2/5}$
43. $2x^{-2/3}$
44. $(3p+q)^{-1/6}$

In Exercises 45–58, simplify each exponential expression. Assume the variables represent positive real numbers. Express the answer without zero or negative exponents.

45. $(4x^2)^{1/2} + (32x^5)^{1/5}$
46. $[16(x+1)^4]^{-1/4}$
47. $\left(\dfrac{x^3}{8}\right)^{-2/3}$
48. $\left(\dfrac{x^5 y^{10}}{32}\right)^{2/5}$
49. $(x+y)^{4/3}(x+y)^{3/2}$
50. $(x^2+1)^{-1/2}(x^2+1)^{3/2}$
51. $(-64x^3 y^{-3})^{-2/3}$
52. $(25x^4 y^{-6})^{3/2}$
53. $\dfrac{x^2(x^2+4)^{1/2}}{x^{2/3}(x^2+4)^{3/4}}$
54. $\dfrac{[36(m+8)^{-2}]^{-1/2}}{(m+8)^{2/3}}$
55. $\dfrac{(2^{-4/3}x^{2/3})^{3/4}}{x^{1/2}y^{-2}}$
56. $\dfrac{(3^{2/3}x^{-2}y^{1/2})^{-3/2}}{25x^{1/3}y^2}$
57. $\dfrac{1}{2}(x^2-1)^{-1/2}(2x)$
58. $\dfrac{1}{3}(3x-6x^3)^{-2/3}(3-18x^2)$
59. Rework Exercise 45 assuming x is *any* real number.
60. Rework Exercise 46 assuming x is *any* real number.

SECTION 1.4 ♦ The Complex Numbers

In Exercises 61–76, write each radical expression in simplified form. Assume the variables represent positive real numbers.

61. $\sqrt{24}$
62. $\sqrt[3]{72}$
63. $\sqrt{(x+1)^2}$
64. $\sqrt[4]{16x^4} + \sqrt[3]{8x^3}$
65. $\sqrt{x^9}$
66. $\sqrt[3]{x^8}$
67. $\sqrt[3]{16x^4 y^3}$
68. $\sqrt{125x^7 y^2 z^3}$
69. $\sqrt{49(a+b)^2(a^2+b^2)}$
70. $\sqrt[3]{24(x^3-8)(x-8)^3}$
71. $\sqrt[3]{\dfrac{54x^3 y^4}{125z^3}}$
72. $\sqrt{\dfrac{b^2-4ac}{4a^2}}$
73. $\sqrt{\sqrt[3]{25x^2 y^4}}$
74. $\sqrt[4]{\sqrt[3]{36x^2}}$
75. $\dfrac{\sqrt{40x^4 yz}}{2\sqrt{5xy^3 z}}$
76. $\sqrt[3]{6x^2 y^2} \cdot \sqrt[3]{9xy^2}$

77. Rework Exercise 63 assuming x can be *any* real number.
78. Rework Exercise 64 assuming x can be *any* real number.

In Exercises 79–82, rationalize the denominator. Assume the variables represent positive real numbers.

79. $\dfrac{12}{\sqrt{2}}$
80. $\dfrac{\sqrt{7}}{6x\sqrt{3x}}$
81. $\sqrt[3]{\dfrac{4}{9xy^2}}$
82. $\dfrac{1}{\sqrt[3]{(x+1)^2}}$

In Exercises 83–86, rationalize the numerator. Assume all variables represent positive real numbers.

83. $\dfrac{7\sqrt{6a}}{5}$
84. $\sqrt{\dfrac{x}{3}}$
85. $\sqrt[3]{\dfrac{(x+2)^2}{3x}}$
86. $\dfrac{\sqrt[4]{2p^2 q^3}}{\sqrt[4]{3}}$

◀ Critical Thinking

In Exercises 87–90, rewrite each product using only one radical symbol. [Hint: Use rational exponents to change the given radicals to radicals that have the same index.]

87. $\sqrt{3} \cdot \sqrt[3]{5}$
88. $\sqrt[3]{2} \cdot \sqrt[4]{3}$
89. $\sqrt{x} \cdot \sqrt[3]{y^2}$
90. $\sqrt{2x} \cdot \sqrt[5]{3x^2}$

91. Perform the indicated operations and write the answer in simplified radical form.
    (a) $\sqrt[3]{4}(\sqrt[3]{2} - \sqrt[3]{4} + \sqrt[3]{10})$
    (b) $\sqrt{abc}(a\sqrt{ab} + b\sqrt{bc} + c\sqrt{ac})$

92. Is $(-1)^{1/n}$ defined when n is an even integer? an odd integer? 0? If it is defined, state its value.

93. The time it takes a pendulum to swing back and forth in completing one cycle is called its *period*. If the length L of a pendulum is known, then its period may be determined by evaluating

$$2\pi\sqrt{\dfrac{L}{32}}.$$

Rationalize the denominator of this expression and then simplify.

94. Which of the following statements is incorrect? Explain.
    (i) $[(-2)^6]^{1/2} = (64)^{1/2} = 8$
    (ii) $[(-2)^6]^{1/2} = (-2)^{6(1/2)} = (-2)^3 = -8$

95. When the depth of water in a circular above-ground swimming pool is 4 ft, the pool holds 6032 gallons of water and the radius (in feet) of the pool may be found by computing

$$0.1825\left(\dfrac{6032}{\pi}\right)^{1/2}.$$

Compute the radius of the pool.

96. To rationalize the denominator of $2/\sqrt{3}$, a student squared both numerator and denominator and obtained $4/3$. Explain why this procedure is wrong.

1.4 The Complex Numbers

◆ Introductory Comments

We know that the square roots of -4 and -9 *cannot* be real numbers, since no real number squared is -4 and no real number squared is -9. One of the basic properties of a real number *a* is that its square is always *nonnegative*:

$$a^2 \geq 0$$

In the 17th century, the French mathematician René Descartes used the word "imaginary" to describe square roots like $\sqrt{-4}$ and $\sqrt{-9}$, and in the 18th century a Swiss mathematician, Leonhard Euler, defined the **imaginary unit *i*.**

◆ **Imaginary Unit *i***

$$i = \sqrt{-1} \quad \text{where} \quad i^2 = -1$$

Thus, if *a* is a real number and $a > 0$, then we can express the *principal square root of* $-a$ in term of *i*:

$$\sqrt{-a} = \sqrt{-1 \cdot a} = \sqrt{-1} \cdot \sqrt{a} = i\sqrt{a}.$$

◆ **Principal Square of a Negative Number**

If *a* is any real number and $a > 0$, then the **principal square root** of $-a$ is

$$\sqrt{-a} = i\sqrt{a}.$$

Thus, for $\sqrt{-4}$ and $\sqrt{-9}$, we have

$$\sqrt{-4} = i\sqrt{4} = 2i \quad \text{and} \quad \sqrt{-9} = i\sqrt{9} = 3i.$$

Note that the squares of $2i$ and $3i$ are -4 and -9, respectively:

$$(2i)^2 = 2^2 i^2 = 4(-1) = -4 \quad \text{and} \quad (3i)^2 = 3^2 i^2 = 9(-1) = -9.$$

In the 19th century, the German mathematician Carl Friedrich Gauss extended the real number system to a larger system that includes the real numbers as well as those numbers whose squares are negative. We refer to this larger system as the **complex number system**. Each number in this system is called a **complex number** and may be written in the form $a + bi$, where *a* and *b* are real numbers.

◆ **Complex Number in Standard Form**

Any number of the form

$$a + bi,$$

where *a* and *b* are real numbers and *i* is the imaginary unit, is called a **complex number in standard form.**

In the complex number $a + bi$, the real number *a* is the *real part* and the real number *b* is the *imaginary part* of the number. If $b = 0$, then $a + bi$

becomes $a + 0i = a$. A complex number of the form $a + 0i$ is a real number. If $b \neq 0$, then $a + bi$ is referred to as an **imaginary number**. Here are some examples of imaginary numbers:

$$2 + 3i \qquad \pi - 6i \qquad 5i \qquad i\sqrt{2}$$

The use of the phrase "imaginary number" indicates the initial uneasiness of mathematicians with nonreal numbers. Today, imaginary numbers often appear in the solutions of quadratic and polynomial equations (see Section 3.4 and Chapter 4) and in applications of electrical circuits and quantum mechanics.

◆ Operations with Complex Numbers

We can add, subtract, multiply, and divide complex numbers by applying the same commutative, associative, and distributive properties that we use with real numbers. Here are some illustrations:

$$\sqrt{-4} + \sqrt{-9} = 2i + 3i = (2 + 3)i = 5i$$
$$\sqrt{-4} - \sqrt{-9} = 2i - 3i = (2 - 3)i = -i$$
$$\sqrt{-4} \cdot \sqrt{-9} = (2i)(3i) = (2 \cdot 3)(i \cdot i) = 6i^2 = 6(-1) = -6$$
$$\frac{\sqrt{-4}}{\sqrt{-9}} = \frac{2i}{3i} = \frac{2}{3}$$

CAUTION The property of radicals $\sqrt{a}\sqrt{b} = \sqrt{ab}$ does not apply when both a and b are negative numbers. For example, to write

$$\sqrt{-4} \cdot \sqrt{-9} = \sqrt{(-4)(-9)} = \sqrt{36} = 6 \quad \text{is WRONG!}$$

When working with the principal square root of a negative number, be sure to apply $\sqrt{-a} = i\sqrt{a}$ before using any of the properties of radicals discussed in Section 1.3.

To find the *sum* of two complex numbers $a + bi$ and $c + di$, we add the real and imaginary parts of the numbers separately. Similarly, to find the *difference* of two complex numbers we subtract the real and imaginary parts of the numbers separately.

EXAMPLE 1 Perform the indicated operations and express each answer in standard form.

(a) $(6 - 3i) - (-4 + 7i)$ **(b)** $\left(3 + 2\sqrt{-3}\right) + \sqrt{-27}$

◀ Solution

(a) $(6 - 3i) - (-4 + 7i) = [6 - (-4)] + (-3 - 7)i$
$$= 10 - 10i$$

(b) $\left(3 + 2\sqrt{-3}\right) + \sqrt{-27} = \left(3 + 2i\sqrt{3}\right) + i\sqrt{27}$
$$= 3 + \left(2i\sqrt{3} + 3i\sqrt{3}\right)$$
$$= 3 + 5i\sqrt{3}$$

PROBLEM 1 Find the difference: $(3 + \sqrt{-25}) - (3 - \sqrt{-36})$

To find the *product* of two complex numbers $a + bi$ and $c + di$, we multiply each part of the first complex number by each part of the second complex number.

EXAMPLE 2 Perform the indicated operations and express each answer in standard form.

(a) $(3 + 4i)(5 - i)$ **(b)** $\sqrt{-2}(\sqrt{-18} + \sqrt{8})$

◀ **Solution**

(a) $(3 + 4i)(5 - i) = 15 - 3i + 20i - 4i^2$
$(3 + 4i)(5 - i) = 15 - 3i + 20i - 4i^2$
$= 15 + 17i - 4(-1)$
$= 19 + 17i$

(b) $\sqrt{-2}(\sqrt{-18} + \sqrt{8}) = i\sqrt{2}(i\sqrt{18} + \sqrt{8})$
$= i^2\sqrt{36} + i\sqrt{16}$
$= 6i^2 + 4i$
$= -6 + 4i$

PROBLEM 2 Find the product: $(1 + 3i)(2 + 5i)$

To find the *quotient* $\dfrac{a + bi}{c + di}$, we multiply numerator and denominator by the **complex conjugate** of the denominator, namely, $c - di$. The product of the complex conjugates is a nonnegative real number:

$(c + di)(c - di) = c^2 - cdi + cdi - d^2i^2 = c^2 + d^2$

　　　　　　　　　　　　　　　　　　　a *nonnegative* real number

The fact that the product of complex conjugates is a nonnegative real number enables us to express the quotient $\dfrac{a + bi}{c + di}$ as a complex number in standard form:

$$\dfrac{a + bi}{c + di} = \dfrac{(a + bi)(c - di)}{(c + di)(c - di)} = \dfrac{(ac + bd) + (bc - ad)i}{c^2 + d^2}$$

$$= \dfrac{ac + bd}{c^2 + d^2} + \dfrac{bc - ad}{c^2 + d^2}i$$

A complex number in standard form

SECTION 1.4 ♦ The Complex Numbers

EXAMPLE 3 Perform the indicated operations and express each answer in standard form.

(a) $\dfrac{3 + 5i}{1 - 5i}$ (b) $\dfrac{8}{-\sqrt{-25}}$

◀ Solution

a) $\dfrac{3 + 5i}{1 - 5i} = \dfrac{(3 + 5i)(1 + 5i)}{(1 - 5i)(1 + 5i)} = \dfrac{3 + 15i + 5i + 25i^2}{1 - 25i^2}$

The complex conjugate of $1 - 5i$

$= \dfrac{-22 + 20i}{26}$

$= \dfrac{-22}{26} + \dfrac{20}{26}i = -\dfrac{11}{13} + \dfrac{10}{13}i$

b) $\dfrac{8}{-\sqrt{-25}} = \dfrac{8}{-5i} = \dfrac{8(5i)}{-5i(5i)} = \dfrac{40i}{-25i^2} = \dfrac{40i}{25} = \dfrac{8}{5}i$

The complex conjugate of $-5i$

PROBLEM 3 Find the quotient: $\dfrac{1 + \sqrt{-9}}{1 - \sqrt{-9}}$

♦ Powers of i

When we raise i to successive positive integer powers, an interesting pattern develops:

$i^1 = i$
$i^2 = -1$
$i^3 = i^2 i = (-1)i = -i$
$i^4 = i^2 i^2 = (-1)(-1) = 1$

$i^5 = i^4 i = 1(i) = i$
$i^6 = i^4 i^2 = 1(-1) = -1$
$i^7 = i^4 i^3 = 1(-i) = -i$
$i^8 = i^4 i^4 = 1(1) = 1$

$i^9 = i^8 i = 1(i) = i$
$i^{10} = i^8 i^2 = 1(-1) = -1$
$i^{11} = i^8 i^3 = 1(-i) = -i$
$i^{12} = i^8 i^4 = 1(1) = 1$

Notice the values repeat in cycles of four according to the pattern $i, -1, -i, 1$. For higher powers of i, we use the fact that $i^4 = 1$ and apply the laws of exponents from Section 1.2. Here are some illustrations of higher powers:

$$i^{13} = (i^4)^3 i = (1)^3 i = i$$

$$i^{100} = (i^4)^{25} = (1)^{25} = 1$$

$$i^{-22} = \dfrac{1}{i^{22}} = \dfrac{1}{(i^4)^5 i^2} = \dfrac{1}{1^5(-1)} = -1$$

EXAMPLE 4 Find the indicated power and express each answer in standard form.

(a) $(\sqrt{-4})^5$ (b) $(1 + 2i)^2$

◆ Solution

(a) $(\sqrt{-4})^5 = (2i)^5 = 32i^5 = 32i^4 i = 32i$

(b) $(1 + 2i)^2 = (1 + 2i)(1 + 2i)$
$= 1 + 2i + 2i + 4i^2$
$= 1 + 4i + 4(-1)$
$= -3 + 4i$

PROBLEM 4 Find the indicated power: $(2 - \sqrt{-9})^2$

Exercises 1.4

Basic Skills

In Exercises 1–30, perform the indicated operations and express each answer in standard form.

1. $(4 + 3i) - (1 - 5i)$
2. $(3 - 2i) + (8 - 3i)$
3. $(9 - 3i) + (6 + 2i) + 4i$
4. $(12 + 6i) - (4 - 3i) + (1 + i)$
5. $8i(2 - 3i)$
6. $-4i(5 + 7i)$
7. $(5 + 2i)(9 + 3i)$
8. $(8 - 2i)(1 - i)$
9. $(\sqrt{2} - 2i)(\sqrt{2} + 2i)$
10. $(3 + i\sqrt{5})(3 - i\sqrt{5})$
11. $\dfrac{4 + 2i}{i}$
12. $\dfrac{1 - 4i}{3i}$
13. $\dfrac{6}{7 - 3i}$
14. $\dfrac{2i}{4i - 1}$
15. $\dfrac{1 - 2i}{1 + 2i}$
16. $\dfrac{4 - i}{2 - 3i}$
17. $\sqrt{-16} + \sqrt{-25}$
18. $\sqrt{-36} - \sqrt{-49}$
19. $\sqrt{3} \cdot \sqrt{-12} \cdot \sqrt{-9}$
20. $\dfrac{\sqrt{-125} \cdot \sqrt{36}}{\sqrt{-5}}$
21. $5\sqrt{-5} + \sqrt{-20} - \frac{4}{3}\sqrt{-45}$
22. $2\sqrt{-3} - 2\sqrt{-27} + 6\sqrt{-48}$
23. $(3 + 2\sqrt{-36}) - (5 - 3\sqrt{-49})$
24. $(8 - 4\sqrt{-9}) + (6 + 5\sqrt{-16})$
25. $\sqrt{-3}(\sqrt{-12} - \sqrt{-27})$
26. $\sqrt{2}(\sqrt{-18} - \sqrt{-50})$
27. $(3 - \sqrt{-4})(4 + 3\sqrt{-9})$
28. $(\sqrt{-25} + 1)(3\sqrt{-4} - 8)$
29. $\dfrac{\sqrt{-4}}{2 - \sqrt{-6}}$
30. $\dfrac{9}{1 + 2\sqrt{-3}}$

In Exercises 31–44, find the indicated power and express each answer in standard form.

31. $5i^{10}$
32. $13i^{36}$
33. $2i^{-5}$
34. $\dfrac{1}{6i^{37}}$
35. $(\sqrt{-9})^3$
36. $(\sqrt{-2})^4$
37. $(3i)^3(2i)^2$
38. $(2i)^4(-3i)^3$
39. $(3 - 2i)^2$
40. $(-1 + 2\sqrt{-25})^2$
41. $\dfrac{1}{(-4 - 5i)^2}$
42. $\dfrac{1}{(2 + \sqrt{-9})^2}$
43. $i + i^2 + i^3 + i^4$
44. $i^{-1} - i^{-2} + i^{-3} - i^{-4}$

◀ Critical Thinking

45. Show that $1 + 0i$ is the *multiplicative identity* for the complex number $a + bi$.

46. Find the *multiplicative inverse*, or *reciprocal*, of the complex number $a + bi$.

47. Evaluate the expression $x^2 - 2x + 2$ when (a) $x = 1 + i$ and (b) $x = 1 - i$.

48. Evaluate the expression $x^2 - 6x + 13$ when (a) $x = 3 - 2i$ and (b) $x = 3 + 2i$.

49. Evaluate the expression $x^2 - 4x + 5$ when (a) $x = 2 - i$ and (b) $x = 2 + i$.

50. Evaluate the expression $x^2 - 2x + 10$ when (a) $x = 1 + 3i$ and (b) $x = 1 - 3i$.

51. What are the two square roots of -9?

52. Show that the two square roots of i are

$$\frac{\sqrt{2} + i\sqrt{2}}{2} \quad \text{and} \quad \frac{-\sqrt{2} - i\sqrt{2}}{2}$$

53. On most calculators, you must enter nonnegative inputs when working with the *square root key*, $\boxed{\sqrt{}}$. (A negative input causes an error message to appear in the display window). Use your calculator in conjunction with the definition of the principal square root of a negative number to evaluate each radical expression to three significant digits. (See Appendix for a review of significant digits.)

(a) $\sqrt{-18}$ (b) $\sqrt{-506}$
(c) $\sqrt{-153.6}$ (d) $\sqrt{-0.8765}$

54. Evaluate the expression

$$\frac{-b \pm \sqrt{b^2 - 4ac}}{2a}$$

for the given values of a, b, and c. Express each answer in standard form.

(a) $a = 1$, $b = -2$, $c = 10$
(b) $a = 2$, $b = 6$, $c = 7$

55. The magnitude of the current (in amperes, A) flowing through the series circuit shown in the sketch may be found by first writing the expression

$$\frac{200}{150 + (188.5 - 76.4)i}$$

in standard form, and then computing $\sqrt{a^2 + b^2}$, where a is the real part and b is the imaginary part of a complex number. Determine the magnitude of the current in this circuit.

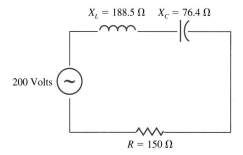

56. The magnitude of the total impedance (in ohms, Ω) for the parallel circuit shown in the sketch may be found by first writing the expression

$$\frac{1}{\dfrac{1}{150} + \dfrac{1}{188.5\,i} - \dfrac{1}{76.4\,i}}$$

in standard form, and then computing $\sqrt{a^2 + b^2}$, where a is the real part and b is the imaginary part of a complex number. Determine the magnitude of the total impedance for this circuit.

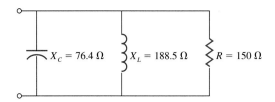

1.5 Polynomials and Special Products

◆ Introductory Comments

A collection of variables and constants formed by using addition, subtraction, multiplication, division, powers, or roots is called an **algebraic expression**.

The building block of algebraic expressions is the **term**—a constant, a variable, or a product or quotient of constants and variables. The algebraic expression

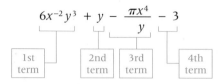

consists of four terms. If a term consists of the product of a real number and one or more variables, then the real number is called the **numerical coefficient** (or simply the *coefficient*) of the term. In the preceding algebraic expression the numerical coefficient of the first term is 6; of the second term, 1; of the third term, $-\pi$. The fourth term is called a *constant term* because it contains no variable.

Terms that differ only in their numerical coefficients are called **like terms**. If two or more like terms appear in an algebraic expression, we usually combine them into a single term by applying the distributive property and adding their numerical coefficients. The procedure is called *combining*, or *collecting*, like terms. In each of the following algebraic expressions, we collect like terms:

$$3x^2y + 5x^2y = (3+5)x^2y = 8x^2y$$

$$3y^{-1} + 4y^{-1} - y^{-1} = (3+4-1)y^{-1} = 6y^{-1}$$

$$3\sqrt{5x} + \sqrt{5x} = (3+1)\sqrt{5x} = 4\sqrt{5x}$$

A *polynomial* is an algebraic expression in which no variable appears in any denominator or in any radicand, and any variable that does appear is raised to a nonnegative integer power.

◀ **Definition of a Polynomial in x of Degree n**

> Any algebraic expression of the form
>
> $$a_nx^n + a_{n-1}x^{n-1} + a_{n-2}x^{n-2} + \cdots + a_2x^2 + a_1x + a_0$$
>
> where n is a nonnegative integer and $a_0, a_1, a_2, \ldots, a_n$ are real numbers with $a_n \neq 0$, is called a **polynomial in x of degree n**.

If, after like terms are combined, a polynomial consists of only *one* term, it is called a **monomial**; if it consists of *two* terms, it is a **binomial**; and if it consists of *three* terms, it is a **trinomial**. Each of the following algebraic expressions is an example of a polynomial in a single variable:

$$3x^3 \quad \text{(a monomial of degree 3)}$$

$$2y^6 - 7y \quad \text{(a binomial of degree 6)}$$

$$z^2 - 3z + 2 \quad \text{(a trinomial of degree 2)}$$

SECTION 1.5 ♦ Polynomials and Special Products

The **degree of a term** of a polynomial is the sum of the exponents in the term. The algebraic expression

$$5x^2y^4 + xy^3 - 7x^2 + 3$$

is a polynomial in two variables, x and y. The degree of the first term of this polynomial is $2 + 4 = 6$; of the second term, $1 + 3 = 4$; and of the third term, 2. The degree of the fourth term, which may be written as $3x^0$, is regarded as 0. The degree of a polynomial in more than one variable is the degree of the term with the *highest* degree in the polynomial. Hence, the degree of the polynomial $5x^2y^4 + xy^3 - 7x^2 + 3$ is 6.

♦ Addition and Subtraction of Polynomials

To *add* two polynomials, we use the commutative and associative properties to group like terms, and then apply the distributive property to combine the like terms. To *subtract* two polynomials, we use the fact that $a - b = a + (-b)$; that is, we change the sign of each term in the polynomial that comes after the subtraction sign, and then combine like terms.

EXAMPLE 1 Perform the indicated operation.

(a) $(3x + 5y) + (5x^2 - 2x + 7y)$

(b) $(x^3 + 4x^2y + xy^2) - (xy^2 - 2yx^2 + 2y^3)$

◀ Solution

(a) $(3x + 5y) + (5x^2 - 2x + 7y)$
$= 5x^2 + (3x - 2x) + (5y + 7y)$ Group like terms
$= 5x^2 + x + 12y$ Combine like terms

(b) $(x^3 + 4x^2y + xy^2) - (xy^2 - 2yx^2 + 2y^3)$
$= x^3 + 4x^2y + xy^2 \ominus xy^2 \oplus 2yx^2 \ominus 2y^3$ Change signs
$= x^3 + (4x^2y + 2yx^2) + (xy^2 - xy^2) - 2y^3$ Group like terms
$= x^3 + 6x^2y - 2y^3$ Combine like terms   ♦

PROBLEM 1 Perform the indicated operations:

$(5x^2 - 3x) + (5x - x^2) - (2x + 4x^2)$ ♦

♦ Multiplication of Polynomials

To multiply two polynomials, both of which do not contain just one term, we use the distributive property. First we multiply each term of the first polynomial by each term of the second polynomial, and then we collect like terms.

EXAMPLE 2 Perform the indicated operations.

(a) $2x^2(3 - 5x - x^3)$ (b) $(3x - 4y)(2x + 5y)$

(c) $(x + 2)(3x^3 - x^2 + 2x - 4)$

◆ Solution

(a) $2x^2(3 - 5x + x^3) = 6x^2 - 10x^3 + 2x^5$

(b) $(3x - 4y)(2x + 5y) = 6x^2 + 15xy - 8xy - 20y^2$
$= 6x^2 + 7xy - 20y^2$

(c) $(x + 2)(3x^3 - x^2 + 2x - 4)$

$= 3x^4 - x^3 + 2x^2 - 4x + 6x^3 - 2x^2 + 4x - 8$
$= 3x^4 + 5x^3 - 8$ ◆

The product in Example 2(c) may also be found by arranging the polynomials *vertically* and multiplying:

$$\begin{array}{r} 3x^3 - x^2 + 2x - 4 \\ x + 2 \\ \hline 3x^4 - x^3 + 2x^2 - 4x \\ 6x^3 - 2x^2 + 4x - 8 \\ \hline 3x^4 + 5x^3 \qquad\qquad - 8 \end{array}$$

Multiply each term of $(3x^3 - x^2 + 2x - 4)$ by x.

Multiply each term of $(3x^3 - x^2 + 2x - 4)$ by 2. Align like terms with the row above.

Add like terms in each row.

The advantage of the vertical method is that like terms are grouped at the same time the multiplication is performed.

PROBLEM 2 Use the vertical method to perform the indicated operation: $(m^3 - m^2 + 3m + 4)(m^2 + 2)$ ◆

◆ **Special Products**

Three *special products* occur frequently when we work with algebraic expressions.

Special Products

For algebraic expressions A and B,

1. **Product of a sum and difference:**
   $(A + B)(A - B) = A^2 - B^2$
2. **Square of a binomial:** $(A + B)^2 = A^2 + 2AB + B^2$
3. **Cube of a binomial:** $(A + B)^3 = A^3 + 3A^2B + 3AB^2 + B^3$

$(a-b)^2 = a^2 - 2ab + b^2$
$(a-b)^3 = a^3 + 3a^2b + 3ab^2 - b^3$

We can verify these special products by multiplying in the conventional manner:

1. $(A + B)(A - B) = A^2 - AB + AB - B^2$
   $= A^2 - B^2$

2. $(A + B)^2 = (A + B)(A + B) = A^2 + AB + AB + B^2$
   $= A^2 + 2AB + B^2$

3. $(A + B)^3 = (A + B)(A + B)^2$
   $= (A + B)(A^2 + 2AB + B^2)$
   $= A^3 + 2A^2B + AB^2 + A^2B + 2AB^2 + B^3$
   $= A^3 + 3A^2B + 3AB^2 + B^3$

The process of squaring or cubing a binomial is called *expanding* the binomial. To expand a binomial in which the power is greater than 3, the best method to use is the *binomial theorem*, which we will discuss in the last chapter of this text.

EXAMPLE 3 Perform the indicated operations:

(a) $(3x + 2)(3x - 2)$ (b) $(4x + 5)^2$ (c) $(3x - 2y)^3$

Solution

(a) Let $A = 3x$ and $B = 2$. Using

$$(A + B)(A - B) = A^2 - B^2,$$
we have $\quad (3x + 2)(3x - 2) = (3x)^2 - (2)^2$
$$= 9x^2 - 4$$

(b) Let $A = 4x$ and $B = 5$. Using

$$(A + B)^2 = A^2 + 2AB + B^2,$$
we have $\quad (4x + 5)^2 = (4x)^2 + 2(4x)(5) + (5)^2$
$$= 16x^2 + 40x + 25$$

(c) By the definition of subtraction, $(3x - 2y)^3 = [3x + (-2y)]^3$. Letting $A = 3x$ and $B = -2y$, we have

$(A + B)^3 = A^3 + 3A^2B + 3AB^2 + B^3$
$(3x - 2y)^3 = (3x)^3 + 3(3x)^2(-2y) + 3(3x)(-2y)^2 + (-2y)^3$
$= 27x^3 - 54x^2y + 36xy^2 - 8y^3$

PROBLEM 3 Perform the indicated operations.

(a) $(2x - 3)(2x + 3)$ 　　(b) $(3x - 5y)^2$ 　　(c) $(x^4 + 4)^3$

◆ Grouping Symbols and the Order of Operation

Grouping symbols such as *parentheses* ( ), *brackets* [ ], and *braces* { } are used to show that the terms contained within them represent a single quantity. Parentheses are commonly used for the *innermost* grouping symbols, then brackets, and then braces.

◀ **Order of Operations**

> To simplify an algebraic expression that contains more than one set of grouping symbols, we use the following **order of operations** to remove each set of grouping symbols, starting with the innermost set.
>
> 1. Raise expressions to powers.
> 2. Perform the multiplications and divisions in order from left to right.
> 3. Perform the additions and subtractions in order from left to right.
>
> Repeat the order of operations until all sets of grouping symbols have been removed.

EXAMPLE 4 Perform the indicated operations.

(a) $2x[(8x + 2) - 2(2x + 1)^2]$

(b) $6x^2 + \{-3[x - (2x - 1)(x + 3)]\}$

◀ **Solution**

(a) [Start with this power: $(2x+1)^2$]

$2x[(8x + 2) - 2(2x + 1)^2] = 2x[(8x + 2) - 2(4x^2 + 4x + 1)]$
$= 2x[(8x + 2) - 8x^2 - 8x - 2]$
$= 2x[-8x^2]$
$= -16x^3$

(b) [Start with this product: $(2x-1)(x+3)$]

$6x^2 + \{-3[x - (2x - 1)(x + 3)]\} = 6x^2 + \{-3[x - (2x^2 + 5x - 3)]\}$

Retain this set of parentheses and subtract all terms of this product.

$= 6x^2 + \{-3[x \ominus 2x^2 \ominus 5x \oplus 3]\}$
$= 6x^2 + \{-3x + 6x^2 + 15x - 9\}$
$= 12x^2 + 12x - 9$

PROBLEM 4 Perform the indicated operations:
$x^2 - [y(x + 2y) - (x + y)(2y - x)]$

◆ Operations with Nonpolynomial Expressions

The procedures for performing operations with polynomials can be extended to algebraic expressions that are not polynomials. For example, to find the product of the nonpolynomial expressions $(3 + 2x^{-1})$ and $(2 - 3x^{-1})$, we multiply as we would two binomials:

$$(3 + 2x^{-1})(2 - 3x^{-1}) = 6 - 9x^{-1} + 4x^{-1} - 6x^{-2}$$
$$= 6 - 5x^{-1} - 6x^{-2}$$

EXAMPLE 5 Perform the indicated operations.

(a) $(4\sqrt{xy} - 3y^{-3})(4\sqrt{xy} + 3y^{-3})$ (b) $(x^{1/2} + 5x^{-1/2})^2$

◀ Solution

(a) Using the special product of a sum and difference with $A = 4\sqrt{xy}$ and $B = 3y^{-3}$, we proceed as follows:

$$(A - B)(A + B) = A^2 - B^2$$
$$(4\sqrt{xy} - 3y^{-3})(4\sqrt{xy} + 3y^{-3}) = (4\sqrt{xy})^2 - (3y^{-3})^2$$
$$= 16xy - 9y^{-6}$$

(b) Using the special product for the square of a binomial with $A = x^{1/2}$ and $B = 5x^{-1/2}$, we proceed as follows:

$$(A + B)^2 = A^2 + 2AB + B^2$$
$$(x^{1/2} + 5x^{-1/2})^2 = (x^{1/2})^2 + 2(x^{1/2})(5x^{-1/2}) + (5x^{-1/2})^2$$
$$= x + 10 + 25x^{-1}$$

PROBLEM 5 Use the special product for the cube of a binomial to find the indicated power: $(x^{-4} + 4)^3$

Exercises 1.5

Basic Skills

In Exercises 1–10, write each algebraic expression as a single term.

1. $5x^2 - 3x^2 + 4x^2 - 7x^2$
2. $2y^2 + 4y^2 - 5y^2 - 3y^2$
3. $x^2y + 3x^2y - 5yx^2 - 4yx^2$
4. $3mn^3 - 2mn^3 + n^3m - 3n^3m + mn^3$
5. $\frac{3}{4}x^{-1} + \frac{1}{2}x^{-1} - \frac{7}{2}x^{-1} + \frac{1}{4}x^{-1}$
6. $\frac{2}{3}x^{1/2} + \frac{4}{3}x^{1/2} + \frac{3}{4}x^{1/2} + \frac{1}{4}x^{1/2}$
7. $6\sqrt{3y} - 3\sqrt{3y} + 7\sqrt{3y}$

8. $1.3\sqrt[3]{x} + 4.6\sqrt[3]{x} - 8.6\sqrt[3]{x}$

9. $\sqrt{10x} + 3\sqrt{40x} - 2\sqrt{90x}$

10. $3\sqrt{64p} - \sqrt{100p} - 5\sqrt{49p}$

In Exercises 11–20, determine whether the expression is a polynomial. If it is a polynomial, give its degree. If it is not a polynomial, explain why.

11. $\pi x^2 y^4$

12. $\dfrac{18x^3 y^2}{z}$

13. $3x - y^{-1}$

14. $7x^2 y^3 - 4z^3$

15. $\sqrt{3}x^2 - 9x + \pi$

16. $5x^{-2}y + z + \sqrt{7}$

17. $9x^3 y^4 - 3xy^7 + 9xy - 3$

18. $3x^{1/2} y^3 - 7x + 4y - 8\pi$

19. $3x^2 + y^3 + z^4 + \sqrt{x^2 + y^2}$

20. $\dfrac{7x}{3} + \dfrac{9x^2}{7} + 2x^3 + 8x^4$

In Exercises 21–60, perform the indicated operations.

21. $(3m^2 + 2m - 9) + (4m^2 - 7m)$

22. $(5x^2 - 3x + 4) + (4 - 3x - 5x^2)$

23. $(x^2 + 3xy) - (x^2 + 2yx) - (x^2 - xy)$

24. $(-xy^2 + 4x^2 y) - (2yx^2 - 4y^2 x)$

25. $-4m^2(8m^3 n + 2m^2 + m - 1)$

26. $(4x^2 y^3 + xy + y^2)5x^2 y^3$

27. $(x^2 - 9y)(3x + 7)$

28. $(2mn + m)(n - m)$

29. $(3x - y)(4x + y)$

30. $(9 - 2x^2)(5 - 3x^2)$

31. $(3x + y)(3x - y)$

32. $(2x + 5)(2x - 5)$

33. $(n - 2)(2n^2 + 5n - 2)$

34. $(p^2 + 4p - 2)(p^2 - p - 3)$

35. $(x + 3)(x^2 - 3x + 9)$

36. $(9x^2 - 6x + 4)(3x + 2)$

37. $(4x^2 + 2x + 1)(2x - 1)$

38. $(2x - 5)(4x^2 + 10x + 25)$

39. $(x^2 + x\sqrt{3} + 3)(x\sqrt{3} - 3)$

40. $(\sqrt{2} + 3a)(\sqrt{2} - 5a)$

41. $(m + 6)^2$

42. $(y - 8)^2$

43. $(2a - 3b^2)^2$

44. $(3x^3 + 4y)^2$

45. $(3x^3 + 5)^2$

46. $(6a - 5b^2)^2$

47. $(x + 3)^3$

48. $(2 - y)^3$

49. $(4x - 3y^2)^3$

50. $(2m^3 + 5)^3$

51. $n - 2(2n^2 + 5n - 2)$

52. $p^2 + 4p - 2(p^2 - p - 3)$

53. $-m^3 - m[8m - 2m(m + 4)]$

54. $a^2 - [a(2a + b) - (2a - b)(a + b)]$

55. $8x(2x - 1)^2$

56. $1 - 3y(3y + 5)^2$

57. $[(a + 4)(a - 4)]^2$

58. $[(4a - 1)(4a + 1)]^2$

59. $-3x[(x - 2) - 2(3x - 1)^3]$

60. $-x^2 - \{-[x + (3x + 2)(3x - 2)]\}$

In Exercises 61–68, perform the indicated operations.

61. $(3x^{-2} + y) + \left(\dfrac{1}{x^2} - y\right)$

62. $(5t - t^{1/2}) - (2t - \sqrt{t})$

63. $(\sqrt{x} + \sqrt{3y})(\sqrt{x} - \sqrt{3y})$

64. $(\sqrt{x} + 3)(\sqrt{x} - 5)$

65. $(2\sqrt{x} - 7y)^2$

66. $(2x^{-1/3} - x)^3$

67. $7x^{1/3} - (x^{1/3} + 4)(x^{1/3} + 2)$

68. $15x + 2\sqrt{x} - (3\sqrt{x} - 2)(5\sqrt{x} + 4)$

Critical Thinking

69. Suppose one polynomial P has degree m and another polynomial Q has degree n, where $m > n$. What is the degree of the sum $P + Q$? of the difference $P - Q$? of the product PQ? of the power P^2?

70. Develop a formula for the expansion of $(A + B)^4$. Then use this formula to expand each binomial:
    (a) $(2x + 1)^4$
    (b) $(x^2 - 2y)^4$

71. The area of a rectangle is the product of its length and width. Find a polynomial that describes the area of the shaded portion of the rectangle shown in the figure.

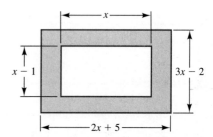

72. The volume of a rectangular solid is the product of its length, width, and height. Find a polynomial that de-

scribes the volume of the rectangular solid shown in the figure.

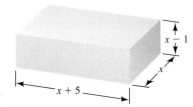

73. The area of a circle is the product of π and its radius squared. Find a polynomial that describes the area of the shaded portion of the circle shown in the figure.

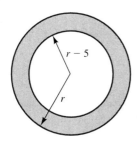

74. The volume of a sphere is the product of $4\pi/3$ and its radius cubed. Find a polynomial that describes the volume of the sphere in the figure.

75. Most calculators are programmed to follow the order of operations discussed in this section. Evaluate each expression by following the order of operations, and then compare the answer to the one you obtain by pressing the keys on your calculator in the order written in the expression. Do your answers agree?
   (a) $28 + 12 \div 2^2 - 4 \cdot 3$
   (b) $(28 + 12) \div (2^2 - 4 \cdot 3)$

76. Use the power key, $\boxed{\wedge}$ or $\boxed{y^x}$, on your calculator to evaluate $(1.01)^3$. Then evaluate $(1 + .01)^3$ by using the special product $(A + B)^3$. Do your answers agree?

77. Each teacher in a school system receives a 6% raise. Find a polynomial that describes a teacher's new salary if x is the previous salary.

78. Suppose a sum of $1000 earns simple interest at the rate of 9% per year, and a sum of $500 earns simple interest at the rate of 6% per year. Find a polynomial that describes the total amount to which these sums will accumulate after t years.

79. A steel beam 36.6 feet long and fixed at both ends carries a uniformly distributed load of 25,000 pounds (lb) as shown in the figure. The deflection of the beam (in inches) at any distance x from one of the fixed ends is given by

$$4.72 \times 10^{-6} x^2 (36.6 - x)^2.$$

(a) Write this expression as a trinomial, recording the numerical coefficients in scientific notation rounded to three significant digits. (See Appendix for a review of significant digits.)

(b) The maximum deflection for this beam occurs at midspan (18.3 ft). Determine the maximum deflection to the nearest hundredth of an inch.

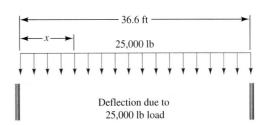

80. A compressor powers the pneumatic tube system used at drive-through bank lanes to carry transactions between a customer and the teller. The horsepower of the compressor designed to carry a tube with a 4-inch diameter a distance of 60 feet is given by

$$0.032 p^{3/2} (0.1815 + 0.00348 p)$$

where p is the air pressure in the system, in pounds per square inch (psi). Find the product, recording the numerical coefficients in scientific notation rounded to three significant digits.

1.6 Factoring Techniques

◆ Introductory Comments

The process of expressing a polynomial as a product of other polynomials is called **factoring** the polynomial. Factoring provides the foundation for simplifying algebraic fractions and for solving several types of equations and inequalities.

In this section our concern is with factoring a polynomial into polynomials that have integer coefficients. This is called *factoring a polynomial over the set of integers*. If a polynomial cannot be written as the product of two polynomials of positive degree that have integer coefficients, it is said to be **prime** over the set of integers. Here are some examples of polynomials that are prime over the set of integers:

$$2x - 3 \qquad x^2 + 1 \qquad x^2 + x - 3 \qquad 3xy^2$$

When a polynomial is written as the product of prime factors, it is said to be **factored completely**. To factor a polynomial completely, we first *factor out* common factors by applying the distributive property from right to left:

$$ab + ac = a(b + c)$$

- a is common to each term
- a is factored out

The process of factoring out common factors is called **common term factoring**. Here are some examples of common term factoring:

$$6x - 9 = 3(2x - 3)$$
$$5x^2(x + 1) + 5(x + 1) = 5(x + 1)(x^2 + 1)$$
$$3x^3y^2 + 3x^2y^2 - 9xy^2 = 3xy^2(x^2 + x - 3)$$

Each of these polynomials is *factored completely over the set of integers*. We can check the factoring process by finding the product of the factors. Monomial factors, such as $3xy^2$, are always considered prime.

◆ Factoring by Grouping Terms

From our knowledge of multiplication of polynomials, we know that

$$(a + b)(c + d) = (c + d)(a + b) = ac + ad + bc + bd.$$

Now, consider factoring $ac + ad + bc + bd$. No factor is common to all four terms; however, by grouping the first two terms and the last two terms, we are able to factor by common term factoring:

SECTION 1.6 ◆ Factoring Techniques

$$ac + ad + bc + bd = (ac + ad) + (bc + bd) \quad \text{Group terms}$$
$$= a(c + d) + b(c + d) \quad \text{Factor out common factors from each group}$$
$$= (a + b)(c + d) \quad \text{Factor out the common binomial } (c + d)$$

We refer to this process as **factoring by grouping terms**. The key to factoring by grouping terms is to develop a common binomial that can be factored out of each group. As illustrated in Example 1(b), this may require factoring out a negative factor from one of the groups.

EXAMPLE 1 Factor completely over the set of integers.

(a) $x^3 - 3x^2 + 5x - 15$ (b) $9 + 2xy - 6x - 3y$

◆ Solution

(a) $x^3 - 3x^2 + 5x - 15 = (x^3 - 3x^2) + (5x - 15)$
$$= x^2(x - 3) + 5(x - 3)$$
$$= (x - 3)(x^2 + 5)$$

(b) $9 + 2xy - 6x - 3y = (9 - 6x) + (2xy - 3y)$
$$= 3(3 - 2x) - y(3 - 2x)$$

Factor out $-y$ to obtain the common binomial $(3 - 2x)$.

$$= (3 - 2x)(3 - y)$$

PROBLEM 1 Terms may usually be grouped in more than one way. Group the first and last terms and the second and third terms of $9 + 2xy - 6x - 3y$, and then factor. You should obtain the same result as in Example 1(b). ◆

◆ Factoring Trinomials with Binomial Factors

Binomials such as $(3x + 4)$ and $(x + 2)$ are called **like binomials**, because their first terms are like terms and their second terms are also like terms. The product of two like binomials is a trinomial:

$$(3x + 4)(x + 2) = 3x^2 + 6x + 4x + 8 = 3x^2 + 10x + 8$$

Now, consider factoring the trinomial $3x^2 + 10x + 8$. Because the first term's coefficient is 3, the middle term's coefficient is *positive*, and the last term's coefficient is *positive*, we begin by writing the like binomials as follows:

inner
$(3x + \blacksquare)(x + \blacksquare)$
outer

The two boxes must be filled in with numbers whose product is 8 such that the sum of the inner and outer terms is $10x$. We have four possibilities:

$(3x + 8)(x + 1)$ sum of inner and outer terms is $11x$
$(3x + 1)(x + 8)$ sum of inner and outer terms is $25x$
$(3x + 2)(x + 4)$ sum of inner and outer terms is $14x$
$(3x + 4)(x + 2)$ sum of inner and outer terms is $10x$

Hence, $(3x + 4)(x + 2)$ is the correct choice, and we conclude that

$$3x^2 + 10x + 8 = (3x + 4)(x + 2).$$

CAUTION Not every trinomial can be factored over the set of integers as the product of two like binomials. For example, the only possible factors of $x^2 + 2x + 3$ are $(x + 1)(x + 3)$. However, since the sum of the inner and outer terms of this product is *not* $2x$, we conclude that $x^2 + 2x + 3$ is prime over the set of integers.

EXAMPLE 2 Factor completely over the set of integers.

(a) $x^2 - 9x + 14$ (b) $x^6 + x^3 - 6$
(c) $9x^2 + 24xy + 16y^2$ (d) $6x^3y - 3x^2y - 18xy$

◆ Solution

(a) For $x^2 - 9x + 14$, we begin with the like binomials

$$(x - \Box)(x - \Box)$$

The two boxes must be filled in with numbers whose product is 14 such that the sum of the inner and outer terms is $-9x$. Checking the possibilities, we find that

$$x^2 - 9x + 14 = (x - 2)(x - 7).$$

(b) For $x^6 + x^3 - 6$, we begin with the like binomials.

$$(x^3 + \Box)(x^3 - \Box)$$

The two boxes must be filled in with numbers whose product is 6 such that the sum of the inner and outer terms is x^3. Checking the possibilities, we find that

$$x^6 + x^3 - 6 = (x^3 + 3)(x^3 - 2).$$

(c) For $9x^2 + 24xy + 16y^2$, we begin with the like binomials

$$(3x + \Box y)(3x + \Box y) \quad \text{or} \quad (9x + \Box y)(x + \Box y)$$

The two boxes must be filled in with numbers whose product is 16 such that the sum of the inner and outer terms is $24xy$. Checking the possibilities, we find that

$$9x^2 + 24xy + 16y^2 = (3x + 4y)(3x + 4y) = (3x + 4y)^2$$

(d) For $6x^3y - 3x^2y - 18xy$, we begin by factoring out the common factor $3xy$:

$$6x^3y - 3x^2y - 18xy = 3xy(2x^2 - x - 6)$$

Now, for $2x^2 - x - 6$, we have the like binomials

$$(2x - \blacksquare)(x + \blacksquare) \quad \text{or} \quad (2x + \blacksquare)(x - \blacksquare)$$

The two boxes must be filled in with numbers whose product is 6 such that the sum of the inner and outer terms is $-x$. Checking the possibilities, we find that

$$2x^2 - x - 6 = (2x + 3)(x - 2)$$

Hence,

$$6x^3y - 3x^2y - 18xy = 3xy(2x + 3)(x - 2) \quad \blacklozenge$$

Referring to Example 2(c), since $9x^2 + 24xy + 16y^2$ may be expressed as the *perfect square* $(3x + 4y)^2$, we refer to $9x^2 + 24xy + 16y^2$ as a **perfect square trinomial**.

PROBLEM 2 Show that $x^2 + 10x + 25$ is a perfect square trinomial. $\quad \blacklozenge$

◆ Factoring Formulas

From our knowledge of multiplication of polynomials, we have the following three products:

1. $(A + B)(A - B) = A^2 - AB + AB - B^2$
$\qquad\qquad\qquad\quad = A^2 - B^2$
2. $(A + B)(A^2 - AB + B^2) = A^3 - A^2B + AB^2 + A^2B - AB^2 + B^3$
$\qquad\qquad\qquad\qquad\qquad\; = A^3 + B^3$
3. $(A - B)(A^2 + AB + B^2) = A^3 + A^2B + AB^2 - A^2B - AB^2 - B^3$
$\qquad\qquad\qquad\qquad\qquad\; = A^3 - B^3$

These results give us three *factoring formulas*.

◄ **Factoring Formulas**

> 1. **Difference of two squares:** $A^2 - B^2 = (A + B)(A - B)$
> 2. **Sum of two cubes:** $A^3 + B^3 = (A + B)(A^2 - AB + B^2)$
> 3. **Difference of two cubes:**
> $A^3 - B^3 = (A - B)(A^2 + AB + B^2)$

EXAMPLE 3 Factor completely over the set of integers.

(a) $9x^2 - 4y^2$
(b) $x^6 + 8$
(c) $x^6 + x^3 - 2$
(d) $2x^4 - 2$

◆ Solution

(a) Applying the difference of two squares formula, we obtain

$$9x^2 - 4y^2 = (3x)^2 - (2y)^2$$
$$= (3x + 2y)(3x - 2y)$$

(b) Applying the sum of two cubes formula, we obtain

$$x^6 + 8 = (x^2)^3 + (2)^3$$
$$= (x^2 + 2)[(x^2)^2 - (x^2)(2) + (2)^2]$$
$$= (x^2 + 2)(x^4 - 2x^2 + 4)$$

(c) First, we factor the trinomial into two like binomials, and then we factor one binomial as the difference of two cubes:

$$x^6 + x^3 - 2 = (x^3 + 2)(x^3 - 1)$$
$$= (x^3 + 2)(x - 1)(x^2 + x + 1)$$

(d) First, we factor out the common factor 2, and then we factor the difference of two squares twice:

$$2x^4 - 2 = 2(x^4 - 1)$$
$$= 2(x^2 + 1)(x^2 - 1)$$
$$= 2(x^2 + 1)(x + 1)(x - 1)$$

The sum of squares $x^2 + 1$ is prime over the set of integers. ◆

PROBLEM 3 Factor completely over the set of integers.

(a) $(x + y)^2 - 25y^4$
(b) $64 - x^3$ ◆

Exercises 1.6

Basic Skills

In Exercises 1–74, factor completely over the set of integers.

1. $3a^2b - ab$
2. $48x^2y - 36xy^2$
3. $6x^3y - 9x^2y + 3xy^2$
4. $6x^2y^3 - 18x^3y^4 + 12x^3y^3$
5. $5(x - 1) - x(x - 1)$
6. $5n(2n - 7) + n(2n - 7)$
7. $6m(m + 3)^3 + m(m + 3)^3$
8. $3p(p + 2)^2 + 9p^2(p + 2)^2$
9. $ac + ad + bc + bd$
10. $xy + 3x + 2y + 6$
11. $x^3 - 3x^2 + 2x - 6$
12. $m^4 - 2m^3 - 3m + 6$
13. $6 - 3x - 2y + xy$

14. $15 + 6y - 20x - 8xy$
15. $15 + 2x^2y - 3x^2 - 10y$
16. $3x^3 - 20y - 15x^2 + 4xy$
17. $a^2 + 3a - 10$
18. $y^2 + 14y + 45$
19. $x^2 - 11xy + 18y^2$
20. $p^2 - pq - 12q^2$
21. $24 + 14x + x^2$
22. $24 - 2m + m^2$
23. $x^4 + 3x^2 - 18$
24. $x^6 + 2x^3 - 63$
25. $2x^2 + 5x + 3$
26. $3x^2 + 8x + 4$
27. $3a^2 + 11ab - 20b^2$
28. $5x^2 + 18xy - 8y^2$
29. $4x^2 - 11x - 3$
30. $6p^2 + 23p - 4$
31. $15 - 16t + 4t^2$
32. $12 - 17t + 6t^2$
33. $12x^2 - 23xy + 10y^2$
34. $18x^2 + 23xy - 6y^2$
35. $2n^4 + 7n^2 + 5$
36. $3y^4 - 13y^2 + 10$
37. $6x^6 - 17x^3y^2 + 10y^4$
38. $10x^8 - x^4y - 3y^2$
39. $8y^{12} + 13y^6 - 6$
40. $12a^8 + 8a^4 - 15$
41. $x^2 - 81$
42. $x^2 - 36$
43. $25x^2 - y^2$
44. $x^2 - 16y^4$
45. $4(x^2 + 5)^2 - 25x^2$
46. $49 - (m + n)^2$
47. $t^3 + 8$
48. $125 + a^3$
49. $27x^3 + 8y^6$
50. $y^3 - 27$
51. $8 - 125n^3$
52. $64n^3 - 1$
53. $t^2 + 8t + 16$
54. $25 - 10x + x^2$
55. $4a^4 - 12a^2b^2 + 9b^4$
56. $9p^6 - 6p^3 + 1$
57. $4x^3y - 28x^2y + 48xy$
58. $3x^4y^2 - 6x^3y^3 - 45x^2y^4$
59. $6x^3 - 8x^2 - 8x$
60. $12x^3 + 3x^2 - 9x$
61. $5a^3 - 45a$
62. $2m^3n - 32mn$
63. $16x^4y + 2xy$
64. $3x^5y^2 - 81x^2y^2$
65. $2n^4 - 32$
66. $81x - 16x^5$
67. $4x^4 - 13x^2 + 9$
68. $9x^4 - 37x^2y^2 + 4y^2$
69. $3a^7 + 6a^4 - 9a$
70. $2x^6y + 26x^3y + 80y$
71. $8x^3y + 8x^2y - 18xy - 18y$
72. $6x^4 + 3x^3 - 24x^2 - 12x$
73. $x^6 - 1$
74. $64 - x^6$

Critical Thinking

In Exercises 75–78, group three terms and factor completely over the integers.

75. $x^2 - 4x + 4 - y^2$
76. $x^2 + 12x + 36 - 9y^2$
77. $4x^2 + 12x - 4y^2 + 9$
78. $9x^2 - 16y^2 - 30x + 25$

In Exercises 79–82, use the factoring formulas given in this section to factor each polynomial over the set of real numbers. [Hint: To factor a polynomial over the set of real numbers, factor the polynomial into polynomials whose coefficients are real numbers.]

79. $x^2 - 10$
80. $3t^2 - 4y^2$
81. $a^3 + 5b^3$
82. $8x^3 - 2$

In Exercises 83–86, use the difference of squares formula to factor each polynomial over the set of complex numbers. [Hint: To factor a polynomial over the set of complex numbers, factor the polynomial into polynomials whose coefficients are complex numbers.]

83. $x^2 + 25$
84. $y^2 + 18$
85. $t^4 - 36$
86. $t^4 - 256$

87. The total amount S of aluminum used to make a beer can is given by the formula

$$S = 2\pi rht + 2\pi r^2 t$$

where r is the radius, h is the height, and t is the thickness of the beer can. Factor the expression on the right-hand side of this formula.

88. Finding the number of households (in units of 1000) that use a certain product after it has been on the market for n months involves working with the expression

$$32n - \tfrac{4}{3}n^2.$$

Factor this expression so that the coefficient of n in the binomial factor is 1.

89. Finding the response time of mice to a certain stimulus involves working with the expression

$$10.24 + 9.24A^2 - A^4$$

where A is the age (in months) of the subject receiving the stimulus. Given that one factor of this expression is $1 + A^2$, find two other factors with rational coefficients.

90. The velocity V of sound in seawater may be approximated by the formula

$$V = 4800 + 14t - 0.12t^2$$

where t is the temperature of the seawater. Factor the expression on the right-hand side of this formula so that the coefficient of t in one of the binomial factors is 1.

1.7 Algebraic Fractions

◆ Introductory Comments

The quotient of two algebraic expressions is called an **algebraic fraction**. If the numerator and denominator of an algebraic fraction are polynomials, then the algebraic fraction is referred to as a **rational expression**. Since division by zero is undefined, we must restrict all the variables in a rational expression to those values that give a nonzero denominator. Thus, when working with a rational expression such as

$$\frac{4y^2 + 5y + 3}{2y - 6},$$

we assume that its denominator $2y - 6 \neq 0$, which implies that $y \neq 3$.

We use the **fundamental property of fractions** to build up an algebraic fraction to higher terms or to reduce an algebraic fraction to lower terms.

◀ Fundamental Property of Fractions

For the fraction $\frac{a}{b}$ (provided $b \neq 0$), and any real number k (except $k = 0$),

$$\frac{a}{b} = \frac{ak}{bk}.$$

Here are examples of building up a rational expression and reducing a rational expression:

$$\frac{2x}{3} = \frac{2x \cdot 5x}{3 \cdot 5x} = \frac{10x^2}{15x}, \, x \neq 0 \qquad \frac{10x^2}{15x} = \frac{2x \cdot 5x}{3 \cdot 5x} = \frac{2x}{3}, \, x \neq 0$$

building up a rational expression to higher terms

reducing a rational expression to lower terms

Two algebraic fractions are said to be **equivalent fractions** if they have the same numerical value for all numbers for which the fractions are defined. The rational expressions $\frac{2x}{3}$ and $\frac{10x^2}{15x}$ are equivalent fractions, provided $x \neq 0$.

◆ Reducing a Rational Expression to Lowest Terms

We say that a rational expression is *reduced to lowest terms* when its numerator and denominator contain no common factor other than ± 1 and we have expressed $-a/(-b)$ as a/b. To reduce a rational expression to lowest terms, we factor its numerator and denominator completely over the set of integers and cancel any common factors. Here is an example of reducing a rational expression to lowest terms:

$$\frac{8x^3y + 12x^2y}{8x^3y} = \frac{\overset{1}{4x^2}\,\overset{}{y}(2x+3)}{\underset{2}{8x^3y}} = \frac{2x+3}{2x}, \quad x \neq y \neq 0$$

CAUTION   We can cancel common *factors* of the numerator and denominator, but never common *terms*. To reduce the fraction as

$$\frac{\overset{1}{\cancel{8x^3y}} + 12x^2y}{\underset{1}{\cancel{8x^3y}}} \quad \text{is WRONG!}$$

EXAMPLE 1   Reduce each rational expression to lowest terms.

(a) $\dfrac{9-x^2}{x^2-2x-15}$ (b) $\dfrac{1-x}{x^3-1}$

◆ **Solution**

(a) $\dfrac{9-x^2}{x^2-2x-15} = \dfrac{(3+x)(3-x)}{(x+3)(x-5)}$ Factor completely

$\phantom{(a) \dfrac{9-x^2}{x^2-2x-15}} = \dfrac{3-x}{x-5}, \quad x \neq -3$ Reduce to lowest terms

(b) $\dfrac{1-x}{x^3-1} = \dfrac{1-x}{(x-1)(x^2+x+1)}$ Factor completely

$\phantom{(b) \dfrac{1-x}{x^3-1}} = \dfrac{-(x-1)}{(x-1)(x^2+x+1)}$ Use $a-b = -(b-a)$

$\phantom{(b) \dfrac{1-x}{x^3-1}} = \dfrac{-1}{x^2+x+1}, \quad x \neq 1$ Reduce to lowest terms ◆

PROBLEM 1   Reduce the rational expression to lowest terms:

$$\frac{16-x^2}{2x^3 - 5x^2 - 12x}$$ ◆

◆ Multiplying and Dividing Rational Expressions

To multiply and divide rational expressions, we use the *multiplication and division properties of fractions.*

Multiplication and Division Properties of Fractions

For fractions $\frac{a}{b}$ and $\frac{c}{d}$ (provided $b \neq 0$ and $d \neq 0$),

Property	Example
Multiplication Property	
$\frac{a}{b} \cdot \frac{c}{d} = \frac{ac}{bd}$	$\frac{2x}{3} \cdot \frac{4x}{5} = \frac{2x \cdot 4x}{3 \cdot 5} = \frac{8x^2}{15}$
Division Property	
$\frac{a}{b} \div \frac{c}{d} = \frac{a}{b} \cdot \frac{d}{c} = \frac{ad}{bc}$	$\frac{2x}{3} \div \frac{4x}{5} = \frac{2x}{3} \cdot \frac{5}{4x} = \frac{\overset{1}{\cancel{2x}} \cdot 5}{3 \cdot \underset{2}{\cancel{4x}}}$
(provided $c \neq 0$)	$= \frac{5}{6},\ x \neq 0$

Before applying these properties to a rational expression, we factor its numerator and denominator completely over the set of integers. In this way, common factors can be cancelled first, yielding a product that is reduced to lowest terms.

EXAMPLE 2 Perform the indicated operations and reduce to lowest terms.

(a) $\dfrac{x-6}{2x^2} \cdot \dfrac{x^2-x}{36-x^2}$

(b) $\dfrac{3x-3}{2x^2+x-6} \div \dfrac{x^2-1}{4x^2-12x+9}$

Solution

(a) $\dfrac{x-6}{2x^2} \cdot \dfrac{x^2-x}{36-x^2} = \dfrac{(x-6) \cdot x(x-1)}{2x^2 \cdot (6+x)(6-x)}$ Factor completely and apply the multiplication property

$= \dfrac{-(6-x) \cdot x(x-1)}{2x^2 \cdot (6+x)(6-x)}$ Use $a - b = -(b - a)$

$= \dfrac{-(x-1)}{2x(6+x)},\ x \neq 0, 6$ Reduce to lowest terms

(b) $\dfrac{3x-3}{2x^2+x-6} \div \dfrac{x^2-1}{4x^2-12x+9}$

$= \dfrac{3x-3}{2x^2+x-6} \cdot \dfrac{4x^2-12x+9}{x^2-1}$ Apply the division property

$= \dfrac{3(x-1) \cdot (2x-3)^2}{(2x-3)(x+2) \cdot (x+1)(x-1)}$ Factor completely and apply the multiplication property

$= \dfrac{3(2x-3)}{(x+2)(x+1)},\ x \neq 1, \tfrac{3}{2}$ Reduce to lowest terms

SECTION 1.7 ◆ Algebraic Fractions

PROBLEM 2 Find the quotient: $\dfrac{x^2}{2x^2 - 5x + 2} \div \dfrac{4x}{4 - x^2}$

◆ **Adding and Subtracting Rational Expressions**

To add and subtract rational expressions, we use the *addition and subtraction properties of fractions*.

◆ **Addition and Subtraction Properties of Fractions**

For fractions $\dfrac{a}{b}$ and $\dfrac{c}{b}$ (provided $b \neq 0$),

Property	Example
Addition Property $\dfrac{a}{b} + \dfrac{c}{b} = \dfrac{a+c}{b}$	$\dfrac{2x}{3} + \dfrac{5x}{3} = \dfrac{2x+5x}{3} = \dfrac{7x}{3}$
Subtraction Property $\dfrac{a}{b} - \dfrac{c}{b} = \dfrac{a-c}{b}$	$\dfrac{2x}{3} - \dfrac{5x}{3} = \dfrac{2x-5x}{3} = \dfrac{-3x}{3} = -x$

After applying these properties, always look for common factors to see if the sum or difference can be reduced to lower terms.

EXAMPLE 3 Perform the indicated operations.

(a) $\dfrac{x+y}{xy} - \dfrac{y-6x}{xy}$ (b) $\dfrac{2x+4}{x-5} - \dfrac{x+3}{5-x}$

◆ Solution

(a) $\dfrac{x+y}{xy} - \dfrac{y-6x}{xy} = \dfrac{(x+y) - (y-6x)}{xy}$ Apply the subtraction property. (Be sure to use parentheses.)

$= \dfrac{x + y \ominus y \oplus 6x}{xy}$ Remove parentheses

$= \dfrac{7\cancel{x}}{\cancel{x}y} = \dfrac{7}{y}, \quad x \neq 0$ Combine like terms and reduce to lowest terms

(b) $\dfrac{2x+4}{x-5} - \dfrac{x+3}{5-x} = \dfrac{2x+4}{x-5} - \dfrac{x+3}{-(x-5)}$ Use $a - b = -(b - a)$

$= \dfrac{2x+4}{x-5} + \dfrac{x+3}{x-5}$ Use $a/(-b) = -(a/b)$

$= \dfrac{(2x+4) + (x+3)}{x-5}$ Apply the addition property

$= \dfrac{3x+7}{x-5}$ Combine like terms

PROBLEM 3 Find the difference: $\dfrac{2x^2 - 9}{2x^2 + 11x + 5} - \dfrac{x^2 + 16}{2x^2 + 11x + 5}$

To add or subtract algebraic fractions with *unlike denominators*, we first find their **least common denominator**, which is denoted as **LCD**. The following two-step procedure may be used to find the LCD.

◀ **Procedure for Determining the LCD**

Example: $\dfrac{x-1}{x^2 + x} - \dfrac{2x}{x^2 + 2x + 1}$

1. Factor completely each denominator, and use exponential notation to represent repeated prime factors that occur in any one of the denominators.

   1. $\dfrac{x-1}{\boxed{x}(x+1)} - \dfrac{2x}{\boxed{(x+1)^2}}$

2. The product of each *different prime factor* to the *highest power* it occurs in any one of the denominators is the LCD.

   2. LCD is $x(x+1)^2$.

After determining the LCD, we apply the fundamental property of fractions and build up the fractions to equivalent fractions, each having the LCD as its new denominator. Once the denominators are the same, we use the addition and subtraction properties of fractions to find the sum or difference of the fractions.

EXAMPLE 4 Perform the indicated operations.

(a) $\dfrac{x-1}{x^2 + x} - \dfrac{2x}{x^2 + 2x + 1}$ (b) $\dfrac{8}{4 - y^2} + \dfrac{y}{2 + y} - \dfrac{1}{y - 2}$

SECTION 1.7 ◆ Algebraic Fractions

◀ Solution

(a) $\dfrac{x-1}{x^2+x} - \dfrac{2x}{x^2+2x+1} = \dfrac{x-1}{x(x+1)} - \dfrac{2x}{(x+1)^2}$

⎫ Factor the denominators. The LCD is $x(x+1)^2$.

$= \dfrac{(x-1)\cdot(x+1)}{x(x+1)\cdot(x+1)} - \dfrac{2x\cdot x}{(x+1)^2\cdot x}$

$= \dfrac{x^2-1}{x(x+1)^2} - \dfrac{2x^2}{x(x+1)^2}$

$= \dfrac{-x^2-1}{x(x+1)^2}$

(b) $\dfrac{8}{4-y^2} + \dfrac{y}{2+y} - \dfrac{1}{y-2}$

$= \dfrac{8}{(2+y)(2-y)} + \dfrac{y}{2+y} \oplus \dfrac{1}{2-y}$

⎫ Factoring and rearranging the last fraction yields an LCD of $(2+y)(2-y)$.

$= \dfrac{8}{(2+y)(2-y)} + \dfrac{y\cdot(2-y)}{(2+y)\cdot(2-y)} + \dfrac{1\cdot(2+y)}{(2-y)\cdot(2+y)}$

$= \dfrac{8+(2y-y^2)+(2+y)}{(2+y)(2-y)}$

$= \dfrac{10+3y-y^2}{(2+y)(2-y)}$

We now reduce this answer to lowest terms:

$\dfrac{10+3y-y^2}{(2+y)(2-y)} = \dfrac{(5-y)(2+y)}{(2+y)(2-y)} = \dfrac{5-y}{2-y}, \quad y \neq -2$ ◆

PROBLEM 4 Find the difference: $\dfrac{x+1}{x^2-2x+1} - \dfrac{1}{x-1}$ ◆

◆ The Algebra of Calculus

We conclude this section by looking at some of the algebraic support that is needed in calculus. We look at three ideas:

1. Rationalizing the numerator of an algebraic fraction.
2. Simplifying a complex fraction.
3. Writing an expression containing negative exponents as an algebraic fraction.

Note: In calculus, the ratio of the change in the variable y to the change in the variable x is denoted by $\Delta y/\Delta x$ (read "delta y to delta x"). In this text, we use the logo $\boxed{\Delta y / \Delta x}$ in the exercise sets to identify problems that are calculus-related.

To rationalize the numerator of an algebraic fraction that contains the form $\sqrt{a} - \sqrt{b}$, we apply the fundamental property of fractions and multiply both numerator and denominator by $\sqrt{a} + \sqrt{b}$. In computing the product $(\sqrt{a} - \sqrt{b})(\sqrt{a} + \sqrt{b})$, we use the fact that $(A - B)(A + B) = A^2 - B^2$:

$$(\sqrt{a} - \sqrt{b})(\sqrt{a} + \sqrt{b}) = (\sqrt{a})^2 - (\sqrt{b})^2 = a - b$$

Each of the factors $\sqrt{a} - \sqrt{b}$ and $\sqrt{a} + \sqrt{b}$ is called the **conjugate** of the other.

EXAMPLE 5 Rationalize the numerator of $\dfrac{\sqrt{x + h} - \sqrt{x}}{h}$ and simplify the result.

Solution The conjugate of the numerator is $\sqrt{x + h} + \sqrt{x}$. Thus, we multiply both numerator and denominator by this conjugate, and then simplify:

$$\dfrac{\sqrt{x + h} - \sqrt{x}}{h}$$

$$= \dfrac{(\sqrt{x + h} - \sqrt{x}) \cdot (\sqrt{x + h} + \sqrt{x})}{h \cdot (\sqrt{x + h} + \sqrt{x})} \quad \text{Apply the fundamental property of fractions}$$

$$= \dfrac{(x + h) - (x)}{h \cdot (\sqrt{x + h} + \sqrt{x})} \quad \text{Multiply}$$

$$= \dfrac{\not{h}}{\not{h} \cdot (\sqrt{x + h} + \sqrt{x})} \quad \text{Simplify the numerator}$$

$$= \dfrac{1}{\sqrt{x + h} + \sqrt{x}} \quad \text{Reduce to lowest terms}$$

PROBLEM 5 Rationalize the numerator and simplify:

$$\dfrac{\sqrt{(x + h) + 1} - \sqrt{x + 1}}{h}$$

An algebraic fraction that contains fractions in its numerator, denominator, or both its numerator and denominator is called a **complex fraction**. One method of simplifying a complex fraction is to first find the LCD for all fractions appearing in its numerator and denominator. After this is accomplished, we multiply both numerator and denominator of the complex fraction by this LCD and then simplify the resulting expression. The procedure is illustrated in the next example.

EXAMPLE 6 Simplify the complex fraction $\dfrac{\dfrac{3}{x + h} - \dfrac{3}{x}}{h}$.

SECTION 1.7 ◆ Algebraic Fractions

◆ **Solution** The LCD for $(x + h)$ and x is $x(x + h)$. Thus, we multiply numerator and denominator of this complex fraction by $x(x + h)$ and then simplify the resulting expression as follows:

$$\frac{\dfrac{3}{x+h} - \dfrac{3}{x}}{h} = \frac{\left(\dfrac{3}{x+h} - \dfrac{3}{x}\right) \cdot [x(x+h)]}{h \cdot [x(x+h)]} \quad \text{Apply the fundamental property of fractions}$$

$$= \frac{3x - 3(x+h)}{h \cdot [x(x+h)]} \quad \text{Multiply and reduce}$$

$$= \frac{-3h}{h \cdot [x(x+h)]} \quad \text{Simplify the numerator}$$

$$= \frac{-3}{x(x+h)} \quad \text{Reduce to lowest terms}$$

◆

PROBLEM 6 As an alternate method of simplifying the complex fraction in Example 6, we can perform the indicated subtraction in the numerator and then divide this result by h. Rework Example 6 using this procedure. Which method to you prefer? ◆

Algebraic expressions containing negative exponents, such as

$$x[-2(1 - 3x)^{-3}(-3)] + (1 - 3x)^{-2},$$

develop frequently in calculus through a process called differentiation. One method of simplifying such expressions is to factor out the common factor with the *smaller exponent*. The procedure is illustrated in the next example.

EXAMPLE 7 Simplify the expression
$x[-2(1 - 3x)^{-3}(-3)] + (1 - 3x)^{-2}$.

◆ **Solution** First we rewrite $(1 - 3x)^{-2}$ as $(1 - 3x)^{-3}(1 - 3x)^{1}$, and then we factor out the common factor $(1 - 3x)^{-3}$ from each term:

$$6x(1 - 3x)^{-3} + (1 - 3x)^{-2}$$
$$= 6x(1 - 3x)^{-3} + (1 - 3x)^{-3}(1 - 3x)^{1}$$
$$= (1 - 3x)^{-3}[6x + (1 - 3x)^{1}]$$
$$= (1 - 3x)^{-3}(3x + 1)$$
$$= \frac{3x + 1}{(1 - 3x)^{3}}$$

◆

PROBLEM 7 As an alternate method for simplifying the expression in Example 7, we use $a^{-n} = 1/a^n$ to form a pair of algebraic fractions, and then add the fractions by finding their LCD. Rework Example 7 using this procedure. Which method do you prefer? ◆

Exercises 1.7

Basic Skills

In Exercises 1–12, reduce the rational expression to lowest terms.

1. $\dfrac{5xy}{5xy - 10x^2y}$

2. $\dfrac{2x - 4x^2}{2x}$

3. $\dfrac{4m - 4n}{12n - 12m}$

4. $\dfrac{9x - 27y}{9y - 3x}$

5. $\dfrac{2x^2 + 11x + 5}{x^2 + 4x - 5}$

6. $\dfrac{16b^2 - a^2}{a^2 + ab - 12b^2}$

7. $\dfrac{n^2 - 5n - 6}{36 - n^2}$

8. $\dfrac{4x^2 - 5x - 6}{8 - x^3}$

9. $\dfrac{x^3 - x}{x^4 + 3x^3 - x^2 - 3x}$

10. $\dfrac{k^4 - 1}{k^8 - 1}$

11. $\dfrac{9 - (x + y)^2}{(x - 3)^2 - y^2}$

12. $\dfrac{x^2 - 2xy + y^2}{x^3 - x^2y - xy^2 + y^3}$

In Exercises 13–38, perform the indicated operations and reduce to lowest terms.

13. $\dfrac{x^2 + 2x + 1}{2xy} \cdot \dfrac{6xy}{x^2 - 1}$

14. $\dfrac{a}{(a - b)^2} \cdot \dfrac{a^2 - b^2}{a^2}$

15. $\dfrac{4 - n^2}{(n - 1)^2} \cdot \dfrac{n^2 + 3n - 4}{n^2 + 2n - 8}$

16. $\dfrac{(x + 2)^3}{x^3 + 8} \cdot \dfrac{x^3 - 2x^2 + 4x}{x^2 + 4x + 4}$

17. $\dfrac{12x^2}{x^2 + 2xy + y^2} \div \dfrac{6x}{x + y}$

18. $\dfrac{m + 3}{5m} \div \dfrac{m^2 - 9}{10m^3}$

19. $\dfrac{3a^3 - 81b^3}{9b^2 - a^2} \div (a^2 + 3ab + 9b^2)$

20. $\dfrac{9 - (x - y)^2}{3xy^2} \div \dfrac{x - y - 3}{3xy}$

21. $\dfrac{3n^2 + 7n + 3}{4n - 1} + \dfrac{5 - n^2}{1 - 4n}$

22. $\dfrac{6x^2 - 12x}{x^2 + 2x + 4} + \dfrac{(x - 2)^3}{x^2 + 2x + 4}$

23. $\dfrac{(x + 1)^2}{3x} - \dfrac{2x + 1}{3x}$

24. $\dfrac{b - a}{2a - b} - \dfrac{a - b}{b - 2a}$

25. $\dfrac{3}{m^2 - 9} + \dfrac{1}{m + 3}$

26. $\dfrac{x + 2}{x - 2} + \dfrac{x - 2}{x + 2}$

27. $\dfrac{1}{(x - 1)^3} + \dfrac{1}{1 - 2x + x^2}$

28. $\dfrac{x + 9}{x^2 - 6x + 9} + \dfrac{2}{x^2 - 9}$

29. $\dfrac{2 - x}{x^2 + 2x} - \dfrac{x}{x^2 + 4x + 4}$

30. $\dfrac{(x - 4)^2}{x + 4} - \dfrac{(x + 4)^2}{x - 4}$

31. $\dfrac{1}{a^2 + a} - \dfrac{1}{a - a^2}$

32. $\dfrac{1}{t - 1} - \dfrac{t^2 + t}{t^3 - 1}$

33. $\dfrac{a + 1}{a^2 + a + 1} + \dfrac{1}{1 - a} + \dfrac{a^2 + 2a + 3}{a^3 - 1}$

34. $\dfrac{x - 3}{3 + x} + \dfrac{x + 3}{3 - x} + \dfrac{36}{x^2 - 9}$

35. $\left(\dfrac{x + y}{x - y} - \dfrac{x - y}{x + y}\right) \div \left(\dfrac{x^2 + y^2}{x^2 - y^2} - \dfrac{x^2 - y^2}{x^2 + y^2}\right)$

36. $\left(x + \dfrac{4}{x}\right)\left(\dfrac{x^2}{4} - 1\right) \div \left(1 - \dfrac{16}{x^4}\right)$

37. $\left(\dfrac{b}{ac} + \dfrac{2}{a} + \dfrac{c}{ab} - \dfrac{a}{bc}\right)\left(1 + \dfrac{b + c}{a - b - c}\right)$

38. $\left(\dfrac{n^2}{25n^2 - 4}\right)\left(5n - 8 - \dfrac{4}{n}\right)\left(5n + 8 - \dfrac{4}{n}\right)$

In Exercises 39–44, rationalize the numerator and reduce to lowest terms.

39. $\dfrac{3 - \sqrt{t + 3}}{t - 6}$

40. $\dfrac{x - \sqrt{x^2 - 1}}{x + \sqrt{x^2 - 1}}$

41. $\dfrac{\sqrt{2(x + h)} - \sqrt{2x}}{h}$

42. $\dfrac{\sqrt{2 + (x + h)} - \sqrt{2 + x}}{h}$

43. $\dfrac{\sqrt{(x + h)^2 + 1} - \sqrt{x^2 + 1}}{h}$

44. $\dfrac{\sqrt{3(x + h) - 1} - \sqrt{3x - 1}}{h}$

In Exercises 45–50, simplify each complex fraction.

45. $\dfrac{\dfrac{1}{a} + 1}{\dfrac{1}{a^2} - 1}$

46. $\dfrac{1 - \dfrac{1}{4n^2}}{2 - \dfrac{1}{n}}$

47. $\dfrac{\dfrac{1}{x+h} - \dfrac{1}{x}}{h}$

48. $\dfrac{\dfrac{2}{(x+h)+1} - \dfrac{2}{x+1}}{h}$

49. $\dfrac{\dfrac{1}{(x+h)^2} - \dfrac{1}{x^2}}{h}$

50. $\dfrac{\dfrac{5}{\sqrt{x+h}} - \dfrac{5}{\sqrt{x}}}{h}$

In Exercises 51–60, simplify each expression.

51. $2x[3(x+3)^2] + (x+3)^3(2)$

52. $4x^3[5(1-2x)^4(-2)] + (1-2x)^5(12x^2)$

53. $t[-3(2-3t)^{-4}(-3)] + (2-3t)^{-3}$

54. $x[-1(3-2x)^{-2}(-2)] + (3-2x)^{-1}$

55. $\tfrac{4}{5}n(1-4n)^{-6/5} + (1-4n)^{-1/5}$

56. $3x[-\tfrac{1}{3}(3-x)^{-4/3}(-1)] + 3(3-x)^{-1/3}$

57. $-t(3-4t)^{-3/4} + (3-4t)^{1/4}$

58. $\tfrac{3}{4}y^2(y^2+1)^{-2/3} - (y^2+1)^{1/3}$

59. $\dfrac{6t^{4/3} - (t^2+1)t^{-2/3}}{t^{2/3}}$

60. $\dfrac{(x^2+4)^{1/2} - x^2(x^2+4)^{-1/2}}{x^2+4}$

Critical Thinking

61. If we reduce $\dfrac{x-1}{x^2-1}$ to lowest terms, we obtain $\dfrac{1}{x+1}$. Are these two rational expressions equal for all real values of x? Explain.

62. Rationalize the numerator of

$$\dfrac{\sqrt[3]{x+h} - \sqrt[3]{x}}{h}$$

by using the rationalizing factor

$$\sqrt[3]{(x+h)^2} + \sqrt[3]{x+h} \cdot \sqrt[3]{x} + \sqrt[3]{x^2}.$$

63. The sum or difference of a polynomial and a rational expression is called a *mixed algebraic expression*. Use the addition and subtraction properties for fractions (reading from right to left) to write each rational expression as a mixed algebraic expression.

    (a) $\dfrac{y^2 + 7y - 5}{y}$ (b) $\dfrac{14 - (t+7)^2}{t+7}$

64. Use the addition and subtraction properties for fractions (reading from right to left) to write each rational expression as a sum or difference of simpler rational expressions, each reduced to lowest terms.

    (a) $\dfrac{12xy^2 - 7x^2y^3 + 6x - 9}{3x^2y}$

    (b) $\dfrac{3a(x+y)^2 + (x+y)^3 - 2(x+y)}{2(x+y)^2}$

65. Rationalize the denominator and simplify.

    (a) $\dfrac{1}{x - \sqrt{x^2 - 1}}$ (b) $\dfrac{a^2}{\sqrt{a^2 + b^2} + b}$

66. A single lens with focal length 11 cm projects an image of an object onto a screen, as shown in the sketch. The distance (in cm) from the lens to the screen is given by

$$\dfrac{1}{\dfrac{1}{11} - \dfrac{1}{12}}.$$

Evaluate this expression by (a) simplifying the complex fraction and (b) by using the reciprocal key, $\boxed{1/x}$ or $\boxed{x^{-1}}$, on your calculator.

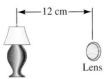

67. If two batteries are connected in parallel, as shown in the figure, and have the same internal resistance r and the same voltage E, then the current i supplied to the external load R is given by the formula

$$i = \dfrac{E}{R + \dfrac{r}{2}}.$$

(*Exercise 67 continues*)

Simplify the complex fraction that appears on the right-hand side of this formula.

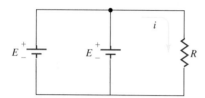

68. If a payment of R dollars is made at the end of each time period for n periods into an annuity that earns interest at the rate of r per period, then the present value A of the annuity is given by the formula

$$A = R\left[\frac{1 - (1 + r)^{-n}}{r}\right].$$

Rewrite the expression on the right-hand side of this formula as a single fraction that contains no negative exponent.

69. Finding the number of crimes committed in a certain part of the city involves working with the expression

$$\frac{60}{2 + \sqrt{n}} - n$$

where n is the number of police that are assigned to the area. Write this expression as a single fraction with a rationalized denominator.

70. The number of items that a machine can produce during an 8-hour day is $t^2 + 16t$, where t is the number of hours that the machine operates. Suppose the total cost to produce these items is $8t + 128$. Find a simplified algebraic expression that represents the *average cost* of production.

1.8 Generating Equivalent Equations and Inequalities

◆ Introductory Comments

An **equation** is a statement declaring that two algebraic expressions are equal. Two examples of equations in one variable are

$$3x + 4x = 7x \quad \text{and} \quad 2x + 1 = 7.$$

The equation $3x + 4x = 7x$ is true no matter what value we choose to replace x. An equation that is true for all permissible values of the variable that it involves is called an **identity**. The equation $2x + 1 = 7$ is true only when we replace x with 3 and is false for all other replacements. An equation that is true for some values of the variable it involves but false for others (or is never true for any value of the variable), is called a **conditional equation**. Any value of the variable that makes an equation a true statement is a **root** of, or **solution** to, the equation. Thus, we say 3 is a root of (or solution to) the equation $2x + 1 = 7$, since 3 is a number that *satisfies* this equation.

An **inequality** is a statement declaring that one algebraic expression is *less than* ($<$), *greater than* ($>$), *less than or equal to* ($\leq$), or *greater than or equal to* ($\geq$) another algebraic expression. Two examples of an inequality in one variable are

$$x + 1 > x \quad \text{and} \quad 2x \geq 6.$$

The inequality $x + 1 > x$ is true no matter what real number we choose to replace x. An inequality that is true for all permissible values of the variable that it involves is called an **absolute inequality**. The inequality $2x \geq 6$ is true only when x is replaced by real numbers greater than or equal to 3. An inequality that is true for some values of the variable it involves but false for others (or is never true for any value of the variable), is called a **conditional inequality**. We refer to values of the variable that make an inequality a true statement as the **solution set** of the inequality. In this text, we use *interval notation* (discussed in Section 1.1) to describe the solution set of an inequality. Hence, we describe the solution set of the inequality $2x \geq 6$ by writing $[3, \infty)$.

◆ Generating Equivalent Equations

Equations that have the same solutions are said to be **equivalent equations**. We can generate equivalent equations from a given equation by performing any of the following manipulations.

◀ **Generating Equivalent Equations**

Performing any of the following manipulations on an equation generates an equivalent equation.

Procedure	Example
1. Simplify algebraic expressions that appear on either side.	$9 = 3x - 5x - 1$ is equivalent to $9 = -2x - 1$
2. Add (or subtract) the same quantity to both sides.	$9 = -2x - 1$ is equivalent to $10 = -2x$
3. Multiply (or divide) both sides by the same *nonzero* quantity.	$10 = -2x$ is equivalent to $-5 = x$
4. Interchange the left-hand and right-hand sides.	$-5 = x$ is equivalent to $x = -5$

To **solve** an equation means to find *all* values of the variable that make the equation become a true statement. The usual strategy in solving an equation is to transform the given equation into an equivalent one whose solutions are apparent.

EXAMPLE 1 Solve each equation.

(a) $2y - 11 = 5y - 5$

(b) $\dfrac{2x + 3}{5} - \dfrac{x - 1}{15} = \dfrac{3x - 1}{3}$

(c) $|4t - 3| - 3 = 2$

(d) $5 - 8x^2 = -11$

◆ **Solution**

(a) To solve the equation $2y - 11 = 5y - 5$, we proceed as follows:

$$2y - 11 = 5y - 5$$

$$-11 = 3y - 5 \qquad \text{Subtract } 2y \text{ from both sides}$$

$$-6 = 3y \qquad \text{Add 5 to both sides}$$

$$y = -2 \qquad \text{Divide both sides by 3 and interchange the sides}$$

Since the solution to the last equation is obviously -2, the solution to the original equation must also be -2. By replacing y with -2 in the original equation, we can verify that -2 is a solution.

(b) For equations involving fractions, we eliminate the fractions by multiplying both sides of the equation by the least common denominator (LCD). The LCD for the denominators 5, 15, and 3 is 15. Thus, we multiply both sides of this equation by 15 and cancel the denominators as follows:

$$\frac{2x+3}{5} - \frac{x-1}{15} = \frac{3x-1}{3}$$

$$15\left(\frac{2x+3}{5} - \frac{x-1}{15}\right) = 15\left(\frac{3x-1}{3}\right) \qquad \text{Multiply both sides by 15}$$

$$3(2x+3) - (x-1) = 5(3x-1) \qquad \text{Simplify by canceling denominators}$$

Remove the parentheses

$$6x + 9 \ominus x \oplus 1 = 15x - 5 \qquad \text{Apply the distributive property}$$

$$5x + 10 = 15x - 5 \qquad \text{Collect like terms on the left-hand side}$$

$$10 = 10x - 5 \qquad \text{Subtract } 5x \text{ from both sides}$$

$$15 = 10x \qquad \text{Add 5 to both sides}$$

$$x = \frac{3}{2} \qquad \text{Divide both sides by 10, interchange sides, and reduce}$$

Thus, the solution of the original equation is also $\frac{3}{2}$.

(c) By adding 3 to both sides of the equation $|4t - 3| - 3 = 2$, we obtain the equivalent equation $|4t - 3| = 5$. By the definition of absolute value (Section 1.1), the equation $|4t - 3| = 5$ is equivalent to the following pair of equations:

$$\begin{array}{ll} 4t - 3 = 5 \qquad \text{or} & -(4t - 3) = 5 \\ 4t = 8 & -4t + 3 = 5 \\ t = 2 & -4t = 2 \\ & t = -\frac{1}{2} \end{array}$$

Thus, the solutions of the original equation are 2 or $-\frac{1}{2}$.

(d) We begin by solving the equation $5 - 8x^2 = -11$ for x^2;

$$5 - 8x^2 = -11$$

$$-8x^2 = -16 \quad \text{Subtract 5 from both sides}$$

$$x^2 = 2 \quad \text{Divide both sides by } -8$$

The solutions of $x^2 = 2$ are the numbers whose squares are 2, namely, the square roots of 2. Hence, we *extract the square roots* from this equation and write the solutions $\sqrt{2}$ and $-\sqrt{2}$, or more compactly, $\pm\sqrt{2}$. ◆

Note: In general, the solutions of the equation $x^2 = a$ are $\pm\sqrt{a}$. Each solution is a real number if $a > 0$, an imaginary number if $a < 0$, or zero if $a = 0$.

PROBLEM 1 Try to solve the equation $|4t - 3| + 3 = 2$. Explain why this equation has no solution. ◆

◆ Generating Equivalent Inequalities

Inequalities that have the same solution set are said to be **equivalent inequalities**. We can generate equivalent inequalities from a given inequality by performing any of the following manipulations.

◀ **Generating Equivalent Inequalities**

Performing any of the following manipulations on an inequality generates an equivalent inequality.

Procedure	Example
1. Simplify algebraic expressions that appear on either side.	$9 < 3x - 5x - 1$ is equivalent to $9 < -2x - 1$
2. Add (or subtract) the same quantity to both sides.	$9 < -2x - 1$ is equivalent to $10 < -2x$
3. Multiply (or divide) both sides by the same *positive* quantity. Multiply (or divide) both sides by the same *negative* quantity and *reverse the direction of the inequality symbol.*	$10 < -2x$ is equivalent to $-5 > x$
4. Interchange the left-hand and right-hand sides and *reverse the direction of the inequality symbol.*	$-5 > x$ is equivalent to $x < -5$

To **solve** an inequality means to find *all* values of the variable that make the inequality become a true statement. The usual strategy in solving an inequality is to transform the given inequality into an equivalent one whose solution set is apparent.

EXAMPLE 2 Solve each inequality. Use interval notation to describe the solution set.

(a) $5(y + 1) < 2(4 - 3y) + 2y$

(b) $\dfrac{2x}{3} - \dfrac{6x - 1}{6} \leq -\dfrac{1}{2}$

(c) $|2x - 3| < 5$

(d) $|2 - t| - 3 \geq 1$

◆ **Solution**

(a) To solve this inequality, we proceed as follows:

$$5(y + 1) < 2(4 - 3y) + 2y$$

$5y + 5 < 8 - 4y$ **Simplify each side**

$9y + 5 < 8$ **Add 4y to both sides**

$9y < 3$ **Subtract 5 from both sides**

$y < \dfrac{1}{3}$ **Divide both sides by 9 and reduce**

Since the solution set of this last inequality is $(-\infty, \frac{1}{3})$, the solution set of the original inequality is also $(-\infty, \frac{1}{3})$.

(b) To solve this inequality, we begin by eliminating fractions as follows:

$$\dfrac{2x}{3} - \dfrac{6x - 1}{6} \leq -\dfrac{1}{2}$$

$4x - (6x - 1) \leq -3$ **Multiply both sides by 6, the LCD**

$-2x + 1 \leq -3$ **Simplify the left-hand side**

$-2x \leq -4$ **Subtract 1 from both sides**

$x \geq 2$ **Divide both sides by -2 and reverse the direction of the inequality symbol**

Since the solution set of this last inequality is $[2, \infty)$, the solution set of the original inequality is also $[2, \infty)$.

(c) From our discussion of absolute value in Section 1.1, we know that the inequality $|2x - 3| < 5$ can be interpreted as the set of all real numbers that are *less than* 5 units from 0. Thus, $|2x - 3| < 5$ is equivalent to the following double inequality:

$-5 < 2x - 3 < 5$

$-2 < 2x < 8$ **Add 3 to each member**

$-1 < x < 4$ **Divide each member by 2**

Hence, the solution set of $|2x - 3| < 5$ is $(-1, 4)$.

(d) Adding 3 to both sides of the inequality $|2 - t| - 3 \geq 1$ gives us the equivalent inequality $|2 - t| \geq 4$. Now, from our discussion of absolute value in Section 1.1, we know that the inequality $|2 - t| \geq 4$ can be interpreted as the set of all real numbers that are *greater than or equal to* 4 units from 0. Thus, $|2 - t| \geq 4$ is equivalent to the following pair of inequalities:

$$2 - t \leq -4 \quad \text{or} \quad 2 - t \geq 4$$
$$-t \leq -6 \quad\quad\quad\quad -t \geq 2$$
$$t \geq 6 \quad\quad\quad\quad t \leq -2$$

Hence, the solution set of the original inequality consists of all real numbers belonging to either of the intervals $(-\infty, -2]$ or $[6, \infty)$. Hence, the solution set of $|2 - t| - 3 \geq 1$ is $(-\infty, -2] \cup [6, \infty)$. ◆

PROBLEM 2 Solve the inequality $\dfrac{y + 1}{2} - \dfrac{5(y - 1)}{8} > 1$. ◆

◆ Literal Equations and Inequalities

An equation (or inequality) that contains letters other than the variable for which we wish to solve is called a **literal equation** (or **literal inequality**). Here are examples of a literal equation and a literal inequality in which we solve for x:

Literal equation:

$$ax + b = c$$
$$ax = c - b$$
$$x = \frac{c - b}{a}, \quad a \neq 0$$

Literal inequality:

$$ax + b < c$$
$$ax < c - b$$
$$x < \frac{c - b}{a}, \quad a > 0$$
$$\text{or} \quad x > \frac{c - b}{a}, \quad a < 0$$

To solve some literal equations (or inequalities) in which the variable appears *more than once*, we use the rules for generating equivalent equations (or inequalities) and apply the following three steps:

1. *Group* all the variable terms for which we wish to solve on one side of the equation (or inequality).
2. *Factor out* the variable for which we wish to solve.
3. *Divide* each side of the equation (or inequality) by the expression being multiplied by the variable.

The procedure is illustrated in the next example.

EXAMPLE 3 Solve the equation $ax + b = cx + d$ for x.

Solution To solve this literal equation for x, we proceed as follows:

$$ax + b = cx + d$$
$$ax - cx = d - b \quad\quad \textit{Group the x terms on one side}$$
$$x(a - c) = d - b \quad\quad \textit{Factor out the variable x}$$
$$x = \frac{d - b}{a - c} \quad\quad \textit{Divide by } (a - c), \textit{ provided } a - c \neq 0$$

◆

PROBLEM 3 In Example 3, we grouped the *x* terms on the left-hand side of the equation. Rework Example 3 by grouping the *x* terms on the right-hand side of the equation and show that the answer is the same as that given in Example 3. ◆

◆ **Application: Common Formulas**

A **formula** is a mathematical or scientific rule in the form of a literal equation that describes a special relationship between two or more variables. Refer to the formulas listed in the following box as you need them to solve applied problems throughout this text.

◀ **Common Formulas**

Uniform Motion

$$d = rt$$

d is distance
r is rate of speed
t is time

Simple Interest

$$A = P(1 + rt)$$

A is amount accrued
P is amount invested
r is rate of interest
t is time

Temperature

$$F = \tfrac{9}{5}C + 32$$

F is degrees Fahrenheit
C is degrees Celsius

Pythagorean Theorem

$$c^2 = a^2 + b^2$$

a and *b* are lengths of legs
c is length of hypotenuse

Geometry (*A* is area, *P* is perimeter, *C* is circumference, *V* is volume, and *S* is total surface area)

Rectangle

$$A = lw$$
$$P = 2l + 2w$$

Triangle

$$A = \tfrac{1}{2}bh$$

Circle

$$A = \pi r^2$$
$$C = 2\pi r$$

Trapezoid

$$A = \tfrac{1}{2}h(b_1 + b_2)$$

SECTION 1.8 ♦ Generating Equivalent Equations and Inequalities

Rectangular Solid

$V = lwh$
$S = 2lw + 2wh + 2lh$

Cylinder

$V = \pi r^2 h$
$S = 2\pi r(r + h)$

Cone

$V = \frac{1}{3}\pi r^2 h$
$S = \pi r(r + s)$

Sphere

$V = \frac{4}{3}\pi r^3$
$S = 4\pi r^2$

EXAMPLE 4

(a) Solve the volume of a sphere formula $V = \frac{4}{3}\pi r^3$ for r.

(b) Use the result of part (a) to find the radius of a sphere whose volume is 36π cubic inches.

◀ **Solution**

(a) To solve this formula for r, we proceed as follows:

$$V = \frac{4}{3}\pi r^3$$

$$3V = 4\pi r^3 \qquad \text{Multiply both sides by 3}$$

$$\frac{3V}{4\pi} = r^3 \qquad \text{Divide both sides by } 4\pi$$

$$r = \sqrt[3]{\frac{3V}{4\pi}} \qquad \text{Extract the real cube root}$$

(b) To find the radius of a sphere whose volume is 36π cubic inches, we replace V with 36π:

$$r = \sqrt[3]{\frac{3V}{4\pi}} = \sqrt[3]{\frac{3(36\pi)}{4\pi}} = \sqrt[3]{27} = 3 \text{ inches} \qquad \blacklozenge$$

PROBLEM 4 Solve the area of a trapezoid formula $A = \frac{1}{2}h(b_1 + b_2)$ for the base b_1. ◆

Exercises 1.8

Basic Skills

In Exercises 1–30, find the solutions of each equation, if they exist.

1. $2x - 7 = -21$
2. $14 = 22 + 3y$
3. $26 = 6 - 5t$
4. $12 - 6x = -18$
5. $3n + 4 = n - 4$
6. $12m + 5 = 2m - 5$
7. $8x + 4 + 3x = x - 11$
8. $7x - 3 + 2x = 4x + 3 + x$
9. $9x - 2(4x + 1) = 3$
10. $16 - (3x + 7) = 3x$
11. $\dfrac{5x}{6} - 1 = \dfrac{x}{3}$
12. $4 + \dfrac{n}{6} = \dfrac{n+4}{2}$
13. $\dfrac{x}{2} + \dfrac{x}{4} = \dfrac{1}{6} + \dfrac{x}{3}$
14. $\dfrac{x}{2} + \dfrac{x}{3} + \dfrac{x}{5} = 1 - \dfrac{x}{6}$
15. $\dfrac{t-1}{6} = \dfrac{2t+3}{12} - \dfrac{7t-4}{10}$
16. $\dfrac{x}{12} - \dfrac{x-3}{8} = \dfrac{2x-1}{3}$
17. $|3x| = 12$
18. $3|-2t| = 15$
19. $6 - |5n - 3| = 4$
20. $|4y - 3| - 5 = 4$
21. $5 - 2|9 - 3y| = -15$
22. $2 - |3x - 11| = -4$
23. $x^2 - 3 = 0$
24. $n^2 - 16 = 9$
25. $\tfrac{1}{2}x^2 + 6 = 14$
26. $2y^2 - 21 = 33$
27. $2x^2 - 5 = 3x^2 - 7$
28. $10 - x^2 = 4 + x^2$
29. $(x - 1)^2 = 49$
30. $3(x + 2)^2 = 24$

In Exercises 31–50, solve each inequality. Use interval notation to describe the solution set.

31. $3x - 2 < 7$
32. $15 - 4x < -3$
33. $2x + 3 \geq 3x + 4$
34. $3m - 4 \leq m - 8$
35. $2(y + 3) > 3 - (4 - y)$
36. $4x - (x - 2) < x - 3$
37. $\dfrac{2p}{3} > \dfrac{p}{9} - \dfrac{5}{6}$
38. $\dfrac{x}{3} + \dfrac{x}{5} < 1 + \dfrac{11x}{15}$
39. $\dfrac{x+5}{2} - 3 \leq \dfrac{3x}{5} - \dfrac{x-4}{4}$
40. $\dfrac{h+1}{2} + \dfrac{2h+3}{4} \geq 2$
41. $|x| < 3$
42. $|2x| - 2 \geq 8$
43. $\left|\dfrac{2x}{3}\right| - 1 > 0$
44. $5\left|\dfrac{-x}{10}\right| + 3 < 13$
45. $|5 - 2x| > 3$
46. $|4x + 3| + 2 < 17$
47. $\left|\dfrac{4-x}{3}\right| \leq \dfrac{1}{2}$
48. $\left|\dfrac{2(4 - 3x)}{3}\right| > 8$
49. $\left|\dfrac{3x}{4} + \dfrac{1 - 3x}{6}\right| \geq 1$
50. $\left|\dfrac{1 - 2x}{4} + \dfrac{1 + 2x}{6}\right| \leq \dfrac{5}{9}$

In Exercises 51–58, solve each literal equation for x.

51. $\dfrac{ax}{b} = c$
52. $\dfrac{a}{bx + c} = 1$
53. $ax = bx + c$
54. $ax - b(c + x) = kx$
55. $\dfrac{a + bx}{k} = 3x + a$
56. $\dfrac{a - x}{b} = \dfrac{c - x}{a}$
57. $|ax + b| = 1$
58. $ax^2 + b = 0$

In Exercises 59–62, solve each literal inequality for x subject to the given conditions.

59. $ax + b \leq cx$, provided $a > c$
60. $ax + b > cx$, provided $a < c$
61. $\dfrac{x}{a} + \dfrac{y}{b} \leq c$, provided $a < 0$ and $b \neq 0$
62. $\dfrac{x}{a} + \dfrac{y}{b} \leq c$, provided $a > 0$ and $b \neq 0$

In Exercises 63–68, use the common formulas given in this section to answer each question.

63. What is the Celsius temperature if the Fahrenheit temperature is 59°F?
64. What time is required for an investment of $10,000 to amount to $20,000 if the interest rate is 5% simple interest?
65. The length of the hypotenuse of an isosceles right triangle is $5\sqrt{2}$ centimeters. What is the length of a leg of this triangle?
66. What is the radius of a circle if its area is 9π square centimeters?

67. What is the height of a cone if its volume is 12π cubic feet and its radius is 2 feet?

68. What is the height of a cylinder if its total surface area is 24π square meters and its radius is 3 meters?

In Exercises 69–78, solve each formula for the indicated variable.

69. Electrical current: $I = \dfrac{E}{R}$
    (a) the voltage E
    (b) the resistance R

70. Tension in an elevator cable: $T = m(g + a)$
    (a) the mass of the elevator m
    (b) the acceleration a

71. Thermal expansion: $L = L_0(1 + \mu \Delta t)$
    (a) the original length L_0
    (b) the temperature change Δt

72. nth term of an arithmetic sequence: $a_n = a_1 + (n - 1)d$
    (a) the common difference d
    (b) the number of terms n

73. Amount of money accrued: $A = P + Prt$
    (a) the interest rate r
    (b) the principal invested P

74. Mechanical advantage of a differential hoist: $M = \dfrac{2R}{R - r}$
    (a) the radius of the small pulley r
    (b) the radius of the large pulley R

75. Thrust of a spaceship's engine: $F = \dfrac{mv - mv_0}{t}$
    (a) the burn time t
    (b) the mass of the spaceship m

76. Sum of a finite geometric series: $S = \dfrac{a_1 - ra_n}{1 - r}$
    (a) the first term a_1
    (b) the common ratio r

77. Total power dissipated by two resistors connected in series: $P = I^2R_1 + I^2R_2$
    (a) the resistance R_1
    (b) the current I

78. Deflection of a cantilever steel beam: $D = \dfrac{8PL^3 + 3WL^3}{24EI}$
    (a) the concentrated load P
    (b) the length of the beam L

Critical Thinking

79. Is the equation $A = B$, where A and B are algebraic expressions, equivalent to the equation $A - B = 0$? to $\dfrac{A}{B} = 1$? to $A^2 = B^2$? Explain.

80. Is the inequality $A < B$, where A and B are algebraic expressions, equivalent to the inequality $A - B < 0$? to $\dfrac{A}{B} < 1$? to $A^2 < B^2$? Explain.

81. Determine the interval for which the expression $\sqrt{3x + 8}$ is undefined in the real number system.

82. Find a value of a such that the given pair of equations are equivalent.
    (a) $a - 4x = 18$ and $2x + 3 = 13$
    (b) $2x - 3a = 5x + 3$ and $3(x + 2) = 6(x + 3)$

83. Given that $x > 5$, find the fallacy in the following argument.

$x > 5$	
$5x > 25$	**Multiply both sides by 5**
$5x - x^2 > 25 - x^2$	**Subtract x^2 from both sides**
$x(5 - x) > (5 + x)(5 - x)$	**Factor each side**
$x > 5 + x$	**Divide both sides by $5 - x$**
$0 > 5$	**Subtract x from both sides**

84. Given that $x = 3$, find the fallacy in the following argument.

$x = 3$	
$x - 3 = 0$	Subtract 3 from both sides
$2x - 6 = 0$	Multiply both sides by 2
$x^2 + 2x - 6 = x^2$	Add x^2 to both sides
$x^2 + 2x - 15 = x^2 - 9$	Subtract 9 from both sides
$(x + 5)(x - 3) = (x + 3)(x - 3)$	Factor each side
$x + 5 = x + 3$	Divide both sides by $(x - 3)$
$5 = 3$	Subtract x from both sides

85. The relationship between the air temperature T_a (in °F), the dew point temperature T_d (in °F), and the height h (in feet) to the base of a cumulus cloud is given by the formula

$$h = \frac{T_a - T_d}{0.0045}.$$

What is the dew point temperature when the base of the cloud is 5200 ft above the ground and the air temperature is 72°F?

86. A pediatrician determines that the weight x in pounds of 95% of the babies born at a certain hospital satisfies the inequality

$$|x - 7.2| \leq 2.3.$$

What is the range of the weights of the other 5% of the babies?

Chapter 1 Review

◆ Questions for Writing or Group Discussion

1. What is the difference between a *rational number* and an *irrational number*?
2. Is the expression $-x$ always negative? Explain.
3. For what values of a does $\frac{|a|}{a} = -1$?
4. What is meant by an *open interval*? a *closed interval*? an *unbounded interval*?
5. List the *laws of exponents*. Illustrate each rule with an example.
6. Look up the meanings of the following prefixes in a dictionary: *pico-, deci-, nano-, giga-,* and *hecto-*. Use these definitions to explain the meanings of picosecond, decimeter, nanofarad, gigawatt, and hectoliter.
7. What is meant by the *principal nth root* of a real number? Is the principal nth root of a real number always a real number? Explain
8. What is a *complex number*? Under what conditions is a complex number a real number? an imaginary number?
9. State the procedures for adding, subtracting, multiplying, and dividing complex numbers.

10. What is a *complex conjugate*? How is it used in finding the quotient of two complex numbers? Illustrate with an example.
11. Does every complex number have a reciprocal? Explain.
12. What is a *polynomial*?
13. What is meant by the *degree of a polynomial*? Illustrate with an example.
14. Explain the procedure for combining *like terms* into a single term.
15. State the procedures for adding, subtracting, and multiplying polynomials.
16. What is meant by *factoring completely* over the set of integers?
17. What are the conditions for a trinomial to be a *perfect square*?
18. From a given fraction, how can an *equivalent fraction* be generated?
19. Describe the process of *reducing* an algebraic fraction to lowest terms.
20. Explain the procedure of determining the *LCD* for a sum or difference of algebraic fractions.
21. What is a *complex fraction*? What is the procedure for simplifying a complex fraction?
22. How are *conjugates* used to rationalize a denominator or numerator?
23. Explain the difference between an *identity* and a *conditional equation*.
24. What is meant by the *solution* of an equation?
25. Does squaring both sides of an equation generate equivalent equations? Explain.
26. When is it necessary to *reverse the direction of the inequality symbol* when generating equivalent inequalities. Illustrate by examples.

Review Exercises

In Exercises 1–6, rewrite each statement using an inequality.

1. a is at least 7.
2. b is at most -9.
3. c is negative and greater than -10.
4. d is nonnegative and at most 6.
5. The distance between a and 3 is at most 2.
6. c is more than 5 units from 0.

In Exercises 7 and 8, find the distance from A to B on the real number line.

7. The coordinate of point A is -9 and the coordinate of point B is 6.
8. The coordinate of point A is 3 and the coordinate of point B is π.

In Exercises 9–16, use interval notation to describe the set of all real numbers designated on each real number line.

9.
10.
11.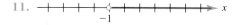
12.
13.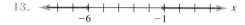
14.
15.
16.

In Exercises 17–26, simplify each exponential expression. Write the answer without zero or negative exponents. Assume all variables represent positive real numbers.

17. $(-2x^4y^5)(5xy^4)^2$
18. $[-2(x + 3)^2]^3$
19. $(2x + 1)^{-3}(2x + 1)^{-1}(2x + 1)^0$
20. $\left(\dfrac{3x^3}{2}\right)^{-2}$
21. $\dfrac{(x - 1)^4}{(x - 1)^3}$
22. $\dfrac{(2m^2n)^{-4}}{8m^{-2}}$
23. $(x + 1)^{-2/3}(x + 1)^{5/3}$
24. $(36x^4y^{-6})^{3/2}y^8$
25. $\left(\dfrac{16x^2}{y^8}\right)^{-3/4}$
26. $\dfrac{[9(x + 4)^{-3}]^{-1/2}}{(x + 4)^{1/2}}$

In Exercises 27 and 28, perform the indicated operations using scientific notation.

27. $\dfrac{0.00000000018}{60,000,000,000,000}$
28. $(30,000,000,000)(0.0000000002)^3$

In Exercises 29 and 30, rewrite each expression in exponential form.

29. $\sqrt[3]{(x + 2)^2}$
30. $\sqrt{x} \cdot \sqrt[4]{x^3}$

In Exercises 31 and 32, rewrite each expression in radical form.

31. $(x + 4)^{-1/6}$
32. $(2ab)^{3/4}$

In Exercises 33–36, write each radical expression in simplified form. Assume all variables represent positive real numbers.

33. $\sqrt{32(x + y)^3}$
34. $\sqrt{\sqrt[3]{36x^4}}$
35. $\sqrt[4]{32x^2y^2} \cdot \sqrt[4]{2xy^2}$
36. $\sqrt{\dfrac{8m^5}{(n + 2)^6}}$

In Exercises 37 and 38, simplify by assuming the variables represent any real number.

37. $\sqrt{9x^2y^2}$
38. $[(-2)^4(y - 1)^8]^{1/4}$

In Exercises 39–46, perform the indicated operations. Express each answer in standard form.

39. $4\sqrt{-45} + 3\sqrt{-80} - \sqrt{-20}$
40. $\sqrt{-24} \cdot \sqrt{6} \cdot \sqrt{-16}$
41. $(2i)^6$
42. $5i^{-33}$
43. $(9 + i) + (3 - 2i)$
44. $(4 - 3\sqrt{-9}) - (9 + 3\sqrt{-25})$
45. $(3 - 5i)(4 + 7i)$
46. $\dfrac{(1 - i)^2}{3 + 4i}$

In Exercises 47–66, perform the indicated operations with the algebraic expressions.

47. $(a^2 - 3a) + (5a - 4a^2) - (a - 5a^2)$
48. $(x^2 - 2x\sqrt{27x}) - (\sqrt{3x^3} - 3x^2)$
49. $(3\sqrt{40x} - 4\sqrt{90x}) + 2\sqrt{160x}$
50. $3x^{1/2} + (2x^{-1} + 4x^{1/2} - 3x^{-1})$
51. $(x^3 + 2x^2 - 3)(2x^3 - 3x - 1)$
52. $(5n^{-1/2} - 3n^{-3/2})(2n^{3/2} - n^{-1/2})$
53. $(5x^2 + 2y)(6x^2 - 5y)$
54. $(3\sqrt{ab} - 2)(4\sqrt{ab} - 5)$
55. $(2t - 5)(2t + 5)$
56. $(\sqrt{2x} - \sqrt{3y})(\sqrt{2x} + \sqrt{3y})$
57. $(x - 4)(x^2 + 4x + 16)$
58. $(4m^{-2} - 6m^{-1} + 9)(2m^{-1} + 3)$
59. $(3x - 7y)^2$
60. $(\sqrt{2n} + 1)^2$
61. $(2x - 5)^3$
62. $(a^{-2} - 2b)^3$
63. $[(x + 3)(x - 3)]^3$
64. $2(1 - a)^2(-5a) + 10a$
65. $y^2 - [x(3x - y) - (3x - 4y)(x + y)]$
66. $x\{x^2 - [(2x + 3)2x - x(3x + 2)]\}$

In Exercises 67 and 68, find a polynomial that represents the area of the shaded region.

67.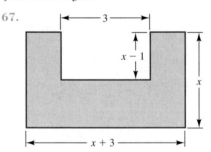

68.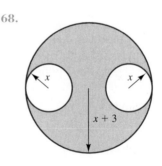

In Exercises 69–88, factor completely over the set of integers.

69. $3y^2(y - 2) + 6y(y - 2)$
70. $x(1 - 3x)^2 + (1 - 3x)^2$
71. $m^3 - 3m^2 + m - 3$
72. $12 - 9y - 8x + 6xy$
73. $6x^2 - xy - 12y^2$
74. $8t^2 + 6t - 5$
75. $2x^4 - 13x^2 + 15$
76. $m^2 + 10mn + 16n^2$
77. $x^2 + 2x - 15$
78. $a^6 - 7a^3b + 12b^2$
79. $25x^2 - 9y^4$
80. $x^3 + 27y^3$
81. $8x^3 - 1$
82. $9m^4 - 12m^2n^2 + 4n^4$
83. $t^3 + 6t^2 + 12t + 8$
84. $x^4 - 4y^4$
85. $8x^3y + 2x^2y - 6xy$
86. $m^6 - 4m^3 - 32$
87. $2xy + 4x^2y - 18x^3y - 36x^4y$
88. $16t^4 - 1$

In Exercises 89 and 90, reduce each fraction to lowest terms.

89. $\dfrac{n^2 + 5n - 36}{3n^2 - 14n + 8}$
90. $\dfrac{16x - x^5}{x^2 - 2x}$

In Exercises 91–106, perform the indicated operations.

91. $\dfrac{-5}{2x} + \dfrac{7}{2x}$
92. $\dfrac{2}{\pi} - \dfrac{9}{\pi}$
93. $\dfrac{-9}{5x} \cdot \dfrac{10x^2}{3}$
94. $\dfrac{3x}{2} \div \dfrac{15x}{8}$
95. $\dfrac{7y}{8x} + \dfrac{-5x^2}{12y}$
96. $\dfrac{5x}{9} - \dfrac{y}{12}$
97. $\dfrac{2x^2 - 3x - 9}{6xy} \cdot \dfrac{2x^2}{9 - x^2}$
98. $\dfrac{2a^2 + ab - 3b^2}{4a^2 - 9b^2} \div \dfrac{b^3 - a^3}{3b - 2a}$
99. $\dfrac{2x + 3}{x^2 - 9} - \dfrac{5x - 6}{x^2 - 9}$
100. $\dfrac{3m(m - n)}{4m - n} + \dfrac{n^2 - m^2}{n - 4m}$
101. $\dfrac{3}{x^2 + 4x} - \dfrac{3}{x^2 + 8x + 16}$
102. $\dfrac{1}{5 + x} + \dfrac{1}{5 - x} - \dfrac{x^2 - 7x}{x^2 - 25}$
103. $\dfrac{1}{x^2 - 4xy + 4y^2} + \dfrac{1}{x^2 - 4y^2}$
104. $\dfrac{y^{-2} - x^{-2}}{x^{-1} - y^{-1}}$
105. $\dfrac{1 + (a - 1)^{-1}}{(a - 1)^{-1}}$
106. $\dfrac{1 - \dfrac{1}{25x^2}}{5 - \dfrac{1}{x}}$

In Exercises 107 and 108, rationalize the numerator and simplify.

107. $\sqrt[4]{\dfrac{(x - 5)^3}{3x}}$
108. $\dfrac{\sqrt{3x} + \sqrt{3y}}{\sqrt{3xy}}$

In Exercises 109 and 110, rationalize the denominator and simplify.

109. $\dfrac{1}{2x - \sqrt{4x^2 - 1}}$
110. $\dfrac{1}{\sqrt[3]{(3 - x)^2}}$

In Exercises 111–118, solve for x.

111. $18 = 24 - 3x$
112. $3x + 4 = 7x - 8$
113. $6x - (3 - 2x) = 2(x - 3)$
114. $\dfrac{x}{8} - \dfrac{3x + 1}{10} = \dfrac{3 - x}{20}$
115. $|4 - 3x| = 16$
116. $\dfrac{1}{x} - \dfrac{1}{a} = \dfrac{1}{b} - \dfrac{1}{x}$
117. $x^2 - 16 = 0$
118. $(2x - 1)^2 = 18$

In Exercises 119 and 120, solve each formula for the indicated variable.

119. Resistance of a conductor: $R = \dfrac{\mu L + \mu L t}{d^2}$

    Solve for the length of the conductor L.

120. Energy output of a battery: $W = I^2 R_1 t + I^2 R_2 t$

    Solve for the current I.

In Exercises 121–126, solve each inequality. Use interval notation to describe the solution set.

121. $4 - x > 7$
122. $3x - 2 < 5x - 4$
123. $\dfrac{x}{4} \leq \dfrac{2x}{5} - \dfrac{3}{10}$
124. $\dfrac{1}{6} - \dfrac{t - 2}{12} \geq \dfrac{t - 1}{4}$
125. $|5 - 3n| > 8$
126. $\left|\dfrac{2x - 3}{5}\right| < \dfrac{2}{3}$

127. The wavelength λ (the Greek letter lambda) of a radio wave of frequency f is given by the formula

$$\lambda = \dfrac{3 \times 10^8}{f}$$

(Exercise 127 continues)

where λ is in meters and f is in hertz. Determine λ if $f = 7.5 \times 10^6$ hertz.

128. The change in length δ (the Greek letter delta) of a concrete slab in a sidewalk due to thermal expansion is given by the formula

$$\delta = 5.5 \times 10^{-6} tL$$

where t is the change in temperature, and L is the original length of the slab. Suppose 30.0-foot concrete slabs are being laid with an expansion gap between the end of one slab and the beginning of the next. What is the minimum expansion gap in *inches* that should be left between the slabs if the temperature of the concrete in the summertime can reach 170°F and the concrete is being laid at a temperature of 44°F?

129. The land area of China is about 3.7×10^6 square miles. In 1980, the population of China was approximately one billion. At that time, what was the amount of land area in *square feet* per person?

130. Presently, the United States natural gas reserve is estimated to be about 180 trillion cubic feet and consumption is averaging approximately 9×10^{11} cubic feet per year. Assuming consumption continues at this rate, how long will the supply last?

131. (a) Solve the volume of a cone formula $V = \tfrac{1}{3}\pi r^2 h$ for r.

(b) Use the result of part (a) to find the radius of a cone whose volume is 24π cubic inches and whose height is 2 inches.

132. A pediatrician determines that the length x in inches of 90% of the babies born at a certain hospital satisfies the inequality

$$|x - 20.6| \le 2.1.$$

What is the range of the lengths of the other 10% of the babies?

The real estate tax T on a property varies directly as its assessed value V.
(a) Express T as a function of V if $T = \$2800$ when $V = \$112,000$.
(b) Generate the graph of the function defined in part (a) for $\$0 \leq V \leq \$500,000$.

For the solution, see Example 3 in Section 2.7.

Functions and Graphs

- **2.1** Working in the Cartesian Plane
- **2.2** Graphs of Equations and the Graphing Calculator
- **2.3** Functions
- **2.4** Graph of a Function
- **2.5** Shifting, Reflecting, and Stretching a Graph
- **2.6** Composite and Inverse Functions
- **2.7** Applied Functions and Variation

2.1 Working in the Cartesian Plane

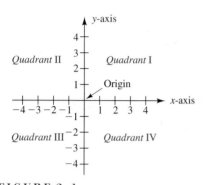

FIGURE 2.1

The coordinate plane, or Cartesian plane

◆ Introductory Comments

Recall from Section 1.1 that for each point on the real number line, there corresponds a real number called its *coordinate*. In this section, we extend this idea by assigning to each point in a plane a pair of real numbers. To do this, we construct horizontal and vertical real number lines that intersect at the zero points of the two lines (see Figure 2.1). The two lines are called *coordinate axes* and the *plane* in which they lie is called the **coordinate plane**, or **Cartesian plane**, after the French mathematician René Descartes (1596–1650). The point where the axes intersect is called the **origin**. The horizontal number line has its positive direction to the right and is usually called the **x-axis**. The vertical number line has its positive direction upward and is usually called the **y-axis**. The coordinate axes divide the plane into four regions, or **quadrants**, which are labeled with Roman numerals, as shown in Figure 2.1.

◆ Plotting Points

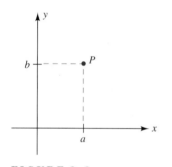

FIGURE 2.2

The coordinates of point P are (a, b).

We can assign to each point P in the coordinate plane a unique pair of numbers, called its **rectangular coordinates** or **Cartesian coordinates**. By drawing horizontal and vertical lines through P, we find that the vertical line intersects the x-axis at some point with coordinate a, and the horizontal line intersects the y-axis at some point with coordinate b, as shown in Figure 2.2. Thus, we assign the pair (a, b) to the point P. In the pair (a, b), the first number a is called the **x-coordinate** or **abscissa** of P, and the second number b is called the **y-coordinate** or **ordinate** of P. Since the x-coordinate is always written first, we refer to (a, b) as an **ordered pair** of numbers.

Conversely, we can locate any ordered pair (a, b) in the coordinate plane by constructing a vertical line through a on the x-axis and a horizontal line through b on the y-axis. The intersection of these lines determines a unique point P, which we designate as $P(a, b)$. Thus, for each point in the coordinate plane there corresponds a unique ordered pair of real numbers, and for each ordered pair of real numbers there is a unique point in the coordinate plane. Hence, we have a *one-to-one correspondence* between pairs of real numbers and points in a coordinate plane.

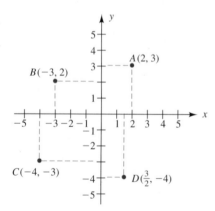

FIGURE 2.3

Plotting points associated with ordered pairs.

EXAMPLE 1 Plot the points $A(2, 3)$, $B(-3, 2)$, $C(-4, -3)$, and $D(\frac{3}{2}, -4)$ on a coordinate plane. Specify the quadrant in which each point lies.

◀ **Solution**   The point $A(2, 3)$ is in quadrant I, $B(-3, 2)$ is in quadrant II, $C(-4, -3)$ is in quadrant III, and $D(\frac{3}{2}, -4)$ is in quadrant IV, as shown in Figure 2.3. ◆

A point on the x-axis, y-axis, or at the origin, is not in any of the quadrants.

PROBLEM 1 Plot the points $A(0, 0)$, $B(0, -4)$, and $C(4, 0)$ on a coordinate plane. ◆

The Distance Formula

By convention, the *line segment* that joins the points $P(x_1, y_1)$ and $Q(x_2, y_2)$ in the coordinate plane is designated as $\overline{PQ}$, and the *length* of $\overline{PQ}$ is designated as PQ. To determine PQ, we begin by constructing a right triangle, as shown in Figure 2.4. The length of the horizontal leg of the triangle is $|x_2 - x_1|$ and the length of the vertical leg is $|y_2 - y_1|$. By the Pythagorean theorem, the sum of the squares of the lengths of these legs is equal to the square of the length of the hypotenuse PQ:

$$(PQ)^2 = |x_2 - x_1|^2 + |y_2 - y_1|^2$$
$$PQ = \sqrt{|x_2 - x_1|^2 + |y_2 - y_1|^2}$$
$$PQ = \sqrt{(x_2 - x_1)^2 + (y_2 - y_1)^2}$$

We refer to this last equation as the **distance formula**.

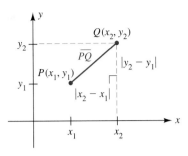

FIGURE 2.4
The length of $\overline{PQ}$, denoted as PQ, may be determined by the Pythagorean theorem.

◀ **Distance Formula**

> The **distance** between two points $P(x_1, y_1)$ and $Q(x_2, y_2)$ in the coordinate plane is given by
> $$PQ = \sqrt{(x_2 - x_1)^2 + (y_2 - y_1)^2}.$$

Note: When we use the distance formula, it does not matter which point is called (x_1, y_1) and which is called (x_2, y_2). This is because $(x_2 - x_1)$ and $(x_1 - x_2)$ are opposites of each other, as are $(y_2 - y_1)$ and $(y_1 - y_2)$, and squaring a number or its opposite gives the same numerical result.

EXAMPLE 2 Find the length of the line segment joining the points $A(-2, 3)$ and $B(4, 1)$.

◀ **Solution** Figure 2.5 shows the points A and B in the coordinate plane. We designate the length of the line segment that joins the points A and B as AB. We use the distance formula with $(x_1, y_1) = (-2, 3)$ and $(x_2, y_2) = (4, 1)$:

$$AB = \sqrt{[4 - (-2)]^2 + (1 - 3)^2}$$
$$= \sqrt{(6)^2 + (-2)^2}$$
$$= \sqrt{40}$$
$$= 2\sqrt{10}$$
$$\approx 6.32$$

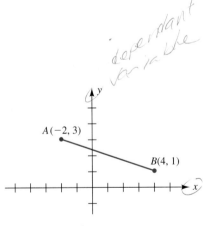

FIGURE 2.5
The length of the line segment joining the points A and B can be found by using the distance formula.

PROBLEM 2 Find the length of the line segment joining the points $P(-3, 1)$ and $Q(1, -2)$.

The Midpoint Formula

Given two points $P(x_1, y_1)$ and $Q(x_2, y_2)$ in the coordinate plane, we can find the coordinates of the midpoint M of $\overline{PQ}$ by finding the average value of the

x-coordinates and the average value of the y-coordinates of the endpoints P and Q. We refer to this fact as the **midpoint formula**.

◆ **Midpoint Formula**

> The coordinates of the **midpoint** M of a line segment joining the points $P(x_1, y_1)$ and $Q(x_2, y_2)$ are
> $$\left(\frac{x_1 + x_2}{2}, \frac{y_1 + y_2}{2} \right).$$

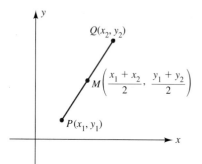

FIGURE 2.6
The midpoint of $\overline{PQ}$ is M, provided $PM = MQ$ and $PM + MQ = PQ$.

To verify the midpoint formula, we use Figure 2.6 and the distance formula to show that $PM = MQ$ and $PM + MQ = PQ$. Note that

$$PM = \sqrt{\left(\frac{x_1 + x_2}{2} - x_1\right)^2 + \left(\frac{y_1 + y_2}{2} - y_1\right)^2}$$

$$= \sqrt{\left(\frac{x_2 - x_1}{2}\right)^2 + \left(\frac{y_2 - y_1}{2}\right)^2}$$

$$= \tfrac{1}{2}\sqrt{(x_2 - x_1)^2 + (y_2 - y_1)^2}$$

and

$$MQ = \sqrt{\left(x_2 - \frac{x_1 + x_2}{2}\right)^2 + \left(y_2 - \frac{y_1 + y_2}{2}\right)^2}$$

$$= \sqrt{\left(\frac{x_2 - x_1}{2}\right)^2 + \left(\frac{y_2 - y_1}{2}\right)^2}$$

$$= \tfrac{1}{2}\sqrt{(x_2 - x_1)^2 + (y_2 - y_1)^2}.$$

Thus $\quad PM + MQ = \sqrt{(x_2 - x_1)^2 + (y_2 - y_1)^2} = PQ.$

Since $PM = MQ$ and the points P, M, and Q are *collinear* ($PM + MQ = PQ$), we conclude that M is the midpoint of $\overline{PQ}$.

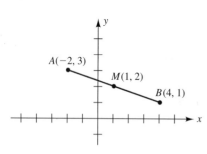

FIGURE 2.7
The coordinates of the midpoint of $\overline{AB}$ are (1, 2).

EXAMPLE 3 Find the coordinates of the midpoint of the line segment joining the points $A(-2, 3)$ and $B(4, 1)$.

◆ **Solution** Using the midpoint formula with $(x_1, y_1) = (-2, 3)$ and $(x_2, y_2) = (4, 1)$, we have

$$\left(\frac{-2 + 4}{2}, \frac{3 + 1}{2}\right) = (1, 2).$$

Thus, as shown in Figure 2.7, the coordinates of the midpoint M are (1, 2). ◆

PROBLEM 3 Find the coordinates of the midpoint of the line segment joining the points $A(-4, -3)$ and $B(5, -1)$. ◆

◆ **Analytic Geometry**

We can place geometric figures on the coordinate plane and solve related problems by using the distance and midpoint formulas. We refer to the

SECTION 2.1 ♦ Working in the Cartesian Plane

branch of mathematics that uses algebra to solve geometric problems as **analytic geometry**.

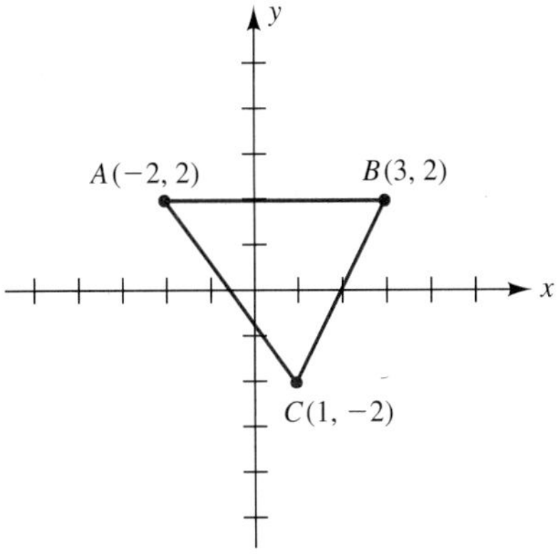

EXAMPLE 4 Show that the triangle with vertices at the points $A(-2, 2)$, $B(3, 2)$, and $C(1, -2)$ is an isosceles triangle—it has two sides of equal length. Then find the area of triangle ABC.

◀ **Solution** We begin by plotting the points and constructing the triangle, as shown in Figure 2.8. Now, using the distance formula, we can find the length of each side:

$$AB = \sqrt{[3 - (-2)]^2 + (2 - 2)^2} = \sqrt{(5)^2 + (0)^2} = 5$$
$$BC = \sqrt{(1 - 3)^2 + (-2 - 2)^2} = \sqrt{(-2)^2 + (-4)^2} = \sqrt{20} = 2\sqrt{5}$$
$$AC = \sqrt{[1 - (-2)]^2 + (-2 - 2)^2} = \sqrt{(3)^2 + (-4)^2} = \sqrt{25} = 5$$

Since $AB = AC$, we conclude the triangle is isosceles with BC the nonequal side.

To find the area of triangle ABC, recall from geometry that the altitude to the nonequal side of an isosceles triangle bisects that side. Thus, if $\overline{AM}$ is the altitude to the base $\overline{BC}$, then M is the midpoint of $\overline{BC}$, as shown in Figure 2.9. By the midpoint formula, the coordinates of the midpoint M are

$$\left(\frac{1 + 3}{2}, \frac{-2 + 2}{2}\right) = (2, 0)$$

and, by the distance formula, the length of the altitude $\overline{AM}$ is

$$AM = \sqrt{[2 - (-2)]^2 + (0 - 2)^2} = \sqrt{(4)^2 + (-2)^2}$$
$$= \sqrt{20} = 2\sqrt{5}.$$

Hence, the area of triangle ABC is

$$\text{Area} = \tfrac{1}{2}(\text{base})(\text{height}) = \tfrac{1}{2}(BC)(AM)$$
$$= \tfrac{1}{2}(2\sqrt{5})(2\sqrt{5})$$
$$= 10 \text{ square units.}$$

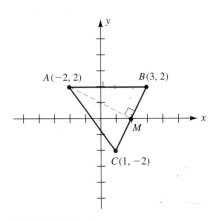

FIGURE 2.8
The triangle formed by the points $A(-2, 2)$, $B(3, 2)$, and $C(1, -2)$.

FIGURE 2.9
The altitude to the nonequal side of an isosceles triangle bisects that side. Thus, M is the midpoint of $\overline{BC}$.

As an alternative method for finding the area of triangle ABC, let $\overline{CN}$ be the altitude to base $\overline{AB}$. Since $\overline{AB}$ is a horizontal line segment, $CN = 4$. Hence,

$$\text{Area} = \tfrac{1}{2}(AB)(CN) = \tfrac{1}{2}(5)(4) = 10 \text{ square units.} \quad ◆$$

The *converse* of the Pythagorean theorem is also true. If the square of the length of the longest side of a triangle is equal to the sum of the squares of the lengths of the two other sides, then the triangle is a right triangle, with the right angle opposite the longest side.

PROBLEM 4 Use the distance formula and the converse of the Pythagorean theorem to show that the triangle with vertices at the points $A(2, 5)$, $B(-2, 3)$, and $C(4, 1)$ is an isosceles right triangle. Then find the area of triangle ABC. ◆

Exercises 2.1

Basic Skills

In Exercises 1–6, fill in the blank to complete the statement.

1. If both coordinates are negative, the point is located in quadrant _____.
2. If the *x*-coordinate is _____ and the *y*-coordinate is _____, the point is located in quadrant II.
3. If the *x*-coordinate is _____ and the *y*-coordinate is _____, the point is located in quadrant IV.
4. If the *x*-coordinate is zero, the point is located on the _____.
5. If the *y*-coordinate is zero, the point is located on the _____.
6. If both coordinates are _____, the point is located at the origin.

In Exercises 7 and 8, name the ordered pair associated with each point.

7.

8.
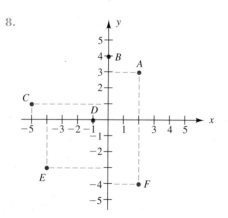

In Exercises 9–18, find
(a) *the length of the line segment joining the points A and B, and*
(b) *the coordinates of the midpoint M of the line segment joining the points A and B.*

9. $A(1, 2)$, $B(4, 6)$
10. $A(3, -6)$, $B(-2, 6)$
11. $A(-2, 3)$, $B(1, 8)$
12. $A(-1, -2)$, $B(-3, 2)$
13. $A(-5, 2)$, $B(3, -5)$
14. $A(-2, -1)$, $B(-6, -5)$
15. $A\left(-\frac{1}{2}, \frac{2}{3}\right)$, $B\left(-\frac{3}{4}, 1\right)$
16. $A\left(\frac{1}{2}, \frac{1}{2}\right)$, $B\left(\frac{2}{3}, \frac{5}{8}\right)$
17. $A(0, \sqrt{2})$, $B(-4, 0)$
18. $A(\sqrt{3}, -1)$, $B(0, 2)$

19. Find a formula for the distance d between the origin and the point (x, y) in the coordinate plane.

20. Find the coordinates of the midpoint M of a line segment joining the origin to the point (x, y) in the coordinate plane.

21. Find the perimeter of a triangle whose vertices are the points $A(-2, 1)$, $B(1, 3)$, and $C(4, -3)$.

22. Find the perimeter of a quadrilateral whose vertices are the points $A(-1, 0)$, $B(2, 4)$, $C(8, -4)$, and $D(4, -12)$.

23. Show that the points $A(1, 4)$, $B(2, -3)$, and $C(-1, -2)$ are the vertices of a right triangle. Then find the area of triangle ABC.

24. Show that the triangle whose vertices are the points $A(-5, 14)$, $B(1, 4)$, and $C(11, 10)$ is isosceles. Then find its area.

25. Show that the triangle whose vertices are $A(1, 1)$, $B(-1, -1)$, and $C(\sqrt{3}, -\sqrt{3})$ is an equilateral triangle. Then find its area.

26. Show that the triangle whose vertices are $A(0, 6)$, $B(2, 0)$, and $C(8, 2)$ is an isosceles right triangle. Then find its area.

27. A line segment joins the points $A(8, -12)$ and $B(-4, 6)$. Use the midpoint formula to find the coordinates of the three points that divide this line segment into four equal parts.

28. Use the distance formula to determine if the points $A(-2, -3)$, $B(1, 3)$, and $C(2, 5)$ are collinear.

SECTION 2.2 ♦ Graphs of Equations and the Graphing Calculator

29. A *median* of a triangle is a line segment that joins a vertex of the triangle to the midpoint of the opposite side. Find the lengths of the medians of a triangle whose vertices are $A(-5, 4)$, $B(5, 2)$, and $C(-1, -4)$.

30. The midpoint of a line segment is $M(3, -2)$. One endpoint of the segment has coordinates $(6, 3)$. Find the coordinates of the other endpoint.

Critical Thinking

31. Find the coordinates of a point on the y-axis that are equidistant from the points $A(-2, -4)$ and $B(3, 5)$.

32. Find the coordinates of a point on the x-axis that are equidistant from the points $A(-2, -4)$ and $B(3, 5)$.

For Exercises 33–36, refer to the right triangle with vertices A, O, and B shown in the figure.

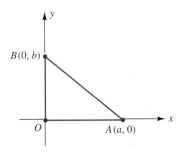

33. Find the coordinates of the midpoint M of the hypotenuse $\overline{AB}$.

34. Find AM and BM.

35. Find OM.

36. What conclusion can you make about the midpoint of the hypotenuse in regard to the three vertices of the right triangle?

For Exercises 37–40, refer to the parallelogram with vertices A, B, C, and O shown in the figure.

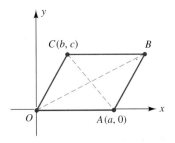

37. Find the coordinates of point B.

38. Find the midpoint of the diagonal $\overline{AC}$.

39. Find the midpoint of the diagonal $\overline{OB}$.

40. What conclusions can you make about the diagonals of a parallelogram?

A point P lies on the line segment joining the points $A(x_1, y_1)$ and $B(x_2, y_2)$ shown in the figure.

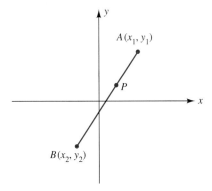

If the ratio $AP/AB = r$, then the point P has the coordinates

$$P\bigl(x_1 + r(x_2 - x_1),\ y_1 + r(y_2 - y_1)\bigr).$$

Use this formula for Exercises 41 and 42.

41. A line segment joins the points $A(6, 4)$ and $B(2, -2)$. Find the coordinates of the two points P_1 and P_2 that divide this line segment into three equal parts.

42. A line segment joins the points $A(5, 6)$ and $B(-3, -4)$. Find the coordinates of the four points P_1, P_2, P_3, and P_4 that divide the line segment into five equal parts.

43. Find the radius and area of a circle that passes through the point $A(2.24, 3.71)$ and has its center at the origin.

44. The line segment that joins the point $A(4.3, 2.5)$ to $B(-2.7, -4.9)$ is the diameter of a circle. Find the center, radius, and area of the circle.

2.2 Graphs of Equations and the Graphing Calculator

♦ Introductory Comments

In Section 1.8, we solved various types of equations containing one unknown. We now turn our attention to equations in *two unknowns*. A solution of an equation in two unknowns x and y is an ordered pair (x_1, y_1) such that when x is replaced by x_1 and y by y_1, the equation becomes true. Consider the equation

$$y = 2x + 1$$

with the two unknowns x and y. We say that the ordered pair $(0, 1)$ is a solution of this equation, since

$$y = 2x + 1$$
$$1 = 2(0) + 1$$
$$1 = 1 \text{ is true.}$$

To find other ordered pairs that are solutions of this equation, we arbitrarily choose values of x and then determine the corresponding values of y. Under these conditions we say that x is the **independent variable** and y the **dependent variable** in the equation. In order to determine other solutions to this equation, it is convenient to set up a **table of values** such as the one shown in Table 2.1.

TABLE 2.1
Table of values for $y = 2x + 1$

x	-2	-1	0	1	2
$y = 2x + 1$	-3	-1	1	3	5

Thus, along with $(0, 1)$ we have $(-2, -3)$, $(-1, -1)$, $(1, 3)$, and $(2, 5)$ as four other solutions of the equation $y = 2x + 1$. By continuing this process, we can generate many other ordered pairs that satisfy this equation. The **graph** of an equation is the set of *all* ordered pairs (x, y) in the coordinate plane that are solutions of the given equation. The graph of the equation $y = 2x + 1$ is a straight line, as shown in Figure 2.10. The arrowhead at each end of the line indicates that the graph continues indefinitely in that direction. Every point on this line is a solution of the equation $y = 2x + 1$.

◆ Point-Plotting Method

We can use the *point-plotting method* to *sketch the graph* of a simple equation.

◀ **Point-Plotting Method**

> To sketch the graph of an equation by the **point-plotting method:**
>
> 1. Set up a table of values and find a few ordered pairs that satisfy the equation.
> 2. Plot and label the corresponding points in the coordinate plane.
> 3. Look for a pattern, and connect the plotted points to form a smooth curve.

Note: Although this method works well for some simple equations in two unknowns, it is inadequate for sketching the graph of more complicated equations. Throughout

SECTION 2.2 ♦ Graphs of Equations and the Graphing Calculator

this chapter, we will introduce graphical aids that are useful for sketching an accurate graph by plotting as few points as possible.

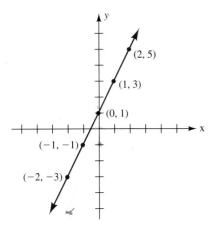

FIGURE 2.10
The graph of the equation $y = 2x + 1$ is a straight line.

EXAMPLE 1 Sketch the graph of each equation by using the point-plotting method.

(a) $y = |x|$ **(b)** $x = y^2$ **(c)** $2y - x^3 = 0$

◀ Solution

(a) We begin by selecting arbitrary values of x and then finding their corresponding values of y. The following table of values organizes our work.

x	-4	-2	-1	0	1	2	4
$y = |x|$	4	2	1	0	1	2	4

We now plot and label the points given by this table and connect them by drawing half-lines through them, according to the suggested pattern. The graph of $y = |x|$ is the **V**-shaped curve shown in Figure 2.11. The arrowheads on the curve indicate that the graph continues upward to the left and upward to the right, according to the pattern suggested by the points we have plotted.

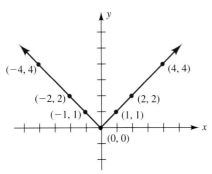

FIGURE 2.11
Graph of $y = |x|$

(b) For the equation $x = y^2$, it is easier to choose arbitrary values of y and then determine the corresponding values of x. In doing so, we are treating y as the independent variable and x as the dependent variable. The following table of values shows several values of y and their corresponding x-values.

y	-3	-2	-1	0	1	2	3
$x = y^2$	9	4	1	0	1	4	9

We plot and label the points given by this table and connect them by drawing a smooth curve, according to the suggested pattern. *Remember that the x-coordinate is always written first in an ordered pair.* The graph of $x = y^2$ is shown in Figure 2.12. It is a cupped-shaped curve called a *parabola*. The arrowheads on the curve indicate that the curve continues indefinitely, according to the pattern suggested by the points we have plotted.

(c) It is best to solve the equation $2y - x^3 = 0$ for either x or y before setting up a table of values. For this equation, solving for y is easier than solving for x, and we obtain

$$y = \tfrac{1}{2}x^3.$$

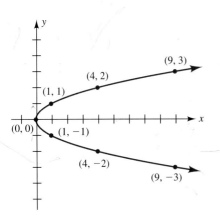

FIGURE 2.12
Graph of $x = y^2$

The following table of values shows several values of x and their corresponding y-values.

84   CHAPTER 2 ◆ *Functions and Graphs*

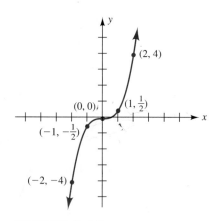

FIGURE 2.13
Graph of $2y - x^3 = 0$

x	-2	-1	0	1	2
$y = \frac{1}{2}x^3$	-4	$-\frac{1}{2}$	0	$\frac{1}{2}$	4

We now plot and label the points given by this table and draw a smooth curve through them. The graph of $2y - x^3 = 0$ is shown in Figure 2.13. The arrowheads on the curve indicate that the graph continues indefinitely, downward to the left and upward to the right. ◆

PROBLEM 1 Sketch the graph of the equation $4x + y = 8$ by using the point-plotting method. ◆

◆ The Graphing Calculator

Current technology enables us to generate the graph of an equation by using a personal computer with graphing software or a hand-held graphing calculator. Either of these graphing utilities is capable of plotting several hundred points in a few seconds. Although you may not have access to a personal computer with graphing software, you can purchase a hand-held graphing calculator at a reasonable cost. Throughout this text we use the graphing calculator to explore, investigate, and check many mathematical problems.

The key to success in using a graphing calculator is to select a viewing screen that shows the portion of the graph you wish to display. The viewing screen on a graphing calculator, which is usually called the *viewing rectangle*, represents a portion of the coordinate plane. On most graphing calculators, the WINDOW key or RANGE key is used to select a viewing rectangle. After pressing this key, you enter several values:

Xmin x-axis minimum value   **Ymin** y-axis minimum value
Xmax x-axis maximum value   **Ymax** y-axis maximum value
Xscl distance between scale marks on x-axis   **Yscl** distance between scale marks on y-axis

These values are shown in Figure 2.14.

EXAMPLE 2 Generate the graph of $2y - x^3 = 0$ on a graphing calculator (see Figure 2.13).

◆ **Solution** We begin by pressing the WINDOW key or RANGE key to choose a viewing rectangle. Although the choice of a viewing rectangle is arbitrary, Figure 2.13 suggests the following viewing rectangle:

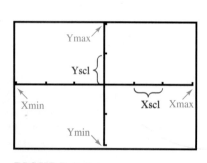

FIGURE 2.14
Viewing rectangle on a graphing calculator

Xmin = −5
Xmax = 5
Xscl = 1
Ymin = −5
Ymax = 5
Yscl = 1

After entering these values, we solve the equation $2y - x^3 = 0$ for y to obtain

$$y = \tfrac{1}{2}x^3,$$

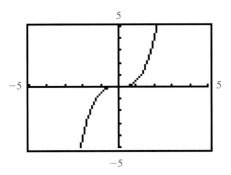

FIGURE 2.15

Graph of $2y - x^3 = 0$ in the viewing rectangle of a graphing calculator

and then enter this form of the equation into the calculator. *An equation must be in the form $y = \blacksquare$ before we can enter the equation into a graphing calculator.* Activating the graphing feature of the calculator, we obtain the graph of $2y - x^3 = 0$, as shown in Figure 2.15. ◆

To generate the graph of $x = y^2$ in the viewing rectangle of a graphing calculator, we first solve the equation for y to obtain $y = \pm\sqrt{x}$. We then enter both equations, $y = \sqrt{x}$ and $y = -\sqrt{x}$, into the calculator.

PROBLEM 2   Generate the graph of $x = y^2$ on a graphing calculator (see Figure 2.12). ◆

Almost every graphing calculator has a ⌈TRACE⌉ key and a ⌈ZOOM⌉ key. The ⌈TRACE⌉ key is used to display the x- and y-coordinates associated with each dot, or *pixel*, along the graph. With the ⌈ZOOM⌉ key, we can magnify the graph n times about any point. These two keys enable us to estimate some special points along the graph, such as the coordinates of the points where the graph intersects the x-axis or y-axis, intersects another graph, or reaches its maximum or minimum value. More advanced graphing calculators have built-in programs for finding these special points. For more detailed explanations about the usage of different types of calculators, consult the manual that came with your calculator or the *Graphing Technology Laboratory Manual* that accompanies this text.

CAUTION   Do not use a graphing calculator to obtain a quick solution to an example or exercise with no understanding of the fundamental mathematical ideas being presented. To become dependent upon any machine for solutions can have a disastrous effect on your learning. Instead, use a calculator to explore, investigate, and check the solution of mathematical problems. In this way, you will enhance your learning and understanding of college mathematics.

◆ Symmetry and Intercepts

Look at Figure 2.11: If we fold the coordinate plane along the y-axis, then the right-hand and left-hand portions of this graph will coincide. Thus, we say the graph of $y = |x|$ is **symmetric with respect to the y-axis**. This means that for each point (x, y) on the graph there corresponds a point $(-x, y)$ that is also on the graph. Now look at Figure 2.12. If we fold this coordinate plane along the x-axis, the upper and lower portions of the graph will coincide. Thus, we say that the graph of $x = y^2$ is **symmetric with respect to the x-axis**. This means that for each point (x, y) on the graph there corresponds a point $(x, -y)$ that is also on the graph. Finally, consider Figure 2.13. If we fold the coordinate plane along the x-axis, and then along the y-axis, the upper and lower portions of the graph will coincide. Thus, we say the graph of $2y - x^3 = 0$ is **symmetric with respect to the origin**. This means that for each point (x, y) on the graph, there corresponds a point $(-x, -y)$ that is also on the graph. The following tests can be used to tell whether an equation has any of these three types of symmetry.

◀ **Tests for Symmetry**

1. The graph of an equation in two unknowns x and y is **symmetric with respect to the y-axis** if replacing x with $-x$ yields an equivalent equation.

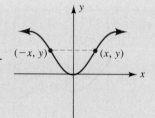

2. The graph of an equation in two unknowns x and y is **symmetric with respect to the x-axis** if replacing y with $-y$ yields an equivalent equation.

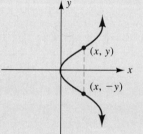

3. The graph of an equation in two unknowns x and y is **symmetric with respect to the origin** if replacing x and y with $-x$ and $-y$, respectively, yields an equivalent equation.

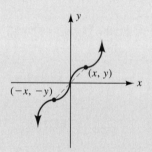

The **intercepts** of a graph are the points where the graph intersects the x-axis and y-axis. An **x-intercept** is the x-coordinate of any point where the graph intersects the x-axis. Since y is zero when the graph intersects the x-axis, the x-intercepts can be found by letting y be zero and solving the equation for x. A **y-intercept** is the y-coordinate of any point where the graph intersects the y-axis. Since x is zero when the graph intersects the y-axis, the y-intercepts can be found by letting x be zero and solving the equation for y.

EXAMPLE 3 Determine if the graph of each equation has symmetry with respect to the x-axis, y-axis, or origin. Then find the x- and y-intercepts, if they exist.

(a) $x^2 + y + 2 = 0$ (b) $xy^2 = 1$ (c) $x^2 + y^2 = 25$

Solution

(a) *Symmetry:*

x-axis (replace y with −y)	y-axis (replace x with −x)	origin (replace x with −x and y with −y)
$x^2 + y + 2 = 0$	$x^2 + y + 2 = 0$	$x^2 + y + 2 = 0$
$x^2 + (-y) + 2 = 0$	$(-x)^2 + y + 2 = 0$	$(-x)^2 + (-y) + 2 = 0$
$x^2 - y + 2 = 0$	$x^2 + y + 2 = 0$	$x^2 - y + 2 = 0$
not an equivalent equation	an equivalent equation	not an equivalent equation

Since replacing x with $-x$ yields an equivalent equation, we conclude that the graph of $x^2 + y + 2 = 0$ is symmetric with respect to the y-axis.

x-intercepts: Letting $y = 0$ in the equation $x^2 + y + 2 = 0$, we obtain

$$x^2 + (0) + 2 = 0, \quad \text{or} \quad x^2 = -2,$$

which has no real solution. Thus, we conclude that the graph has no x-intercept.

y-intercepts: Letting $x = 0$ in the equation $x^2 + y + 2 = 0$, we obtain

$$(0)^2 + y + 2 = 0, \quad \text{or} \quad y = -2.$$

Hence the y-intercept is -2.

We can use a graphing calculator to verify our work. Solving the equation $x^2 + y + 2 = 0$ for y, we obtain

$$y = -x^2 - 2.$$

Entering this form of the equation into the calculator gives us the graph in Figure 2.16. Note that the graph has symmetry with respect to the y-axis, no x-intercept, and y-intercept -2.

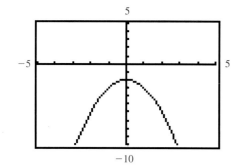

FIGURE 2.16
Graph of $x^2 + y + 2 = 0$

(b) *Symmetry:*

x-axis (replace y with −y)	y-axis (replace x with −x)	origin (replace x with −x and y with −y)
$xy^2 = 1$	$xy^2 = 1$	$xy^2 = 1$
$x(-y)^2 = 1$	$(-x)y^2 = 1$	$(-x)(-y)^2 = 1$
$xy^2 = 1$	$-xy^2 = 1$	$-xy^2 = 1$
an equivalent equation	not an equivalent equation	not an equivalent equation

Since replacing y with $-y$ yields an equivalent equation, we conclude that the graph of $xy^2 = 1$ is symmetric with respect to the x-axis.

x-intercepts: Letting $y = 0$ in the equation $xy^2 = 1$, we obtain

$$x(0)^2 = 1, \quad \text{or} \quad 0x = 1.$$

However, since no number times zero is one, we conclude that the graph has no x-intercept.

y-intercepts: Letting $x = 0$ in the equation $xy^2 = 1$, we obtain

$$(0)y^2 = 1.$$

However, since no number squared times zero is one, we conclude that the graph has no y-intercept.

We can use a graphing calculator to verify our work. Solving the equation $xy^2 = 1$ for y, we obtain

$$y = \pm\sqrt{\frac{1}{x}}.$$

Entering *both* of these equations into the calculator gives us the graph in Figure 2.17. Note that the graph has symmetry with respect to the x-axis, no x-intercept, and no y-intercept.

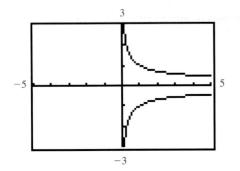

FIGURE 2.17
Graph of $xy^2 = 1$

(c) *Symmetry:*

x-axis (replace y with $-y$)	y-axis (replace x with $-x$)	origin (replace x with $-x$ and y with $-y$)
$x^2 + y^2 = 25$	$x^2 + y^2 = 25$	$x^2 + y^2 = 25$
$x^2 + (-y)^2 = 25$	$(-x)^2 + y^2 = 25$	$(-x)^2 + (-y)^2 = 25$
$x^2 + y^2 = 25$	$x^2 + y^2 = 25$	$x^2 + y^2 = 25$
an equivalent equation	an equivalent equation	an equivalent equation

Since an equivalent equation is generated in each case, we conclude that the graph of $x^2 + y^2 = 25$ is symmetric with respect to the x-axis, to the y-axis, and to the origin.

x-intercepts: Letting $y = 0$ in the equation $x^2 + y^2 = 25$, we obtain

$$x^2 + (0)^2 = 25, \quad \text{or} \quad x^2 = 25,$$

which implies $x = \pm\sqrt{25}$, or $x = \pm 5$. Hence, the x-intercepts of the graph of $x^2 + y^2 = 25$ are ± 5.

y-intercepts: Letting $x = 0$ in the equation $x^2 + y^2 = 25$, we obtain

$$(0)^2 + y^2 = 25, \quad \text{or} \quad y^2 = 25,$$

which implies $y = \pm\sqrt{25}$ or $y = \pm 5$. Hence, the y-intercepts of the graph of $x^2 + y^2 = 25$ are also ± 5.

We can use a graphing calculator to verify our work. Solving the equation $x^2 + y^2 = 25$ for y, we obtain

$$y = \pm\sqrt{25 - x^2}.$$

Entering *both* of these equations into the calculator gives us the graph in Figure 2.18. To show the true geometric perpective of this graph, which is a circle, we use a *square set viewing rectangle*—one in which the vertical and horizontal tick marks have equal spacing. The graph has symmetry with respect to the x-axis, to the y-axis, and to the origin, and has x-intercepts ± 5, and y-intercepts ± 5. ◆

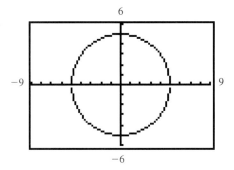

FIGURE 2.18
Graph of $x^2 + y^2 = 25$

PROBLEM 3   Test the equation $y^2 + x = 4$ for symmetry. Then find the x- and y-intercepts, if they exist. ◆

To find the x- and y-intercepts for some equations, we factor to form a product and then apply the *zero product property*.

◆ Zero Product Property

For real numbers p and q,

if $pq = 0$,   then either $p = 0$ or $q = 0$ (or both p and q are zero).

The procedure is illustrated in the next example.

EXAMPLE 4   Find the x- and y-intercepts for the graph of each equation.

(a) $y = x^3 - 3x$ **(b)** $x^2 + 3y^2 - 4x - 12 = 0$

◆ **Solution**

(a) *x-intercepts:* Letting $y = 0$ in the equation $y = x^3 - 3x$, we obtain

$$0 = x^3 - 3x$$
$$0 = x(x^2 - 3) \qquad \text{Factor}$$
$$x = 0 \quad \text{or} \quad x^2 - 3 = 0 \qquad \text{Apply the zero product property}$$
$$x = \pm\sqrt{3} \qquad \text{Solve for } x$$

Hence, we conclude that the x-intercepts are 0, $\sqrt{3} \approx 1.73$, and $-\sqrt{3} \approx -1.73$.

y-intercepts: Letting $x = 0$ in the equation $y = x^3 - 3x$, we obtain

$$y = (0)^3 - 3(0), \quad \text{or} \quad y = 0.$$

Hence, we conclude the y-intercept is 0.

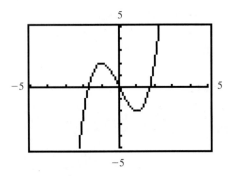

FIGURE 2.19
The graph of $y = x^3 - 3x$ has x-intercepts 0 and $\pm\sqrt{3}$ and y-intercept 0.

Figure 2.19 shows the graph of $y = x^3 - 3x$ in the viewing rectangle of a graphing calculator. We can use the $\boxed{\text{TRACE}}$ key to verify the intercepts.

(b) *x-intercepts:* Letting $y = 0$ in the equation $x^2 + 3y^2 - 4x - 12 = 0$, we obtain

$$x^2 - 4x - 12 = 0$$

$$(x + 2)(x - 6) = 0 \qquad \text{Factor}$$

$$x + 2 = 0 \quad \text{or} \quad x - 6 = 0 \qquad \text{Apply the zero product property}$$

$$x = -2 \qquad\qquad x = 6 \qquad \text{Solve for } x$$

Hence, we conclude that the x-intercepts are -2 and 6.

y-intercepts: Letting $x = 0$ in the equation $x^2 + 3y^2 - 4x - 12 = 0$, we obtain

$$3y^2 - 12 = 0$$

$$y^2 - 4 = 0 \qquad \text{Divide both sides by 3}$$

$$(y + 2)(y - 2) = 0 \qquad \text{Factor}$$

$$y + 2 = 0 \quad \text{or} \quad y - 2 = 0 \qquad \text{Apply zero product property}$$

$$y = -2 \qquad\qquad y = 2 \qquad \text{Solve for } y$$

Hence, we conclude that the y-intercepts are ± 2.

Figure 2.20 shows the graph of $x^2 + 3y^2 - 4x - 12 = 0$, which is equivalent to

$$y = \pm\sqrt{\frac{12 + 4x - x^2}{3}},$$

in the viewing rectangle of a graphing calculator. Note that the x-intercepts are -2 and 6 and the y-intercepts are ± 2. The graph of this equation is called an *ellipse*. ◆

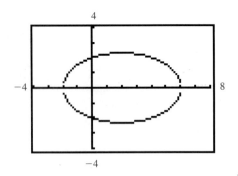

FIGURE 2.20
The graph of $x^2 + 3y^2 - 4x - 12 = 0$ has x-intercepts -2 and 6 and y-intercepts ± 2.

$\boxed{\text{PROBLEM 4}}$ The graph of the equation in Figure 2.19 appears to be symmetric with respect to the origin, and the graph of the equation in Figure 2.20 appears to be symmetric with respect to the x-axis. Algebraically verify these observations. ◆

◆ Application: Graphs from Experimental Data

In science and engineering, the relationship between two variables is often given by a table of values obtained from experimentation. Although we may

SECTION 2.2 ♦ Graphs of Equations and the Graphing Calculator

not know a formula that equates the variables in the experiment, we can use the experimental data from the table to draw a graph and to answer questions concerning the relationship between the variables.

EXAMPLE 5 A pizza baked at 400°F was removed from the oven and placed in a room with a constant temperature. The temperature of the pizza was then recorded each minute for the next 6 minutes, as shown in the table of values that follows.

Time t (in minutes)	0	1	2	3	4	5	6
Temperature T (in °F)	400	345	300	262	230	203	181

(a) Plot the graph of this data using t as the horizontal axis and T as the vertical axis.

(b) Estimate the temperature of the pizza for $t = 2.3$ minutes.

◀ Solution

(a) We observe from the data that the temperature T decreases as time t increases. Plotting the points (t, T) in the coordinate plane and connecting them to form a smooth curve gives us the graph in Figure 2.21.

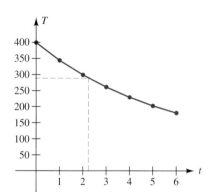

FIGURE 2.21

Graph of the experimental data in Example 5

(b) We can use the graph in Figure 2.21 to estimate the temperature T when $t = 2.3$ minutes. First we estimate 2.3 in the interval [2, 3] on the t-axis and draw a vertical line from the axis to the graph. Then we draw a horizontal line from this point on the graph to the vertical axis. Reading the value of T on the vertical axis, we find $T \approx 290$ °F when $t = 2.3$ minutes, as shown in Figure 2.21.

We can also estimate the temperature T by using the **method of linear interpolation**, which assumes that any two consecutive points in the table are connected by a straight line. If values in the table are sufficiently close together, this assumption yields an accurate approximation. Figure 2.22 illustrates that we can set up a direct proportion to find the value of x as follows:

$$\frac{1}{0.3} = \frac{38}{x}$$

$$x = (0.3)(38)$$

$$x = 11.4$$

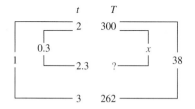

FIGURE 2.22

The value of x may be found by setting up a direct proportion.

Therefore, when $t = 2.3$ minutes,

$$T \approx 300 - 11.4 = 288.6 \text{ °F}.$$ ◆

PROBLEM 5 Use linear interpolation to estimate the temperature of the pizza for $t = 3.7$ minutes. Use the graph in Figure 2.21 to check your answer. ◆

Exercises 2.2

Basic Skills

In Exercises 1–12, use the point-plotting method to sketch the graph of each equation. Check graphically with a graphing calculator.

1. $y = x$
2. $y = x^2$
3. $y = 2\sqrt{x}$
4. $y = -3\sqrt[3]{x}$
5. $x = y^2 + 2$
6. $x = y^3 - 4$
7. $y - x^3 = 1$
8. $y + |x| = 3$
9. $xy = 1$
10. $y - 3x = 0$
11. $3x + 2y = 12$
12. $4x - 3y = 12$

In Exercises 13–40,
(a) *determine algebraically if the graph of each equation has symmetry with respect to the x-axis, y-axis, or origin.*
(b) *determine algebraically the x-intercept(s) and y-intercept(s) for the graph of each equation.*

Check each answer graphically with a graphing calculator.

13. $x^2 + 4y = 9$
14. $x - y^4 = 4$
15. $xy = 2$
16. $y = x^{2/3}$
17. $y = 3 - |x|$
18. $x^2 + 4y^2 = 16$
19. $x^2 + 1 = xy$
20. $x^2y + y = 1$
21. $y^2 = \sqrt{x - 2}$
22. $y = |2x - 5|$
23. $6x + 5y = 10$
24. $y^5 - 2x = 32$
25. $x^2 + 3y - 36 = 0$
26. $y = 2x^2 + 5x - 3$
27. $y = 8x - x^4$
28. $y = 12x - x^3$
29. $x^2 + y^2 - 6x - 16 = 0$
30. $4x^2 + 4x - 2y + 1 = 0$
31. $y = \dfrac{\sqrt{x^2 - 2}}{x}$
32. $y = \dfrac{x - 4}{2x - 3}$
33. $3y + 6 = |2x + 9|$
34. $2y - 3x^3 - 2x^2 = 0$
35. $y = 3x - 2x^3 - x^5$
36. $y = x^4 - 4x^2$
37. $x = -\sqrt{4 - y^2}$
38. $3y - \sqrt{x^2 - 9} = 0$
39. $y = \dfrac{1}{x^2 + 1}$
40. $y = \dfrac{x^2 + 1}{x^2}$

41. A bottle of wine at 70 °F is placed in a refrigerator with a constant temperature. The temperature T of the wine is then recorded every 5 minutes for the next 30 minutes, as shown in the table of values that follows.

t (in min)	0	5	10	15	20	25	30
T (in °F)	70	65	61	58	55	52	50

(a) Plot the graph of this data using time t as the horizontal axis and T as the vertical axis.

(b) Use linear interpolation to estimate the temperature of the wine for $t = 12$ minutes. Check your answer by using the graph in part (a).

42. Initially, a culture contains 18 bacteria per microliter. They start to multiply and the total number N present in a microliter is recorded every 3 minutes for the next 15 minutes, as shown in the table of values that follows.

t (in min)	0	3	6	9	12	15
N	18	27	41	61	91	137

(a) Plot the graph of this data using time t as the horizontal axis and N as the vertical axis.

(b) Use linear interpolation to estimate the number of bacteria present in the culture for $t = 7$ minutes. Check your answer by using the graph in part (a).

43. A stone is dropped from the top of a 400-foot building and its height h above the ground is recorded each second for the next 5 seconds, as shown in the table of values that follows.

t (in s)	0	1	2	3	4	5
h (in ft)	400	384	336	256	144	0

(a) Plot the graph of this data using time t as the horizontal axis and h as the vertical axis.

(b) Use linear interpolation to estimate the height of the stone above the ground for $t = 3.4$ seconds. Check your answer by using the graph in part (a).

44. In an electrical circuit with a constant voltage source, the current i (in amperes) is measured through a vari-

able resistor R (in ohms). The data is shown in the table of values that follows:

Resistance R (in ohms)	10	20	30	40	50
Current i (in amperes)	11.7	5.85	3.90	2.93	2.34

(a) Plot the graph of this data using R as the horizontal axis and i as the vertical axis.

(b) Use linear interpolation to estimate the current in the circuit for $R = 38$ ohms. Check your answer by using the graph in part (a).

Critical Thinking

45. Generate the graph of the equation $y = |x|$ in the viewing rectangle of a graphing calculator. Then, in the same viewing rectangle, generate the graphs of the equations

$$y = |x| + 1, \quad y = |x| + 2, \quad \text{and} \quad y = |x| + 3.$$

Compare each of these graphs to the graph of $y = |x|$. Describe the graph of $y = |x| + c$ for any constant $c > 0$.

46. Generate the graph of the equation $y = x^3$ in the viewing rectangle of a graphing calculator. Then, in the same viewing rectangle, generate the graphs of the equations

$$y = x^3 - 1, \quad y = x^3 - 2, \quad \text{and} \quad y = x^3 - 3.$$

Compare each of these graphs to the graph of $y = x^3$. Describe the graph of $y = x^3 - c$ for any constant $c > 0$.

47. Generate the graph of the equation $y = \sqrt{x}$ in the viewing rectangle of a graphing calculator. Then, in the same viewing rectangle, generate the graphs of the equations

$$y = \sqrt{x - 1}, \; y = \sqrt{x - 2}, \; \text{and} \; y = \sqrt{x - 3}.$$

Compare each of these graphs to the graph of $y = \sqrt{x}$. Describe the graph of $y = \sqrt{x - c}$ for any constant $c > 0$.

48. Generate the graph of the equation $y = x^2$ in the viewing rectangle of a graphing calculator. Then, in the same viewing rectangle, generate the graphs of the equations

$$y = (x + 1)^2, \quad y = (x + 2)^2, \quad \text{and} \quad y = (x + 3)^2.$$

Compare each of these graphs to the graph of $y = x^2$. Describe the graph of $y = (x + c)^2$ for any constant $c > 0$.

49. When a meteorologist gives the outside temperature in degrees Celsius C and in degrees Fahrenheit F, it is observed that $5F - 9C = 160$.

(a) Complete the following table of values for this relationship between C and F. Record each answer to the nearest tenth of a degree.

C (in °C)			12	31	-6
F (in °F)	39	74	-3		

(b) Use the table of values in part (a) to sketch the graph of this relationship between C and F. Use F as the horizontal axis and C as the vertical axis.

(c) Find the F-intercept and C-intercept for the graph in part (b). Interpret their meanings.

(d) Generate the graph of the equation $5F - 9C = 160$ on a graphing calculator. Use the $\boxed{\text{TRACE}}$ key to find the point on the graph where the Celsius and Fahrenheit temperatures are exactly the same. Explain how to find this point algebraically.

50. In a chemistry experiment, it is observed that when the temperature of a confined gas remains constant, the product of its pressure P (in pounds per square inch, psi) and its volume V (in cubic inches, in^3) is always equal to 125.

(a) Complete the following table of values for this relationship between P and V. Record each answer to two significant digits.

P (in psi)	3.4	11	41			
V (in in^3)				22	6.8	4.0

(*Exercise 50 continues*)

(b) Use the table of values in part (a) to sketch the graph of this relationship between P and V. Use P as the horizontal axis and V as the vertical axis.

(c) Does the graph in part (b) have axis intercepts? Explain.

2.3 Functions

◆ Introductory Comments

Given an equation in two unknowns x and y, suppose we *choose* values for x and *obtain* corresponding values for y. Hence, we are using x as the independent variable and y as the dependent variable. If for each value of x we choose there corresponds *exactly one* value of y, then we say that the equation defines y as *a function of x*.

◆ Definition of a Function

> A **function** from a set X to a set Y is a rule of correspondence that assigns to each element x in X exactly one element y in Y. Set X is called the **domain** of the function and set Y contains the **range** of the function.

A function may be represented by a list or table, by a graph, or by a formula or equation. Many functions in this text are specified by equations. For example, consider the equation $y = x^2$. It is useful to think of the x-values as **inputs** and their corresponding y-values as **outputs**. We can think of this equation as a rule that says "Square the input." A *function machine* for this equation is shown in Figure 2.23. If we place $x = 3$ in the input hopper, the machine follows the rule "square the input" and gives us an output of $y = 9$. The rule "square the input" assigns to each input value x one and only one output value y. Thus we say the equation $y = x^2$ *defines y as a function of x*.

Not every equation in two unknowns x and y defines y as a function of x. Consider the equation $y^2 = x$. If we choose the input $x = 9$, the equation becomes $y^2 = 9$ and, consequently, $y = \pm 3$. In this equation we obtain two outputs for the input $x = 9$. Since our definition of a function requires that there be exactly one output for each input, we know that the equation $y^2 = x$ does not define y as a function of x.

◆ Vertical Line Test

Suppose we are given the graph of an equation in two unknowns x and y. If every vertical line that can be drawn in the coordinate plane intersects the graph *at most once*, then we can say the equation has exactly one output y for each input value x that we can assign. Hence, the equation defines y as a function of x. We refer to this graphical method of determining whether y is a function of x as the **vertical line test**.

FIGURE 2.23
Function machine for $y = x^2$

Vertical Line Test

An equation defines *y* as a function of *x* if and only if every vertical line in the coordinate plane intersects the graph of the equation at most once.

The graph of $y = x^2$ (in Figure 2.24) passes the vertical line test, whereas the graph of $y^2 = x$ (in Figure 2.25) fails this test.

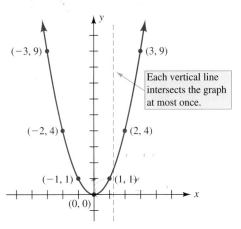

FIGURE 2.24
The graph of $y = x^2$ passes the vertical line text.

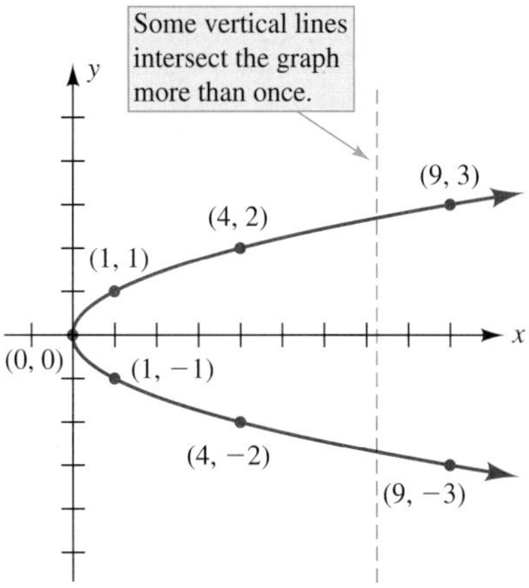

FIGURE 2.25
The graph of $y^2 = x$ fails the vertical line test.

EXAMPLE 1 Determine if each equation defines *y* as a function of *x*.

(a) $x^2 + y - 4 = 0$ (b) $x^2 + y^2 - 4 = 0$

Solution

(a) If we solve the equation $x^2 + y - 4 = 0$ for *y*, we obtain

$$y = 4 - x^2.$$

From this form of the equation we can see that for each input value *x* we choose, one—and only one—output value *y* corresponds to it. Hence, the equation $x^2 + y - 4 = 0$ defines *y* as a function of *x*. Figure 2.26 shows the graph of this equation in the viewing rectangle of a graphing calculator. The equation $x^2 + y - 4 = 0$ passes the vertical line test. Note that every vertical line intersects this graph *at most once*.

(b) If we solve the equation $x^2 + y^2 - 4 = 0$ for *y*, we obtain

$$y = \pm\sqrt{4 - x^2}.$$

This form of the equation makes it obvious that we have *two* outputs for each input value *x* in the interval $(-2, 2)$. Hence, the equation

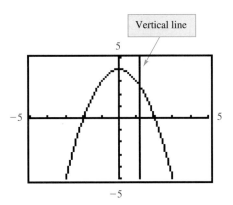

FIGURE 2.26
The graph of $x^2 + y - 4 = 0$ passes the vertical line text. Hence, this equation defines *y* as a function of *x*.

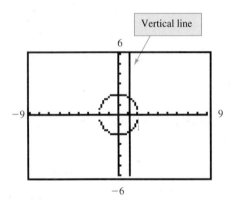

FIGURE 2.27

The graph of $x^2 + y^2 - 4 = 0$ fails the vertical line test. Hence, this equation does not define y as a function of x.

$x^2 + y^2 - 4 = 0$ does *not* define y as a function of x. Figure 2.27 shows the graph of this equation in the viewing rectangle of a graphing calculator. The equation $x^2 + y^2 - 4 = 0$ fails the vertical line test. Note that every vertical line in the interval $(-2, 2)$ intersects the graph twice. ◆

Note: Taken separately, the equations $y = \sqrt{4 - x^2}$ and $y = -\sqrt{4 - x^2}$ each define y as a function of x, and the graph of each function is a semicircle. Referring to Figure 2.27, we see that the graph of $y = \sqrt{4 - x^2}$ is the semicircle above the x-axis, and the graph of $y = -\sqrt{4 - x^2}$ is the semicircle below the x-axis. Each semicircle passes the vertical line test.

PROBLEM 1 Determine if the equation $2x + y^2 = 4$ defines y as a function of x. ◆

◆ Functional Notation

The letters f, F, g, G, h, and H are often used to represent functions. We use the notation $f(x)$, read "f of x," to denote the **value of the function f at x**.

◀ **Functional Notation**

> If f is a function and x an element of the domain of f, then the **functional notation**
>
> $$f(x), \quad \text{read "} f \text{ of } x,\text{"}$$
>
> denotes the *value of the function at x.*

CAUTION The notation $f(x)$ does *not* mean "f times x." Also, the letter f is the *name* of the function, whereas the notation $f(x)$ is the *value* of the function at x. Do not confuse these symbols.

The function f defined by the equation $y = x^2$ may be written using functional notation:

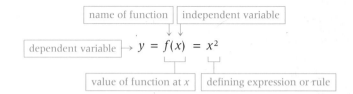

We refer to the process of determining the value of $f(x)$ for a given value of x as *computing the functional value*. To compute the functional value at some real number a, we replace x with a in the defining expression for f and perform the indicated operations. Here are examples of computing the functional value of $f(x) = x^2$ at two values of x, -3 and 10:

$$f(-3) = (-3)^2 = 9 \qquad f(10) = (10)^2 = 100$$

"f of -3" "f of 10"

SECTION 2.3 ◆ Functions

Note: The letters we use to represent the function and the independent variable are immaterial. The function g defined by $g(t) = t^2$ describes the same function as the one defined by $f(x) = x^2$.

EXAMPLE 2 Let f be a function defined by $f(x) = x^2 - 4x + 3$. Compute each functional value.

(a) $f(-2)$ **(b)** $f(3)$ **(c)** $f(2a)$ **(d)** $f(x-1)$

◀ Solution

(a) The notation $f(-2)$ denotes the value of the function at -2. By replacing x with -2 in the defining expression $x^2 - 4x + 3$, we obtain the functional value at -2:

$$f(-2) = (-2)^2 - 4(-2) + 3$$
$$= 4 + 8 + 3$$
$$= 15$$

(b) The notation $f(3)$ denotes the value of the function at 3. By replacing x with 3 in the defining expression $x^2 - 4x + 3$, we obtain the functional value at 3:

$$f(3) = (3)^2 - 4(3) + 3$$
$$= 9 - 12 + 3$$
$$= 0$$

(c) The notation $f(2a)$ denotes the value of the function at $2a$. By replacing x with $2a$ in the defining expression $x^2 - 4x + 3$, we obtain the functional value at $2a$:

$$f(2a) = (2a)^2 - 4(2a) + 3$$
$$= 4a^2 - 8a + 3$$

(d) The notation $f(x-1)$ denotes the value of the function at $x - 1$. By replacing x with $x - 1$ in the defining expression $x^2 - 4x + 3$, we obtain the functional value at $x - 1$:

$$f(x-1) = (x-1)^2 - 4(x-1) + 3$$
$$= x^2 - 2x + 1 - 4x + 4 + 3$$
$$= x^2 - 6x + 8$$

◆

PROBLEM 2 For the function f defined by $f(x) = x^2 - 4x + 3$, find $f(\pi)$.

◆

In calculus, it is often necessary to evaluate the **difference quotient** for a given function f:

$$\frac{f(x+h) - f(x)}{h}, \quad h \neq 0$$

The notations $f(x + h)$ and $f(x)$ denote the values of the function f at $x + h$ and at x, respectively.

EXAMPLE 3 Find the difference quotient for the function f defined by $f(x) = x^2 - 4x + 3$.

◆ **Solution** To find the difference quotient of a function, we proceed in three steps.

Step 1: First, we compute $f(x + h)$:

$$f(x + h) = (x + h)^2 - 4(x + h) + 3$$
$$= x^2 + 2xh + h^2 - 4x - 4h + 3$$

Step 2: Next we subtract $f(x)$ from the result in step 1 and simplify:

$$f(x + h) - f(x) = x^2 + 2xh + h^2 - 4x - 4h + 3 - (x^2 - 4x + 3)$$
$$= x^2 + 2xh + h^2 - 4x - 4h + 3 \ominus x^2 \oplus 4x \ominus 3$$
$$= 2xh + h^2 - 4h$$

Step 3: Finally, we divide the result in step 2 by h:

$$\frac{f(x + h) - f(x)}{h} = \frac{2xh + h^2 - 4h}{h}$$
$$= \frac{h(2x + h - 4)}{h} \quad \text{Factor the numerator}$$
$$= 2x + h - 4 \quad \text{Reduce to lowest terms} \quad ◆$$

PROBLEM 3 Find the difference quotient for the function g defined by $g(x) = 4 - 3x$. ◆

As illustrated in the next example, a function can be defined by a multipart rule. We refer to such a function as a **piecewise-defined function**. When we work with a piecewise-defined function, it is important to observe how the rule changes over different subsets of the domain.

EXAMPLE 4 Let h be a function defined by

$$h(x) = \begin{cases} 1 - x^2 & \text{if } x < 1 \\ x - 1 & \text{if } 1 \leq x \leq 3 \\ 4 & \text{if } x > 3. \end{cases}$$

Compute each functional value.

(a) $h(-2)$ **(b)** $h(3)$ **(c)** $h(\pi)$

Solution

(a) If $x < 1$, then the function h is defined by $h(x) = 1 - x^2$. Thus, for $x = -2$, we have

$$h(-2) = 1 - (-2)^2 = 1 - 4 = -3.$$

(b) If $1 \leq x \leq 3$, then the function h is defined by $h(x) = x - 1$. Thus, for $x = 3$, we have

$$h(3) = (3) - 1 = 2.$$

(c) If $x > 3$, then the function h is defined by $h(x) = 4$. Thus, for $x = \pi \approx 3.14$, we have

$$h(\pi) = 4.$$

PROBLEM 4 For the function h defined in Example 4, find $h(1)$.

◆ Finding the Domain of a Function

We can think of the *domain* of a function as the set of all possible inputs. When a function is defined by an equation, the domain is assumed to be the set of all real numbers that give us real-number outputs. For the function f defined by $f(x) = x^2$, every real number assigned to x leads to a corresponding real number for $f(x)$. Thus, the domain of the function f is the set of all real numbers. In this text, we express the domain of a function by using interval notation (see Section 1.1). Thus, we express the domain of f as $(-\infty, \infty)$.

The *range* of a function is the set of all function values that we obtain from the elements in the domain. Consider the function f defined by $f(x) = x^2$. When real numbers are squared, we obtain *nonnegative* real numbers. Thus, using interval notation, we conclude that the range of f is $[0, \infty)$. For most functions, however, the range is quite difficult to find. The best method for obtaining the range is usually to read it from the graph of the function. We will discuss this method in Section 2.4.

As illustrated in the next example, we must be aware of two properties when determining the domain of a function:

1. Division by zero is undefined.
2. Even roots of negative numbers are *not* real numbers.

EXAMPLE 5 State the domain of each function using interval notation.

(a) $f(x) = \dfrac{1}{x + 4}$

(b) $F(x) = \dfrac{2x + 5}{x^3 - 3x^2 + 2x - 6}$

(c) $G(x) = \sqrt[3]{3 - x}$

(d) $g(x) = \dfrac{\sqrt{2 - x}}{x + 1}$

Solution

(a) Division by zero is undefined. If $x = -4$, then the denominator becomes zero. Thus, the domain of the function f defined by $f(x) = 1/(x + 4)$ is all real numbers *except* $x = -4$. This domain is shown on the real number line in Figure 2.28. Using interval notation, we express the domain of this function as

$$(-\infty, -4) \cup (-4, \infty).$$

FIGURE 2.28

The domain of $f(x) = \dfrac{1}{x + 4}$ is $(-\infty, -4) \cup (-4, \infty)$.

(b) Division by zero is undefined. Thus, the domain of the function F defined by

$$F(x) = \frac{2x + 5}{x^3 - 3x^2 + 2x - 6}$$

is all real numbers x except those for which the denominator is zero:

$$\begin{aligned} x^3 - 3x^2 + 2x - 6 &= 0 \\ x^2(x - 3) + 2(x - 3) &= 0 \quad &\text{Factor by grouping terms} \\ (x - 3)(x^2 + 2) &= 0 \\ x - 3 = 0 \quad \text{or} \quad x^2 + 2 &= 0 \quad &\text{Apply the zero product property} \\ x = 3 \end{aligned}$$

(no real solution)

Hence, the domain of this function is all real numbers *except* $x = 3$. The domain is shown on the real number line in Figure 2.29. Using interval notation, we can express the domain of this function as

$$(-\infty, 3) \cup (3, \infty).$$

FIGURE 2.29

The domain of

$$F(x) = \frac{2x + 5}{x^3 - 3x^2 + 2x - 6}$$

is $(-\infty, 3) \cup (3, \infty)$.

(c) Odd roots of real numbers *are* real numbers. Hence, the domain of the function G defined by $G(x) = \sqrt[3]{3 - x}$ is the set of all real numbers. This domain is shown on the real number line in Figure 2.30. Using interval notation, we can express the domain of this function as

$$(-\infty, \infty).$$

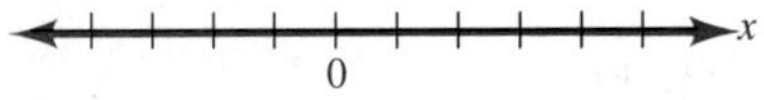

FIGURE 2.30

The domain of $G(x) = \sqrt[3]{3 - x}$ is $(-\infty, \infty)$.

(d) Even roots of negative numbers are *not* real numbers. Thus, for the function

$$g(x) = \frac{\sqrt{2 - x}}{x + 1},$$

the radicand $2 - x$ must be *nonnegative*, that is,

$$2 - x \geq 0$$

$$-x \geq -2 \quad \text{Add } -2 \text{ to both sides}$$

$$x \leq 2 \quad \text{Multiply both sides by } -1 \text{ and reverse the direction of the inequality symbol}$$

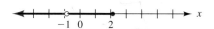

FIGURE 2.31

The domain of $g(x) = \dfrac{\sqrt{2-x}}{x+1}$ is $(-\infty, -1) \cup (-1, 2]$.

Also, we must exclude $x = -1$ from the domain, since this value of x makes the denominator zero. Hence, the domain of this function is all real numbers less than or equal to 2 except $x = -1$. We can show the domain on the real number line as in Figure 2.31, or write it in interval notation as

$$(-\infty, -1) \cup (-1, 2].$$

PROBLEM 5 Determine the domain of the function H defined by
$$H(x) = \dfrac{x+1}{\sqrt{2-x}}.$$

Exercises 2.3

Basic Skills

In Exercises 1–14, generate the graph of each equation in the viewing rectangle of a graphing calculator. Then use the vertical line test to determine if the equation defines y as a function of x.

1. $x^2 + y - 1 = 0$
2. $x^2 + y - 2x = 0$
3. $x^3 + 7y = 1$
4. $3x + 2y = 6$
5. $y = \sqrt{x^2 - 4}$
6. $y = |x - 5|$
7. $x^2 + y^2 - 4x = 0$
8. $x^2 - y^2 - 4 = 0$
9. $|x||y| = 1$
10. $|xy| = 1$
11. $y^3 + x - 1 = 0$
12. $x^3 + y^3 = 1$
13. $xy - x^2 = y$
14. $x^2y + y = 1$

Given the functions f, g, and h, defined by

$$f(x) = 4x^2 - 2x + 1, \quad g(x) = \sqrt{x - 4}, \quad h(x) = |2x + 3|,$$

compute the functional values given in Exercises 15–34.

15. $f(2)$
16. $g(9)$
17. $h(-4)$
18. $g(\tfrac{1}{4}) + g(1)$
19. $f(\sqrt{2})$
20. $-h(\tfrac{3}{2})$
21. $h(ab)$
22. $f(3p)$
23. $g(t^2)$, $t \geq 0$
24. $f(-x)$
25. $f\left(\dfrac{n}{2}\right)$
26. $h(x^2 - 3)$
27. $g(x) - g(0)$
28. $f(x + 2)$
29. $g(1 + x^2)$
30. $f(x + h) - f(x)$
31. $f(\sqrt{x - 2})$
32. $h\left(-\dfrac{x}{2}\right)$
33. $f(\sqrt{x} - 4)$
34. $g(4x^2 - 2x + 1)$

For the piecewise-defined functions in Exercises 35–38, compute each functional value.

35. $f(x) = \begin{cases} 2x - 5 & \text{if } x < 0 \\ x^2 & \text{if } x \geq 0 \end{cases}$
   (a) $f(3)$ (b) $f(-3)$
   (c) $f(2)$ (d) $f(0)$

36. $g(x) = \begin{cases} \sqrt{3 - x} & \text{if } x \leq 3 \\ 2 - x & \text{if } 3 < x < 6 \\ 1 & \text{if } x \geq 6 \end{cases}$
   (a) $g(0)$ (b) $g(4)$
   (c) $g(3)$ (d) $g(7)$

37. $h(x) = \begin{cases} x & \text{if } x > 0 \\ 0 & \text{if } x = 0 \\ -x & \text{if } x < 0 \end{cases}$
   (a) $h(7)$ (b) $h(-3)$
   (c) $h(5) - h(0)$ (d) $h(x + 1)$

38. $f(x) = \begin{cases} 1 & \text{if } x \text{ is a rational number} \\ -1 & \text{if } x \text{ is an irrational number} \end{cases}$
   (a) $f(0)$ (b) $f(6)$
   (c) $f(\pi)$ (d) $f(\sqrt{2})$

 In Exercises 39–46, find the difference quotient for the given function f, and simplify the result.

39. $f(x) = 2x + 1$
40. $f(x) = 2 - 3x$
41. $f(x) = x^2$
42. $f(x) = x^3$
43. $f(x) = x^2 + 2x - 3$
44. $f(x) = 2x^2 - 3x$
45. $f(x) = \dfrac{1}{x}$ [*Hint:* Subtract fractions in the numerator.]
46. $f(x) = \sqrt{x}$ [*Hint:* Rationalize the numerator.]

In Exercises 47–62, state the domain of each function using interval notation.

47. $f(x) = 3x + 2$
48. $g(x) = 1 - x^2$
49. $f(x) = x^3 + 2$
50. $f(x) = \sqrt[3]{2x - 1}$
51. $H(x) = \sqrt{4 - x}$
52. $h(x) = \sqrt{2x - 3}$
53. $g(x) = -\sqrt{16 - x^2}$
54. $F(x) = \sqrt{x^2 - 4}$
55. $f(x) = \dfrac{x}{x + 2}$
56. $f(x) = \dfrac{-3}{\sqrt{2 - x}}$
57. $G(x) = \dfrac{1}{x^2 - 4}$
58. $f(x) = \dfrac{2x}{x^2 + 3}$
59. $f(x) = \dfrac{-3}{x^2 + 3x - 10}$
60. $f(x) = \dfrac{2x - 3}{x^3 - 4x^2 + x - 4}$
61. $F(x) = \dfrac{\sqrt{2x - 1}}{4 - x}$
62. $H(x) = \sqrt{x^3 - 2x^2 - 8x}$

63. The radius r (in meters) of an oil spill is a function of the time t (in minutes) that the oil has been leaking. This function is given by

$$r(t) = \sqrt{2t - 1}, \quad \text{for } t \geq 1.$$

Find $r(25)$ and describe what this value represents.

64. The bending moment M (in pound-feet, lb-ft) along a simply supported 9-ft steel beam that carries a uniformly distributed load of 800 pounds per foot (lb/ft) is a function of the distance x (in feet) from one end of the beam and is given by

$$M(x) = \tfrac{1}{2}(9)(800)x - \tfrac{1}{2}(800)x^2$$
$$= 3600x - 400x^2, \quad \text{for } 0 \leq x \leq 9.$$

Find $M(1)$ and describe what this value represents.

65. The cost C (in dollars) for a taxicab fare is a function of the distance x driven (in miles), and is given by

$$C(x) = 1.50 + 1.45x, \quad \text{for } x \geq 0.$$

Find $C(12.4)$ and describe what this value represents.

66. If $5000 is borrowed at $10\tfrac{1}{2}\%$ simple interest per year, the amount A (in dollars) that must be paid back is a function of the time t (in years) over which it is repaid and is given by

$$A(t) = 5000(1 + 0.105t), \quad \text{for } t \geq 0.$$

Find $A(12)$ and describe what this value represents.

 Critical Thinking

67. Does the inequality $y < x$ define y as a function of x? Explain.

68. Does the equation $y^n = x$ define y as a function of x when n is an odd integer? an even integer? zero? Explain.

69. Let $X = \{1, 2, 3\}$ and $Y = \{a, b, c, d\}$. Use the definition of a function to determine whether each of the following sets of ordered pairs or figures defines a function from set X to set Y. Explain.

    (a) $\{(1, a), (2, b), (3, d)\}$
    (b) $\{(1, c), (3, a)\}$
    (c)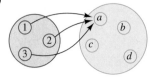
    (d)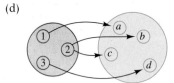

70. Does $f(a + b) = f(a) + f(b)$ for a function f? Use the function f defined by $f(x) = x^2$ to support your claim.

71. For a function f defined as follows, determine the value of A if $f(2) = 12$.

    (a) $f(x) = x^2 + Ax$
    (b) $f(x) = \sqrt{5x + A}$

72. Give an example of a function f with the property that $f(-x) = f(x)$ for all real numbers x and $-x$ in the domain of f. [Answers are not unique.]

73. The population size P of a certain organism is a function of the temperature t (in degrees Celsius) of the medium in which the organism exists and is given by

$P(t) = 3t^2$, for $0 \le t \le 40$.

If t_1 is the present temperature of the medium, compare $P(t_1)$ with $P(\tfrac{1}{2}t_1)$, and state the effect of halving the temperature on the population size.

74. In a psychological experiment, a student is given electrical shocks of varying intensities and is asked to rate the intensity of each shock in relation to an initial shock s_0, which is given a rating of 10. It is found that the response number R given by the student is a function of the magnitude (in milliamperes) of the shock s and is defined by

$$R(s) = \frac{s^{3/2}}{40}.$$

Compare $R(s_0)$ with $R(4s_0)$, and state the effect of quadrupling the intensity on the response number.

2.4 Graph of a Function

◆ Introductory Comments

The *graph of a function f* is the same as the graph of the equation $y = f(x)$ and is defined as follows.

◀ **Graph of a Function**

> The **graph of a function f** is the set of all points (x, y) in the coordinate plane such that x is in the domain of f and $y = f(x)$.

The point-plotting method discussed in Section 2.2 may be used to sketch the graph of a simple function. First, we select various values of x from the domain of f and compute the functional values $f(x)$. Next, we plot the points $(x, f(x))$ in the coordinate plane, where x is the directed distance from the y-axis and $f(x)$ is the directed distance from the x-axis. Finally, we connect these points to form a smooth curve, which is the graph of f. The graph of the function f defined by $f(x) = x^2$ is shown in Figure 2.32. The graph of this function is the same as the graph of the equation $y = x^2$ (see Figure 2.24).

To generate the graph of a function f in the viewing rectangle of a graphing calculator, we enter the equation $y = f(x)$. For example, to generate the graph of the function f defined by $f(x) = x^2$ on a graphing calculator, we simply enter the equation $y = x^2$.

◆ Finding the Domain and Range from the Graph of a Function

Observe in Figure 2.32 that the domain and range of the function f defined by $f(x) = x^2$ are evident from its graph. For the domain, we look along the x-axis and note the interval(s) over which the graph is drawn. For the range, we look along the y-axis and note the interval(s) over which the graph is drawn.

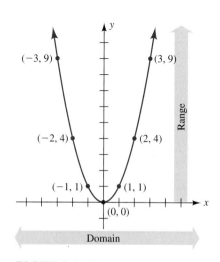

FIGURE 2.32
Graph of $f(x) = x^2$.
Domain: $(-\infty, \infty)$.   Range: $[0, \infty)$.

EXAMPLE 1 The graph of the function F defined by

$$F(x) = -1 + \sqrt{15 + 2x - x^2}$$

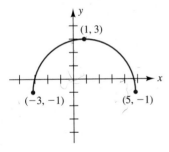

FIGURE 2.33

is shown in Figure 2.33. Use this graph to determine the domain and range of F.

◆ **Solution**   As we look along the x-axis, we see that the graph extends from $x = -3$ to $x = 5$. Hence, we conclude that the domain of this function is all real numbers in the closed interval $[-3, 5]$. As we look along the y-axis, we see that the graph extends from $F(-3) = -1$ [or $F(5) = -1$] to $F(1) = 3$. Hence, the range of this function is all real numbers in the closed interval $[-1, 3]$. ◆

PROBLEM 1 Referring to the function F in Example 1, compute each functional value, if possible. Check each answer with the graph in Figure 2.33.
(a) $F(0)$ **(b)** $F(6)$ ◆

◆ **Symmetry and Real Zeros of Functions**

We say that a function is an *even function* if its graph is symmetric with respect to the y-axis or that it is an *odd function* if its graph is symmetric with respect to the origin. The following tests can be used to determine if a function is even or odd.

◆ **Tests for Even and Odd Functions**

1. A function f is an **even function** if for every x in the domain of f, $-x$ is also in the domain of f, and $f(-x) = f(x)$.

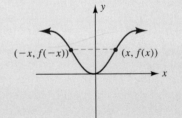

2. A function f is an **odd function** if for every x in the domain of f, $-x$ is also in the domain of f, and $f(-x) = -f(x)$.

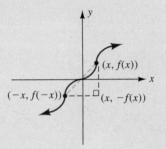

The **real zeros** of a function f are the real numbers x for which $f(x) = 0$. When $f(x) = 0$, the graph of the function f intersects the x-axis. Hence, the real zeros of a function f are the x-intercepts of the graph of f.

SECTION 2.4 ◆ Graph of a Function

EXAMPLE 2 Determine if the function is even, odd, or neither. Then find the real zero(s) of the function.

(a) $g(x) = \sqrt{x^2 - 4}$ **(b)** $h(x) = x - x^3$

◀ **Solution**

(a) *Even, odd, or neither:* The function g is *even* because

$$g(-x) = \sqrt{(-x)^2 - 4} = \sqrt{x^2 - 4} = g(x).$$

Real zeros: The zeros of g are the real roots of the equation $g(x) = \sqrt{x^2 - 4} = 0$. Since the square root of zero is zero, we have

$$x^2 - 4 = 0 \quad \text{or} \quad x = \pm 2.$$

Hence, the real zeros of the function g are ± 2.

We can use a graphing calculator to verify our work. The graph shown in Figure 2.34 has x-intercepts ± 2 and symmetry with the respect to the y-axis. Thus, the function g is an even function with real zeros ± 2.

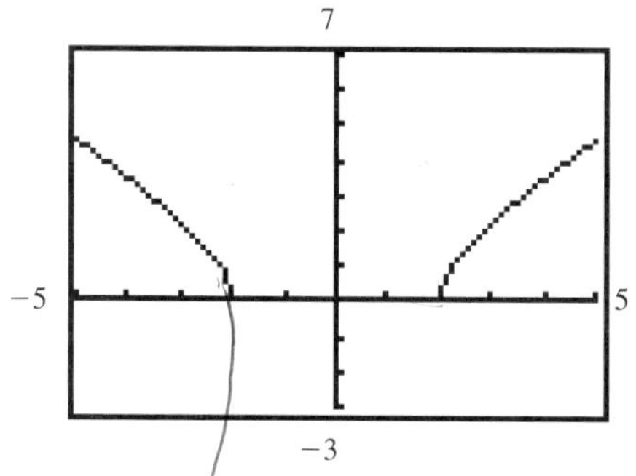

(b) *Even, odd, or neither:* The function h is *odd* because

$$h(-x) = (-x) - (-x)^3 = -x + x^3 = -(x - x^3) = -h(x).$$

Real zeros: The zeros of h are the real roots of the equation $h(x) = x - x^3 = 0$. By factoring and then applying the zero product property, we can find the roots:

$$x - x^3 = 0$$
$$x(1 - x^2) = 0$$
$$x = 0 \quad \text{or} \quad 1 - x^2 = 0$$
$$x = \pm 1$$

Hence, the real zeros of the function h are 0 and ± 1.

We can use a graphing calculator to verify our work. The graph in Figure 2.35 has x-intercepts 0 and ± 1 and symmetry with respect to the origin. Thus, the function h is an odd function with real zeros 0 and ± 1. ◆

PROBLEM 2 Use the graphs of the functions g and h in Figures 2.34 and 2.35 to determine their domains and ranges. ◆

◆ **Solving Equations and Inequalities Graphically**

Example 2 illustrates that the concepts of real zeros, x-intercepts, and real roots of equations are closely related. In general, if f is a function and k is a real number, then the following statements are equivalent:

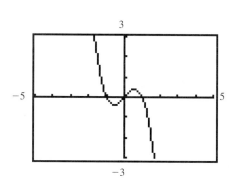

FIGURE 2.34
Graph of $g(x) = \sqrt{x^2 - 4}$, an even function with zeros ± 2

FIGURE 2.35
Graph of $h(x) = x - x^3$, an odd function with zeros 0 and ± 1.

1. k is an x-intercept for the graph of f.
2. k is a real zero of f.
3. k is a real root of the equation $f(x) = 0$.

In Example 2, we found the real roots of the equation $f(x) = 0$ to get the real zeros of f, which are the x-intercepts of the graph of f. If we reverse the order of these steps, we have a method for *solving an equation graphically*: find the x-intercepts of the graph of f to get the real zeros of f, which are the real roots of the equation $f(x) = 0$. We refer to this procedure as the **x-intercept method for solving an equation**.

Many graphing calculators have a built-in program for finding an x-intercept. If your calculator does not have a built-in program for finding the x-intercept, use the TRACE key and ZOOM key on the calculator to estimate the x-intercept.

EXAMPLE 3 Solve each equation by using the x-intercept method.

(a) $5 - 8x^2 = -11$ **(b)** $\sqrt{2x - 1} + 2 = x$

◆ Solution

(a) We begin by writing the equation in the form $f(x) = 0$:

$$5 - 8x^2 = -11$$
$$\underbrace{16 - 8x^2}_{f(x)} = 0 \quad \text{Add 11 to both sides}$$

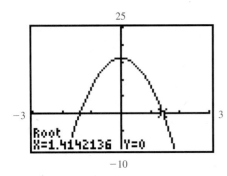

FIGURE 2.36

The roots of the equation $5 - 8x^2 = -11$ are approximately ± 1.414.

Note that for $f(x) = 16 - 8x^2$, we have $f(x) = f(-x)$ for all x in the domain of f. This indicates that f is an even function and its graph is symmetric with respect to the y-axis. We now generate the graph of $y = f(x)$ in the viewing rectangle of a graphing calculator. As shown in Figure 2.36, the x-intercept on the right is approximately 1.414. Because this function is an even function, the x-intercept on the left must be -1.414. Hence, the roots of the original equation are approximately ± 1.414. For the algebraic solution of this equation, see Example 1(d) in Section 1.8.

(b) We begin by writing the equation in the form $f(x) = 0$:

$$\sqrt{2x - 1} + 2 = x$$
$$\underbrace{\sqrt{2x - 1} + 2 - x}_{f(x)} = 0 \quad \text{Subtract } x \text{ from both sides}$$

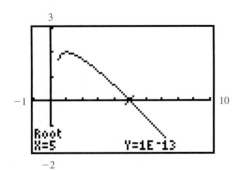

FIGURE 2.37

The root of the equation $\sqrt{2x - 1} + 2 = x$ is 5.

We now generate the graph of $y = f(x)$ in the viewing rectangle of a graphing calculator. As shown in Figure 2.37, the x-intercept of this graph is 5. Hence, the root of the original equation is also 5. We can check this solution by replacing x with 5 in the original equation:

SECTION 2.4 ◆ Graph of a Function

$$\sqrt{2(5) - 1} + 2 \stackrel{?}{=} (5)$$
$$\sqrt{9} + 2 \stackrel{?}{=} 5$$
$$3 + 2 \stackrel{?}{=} 5$$
$$5 = 5$$
◆

Equations that contain an unknown in the radicand, as in Example 3(b), are called **radical equations**. Algebraically, we can eliminate the radical sign in a radical equation that contains one radical expression by isolating the radical on one side of the equation and then raising both sides of the equation to the power that is equal to the index of the radical. But does raising both sides of an equation to a positive integer power always generate equivalent equation? Consider the equation.

$$x = 5$$

Of course, the solution of this equation is 5. If we square both sides of this equation, we obtain

$$x^2 = 25,$$

which implies that $x = \pm 5$. Note that the solution of the original equation, 5, appears as one of the solutions of the equation $x^2 = 25$. However, this equation also has the root -5, which is *not* a solution of the original equation. We refer to a root such as -5 that develops through this algebraic process, but does not satisfy the original equation, as an **extraneous root**.

When raising both sides of an equation to a positive *even* integer power, it is essential to check all apparent solutions which develop in order to make certain that no extraneous root is recorded as a solution to the original equation.

PROBLEM 3 Solve the equation $\sqrt{2x - 1} + 2 = x$ algebraically by isolating the radical expression on one side of the equation and then squaring both sides. What extraneous root develops through this process? ◆

Graphically, we can solve the inequality $f(x) > 0$ by determining the interval(s) on which the graph of f is above the x-axis. To solve $f(x) < 0$, we determine the interval(s) on which the graph of f is below the x-axis. For inequalities containing $\leq$ or $\geq$, we include the real zeros (x-intercepts) of f as part of the solution set. We refer to this procedure as the ***x*-intercept method for solving an inequality**.

EXAMPLE 4 Solve each inequality by using the x-intercept method.

(a) $|2x - 3| < 5$ (b) $\dfrac{x + 1}{2} - \dfrac{5(x - 1)}{8} \geq 1$

Solution

(a) We begin by writing the inequality in the form $f(x) < 0$:

$$|2x - 3| < 5$$
$$\underbrace{|2x - 3| - 5}_{f(x)} < 0 \quad \text{Subtract 5 from both sides}$$

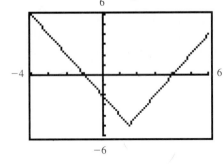

FIGURE 2.38
The solution set of the inequality $|2x - 3| < 5$ is $(-1, 4)$.

We now generate the graph of $y = f(x)$ in the viewing rectangle of a graphing calculator. Since the inequality symbol is *less than* ($<$), we look for the interval on which the graph is *below* the x-axis. As shown in Figure 2.38, the graph is below the x-axis when $-1 < x < 4$. Hence, the solution set of the original inequality is $(-1, 4)$. For the algebraic solution of this inequality, see Example 2(c) in Section 1.8.

(b) We begin by writing the inequality in the form $f(x) \geq 0$:

$$\frac{x + 1}{2} - \frac{5(x - 1)}{8} \geq 1$$

$$\underbrace{\frac{x + 1}{2} - \frac{5(x - 1)}{8} - 1}_{f(x)} \geq 0 \quad \text{Subtract 1 from both sides}$$

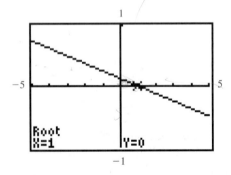

FIGURE 2.39
The solution set of the inequality $\frac{x + 1}{2} - \frac{5(x - 1)}{8} \geq 1$ is $(-\infty, 1]$.

We now generate the graph of $y = f(x)$ in the viewing rectangle of a graphing calculator. Since the inequality symbol is *greater than or equal to* ($\geq$), we look for the interval(s) on which the graph intersects the x-axis or is *above* the x-axis. As shown in Figure 2.39, the graph intersects the x-axis at $x = 1$ and is above the x-axis when $x < 1$. Hence, the solution set of the original inequality is $(-\infty, 1]$. ◆

PROBLEM 4 Solve the inequality $\frac{x + 1}{2} - \frac{5(x - 1)}{8} \geq 1$ algebraically. Check your answer with the graphical result in Example 4(b). ◆

◆ Application: Solving Word Problems

An essential goal of our study of algebra is to become skillful in reading word problems and changing the words into algebraic equations (or inequalities) that can be solved for the *unknowns*. Unfortunately, no one single procedure can be given to solve all word problems. In general, however, we assign one of the unknowns the variable x and express each of the other unknowns in terms of x. We then develop an equation (or inequality) that relates the known and unknown quantities and solve for x. The procedure is shown in the next example.

SECTION 2.4 ◆ Graph of a Function

EXAMPLE 5 A length of rope hangs from the top of a flagpole, with 6 feet coiled up on the ground at the base of the pole. When the free end of the rope is brought out 18 feet from the base of the pole, it just touches the ground. Assuming the flagpole is on level ground, what is the height of the flagpole?

◆ **Solution** Referring to Figure 2.40, we let

$$x = \text{height of the flagpole (in feet)}.$$

Since the rope is 6 feet longer than the height of the flagpole, we have

$$x + 6 = \text{length of the rope (in feet)}.$$

From the sketch in Figure 2.40, we can see that x, $x + 6$, and 18 are related by the Pythagorean theorem. Thus,

$$\boxed{(\text{hypotenuse})^2} = \boxed{(\text{leg})^2} + \boxed{(\text{leg})^2}$$

$$(x + 6)^2 = x^2 + 18^2$$

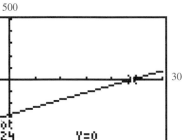

FIGURE 2.40

The quantities x, $x + 6$, and 18 are related by the Pythagorean theorem.

We can solve this equation algebraically or graphically.

Choosing the graphical approach, we write the equation in the form $f(x) = 0$:

$$(x + 6)^2 = x^2 + 18^2$$
$$\underbrace{(x + 6)^2 - x^2 - 18^2}_{f(x)} = 0 \quad \text{Subtract } x^2 \text{ and } 18^2 \text{ from both sides}$$

We now generate the graph of $y = f(x)$ in the viewing rectangle of a graphing calculator. As shown in Figure 2.41, the x-intercept of this graph is 24. Hence, the root of the original equation is also 24. Therefore, the height of the flagpole is 24 ft.

In Problem 5 you are asked to verify the solution to this problem by solving the equation $(x + 6)^2 = x^2 + 18^2$ algebraically. ◆

FIGURE 2.41

The root of the equation $(x + 6)^2 - x^2 - 18^2 = 0$ is 24.

PROBLEM 5 Check the result of Example 5 by solving the equation $(x + 6)^2 = x^2 + 18^2$ algebraically. ◆

Exercises 2.4

Basic Skills

In Exercises 1–10, a function and its graph are given.
(a) Determine the domain of the function.
(b) Determine the range of the function.
(c) Find any real zero(s) of the function.
(d) State whether the function is even, odd, or neither.

1. $f(x) = |x| - 2$
2. $g(x) = 3x - 2$

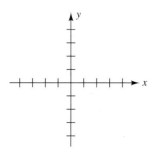
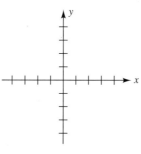

3. $g(x) = x^2 + 2x$
4. $H(x) = 2x^2 - x^4$

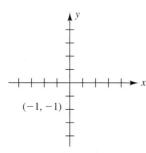
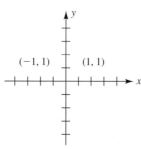

5. $f(x) = \sqrt{5 - x^2}$
6. $g(x) = -\sqrt{x^2 - 3}$

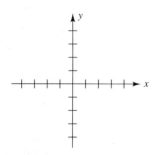
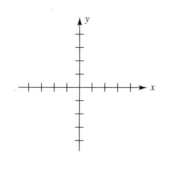

7. $f(x) = \dfrac{-2}{x^2 + 1}$
8. $g(x) = \dfrac{x^2 + 1}{x}$

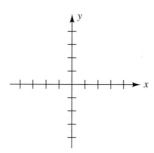
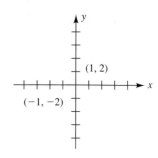

9. $g(x) = x^5 - 5x$, for $-2 \le x \le 2$

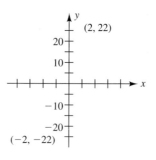

10. $f(x) = \begin{cases} x^2 & \text{if } -2 \le x < 1 \\ 3 - 2x & \text{if } 1 \le x \le 3 \end{cases}$

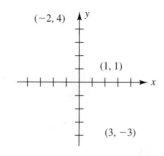

In Exercises 11–20, determine whether the function is even, odd, or neither. Check graphically with a graphing calculator.

11. $F(x) = x$
12. $g(x) = x^3$
13. $f(x) = 3 - |x|$
14. $f(x) = 2x^2 - 4$
15. $f(x) = 4x - x^5$
16. $f(x) = x^4 + 2x^2 + 3$

17. $g(x) = \dfrac{3 - x^2}{2 + x^6}$

18. $H(x) = \dfrac{x + x^3}{2|x|}$

19. $h(x) = x^3 - 3x + 1$

20. $g(x) = (3x^2 - 2x)^2$

In Exercises 21–30, find the real zero(s) of each function. Check graphically with a graphing calculator.

21. $g(x) = 3x - 5$
22. $f(x) = 4 + 5x$
23. $F(x) = x^2 - 3x - 40$
24. $g(x) = x^2 - 8x + 12$
25. $h(x) = 2x^2 - 7x + 3$
26. $f(x) = 2x^4 - 5x^2 - 12$
27. $g(x) = \dfrac{2x - 3}{x^2 + 4}$
28. $G(x) = \dfrac{9 - x^2}{2x + 3}$
29. $H(x) = 5 - \sqrt{x^4 + 9}$
30. $h(x) = x - \sqrt{4x^2 - x - 2}$

In Exercises 31–50, solve each equation or inequality by using the x-intercept method in conjunction with a graphing calculator. Check algebraically.

31. $9x - 2(4x + 1) = 3$
32. $\dfrac{x}{6} = \dfrac{x + 4}{2}$
33. $6x^2 = x + 2$
34. $x^3 - 3x^2 - 2x + 6 = 0$
35. $\left|\dfrac{x + 4}{3}\right| = 7$
36. $2 - |3x - 11| = -4$
37. $3\sqrt{x + 1} = 4$
38. $4\sqrt{3x + 1} = 9$
39. $\sqrt{1 - x} - x = 5$
40. $\sqrt{5x - 1} + 3 = x$
41. $\sqrt[3]{x^3 + 8} - x = 2$
42. $\sqrt[4]{x^4 - x + 3} = x$
43. $2x + 6 \geq 3x + 4$
44. $4x - (x - 2) < x - 3$
45. $\dfrac{x + 5}{2} - 3 \leq \dfrac{3x}{5} - \dfrac{x - 4}{4}$
46. $\dfrac{x + 1}{2} + \dfrac{2x + 3}{4} \geq 2$
47. $|5 - 2x| > 3$
48. $|4x + 3| + 2 < 17$
49. $|x^2 - 6| \geq 3$
50. $|2 - x^3| < 6$

In Exercises 51–60, set up an equation or inequality and solve by the x-intercept method. Check each result algebraically.

51. The hypotenuse of a right triangle is 1 inch longer than one of its legs. The other leg is 5 inches long. What is the length of the hypotenuse?

52. The length of a rectangle is twice its width. If each side of the rectangle were increased by 3 centimeters, the new perimeter would be 48 centimeters. What are the dimensions of the original rectangle?

53. At a Red Sox baseball game, the price for a box seat is $9 and for grandstand seat, $6. A total of 24,320 seats are sold for a particular game, and the receipts from the box seats equal those from the grandstand seats. How many box seats are sold?

54. A developer purchased 22 acres of land for $900,000. He agreed to pay $50,000 per acre for buildable land and $10,000 per acre for marshland. How many acres of buildable land were purchased?

55. Oak flooring is installed in a square room and a rectangular room of a new home. The square room requires 18 square feet (sq ft) less flooring than is used in the rectangular room. The length of the rectangular room is 6 feet more than a side of the square room, and the width of the rectangular room is 4 feet less than a side of the square room. Find the amount of oak flooring used in each room.

56. How many pounds of cashews, worth $6.40 per pound, must be added to 12 pounds of peanuts, worth $2.40 per pound, to form a mixture that is worth $4.00 per pound?

57. A campsite is set up on the bank of a river that is 40 ft wide. On the other side of the river and 200 ft downstream is another campsite. A newly constructed footbridge and walkway join the two campsites, as shown in the sketch. The walkway, which runs along the edge of the straight river, cost $2 per foot to construct. The footbridge, which is built diagonally over the river, costs $10 per foot. If the total cost of this project is $840, find the length of the walkway.

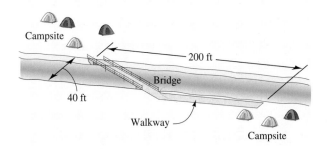

58. In order to widen a highway, the state takes by eminent domain a strip of land from Lot A, which is shown in the sketch at the top of the next page. What is the largest width for this strip of land that can be taken if Lot A must remain at least one acre after removing the strip? [*Hint:* 1 acre = 43,560 square feet.]

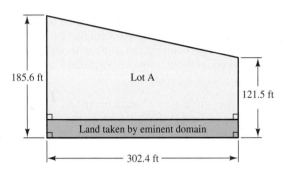

59. Find the set of three consecutive positive odd integers such that the largest is greater than twice the smallest.

60. Find all sets of four consecutive positive even integers such that the largest is greater than twice the smallest.

 Critical Thinking

61. Suppose that f is an even function and one point on its graph is $(-2, 3)$. Find the coordinates of a second point on its graph.

62. Suppose that f is an odd function and one point on its graph is $(-2, 3)$. Find the coordinates of a second point on its graph.

63. Suppose the graph of a function f does not intersect the x-axis. What can we conclude about the solution of the equation $f(x) = 0$?

64. Under what conditions is the function f defined by $f(x) = x^n$ an even function? an odd function?

65. How many y-intercepts can the graph of a function have? How many x-intercepts?

66. Use a graphing calculator to generate the graphs of the functions f and g defined by

$$f(x) = |x| \text{ and } g(x) = \sqrt{x^2}.$$

What can we conclude about $|x|$ and $\sqrt{x^2}$?

67. Can the graph of a nonzero function be symmetric with respect to the x-axis? Explain.

68. Is the function g defined by $g(x) = xf(x)$ even, odd, or neither when the function f is an even function? when the function f is an odd function?

69. Biologists estimate that the population P of the endangered red-spotted turtle is a function of time t (in years) and is given by

$$P(t) = 90 - 10(34t + 22)^{2/5}.$$

In how many years will this species become extinct?

70. A company determines that when it produces q units of a product, its total revenue from sales is $200q^{1/2}$ dollars. Suppose the company can produce no more than 500 units and that the variable cost per unit is $4 and the fixed cost is $1600. How many units must the company produce to break even (no profit). [*Hint:* Revenue $-$ cost = profit.]

2.5 Shifting, Reflecting, and Stretching a Graph

◆ Introductory Comments

Figure 2.42 shows the graphs of eight basic functions that occur frequently in mathematics: *constant function, identity function, absolute value function, squaring function, cubing function, reciprocal function, square root function,* and *cube root function*. By plotting points or using a graphing calculator, we can verify the graph of each of these functions.

SECTION 2.5 ♦ *Shifting, Reflecting, and Stretching a Graph* 113

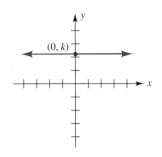

(a) Constant function
$f(x) = k$
Domain: $(-\infty, \infty)$
Range: $\{k\}$

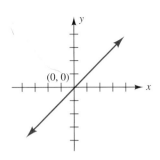

(b) Identity function
$f(x) = x$
Domain: $(-\infty, \infty)$
Range: $(-\infty, \infty)$

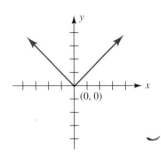

(c) Absolute value function
$f(x) = |x|$
Domain: $(-\infty, \infty)$
Range: $[0, \infty)$

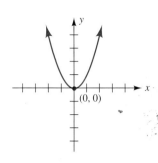

(d) Squaring function
$f(x) = x^2$
Domain: $(-\infty, \infty)$
Range: $[0, \infty)$

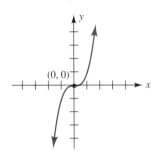

(e) Cubing function
$f(x) = x^3$
Domain: $(-\infty, \infty)$
Range: $(-\infty, \infty)$

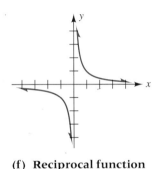

(f) Reciprocal function
$f(x) = 1/x$
Domain: $(-\infty, 0) \cup (0, \infty)$
Range: $(-\infty, 0) \cup (0, \infty)$

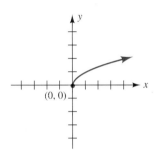

(g) Square root function
$f(x) = \sqrt{x}$
Domain: $[0, \infty)$
Range: $[0, \infty)$

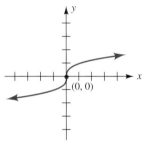

(h) Cube root function
$f(x) = \sqrt[3]{x}$
Domain: $(-\infty, \infty)$
Range: $(-\infty, \infty)$

FIGURE 2.42
Graphs of eight basic functions

In this section, we discuss techniques of graphing a function without using a graphing calculator or plotting points. We use the graphs of these eight basic functions to sketch the graphs of many other related functions by using one or more procedures:

1. *Shifting* a graph vertically or horizontally.
2. *Reflecting* a graph about the *x*-axis or *y*-axis.
3. *Stretching* or *compressing* a graph.

♦ **Vertical Shift Rule**

Figure 2.43 shows the graphs of the absolute value function $f(x) = |x|$ and the function $F(x) = |x| + 2$ in the same viewing rectangle of a graphing calculator. The graph of $F(x) = |x| + 2$ appears to be the same as the graph of $f(x) = |x|$, but *shifted vertically upward* 2 units. When we substitute the same input into the functions F and f, the output $F(x)$ is always 2 more than the output $f(x)$. Without using a graphing calculator or plotting points, we can sketch the graph of $F(x) = |x| + 2$ by shifting the graph of $f(x) = |x|$ vertically upward 2 units. We can apply this rule to any function. In summary, we state this property as the *vertical shift rule*.

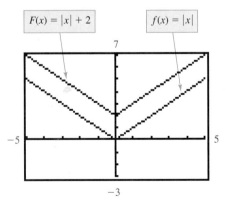

FIGURE 2.43
Comparison of the graphs of
$F(x) = |x| + 2$ and $f(x) = |x|$

Vertical Shift Rule

If f is a function and c is a constant, then the graph of the function F defined by

$$F(x) = f(x) + c$$

is the same as the graph of f shifted *vertically upward* $|c|$ units if $c > 0$ or shifted *vertically downward* $|c|$ units if $c < 0$.

To obtain an accurate sketch, it is good practice to determine and label the *intercepts* of the graph. Recall that the *x*-intercepts of the graph of a function f are the real zeros of the function and may be found by solving the equation $f(x) = 0$. The *y*-intercept of the graph of a function f may be found by evaluating $f(0)$. The graph of a function can have at most one *y*-intercept.

EXAMPLE 1 Sketch the graph of each function and label the intercepts, if they exist.

(a) $F(x) = x^2 + 1$ (b) $G(x) = \sqrt{x} - 2$

◆ Solution

(a) We can think of this function as $F(x) = f(x) + 1$, where $f(x) = x^2$ is the squaring function in part (d) of Figure 2.42. Thus, by the vertical shift rule, the graph of $F(x) = x^2 + 1$ is the same as the graph of $f(x) = x^2$, but shifted vertically *upward* 1 unit. The graph of the function F is shown in Figure 2.44.

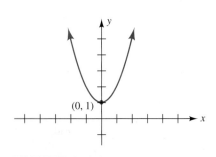

FIGURE 2.44
Graph of $F(x) = x^2 + 1$

x-intercept:

$F(x) = 0$
$x^2 + 1 = 0$
$x^2 = -1$

y-intercept:

$F(0) = (0)^2 + 1 = 1$

(b) We can think of this function as $G(x) = f(x) - 2$, where $f(x) = \sqrt{x}$ is the square root function in part (g) of Figure 2.42. Thus, by the vertical shift rule, the graph of $G(x) = \sqrt{x} - 2$ is the same as the graph of $f(x) = \sqrt{x}$, but shifted vertically *downward* 2 units. The graph of the function G is shown in Figure 2.45.

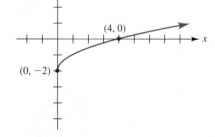

FIGURE 2.45
Graph of $G(x) = \sqrt{x} - 2$

x-intercept:

$G(x) = 0$
$\sqrt{x} - 2 = 0$
$\sqrt{x} = 2$
$x = 4$ Square both sides

y-intercept:

$G(0) = \sqrt{0} - 2 = -2$

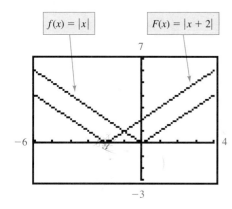

FIGURE 2.46

Comparison of the graphs of $F(x) = |x + 2|$ and $f(x) = |x|$

PROBLEM 1 Sketch the graph of $h(x) = x^3 - 1$ and label the intercepts.

◆ Horizontal Shift Rule

Figure 2.46 shows the graphs of the absolute value function $f(x) = |x|$ and the function $F(x) = |x + 2|$ in the same viewing rectangle of a graphing calculator. The graph of $F(x) = |x + 2|$ appears to be the same as the graph of $f(x) = |x|$, but *shifted horizontally to the left* 2 units. When we substitute an input into F that is 2 *less than* the input we use in f, the values of the outputs $F(x)$ and $f(x)$ are equal. Without using a graphing calculator or plotting points, we can sketch the graph of $F(x) = |x + 2|$ by shifting the graph of $f(x) = |x|$ horizontally to the left 2 units. Like the vertical shift rule, we can apply this property to any function. In summary, we state the *horizontal shift rule*.

◀ Horizontal Shift Rule

> If f is a function and c is a constant, then the graph of the function F defined by
>
> $$F(x) = f(x + c)$$
>
> is the same as the graph of f shifted *horizontally to the left* $|c|$ units if $c > 0$ or shifted *horizontally to the right* $|c|$ units if $c < 0$.

EXAMPLE 2 Sketch the graph of each function and label the intercepts.

(a) $F(x) = (x + 1)^2$ **(b)** $G(x) = \sqrt[3]{x - 2}$

◀ Solution

(a) We can think of this function as $F(x) = f(x + 1)$, where $f(x) = x^2$ is the squaring function in part (d) of Figure 2.42. Thus, by the horizontal shift rule, the graph of $F(x) = (x + 1)^2$ is the same as the graph of $f(x) = x^2$, but shifted horizontally *to the left* 1 unit. The graph of the function F is shown in Figure 2.47.

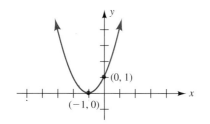

FIGURE 2.47

Graph of $F(x) = (x + 1)^2$

$$\begin{array}{ll} \text{x-intercept:} & \text{y-intercept:} \\ F(x) = 0 & F(0) = (0 + 1)^2 = 1 \\ (x + 1)^2 = 0 & \\ x + 1 = 0 \quad \text{Take the square root} & \\ \text{of both sides} & \\ x = -1 & \end{array}$$

(b) We can think of this function as $G(x) = f(x - 2)$, where $f(x) = \sqrt[3]{x}$ is the cube root function in part (h) of Figure 2.42. Thus, by the horizontal shift rule, the graph of $G(x) = \sqrt[3]{x - 2}$ is the same as the graph of $f(x) = \sqrt[3]{x}$, but shifted horizontally *to the right* 2 units. The graph of the function G is shown in Figure 2.48.

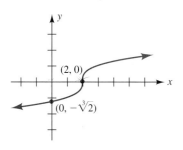

FIGURE 2.48

Graph of $G(x) = \sqrt[3]{x - 2}$

CHAPTER 2 ◆ Functions and Graphs

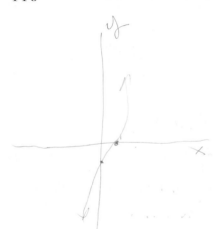

x-intercept:
$$G(x) = 0$$
$$\sqrt[3]{x-2} = 0$$
$$x - 2 = 0 \quad \text{Cube both sides}$$
$$x = 2$$

y-intercept:
$$G(0) = \sqrt[3]{0-2} = \sqrt[3]{-2}$$
$$\approx -1.26$$

◆

PROBLEM 2 Sketch the graph of $h(x) = (x-1)^3$ and label the intercepts.

◆

As illustrated in the next example, we can use both the vertical and horizontal shift rules to help sketch the graph of a function.

EXAMPLE 3 Sketch the graph of $F(x) = \dfrac{1}{x-2} + 1$ and label the intercepts.

◆ **Solution** We can think of this function as $F(x) = f(x-2) + 1$, where $f(x) = 1/x$ is the reciprocal function in part (f) of Figure 2.42. Thus, by the horizontal and vertical shift rules, we obtain the graph of F by shifting the graph of $f(x) = 1/x$ to the right 2 units and upward 1 unit, as shown in Figure 2.49.

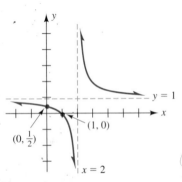

FIGURE 2.49

Graph of $F(x) = \dfrac{1}{x-2} + 1$

x-intercept:
$$F(x) = 0$$
$$\frac{1}{x-2} + 1 = 0$$
$$1 + (x-2) = 0 \quad \begin{array}{l}\text{Multiply both sides by}\\ (x-2), \text{ provided}\\ x - 2 \neq 0\end{array}$$
$$x = 1$$

y-intercept:
$$F(0) = \frac{1}{0-2} + 1 = \frac{1}{2}$$

The vertical dashed line that passes through (2, 0) is called a *vertical asymptote* for the graph of F. The horizontal dashed line that passes through (0, 1) is called a *horizontal asymptote* for the graph of F. In Chapter 4 we will discuss asymptotes in more detail.

◆

PROBLEM 3 Sketch the graph of $F(x) = (x+1)^2 - 4$ and label the intercepts.

◆

◆ x-Axis and y-Axis Reflection Rules

Figure 2.50 shows the graphs of the square root function $f(x) = \sqrt{x}$ and the functions $F(x) = -\sqrt{x}$ and $G(x) = \sqrt{-x}$ in the same viewing rectangle of a graphing calculator. It appears the graph of $F(x) = -\sqrt{x}$ is the same as the graph of $f(x) = \sqrt{x}$, but reflected about the x-axis. When we substitute the same input into the functions f and F, the output $F(x)$ is always the *opposite* of the output $f(x)$. This property can be applied to any function. In summary, we state the *x-axis reflection rule*.

FIGURE 2.50

Comparison of the graphs of $f(x) = \sqrt{x}$, $F(x) = -\sqrt{x}$, and $G(x) = \sqrt{-x}$

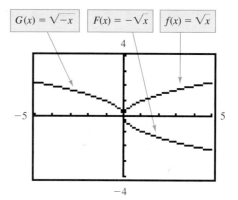

◆ **x-Axis Reflection Rule**

If f is a function, then the graph of the function F defined by

$$F(x) = -f(x)$$

is the same as the graph of f, but *reflected about the x-axis*.

Figure 2.50 also shows that the graph of $G(x) = \sqrt{-x}$ is the same as the graph of $f(x) = \sqrt{x}$, but reflected about the *y*-axis. When we substitute an input into the function G that is the *opposite* of the input we use for f, the values of the outputs $G(x)$ and $f(x)$ are equal. Like the other shift and reflect properties, we can apply this property to any function. In summary, we state the *y-axis reflection rule*.

◆ **y-Axis Reflection Rule**

If f is a function, then the graph of the function F defined by

$$F(x) = f(-x)$$

is the same as the graph of f, but *reflected about the y-axis*.

EXAMPLE 4 Sketch the graph of each function and label the intercepts.

(a) $F(x) = 2 - x^3$ (b) $h(x) = \sqrt{2 - x}$

◆ Solution

(a) We can think of this function as $F(x) = -f(x) + 2$, where $f(x) = x^3$ is the cubing function in part (e) of Figure 2.42. Thus, by the *x*-axis reflec-

118   CHAPTER 2 ♦ *Functions and Graphs*

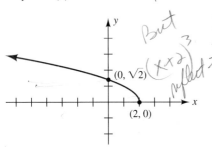

FIGURE 2.51
Graph of $F(x) = 2 - x^3$

tion rule and the vertical shift rule, the graph of $F(x) = 2 - x^3$ is obtained by reflecting the graph of $f(x) = x^3$ *about the x-axis* and then shifting this graph *upward* 2 units, as shown in Figure 2.51.

x-intercept:	y-intercept:
$F(x) = 0$	$F(0) = 2 - (0)^3 = 2$
$2 - x^3 = 0$	
$x^3 = 2$	
$x = \sqrt[3]{2} \approx 1.26$ **Extract the cube root**	

(b) The graph of $h(x) = \sqrt{2 - x}$ is the same as the graph of $h(-x) = \sqrt{2 + x}$, but reflected about the y-axis. To sketch $h(-x) = \sqrt{2 + x}$, we shift the graph of the square root function $f(x) = \sqrt{x}$ to the left 2 units. Now, reflecting this graph *about the y-axis* gives us the graph of $h(x) = \sqrt{2 - x}$, as shown in Figure 2.52.

FIGURE 2.52
Graph of $h(x) = \sqrt{2 - x}$

x-intercept:	y-intercept:
$h(x) = 0$	$h(0) = \sqrt{2 + 0} = \sqrt{2}$
$\sqrt{2 - x} = 0$	≈ 1.41
$2 - x = 0$ **Square both sides**	
$x = 2$	

CAUTION   The order in which we apply the shift rules and the axis reflection rules is important. The graph of $F(x) = 2 - x^3$ [in Example 4(a)] is *not* formed by shifting the graph of $f(x) = x^3$ vertically upward 2 units and then reflecting this graph about the x-axis. This sequence of steps would produce the graph of the function $G(x) = -(x^3 + 2)$.

PROBLEM 4   Sketch the graph of $F(x) = -|x + 2|$ and label the intercepts.

♦ Vertical Stretch and Compress Rule

Figure 2.53 shows the graph of the absolute value function $f(x) = |x|$ and the functions $F(x) = 2|x|$ and $G(x) = \frac{1}{2}|x|$ in the viewing rectangle of

FIGURE 2.53
Comparison of the graphs of $f(x) = |x|$, $F(x) = 2|x|$, and $G(x) = \frac{1}{2}|x|$

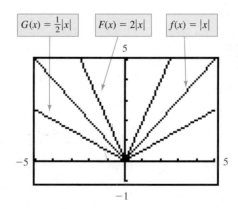

a graphing calculator. The graph of $F(x) = 2|x|$ appears to be a steeper V-shape than the graph of $f(x) = |x|$. The graph of $G(x) = \frac{1}{2}|x|$ appears to be a flatter V-shape than the graph of $f(x) = |x|$. When we substitute the same input into the functions f, F, and G, the output $F(x)$ is *twice* the output $f(x)$, and the output $G(x)$ is *half* the output $f(x)$. The graph of F is *stretched vertically* by a factor of 2, and the graph of G is *compressed vertically* by a factor of 2. In summary, we state the *vertical stretch and compress rules*.

◀ **Vertical Stretch and Compress Rules**

If f is a function and c is a real number with $c > 1$, then the graph of the function F defined by

$$F(x) = cf(x)$$

is similar to the graph of f, but is *stretched vertically* by a factor of c, and the graph of the function G defined by

$$G(x) = \frac{1}{c}f(x)$$

is similar to the graph of f, but is *compressed vertically* by a factor of c.

EXAMPLE 5   Sketch the graph of each function and label the intercepts.

(a) $G(x) = -\dfrac{|x|}{4}$ (b) $F(x) = 4(x - 1)$

◀ **Solution**

(a) We can think of this function as $G(x) = -\frac{1}{4}f(x)$, where $f(x) = |x|$. Thus, by the *x*-axis reflection rule and the vertical compress rule, we obtain the graph of $G(x) = -|x|/4$ by reflecting the graph of the absolute value function about the *x*-axis and then compressing vertically by a factor of 4, as shown in Figure 2.54.

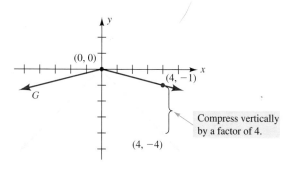

FIGURE 2.54

The graph of $G(x) = -|x|/4$ is formed from the graph of $y = -|x|$ by compressing vertically by a factor of 4.

(b) We can think of this function as $F(x) = 4f(x - 1)$, where $f(x) = x$ is the identity function in part (b) of Figure 2.42. By the horizontal shift rule and vertical stretch rule, we obtain the graph of $F(x) = 4(x - 1)$ by shifting the graph of $f(x) = x$ to the right 1 unit and then stretching vertically by a factor of 4, as shown in Figure 2.55.

x-intercept:

$$F(x) = 0$$
$$4(x - 1) = 0$$
$$x - 1 = 0$$
$$x = 1$$

y-intercept:

$$F(0) = 4(0 - 1) = -4$$

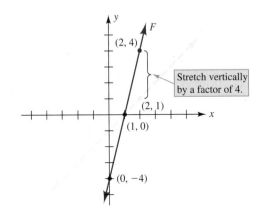

FIGURE 2.55
The graph of $F(x) = 4(x - 1)$ is formed from the graph of $y = x - 1$ by stretching vertically by a factor of 4.

PROBLEM 5 Sketch the graph of $F(x) = 4x - 1$ and label the intercepts.

◆ Increasing, Decreasing, and Constant Functions

A function f is said to be an *increasing function* if as x increases, the value of $f(x)$ also *increases*. A function f is said to be a *decreasing function* if as x increases, $f(x)$ *decreases*. A function f is said to be a *constant function* if as x increases, $f(x)$ *remains the same* [see part (a) of Figure 2.42.] More precisely, we state the following definitions.

◀ **Increasing, Decreasing, and Constant Functions**

1. A function f is an **increasing function** if for all a and b in the domain of f, $f(a) < f(b)$ whenever $a < b$.

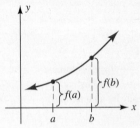

2. A function f is a **decreasing function** if for all a and b in the domain of f, $f(a) > f(b)$ whenever $a < b$.

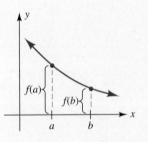

3. A function f is a **constant function** if for all a and b in the domain of f, $f(a) = f(b)$.

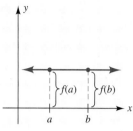

From the graph of a function, we may determine whether the function is increasing or decreasing. The graph of $F(x) = 4(x - 1)$, shown in Figure 2.55, always *rises* as we move from left to right along the x-axis. Thus, as x increases, $F(x)$ also increases, and we conclude the function F is an increasing function. The graph of $h(x) = \sqrt{2 - x}$, shown in Figure 2.52, always *falls* as we move from left to right along the x-axis. Hence, as x increases, $h(x)$ decreases, and we conclude the function h is a decreasing function.

The graph of $G(x) = -|x|/4$, shown in Figure 2.54, rises as we move from left to right until we reach the y-axis. The graph then falls as we continue to move along the x-axis from the y-axis toward the right. Thus, this function is *neither increasing nor decreasing* over its entire domain. This situation is typical of many functions. A function may be increasing over some intervals in its domain and decreasing over other intervals. For $G(x) = -|x|/4$, we say that G is *increasing on the interval* $(-\infty, 0)$ and *decreasing on the interval* $(0, \infty)$.

FIGURE 2.56

Graphs of $y = 1 - x^2$, $y = x - 1$, and $y = 4$, with dashed lines

EXAMPLE 6

(a) Sketch the graph of the piecewise-defined function g:

$$g(x) = \begin{cases} 1 - x^2 & \text{if } x < 1 \\ x - 1 & \text{if } 1 \leq x \leq 3 \\ 4 & \text{if } x > 3. \end{cases}$$

(b) Use the graph of function g to determine the open intervals on which the function is increasing, decreasing, or constant.

◀ Solution

(a) First, we apply the graphing techniques from this section to sketch the graphs of $y = 1 - x^2$, $y = x - 1$, and $y = 4$ with dashed lines, as shown in Figure 2.56. To form the graph of g, we darken the graphs of

$$y = 1 - x^2 \quad \text{for } x < 1$$

and

$$y = x - 1 \quad \text{for } 1 \leq x \leq 3$$

$$y = 4 \quad \text{for } x > 3$$

as shown in Figure 2.57. We place a *solid dot* at the point $(3, 2)$ to indicate that this point is part of the graph of g, and we place an *open circle* at the point $(3, 4)$ to indicate that this point is *not* part of the graph of g.

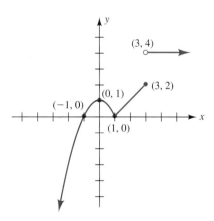

FIGURE 2.57

Graph of

$$g(x) = \begin{cases} 1 - x^2 & \text{if } x < 1 \\ x - 1 & \text{if } 1 \leq x \leq 3 \\ 4 & \text{if } x > 3 \end{cases}$$

Increasing on intervals $(-\infty, 0)$ and $(1, 3)$
Decreasing on the interval $(0, 1)$
Constant on the interval $(3, \infty)$

(b) As we move along the *x*-axis from left to right, the graph *rises* until we reach the *y*-axis. From the *y*-axis to $x = 1$, the graph *falls*. From $x = 1$ to $x = 3$, the graph again *rises*. Finally, to the right of $x = 3$, the graph remains the same. Thus, we conclude the function is *increasing* on the intervals $(-\infty, 0)$ and $(1, 3)$, *decreasing* on the interval $(0, 1)$, and *constant* on the interval $(3, \infty)$. ◆

To generate the graph of a piecewise-defined function on a graphing calculator, we enter each part of the function separately and include with each part of the function the interval over which that part is defined. Here is a typical display showing the entries for the piecewise-defined function in Example 6:

$$\boxed{y_1=} \ (1 - x^2)(x < 1)$$
$$\boxed{y_2=} \ (x - 1)(x \geq 1)(x \leq 3)$$
$$\boxed{y_3=} \ (4)(x > 3)$$

Also, when graphing a piecewise-defined function on a graphing calculator it is better to use the "dot" mode instead of the "connected" mode. In connected mode, the calculator may erroneously connect one part of the graph to the next part.

PROBLEM 6 Consult the manual that came with your calculator or the *Graphing Technology Laboratory Manual* that accompanies this text regarding piecewise-defined functions. Then use a graphing calculator to generate the graph of the function *g* in Example 6. ◆

Exercises 2.5

Basic Skills

In Exercises 1–42, use the shift rules, axis reflection rules, and stretch and compress rules to sketch the graph of each function. Label the axis intercepts, if they exist. Use a graphing calculator to verify each graph.

1. $f(x) = x + 3$
2. $g(x) = x^2 - 1$
3. $G(x) = x^3 - 2$
4. $H(x) = \dfrac{1}{x} + 2$
5. $F(x) = 3 + \sqrt{x}$
6. $f(x) = -2 + |x|$
7. $f(x) = |x - 1|$
8. $g(x) = \sqrt{x + 3}$
9. $H(x) = (x + 2)^2$
10. $F(x) = (x + 1)^3$
11. $F(x) = \dfrac{1}{x + 2}$
12. $f(x) = \sqrt[3]{x - 1}$
13. $h(x) = -\dfrac{1}{x}$
14. $g(x) = -x^3$
15. $f(x) = -x^2$
16. $H(x) = \sqrt[3]{-x}$
17. $g(x) = \sqrt{1 - x}$
18. $f(x) = -\sqrt{3 - x}$
19. $F(x) = (x - 1)^2 + 3$
20. $G(x) = |x - 4| - 1$
21. $f(x) = \sqrt{x + 2} - 1$
22. $h(x) = \dfrac{1}{x + 1} - 2$
23. $h(x) = -|x - 3|$
24. $f(x) = -(x + 1)$
25. $G(x) = 3 - x^2$
26. $H(x) = 2 - \sqrt{x}$
27. $F(x) = 2 - (x + 4)^3$
28. $h(x) = 1 - |x - 1|$
29. $f(x) = \sqrt[3]{x + 2} - 3$
30. $f(x) = 2 - \sqrt{4 - x}$

31. $H(x) = \dfrac{1}{2 - x}$

32. $g(x) = 3 + (2 - x)^2$

33. $g(x) = 3|x|$

34. $f(x) = -2x$

35. $f(x) = -\tfrac{1}{3}x^2$

36. $G(x) = \dfrac{x^3}{4}$

37. $G(x) = 2x - 1$

38. $F(x) = 3 - \tfrac{1}{2}x$

39. $F(x) = 2\sqrt{x - 4}$

40. $g(x) = 3 - 2|x|$

41. $F(x) = \dfrac{5}{x + 2}$

42. $h(x) = \dfrac{(x - 1)^2}{4}$

43. Which of the eight basic functions in Figure 2.42 are increasing functions? decreasing functions?

44. For each of the eight basic functions in Figure 2.42 that is neither an increasing nor decreasing function, determine the open intervals over which the function is increasing, decreasing, or constant.

In Exercises 45–52, determine whether the function is increasing, decreasing, or neither.

45. The function G in Exercise 3
46. The function g in Exercise 8
47. The function g in Exercise 17
48. The function g in Exercise 14
49. The function G in Exercise 25
50. The function G in Exercise 20
51. The function f in Exercise 29
52. The function f in Exercise 34

In Exercises 53–58, use the shift rules, axis reflection rules, and stretch and compress rules to sketch the graph of each piecewise-defined function. Label the axis intercepts, if they exist. Use the graph to determine the open intervals over which the function is increasing, decreasing, or constant. Use a graphing calculator to verify each graph.

53. $g(x) = \begin{cases} x + 1 & \text{if } x < 2 \\ 3 & \text{if } x \geq 2 \end{cases}$

54. $f(x) = \begin{cases} -x & \text{if } x \leq 4 \\ \sqrt{x} & \text{if } x > 4 \end{cases}$

55. $f(x) = \begin{cases} x + 2 & \text{if } x < 0 \\ 2 & \text{if } 0 \leq x < 3 \\ (x - 3)^2 & \text{if } x \geq 3 \end{cases}$

56. $h(x) = \begin{cases} x^3 & \text{if } x < 1 \\ 2 - x^2 & \text{if } 1 \leq x < 3 \\ -7 & \text{if } x \geq 3 \end{cases}$

57. $h(x) = \begin{cases} |x| & \text{if } x < 1 \\ -x^2 & \text{if } 1 \leq x \leq 3 \\ 1/x & \text{if } x > 3 \end{cases}$

58. $g(x) = \begin{cases} x^2 - 1 & \text{if } x \leq 1 \\ x - 1 & \text{if } 1 < x < 3 \\ 3 & \text{if } 3 \leq x \leq 5 \\ 13 - 2x & \text{if } x > 5 \end{cases}$

Critical Thinking

The graph of $y = f(x)$ is shown in the figure. Use this graph and the techniques of shifting and reflecting to sketch the graph of each of the equations in Exercises 59–62.

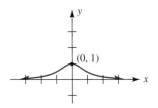

59. $y = -f(x)$
60. $y = f(x) - 3$
61. $y = f(x - 2)$
62. $y = 2 - f(x + 1)$

The graph of $y = \sqrt{4 - x^2}$ is shown in the figure. Use this graph to write an equation for each of the functions whose graphs are shown in Exercises 63–66. Check the equation by generating its graph on a graphing calculator to determine if it matches the given graph.

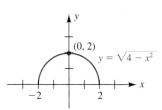

63.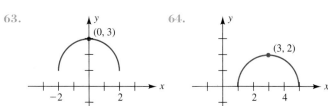

64.

65.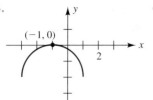

66.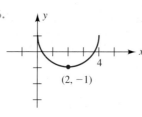

In Exercises 67–70, use a graphing calculator to generate the graphs of the given equations in the same viewing rectangle. What rule in this section verifies the picture you observe?

67. $y = |x|$, $y = |x| + 1$, $y = |x| + 2$, $y = |x| - 1$, $y = |x| - 2$

68. $y = x^3$, $y = (x + 1)^3$, $y = (x + 2)^3$, $y = (x - 1)^3$, $y = (x - 2)^3$

69. $y = \sqrt{x}$, $y = -\sqrt{x}$, $y = \sqrt{-x}$, $y = -\sqrt{-x}$

70. $y = x^2$, $y = 0.25x^2$, $y = 0.5x^2$, $y = 2x^2$, $y = 4x^2$

71. Use your observations in Exercises 67–70 to fill in the following table:

To graph:	Draw the graph of $y = f(x)$ and then
$y = f(x) + c$ for $c > 0$	
$y = f(x) - c$ for $c > 0$	
$y = f(x + c)$ for $c > 0$	
$y = f(x - c)$ for $c > 0$	
$y = -f(x)$	
$y = f(-x)$	
$y = cf(x)$ for $c > 1$	
$y = cf(x)$ for $0 < c < 1$	

72. In mathematics, the notation $[\![x]\!]$ denotes the largest integer n such that $n \le x$. For example,

$$[\![8.9]\!] = 8, \quad [\![5]\!] = 5, \quad [\![\pi]\!] = 3,$$
$$[\![-8.9]\!] = -9, \quad [\![-\sqrt{3}]\!] = -2, \quad \text{and so on.}$$

The **greatest integer function** f is defined by $f(x) = [\![x]\!]$. Sketch the graph of this function and explain why the function is classified as a *step function*.

73. The cost C (in cents) to mail a letter with first-class postage is a function of its weight w (in ounces) and is given by

$$C(w) = \begin{cases} 32 & \text{if } 0 < w \le 1 \\ 55 & \text{if } 1 < w \le 2 \\ 78 & \text{if } 2 < w \le 3 \\ 101 & \text{if } 3 < w \le 4 \\ 124 & \text{if } 4 < w \le 5 \end{cases}$$

This function is another example of a step function (see Exercise 72).

(a) Sketch the graph of this function.

(b) Find $C(2.4)$ and describe what this value represents.

74. The voltage v (in volts) that is applied to an electrical circuit is a function of time t (in milliseconds) and is given by

$$v(t) = \begin{cases} t^2 & \text{if } 0 \le t \le 2 \\ 4 & \text{if } 2 < t \le 6 \\ (t - 8)^2 & \text{if } 6 < t \le 8 \end{cases}$$

(a) Sketch the graph of this function.

(b) Find $v(7)$ and describe what this value represents.

2.6 Composite and Inverse Functions

◆ Introductory Comments

If we select a number, cube it, and then take the cube root of the result, we obtain the number which we selected to cube. Reversing the procedure, that is, taking the cube root first and then cubing the result, also results in the original number. Since one operation "undoes" the other, we say that cubing and taking a cube root are *inverse operations*. In this section, we discuss functions that behave in a similar manner. We begin by discussing a way in which we may combine two functions.

◆ Composition of Functions

One method of combining two functions is called the *composition* of functions. If f and g are functions, then f composed with g, denoted $f \circ g$, is formed by using the output of g as the input of f.

◀ **Composition of Two Functions**

If f and g are functions, then the **composite function** $f \circ g$ is defined by

$$(f \circ g)(x) = f(g(x)).$$

The domain of $f \circ g$ is the set of all x in the domain of g such that $g(x)$ is in the domain of f.

EXAMPLE 1 Use the graphs of f and g in Figure 2.58 to compute each functional value.

(a) $(f \circ g)(4)$

(b) $(g \circ f)(0)$

FIGURE 2.58
The graphs of f and g for Example 1.

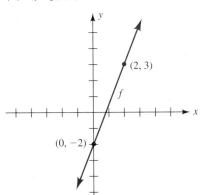
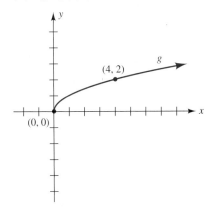

◀ **Solution**

(a) From the graph of g in Figure 2.58, we observe that $g(4) = 2$. Thus,

$$(f \circ g)(4) = f(g(4)) = f(2).$$

Now, from the graph of f in Figure 2.58, we observe that $f(2) = 3$. Hence,

$$(f \circ g)(4) = 3.$$

(b) The composite function $g \circ f$ is defined by $(g \circ f)(x) = g(f(x))$. From the graph of f, we observe that $f(0) = -2$. Thus,

$$(g \circ f)(0) = g(f(0)) = g(-2).$$

However, as shown in Figure 2.58, -2 is *not* in the domain of g. Hence, we conclude that

$$(g \circ f)(0) \text{ is } \textit{undefined}.$$

For $g(f(x))$ to make sense, the value of $f(x)$ must be in the domain of the function g. ◆

PROBLEM 1 Using the graphs of f and g in Figure 2.58, find $(f \circ g)(0)$. ◆

As illustrated in the next example, the composite functions $f \circ g$ and $g \circ f$ are *not* necessarily the same function.

EXAMPLE 2 Given the functions f and g defined by $f(x) = \sqrt{x}$ and $g(x) = x^2 - 4$, find each of the following functions. Then determine the domain of each function.

(a) $f \circ g$ **(b)** $g \circ f$

◆ Solution

(a) The function $f \circ g$ is defined by

$$(f \circ g)(x) = f(g(x)) = f(x^2 - 4) = \sqrt{x^2 - 4}.$$

The domain of $f \circ g$ is all real numbers in the domain of g such that $g(x)$ is in the domain of f. The domain of g is $(-\infty, \infty)$, and the domain of f is $[0, \infty)$. Thus, from the interval $(-\infty, \infty)$ we can select only inputs x such that the output $g(x)$ is *nonnegative*:

$$x^2 - 4 \geq 0$$

We can solve this inequality by generating the graph of $y = x^2 - 4$ in the viewing rectangle of a graphing calculator. As shown in Figure 2.59, the graph intersects the x-axis at $x = \pm 2$ and is above the x-axis when $x < -2$ or $x > 2$. Hence, the solution set of this inequality is

$$x \leq -2 \quad \text{or} \quad x \geq 2.$$

Therefore, the domain of $f \circ g$ is $(-\infty, -2] \cup [2, \infty)$.

(b) The function $g \circ f$ is defined by

$$(g \circ f)(x) = g(f(x)) = g(\sqrt{x}) = (\sqrt{x})^2 - 4 = x - 4.$$

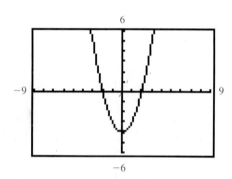

FIGURE 2.59

The solution set of the inequality $x^2 - 4 \geq 0$ is $(-\infty, -2] \cup [2, \infty)$.

The domain of $g \circ f$ is all real numbers in the domain of f such that $f(x)$ is in the domain of g. Since the domain of f is $[0, \infty)$, we can choose inputs only from this interval. Since each *nonnegative* input x gives us an output $f(x)$ that is in the domain of g, the domain of $g \circ f$ is also $[0, \infty)$. Hence, we write

$$(g \circ f)(x) = x - 4, \quad \text{for } x \geq 0.$$

Note that in this example, $(g \circ f)(x) \neq (f \circ g)(x)$. ◆

PROBLEM 2 For the functions f and g given in Example 2, find $(g \circ f)(9)$ by each of the following methods.

(a) First find $f(9)$, and then evaluate $g(f(9))$.

(b) Use the result of Example 2(b), $(g \circ f)(x) = x - 4$ for $x \geq 0$. ◆

◆ One-to-One Functions

A function f is said to be *one-to-one* if each element in the range of f is associated with only one element in its domain. More precisely, we state the following definition.

◀ **One-to-One Function**

> A function f is a **one-to-one function** if, for all a and b in the domain of f,
>
> $$f(a) = f(b) \quad \text{implies} \quad a = b.$$

A simple method for determining whether a function is one-to-one is to look at its graph. If every horizontal line that can be drawn in the coordinate plane intersects the graph at most once, then each output of the function is associated with only one input. Hence, the function is one-to-one. We refer to this graphical method of determining whether a function is one-to-one as the **horizontal line test**.

◀ **Horizontal Line Test**

> A function f is one-to-one if no horizontal line in the coordinate plane intersects the graph of the function in more than one point.

Every function that is either an increasing function or a decreasing function (see Section 2.5) passes the horizontal line test and, therefore, is also a one-to-one function.

EXAMPLE 3 Determine if each function is a one-to-one function.

(a) $f(x) = (x - 1)^3$ **(b)** $h(x) = x^2 - 1$

128   CHAPTER 2 ♦ *Functions and Graphs*

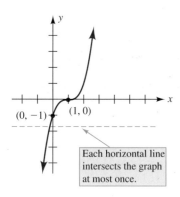

FIGURE 2.60
The graph of $f(x) = (x - 1)^3$ passes the horizontal line test. Hence, f is a one-to-one function.

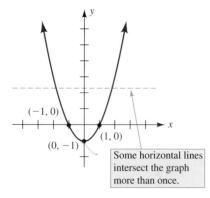

FIGURE 2.61
The graph of $h(x) = x^2 - 1$ fails the horizontal line test. Hence, h is *not* a one-to-one function.

♦ **Solution**

(a) Using a graphing calculator or the techniques of graphing a function that we discussed in Section 2.5, we generate the graph of $f(x) = (x - 1)^3$, as shown in Figure 2.60. Since the function f is an increasing function, we know that each horizontal line drawn in the coordinate plane must intersect the graph at most once. Thus, by the horizontal line test, the function f is a one-to-one function. Also, note that

$$f(a) = f(b)$$

implies that $\quad (a - 1)^3 = (b - 1)^3$

$\qquad\qquad\qquad a - 1 = b - 1 \qquad$ **Take the cube root of both sides**

$\qquad\qquad\qquad\quad\; a = b \qquad\qquad$ **Add 1 to both sides**

Since $f(a) = f(b)$ implies $a = b$, we conclude that f is a one-to-one function.

(b) Using a graphing calculator or the techniques of graphing a function that we discussed in Section 2.5, we generate the graph of $h(x) = x^2 - 1$, as shown in Figure 2.61. Notice that some horizontal lines intersect the graph more than once. Thus, by the horizontal line test, the function h is not a one-to-one function. Also, note that

$$h(a) = h(b)$$

implies that $\quad a^2 - 1 = b^2 - 1$

$\qquad\qquad\qquad\quad a^2 = b^2 \qquad\quad$ **Add 1 to both sides**

$\qquad\qquad\qquad\quad\; a = \pm b \qquad\quad$ **Take the square root of both sides**

Since $h(a) = h(b)$ does not imply $a = b$, we may conclude that h is not a one-to-one function. ♦

If a function is *not* one-to-one, we can often find a suitable restriction on its domain in order to form a new function that *is* one-to-one. For example, if we restrict the domain of the function h defined in Example 3(b) to *nonnegative* numbers, the new function that is formed is one-to-one.

PROBLEM 3   Show that the function H defined by $H(x) = x^2 - 1$ for $x \geq 0$ is a one-to-one function. ♦

♦ **Inverse Functions**

Suppose f is a one-to-one function, and for each element x in its domain X there corresponds an element $f(x)$ in its range Y. Because f if a one-to-one function, we can assign to each output $f(x)$ the input x from which it came. Hence, we can define a function g from Y to X in which

$$g(f(x)) = x \quad \text{for every } x \text{ in } X$$

We refer to g as the **inverse function** of f, as shown in Figure 2.62.

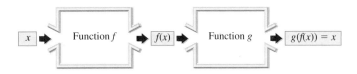

FIGURE 2.62
If g is the inverse function of f, then $g(f(x)) = x$ for every x in the domain of f.

If we let $f(x) = y$, then $g(f(x)) = x$ implies $g(y) = x$. Hence, we also have the following composition:

$$f(g(y)) = y \quad \text{for every } y \text{ in } Y.$$

For notation purposes, we replace y with x and write

$$f(g(x)) = x \quad \text{for every } x \text{ in } Y.$$

◀ **Inverse Function**

Let f be a one-to-one function with domain X and range Y. The **inverse function** of f is a function g with domain Y and range X such that

$$g(f(x)) = x \quad \text{for every } x \text{ in } X$$
and
$$f(g(x)) = x \quad \text{for every } x \text{ in } Y$$

If g is the inverse of f, then f is the inverse of g; that is, inverse functions always come in pairs. Consider the cubing function f defined by $f(x) = x^3$ and the cube root function g defined by $g(x) = \sqrt[3]{x}$. Both of these functions are one-to-one functions and the domain and range of each is the set of all real numbers. [See parts (e) and (h) of Figure 2.42 in Section 2.5.] Note that

$$g(f(x)) = g(x^3) = \sqrt[3]{x^3} = x$$
and
$$f(g(x)) = f(\sqrt[3]{x}) = (\sqrt[3]{x})^3 = x.$$

In summary, the cubing function and the cube root function are *inverses of each other*.

The graphs of the cubing function f and the cube root function g are shown in Figure 2.63. If we fold the coordinate plane along the dotted line $y = x$, the graphs of f and g will coincide. In other words, the graphs of f and g are *reflections of one another in the line $y = x$*. For every point (a, b) on the graph of f there corresponds a point (b, a) on the graph of g. This special relationship between the graphs of a pair of inverse functions gives us a graphical method for verifying whether one function is the inverse of another.

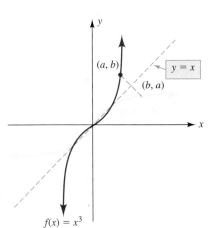

FIGURE 2.63
The graphs of a pair of inverse functions are reflections of one another in the line $y = x$.

EXAMPLE 4 Determine whether the functions f and g are inverses of each other:

$$f(x) = 2x - 8 \qquad g(x) = \tfrac{1}{2}x + 4$$

130 CHAPTER 2 ◆ *Functions and Graphs*

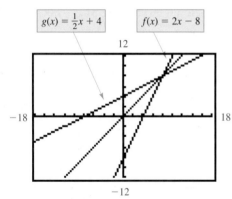

FIGURE 2.64
The functions f and g defined by $f(x) = 2x - 8$ and $g(x) = \frac{1}{2}x + 4$ appear to be inverses of each other.

◆ **Solution** Figure 2.64 shows the graphs of $f(x) = 2x - 8$ and $g(x) = \frac{1}{2}x + 4$ in a square-set viewing rectangle of a graphing calculator. The functions f and g are one-to-one functions and their domains and ranges are the set of all real number. It appears that the graphs of f and g are reflections of one another in the line $y = x$, which indicates that these functions are inverses of each other. We can verify this conjecture algebraically:

$$\begin{aligned} f(g(x)) &= f(\tfrac{1}{2}x + 4) \\ &= 2(\tfrac{1}{2}x + 4) - 8 \\ &= x + 8 - 8 \\ &= x \end{aligned} \quad \text{and} \quad \begin{aligned} g(f(x)) &= g(2x - 8) \\ &= \tfrac{1}{2}(2x - 8) + 4 \\ &= x - 4 + 4 \\ &= x \end{aligned}$$

Since $f(g(x)) = g(f(x)) = x$, we conclude that the functions f and g are inverses of each other. ◆

PROBLEM 4 Determine whether $f(x) = (x - 2)^5$ and $g(x) = \sqrt[5]{x} + 2$ are inverses of each other. Check graphically with a graphing calculator. ◆

◆ Finding the Inverse of a Function

We usually denote the inverse function of the function f by using the notation f^{-1} (read "f inverse"). Thus, if f and f^{-1} are inverses of each other, then we write

$$\boxed{f(f^{-1}(x)) = f^{-1}(f(x)) = x}$$

CAUTION In the notation f^{-1}, the superscript -1 denotes an inverse, *not* an exponent; that is, f^{-1} does not mean $1/f$.

We now discuss a procedure that we can use to find the inverse function of a one-to-one function defined by the equation $y = f(x)$. If we apply f^{-1} to both sides of the equation $y = f(x)$, we obtain

$$\begin{aligned} f^{-1}(y) &= f^{-1}(f(x)) \\ &= x \end{aligned}$$

The statement $x = f^{-1}(y)$ suggests that we can find an equation that defines f^{-1} by solving the equation $y = f(x)$ for x. For example, to find the inverse of the function f defined by $f(x) = 2x - 8$, we solve the equation $y = f(x)$ for x:

$$y = 2x - 8$$

$$y + 8 = 2x \qquad \text{Add 8 to both sides}$$

$$x = \tfrac{1}{2}y + 4 \qquad \text{Divide both sides by 2}$$

Hence, the equation $x = f^{-1}(y) = \tfrac{1}{2}y + 4$ defines the inverse function of f. The letter we use in the defining expression for f^{-1} is immaterial; however, when we work with functions, it is customary to use the letter x as the independent variable and y as the dependent variable. Thus, replacing x with y, we write

$$f^{-1}(x) = \tfrac{1}{2}x + 4,$$

which agrees with our work in Example 4.

The following procedure may be used to find the inverse of a one-to-one function.

◀ **Procedure for Finding the Inverse of a One-to-One Function**

If f is a one-to-one function, then its inverse function f^{-1} may be found by the following procedure:

1. Set $y = f(x)$.
2. Solve the equation $y = f(x)$ for x in terms of y.
3. Interchange x and y.
4. Replace y with $f^{-1}(x)$ and restrict the domain of f^{-1} (if necessary) so that it equals the range of f.

This procedure is illustrated in the next example.

EXAMPLE 5   Find the inverse of each one-to-one function.

(a) $f(x) = x^3 - 1$ (b) $g(x) = \sqrt{x - 1}$

◀ **Solution**

(a) To find the inverse of the function f defined by $f(x) = x^3 - 1$, we proceed as follows:

$$y = x^3 - 1 \quad \text{Replace } f(x) \text{ with } y$$
$$y + 1 = x^3 \quad \text{Solve for } x$$
$$x = \sqrt[3]{y + 1}$$
$$y = \sqrt[3]{x + 1} \quad \text{Interchange } x \text{ and } y$$
$$f^{-1}(x) = \sqrt[3]{x + 1} \quad \text{Replace } y \text{ with } f^{-1}(x)$$

Since the domains and ranges of both f and f^{-1} are the set of all real numbers, we conclude that the inverse function of f is defined by

$$f^{-1}(x) = \sqrt[3]{x + 1}.$$

Figure 2.65 shows the graphs of f and f^{-1} in the same viewing rectangle of a graphing calculator. Note that the graphs are reflections of one an-

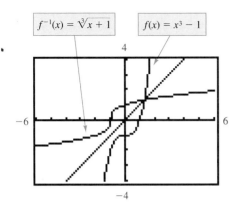

FIGURE 2.65
The graphs of $f(x) = x^3 - 1$ and $f^{-1}(x) = \sqrt[3]{x + 1}$ are reflections of one another in the line $y = x$.

other in the line $y = x$. This relationship between the graphs verifies our work.

(b) To find the inverse of the function g defined by $g(x) = \sqrt{x - 1}$, we proceed as follows:

$$y = \sqrt{x - 1} \qquad \text{Replace } g(x) \text{ with } y$$

$$y^2 = x - 1 \qquad \text{Solve for } x$$

$$x = y^2 + 1$$

$$y = x^2 + 1 \qquad \text{Interchange } x \text{ and } y$$

$$g^{-1}(x) = x^2 + 1 \qquad \text{Replace } y \text{ with } g^{-1}(x)$$

Since the domain of g^{-1} must be equal to the range of g, we must restrict the domain of g^{-1} to the interval $[0, \infty)$. Hence, we conclude that g^{-1} is defined by

$$g^{-1}(x) = x^2 + 1 \quad \text{for } x \geq 0.$$

Figure 2.66 shows the graphs of g and g^{-1} in the same viewing rectangle of a graphing calculator. The graphs are reflections of one another in the line $y = x$. This relationship between the graphs verifies our work. ◆

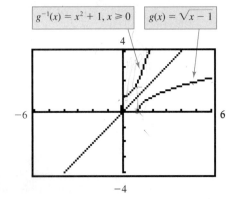

FIGURE 2.66
The graphs of $g(x) = \sqrt{x - 1}$ and $g^{-1}(x) = x^2 + 1$, for $x \geq 0$, are reflections of one another in the line $y = x$.

PROBLEM 5 Find the inverse of the one-to-one function f defined by $f(x) = \dfrac{1}{x + 2}$. ◆

Exercises 2.6

 Basic Skills

In Exercises 1–8, use the graphs of f, g, and h to compute (if possible) the given functional value.

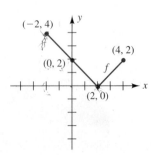 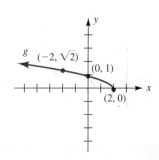 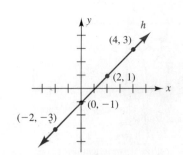

SECTION 2.6 ◆ Exercises

1. $(g \circ f)(0)$
2. $(h \circ f)(-2)$
3. $(f \circ h)(-2)$
4. $(h \circ g)(3)$
5. $(f \circ g)(2)$
6. $(f \circ f)(-2)$
7. $[h \circ (f \circ g)](2)$
8. $[h \circ (g \circ f)](0)$

In Exercises 9–20, find each composite function, given that f, g, h, F, G, and H are defined as follows:

$$f(x) = 2x + 1 \qquad F(x) = \sqrt{x^2 - 4}$$
$$g(x) = x^2 + 1 \qquad G(x) = \frac{1}{x}$$
$$h(x) = \sqrt{x} \qquad H(x) = \frac{2}{x - 1}$$

Then find the domain of the composite function.

9. $f \circ g$
10. $g \circ f$
11. $G \circ f$
12. $G \circ F$
13. $F \circ h$
14. $g \circ h$
15. $G \circ H$
16. $H \circ G$
17. $H \circ f$
18. $G \circ G$
19. $f \circ f$
20. $H \circ H$

21. Use the result of Exercise 15 to evaluate $(G \circ H)(2)$.
22. Use the result of Exercise 16 to evaluate $(H \circ G)(2)$.

In Exercises 23–32,
(a) determine whether the function is one-to-one.
(b) if the function is not one-to-one, find a suitable restriction on its domain in order to form a new function that is one-to-one. (More than one restriction is possible.)

23. $f(x) = 2x$
24. $f(x) = \frac{x^3}{3}$
25. $f(x) = x^2$
26. $f(x) = |x| - 3$
27. $f(x) = \sqrt{9 - x}$
28. $f(x) = \frac{1}{x - 3}$
29. $f(x) = |x + 2|$
30. $f(x) = 1 - \sqrt[3]{x}$
31. $f(x) = (x - 2)^2 - 1$
32. $f(x) = (x - 1)^{2/3}$

In Exercises 33–44, use a graphing calculator to determine whether the functions f and g appear to be inverses of each other. If the functions appear to be inverses, verify this fact algebraically.

33. $f(x) = 3x - 1$
    $g(x) = \frac{x + 1}{3}$
34. $f(x) = 5 - 2x$
    $g(x) = \frac{5 - x}{2}$
35. $f(x) = \sqrt{x}$
    $g(x) = x^2$
36. $f(x) = x$
    $g(x) = \frac{1}{x}$
37. $f(x) = x^3 - 8$
    $g(x) = \sqrt[3]{x + 8}$
38. $f(x) = (x - 2)^3$
    $g(x) = \sqrt[3]{x} + 2$
39. $f(x) = \frac{1}{x - 1}$
    $g(x) = \frac{1}{x} + 1$
40. $f(x) = \frac{x}{x + 3}$
    $g(x) = \frac{-3x}{x - 1}$
41. $f(x) = \frac{3x + 1}{x}$
    $g(x) = \frac{1}{x - 3}$
42. $f(x) = \frac{x}{x - 2}$
    $g(x) = \frac{2x}{x - 1}$
43. $f(x) = (x - 1)^2$ for $x \geq 1$
    $g(x) = \sqrt{x} + 1$
44. $f(x) = x^2 + 2$ for $x \geq 0$
    $g(x) = \sqrt{x - 2}$

Each function f in Exercises 45–62 is a one-to-one function.
(a) Find the inverse function f^{-1}.
(b) Verify the answer to part (a) by showing that the graphs of f and f^{-1} are reflections of one another in the line $y = x$.

45. $f(x) = 2x$
46. $f(x) = \frac{-x}{4}$
47. $f(x) = 3 - x$
48. $f(x) = x + 1$
49. $f(x) = 2 - x^3$
50. $f(x) = (x + 1)^3$
51. $f(x) = \sqrt[3]{x - 4}$
52. $f(x) = 3 + \sqrt[3]{x}$
53. $f(x) = \frac{1}{x + 1}$
54. $f(x) = \frac{1}{x} - 2$
55. $f(x) = \sqrt{x - 4} + 1$
56. $f(x) = \sqrt{x} + 3$
57. $f(x) = (x - 1)^2, x \geq 1$
58. $f(x) = (x - 3)^2 + 2, x \geq 3$
59. $f(x) = \sqrt{9 - x^2}, 0 \leq x \leq 3$
60. $f(x) = \sqrt{x^2 - 9}, x \geq 3$
61. $f(x) = x^2 + 3, x \leq 0$
62. $f(x) = 2 - x^2, x \leq 0$

Critical Thinking

For Exercises 63–68, refer to the following table of values.

x	$f(x)$	$g(x)$
-1	4	3
0	7	2
2	3	-1
3	-1	-5

63. Given that $h(x) = (f \circ g)(x)$, find $h(2)$.
64. Given that $h(x) = (g \circ f)(x)$, find $h(2)$.
65. Given that $g(x) = (f \circ h)(x)$, find $h(2)$.
66. Given that $f(x) = (g \circ h)(x)$, find $h(2)$.
67. Given that $f(x) = (g \circ h)(x)$, find $h(3)$.
68. Given that $g(x) = (f \circ h)(x)$, find $h(-1)$.

In Exercises 69–72, the graph of a function f is shown. Determine whether the function f has an inverse. If so, sketch the graph of f^{-1} and complete the following table of values.

x	-2	0	3
$f^{-1}(x)$			

69.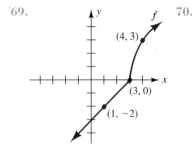

70.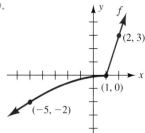

71.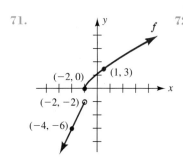

72.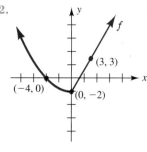

For Exercises 73 and 74, assume the functions f and g are defined as follows:

$$f(x) = \frac{x-2}{3} \quad \text{and} \quad g(x) = 3x + 1$$

73. Compare $(f \circ g)^{-1}(x)$ with $(g^{-1} \circ f^{-1})(x)$. What can you conclude?

74. Compare $(g \circ f)^{-1}(x)$ with $(f^{-1} \circ g^{-1})(x)$. What can you conclude?

In Exercises 75 and 78, determine the range of the given function by examining the domain of its inverse function.

75. $f(x) = \dfrac{1}{2x-1}$

76. $f(x) = \dfrac{x}{6-3x}$

77. $f(x) = \dfrac{2x+1}{3x-2}$

78. $f(x) = \dfrac{9x-3}{3x+1}$

79. Find $f(x)$ if $(g \circ f)(x) = 2x - 8$ and $g(x) = 6x + 4$.

80. (a) Determine whether each of the eight basic functions in Figure 2.42 (in Section 2.5) is a one-to-one function.
    (b) Which of the one-to-one functions in part (a) is its own inverse?

81. A spherical balloon is being inflated with air. The volume V of the balloon is a function of its radius r and is given by $V(r) = \frac{4}{3}\pi r^3$. In turn, the radius r (in inches) is a function of the time t (in seconds) after the inflation process begins and is given by $r(t) = 3(t+1)$.
    (a) Determine $(V \circ r)(t)$.
    (b) What does the composition function $V \circ r$ describe?
    (c) Find the volume of the balloon after 3 seconds of inflation.
    (d) Find the time at which the volume is $36{,}000\pi$ cubic inches.

82. In a ski factory, the daily cost C (in dollars) of producing n pairs of skis is given by

$$C(n) = 225n - 0.8n^2 \quad \text{for } 0 \leq n \leq 40$$

and the number of pairs produced in t hours is given by

$$n(t) = 4.5t \quad \text{for } t \leq 10.$$

   (a) Determine $(C \circ n)(t)$.
   (b) What does the composition function $C \circ n$ describe?
   (c) Find the daily production cost (to the nearest dollar) if the factory runs for 8.2 hours per day.
   (d) How many hours does the factory operate if the daily production cost is $4228?

2.7 Applied Functions and Variation

◆ Introductory Comments

To solve many types of word problems—especially those that require the techniques of calculus—we must begin by setting up an equation that defines a function. We may then analyze the function, draw its graph, and answer questions concerning the functional relationship between the quantities. In this section we practice setting up functions from words.

◆ Applied Functions

For many applied problems, we begin with an established formula and then use *substitution* to obtain a functional relationship between the desired variables. The procedure is illustrated in the next two examples.

EXAMPLE 1 Suppose 200 feet of fencing is needed to enclose a rectangular garden.

(a) Express the area A of the rectangular garden as a function of its length x.

(b) State the domain of the function in part (a).

(c) Generate the graph of the function in part (a). Use the graph to find the length x of the garden for which the area A is the *largest* possible.

◀ Solution

(a) Drawing a rectangular garden, we let

$$x = \text{the length} \quad \text{and} \quad y = \text{the width}$$

as shown in Figure 2.67. Recall that the area A of a rectangle is the product of its length and width:

$$A = xy$$

For the area A to be a function of its length x, we need to write A in terms of only x. Since 200 feet of fencing are needed to enclose the garden, we have

$$200 = 2x + 2y.$$

Solving this equation for y gives us

$$y = 100 - x.$$

Now, we substitute $100 - x$ for y in the area formula, $A = xy$, and obtain

$$A = x(100 - x), \quad \text{or} \quad A = 100x - x^2.$$

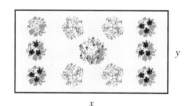

FIGURE 2.67

Rectangular garden of length x and width y

The equation $A = 100x - x^2$ defines the area A of this rectangle as a function of its length x. If we wish to emphasize this functional relationship, we may use functional notation to write

$$A(x) = 100x - x^2.$$

(b) Algebraically, we know that the domain of this function is the set of all real numbers but, geometrically, this does not make sense. This is because x represents the length of the rectangle, and a length can only be *positive*. Thus, we must have $x > 0$. Also, since the perimeter is 200 feet, we must have $x < 100$, otherwise, the width y would be zero or negative. Because of the geometric restrictions required for this function, the domain is (0, 100).

(c) Figure 2.68 shows the graph of $A = 100x - x^2$, with $0 < x < 100$, in the viewing rectangle of a graphing calculator. The x-axis denotes the length x of the rectangle and the y-axis denotes its area A. The coordinates (x, A) at the peak of the curve tell us the length x for which the area A is the *largest* possible. We refer to the value of A at the peak of the curve as the *maximum functional value*. Most graphing calculators have a built-in program for finding the maximum value of a function. If your calculator does not have a built-in program for finding the maximum value, use the TRACE key and ZOOM key to estimate the maximum value. Using either method, we find that the maximum area is 2500 square feet and this occurs when $x = 50$ ft. ◆

FIGURE 2.68

The coordinates at the peak of the curve tell us the length x of the rectangle for which the area A is the largest possible.

PROBLEM 1 Determine the width y of the rectangle in Example 1 for which the area is the *largest* possible. What special type of rectangle gives the maximum area for a fixed perimeter? ◆

EXAMPLE 2 Water is flowing into a conical funnel at the rate of 8 cubic inches per minute. The diameter of the base of the funnel is 8 inches and its height is 12 inches.

(a) Express the volume V of water in the funnel as a function of the height h of the water in the funnel.

(b) State the domain and range of the function in part (a). Then generate its graph.

(c) Determine the time t (in minutes) it takes for the water to rise to a height of 6 inches.

◀ Solution

(a) Drawing a conical funnel as in Figure 2.69, we let

$h =$ height of the water in the funnel

and $r =$ radius of the surface of the water.

From elementary geometry, we know that the volume of water in the funnel is given by

$$V = \tfrac{1}{3}\pi r^2 h.$$

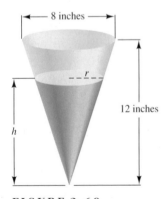

FIGURE 2.69

A conical funnel 8 inches wide and 12 inches deep

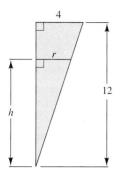

FIGURE 2.70

Similar triangles formed from the cross-section of the funnel

For the volume V to be a function of h, we must write the volume in terms of only h. To do this, we use similar triangles from the cross-section of the funnel shown in Figure 2.70. Since corresponding sides of similar triangles are proportional, we have

$$\frac{4}{12} = \frac{r}{h}, \quad \text{or} \quad r = \frac{h}{3}.$$

Now, we substitute $h/3$ for r in the volume formula, $V = \frac{1}{3}\pi r^2 h$, and obtain

$$V = \frac{1}{3}\pi \left(\frac{h}{3}\right)^2 h, \quad \text{or} \quad V = \frac{1}{27}\pi h^3.$$

If we wish to emphasize this functional relationship, we use functional notation to write

$$V(h) = \frac{1}{27}\pi h^3.$$

(b) Since the height of the funnel is 12 inches, the water level must be between 0 and 12 inches. Thus, the domain of this function is [0, 12]. The function defined by the equation $V = \frac{1}{27}\pi h^3$ is an increasing function. By the vertical compress rule (Section 2.5), we know that the graph of this function is similar to the graph of the cubing function. For $h = 0$ inch, we have $V = 0$ cubic inch and for $h = 12$ inches, we have $V = 64\pi \approx 201$ cubic inches. Hence, the range of this function is $[0, 64\pi] \approx [0, 201]$. This information suggests that an appropriate viewing rectangle for observing the graph of $V = \frac{1}{27}\pi h^3$ is

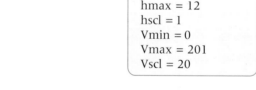

Figure 2.71 shows the graph of $V = \frac{1}{27}\pi h^3$ in this viewing rectangle. The x-axis denotes the height of water h, and the y-axis denotes the volume of water V.

(c) When $h = 6$ inches, the volume of water in the funnel is

$$V(6) = \frac{1}{27}\pi(6)^3 = 8\pi \approx 25.1 \text{ cubic inches}$$

Figure 2.71 verifies this fact. Since the water is flowing into the funnel at the rate of 8 cubic inches per minute, it takes

$$\frac{8\pi}{8} = \pi \approx 3.14 \text{ minutes}$$

for the water to reach a height of 6 inches. ◆

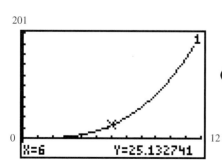

FIGURE 2.71

Graph of $V = \frac{1}{27}\pi h^3$
Domain: [0, 12]
Range: $[0, 64\pi] \approx [0, 201]$

PROBLEM 2 Referring to Example 2, express the volume V of water in the funnel as a function of the radius r of water in the funnel. ◆

◆ Variation

In business, engineering, and the sciences, the functional relationship between two quantities is often given in terms of *variation*. When we state that **y varies directly as x** or that **y is directly proportional to x**, we mean that the ratio of y to x is always the same. In other words,

$$\frac{y}{x} = k \quad \text{or} \quad y = kx$$

where k is called the *variation constant*.

EXAMPLE 3 *Chapter Opening Problem*
The real estate tax T on a property varies directly as its assessed value V.

(a) Express T as a function of V if $T = \$2800$ when $V = \$112{,}000$.

(b) Generate the graph of the function defined in part (a) for $\$0 \leq V \leq \$500{,}000$.

◀ Solution

(a) Since T varies directly as V, we write

$$T = kV$$

where k is the variation constant to be determined. To find k, we replace T with 2800 and V with 112,000:

$$T = kV$$
$$2800 = k(112{,}000)$$
$$k = \frac{2800}{112{,}000} = \frac{1}{40}$$

Thus, $\quad T = \dfrac{1}{40}V, \quad$ or $\quad T = \dfrac{V}{40} \quad$ with $V \geq 0$.

(b) The function defined by the equation $T = V/40$ for $V \geq 0$ is an increasing function. By the vertical compress rule (Section 2.5), we know that the graph of this function is similar to the graph of the identity function. For $V = \$0$, we have $T = \$0$ and for $V = \$500{,}000$, we have $T = \$500{,}000/40$, or $T = \$12{,}500$. This information suggests that an appropriate viewing rectangle for observing the graph of $T = V/40$ with $0 \leq V \leq 500{,}000$ is

SECTION 2.7 ◆ *Applied Functions and Variation* 139

```
Vmin = 0
Vmax = 500000
Vscl = 50000
Tmin = 0
Tmax = 12500
Tscl = 1000
```

Figure 2.72 shows the graph of $T = V/40$ in this viewing rectangle. The *x*-axis denotes the assessed value V, and the *y*-axis denotes the tax T. Note that $(112000, 2800)$ is a point on this graph.

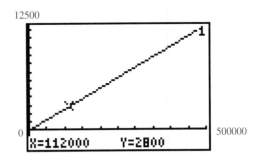

FIGURE 2.72
Graph of the function defined by the equation $T = V/40$ with $0 \leq V \leq 500,000$

PROBLEM 3 Use the result from Example 3(a) to find the tax on a property with an assessed value of $168,000. Check your answer by using the graph in Figure 2.72. ◆

When we state that *y* **varies inversely as** *x* or that *y* **is inversely proportional to** *x*, we mean that the product of *y* and *x* is always the same. In other words,

$$yx = k \quad \text{or} \quad y = \frac{k}{x}$$

where *k* is the *variation constant*.

EXAMPLE 4 The time *t* it takes a person to travel a fixed distance varies inversely as the rate of speed *r* at which the person travels.

(a) Express *t* as a function of *r* if $t = 4$ hours when $r = 60$ mi/h.

(b) Generate the graph of the function in part (a) for $0 \text{ mi/h} < r \leq 120 \text{ mi/h}$.

◆ Solution

(a) Since *t* varies inversely as *r*, we write

$$t = \frac{k}{r}$$

where k is the variation constant to be determined. To find k, we replace t with 4 and r with 60 as follows:

$$t = \frac{k}{r}$$

$$4 = \frac{k}{60}$$

$$k = 240$$

Thus, $\quad t = \dfrac{240}{r} \quad$ for $r > 0$.

(b) The function defined by the equation $t = 240/r$ for $r > 0$ is a decreasing function. By the vertical stretch rule (Section 2.5), we know that the graph of this function is similar to the graph of the reciprocal function. Values of r that are close to 0 yield large values of t. For example, if $r = 5$ miles per hour (mi/h), then $t = \frac{240}{5}$, or $t = 48$ h. For $r = 120$ mi/h, we have $t = \frac{240}{120}$, or $t = 2$ h. An appropriate viewing rectangle that shows this information as well as the axes for this graph is

```
rmin = 0
rmax = 120
rscl = 10
tmin = 0
tmax = 50
tscl = 10
```

Figure 2.73 shows the graph of $t = 240/r$ in this viewing rectangle. The x-axis denotes the rate of speed r, and the y-axis denotes the time t. Note that (60, 4) is a point on this graph.

FIGURE 2.73

Graph of the function defined by the equation $t = 240/r$ for $r > 0$

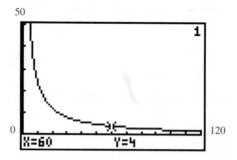

PROBLEM 4 Use the results of Example 4(a) to find the time t it takes to travel a fixed distance if the rate of speed is 50 mi/h. Check your answer by using the graph in Figure 2.73.

The following table gives several other types of variation that occur frequently. In each case, k denotes the variation constant.

Statement	Formula
1. y varies directly as the nth power of x.	$y = kx^n$
2. y varies inversely as the nth power of x.	$y = \dfrac{k}{x^n}$
3. y varies directly as x and inversely as z.	$y = \dfrac{kx}{z}$
4. y varies jointly as x and z.	$y = kxz$

FIGURE 2.74
A right circular cylinder

EXAMPLE 5 The volume V of a right circular cylinder (see Figure 2.74) varies jointly as its height h and the square of its radius r. Describe what happens to the volume of a cylinder if its height and radius are doubled.

◆ **Solution** Since V varies jointly as h and the square of r, we write

$$V = kr^2 h$$

where k is the variation constant. If we double both the height and the radius, we obtain

$$V = k(2r)^2(2h) = 8kr^2 h.$$

From this equation we can see that the volume of the cylinder is *eight times the original volume*. Thus, when the height and radius of a cylinder are doubled, its volume becomes eight times as large. ◆

PROBLEM 5 The electric resistance R of a wire varies directly as its length L and inversely as the square of its radius r. Describe what happens to the resistance of a wire if its length and radius are doubled. ◆

Exercises 2.7

◆ Basic Skills

1. Express the diameter d of a circle as a function of its circumference C.

2. The length of a rectangle is 5 cm. Express its width w as a function of its perimeter P.

3. Suppose an automobile leasing company charges $40 per day plus $0.20 per mile to rent a car. Express the daily cost C (in dollars) of renting a car as a function of the distance driven n (in miles).

4. Suppose a salesperson earns $200 per week plus 25% commision on all sales. Express the weekly earnings E (in dollars) of a salesperson as a function of the amount A of merchandise (in units) that she sells.

5. A square with side s and diagonal d is shown in the figure.

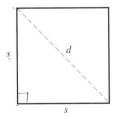

(a) Express s as a function of d.
(b) Express the area A of the square as a function of d.
(Exercise 5 continues)

(c) Express the perimeter P of the square as a function of d.

6. An equilateral triangle with side s and height h is shown in the figure.

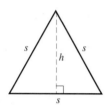

(a) Express h as a function of s.
(b) Express the area A of the triangle as a function of s.
(c) Express the perimeter P of the triangle as a function of h.

7. A small company has a present net worth of $125,000. It is estimated that the future weekly income and expenses for the company will be $30,000 and $26,000, respectively. Express the net worth W of the company at the end of t weeks as a function of time, t.

8. An automobile that was bought for $15,000 depreciates 20% of its original value each year. Express the value V of the automobile at the end of t years as a function of time, t.

9. The radius r of a pile of sand in the shape of a cone is twice its height h.
    (a) Express the volume V of sand as a function of h.
    (b) Express the volume V of sand as a function of r.

10. The height h of a tin can in the shape of a cylinder is equal to its diameter d.
    (a) Express the volume V of the tin can as a function of h.
    (b) Express the surface area S of the tin can as a function of d.

11. The point $P(x, y)$ lies on the graph of the circle $x^2 + y^2 = 1$ as shown in the figure.

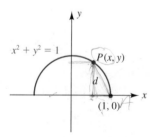

(a) Express the distance d from point P to the point $(1, 0)$ as a function of the x-coordinate of P.
(b) State the domain of the function in part (a).

12. The point $P(x, y)$ lies on the graph of the parabola $2y = x^2$ shown in the figure.
    (a) Express the distance d from the point P to the point $(0, 1)$ as a function of the y-coordinate of P.
    (b) State the domain of the function in part (a).

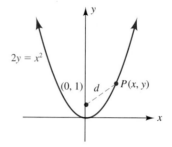

13. The point $Q(x, y)$ lies on the graph of $y = \sqrt{x}$ shown in the figure.

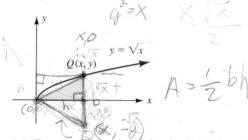

(a) Express the area A of the shaded right triangle as a function of the x-coordinate of Q.
(b) Express the perimeter P of the shaded right triangle as a function of the x-coordinate of Q.

14. The point $Q(x, y)$ lies on the graph of the semicircle $y = \sqrt{25 - x^2}$ shown in the figure.

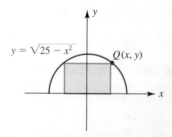

(a) Express the area A of the shaded rectangle as a function of the x-coordinate of Q.
(b) Express the perimeter P of the shaded rectangle as a function of the x-coordinate of Q.

15. The volume of a rectangular box with a square base is 64 cubic inches. Express its surface area S as a function of the width x of its base.

16. The volume of a rectangular box with a square base and an open top is 100 cubic meters. The material used to

construct the base costs $4 per square meter, and the material for the sides cost $2.50 per square meter. Express the cost C to construct the rectangular box as a function of the width x of its base.

17. The ends of a water trough are isosceles triangles, as shown in the figure.

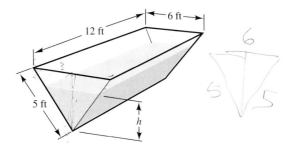

(a) Express the volume V of water in the trough as a function of the height h of the water in the trough.
(b) Give the domain of the function in part (a), and
(c) Sketch the graph of the function in part (a).

18. Suppose the trough described in Exercise 17 is initially empty and then begins to fill with water at the rate of 1 cubic foot per minute (ft³/min). Express the height h of the water in the trough as a function of the time t (in minutes) that the water flows into the trough.

19. A baseball diamond is a square with 90-foot base paths, as shown in the figure. Suppose a runner on first base runs toward second base at the rate of 30 feet per second (ft/s) as soon as the pitcher throws the ball to home plate.

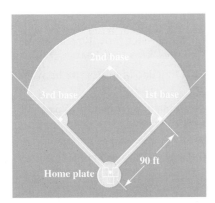

(a) Express the runner's distance d from home plate as a function of the time t (in seconds) after the pitcher throws the ball.
(b) Describe the domain and range of the function in part (a).

20. The fixed cost to run a wood-stove company is $25,000 per month, and the variable cost to produce each stove is $200 per unit. If each stove sells for x dollars, with $200 \le x \le 500$, then the number of stoves sold per month is estimated to be $1000 - 2x$. Express the monthly profit of the company as a function of x. [*Hint*: Profit = total revenue − total cost.]

21. The length of the hypotenuse of a right triangle is 16.2 inches.
(a) Express the area A of the triangle as a function of the length x of one of the legs.
(b) Find the area when $x = 12.4$ inches.

22. Two joggers start from the same place at the same time. One runs due east at 8.2 miles per hour (mi/h) and the other runs due north at 6.3 mi/h.
(a) Determine the distance d between the joggers as a function of the time t (in hours) that they have been running.
(b) Find the distance between the joggers after 1 hour 24 minutes.

In Exercises 23–36, the functional relationship between the quantities is given in terms of variation.

23. Hooke's law states that the distance d a spring stretches varies directly as the force F applied to the spring, as long as the elastic limit of the spring is not exceeded. Assume that the elastic limit of a spring occurs at 2400 newtons (N).
(a) Express d as a function of F if $d = 8$ cm when $F = 400$ N.
(b) Generate the graph of the function in part (a) for $0 \text{ N} \le F \le 2400 \text{ N}$.
(c) Find the distance the spring stretches when $F = 1800$ N. Check your answer by using the graph in part (b).

24. The electrical resistance R of a wire with a constant diameter is directly proportional to the length l of the wire.
(a) Express R as a function of l if $R = 2$ ohms when $l = 30$ meters (m).
(b) Generate the graph of the function in part (a) for $0 \text{ m} \le l \le 100 \text{ m}$.
(c) Find the resistance of the wire when $l = 75$ m. Check your answer by using the graph in part (b).

25. The weight W (in pounds, lb) of a steel beam with a constant cross-sectional area varies directly as its length l (in feet).
(a) Express W as a function of l if $W = 1500$ lb when $l = 12$ ft.
(b) Generate the graph of the function in part (a) for $0 \text{ ft} \le l \le 48 \text{ ft}$.

(*Exercise 25 continues*)

(c) Find the weight of the beam when $l = 32$ ft. Check your answer by using the graph in part (b).

26. The weekly payroll P (in dollars) of a company varies directly as the number n of workers assigned to the job.
    (a) Express P as a function of n if $P = \$5200$ when $n = 13$ workers.
    (b) Generate the graph of the function in part (a) for $0 \leq n \leq 50$.
    (c) Find the weekly payroll when $n = 41$. Check your answer by using the graph in part (b).

27. The number N of long-distance phone calls per day between two towns, each with a population of approximately 25,000 people, varies inversely as the distance d (in miles) between the towns.
    (a) Express N as a function of d if $N = 20$ calls when $d = 500$ miles.
    (b) Generate the graph of the function in part (a) for $0 \text{ mi} < d \leq 2500 \text{ mi}$.
    (c) Find the number of calls made per day between two towns when $d = 800$ miles. Check your answer by using the graph in part (b).

28. *Boyle's law* states that when the temperature of a confined gas remains constant, the pressure P (in pounds per square inch, psi) it exerts varies inversely as the volume V (in cubic inches, in³) it occupies.
    (a) Express P as a function of V if $P = 4$ psi when $V = 30$ in³.
    (b) Generate the graph of the function in part (a) for $0 \text{ in}^3 < V \leq 120 \text{ in}^3$.
    (c) Find the pressure exerted when $V = 24$ in³. Check your answer by using the graph in part (b).

29. The wavelength W (in meters, m) of a radio wave varies inversely as its frequency f (in megahertz, MHz).
    (a) Express W as a function of f if $W = 40$ m when $f = 7.5$ MHz.
    (b) Generate the graph of the function in part (a) for $0 \text{ MHz} < f \leq 50 \text{ MHz}$.
    (c) Find the wavelength when $f = 30$ MHz. Check your answer by using the graph in part (b).

30. If the voltage in an electric circuit is constant, the current I (in milliamperes, mA) through a resistor varies inversely as its resistance R (in ohms, Ω).
    (a) Express I as a function of R if $I = 2$ mA when $R = 100$ Ω.
    (b) Generate the graph of the function in part (a) for $0 \Omega < R \leq 200 \Omega$.
    (c) Find the current generated when $R = 25$ Ω. Check your answer by using the graph in part (b).

31. The gravitational force F (in newtons, N) between two objects varies inversely as the square of the distance d (in meters, m) between them.
    (a) Express F as a function of d if $F = 100$ N when $d = 40$ m.
    (b) Generate the graph of the function in part (a) for $0 \text{ m} < d \leq 400 \text{ m}$.
    (c) Find the force exerted when $d = 125$ m. Check your answer by using the graph in part (b).

32. The lift L (in pounds per square foot, lb/ft²) of an airplane wing varies directly as the square of the speed v (in miles per hour, mi/h) of air flowing over it.
    (a) Express L as a function of v if $L = 225$ lb/ft² when $v = 150$ mi/h.
    (b) Generate the graph of the function in part (a) for $0 \text{ mi/h} \leq v \leq 500 \text{ mi/h}$.
    (c) Find the lift required when $v = 300$ mi/h. Check your answer by using the graph in part (b).

33. The period T (in seconds, s) of a pendulum is directly proportional to the square root of the length l (in meters, m) of the pendulum.
    (a) Express T as a function of l if $T = 2.1$ s when $l = 1.095$ m.
    (b) Generate the graph of the function in part (a) for $0 \text{ m} \leq l \leq 9 \text{ m}$.
    (c) Find the period when $l = 2.405$ m. Check your answer by using the graph in part (b).

34. The power H (in horsepowers, hp) required to propel a motorboat through the water varies directly as the cube of the speed s (in knots, kn) of the boat.
    (a) Express H as a function of s if $H = 48$ hp when $s = 11.2$ kn.
    (b) Generate the graph of the function in part (a) for $0 \text{ kn} \leq s \leq 20 \text{ kn}$.
    (c) Find the power required when $s = 15.6$ kn. Check your answer by using the graph in part (b).

35. The centripetal force acting on an object traveling in a circular path varies directly as the square of its velocity and inversely as the radius of the circle. Describe what happens to the centripetal force acting on the object if the velocity and radius are both tripled.

36. The power produced by an electric generator varies jointly as the load resistance and the square of the current. Describe what happens to the power produced by the generator if the load resistance is doubled and the current is halved.

Critical Thinking

37. A rectangular pasture along a river is to be enclosed using 1000 feet of fencing, as shown in the figure. No fencing is needed along the river.

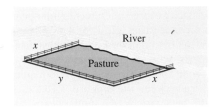

(a) Express the area A of the pasture as a function of the length of a side, x.

(b) State the domain of the function defined in part (a).

(c) Use a graphing calculator to generate the graph of the function defined in part (a). Trace to the peak of this curve, then state the dimensions of the rectangular pasture whose area is the *largest* possible.

(d) Use the result from part (c) to compare the sum of the two lengths x to the length y. What do you observe? Make a conjecture about the dimensions of x and y if a fixed amount of fencing k is used to enclose three sides of a rectangular pasture so that its area is the largest possible.

38. Three adjacent rectangular corrals are to be built with 120 feet of fencing, as shown in the figure.

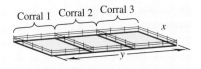

(a) Express the total enclosed area A as a function of the common length x.

(b) State the domain of the function defined in part (a).

(c) Use a graphing calculator to generate the graph of the function defined in part (a). Trace to the peak of this curve, then state the overall dimensions of the corrals so that the total enclosed area is the *largest* possible.

(d) Use the result from part (c) to compare the sum of the four lengths x to the sum of the two lengths y. What do you observe? Make a conjecture about the dimensions of x and y if a fixed amount of fencing k is used to enclose n adjacent rectangular corrals so that the area is the largest possible.

39. A church window has the shape of a rectangle surmounted by a semicircle, as shown in the figure.

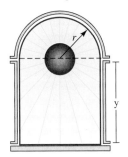

(a) Express the area A of the window as a function of the radius r of the semicircle if the perimeter of the window is 30 ft.

(b) Use a graphing calculator to generate the graph of the function defined in part (a). Trace to the peak of this curve, then state the value of r that makes the area of the window the *largest* possible.

(c) Use the value of r from part (b) to estimate the length y when the area of the window is the largest possible. Make a conjecture about the values of r and y when the area of the window is the largest possible.

40. A football stadium has the shape of a rectangle with semicircular ends as shown in the figure.

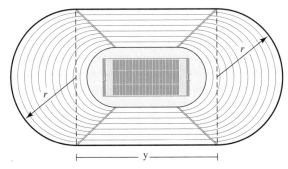

(a) Express the area A of the stadium as a function of the radius r of the semicircle if the perimeter of the stadium is 1 mile.

(b) Use a graphing calculator to generate the graph of the function defined in part (a). Trace to the peak of this curve, then estimate (to three significant digits) the value of r that makes the area of the stadium the *largest* possible.

(c) Use the value of r from part (b) to estimate the distance y when the area of the stadium is the largest possible. Make a conjecture about the values of r and y when the area of the stadium is the largest possible. What is the actual shape of the stadium?

41. A baseball player hits the ball from home plate to the outfield and reaches second base safely. Suppose the player runs at the rate of 30 ft/s, directly on the base paths. (See the figure in Exercise 19.) Express the straight-line distance d from home plate to the runner as a function of the time t (in seconds) after he hits the ball. [*Hint:* Use a piecewise-defined function.]

42. An author is paid a royalty of $3.00 per book for the first 1000 books sold. The royalty increases to $3.05 per book for each book sold in excess of 1000. Express the author's royalty R as a function of the number n of books sold. [*Hint:* Use a piecewise-defined function.]

43. The time required for an elevator to lift its passengers varies jointly as the weight of the passengers and the distance they are lifted and inversely as the power of the motor used. Suppose it takes 10 seconds for an elevator to lift 800 pounds to a height of 40 feet with a 20-horsepower motor. Determine the time it takes this elevator to lift 1000 pounds to a height of 80 feet with the same motor.

44. The safe-load capacity of a wooden rectangular beam supported at both ends varies jointly as its width and the square of its depth and inversely as the distance between the supports. Suppose the safe-load capacity of a beam 4 inches wide and 12 inches deep is 3600 pounds when the distance between the supports is 16 feet. Determine the safe-load capacity of a similar beam that is 8 inches wide and 8 inches deep if the distance between the supports remains 16 feet.

Chapter 2 Review

◆ Questions for Writing or Group Discussion

1. Explain the difference between the ordered pair (a, b) and the interval (a, b).
2. Explain in words how to find the *midpoint* of a line segment.
3. Explain in words how to find the *distance* between two points in the *Cartesian plane*.
4. What is the difference between the *independent variable* and the *dependent variable* in an equation?
5. Describe the *point-plotting method* of sketching the graph of a simple equation.
6. What is meant by the *x-intercept* and *y-intercept* of a graph? How can the intercepts be determined from a given equation?
7. How can the *vertical line test* be used to determine if an equation defines y as a function of x?
8. What is a *function*? What is meant by its *domain* and *range*?
9. How can the *zeros* of a function be determined? Graphically, what do the real zeros represent?
10. What is the difference between an *even function* and an *odd function*? Give an example of each.
11. Which of the eight basic functions in Figure 2.42 are *increasing functions*? *decreasing functions*?
12. Does the *squaring function* pass the *horizontal line test*? Explain. What does this tell us about the squaring function?
13. Does the *cubing function* pass the *horizontal line test*? Explain. What does this tell us about the cubing function?
14. How does the graph of a function relate to the graph of its inverse?
15. Explain the algebraic procedure for finding the inverse of a one-to-one function.

16. Can the graph of a function be symmetric with respect to the x-axis? Explain.

17. How are the graphs of each of the following functions related to the graph of $y = x^2$?
    (a) $y = x^2 + 3$
    (b) $y = (x - 2)^2$
    (c) $y = -x^2$
    (d) $y = (x + 2)^2 - 1$

18. What is the difference between $f \circ g$ and $g \circ f$? Illustrate with examples.

19. How may *composition* be used to show that the functions f and g are *inverse functions*?

20. If f is an even function, what can you conclude about f^{-1}? Explain.

21. Can a function be its own inverse? If so, give an example.

22. Explain how the *distance formula* may be used to determine if three points are *collinear*.

23. Suppose y varies inversely as x and x varies inversely as t. What can you conclude about y and t?

24. The volume V of helium in a balloon is a function of the radius r of the balloon, and the radius r of the balloon is a function of time t. What does the composite function $V \circ r$ describe?

Review Exercises

In Exercises 1–6, find
(a) *the length of the line segment joining the points A and B, and*
(b) *the coordinates of the midpoint M of the line segment joining the points A and B.*

1. $A(0, 3)$; $B(4, 6)$
2. $A(2, 0)$; $B(7, 12)$
3. $A(-2, 1)$; $B(2, 3)$
4. $A(4, -1)$; $B(1, -4)$
5. $A(-3, -5)$; $B(6, -1)$
6. $A(7, -3)$; $B(-3, -5)$

7. Show that the quadrilateral that joins the points $A(2, 1)$, $B(4, 3)$, $C(2, 5)$, and $D(0, 3)$ is a square. Then find its perimeter and area.

8. Show that the triangle that joins the points $A(-1, 8)$, $B(5, -2)$, and $C(15, 4)$ is an isosceles triangle. Then find its perimeter and area.

For Exercises 9–12, refer to the triangle with vertices A, O, and B shown in the figure.

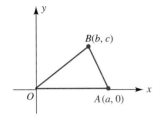

9. Find the length of the line segment joining the midpoints of the sides $\overline{OB}$ and $\overline{AB}$. Compare this length to OA, the length of the third side of the triangle.

10. Find the length of the line segment joining the midpoints of the sides $\overline{OA}$ and $\overline{AB}$. Compare this length to OB, the length of the third side of the triangle.

11. Find the length of the line segment joining the midpoints of the sides $\overline{OA}$ and $\overline{OB}$. Compare this length to AB, the length of the third side of the triangle.

12. What conclusion can you make about the length of a line segment joining the midpoints to any two sides of a triangle with respect to the length of the third side of the triangle?

For each of the equations in Exercises 13–24,
(a) *determine any intercepts,*
(b) *check for symmetry with respect to the x-axis, y-axis, and origin,*
(c) *generate the graph by plotting points or by using a graphing calculator, and*
(d) *determine if the equation defines y as a function of x.*

13. $y = \frac{1}{2}x^3$
14. $4x = y^2$
15. $3x - |y| = 0$
16. $y - 2\sqrt{x} = 0$
17. $x^2 y = 1$
18. $xy = 2$
19. $2x + 3y = 6$
20. $3x - y = 9$

21. $x^2 - 3y = 9$
22. $x + |y| = 3$
23. $|x| + 2|y| = 2$
24. $|x + y| = 1$

Given the function f, g, and h, defined by

$$f(x) = 3x + 4, \quad g(x) = 2x^2 - x, \quad h(x) = \frac{x+1}{x-3}$$

compute the functional values given in Exercises 25–44.

25. $f(-1)$
26. $g(3)$
27. $h(4)$
28. $f(2x)$
29. $g(x-3)$
30. $h\left(\dfrac{a}{2}\right)$
31. $f(\tfrac{1}{3}) - f(2)$
32. $g(2) \cdot f(2)$
33. $\dfrac{f(7)}{h(4)}$
34. $(f \circ h)(2)$
35. $(g \circ f)(x^2)$
36. $(f \circ f)(-5)$
37. $h^{-1}(x)$
38. $f(x) + f^{-1}(x)$
39. $f^{-1}(x+2)$
40. $h^{-1}(-x)$
41. $(h \circ h^{-1})(2)$
42. $(g \circ f^{-1})(x)$
43. $\dfrac{g(x+h) - g(x)}{h}$
44. $\dfrac{f^{-1}(a+h) - f^{-1}(a)}{h}$

Answer questions (a)–(g) for each of the functions in Exercises 45–64.
(a) Find the domain of the function.
(b) Find any real zeros of the function.
(c) Determine if the function is even, odd, or neither.
(d) Sketch the graph of the function (if it is not given) by using the techniques of shifting, reflecting, stretching, and compressing.
(e) Determine the open intervals where the function is increasing, decreasing, or constant.
(f) Determine if the function is a one-to-one function.
(g) Use the graph to find the range of the function.

45. $f(x) = x + 3$
46. $g(x) = -2x$
47. $h(x) = \sqrt[3]{2x}$
48. $F(x) = \sqrt[3]{x - 1}$
49. $G(x) = x^2 - 9$
50. $H(x) = |x| + 5$
51. $g(x) = (x + 3)^2$
52. $F(x) = \sqrt{3 - x}$
53. $G(x) = x^3 + 8$
54. $h(x) = \dfrac{1}{x - 4}$
55. $F(x) = |x - 3| - 1$
56. $H(x) = 8 - (x + 1)^3$

57. $H(x) = 6x - x^2$

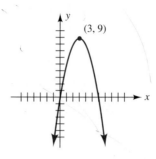

58. $f(x) = |x + 1| + |x - 1|$

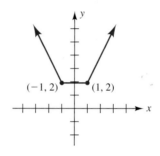

59. $f(x) = \sqrt{100 - x^2}$

60. $g(x) = \sqrt{x^2 - 49}$

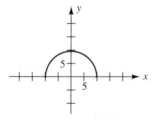 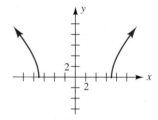

61. $g(x) = 2x^2 - x^4$

62. $H(x) = 2x^3 - 6x$

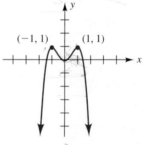 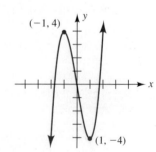

63. $F(x) = \begin{cases} x^2 + 1 & \text{if } x \le 0 \\ 1 & \text{if } 0 < x \le 3 \end{cases}$

64. $h(x) = \begin{cases} \sqrt{x - 1} & \text{if } 1 \le x \le 5 \\ x - 3 & \text{if } x > 5 \end{cases}$

In Exercises 65–70, use the graph of $y = x^{2/3}$ in the figure and the techniques of shifting, reflecting, stretching, and compressing to sketch the graph of the function defined by each equation.

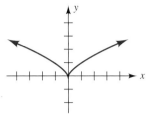

65. $y = x^{2/3} + 2$
66. $y = (x - 3)^{2/3}$
67. $y = (x + 1)^{2/3} - 4$
68. $y = 3 - x^{2/3}$
69. $y = -3x^{2/3}$
70. $y = \dfrac{x^{2/3}}{3}$

In Exercises 71–78, each function f is a one-to-one function.
(a) Find the inverse function f^{-1}.
(b) Check your answer in part (a) by generating the graphs of f and f^{-1} to determine if they are reflections of one another in the line $y = x$.

71. $f(x) = x^2, x \leq 0$
72. $f(x) = 7x - 8$
73. $f(x) = 3 - x^3$
74. $f(x) = \dfrac{1}{x} - 3$
75. $f(x) = \sqrt{x - 2}$
76. $f(x) = \sqrt{x} + 3$
77. $f(x) = \begin{cases} x^3 & \text{if } x < 1 \\ \dfrac{x + 1}{2} & \text{if } x \geq 1 \end{cases}$
78. $f(x) = \begin{cases} x^{2/3} & \text{if } x \geq 0 \\ -(x^2 + 2) & \text{if } x < 0 \end{cases}$

In Exercises 79–86, refer to the graph of the function f shown in the figure.

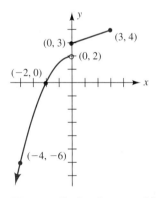

79. Specify the domain of f.
80. Specify the range of f.
81. Find $f(0)$.
82. Find $f(-2)$.
83. Find $f(f(0))$.
84. Find $f(f(-2))$.
85. Draw the graph of f^{-1}.
86. Compute each functional value.
    (a) $f^{-1}(0)$ (b) $f^{-1}(f(4))$ (c) $f(f^{-1}(-6))$

In Exercises 87–92, solve each equation or inequality by using the x-intercept method in conjunction with a graphing calculator. Check algebraically.

87. $6x - (3 - 2x) = 2(x - 3)$
88. $\dfrac{x}{8} - \dfrac{3x + 1}{10} = \dfrac{3 - x}{20}$
89. $\sqrt{2 - 2x} = 3x + 1$
90. $4(2x - 1)^2 = 25$
91. $\dfrac{1}{6} - \dfrac{t - 2}{12} \geq \dfrac{t - 1}{4}$
92. $\left| \dfrac{2x - 3}{5} \right| < \dfrac{2}{3}$

In Exercises 93 and 94, set up an equation and solve by the x-intercept method. Check algebraically.

93. If $12,500 is invested, a portion at 8% simple interest and the remainder at 9%, the interest earned yearly is $41 more than if all the funds had been invested at 8%. How much is invested at each rate?

94. If the radius of a circle is increased by 1 meter, the area of the circle is increased by 19π square meters. Find the radius of the circle.

95. The cost C (in dollars) for a daily truck rental is a function of the distance x driven (in miles) and is given by

$$C(x) = 32 + 0.25x.$$

(a) Evaluate $C(125.7)$, rounding to the nearest cent.
(b) Determine the distance driven when $C(x) = \$138.40$.

96. The volume of a rectangular box with a square base and open top is 9 cubic feet. Express the total surface area S as a function of the width x of its base.

97. The monthly charge for water in a small town is $0.015 per gallon for the first 1000 gallons used and $0.02 per gallon for each gallon used in excess of 1000 gallons. Express the charge C for the water as a function of the number n of gallons used.

98. The fixed cost to run a company that manufactures picnic tables is $10,000 per month, and the variable cost to produce each table is $100 per unit. If each table sells for x dollars, then the number of tables sold per month is estimated to be $800 - 2x$. Express the monthly profit P of the company as a function of x. [Hint: Profit = total revenue − total cost.]

99. The point $P(x, y)$ lies on the right-hand portion of the parabola $y = 3 - x^2$ and is in quadrant I, as shown in the figure.

(a) Express the area A of the shaded isosceles triangle as a function of the x-coordinate of P.
(b) State the domain of the function defined in part (a).

(Exercise 99 continues)

(c) Use a graphing calculator to generate the graph of the function defined in part (a). Trace to the peak of this curve, then state the coordinates of point P that make the area of the triangle the *largest* possible.

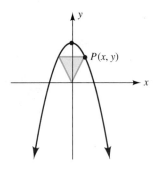

100. A street light 30 ft above level ground casts the shadow of a man 6 ft tall.
   (a) Express the shadow length s as a function of the man's distance d from the base of the light pole.
   (b) If the man stands 8 ft from the base of the light pole and then walks away from the pole at the rate of 4 ft/s, express the shadow length s (in feet) as a function of the time t (in seconds).

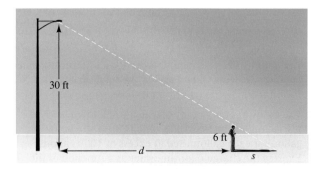

101. The excise tax T (in dollars) on an automobile varies directly as the vehicle's actual value V (in dollars).
   (a) Express T as a function of V if $T = \$200$ when $V = \$5000$.
   (b) Generate the graph of the function in part (a) for $\$0 \leq V \leq \$20,000$.
   (c) Find the tax when $V = \$12,000$. Check your answer by using the graph in part (b).

102. The intensity I (in candelas, cd) of a light source varies inversely as the square of the distance d (in feet) from the light source.
   (a) Express I as a function of d if $I = 4$ cd when d is 5 ft.
   (b) Generate the graph of the function in part (a) for 0 ft $< d \leq$ 20 ft.
   (c) Find the intensity when $d = 10$ ft. Check your answer by using the graph in part (b).

103. A large meadow is capable of supporting life for a maximum of 2000 field mice. Suppose the rate of growth G on the mouse population varies jointly as the number n of mice present and the difference between the maximum number supportable by the field and the number present.
   (a) Express G as a function of n if $G = 50$ field mice per week when $n = 1200$ mice.
   (b) Find the rate of growth when 1500 field mice are living in the field.

104. The time t required for the excavation of a sewer line varies jointly as its length l, width w, and square of its depth d and inversely as the number n of backhoes used. Describe what happens to the time for excavation if the depth is halved and the number of backhoes is doubled.

A four-year-old car has a value of $3000. When one year old, its value was $9000.
(a) Assuming the car's depreciation is linear, express the value V of the car as a function of its age x (in years).
(b) What is the age of the car when its value is fully depreciated?

For the solution, see Example 5 in Section 3.2.

Linear and Quadratic Functions

3.1 Linear Functions
3.2 Determining the Equation of a Line
3.3 Quadratic Functions
3.4 Quadratic Equations and Inequalities
3.5 Intersection Points of Two Graphs

3.1 Linear Functions

◆ Introductory Comments

Recall from Section 2.5 that the graph of the identity function $f(x) = x$ is a straight line that passes through the origin and splits quadrants I and III in half [see part (b) of Figure 2.42 in Section 2.5]. By the vertical stretch and compress rule (Section 2.5), we know that the graph of $g(x) = ax$ is also a straight line, but with a *steeper* incline than the graph of $f(x) = x$ if $a > 1$, or a *flatter* incline if $0 < a < 1$. By the x-axis reflection rule (Section 2.5), the line slants *upward* if $a > 0$, or *downward* if $a < 0$, as we move from left to right along the x-axis. In summary, the value of a determines the steepness and direction of the line. In Figure 3.1 we use a graphing calculator to illustrate this fact by generating the graph of $g(x) = ax$ for $a = 2, -5$, and $\frac{1}{4}$.

By the vertical shift rule (Section 2.5), we know that the graph of $F(x) = ax + b$ is the same as the graph of $g(x) = ax$, but shifted vertically upward b units if $b > 0$, or downward $|b|$ units if $b < 0$. Hence, the value of b is the y-intercept of the graph of F. In Figure 3.2 we use a graphing calculator to illustrate this fact by generating the graph of $F(x) = ax + b$ for several values of a and b.

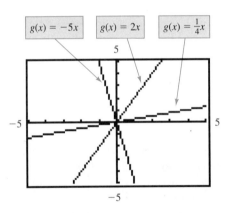

FIGURE 3.1

In the graph of $g(x) = ax$, the value of a determines the steepness and direction of the line.

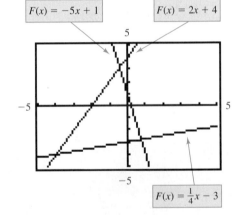

FIGURE 3.2

In the graph of $F(x) = ax + b$, the value of b is the y-intercept of the line.

We refer to the function F defined by $F(x) = ax + b$ as a *linear function:* its graph is a straight line with y-intercept b.

◆ Linear Function

If a and b are any real numbers with $a \neq 0$, then the function F defined by

$$F(x) = ax + b$$

is a **linear function.** The graph of this function is a **straight line** with y-intercept b.

SECTION 3.1 ♦ Linear Functions

In this section, we study linear functions and related equations whose graphs are straight lines. We begin our discussion by defining the *slope* of a line.

◆ The Slope of a Line

The slope of a nonvertical line is a measure of the line's steepness and is defined as the ratio of the vertical rise to the horizontal run between any two distinct points on the line:

$$\text{Slope} = \frac{\text{vertical rise}}{\text{horizontal run}}$$

Consider the straight line through the points $P(x_1, y_1)$ and $Q(x_2, y_2)$ shown in Figure 3.3. The vertical rise between the points P and Q is $y_2 - y_1$, and the horizontal run between these points is $x_2 - x_1$. Thus, the ratio of $(y_2 - y_1)$ to $(x_2 - x_1)$ defines the slope of this line. It is common practice to designate the slope of a line with the lowercase letter m.

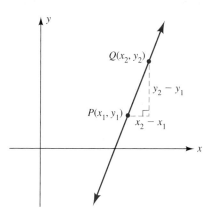

FIGURE 3.3

The slope of the line through the points P and Q is the ratio of the vertical rise $(y_2 - y_1)$ to the horizontal run $(x_2 - x_1)$.

◆ Slope Formula

The **slope** m of a nonvertical line through the distinct points $P(x_1, y_1)$ and $Q(x_2, y_2)$, provided $x_1 \neq x_2$, is

$$m = \frac{y_2 - y_1}{x_2 - x_1}.$$

If $x_1 = x_2$, the line is vertical and its slope is *undefined*.

Regardless of which two distinct points we choose on the line, the slope of the line is always the same. To show this, consider the straight line that passes through the points $P(x_1, y_1)$ and $Q(x_2, y_2)$ in Figure 3.3 and also through two other distinct points $R(x_3, y_3)$ and $S(x_4, y_4)$ as shown in Figure 3.4. Notice that the triangles *PAQ* and *RBS* are similar, because their corresponding angles are equal. Thus, the corresponding sides of these triangles are proportional, and we have

$$\frac{y_2 - y_1}{x_2 - x_1} = \frac{y_4 - y_3}{x_4 - x_3}.$$

Since these two ratios are equal, and we have already defined the slope of this line as $m = \frac{y_2 - y_1}{x_2 - x_1}$, we know that $\frac{y_4 - y_3}{x_4 - x_3}$ also defines the slope of this line.

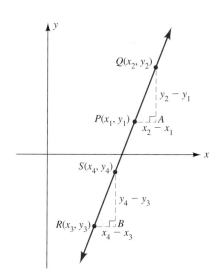

FIGURE 3.4

Since the triangles *PAQ* and *RBS* are similar, the slope of the line is the same value, regardless of which two distinct points we choose on the line.

EXAMPLE 1 Determine the slope of the line that passes through the given points.

(a) $(-1, 2)$ and $(3, 4)$ **(b)** $(2, -3)$ and $(-2, 4)$

(c) $(-2, 3)$ and $(3, 3)$ **(d)** $(-1, 1)$ and $(-1, -4)$

Solution

(a) Let $(x_1, y_1) = (-1, 2)$ and $(x_2, y_2) = (3, 4)$. Then

$$m = \frac{y_2 - y_1}{x_2 - x_1} = \frac{4 - 2}{3 - (-1)} = \frac{2}{4} = \frac{1}{2} \qquad \text{(See Figure 3.5.)}$$

(b) Let $(x_1, y_1) = (2, -3)$ and $(x_2, y_2) = (-2, 4)$. Then

$$m = \frac{y_2 - y_1}{x_2 - x_1} = \frac{4 - (-3)}{-2 - 2} = -\frac{7}{4} \qquad \text{(See Figure 3.6.)}$$

(c) Let $(x_1, y_1) = (-2, 3)$ and $(x_2, y_2) = (3, 3)$. Then

$$m = \frac{y_2 - y_1}{x_2 - x_1} = \frac{3 - 3}{3 - (-2)} = \frac{0}{5} = 0 \qquad \text{(See Figure 3.7.)}$$

(d) Since the x-coordinates of the points are the same, $x_1 = x_2 = -1$, the slope is undefined (see Figure 3.8).

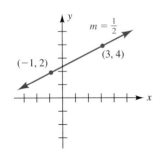

FIGURE 3.5
A line with a *positive* slope

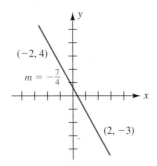

FIGURE 3.6
A line with a *negative* slope

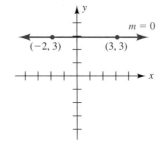

FIGURE 3.7
A line with a *zero* slope

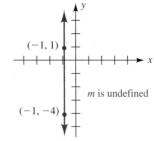

FIGURE 3.8
A line with an *undefined* slope

PROBLEM 1 When determining the slope, does it matter which point you label (x_1, y_1) or (x_2, y_2)? Interchange the labels of the points (x_1, y_1) and (x_2, y_2) in parts (a)–(c) of Example 1, and then determine the slope. What do you conclude?

We can use Example 1 to make some observations about the slope of a line:

1. If the line slants *upward* (increases) as we move from left to right along the x-axis, as in Figure 3.5, then the slope of the line is *positive*.
2. If the line slants *downward* (decreases) as we move from left to right along the x-axis, as in Figure 3.6, then the slope of the line is *negative*.
3. If the line is *horizontal* (parallel to the x-axis), as in Figure 3.7, then $y_1 = y_2$, and therefore the slope of the line is *zero*.

4. If the line is *vertical* (perpendicular to the *x*-axis), as in Figure 3.8, then $x_1 = x_2$, and therefore the slope of the line is *undefined*.

EXAMPLE 2 Sketch the line that passes through the point $(2, -1)$ and has the given slope.

(a) $m = 3$ **(b)** $m = -\frac{2}{3}$

◆ Solution

(a) A slope of $3 = \frac{3}{1}$ represents a vertical rise of 3 to a horizontal run of 1 (or a vertical rise of 6 to a horizontal run of 2, and so on). Thus, starting at the point $(2, -1)$, we can go *up 3 units* and run *right 1 unit* to locate another point on the line. Connecting these points with a straightedge gives us a sketch of this straight line, as shown in Figure 3.9.

(b) A slope of $-\frac{2}{3}$ represents a vertical drop of 2 to a horizontal run of 3 (or a vertical drop of 4 to a horizontal run of 6, and so on.) Thus, starting at the point $(2, -1)$, we can go *down 2 units* and run *right 3 units* to locate another point on the line. Connecting these points with a straightedge gives us a sketch of this straight line, as shown in Figure 3.10.

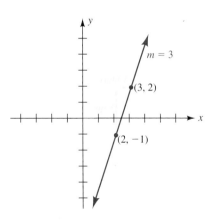

FIGURE 3.9
A line passing through the point $(2, -1)$ with a slope of 3

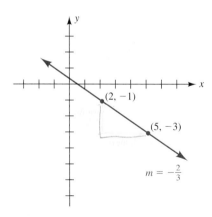

FIGURE 3.10
A line passing through the point $(2, -1)$ with a slope of $-\frac{2}{3}$

PROBLEM 2 Sketch the line that passes through the point $(-3, 0)$ and has slope $\frac{5}{2}$.

◆ Sketching the Graph of a Linear Function

The slope *m* of the line with the equation $y = ax + b$ is the number *a*. To show this, we let $P(x_1, y_1)$ and $Q(x_2, y_2)$ be any two distinct points on this line. Since these two points must satisfy the equation $y = ax + b$, we have

$$y_1 = ax_1 + b \quad \text{and} \quad y_2 = ax_2 + b.$$

Thus,

$$m = \frac{y_2 - y_1}{x_2 - x_1} = \frac{(ax_2 + b) - (ax_1 + b)}{x_2 - x_1}$$

$$= \frac{a(x_2 - x_1)}{x_2 - x_1}$$

$$= a$$

When we replace a with m in the equation $y = ax + b$, we obtain

$$y = mx + b$$

which is called the **slope-intercept form** for the equation of a straight line. As illustrated in the next example, we can quickly sketch the graph of a linear function simply by noting its slope and y-intercept.

EXAMPLE 3 Identify the slope and y-intercept, and then sketch the graph of each linear function.

(a) $f(x) = \frac{4}{3}x - 2$ **(b)** $y = 3 - 5x$

◆ Solution

(a) The linear function f defined by

$$f(x) = \frac{4}{3}x + (-2),$$

has slope $\frac{4}{3}$ and y-intercept -2. To graph this linear function, we begin by plotting the point $(0, -2)$. Next, remember that the slope is the ratio of the rise to the run. Thus, from the point $(0, -2)$, we go *up 4 units* and run *right 3 units* to locate another point on the line. Connect these two points to form the graph of $f(x) = \frac{4}{3}x - 2$, as shown in Figure 3.11.

(b) The linear function defined by the equation

$$y = 3 - 5x = -5x + 3$$

has slope -5, or $-\frac{5}{1}$, and y-intercept 3. Since the slope is negative, we start at the point $(0, 3)$ and go *down 5 units*, then run *right 1 unit* to locate another point on the line. Connecting these two points, we have the graph of $y = 3 - 5x$, as shown in Figure 3.12. ◆

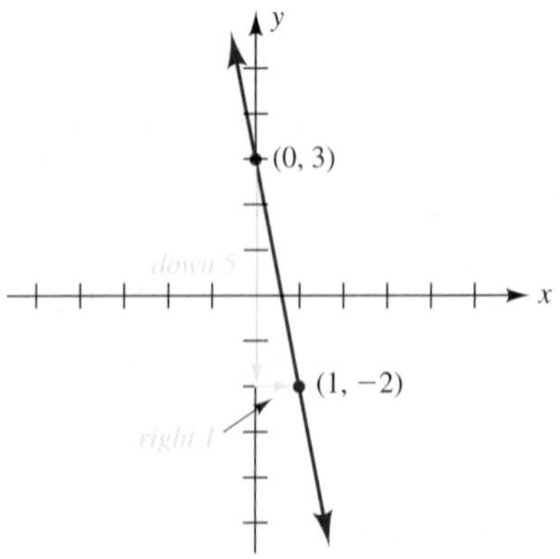

PROBLEM 3 Identify the slope and y-intercept of the function g defined by $g(x) = 3x - \frac{1}{2}$. Then sketch the graph of this linear function. ◆

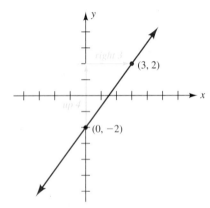

FIGURE 3.11
The graph of $f(x) = \frac{4}{3}x - 2$ has slope $\frac{4}{3}$ and y-intercept -2.

FIGURE 3.12
The graph of $y = 3 - 5x$ has slope -5 and y-intercept 3.

◆ General Form for the Equation of a Straight Line

Consider the equation $Ax + By + C = 0$ with constants A, B, and C, provided $B \neq 0$. Solving this equation for y, we obtain

$$y = \left(-\frac{A}{B}\right)x + \left(-\frac{C}{B}\right).$$

where $-\frac{A}{B}$ is the slope and $-\frac{C}{B}$ is the y-intercept.

We know that this is the *slope-intercept form* for the equation of a straight line. Thus, we also know that the graph of the equation $Ax + By + C = 0$ is a straight line with slope $-A/B$ and y-intercept $-C/B$, provided $B \neq 0$. We refer to the equation

$$\boxed{Ax + By + C = 0}$$

as the **general form** for the equation of a straight line.

Suppose $A = 0$ in the equation $Ax + By + C = 0$ with $B \neq 0$. Solving for y, we obtain

$$y = 0x + \left(-\frac{C}{B}\right).$$

where 0 is the slope and $-\frac{C}{B}$ is the y-intercept.

Since the slope of the line is zero, we conclude that $0x + By + C = 0$, or

$$\boxed{By + C = 0}$$

is the equation of a *horizontal line* with y-intercept $-C/B$. Actually, the equation $By + C = 0$ defines a *constant function*.

If $B = 0$ in the equation $Ax + By + C = 0$ with $A \neq 0$, then we cannot write the equation in the slope-intercept form $y = mx + b$, because division by zero is not defined. However, regardless of what value we choose for y in the equation $Ax + 0y + C = 0$, the quantity $0y$ is always zero. This means that ordered pairs that satisfy $Ax + 0y + C = 0$, or $Ax + C = 0$, can have any y-value as long as the x-value is $-C/A$. Thus, we conclude that

$$\boxed{Ax + C = 0}$$

is the equation of a *vertical line* with x-intercept $-C/A$. Although the graph of $Ax + C = 0$ is a straight line, the equation $Ax + C = 0$ does *not* define a linear function, because this graph fails the vertical line test (see Section 2.3).

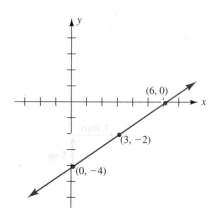

FIGURE 3.13
Graph of $2x - 3y = 12$

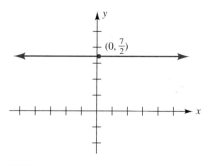

FIGURE 3.14
Graph of $2y - 7 = 0$

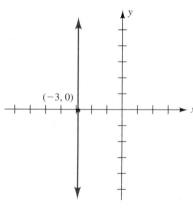

FIGURE 3.15
Graph of $2x + 6 = 0$

EXAMPLE 4 Sketch the graph of each equation.

(a) $2x - 3y = 12$ **(b)** $2y - 7 = 0$ **(c)** $2x + 6 = 0$

◆ **Solution**

(a) Solving for y, we obtain

$$2x - 3y = 12$$
$$-3y = -2x + 12 \quad \text{Subtract } 2x \text{ from both sides}$$
$$y = \tfrac{2}{3}x - 4 \quad \text{Divide both sides by } -3$$

This last equation is the slope-intercept form for the equation of a line. The graph of this equation is a straight line with slope $\tfrac{2}{3}$ and y-intercept -4, as shown in Figure 3.13.

(b) The equation $2y - 7 = 0$ can be written in slope-intercept form as

$$y = 0x + \tfrac{7}{2}.$$

where 0 is the slope and $\tfrac{7}{2}$ is the y-intercept.

Since the slope is zero, the graph of this equation is a horizontal line with y-intercept $\tfrac{7}{2}$, as shown in Figure 3.14.

(c) Although the equation $2x + 6 = 0$ cannot be written in slope-intercept form, we can write it in general form as

$$2x + 0y + 6 = 0.$$

Ordered pairs that satisfy this equation can have any y-value provided that $x = -3$. Thus, the graph of this equation is a vertical line with x-intercept -3, as shown in Figure 3.15. ◆

To graph the equation $Ax + By + C = 0$, where A, B, and C are nonzero numbers, we can use an alternate method: find the x-intercept and y-intercept. Connecting these intercepts with a straightedge also gives us the graph of the straight line. For the equation $2x - 3y = 12$ in Example 4(a), we find the intercepts as follows.

Find the x-intercept:	Find the y-intercept:
$2x - 3y = 12$	$2x - 3y = 12$
Let $y = 0$: $2x - 3(0) = 12$	Let $x = 0$: $2(0) - 3y = 12$
$2x = 12$	$-3y = 12$
$x = 6$	$y = -4$
Coordinates of	Coordinates of
x-intercept: $(6, 0)$	y-intercept: $(0, -4)$

Connecting the points $(6, 0)$ and $(0, -4)$ also gives us the graph of the straight line illustrated in Figure 3.13.

To display the graph of a nonvertical line $Ax + By = C$ on a graphing calculator, we must enter the equation in the form $y = f(x)$. To show a true geometric perspective of the slope of the line, we use a square set viewing rectangle—one in which the vertical and horizontal tick marks have equal spacing. Figure 3.16 shows how the graph of the equation $2x - 3y = 12$ (or $y = \frac{2}{3}x - 4$) looks in different viewing rectangles. Only the line in the square set viewing rectangle in part (a) visually appears to have a slope of $\frac{2}{3}$.

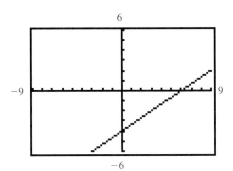
(a) Equal spacing between tick marks (square set viewing rectangle)

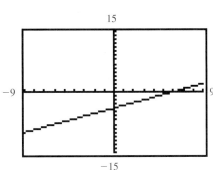
(b) Smaller spacing between tick marks on y-axis than on x-axis

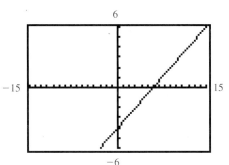
(c) Smaller spacing between tick marks on x-axis than on y-axis

FIGURE 3.16
Appearance of $2x - 3y = 12$ in different viewing rectangles.

PROBLEM 4 Find the *x*-intercept and *y*-intercept for the equation $3x + 8y = 12$. Then sketch its graph. Use a graphing calculator to verify the sketch.

◆ Application: Tax Rate Schedule

The amount of federal income tax you pay each year is a function of your taxable income for that year. This amount is usually given by the Internal Revenue Service (IRS) in terms of a *tax rate schedule*. Each line in the tax rate schedule can be expressed as a different linear function. If a function is defined by different linear functions on distinct subsets of its domain, it is called a **piecewise linear function**.

EXAMPLE 5 A recent tax rate schedule for a single person with a taxable income up to $93,130 is given by the IRS as follows.

Taxable income	Tax
$0 to $18,550	15% of the amount over $0
$18,550 to $44,900	$2782.50 + 28% of the amount over $18,550
$44,900 to $93,130	$10,160.50 + 33% of the amount over $44,900

(a) Find the piecewise linear function defined by this tax rate schedule.

(b) Use a graphing calculator to generate the graph of the function in part (a).

(c) Use the piecewise linear function in part (a) to determine the amount of tax a single person is required to pay for a taxable income of $39,440. Then use the tax rate schedule to find this tax.

◆ Solution

(a) Using this tax rate schedule, we can write three equations:

1. If the taxable income x is $0 to $18,550, then the required tax T (in dollars) is

$$T(x) = 0.15x$$

2. If the taxable income x is more than $18,550 but not more than $44,900, the required tax T (in dollars) is

$$T(x) = 2782.50 + 0.28(x - 18{,}550)$$
$$= 0.28x - 2411.50$$

3. If the taxable income x is more than $44,900 but not more than $93,130, the required tax T (in dollars) is

$$T(x) = 10{,}160.50 + 0.33(x - 44{,}900)$$
$$= 0.33x - 4656.50$$

Thus, the piecewise linear function defined by this tax rate schedule has three parts:

$$T(x) = \begin{cases} 0.15x & \text{if } 0 \le x \le 18550 \\ 0.28x - 2411.50 & \text{if } 18{,}550 < x \le 44{,}900 \\ 0.33x - 4656.50 & \text{if } 44{,}900 < x \le 93{,}130 \end{cases}$$

(b) To generate the graph of this piecewise linear function on a graphing calculator, we enter each piece of the function separately, specifying the interval over which that piece is defined. Figure 3.17 shows the graph of the function T defined in part (a). The x-axis denotes the taxable income x, and the y-axis denotes the required tax T.

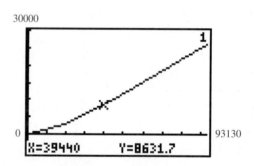

FIGURE 3.17

Graph of the piecewise linear function defined in Example 5(a)

(c) Since a taxable income of $39,440 is more than $18,550 but not more than $44,900, we use the second part of the function to compute the tax:

$$T(x) = 0.28x - 2411.50$$

$$T(39,440) = 0.28(39,440) - 2411.50 = \$8631.70$$

Also, using the second line of the tax rate schedule, we find that the tax is

$$\$2782.50 + 0.28 \underbrace{(\$39,440 - \$18,550)}_{\text{The amount over \$18,550}} = \$8631.70.$$

Referring to Figure 3.17, we note that (39440, 8631.70) is a point on the graph.

PROBLEM 5 Use the graph in Figure 3.17 to estimate the taxable income of a single person who is required to pay an income tax of $15,000. Then use the piecewise linear function in Example 5(a) to find the taxable income to the nearest cent.

Exercises 3.1

◆ Basic Skills

In Exercises 1–12, find the slope (if it exists) of the line that passes through the given points.

1. (2, 0) and (3, 2)
2. (4, 1) and (−1, −1)
3. (−3, 4) and (6, −2)
4. (3, −2) and (4, −3)
5. (−4, 3) and (−4, 1)
6. (−2, 5) and (4, 5)
7. $\left(\frac{3}{2}, -1\right)$ and $\left(-\frac{1}{2}, 7\right)$
8. $\left(\frac{3}{4}, \frac{2}{3}\right)$ and $\left(-1, \frac{1}{6}\right)$
9. (4, 3) and (0, b)
10. (a, 0) and (0, b)
11. (a, b) and (b, a)
12. $\left(a, \frac{a}{b}\right)$ and $\left(b, \frac{b}{a}\right)$

In Exercises 13–18, sketch the line that passes through the given point and has the given slope.

13. Through (3, 1) with slope 3
14. Through (−3, −1) with slope $\frac{4}{3}$
15. Through $\left(0, -\frac{2}{3}\right)$ with slope $-\frac{1}{3}$
16. Through $\left(\frac{1}{2}, \frac{3}{2}\right)$ with slope −2
17. Through (−1, 2) with slope 0
18. Through (2, 3) with undefined slope

In Exercises 19–30, identify the slope and y-intercept, and then sketch the graph of the linear function. Use a graphing calculator to verify each graph.

19. $y = 3x$
20. $y = -2x$
21. $f(x) = -\frac{1}{3}x$
22. $F(x) = \frac{3x}{4}$
23. $y = 2x + 3$
24. $y = 3x - 4$
25. $g(x) = -4x + 1$
26. $G(x) = 5 - 2x$
27. $y = \frac{x}{2} - 1$
28. $y = \frac{2}{3}x - 2$
29. $h(x) = \frac{1 - 3x}{4}$
30. $H(x) = \frac{-x + 3}{2}$

In Exercises 31–42, sketch the graph of each equation, state the slope, and label the x-intercept and y-intercept (if they exist). Use a graphing calculator to verify the graph of each nonvertical line.

31. $x + y = 2$
32. $x - y = -3$
33. $2x + 5y = 10$
34. $3x + y = -6$
35. $21x - 28y = -56$
36. $9x - 18y = 60$

37. $x = 2y$
38. $4x - 3y = 0$
39. $x = -2$
40. $y = 4$
41. $\frac{4}{3}y - 3 = 0$
42. $2.5 - 0.2x = 0$

In Exercises 43–48, sketch the graph of each piecewise linear function. Use a graphing calculator to verify each graph.

43. $f(x) = \begin{cases} x & \text{if } x \leq 2 \\ 2x - 2 & \text{if } x > 2 \end{cases}$

44. $F(x) = \begin{cases} 2x & \text{if } x \leq 1 \\ 3 - x & \text{if } x > 1 \end{cases}$

45. $h(x) = \begin{cases} 2x + 5 & \text{if } x \leq -2 \\ x + 3 & \text{if } x > -2 \end{cases}$

46. $H(x) = \begin{cases} -(3x + 4) & \text{if } x \leq -1 \\ \dfrac{x - 2}{3} & \text{if } x > -1 \end{cases}$

47. $g(x) = \begin{cases} 3x - 1 & \text{if } x \leq 0 \\ x - 1 & \text{if } 0 < x \leq 2 \\ 6 - x & \text{if } x > 2 \end{cases}$

48. $G(x) = \begin{cases} x + 3 & \text{if } x \leq 2 \\ 7 - 2x & \text{if } 2 < x \leq 4 \\ x - 5 & \text{if } x > 4 \end{cases}$

In Exercises 49–56, use a graphing calculator to generate the graph of each equation in the viewing rectangle. Then estimate the x-intercept and y-intercept by tracing to these points. Round each estimate to three significant digits. Check each estimate algebraically.

49. $7x + 4y = 24$
50. $16y - 18x = -33$
51. $14x - 18y + 57 = 0$
52. $124x = 234y - 167$
53. $6.45x - 8.76y = 34.5$
54. $345.6x - 124.8y + 187.4 = 0$
55. $7.89y - 1.23x = 14.8$
56. $18.4x = 67.3 - 18.7y$

57. The linear equation $F = \frac{9}{5}C + 32$ expresses the Fahrenheit temperature F as a function of the Celsius temperature C.

(a) Identify the slope and F-intercept.

(b) Sketch the graph of the linear equation.

58. The linear equation $C = \frac{5}{9}(F - 32)$ expresses the Celsius temperature C as a function of the Fahrenheit temperature F.

(a) Identify the slope and C-intercept.

(b) Sketch the graph of the linear equation.

59. The linear equation $v = 44 - 3t$ expresses the velocity v (in feet per second, ft/s) of an object as a function of the time t (in seconds) the object travels at a constant deceleration of 3 ft/s². [*Note:* The unit ft/s² is a more compact way of writing *feet per second per second*, or $\dfrac{\text{ft/s}}{\text{s}}$.]

(a) Identify the slope and v-intercept.

(b) Sketch the graph of the linear equation.

(c) What is the significance of the t-intercept?

60. The linear equation $A = 500 + 3.75t$ expresses the amount A (in dollars) accumulated in a certain savings account after t months ($0 \leq t \leq 12$) if the money earns 9% simple interest per year.

(a) Identify the slope and A-intercept.

(b) Sketch the graph of the linear equation for $0 \leq t \leq 12$.

(c) What is the significance of the slope and the A-intercept?

61. The cost C of electricity is a function of the amount x of electricity consumed (in kilowatt-hours, kwh). Suppose the cost of electricity is given by the following rate schedule.

Usage	Cost
First 30 kwh or less	$3.00
Next 70 kwh	$0.06 per kwh
Next 200 kwh	$0.04 per kwh
All kilowatt-hours over 300	$0.01 per kwh

(a) Find the piecewise linear function that describes this rate schedule.

(b) Use a graphing calculator to generate the graph of the function in part (a).

(c) Use the rate table to find the cost of using 240 kwh of electricity. Then use the piecewise linear function defined in part (a) to find the cost.

(d) Use the graph from part (b) to estimate the number of kilowatts-hours consumed if the cost is $14.08. Then use the piecewise linear function from part (a) to find the exact number of kilowatt-hours consumed.

62. The amount of federal income tax T you pay is a function of your taxable income x. A recent tax rate schedule for a single person is given:

Taxable income	Tax
$0 to $1800	11% of the amount over $0
$1800 to $16,800	$198 + 15% of the amount over $1800
$16,800 to $27,000	$2448 + 28% of the amount over $16,800
$27,000 to $54,000	$5304 + 35% of the amount over $27,000
$54,000 and up	$14,754 + 38.5% of the amount over $54,000

(a) Find the piecewise linear function defined by this rate schedule.

(b) Use a graphing calculator to generate the graph of the function in part (a).

(c) Use the tax rate table to determine the amount of tax a single person is required to pay for a taxable income of $35,000. Then use the piecewise linear function from part (a) to find the tax.

(d) Use the graph from part (b) to estimate the taxable income of a single person whose income tax is $7,000. Then use the piecewise linear function defined in part (a) to find this taxable income to the nearest cent.

◆ Critical Thinking

63. The center of a circle of radius 5 has coordinates $(-1, 2)$. Determine the coordinates of the endpoints of a diameter whose slope is $-\frac{3}{4}$.

64. The midpoint of a line segment 20 units long is $(2, -1)$. If the slope of the line segment is $\frac{4}{3}$, what are the coordinates of its endpoints?

65. For what value of a will the line through the points $(3, -2)$ and $(a, 2)$ have the given slope?
    (a) $m = -2$ (b) $m = \frac{2}{3}$ (c) $m = -\frac{1}{2}$

66. For what value of a will the points $(-1, 3)$, $(2, a)$, and $(3, -3)$ be collinear (lie on the same line)?

67. Find the inverse function f^{-1} for the linear function $f(x) = mx + b$ with $m \neq 0$.

68. Suppose f and g are linear functions defined by $f(x) = m_1 x + b_1$ and $g(x) = m_2 x + b_2$, where $m_1 \neq 0$ and $m_2 \neq 0$.
    (a) Find $f \circ g$.
    (b) Find $g \circ f$.

(c) Is $f \circ g$ a linear function? If so, give its slope and y-intercept.

(d) Is $g \circ f$ a linear function? If so, give its slope and y-intercept.

69. Consider the equation $Ax + By = C$ with $A \neq 0$, $B \neq 0$, and $C \neq 0$. Divide both sides of this equation by C and write the equation in the form

$$\frac{x}{a} + \frac{y}{b} = 1.$$

Explain why this equation is called the *equation of a line in intercept form*.

70. Use the results of Exercise 69 to sketch the graph of each equation. Use a graphing calculator to verify each sketch.

(a) $\dfrac{x}{2} + \dfrac{y}{3} = 1$ (b) $\dfrac{x}{5} - \dfrac{y}{2} = 1$

(c) $\dfrac{y}{4} - \dfrac{x}{6} = 1$ (d) $-\dfrac{x}{\frac{3}{2}} - \dfrac{y}{\frac{8}{3}} = 1$

3.2 Determining the Equation of a Line

◆ Introductory Comments

In Section 3.1, we were given the equation of a line and then found some information about the line, such as its slope or x- and y-intercepts. In this section, we reverse the procedure—from some given information about a line, we determine its equation. We begin by developing the *point-slope form* for the equation of a straight line. This form is useful for determining the equation of a line when we are given the slope of the line and the coordinates of a fixed point on the line.

◆ Point-Slope Form for the Equation of a Line

Let m be the given slope of a line and $P(x_1, y_1)$ be a given point on the line. Also, let $Q(x, y)$ be *any* point other than P on the given line, as shown in

Figure 3.18. By the slope formula (in Section 3.1), we have

$$m = \frac{y - y_1}{x - x_1}$$

$$y - y_1 = m(x - x_1) \quad \text{Multiply both sides by } (x - x_1)$$

This last equation is called the **point-slope form** for the equation of a line.

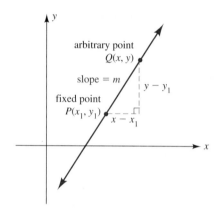

FIGURE 3.18
The slope m of the line through a fixed point $P(x_1, y_1)$ and an arbitrary point $Q(x, y)$ on the line is $(y - y_1)/(x - x_1)$.

◆ Point-Slope Form

> The equation of a line having a given slope m and passing through the fixed point $P(x_1, y_1)$ and arbitrary point $Q(x, y)$ is
>
> $$y - y_1 = m(x - x_1).$$

EXAMPLE 1 Find the equation of a line that passes through the point $(2, -1)$ and has a slope of 3. Write this equation in slope-intercept form (see Section 3.1).

◆ Solution Using the point-slope form for the equation of a line with $(x_1, y_1) = (2, -1)$ and $m = 3$, we obtain the following equation:

$$y - y_1 = m(x - x_1)$$
$$y - (-1) = 3(x - 2) \quad \text{Substitute}$$
$$y + 1 = 3x - 6 \quad \text{Multiply}$$
$$y = 3x - 7 \quad \text{Write in slope-intercept form}$$

Figure 3.19 shows the graph of the equation $y = 3x - 7$ in the viewing rectangle of a graphing calculator. Note that this line passes through the given point $(2, -1)$ and has slope 3. ◆

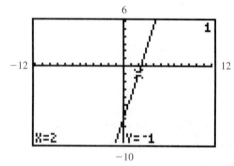

FIGURE 3.19
The graph of $y = 3x - 7$ is a line that passes through the point $(2, -1)$ and has slope 3.

PROBLEM 1 Find the equation of a line with slope $\frac{1}{2}$ and x-intercept -3. Write this equation in slope-intercept form. ◆

SECTION 3.2 ♦ Determining the Equation of a Line

If two points on a line are given, then the equation of the line can be determined by using the slope formula (from Section 3.1) together with the point-slope form for the equation of a line.

EXAMPLE 2 Find the equation of the line that passes through the points $(-2, -3)$ and $(2, 1)$. Write the equation in slope-intercept form.

◆ Solution   The slope of the line that passes through the points $(-2, -3)$ and $(2, 1)$ is

$$m = \frac{1 - (-3)}{2 - (-2)} = \frac{4}{4} = 1.$$

Using $(x_1, y_1) = (-2, -3)$ as the fixed point and $m = 1$ as the slope, we have

$$y - y_1 = m(x - x_1)$$
$$y - (-3) = 1(x - (-2)) \quad \text{Substitute}$$
$$y + 3 = x + 2 \quad \text{Simplify}$$
$$y = x - 1 \quad \text{Write in slope-intercept form}$$

Figure 3.20 shows the graph of the equation $y = x - 1$ in the viewing rectangle of a graphing calculator. Note that this line passes through the given points $(-2, -3)$ and $(2, 1)$. ◆

PROBLEM 2 Use $(x_1, y_1) = (2, 1)$ as a fixed point and $m = 1$ as the slope to determine the equation of a line. Write the equation in slope-intercept form. Your answer should agree with the result in Example 2. ◆

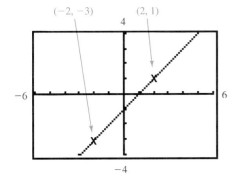

FIGURE 3.20

The graph of $y = x - 1$ is a line that passes through the points $(-2, -3)$ and $(2, 1)$.

◆ Parallel Lines

Suppose two distinct, nonvertical lines $y = m_1 x + b_1$ and $y = m_2 x + b_2$ are parallel, as shown in Figure 3.21.

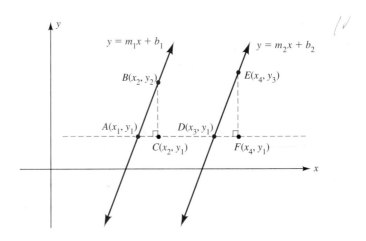

FIGURE 3.21

If the lines $y = m_1 x + b_1$ and $y = m_2 x + b_2$ are *parallel*, then the triangles ABC and DEF are similar and their corresponding sides are proportional.

The triangles ABC and DEF are similar, since their corresponding angles are equal. Thus, their corresponding sides are proportional and we have

$$\frac{y_2 - y_1}{x_2 - x_1} = \frac{y_3 - y_1}{x_4 - x_3}.$$

However, $\quad \dfrac{y_2 - y_1}{x_2 - x_1} = m_1 \quad$ and $\quad \dfrac{y_3 - y_1}{x_4 - x_3} = m_2.$

Therefore, $m_1 = m_2$. Thus, if two nonvertical lines are parallel, they have the same slope. The converse of this statement is also true: If two distinct nonvertical lines have the same slope, then the lines are parallel.

◆ Parallel Lines

Two distinct nonvertical lines with slopes m_1 and m_2 are **parallel** if and only if

$$m_1 = m_2.$$

EXAMPLE 3 Find the equation of the line that passes through the point $(1, 3)$ and is parallel to the line that passes through the points $(-2, -4)$ and $(2, -1)$. Write this equation in slope-intercept form.

◆ Solution

The slope of the line that passes through the points $(-2, -4)$ and $(2, -1)$ is

$$m = \frac{-1 - (-4)}{2 - (-2)} = \frac{3}{4}.$$

Since two parallel lines have the same slope, we know that the slope of the line that passes through the point $(1, 3)$ is also $\frac{3}{4}$. Now using the point-slope form for the equation of a line with fixed point $(x_1, y_1) = (1, 3)$ and slope $m = \frac{3}{4}$, we can find the equation of the desired line as follows:

$$y - y_1 = m(x - x_1)$$
$$y - 3 = \tfrac{3}{4}(x - 1) \qquad \text{Substitute}$$
$$y - 3 = \tfrac{3}{4}x - \tfrac{3}{4} \qquad \text{Multiply}$$
$$y = \tfrac{3}{4}x + \tfrac{9}{4} \qquad \text{Write in slope-intercept form}$$

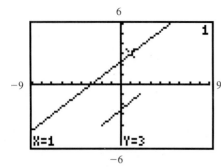

FIGURE 3.22

The graph of $y = \tfrac{3}{4}x + \tfrac{9}{4}$ is a line that passes through $(1, 3)$. This line is parallel to the line that joins the points $(-2, -4)$ and $(2, -1)$.

Figure 3.22 shows the graph of the equation $y = \tfrac{3}{4}x + \tfrac{9}{4}$ in the viewing rectangle of a graphing calculator. Note that the line passes through the given point $(1, 3)$. In this viewing rectangle, we have also constructed a line segment between the given points $(-2, -4)$ and $(2, -1)$ by using the [DRAW] key on the calculator. Observe that the lines appear to be parallel. ◆

SECTION 3.2 ♦ *Determining the Equation of a Line* 167

PROBLEM 3 Find the equation of a line that has x-intercept -3 and is parallel to a line whose slope is $\frac{2}{3}$. Write this equation in slope-intercept form.

♦ Perpendicular Lines

Suppose two nonvertical lines $y = m_1x + b_1$ and $y = m_2x + b_2$ are perpendicular, as shown in Figure 3.23.

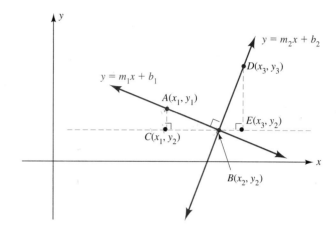

FIGURE 3.23
If the lines $y = m_1x + b_1$ and $y = m_2x + b_2$ are *perpendicular*, then the triangles *BAC* and *DBE* are similar and their corresponding sides are proportional.

Angle *BAC* is complementary to angle *ABC*, and angle *ABC* is complementary to angle *DBE*. Thus, angles *BAC* and *DBE* are equal and so triangles *BAC* and *DBE* are similar. Using the fact that the corresponding sides of these triangles are proportional, we have

$$\frac{y_3 - y_2}{x_3 - x_2} = \frac{x_2 - x_1}{y_1 - y_2}$$

However,

$$\frac{y_3 - y_2}{x_3 - x_2} = m_2,$$

and

$$\frac{x_2 - x_1}{y_1 - y_2} = -\frac{x_1 - x_2}{y_1 - y_2} = -\frac{1}{\dfrac{y_1 - y_2}{x_1 - x_2}} = -\frac{1}{m_1}.$$

Therefore, $m_2 = -\dfrac{1}{m_1}$. Thus, if two nonvertical lines are perpendicular, the slope of one line is the negative reciprocal of the slope of the other. The converse of this statement is also true: If the slope of one line is the negative reciprocal of the slope of another, then the lines are perpendicular.

Perpendicular Lines

Two nonvertical lines with slopes m_1 and m_2 are **perpendicular** if and only if

$$m_2 = -\frac{1}{m_1}.$$

EXAMPLE 4 Find the equation of a line that passes through the point $(2, 1)$ and is perpendicular to the line $3x - 5y = 10$.

Solution We can determine the slope of the line $3x - 5y = 10$ by writing the equation in slope-intercept form:

$$3x - 5y = 10$$
$$-5y = -3x + 10$$
$$y = \underbrace{\frac{3}{5}}_{\text{slope}} x - 2$$

Since the slope of the line $3x - 5y = 10$ is $\frac{3}{5}$, the slope m of the perpendicular line is

$$m = -\frac{1}{\frac{3}{5}} = \underbrace{-\frac{5}{3}}_{\text{Negative reciprocal of } \frac{3}{5}}.$$

Now, using the point-slope form for the equation of a line with fixed point $(x_1, y_1) = (2, 1)$ and $m = -\frac{5}{3}$, we can find the desired equation:

$$y - y_1 = m(x - x_1)$$
$$y - 1 = -\tfrac{5}{3}(x - 2) \quad \text{Substitute}$$
$$y - 1 = -\tfrac{5}{3}x + \tfrac{10}{3} \quad \text{Multiply}$$
$$y = -\tfrac{5}{3}x + \tfrac{13}{3} \quad \text{Write in slope-intercept form}$$

Figure 3.24 shows the graph of the equation $y = -\frac{5}{3}x + \frac{13}{3}$ in the viewing rectangle of a graphing calculator. Note that the line passes through the given point $(2, 1)$. In this viewing rectangle, we also have generated the graph of the equation $3x - 5y = 10$. Observe that the lines appear to be perpendicular. It is important to use a square set viewing rectangle to show the true geometric perspective of the slopes and, hence, the perpendicularity of the lines. ◆

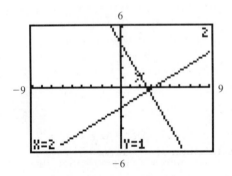

FIGURE 3.24
The graph of $y = -\frac{5}{3}x + \frac{13}{3}$ is a line that passes through $(2, 1)$. This line is perpendicular to the line $3x - 5y = 10$.

PROBLEM 4
Find the equation of a line that passes through the point $(3, -4)$ and is perpendicular to a line that passes through the points $(3, 0)$ and $(2, -3)$. ◆

SECTION 3.2 ◆ Determining the Equation of a Line

◆ **Application: Linear Depreciation**

One type of applied problem that occurs frequently in business is *linear depreciation*. In linear depreciation, the value of an asset *decreases* linearly over time.

EXAMPLE 5 *Chapter Opening Problem*

A four-year-old car has a value of $3000. When one year old, its value was $9000.

(a) Assuming the car's depreciation is linear, express the value V of the car as a function of its age x (in years).

(b) What is the age of the car when its value is fully depreciated?

◆ Solution

(a) To illustrate the linear depreciation of the car between the ages of 1 and 4 years, we use the [DRAW] key on a graphing calculator to construct a line segment between the points (1, 9000) and (4, 3000), as shown in Figure 3.25. The slope of this line segment is

$$m = \frac{\$9000 - \$3000}{1 \text{ yr} - 4 \text{ yr}} = -\$2000 \text{ per year.}$$

Note that the slope tells us the amount of *depreciation per year*. Now, to express the value V of the car as a function of its age x, we use the point-slope form for the equation of a line with fixed point $(x_1, V_1) = (1, 9000)$ and slope $m = -2000$. Thus,

$$V - V_1 = m(x - x_1)$$
$$V - 9000 = -2000(x - 1) \quad \textbf{Substitute}$$
$$V - 9000 = -2000x + 2000 \quad \textbf{Multiply}$$
$$V = 11{,}000 - 2000x \quad \textbf{Write in slope-intercept form}$$

FIGURE 3.25

A graph of the linear depreciation of the car between the ages of 1 and 4 years

(b) The car's value is fully depreciated when $V = 0$. Assuming the car continues to depreciate according to the equation $V = 11{,}000 - 2000x$, we have

$$0 = 11{,}000 - 2000x$$
$$2000x = 11{,}000$$
$$x = 5\tfrac{1}{2}$$

Thus, the car's value is fully depreciated when it is $5\tfrac{1}{2}$ years old. Figure 3.26 shows that $x = 5\tfrac{1}{2}$ is the x-intercept of the graph of $V = 11{,}000 - 2000x$, which confirms our answer. ◆

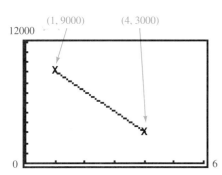

FIGURE 3.26

The x-intercept of the graph of $V = 11{,}000 - 2000x$ denotes the time when the car's value is fully depreciated.

PROBLEM 5 Referring to Example 5, what is the significance of the V-intercept in the equation $V = 11{,}000 - 2000x$? ◆

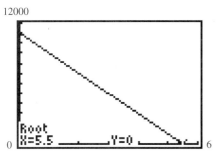

Exercises 3.2

Basic Skills

In Exercises 1–38, find the equation of a line that satisfies the given conditions. For nonvertical lines, write the equation in slope-intercept form. For vertical lines, write the equation in general form with integer coefficients.

1. y-intercept 3 and slope $\frac{3}{4}$
2. y-intercept -2 and slope $-\frac{1}{2}$
3. Through $(-1, 2)$ with slope 3
4. Through $(3, 4)$ with slope -2
5. x-intercept 4 and slope $-\frac{7}{3}$
6. x-intercept -3 and slope $\frac{2}{5}$
7. Through the origin with slope -3
8. Through the origin and $(6, 5)$
9. Through $(4, -1)$ and $(2, 5)$
10. Through $(-1, -2)$ and $(4, -3)$
11. Through $\left(-2, \frac{4}{3}\right)$ and $(3, -1)$
12. Through $\left(\frac{6}{5}, -2\right)$ and $\left(-\frac{3}{2}, 1\right)$
13. Through $(-3, 4)$ with y-intercept $\frac{1}{6}$
14. Through $(5, 3)$ with x-intercept -3
15. x-intercept 3 and y-intercept -2
16. x-intercept -4 and y-intercept 6
17. x-intercept $\frac{1}{4}$ and y-intercept 6
18. x-intercept -3 and y-intercept $-\frac{5}{3}$
19. y-intercept $\frac{3}{2}$ and slope 0
20. Through $(4, -3)$ and $(-1, -3)$
21. Through $(3, -2)$ and $(3, 4)$
22. x-intercept $-\frac{2}{3}$ and undefined slope
23. Through $(5, -1)$ with same y-intercept as $3x - 2y = 4$
24. Through $\left(\frac{1}{2}, -1\right)$ with same x-intercept as $4x - y + 4 = 0$
25. Through $(2, 3)$ with x-intercept a and y-intercept a, provided $a \neq 0$
26. Through $\left(\frac{1}{3}, -5\right)$ with x-intercept a and y-intercept $-a$, provided $a \neq 0$
27. x-intercept -2 and parallel to a line with slope $-\frac{1}{2}$
28. y-intercept $\frac{4}{3}$ and parallel to a line with slope -4
29. Through $(-3, -1)$ and parallel to the line that passes through $(1, 0)$ and $(2, 3)$
30. Through $(4, 5)$ and parallel to the line with x-intercept 3 and y-intercept -2
31. Through $(2, -3)$ and parallel to the line $2x - 3y + 6 = 0$
32. Through the origin and parallel to the line $x - 2y - 3 = 0$
33. y-intercept 3 and perpendicular to a line with slope -4
34. x-intercept $\frac{5}{2}$ and perpendicular to a line with slope $\frac{8}{5}$
35. Through $(-1, 0)$ and perpendicular to the line that passes through $(2, 0)$ and $(-1, 4)$
36. Through $(5, -2)$ and perpendicular to the y-axis
37. Through the origin and perpendicular to the line $2x + 4y = 5$
38. Through $(2, -7)$ and perpendicular to the line $3x - 2y - 9 = 0$
39. Use a graphing calculator to generate the graphs of the following equations in the same viewing rectangle:

    $$2.795x - 1.625y = 8.325$$
    $$6.622x - 3.850y = -5.765$$

    Use the graphs to determine if the pair of lines appear to be parallel. Verify your answer algebraically.

40. Use a graphing calculator to generate the graphs of the following equations in the same viewing rectangle:

    $$2.740x + 0.8768y = 10.65$$
    $$0.6672x - 2.085y = 15.76$$

    Use the graphs to determine if the pair of lines appear to be perpendicular. Verify your answer algebraically.

41. At $60°$F a cricket chirps 50 times per minute and at $80°$F it chirps 100 times per minute.
    (a) Assuming that the rate of chirping varies linearly with the temperature, express the cricket's chirping rate r as a function of the temperature T.
    (b) Determine the cricket's chirping rate when $T = 40°$F.

42. A salesperson's weekly pay is $450 for weekly sales of $3000 and $500 for weekly sales of $4000.
    (a) Assuming that weekly pay varies linearly with the amount of weekly sales, express the weekly pay P as a function of the amount of weekly sales S.

(b) Determine the base pay and the rate of commission for this salesperson.

43. A new dump truck cost $60,000 and depreciates linearly to a value of $6000 after 6 years.
    (a) Express the value V of the truck as a function of its age x (in years).
    (b) What is the value of the truck after 4 years?

44. A factory owner buys a new machine for $25,000. The value of the machine depreciates linearly, and the machine has no resale value after 10 years.
    (a) Express the value V of the machine as a function of its age x (in years).
    (b) What is the resale value of the machine after 6 years?

45. A house cost $50,000 to build 8 years ago. Its value has appreciated linearly and today the house is worth $170,000.
    (a) Express the value V of the house as a function of its age x (in years).
    (b) What is the projected value of the house 4 years from now?

46. A diamond ring was purchased for $600. Fifteen years later the ring was appraised for $2400. Assume the ring appreciated linearly.
    (a) Express the value V of the ring as a function of its age x (in years).
    (b) Find the value of the ring 20 years after it was purchased.

47. Trolley tracks are laid with expansion gaps between the steel rails so that the rails can expand without distortion. As the temperature increases, the width of the gap between the rails decreases linearly. Suppose the gap is 2.5 mm wide when the temperature is $51\,°F$ and 1.2 mm wide when the temperature is $83\,°F$.
    (a) Express the width W of the expansion gap as a function of the temperature T.
    (b) What is the approximate width of the expansion gap when the temperature is $35\,°F$?
    (c) At what temperature will the rails just touch?

48. The fuel tank of an automobile is filled with 22.3 gallons of gasoline. When the auto is traveling at a constant rate of speed, the amount of fuel remaining in the tank decreases linearly. After 1 hour 12 minutes of travel, the tank contains 18.1 gallons of fuel.
    (a) Express the number N of gallons of gasoline in the tank as a function of the amount of time t traveled.
    (b) Approximately how many gallons of gasoline remain after traveling for 2 hours 21 minutes?
    (c) If the auto continues to travel at this constant rate, when is the fuel tank empty?

Critical Thinking

49. The *perpendicular bisector* of a line segment is a line that is perpendicular to the segment at its midpoint. Find the equation of the perpendicular bisector of the line segment that joins the given points.
    (a) $(4, 0)$ and $(0, -12)$ (b) $(3, -2)$ and $(-1, 1)$

50. Use slopes to show that the triangle with vertices $A(-2, 1)$, $B(4, -8)$, and $C(7, -6)$ is a right triangle. Then find the area of this right triangle.

In Exercises 51–54, find the linear function that satisfies the given conditions.

51. $f(2) = 7$ and $f(-1) = 3$
52. $h(3) = 1$ and $h(0) = 2$
53. $g(2) = -1$ and the graph of g has slope 3
54. $F(-1) = -3$ and the graph of F has slope -2
55. Find $H(3)$ given that H is a linear function with $H(5) = 4$ and $H(0) = 0$
56. Find $G(-4)$ given that G is a linear function with $G(-7) = 3$ and $G(5) = 1$

57. Find the value of k such that the line $kx + 2y + 4 = 0$ satisfies the given condition.
    (a) passes through $(3, 1)$
    (b) is parallel to the x-axis
    (c) is parallel to the line $y = 3x - 4$
    (d) is perpendicular to the line $2x - 4y = 5$

58. Find the value of k such that the line $3x + ky = 6$ satisfies the given condition.
    (a) passes through $(-2, 4)$
    (b) is perpendicular to the x-axis
    (c) is parallel to the line $y = 5 - 2x$
    (d) is perpendicular to the line $x - y = 0$

59. A *tangent line* to a curve is a line that just touches the curve at a single point and is closer to the curve in the vicinity of the point than any other line. Using calculus, it can be shown that the *slope of the tangent line* (denoted $m_{\tan}$) to the curve $y = 1 - 3x^2$ at any point (x, y) on the curve is given by $m_{\tan} = -6x$. Find the equation of the

tangent line that touches the point (1, −2), as shown in the sketch.

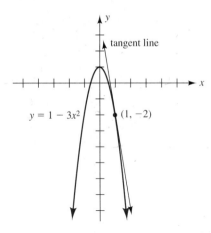

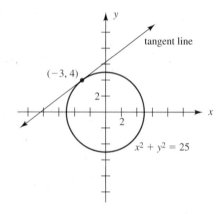

60. Refer to Exercise 59. Using calculus, it can be shown that the slope of the tangent line to the circle $x^2 + y^2 = 25$ at any point (x, y) on the circle is given by $m_{\tan} = -x/y$. Find the equation of the tangent line that touches the point $(-3, 4)$ as shown in the sketch.

61. Find the equation of the line that passes through the point $P(x_1, y_1)$ and is parallel to the line $Ax + By + C = 0$. Write the equation in slope-intercept form.

62. Find the equation of the line that passes through the origin and is perpendicular to the line $Ax + By + C = 0$. Write the equation in general form.

3.3 Quadratic Functions

◆ Introductory Comments

Recall from Section 2.5 that the graph of the squaring function $f(x) = x^2$ is a cupped-shaped curve that is symmetric with respect to the y-axis [see part (d) of Figure 2.42 in Section 2.5]. We refer to this curve as a **parabola** and the vertical line $x = 0$ (the y-axis) as its **axis of symmetry**. The point at which the parabola intersects its axis of symmetry is called the **vertex** of the parabola (see Figure 3.27).

By the vertical stretch and compress rule (in Section 2.5), we know that the graph of $g(x) = ax^2$ is also a parabola, but with a *narrower* opening than the graph of $f(x) = x^2$ if $|a| > 1$, or a *wider* opening if $0 < |a| < 1$. By the x-axis reflection rule (Section 2.5), the parabola opens *upward* if $a > 0$, or *downward* if $a < 0$. In summary, the value of a determines the width and direction of the opening of the parabola. In Figure 3.28, we use a graphing calculator to illustrate this fact by generating the graph of $g(x) = ax^2$ for $a = 2, -5,$ and $\frac{1}{4}$.

By the horizontal shift rule and the vertical shift rule (in Section 2.5), we know that the graph of $F(x) = a(x - h)^2 + k$ is the same as the graph of $g(x) = ax^2$, but shifted horizontally to the *right* h units if $h > 0$, or to the *left* $|h|$ units if $h < 0$, and shifted vertically *upward* k units if $k > 0$, or *downward* $|k|$ units if $k < 0$. Hence, the coordinates of the vertex of the parabola are now (h, k) and its axis of symmetry is the vertical line $x = h$. In Figure 3.29, we use a graphing calculator to illustrate this fact by generating the graph of $F(x) = a(x - h)^2 + k$ for several values of a, h, and k.

FIGURE 3.27

The graph of $f(x) = x^2$ is a parabola with vertex $(0, 0)$ and axis of symmetry $x = 0$.

SECTION 3.3 ◆ *Quadratic Functions* 173

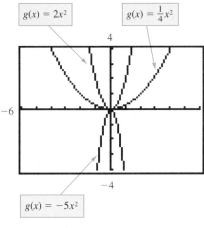

FIGURE 3.28
In the graph of $g(x) = ax^2$, the value of a determines the width and direction of the opening of the parabola.

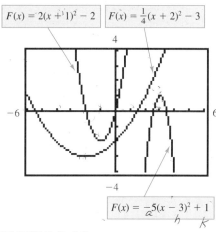

FIGURE 3.29
In the graph of $F(x) = a(x - h)^2 + k$, the coordinates of the vertex of the parabola are (h, k).

We refer to the function F defined by $F(x) = a(x - h)^2 + k$ as a *quadratic function in standard form*. Its graph is a parabola with vertex (h, k) and axis of symmetry $x = h$.

◆ **Quadratic Function in Standard Form**

> If a, h, and k are real numbers with $a \neq 0$, then the function F defined by
>
> $$F(x) = a(x - h)^2 + k$$
>
> is called a **quadratic function in standard form**. The graph of this function is a **parabola** with vertex (h, k) and axis of symmetry $x = h$.

In this section, we study quadratic functions and their applications.

◆ **Working with Quadratic Functions in Standard Form**

To sketch the graph of the quadratic function $F(x) = a(x - h)^2 + k$ without plotting points or using a graphing calculator, we need to locate the vertex (h, k) and determine the y-intercept.

EXAMPLE 1 Determine the vertex and y-intercept for the graph of the quadratic function $f(x) = -2(x + 1)^2 + 8$. Then sketch the graph of this function.

◆ **Solution** We begin by writing this quadratic function in standard form:

$$f(x) = \underbrace{-2}_{a}[x - \underbrace{(-1)}_{h}]^2 + \underbrace{8}_{k}$$

CHAPTER 3 ♦ Linear and Quadratic Functions

Since $a = -2$ (so $a < 0$), we know the graph is a parabola that opens *downward*. Since $h = -1$ and $k = 8$, its vertex is $(-1, 8)$.

To find the *y*-intercept, we let $x = 0$ and determine $f(0)$ as follows:

$$f(0) = -2(0 + 1)^2 + 8 = -2 + 8 = 6$$

y-intercept

The graph of $f(x) = -2(x + 1)^2 + 8$ is the parabola shown in Figure 3.30. We draw the right-hand side of the parabola by using the vertex and the *y*-intercept. We then use the fact that the parabola is symmetric with respect to the vertical line $x = -1$ to draw the left-hand side of this curve. ♦

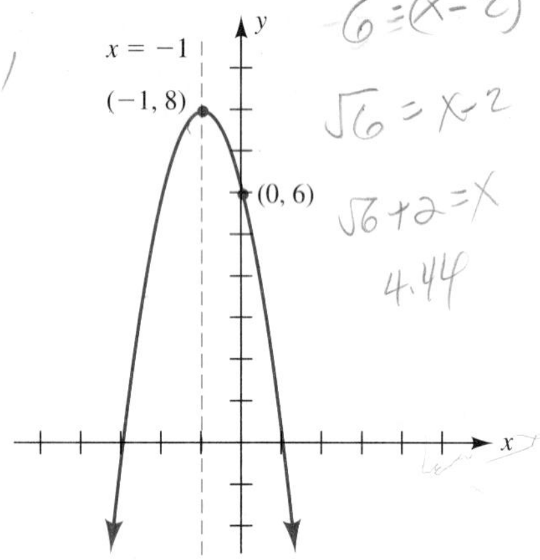

PROBLEM 1 Determine the vertex and intercepts for the quadratic function g defined by $g(x) = (x - 3)^2 - 1$. ♦

As illustrated in the next example, we can determine the equation of a quadratic function if we know the vertex of the parabola and the coordinates of just one other point on the parabola.

EXAMPLE 2 Determine the equation of the quadratic function whose graph passes through the point $(4, 1)$ and has its vertex at $(2, 3)$.

♦ **Solution** Since the vertex is $(2, 3)$, the quadratic function must have the form

$$y = a(x - 2)^2 + 3.$$

To determine the value of a, we use the fact that the graph passes through the point $(4, 1)$:

$$y = a(x - 2)^2 + 3$$
$$1 = a(4 - 2)^2 + 3 \quad \text{Substitute } x = 4 \text{ and } y = 1$$
$$1 = 4a + 3 \quad \text{Solve for } a$$
$$-2 = 4a$$
$$a = -\tfrac{1}{2}$$

Thus, the required quadratic function is $y = -\tfrac{1}{2}(x - 2)^2 + 3$. Figure 3.31 shows the graph of this quadratic function in the viewing rectangle of a graphing calculator. The parabola passes through the point $(4, 1)$ and has its vertex at $(2, 3)$; these facts verify our work. ♦

PROBLEM 2 Determine the equation of the quadratic function whose graph passes through the point $(2, 3)$ and has its vertex at $(4, 1)$. ♦

FIGURE 3.30
The graph of $f(x) = -2(x + 1)^2 + 8$ is a parabola that opens *downward* with vertex $(-1, 8)$. The graph is symmetric with respect to the vertical line $x = -1$.

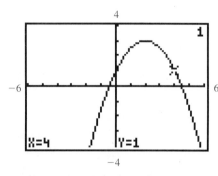

FIGURE 3.31
The graph of the quadratic function $y = -\tfrac{1}{2}(x - 2)^2 + 3$ passes through the point $(4, 1)$ and has its vertex at $(2, 3)$.

◆ Quadratic Functions in General Form

If we square the expression $(x - h)$ in $F(x) = a(x - h)^2 + k$, we obtain

$$F(x) = a(x^2 - 2hx + h^2) + k$$
$$= ax^2 + (-2ah)x + (ah^2 + k)$$

Replace the constant $-2ah$ with b and the constant $ah^2 + k$ with c.

$$= ax^2 + bx + c$$

where a, b, and c are constants and $a \neq 0$. We refer to a function of this form as a *quadratic function in general form*.

◆ **Quadratic Function in General Form**

If a, b, and c are real numbers and $a \neq 0$, then the function F defined by

$$F(x) = ax^2 + bx + c$$

is a **quadratic function in general form**.

If a quadratic function is written in the general form $F(x) = ax^2 + bx + c$, then we know that its graph is a parabola. We can find the x-coordinate of its vertex by writing the first two terms, $ax^2 + bx$, in the form of $a(x - h)^2$ and noting the value of h. To accomplish this task, we use an algebraic method called **completing the square**. First, we factor a from the first two terms:

$$ax^2 + bx = a\left(x^2 + \frac{b}{a}x\right).$$

Next, we add a special constant to the expression $x^2 + \frac{b}{a}x$ to form a perfect square trinomial that factors to the form $(x - h)^2$. If we add *the square of half the coefficient of x*, or $\left(\frac{b}{2a}\right)^2$, we obtain

$$x^2 + \frac{b}{a}x + \left(\frac{b}{2a}\right)^2 \quad \text{or} \quad \left(x + \frac{b}{2a}\right)^2.$$

The expression $\left(x + \frac{b}{2a}\right)^2$ is of the form $(x - h)^2$ with $h = -\frac{b}{2a}$. Hence, we conclude that $-\frac{b}{2a}$ is the x-coordinate of the vertex of the parabola defined by the function $F(x) = ax^2 + bx + c$. The y-coordinate of the vertex of the parabola may be found by evaluating $F\left(-\frac{b}{2a}\right)$. We refer to this fact as the **vertex formula for a parabola**.

Vertex Formula for a Parabola

> The graph of the quadratic function $F(x) = ax^2 + bx + c$ is a parabola with
>
> $$\text{axis of symmetry } x = -\frac{b}{2a}$$
>
> $$\text{vertex } \left(-\frac{b}{2a}, F\left(-\frac{b}{2a}\right)\right).$$

EXAMPLE 3 Determine the vertex and y-intercept for the graph of each quadratic function. Then sketch the graph of the function.

(a) $f(x) = x^2 - 3x - 4$ (b) $y = -2x^2 + 4x - 5$

Solution

(a) For $f(x) = x^2 - 3x - 4$, we have $a = 1$, $b = -3$, and $c = -4$. Thus, the x-coordinate of the vertex of the parabola is

$$x = -\frac{b}{2a} = -\frac{-3}{2(1)} = \frac{3}{2}$$

and the y-coordinate is

$$y = f\left(-\frac{b}{2a}\right) = f\left(\frac{3}{2}\right) = \left(\frac{3}{2}\right)^2 - 3\left(\frac{3}{2}\right) - 4 = -\frac{25}{4}.$$

Hence, the vertex of the parabola is $\left(\frac{3}{2}, -\frac{25}{4}\right)$. Since $a = 1$ (so $a > 0$), we know the parabola opens upward from the point $\left(\frac{3}{2}, -\frac{25}{4}\right)$.

We find the y-intercept by letting $x = 0$ and evaluating $f(0)$:

$$f(0) = (0)^2 - 3(0) - 4 = -4$$

                                                y-intercept

The graph of $f(x) = x^2 - 3x - 4$ is shown in Figure 3.32. We draw the left-hand side of the parabola by using the vertex and the y-intercept. We then use the fact that the parabola is symmetric with respect to the line $x = \frac{3}{2}$ to draw the right-hand side of the curve.

(b) For $y = -2x^2 + 4x - 5$, we have $a = -2$, $b = 4$, and $c = -5$. Thus, the x-coordinate for the vertex of the parabola is

$$x = -\frac{b}{2a} = -\frac{4}{2(-2)} = 1.$$

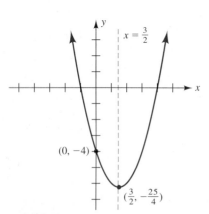

FIGURE 3.32
The graph of $f(x) = x^2 - 3x - 4$ is a parabola that opens upward and has vertex $\left(\frac{3}{2}, -\frac{25}{4}\right)$.

To find the y-coordinate of the vertex, we replace x with 1 in the equation $y = -2x^2 + 4x - 5$ as follows:

$$y = -2(1)^2 + 4(1) - 5 = -3$$

Hence, the vertex has coordinates $(1, -3)$ and, since $a = -2$ (so $a < 0$), the parabola opens downward from the point $(1, -3)$.

We find the y-intercept by letting $x = 0$ and solving for y as follows:

$$y = -2(0)^2 + 4(0) - 5 = \underbrace{-5}_{y\text{-intercept}}$$

The graph of $y = -2x^2 + 4x - 5$ is shown in Figure 3.33. We draw the left-hand side of the parabola by using the vertex and the y-intercept. We then use the fact that the parabola is symmetric with respect to the line $x = 1$ to draw the right-hand side of the curve. ◆

PROBLEM 3   Use factoring and the zero product property (in Section 2.2) to find the x-intercepts of the parabola in Example 3(a). Check your answers with the graph in Figure 3.32. ◆

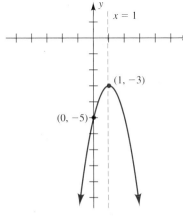

FIGURE 3.33

The graph of $y = -2x^2 + 4x - 5$ is a parabola that opens downward and has vertex $(1, -3)$.

◆ Maximum and Minimum Values of Quadratic Functions

If $a > 0$, then the vertex (h, k) is the *lowest point* on the graph of the function $F(x) = a(x - h)^2 + k$. Hence, we say that the function F has a **minimum value** of k when $x = h$. For example, Figure 3.32 indicates that the lowest point on the graph of $f(x) = x^2 - 3x - 4$ is its vertex, $\left(\frac{3}{2}, -\frac{25}{4}\right)$. We refer to $-\frac{25}{4}$ as the *minimum value* for this quadratic function and say that it occurs when $x = \frac{3}{2}$.

If $a < 0$, then the vertex (h, k) is the *highest point* on the graph of the function $F(x) = a(x - h)^2 + k$. Hence, we say that the function F has a **maximum value** of k when $x = h$. For example, Figure 3.33 indicates that the highest point on the graph of $y = -2x^2 + 4x - 5$ is its vertex, $(1, -3)$. We refer to -3 as the *maximum value* for this quadratic function and say that it occurs when $x = 1$.

In general, calculus is needed to find the maximum and minimum values of a function. However, if the function is quadratic, we may use the vertex formula for a parabola to find the maximum or minimum value of the function.

EXAMPLE 4   Determine if the function $g(x) = 3x^2 + 6x - 8$ has a maximum or minimum value. Then find this value.

◀ Solution   Since the function g is quadratic, we can use the vertex formula for a parabola to determine the maximum or minimum value.

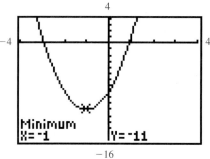

FIGURE 3.34

The quadratic function $g(x) = 3x^2 + 6x - 8$ has a minimum value of -11, and this occurs when $x = -1$.

Since $a = 3$ (so $a > 0$), we know the parabola opens upward. Thus, g has a *minimum value* at

$$x = -\frac{b}{2a} = -\frac{6}{2(3)} = -1.$$

To find the minimum value, we replace x with -1 and evaluate $g(-1)$:

$$g(-1) = 3(-1)^2 + 6(-1) - 8 = -11$$

Thus, the minimum value for the quadratic function $g(x) = 3x^2 + 6x - 8$ is -11, and it occurs when $x = -1$. Figure 3.34 shows the graph of the function g in the viewing rectangle of a graphing calculator. The coordinates of the vertex of the parabola are $(-1, -11)$; this fact confirms our work. ◆

PROBLEM 4 Determine whether the function defined by the equation $y = 3 - 3x^2$ has a minimum or maximum value. Then find this value. ◆

◆ Application: Maximizing the Area of a Garden

The idea of maximum and minimum values occurs in many applied problems. We illustrate one such application in our next example.

EXAMPLE 5 One hundred feet of fencing is to be used to enclose a rectangular garden that abuts a barn. No fencing is needed along the barn. What is the *largest* possible area that can be enclosed?

◆ **Solution** In this problem we want the *area* to be the *largest* possible. Thus, we express the area as a quadratic function and then find its *maximum value* by using the vertex formula for a parabola. We begin by letting

w = the width of the rectangle   and l = the length of the rectangle

as shown in Figure 3.35. Then the area A of the rectangle is

$$A = lw.$$

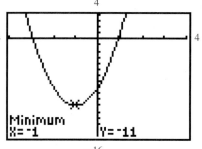

Since 100 ft of fencing is available to enclose the rectangular garden, we have

$$100 = l + 2w \quad \text{or} \quad l = 100 - 2w.$$

FIGURE 3.35

A rectangular garden of length l and width w with no fencing along the barn

Substituting $100 - 2w$ for l in the area formula $A = lw$, we obtain

$$A = (100 - 2w)w$$
$$A = 100w - 2w^2$$

A quadratic function with $a = -2$, $b = 100$, and $c = 0$

SECTION 3.3 ◆ Quadratic Functions

Since $a = -2$ (so $a < 0$), we know that the graph of $A = 100w - 2w^2$ is a parabola that opens downward. Thus, this quadratic function has a maximum value that occurs at

$$w = -\frac{b}{2a} = -\frac{100}{2(-2)} = 25.$$

Replacing w with 25 in $A = 100w - 2w^2$, we obtain

$$A = 100(25) - 2(25)^2 = \underbrace{1250}_{\text{maximum value}}.$$

Thus, the largest possible area that can be enclosed is 1250 square feet. This occurs when the width w is 25 ft and the length l is $100 - 2(25)$, or 50 ft. Figure 3.36 shows the graph of $A = 100w - 2w^2$ for $0 < w < 50$ in the viewing rectangle of a graphing calculator. The x-axis denotes the width w of the rectangular garden and the y-axis denotes the area A of the garden. Note that the coordinates of the vertex of the parabola are (25, 1250); this fact confirms our work. ◆

FIGURE 3.36
The quadratic function $A = 100w - 2w^2$ has a maximum value of 1250, and this occurs when $w = 25$.

PROBLEM 5 Suppose 100 ft of fencing is to be used to enclose a rectangular garden on all four sides. What is the *largest* possible area that can be enclosed? ◆

◆ Application: Law of Motion for Freely Falling Objects

Nearly four hundred years ago, Galileo discovered the *law of motion for freely falling objects:* If air resistance is not considered for a projectile that is fired straight upward with an initial velocity v_0 from an initial height s_0, then its distance s above the ground t seconds later is given by

$$s = -\tfrac{1}{2}gt^2 + v_0 t + s_0$$

where g is the acceleration due to gravity (approximately 32 ft/s², or 9.8 m/s²). This equation defines a quadratic function, and we can use it to find maximum and minimum values.

EXAMPLE 6 The roof in an enclosed baseball stadium is 150 ft above home plate. A baseball player standing at home plate hits a pop fly straight upward with an initial velocity of 96 ft/s from an initial height of 4 ft. Does the ball hit the roof?

◆ **Solution** Since the measurements are given in feet, we use $g = 32$ ft/s² with $v_0 = 96$ ft/s and $s_0 = 4$ ft. Hence, the quadratic function that defines the distance s of the ball above the ground is

$$s = -16t^2 + 96t + 4.$$

Since this equation defines a quadratic function, we can use the vertex formula for a parabola to determine the maximum distance above the ground that the ball travels. Since $a = -16$ (so $a < 0$), we know the parabola opens downward and, therefore, has a maximum value. The maximum value occurs when

$$t = -\frac{b}{2a} = -\frac{96}{2(-16)} = 3 \text{ seconds}.$$

To find the maximum distance above the ground, we replace t with 3 in the equation $s = -16t^2 + 96t + 4$ and evaluate s:

$$s = -16(3)^2 + 96(3) + 4 = 148 \text{ ft}$$

Hence, the maximum distance above the ground that the ball travels is 148 ft. Since the roof is 150 ft above home plate, the ball misses hitting the roof by 2 ft. The graph in Figure 3.37 verifies our work. This graph is not the actual path of the ball (which is hit straight upward), but rather the relationship between time and height. ◆

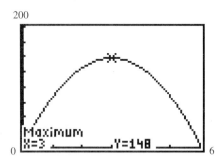

FIGURE 3.37
The quadratic function $s = -16t^2 + 96t + 4$ has a maximum value of 148, and this occurs when $t = 3$.

PROBLEM 6 Refer to Example 6. Suppose the catcher catches the ball 4 feet above the ground. How long is the ball in flight? ◆

Exercises 3.3

◆ Basic Skills

In Exercises 1–24, determine the vertex and y-intercept for the graph of each quadratic function. Then sketch the graph of the function. Use a graphing calculator to confirm the graph.

1. $y = (x - 1)^2 + 3$
2. $y = (x + 2)^2 - 1$
3. $f(x) = -(x + 3)^2 + 4$
4. $F(x) = -2(x - 1)^2 - 2$
5. $y = 3 - 3(x + 1)^2$
6. $y = \frac{1}{4}(x - 4)^2$
7. $G(x) = \frac{1}{2}(x - 4)^2 - 1$
8. $y = 2(x - \frac{1}{2})^2 - \frac{3}{2}$
9. $y = x^2 - 1$
10. $F(x) = 2 - \frac{1}{2}x^2$
11. $f(x) = 4x - x^2$
12. $g(x) = x^2 - 6x$
13. $y = x^2 + 2x - 3$
14. $G(x) = x^2 + 8x + 12$
15. $f(x) = 5 + 4x - x^2$
16. $y = -x^2 - 6x + 7$
17. $y = x^2 + x - 6$
18. $f(x) = 2x^2 - 3x - 5$
19. $h(x) = -2x^2 + 8x - 3$
20. $y = 3x^2 - 6x + 1$
21. $y = 3x^2 - 2x + 4$
22. $y = 3 + 5x - 5x^2$
23. $y = 1 + 4x + 4x^2$
24. $H(x) = 4x^2 - 12x + 9$

In Exercises 25–32, determine the equation of the quadratic function in general form whose graph satisfies the given conditions. Then generate the graph of the equation on a graphing calculator and verify that these conditions have been satisfied.

25. Vertex at (0, 0) and passing through (2, 1)
26. Vertex at the origin and passing through $(-1, -3)$
27. Vertex (0, 2) and x-intercepts ± 2
28. Vertex $(0, -4)$ and passing through (1, 4)
29. Vertex (3, 2) and passing through the origin
30. Vertex $(2, -1)$ and passing through $(-1, -2)$

1-55 every other odd # 1, 5, 9

31. Vertex $\left(-1, \frac{4}{3}\right)$ and y-intercept 2
32. Vertex $\left(-\frac{3}{4}, -\frac{49}{8}\right)$ and passing through $(-1, 0)$

In Exercises 33–38, determine if the quadratic function has a maximum or minimum value. Then find this value. Use a graphing calculator to verify your answer.

33. $y = 8x - x^2$
34. $f(x) = x^2 + 5x$
35. $G(x) = 4x^2 + 6x + 3$
36. $y = -2x^2 - 4x + 8$
37. $h(x) = -3(x - 7)^2 + 9$
38. $F(x) = 5(x + 4)^2 - 11$

In Exercises 39–42, use a graphing calculator to find the coordinates of the vertex for the graph of each quadratic function. Round each coordinate to three significant digits. Check algebraically by using the vertex formula for a parabola.

39. $f(x) = 12.5x^2 - 18.6x + 10.7$
40. $f(x) = 1.24x^2 + 4.56x - 9.55$
41. $f(x) = -0.024x^2 + 0.126x + 1.344$
42. $f(x) = 107 - 425x - 121x^2$

In Exercises 43 and 44, use the law of motion for freely falling objects given on page 179.

43. A ball is thrown vertically upward with an initial velocity of 64 ft/s from a height of 80 ft above the ground.
    (a) Use the law of motion to express the distance s (in feet) of the ball above the ground after t seconds.
    (b) When does the ball reach its maximum height?
    (c) What is the ball's maximum height?
    (d) How long is the ball in flight?

44. A model rocket (without its parachute) is fired directly upward with an initial velocity of 147 m/s from ground level.
    (a) Use the law of motion to express the distance s (in meters) of the rocket above the ground after t seconds.
    (b) When does the rocket reach its maximum height?
    (c) What is the rocket's maximum height?
    (d) How long is the rocket in flight?

45. The daily manufacturing cost C (in dollars) for a company that makes skis is given by

$$C(x) = x^2 - 100x + 4800,$$

where x is the number of pairs of skis produced per day.
    (a) How many pairs of skis should be produced to minimize the daily manufacturing cost?
    (b) What is the minimum daily manufacturing cost?

46. The daily profit P (in dollars) for a company that makes tennis rackets is given by

$$P(x) = -x^2 + 240x - 5400,$$

where x is the number of rackets produced per day.
    (a) How many tennis rackets should be produced to maximize the daily profit?
    (b) What is the maximum daily profit?

47. Find two numbers such that their sum is 124 and their product is the largest possible.

48. Find two numbers such that their sum is -30 and the sum of their squares is a minimum.

49. Of all rectangles with a perimeter of 44 inches, what are the dimensions of the one with the maximum area?

50. A long piece of sheet metal 12 inches wide is to be made into a rain gutter by bending up two long edges to make straight sides. The gutter will have a rectangular cross-sectional area, as shown in the figure. What depth will give the greatest water carrying capacity?

51. Two hundred feet of fencing is to be used to enclose two identical rectangular animal pens that abut a barn, as shown in the figure. No fencing is needed along the barn. What dimensions for the total enclosure make the area of the pens the largest possible?

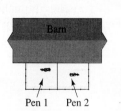

52. A football stadium is to be built in the shape of a rectangle with semicircular ends, as shown in the figure at the top of the next page. If the perimeter of the stadium is 1 mile, find the dimensions of x and r so that the area of the rectangular part of the stadium is the largest possible.

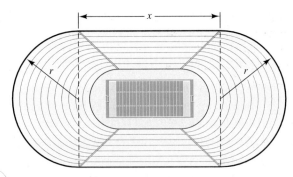

53. For the electrical circuit shown in the sketch, the power P (in watts) delivered to the load resistance R_L (in ohms) is given by

$$P = 21.6i - 0.34i^2$$

where i is the current (in amperes) flowing through the circuit. What is the maximum power that can be delivered to the load resistance?

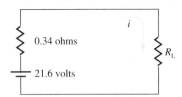

54. A wire 21.6 cm long is cut into two pieces. One of the pieces is bent into a square and the other piece is bent into a circle. Where should the wire be cut if the total area of the square and circle is to be a minimum?

55. A bus service between Plymouth and Boston charges $8 per person for a one-way trip and carries 400 passengers per day. Research shows that for each $1 increase in fare the company loses 25 passengers per day.
    (a) What daily rate gives the company the maximum revenue?
    (b) What is this maximum revenue?

56. A tutor charges $30 per hour for math tutoring and averages 21 students per week in one-hour sessions. The tutor has found that for each $2 decrease in the hourly rate, 3 more students per week request tutoring services.
    (a) What hourly rate should the tutor charge for maximum revenue?
    (b) What is this maximum revenue?

 Critical Thinking

57. If a parabola has two distinct x-intercepts a and b, what is the x-coordinate of its vertex?

58. Find all values of k such that the vertex of the graph of $y = x^2 + kx + 9$ lies on the x-axis.

59. Determine the quadratic function f with zeros -1 and 3 and range $[-2, \infty)$.

60. If the vertex of the graph of $y = 2x^2 + bx + c$ has coordinates $(3, 8)$, find the values of b and c.

61. If the graph of $f(x) = ax^2 + bx + c$ passes through the origin, what is the value of c?

62. Consider the linear function $f(x) = mx + b$ and the quadratic function $g(x) = a(x - h)^2 + k$ with $m \neq 0$ and $a \neq 0$.
    (a) Find $f \circ g$.
    (b) Find $g \circ f$.
    (c) Classify the functions $f \circ g$ and $g \circ f$ as linear, quadratic, or neither.

63. Use a graphing calculator to generate the five graphs of $y = x^2 + 4x + c$ for $c = -2, -1, 0, 1,$ and 2 in the same viewing rectangle. What effect does the constant c have on the axis of symmetry of these parabolas?

64. Use a graphing calculator to generate the four graphs of $y = a(x - 2)^2 + 1$ for $a = -2, -1, 1,$ and 2 in the same viewing rectangle. What effect does the constant a have on the vertex of these parabolas?

3.4 Quadratic Equations and Inequalities

◆ Introductory Comments

Any equation of the form

$$ax^2 + bx + c = 0$$

where x is the variable and a, b, and c are any real numbers such that $a \neq 0$, is called a **quadratic equation in standard form**. If we replace the equal sign with one of the inequality symbols ($<$, $>$, $\leq$, or $\geq$), we have a **quadratic inequality in standard form**. In this section, we discuss methods for solving quadratic equations and quadratic inequalities.

◆ Solving Quadratic Equations

Some quadratic equations can be solved easily by factoring. This method relies on the zero product property (in Section 2.2). For example, to solve the quadratic equation $x^2 - x = 12$, we may proceed as follows:

$$x^2 - x = 12$$
$$x^2 - x - 12 = 0 \quad \text{Write in standard form}$$
$$(x - 4)(x + 3) = 0 \quad \text{Factor the trinomial}$$
$$x - 4 = 0 \quad \text{or} \quad x + 3 = 0 \quad \text{Apply the zero product property}$$
$$x = 4 \qquad\qquad x = -3 \quad \text{Solve for } x$$

Thus, the roots of $x^2 - x = 12$ are 4 and -3.

CAUTION The quadratic equation must be in *standard form* before we factor and apply the zero product property. To write

$$x^2 - x = 12$$
$$x(x - 1) = 12$$
$$x = 12 \quad \text{or} \quad x - 1 = 12$$
$$x = 13 \quad \text{is WRONG!}$$

From our preceding work, we know that the solutions are 4 and -3, *not* 12 and 13.

Recall from Section 2.4 that the concepts of real roots of equations, x-intercepts of graphs, and real zeros of functions are all closely related. In the preceding example, we see that the factors of $x^2 - x - 12$, namely $(x - 4)$ and $(x + 3)$, are also closely related to these concepts. The roots of the equation $x^2 - x = 12$ are 4 and -3, the x-intercepts of the graph of $f(x) = x^2 - x - 12$ are 4 and -3 (see Figure 3.38), the real zeros of the function f are 4 and -3, and the factors of $x^2 - x - 12$ are $(x - 4)$ and $(x + 3)$. In general, if f is a quadratic function and k is a real number, then the following statements are equivalent:

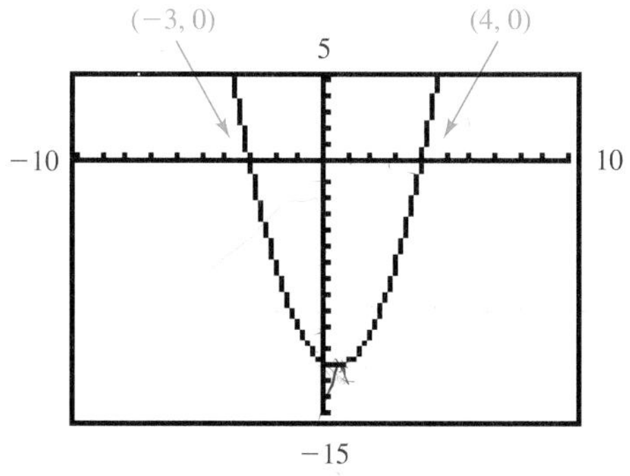

1. k is a real root of the equation $f(x) = 0$.
2. k is an x-intercept of the graph of the function f.
3. k is a real zero of f.
4. $x - k$ is a factor of $f(x)$.

FIGURE 3.38
The roots of the equation $x^2 - x = 12$ are the x-intercepts of the graph of $f(x) = x^2 - x - 12$.

EXAMPLE 1 Write each equation in the standard quadratic form $ax^2 + bx + c = 0$. Then factor and solve by using the zero product property.

(a) $(x - 3)(x - 10) = x - 10$ **(b)** $\dfrac{x - 2}{8} + 16 = \dfrac{x^2 - 3}{2}$

◆ **Solution**

(a) To solve the equation $(x - 3)(x - 10) = x - 10$, we proceed as follows:

$$(x - 3)(x - 10) = x - 10$$
$$x^2 - 13x + 30 = x - 10 \quad \text{Multiply}$$
$$x^2 - 14x + 40 = 0 \quad \text{Write in standard form}$$
$$(x - 4)(x - 10) = 0 \quad \text{Factor the trinomial}$$
$$x - 4 = 0 \quad \text{or} \quad x - 10 = 0 \quad \text{Apply the zero product property}$$
$$x = 4 \qquad\qquad x = 10$$

Thus, the roots of the original equation are 4 and 10. Figure 3.39 shows the graph of $f(x) = (x - 3)(x - 10) - x + 10$ in the viewing rectangle of a calculator. Note that the graph is a parabola with x-intercepts 4 and 10.

(b) To solve the equation $\dfrac{x - 2}{8} + 16 = \dfrac{x^2 - 3}{2}$, we multiply both sides of the equation by 8 (the least common denominator), cancel the denominators, and then write the equation in standard form:

$$\dfrac{x - 2}{8} + 16 = \dfrac{x^2 - 3}{2}$$

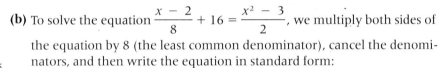

$$8\left(\dfrac{x - 2}{8} + 16\right) = 8\left(\dfrac{x^2 - 3}{2}\right) \quad \text{Multiply both sides by 8}$$
$$(x - 2) + 128 = 4(x^2 - 3) \quad \text{Simplify by canceling denominators}$$
$$4x^2 - x - 138 = 0 \quad \text{Write in standard form}$$

To help factor $4x^2 - x - 138$, we can look at the x-intercepts of the graph of $f(x) = 4x^2 - x - 138$. The graph in Figure 3.40 indicates that one of the x-intercepts is 6. If 6 is an x-intercept, then $(x - 6)$ is one of the factors, and we continue as follows:

$$4x^2 - x - 138 = 0$$
$$(x - 6)(4x + 23) = 0 \quad \text{Factor the trinomial}$$
$$x - 6 = 0 \quad \text{or} \quad 4x + 23 = 0 \quad \text{Apply the zero product property}$$
$$x = 6 \qquad\qquad x = -\dfrac{23}{4}$$

Thus, the roots of the original equation are 6 and $-\dfrac{23}{4}$.   ◆

PROBLEM 1 Write the equation $x^2 - x - 12 = 3(x + 2)(x - 2)$ in standard quadratic form. Then factor and solve by using the zero product property.   ◆

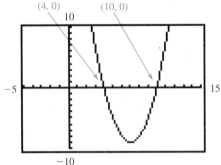

FIGURE 3.39
The roots of the equation $(x - 3)(x - 10) = x - 10$ are the x-intercepts of the graph of $f(x) = (x - 3)(x - 10) - x + 10$.

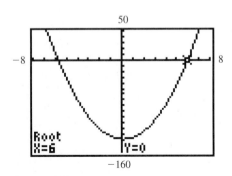

FIGURE 3.40
If 6 is an x-intercept of the graph of $f(x) = 4x^2 - x - 138$, then $(x - 6)$ is a factor of $4x^2 - x - 138$.

SECTION 3.4 ♦ Quadratic Equations and Inequalities

We can solve the quadratic equation $ax^2 + bx + c = 0$ by the zero product property only if we recognize the factors of the trinomial $ax^2 + bx + c$. The method of completing the square (in Section 3.3) enables us to derive a formula that can be used to solve a quadratic equation whether or not we recognize the factors.

Beginning with $ax^2 + bx + c = 0$, we proceed as follows:

$$ax^2 + bx + c = 0$$

$$ax^2 + bx = -c \qquad \text{Subtract } c \text{ from both sides}$$

$$x^2 + \frac{b}{a}x = -\frac{c}{a} \qquad \text{Divide both sides by } a$$

$$x^2 + \underbrace{\frac{b}{a}x}_{\text{(half of } b/a)^2} + \left(\frac{b}{2a}\right)^2 = \left(\frac{b}{2a}\right)^2 - \frac{c}{a} \qquad \text{Complete the square by adding } (b/2a)^2 \text{ to both sides}$$

$$\left(x + \frac{b}{2a}\right)^2 = \frac{b^2 - 4ac}{4a^2} \qquad \text{Factor the perfect square trinomial and add fractions on the right-hand side}$$

$$x + \frac{b}{2a} = \pm\sqrt{\frac{b^2 - 4ac}{4a^2}} \qquad \text{Extract the square roots}$$

$$x + \frac{b}{2a} = \pm\frac{\sqrt{b^2 - 4ac}}{2a} \qquad \text{Simplify the radical expression}$$

$$x = \frac{-b \pm \sqrt{b^2 - 4ac}}{2a} \qquad \text{Add } -b/2a \text{ to both sides}$$

This last equation represents a formula that can be used to find the roots of *any* quadratic equation that is written in standard form. We refer to this formula as the *quadratic formula*.

◆ **Quadratic Formula**

> If $ax^2 + bx + c = 0$, provided $a \neq 0$, then
>
> $$x = \frac{-b \pm \sqrt{b^2 - 4ac}}{2a}.$$

In the quadratic formula, the quantity $b^2 - 4ac$ is called the **discriminant**. It indicates the *nature of the roots* of the quadratic equation $ax^2 + bx + c = 0$ whenever a, b, and c are real numbers. We consider three cases.

Case 1 If $b^2 - 4ac > 0$, then $\sqrt{b^2 - 4ac}$ is a *real number* and the quadratic formula gives two distinct real roots:

$$\frac{-b + \sqrt{b^2 - 4ac}}{2a} \qquad \text{and} \qquad \frac{-b - \sqrt{b^2 - 4ac}}{2a}$$

If a, b, and c are rational numbers and $b^2 - 4ac$ is the square of a rational number, then these two distinct real roots are rational numbers. Otherwise, the roots are irrational numbers.

Case 2 If $b^2 - 4ac < 0$, then $\sqrt{b^2 - 4ac}$ is an *imaginary number* (see Section 1.4) and the quadratic formula gives two complex conjugate roots:

$$-\frac{b}{2a} + \frac{\sqrt{|b^2 - 4ac|}}{2a}i \quad \text{and} \quad -\frac{b}{2a} - \frac{\sqrt{|b^2 - 4ac|}}{2a}i$$

Case 3 If $b^2 - 4ac = 0$, then $\sqrt{b^2 - 4ac}$ is *zero* and the quadratic formula gives only one real root:

$$-\frac{b}{2a}$$

The quantity $-\dfrac{b}{2a}$ is called a **repeated root of multiplicity two**.

We now summarize the nature of the roots of the quadratic equation $ax^2 + bx + c = 0$ with real numbers a, b, and c:

Discriminant	Nature of the roots	Typical Graph
$b^2 - 4ac > 0$	Two distinct real roots	
$b^2 - 4ac < 0$	Two complex conjugate roots	
$b^2 - 4ac = 0$	One real root of multiplicity two	

EXAMPLE 2 Solve each quadratic equation by using the quadratic formula.

(a) $x^2 + 6x = 18$ (b) $x^2 + 5 = 2x$ (c) $2x^2 - 2\sqrt{10}\,x + 5 = 0$

◆ Solution

(a) Before identifying a, b, and c, we write $x^2 + 6x = 18$ in standard form:

$$x^2 + 6x - 18 = 0$$

SECTION 3.4 ♦ Quadratic Equations and Inequalities

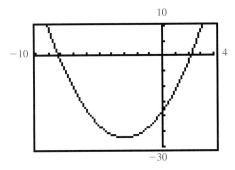

FIGURE 3.41
The x-intercepts of the graph of $f(x) = x^2 + 6x - 18$ are the roots of the equation $x^2 + 6x = 18$.

Now, $a = 1$, $b = 6$, and $c = -18$. Using the quadratic formula, we have

$$x = \frac{-6 \pm \sqrt{6^2 - 4(1)(-18)}}{2(1)} = \frac{-6 \pm \sqrt{108}}{2}$$

$$= \frac{-6 \pm 6\sqrt{3}}{2} \quad \text{Simplify the radical}$$

$$= \frac{2(-3 \pm 3\sqrt{3})}{2} \quad \text{Factor}$$

$$= -3 \pm 3\sqrt{3} \quad \text{Reduce}$$

Thus, the two real roots are $-3 + 3\sqrt{3} \approx 2.20$ and $-3 - 3\sqrt{3} \approx -8.20$. Figure 3.41 shows the graph of $f(x) = x^2 + 6x - 18$ in the viewing rectangle of a graphing calculator. Note that the x-intercepts of the parabola are approximately 2.20 and -8.20.

(b) We begin by writing $x^2 + 5 = 2x$ in standard form:

$$x^2 - 2x + 5 = 0$$

Therefore, $a = 1$, $b = -2$, and $c = 5$. Using the quadratic formula, we have

$$x = \frac{-(-2) \pm \sqrt{(-2)^2 - 4(1)(5)}}{2(1)} = \frac{2 \pm \sqrt{-16}}{2}$$

$$= \frac{2 + 4i}{2} \quad \text{Simplify the radical}$$

$$= \frac{2(1 \pm 2i)}{2} \quad \text{Factor}$$

$$= 1 \pm 2i \quad \text{Reduce}$$

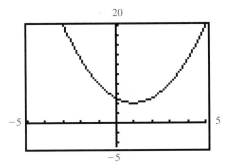

FIGURE 3.42
The graph of $f(x) = x^2 - 2x + 5$ has no x-intercept. This indicates that the roots of the equation $x^2 + 5 = 2x$ are not real numbers.

Thus, the roots are the complex conjugates $1 + 2i$ and $1 - 2i$. Figure 3.42 shows the graph of $f(x) = x^2 - 2x + 5$ in the viewing rectangle of a graphing calculator. Note that the parabola has no x-intercept. This indicates that the roots of the quadratic equation are not real numbers.

(c) For $2x^2 - 2\sqrt{10}\,x + 5 = 0$, we have $a = 2$, $b = -2\sqrt{10}$, and $c = 5$. Using the quadratic formula, we find

$$x = \frac{-(-2\sqrt{10}) \pm \sqrt{(-2\sqrt{10})^2 - 4(2)(5)}}{2(2)} = \frac{2\sqrt{10} \pm \sqrt{0}}{4}$$

$$= \frac{2\sqrt{10}}{4}$$

$$= \frac{\sqrt{10}}{2} \approx 1.58$$

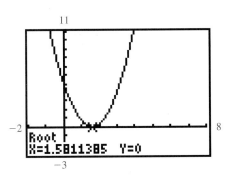

FIGURE 3.43
The graph of $f(x) = 2x^2 - 2\sqrt{10}\,x + 5$ is a parabola with its vertex on the x-axis. This indicates that $2x^2 - 2\sqrt{10}\,x + 5 = 0$ has one root of multiplicity two.

We refer to $\sqrt{10}/2$ as a repeated root of multiplicity two. Figure 3.43 shows the graph of $f(x) = 2x^2 - 2\sqrt{10}\,x + 5$ in the viewing rectangle of a graphing calculator. Note that the vertex of the parabola is the only x-intercept of the graph. This indicates that the quadratic equation has one real root of multiplicity two.

To check an imaginary root, like either of the complex conjugates in Example 2(b), we replace x in the original equation with the imaginary root and perform the indicated operations using the fact that $i^2 = -1$.

PROBLEM 2 Check the solutions to the quadratic equation in Example 2(b).

◆ Solving Quadratic Inequalities

The inequalities

$$x^2 - x - 6 < 0 \quad \text{and} \quad x^2 - x - 6 > 0$$

are examples of *quadratic inequalities in standard form*. We can use the graph of the related quadratic function $f(x) = x^2 - x - 6$ to help us find the solution set of each of these inequalities. We know that the graph of the quadratic function f is a parabola. The parabola opens upward and the coordinates of its vertex are

$$\left(-\frac{b}{2a}, f\left(-\frac{b}{2a}\right)\right) = \left(-\frac{1}{2}, -\frac{25}{4}\right).$$

The y-intercept is $f(0) = -6$, and the x-intercepts may be found by factoring and applying the zero product property as follows:

$$x^2 - x - 6 = 0$$
$$(x + 2)(x - 3) = 0$$
$$x + 2 = 0 \quad \text{or} \quad x - 3 = 0$$
$$x = -2 \quad \quad x = 3$$

The graph of $f(x) = x^2 - x - 6$ is shown in Figure 3.44.

We refer to the x-intercepts, -2 and 3, as the **boundary numbers** of the inequality $x^2 - x - 6 > 0$, since these are the values of x for which $x^2 - x - 6$ is *neither positive nor negative*. As illustrated in Figure 3.44, these two boundary numbers divide the x-axis into three intervals:

$$(-\infty, -2) \quad (-2, 3) \quad (3, \infty)$$

The graph of $f(x) = x^2 - x - 6$ is *below* the x-axis in the interval $(-2, 3)$. Hence, $f(x) < 0$ in this interval. Thus, the solution set of the inequality $x^2 - x - 6 < 0$ is

$$(-2, 3)$$

The graph of $f(x) = x^2 - x - 6$ is *above* the x-axis in the interval $(-\infty, -2)$

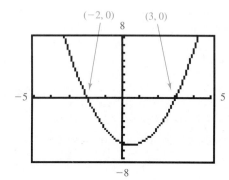

FIGURE 3.44

The x-intercepts of the graph of $f(x) = x^2 - x - 6$ divide the x-axis into three intervals: $(-\infty, -2)$, $(-2, 3)$, and $(3, \infty)$.

SECTION 3.4 ◆ Quadratic Equations and Inequalities

or $(3, \infty)$. Hence, $f(x) > 0$ in these intervals. Thus, the solution set of the inequality $x^2 - x - 6 > 0$ is

$$(-\infty, -2) \cup (3, \infty)$$

If an inequality contains the symbols $\leq$ or $\geq$, we include its boundary numbers as part of the solution set. For example, the solution set of $x^2 - x - 6 \leq 0$ is $[-2, 3]$, and the solution set of $x^2 - x - 6 \geq 0$ is $(-\infty, -2] \cup [3, \infty)$.

EXAMPLE 3 Solve each quadratic inequality.

(a) $2x^2 + 3x < 20$ **(b)** $2x + 9 \leq x^2$

◆ Solution

(a) We begin by subtracting 20 from both sides of the inequality and writing it in standard form:

$$2x^2 + 3x - 20 < 0$$

The boundary numbers of this inequality are the x-intercepts of the graph of $f(x) = 2x^2 + 3x - 20$. We can find these boundary numbers by solving the equation $f(x) = 0$:

$2x^2 + 3x - 20 = 0$

$(2x - 5)(x + 4) = 0$ Factor

$2x - 5 = 0$ or $x + 4 = 0$ Apply the zero product property

$x = \frac{5}{2}$ $x = -4$

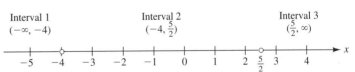

These two boundary numbers divide the x-axis into three intervals:

Interval 1   Interval 2   Interval 3
$(-\infty, -4)$ $(-4, \frac{5}{2})$ $(\frac{5}{2}, \infty)$

We can confirm these intervals by generating the graph of $f(x) = 2x^2 + 3x - 20$ in the viewing rectangle of a graphing calculator. As shown in Figure 3.45, the graph of f is *below* the x-axis in the interval $(-4, \frac{5}{2})$. Hence, $f(x) < 0$ in this interval. Thus, the solution set of the inequality $2x^2 + 3x - 20 < 0$ is $(-4, \frac{5}{2})$.

(b) We begin writing the inequality in standard form with a nonnegative x^2 term:

$$2x + 9 \leq x^2$$

$-x^2 + 2x + 9 \leq 0$ Subtract x^2 from both sides

$x^2 - 2x - 9 \geq 0$ Multiply both sides by -1 and reverse the inequality

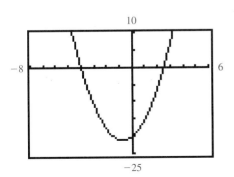

FIGURE 3.45
The graph of $f(x) = 2x^2 + 3x - 20$ is below the x-axis in the interval $(-4, \frac{5}{2})$.

The boundary numbers of this inequality are the *x*-intercepts of the graph of $f(x) = x^2 - 2x - 9$. We can find these boundary numbers by solving the equation $x^2 - 2x - 9 = 0$. Since $x^2 - 2x - 9$ is not factorable over the set of integers, we solve this equation by using the quadratic formula:

$$x = \frac{-(-2) \pm \sqrt{(-2)^2 - 4(1)(-9)}}{2(1)} = 1 \pm \sqrt{10}$$

Hence, the two boundary numbers are $1 + \sqrt{10} \approx 4.2$ and $1 - \sqrt{10} \approx -2.2$. These two boundary numbers divide the *x*-axis into three intervals:

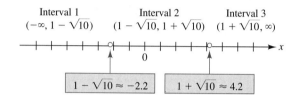

We can confirm these intervals by generating the graph of $f(x) = x^2 - 2x - 9$ in the viewing rectangle of a graphing calculator. As shown in Figure 3.46, the graph of f is *above* the *x*-axis in the interval $(-\infty, 1 - \sqrt{10})$ or $(1 + \sqrt{10}, \infty)$. Hence, $f(x) \geq 0$ in the interval $(-\infty, 1 - \sqrt{10}]$ or $[1 + \sqrt{10}, \infty)$. Thus, the solution set of the inequality $x^2 - 2x - 9 \geq 0$ is

$$(-\infty, 1 - \sqrt{10}] \cup [1 + \sqrt{10}, \infty). \qquad \blacklozenge$$

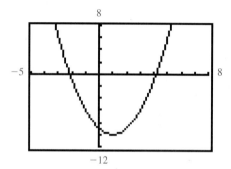

FIGURE 3.46

The graph of $f(x) = x^2 - 2x - 9$ is above the *x*-axis in the intervals $(-\infty, 1 - \sqrt{10})$ and $(1 + \sqrt{10}, \infty)$.

Look at the graphs in Figures 3.45 and 3.46: In each of the intervals formed by the boundary numbers, the graph of $f(x) = ax^2 + bx + c$ is entirely above the *x*-axis or entirely below the *x*-axis. We can use this fact to help solve a quadratic inequality in standard form without referring to a graph. We simply select an arbitrary *test number* from each interval formed by the boundary numbers, and then determine the algebraic sign of the quadratic expression $ax^2 + bx + c$ at this number. If the quadratic expression is negative for a particular test number in an interval, then it is negative for every value of *x* in that interval. If the quadratic expression is positive for a particular test number in an interval, then it is positive for every value of *x* in that interval. We refer to this algebraic procedure of solving a quadratic inequality as the **test number method**.

For example, to solve the quadratic inequality $2x^2 + 3x - 20 < 0$ in Example 3(a) by using the test number method, we select an arbitrary test number from each of the intervals $(-\infty, -4)$, $(-4, \frac{5}{2})$, and $(\frac{5}{2}, \infty)$, and determine the algebraic sign of the quadratic expression $2x^2 + 3x - 20$ at each test number. The following table organizes our work:

Interval	Test Number	Sign of $2x^2 + 3x - 20$
$(-\infty, -4)$	-5	$2(-5)^2 + 3(-5) - 20 = 15 > 0$
$(-4, \frac{5}{2})$	1	$2(1)^2 + 3(1) - 20 = -15 < 0$
$(\frac{5}{2}, \infty)$	3	$2(3)^2 + 3(3) - 20 = 7 > 0$

SECTION 3.4 ♦ Quadratic Equations and Inequalities

Since $2x^2 + 3x - 20 < 0$ when $x = 1$, we know that $2x^2 + 3x - 20 < 0$ for every value of x in the interval $\left(-4, \frac{5}{2}\right)$. Hence, we conclude that the solution set of this inequality is $\left(-4, \frac{5}{2}\right)$, which agrees with our work in Example 3(a).

PROBLEM 3 Solve the quadratic inequality $x^2 < 16x$ by using the test number method. ♦

♦ Application: Parking Lot Addition

Many applied problems lead to quadratic equations. In the next example, we illustrate how a construction company uses such equations to determine a uniform width that doubles the size of an existing rectangular parking lot.

EXAMPLE 4 An existing rectangular parking lot is 100 ft by 220 ft. A construction company is asked to double the size of this parking lot by adding an area of uniform width, as shown in Figure 3.47.

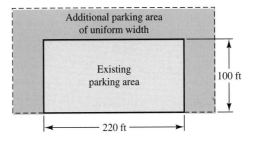

FIGURE 3.47

(a) What is the uniform width of the additional parking area?

(b) What are the dimensions and area of the new parking lot?

◆ Solution

(a) Suppose we let

x = the uniform width (in feet) of the additional parking area.

Then

$220 + 2x$ = the overall length (in feet) of the new parking lot

and

$100 + x$ = the overall width (in feet) of the new parking lot

as shown in Figure 3.48. Since the area of a rectangle is the product of its length and width, we have

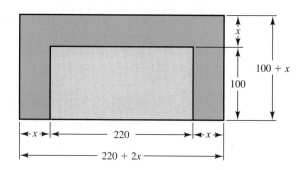

FIGURE 3.48

If x is the uniform width, then the dimensions of the new parking lot are $(220 + 2x)$ ft by $(100 + x)$ ft.

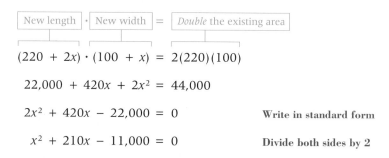

$$(220 + 2x) \cdot (100 + x) = 2(220)(100)$$

$$22{,}000 + 420x + 2x^2 = 44{,}000$$

$$2x^2 + 420x - 22{,}000 = 0 \qquad \text{Write in standard form}$$

$$x^2 + 210x - 11{,}000 = 0 \qquad \text{Divide both sides by 2}$$

To solve this quadratic equation, we apply the quadratic formula:

$$x = \frac{-210 \pm \sqrt{(210)^2 - 4(1)(-11{,}000)}}{2(1)} = \frac{-210 \pm \sqrt{88{,}100}}{2}$$

Using a calculator, we find that the approximate roots are

$$\frac{-210 + \sqrt{88{,}100}}{2} \approx 43.4 \quad \text{and} \quad \frac{-210 - \sqrt{88{,}100}}{2} \approx -253.4.$$

Since a width cannot be negative, we discard the negative solution, and we conclude that the uniform width of the additional parking area is 43.4 ft. Figure 3.49 shows the graph of $f(x) = x^2 + 210x - 11{,}000$ in the viewing rectangle of a graphing calculator. Note that the positive x-intercept of the parabola is approximately 43.4.

(b) The dimensions of the new parking lot (to the nearest tenth of a foot) are

$$\text{Width:} \quad 100 + x = 100 + 43.4 = 143.4 \text{ ft}$$

$$\text{Length:} \quad 220 + 2x = 220 + 2(43.4) = 306.8 \text{ ft}$$

Hence, the area of the new lot is

$$(143.4 \text{ ft})(306.8 \text{ ft}) = 43{,}995.12 \text{ sq ft} \approx 44{,}000 \text{ sq ft},$$

which is double the original area ($100 \times 220 = 22{,}000$ sq ft). ◆

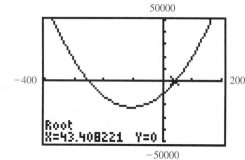

FIGURE 3.49

The positive x-intercept of the graph of $f(x) = x^2 + 210x - 11000$ is the uniform width of the additional parking area, 43.4 ft.

SECTION 3.4 ◆ *Quadratic Equations and Inequalities*   193

PROBLEM 4   Repeat Example 4 if the construction company is asked to double the size of the existing parking lot by adding an area of uniform width as shown in Figure 3.50.

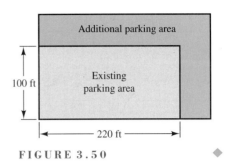

FIGURE 3.50

◆ Application: Hitting a Pop Fly

Many applied problems lead to quadratic inequalities rather than quadratic equations. In the next example, we use the law of motion for freely falling objects (see Section 3.3) to determine the time interval when a pop fly is more than a given distance above the ground.

EXAMPLE 5   A baseball player hits a pop fly straight upward with an initial velocity of 96 ft/s from an initial height of 4 ft. What is the time interval when the baseball is more than 84 feet above the ground?

◆ Solution   From Example 6 in Section 3.3, the quadratic function that defines the distance s of the ball above the ground is

$$s = -16t^2 + 96t + 4.$$

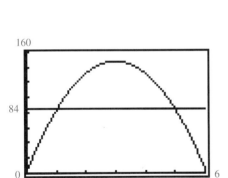

FIGURE 3.51

Graphs of $s = -16t^2 + 96t + 4$ and $s = 84$ in the viewing rectangle of a graphing calculator

Figure 3.51 shows the graph of this quadratic function, along with the graph of the horizontal line $s = 84$. The time interval where the parabola is *above* the horizontal line $s = 84$ represents the time interval when the baseball is *more than* 84 ft above the ground. To determine this time interval algebraically, we solve the quadratic inequality

$$-16t^2 + 96t + 4 > 84.$$

Writing this inequality in standard form, we obtain

$-16t^2 + 96t - 80 > 0$ Subtract 84 from both sides

$\underbrace{t^2 - 6t + 5}_{f(t)} < 0$ Divide both sides by -16 and *reverse the inequality*

We find the boundary numbers of this inequality by solving the equation $f(t) = 0$:

$$t^2 - 6t + 5 = 0$$

$(t - 1)(t - 5) = 0$ Factor

$t - 1 = 0$ or $t - 5 = 0$ Apply the zero product property

$t = 1$ $t = 5$

Hence, the baseball is more than 84 feet above the ground during the time interval from 1 second to 5 seconds.

PROBLEM 5 Generate the graphs in Figure 3.51 on a graphing calculator and trace to the points of intersection of the parabola and the line. Show that this graphical approach also gives the time interval (1, 5).

Exercises 3.4

Basic Skills

In Exercises 1–12, write each quadratic equation in the standard form $ax^2 + bx + c = 0$. Then factor and solve by using the zero product property. Check graphically.

1. $x^2 + 2x = 8$
2. $x^2 - 48 = 13x$
3. $6y^2 = 12y$
4. $3t^2 = t$
5. $x^2 - 36 = 0$
6. $100 - 9n^2 = 0$
7. $n(2n - 3) = 5$
8. $(x - 1)(3x + 2) = 2$
9. $(x + 4)(x + 6) = x + 6$
10. $(m - 7)(m + 4) = 2(m - 7)$
11. $y^2 + (y - 2)(y + 8) = 2y$
12. $(x + 3)^2 = x(3x - 1)$

In Exercises 13–24, write each quadratic equation in the standard form $ax^2 + bx + c = 0$. Then solve by using the quadratic formula. Check graphically.

13. $x^2 = 11x + 1$
14. $m^2 - 11 = 2m$
15. $20y^2 + y = 12$
16. $18x^2 + 3x = 10$
17. $5x^2 + 1 = 4x$
18. $2x^2 + 3 = 4x$
19. $(2x - 1)(x + 4) = (x + 1)^2$
20. $y(y - 3) = (2y + 3)(y - 1)$
21. $2(x^2 + 2) = \sqrt{7}x$
22. $3x^2 + 2(1 + \sqrt{6}x) = 0$
23. $1 - \dfrac{2n^2}{5} = \dfrac{1}{2}$
24. $\dfrac{2x}{3} = x^2 + \dfrac{1}{2}$

In Exercises 25–40, solve each quadratic equation by using any convenient method.

25. $4p^2 - 15 = 0$
26. $3x^2 - 25 = 0$
27. $4x^2 + 49 = 0$
28. $16x^2 + 75 = 0$
29. $(4y + 3)^2 = 32$
30. $(2m - 1)^2 + 24 = 0$
31. $n^2 - 10n + 25 = -18$
32. $4t^2 + 4t + 1 = 28$
33. $x^2 + 12x = 12$
34. $x^2 = 10x + 15$
35. $n^2 - 4n - 12 = 0$
36. $y^2 + 14y + 24 = 0$
37. $x^2 = 3x + 9$
38. $m^2 + 5m = 5$
39. $p^2 - \frac{4}{5}p = \frac{21}{25}$
40. $x^2 + \frac{3}{2}x = \frac{7}{16}$

In Exercises 41–52, solve each quadratic inequality. Write the solution set using interval notation.

41. $x^2 - 9 < 0$
42. $p^2 > 25$
43. $y^2 \geq 12y$
44. $x^2 + 20x \leq 0$
45. $x^2 - 8x \leq 9$
46. $y(y + 10) \geq 24$
47. $4x(x + 3) > 7$
48. $4x^2 + 9 < 20x$
49. $n^2 < 11n + 1$
50. $x^2 + 3x > 8 - 3x$
51. $(x - 1)^2 \geq (2x + 3)^2$
52. $(4n - 1)(2n + 5) \geq (3n + 1)^2$

In Exercises 53–56, use a graphing calculator to solve each equation. Write each solution as a decimal number rounded to three significant digits.

53. $3.81x^2 - 8.94x = 12.6$
54. $0.238x^2 + 0.342x - 1.14 = 0$
55. $872x^2 = 169x + 906$
56. $-968x^2 + 543 = 272x$

In Exercises 57–60, use a graphing calculator to solve each inequality. Describe each solution set using interval notation, and round each decimal number to three significant digits.

57. $27.93x^2 - 92.24 > 0$
58. $2.6x^2 - 5.4x \leq 9.2$
59. $42.6x^2 \leq 18.3x + 92.4$
60. $298x^2 < 987x - 500$

In Exercises 61–78, set up an equation or inequaltity, then solve.

61. Find two integers whose sum is 32 and whose product is 252.

62. Find two consecutive, positive even integers such that the sum of their squares is 340.

63. The area of a triangle is 54 square inches and its base is 3 inches shorter than its height. Find the base and height of the triangle.

64. A rectangle has a perimeter of 84 ft and a diagonal of 30 ft. Find the length and width of the rectangle.

65. After purchasing a coal-burning stove, a homeowner constructs a coal bin with rectangular sides to hold four tons of coal. The height of the coal bin is 4 feet and the length is 2 feet more than the width. Allowing 35 cubic feet for each ton of coal, find the dimensions of the coal bin.

66. A picture is encased in a wood frame of uniform width as shown in the sketch. The dimensions of the picture are 8 inches by 10 inches and the total area of both picture and frame is $131\frac{1}{4}$ square inches. Find the width of the frame.

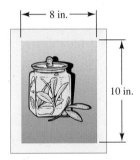

67. A rectangular flower garden, 16.6 ft by 32.2 ft, is to be surrounded by a brick walkway of uniform width, as shown in the sketch. The contractor has enough brick to cover an area of 400 square feet. How wide should the walk be if all the bricks are to be used?

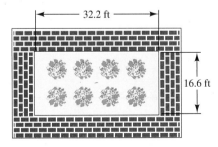

68. A 4-ft walkway surrounds a circular flower garden, as shown in the sketch. The area of the walk is 44% of the area of the garden. Find the radius of the garden.

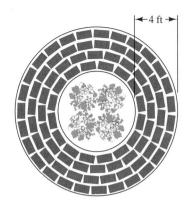

69. When the owners of an apartment complex charge rent of $600 per month for each unit, all 120 apartments are occupied. For each $20 increase in the monthly rent, 3 apartments become vacant. What should the owners of the complex charge for monthly rent if they wish to earn $73,500 per month from rental income?

70. A campus bookstore can order sweatshirts stamped with the university emblem for $12 per sweatshirt when less than 20 shirts are ordered from the supplier. If more than 20 but less than 200 sweatshirts are ordered, the price per shirt is reduced by 3 cents times the number of shirts ordered. How many sweatshirts can the bookstore order for $900?

71. The length of a rectangle is 2 meters more than its width, and its area is less than 99 square meters. What is the range of values for the length of the rectangle?

72. One leg of a right triangle is 3 inches longer than the other leg. If the hypotenuse is less than $\sqrt{39}$ inches, what is the range of values for the shorter leg?

73. Find all sets of three consecutive, even integers such that the square of the largest integer is greater than the sum of the squares of the other two integers.

74. Find all sets of two consecutive integers such that the square of the smaller integer is less than four times the larger integer.

75. When the selling price of a product is p dollars, a manufacturer will *supply* the market with $4p^2 - 5p$ units of their product and consumers will *demand* to buy $30 - p^2$ units. Under what conditions is supply greater than demand?

76. A highway safety group recommends the formula

$$d = 0.045v^2 + v$$

for determining the minimum safe distance d (in feet) that should be maintained between two cars traveling along a highway at a speed of v miles per hour (mi/h).

Exercise 76 continues

If the distance between two cars does not exceed 200 ft, find the interval of safe speeds.

77. If a rocket is fired straight upward from the ground with an initial velocity of 320 ft/s, what is the time interval during which the rocket is more than 1200 ft above the ground? (For the law of motion of freely falling objects, see Section 3.3).

78. If an arrow is shot straight upward from an initial height of 1.2 meters (m) with an initial velocity of 32.0 m/s, what are the time intervals when the arrow is less than 19.2 m above the ground? (For the law of motion of freely falling objects, see Section 3.3).

  Critical Thinking

In Exercises 79–84, solve each literal equation for x.

79. $ax^2 + bx = 0$
80. $a^2x^2 - b^2 = 0$
81. $x^2 + 2ax + a^2 = b^2$
82. $a^2x^2 + b^2 = 2abx$
83. $x^2 + 2bx - a^2 = 2ab$
84. $x^2 + x = a^2 + a$

In Exercises 85–90, find the value(s) of k such that each quadratic equation has exactly one root of multiplicity two.

85. $x^2 + 16x + k = 0$
86. $kx^2 - 4x + 2 = 0$
87. $2x^2 - kx + 3 = 0$
88. $5x^2 - (\sqrt{3}k)x + 3 = 0$
89. $kx^2 - kx + 2 = 0$
90. $3x^2 - kx + k = 0$

In Exercises 91 and 92, solve each double inequality. Use interval notation to describe the solution set.

91. $6 < x^2 - x < 20$
92. $10 - x < x^2 - 4x \leq 2x + 16$

93. Find the range of values for *x* if the area of the region shaded in the sketch is any size from 42 to 127, inclusive.

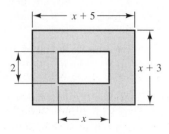

94. Find the range of values for *x* if the area of the region shaded in the sketch is any size from 21π and 33π, inclusive.

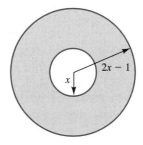

In Exercises 95 and 96, use interval notation to state the domain of each function. Check graphically.

95. $f(x) = \sqrt{x^2 - 2x - 8}$
96. $f(x) = \sqrt[4]{x^2 + 7x + 10}$

In Exercises 97–100, determine the values of k for which each quadratic equation has real roots.

97. $x^2 + kx + 6 = 0$
98. $2x^2 + kx + 3 = 0$
99. $x^2 + kx + k = 0$
100. $kx^2 + x + k = 0$

3.5 Intersection Points of Two Graphs

◆ Introductory Comments

In Figure 3.52, we show the graphs of the equations

$$y = f(x) \quad \text{and} \quad y = g(x)$$

SECTION 3.5 ♦ Intersection Points of Two Graphs

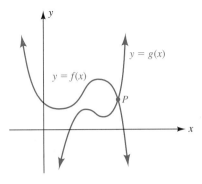

FIGURE 3.52
If the graphs of two equations intersect at point P, then P is called an *intersection point*.

in the same coordinate plane. We refer to the point P as an **intersection point** of the graphs of the two equations. If P is an intersection point of the graphs of the two equations, then the coordinates of point P must satisfy *both* equations. One method of finding the x-coordinates of point P is to substitute $f(x)$ for y in the equation $y = g(x)$ and solve the equation

$$f(x) = g(x)$$

for x. We refer to this procedure of solving two equations simultaneously as the **substitution method**. We begin by discussing the intersection of two lines.

♦ Intersection of Two Lines

Figure 3.53 illustrates the three possibilities for the intersection of two lines graphed on the same coordinate plane.

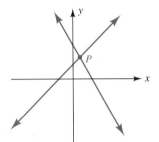

(a) The lines intersect at one point.

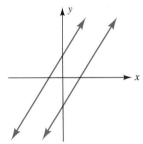

(b) The lines are parallel and do not intersect.

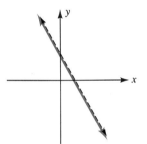

(c) The lines coincide and intersect at an infinite number of points.

FIGURE 3.53
The three possibilities for the intersection of two lines graphed on the same coordinate plane.

EXAMPLE 1 Find the coordinates of the intersection point for the lines with the given equations.

(a) $y = 3 - 5x$ and $y = \frac{3}{2}x - 4$ **(b)** $y = 2x + 4$ and $y = 2x - 1$

♦ **Solution**

(a) Figure 3.54 shows the graphs of the lines $y = 3 - 5x$ and $y = \frac{3}{2}x - 4$ in the viewing rectangle of a graphing calculator. The x-coordinate of the intersection point appears to be about 1, and the y-coordinate appears to be between -2 and -3. One method of finding the x-coordinate of the intersection point is to solve the two equations simultaneously by using substitution. We substitute $3 - 5x$ for y in the equation $y = \frac{3}{2}x - 4$ and solve for x as follows:

$$3 - 5x = \tfrac{3}{2}x - 4$$
$$6 - 10x = 3x - 8 \quad \text{Multiply both sides by 2}$$
$$14 = 13x \quad \text{Add } 10x \text{ and 8 to both sides}$$
$$x = \tfrac{14}{13} \quad \text{Divide both sides by 13}$$

FIGURE 3.54
The lines with equations $y = 3 - 5x$ and $y = \frac{3}{2}x - 4$ intersect at one point.

We can find the y-coordinate of the intersection point by substituting $\frac{14}{13}$ for x in one of the original equations. Choosing the equation $y = 3 - 5x$, we find

$$y = 3 - 5\left(\tfrac{14}{13}\right) = 3 - \tfrac{70}{13} = -\tfrac{31}{13}.$$

Thus, the lines intersect at the point $\left(\tfrac{14}{13}, -\tfrac{31}{13}\right)$.

(b) The graphs of $y = 2x + 4$ and $y = 2x - 1$ are lines with the same slope, 2, but different y-intercepts (4 and -1). Thus, we conclude these lines are *parallel* and hence have *no intersection point* (see Figure 3.55). If we solve the equations simultaneously by substituting $2x + 4$ for y in the equation $y = 2x - 1$, we find that

$$y = 2x - 1$$

$$2x + 4 = 2x - 1 \qquad \textbf{Substitute } 2x + 4 \textbf{ for } y$$

$$4 = -1 \qquad \textbf{Subtract } 2x \textbf{ on both sides}$$

a false statement

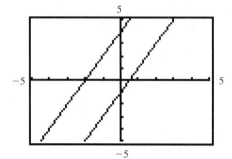

FIGURE 3.55
The lines with equations $y = 2x + 4$ and $y = 2x - 1$ are parallel. Hence there is no intersection point.

When the variables drop out and the resulting statement is *false*, we conclude that the lines are *parallel* and have *no intersection point*. ◆

PROBLEM 1 Verify that the point with coordinates $\left(\tfrac{14}{13}, -\tfrac{31}{13}\right)$ satisfies the equations of both lines in Example 1(a). ◆

We now discuss an alternative method for solving two equations simultaneously. Consider the two equations $A = B$ and $C = D$. The equation $C = D$ states that C is the same as D. Thus, if we add C to the left-hand side and D to the right-hand side of the equation $A = B$, we are actually adding the *same* quantity to both sides of the equation $A = B$. Therefore, we can state the following:

If $\qquad A = B$
and $\qquad C = D$
then $\quad A + C = B + D \qquad$ **Add the left-hand and right-hand sides**

This idea gives us an alternative method for finding the intersection point of two lines. We solve the equations simultaneously by adding the left-hand and right-hand sides of the equations in order to eliminate one of the variables. To accomplish this task, it may be necessary to first multiply one of the equations by a constant so that the coefficient of a variable (either x or y) in one equation is the opposite of the same variable's coefficient in the other equation. In fact, we must sometimes multiply both equations (by different constants) to obtain coefficients of a variable that can be eliminated by adding the equations. We refer to this method as the **elimination method** and illustrate the procedure in the next example.

SECTION 3.5 ♦ Intersection Points of Two Graphs

EXAMPLE 2 Find the coordinates of the intersection point for the lines with the given equations.

(a) $3x - 3y = 10$ and $5x + 6y = 2$
(b) $-6x + 2y = 1$ and $9x + 3y = -\frac{3}{2}$

◀ Solution

(a) Figure 3.56 shows the graphs of the equations $3x - 3y = 10$ (or $y = x - \frac{10}{3}$) and $5x + 6y = 2$ (or $y = -\frac{5}{6}x + \frac{1}{3}$) in the viewing rectangle of a graphing calculator. The x-coordinate of the intersection point appears to be about 2 and the y-coordinate appears to be between -1 and -2. One method of finding the x-coordinates of the intersection point is to solve the equations simultaneously by using the elimination method:

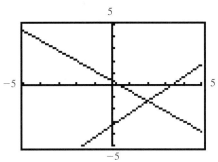

FIGURE 3.56
The lines with equations $3x - 3y = 10$ and $5x + 6y = 2$ intersect at one point.

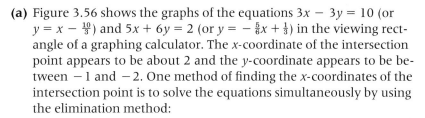

To find the y-coordinate of the intersection point, we substitute 2 for x in either of the original equations and solve for y. Replacing x with 2 in the equation $3x - 3y = 10$, we find $y = -\frac{4}{3}$. Thus, the lines intersect at the point $(2, -\frac{4}{3})$.

(b) We first write both equations in slope-intercept form, $y = mx + b$:

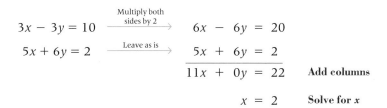

Since these equations have the same slope and y-intercept, we conclude that the lines *coincide*. Thus, every point on each line satisfies the equation of the other line, and we conclude that the lines have an *infinite number of intersection points*. If we solve the equations simultaneously by using the elimination method, we find that

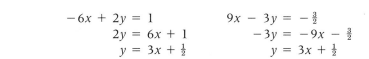

When the variables drop out and the resulting statement is true, we conclude that the lines coincide and, therefore, have an infinite number of intersection points. ◆

PROBLEM 2 Find the coordinates of the intersection point for the lines with the equations $2x - y = 3$ and $y = 4x + 5$ by using each method.

(a) Substitution method   **(b)** Elimination method

◆ Intersection of a Line and a Parabola

Figure 3.57 illustrates the three possibilities for the intersection of a nonvertical line and a parabola graphed on the same coordinate plane.

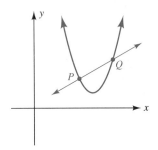
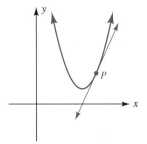
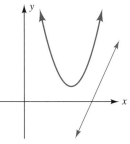

(a) The line and the parabola intersect at two points.
(b) The line and the parabola intersect at one point.
(c) The line and the parabola do not intersect.

FIGURE 3.57
The three possibilities for the intersection of a nonvertical line and a parabola graphed on the same coordinate plane.

Although the notion of *tangency* is studied in detail in calculus, we mention the idea at this time. If a line is not parallel to the axis of a parabola and touches a parabola at only one point, as shown in Figure 3.57(b), then the line is said to be **tangent** to the parabola at that point.

We may find the coordinates of the intersection points of a line and a parabola by solving their equations simultaneously by using the substitution method. The coordinates of the intersection points must satisfy *both* the equation of the line and the equation of the parabola.

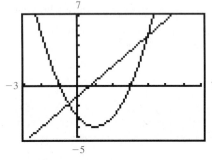

FIGURE 3.58
The line with the equation $3x - 2y = 2$ and the parabola with the equation $y = x^2 - 2x - 3$ intersect at two points.

EXAMPLE 3 Find the coordinates of the intersection points for the line and the parabola with the given equations.

(a) $3x - 2y = 2$ and $y = x^2 - 2x - 3$

(b) $y = x - 1$ and $y = x^2 - 2x + 3$

◆ Solution

(a) Figure 3.58 shows that the graphs of $3x - 2y = 2$ (or $y = \frac{3}{2}x - 1$) and $y = x^2 - 2x - 3$ intersect at two points. To find the *x*-coordinates of these intersection points, we solve the equations simultaneously by using the substitution method:

SECTION 3.5 ♦ Intersection Points of Two Graphs

$$3x - 2y = 2$$

$$3x - 2(x^2 - 2x - 3) = 2 \quad \text{Substitute } x^2 - 2x - 3 \text{ for } y$$

$$3x - 2x^2 + 4x + 6 = 2 \quad \text{Simplify}$$

$$2x^2 - 7x - 4 = 0 \quad \text{Write the quadratic equation in standard form}$$

$$(2x + 1)(x - 4) = 0 \quad \text{Factor}$$

$$2x + 1 = 0 \quad \text{or} \quad x - 4 = 0 \quad \text{Apply the zero product property}$$

$$x = -\tfrac{1}{2} \qquad\qquad x = 4 \quad \text{Solve for } x$$

⎿ x-coordinates of the intersection points ⏌

To find the y-coordinates of the intersection points that correspond to these x-coordinates, we substitute the x-values of $-\tfrac{1}{2}$ and 4 in either one of the original equations and solve for y. Substituting $x = -\tfrac{1}{2}$ and $x = 4$ into the equation $y = x^2 - 2x - 3$, we have

$$y = (-\tfrac{1}{2})^2 - 2(-\tfrac{1}{2}) - 3 = -\tfrac{7}{4} \quad \text{and} \quad y = (4)^2 - 2(4) - 3 = 5.$$

⎿ y-coordinates of the intersection points ⏌

Thus, the coordinates of the intersection points are $\left(-\tfrac{1}{2}, -\tfrac{7}{4}\right)$ and $(4, 5)$.

(b) The graphs of $y = x - 1$ and $y = x^2 - 2x + 3$ are shown in Figure 3.59. These graphs appear to have no intersection. If we solve the equations simultaneously by the substitution method, we find

$$x - 1 = x^2 - 2x + 3$$

$$x^2 - 3x + 4 = 0 \quad \text{Write the quadratic equation in standard form}$$

$$x = \frac{-(-3) \pm \sqrt{(-3)^2 - 4(1)(4)}}{2(1)} \quad \text{Apply the quadratic formula}$$

$$x = \frac{3 \pm \sqrt{-7}}{2} \quad \text{Simplify}$$

not a real solution

If the equations of a line and a parabola have no real solution when solved simultaneously, then the line and the parabola do not intersect. ◆

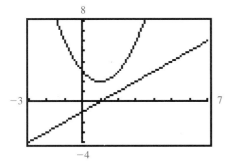

FIGURE 3.59
The line with the equation $y = x - 1$ and the parabola with the equation $y = x^2 - 2x + 3$ do *not* intersect.

PROBLEM 3 Find the coordinates of the intersection points for the line and the parabola with equations $4x - y = 6$ and $y = x^2 - 2x + 3$, respectively. ◆

Intersection Points of Other Curves

As illustrated in the next example, the substitution and elimination methods can be used to find the intersection points of other curves as well.

EXAMPLE 4 Find the coordinates of the intersection points of the curves with the given equations.

(a) $y = \sqrt{x}$ and $x + y = 6$ (b) $x^2 + y^2 = 4$ and $x^2 - 3y = 0$

◆ Solution

(a) Figure 3.60 shows the graphs of the equations $y = \sqrt{x}$ and $x + y = 6$ (or $y = 6 - x$) in the viewing rectangle of a graphing calculator. To find the x-coordinate of the intersection point, we solve the equations simultaneously by substituting $\sqrt{x}$ for y in the equation $x + y = 6$ as follows:

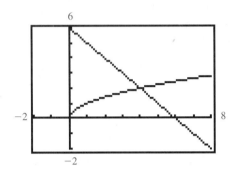

FIGURE 3.60

The graphs of the equations $y = \sqrt{x}$ and $x + y = 6$ intersect at one point.

$x + y = 6$	
$x + \sqrt{x} = 6$	Substitute $\sqrt{x}$ for y
$\sqrt{x} = 6 - x$	Isolate the radical
$x = 36 - 12x + x^2$	Square both sides

Squaring both sides may introduce extraneous roots.

$0 = 36 - 13x + x^2$	Solve the quadratic equation
$0 = (9 - x)(4 - x)$	
$9 - x = 0$ or $4 - x = 0$	
$x = 9$ or $x = 4$	

We have found two x-coordinates, but Figure 3.60 indicates that the curves have only one intersection point. Substituting $x = 9$ into the equation $y = \sqrt{x}$, we find

$$y = \sqrt{9} = 3.$$

However, the coordinates $(9, 3)$ do not satisfy the equation $x + y = 6$. Thus, $(9, 3)$ is *not* an intersection point.

Substituting $x = 4$ into the equation $y = \sqrt{x}$, we find

$$y = \sqrt{4} = 2.$$

Since the coordinates $(4, 2)$ also satisfy the equation $x + y = 6$, we conclude that the only intersection point is $(4, 2)$.

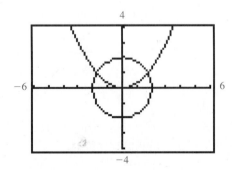

FIGURE 3.61

The graphs of the equations $x^2 + y^2 = 4$ (circle) and $x^2 - 3y = 0$ (parabola) intersect at two points.

(b) Figure 3.61 shows the graphs of the equations $x^2 + y^2 = 4$ (or $y = \pm\sqrt{4 - x^2}$) and $x^2 - 3y = 0$ (or $y = \frac{1}{3}x^2$) in the viewing rectangle of a graphing calculator. To find the y-coordinates of the intersection points, we eliminate the x^2 terms by using the elimination method:

$$x^2 + y^2 = 4 \quad \xrightarrow{\text{No change}} \quad x^2 + y^2 = 4$$

$$x^2 - 3y = 0 \quad \xrightarrow{\substack{\text{Multiply both} \\ \text{sides by } -1}} \quad -x^2 + 3y = 0$$

$$\overline{ y^2 + 3y = 4}$$

Add and solve the quadratic equation for y.

$$y^2 + 3y - 4 = 0$$
$$(y - 1)(y + 4) = 0$$
$$y - 1 = 0 \quad \text{or} \quad y + 4 = 0$$
$$y = 1 \qquad\qquad y = -4$$

Substituting $y = 1$ into the equation $x^2 - 3y = 0$ gives us

$$x^2 = 3 \quad \text{or} \quad x = \pm\sqrt{3}.$$

Substituting $y = -4$ into the equation $x^2 - 3y = 0$ gives us

$$x^2 = -12.$$

This equation has no real solution, so the only intersection points are $(\sqrt{3}, 1)$ and $(-\sqrt{3}, 1)$. ◆

PROBLEM 4 Find the coordinates of the intersection points for the parabolas with equations $y = x^2$ and $y = 4x - x^2$. ◆

◆ Approximating the Intersection Points of Two Curves

The substitution procedure sometimes yields an equation in one unknown that may be challenging to solve. For problems of this nature, we can use a graphing calculator to find the approximate points of intersection.

EXAMPLE 5 Find the approximate coordinates of the intersection points for the curves with equations $y = x^3$ and $4x^2 + y^2 = 16$.

◆ **Solution** To determine the x-coordinates of the intersection points, we substitute x^3 for y in the equation $4x^2 + y^2 = 16$ and obtain

$$4x^2 + (x^3)^2 = 16 \quad \text{or} \quad x^6 + 4x^2 - 16 = 0.$$

Note that the equation $x^6 + 4x^2 - 16 = 0$ is *not* of quadratic type and, hence, the chance of solving this equation by any of the algebraic methods that we have discussed in this chapter seems bleak. Instead, we approximate the coordinates of the intersection points by using a graphing calculator.

Figure 3.62 shows the graphs of the equations $y = x^3$ and $4x^2 + y^2 = 16$ (or $y = \pm 2\sqrt{4 - x^2}$) in the viewing rectangle of a graphing calculator.

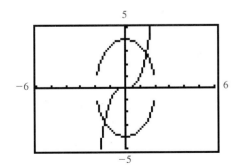

FIGURE 3.62

The graphs of the equations $4x^2 + y^2 = 16$ (ellipse) and $y = x^3$ intersect at two points.

Many graphing calculators have a built-in program for finding an intersection point. If your calculator does not have a built-in program for finding an intersection point, use the [TRACE] key and [ZOOM] key on the calculator to estimate the coordinates of these points. Either way, we find that the coordinates of the intersection points are $(-1.414, -2.828)$ and $(1.414, 2.828)$ when rounded to the nearest thousandth. ◆

PROBLEM 5 Referring to Example 5, when we substitute x^3 for y in the equation $4x^2 + y^2 = 16$, we obtain $x^6 + 4x^2 - 16 = 0$. With some *creative* factoring, we can find the *exact* solutions of this equation. First, rewrite this equation as

$$(x^6 - 8) + 4x^2 - 8 = 0.$$

Now, factor $x^6 - 8$ as the difference of cubes (Section 1.6) and then look for a common binomial factor between this expression and $4x^2 - 8$. State the exact coordinates of the intersection points. ◆

◆ **Solving Equations by Finding Intersection Points**

To determine the intersection points of the graphs of the equations $y = f(x)$ and $y = g(x)$, we find the real roots of the equation $f(x) = g(x)$. Conversely, we can find the real roots of an equation of the form $f(x) = g(x)$ by graphing $y = f(x)$ and $y = g(x)$ in the same coordinate plane and noting the x-coordinates of their intersection points. We refer to this graphical procedure for solving an equation as the **intersection point method**. Hence, we have two graphical methods for solving an equation of the form $f(x) = g(x)$.

1. ***x*-intercept method**: Write the equation in the form $f(x) - g(x) = 0$, then generate the graph of $y = f(x) - g(x)$. The x-intercepts of this graph are the real solutions of the equation $f(x) = g(x)$. This procedure is outlined in Section 2.4.

2. **Intersection point method**: Generate the graphs of $y = f(x)$ and $y = g(x)$ in the same coordinate plane. The x-coordinates of the intersection points are the real solutions of the equation $f(x) = g(x)$.

EXAMPLE 6 Solve graphically by using the intersection point method.

(a) $2x - 11 = 5x - 5$ (b) $(x - 3)(x - 10) = x - 10$

◆ Solution

(a) We begin by generating the graphs of $y = 2x - 11$ and $y = 5x - 5$ in the same viewing rectangle of a graphing calculator. As shown in Figure 3.63, the lines intersect at one point. Using either the calculator's built-in program for finding the intersection point or the [TRACE] key and [ZOOM] key, we find that the x-coordinate of the intersection point is -2. Hence, the real root of the original equation is also -2. For the algebraic solution of this equation, see Example 1(a) in Section 1.8.

(b) We begin by generating the graphs of $y = (x - 3)(x - 10)$ and $y = x - 10$ in the viewing rectangle of a graphing calculator. As shown in Figure 3.64, the parabola and the line intersect at two points. Using either the calculator's built-in program for finding the intersection points or the TRACE key and ZOOM key, we find that the x-coordinates of the intersection points are 4 and 10. Hence, the real roots of the original equation are also 4 and 10. For the algebraic solution of this equation, see Example 1(a) in Section 3.4.

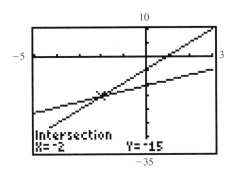

FIGURE 3.63
The real root of the equation $2x - 11 = 5x - 5$ is the x-coordinate of the intersection points of the lines $y = 2x - 11$ and $y = 5x - 5$.

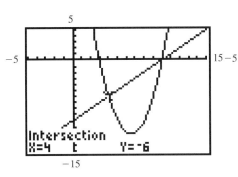

FIGURE 3.64
The real roots of the equation $(x - 3)(x - 10) = x - 10$ are the x-coordinates of the intersection points of the parabola $y = (x - 3)(x - 10)$ and the line $y = x - 10$.

PROBLEM 6 Solve graphically: $x^2 - x - 12 = 3(x + 2)(x - 2)$. (For the algebraic solution, see Problem 1 in Section 3.4.)

Exercises 3.5

◆ Basic Skills

In Exercises 1–12, use the substitution method or the elimination method to find the coordinates of any intersection point for the lines with the given equations. Check graphically.

1. $3x + 5y = 8$ and $x = 5y$
2. $y = x + 6$ and $3x + 5y = 4$
3. $y = 8x + 3$ and $2x - 3y = 2$
4. $y = 3x - 5$ and $y = 5 - 2x$
5. $y = 4 - 3x$ and $y = 3 - 5x$
6. $y + 2x = 0$ and $3x - 2y = 14$
7. $7x - y = 2$ and $5x - y = 8$
8. $2x + 3y = 7$ and $3x - 2y = 4$
9. $3x - 3y = 10$ and $x - y = 1$
10. $4x - 3y = 8$ and $4x - 2y = 7$
11. $5x - 2y = 10$ and $3x + 6y = 12$
12. $4y = x + 6$ and $3x - 12y = -18$

In Exercises 13–24, use the substitution method to find the coordinates of any intersection point for the line and parabola with the given equations. Check graphically.

13. $y = 2x$ and $y = 3 - x^2$
14. $x + y = 6$ and $y = x^2$
15. $y = x + 3$ and $y = 2x^2 - 3$
16. $6x - 2y = 31$ and $x^2 - 2y - 6x + 5 = 0$

17. $3x - 2y = 5$ and $y = x^2 - 4x$
18. $x + y = -6$ and $y = x^2 - 2x - 8$
19. $y = \frac{5}{2}x + 2$ and $y = 3 - x - 2x^2$
20. $y = \frac{1}{4}x + 2$ and $y = \frac{1}{4}x^2 - x + 3$
21. $2y = x$ and $x^2 - y + 6x + 5 = 0$
22. $5x + 2y = 10$ and $y = 4 - 2x^2$
23. $y = \frac{1}{4}x^2$ and $5x - 2y = 6$
24. $y = x - 2$ and $x^2 - 4x + 3y - 6 = 0$

In Exercises 25–50, use the substitution method or the elimination method to find the coordinates of the intersection points for the graphs with the given equations. Check graphically with a graphing calculator.

25. $y = x^2 - 4x + 3$ and $y = 3 + 2x - x^2$
26. $y = x^2 - 4x$ and $y = 6x - x^2$
27. $x^2 + y^2 = 36$ and $y^2 - 5x = 0$
28. $x^2 + y^2 = 1$ and $y = 2x^2 - 1$
29. $x^2 + y^2 = 25$ and $y = \frac{1}{3}x + 3$
30. $x^2 + y^2 = 4$ and $y = \frac{2}{x}$
31. $y = \frac{1}{x}$ and $2x - y = 1$
32. $y = \sqrt{x}$ and $y = x - 2$
33. $x - 2y = 4$ and $y = \sqrt{x - 4}$
34. $y = \sqrt{x + 4}$ and $x = -y^2$
35. $y = \frac{1}{3}x + 1$ and $y = \sqrt{x}$
36. $x + y = 3$ and $xy = 3$
37. $9x^2 + y^2 = 36$ and $x^2 + y^2 = 28$
38. $x^2 + 2y^2 = 18$ and $2x^2 + y^2 = 15$
39. $y = x + 1$ and $x^2 + 4y^2 = 4$
40. $y = x^2 - 1$ and $4x^2 + y^2 = 16$
41. $x^2 + 2y^2 = 8$ and $y = \sqrt{x}$
42. $9x^2 + 4y^2 = 25$ and $y = \frac{2}{x}$
43. $9y^2 - 4x^2 = 36$ and $2x^2 + y^2 = 4$
44. $4x^2 - y^2 = 16$ and $x^2 + y^2 = 9$
45. $y = x - 2$ and $x^2 - y^2 = 1$
46. $y = x^2 - 1$ and $x^2 - 4y^2 = 1$
47. $x^2 - 5y^2 = 36$ and $x = y^2$
48. $10y^2 - x^2 = 16$ and $y = \sqrt{x}$
49. $4x^2 + y^2 - 8x = 0$ and $y = 6x - 4$
50. $x^2 + 4y^2 - 6x + 5 = 0$ and $y = 2x - 5$

In Exercises 51–58, use a graphing calculator to approximate the coordinates of the intersection points of the graphs with the given equations. Round each coordinate to the nearest thousandth.

51. $y = 3x - 1$ and $y = x^3$
52. $y = 2x^2 - 3x - 5$ and $y = x^3$
53. $x^2 + y^2 = 16$ and $y = x^2 - 2x$
54. $x^2 - y^2 = 4$ and $y = 4x - x^2$
55. $x^2 + 4y^2 - 4x = 0$ and $y = x - 1$
56. $y^2 + 16x = 0$ and $y^2 - 9x^2 - 18x = 18$
57. $y = 4 - 3x - x^2$ and $y = 2x^4$
58. $y = \sqrt[3]{x}$ and $y = 1 - x^2$

In Exercises 59–74, solve each equation graphically by using the intercept method in conjunction with a graphing calculator. Check algebraically.

59. $3n + 4 = n - 4$
60. $12m + 5 = 2m - 5$
61. $\frac{5x}{6} - 1 = \frac{x}{3}$
62. $4 + \frac{n}{6} = \frac{n + 4}{2}$
63. $n(2n - 3) = 5$
64. $(x - 1)(3x + 2) = 2$
65. $(x + 4)(x + 6) = x + 6$
66. $(m - 7)(m + 4) = 2(m - 7)$
67. $y^2 + (y - 2)(y + 8) = 2y$
68. $(x + 3)^2 = x(3x - 1)$
69. $5x^2 + 1 = 4x$
70. $2x^2 + 3 = 4x$
71. $(2x - 1)(x + 4) = (x + 1)^2$
72. $y(y - 3) = (2y + 3)(y - 1)$
73. $2(x^2 + 2) = \sqrt{7}x$
74. $\frac{2x}{3} = x^2 + \frac{1}{2}$

Critical Thinking

75. Find the coordinates of the vertices of a triangle whose sides lie on the lines $x + y = 4$, $3x - y = 4$, and $x - y = 6$.

76. Show that the triangle in Exercise 75 is a right triangle. Then find its area.

77. Two sides of a parallelogram lie on the lines $2x - y = -3$ and $x + 2y = 11$, and one of its vertices is $(7, -3)$. Find the coordinates of the other three vertices of the parallelogram.

78. Show that the parallelogram in Exercise 77 is a rectangle. Then find its area.

79. Find the length of the line segment that joins the intersection points of the line $y = 2x + 3$ and the parabola opening upward with vertex $(-2, -4)$ and y-intercept 0.

80. Find the coordinates of the midpoint of the line segment that joins the intersection points of the parabola $y = 3 - x^2$ and the line with slope 1 and y-intercept 1.

$\boxed{\frac{\Delta y}{\Delta x}}$ *Calculus can be used to find the area of a region bounded by two or more curves. In order to find the area, we must first find the intersection points of the curves. In Exercises 81–86, find the coordinates of the intersection points P, Q, R, and (when applicable) S.*

81.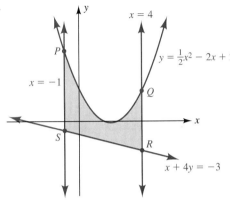

82.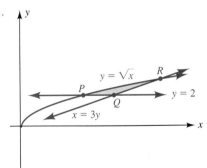

83.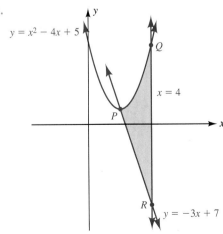

84.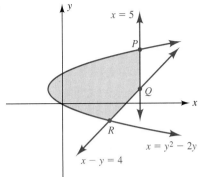

85.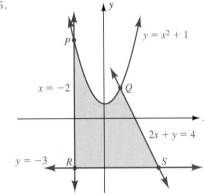

86.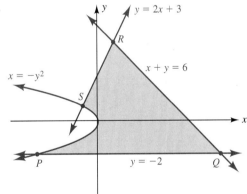

Chapter 3 Review

◆ Questions for Writing or Group Discussion

1. What is meant by the *slope* of a straight line? Does every line have a slope? Explain.
2. How can you find the slope of a straight line from its equation? from its graph?
3. Describe the graph of a *linear function*. Explain the procedure for graphing a linear function.
4. Explain the procedure for finding the equation of a straight line when two points on the line are given.
5. On a coordinate plane, sketch two lines that have the given conditions:
   (a) negative reciprocal slopes and the same y-intercept
   (b) the same slope and different y-intercepts
6. Describe the graph of a *quadratic function*. Explain the procedure for graphing a quadratic function.
7. How can we determine whether a quadratic function has a *maximum* or *minimum value*? Explain how the maximum or minimum value can be found.
8. State the *quadratic formula*. Illustrate its use with an example.
9. Describe the method of *completing the square*.
10. By using the *discriminant*, explain how you can determine whether the roots of a quadratic equation are (a) real, (b) imaginary, or (c) equal.
11. Given that a is a real number, are the roots of the equation $5x^2 + 2ax + a^2 = 0$ real or imaginary? Explain your reasoning.
12. Is it possible to find three consecutive integers whose sum equals the product of the smallest and largest of these integers? Explain.
13. Illustrate with sketches some possibilities of intersection for two lines graphed in the same coordinate plane.
14. Describe two methods that can be used to determine the coordinates of the intersection points of two curves.
15. What can we conclude if the equations of two curves have no real solution when solved simultaneously?
16. Describe two graphical methods that can be used to solve an equation of the form $f(x) = g(x)$. Illustrate with an example.

◆ Review Exercises

In Exercises 1–14, identify the graph of each equation as a line or a parabola. If the graph is a line, give its slope and find any x-intercept and y-intercept. If the graph is a parabola, give the coordinates of its vertex and find any x- and y-intercept(s).

1. $y = 2x - 3$
2. $y = 1 - 4x$
3. $y = -\frac{3}{4}x - 1$
4. $y = \frac{2}{3}x + 2$
5. $y = (x - 2)^2 - 3$
6. $y = -2(x + 1)^2 + 2$
7. $y = 15 + 2x - x^2$
8. $y = 2x^2 + x - 10$
9. $6x - 2y = 8$
10. $3x - 4y + 18 = 0$
11. $3y - 18 = 0$
12. $2x + 5 = 0$
13. $x^2 + 2x - 2y + 5 = 0$
14. $2x^2 - 8x - 3y + 5 = 0$

In Exercises 15–24, determine the equation from the given information.

15. Line with slope $\frac{3}{4}$ and y-intercept -2
16. Line with x-intercept 4 and y-intercept -2
17. Line through $(-2, 4)$ with slope 3
18. Line through the origin and parallel to $x - 3y = 4$
19. Line through $(-2, 1)$ and perpendicular to the line through $(1, 1)$ and $(4, 3)$
20. Line through $(-3, 2)$ and perpendicular to the x-axis
21. Parabola with vertex $(0, -4)$ and x-intercepts ± 2
22. Parabola with vertex $(1, 2)$ and x-intercepts -1 and 3
23. Parabola passing through the origin, with vertex $(-2, -3)$ and symmetric with respect to $x = -2$
24. Parabola passing through $(-1, 0)$, with vertex $(2, 3)$ and symmetric with respect to $x = 2$

In Exercises 25–34, solve each quadratic equation.

25. $x^2 - 6x = 16$
26. $x(3x - 4) = 4$
27. $3x^2 + 48 = 0$
28. $16 + 2x^2 = 0$
29. $x^2 = 4x + 10$
30. $2x^2 + 3x = 1$
31. $x^2 + 5 = 2x$
32. $x^2 + 8x + 20 = 0$
33. $2(x^2 - 3) = \sqrt{6}x$
34. $2x^2 = 2\sqrt{2}x - 3$

In Exercises 35–40, solve each quadratic inequality. Use interval notation to write the solution set.

35. $3x^2 + 7x \le 6$
36. $25 - x^2 > 16$
37. $x^2 + 2x > 4$
38. $x^2 - 5x + 4 < 0$
39. $x + 4 < 3x(x - 1)$
40. $(2y + 1)(y - 3) \ge 2y - 1$

In Exercises 41–46, solve each equation graphically by using the intersection point method and a graphing calculator.

41. $7x - 6 = 4(3x + 1)$
42. $3x + 4 = 7x - 8$
43. $2 - 2x = 9x^2 + 6x + 1$
44. $4x^2 = (2x - 3)^2 + 9$
45. $x(x - 2) = 4(x - 1)$
46. $2(x - 1)^2 = 3x^2 - 5x$

In Exercises 47–56, use the substitution method or the elimination method to find the intersection points of the graphs with the given equations. Check graphically with a graphing calculator.

47. $y = 4x - 5$ and $2x - 3y = 10$
48. $3x - 4y = 6$ and $6x - 3y = 7$
49. $y = x^2 - 4x + 6$ and $y = x + 2$
50. $x = 4y^2 - 8y$ and $x - 8y = -12$
51. $x^2 + y^2 = 36$ and $y = x^2 - 6$
52. $y = 2 - 2x$ and $4x^2 + 9y^2 = 36$
53. $y^2 - x^2 = 1$ and $x^2 + 4y^2 = 9$
54. $y = \sqrt{x + 1}$ and $x^2 + 4y^2 = 9$
55. $x^2 - 4y^2 - 2x - 15 = 0$ and $x - 4y = 0$
56. $y = x^2 - 2x - 3$ and $4x^2 - 12x + 3y = 0$

In Exercises 57 and 58, use a graphing calculator to approximate the coordinates of the intersection points of the given curves. Round each coordinate to the nearest thousandth.

57. $3x + 2y = 6$ and $y = x^3$
58. $4x^2 + y^2 = 16$ and $y = x^2 - 4x$

59. Given the equations $A_1 x + B_1 y = C_1$ and $A_2 x + B_2 y = C_2$, solve the equations simultaneously (a) for x and (b) for y.

60. Given the equations $A_1 x^2 + B_1 y^2 = C_1$ and $A_2 x^2 + B_2 y^2 = C_2$, solve the equations simultaneously (a) for x and (b) for y.

61. The graph of the line $Ax + By + C = 0$ with $A \ne 0$ and $B \ne 0$ is sketched in the figure.

    (a) Determine the equation of the line that passes through the origin and is perpendicular to the line $Ax + By + C = 0$.

    (b) Determine the coordinates of the intersection point P.

    (c) Show that the distance d from the line $Ax + By + C = 0$ to the origin is given by

    $$d \quad \frac{|C|}{\sqrt{A^2 + B^2}}$$

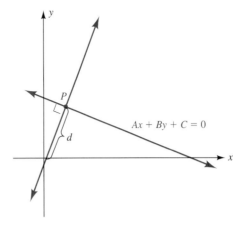

62. Use the formula from Exercise 61(c) to find each distance:

    (a) the distance from the origin to the line $3x + 4y - 12 = 0$

    Exercise 62 continues

(b) the distance between the parallel lines $3x + y = 4$ and $3x + y = 1$

63. Two sides of a parallelogram lie along the lines $y = x + 4$ and $x + 2y = 2$. One diagonal of the parallelogram is on the line $5x - 2y = -4$. Determine the coordinates of the vertices of the parallelogram.

64. Find the area of the triangle bounded by the lines $2x - y = 6$, $x + 2y = 18$, and $3x + y = 4$.

65. Determine whether the quadratic function $f(x) = 3x^2 - 4x + 5$ has a maximum or minimum value. Then find this value.

66. Determine whether the quadratic function $f(x) = 5x - 2x^2$ has a maximum or minimum value. Then find this value.

67. Determine the value of a such that the quadratic function $f(x) = ax^2 - 4x + 3$ has a maximum value of 5.

68. Let $f(x) = 2x + 3$ and $g(x) = x^2 - 3x + 4$.
    (a) Determine the minimum value of $f \circ g$.
    (b) Determine the minimum value of $g \circ f$.

69. The value of a personal computer depreciates linearly from $2400 at time of purchase to $400 after 5 years.
    (a) Express the value V (in dollars) of the computer as a function of its age x (in years).
    (b) What is the age of the computer when its value is fully depreciated?

70. Each week a salesperson gets paid a base salary plus a commission, which is a fixed percentage of the week's sales. For weekly sales of $1000, total pay is $280, and for weekly sales of $3000, the total pay is $440.
    (a) Express the salesperson's total pay P as a function of weekly sales x.
    (b) What is the base salary?
    (c) What is the significance of the slope of the graph of the function defined in part (a)?

71. In a certain city, the yearly cost C for water that is supplied to a household is a function of the number x of gallons that are used. Suppose the cost of water is given by the following rate schedule.

Usage	Cost
First 5000 gallons or less	$120
Next 10,000 gallons	$0.018 per gallon
Next 15,000 gallons	$0.011 per gallon
All gallons over 30,000	$0.003 per gallon

Find the piecewise linear function that describes this usage schedule. Then sketch its graph.

72. Five hundred feet of fencing is to be used to build three adjacent rectangular corrals (see figure). Find the overall dimensions of the corrals if the total enclosed area is to be the *largest* possible.

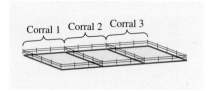

73. A triangular piece of land has 400 feet of frontage on a tar road and 250 feet of frontage on a gravel road that is perpendicular to the tar road. What are the dimensions of the *largest* rectangular building lot that can be inscribed in this triangular piece of land if the sides of the rectangular lot lie along the tar and gravel roads?

74. When the owner of an apartment complex charges rent of $600 per month for each unit, all 60 apartments are occupied. Research shows that for each $50 increase in the monthly rent, 3 apartments become vacant. What monthly charge for rent gives the owner maximum revenue?

75. The shape of a cable that supports the weight of a suspension bridge approximates a parabola with its vertex halfway between the supports, as shown in the sketch. Determine the length x of the vertical cable that is 30 meters from the end of a support.

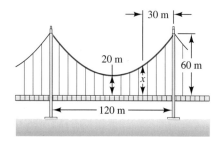

76. The dimensions of a picture are 12 in. by 15 in., as shown in the sketch. Encasing the picture in a frame of uniform width increases its area by 50%. What is the width of the frame?

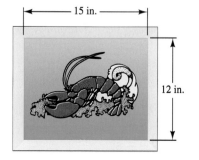

77. An open box is made from a 6-in. by 9-in. piece of metal by cutting out a square from each corner and then turning up the sides along the dashed lines, as shown in the figure. If the area of the bottom of the box is 18 sq in., what is the height of the box?

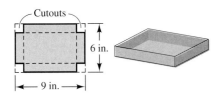

78. Find the length and width of a rectangle if its diagonal is 2 centimeters more than the longer side, which in turn is twice the shorter side.

79. Find k such that one root of $25x^2 + kx + 1 = 0$ is two more than the other root.

80. The base of a triangle is 4 ft more than its height, and the area of the triangle is less than 96 sq ft (or ft²). What is the range of values for the base of the triangle?

Cumulative Review Exercises for Chapters 1, 2, and 3

1. Factor completely over the set of integers.
   (a) $x(x + 1)(2x - 5) - 12(x + 1)$
   (b) $x^2 - 4xy - 4a^2 + 8ay$

2. Simplify each fractional expression.
   (a) $\dfrac{(x - 3)^{1/2} - 4(x - 3)^{-1/2}}{x - 7}$
   (b) $\dfrac{x^{-1} + y^{-1}}{(x + y)^{-1}}$

3. Express $8^0 + 3\tfrac{1}{2} - 6^{-1} + \dfrac{1}{\sqrt[3]{-27}} + 8^{2/3}$ as an integer.

4. If $\dfrac{\sqrt{19} - \sqrt{10}}{x} = \dfrac{x}{\sqrt{19} + \sqrt{10}}$, find x.

5. For what values of k does the equation $4x^2 - 12x + k = 0$ have the given roots?
   (a) one root that is zero   (b) equal roots
   (c) real roots   (d) imaginary roots

6. For the equation $ax^2 - 5x + 2 = 0$, find the value of a if one root is to be the reciprocal of the other.

7. Expand and simplify $(1 + i)^2$, where $i = \sqrt{-1}$.

8. Arrange the given numbers in order from smallest to largest:

   $4^{-3/2}$ $\left(\dfrac{1}{4}\right)^{-2}$ $(4^{-1/3})^6$ $\left(\dfrac{1}{4}\right)^{-1/2}$ $\dfrac{1}{4^0}$

9. A rectangular field is twice as long as it is wide and is enclosed by x feet of fencing. Express the area A of the field as a function of x.

10. Simplify $(x^2 + 10x + 25)^{1/2} - (x^2 - 10x + 25)^{1/2}$ given the indicated conditions.
    (a) x is any real number   (b) $x \geq 5$

11. Perform the indicated operations and reduce to lowest terms.
    (a) $\dfrac{a}{(a - b)(a - c)} + \dfrac{b}{(b - c)(b - a)} + \dfrac{a}{(c - a)(c - b)}$
    (b) $\dfrac{1 + \dfrac{1}{x^2} + \dfrac{1}{x^4}}{1 + \dfrac{1}{x} + \dfrac{1}{x^2}}$

12. Find the x- and y-intercepts for the graph of the equation $y^2 = x(x - 2)^2$.

13. If the acceleration of an object is constant, its velocity increases linearly with time. Suppose that after 1 second a certain object with a constant acceleration has a velocity of 12 feet/second (ft/s), and after 5 seconds it has a velocity of 20 ft/s.
    (a) Express the velocity v of the object as a function of time t.
    (b) Sketch the graph of the function defined in part (a) and state the significance of the y-intercept.

14. Find the perimeter and the area of the triangle formed by the intersecting lines $x + y = -3$, $3x - 4y = -9$, and $4x - 3y = 2$.

15. For the graph of the equation $y = f(x)$, what is the y-intercept?

16. Sketch the graph of each quadratic function. Label the vertex and the x- and y-intercept(s).
    (a) $f(x) = x^2 - 4x + 3$ (b) $g(x) = 1 - 4x - x^2$

17. A car radiator contains 15 quarts of a 20% antifreeze solution. How many quarts of 20% solution must be drained and replaced with pure antifreeze (100% solution) to achieve a solution of at least 60% antifreeze?

18. Determine k so that the line $x - 3y + k = 0$ passes through the point $(3, -1)$.

19. Find the area of a triangle whose vertices are the points $(8, 8)$, $(-7, -2)$, and $(1, -6)$.

20. Given that $f(x) = \sqrt{x}$, express
    $$\dfrac{f(x + h) - f(x)}{h}$$
    in simplest form with a rationalized numerator. Then evaluate this expression as h approaches 0.

21. Determine whether the graph of each equation has symmetry with respect to the x-axis, y-axis, or origin.
    (a) $x^2y = 9$ (b) $x + y = x^3$
    (c) $x^2 + y^2 - 6x = 16$ (d) $y = \pm\sqrt{9 - x^2}$

22. Determine what two amounts $22,400 can be split into so that the annual interest earned from one amount, which is invested at 8%, is double the interest from the other amount, which is invested at 10%.

23. Find the equation of a line that passes through the point of intersection of the lines $3x - 2y = 7$ and $y = 7 - 2x$ and also satifies the given condition.
    (a) is parallel to the x-axis
    (b) is perpendicular to $x + y = 8$
    (c) passes through $(5, -2)$
    (d) has a slope of -3

24. A business determines that consumers will purchase x units of a product each month when the price (in dollars) of the unit is $(60 - x)/3$. How many units are sold if the total monthly revenue from sales is $225?

25. Sketch the graph of the equation $y = 1/x^2$. Then use the ideas of shifting and reflecting to help sketch the graphs

of the following functions. Label all intercepts and asymptotes.

(a) $f(x) = 2 - \dfrac{1}{x^2}$

(b) $g(x) = \dfrac{1}{(x-1)^2}$

26. A point is equidistant from (2, 1) and (−4, 3), and the slope of the straight line joining this point to the origin is 2. Determine the coordinates of the point.

27. Determine the value of m so that the line $y = mx - 2$ passes through the point of intersection of the lines $y = 3x + 2$ and $y = x - 4$.

28. Determine the quadratic function with real coefficients that has $2 \pm 3i$ as its zeros.

29. A square piece of sheet metal is to be made into a rectangular box with an open top by cutting out a 2-in. by 2-in. square from each corner and bending along the dashed lines, as shown in the sketch. Determine the range of dimensions of the original piece of sheet metal that results in a box volume from 50 to 98 cubic inches.

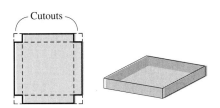

30. The power P (in watts) dissipated in a certain electrical circuit is directly proportional to the square of the current i (in amperes) flowing through the circuit. Express P as a function of i if $P = 5$ watts when $i = 20$ milliamperes.

31. If a farmer picks his pumpkin crop today, he will have 30,000 pounds of pumpkins worth 20¢ per pound. If he waits, the crop will grow at 2,000 pounds per week, but the price will drop by 1¢ per pound per week.
    (a) Express the amount A of money in dollars that the farmer receives at the end of x weeks as a function of x.
    (b) When is the most profitable time for the farmer to pick and sell the pumpkin crop?

32. A tank contains 1000 gallons of water. The spigot is opened and water flows out of the tank at the uniform rate of $\tfrac{2}{3}$ gallon per minute.
    (a) Express the amount A of water (in gallons) in the tank as a function of the time t (in minutes) that the spigot is opened.
    (b) State the domain and range of the function defined in part (a).

A silo has the shape of a right circular cylinder surmounted by a hemisphere and holds 3000 cubic feet of grain. If the height of the silo is 60 ft, what is the radius of its base?

For the solution, see Example 5 in Section 4.4.

Polynomial and Rational Functions

- 4.1 Polynomial Functions and Their Graphs
- 4.2 Polynomial Division
- 4.3 Factors and Zeros of Polynomial Functions
- 4.4 Polynomial Equations and Inequalities
- 4.5 Rational Functions and Their Graphs
- 4.6 Rational Equations and Inequalities

4.1 Polynomial Functions and Their Graphs

◆ Introductory Comments

Any function P defined by

$$P(x) = a_n x^n + a_{n-1} x^{n-1} + a_{n-2} x^{n-2} + \cdots + a_2 x^2 + a_1 x + a_0$$

where n is a nonnegative integer and $a_n \neq 0$, is called a **polynomial function of degree n**. In this text, we restrict our discussion to polynomial functions with *real coefficients*, that is, $a_0, a_1, a_2, \ldots, a_n$ are real numbers with $a_n \neq 0$. The domain of any polynomial function is $(-\infty, \infty)$.

A polynomial function of *degree 0* is a constant function $f(x) = a_0$, of degree 1 is a linear function $g(x) = a_1 x + a_0$, and of degree 2 is a quadratic function $h(x) = a_2 x^2 + a_1 x + a_0$. Since we have discussed constant functions, linear functions, and quadratic functions in previous chapters, we devote this chapter to polynomial functions of degree 3 or greater, such as the following:

1. Cubic function: $F(x) = a_3 x^3 + a_2 x^2 + a_1 x + a_0$
2. Quartic function: $G(x) = a_4 x^4 + a_3 x^3 + a_2 x^2 + a_1 x + a_0$
3. Quintic function: $H(x) = a_5 x^5 + a_4 x^4 + a_3 x^3 + a_2 x^2 + a_1 x + a_0$

The graph of a polynomial function of degree 3 or greater is *continuous* with smooth, rounded turns—it contains no gap, break, jump, or sharp turn. The sketches in Figure 4.1. illustrate the differences between the graphs of polynomial and nonpolynomial functions. In this section, we discuss the important characteristics of the graph of a polynomial function.

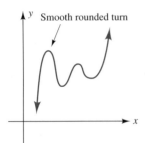

(a) Polynomial function: Smooth, rounded turns; no gap, break, jump, or sharp turn

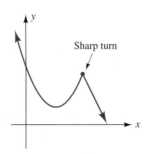

(b) Nonpolynomial function: Sharp turn

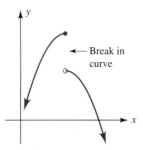

(c) Nonpolynomial function: Break in curve (discontinuous)

FIGURE 4.1

Graphs of polynomial and nonpolynomial functions

◆ Power Functions

The simplest polynomial function contains a single term and has the form

$$P(x) = x^n$$

with n a positive integer. In this text, we refer to this type of polynomial function as a **power function**.

When n is an *even* integer with $n \geq 4$, the graph of $P(x) = x^n$ is similar to the graph of the squaring function $f(x) = x^2$ (Section 2.5). When n is an *odd* integer with $n \geq 5$, the graph of $P(x) = x^n$ is similar to the graph of the cubing function $f(x) = x^3$ (Section 2.5). The greater the value of n (odd or even), the flatter the graph of P in the interval $(-1, 1)$ and the steeper the graph of P in the intervals $(-\infty, -1)$ and $(1, \infty)$. Figure 4.2 shows the graphs of $P(x) = x^4$ and $f(x) = x^2$ in the same viewing rectangle of a graphing calculator, and Figure 4.3 shows the graphs of $P(x) = x^5$ and $f(x) = x^3$ in the same viewing rectangle of a graphing calculator. Note that in both figures, in the interval $(-1, 1)$, the graph of P is flatter than the graph of f, but in the intervals $(-\infty, -1)$ and $(1, \infty)$ the graph of P is steeper than the graph of f.

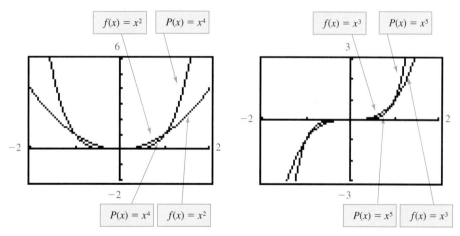

FIGURE 4.2

The graph of $P(x) = x^4$ resembles the graph of $f(x) = x^2$.

FIGURE 4.3

The graph of $P(x) = x^5$ resembles the graph of $f(x) = x^3$.

We can use the graph of a power function in conjunction with the shift rules, reflection rules, and stretch and compress rules (Section 2.5) to sketch the graphs of some polynomial functions.

EXAMPLE 1 Sketch the graph of each function and label the intercepts.

(a) $P(x) = \frac{1}{2}(x - 1)^5$ **(b)** $P(x) = 16 - x^4$

◆ Solution

(a) We can think of this function as $P(x) = \frac{1}{2}f(x - 1)$ where $f(x) = x^5$. Thus, by the horizontal shift rule and the vertical compress rule, the graph of $P(x) = \frac{1}{2}(x - 1)^5$ is obtained by shifting the graph of $f(x) = x^5$ right 1 unit, and then compressing this graph by a factor of 2. The graph of the function P is shown in Figure 4.4.

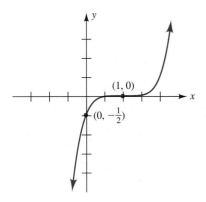

FIGURE 4.4
Graph of $P(x) = \frac{1}{2}(x - 1)^5$

x-intercept:
$$P(x) = 0$$
$$\tfrac{1}{2}(x - 1)^5 = 0$$
$$(x - 1)^5 = 0 \quad \text{Multiply both sides by 2}$$
$$x - 1 = 0 \quad \text{Take the fifth root of both sides}$$
$$x = 1$$

y-intercept:
$$P(0) = \tfrac{1}{2}(0 - 1)^5 = -\tfrac{1}{2}$$

(b) We can think of this function as $P(x) = -f(x) + 16$ where $f(x) = x^4$. Thus, by the x-axis reflection rule and the vertical shift rule, the graph of $P(x) = 16 - x^4$ is obtained by reflecting the graph of $f(x) = x^4$ about the x-axis and then shifting this graph upward 16 units. The graph of the function P is shown in Figure 4.5.

x-intercept:
$$P(x) = 0$$
$$16 - x^4 = 0$$
$$x^4 = 16$$
$$x = \pm \sqrt[4]{16} \quad \text{Extract the two real fourth roots}$$
$$x = \pm 2$$

y-intercept:
$$P(0) = 16 - (0)^4 = 16$$

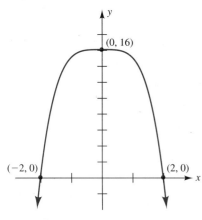

FIGURE 4.5
Graph of $P(x) = 16 - x^4$

PROBLEM 1 Sketch the graph of $P(x) = 1 - (x + 2)^6$ and label the intercepts.

◆ Left and Right Behavior and Relative Extrema

When a polynomial function is written in descending powers of x, the coefficient of the term with the highest degree is the **leading coefficient**. In the polynomial function

$$P(x) = a_n x^n + a_{n-1} x^{n-1} + \cdots + a_1 x + a_0$$

a_n is the leading coefficient. For very large values of x (either negative or positive), the highest-degree term of the polynomial function, $a_n x^n$, becomes dominant in relation to the other terms of the polynomial. Because of this, the graph of a polynomial function P eventually rises or falls, depending on

  (i) the sign of its leading coefficient a_n, and
  (ii) whether n is even or odd.

Figure 4.6 illustrates the four possibilities for the left behavior of the graph of a polynomial function P as x decreases without bound ($x \to -\infty$) and for the

SECTION 4.1 ♦ Polynomial Functions and Their Graphs

right behavior as x increases without bound ($x \to \infty$). For convenience, we use quadrants I and IV to illustrate the possible shapes.

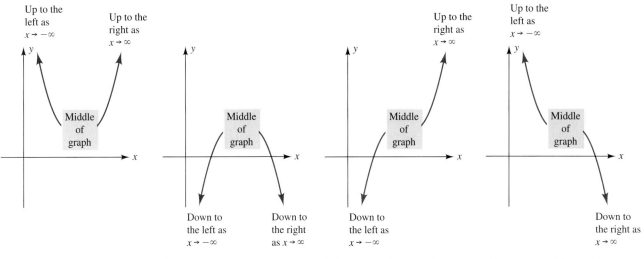

(a) $a_n > 0$ with n even   **(b)** $a_n < 0$ with n even   **(c)** $a_n > 0$ with n odd   **(d)** $a_n < 0$ with n odd

FIGURE 4.6
Left and right behavior of the graph of $P(x) = a_n x^n + a_{n-1} x^{n-1} + \cdots + a_1 x + a_0$

On the graph of a polynomial function, any point at which the function changes from increasing to decreasing or from decreasing to increasing is called a **relative extremum**. The graph of the polynomial function in Figure 4.7 has four such points, or extrema, labeled A, B, C, and D. The points A and C are called **relative maxima** (peaks), and the points B and D are called **relative minima** (valleys) of the function. A single extremum point is called a relative *maximum* or *minimum*.

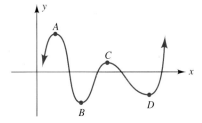

FIGURE 4.7
The graph of this polynomial function has relative maxima at A and C and relative minima at B and D.

Using calculus, it can be shown that if P is a polynomial function of degree n, then the number of relative extrema either is equal to $n - 1$ or is less than this number ($n - 1$) by an even number. This fact enables us to determine the possible shapes for the shaded portions of the graphs in Figure 4.6.

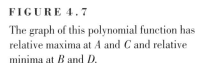 **Relative Extrema Rule**

If P is a polynomial function of degree n with real coefficients, then the number of relative extrema either is equal to $n - 1$ or is less than this number by an even number.

(a) Up to the left and up to the right, with three relative extrema

(b) Up to the left and up to the right, with one relative extremum

FIGURE 4.8

Sketches of two possible shapes for the graph of a polynomial function of degree 4 with a positive leading coefficient

◆ **Algebraic Procedure for Graphing a Polynomial Function**

EXAMPLE 2 For the polynomial function P defined by $P(x) = 3x^4 - 9x^2 - 6x - 1$, determine the left and right behavior of its graph and the possible number of relative extrema.

◆ **Solution** For $P(x) = 3x^4 - 9x^2 - 6x - 1$, we have $a_n = 3$ and $n = 4$. Since $a_n > 0$ and n is even, we know that the graph of P goes up to the left as $x \to -\infty$ and up to the right as $x \to \infty$, as shown in Figure 4.6(a). Since the degree of P is 4, its graph has either $4 - 1 = 3$ relative extrema or one relative extremum for the middle portion of the graph of P. Sketches of two possible shapes for the graph of P are shown in Figure 4.8. ◆

When you use a graphing calculator to generate the graph of a polynomial function, it is important to choose a viewing rectangle that shows the *complete graph*, that is, a viewing rectangle that shows the left and right behavior of the graph as well as any relative extrema in the middle of the graph.

PROBLEM 2 Use a graphing calculator to generate a complete graph of the polynomial function $P(x) = 3x^4 - 9x^2 - 6x - 1$. Compare your graph with the sketches in Figure 4.8. ◆

◆ **Sketching the Graph of a Polynomial Function**

The most efficient method of sketching the graph of a polynomial function of degree $n \geq 3$ is to use calculus. The methods of calculus give the relative extrema of a polynomial function, and joining its relative extrema to form a smooth curve gives us its graph. However, if we can write a polynomial function as a product of linear factors $ax + b$, with a and b real numbers, then we can draw a rough sketch of its graph by using the following algebraic procedure.

To sketch the graph of a polynomial function $y = P(x)$ of degree $n \geq 3$:

Step 1 Express $P(x)$ as the product of linear factors $ax + b$, with a and b real numbers. Then determine the x-intercept(s) by solving the equation $P(x) = 0$.

Step 2 Determine the left and right behavior of the graph and the shape of the graph near each x-intercept.

Step 3 Sketch the graph of the function using the information in step 2 as a guide.

(*Note:* To sketch a more accurate graph, determine the y-intercept and plot a few additional points.)

EXAMPLE 3 Sketch the graph of each polynomial function. Label the intercepts.

(a) $P(x) = 2x^3 - x^4$ **(b)** $P(x) = x^3 - 2x^2 - 4x + 8$

SECTION 4.1 ♦ Polynomial Functions and Their Graphs

Solution

(a) Step 1 We use the techniques of factoring from Section 1.6:

$$P(x) = 2x^3 - x^4$$
$$= x^3(2-x) \quad \text{Factor out } x^3$$
$$= (x-0)^3(2-x) \quad \text{Write as the product of linear factors}$$

To find the x-intercepts of the graph of P, we solve the equation $P(x) = 0$ by applying the zero product property (Section 2.2):

$$(x-0)^3(2-x) = 0$$
$$(x-0)^3 = 0 \quad \text{or} \quad 2-x = 0$$
$$x = 0 \quad\quad\quad\quad x = 2$$

x-intercepts

Step 2 For $P(x) = 2x^3 - x^4$, we have $a_n = -1$ and $n = 4$. Since $a_n < 0$ and n is even, we know that the graph of P goes down to the left as $x \to -\infty$ and down to the right as $x \to \infty$. To determine the shape of the graph near an x-intercept, we retain the factor corresponding to the intercept and evaluate the other factor(s) at the x-intercept:

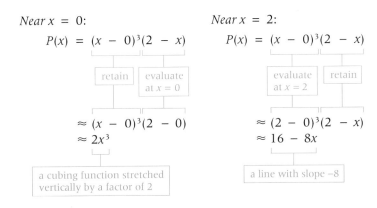

Near $x = 0$:
$$P(x) = (x-0)^3(2-x)$$
retain | evaluate at $x = 0$
$$\approx (x-0)^3(2-0)$$
$$\approx 2x^3$$
a cubing function stretched vertically by a factor of 2

Near $x = 2$:
$$P(x) = (x-0)^3(2-x)$$
evaluate at $x=2$ | retain
$$\approx (2-0)^3(2-x)$$
$$\approx 16 - 8x$$
a line with slope -8

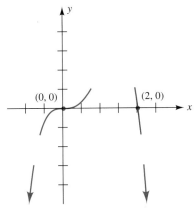

FIGURE 4.9
The left and right behavior of the graph of $P(x) = 2x^3 - x^4$ and its shape near each x-intercept

Figure 4.9 summarizes the information in Step 2.

Step 3 The information in Figure 4.9 enables us to sketch a rough graph of this function. For a more accurate sketch, we also plot the point $(1, 1)$, as shown in Figure 4.10. The graph of this polynomial function has one relative extremum, which satisfies the relative extrema rule for $n = 4$. Using calculus, it can be shown that this relative maximum point has coordinates $\left(\frac{3}{2}, \frac{27}{16}\right)$.

(b) Step 1 We again use the techniques of factoring from Section 1.6:

$$P(x) = x^3 - 2x^2 - 4x + 8$$
$$= x^2(x-2) - 4(x-2) \quad \text{Factor by grouping terms}$$
$$= (x-2)(x^2-4)$$
$$= (x-2)^2(x+2) \quad \text{Factor the difference of squares}$$

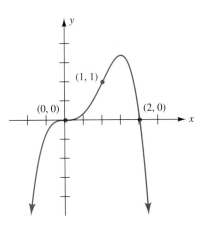

FIGURE 4.10
Graph of $P(x) = 2x^3 - x^4$

To find the x-intercepts of the graph of P, we solve the equation $P(x) = 0$ by applying the zero product property (Section 2.2):

$$(x - 2)^2(x + 2) = 0$$
$$(x - 2)^2 = 0 \quad \text{or} \quad x + 2 = 0$$
$$x = 2 \qquad\qquad\qquad x = -2$$

x – intercepts

Step 2 For $P(x) = x^3 - 2x^2 - 4x + 8$, we have $a_n = 1$ and $n = 3$. Since $a_n > 0$ and n is odd, we know that the graph of P goes down to the left as $x \to -\infty$ and up to the right as $x \to \infty$. To determine the shape of the graph near an x-intercept, we retain the factor corresponding to the intercept and evaluate the other factor(s) at the x-intercept:

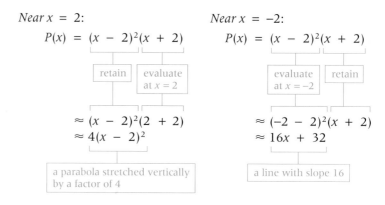

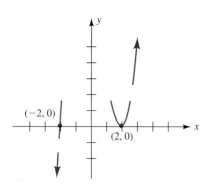

FIGURE 4.11

The left and right behavior of the graph of $P(x) = x^3 - 2x^2 - 4x + 8$ and its shape near each x-intercept

Figure 4.11 summarizes the information in Step 2.

Step 3 The information in Figure 4.11 enables us to sketch a rough graph of this function. For a more accurate sketch, we also plot the point $(-1, 9)$ and the y-intercept, as shown in Figure 4.12. Note that the graph of this polynomial function has two relative extrema, which satisfies the relative extrema rule for $n = 3$. The relative minimum point has coordinates $(2, 0)$. Using calculus, it can be shown that the relative maximum point has coordinates $\left(-\frac{2}{3}, \frac{256}{27}\right)$. ◆

Most graphing calculators have built-in programs for determining the relative extrema of a polynomial function. If a calculator does not have these built-in programs, we use the $\boxed{\text{TRACE}}$ key and $\boxed{\text{ZOOM}}$ key to estimate the coordinates of any relative extrema.

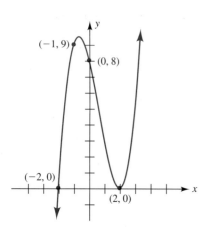

FIGURE 4.12

Graph of $P(x) = x^3 - 2x^2 - 4x + 8$

$\boxed{\text{PROBLEM 3}}$ Generate the graph of each polynomial function in Example 3 in the viewing rectangle of a graphing calculator. Estimate the coordinates of any relative extrema and compare the answers to those given in the solution. ◆

◆ Application: Bending Moment of a Beam

When a load (or weight) is applied to a steel or wood beam, the force produced by the load creates a *bending moment*. Due to this bending moment, the beam will deform slightly, resulting in a curvature of the beam. Structural engineers must investigate the bending moment at various points along a beam in order to select a beam of proper size to safely carry (support) a particular load. Bending moments are described frequently by polynomial functions of degree $n \geq 3$.

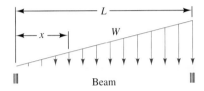

FIGURE 4.13
Simple beam with uniformly increasing load

EXAMPLE 4 For a simply supported beam of length L that carries a uniformly increasing load of weight W, as shown in Figure 4.13, the bending moment M at any distance x from the left end of the beam is given by

$$M(x) = \frac{Wx}{3L^2}(L^2 - x^2).$$

(a) If $W = 300$ lb and $L = 10$ ft, express the function M as a product of linear factors. Then sketch the graph of this function.

(b) Generate the graph of the function in part (a) in the viewing rectangle of a graphing calculator. Then approximate the distance x (from the left end of the beam) at which the bending moment is greatest.

◀ Solution

(a) Replacing W with 300 and L with 10 gives us

$$\begin{aligned} M(x) &= \frac{300x}{3(10)^2}(10^2 - x^2) \\ &= 100x - x^3 \\ &= x(10 + x)(10 - x) \quad \text{Write as the product of linear factors} \end{aligned}$$

Observe that M is a polynomial function of degree 3. Since the coefficient of x^3 is negative, the graph of M goes up to the left as $x \to -\infty$ and down to the right as $x \to \infty$. Also, by the relative extrema rule, the graph of M has either no relative extremum or two relative extrema.

To find the x-intercepts of the graph of M, we solve the equation

$$x(10 + x)(10 - x) = 0.$$

By the zero product property (Section 2.2), the x-intercepts are

$$x = 0 \qquad x = -10 \qquad x = 10$$

Now we investigate the shape of the graph near each x-intercept.

Near x = 0:
$$M(x) = x(10 + x)(10 - x)$$
$$\approx x(10)(10)$$
$$\approx 100x$$

a line with slope 100

Near x = −10:
$$M(x) = x(10 + x)(10 - x)$$
$$\approx -10(10 + x)(20)$$
$$\approx -2000 - 200x$$

a line with slope −200

Near x = 10:
$$M(x) = x(10 + x)(10 - x)$$
$$\approx 10(20)(10 - x)$$
$$\approx 2000 - 200x$$

Figure 4.14 summarizes the information we have gathered. The large slopes near the *x*-intercepts indicate that $M(x)$ may be quite large for certain values of *x*. For example, when $x = 5$, we obtain $M(5) = 375$. Graphically, we can compensate for these large values of $M(x)$ by selecting different scales for the *x*-axis and *M*-axis. Although the choice of scales is arbitrary, a reasonable choice is to choose 5 units between tick marks on the *x*-axis and 100 units between tick marks on the *M*-axis. The graph of *M* is shown in Figure 4.15. Since the length of the beam is 10 ft, we must have $0 \leq x \leq 10$. We indicate this restriction on the domain of this function by showing a solid curve in the interval [0, 10] and a dashed curve elsewhere.

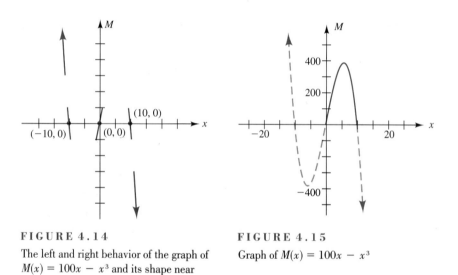

FIGURE 4.14

The left and right behavior of the graph of $M(x) = 100x - x^3$ and its shape near each *x*-intercept

FIGURE 4.15

Graph of $M(x) = 100x - x^3$

(b) To generate the graph of $y = M(x) = 100x - x^3$ for $0 \leq x \leq 10$ on a graphing calculator, we begin by choosing a viewing rectangle. We use our work in part (a) to make a reasonable choice:

```
Xmin = 0
Xmax = 10
Xscl = 1
Ymin = 0
Ymax = 400
Yscl = 100
```

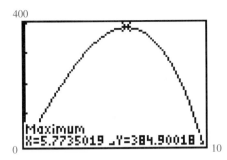

FIGURE 4.16
The graph of $M(x) = 100x - x^3$, $0 \leq x \leq 10$, in the viewing rectangle of a graphing calculator

Now, by entering the equation $y = 100x - x^3$, we obtain the desired portion of the graph in the viewing rectangle (see Figure 4.16). The point where the bending moment is greatest is at the peak of this curve. Using either the calculator's built-in program for finding a relative maximum or the [TRACE] key and [ZOOM] key, we find that the coordinates of the relative maximum are (5.774, 384.900) to the nearest thousandths. In summary, the maximum bending moment along the beam is approximately 384.900 pound-feet (lb-ft), and this occurs when $x \approx 5.774$ ft. Using calculus, it can be shown that the *exact* maximum bending moment is $\dfrac{2000\sqrt{3}}{9}$ lb-ft, and this occurs when $x = \dfrac{10\sqrt{3}}{3}$ ft. ◆

[PROBLEM 4] Is $M(x) = 100x - x^3$ an even function, an odd function, or neither? Explain. ◆

Exercises 4.1

 Basic Skills

In Exercises 1–12, use the shift rules, axis reflection rules, and stretch and compress rules (from Section 2.5) to sketch the graph of each polynomial function. Label any intercepts. Use a graphing calculator to verify each graph.

1. $f(x) = -x^5$
2. $f(x) = -x^4$
3. $f(x) = x^4 + 1$
4. $f(x) = x^5 - 1$
5. $f(x) = \frac{1}{4}(x - 1)^7$
6. $f(x) = 2(x + 1)^6$
7. $f(x) = 64 - x^6$
8. $f(x) = 3 - 3x^7$
9. $f(x) = (x - 3)^4 - 16$
10. $f(x) = (x + 1)^5 + 32$
11. $f(x) = 1 - (2 - x)^5$
12. $f(x) = 16 - (1 - x)^4$

For each polynomial function in Exercises 13–20, determine the following features of its graph: (a) the left and right behavior and (b) the possible number of relative extrema.

13. $P(x) = 2x^3 - 7x^2 + 3x - 2$
14. $P(x) = 7 - 5x - 4x^3$
15. $F(x) = 6 - 2x + 8x^2 - 5x^4$
16. $f(x) = 3x^8 - 2x^7 + x^6 - 3x + 1$
17. $P(x) = -x^7 - 2x^5 + 4x - 3$
18. $P(x) = 2x^5 - 20x^4 + 3x - 9$
19. $g(x) = 3x^6 - 2x^4 + 3x^2 + 1$
20. $G(x) = -5x^6 + 3x^5 + 18x^4 - 1$

In Exercises 21–28, generate a complete graph of each polynomial function in the viewing rectangle of a graphing calculator. Then determine the coordinates of any relative extrema. Round each coordinate to three significant digits.

21. The function P in Exercise 13
22. The function P in Exercise 14
23. The function F in Exercise 15
24. The function f in Exercise 16
25. The function P in Exercise 17
26. The function P in Exercise 18
27. The function g in Exercise 19
28. The function G in Exercise 20

In Exercises 29–46, sketch the graph of each polynomial function by using the three-step procedure outlined in this section. Label any intercepts. Use a graphing calculator to verify each graph.

29. $P(x) = x^3 + x^2 - 2x$
30. $P(x) = 12 + 4x - 3x^2 - x^3$
31. $G(x) = x^4 - 10x^2 + 9$
32. $P(x) = 3 - 2x^2 - x^4$
33. $h(x) = 3x^3 - x^4$
34. $f(x) = 4x^3 - 4x^2 + x$
35. $P(x) = x^5 - x^4$
36. $F(x) = 9x^3 - 4x^5$
37. $P(x) = 3x^4 - 12x^3 + 12x^2$
38. $Q(x) = 2x^5 - 12x^4 + 18x^3$
39. $f(x) = x^4 - 2x^2 + 1$

40. $h(x) = 16x^2 - 8x^4 + x^6$
41. $D(x) = x^3 + 2x^2 - 4x - 8$
42. $P(x) = 8x^3 + 4x^2 - 2x - 1$
43. $P(x) = x^2(x - 2)^3$
44. $P(x) = 2x(x + 3)^4$
45. $f(x) = 3x^3(x + 1)^2(x - 4)^3$
46. $G(x) = 4x^4(x - \frac{1}{2})^2(x + 6)^2$

47. The response time T (in microseconds) to a certain stimulus is a function of one's age x (in years) and is given by

$$T(x) = 20 + 79x + 76x^2 - 4x^3 \text{ for } 10 \leq x \leq 18.$$

Generate the graph of this function in the viewing rectangle of a graphing calculator. Then use this graph to estimate the age (to the nearest year) when the response time is greatest.

48. The profit P (in thousands of dollars) that a manufacturer makes is a function of the number of units x (in hundreds) that are produced and is given by

$$P(x) = 15 + 4x^2 - x^3 \text{ for } 0 \leq x \leq 6.$$

Generate the graph of this function in the viewing rectangle of a graphing calculator. Then use this graph to estimate the number of units (to the nearest whole number) that should be produced to maximize the manufacturer's profit.

49. A rectangular storage bin, with a square base of dimension x and an open top, is made from 36 square feet of material.
    (a) Show that the volume V of the box can be written as the product of linear factors:

    $$V = \tfrac{1}{4}x(6 - x)(6 + x)$$

    Then state the domain of the function.
    (b) Use the three-step procedure outlined in this section to sketch the graph of the polynomial function defined in part (a).
    (c) Generate the graph of the function defined in part (a) on a graphing calculator. Then find the dimension x for the storage bin with the largest volume.

50. A rectangle has its base on the x-axis and its two upper vertices touch the parabola $y = 3 - x^2$ at the points $P(x, y)$ and $Q(-x, y)$.
    (a) Show that the area A of the rectangle can be written as the product of linear factors:

    $$A = 2x(\sqrt{3} - x)(\sqrt{3} + x)$$

    Then state the domain of the function.

    (b) Use the three-step procedure that is outlined in this section to sketch the graph of the polynomial function defined in part (a).
    (c) Generate the graph of the function defined in part (a) on a graphing calculator. Then find the coordinates of points P and Q for the rectangle with the largest area.

A trucking company will accept a package for shipment only if the sum of the length and girth (the distance around the package) does not exceed 180 inches. For the packages shown in Exercises 51 and 52, answer the following questions.
(a) *Express the volume V of the package as a function of x. Then state the domain of the function.*
(b) *Use the three-step procedure outlined in this section to sketch the graph of the polynomial function defined in part (a).*
(c) *Generate the graph of the function defined in part (a) on a graphing calculator. Then find the dimension x for the package with the largest volume that can be accepted for shipping.*

51.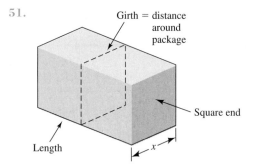

52.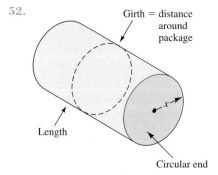

53. A piece of cardboard, 8 cm by 16 cm, is cut and folded as shown in the sketch to form a rectangular box with a closed top and flap.
    (a) Express the volume V of this box as a function of x.
    (b) Use the three-step procedure outlined in this section to sketch the graph of the polynomial function defined in part (a).
    (c) Generate the graph of the function in part (a) on a graphing calculator. Then find the approximate

value of x for which the volume of the box is greatest.

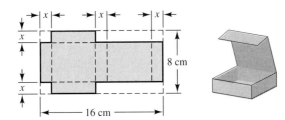

54. A beam of length L is fixed at one end and is simply supported at the other end. It carries a uniformly increasing load of weight W, as shown in the figure. The bending moment M at any distance x from the left end of this beam is given by

$$M(x) = \frac{Wx}{15L^2}(3L^2 - 5x^2).$$

(a) Determine the bending moment function when $W = 1500$ lb and $L = 10$ ft.

(b) Use the three-step procedure outlined in this section to sketch the graph of the polynomial function defined in part (a).

(c) Generate the graph of the function in part (a) on a graphing calculator. Then find the approximate distance x (from the left end of the beam) at which the bending moment is greatest.

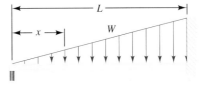

Critical Thinking

In Exercises 55–62,

(a) *sketch the possible shapes of the graph of a polynomial function with the given characteristics.*

(b) *use a graphing calculator to find a particular polynomial function whose graph has the shape given in part (a). Answers are not unique.*

55. Degree 3 with a negative leading coefficient
56. Degree 2 with a positive leading coefficient
57. Degree 4 with a positive leading coefficient
58. Degree 4 with a negative leading coefficient
59. Degree 5 with a positive leading coefficient
60. Degree 7 with a negative leading coefficient
61. Degree 6 with a negative leading coefficient
62. Degree 6 with a positive leading coefficient
63. By solving the equations simultaneously, determine the coordinates of the point(s) where the graphs of the equations intersect. Check graphically.
    (a) $y = x^3 - x^2$ and $y = 3x - 3$
    (b) $y = x^3 - 3x^2 - 4x$ and $x - y = 15$
64. Determine the maximum number of times the graphs of two polynomial functions P and Q may intersect if the degree of P is m and the degree of Q is n, provided that $m > n$? Explain.

In Exercises 65 and 66, sketch the graph of a polynomial function P with the given characteristics.

65. Degree 3; an odd function; $P(-2) = 0$; $P(4) = -12$
66. Degree 4; an even function; $P(0) = -27$; graph is tangent to the x-axis at $(3, 0)$
67. May a polynomial function of degree 5 be an even function? If so, give an example. If not, explain why.
68. Shown in the sketch are the graphs of several power functions $y = x^n$ with $n \geq 2$ and $0 \leq x \leq 1$. Which graph represents the function with the largest exponent n? Explain.

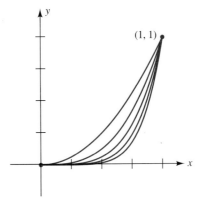

4.2 Polynomial Division

◆ Introductory Comments

In order to work with polynomial functions of degree 3 or greater, we must be able to divide one polynomial by another. The procedure that we use is similar to the one for dividing whole numbers. For example, to divide 349 by 15, we can use the long division scheme from arithmetic:

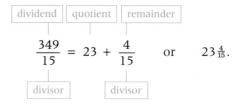

Since 23 is the quotient and 4 is the remainder, we write

$$\frac{349}{15} = 23 + \frac{4}{15} \quad \text{or} \quad 23\frac{4}{15}.$$

In a similar manner, we can divide one polynomial $P(x)$ by another polynomial $D(x)$, provided that $D(x) \neq 0$. This long division scheme, which is called the **division algorithm for polynomials**, yields a quotient polynomial $Q(x)$ and a remainder polynomial $R(x)$ whose degree is less than that of the divisor $D(x)$ such that

$$\frac{P(x)}{D(x)} = Q(x) + \frac{R(x)}{D(x)}$$

◆ Polynomial Long Division

We refer to the procedure of dividing one polynomial by another polynomial as *polynomial long division*.

EXAMPLE 1 Divide $P(x) = 2x^3 - 5x^2 + 3x - 4$ by $D(x) = x - 2$ using polynomial long division.

◆ **Solution** We begin by dividing the first term of the divisor, x, into the first term of the dividend, $2x^3$, to obtain the first term of the quotient:

SECTION 4.2 ◆ Polynomial Division

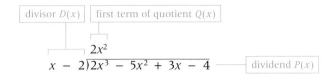

Next, we multiply the first term of the quotient by the divisor and *subtract* this product from the dividend to obtain the first remainder:

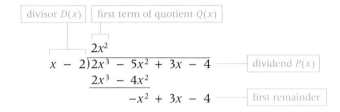

Now, using the first remainder as the *new* dividend, we continue this process until we obtain a remainder that is either zero or a polynomial whose degree is lower than that of the divisor:

$$\begin{array}{r}
2x^2 - x + 1 \\
x - 2 \overline{\smash{\big)}\, 2x^3 - 5x^2 + 3x - 4} \\
\underline{2x^3 - 4x^2} \\
-x^2 + 3x - 4 \\
\underline{-x^2 + 2x} \\
x - 4 \\
\underline{x - 2} \\
-2
\end{array}$$

Thus, we write

$$\frac{2x^3 - 5x^2 + 3x - 4}{x - 2} = 2x^2 - x + 1 + \frac{-2}{x - 2}. \quad \blacklozenge$$

To check polynomial long division, we use the same method of checking as for arithmetic division, and show that

$$P(x) = D(x)Q(x) + R(x).$$

PROBLEM 1 Check the polynomial long division in Example 1. ◆

Before beginning the polynomial long division process, we usually write the dividend and divisor in descending powers of x, using 0 as a coefficient of any term that is missing from either the dividend or divisor. This helps us align like terms in the same column when we carry out the division.

EXAMPLE 2 Divide $P(x) = 4x^4 + 3x^5 - 2x^2 - 3x - 2$ by $D(x) = x^2 - 1$ using polynomial long division.

◆ **Solution** We first arrange the dividend and divisor in descending powers of x, using 0 as a coefficient of any missing term. Then carry out the long division as follows:

$$
\begin{array}{r}
3x^3 + 4x^2 + 3x + 2 \\
x^2 + 0x - 1 \overline{)\,3x^5 + 4x^4 + 0x^3 - 2x^2 - 3x - 2\,} \\
\underline{3x^5 + 0x^4 - 3x^3} \\
4x^4 + 3x^3 - 2x^2 - 3x - 2 \\
\underline{4x^4 + 0x^3 - 4x^2} \\
3x^3 + 2x^2 - 3x - 2 \\
\underline{3x^3 + 0x^2 - 3x} \\
2x^2 + 0x - 2 \\
\underline{2x^2 + 0x - 2} \\
0
\end{array}
$$

where $D(x)$ is the divisor, $Q(x)$ is the quotient, $P(x)$ is the dividend, and $R(x)$ is the remainder.

Since the remainder is 0, we write

$$\frac{3x^5 + 4x^4 - 2x^2 - 3x - 2}{x^2 - 1} = 3x^3 + 4x^2 + 3x + 2.$$ ◆

PROBLEM 2 Divide $P(x) = 6x^3 - 3x^2$ by $D(x) = 3x^2 - 2$ using polynomial long division. ◆

◆ Synthetic Division

As we illustrated in Example 1, when a polynomial $P(x)$ of degree $n \geq 1$ is divided by a first-degree polynomial $D(x) = x - r$, the remainder $R(x)$ is a constant k and the degree of the quotient polynomial $Q(x)$ is one less than the degree of $P(x)$. That is,

$$\frac{P(x)}{x - r} = Q(x) + \frac{k}{x - r}.$$

A shortcut called *synthetic division* may be used to divide a polynomial of degree $n \geq 1$ by a first-degree polynomial $x - r$. In **synthetic division**, we use only the essential data from the long division process by eliminating any power of x or coefficient that is a duplicate of that directly above it (in the same column). Also, we replace the divisor $x - r$ by the *synthetic divisor* r so that we may *add* rather than subtract columns. In the next example, we rework Example 1 using synthetic division.

SECTION 4.2 ◆ Polynomial Division

EXAMPLE 3 Divide $P(x) = 2x^3 - 5x^2 + 3x - 4$ by $D(x) = x - 2$ using synthetic division.

◀ **Solution**   We begin by replacing the divisor $x - 2$ with the synthetic divisor 2 and writing the coefficients of the dividend:

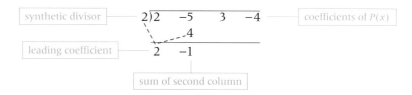

We bring down the leading coefficient two rows, multiply the synthetic divisor by the leading coefficient, and record this product in the middle row under the second coefficient of $P(x)$. Then we add to obtain the sum of the second column:

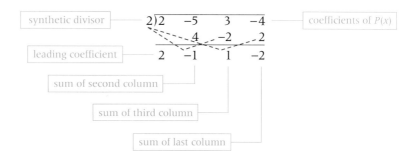

Now, we multiply the synthetic divisor by the sum of the second column; record this product in the middle row under the third coefficient of $P(x)$; then add to obtain the sum of the third column. We continue in this manner until all columns have a sum:

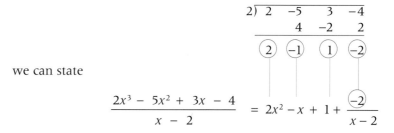

The sum of the last column is the constant remainder k. The preceding sums, to the left of the last sum, are the coefficients of the successive terms in the quotient polynomial $Q(x)$, whose degree is one less than that of $P(x)$. Thus, from the synthetic division

$$\begin{array}{r|rrrr} 2) & 2 & -5 & 3 & -4 \\ & & 4 & -2 & 2 \\ \hline & 2 & -1 & 1 & -2 \end{array}$$

we can state

$$\frac{2x^3 - 5x^2 + 3x - 4}{x - 2} = 2x^2 - x + 1 + \frac{-2}{x - 2}$$

This result agrees with the quotient and remainder polynomials we obtained in Example 1.   ◆

PROBLEM 3 Divide $P(x) = x^3 - 5x^2 + 7x - 3$ by $D(x) = x - 1$ using synthetic division. ◆

CAUTION The shortcut of synthetic division works only when the divisor $D(x)$ is a first-degree polynomial of the form $x - r$. For division problems such as Example 2, where the divisor $D(x)$ is quadratic, synthetic division does *not* apply. See Exercises 71–76 for the procedure to use when the divisor $D(x)$ is a first-degree polynomial of the form $ax + b$.

In synthetic division, we must record the coefficients of the dividend $P(x)$ in the order that they appear when $P(x)$ is written in descending powers of x, and if a power of x is missing, we must record 0 as its coefficient.

EXAMPLE 4 Divide $P(x) = 2x^4 - 15x^2 + 9x + 1$ by $D(x) = x + 3$ using synthetic division.

◆ **Solution** Since the divisor $D(x) = x + 3 = x - (-3)$, we use -3 as the synthetic divisor. Now, after inserting 0 for the coefficient of the x^3 term in the dividend, we carry out the synthetic division as follows:

```
synthetic divisor ─── -3)2    0   -15    9    1  ─── coefficients of P(x)
                         -6    18   -9    0
                      ─────────────────────────
                       2   -6    3    0   (1) ─── constant remainder k
                      ─────────────────
                      coefficients of Q(x), whose
                      degree is 1 less than P(x)
```

Thus,

$$\frac{2x^4 - 15x^2 + 9x + 1}{x + 3} = 2x^3 - 6x^2 + 3x + \frac{1}{x + 3}. \quad ◆$$

We can check the synthetic division process by showing that

$$P(x) = (x - r)Q(x) + k.$$

PROBLEM 4 Check the synthetic division in Example 4. ◆

◆ The Remainder Theorem

If $Q(x)$ is the quotient polynomial and k is the constant remainder obtained when a polynomial $P(x)$ is divided by a polynomial $x - r$, then we have

$$\frac{P(x)}{x - r} = Q(x) + \frac{k}{x - r}.$$

Multiplying both sides of this equation by $x - r$, we obtain

$$P(x) = (x - r)Q(x) + k.$$

Now, replacing x with r, we find

$$P(r) = (r - r)Q(r) + k$$
$$= 0 \cdot Q(r) + k$$
$$= k$$

In other words, $P(r)$ is the same number as the remainder k obtained by dividing $P(x)$ by $x - r$. We refer to this fact about polynomial division as the **remainder theorem**.

◀ **Remainder Theorem**

> When a polynomial $P(x)$ is divided by $x - r$, the remainder is $P(r)$.

EXAMPLE 5 Use the remainder theorem to determine the remainder when

$$P(x) = x^4 - 3x^3 - 4x^2 + 13x - 8 \quad \text{is divided by} \quad x - 3.$$

◆ **Solution** The remainder theorem tells us that when $P(x)$ is divided by $x - 3$, the remainder is $P(3)$. Evaluating $P(3)$, we find

$$P(3) = (3)^4 - 3(3)^3 - 4(3)^2 + 13(3) - 8$$
$$= 81 - 81 - 36 + 39 - 8$$
$$= -5$$

Thus, when $P(x)$ is divided by $x - 3$, the remainder is -5. ◆

PROBLEM 5 Verify the remainder in Example 5 by dividing $P(x)$ by $x - 3$ using synthetic division. ◆

As illustrated in the following example, we can evaluate $P(r)$ by using the remainder theorem and synthetically dividing $P(x)$ by $x - r$. This approach is often easier than finding $P(r)$ by direct substitution.

EXAMPLE 6 Given that $P(x) = 4x^4 - 6x^3 + 4x - 3$, find $P(-\frac{1}{2})$ by using synthetic division and the remainder theorem.

◆ **Solution** According to the remainder theorem, $P(-\frac{1}{2})$ is the remainder obtained when $P(x)$ is divided by $x - (-\frac{1}{2}) = x + \frac{1}{2}$. Using synthetic division, we find the remainder:

$$\begin{array}{r|rrrrr} -\frac{1}{2} & 4 & -6 & 0 & 4 & -3 \\ & & -2 & 4 & -2 & -1 \\ \hline & 4 & -8 & 4 & 2 & \boxed{-4} \end{array} \quad \text{remainder}$$

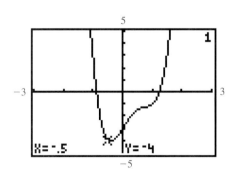

FIGURE 4.17
For the polynomial function $P(x) = 4x^4 - 6x^3 + 4x - 3$, the graph confirms that $P(-\frac{1}{2}) = -4$ is a functional value.

Thus, $P(-\frac{1}{2}) = -4$. Figure 4.17 shows the graph of this polynomial function in the viewing rectangle of a graphing calculator. Note that $(-\frac{1}{2}, -4)$ is a point on this graph, which confirms our answer. ◆

PROBLEM 6 Given that $P(x) = 4x^4 - 6x^3 + 4x - 3$, find $P(\frac{1}{2})$ by using synthetic division and the remainder theorem. Check graphically, and compare your result with Figure 4.17.

If $P(x)$ is a polynomial and r is a complex number (see Section 1.4 for a discussion of complex numbers) such that $P(r) = 0$, we say that r is a **zero** of the polynomial function P or a **root** of the polynomial equation $P(x) = 0$. Also, recall from Section 2.4 that any *real* zero of a function P represents an x-intercept of the graph of P.

EXAMPLE 7 Given the function P defined by $P(x) = 3x^3 - 8x^2 + 10x - 4$, use synthetic division and the remainder theorem to determine whether each number is a zero of P:

(a) $\frac{2}{3}$ (b) $1 + i$

◆ Solution

(a) If $\frac{2}{3}$ is a zero of the function P, then $P(\frac{2}{3}) = 0$. Using synthetic division, we find

$$\frac{2}{3} \overline{)\begin{array}{cccc} 3 & -8 & 10 & -4 \\ & 2 & -4 & 4 \\ \hline 3 & -6 & 6 & 0 \end{array}} \text{ remainder}$$

Since the remainder is 0, we know that $P(\frac{2}{3}) = 0$, and so $\frac{2}{3}$ is a zero of the polynomial function P defined by $P(x) = 3x^3 - 8x^2 + 10x - 4$. Figure 4.18 shows the graph of P in the viewing rectangle of a graphing calculator. Note that $\frac{2}{3}$ is an x-intercept of this graph.

(b) If $1 + i$ is a zero of the function P, then $P(1 + i) = 0$. Using synthetic division and recalling from Section 1.4 that $i^2 = -1$, we find

$$1 + i \overline{)\begin{array}{cccc} 3 & -8 & 10 & -4 \\ & 3 + 3i & -8 - 2i & 4 \\ \hline 3 & -5 + 3i & 2 - 2i & 0 \end{array}} \text{ remainder}$$

Since the remainder is 0, we know that $P(1 + i) = 0$, and so $1 + i$ is a zero of the polynomial function P defined by $P(x) = 3x^3 - 8x^2 + 10x - 4$. Although the imaginary number $1 + i$ is a zero of P, it is not an x-intercept of the graph of P. Remember, only real zeros of a function P represent x-intercepts of the graph of P. ◆

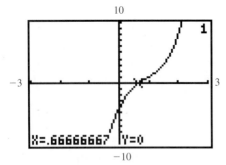

FIGURE 4.18

If $\frac{2}{3}$ is a real zero of $P(x) = 3x^3 - 8x^2 + 10x - 4$, then $\frac{2}{3}$ is an x-intercept of the graph of P.

PROBLEM 7 Use synthetic division and the remainder theorem to determine if $1 - i$ is a zero of the polynomial function P defined in Example 7. ◆

Exercises 4.2

Basic Skills

In Exercises 1–16, divide P(x) by D(x) using polynomial long division.

1. $P(x) = 3x^3 - 2x^2 + x - 3$
   $D(x) = x - 1$

2. $P(x) = 2x^4 + 3x^3 - 4x^2 - 2x + 1$
   $D(x) = x + 2$

3. $P(x) = 3x^4 - 10x^3 + 16x^2 - 14x + 3$
   $D(x) = x^2 - 2x + 3$

4. $P(x) = 2x^3 - 7x^2 + 11x - 4$
   $D(x) = x^2 - 3x + 4$

5. $P(x) = 4x^3 + 3x - 3$
   $D(x) = 2x - 1$

6. $P(x) = 9x^4 + 6x^3 + 3x - 2$
   $D(x) = 3x + 2$

7. $P(x) = 2x^5 + 2x^3 - 2x^2 + 5$
   $D(x) = 2x^3 + 1$

8. $P(x) = x^6 - 2x^2 - 3x + 4$
   $D(x) = x^3 - 3x$

9. $P(x) = x^3$
   $D(x) = x + 1$

10. $P(x) = x^4$
    $D(x) = x^2 + 1$

11. $P(x) = 6x^5 - x^3 + 2x^2 - x - 1$
    $D(x) = 1 - 2x^2$

12. $P(x) = 3x^4 + 4x^3 - 15x^2 - 4x - 12$
    $D(x) = 6 - x - x^2$

13. $P(x) = 3 - 6x$
    $D(x) = 2x + 1$

14. $P(x) = 8 - 2x - x^2$
    $D(x) = x^2 + 1$

15. $P(x) = 8x^3$
    $D(x) = (2x - 1)^2$

16. $P(x) = (x + 1)^4$
    $D(x) = x^2 + 1$

In Exercises 17–34, divide P(x) by D(x) using synthetic division. In each case, verify the remainder by using the remainder theorem.

17. $P(x) = 2x^2 - 3x + 4$
    $D(x) = x - 1$

18. $P(x) = 3x^2 - 6x - 5$
    $D(x) = x + 3$

19. $P(x) = x^3 + 5x^2 - 2x + 4$
    $D(x) = x + 2$

20. $P(x) = x^4 - 6x^3 + 4x^2 + 15x + 2$
    $D(x) = x - 4$

21. $P(x) = 2x^4 - 7x^3 - 3x^2 - 16$
    $D(x) = x - 4$

22. $P(x) = 3x^3 - 70x + 25$
    $D(x) = x + 5$

23. $P(x) = 4x^4 + 25x^3 - 32x$
    $D(x) = x + 6$

24. $P(x) = x^5 - 6x^3 + 7x^2 - 3$
    $D(x) = x - 2$

25. $P(x) = 3 - x^5$
    $D(x) = x - 2$

26. $P(x) = 7 + x - 2x^3$
    $D(x) = x + 1$

27. $P(x) = 3x^3 - 8x^2 + 19x - 10$
    $D(x) = x - \frac{2}{3}$

28. $P(x) = 4x^3 - 8x^2 + x + 3$
    $D(x) = x + \frac{1}{2}$

29. $P(x) = 2x^4 + x^3 - 19x + 8$
    $D(x) = x + \frac{5}{2}$

30. $P(x) = 4x^4 + 3x^2 - 27$
    $D(x) = x - \frac{3}{2}$

31. $P(x) = 14.5x^3 + 266.4x^2 - 828.1$
    $D(x) = x + 18.2$

32. $P(x) = 2.5x^3 - 6.0x^2 - 22.3x + 41.4$
    $D(x) = x - 3.6$

33. $P(x) = 275x^4 - 126x^2 - 627x - 1350$
    $D(x) = x - 1.8$

34. $P(x) = 125x^4 + 2680x^3 - 1625x^2 - 1824x + 1012$
    $D(x) = x + 22$

In Exercises 35–50, use synthetic division and the remainder theorem to find the indicated functional value. Check graphically whenever possible.

35. $P(x) = 2x^3 - 3x^2 + 4x + 5$; $P(1)$
36. $Q(x) = 3x^3 - 4x^2 + x - 9$; $Q(-1)$
37. $F(x) = x^4 + 2x^3 - 5x - 3$; $F(-2)$
38. $g(x) = x^3 - 2x^2 - 5x + 6$; $g(3)$
39. $P(x) = 2x^3 - 7x^2 + 2x + 1$; $P(4)$
40. $Q(x) = -x^3 + 4x^2 + 12x + 2$; $Q(6)$
41. $Q(x) = 4x^3 + 2x^2 - 1$; $Q(\frac{1}{2})$
42. $P(x) = 3x^4 - x^3 + x - 1$; $P(-\frac{2}{3})$
43. $f(x) = 3x^4 - 5x^3 + 7x^2 + 9x + 2$; $f(-\frac{1}{3})$
44. $H(x) = 5x^4 - 7x^2 + 12x - 4$; $H(\frac{2}{5})$
45. $h(x) = 2x^3 - 2x^2 + 4x + 7$; $h(1 - i)$

46. $G(x) = 2x^4 + x^3 - 8x^2 - 1;\quad G(-2i)$
47. $P(x) = x^4 - x^3 + 3x^2 + 2x + 2;\quad P(\sqrt{2})$
48. $P(x) = 3x^3 - 21x - 4;\quad P(2 + \sqrt{3})$
49. $P(x) = 2.6x^3 - 9.8x^2 + 3.7x + 4.2;\quad P(1.7)$
50. $P(x) = 0.035x^4 + 1.025x^3 - 2134;\quad P(-32.2)$

In Exercises 51–58, use synthetic division to determine whether the given value of x is a zero of the polynomial function P. Check graphically whenever possible.

51. $P(x) = 2x^3 + x^2 - 14x - 3;\quad x = -3$
52. $P(x) = 4x^3 + 2x^2 - 11x - 18;\quad x = 2$
53. $P(x) = -2x^4 + 3x^3 + 8x^2 - 18;\quad x = \frac{3}{2}$
54. $P(x) = -6x^4 + x^3 + 2;\quad x = \frac{2}{3}$
55. $P(x) = x^4 - 3x^3 + 13x^2 - 27x + 36;\quad x = 3i$
56. $P(x) = 4x^3 - 17x^2 + 24x - 5;\quad x = 2 - i$
57. $P(x) = 3x^3 - 10x^2 + 5x + 4;\quad x = 1 - \sqrt{2}$
58. $P(x) = x^4 - 2x^3 + x^2 + 6x - 12;\quad x = \sqrt{3}$

In Exercises 59–66, use synthetic division to determine whether the given value of x is a root of the polynomial equation. Check graphically whenever possible.

59. $3x^4 - 5x^3 - 12x^2 + 19x + 2 = 0;\quad x = -2$
60. $-6x^3 + 2x^2 + 5x + 21 = 0;\quad x = 3$
61. $6x^3 + 2x^2 + 3x + 1 = 0;\quad x = -\frac{1}{3}$
62. $25x^5 + x^3 - 2x^2 + 5x + 2 = 0;\quad x = \frac{2}{5}$
63. $2x^3 - 3x^2 + 8x - 12 = 0;\quad x = -2i$
64. $x^4 + 2x^3 + x^2 - 2x - 2 = 0;\quad x = -1 + i$
65. $3x^4 - 2x^3 - 8x^2 + 4x + 4 = 0;\quad x = \sqrt{2}$
66. $x^4 - 2x^3 - 6x^2 + 8x + 8 = 0;\quad x = 1 + \sqrt{3}$

67. When a Patriot missile is fired from its launchpad, the height h (in meters) of the missile above the ground is a function of the horizontal distance x (in kilometers) from the launchpad and is given by

$$h(x) = 224x + 35x^2 - 24x^3 \quad \text{for } 0 \leq x \leq 4$$

Use synthetic division and the remainder theorem to find $h(2.45)$. Round your answer to three significant digits.

68. The current i (in milliamperes) flowing through an electric circuit is a function of time t (in seconds) and is given by

$$i(t) = 545 + 8.2t^2 - 1.2t^4 \quad \text{for } 0 \leq t \leq 5$$

Use synthetic division and the remainder theorem to find $i(3.8)$. Round your answer to three significant digits.

Critical Thinking

69. When the polynomial $P(x)$ is divided by $x + 6$, the quotient is $x^3 + x^2 + x + 5$ and the remainder is -12. Find $P(x)$.

70. When $P(x) = 2x^3 - 5x^2 + 8x - 11$ is divided by the polynomial $D(x)$, the quotient is $x^2 - 2x + 3$ and the remainder is -8. Find $D(x)$.

Synthetic division may be used to divide a polynomial $P(x)$ of degree $n \geq 1$ by a first-degree polynomial $D(x) = ax + b$. Since

$$\frac{P(x)}{ax + b} = \frac{P(x)}{a\left(x + \frac{b}{a}\right)} = \frac{\frac{1}{a}P(x)}{x + \frac{b}{a}},$$

we can obtain the quotient and remainder by using a synthetic divisor $-b/a$ with dividend $(1/a)P(x)$. In Exercises 71–76, divide $P(x)$ by $D(x)$ using synthetic division.

71. $P(x) = 6x^3 + 7x^2 - 18x + 16$
    $D(x) = 3x + 8$
72. $P(x) = 8x^4 + 2x^3 - 3x^2 + 2x - 1$
    $D(x) = 2x - 1$
73. $P(x) = 32x^4 - 4x^3 + 7x^2 - 10x + 2$
    $D(x) = 4x - 1$
74. $P(x) = 3x^3 + x^2 + 2x + 8$
    $D(x) = 3x + 4$
75. $P(x) = x^4 - 2x^3 + 2x^2 - 7x + 6$
    $D(x) = 2 - x$
76. $P(x) = 24 - 20x - 12x^2 + 28x^3 - 15x^4$
    $D(x) = 6 - 5x$

In Exercises 77 and 78, use synthetic division to determine a value of k such that -2 is a zero of the given polynomial function. Check graphically.

77. $P(x) = x^4 + 2x^3 + 6x^2 + kx + 8$
78. $P(x) = kx^3 + 3x^2 + 6x - 16$

In Exercises 79 and 80, use synthetic division to determine a value of k such that 4 is a root of the given polynomial equation. Check graphically.

79. $kx^3 - 10x^2 + kx + 24 = 0$
80. $2x^3 + k^2x^2 - 2x + 88k = 0$

4.3 Factors and Zeros of Polynomial Functions

◆ Introductory Comments

Finding the zeros of a polynomial function of degree 2 is simple: apply the quadratic formula. Finding the zeros of a polynomial function of degree 3 or greater is usually much more difficult, but mathematicians have developed formulas for this purpose. Unlike the quadratic formula, however, formulas for finding the zeros of a polynomial function of degree 3 or 4 are extremely complicated. Mathematicians have also shown that no such general formulas exist for finding the zeros of polynomial functions of degree 5 or greater.

In this section, we discuss some theorems that we can use to help find the zeros of a polynomial function. We begin by proving the factor theorem, which shows a relationship between the zeros and factors of a polynomial function.

◆ The Factor Theorem

If $Q(x)$ is the quotient polynomial and k the constant remainder obtained when a polynomial $P(x)$ is divided by $x - r$, then we have

$$\frac{P(x)}{x - r} = Q(x) + \frac{k}{x - r}.$$

Multiplying both sides of this equation by $x - r$, we obtain

$$P(x) = (x - r)Q(x) + k.$$

Now by the remainder theorem (Section 4.2), we know that $P(r) = k$. So we can also write

$$P(x) = (x - r)Q(x) + P(r).$$

However, if the remainder $P(r) = 0$, then

$$P(x) = (x - r)Q(x)$$

and we conclude that $(x - r)$ is a factor of $P(x)$. Conversely, if $x - r$ is a factor of $P(x)$, then

$$P(x) = (x - r)Q(x)$$

and it follows that

$$\begin{aligned} P(r) &= (r - r)Q(r) \\ &= 0 \cdot Q(r) \\ &= 0. \end{aligned}$$

We refer to this consequence of the remainder theorem as the **factor theorem**.

◀ **Factor Theorem**

> A first-degree polynomial $x - r$ is a factor of a polynomial $P(x)$ if and only if $P(r) = 0$ or, equivalently, if and only if r is a zero of P.

It is important to recognize that the concepts of factors, roots, and zeros are closely related. In general, if P is a polynomial function and r is a complex number, then the following statements are equivalent:

1. $x - r$ is a factor of the polynomial $P(x)$.
2. r is a zero of the function P.
3. r is a root of the equation $P(x) = 0$.

Also, if r is a real number, then

4. r is an x-intercept of the graph of P.

EXAMPLE 1 Given that $P(x) = 2x^3 - 3x^2 + 8x - 12$, determine whether each expression is a factor of $P(x)$.

(a) $x - 2$ (b) $x - \frac{3}{2}$ (c) $x + 2i$

◀ **Solution**

(a) According to the factor theorem, $x - 2$ is a factor of $P(x)$ if $P(2) = 0$. To determine $P(2)$, we can use either direct substitution or synthetic division. Choosing synthetic division, we find

$$\begin{array}{r|rrrr} 2 & 2 & -3 & 8 & -12 \\ & & 4 & 2 & 20 \\ \hline & 2 & 1 & 10 & \boxed{8} \end{array} \longrightarrow P(2)$$

Since $P(2) = 8 \neq 0$, we conclude that $x - 2$ is not a factor of $P(x)$.

(b) According to the factor theorem, $x - \frac{3}{2}$ is a factor of $P(x)$ if $P(\frac{3}{2}) = 0$. To determine $P(\frac{3}{2})$, we can use either direct substitution or synthetic division. Choosing synthetic division, we find

$$\begin{array}{r|rrrr} \frac{3}{2} & 2 & -3 & 8 & -12 \\ & & 3 & 0 & 12 \\ \hline & 2 & 0 & 8 & \boxed{0} \end{array} \longrightarrow P(\tfrac{3}{2})$$

Since $P(\frac{3}{2}) = 0$, we conclude that $x - \frac{3}{2}$ is a factor of $P(x)$. The graph in Figure 4.19 also illustrates this fact.

(c) According to the factor theorem, $x + 2i$ is a factor of $P(x)$ if $P(-2i) = 0$. To determine $P(-2i)$, we can use direct substitution or

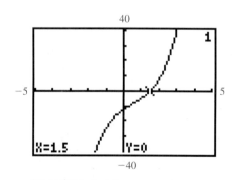

FIGURE 4.19
Since $\frac{3}{2}$ is an x-intercept of the graph of $P(x) = 2x^3 - 3x^2 + 8x - 12$, we know that $x - \frac{3}{2}$ is a factor of $P(x)$.

synthetic division. Choosing synthetic division and recalling from Section 1.4 that $i^2 = -1$, we find

$$
\begin{array}{r|rrrr}
-2i & 2 & -3 & 8 & -12 \\
 & & -4i & 6i-8 & 12 \\
\hline
 & 2 & -3-4i & 6i & 0 \quad \leftarrow P(-2i)
\end{array}
$$

Since $P(-2i) = 0$, we conclude that $x + 2i$ is a factor of $P(x)$. ◆

PROBLEM 1 Determine whether $x - 2i$ is a factor of $P(x) = 2x^3 - 3x^2 + 8x - 12$. ◆

◆ Fundamental Theorem of Algebra

A linear function $f(x) = ax + b$, with $a \neq 0$, has just one zero, $-b/a$. A quadratic function $g(x) = ax^2 + bx + c$, with $a \neq 0$, has two zeros:

$$\frac{-b + \sqrt{b^2 - 4ac}}{2a} \quad \text{and} \quad \frac{-b - \sqrt{b^2 - 4ac}}{2a}$$

How many zeros for a cubic function? a quartic function? a quintic function? The **fundamental theorem of algebra** begins to answer these questions.

◀ **Fundamental Theorem of Algebra**

> Every polynomial function P of degree $n \geq 1$ has at least one zero in the set of complex numbers.

The proof of this theorem was provided by Carl Friedrich Gauss in 1799, but is beyond the scope of this book. Thus, we shall simply accept its validity and use its results.

Consider the polynomial function P of degree $n \geq 1$ defined by

$$P(x) = a_n x^n + a_{n-1} x^{n-1} + \cdots + a_1 x + a_0 \quad \text{for } a_n \neq 0.$$

According to the fundamental theorem of algebra, the function P has at least one zero. If this zero is r_1, then, by the factor theorem, $x - r_1$ is a factor of the polynomial $P(x)$. Thus, we can write

$$P(x) = (x - r_1) Q_1(x),$$

where the polynomial $Q_1(x)$ is of degree $n - 1$. Now if $n - 1 \geq 1$, then, by the fundamental theorem of algebra, the function Q_1 has at least one zero. If this zero is r_2, then, by the factor theorem, $x - r_2$ is a factor of the polynomial $Q_1(x)$ and we have

$$Q_1(x) = (x - r_2) Q_2(x),$$

where the polynomial $Q_2(x)$ is of degree $n - 2$. Thus, using substitution, we obtain

$$P(x) = (x - r_1)(x - r_2)Q_2(x).$$

If $n - 2 \geq 1$, we continue this procedure until we obtain a quotient polynomial $Q_n(x)$ of degree 0, that is, until $Q_n(x)$ is the constant a_n. We can then write

$$P(x) = \underbrace{(x - r_1)(x - r_2)\cdots(x - r_n)}_{n \text{ linear factors}} a_n,$$

and conclude that P has exactly n linear factors. Hence, P has exactly n zeros, namely $r_1, r_2, \ldots, r_n$.

◀ **Extension of the Fundamental Theorem**

> Every polynomial function P of degree $n \geq 1$ can be expressed as the product of n linear factors and, consequently, P has exactly n zeros in the set of complex numbers.

Thus, a cubic function can be expressed as the product of three linear factors, and it has three zeros; a quartic function can be expressed as the product of four linear factors, and it has four zeros; a quintic function can be expressed as the product of five linear factors, and it has five zeros; and so on.

The n zeros of a polynomial function of degree $n \geq 2$ are not necessarily distinct. For example, consider the polynomial function P defined by

$$P(x) = 2x^5 - 4x^4 + 2x^3.$$

Using the factoring techniques from Section 1.6, we can write $P(x)$ as the product of five linear factors:

$$\begin{aligned}
P(x) = 2x^5 - 4x^4 + 2x^3 &= 2x^3(x^2 - 2x + 1) &&\textbf{Factor out } 2x^3 \\
&= 2x^3(x - 1)^2 &&\textbf{Factor the trinomial} \\
&= 2\underbrace{(x - 0)^3(x - 1)^2}_{\text{five linear factors}} &&\textbf{Rewrite}
\end{aligned}$$

Since the linear factor $(x - 0)$ appears three times and the linear factor $(x - 1)$ appears twice, we say that 0 is a zero of *multiplicity three* and 1 is a zero of *multiplicity two*. In general, if a linear factor $(x - r)$ appears k times, then r is a zero of multiplicity k.

Figure 4.20 shows the graph of

$$P(x) = 2x^5 - 4x^4 + 2x^3 = 2(x - 0)^3(x - 1)^2$$

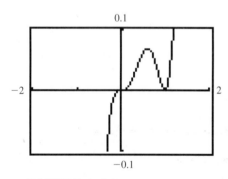

FIGURE 4.20

The graph of $P(x) = 2x^5 - 4x^4 + 2x^3$ is tangent to the x-axis at $x = 0$ and $x = 1$.

in the viewing rectangle of a graphing calculator. Note the appearance of the graph near the zeros of the function. In general, if r is a zero of multiplicity k, provided $k \geq 2$ for a polynomial function P, then the graph of P is *tangent* to the x-axis at $x = r$. If k is even, then the tangency looks similar to either one of the sketches shown in Figure 4.21, that is, the graph "bounces off" the x-axis. If k is odd, then the tangency looks similar to either one of the sketches shown in Figure 4.22, that is, the graph goes through the x-axis in an S-turn.

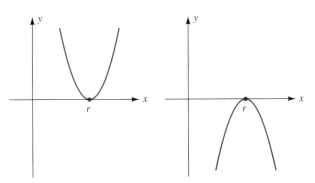

FIGURE 4.21

Appearance of tangency when r is a zero of multiplicity k and k is even

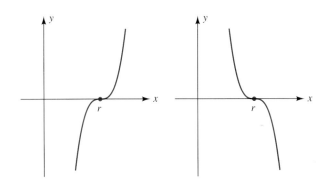

FIGURE 4.22

Appearance of tangency when r is a zero of multiplicity k and k is odd

If we know the n zeros of a polynomial function P of degree n and the coordinates of just one other point on its graph, then we can find the equation that defines the polynomial function. This procedure is illustrated in the next example.

EXAMPLE 2 Find the equation of a polynomial function P of degree 4 if the graph passes through the point $(2, 27)$ and the zeros of P are 1 of multiplicity two and -1 of multiplicity two.

◆ Solution By the factor theorem, since the zeros of P are 1 of multiplicity two and -1 of multiplicity two, the general form of the polynomial function is

$$P(x) = a(x - 1)^2(x + 1)^2.$$

To determine the value of a, we use the fact that the graph passes through the point $(2, 27)$:

$$P(x) = a(x - 1)^2(x + 1)^2$$

$$27 = a(2 - 1)^2(2 + 1)^2 \quad \text{Substitute } x = 2 \text{ and } P(x) = 27$$

$$27 = 9a \quad \text{Solve for } a$$

$$a = 3$$

Hence, the polynomial function with the given characteristics is

$$P(x) = 3(x - 1)^2(x + 1)^2$$
$$= 3[(x - 1)(x + 1)]^2$$
$$= 3(x^2 - 1)^2$$
$$= 3(x^4 - 2x^2 + 1)$$
$$= 3x^4 - 6x^2 + 3$$

Figure 4.23 shows the graph of $P(x) = 3x^4 - 6x^2 + 3$ in the viewing rectangle of a graphing calculator. Note that the graph satisfies the given conditions. ◆

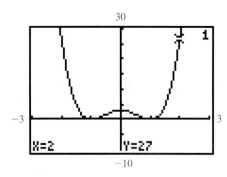

FIGURE 4.23
The graph of $P(x) = 3x^4 - 6x^2 + 3$ is tangent to the x-axis at $x = \pm 1$ and passes through the point (2, 27).

PROBLEM 2 Find the equation of a polynomial function P of degree 4 if its graph passes through the point (1, 8) and the zeros of P are 0 of multiplicity three and -3. ◆

◆ Rational Zero Theorem

Unfortunately, the zeros of a polynomial function are not always as easily attainable as those found for

$$P(x) = 2x^5 - 4x^4 + 2x^3 = 2(x - 0)^3(x - 1)^2$$

because not every polynomial can be factored by the methods discussed in Section 1.6. For example, consider finding the zeros of the polynomial function P defined by

$$P(x) = 2x^3 - 11x^2 + 13x - 4.$$

We can't factor out any common term, and factoring by grouping terms doesn't work for this polynomial. Then how do we factor this polynomial? To help us find some of the linear factors of a polynomial function, we use the **rational zero theorem**.

◀ **Rational Zero Theorem**

Let
$$P(x) = a_n x^n + a_{n-1} x^{n-1} + \cdots + a_1 x + a_0$$

be a polynomial function with integer coefficients and with $a_n \neq 0$ and $a_0 \neq 0$. If the rational number b/c, in lowest terms, is a zero of P, then b must be a factor of the constant term a_0 and c must be a factor of the coefficient of the highest-degree term a_n.

CAUTION The rational zero theorem does not say that a polynomial function with integer coefficients *has* rational zeros. It simply enables us to develop a list of all *possible* rational zeros of a polynomial function P with integer coefficients. By using direct substitution or synthetic division, we can then determine if any of the rational numbers in this list *are* zeros of P.

In the next example, we illustrate that if we can express a polynomial function in terms of linear and quadratic factors, we can then find *all* zeros of the function.

EXAMPLE 3 Find all zeros of $P(x) = 2x^3 - 11x^2 + 13x - 4$.

Solution For $P(x) = 2x^3 - 11x^2 + 13x - 4$, the constant term is -4 and the coefficient of the highest-degree term is 2. Now the factors of -4 are ± 1, ± 2, and ± 4, and the factors of 2 are ± 1 and ± 2. Thus, according to the rational zero theorem, the possible rational zeros of P are

$$\frac{\pm 1}{\pm 1}, \quad \frac{\pm 2}{\pm 1}, \quad \frac{\pm 4}{\pm 1} \quad \text{and} \quad \frac{\pm 1}{\pm 2}, \quad \frac{\pm 2}{\pm 2}, \quad \frac{\pm 4}{\pm 2},$$

which simplify to ± 1, ± 2, ± 4, and $\pm \frac{1}{2}$. Using this list of numbers and the graph of P in Figure 4.24, we note that 4, 1, and $\frac{1}{2}$ may be rational zeros of P. Testing 4 using synthetic division, we find

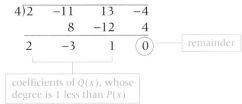

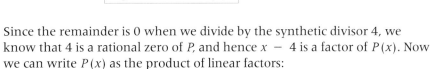

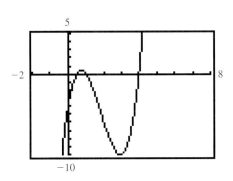

FIGURE 4.24

The polynomial function $P(x) = 2x^3 - 11x^2 + 13x - 4$ appears to have rational zeros of $x = 4, 1,$ and $\frac{1}{2}$.

Since the remainder is 0 when we divide by the synthetic divisor 4, we know that 4 is a rational zero of P, and hence $x - 4$ is a factor of $P(x)$. Now we can write $P(x)$ as the product of linear factors:

$$P(x) = (x - 4)Q(x) = (x - 4)(2x^2 - 3x + 1)$$
$$= (x - 4)(x - 1)(2x - 1)$$

Hence, by the factor theorem, the three zeros of P are 4, 1, and $\frac{1}{2}$. ◆

PROBLEM 3 Find all zeros of $P(x) = 3x^3 + 22x^2 + 25x + 6$. ◆

We can find repeated rational zeros of a polynomial function by inspecting the tangency of the graph to the x-axis.

EXAMPLE 4 Find all zeros of $P(x) = x^4 - 8x^3 + 26x^2 - 48x + 45$.

Solution For $P(x) = x^4 - 8x^3 + 26x^2 - 48x + 45$, the constant term is 45 and the coefficient of the highest-degree term is 1. When the coefficient of the highest-degree term is 1, the possible rational zeros of P are simply the factors of the constant term:

$$\pm 1, \quad \pm 3, \quad \pm 5, \quad \pm 9, \quad \pm 15, \quad \text{and} \quad \pm 45.$$

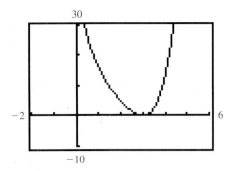

FIGURE 4.25
The polynomial function
$P(x) = x^4 - 8x^3 + 26x^2 - 48x + 45$
appears to have a rational zero of multiplicity two at $x = 3$.

Using this list of numbers and the graph of P in Figure 4.25, we note that 3 may be a rational zero of even multiplicity. Testing 3 twice using synthetic division, we find the coefficients of $Q(x)$:

$$
\begin{array}{r|rrrrr}
3 & 1 & -8 & 26 & -48 & 45 \\
  &   & 3 & -15 & 33 & -45 \\ \hline
3 & 1 & -5 & 11 & -15 & \boxed{0} \quad \text{— remainder}\\
  &   & 3 & -6 & 15 & \\ \hline
  & 1 & -2 & 5 & \boxed{0} & \quad \text{— remainder}
\end{array}
$$

coefficients of $Q(x)$ whose degree is 2 less than $P(x)$

Since the remainder is 0 each time we divide by the synthetic divisor 3, we know that 3 is a rational zero of multiplicity two, and hence $(x - 3)^2$ is a factor of $P(x)$. Therefore,

$$P(x) = (x - 3)^2 Q(x) = (x - 3)^2(x^2 - 2x + 5).$$

We can find the zeros of the quadratic function $Q(x) = x^2 - 2x + 5$ by applying the quadratic formula:

$$x = \frac{-(-2) \pm \sqrt{(-2)^2 - 4(1)(5)}}{2(1)} = \frac{2 \pm \sqrt{-16}}{2} = \frac{2 \pm 4i}{2} = 1 \pm 2i.$$

Thus, the four zeros of the function P are 3 of multiplicity two, $1 + 2i$, and $1 - 2i$. ◆

Using the factor theorem, we can express the polynomial function in Example 4 as the product of four linear factors:

$$P(x) = (x - 3)(x - 3)[x - (1 + 2i)][x - (1 - 2i)].$$

PROBLEM 4 Find the product $(x - 3)^2[x - (1 + 2i)][x - (1 - 2i)]$. You should obtain the original form of the polynomial given in Example 4. ◆

As illustrated in the next example, we can use the rational zero theorem to factor completely a polynomial over the set of integers.

EXAMPLE 5 Factor completely
$P(x) = 2x^5 + x^4 + x^3 + 13x^2 + 5x - 6$ over the set of integers.

◆ **Solution** For $P(x) = 2x^5 + x^4 + x^3 + 13x^2 + 5x - 6$, the constant term is -6 and the coefficient of the highest-degree term is 2. The factors of -6 are ± 1, ± 2, ± 3, and ± 6, and the factors of 2 are ± 1, ± 2. According to the rational zero theorem, the possible rational zeros of P are

$$\frac{\pm 1}{\pm 1}, \frac{\pm 2}{\pm 1}, \frac{\pm 3}{\pm 1}, \frac{\pm 6}{\pm 1} \quad \text{and} \quad \frac{\pm 1}{\pm 2}, \frac{\pm 2}{\pm 2}, \frac{\pm 3}{\pm 2}, \frac{\pm 6}{\pm 2},$$

SECTION 4.3 ♦ Factors and Zeros of Polynomial Functions

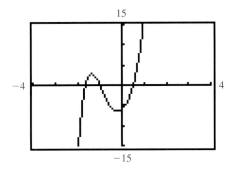

FIGURE 4.26

The polynomial function $P(x) = 2x^5 + x^4 + x^3 + 13x^2 + 5x - 6$ might have rational zeros at $x = -\frac{3}{2}, -1,$ and $\frac{1}{2}$.

which simplify to $\pm 1, \pm 2, \pm 3, \pm 6, \pm \frac{1}{2},$ and $\pm \frac{3}{2}$. Using this list of numbers and the graph of P in Figure 4.26, we note that $-\frac{3}{2}, -1,$ and $\frac{1}{2}$ may be rational zeros of P. Testing $-\frac{3}{2}$ using synthetic division, we find

$$-\tfrac{3}{2}\overline{)\,\begin{array}{rrrrrr} 2 & 1 & 1 & 13 & 5 & -6 \\ & -3 & 3 & -6 & -\tfrac{21}{2} & \tfrac{33}{4} \\ \hline 2 & -2 & 4 & 7 & -\tfrac{11}{2} & \boxed{\tfrac{9}{4}} \end{array}} \quad \text{remainder}$$

Since the remainder is *not* 0 when we divide by the synthetic divisor $-\frac{3}{2}$, we know that $-\frac{3}{2}$ is *not* a rational zero of P. Since the real zero between -2 and -1 isn't a rational number, we conclude that it must be an irrational number.

We now test our other possible rational zeros, $-1,$ and $\frac{1}{2}$. The order in which we test these numbers is immaterial:

$$\begin{array}{r} -1\overline{)\,\begin{array}{rrrrrr} 2 & 1 & 1 & 13 & 5 & -6 \\ & -2 & 1 & -2 & -11 & 6 \\ \end{array}} \\ \tfrac{1}{2}\overline{)\,\begin{array}{rrrrrr} 2 & -1 & 2 & 11 & -6 & \boxed{0} \\ & 1 & 0 & 1 & 6 & \\ \hline 2 & 0 & 2 & 12 & \boxed{0} & \end{array}} \end{array}$$

coefficients of $Q(x)$ whose degree is 2 less than $P(x)$

Since the remainder is 0 when we divide by the synthetic divisors -1 and $\frac{1}{2}$, we conclude that -1 and $\frac{1}{2}$ are both rational zeros of P. Therefore,

$$P(x) = (x+1)(x-\tfrac{1}{2})Q(x)$$
$$= (x+1)(x-\tfrac{1}{2})(2x^3 + 2x + 12).$$

Thus, we can factor completely the polynomial P over the set of integers:

$$P(x) = (x+1)(2x-1)(x^3 + x + 6) \qquad \blacklozenge$$

We know that the polynomial function in Example 5 has an irrational zero in the interval $(-2, -1)$ and two rational zeros, -1 and $\frac{1}{2}$. Since the graph of this function intersects the x-axis only three times, the other two zeros of this fifth-degree polynomial function must be imaginary numbers. Although we cannot find these imaginary zeros by the methods discussed in this section, we can estimate the irrational zero by using the TRACE key and ZOOM key on a graphing calculator or the calculator's built-in program for finding an x-intercept.

PROBLEM 5 Use a graphing calculator to estimate (to the nearest thousandth) the irrational zero in the interval $(-2, -1)$ for the polynomial P defined in Example 5. ♦

◆ Conjugate Pair Theorem

In Example 4, we found that two zeros of the polynomial function

$$P(x) = x^4 - 8x^3 + 26x^2 - 48x + 45$$

are the imaginary numbers $1 + 2i$ and $1 - 2i$. Together, these imaginary numbers are called a *conjugate pair*. If a polynomial function with real coefficients has imaginary zeros, then these zeros always occur in conjugate pairs. We refer to this fact as the **conjugate pair theorem**.

◆ **Conjugate Pair Theorem**

> If P is a polynomial function of degree $n \geq 1$ with real coefficients, and if the imaginary number $a + bi$ is a zero of P, then its conjugate $a - bi$ is also a zero of P.

We can state two consequences of the conjugate pair theorem:

1. A polynomial function with real coefficients of *odd* degree must have at least one real zero. In fact, such polynomial functions must have an odd number of real zeros (see Examples 3 and 5).

2. A polynomial function with real coefficients of *even* degree must have either an even number of real zeros or no real zero at all (see Examples 4 and 6).

EXAMPLE 6 Find all zeros of $P(x) = x^4 - 6x^3 + 15x^2 - 18x + 10$, given that $2 - i$ is a zero of P.

◆ **Solution** This fourth-degree polynomial function must have four zeros in the set of complex numbers. By the conjugate pair theorem, since $2 - i$ is a zero of P and P has real coefficients, we know that $2 + i$ is also a zero of P. Thus, $x - (2 - i)$ and $x - (2 + i)$ are factors of the polynomial $P(x)$. Hence, we can write

$$P(x) = [x - (2 - i)][x - (2 + i)]Q(x)$$

Now we can find $Q(x)$ by dividing $P(x)$ by $x - (2 - i)$ using synthetic division, and then dividing the result of this division by $x - (2 + i)$, using synthetic division again:

$$
\begin{array}{r|rrrrr}
2-i & 1 & -6 & 15 & -18 & 10 \\
 & & 2-i & -9+2i & 14-2i & -10 \\
\hline
2+i & 1 & -4-i & 6+2i & -4-2i & \boxed{0} \\
 & & 2+i & -4-2i & 4+2i & \\
\hline
 & 1 & -2 & 2 & \boxed{0} & \\
\end{array}
$$

— remainder

— remainder

coefficients of $Q(x)$ whose degree is 2

Thus, $Q(x) = x^2 - 2x + 2$. We can determine the zeros of this quadratic function by applying the quadratic formula:

$$x = \frac{-(-2) \pm \sqrt{(-2)^2 - 4(1)(2)}}{2(1)} = \frac{2 \pm \sqrt{-4}}{2} = 1 \pm i.$$

Thus, the four zeros of P are the (nonreal) conjugate pairs $2 \pm i$ and $1 \pm i$. ◆

PROBLEM 6 Use a graphing calculator to generate the graph of the polynomial function in Example 6. How can we tell from the graph that this function has no real zero? ◆

Exercises 4.3

◆ Basic Skills

In Exercises 1–12, use the factor theorem to determine if the first polynomial is a factor of the second polynomial.

1. $x - 1$; $x^3 - 4x^2 + x - 4$
2. $x + 2$; $x^3 - 4x^2 + x - 4$
3. $x + 3$; $x^4 - 7x^3 + 10x^2 + 3x - 15$
4. $x - 5$; $x^4 - 7x^3 + 10x^2 + 3x - 15$
5. $x - 4$; $x^4 - 3x^3 - 12x - 16$
6. $x + 1$; $x^4 - 3x^3 - 12x - 16$
7. $x + \frac{2}{3}$; $6x^3 + x^2 + 7x + 6$
8. $x - \frac{1}{3}$; $6x^3 + x^2 + 7x + 6$
9. $x - i$; $x^4 - 2x^2 + 1$
10. $x + i$; $x^4 - 2x^2 + 1$
11. $x - (1 + i)$; $x^3 - 2x + 4$
12. $x - (1 - i)$; $x^3 - 2x + 4$

In Exercises 13–20, determine the polynomial function P with the given characteristics.

13. Degree 3; zeros are 1, −1, and 2; graph passes through the point (3, 16)
14. Degree 3; zeros are 0, −2, and 4; graph passes through the point (−1, −15)
15. Degree 4; zeros are −1, 1, and 2 of multiplicity two; graph passes through the point (0, 16)
16. Degree 3; zeros are 2 of multiplicity three; graph passes through the point (1, 7)
17. Degree 4; zeros are ±1 and ±i; graph passes through the point (3, 80)
18. Degree 4; zeros are $1 \pm i$ of multiplicity two; graph passes through the point (0, −12)
19. Degree 5; zeros are 0 of multiplicity three and $\pm\sqrt{2}$; graph passes through the point (1, 5)
20. Degree 6; zeros are 0 of multiplicity two, 1 of multiplicity two, and $1 \pm \sqrt{3}$; graph passes through the point (2, −24)

For the polynomial function P defined in Exercises 21–32, (a) write $P(x)$ as the product of linear factors, and (b) find all the zeros of P.

21. $P(x) = x^3 + x^2 - 5x + 3$
22. $P(x) = 2x^3 + 9x^2 + x - 12$
23. $P(x) = 6x^3 + x^2 - 9x + 4$
24. $P(x) = 18x^3 - 33x^2 + 20x - 4$
25. $P(x) = x^4 - 4x^3 + 5x^2 - 4x + 4$
26. $P(x) = x^5 - 5x^4 + 10x^3 - 10x^2 + 5x - 1$
27. $P(x) = 2x^4 - 17x^3 + 53x^2 - 72x + 36$
28. $P(x) = 8x^4 - 14x^3 - 71x^2 - 10x + 24$
29. $P(x) = x^4 - 3x^3 - 12x - 16$
30. $P(x) = 2x^4 - 3x^3 - 3x - 2$
31. $P(x) = x^4 - 4x^3 - 7x^2 + 50x - 50$; $3 - i$ is a zero

32. $P(x) = x^4 - 6x^3 + 18x^2 - 30x + 25$; $[x - (1 + 2i)]$ is a factor

In Exercises 33–50, (a) find all rational zeros of each polynomial function and (b) factor the polynomial completely over the set of integers.

33. $P(x) = x^3 - 6x^2 + 11x - 6$
34. $P(x) = x^3 + 7x^2 + 16x + 12$
35. $f(x) = 2x^3 - 5x^2 - 14x + 8$
36. $F(x) = 6x^3 + 35x^2 - 8x - 12$
37. $H(x) = 4x^3 + 9x^2 + 26x + 6$
38. $G(x) = 2x^3 - 9x^2 + 14x - 10$
39. $P(x) = 12x^3 + 4x^2 - 17x + 6$
40. $Q(x) = 24x^3 - 70x^2 + 63x - 18$
41. $f(x) = 2x^4 + x^3 + 3x - 18$
42. $F(x) = x^4 - 11x^3 - 14x - 26$
43. $g(x) = 3x^4 - 14x^3 + 16x^2 - 3x + 8$
44. $h(x) = 4x^4 - 3x^3 - 12x^2 + 25x - 12$
45. $P(x) = 2x^4 + 9x^3 + 14x^2 + 9x + 2$
46. $f(x) = 4x^4 - 28x^3 + 67x^2 - 63x + 18$
47. $f(x) = x^5 + x^4 - 6x^3 + 8x - 16$
48. $P(x) = 6x^5 - 8x^4 + 5x^3 - 16x^2 - 25x + 10$
49. $F(x) = 2x^5 + 5x^4 - 4x^3 - 11x^2 + 4x + 4$
50. $D(x) = 16x^5 - 80x^4 + 127x^3 - 59x^2 - 8x + 4$

Critical Thinking

51. Is $x + a$, with $a > 0$, a factor of $x^n - a^n$ when n is even? when n is odd? Explain.
52. Is $x + a$, with $a > 0$, a factor of $x^n + a^n$ when n is even? when n is odd? Explain.
53. Determine the value of k if $x - 2$ is a factor of $x^4 + 3x^3 + kx^2 - 5x - 2$.
54. Determine the value of k if $x + 3$ is a factor of $2x^4 + 4x^3 - 5x^2 + 2k + k^2$.
55. Find all integers k for which the polynomial function $P(x) = x^3 - 2x^2 + kx - 3$ has one rational zero and two imaginary zeros.
56. What are the possible rational zeros of

$$P(x) = x^n + a_{n-1}x^{n-1} + a_{n-2}x^{n-2} + \cdots + a_1x + a_0$$

if each coefficient is an integer and a_0 is a prime number?

In Exercises 57–62, find all rational zeros of each polynomial function.
[*Hint:* *To extend the rational zero theorem to include a polynomial function with rational coefficients, multiply both sides of the function by the LCD to eliminate all fractions. The new function that is formed has the same zeros as the original function.*]

57. $P(x) = \tfrac{1}{2}x^3 + \tfrac{5}{6}x^2 + \tfrac{8}{3}x - 1$
58. $f(x) = \tfrac{1}{2}x^3 + x^2 - \tfrac{1}{8}x - \tfrac{1}{4}$
59. $g(x) = \tfrac{3}{4}x^4 + x^3 - \tfrac{1}{4}x^2 + x - 1$
60. $G(x) = \tfrac{1}{10}x^4 - \tfrac{1}{2}x^2 - x - \tfrac{3}{5}$
61. $f(x) = 0.2x^3 - 2.3x^2 + 3.1x + 1.4$
62. $P(x) = 0.6x^4 - 1.1x^3 + 0.7x^2 - 2.2x + 2.4$

The graphs of cubic functions are shown in Exercises 63 and 64. Determine the equation of each function.

63.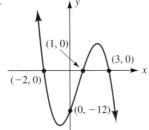

64.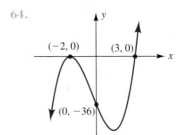

The graphs of quartic functions are shown in Exercises 65 and 66. Determine the equation of each function.

65.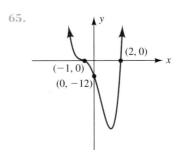

66.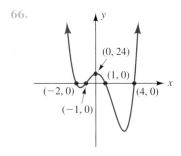

67. The velocity v (in feet per second, or ft/s) of the current at the mouth of a tidal river is a function of the time t (in hours) after high tide and is given by

$$v(t) = -0.240t^3 + 0.064t^2 + 8.527t - 2.982$$
for $0 \leq t \leq 6$

Find, to the nearest minute, the time(s) t when the velocity of the current is 0 ft/s.

68. The deflection d (in millimeters, mm) of a 34.0-meter steel beam is a function of the distance x (in meters, m) from the left end of the beam and is given by

$$d(x) = 0.002x^4 - 0.129x^3 + 2.480x^2 - 13.804x$$
for $0 \leq x \leq 34$

Find the distance(s) x for which the deflection of the beam is 0 mm.

4.4 Polynomial Equations and Inequalities

◆ Introductory Comments

Any equation of the form

$$a_n x^n + a_{n-1} x^{n-1} + a_{n-2} x^{n-2} + \cdots + a_2 x^2 + a_1 x + a_0 = 0$$

where n is a nonnegative integer and $a_n, a_{n-1}, a_{n-2}, \ldots, a_2, a_1,$ and a_0 are real numbers with $a_n \neq 0$, is called a **polynomial equation of degree n in standard form**. If we replace the equal sign with one of the inequality symbols ($<$, $>$, $\leq$, or $\geq$) we have a **polynomial inequality of degree n in standard form**. In this section, we discuss methods for solving polynomial equations and polynomial inequalities.

◆ Solving Polynomial Equations

We can solve some polynomial equations by using the factoring techniques introduced in Section 1.6.

EXAMPLE 1 Find all roots of each polynomial equation.

(a) $(x - 2)^2(x + 1) = 2x - 2$
(b) $\dfrac{4x^2(x - 3)^2}{5} + \dfrac{x(x - 3)^3}{3} = 0$

◆ Solution

(a) We begin by writing the equation in standard form:

$(x - 2)^2(x + 1) = 2x - 2$

$(x^2 - 4x + 4)(x + 1) = 2x - 2$ Expand $(x - 2)^2$

$x^3 - 3x^2 + 4 = 2x - 2$ Multiply polynomials

$x^3 - 3x^2 - 2x + 6 = 0$ Write in standard form

This third-degree polynomial equation must have three roots in the set of complex numbers. We can find these three roots as follows:

$$x^3 - 3x^2 - 2x + 6 = 0$$
$$x^2(x - 3) - 2(x - 3) = 0 \quad \text{Factor by grouping terms}$$
$$(x - 3)(x^2 - 2) = 0$$
$$x = 3, \quad x = \pm\sqrt{2}$$

Hence, the three roots of the original equation are 3 and $\pm\sqrt{2} \approx \pm 1.414$.

Figure 4.27 shows the graph of $P(x) = (x - 2)^2(x + 1) - 2x + 2$ in the viewing rectangle of a graphing calculator. The graph indicates that the x-intercepts are 3 and approximately ± 1.414; these facts confirm our answer.

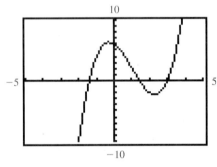

FIGURE 4.27
The x-intercepts of the graph of $P(x) = (x - 2)^2(x + 1) - 2x + 2$ are 3 and $\pm\sqrt{2} \approx \pm 1.414$.

(b) We begin by eliminating the fractions:

$$\frac{4x^2(x - 3)^2}{5} + \frac{x(x - 3)^3}{3} = 0$$

$$15\left[\frac{4x^2(x - 3)^2}{5} + \frac{x(x - 3)^3}{3}\right] = 15(0) \quad \text{Multiply both sides by the LCD}$$

$$12x^2(x - 3)^2 + 5x(x - 3)^3 = 0 \quad \text{Simplify}$$

It is best *not* to write this equation in standard form. Instead, we factor out the common factors from each term:

$$12x^2(x - 3)^2 + 5x(x - 3)^3 = 0$$
$$x(x - 3)^2[12x + 5(x - 3)] = 0 \quad \text{Factor out } x(x-3)^2$$
$$x(x - 3)^2(17x - 15) = 0 \quad \text{Simplify the third factor}$$
$$x = 0, \quad x = 3, \quad x = \tfrac{15}{17}$$

Thus, the four roots of the original fourth-degree polynomial equation are 0, 3 of multiplicity two, and $\tfrac{15}{17} \approx 0.882$. The graph in Figure 4.28 supports our work. ◆

PROBLEM 1 Find all roots of the equation $(x^2 + 4)(x^2 - 3x - 4) = 0$. ◆

If a polynomial equation is not factorable by the techniques discussed in Section 1.6, then we apply the rational zero theorem (Section 4.3) and write the equation as the product of linear and quadratic factors.

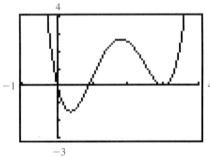

FIGURE 4.28
The x-intercepts of the graph $P(x) = \dfrac{4x^2(x - 3)^2}{5} + \dfrac{x(x - 3)^3}{3}$ are 0, 3, and $\tfrac{15}{17} \approx 0.882$.

EXAMPLE 2 Find all roots of each polynomial equation.

(a) $\dfrac{x^3 + 5x}{3} - \dfrac{x - 3x^2}{2} = 1$ **(b)** $x^2(x - 1)^2 = 4(3x - 5)$

Solution

(a) We begin by eliminating the fractions:

$$\frac{x^3 + 5x}{3} - \frac{x - 3x^2}{2} = 1$$

$$6\left[\frac{x^3 + 5x}{3} - \frac{x - 3x^2}{2}\right] = 6(1) \quad \text{Multiply both sides by the LCD}$$

$$2(x^3 + 5x) - 3(x - 3x^2) = 6 \quad \text{Simplify}$$

$$2x^3 + 9x^2 + 7x - 6 = 0 \quad \text{Write in standard form}$$

This third-degree polynomial equation must have three roots in the set of complex numbers. By the rational zero theorem (Section 4.3), the possible rational roots of this equation are

$$\pm 1, \quad \pm 2, \quad \pm 3, \quad \pm 6, \quad \pm \frac{1}{2}, \quad \text{and} \quad \pm \frac{3}{2}.$$

The graph in Figure 4.29 indicates that -3, -2, and $\frac{1}{2}$ may be possible roots. Testing -3, we find

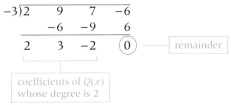

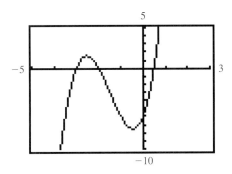

FIGURE 4.29

The graph of $P(x) = 2x^3 + 9x^2 + 7x - 6$ indicates that -3, -2, and $\frac{1}{2}$ may be rational roots of $2x^3 + 9x^2 + 7x - 6 = 0$.

Since the remainder is zero, we know that -3 is a root of the equation. Hence, $x + 3$ is a factor of the polynomial $2x^3 + 9x^2 + 7x - 6$. We can now write the equation as the product of three linear factors:

$$2x^3 + 9x^2 + 7x - 6 = 0$$
$$(x + 3)(2x^2 + 3x - 2) = 0$$
$$(x + 3)(x + 2)(2x - 1) = 0$$

$$x = -3, \; x = -2, \; x = \tfrac{1}{2}$$

Therefore, the three roots of the original equation are -3, -2, and $\frac{1}{2}$.

(b) We begin by writing the equation in standard form:

$$x^2(x - 1)^2 = 4(3x - 5)$$
$$x^2(x^2 - 2x + 1) = 12x - 20$$
$$x^4 - 2x^3 + x^2 - 12x + 20 = 0$$

This fourth-degree polynomial equation must have four roots in the set of complex numbers. By the rational zero theorem (Section 4.3), the possible rational roots of this equation are

$$\pm 1, \quad \pm 2, \quad \pm 4, \quad \pm 5, \quad \pm 10, \quad \text{and} \quad \pm 20$$

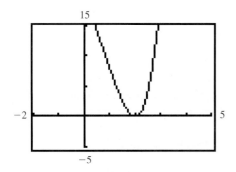

FIGURE 4.30
The graph of
$P(x) = x^4 - 2x^3 + x^2 - 12x + 20$
indicates that 2 may be a rational root of even multiplicity of
$x^4 - 2x^3 + x^2 - 12x + 20 = 0$.

The graph in Figure 4.30 indicates that 2 may be a root with even multiplicity. Checking 2, we find

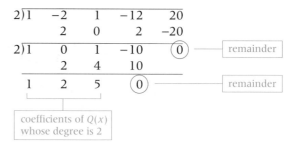

Since the remainder is 0 each time we divide by the synthetic divisor 2, we know that 2 is a rational root of multiplicity two and, hence, $(x - 2)^2$ is a factor of $x^4 - 2x^3 + x^2 - 12x + 20$. We can now write the equation as the product of linear and quadratic factors:

$$x^4 - 2x^3 + x^2 - 12x + 20 = 0$$
$$(x - 2)^2(x^2 + 2x + 5) = 0$$

Apply the quadratic formula.

$$x = 2 \qquad x = 1 \pm 2i$$

Therefore, the four roots of the original equation are 2 of multiplicity two and the complex conjugates $1 + 2i$ and $1 - 2i$. ◆

PROBLEM 2 Find all roots of the equation $x^2(x - 1) = 2(5x + 4)$. ◆

Although we may not be able to find all complex roots of a polynomial equation $P(x) = 0$ by using the techniques discussed in this chapter, we can find all its real roots by locating the x-intercepts of the graph of P.

EXAMPLE 3 Find all real roots of each polynomial equation. Record any rational root as a reduced fraction and any irrational root as a decimal number rounded to the nearest thousandth.

(a) $x^3 = x - 3$ **(b)** $4(x^2 - 1)(x^2 + 1) = 2x^3 - 2x^2 - 7x$

◆ **Solution**

(a) Writing the equation in standard form, we obtain

$$x^3 - x + 3 = 0.$$

As a consequence of the conjugate pair theorem (Section 4.3), this third-degree polynomial equation has either 1 or 3 real roots. Figure 4.31 shows a complete graph of $P(x) = x^3 - x + 3$ in the viewing rectangle of a graphing calculator. Note that the graph of P intersects the

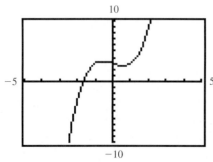

FIGURE 4.31
The equation $x^3 - x + 3 = 0$ has one real root in the interval $(-2, -1)$.

x-axis once. Hence, the equation $x^3 - x + 3 = 0$ has only one real root, and this root is in the interval $(-2, -1)$.

By the rational zero theorem (Section 4.3), the only possible rational roots of the equation $x^3 - x + 3 = 0$ are ± 1 and ± 3. Therefore, the root in the interval $(-2, -1)$ must be irrational. To find this irrational root, we use a graphing calculator to estimate the x-intercept of the graph. Using the calculator's built-in program for finding an x-intercept or the [TRACE] key and [ZOOM] key, we find that $x \approx -1.672$, rounded to the nearest thousandth. Hence, the only real root of the original equation is approximately -1.672.

(b) We begin by writing the equation in standard form:

$$4(x^2 - 1)(x^2 + 1) = 2x^3 - 2x^2 - 7x$$
$$4(x^4 - 1) = 2x^3 - 2x^2 - 7x$$
$$4x^4 - 2x^3 + 2x^2 + 7x - 4 = 0$$

As a consequence of the conjugate pair theorem (Section 4.3), this fourth-degree polynomial equation has either 0, 2, or 4 real roots. Figure 4.32 shows a complete graph of $P(x) = 4x^4 - 2x^3 + 2x^2 + 7x - 4$ in the viewing rectangle of a graphing calculator. Since the graph of P intersects the x-axis twice, the equation $4x^4 - 2x^3 + 2x^2 + 7x - 4 = 0$ must have two real roots.

By the rational zero theorem (Section 4.3), the possible rational roots of the equation are ± 1, ± 2, ± 4, $\pm \frac{1}{2}$, and $\pm \frac{1}{4}$. Using this list of possibilities and the graph of P in Figure 4.32, we note that $\frac{1}{2}$ and -1 may be rational roots. Testing $\frac{1}{2}$, we find

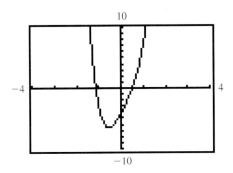

FIGURE 4.32
The equation
$4x^4 - 2x^3 + 2x^2 + 7x - 4 = 0$ has two real roots.

$$\frac{1}{2}\overline{)\begin{array}{ccccc} 4 & -2 & 2 & 7 & -4 \\ & 2 & 0 & 1 & 4 \\ \hline 4 & 0 & 2 & 8 & \boxed{0} \end{array}} \longleftarrow \text{remainder}$$

Since the remainder is zero, we conclude that $\frac{1}{2}$ is a rational root of the original equation. Testing -1 as a possible rational root, we find

$$-1\overline{)\begin{array}{ccccc} 4 & -2 & 2 & 7 & -4 \\ & -4 & 6 & -8 & 1 \\ \hline 4 & -6 & 8 & -1 & \boxed{-3} \end{array}} \longleftarrow \text{remainder}$$

Since the remainder is *not* zero, we conclude that -1 is *not* a rational root. Hence, the root near $x = -1$ must be irrational. Using the calculator's built-in program for finding an x-intercept or the [TRACE] key and [ZOOM] key, we find that $x \approx -1.128$, rounded to the nearest thousandth. In summary, the real roots of the original equation are $\frac{1}{2}$ and approximately -1.128. ◆

PROBLEM 3 Find all roots of the equation $x^2(x + 1)^2 = 2x + 8$. ◆

◆ Solving Polynomial Inequalities

We can use the procedure for solving a quadratic inequality (Section 3.4) to help solve a polynomial inequality. To solve the polynomial inequality $P(x) > 0$, we find the *boundary numbers* of the inequality—the real values of x for which the inequality is neither positive nor negative. These boundary numbers are the x-intercepts of the graph of P, and they divide the x-axis into intervals. The solution set of the inequality $P(x) > 0$ includes all the intervals on which the graph of P is *above* the x-axis. For the inequality $P(x) < 0$, the solution set of the inequality is all the intervals on which the graph of P is *below* the x-axis. If the inequality symbol is $\leq$ or $\geq$, we include the boundary numbers as part of the solution set.

EXAMPLE 4 Solve each polynomial inequality.

(a) $9 > x^4 - 8x^2$ (b) $x^3 - 2x^2 + 8 \geq 10x$

◆ Solution

(a) We begin by writing the inequality in standard form:

$$9 > x^4 - 8x^2$$
$$0 > x^4 - 8x^2 - 9$$
$$x^4 - 8x^2 - 9 < 0 \qquad \text{Interchange sides and reverse the direction of the inequality}$$

The boundary numbers of this inequality are the x-intercepts of the graph of $P(x) = x^4 - 8x^2 - 9$. We can find these boundary numbers by solving the equation $P(x) = 0$:

$$x^4 - 8x^2 - 9 = 0$$
$$(x^2 - 9)(x^2 + 1) = 0 \qquad \text{Factor the trinomial}$$
$$(x + 3)(x - 3)(x^2 + 1) = 0 \qquad \text{Factor the difference of squares}$$

$x = -3 \quad x = 3 \quad$ no real solution

boundary numbers

These two boundary numbers divide the x-axis into three intervals, as shown:

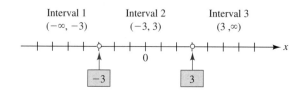

We can confirm these intervals by generating the graph of $P(x) = x^4 - 8x^2 - 9$ in the viewing rectangle of a graphing calculator. As shown in Figure 4.33, the graph of $P(x) = x^4 - 8x^2 - 9$ is *below* the x-axis in the interval $(-3, 3)$. Hence, $P(x) < 0$ in this interval. Thus, the solution set of the original inequality is $(-3, 3)$.

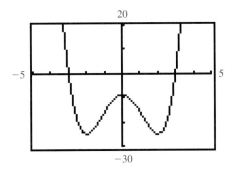

FIGURE 4.33
The graph of $P(x) = x^4 - 8x^2 - 9$ is below the x-axis in the interval $(-3, 3)$.

(b) We begin by subtracting $10x$ from both sides of the inequality and then writing the inequality in standard form:

$$x^3 - 2x^2 - 10x + 8 \geq 0$$

The boundary numbers of this inequality are the x-intercepts of the graph of $P(x) = x^3 - 2x^2 - 10x + 8$. We can find these boundary numbers by solving the equation $x^3 - 2x^2 - 10x + 8 = 0$. By the rational zero theorem (Section 4.3), the possible rational roots of this equation are $\pm 1, \pm 2, \pm 4,$ and ± 8. The graph in Figure 4.34 indicates that 4 may be a possible root, so we test the number:

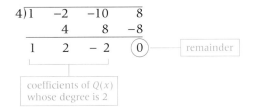

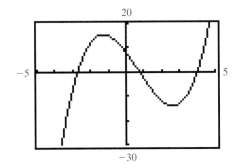

FIGURE 4.34
The graph of $P(x) = x^3 - 2x^2 - 10x + 8$ indicates that 4 may be a rational root of the equation $x^3 - 2x^2 - 10x + 8 = 0$.

If 4 is a root of the equation, then we have

$$x^3 - 2x^2 - 10x + 8 = 0$$
$$(x - 4)(x^2 + 2x - 2) = 0$$

Apply the quadratic formula.

$$x = 4 \quad x = -1 \pm \sqrt{3}$$

boundary numbers

These three boundary numbers divide the x-axis into four intervals:

Interval 1: $(-\infty, -1 - \sqrt{3})$
Interval 2: $(-1 - \sqrt{3}, -1 + \sqrt{3})$
Interval 3: $(-1 + \sqrt{3}, 4)$
Interval 4: $(4, \infty)$

$-1 - \sqrt{3} \approx -2.73$
$-1 + \sqrt{3} \approx 0.73$

We can confirm these intervals by generating the graph of $P(x) = x^3 - 2x^2 - 10x + 8$ in the viewing rectangle of a graphing calculator. As shown in Figure 4.34, the graph of P is *above* the x-axis in the interval $(-1 - \sqrt{3}, -1 + \sqrt{3})$ or $(4, \infty)$. Hence, $P(x) \geq 0$ in the interval $[-1 - \sqrt{3}, -1 + \sqrt{3}]$ or $[4, \infty]$. Thus, the solution set of the original inequality is $[-1 - \sqrt{3}, -1 + \sqrt{3}] \cup [4, \infty)$. ◆

Observe in Figures 4.33 and 4.34 that in each of the intervals formed by the boundary numbers, the graph of the polynomial function is entirely above the x-axis or entirely below the x-axis. Hence, we can use the test number method (Section 3.4) to help solve a polynomial inequality in standard form without referring to a graph. We simply select an arbitrary test number from each interval formed by the boundary numbers and then determine the algebraic sign of the polynomial. If the polynomial is negative for a particular test number in an interval, then it is negative for every value of x in that interval. If the polynomial is positive for a particular test number in an interval, then it is positive for every value of x in that interval.

For example, to solve the polynomial inequality $x^4 - 8x^2 - 9 < 0$ in Example 4(a) by using the test number method, we select an arbitrary test number from each of the intervals $(-\infty, -3)$, $(-3, 3)$, and $(3, \infty)$ and then determine the algebraic sign of the polynomial $x^4 - 8x^2 - 9$ for each test number. The following table organizes our work:

Interval	Test Number	Sign of $x^4 - 8x^2 - 9$
$(-\infty, -3)$	-4	$(-4)^4 - 8(-4)^2 - 9 = 119 > 0$
$(-3, 3)$	1	$(1)^4 - 8(1)^2 - 9 = -16 < 0$
$(3, \infty)$	4	$(4)^4 - 8(4)^2 - 9 = 119 > 0$

Since $x^4 - 8x^2 - 9 < 0$ when $x = 1$, then we know that $x^4 - 8x^2 - 9 < 0$ for every value of x in the interval $(-3, 3)$. Hence, we conclude that the solution set of this inequality is $(-3, 3)$, which agrees with our work in Example 4(a).

PROBLEM 4 Solve the polynomial inequality $x^3 + 6 \geq 7x$ by using the test number method. ◆

◆ Application: Silo for Storing Grain

Many applied problems involve the solution of a polynomial equation of the form

$$a_n x^n + a_{n-1} x^{n-1} + \cdots + a_1 x + a_0 = 0,$$

where $n \geq 3$. Such equations often have solutions that are irrational numbers. We illustrate one such application in the next example.

EXAMPLE 5 *Chapter Opening Problem*

A silo has the shape of a right circular cylinder surmounted by a hemisphere and holds 3000 cubic feet of grain. If the height of the silo is 60 ft, what is the radius of its base?

SECTION 4.4 ◆ *Polynomial Equations and Inequalities* 257

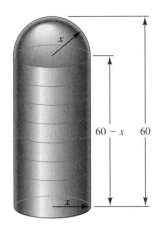

FIGURE 4.35
A silo with height 60 ft and radius x

◆ Solution   Let

$$x = \text{the radius of the cylindrical part of the silo.}$$

Since the radius of the hemisphere is the same as the radius of the cylinder, the height of the cylindrical part of the silo is $60 - x$, as shown in Figure 4.35. Now the volume V of the silo is the sum of the volumes of the cylinder and hemisphere:

$$V = \underbrace{\pi r^2 h}_{\text{cylinder}} + \underbrace{\frac{2}{3}\pi r^3}_{\text{hemisphere}}$$

$$3000 = \pi x^2(60 - x) + \frac{2}{3}\pi x^3$$

$$9000 = 3\pi x^2(60 - x) + 2\pi x^3 \qquad \textbf{Multiply both sides by 3}$$

$$\pi x^3 - 180\pi x^2 + 9000 = 0 \qquad \textbf{Write in standard form}$$

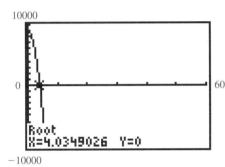

FIGURE 4.36
The equation
$\pi x^3 - 180\pi x^2 + 9000 = 0$ has only one real root in the interval $(0, 60)$.

Because the irrational number π is a coefficient of only some terms of this polynomial equation, the equation cannot have rational roots. Also, since the height of the cylindrical part of the silo is $60 - x$, we must have $x < 60$. As shown in Figure 4.36, the graph of $P(x) = \pi x^3 - 180\pi x^2 + 9000$ intersects the x-axis only once in the interval $(0, 60)$. Hence, the equation $\pi x^3 - 180\pi x^2 + 9000 = 0$ has only one real root in this interval. Using the calculator's built-in program for finding an x-intercept or the TRACE key and ZOOM key, we find that $x \approx 4.035$, rounded to the nearest thousandth. In summary, the radius of the base of the silo is approximately 4.035 ft. ◆

PROBLEM 5   If the silo described in Example 5 holds 4000 cubic feet of grain, what is the radius of its base? ◆

◆ **Application: Ash Pan for a Wood Stove**

For some applied problems, it is necessary to solve a polynomial inequality of degree $n \geq 3$. We illustrate one such application in the next example.

EXAMPLE 6   An ash pan for a wood stove is made from a rectangular piece of sheet metal, 8 inches by 15 inches, by cutting out a square from each corner and turning up the resulting sides. For safety reasons, the volume of the pan that is formed must be greater than 90 cubic inches. Find an interval for the acceptable dimensions of the square cut-out.

◆ Solution   Recall from elementary geometry that the volume V of a rectangular box is the product of its length, width, and height. If we let

$$x = \text{the length of the side of the square cut-out,}$$

then the dimensions of the ash pan are

$$\text{height} = x \qquad \text{length} = 15 - 2x \qquad \text{width} = 8 - 2x$$

as shown in Figure 4.37. Thus,

$$V = (15 - 2x)(8 - 2x)x.$$

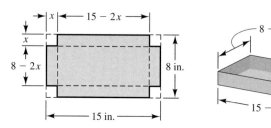

FIGURE 4.37
The dimensions of the ash pan described in Example 6

Now, to find the interval when $V > 90$ cubic inches we solve the corresponding inequality:

$$(15 - 2x)(8 - 2x)x > 90$$

$$(120 - 46x + 4x^2)x > 90 \qquad \text{Multiply the binomials}$$

$$4x^3 - 46x^2 + 120x - 90 > 0 \qquad \text{Write in standard form}$$

$$2x^3 - 23x^2 + 60x - 45 > 0 \qquad \text{Divide both sides by 2}$$

By the rational zero theorem, the possible rational boundary numbers of this inequality are:

$$\pm 1, \quad \pm 3, \quad \pm 5, \quad \pm 9, \quad \pm 15, \quad \pm 45, \quad \pm \tfrac{1}{2}, \quad \pm \tfrac{3}{2}, \quad \pm \tfrac{5}{2}, \quad \pm \tfrac{9}{2}, \quad \pm \tfrac{15}{2}, \quad \pm \tfrac{45}{2}$$

However, the length of the side of the cut-out cannot be a negative number, nor can it be greater than 4 inches (since the piece of metal is 8 inches wide). Since $0 \leq x \leq 4$, we check only $1, 3, \tfrac{1}{2}, \tfrac{3}{2}$, and $\tfrac{5}{2}$ as possible rational boundary numbers. The graph in Figure 4.38 indicates that $\tfrac{3}{2}$ may be a boundary number of the inequality. Testing $\tfrac{3}{2}$, we find

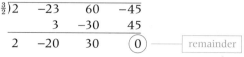

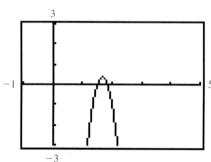

FIGURE 4.38
The graph of
$P(x) = 2x^3 - 23x^2 + 60x - 45$
indicates that $\tfrac{3}{2}$ may be a critical value of the inequality
$2x^3 - 23x^2 + 60x - 45 > 0$.

Since the remainder is zero, we know that $x = \tfrac{3}{2}$ is a boundary number of the inequality. The other boundary number in the interval (1, 2) must be irrational. Using the calculator's built-in program for finding an x-intercept or the TRACE key and ZOOM key, we find that $x \approx 1.838$, rounded to the nearest thousandth. In summary, the volume of the pan is greater than 90 cubic inches when the dimensions of the square cut-out are in the interval between 1.5 inches and approximately 1.838 inches. ◆

PROBLEM 6 Factor the polynomial in Example 6, $2x^3 - 23x^2 + 60x - 45$, completely over the set of integers. Then show that the exact value of the irrational boundary number in the interval $(1, 2)$ is $5 - \sqrt{10}$.

Exercises 4.4

◆ Basic Skills

In Exercises 1–20, find all roots of each equation and give the multiplicity of any repeated root.

1. $y^3 - y = 0$
2. $2x^4 = 16x^2$
3. $x^4 - 3x^3 - 2x^2 = 0$
4. $10x^2 + 3x = 8x^2$
5. $x^4(2x - 3) = x^3$
6. $x^2(5x - 8) = 2x$
7. $\dfrac{x(x^2 + 2)}{3} = \dfrac{x^2 + 2}{2}$
8. $\dfrac{n^5(n - 2)}{6} - \dfrac{n(n - 2)}{2} = 0$
9. $3x^2(x + 5)^2 + 2x(x + 5)^3 = 0$
10. $4x^2(x + 1)^3 + 2x(x + 1)^4 = 0$
11. $6x^3 + x^2 - 31x + 24 = 0$
12. $6x^3 + x^2 - 29x = 14$
13. $x^2(9x^2 + 10) - 6x(x^2 + 1) = -1$
14. $12(4x - 3) - 25x^2 = 4x^3(x - 3)$
15. $x^4 + 27x = 0$
16. $y^3 = 8y^6$
17. $x^2(x + 1)^2 = 4(2x + 5)$
18. $x^2(2x^2 - 17x + 57) = 97x + 65$
19. $\dfrac{x(43x^2 + x - 2)}{6} + 4 = x^2(x^2 + 13)$
20. $\dfrac{x^6 - x^2}{10} - \dfrac{2x^4 - 2}{5} = \dfrac{x^5 - 5x^3 + 4x}{3}$

In Exercises 21–40, find the real roots of each equation. Record any rational root as a reduced fraction and any irrational root as a decimal number rounded to the nearest hundredth.

21. $x^3 + x = 4$
22. $2x^3 + x + 9 = 0$
23. $x^2(2x + 1) = -5$
24. $x^2(x - 6) = 5$
25. $\dfrac{x^2(x - 3)}{6} = x - 1$
26. $\dfrac{2x^3 - 5}{4} = x - x^2$
27. $x(3x + 4)(x - 2) = 2$
28. $(3 - 5x)(1 + x) = x^3$
29. $\dfrac{x(x^3 - 4)}{3} - 1 = 0$
30. $\dfrac{x^3(x + 3)}{2} + 1 = 0$
31. $2x^4 + 2x^3 + x^2 - 3x = 4$
32. $4x^4 - 4x^2 + x = 1$
33. $x[x^2(2x + 3) + 10x + 17] + 5 = 2$
34. $x[x^2(2x - 7) + 3x - 4] + 8 = 6$
35. $x[x^2(3x^2 - 5) - 2x(2x^2 + 1) + 4] = -8$
36. $x[2x^2(x + 1)(x - 1) - x(7x^2 + 5) - 11] = 4$
37. $\dfrac{x^4(x + 3)}{5} - \dfrac{x^2(2x + 5)}{2} + \dfrac{4(2x + 1)}{10} = 0$
38. $\dfrac{x^3(x^2 + 1)}{11} + \dfrac{x^2(x^2 - 2)}{3} = \dfrac{4(3x - 2)}{33}$
39. $\dfrac{x(x^5 - x^4 + 2)}{2} + \dfrac{x^2(x - 1)}{3} = 1$
40. $x^2(x^4 - 3x^3 - 2) - \dfrac{x^3(x - 3)}{4} + \dfrac{1}{2} = 0$

In Exercises 41–56, solve each inequality.

41. $x^3 \le 4x^2 + 8x$
42. $2x^4 + 3x^3 > 2x^2$
43. $t^4 + 3t^3 > 2t^2 + 6t$
44. $\dfrac{2h^3 - 9}{3} \le h(h - 2)$
45. $\dfrac{n^2(n^2 + 4)}{5} < 9$
46. $x^6 \ge x^3 + 2$
47. $2x^2(3x - 2)^3 + (3x - 2)^4 \ge 0$
48. $t(2t + 1)^2(t - 3)^3(t + 2) < 0$
49. $2x^4 - 3x^3 - 3x^2 - 3x \le 5$
50. $3x^4 - 4x^3 + 8x^2 - 16x > 16$
51. $3 - 7x > x^2(x - 5)$
52. $x^2(2x^2 - 11x + 18) \le 4(x + 2)$
53. $6x^4 - 41x^3 \le 19(4x - 1) - 5(23x^2 - 1)$
54. $3x^2(x + 2)(x - 3) + 11x(x^2 - 3) < 2(x - 15)$

55. $\dfrac{(x^2 + 5)(x^2 - 3)}{6} \geq x^3 - 3x$

56. $\dfrac{x^5 - x^2 + 8}{4} - \dfrac{x^4 - x}{2} > 2x^3$

For the applied problems in Exercises 57–68, express any rational solution as a reduced fraction and any irrational solution as a decimal number rounded to the nearest thousandth.

57. As a result of poaching, the number of elephants in a certain herd is declining rapidly. It is estimated that the population P of the herd is a function of time t (in years) and is given by

$$P(t) = -t^3 - t^2 + t + 150.$$

In approximately how many years will the herd be extinct?

58. The charge q (in coulombs) on a certain electrical capacitor is a function of the amount of time t (in seconds) that the capacitor has been charging and is given by

$$q(t) = 3t^3 + 4t.$$

After how many seconds of charging does the capacitor reach a charge of 10 coulombs?

59. The total profit P (in dollars) a company makes is a function of the number of units x it sells and is given by

$$P(x) = x^3 - x^2 - 12x - 210 \quad \text{for } x \geq 0.$$

How many units must the company sell to make a profit?

60. Under certain conditions, the velocity v (in feet per second, or ft/s) of an object as a function of time t (in seconds) is given by

$$v(t) = 4t^3 - 11t^2 - 6t + 9 \quad \text{for } 0 \leq t \leq 4.$$

Determine the time interval(s) in which $v(t) \leq 0$ ft/s.

61. The rectangular jewelry box shown in the figure has a square base, and its height is 2 inches less than the length of its base. Find the dimensions of the box if the volume of the box is 48 cubic inches.

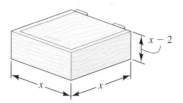

62. The allergy medication capsule shown in the figure has the shape of a cylinder surmounted by a hemisphere on each end. The total length of the capsule is 10 millimeters and it holds 25π cubic millimeters of medicine. Find the diameter d of the capsule.

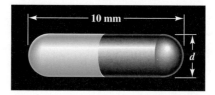

63. The width of a rectangular toy box is twice its height and the length of the box is 1 foot more than its width, as shown in the figure. The volume of the box is 18 cubic feet. What are the dimensions of the box?

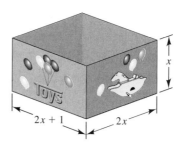

64. Recall from elementary geometry that the volume V of a cone with radius r and height h is $V = \tfrac{1}{3}\pi r^2 h$. If a conical sandpile has a slant height of 10 meters, as shown in the sketch, for what value(s) of h is $V = 128\pi$ cubic meters?

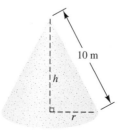

65. When the length of one side of a cube is reduced by 2 inches, the volume of the remaining rectangular solid is 75 cubic inches. What is the length of the side of the original cube?

66. A rectangular box has length 2 inches, width 2 inches, and height 3 inches. Each of these dimensions is increased uniformly until a new box is formed whose volume is triple that of the original box. What are the dimensions of the new box?

67. A piece of cardboard, 4 cm by 5 cm, is made into an open rectangular box by cutting out a square from each corner and turning up the resulting sides, as shown in the figure. The volume of the box that is formed is less than 3 cubic centimeters. Find an interval(s) for the acceptable dimensions of the square cut-out.

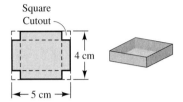

68. Recall from elementary geometry that the volume V of a cylinder with radius r and height h is $V = \pi r^2 h$. Suppose the slant distance from the top of an oil drum to the bottom is 12 feet, as shown in the figure. Find an interval for the height h of the drum if its volume is greater than 160π cubic feet.

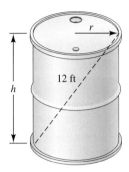

Critical Thinking

A **cubic formula** may be used to solve cubic equations of the form $x^3 + ax + b = 0$, where a and b are real numbers. The formula states that the three roots of this equation are

$$A + B, \quad -\frac{A+B}{2} + \frac{A-B}{2}\sqrt{-3},$$

and

$$-\frac{A+B}{2} - \frac{A-B}{2}\sqrt{-3}$$

where

$$A = \sqrt[3]{-\frac{b}{2} + \sqrt{\frac{b^2}{4} + \frac{a^3}{27}}}$$

and

$$B = \sqrt[3]{-\frac{b}{2} - \sqrt{\frac{b^2}{4} + \frac{a^3}{27}}}.$$

In Exercises 69 and 70, use the cubic formula to approximate the three roots of each equation.

69. $x^3 + x - 4 = 0$ 70. $2x^3 + x + 9 = 0$

The cubic equation $y^3 + py^2 + qy + r = 0$ can be written in the form $x^3 + ax + b = 0$ by replacing y with $x - \frac{p}{3}$. In Exercises 71 and 72,

(a) use this substitution to write the given equation in the form $x^3 + ax + b = 0$.

(b) use the cubic formula (see Exercises 69 and 70) to approximate the three roots of the given equation.

71. $y^3 - 3y^2 - 9y + 27 = 0$
72. $y^3 - 6y^2 - 5 = 0$

In Exercises 73 and 74, find the coordinates of the intersection point(s) of the graphs of the two equations by solving the two equations simultaneously. Check graphically with a graphing calculator.

73. $y = x^3$ and $21x - 4y = 10$
74. $y = x^4$ and $20x + 27y = 7$

75. Use a graphing calculator to generate the graph of $P(x) = x^3 + ax^2 + b$ for several positive values of a and b. Then discuss the nature of the roots of the equation $x^3 + ax^2 + b = 0$.

76. Use a graphing calculator to generate the graph of $P(x) = x^3 + ax^2 + b$ for several negative values of a and b. Then discuss the nature of the roots of the equation $x^3 + ax^2 + b = 0$.

77. Use a graphing calculator to generate the graph of $P(x) = x^4 + ax + b$ for several negative values of a and b. Then discuss the nature of the roots of the equation $x^4 + ax + b = 0$.

78. Use a graphing calculator to generate the graph of $P(x) = x^4 + ax + b$ for several positive values of a and negative values b. Then discuss the nature of the roots of the equation $x^4 + ax + b = 0$.

4.5 Rational Functions and Their Graphs

◆ Introductory Comments

Any function f of the form

$$f(x) = \frac{P(x)}{D(x)},$$

where $P(x)$ and $D(x)$ are polynomials and $D(x) \neq 0$, is called a **rational function**. Since division by zero is undefined, the domain of a rational function f is the set of all real numbers other than those numbers that make its denominator zero, that is, all real numbers except the real zeros of D.

When working with rational functions, we often need to evaluate expressions of the form $1/c$, where $|c|$ is a very large number or a very small number. If $|c|$ is a very large number, then $1/c$ is close to 0, and if $|c|$ is a very small number, then $1/c$ is very large in absolute value. Here are some illustrations:

$$\underbrace{\frac{1}{1000} = 0.001 \qquad \frac{1}{-1000} = -0.001}_{\text{close to zero}} \qquad \underbrace{\frac{1}{0.001} = 1000 \qquad \frac{1}{-0.001} = -1000}_{\text{very large in absolute value}}$$

◆ Vertical Asymptotes

The reciprocal function $f(x) = 1/x$ (Section 2.5) is one of the simplest rational functions. Its domain is the set of all real numbers except 0. Using interval notation, we write the domain as $(-\infty, 0) \cup (0, \infty)$. Figure 4.39 shows the graph of the reciprocal function. We say that the y-axis (the line $x = 0$) is a *vertical asymptote* of the graph of f because the graph approaches the y-axis as x gets closer and closer to zero.

As x approaches zero from the *left* ($x \to 0^-$), the value of $f(x)$ *decreases* without bound [$f(x) \to -\infty$]:

$$\text{As } x \to 0^-, \quad f(x) = \frac{1}{x} \to \frac{1}{\underbrace{-0}_{\text{a negative number close to 0}}} \to -\infty$$

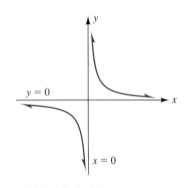

FIGURE 4.39

Graph of $f(x) = \dfrac{1}{x}$

As x approaches zero from the *right* ($x \to 0^+$), the value of $f(x)$ *increases* without bound [$f(x) \to +\infty$]:

$$\text{As } x \to 0^+, \quad f(x) = \frac{1}{x} \to \frac{1}{\underbrace{+0}_{\text{a positive number close to 0}}} \to +\infty$$

We define a vertical asymptote as follows.

Vertical Asymptotes

The line $x = a$ is a **vertical asymptote** of the graph of a rational function f if at least one of the following statements is true.

(a) As $x \to a^-$, $f(x) \to +\infty$.   **(b)** As $x \to a^+$, $f(x) \to +\infty$.

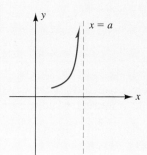 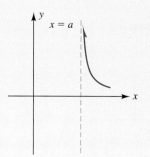

(c) As $x \to a^-$, $f(x) \to -\infty$.   **(d)** As $x \to a^+$, $f(x) \to -\infty$.

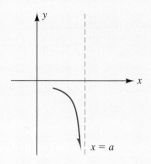 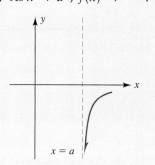

To determine the vertical asymptote(s) of a rational function f defined by $f(x) = P(x)/D(x)$, where $P(x)/D(x)$ is reduced to lowest terms, we find the real zeros of D. If a is a real zero of D, then the vertical line $x = a$ is a vertical asymptote of the graph of f.

EXAMPLE 1 Find the vertical asymptote(s) of the graph of each function, and then determine the appearance of its graph near each vertical asymptote.

(a) $F(x) = \dfrac{3x - 2}{x - 1}$ **(b)** $f(x) = \dfrac{x + 3}{x^2 - 2x - 8}$

Solution

(a) The zero of the denominator $D(x) = x - 1$ is 1. Therefore, since $P(x)/D(x)$ is reduced to lowest terms, we know that $x = 1$ is the only vertical asymptote of the graph of F.

To determine whether $F(x) \to +\infty$ or $F(x) \to -\infty$ as x approaches the vertical asymptote $x = 1$, we check the algebraic sign of $F(x)$ near the asymptote as follows:

$$\text{As } x \to 1^-, \quad F(x) = \frac{3x-2}{x-1} \to \frac{1}{-0} \to -\infty$$

— a negative number close to 0

$$\text{As } x \to 1^+, \quad F(x) = \frac{3x-2}{x-1} \to \frac{1}{+0} \to +\infty$$

— a positive number close to 0

Thus, we conclude that $F(x) \to -\infty$ as $x \to 1^-$ and that $F(x) \to +\infty$ as $x \to 1^+$. Figure 4.40 summarizes this information.

(b) Factoring the denominator, we have

$$f(x) = \frac{x+3}{x^2 - 2x - 8} = \frac{x+3}{(x+2)(x-4)}.$$

Thus, the zeros of $D(x) = x^2 - 2x - 8$ are -2 and 4. Since $P(x)/D(x)$ is in reduced form, we conclude that $x = -2$ and $x = 4$ are vertical asymptotes of the graph of f.

To determine whether $f(x) \to +\infty$ or $f(x) \to -\infty$ as x approaches each vertical asymptote, we check the algebraic sign of $f(x)$ near each asymptote as follows:

$$\text{As } x \to -2^-, \quad f(x) = \frac{x+3}{(x+2)(x-4)} \to \frac{1}{(-0)(-6)} \to +\infty$$

— a negative number close to 0

$$\text{As } x \to -2^+, \quad f(x) = \frac{x+3}{(x+2)(x-4)} \to \frac{1}{(+0)(-6)} \to -\infty$$

— a positive number close to 0

$$\text{As } x \to 4^-, \quad f(x) = \frac{x+3}{(x+2)(x-4)} \to \frac{7}{(6)(-0)} \to -\infty$$

— a negative number close to 0

$$\text{As } x \to 4^+, \quad f(x) = \frac{x+3}{(x+2)(x-4)} \to \frac{7}{(6)(+0)} \to +\infty$$

— a positive number close to 0

Thus, we conclude that $f(x) \to +\infty$ as $x \to -2^-$, $f(x) \to -\infty$ as $x \to -2^+$, $f(x) \to -\infty$ as $x \to 4^-$, and $f(x) \to +\infty$ as $x \to 4^+$. Figure 4.41 summarizes this information. ◆

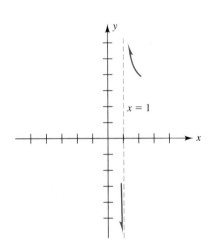

FIGURE 4.40
Appearance of the graph of $F(x) = \dfrac{3x-2}{x-1}$ as x approaches the vertical asymptote $x = 1$

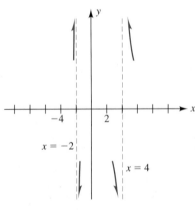

FIGURE 4.41
Appearance of the graph of $f(x) = \dfrac{x+3}{x^2 - 2x - 8}$ as x approaches each vertical asymptote

PROBLEM 1 State the domain of each of the rational functions defined in Example 1. ◆

◆ Horizontal Asymptotes

Recall the graph of the reciprocal function, $f(x) = 1/x$, sketched in Figure 4.39. We say that the x-axis (the line $y = 0$) is a *horizontal asymptote* of the graph of f because the graph approaches the x-axis for very large values of x (either positive or negative). As x increases without bound ($x \to +\infty$), $f(x)$ approaches zero through values greater than zero [$f(x) \to 0^+$], and as x decreases without bound ($x \to -\infty$), $f(x)$ approaches zero through values less than zero [$f(x) \to 0^-$]. We define a horizontal asymptote as follows.

◀ **Horizontal Asymptotes**

If $y = b$ is a **horizontal asymptote** of the graph of a rational function f, then at least one of the following statements is true.

(a) As $x \to +\infty$, $f(x) \to b^+$.　　　**(b)** As $x \to +\infty$, $f(x) \to b^-$.

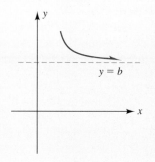
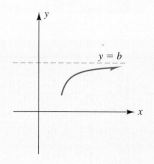

(c) As $x \to -\infty$, $f(x) \to b^+$.　　　**(d)** As $x \to -\infty$, $f(x) \to b^-$.

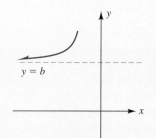
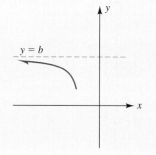

To determine whether a rational function has a horizontal asymptote, we begin by dividing each term of the numerator and denominator by the highest power of x that appears in the rational expression. In doing so, we develop fractions of the form k/x^n, where k is a constant and n is a positive integer. Now, for very large values of x (either positive or negative), x^n become very large and, consequently, expressions of the form k/x^n become very small. We indicate this by writing

$$\frac{k}{x^n} \to 0 \quad \text{as} \quad |x| \to \infty.$$

Although a rational function may have many vertical asymptotes, it can have at most one horizontal asymptote. By using the method described above, it can be shown that the rational function $f(x) = P(x)/D(x)$, where $P(x)/D(x)$ is reduced to lowest terms, follows these rules for the existence of a horizontal asymptote:

1. one horizontal asymptote if the degree of $P(x)$ is less than or equal to the degree of $D(x)$;
2. no horizontal asymptote if the degree of $P(x)$ is greater than the degree of $D(x)$.

EXAMPLE 2 Find the horizontal asymptote of the graph of each function. Then determine the appearance of its graph as $|x| \to \infty$.

(a) $F(x) = \dfrac{3x - 2}{x - 1}$ (b) $f(x) = \dfrac{x + 3}{x^2 - 2x - 8}$

◆ **Solution**

(a) Dividing each term of the numerator and denominator by x, we obtain

$$F(x) = \frac{3x - 2}{x - 1} = \frac{3 - \dfrac{2}{x}}{1 - \dfrac{1}{x}}.$$

Now, as $|x| \to \infty$, each of the fractions $\dfrac{2}{x}$ and $\dfrac{1}{x}$ approaches zero. Thus,

$$\text{as } |x| \to \infty, \quad F(x) \to \frac{3 - 0}{1 - 0} = 3.$$

Hence, the line $y = 3$ is a horizontal asymptote of the graph of F.

To determine if $F(x)$ approaches this horizontal asymptote through values greater than three $[F(x) \to 3^+]$ or through values less than three $[F(x) \to 3^-]$ as $|x| \to \infty$, we choose an arbitrary large value of $|x|$ and determine if the approach is from above or from below the horizontal line $y = 3$. Here we choose $|x| = 100$:

$$F(100) = \frac{3(100) - 2}{(100) - 1} = \frac{298}{99} \approx 3.01$$

approaching $y = 3$ through values greater than 3

$$F(-100) = \frac{3(-100) - 2}{(-100) - 1} = \frac{-302}{-101} \approx 2.99$$

approaching $y = 3$ through values less than 3

Thus, we conclude $F(x) \to 3^+$ as $x \to +\infty$ and $F(x) \to 3^-$ as $x \to -\infty$. The appearance of the graph of F as $|x| \to \infty$ is shown in Figure 4.42.

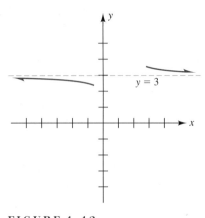

FIGURE 4.42

Appearance of the graph of

$F(x) = \dfrac{3x - 2}{x - 1}$ as $|x| \to \infty$

(b) The highest power of x that appears in the rational expression is x^2. Dividing each term of the numerator and denominator by x^2, we obtain

$$f(x) = \frac{x+3}{x^2 - 2x - 8} = \frac{\dfrac{1}{x} + \dfrac{3}{x^2}}{1 - \dfrac{2}{x} + \dfrac{8}{x^2}}.$$

Now, as $|x| \to \infty$, each of the fractions $\dfrac{1}{x}, \dfrac{3}{x^2}, \dfrac{2}{x}$, and $\dfrac{8}{x^2}$ approaches zero. Thus,

$$\text{as } |x| \to \infty, \quad f(x) \to \frac{0 + 0}{1 - 0 - 0} = 0.$$

Hence, the line $y = 0$ (the x-axis) is a horizontal asymptote of the graph of f. To determine whether

$$f(x) = \frac{x+3}{x^2 - 2x - 8} = \frac{x+3}{(x+2)(x-4)}$$

approaches the x-axis through values greater than zero [$f(x) \to 0^+$] or through values less than zero [$f(x) \to 0^-$] as $|x| \to \infty$, we choose an arbitrary large value of $|x|$ and determine if the approach is from above or from below the x-axis. We choose $|x| = 100$:

$$f(100) = \frac{103}{(102)(96)} \approx +0.01$$

$$f(-100) = \frac{-97}{(-98)(-104)} \approx -0.01$$

Thus, we conclude $f(x) \to 0^+$ as $x \to +\infty$ and $f(x) \to 0^-$ as $x \to -\infty$. The appearance of the graph of f as $|x| \to \infty$ is shown in Figure 4.43. ◆

Although the graph of a rational function can not intersect any of its vertical asymptotes, it can intersect its horizontal asymptote. To determine if the rational function

$$f(x) = \frac{x+3}{x^2 - 2x - 8}$$

in Example 2(b) intersects its horizontal asymptote $y = 0$, we solve the equation $f(x) = 0$ as follows:

$$\frac{x+3}{x^2 - 2x - 8} = 0$$

$$x + 3 = 0 \qquad \text{Multiply both sides by } x^2 - 2x - 8 \neq 0$$

$$x = -3$$

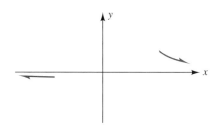

FIGURE 4.43
Appearance of the graph of
$f(x) = \dfrac{x+3}{x^2 - 2x - 8}$ as $|x| \to \infty$

Hence, we conclude that the graph of this rational function intersects its horizontal asymptote at $x = -3$.

PROBLEM 2 Show that the rational function defined in Example 2(a) does *not* intersect its horizontal asymptote $y = 3$ by showing that the equation $F(x) = 3$ leads to a contradiction. ◆

◆ **Oblique Asymptotes**

If the degree of the numerator $P(x)$ of a rational function is one more than the degree of the denominator $D(x)$ and $P(x)/D(x)$ is reduced to lowest terms, then the graph of the rational function $f(x) = P(x)/D(x)$ has an **oblique asymptote**—one that is neither vertical nor horizontal. To determine the oblique asymptote of the rational function $f(x) = P(x)/D(x)$, where the degree of $P(x)$ is one more than the degree of $D(x)$, we use polynomial long division to write the rational function in the form

$$f(x) = (mx + b) + \frac{R(x)}{D(x)}$$

where the degree of $R(x)$ is less than the degree of $D(x)$. Now,

$$\text{as } |x| \to \infty, \quad \frac{R(x)}{D(x)} \to 0.$$

Hence, $f(x) \to mx + b$, which we recognize as the equation of a straight line with slope m and y-intercept b.

EXAMPLE 3 Find the oblique asymptote for the graph of the rational function g defined by $g(x) = \dfrac{2x^3}{x^2 + 3}$. Then determine the appearance of the graph of g as $|x| \to \infty$.

◆ **Solution** Using polynomial long division, we divide $2x^3$ by $x^2 + 3$ as follows:

$$\begin{array}{r} 2x \\ x^2 + 3 \overline{) 2x^3 } \\ 2x^3 + 6x \\ \hline -6x \end{array}$$

Hence, we can write

$$g(x) = 2x + \frac{-6x}{x^2 + 3} = 2x + \frac{\dfrac{-6}{x}}{1 + \dfrac{3}{x^2}}$$

As $|x| \to \infty$, each of the fractions $\dfrac{-6}{x}$ and $\dfrac{3}{x^2}$ approaches zero. Thus,

$$\text{as } |x| \to \infty, \qquad g(x) \to 2x + \dfrac{0}{1+0} = 2x.$$

In summary, the line $y = 2x$ is an oblique asymptote of the graph of g.

To determine the appearance of the graph of g as $|x| \to \infty$, we choose an arbitrary large value of $|x|$ and determine if the approach is from above the line $y = 2x$ or from below this line. Choosing $x = 100$, we have $y = 2(100) = 200$, and

$$g(100) = \dfrac{2(100)^3}{(100)^2 + 3} \approx \underbrace{199.94}_{\text{approaching } y = 2x \text{ from below the line}}$$

Choosing $x = -100$, we have $y = 2(-100) = -200$, and

$$g(-100) = \dfrac{2(-100)^3}{(-100)^2 + 3} \approx \underbrace{-199.94}_{\text{approaching } y = 2x \text{ from above the line}}$$

The appearance of the graph of g as $|x| \to \infty$ is shown in Figure 4.44. ◆

Although the graph of a rational function cannot intersect its vertical asymptotes, the graph of a rational function *may* intersect its oblique asymptote. To determine whether the graph of a rational function f intersects its oblique asymptote $y = mx + b$, we solve the equation $f(x) = mx + b$ for x.

PROBLEM 3 Show that the rational function defined in Example 3 intersects its oblique asymptote $y = 2x$ at the origin. ◆

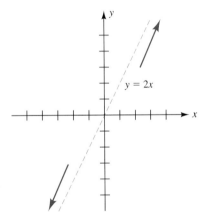

FIGURE 4.44

Appearance of the graph of $g(x) = \dfrac{2x^3}{x^2 + 3}$ as $|x| \to \infty$

◆ Sketching the Graph of a Rational Function

We can usually determine the basic shape of the graph of a rational function by noting the manner in which the graph approaches its asymptotes. To complete the graph of the function, we determine the intercepts, observe any symmetry and, if necessary, plot a few additional points. We now state a general procedure for graphing a rational function

$$f(x) = \dfrac{P(x)}{D(x)},$$

where $P(x)/D(x)$ is reduced to lowest terms.

General Procedure for Graphing a Rational Function

1. Find the vertical asymptotes of the graph of f, and determine the appearance of the graph as x approaches each vertical asymptote from the left and from the right.

2. Find the horizontal or oblique asymptote of the graph of f, and determine the appearance of the graph of f as $|x| \to \infty$.

3. (a) Determine whether the function f is an even function, an odd function, or neither.

   (b) Find the x-intercepts for the graph of f. Since $f(x) = 0$ only when its numerator $P(x) = 0$, the x-intercepts of the graph of f are the real zeros of P.

   (c) Find the y-intercept for the graph of f. This is obtained by evaluating $f(0)$.

   (d) Determine whether the graph crosses its horizontal or oblique asymptote.

4. Plot a few additional points, if necessary, and sketch the graph.

EXAMPLE 4 Sketch the graph of each rational function. Label the asymptotes and the intercepts.

(a) $F(x) = \dfrac{3x - 2}{x - 1}$ (b) $f(x) = \dfrac{x + 3}{x^2 - 2x - 8}$ (c) $g(x) = \dfrac{2x^3}{x^2 + 3}$

◆ **Solution**

(a) By combining the information in Figures 4.40 and 4.42, we begin to see the shape of the graph of $F(x) = \dfrac{3x - 2}{x - 1}$. To complete the graph of this function, we determine the x- and y-intercepts.

x-intercept: $3x - 2 = 0$ y-intercept: $F(0) = \dfrac{3(0) - 2}{(0) - 1} = 2$

$x = \dfrac{2}{3}$

Plotting the intercepts and noting the manner in which the graph approaches its asymptotes (Figures 4.40 and 4.42), we sketch the graph of this function, as shown in Figure 4.45.

(b) By combining the information in Figures 4.41 and 4.43, we begin to see the shape of the graph of $f(x) = \dfrac{x + 3}{x^2 - 2x - 8}$. To complete the graph of this function, we determine the x- and y-intercepts:

x-intercept: $x + 3 = 0$ y-intercept: $f(0) = \dfrac{(0) + 3}{(0)^2 - 2(0) - 8}$

$x = -3$

$= -\dfrac{3}{8}$

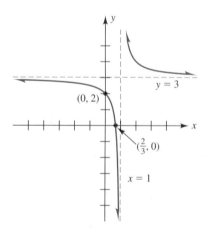

FIGURE 4.45

Graph of $F(x) = \dfrac{3x - 2}{x - 1}$

SECTION 4.5 ◆ *Rational Functions and Their Graphs* 271

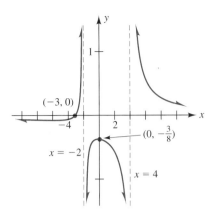

FIGURE 4.46

Graph of $f(x) = \dfrac{x+3}{x^2 - 2x - 8}$

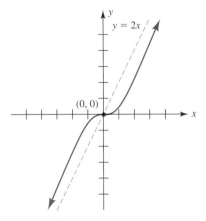

FIGURE 4.47

Graph of $g(x) = \dfrac{2x^3}{x^2 + 3}$

FIGURE 4.48

Graph of $f(x) = \dfrac{x+3}{x^2 - 2x - 8}$ as shown in two viewing rectangles

Plotting the intercepts and noting the manner in which the graph approaches its asymptotes (see Figures 4.41 and 4.43), we sketch the graph of $f(x) = \dfrac{x+3}{x^2 - 2x - 8}$, as shown in Figure 4.46.

(c) Using the information in Figure 4.44 and the fact that the graph of g has no vertical asymptote (why?), we begin to see the shape of the graph of $g(x) = \dfrac{2x^3}{x^2 + 3}$. Also, we note that $g(-x) = -g(x)$ for all x, so we know that g is an odd function. Hence, its graph is symmetric with respect to the origin. To complete the graph of this function, we determine the x- and y-intercepts.

$x\text{-intercept:}\quad 2x^3 = 0 \qquad y\text{-intercept:}\quad g(0) = \dfrac{2(0)^3}{(0)^2 + 3}$
$\phantom{x\text{-intercept:}\quad} x = 0 = 0$

Using the intercepts, the manner in which the graph approaches its oblique asymptote, and the fact that the graph is symmetric with respect to the origin, we sketch the graph of this function, as shown in Figure 4.47. ◆

Note that the graph of the rational function f in Figure 4.46 has a relative maximum point in the interval $(-2, 4)$. Although the methods of calculus are needed to determine the exact coordinates of this relative maximum point, we can estimate its coordinates by using a graphing calculator. When generating the graph of a rational function on a graphing calculator, we may obtain a distorted picture in the viewing rectangle. Figure 4.48 shows how the graph of $f(x) = \dfrac{x+3}{x^2 - 2x - 8}$ looks in two different viewing rectangles.

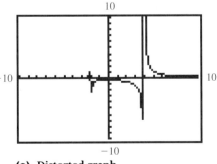

(a) Distorted graph

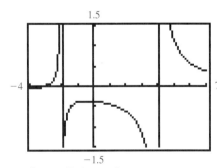

(b) Realistic graph

Note that the graph in part (a) of Figure 4.48 contains spikes and seems distorted, whereas the graph in part (b) is a more realistic picture. The vertical line segments through $x = -2$ and $x = 4$ approximate the vertical asymptotes of the graph and are not actually part of the graph. They occur in the viewing rectangle because the calculator evaluates the function only for a finite number of values, plots these values, and then attempts to connect these values to make a "continuous" curve. To generate a graph without these vertical lines, use the "dot" mode instead of the "connected" mode.

PROBLEM 4 Use a graphing calculator to estimate the coordinates of the relative maximum point for $f(x) = \dfrac{x + 3}{x^2 - 2x - 8}$. Record the coordinates to three significant digits. ◆

◆ Rational Functions with Common Factors

To sketch the graph of a rational function $f(x) = P(x)/D(x)$, where $P(x)$ and $D(x)$ have common factors, we begin by reducing $P(x)/D(x)$ to lowest terms. The graph of f is the same as the graph of the new function that is formed by the reducing process, except it may have one or more "holes," which indicates that f is not a continuous function.

EXAMPLE 5 Sketch the graph of $f(x) = \dfrac{x^2 + x - 2}{x - 1}$.

Solution Factoring the numerator, we obtain

$$f(x) = \frac{(x + 2)(x - 1)}{x - 1}$$

and reducing to lowest terms gives us the new function

$$F(x) = x + 2.$$

Note that the functions f and F are *not* the same functions. The domain of f is the set of all real numbers except 1, whereas the domain of F is the set of all real numbers. The graph of F is a straight line with slope 1 and y-intercept 2. The graph of f is the same as the graph of F, except for a hole at $(1, 3)$, which indicates that f is undefined at this point. The graph of f is shown in Figure 4.49. ◆

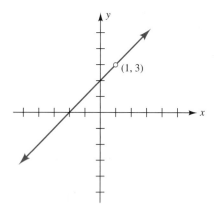

FIGURE 4.49

The graph of $f(x) = \dfrac{x^2 + x - 2}{x - 1}$ is discontinuous at $(1, 3)$.

PROBLEM 5 Generate the graph of the rational function defined in Example 5 in the viewing rectangle of a graphing calculator. Then try to evaluate the function at $x = 1$. What does the calculator display suggest? ◆

◆ Application: Average Cost Function

The *total cost* to run a business is the sum of the *fixed cost* (costs that do not depend on the level of production, such as mortgage payment, insurance, utilities, and so on) and *variable cost* (costs that vary with the level of production, such as maintenance of equipment, cost of materials, labor, and so on). If $T(x)$ is the total cost of producing x units of a product, then the **average cost function** C is defined by

SECTION 4.5 ♦ Rational Functions and Their Graphs

$$C(x) = \frac{T(x)}{x}$$

EXAMPLE 6 A certain manufacturer produces a kit for building a log home. The fixed cost for the company is $141,120 per year and the variable cost (in dollars) of producing x log-home kits per year is $12,000x + 20x^2$ for $0 \leq x \leq 300$.

(a) Define a rational function that represents the average cost per kit.

(b) Generate the graph of the function in part (a) in the viewing rectangle of a graphing calculator. Then find the approximate number of kits that should be produced in order to minimize the average cost per kit.

◀ Solution

(a) If $T(x)$ is the total cost of producing x log-home kits per year, then

$$T(x) = \underbrace{141,120}_{\text{fixed cost}} + \underbrace{(12,000x + 20x^2)}_{\text{variable cost}} \quad \text{for } 0 \leq x \leq 300.$$

Hence, the average cost per kit is

$$C(x) = \frac{141,120 + 12,000x + 20x^2}{x} \quad \text{for } 0 < x \leq 300.$$

(b) Observe the following characteristics of the graph of this function:

Vertical asymptote: $x = 0$

As $x \to 0^+$, $C(x) = \dfrac{141,120 + 12,000x + 20x^2}{x} \to +\infty$

Oblique asymptote: Dividing numerator and denominator by x, we obtain

$$C(x) = \frac{141,120}{x} + 12,000 + 20x.$$

Now,

$$\text{as } |x| \to \infty, \quad \frac{141,120}{x} \to 0.$$

Hence, $C(x) \to 12,000 + 20x$ as $|x| \to \infty$. Thus, the line $y = 20x + 12,000$ is an oblique asymptote of the graph of C. This information suggests that an appropriate viewing rectangle for observing the graph of this average cost function is as follows:

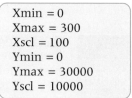

Figure 4.50 shows the graphs of the average cost function and its oblique asymptote in this viewing rectangle. The *x*-coordinate of the lowest point on this curve represents the number of kits that should be produced per year in order to minimize the average cost per kit. Using either the calculator's built-in program for finding a relative minimum point or the TRACE key and ZOOM key, we find that the coordinates of the relative minimum point are (84, 15360), rounded to the nearest whole number. In summary, the minimum average cost per kit is $15,360 and this occurs when 84 kits are produced per year. ◆

FIGURE 4.50
Graph of
$$C(x) = \frac{141{,}120 + 12{,}000x + 20x^2}{x}$$
for $0 < x \leq 300$ and its oblique asymptote $y = 20x + 12{,}000$

PROBLEM 6 For the problem described in Example 6, determine how many kits are produced per year if the average cost per kit is $16,640. ◆

Exercises 4.5

◆ Basic Skills

In Exercises 1–10,
(a) *find the vertical asymptote(s) of the graph of f, if any exist.*
(b) *if the graph of f has vertical asymptotes, determine the appearance of the graph near each vertical asymptote. Verify the appearance by using a graphing calculator.*

1. $f(x) = \dfrac{x}{x + 2}$

2. $f(x) = \dfrac{x + 1}{2x - 3}$

3. $f(x) = \dfrac{1}{x^2 + 4}$

4. $f(x) = \dfrac{x + 3}{x^2 - 4}$

5. $f(x) = \dfrac{x^3 - 1}{2x^2 + 5x - 3}$

6. $f(x) = \dfrac{-1}{x^2 + x + 4}$

7. $f(x) = \dfrac{x + 4}{x^3 - 2x - 4}$

8. $f(x) = \dfrac{x^2 + 2}{x^3 + x^2 - 2}$

9. $f(x) = \dfrac{2 - 3x}{x^3 - 5x^2 + 2x + 8}$

10. $f(x) = \dfrac{2x + 1}{2x^4 - x^3 - 18x^2 + 9x}$

In Exercises 11–20,
(a) *find the horizontal asymptote of the graph of f, if one exists.*
(b) *if the graph of f has a horizontal asymptote, determine the appearance of the graph as $|x| \to \infty$. Verify the appearance by using a graphing calculator.*
(c) *if the graph of f has a horizontal asymptote, determine whether the graph crosses the horizontal asymptote and, if so, where. Verify the coordinates by using a graphing calculator.*

11. $f(x) = \dfrac{x + 2}{x - 3}$

12. $f(x) = \dfrac{x^2 - 4}{2x + 1}$

13. $f(x) = \dfrac{x^2 - 1}{2x^2 + 7x - 15}$

14. $f(x) = \dfrac{2x^2 - 3x + 4}{x^3 + x - 1}$

15. $f(x) = \dfrac{x^3 + 3}{1 - 3x + 2x^2}$

16. $f(x) = \dfrac{x^3 + 2x - 1}{x^3 + 8}$

17. $f(x) = \dfrac{3x^2 - 2x + 1}{4x - x^2}$

18. $f(x) = \dfrac{1 - 4x^2}{x^2}$

19. $f(x) = \dfrac{3x - 1}{x^3 - 27}$

20. $f(x) = \dfrac{2x^3 - 3x + 4}{1 + x - 3x^3}$

In Exercises 21–36, sketch the graph of each rational function. Label the x- and y-intercepts and the vertical and horizontal asymptote(s), if they exist. Use a graphing calculator to verify each sketch.

21. $f(x) = \dfrac{3x}{x - 3}$

22. $f(x) = \dfrac{2x + 1}{x + 2}$

23. $F(x) = \dfrac{x + 6}{x^2 - 9}$

24. $F(x) = \dfrac{2}{4 - x^2}$

25. $g(x) = \dfrac{-1}{x^2 + 1}$

26. $g(x) = \dfrac{3x}{x^2 + 9}$

27. $G(x) = \dfrac{x}{x^2 - 4x - 12}$

28. $G(x) = \dfrac{10 - x}{x^2 + 3x - 10}$

29. $h(x) = \dfrac{x - 2}{2x^2 - x - 3}$

30. $h(x) = \dfrac{2x - 1}{x^2 + 2x + 1}$

31. $H(x) = \dfrac{2x^2 + 2x + 2}{x^2 - 4x}$

32. $H(x) = \dfrac{x^2 + 6x + 9}{2x^2 - 6x}$

33. $f(x) = \dfrac{6}{2x^3 + 9x^2 + 3x - 4}$

34. $F(x) = \dfrac{x^2}{x^3 + 3x^2 - x - 3}$

35. $G(x) = \dfrac{x + 2}{x^4 - 2x^3 + 2x - 1}$

36. $g(x) = \dfrac{-x^4}{x^4 + 2x^3 - 7x^2 + 2x - 8}$

In Exercises 37–44,
(a) find the oblique asymptote of the graph of f.
(b) determine whether the graph of f crosses the oblique asymptote and, if so, where.
(c) sketch the graph of f and label the x- and y-intercepts and any asymptote(s). Use a graphing calculator to verify each sketch.

37. $f(x) = \dfrac{x^2 - 4}{x}$

38. $f(x) = \dfrac{x^3 + 1}{x^2}$

39. $f(x) = \dfrac{x^2}{x - 3}$

40. $f(x) = \dfrac{x^2 + 1}{x + 1}$

41. $f(x) = \dfrac{x^2 - 3x - 4}{x + 3}$

42. $f(x) = \dfrac{x^2 - x}{x - 2}$

43. $f(x) = \dfrac{x^3 + 8}{x^2 - 3x - 4}$

44. $f(x) = \dfrac{x^3 - 2x - 4}{x^2 - 1}$

Each of the rational functions P(x)/D(x) in Exercises 45–50 is not reduced to lowest terms. Sketch the graph of the function and label the x- and y-intercepts and any asymptote or "hole." Use a graphing calculator to verify each sketch and to confirm any "hole" in the graph.

45. $F(x) = \dfrac{x^2 + 3x - 10}{x - 2}$

46. $f(x) = \dfrac{x^3 + 2x^2}{x}$

47. $g(x) = \dfrac{x^2 - x - 6}{x^2 + 4x + 4}$

48. $R(x) = \dfrac{2x^2 - 5x - 3}{x^2 - 4x + 3}$

49. $f(x) = \dfrac{x^4 - 2x^3 - x + 2}{x^3 - 2x^2}$

50. $h(x) = \dfrac{2x^4 + x^3 - 3x^2 + 5x - 5}{x^3 + x^2 + x - 3}$

51. The margins at the top and bottom of a poster are 2 inches each, and the margins at the sides are 4 inches each, as shown in the figure. The poster contains 50 square inches of printed matter.

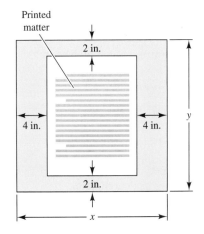

(a) Show that the area A of the entire poster is given by $A = \dfrac{4x^2 + 18x}{x - 8}$.

(b) Generate the graph of the rational function defined in part (a) on a graphing calculator. Then find the overall dimensions of the poster whose area is the smallest possible.

52. A rectangular box with an open top has a square base, as sketched in the figure. The volume of the box is 64 cubic inches.

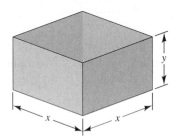

(a) Show that the surface area S of the box is given by $S = \dfrac{x^3 + 256}{x}$.

(b) Generate the graph of the rational function defined in part (a) on a graphing calculator. Then find the dimensions of the box whose surface area is the smallest possible.

53. A certain manufacturer produces wood-burning stoves. The fixed cost for the company is $81,000 per year. The variable cost (in dollars) of producing x stoves per year is $400x + 0.1x^2$ for $0 \le x \le 2000$.

(a) Define a function that represents the average cost per stove.

(b) Generate the graph of the function in part (a) on a graphing calculator. Then find the number of stoves that should be produced per year in order to minimize the average cost per stove.

54. A trucking company determines that the cost of operating one of its trucks is $(7 + 0.008x)$ dollars per mile when the truck is driven at x miles per hour ($x \le 75$). In addition, the company pays the driver $20 per hour.

(a) Express the total cost C per mile as a function of x.

(b) Generate the graph of the function in part (a) on a graphing calculator. Then find the average speed at which the trucks should be driven in order to minimize the total cost per mile.

 Critical Thinking

55. Can the graph of a rational function have two or more horizontal asymptotes? Explain.

56. Can the graph of a rational function have both a horizontal and oblique asymptote? Explain.

57. Give an example of a rational function whose graph crosses its horizontal asymptote more than once.

58. Give an example of a rational function whose graph crosses its oblique asymptote more than once.

59. Is $x = 1$ always a vertical asymptote of the graph of the rational function

$$f(x) = \dfrac{P(x)}{x - 1}$$

Explain.

60. If the degree of the numerator of a rational function is 2 more than the degree of its denominator, then the rational function has a *parabolic asymptote*. Find the parabolic asymptote of each function. Then use the parabolic asymptote and any vertical asymptotes and intercepts to help sketch the graph of each function. Use a graphing calculator to verify each sketch.

(a) $f(x) = \dfrac{x^3 + 1}{x}$

(b) $f(x) = \dfrac{x^4 + 6x^2 + 9}{x^2 - 1}$

61. Sketch the graph of each equation.

(a) $x + x^2y - y = 0$

(b) $2x^2 - xy + 3y - 1 = 0$

62. Sketch the graphs of the two given equations on the same coordinate plane. Then, by solving the equations simultaneously, determine the coordinates of the points where their graphs intersect. Use a graphing calculator to verify the coordinates.

(a) $y = \dfrac{1}{x^2 - 4}$ and $3x - 5y = 8$

(b) $y = \dfrac{x^3 - 1}{x^2}$ and $y = -2x - 4$

63. It is estimated that the cost C (in millions of dollars) to remove x percent of the pollution from Boston Harbor is given by $C(x) = \dfrac{500x}{100 - x}$. What can be said about the cost of cleaning up the harbor as $x \to 100^-$?

64. When a 100-ohm resistor is connected in parallel with a variable resistor x, the total resistance R is given by $R(x) = \dfrac{100x}{x + 100}$. What is the significance of the horizontal asymptote for the graph of R?

4.6 Rational Equations and Inequalities

◆ Introductory Comments

Recall from Section 1.7 that if the numerator and denominator of a fraction are polynomials of degree $n \geq 0$, then the fraction is called a *rational expression*. Since division by zero is undefined, the variables in a rational expression must be restricted to those values that give a nonzero denominator. For example, when working with the rational expression

$$\frac{4x^2 + 5x + 3}{2x - 6},$$

we assume that its denominator $2x - 6 \neq 0$, which implies that $x \neq 3$.

If both sides of an equation are composed of rational expressions, the equation is called a **rational equation**, and if both sides of an inequality are composed of rational expressions, the inequality is called a **rational inequality**. In this section, we discuss methods of solving rational equations and rational inequalities in one unknown.

◆ Solving Rational Equations

A rational equation is reducible to a polynomial equation (Section 4.4) after the denominators in the rational equation have been eliminated. To eliminate the denominators, we multiply both sides of the rational equation by the least common denominator (LCD) of all the fractions in the equation. When we multiply both sides of a rational equation by an LCD that contains a variable, we may introduce an *extraneous root*—a root that develops through the algebraic process but does not satisfy the original equation. This will occur when we multiply both sides of an equation by an LCD whose value is zero for certain values of the variable. Remember, multiplying both sides of an equation by zero does *not* generate an equivalent equation.

EXAMPLE 1 Solve each rational equation.

(a) $\dfrac{9}{x - 1} - 4 = \dfrac{1}{2}$ (b) $\dfrac{3}{2} - \dfrac{1}{m + 2} = \dfrac{m}{2m + 4}$

◀ Solution

(a) The LCD for the denominators $(x - 1)$ and 2 is $2(x - 1)$. We multiply both sides of this rational equation by $2(x - 1)$, assuming that $2(x - 1) \neq 0$:

$$2(x-1)\left[\frac{9}{x-1} - 4\right] = 2(x-1)\left[\frac{1}{2}\right] \quad \text{Multiply both sides by } 2(x-1)$$

$$18 - 8(x-1) = x - 1 \quad \text{Simplify}$$

$$26 - 8x = x - 1 \quad \text{Collect like terms on the left-hand side}$$

$$27 = 9x \quad \text{Add } (8x+1) \text{ to both sides}$$

$$x = 3 \quad \text{Divide both sides by 9 and interchange sides}$$

Note that if $x = 3$, then the LCD $2(x - 1) \neq 0$. Hence, we have generated equivalent equations and, therefore, the solution of the original equation must also be 3. The graph in Figure 4.51 confirms our work.

(b) The LCD for the denominators 2, $m + 2$, and $2m + 4$ [which factors to $2(m + 2)$] is $2(m + 2)$. We multiply both sides of this rational equation by $2(m + 2)$, assuming that $2(m + 2) \neq 0$:

$$2(m+2)\left[\frac{3}{2} - \frac{1}{m+2}\right] = 2(m+2)\left[\frac{m}{2(m+2)}\right] \quad \text{Multiply both sides by } 2(m+2)$$

$$3(m+2) - 2 = m \quad \text{Simplify}$$

$$3m + 4 = m \quad \text{Collect like terms on left-hand side}$$

$$4 = -2m \quad \text{Subtract } 3m \text{ from both sides}$$

$$m = -2 \quad \text{Divide both sides by } -2 \text{ and interchange sides}$$

However, if $m = -2$, then the LCD $2(m + 2) = 0$. Since multiplying both sides of an equation by zero does *not* generate an equivalent equation, the apparent solution, -2, must be discarded. Therefore, we conclude this rational equation has *no solution*. Also, note that the graph in Figure 4.52 does not intersect the x-axis, which indicates that the original equation has no solution. ◆

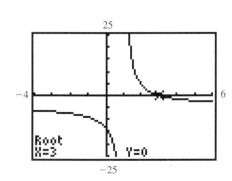

FIGURE 4.51
The x-intercept of the graph of $f(x) = \dfrac{9}{x-1} - \dfrac{9}{2}$ is the solution of the equation $\dfrac{9}{x-1} - 4 = \dfrac{1}{2}$.

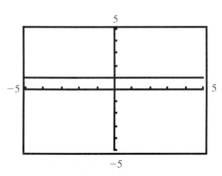

FIGURE 4.52
The graph of $f(m) = \dfrac{3}{2} - \dfrac{1}{m+2} - \dfrac{m}{2m+4}$ does not intersect the m-axis.

PROBLEM 1 Rewrite the function

$$f(m) = \frac{3}{2} - \frac{1}{m+2} - \frac{m}{2m+4}$$

with a single fraction on the right-hand side. Then show that the new function that is formed is $F(m) = 1$ provided $m = -2$. Compare your graph with Figure 4.52. ◆

SECTION 4.6 ♦ Rational Equations and Inequalities

The rational equations in Example 1 are reducible to polynomial equations of degree 1 (linear equations). In the next example we solve rational equations that are reducible to polynomial equations of degree $n \geq 2$.

EXAMPLE 2 Solve each rational equation.

(a) $\dfrac{1}{n-3} - \dfrac{6}{n^2 - 9} = 2$ (b) $x + \dfrac{3x+1}{x-1} + \dfrac{6}{x} = 12$

◀ **Solution**

(a) The LCD for $n - 3$ and $n^2 - 9$ [which factors to $(n - 3)(n + 3)$] is $(n - 3)(n + 3)$. We can eliminate fractions by multiplying both sides of this rational equation by the LCD, provided that $(n - 3)(n + 3) \neq 0$:

$$\dfrac{1}{n-3} - \dfrac{6}{n^2 - 9} = 2$$

$(n + 3) - 6 = 2(n - 3)(n + 3)$ **Multiply both sides by $(n-3)(n+3)$**

$n - 3 = 2n^2 - 18$ **Simplify both sides**

$2n^2 - n - 15 = 0$ **Write in standard form**

$(2n + 5)(n - 3) = 0$ **Factor**

$n = -\tfrac{5}{2} \quad n = 3$ **Solve for n**

Note that if $n = -\tfrac{5}{2}$, then the LCD $(n - 3)(n + 3) \neq 0$. Hence, we have generated equivalent equations and, therefore, a solution of the original equation must also be $-\tfrac{5}{2}$. However, if $n = 3$, then the LCD $(n - 3)(n + 3) = 0$. Since multiplying both sides of an equation by zero does not generate an equivalent equation, the apparent solution, 3, must be discarded. Therefore, we conclude this rational equation has only one solution, namely, $-\tfrac{5}{2}$, as illustrated in Figure 4.53.

(b) The LCD for $x - 1$ and x is $x(x - 1)$. We can eliminate fractions by multiplying both sides of this rational equation by the LCD, provided that $x(x - 1) \neq 0$:

$$x + \dfrac{3x+1}{x-1} + \dfrac{6}{x} = 12$$

$x[x(x-1)] + (3x+1)x + 6(x-1) = 12[x(x-1)]$ **Multiply both sides by $x(x-1)$**

$x^3 - 10x^2 + 19x - 6 = 0$ **Write in standard form**

By the rational zero theorem (Section 4.3), the possible rational roots of this polynomial equation are ± 1, ± 2, ± 3, and ± 6. The graph in

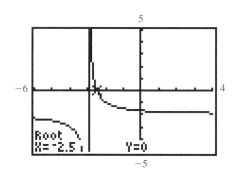

FIGURE 4.53
The n-intercept of the graph of
$f(n) = \dfrac{1}{n-3} - \dfrac{6}{n^2-9} - 2$
is the solution of the equation
$\dfrac{1}{n-3} - \dfrac{6}{n^2-9} = 2.$

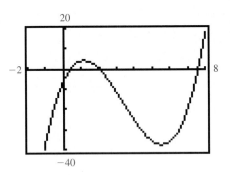

FIGURE 4.54
The graph of
$P(x) = x^3 - 10x^2 + 19x - 6$ indicates that 2 might be a rational root of the equation $x^3 - 10x^2 + 19x - 6 = 0$.

Figure 4.54 indicates that 2 may be a rational root, so we test 2 using synthetic division:

$$\begin{array}{r|rrrr} 2) & 1 & -10 & 19 & -6 \\ & & 2 & -16 & 6 \\ \hline & 1 & -8 & 3 & \boxed{0} \end{array}$$ ⎯⎯ remainder

Since the remainder is zero, we continue as follows:

$$x^3 - 10x^2 + 19x - 6 = 0$$
$$(x - 2)(x^2 - 8x + 3) = 0$$

  Apply the quadratic formula.

$$x = 2 \qquad x = 4 \pm \sqrt{13}$$

Note that if $x = 2$ or $x = 4 \pm \sqrt{13}$, then the LCD $x(x - 1) \neq 0$. Hence, we have generated an equivalent equation and, therefore, the solution of the original equation must also be

$$x = 2, \quad 4 + \sqrt{13} \approx 7.61, \quad \text{and} \quad 4 - \sqrt{13} \approx 0.394. \qquad \blacklozenge$$

PROBLEM 2 Generate the graph of

$$f(x) = x + \frac{3x + 1}{x - 1} + \frac{6}{x} - 12$$

in the viewing rectangle of a graphing calculator, and confirm that the roots are $x = 2$, $4 + \sqrt{13} \approx 7.61$, and $4 - \sqrt{13} \approx 0.394$, as we found in Example 2(b). $\qquad \blacklozenge$

◆ Solving Rational Inequalities

An inequality that can be written with a rational expression on the left-hand side and zero on the right-hand side is called a **rational inequality in standard form**. The rational inequality

$$\frac{5x - 2}{x - 1} \leq 4,$$

can be written in standard form by proceeding as follows:

$$\frac{5x - 2}{x - 1} - 4 \leq 0 \qquad \textbf{Subtract 4 from both sides}$$

$$\frac{(5x - 2) - 4(x - 1)}{x - 1} \leq 0 \qquad \textbf{Write as a single fraction}$$

$$\frac{x + 2}{x - 1} \leq 0 \qquad \textbf{Simplify}$$

CAUTION When writing a rational inequality in standard form, we cannot multiply both sides of the inequality by an expression that contains the unknown. Since we don't know whether the unknown quantity is positive or negative, we don't know whether to change the direction of the inequality symbol. For example, to eliminate the fraction by writing

$$\frac{5x-2}{x-1} \leq 4 \quad \text{as} \quad 5x - 2 \leq 4(x-1) \quad \text{is WRONG!}$$

This procedure assumes that $x - 1 > 0$, which may not be true.

To solve the rational inequality in standard form $f(x) > 0$, we begin by finding the values of the unknown that make the numerator or denominator of the rational expression zero. We refer to these values of the unknown as the *boundary numbers* of the rational inequality. These boundary numbers are the x-intercepts and vertical asymptotes of the graph of f, and they divide the x-axis into intervals. The solution set of the rational inequality $f(x) > 0$ is all the intervals on which the graph of f is *above* the x-axis. For the rational inequality $f(x) < 0$, the solution set is all the intervals on which the graph of f is *below* the x-axis. If the inequality symbol is $\leq$ or $\geq$, we include the boundary numbers that represent the x-intercepts. Since division by zero is undefined, the solution set of a rational inequality cannot include any boundary numbers that represents a vertical asymptote.

EXAMPLE 3 Solve each rational inequality.

(a) $\dfrac{5x-2}{x-1} \leq 4$ (b) $\dfrac{2x+3}{x-3} > \dfrac{x+1}{x-1}$

◆ Solution

(a) As we have found in the preceding discussion, the rational inequality $\dfrac{5x-2}{x-1} \leq 4$ is equivalent to the inequality

$$\frac{x+2}{x-1} \leq 0.$$

We find the boundary numbers of a rational inequality in standard form by setting its numerator and denominator equal to zero.

For numerator: $x + 2 = 0$ For denominator: $x - 1 = 0$
                 $x = -2$ $x = 1$

These two boundary numbers divide the real number line into three intervals:

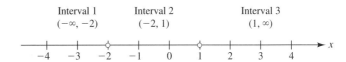

We can confirm these intervals by generating the graph of $f(x) = \dfrac{x + 2}{x - 1}$ in the viewing rectangle of a graphing calculator. As shown in Figure 4.55, the graph of f is *below* the x-axis in the interval $(-2, 1)$. Hence, $f(x) \leq 0$ in the interval $[-2, 1)$. Thus, the solution set of the original inequality is $[-2, 1)$.

(b) We begin by writing the inequality in standard form:

$$\frac{2x + 3}{x - 3} > \frac{x + 1}{x - 1}$$

$$\frac{2x + 3}{x - 3} - \frac{x + 1}{x - 1} > 0 \qquad \text{Subtract } \frac{x + 1}{x - 1} \text{ from both sides}$$

$$\frac{(2x + 3)(x - 1) - (x + 1)(x - 3)}{(x - 3)(x - 1)} > 0 \qquad \text{Subtract fractions}$$

$$\frac{x^2 + 3x}{(x - 3)(x - 1)} > 0 \qquad \text{Simplify the numerator}$$

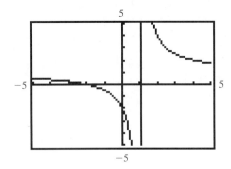

FIGURE 4.55

The graph of $f(x) = \dfrac{x + 2}{x - 1}$ is below the x-axis in the interval $(-2, 1)$.

We then determine the boundary numbers:

For numerator:

$x^2 + 3x = 0$

$x(x + 3) = 0$

$x = 0 \quad x = -3$

For denominator:

$(x - 3)(x - 1) = 0$

$x = 3 \quad x = 1$

Thus, the four boundary numbers are $-3, 0, 1,$ and 3. These four boundary numbers divide the real number line into five intervals:

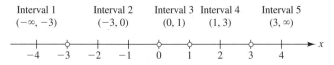

We can confirm these intervals by generating the graph of $f(x) = \dfrac{x^2 + 3x}{(x - 3)(x - 1)}$ in the viewing rectangle of a graphing calculator. As shown in Figure 4.56, the graph of f is above the x-axis in the interval $(-\infty, -3), (0, 1),$ or $(3, \infty)$. Hence, $f(x) > 0$ in the interval $(-\infty, -3), (0, 1),$ or $(3, \infty)$. Thus, the solution set of the original inequality is

$$(-\infty, -3) \cup (0, 1) \cup (3, \infty).$$

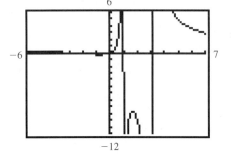

FIGURE 4.56

The graph of $f(x) = \dfrac{x^2 + 3x}{(x - 3)(x - 1)}$ is above the x-axis in the interval $(-\infty, -3), (0, 1),$ or $(3, \infty)$.

Observe in Figures 4.55 and 4.56 that in each of the intervals formed by the boundary numbers, the graph of the rational function is entirely above the x-axis or entirely below the x-axis. Hence, we can use the test number method (Section 3.4) to help solve a rational inequality in standard form without referring to a graph. We simply select an arbitrary test number from each interval formed by the boundary numbers, and then determine the algebraic sign of the rational expression. If the rational expression is negative for a particular test number in an interval, then it is negative for every value of x in that interval. If the rational expression is positive for a particular test number in an interval, then it is positive for every value of x in that interval.

For example, to solve the rational inequality

$$\frac{x + 2}{x - 1} \leq 0$$

in Example 3(a) by using the test number method, we select an arbitrary test number from each of the intervals $(-\infty, -2)$, $(-2, 1)$, and $(1, \infty)$, and then determine the algebraic sign of the rational expression $\dfrac{x + 2}{x - 1}$ for each of the test numbers. The following table organizes our work:

Interval	Test Number	Sign of $\dfrac{x + 2}{x - 1}$
$(-\infty, -2)$	-3	$\dfrac{-3 + 2}{-3 - 1} = \dfrac{1}{4} > 0$
$(-2, 1)$	-1	$\dfrac{-1 + 2}{-1 - 1} = -\dfrac{1}{2} < 0$
$(1, \infty)$	2	$\dfrac{2 + 2}{2 - 1} = 4 > 0$

Since $\dfrac{x + 2}{x - 1} < 0$ when $x = -1$, we know that $\dfrac{x + 2}{x - 1} < 0$ for every value of x in the interval $(-2, 1)$. Hence, we conclude that the solution set of $\dfrac{x + 2}{x - 1} \leq 0$ is

$$[-2, 1)$$

Never include the value of the asymptote in a solution set.

This answer agrees with our work in Example 3(a).

PROBLEM 3 Solve the rational inequality $\dfrac{x^2 + 2x - 4}{x + 2} \geq 1$ by using the test number method. ◆

◆ **Application: Uniform Motion**

Objects that move at a constant rate of speed are in *uniform motion*. The formula

$$d = rt$$

is used to find the distance d traveled when given the constant rate of speed r over a given time t. We may also write this formula as follows:

$$r = \frac{d}{t} \quad \text{and} \quad t = \frac{d}{r}$$

As illustrated in the next example, uniform motion problems often involve the solution of a rational equation.

EXAMPLE 4 A college rowing team practices on the Charles River in Boston. In order to qualify for the NCAA championship trials, the crew must be able to row at a constant rate of 18 mi/h in still water. Their coach observes that while rowing at a constant rate of speed, the crew travels 3 miles downstream in the same time it takes them to row 2 miles upstream. He knows that the Charles River flows at a constant rate of 3 mi/h. Will his crew qualify for the NCAA championship trials?

◆ **Solution**   Let

   $x =$ the crew's rate of rowing (in mi/h) in still water.

Since the river flows at a constant rate of 3 mi/h, we can state that

   $x + 3 =$ the crew's rate of rowing (in mi/h) *downstream*

and

   $x - 3 =$ the crew's rate of rowing (in mi/h) *upstream*.

We can arrange these facts as a table:

	Rate, r	Distance, d	Time, $t = \dfrac{d}{r}$
Downstream	$x + 3$	3	$\dfrac{3}{x + 3}$
Upstream	$x - 3$	2	$\dfrac{2}{x - 3}$

Since the problem states that the time to travel 3 miles downstream *is the same as* the time to travel 2 miles upstream, we can write the following equation:

SECTION 4.6 ♦ *Rational Equations and Inequalities*

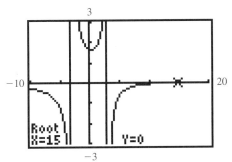

FIGURE 4.57
The x-intercept of the graph of
$f(x) = \dfrac{3}{x+3} - \dfrac{2}{x-3}$ is 15.

$$\text{time downstream} = \text{time upstream}$$

$$\frac{3}{x+3} = \frac{2}{x-3}$$

$$3(x-3) = 2(x+3) \qquad \text{Multiply both sides by the LCD, } (x+3)(x-3)$$

$$3x - 9 = 2x + 6$$

$$x = 15$$

Thus, the crew's rate in still water is 15 mi/h. At this rate, the team will not qualify for the NCAA championship trials. The graph in Figure 4.57 supports our work. ◆

PROBLEM 4 Refer to Example 4: If the speed of the river is 3 mi/h and the crew can row 4 miles downstream in the same time it takes them to row 3 miles upstream, will the team qualify for the NCAA trials? ◆

♦ **Application: Rate of Work**

Another type of applied problem that involves solving a rational equation is the *rate of work problem*. Suppose it takes t hours to complete a certain job when working at a constant rate of speed. Then, $1/t$ of the job is done in 1 hour. For example, if it takes 3 hours to mow a lawn, then $\frac{1}{3}$ of the job is done in 1 hour. We use this idea to solve the next example.

EXAMPLE 5 Two spillways from a reservoir can be used to flood a cranberry bog. If both spillways are used together, the bog floods in 24 hours. If only the larger spillway is used, the bog will flood in 10 hours less time than if only the smaller spillway is used. What time is required using each spillway alone to flood the cranberry bog?

◆ **Solution** Let

$$x = \text{amount of time (in hours) to flood the bog using only the smaller spillway}$$

and

$$x - 10 = \text{amount of time (in hours) to flood the bog using only the larger spillway.}$$

Hence, we have the following equation:

$$\underbrace{\frac{1}{x}}_{\substack{\text{part of job done}\\\text{by smaller spill-}\\\text{way in 1 hour}}} + \underbrace{\frac{1}{x-10}}_{\substack{\text{part of job done}\\\text{by larger spill-}\\\text{way in 1 hour}}} = \underbrace{\frac{1}{24}}_{\substack{\text{part of job done}\\\text{by both spillways}\\\text{together in 1 hour}}}$$

$$24(x - 10) + 24x = x(x - 10) \qquad \text{Multiply both sides by the LCD}$$

$$x^2 - 58x + 240 = 0 \qquad \text{Write in standard form}$$

Now, to solve this quadratic equation in standard form, we apply the quadratic formula:

$$x = \frac{-(-58) \pm \sqrt{(-58)^2 - 4(1)(240)}}{2(1)} \approx 4.5 \text{ or } 53.5$$

We disregard the solution of 4.5, since this value yields a negative time for the larger spillway to flood the bog. Hence, the only acceptable solution for this quadratic equation is 53.5. In summary, the bog floods in 53.5 hours using only the smaller spillway and in 43.5 hours using only the larger spillway. The graph in Figure 4.58 supports our work. ◆

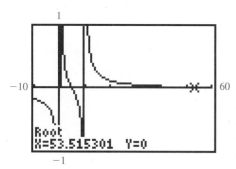

FIGURE 4.58
An x-intercept of the graph of $f(x) = \dfrac{1}{x} + \dfrac{1}{x-10} - \dfrac{1}{24}$ is approximately 53.5.

PROBLEM 5 Suppose that a cranberry bog floods in 48 hours when a smaller spillway is used alone and it floods in 36 hours when a larger spillway is used alone. How long does it take to flood the cranberry bog if both spillways operate simultaneously? ◆

Exercises 4.6

◆ Basic Skills

In Exercises 1–28, solve each rational equation. Check graphically with a graphing calculator.

1. $\dfrac{3}{x+2} - \dfrac{1}{x-2} = \dfrac{x}{x^2-4}$

2. $\dfrac{1}{x+1} - \dfrac{1}{x-1} = \dfrac{2x}{x^2-1}$

3. $\dfrac{2x-1}{4x+1} - \dfrac{3x+2}{6x+1} = 0$

4. $\dfrac{2x+3}{3x} + \dfrac{1}{3} = \dfrac{x-3}{x+3}$

5. $\dfrac{3}{x-2} - \dfrac{1}{2-3x} = \dfrac{2}{3x^2-8x+4}$

6. $\dfrac{2}{x^2+5x+4} + \dfrac{1}{x^2+x} = \dfrac{2}{x^2+4x}$

7. $\dfrac{3}{x^3+1} + \dfrac{x}{x^2-x+1} = \dfrac{1}{x+1}$

8. $\dfrac{1}{8x^3-1} = \dfrac{2x}{4x^2+2x+1} - \dfrac{1}{2x-1}$

9. $\dfrac{n-1}{5} = \dfrac{4}{n}$

10. $\dfrac{x}{27} - \dfrac{1}{x} = \dfrac{2}{9}$

11. $\dfrac{1}{x} - \dfrac{1}{6} = \dfrac{x-1}{x+1}$

12. $\dfrac{1}{p} + \dfrac{3}{4} = \dfrac{2p+3}{3p-1}$

13. $\dfrac{5-x}{x-3} = \dfrac{3}{x^2-4x+3}$

14. $\dfrac{7}{x^2-4} - \dfrac{2}{x+2} = 1$

15. $\dfrac{x+2}{x+1} - \dfrac{x-3}{x-1} = \dfrac{x^2+2x-1}{x^2-1}$

16. $\dfrac{x}{2x+1} - \dfrac{3}{3-x} = \dfrac{13}{2x^2-5x-3}$

17. $\dfrac{n-1}{n-3} - 2 = \dfrac{1}{2} - \dfrac{n-5}{n^2-5n+6}$

18. $\dfrac{2}{m-1} + \dfrac{1}{m^2-4m+3} + 2 = \dfrac{3}{m-3}$

19. $\dfrac{1}{n} - \dfrac{2n}{5} = \dfrac{1}{2n}$

20. $\dfrac{2x}{3} = x^2 + \dfrac{1}{2}$

21. $\dfrac{1}{p} - \dfrac{1}{3} = \dfrac{4}{p-3}$

22. $\dfrac{x}{x+1} - \dfrac{x+1}{x} = \dfrac{1}{2}$

23. $\dfrac{x}{x^2-4} + \dfrac{1}{x+2} = 1$

24. $\dfrac{2}{3t+2} + 1 = \dfrac{t}{3t-2} + \dfrac{8}{4-9t^2}$

25. $\dfrac{4}{x} + \dfrac{2}{x-1} = \dfrac{6x}{x+1}$

26. $\dfrac{5}{x+2} - \dfrac{2x}{3} = \dfrac{x-6}{x}$

27. $\dfrac{x}{x+3} + \dfrac{16x}{x^2-9} = \dfrac{3x^2}{2}$

28. $\dfrac{x+1}{x-2} + \dfrac{x-1}{x+2} = \dfrac{3x}{x^2-4} + \dfrac{3}{x}$

In Exercises 29–48, solve each rational inequality. Write each solution set using interval notation.

29. $\dfrac{1}{x} > 2$

30. $\dfrac{2}{3x} < 1$

31. $\dfrac{x+3}{x-1} < 0$

32. $\dfrac{3-2x}{5x} > 0$

33. $\dfrac{x-4}{2x+1} \geq 3$

34. $\dfrac{3x+2}{x-1} \leq 1$

35. $\dfrac{x^2+3x-4}{x-2} \leq 0$

36. $\dfrac{x^2+5x-14}{x^2+5x-24} \geq 0$

37. $\dfrac{x^2+6x+2}{x^2-25} < 0$

38. $\dfrac{3x-4}{x^2+4x+1} < 0$

39. $\dfrac{1}{x} > \dfrac{2x+1}{x+8}$

40. $\dfrac{2x-3}{x+4} < \dfrac{x-1}{x+3}$

41. $\dfrac{1}{x-2} + \dfrac{x}{x+3} < \dfrac{1}{2}$

42. $x + \dfrac{x-3}{2x+5} < 3$

43. $\dfrac{7}{x^2+7x+12} \geq \dfrac{x-3}{x+4}$

44. $\dfrac{x}{x^2+2x-15} \geq \dfrac{2}{x^2-25}$

45. $\dfrac{x-2}{x^3+2x^2+5x+10} < 0$

46. $\dfrac{x^3-3x+2}{x^2+1} > 0$

47. $\dfrac{5}{x^2-4} \geq \dfrac{4x}{3x+3}$

48. $\dfrac{5x+3}{x^2+2x+5} \leq \dfrac{3x}{2x^2-2}$

In Exercises 49–60, set up a rational equation, and then solve it.

49. One number is three times another, and the sum of their reciprocals is $\frac{1}{3}$. Find the two numbers.

50. The denominator of a fraction exceeds its numerator by 18. The reduced value of the fraction is $\frac{4}{7}$. What is the fraction?

51. A girl, starting from her home, leaves on a bicycle trip. After riding 2 miles at the rate of 12 mi/h, the chain on her bicycle becomes disengaged. She immediately jogs back home along the same route. Find the rate of speed (in mi/h) at which she jogs if she returns home 30 minutes after she left.

52. A canoeist paddles 8 miles upstream in twice the time it takes to paddle the same distance downstream. Find the rate of the current in the river if the canoeist can paddle 3 mi/h in still water.

53. Two ferries leave Woods Hole for Nantucket. One ferry travels 2.2 knots faster than the other and requires 24 minutes less time to make the trip. Find the speed (in knots) of each ferry if the distance between ports is 24 nautical miles.

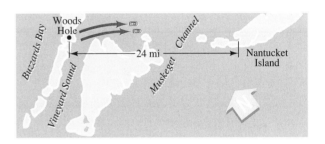

54. A college rowing team rows 6 miles down river and back again in 1 hour 15 minutes. If the river flows at 2 mi/h, find the crew's rate of speed in still water.

55. A tank can be filled by an inlet pipe in 15 hours, and emptied by an outlet pipe in 25 hours. The tank is initially empty, and then both pipes are opened. How long will it take to fill the tank?

56. A man can paint his house in 8 days. His son alone could paint it in 12 days. How long will it take to paint the house if they work together?

57. Using a large sump pump and small sump pump together, a homeowner can drain all the water from a cellar in 24 minutes. Using only one sump pump to drain the water requires 20 minutes longer with the smaller pump than with the larger one. Find the time required to drain this amount of water using each pump alone.

58. A swimming pool can be filled with two hoses in 8 hours. Using the smaller hose alone, the pool can be filled in 20 hours. How long would it take to fill the pool using only the larger hose?

59. A group of college students rents a bus for a ski trip. The cost for the bus is $480, and the students divide this expense evenly. Because of sickness, four students are unable to travel. This increases the share that each

remaining person must pay by $10. How many students go on the ski trip?

60. A real estate developer bought two parcels of land for $600,000 each. The larger parcel contains 10 acres more than the smaller parcel yet costs $10,000 per acre less. How many acres are in each parcel?

Critical Thinking

In Exercises 61–68, solve each literal equation for x.

61. $\dfrac{1}{m} - \dfrac{1}{x} = \dfrac{1}{n} + \dfrac{1}{x}$

62. $\dfrac{1}{a} - \dfrac{1}{x} = \dfrac{1}{b}$

63. $\dfrac{2x + k}{x - k} - \dfrac{x - k}{x + k} = 1$

64. $\dfrac{a + x}{a - x} = \dfrac{x^2}{a^2 - x^2}$

65. $\dfrac{a}{x - 1} + a = bx$

66. $\dfrac{a}{x} + x = a + \dfrac{1}{x}$

67. $x + \dfrac{1}{a} = a + \dfrac{1}{x}$

68. $\dfrac{x}{a} + \dfrac{a}{x} = \dfrac{a}{b} + \dfrac{b}{a}$

In Exercises 69 and 70, solve each double inequality. Use interval notation to describe the solution set

69. $-3 \leq \dfrac{2x - 1}{x + 2} \leq 7$

70. $\dfrac{1}{x} \leq \dfrac{2 - 6x}{x - 5} < 10$

In Exercises 71 and 72, use interval notation to describe the domain of each function.

71. $f(x) = \sqrt{\dfrac{2x + 3}{x - 1}}$

72. $f(x) = \sqrt{\dfrac{x^2 - 4}{x + 1}}$

73. On a math test, a student multiplied both sides of the inequality $\dfrac{x}{x + 1} < 0$ by $x + 1$ to obtain $x < 0$, and wrote $(-\infty, 0)$ as the solution set. Explain why this procedure is incorrect, and then show the correct method of solving this inequality.

74. Find a rational inequality whose solution set is $(-\infty, -2) \cup [0, \infty)$. Several answers are correct.

75. If two resistances, R_1 and R_2, are connected in parallel in an electric circuit, then the total effective resistance R can be found by using the formula

$$R = \dfrac{R_1 R_2}{R_1 + R_2}$$

If R_1 is 10.6 ohms (Ω), what is the range of values for R_2 if the total effective resistance is positive but not more than 8.2 Ω?

76. In an ant colony, the relationship between the number of adult ants A and the number of larvae L that survive to become adults is given by

$$L = \dfrac{6000A}{A + 1200}.$$

Under what conditions is the number of larvae that survive to become adults less than the number of adult ants in the colony?

Chapter 4 Review

◆ Questions for Writing or Group Discussion

1. What are *relative extrema* of a polynomial function? How are extrema used as an aid in graphing a polynomial function?
2. What can be said about the zeros of a polynomial function P of degree n if the graph of P
   (a) does not cross or touch the x-axis.
   (b) crosses the x-axis n times.

3. Under what conditions is the graph of a polynomial function *tangent* to the *x*-axis? Discuss the types of tangency that can occur for the graph of a polynomial function.

4. Does each of the following sketches represent the graph of a *polynomial function*? Explain.

(a)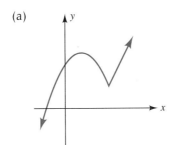

(b)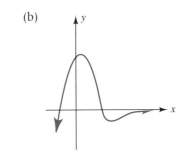

5. Explain the method for checking *polynomial long division*. Illustrate with an example.

6. Under what conditions is it possible to divide one polynomial by another using *synthetic division*?

7. Explain how synthetic division and the *remainder theorem* can be used to determine if r is a zero of a polynomial function P.

8. Explain how synthetic division and the *factor theorem* can be used to determine if $x - r$ is a factor of a polynomial $P(x)$.

9. Determine the possible combinations of real and imaginary zeros of a polynomial function P if
   (a) P is a *quartic function* with real coefficients.
   (b) P is a *quintic function* with real coefficients.

10. If P is a *cubic function* with real coefficients, find
    (a) the maximum number of real zeros of P.
    (b) the minimum number of real zeros of P.

11. Explain the procedure for determining the possible *rational zeros* of a polynomial function.

12. Does the *fundamental theorem of algebra* tell us how to find a zero of a polynomial function? If not, what does it tell us?

13. Is it possible for a *polynomial equation* with real coefficients to have three imaginary roots? Explain.

14. How is the domain of a *rational function* determined? Give an example of a rational function whose domain is (a) $(-\infty, a) \cup (a, \infty)$ or (b) all real numbers.

15. Is it possible for the graph of a rational function to cross its *vertical asymptote*? Explain.

16. Discuss the procedure for finding the *horizontal asymptote* of a rational function.

17. Discuss the procedure for finding the *oblique asymptote* of a rational function.

18. Does each of the following sketches represent the graph of a rational function? Explain.

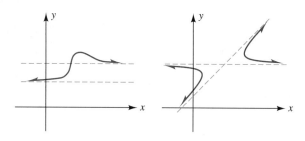

19. Explain how an extraneous root may develop when solving a *rational equation*. Illustrate with an example.

20. What is meant by the *boundary numbers* of a polynomial inequality? of a rational inequality?

Review Exercises

In Exercises 1–8, divide the first polynomial by the second polynomial. Use synthetic division whenever possible.

1. $3x^3 + 2x^2 - 3x + 4$; $x^2 + 1$
2. $4x^3 + 3x^2 - x + 6$; $x - 1$
3. $2x^4 + 2x^3 + 3x - 1$; $x + 2$
4. $3x^4 + 8x^2 + 3x - 1$; $x^2 + 2x - 2$
5. $6x^4 - 2x^3 + 9x^2 - 4$; $x - \frac{1}{3}$
6. $4x^4 - x^3 + 5x^2 - 2x + 1$; $x + \frac{3}{4}$
7. $x^6 + 1$; $x^3 + x + 1$
8. $x^5 - 8x^3 - 7x + 6$; $(x + 3)^2$

In Exercises 9–12, use synthetic division and the remainder theorem to find the indicated functional value. Check graphically with a graphing calculator.

9. $P(x) = 2x^3 - 3x^2 + 5x - 8$; $P(2)$
10. $P(x) = 4x^3 + 7x^2 - 8x - 6$; $P(-3)$
11. $f(x) = 3x^4 - x^3 + 4x^2 - 5x - 6$; $f\left(-\frac{2}{3}\right)$
12. $F(x) = 6x^4 - x^3 - 10x^2 - 5x + 3$; $F\left(\frac{3}{2}\right)$

In Exercises 13–18, use the factor theorem to determine whether the first polynomial is a factor of the second polynomial.

13. $x - 2$; $4x^3 - 7x^2 - 4$
14. $x + 3$; $2x^3 + 3x^2 - 7x - 9$
15. $x + \frac{4}{3}$; $3x^4 + x^3 + 2x^2 + 8x + 4$
16. $x - \frac{1}{4}$; $4x^4 + 3x^3 - x^2 - 16x + 4$
17. $x - 2i$; $x^5 + 3x^3 + 2x^2 - 4x + 8$
18. $x - (2 + i)$; $2x^4 + 7x^3 + 5x^2 - 9x - 5$

In Exercises 19–30, (a) write P(x) as the product of linear factors, and (b) find all zeros of P.

19. $P(x) = x^3 - 5x^2 + 2x + 8$
20. $P(x) = 2x^3 + x^2 - 13x + 6$
21. $P(x) = 3x^3 - 10x^2 - 7x + 20$
22. $P(x) = x^3 - 10x^2 + 34x - 40$
23. $P(x) = 2x^4 + 5x^3 + 5x^2 + 20x - 12$
24. $P(x) = 2x^4 - 9x^3 + 8x^2 + 19x - 30$
25. $P(x) = 4x^5 - 4x^4 - 15x^3 + 10x^2 + 11x - 6$
26. $P(x) = 4x^5 + 16x^4 - 23x^3 - 13x^2 + 11x + 5$
27. $P(x) = x^4 - 4x^3 + 9x^2 - 4x + 8$; $x + i$ is a factor
28. $P(x) = x^4 - 6x^3 + 11x^2 - 10x + 2$; $x - (1 - i)$ is a factor
29. $P(x) = 2x^5 - 9x^4 + 22x^3 - 41x^2 + 56x - 20$; $2 + i$ is a zero
30. $P(x) = x^6 + 2x^5 + 5x^4 - 12x^3 - 29x^2 - 70x - 25$; $2i - 1$ is a zero of multiplicity two

In Exercises 31–44, find all real roots of each equation. Record irrational roots to the nearest hundredth.

31. $x^3 + 2x - 8 = 0$
32. $2x^3 + 5x^2 + 3 = 0$
33. $2x^4 - 5x^3 + 6x^2 - 5x = 25$
34. $x^3 + 8x^2 + 2x = 3$
35. $x^4 + 6x^3 + 4 = 0$
36. $x^5 - 3x^4 - 2x^2 + x + 15 = 0$
37. $x^5 - 8x^4 + 20x^3 - 19x^2 + 12x = 12$
38. $3x^6 + 9x^5 + 10x^4 - 15x^2 - 17x = 6$

39. $\dfrac{x}{2x-1} + \dfrac{1}{2} = \dfrac{3x-1}{6x-3}$

40. $\dfrac{3}{x+3} - \dfrac{1}{x-3} = \dfrac{x}{x^2-9}$

41. $\dfrac{x}{x+4} - \dfrac{2}{x-1} = \dfrac{20}{x^2+3x-4}$

42. $\dfrac{x-1}{x} = \dfrac{5}{2} - \dfrac{x}{x-1}$

43. $\dfrac{1}{2} - \dfrac{1}{x} = \dfrac{1}{x-2}$

44. $x + \dfrac{1}{x} = \dfrac{1}{3} + 3$

In Exercises 45–52, solve each inequality. Use interval notation to describe the solution set.

45. $x^3 + 2x^2 \geq 4x$

46. $x^4 - 5x^2 + 4 < 0$

47. $x^5 - 6x^4 + 16x^3 - 32x^2 + 48x > 32$

48. $x^6 - 2x^5 - 7x^4 + 16x^3 - 17x^2 + 18x \leq 9$

49. $\dfrac{p-4}{p+2} \geq 0$

50. $\dfrac{x^2 - 2x + 3}{x+1} \leq 0$

51. $\dfrac{1}{x} < \dfrac{3(x-1)}{x+4}$

52. $m + \dfrac{3m}{2-m} > 4$

In Exercises 53–66, sketch the graph of each function. Label the x- and y-intercepts and any asymptote(s). Check each sketch with a graphing calculator.

53. $g(x) = 2x^2 - x^4$

54. $h(x) = 4x^3 - 3x^4$

55. $f(x) = x^3 + 3x^2 - 6x - 8$

56. $F(x) = 10 - 11x - 4x^2 - 3x^3$

57. $G(x) = 4x^4 - 12x^3 - 39x^2 - 3x - 10$

58. $P(x) = x^4 - x^3 - 9x^2 + 9x$

59. $f(x) = \dfrac{3x-2}{x+1}$

60. $F(x) = \dfrac{1}{x^2+4}$

61. $G(x) = \dfrac{x}{x^2 - 3x - 10}$

62. $g(x) = \dfrac{x+4}{x^3 - 4x^2 + 4x}$

63. $F(x) = \dfrac{x^2 - 1}{2x}$

64. $f(x) = \dfrac{x^2 + 4}{x - 1}$

65. $H(x) = \dfrac{x^2 - 8x + 15}{x - 5}$

66. $R(x) = \dfrac{x^2 - x - 12}{x^2 - 9}$

67. Determine a polynomial function P of degree 3 with zeros 1, -2, and 3 such that
    (a) $P(0) = 3$ (b) $P(2) = -8$

68. Determine a polynomial function P of degree 4 with zeros -1, -1, $\tfrac{1}{2}$, and $-\tfrac{3}{2}$ such that
    (a) $P(0) = -3$ (b) $P(1) = 24$

69. Sketch the graphs of the equations $y = x^3$ and $7x - 4y = -3$ on the same coordinate plane. Then determine the coordinates of the points where their graphs intersect.

70. Sketch the graphs of the functions
$$f(x) = \dfrac{4}{x} \quad \text{and} \quad g(x) = x^2 - 2x - 8$$
on the same coordinate plane. Then determine to the nearest hundredth the coordinates of the points where their graphs intersect.

71. Determine the horizontal asymptote for
$$f(x) = \dfrac{2x^2 + 5x - 7}{3x^2 + 2}.$$
Does the graph of this function cross the horizontal asymptote? If so, find the coordinates of the point where it does cross.

72. Determine the oblique asymptote for
$$f(x) = \dfrac{x^3 - 2x^2 + 3x}{x^2 - 4x + 3}.$$
Does the graph of this function cross the oblique asymptote? If so, find the coordinates of the point where it does cross.

73. A forester estimates that the number N of 2 in. by 3 in. by 8 ft construction-grade studs that can be cut from a fir tree is given by $N(x) = 4x^3 - 8x^2 + 5x - 1$, where x is the circumference (in feet) of the base of the tree trunk and $x \geq 1$. Determine the circumference of a tree that can yield 24 studs.

74. When a 20-microfarad capacitor is connected in series with a variable capacitor x, the total capacitance C is given by $C(x) = \dfrac{20x}{x+20}$.
    (a) Sketch the graph of this function for $x \geq 0$.
    (b) What is the significance of the horizontal asymptote?

75. The height of a tin can is 3 cm more than its radius. Determine the radius of the can if its volume is 112π cubic cm.

76. An open rectangular box has four sides and a bottom of uniform thickness. The outside dimensions of the box are 3 ft by 3 ft by 2 ft, as shown in the sketch at the top of the next page. Determine the uniform thickness (to the nearest hundredth) if the box holds 12 cubic feet of water when filled to the top.

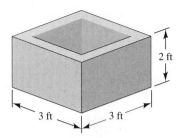

77. A piece of cardboard, 15 cm by 20 cm, is made into an open rectangular box by cutting out a square from each corner and turning up the sides, as shown in the figure.

(a) Express the volume V of this box as a function of x.

(b) Find *two* possible lengths of the side of the square cut-out if the volume of the box is 250 cubic centimeters.

(c) Generate the graph of the function in part (a) on a graphing calculator. Then approximate the value of x for which the volume of the box is greatest.

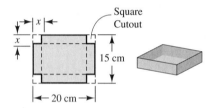

78. A beam of length L is simply supported at both ends and carries a load of weight W that decreases uniformly from the center, as shown in the figure. The bending moment M at any distance x ($x \leq L/2$) from the left end of this beam is given by

$$M(x) = \frac{Wx}{6L^2}(3L^2 - 4x^2).$$

(a) Determine the bending moment function when $W = 2400$ lb and $L = 20$ ft. Then generate the graph of this function on a graphing calculator.

(b) At what distance x from the left end of the beam is the bending moment 6336 lb-ft?

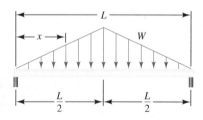

79. A rectangular box has a square base and open top, as shown in the sketch. The volume of the box is 100 cubic meters. The material used to construct the base costs $4 per square meter and the material for the sides cost $2.50 per square meter.

(a) Show that the cost C of materials to construct the box is given by

$$C = \frac{4x^3 + 1000}{x} \quad \text{for } x > 0.$$

(b) Generate the graph of the rational function defined in part (a) on a graphing calculator. Then approximate the dimensions of the box for which the cost is the smallest possible.

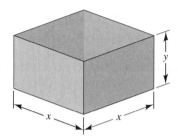

80. A certain manufacturer produces wood-burning stoves. The fixed cost for the company is $81,000 per year and the variable cost (in dollars) of producing x stoves is $400x + 0.1x^2$.

(a) Define a function that represents the average cost per stove.

(b) Generate the graph of the function in part (a) on a graphing calculator. Then approximate the number of stoves that should be produced per year in order to minimize the average cost per stove.

81. A triathlete in training runs 2 miles at the rate of 8 mi/h and returns home, along the same route, riding a bicycle. The entire trip takes 20 minutes. Find the rate of speed at which the triathlete pedals.

82. A gas tank is filled in $3\frac{3}{4}$ hours when two pipelines are used. The time required to fill the tank is four hours less when using the larger pipeline than when using the smaller one. What time is required by each pipeline alone to fill the tank?

Seismologists estimate that the San Francisco earthquake of 1906 measured 8.3 on the Richter scale. How many times more intense was this earthquake than the Loma Prieta quake, which occurred during the 1989 World Series and measured 7.1 on the Richter scale?

For the solution, see Example 5 in Section 5.3.

Exponential and Logarithmic Functions

5.1 Exponential Functions and Their Graphs

5.2 Logarithmic Functions and Their Graphs

5.3 Properties of Logarithms

5.4 Exponential and Logarithmic Equations

5.1 Exponential Functions and Their Graphs

◆ Introductory Comments

Any function that can be expressed as sums, differences, products, quotients, powers, or roots of polynomials is classified as an **algebraic function**. All of the functions that we have discussed in Chapters 2, 3, and 4, are algebraic functions. Any function that goes beyond the limits of, or *transcends,* an algebraic function is called a **transcendental function**. In this section we study a class of transcendental functions in which the independent variable appears as an exponent.

◆ **Exponential Function**

> If b is a real number such that $b > 0$ and $b \neq 1$, then the function f defined by
> $$f(x) = b^x$$
> is called an **exponential function** with base b.

Note: We exclude 1 as a base for exponential functions because $f(x) = 1^x = 1$ is a constant function. We exclude zero as a base because 0^x is undefined when $x \leq 0$. We exclude negative numbers as bases because b^x for $b < 0$ is an imaginary number for infinitely many values of x, such as $x = \frac{1}{2}, \frac{1}{4}, \frac{1}{6}$, and so on.

In Sections 1.2 and 1.3, we evaluated expressions of the form b^x, where b is a *positive real number* and x is a *rational number*. Whether x is a positive integer, zero, a negative integer, a common fraction, or a decimal fraction, we can evaluate b^x as a positive real number, as illustrated in the following examples.

$$4^2 = 16$$
$$4^0 = 1$$
$$4^{-2} = \frac{1}{4^2} = \frac{1}{16}$$
$$4^{3/2} = (4^{1/2})^3 = (\sqrt{4})^3 = (2)^3 = 8$$
$$4^{1.4} = 4^{14/10} = 4^{7/5} \approx 6.9644 \quad \leftarrow \text{We can use a calculator to express the fifth root of 4 to the seventh power as an approximate decimal number.}$$

If b is a positive real number and x is an *irrational* number, then we define b^x as the number approached by b^r as r takes on rational values that get closer and closer to x. For example, we define $4^{\sqrt{2}}$ as the real number approached by 4^r as r takes on rational values that get closer and closer to $\sqrt{2}$:

r	1	1.4	1.41	1.414	1.4142	1.41421
4^r	4	6.964405...	7.061624...	7.100891...	7.102860...	7.102958...

SECTION 5.1 ◆ Exponential Functions and Their Graphs

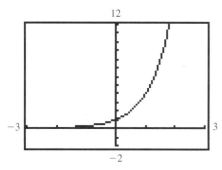

FIGURE 5.1

The graph of $g(x) = 4^x$ rises from left to right. Its y-intercept is 1 and the x-axis is a horizontal asymptote.

From these values we see that $4^{\sqrt{2}} \approx 7.103$, rounded to four significant digits. *For each real number x, b^x is a unique positive real number.* Hence, we conclude that the domain of an exponential function is $(-\infty, \infty)$ and its range is $(0, \infty)$.

◆ Graph of an Exponential Function

Figure 5.1 shows the graph of $g(x) = 4^x$ in the viewing rectangle of a graphing calculator. Notice from the graph that this function is always increasing and its y-intercept is 1. Since $g(x)$ approaches zero [$g(x) \to 0$] as x decreases without bound ($x \to -\infty$), the x-axis is a horizontal asymptote. The graph of an exponential function $f(x) = b^x$ with $b > 4$ rises *more* rapidly than the graph of $g(x) = 4^x$, and the graph of $f(x) = b^x$ with $1 < b < 4$ rises *less* rapidly than the graph of g.

The basic characteristics of the graph of $f(x) = b^x$ for $b > 1$ are as follows.

◆ **Characteristics of the Graph of $f(x) = b^x$ for $b > 1$**

1. The y-intercept is 1, and the graph has no x-intercept.
2. The x-axis is a horizontal asymptote [$f(x) \to 0$ as $x \to -\infty$].
3. The graph is always rising.

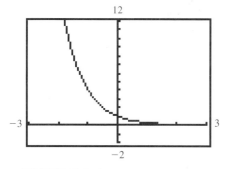

FIGURE 5.2

The graph of $h(x) = (\frac{1}{4})^x = 4^{-x}$ falls from left to right. Its y-intercept is 1 and the x-axis is a horizontal asymptote.

If the base of an exponential function is a fraction $\frac{1}{b}$ between 0 and 1, we usually write the function in terms of a negative exponent:

$$f(x) = \left(\frac{1}{b}\right)^x = (b^{-1})^x = b^{-x}$$

Figure 5.2 shows the graph of $h(x) = (\frac{1}{4})^x = 4^{-x}$ in the viewing rectangle of a graphing calculator. By the y-axis reflection rule (Section 2.5), the graph of h is the same as the graph of $g(x) = 4^x$ (see Figure 5.1) reflected about the y-axis. The graph of $h(x) = (\frac{1}{4})^x = 4^{-x}$ is always falling and its y-intercept is 1. Since $h(x)$ approaches zero [$h(x) \to 0$] as x increases without bound ($x \to +\infty$), the x-axis is a horizontal asymptote.

The basic characteristics of the graph of $f(x) = b^{-x}$ for $b > 1$ are as follows.

◆ **Characteristics of the Graph of $f(x) = b^{-x}$ for $b > 1$**

1. The y-intercept is 1, and the graph has no x-intercept.
2. The x-axis is a horizontal asymptote [$f(x) \to 0$ as $x \to +\infty$].
3. The graph is always falling.

Knowing the characteristics of the graph of an exponential function enables us to graph several related functions by applying the shift rules and axis reflection rules (see Section 2.5).

EXAMPLE 1 Sketch the graph of each function. Label the horizontal asymptote and the y-intercept.

(a) $G(x) = 4^{x+2}$ (b) $H(x) = 3 - 4^{-x}$

Solution

(a) By the horizontal shift rule (Section 2.5), the graph of $G(x) = 4^{x+2}$ is the same as the graph of $g(x) = 4^x$ (see Figure 5.1) shifted horizontally to the left 2 units. The y-intercept is

$$G(0) = 4^{0+2} = 4^2 = 16$$

and the horizontal asymptote remains the x-axis. The graph of $G(x) = 4^{x+2}$ is shown in Figure 5.3.

(b) By the x-axis reflection rule and the vertical shift rule (Section 2.5), the graph of $H(x) = 3 - 4^{-x}$ is the same as the graph of $h(x) = 4^{-x}$ (see Figure 5.2) reflected about the x-axis and then shifted vertically upward 3 units. The y-intercept is

$$H(0) = 3 - 4^0 = 3 - 1 = 2.$$

Note that $4^{-x} \to 0$ as $x \to +\infty$. Hence, $H(x) \to 3$ as $x \to +\infty$. Thus, $y = 3$ is the horizontal asymptote. The graph of $H(x) = 3 - 4^{-x}$ is shown in Figure 5.4. ◆

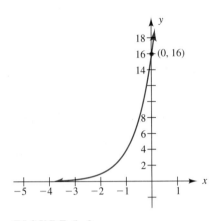

FIGURE 5.3
Graph of $G(x) = 4^{x+2}$

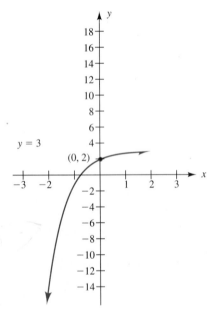

FIGURE 5.4
Graph of $H(x) = 3 - 4^{-x}$

Note: Observe that the graph of H in Figure 5.4 has an x-intercept between 0 and -1. To find the x-intercept, we solve the equation

$$3 - 4^{-x} = 0 \quad \text{or} \quad 4^{-x} = 3$$

for x. However, none of our previous methods can be used to solve an equation in which the unknown appears as an exponent. In Section 5.2, we introduce the *logarithmic function,* which we can use to help solve this type of equation.

PROBLEM 1 State the domain and range of the function H defined in Example 1(b). ◆

◆ Base e

In applications of exponential functions, one particular irrational number occurs frequently as the base. This irrational number is denoted by the letter e, and its value is

$$e \approx 2.71828 \ldots$$

The function f defined by

$$f(x) = e^x$$

is called the **exponential function with base** e. With a calculator, we can evaluate $f(x) = e^x$ for $x = 0, \pm 1, \pm 2, \pm 3, \ldots$ by using the $\boxed{e^x}$ key. These values are displayed in the following table.

SECTION 5.1 ♦ Exponential Functions and Their Graphs

x	-3	-2	-1	0	1	2	3
e^x	$e^{-3} \approx 0.05$	$e^{-2} \approx 0.14$	$e^{-1} \approx 0.37$	1	$e \approx 2.72$	$e^2 \approx 7.39$	$e^3 \approx 20.1$

Figure 5.5 shows the graphs of $f(x) = e^x$ and $g(x) = e^{-x}$ in the same viewing rectangle of a graphing calculator. The graph of $f(x) = e^x$ rises more rapidly than the graph of $y = 2^x$, but less rapidly than the graph of $y = 3^x$.

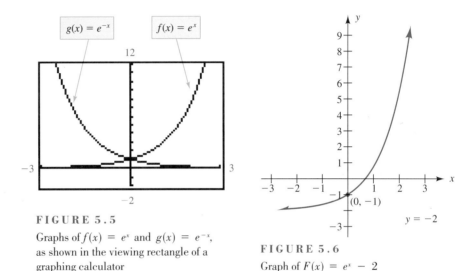

FIGURE 5.5
Graphs of $f(x) = e^x$ and $g(x) = e^{-x}$, as shown in the viewing rectangle of a graphing calculator

FIGURE 5.6
Graph of $F(x) = e^x - 2$

EXAMPLE 2 Sketch the graph of the function F defined by $F(x) = e^x - 2$. Label the horizontal asymptote and the y-intercept.

◀ **Solution** By the vertical shift rule (Section 2.5), the graph of $F(x) = e^x - 2$ is the same as the graph of $f(x) = e^x$ (see Figure 5.5) shifted vertically downward 2 units, as shown in Figure 5.6. The y-intercept is

$$F(0) = e^0 - 2 = 1 - 2 = -1.$$

Note that $F(x) \to -2$ as $x \to -\infty$. Thus, $y = -2$ is the horizontal asymptote. ◆

PROBLEM 2 Use a graphing calculator to verify the graph of the function shown in Figure 5.6. ◆

♦ Properties of Real Exponents

The laws of exponents given in Section 1.2 apply to all real exponents. We restate them here and refer to them as the **properties of real exponents**.

Properties of Real Exponents

If the bases a and b are positive real numbers and the exponents x and y represent real numbers, then

1. $b^0 = 1$
2. $b^{-x} = \dfrac{1}{b^x}$
3. $b^x b^y = b^{x+y}$
4. $(b^x)^y = b^{xy}$
5. $\dfrac{b^x}{b^y} = b^{x-y}$
6. $(ab)^x = a^x b^x$
7. $\left(\dfrac{a}{b}\right)^x = \dfrac{a^x}{b^x}$
8. $\dfrac{a^{-x}}{b^{-y}} = \dfrac{b^y}{a^x}$

EXAMPLE 3 Use the properties of real exponents to write each expression as a constant or as an expression in the form b^x or b^{-x} with $b > 1$.

(a) $\dfrac{5^\pi}{5^{2x+\pi}}$ (b) $4^{-3x} \cdot 8^{2x+1}$

Solution

(a) Since the bases are the same, we begin by applying property 5:

$\dfrac{5^\pi}{5^{2x+\pi}} = 5^{\pi-(2x+\pi)}$ **Subtract exponents**

$\qquad = 5^{-2x}$ **Simplify**

$\qquad = (5^2)^{-x}$ or 25^{-x} **Rewrite using property 4 and simplify**

Figure 5.7 shows that the graphs of

$$f(x) = \dfrac{5^\pi}{5^{2x+\pi}} \quad \text{and} \quad g(x) = 25^{-x}$$

are identical. This fact confirms our work.

(b) First, we write each factor with the same base so that we can apply the properties of real exponents:

$$4^{-3x} \cdot 8^{2x+1} = (2^2)^{-3x} \cdot (2^3)^{2x+1}$$

Now we apply property 4, then property 3:

$4^{-3x} \cdot 8^{2x+1} = (2^2)^{-3x} \cdot (2^3)^{2x+1}$

$\qquad = 2^{-6x} \cdot 2^{6x+3}$ **Multiply the exponents**

$\qquad = 2^3$ or 8 **Add the exponents and simplify**

Figure 5.8 confirms our work.

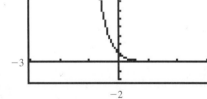

FIGURE 5.7

The graphs of $f(x) = \dfrac{5^\pi}{5^{2x+\pi}}$ and $g(x) = 25^{-x}$ are identical.

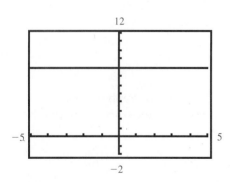

FIGURE 5.8

The graphs of $f(x) = 4^{-3x} \cdot 8^{2x+1}$ and $g(x) = 8$ are identical.

PROBLEM 3 Use the properties of real exponents to write the expression $\dfrac{125^{x-2}}{25^{2x-3}}$ as an expression in the form b^x or b^{-x}. ◆

◆ **Application: Compound Interest**

Exponential functions occur in several types of applied problems. One of the most familiar types involves the interest that we earn on a savings account. The amount of simple interest i earned on a principal P invested at a certain rate of interest r per year over a time of t years is given by the formula

$$i = Prt.$$

Thus, the amount A in the account after t years is

$$A = P + i = P + Prt = P(1 + rt).$$

Compound interest is interest paid both on the principal and on any interest earned previously. If a certain principal P is deposited in a savings account at an interest rate r per year and the interest is *compounded n times per year*, then interest is paid on both principal and any interest earned previously every $(1/n)$th of a year. The interest i earned after the first compounding period is

$$i = Prt = Pr\frac{1}{n}.$$

Hence the amount A_1 in the account after the first compounding period is

$$\begin{aligned} A_1 &= P + i \\ &= P + Pr\frac{1}{n} \\ &= P\left(1 + \frac{r}{n}\right) \quad \text{Factor out } P \end{aligned}$$

The interest i earned after the second compounding period is

$$i = A_1 r \frac{1}{n},$$

and the amount A_2 in the account after the second compounding period is

$$A_2 = A_1 + i = A_1 + A_1 r \frac{1}{n}.$$

Now, replacing A_1 with $P\left(1 + \dfrac{r}{n}\right)$, we have

$$A_2 = P\left(1 + \frac{r}{n}\right) + P\left(1 + \frac{r}{n}\right)r\frac{1}{n}$$

$$= P\left(1 + \frac{r}{n}\right)\left[1 + \frac{r}{n}\right] \qquad \text{Factor out } P\left(1 + \frac{r}{n}\right) \text{ from each term}$$

$$= P\left(1 + \frac{r}{n}\right)^2 \qquad \text{Simplify}$$

Continuing in this manner, we can show that after one year (with n compounding periods in a year) the amount A in the account is

$$A = P\left(1 + \frac{r}{n}\right)^n,$$

and after t years the amount in the account is

$$A = P\left(1 + \frac{r}{n}\right)^{nt}.$$

◆ **Compound Interest Formula for n Compoundings Per Year**

If a certain principal P is deposited in a savings account at an interest rate r per year and interest is compounded n times per year, then the amount A in the account after t years is given by the formula

$$A = P\left(1 + \frac{r}{n}\right)^{nt}.$$

EXAMPLE 4 Suppose that $1000 is deposited in a bank account at an interest rate of 8% per year compounded quarterly. Assume the depositor makes no subsequent deposit or withdrawal.

(a) Express the amount in the account as a function of time t.

(b) Find the amount in the account after five years.

◆ **Solution**

(a) If the interest is compounded quarterly, then it is compounded four times per year. Using the formula $A = P\left(1 + \frac{r}{n}\right)^{nt}$ with $P = 1000$, $r = 0.08$, and $n = 4$, we have

$$A = 1000\left(1 + \frac{0.08}{4}\right)^{4t} = 1000(1.02)^{4t}.$$

(b) To find the amount in the account after five years, we substitute $t = 5$ in the formula from part (a), and then use a calculator to evaluate this expression:

$$A = 1000(1.02)^{4 \cdot 5} \approx 1485.95$$

Thus, the amount in the account after 5 years is $1485.95. The graph in Figure 5.9 confirms our work.

PROBLEM 4 Use the graph in Figure 5.9 to estimate the time it takes for the amount in the account to reach $2000.

If the number of compoundings per year increases *without bond* ($n \to +\infty$), then we have **continuous compounding**. In the formula

$$A = P\left(1 + \frac{r}{n}\right)^{nt}$$

suppose we let $n/r = m$. Then $n = mr$, and $m \to +\infty$ implies that $n \to +\infty$. Now, by direct substitution and the properties of real exponents, we have

$$A = P\left(1 + \frac{r}{n}\right)^{nt} = P\left(1 + \frac{1}{m}\right)^{mrt} = P\left[\left(1 + \frac{1}{m}\right)^{m}\right]^{rt}$$

Using a calculator, we can show that as $m \to +\infty$, the expression $\left(1 + \frac{1}{m}\right)^{m}$ seems to approach the number $e \approx 2.71828\ldots$. The following table displays some calculated values for this expression.

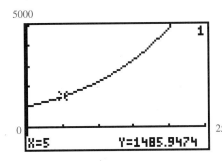

FIGURE 5.9
The graph of $A = 1000(1.02)^{4t}$ for $t \geq 0$ indicates that for $t = 5$ years, $A \approx \$1485.95$.

m	1	10	100	1000	10,000	100,000
$\left(1 + \frac{1}{m}\right)^m$	2	2.59374...	2.70481...	2.71692...	2.71814...	2.71826...

Hence,

$$\text{as} \quad m \to +\infty, \quad A = P\left[\left(1 + \frac{1}{m}\right)^{m}\right]^{rt} \to Pe^{rt}$$

◀ **Compound Interest Formula for Continuous Compounding**

If a certain principal P is deposited in a savings account at an interest rate r per year and the interest is compounded continuously, then the amount A in the account after t years is given by the formula

$$A = Pe^{rt}.$$

EXAMPLE 5 Suppose that $1000 is deposited in a bank account at an interest rate of 8% per year compounded continuously. Assume the depositor makes no subsequent deposit or withdrawal.

(a) Express the amount in the account as a function of time t.

(b) Find the amount in the account after five years

Solution

(a) If the interest is compounded continuously, then we use the formula $A = Pe^{rt}$ with $P = 1000$ and $r = 0.08$ to obtain

$$A = 1000e^{0.08t}.$$

(b) To find the amount in the account after five years, we substitute $t = 5$ in the formula from part (a), and then use a calculator to evaluate this expression:

$$A = 1000e^{(0.08)(5)} \approx 1491.82$$

Thus, if the interest is compounded continuously, the amount in the account after five years is $1491.82. The graph in Figure 5.10 confirms our work.

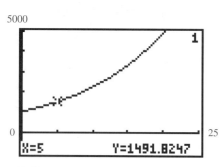

FIGURE 5.10
The graph of $A = 1000e^{0.08t}$ for $t \geq 0$ indicates that for $t = 5$ years, $A \approx \$1491.82$.

PROBLEM 5 Compare the answers in Example 4(b) and Example 5(b). Which type of compounding is better for the investor?

Exercises 5.1

Basic Skills

1. Generate the graphs of the exponential functions defined by the equations $y = 2^x$, $y = 3^x$, and $y = e^x$ in the same viewing rectangle of a graphing calculator. Then answer the following questions.
   (a) What is the point of intersection of these graphs?
   (b) For an x-value on the interval $(-\infty, 0)$, which equation has the largest y-value? which has the smallest y-value?
   (c) For an x-value on the interval $(0, \infty)$, which equation has the largest y-value? which has the smallest y-value?

2. Generate the graphs of the exponential functions defined by the equations $y = 2^{-x}$, $y = 3^{-x}$, and $y = e^{-x}$ in the same viewing rectangle of a graphing calculator. Then answer the following questions.
   (a) What is the point of intersection of these graphs?
   (b) For an x-value on the interval $(-\infty, 0)$, which equation has the largest y-value? which has the smallest y-value?
   (c) For an x-value on the interval $(0, \infty)$, which equation has the largest y-value? which has the smallest y-value?

In Exercises 3–16, use the shift rules and axis reflection rules (Section 2.5) to sketch the graph of each function. For each graph, label the y-intercept and horizontal asymptote. Use a graphing calculator to verify each sketch.

3. $f(x) = -2^x$
4. $f(x) = -3^{-x}$
5. $h(x) = 2^{x-2}$
6. $h(x) = e^{x+1}$
7. $F(x) = -(3^x + 2)$
8. $F(x) = 2^x + 1$
9. $G(x) = 2(\frac{1}{2})^x$
10. $G(x) = 9 \cdot 3^{-x}$
11. $H(x) = 1 - e^{-x}$
12. $H(x) = 2 - (\frac{1}{3})^{-x}$
13. $A(x) = 2 + 3^{2-x}$
14. $A(x) = 2^{x+1} - 3$
15. $h(x) = \dfrac{e}{e^x}$
16. $h(x) = 2 - e^{1-x}$

In Exercises 17–30, use the properties of real exponents to write each expression as a constant or as an expression in the form b^x or b^{-x} with $b > 1$. Check graphically with a graphing calculator.

17. 1^{5x}
18. $1^{x-\pi}$
19. $(6^x)^{-2}$
20. $(\frac{1}{3})^{-2x}$
21. $(4^{x-1})^0$
22. $(2^{2x})^{-2}$
23. $e(e^{x-1})$
24. $e^x \cdot e^{-x}$
25. $(3)(3^{2x-1})$
26. $\dfrac{36}{6^{2-x}}$
27. $(9^{3x-2})(27^{1-2x})$
28. $\dfrac{8^{2-x}}{4^{x+3}}$

$(3^2(3x-2))(3^3(1-2x)) = (3^{6x-4})(3^{3-6x})$
$3^{6x-4+3-6x} = 3^{-1} = \dfrac{1}{3}$

SECTION 5.1 ♦ Exercises

29. $\left(\dfrac{5^{2x}}{25^{x-1}}\right)^{1/2}$

30. $(2^{-x} \cdot 4^x)^2$

In Exercises 31–38,
(a) express the amount A in the account as a function of time t.
(b) find the amount in the account after the indicated number of years. Check graphically with a graphing calculator.

31. $5000 invested at 9% per year compounded annually; 4 years

32. $3000 invested at 7% per year compounded semiannually; 8 years

33. $10,000 invested at $8\frac{1}{2}$% per year compounded quarterly; 12 years

34. $800 invested at $9\frac{3}{4}$% per year compounded daily; 10 years

35. $2000 invested at 6% compounded continuously; 9 years

36. $15,000 invested at 10% compounded continuously; 6 years

37. $120,000 invested at 8% compounded continuously; 6 months

38. $250,000 invested at $12\frac{3}{4}$% compounded continuously; 8 months

39. Suppose P is the air pressure (in pounds per square inch, or psi) at a certain height h (in feet) above sea level, and $P = ke^{-0.00004h}$ where k is a constant.
    (a) Determine k if the air pressure at sea level is 15 psi.
    (b) What is the pressure outside the cabin of an airplane whose altitude is 20,000 feet?

40. Suppose A is the number of moose living in a protected forest in Maine after a certain time t (in years since 1968), and $A = ke^{0.15t}$ where k is a constant.
    (a) Determine k if 201 moose were present in 1988.
    (b) What is the expected moose population in 1998?

41. The number A of bacteria in a certain culture is a function of time t (in minutes) and is described by $A(t) = 18e^{0.2t}$.

(a) Find the number of bacteria initially present in the culture.
(b) Find the number of bacteria present in the culture after 10 minutes. Use a graphing calculator to verify the answer.

42. The voltage V (in volts) in an electrical circuit is a function of time t (in milliseconds, or ms) and is given by

$$V(t) = 400e^{-t/10}.$$

(a) Find the initial voltage in the circuit.
(b) Find the voltage in the circuit after 30 ms.
(c) Use a graphing calculator to generate the graph of the function V for $t \geq 0$. Then discuss the voltage in the circuit as t increases without bound ($t \to +\infty$).

43. The value V (in dollars) of a certain piece of machinery is a function of its age t (in years) and is given by

$$V(t) = 2000 + 10{,}000e^{-0.35t}.$$

(a) What is the value of the machine when it is purchased?
(b) What is the value of the machine 4 years after it is purchased?
(c) Use a graphing calculator to generate the graph of the function V for $t \geq 0$. Then discuss the machine's scrap value, that is, the value of the machinery as t increases without bound ($t \to +\infty$).

44. For a certain secretarial student, the number N of words typed (correctly) per minute on a new word processor is a function of the time t (in hours) that the student has practiced on the machine and is given by

$$N(t) = 60(1 - e^{-0.05t}).$$

(a) What is the student's rate of typing after 10 hours of practice?
(b) Use a graphing calculator to generate the graph of the function N for $t \geq 0$. What is the best rate of typing this student can hope to achieve on this word processor?

Critical Thinking

45. Perform the indicated operations and simplify.
    (a) $(2^x + 2^{-x})(2^x - 2^{-x})$
    (b) $(3^x + 3^{-x})^2 - (3^x - 3^{-x})^2$

46. Given the function f defined by $f(x) = e^x$, determine whether f is a one-to-one function and whether there exists an inverse function f^{-1}. If f^{-1} exists, sketch its graph.

In Exercises 47–52, find the real zeros of each function. Check graphically with a graphing calculator.

47. $F(x) = 2e^{-2x} - xe^{-2x}$

48. $G(x) = 2x^2e^{5x} - xe^{5x} - e^{5x}$

49. $g(x) = \dfrac{(x^2 - 4x + 4)e^x}{x + 1}$

50. $h(x) = \dfrac{x^2 e^x + 2xe^x}{x - 3}$

51. $H(x) = x^2 3^{2x} - 9^x$

52. $f(x) = x^2 (\frac{1}{2})^{3x} - x 8^{-x}$

53. The graph of the exponential function $y = a^x$ passes through the point $(-2, 16)$. Determine the value of the base a.

54. Given that f is an exponential function defined by $f(x) = e^x$, show that each of the following statements is true.
    (a) $f(a)f(b) = f(a + b)$
    (b) $\dfrac{f(a)}{f(b)} = f(a - b)$
    (c) $[f(a)]^n = f(na)$

55. Complete the table, given that the function f is defined by $f(x) = (-2)^x$.

x	0	1	2	3	4	-1	-2	-3	-4
$f(x)$									

Explain why we cannot sketch the graph of f by plotting the points in this table and connecting them with a smooth curve.

56. The *hyperbolic sine function* f and the *hyperbolic cosine function* g are defined by

$$f(x) = \dfrac{e^x - e^{-x}}{2} \quad \text{and} \quad g(x) = \dfrac{e^x + e^{-x}}{2}.$$

   (a) Determine whether f and g are even or odd functions.
   (b) Generate the graphs of f and g in the viewing rectangle of a graphing calculator.

57. Over a period of 10 years, which interest rate is the best investment?
   (i) 8.1% compounded semiannually
   (ii) 8.0% compounded monthly
   (iii) 7.9% compounded continuously

58. In calculus, it is shown that if $|x| \leq 1$, then

$$e^x \approx 1 + x + \dfrac{x^2}{2} + \dfrac{x^3}{6} + \dfrac{x^4}{24} + \dfrac{x^5}{120}.$$

Use this formula and a calculator to evaluate each expression.
   (a) $e^{0.5}$
   (b) $e^{-0.4}$
   (c) e
   (d) e^{-1}

59. A function is defined as $f(x) = e^x$.
   (a) Show that the difference quotient

$$\dfrac{f(x + h) - f(x)}{h} = e^x \left(\dfrac{e^h - 1}{h} \right).$$

   (b) Use a calculator to evaluate $\dfrac{e^h - 1}{h}$ for the values of h given in the following table.

h	1	0.1	0.01	0.001	0.0001
$\dfrac{e^h - 1}{h}$					

   (c) Use your table entries from part (b) to determine the number that $\dfrac{e^h - 1}{h}$ seems to approach as $h \to 0^+$.

60. In statistics, the *normal probability distribution function* is given by

$$f(x) = \dfrac{1}{\sigma \sqrt{2\pi}} e^{-(x-\mu)^2 / 2\sigma^2}$$

where μ is the mean and σ the standard deviation of the distribution. The graph of this function is usually referred to as a "bell-shaped" curve. Use a graphing calculator to generate the graph of the bell-shaped curve for $\mu = 0$ and $\sigma = 1$.

5.2 Logarithmic Functions and Their Graphs

◆ Introductory Comments

As we discussed in Section 5.1, the exponential function $f(x) = b^x$ is always increasing when $b > 1$ and is always decreasing when $0 < b < 1$. Recall from Section 2.6 that every function f that is either an increasing function or a decreasing function is also a one-to-one function and has an inverse function f^{-1}. Using the method suggested in Section 2.6, we can attempt to find the inverse function of $f(x) = b^x$ by solving the equation

$$y = b^x$$

for x. However, we have no algebraic procedure that we can use to solve this last equation for x. By convention, we solve this equation for x by writing

$$x = \log_b y \quad \text{—— read "log base } b \text{ of } y\text{"}$$

The expression $\log_b y$ represents *the power to which b must be raised in order to obtain y.*

If we interchange x and y in the expression $x = \log_b y$, and then replace y with $f^{-1}(x)$, we obtain the inverse of the exponential function $f(x) = b^x$:

$$f^{-1}(x) = \log_b x$$

The function f^{-1} is called the *logarithmic function with base b*. Since the exponential function $f(x) = b^x$ and logarithmic function $f^{-1}(x) = \log_b x$ are inverses of each other, the domain of the logarithmic function must be the range of the exponential function, namely, $(0, \infty)$. In other words, $\log_b x$ is defined only when x is positive.

◀ **Logarithmic Function with Base b**

> For $b > 0$ and $b \neq 1$, the **logarithmic function with base b** is defined as
>
> $$y = \log_b x \quad \text{if and only if} \quad b^y = x \quad \text{for } x > 0.$$

From our definition of the logarithmic function, we can state that the equation

$$\log_b u = v \quad \text{is equivalent to} \quad b^v = u.$$

The equation $\log_b u = v$ is said to be in **logarithmic form** and the equation $b^v = u$ is said to be in **exponential form**. If we change an equation from logarithmic form to exponential form, or vice versa, the base b remains the same in each case; that is, in logarithmic form the base is the subscript b, and in exponential form the base is the number b being raised to a power v. Here are some illustrations of equivalent equations:

Logarithmic form		Exponential form
$\log_4 16 = 2$	is equivalent to	$4^2 = 16$
$\log_{16} 4 = \frac{1}{2}$	is equivalent to	$16^{1/2} = 4$
$\log_2 \frac{1}{16} = -4$	is equivalent to	$2^{-4} = \frac{1}{16}$

Two logarithmic bases occur in most applied problems—base 10 and base e. We refer to a base 10 logarithm ($\log_{10}$) as a **common logarithm** and a base e logarithm ($\log_e$) as a **natural logarithm**, or Napierian logarithm (named after the Scottish mathematician John Napier, 1550–1617). The common logarithm of a positive number a is usually written **log a** (read "log of a"), and the natural logarithm of a positive number a is usually written **ln a** (read "el-en of a").

$$\boxed{\log a \text{ means } \log_{10} a} \quad \text{and} \quad \boxed{\ln a \text{ means } \log_e a}$$

The function F defined by

$$F(x) = \log x \quad \text{for } x > 0$$

is called the **common logarithmic function**, and the function G defined by

$$G(x) = \ln x \quad \text{for } x > 0$$

is called the **natural logarithmic function**.

◆ Evaluating Logarithms

To evaluate the logarithmic expression $\log_b a$, we find the power to which b can be raised to obtain a. The procedure is illustrated in the next example.

EXAMPLE 1 Find the value of each logarithm, if it is defined.

(a) $\log_5 25$ (b) $\log \frac{1}{100}$
(c) $\log_8 16$ (d) $\ln(-1)$

◆ Solution

(a) The logarithm $\log_5 25$ represents the power y such that $5^y = 25$. Since $5^2 = 25$, we conclude that $y = 2$. Hence, $\log_5 25 = 2$.

(b) The common logarithm $\log \frac{1}{100} = \log_{10} \frac{1}{100}$ represents the power y such that $10^y = \frac{1}{100}$. Since $10^{-2} = \frac{1}{100}$, we conclude that $\log \frac{1}{100} = -2$.

(c) The logarithm $\log_8 16$ represents the power y such that $8^y = 16$. We know that $8^1 = 8$ and $8^2 = 64$. Since 16 is between 8 and 64 and the base 8 is a perfect cube, we try a fractional exponent between 1 and 2 whose denominator is 3. Note that $8^{4/3} = 16$. Hence, $\log_8 16 = \frac{4}{3}$.

(d) The domain of the natural logarithmic function $y = \ln x$ is $(0, \infty)$. Thus, for $x \leq 0$ the natural logarithmic function is undefined. Hence, we conclude

$$\ln(-1) = \log_e(-1) \quad \text{is undefined.}$$

In general, $\log_b a$ is undefined whenever $a \leq 0$. ◆

SECTION 5.2 ♦ Logarithmic Functions and Their Graphs

As a direct consequence of the definition of the logarithmic function, we have two identities:

$$\boxed{\log_b 1 = 0} \quad \text{and} \quad \boxed{\log_b b = 1}$$

In each identity, we assume that $b > 0$ and $b \neq 1$.

PROBLEM 1 Find the value of each logarithm: **(a)** $\log 1$ **(b)** $\ln e$ ♦

Most calculators have [LOG] and [LN] keys, which we can use to evaluate common logarithms and natural logarithms, respectively. Also, we can use the [LN] key to evaluate $\log_b a$ by applying the **base e conversion formula**. This formula states that we can express all logarithms in terms of the natural logarithm.

◀ **Base e Conversion Formula**

If $\log_b a$ is defined, then

$$\log_b a = \frac{\ln a}{\ln b}$$

Note: The base e conversion formula is a consequence of the more general *change of base formula*, which we develop in Section 5.3. For now, we shall accept the validity of this formula and use its results.

EXAMPLE 2 Use a calculator to find the approximate value of each logarithm.

(a) $\log 25$ **(b)** $\ln \frac{1}{2}$ **(c)** $\log_4 24$

◀ Solution

(a) The common logarithm $\log 25$ represents the power y such that $10^y = 25$. We know that $10^1 = 10$ and $10^2 = 100$. Since 25 is between 10 and 100, we know that $\log 25$ must be some real number between 1 and 2. Using the [LOG] key on a calculator, we approximate the value to four significant digits:

$$\log 25 \approx 1.398.$$

In other words, $25 \approx 10^{1.398}$.

(b) The natural logarithm $\ln \frac{1}{2}$ represents the power y such that $e^y = \frac{1}{2}$, or 0.5. Since $e^0 = 1$ and $e^{-1} = 1/e \approx 0.37$, we know that $\ln \frac{1}{2}$ must be some real number between 0 and -1. Using the [LN] key on a calculator, we approximate the value to four significant digits:

$$\ln \tfrac{1}{2} \approx -0.6931.$$

(c) The expression $\log_4 24$ represents the power to which 4 must be raised to get 24. We know that $4^2 = 16$ and $4^3 = 64$. Since 24 is between 16 and 64, $\log_4 24$ must be a real number between 2 and 3. Using the base e conversion formula, we write

$$\log_4 24 = \frac{\ln 24}{\ln 4}.$$

Now, using the $\boxed{\text{LN}}$ key on a calculator and rounding to four significant digits, we obtain

$$\log_4 24 = \frac{\ln 24}{\ln 4} \approx 2.292.$$

PROBLEM 2 Use a calculator to verify the results obtained in Problem 1.

◆ Graph of a Logarithmic Function

Recall from Section 2.6 that the graph of a function and its inverse are symmetric with respect to the line $y = x$. Hence, the graph of $f(x) = \log_b x$ may be obtained from the graph of $g(x) = b^x$ by reflecting the graph of g in the line $y = x$, as shown in Figure 5.11.

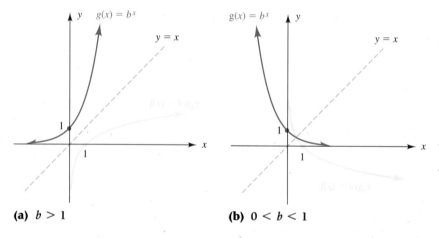

FIGURE 5.11
Like the graphs of any pair of inverse functions, the graphs of $f(x) = \log_b x$ and $g(x) = b^x$ are symmetric with respect to the line $y = x$.

(a) $b > 1$ (b) $0 < b < 1$

In Figure 5.11(a) the graph of $f(x) = \log_b x$ for $b > 1$ is always rising and its x-intercept is 1. Also, for $b > 1$, observe that $f(x)$ decreases without bound $[f(x) \to -\infty]$ as x approaches 0 from the right $(x \to 0^+)$. Hence, the y-axis is a vertical asymptote.

In Figure 5.11(b) the graph of $f(x) = \log_b x$ for $0 < b < 1$ is always falling, although its x-intercept remains 1. Also, for $0 < b < 1$, observe that $f(x)$ increases without bound $[f(x) \to +\infty]$ as x approaches 0 from the right $(x \to 0^+)$. Again, the y-axis is a vertical asymptote. We can summarize these features of the graph of a logarithmic function as follows.

Characteristics of the Graph of $f(x) = \log_b x$

1. The x-intercept is 1, and the graph has no y-intercept.
2. The y-axis is a vertical asymptote.
3. If $b > 1$, the graph of $f(x) = \log_b x$ is always rising.
4. If $0 < b < 1$, the graph of $f(x) = \log_b x$ is always falling.

To sketch the graph of a logarithmic function, we may simply use the basic characteristics of the general graph and plot a few points.

EXAMPLE 3 Sketch the graph of each logarithmic function.

(a) $f(x) = \log_4 x$ (b) $g(x) = \log_{1/4} x$

Solution

(a) Since the base b is 4 (and $4 > 1$), the graph of $f(x) = \log_4 x$ is always rising. The x-intercept is 1 and, since $f(x) \to -\infty$ as $x \to 0^+$, the y-axis is a vertical asymptote. Selecting a few convenient inputs for x, we find their corresponding outputs $f(x)$, as shown in the table:

x	$\frac{1}{4}$	1	4
$f(x)$	-1	0	1

Plotting the points associated with these ordered pairs and connecting them to form a smooth curve gives us the graph of $f(x) = \log_4 x$, as shown in Figure 5.12.

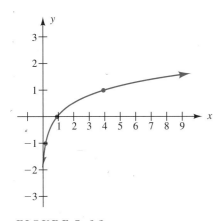

FIGURE 5.12
The function $f(x) = \log_4 x$ is an increasing function.

(b) Since the base b is $\frac{1}{4}$ (and $0 < \frac{1}{4} < 1$), the graph of $g(x) = \log_{1/4} x$ is always falling. The x-intercept is 1 and, since $g(x) \to +\infty$ as $x \to 0^+$, the y-axis is a vertical asymptote. Selecting a few convenient inputs for x, we find their corresponding outputs $g(x)$ as follows:

x	$\frac{1}{4}$	1	4
$g(x)$	1	0	-1

Plotting the points associated with these ordered pairs and connecting them to form a smooth curve gives us the graph of $g(x) = \log_{1/4} x$, as shown in Figure 5.13. ◆

Observe that the graph of $g(x) = \log_{1/4} x$ is the same as the graph of $f(x) = \log_4 x$ reflected about the x-axis. In general, if b is a real number such that $b > 0$ and $b \neq 1$, then the graph of $g(x) = \log_{1/b} x$ is the same as the graph of $f(x) = \log_b x$ reflected about the x-axis.

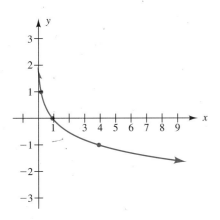

FIGURE 5.13
The function $g(x) = \log_{1/4} x$ is a decreasing function.

To generate the graph of the common logarithmic function or the natural logarithmic function on a graphing calculator, we use the [LOG] key or [LN] key, respectively. For bases other than 10 or e, we apply the base e conversion formula:

$$f(x) = \log_b x = \frac{\ln x}{\ln b}$$

For example, to generate the graph of $f(x) = \log_4 x$ in the viewing rectangle of a graphing calculator, we enter the function

$$f(x) = \frac{\ln x}{\ln 4}.$$

The graph is shown in Figure 5.14.

PROBLEM 3 Generate the graph of $g(x) = \log_{1/4} x$ in the viewing rectangle of a graphing calculator. Compare this graph with the sketch in Figure 5.13. ◆

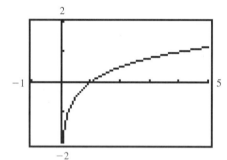

FIGURE 5.14
Graph of $f(x) = \log_4 x$ shown in the viewing rectangle of a graphing calculator

Knowing the basic shape of the graph of the logarithmic function $f(x) = \log_b x$, enables us to graph related functions by applying shift rules and axis reflection rules, which we discussed in Section 2.5.

EXAMPLE 4 Sketch the graph of each function. Label the vertical asymptote and any x- and y-intercepts.

(a) $F(x) = \log_4(x + 2)$ (b) $G(x) = 1 + \log_4(-x)$

◀ Solution

(a) By the horizontal shift rule (Section 2.5), the graph of $F(x) = \log_4(x + 2)$ is the same as the graph of $f(x) = \log_4 x$ (see Figure 5.12) shifted to the left 2 units, as shown in Figure 5.15. We find the y-intercept by evaluating $F(0)$:

$$F(0) = \log_4(0 + 2) = \log_4 2 = \tfrac{1}{2}$$

We find the x-intercept by solving the equation $\log_4(x + 2) = 0$. To solve this equation, we change from logarithmic form to exponential form:

$$\log_4(x + 2) = 0 \quad \text{is equivalent to} \quad 4^0 = x + 2$$
$$1 = x + 2$$
$$x = -1$$

Hence, the x-intercept of the graph of F is -1. Note that $F(x) \to -\infty$ as x approaches -2 from the right ($x \to -2^+$). Thus, the vertical asymptote is the line $x = -2$.

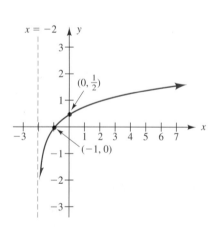

FIGURE 5.15
Graph of $F(x) = \log_4(x + 2)$

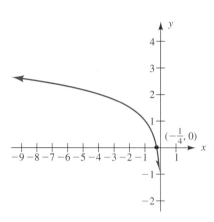

FIGURE 5.16
Graph of $G(x) = 1 + \log_4(-x)$

(b) By the y-axis reflection rule and the vertical shift rule (Section 2.5), the graph of $G(x) = 1 + \log_4(-x)$ is the same as the graph of $f(x) = \log_4 x$ (see Figure 5.12) reflected about the y-axis and then shifted vertically upward 1 unit, as shown in Figure 5.16. Since $G(0)$ is undefined, the graph has no y-intercept. The x-intercept is found by solving the equation $1 + \log_4(-x) = 0$. To solve this equation, we write it in logarithmic form and then change to exponential form:

$$1 + \log_4(-x) = 0$$

$\log_4(-x) = -1 \quad$ is equivalent to $\quad 4^{-1} = -x$
$$\tfrac{1}{4} = -x$$
$$x = -\tfrac{1}{4}$$

Thus, the x-intercept of the graph of G is $-\tfrac{1}{4}$. Note that $G(x) \to -\infty$ as x approaches 0 from the left ($x \to 0^-$). Hence, the y-axis is a vertical asymptote.

PROBLEM 4 Use a graphing calculator to confirm the sketches in Figures 5.15 and 5.16.

◆ Simplifying Logarithmic Expressions

We know that the logarithmic function with base b and the exponential function with base b are inverses of each other. If

$$f(u) = b^u \quad \text{and} \quad g(u) = \log_b u \quad \text{for } u > 0$$

then composing f with g in either order must yield the identity function:

$$f(g(u)) = b^{(\log_b u)} = u \quad \text{for } u > 0$$

and
$$g(f(u)) = \log_b(b^u) = u.$$

In summary, we have

$$\boxed{b^{\log_b u} = u \quad \text{for } u > 0} \quad \text{and} \quad \boxed{\log_b b^u = u}$$

We can illustrate each of these identities:

$$e^{\ln 6} = e^{\log_e 6} = 6 \quad \text{and} \quad \log \sqrt{10} = \log_{10} 10^{1/2} = \tfrac{1}{2}$$

In the next example, we apply these logarithmic identities to simplify expressions containing logarithms. This procedure is often employed in calculus.

EXAMPLE 5 Simplify each expression.

(a) $e^{\ln(x+1)}$ **(b)** $\log_2 16^{x-2}$

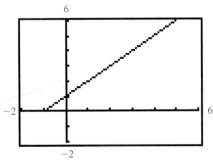

FIGURE 5.17
The graph of $f(x) = e^{\ln(x+1)}$ is identical to the graph of $y = x + 1$ for $x > -1$.

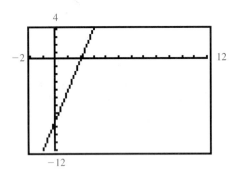

FIGURE 5.18
The graph of $f(x) = \log_2 16^{x-2}$ is identical to the graph of $y = 4x - 8$.

Solution

(a) Since $b^{\log_b u} = u$ for $u > 0$, we have

$$e^{\ln(x+1)} = e^{\log_e(x+1)} = x + 1$$

provided that $x + 1 > 0$, which implies that $x > -1$. Figure 5.17 shows the graph of $f(x) = e^{\ln(x+1)}$ in the viewing rectangle of a graphing calculator. The graph of this function is the same as the graph of the line $y = x + 1$ for $x > -1$, which confirms our answer.

(b) We begin by applying the properties of real exponents (Section 5.1) and matching the base of the exponential expression to the base of the logarithm:

$$\log_2 16^{x-2} = \log_2 (2^4)^{x-2} = \log_2 2^{4x-8}$$

Since $\log_b b^u = u$ for any permissible base b and any real number u, we have

$$\log_2 16^{x-2} = \log_2 2^{4x-8} = 4x - 8.$$

Figure 5.18 shows the graph of $f(x) = \log_2 16^{x-2}$ in the viewing rectangle of a graphing calculator. The graph of this function is the same as the graph of the line $y = 4x - 8$, which confirms our answer. ◆

PROBLEM 5 Simplify $\log_3 9^{2x-3}$. Use a graphing calculator to verify your answer. ◆

◆ Application: The Retention Function

In psychological experiments it has been shown that most of what we learn we forget rapidly and the remainder of the learned material recedes slowly from our memory. The function R that describes the percentage of the learned material we retain over a period of time t is called the **retention function**. The retention function is often defined in terms of logarithms, and its graph is called the **retention curve**.

EXAMPLE 6 In a psychological experiment, a student is asked to memorize a list of nonsense syllables by studying the list until one perfect repetition is performed from memory. At various times during the following week, the student is asked to recall the list of syllables. It is found that the retention function R associated with this experiment is given by

$$R(t) = 100 - 29 \log(27t + 1) \quad \text{for } 0 \leq t \leq 7,$$

where t is time in days.

(a) What percentage of the learned material does the student retain initially (at $t = 0$)?

(b) What percentage of the learned material does the student retain after 8 hours?

◀ Solution

(a) The percentage of the learned material that is retained initially is

$$R(0) = 100 - 29 \log[27(0) + 1]$$
$$= 100 - 29 \log 1$$
$$= 100 - 29(0)$$
$$= 100\%$$

(b) The percentage of the learned material retained after 8 hours $(t = \frac{1}{3}$ day) is

$$R(\tfrac{1}{3}) = 100 - 29 \log[27(\tfrac{1}{3}) + 1]$$
$$= 100 - 29 \log 10$$
$$= 100 - 29(1)$$
$$= 71\%$$

Figure 5.19 shows the retention curve associated with this experiment in the viewing rectangle of a graphing calculator. The point $(\frac{1}{3}, 71)$ lies on this curve, which confirms our work.

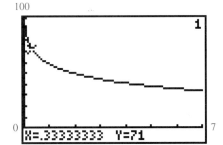

FIGURE 5.19
Graph of $R(t) = 100 - 29 \log(27t + 1)$ with $0 \le t \le 7$

PROBLEM 6 For the retention function given in Example 6, determine the percentage of the learned material that is retained after 5 days. Check this value with the retention curve in Figure 5.19.

Exercises 5.2

◆ Basic Skills

In Exercises 1–30, find the value of each expression, if it is defined.

1. $\log_6 36$
2. $\log_2 8$
3. $\log_5 125$
4. $\log_3 81$
5. $\log_3 3$
6. $\log_7(-7)$
7. $\log_2 \frac{1}{16}$
8. $\log_3 \frac{1}{27}$
9. $\log_{1/3} 9$
10. $\log_{1/2} 16$
11. $\log 100$
12. $\log \frac{1}{1000}$
13. $\ln 1$
14. $\log 1$
15. $\ln e^2$
16. $\ln \sqrt{e}$
17. $\log_{64}(-4)$
18. $\log_9 3$
19. $\log_4 8$
20. $\log_8 32$
21. $\log_{1/8} 16$
22. $\log_{27} \frac{1}{9}$
23. $4^{\log_4 3}$
24. $3^{\log_3 10}$
25. $e^{\ln 5}$
26. $10^{\log 2}$
27. $10^{\ln e}$
28. $e^{\log 10}$
29. $\log(\ln e)$
30. $\ln(\log 10)$

In Exercises 31–42, use a calculator to find the approximate value of each logarithm. Round each answer to four significant digits.

31. $\log 28$ 1.447
32. $\log 750$
33. $\ln 8$ 2.079
34. $\ln 56$
35. $\log_2 12$ 3.585
36. $\log_4 9$
37. $\log_{12} 945$ 2.757
38. $\log_{20} 1250$
39. $\log_3 \frac{2}{3}$ -0.3691
40. $\log_5 \frac{17}{5}$
41. $\log_{2/3} 34$ -8.697
42. $\log_{1/2} \frac{3}{4}$

43. Generate the graphs of the logarithmic functions defined by the equations $y = \log_2 x$, $y = \log_3 x$,

$y = \log x$, and $y = \ln x$ in the same viewing rectangle of a graphing calculator. Then answer the questions that follow:

(a) What is the point of intersection of these graphs?

(b) For an x-value on the interval $(0, 1)$, which equation has the largest y-value? which has the smallest y-value?

(c) For an x-value on the interval $(1, \infty)$, which equation has the largest y-value? which has the smallest y-value?

44. Generate the graphs of the logarithmic functions defined by the equations $y = \log_{1/2} x$, $y = \log_{1/3} x$, $y = \log_{1/10} x$, and $y = \log_{1/e} x$ in the same viewing rectangle of a graphing calculator. Then answer the questions that follow:

(a) What is the point of intersection of these graphs?

(b) For an x-value on the interval $(0, 1)$, which equation has the largest y-value? which has the smallest y-value?

(c) For an x-value on the interval $(1, \infty)$, which equation has the largest y-value? which has the smallest y-value?

In Exercises 45–58, use the shift rules and axis reflection rules (Section 2.5) to sketch the graph of each function. In each case, label the vertical asymptote and any x- and y-intercepts. Use a graphing calculator to verify each sketch.

45. $F(x) = -\log_2 x$
46. $g(x) = -\log_{1/3} x$
47. $h(x) = \log_{1/10}(-x)$
48. $f(x) = -\ln(-x)$
49. $g(x) = \log_3 x + 2$
50. $G(x) = \log_2 x - 1$
51. $G(x) = \log_3(x + 2)$
52. $H(x) = \log_2(x - 1)$
53. $f(x) = -3 + \ln(x - 1)$
54. $f(x) = 2 + \log(x + 2)$
55. $H(x) = 2 - \log_2(x + 3)$
56. $F(x) = 1 - \log(x - 4)$
57. $f(x) = \ln(3 - x) + 2$
58. $f(x) = 2 - \log_3(2 - x)$

In Exercises 59–74, simplify each expression. Use a graphing calculator to confirm your answer. Assume any variable is restricted to those real numbers that allow the logarithmic expression to be defined.

59. $\log_4 4^{x-3}$
60. $\log_3 3^{x+1}$
61. $6^{\log_6(x+2)}$
62. $5^{\log_5(x-4)}$
63. $e^{\ln(x^2+4)}$
64. $10^{\log(1/x)}$
65. $\ln e^{x^2+2x}$
66. $(\ln e)^{x^2+2x}$
67. $(\log 10)^{2-x^2}$
68. $\log 10^{2-x^2}$
69. $\log_2 8^{x+2}$
70. $\log_9 3^{2x-2}$

71. $e^{x+\ln x}$ [*Hint:* $a^{x+y} = a^x a^y$]
72. $e^{x-\ln x}$ [*Hint:* $a^{x-y} = a^x/a^y$]
73. $e^{2\ln x}$ [*Hint:* $a^{xy} = (a^x)^y$]
74. $e^{-\ln x}$ [*Hint:* $a^{xy} = (a^x)^y$]

Chemists describe the acidity or alkalinity of a liquid by denoting its pH. An acid has pH < 7 and an alkaline has pH > 7. By definition,

$$\text{pH} = -\log[\text{H}^+],$$

where $[\text{H}^+]$ is the liquid's concentration of hydrogen ions, measured in moles per liter (m/L). In Exercises 75 and 76, determine the pH of each liquid.

75. Milk: $[\text{H}^+] \approx 4.1 \times 10^{-7}$ m/L
76. Apple juice: $[\text{H}^+] \approx 6.2 \times 10^{-4}$ m/L

77. In a psychological experiment, a student is asked to memorize a list of telephone numbers by studying the list until one perfect repetition is performed from memory. At various times during the following week, the student is asked to recall the list of numbers. It is found that the retention function R associated with this experiment is given by

$$R(t) = 100 - 31 \log(36t + 1) \quad \text{for } 0 \le t \le 7$$

where t is time in days.

(a) What percentage of the learned material is retained initially (at $t = 0$)?

(b) What percentage of the learned material is retained after 6 hours?

(c) Use a graphing calculator to generate the graph of the retention curve associated with this function. From this graph, estimate how much time has elapsed when only 33% of the learned material is still retained.

78. Students taking their last calculus course are given a final exam, and the average score of the group is recorded. Each month thereafter the students are given an equivalent exam, and for each exam their average score is recorded. It is found that the average score S is a function of time t (in months) and is defined by

$$S(t) = 72 - 22 \log(t + 1) \quad \text{for } 0 \le t \le 12.$$

(a) What is the average score on the original final exam?

(b) What is the average score on an exam given six months later?

(c) Use a graphing calculator to generate the graph of the retention curve associated with this function. From this graph, estimate how much time has elapsed when the average score is 50.

79. The number N of computers sold by a certain company in a month is a function of the amount x (in thousand of dollars) that is spent on advertising and is given by

$$N(x) = 1000[1 + \ln(x + 1)].$$

(a) If no money is spent on advertising, how many computers are sold in a month?

(b) If $100,000 is spent on advertising, approximately how many computers are sold in a month?

(c) Use a graphing calculator to generate the graph of the curve associated with this function. From this graph, estimate how much money on advertising must be spent to sell 4000 computers in a month.

80. The time t (in minutes) required for a cup of hot tea to cool to a temperature of T degrees Fahrenheit when placed in a room whose temperature is maintained at 70°F is given by

$$t = -8 \ln \frac{T - 70}{100} \quad \text{for } 70 < T \leq 170.$$

(a) How long does it take the tea to cool to 140°F?

(b) How long does it take the tea to cool to 100°F?

(c) Use a graphing calculator to generate the graph of the curve associated with this function. From this graph, estimate the temperature of the tea when $t = 4$ minutes.

Critical Thinking

81. Find the base b of the logarithmic function $f(x) = \log_b x$ if the graph of this function passes through the given point.

(a) $(81, 4)$ (b) $(\frac{1}{8}, -\frac{3}{2})$

82. Without using a calculator, determine between which two consecutive integers the value of each logarithm lies.

(a) $\log_3 99$ (b) $\log_5 150$
(c) $\log_8 7$ (d) $\log_{1/2} 40$

83. Use interval notation to describe the domain of each function.

(a) $f(x) = \dfrac{1}{1 - \ln x}$ (b) $F(x) = \sqrt{10 - \log x}$

84. Given that $f(x) = e^x$ and $g(x) = \ln x$, find $f(g(x))$ and $g(f(x))$. What do these results tell us about the functions f and g?

In Exercises 85–88,
(a) find the inverse of each function.
(b) sketch the graphs of the function and its inverse on the same coordinate plane.

85. $f(x) = 6^x$ **86.** $g(x) = (\frac{1}{6})^x$

87. $f(x) = \log_8 x$ **88.** $g(x) = \log_{1/8} x$

89. Use the LN key on your calculator to help complete the following table. Then answer the questions that follow.

x	y	$\ln(xy)$	$\ln x + \ln y$	$\ln \frac{x}{y}$	$\ln x - \ln y$	$\ln x^y$	$y \ln x$
2	3						
18	4						
24	15						
π	10						

(a) Compare the values in the table for $\ln(xy)$ and $\ln x + \ln y$. What do you discover?

(b) Compare the values in the table for $\ln \dfrac{x}{y}$ and $\ln x - \ln y$. What do you discover?

(c) Compare the values in the table for $\ln x^y$ and $y \ln x$. What do you discover?

90. In calculus, it is shown that if $|x| < 1$, then

$$\ln(x + 1) \approx x - \frac{x^2}{2} + \frac{x^3}{3} - \frac{x^4}{4} + \frac{x^5}{5} - \frac{x^5}{6}.$$

Use this formula and a calculator to evaluate each logarithm.

(a) $\ln 1.8$ (b) $\ln 0.2$
(c) $\ln 1.25$ (d) $\ln 0.75$

Given the functions f and g defined by

$$f(x) = \ln(x^2 - 2x + 3) \quad \text{and} \quad g(x) = \log \frac{1}{x^2 + 2},$$

compute the functional values given in Exercises 91 and 92. Round each value to three significant digits.

91. $(f \circ g)(4.4)$ **92.** $(g \circ f)(21.8)$

93. The function f is defined as $f(x) = \dfrac{\ln x}{x}$.

   (a) Determine the domain of the function f.
   (b) Determine any zeros of f.
   (c) Explain the behavior of $f(x)$ as $x \to +\infty$.
   (d) Explain the behavior of $f(x)$ as $x \to 0^+$.
   (e) Generate the graph of the function f in the viewing rectangle of a graphing calculator. Then estimate the value of x at which the maximum value of this function seems to occur.

94. The function f is defined as $f(x) = \dfrac{1}{x \ln x}$.

   (a) Determine the domain of the function f.
   (b) Determine any zeros of f.
   (c) Explain the behavior of $f(x)$ as $x \to +\infty$ and as $x \to 1^+$.
   (d) Explain the behavior of $f(x)$ as $x \to 0^+$ and as $x \to 1^-$.
   (e) Generate the graph of this function in the viewing rectangle of a graphing calculator. Then estimate the maximum value of the function f in the interval $(0, 1)$.

5.3 Properties of Logarithms

◆ Introductory Comments

Recall from Section 5.1 three properties of real exponents:

1. $b^x b^y = b^{x+y}$ 2. $\dfrac{b^x}{b^y} = b^{x-y}$ 3. $(b^x)^y = b^{xy}$

By using the log identity $b^{\log_b u} = u$ for $u > 0$ and these three properties of real exponents, we can obtain three corresponding **properties of logarithms**:

1. For $x > 0$ and $y > 0$,

$$b^{\log_b xy} = xy$$
$$= b^{\log_b x} \cdot b^{\log_b y} \quad \text{Rewrite } x \text{ and } y \text{ using } u = b^{\log_b u}$$
$$= b^{\log_b x + \log_b y} \quad \text{Add exponents}$$

Since the exponential function is a one-to-one function, we conclude that the exponents $\log_b xy$ and $\log_b x + \log_b y$ must be equal, that is,

$$\boxed{\log_b xy = \log_b x + \log_b y}$$

The logarithm of a product of two positive real numbers is the sum of the logarithms of its factors.

2. For $x > 0$ and $y > 0$,

$$b^{\log_b (x/y)} = \dfrac{x}{y}$$
$$= \dfrac{b^{\log_b x}}{b^{\log_b y}} \quad \text{Rewrite } x \text{ and } y \text{ using } u = b^{\log_b u}$$
$$= b^{\log_b x - \log_b y} \quad \text{Subtract exponents}$$

Since the exponential function is a one-to-one function, we conclude that the exponents $\log_b(x/y)$ and $\log_b x - \log_b y$ must be equal, that is,

$$\log_b \frac{x}{y} = \log_b x - \log_b y$$

The logarithm of a quotient of two positive real numbers is the logarithm of its numerator minus the logarithm of its denominator.

3. For $x > 0$ and any real number n,

$$b^{\log_b x^n} = x^n$$
$$= (b^{\log_b x})^n \quad \text{Rewrite } x \text{ using } u = b^{\log_b u}$$
$$= b^{n \log_b x} \quad \text{Multiply exponents}$$

Since the exponential function is a one-to-one function, we conclude that the exponents $\log_b x^n$ and $n \log_b x$ must be equal, that is,

$$\log_b x^n = n \log_b x$$

The logarithm of a positive real number raised to a power is the product of the power and the logarithm of that real number.

We now summarize these properties.

◀ **Properties of Logarithms**

If $b > 0$ and $b \neq 1$, then for any positive real numbers x and y,

1. $\log_b xy = \log_b x + \log_b y$
2. $\log_b \dfrac{x}{y} = \log_b x - \log_b y$
3. $\log_b x^n = n \log_b x$ where n is a real number

These properties of logarithms enable us to rewrite certain functions that contain logarithms so that we may sketch the graphs of these functions. Also, these properties enable us to rewrite certain logarithmic equations so that we may solve these equations, as we will show in Section 5.4.

◆ **Expanding Logarithmic Expressions**

We can use the properties of logarithms to write a single logarithmic expression as the sum and difference of simpler logarithmic expressions without logarithms of products, quotients, and powers. This procedure, which is used frequently in calculus, is called *expanding* a logarithmic expression.

EXAMPLE 1 Expand each logarithmic expression.

(a) $\log \sqrt[3]{10x^2}$ (b) $\ln \dfrac{\sqrt{x^2+4}}{xe^{3x}}$

◆ **Solution**

(a) Assuming that $x > 0$, we have

$$\begin{aligned}\log \sqrt[3]{10x^2} &= \log(10x^2)^{1/3} &&\text{Change radical to a rational exponent}\\ &= \tfrac{1}{3}\log(10x^2) &&\text{Apply log property 3}\\ &= \tfrac{1}{3}(\log 10 + \log x^2) &&\text{Apply log property 1}\\ &= \tfrac{1}{3}(\log 10 + 2 \log x) &&\text{Apply log property 3}\\ &= \tfrac{1}{3}(1 + 2 \log x) &&\text{Evaluate log 10}\end{aligned}$$

Figure 5.20 shows that the graphs of

$$f(x) = \log \sqrt[3]{10x^2} \text{ for } x > 0 \quad \text{and} \quad g(x) = \tfrac{1}{3}(1 + 2 \log x)$$

are identical. This fact confirms our work.

(b) Assuming that $x > 0$, we have

$$\begin{aligned}\ln \dfrac{\sqrt{x^2+4}}{xe^{3x}} &= \ln \dfrac{(x^2+4)^{1/2}}{xe^{3x}} &&\text{Change radical to a rational exponent}\\ &= \ln(x^2+4)^{1/2} - \ln xe^{3x} &&\text{Apply log property 2}\\ &= \ln(x^2+4)^{1/2} - (\ln x + \ln e^{3x}) &&\text{Apply log property 1}\\ &&&\boxed{\text{Keep the parentheses.}}\\ &= \tfrac{1}{2}\ln(x^2+4) - (\ln x + 3x \ln e) &&\text{Apply log property 3}\\ &= \tfrac{1}{2}\ln(x^2+4) - \ln x - 3x &&\text{Evaluate } \ln e = 1\end{aligned}$$

Figure 5.21 confirms our work. ◆

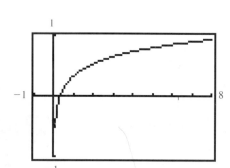

FIGURE 5.20

The graphs of $f(x) = \log \sqrt[3]{10x^2}$ for $x > 0$ and $g(x) = \tfrac{1}{3}(1 + 2 \log x)$ are identical.

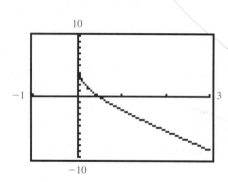

FIGURE 5.21

The graphs of $f(x) = \ln \dfrac{\sqrt{x^2+4}}{xe^{3x}}$ and $g(x) = \tfrac{1}{2}\ln(x^2+4) - \ln x - 3x$ are identical.

Note: The logarithmic expression in Example 1(a) is defined for *all* real numbers except 0. We make the assumption that $x > 0$ so that the properties of logarithms may be applied. If we allow that $x < 0$, then we use the absolute value of x to write

$$\log \sqrt[3]{10x^2} = \tfrac{1}{3}(1 + 2 \log |x|).$$

CAUTION Do not misapply the three properties of logarithms. In Example 1(a), to apply log property 3 and write

$$\ln xe^{3x} \quad \text{as} \quad 3x \ln xe \quad \text{is WRONG!}$$

The expression $3x \ln xe$ is the simplification of $\ln(xe)^{3x}$, not of $\ln xe^{3x}$. Several other errors that are often made when working with the properties of logarithms are illustrated in Exercises 1–4.

PROBLEM 1 Show, by using a specific example, that $(\log_b x)^n$ and $n \log_b x$ are *not* equivalent logarithmic expressions. ◆

We often expand a logarithmic function in order to obtain an equivalent function whose graph is easily plotted. When applying the properties of logarithms to a logarithmic function, we must always preserve the domain of the original function. For example, consider the functions F and G defined by

$$F(x) = \log_4 x^2 \quad \text{and} \quad G(x) = 2 \log_4 x$$

The domain of the function F is the set of all real numbers x with $x \neq 0$. However, the domain of the function G is the set of all positive real numbers x. Since the functions F and G have different domains, they are not equivalent functions. To apply logarithmic property 3 to function F and preserve the same domain, we must use the absolute value of x and write

$$F(x) = \log_4 x^2 = \log_4 |x|^2 = 2 \log_4 |x|.$$

EXAMPLE 2 Sketch the graph of $H(x) = \log_4 \dfrac{1}{x-1}$. Label the vertical asymptote and any x- and y-intercepts.

◀ **Solution** We begin by expanding this logarithmic function:

$$\begin{aligned}
H(x) = \log_4 \frac{1}{x-1} &= \log_4 1 - \log_4(x-1) & &\text{Apply log property 2} \\
&= 0 - \log_4(x-1) & &\text{Evaluate } \log_4 1 = 0 \\
&= -\log_4(x-1) & &\text{Simplify}
\end{aligned}$$

By the horizontal shift rule and x-axis reflection rule (Section 2.5), the graph of $H(x) = -\log_4(x-1)$ is the same as the graph of $f(x) = \log_4 x$ (see Figure 5.12 in Section 5.2) shifted right 1 unit and then reflected about the x-axis, as shown in Figure 5.22. We find the x-intercept by solving the equation $H(x) = 0$:

$$\log_4 \frac{1}{x-1} = 0 \quad \text{is equivalent to} \quad 4^0 = \frac{1}{x-1}$$

$$1 = \frac{1}{x-1}$$

$$x - 1 = 1$$

$$x = 2$$

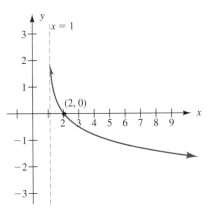

FIGURE 5.22
Graph of $H(x) = \log_4 \dfrac{1}{x-1}$

Hence, the x-intercept of the graph of H is 2. Note that $H(x) \to +\infty$ as $x \to 1^+$. Hence $x = 1$ is a vertical asymptote. ◆

PROBLEM 2 Use a graphing calculator to confirm the sketch in Figure 5.22.

◆ Condensing Logarithmic Expressions

The properties of logarithms can also be used to write sums and differences of logarithmic expressions as a single logarithmic expression with a coefficient of 1. (This is a reversal of the *expansion* process.) For problems of this type, we begin by applying logarithmic property 3 and then changing any numerical coefficient to an exponent. Once we have removed numerical coefficients, we can apply properties 1 and 2. This procedure, which is used frequently to solve logarithmic equations (see Section 5.4), is called *condensing* a logarithmic expression.

EXAMPLE 3 Condense each logarithmic expression.

(a) $\log 50 + 2 \log 4 - 3 \log 2$ (b) $\ln(x^2 - 1) - 2 \ln(x + 1)$

◆ Solution

(a)

$$\begin{aligned}
\log 50 + 2 \log 4 - 3 \log 2 &= \log 50 + \log 4^2 - \log 2^3 && \text{Apply log property 3} \\
&= \log 50 + \log 16 - \log 8 && \text{Simplify} \\
&= \log(50 \cdot 16) - \log 8 && \text{Apply log property 1} \\
&= \log \frac{50 \cdot 16}{8} && \text{Apply log property 2} \\
&= \log 100 = 2 && \text{Reduce and evaluate}
\end{aligned}$$

(b) The domain of $\ln(x^2 - 1) - 2 \ln(x + 1)$ is $(1, \infty)$. Hence,

$$\begin{aligned}
\ln(x^2 - 1) - 2 \ln(x + 1) &= \ln(x^2 - 1) - \ln(x + 1)^2 && \text{Apply log property 3} \\
&= \ln \frac{x^2 - 1}{(x + 1)^2} && \text{Apply log property 2} \\
&= \ln \frac{(x + 1)(x - 1)}{(x + 1)^2} && \text{Factor the numerator} \\
&= \ln \frac{x - 1}{x + 1} \quad \text{for } x > 1 && \text{Reduce}
\end{aligned}$$

Figure 5.23 shows that the graphs of $f(x) = \ln(x^2 - 1) - 2 \ln(x + 1)$ and $g(x) = \ln \frac{x - 1}{x + 1}$ for $x > 1$ are identical. This fact confirms our work.

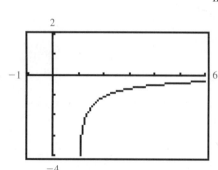

FIGURE 5.23

The graphs of
$f(x) = \ln(x^2 - 1) - 2 \ln(x + 1)$
and $g(x) = \ln \frac{x - 1}{x + 1}$ for $x > 1$ are identical.

PROBLEM 3 Use the [LOG] key on a calculator to evaluate $\log 50 + 2 \log 4 - 3 \log 2$. Your answer should agree with the result in Example 3(a).

◆ Change of Base Formula

In Section 5.2 we used the base e conversion formula to evaluate $\log_b a$ and to generate the graph of $y = \log_b x$ in the viewing rectangle of a graphing calculator. The base e conversion formula is a consequence of the more general *change of base formula*, which we now develop.

By the definition of the logarithmic function with base b, we know that

$$y = \log_b x \quad \text{is equivalent to} \quad b^y = x.$$

Taking the base a logarithm of both sides of the equation $b^y = x$, we obtain

$$\log_a b^y = \log_a x$$
$$y \log_a b = \log_a x \qquad \text{Apply log property 3}$$
$$y = \frac{\log_a x}{\log_a b} \qquad \text{Divide both sides by } \log_a b$$

However, since $y = \log_b x$, we substitute and obtain

$$\log_b x = \frac{\log_a x}{\log_a b},$$

which is the **change of base formula**.

◁ **Change of Base Formula**

If $\log_b x$ is defined, then

$$\log_b x = \frac{\log_a x}{\log_a b} \quad \text{for } a > 0 \text{ and } a \neq 1.$$

If we let $a = e$ in this formula, we obtain the base e conversion formula:

$$\log_b a = \frac{\ln a}{\ln b}$$

If we let $a = 10$ in this formula, we obtain the *base 10 conversion formula*:

$$\log_b a = \frac{\log a}{\log b}$$

Hence, to evaluate $\log_4 24$ on a calculator, we can use the $\boxed{\text{LN}}$ key (see Example 2(c) in Section 5.2) or the $\boxed{\text{LOG}}$ key:

$$\log_4 24 = \frac{\log 24}{\log 4} \approx 2.292.$$

If we replace x with a in the change of base formula, we obtain

$$\log_b a = \frac{1}{\log_a b}$$

or, equivalently,

$$(\log_b a)(\log_a b) = 1$$

EXAMPLE 4 Evaluate each logarithmic expression.

(a) $\dfrac{1}{\log_2 12} + \dfrac{1}{\log_6 12}$ (b) $(\log_3 16)(\log_2 27)$

◆ **Solution**

(a) $\dfrac{1}{\log_2 12} + \dfrac{1}{\log_6 12} = \log_{12} 2 + \log_{12} 6$ Apply $\dfrac{1}{\log_a b} = \log_b a$

$\phantom{\dfrac{1}{\log_2 12} + \dfrac{1}{\log_6 12}} = \log_{12}(2 \cdot 6)$ Apply log property 1

$\phantom{\dfrac{1}{\log_2 12} + \dfrac{1}{\log_6 12}} = \log_{12} 12$ Simplify

$\phantom{\dfrac{1}{\log_2 12} + \dfrac{1}{\log_6 12}} = 1$ Apply $\log_b b = 1$

(b)

$(\log_3 16)(\log_2 27) = (\log_3 2^4)(\log_2 3^3)$ Rewrite

$ = (4 \log_3 2)(3 \log_2 3)$ Apply log property 3

$ = (4 \cdot 3)[(\log_3 2)(\log_2 3)]$ Rearrange the factors

$ = 12[1]$ Apply $(\log_b a)(\log_a b) = 1$

$ = 12$ Simplify ◆

PROBLEM 4 Use the [LN] key or the [LOG] key on a calculator to evaluate each logarithmic expression in Example 4. Your answers should agree with the results obtained in the example. ◆

◆ **Application: Richter Scale**

When a physical quantity varies over a large range of values, it is convenient to work with a *logarithmic scale* in order to obtain a more manageable set of numbers. Some examples of logarithmic scales are the **brightness scale**, used for measuring the magnitude of a star (see Exercises 49 and 50); the **decibel scale**, used to measure the magnitude of sound (see Exercises 51 and 52), and the **Richter scale** (named after the seismologist Charles F. Richter), used for measuring the magnitude of an earthquake.

On the Richter scale, the magnitude R of an earthquake is given by

$$R = \log \dfrac{I}{I_0}$$

where I is the intensity of the earthquake and I_0 is the intensity of a zero-level earthquake having magnitude $R = 0$.

EXAMPLE 5 *Chapter Opening Problem*

Seismologists estimate that the San Francisco earthquake of 1906 measured 8.3 on the Richter scale. How many times more intense was this earthquake than the Loma Prieta quake, which occurred during the 1989 World Series and measured 7.1 on the Richter scale?

◆ **Solution**   Let

$$I_a = \text{the intensity of the 1906 earthquake}$$

and

$$I_b = \text{the intensity of the 1989 earthquake.}$$

We use the formula $R = \log(I/I_0)$ and log property 2:

$$8.3 = \log \frac{I_a}{I_0} \qquad \text{and} \qquad 7.1 = \log \frac{I_b}{I_0}$$
$$8.3 = \log I_a - \log I_0 \qquad\qquad 7.1 = \log I_b - \log I_0$$

Solving these equations simultaneously by using the elimination method (Section 3.5), we find

$$\begin{aligned} 8.3 &= \log I_a - \log I_0 \\ 7.1 &= \log I_b - \log I_0 \\ \hline 1.2 &= \log I_a - \log I_b \quad \text{Subtract} \\ 1.2 &= \log \frac{I_a}{I_b} \quad\quad\quad\;\; \text{Apply log property 2} \end{aligned}$$

Now, to find the ratio I_a/I_b, we change logarithmic form to exponential form:

$$1.2 = \log \frac{I_a}{I_b} \quad \text{is equivalent to} \quad \frac{I_a}{I_b} = 10^{1.2} \approx 16$$

Hence, the 1906 earthquake was about 16 times as intense as the 1989 quake. ◆

PROBLEM 5   One of the strongest earthquakes ever recorded occurred in Japan in 1933 and measured 8.9 on the Richter scale. How many times more intense was this quake than the 1989 Loma Prieta quake in California? ◆

Exercises 5.3

Basic Skills

In Exercises 1–4, show by using specific examples that the given logarithmic expressions are not equivalent.

1. $\log_b(x+y)$ and $\log_b x + \log_b y$
2. $\log_b(x-y)$ and $\log_b x - \log_b y$
3. $\dfrac{\log_b x}{\log_b y}$ and $\log_b \dfrac{x}{y}$
4. $(\log_b x)(\log_b y)$ and $\log_b xy$

In Exercises 5–18, expand each logarithmic expression to eliminate logarithms of products, quotients, and powers. Assume that each variable is restricted only to real numbers for which the properties of logarithms can be applied.

5. $\log_3(27^2 \cdot 81)$
6. $\log 100x^2$
7. $\log_2[8x(x+2)]$
8. $\log_5 \sqrt[4]{25x}$
9. $\ln \dfrac{xe^2}{10}$
10. $\log \dfrac{(x+4)^2}{2x}$
11. $\log_3 \sqrt[3]{\dfrac{x^2}{27}}$
12. $\log_b \left(\dfrac{x^2+2}{x+3}\right)^3$
13. $\log_b \dfrac{x^2}{\sqrt[3]{(x+1)^2}}$
14. $\ln \dfrac{1}{2\sqrt{3x^2+2}}$
15. $\log_3 \dfrac{1}{9\sqrt{x^3 y^{2/3}}}$
16. $\log \dfrac{\sqrt[3]{1+y^2}}{(x-2)(x+2)^{1/2}}$
17. $\ln\left(\dfrac{e^{x^2}}{e^x+1}\right)^2$
18. $\ln \dfrac{xe^{-2x}}{\sqrt{2e^x-1}}$

In Exercises 19–30, condense each expression to a single logarithm with a coefficient of 1. Assume that each variable is restricted only to real numbers for which the properties of logarithms can be applied.

19. $\ln 2 + \ln 3 + \ln 4$
20. $\ln 6 - \ln 3 + \ln 5$
21. $\log 40 - (3 \log 2 - \log 20)$
22. $2 \log 6 - (2 \log 3 + \log 4)$
23. $-2 \log_3 6 + 4 \log_3 2 - \tfrac{1}{2} \log_3 16$
24. $5 \log_6 2 + \tfrac{2}{3} \log_6 27 + \tfrac{3}{2} \log_6 4$
25. $\log_5(x-1) - 2 \log_5 x + \log_5(x+3)$
26. $3 \ln x + \tfrac{2}{3} \ln(x-1) + \ln 2$
27. $2 \ln(x-2) - \ln(x^2-4)$
28. $2 \ln(x^3+1) - \ln(x^2-x+1)$
29. $\tfrac{1}{2}[\ln(x-3) + \ln(x+3)] - 2(\ln x - \ln 3)$
30. $3\left[\log\left(x+\sqrt{x^2-1}\right) + \log\left(x-\sqrt{x^2-1}\right)\right] - \log 2$

In Exercises 31–38, use the properties of logarithms to help sketch the graph of each function. Label any x- and y-intercept(s) and the vertical asymptote.

31. $f(x) = \dfrac{1}{2} \log x^4$
32. $f(x) = \ln x^3$
33. $f(x) = \ln \dfrac{1}{x+3}$
34. $f(x) = \log_3 \dfrac{9}{x}$
35. $f(x) = \log_{1/2}[8(x-1)]$
36. $f(x) = \log_2(16x + 32)$
37. $f(x) = \log_3 \dfrac{1}{18-9x}$
38. $f(x) = \log \dfrac{10}{(x-1)^2}$

In Exercises 39–46, use $\log_b a = \dfrac{1}{\log_a b}$ to help express each logarithmic expression as a whole number.

39. $(\log_4 5)(\log_5 4)$
40. $(\log 7)(\log_7 10)$
41. $(\log_3 4)(\log_4 81)$
42. $(\log_5 16)(\log_2 125)$
43. $\log 20 + \dfrac{1}{\log_5 10}$
44. $\dfrac{1}{\log_4 10} + \dfrac{1}{\log_{25} 10}$
45. $\dfrac{1}{\log_a ab} + \dfrac{1}{\log_b ab}$
46. $\dfrac{\log_b x}{\log_{ab} x} - \dfrac{\log_b x}{\log_a x}$

47. The strongest earthquake ever recorded in the United States occurred in Alaska on Good Friday, March 27, 1964, and measured 8.6 on the Richter scale. How many times more intensive was this earthquake than the 1989 Loma Prieta quake in California, which measured 7.1?

48. One of the aftershocks from the 1989 Loma Prieta earthquake was 60 times less intense than the major earthquake (magnitude 7.1). What was the magnitude of the aftershock?

The magnitude m of a star is given by

$$m = -2.5 \log \dfrac{B}{B_0},$$

where B is the brightness of the star and B_0 is the brightness of a zero-level star having magnitude $m = 0$. Use this formula to answer Exercises 49 and 50.

49. The magnitude of our Sun is -26.8 and the magnitude of Sirius, the brightest star in the heavens, is -1.5. How many times brighter is the sun than Sirius?

50. What is the difference in magnitude between two stars if the brightness of one is 50 times the brightness of the other?

The loudness L (in decibels, db) of a sound is given by

$$L = 10 \log \frac{I}{I_0}$$

where I is the intensity of the sound and I_0 is the intensity of the faintest sound that can be heard. Use this formula to answer Exercises 51 and 52.

51. The loudness of a whisper is 30 db and the loudness of a rock concert is 120 db. How many times more intense is the rock concert than the whisper?

52. If the loudness of two sounds differs by 20 db, how many times more intense is the more audible sound than the other sound?

Critical Thinking

53. Suppose the functions f and g are defined by

$$f(x) = \ln x^2 \quad \text{and} \quad g(x) = 2 \ln x.$$

Are f and g the same function? Explain.

54. Show that if $f(x) = \ln x$, then $f\left(\dfrac{1}{x}\right) = -f(x)$.

55. Given that $\log_b A = 2$ and $\log_b B = 3$, find the value of each logarithm.
   (a) $\log_A b$ (b) $\log_B b$
   (c) $\log_A b^2$ (d) $\log_B \sqrt{b}$
   (e) $\log_{AB} b$ (f) $\log_{AB}(1/b)$
   (g) $\log_{A/B} b$ (h) $\log_{B/A} b$

56. Explain the fallacy in the following argument:

$$\begin{aligned} 3 &= \log_2 8 \\ &= \log_2(4 + 4) \\ &= \log_2 4 + \log_2 4 \\ &= 2 + 2 \\ &= 4 \end{aligned}$$

57. Use the LOG key on your calculator to find the approximate value of each of the following logarithms:

   log 6.2   log 62   log 620   log 6200

   Compare these values. Do you observe a pattern? Show why this pattern develops by expressing 62, 620, and 6200 in scientific notation and then finding the common logarithms of these expressions by using the properties of logarithms.

58. Use the LOG key on your calculator to find the approximate value of each of the following logarithms:

   log 3.5   log 0.35   log 0.035   log 0.0035

   Compare these values. Do you observe a pattern? Show why this pattern develops by expressing 0.35, 0.035, and 0.0035 in scientific notation and then finding the common logarithms of these expressions by using the properties of logarithms.

59. Evaluate the following expressions by each of the given procedures:

$$\ln \pi + \ln\left(\frac{\sqrt{2}}{\pi}\right) + \frac{1}{2}\ln\left(\frac{3}{2}\right) - \ln\left(\frac{\sqrt{3}}{e}\right)$$

   (a) Use the LN key on your calculator.
   (b) Use the properties of logarithms.

60. Use the change of base formula and a calculator to complete the following table.

a	b	$\log_b \dfrac{1}{a}$	$\log_{1/b} a$
2	3		
18	4		
24	15		
π	10		

Compare the values of $\log_b(1/a)$ and $\log_{1/b} a$. Prove the relationship that you observe.

5.4 Exponential and Logarithmic Equations

◆ **Introductory Comments**

An equation in which the variable appears in an exponent is referred to as an **exponential equation**, and an equation that contains logarithmic expressions is referred to as a **logarithmic equation**. In this section, we discuss procedures for solving these two types of equations. We then look at some applied problems that involve the solution techniques introduced in this section.

◆ Solving Exponential Equations

If both sides of an exponential equation can be written as powers of the same base, then we can solve the equation by equating the powers; that is,

$$\text{if } b^x = b^y, \text{ then } x = y.$$

For example, we can solve the exponential equation $27^{x-2} = 9$ by writing both sides of the equation in terms of the same base, 3:

$$27^{x-2} = 9$$

$$(3^3)^{x-2} = 3^2 \quad \text{Write both sides with the same base}$$

$$3^{3(x-2)} = 3^2 \quad \text{Multiply exponents}$$

$$3(x - 2) = 2 \quad \text{Equate exponents and solve for } x$$

$$3x = 8$$

$$x = \tfrac{8}{3}$$

Check: $27^{[(8/3)-2]} = 27^{2/3} = 9$

An alternative method for solving this exponential equation is to take the natural (or common) logarithm of both sides of the equation:

$$27^{x-2} = 9$$

$$\ln 27^{x-2} = \ln 9 \quad \text{Take the natural logarithm of both sides}$$

$$(x - 2) \ln 27 = \ln 9 \quad \text{Apply log property 3}$$

$$x - 2 = \frac{\ln 9}{\ln 27} \quad \text{Divide both sides by } \ln 27$$

$$x = 2 + \frac{\ln 9}{\ln 27} \quad \text{Add 2 to both sides}$$

Now, since $9 = 3^2$ and $27 = 3^3$, we apply logarithmic property 3 and write

$$x = 2 + \frac{\ln 9}{\ln 27} = 2 + \frac{2 \ln 3}{3 \ln 3} = 2 + \frac{2}{3} = \frac{8}{3},$$

This result agrees with the answer we obtained previously. This alternative procedure is particularly useful when both sides of an exponential equation *cannot* be written as powers of the same base.

SECTION 5.4 ◆ Exponential and Logarithmic Equations

◆ **Procedure for Solving an Exponential Equation**

1. Isolate the exponential expression on one side of the equation.
2. Take the natural (or common) logarithm of both sides of this equation, apply the properties of logarithms, and solve for the unknown.

EXAMPLE 1 Solve each exponential equation.

(a) $6^x = 50$ (b) $4e^{2x-1} = 100$

◆ **Solution**

(a) Since both sides of this equation cannot be written as powers of the same base, we apply the given procedure:

$$6^x = 50$$
$$\ln 6^x = \ln 50 \quad \text{Take the natural logarithm of both sides}$$
$$x \ln 6 = \ln 50 \quad \text{Apply log property 3}$$
$$x = \frac{\ln 50}{\ln 6} \approx 2.183 \quad \text{Solve for } x$$

Figure 5.24 shows the graph of $f(x) = 6^x - 50$ in the viewing rectangle of a graphing calculator. The x-intercept is approximately 2.183, which confirms our answer.

(b) Since both sides of this equation cannot be written as powers of the same base, we apply the given procedure:

$$4e^{2x-1} = 100$$
$$e^{2x-1} = 25 \quad \text{Isolate the exponential expression}$$
$$\ln e^{2x-1} = \ln 25 \quad \text{Take the natural logarithm of both sides}$$
$$2x - 1 = \ln 25 \quad \text{Simplify the left-hand side}$$
$$x = \frac{\ln 25 + 1}{2} \approx 2.109 \quad \text{Solve for } x$$

Figure 5.25 shows the graph of $f(x) = 4e^{2x-1} - 100$ in the viewing rectangle of a graphing calculator. The x-intercept is approximately 2.109, which confirms our answer. ◆

Note: To solve the equation in Example 1(a) we may also change the exponential form $6^x = 50$ to its equivalent logarithmic form:

$$x = \log_6 50$$

Now to evaluate $\log_6 50$ we use the base e conversion formula (Section 5.2):

$$x = \log_6 50 = \frac{\ln 50}{\ln 6} \approx 2.183$$

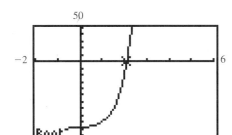

FIGURE 5.24

The x-intercept of the graph of $f(x) = 6^x - 50$ is the root of the equation $6^x = 50$.

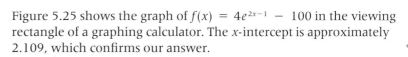

FIGURE 5.25

The x-intercept of the graph of $f(x) = 4e^{2x-1} - 100$ is the root of the equation $4e^{2x-1} = 100$.

PROBLEM 1 Solve the equation $5^{x-1} = 325$.

As illustrated in the next example, we can solve an exponential equation that involves more than one exponential expression by using a similar procedure.

EXAMPLE 2 Solve the equation $3^{x-2} = 2^{-x}$.

◆ **Solution** To solve this equation, we proceed as follows:

$$3^{x-2} = 2^{-x}$$
$$\ln 3^{x-2} = \ln 2^{-x} \quad \text{Take the natural logarithm of both sides}$$
$$(x - 2) \ln 3 = -x \ln 2 \quad \text{Apply log property 3}$$

Keep the parentheses.

$$x \ln 3 - 2 \ln 3 = -x \ln 2 \quad \text{Multiply}$$
$$x \ln 3 + x \ln 2 = 2 \ln 3 \quad \text{Group the } x \text{ terms on one side}$$
$$x (\ln 3 + \ln 2) = 2 \ln 3 \quad \text{Factor out } x$$
$$x \ln 6 = \ln 9 \quad \text{Apply log properties 1 and 3}$$
$$x = \frac{\ln 9}{\ln 6} \approx 1.226 \quad \text{Divide both sides by } \ln 6$$

Figure 5.26 shows the graph of $f(x) = 3^{x-2} - 2^{-x}$ in the viewing rectangle of a graphing calculator. The x-intercept is approximately 1.226, which confirms our answer. ◆

PROBLEM 2 Solve the equation $e^{x-1} = 3^x$.

◆ Solving Logarithmic Equations

We can solve a logarithmic equation in logarithmic form by changing to exponential form. For example,

logarithmic form   exponential form

$\log_3(x - 12) = 2$ is equivalent to $3^2 = x - 12$
$\qquad\qquad\qquad\qquad\qquad\qquad\qquad x = 3^2 + 12$
$\qquad\qquad\qquad\qquad\qquad\qquad\qquad x = 21$

Check: $\log_3[(21) - 12] = \log_3 9 = 2$

To solve many logarithmic equations that are not initially in logarithmic form, we use the following procedure:

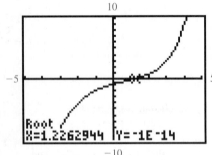

FIGURE 5.26

The x-intercept of the graph of $f(x) = 3^{x-2} - 2^{-x}$ is the root of the equation $3^{x-2} = 2^{-x}$.

◀ **Procedure for Solving Logarithmic Equations**

1. Isolate the logarithmic expressions on one side of the equation.
2. Apply the properties of logarithms, and write the equation in logarithmic form.
3. Change to exponential form, and solve for the unknown.

Note: This procedure may produce extraneous roots.

EXAMPLE 3 Solve each logarithmic equation.

(a) $2 - \log x = \log 3$ (b) $\log_4(x - 2) + 2\log_4 x = 1 + \log_4 2x$

◀ **Solution**

(a) Using the given procedure, we have

$$2 - \log x = \log 3$$
$$2 = \log 3 + \log x \quad \text{Isolate the logarithms on one side}$$
$$2 = \log 3x \quad \text{Apply log property 1}$$
$$10^2 = 3x \quad \text{Change to exponential form}$$
$$x = \frac{100}{3} \quad \text{Solve for } x$$

Figure 5.27 shows the graph of $f(x) = 2 - \log x - \log 3$ in the viewing rectangle of a graphing calculator. The x-intercept is $\frac{100}{3}$, which confirms our answer.

(b) Using the given procedure, we have

$$\log_4(x - 2) + 2\log_4 x = 1 + \log_4 2x$$

$\log_4(x - 2) + 2\log_4 x - \log_4 2x = 1$ Isolate the logarithmic expressions on one side

$\log_4 \dfrac{x^2(x - 2)}{2x} = 1$ Apply the log properties and write in logarithmic form

$\log_4 \dfrac{x(x - 2)}{2} = 1$ Reduce, assuming $x \neq 0$

$4^1 = \dfrac{x(x - 2)}{2}$ Change to exponential form

$x^2 - 2x - 8 = 0$ Write in quadratic form and solve for x

$(x - 4)(x + 2) = 0$

$x = 4 \quad x = -2$

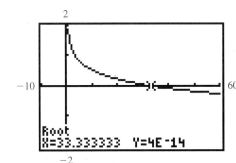

FIGURE 5.27
The x-intercept of the graph of $f(x) = 2 - \log x - \log 3$ is the root of the equation $2 - \log x = \log 3$.

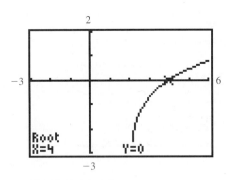

FIGURE 5.28

The graph of $f(x) = \log_4(x - 2) + 2\log_4 x - 1 - \log_4 2x$ has only one x-intercept.

Figure 5.28 shows the graph of

$$f(x) = \log_4(x - 2) + 2 \log_4 x - 1 - \log_4 2x$$

in the viewing rectangle of a graphing calculator. Note that 4 is the only x-intercept. This indicates that the other value, -2, is an extraneous root. Observe that $f(-2)$ is undefined. Hence, $x = 4$ is the only solution. ◆

PROBLEM 3 Solve the logarithmic equation $2 \ln x - \ln 9 = 4$. ◆

Many literal equations and formulas that contain logarithmic expressions may be solved for one of the unknowns by using this procedure.

EXAMPLE 4 Solve the literal equation $\ln(y + a) - 2 \ln(x + b) = c$ for y. Assume that a, b, and c are constants.

◆ **Solution** We solve for y as follows:

$$\ln(y + a) - 2 \ln(x + b) = c$$

$$\ln \frac{y + a}{(x + b)^2} = c \quad \text{Apply the log properties and write in logarithmic form}$$

$$e^c = \frac{y + a}{(x + b)^2} \quad \text{Change to exponential form}$$

$$y = e^c(x + b)^2 - a \quad \text{Solve for } y$$

Since c is a constant, e^c is also a constant. Relabeling e^c as the constant k gives us

$$y = k(x + b)^2 - a \quad \text{where } k = e^c.$$

◆

PROBLEM 4 Solve the following literal equation for y:

$$x + \ln y = \ln c \quad \text{where } c \text{ is a constant} \quad ◆$$

◆ Solving Factorable Equations

We can solve other exponential and logarithmic equations by factoring and applying the zero product property (Section 2.2).

EXAMPLE 5 Solve each equation.

(a) $(\log x)^2 = \log x^2$ **(b)** $e^x + 3e^{-x} = 4$

Solution

(a) We begin by rewriting the equation to obtain an equivalent equation that we can solve by the zero product property:

$$(\log x)^2 = \log x^2$$

$$(\log x)^2 = 2 \log x \qquad \text{Apply log property 3}$$

$$(\log x)^2 - 2 \log x = 0 \qquad \text{Subtract } 2 \log x \text{ from both sides}$$

Now we can factor out $\log x$ from this last equation and solve for x by applying the zero product property:

$$(\log x)^2 - 2 \log x = 0$$

$$\log x (\log x - 2) = 0 \qquad \text{Factor out } \log x$$

$$\log x = 0 \quad \text{or} \quad \log x - 2 = 0 \qquad \text{Apply the zero product property}$$

$$x = 10^0 \qquad\qquad\qquad x = 10^2 \qquad \text{Change to exponential form}$$

$$x = 1 \qquad\qquad\qquad\quad x = 100 \qquad \text{Simplify}$$

The graph in Figure 5.29 confirms our answers.

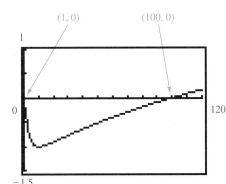

FIGURE 5.29

The graph of $f(x) = (\log x)^2 - \log x^2$ has two x-intercepts.

(b) We begin by rewriting the equation to obtain an equivalent equation that we can solve by the zero product property:

$$e^x + 3e^{-x} = 4$$

$$e^{2x} + 3 = 4e^x \qquad \text{Multiply both sides by } e^x$$

$$e^{2x} - 4e^x + 3 = 0 \qquad \text{Subtract } 4e^x \text{ from both sides}$$

This last equation is of the quadratic form $u^2 - 4u + 3 = 0$ with $u = e^x$:

$$u^2 - 4u + 3 = 0$$

$$(u - 1)(u - 3) = 0 \qquad \text{Factor}$$

$$u = 1 \quad \text{or} \quad u = 3 \qquad \text{Solve for } u$$

$$e^x = 1 \qquad\qquad e^x = 3 \qquad \text{Replace } u \text{ with } e^x$$

$$x = \ln 1 \qquad\quad x = \ln 3 \qquad \text{Take the natural logarithm of both sides}$$

$$x = 0 \qquad\qquad x \approx 1.099 \qquad \text{Evaluate}$$

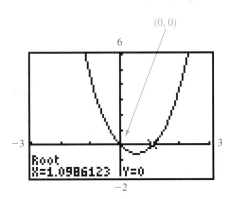

FIGURE 5.30

The x-intercepts of the graph of $f(x) = e^x + 3e^{-x} - 4$ are $x = 0$ and $x \approx 1.099$.

Figure 5.30 shows the graph of $f(x) = e^x + 3e^{-x} - 4$ in the viewing rectangle of a graphing calculator. The x-intercepts are 0 and approximately 1.099; these values confirm our answer.

PROBLEM 5 Solve the equation $\log \sqrt{x} = \sqrt{\log x}$ by squaring both sides. ◆

◆ Graphical Methods of Solutions

If the techniques that we have discussed in this section are insufficient to solve a particular logarithmic or exponential equation, we may use a calculator or computer with graphing capabilities to find the approximate solutions. For example, we cannot solve the equation $2^x + 3^x = 9$ by taking the natural logarithm of both sides. Hence, we use the graphical approach illustrated in the next example.

EXAMPLE 6 Use a graphing calculator to find the approximate solution(s) of the equation $2^x + 3^x = 9$.

◆ **Solution** One method of solving this equation graphically is to generate the graphs of the equations

$$y = 2^x + 3^x \quad \text{and} \quad y = 9$$

in the same viewing rectangle of a graphing calculator and then determine the x-coordinate of the intersection point(s). As shown in Figure 5.31, the graphs intersect once, and the x-coordinate of the intersection point appears to be between $x = 1$ and $x = 2$.

Most graphing calculators have a built-in program for finding this intersection point. If a calculator does not have this feature, we can use the $\boxed{\text{TRACE}}$ key and $\boxed{\text{ZOOM}}$ key to approximate the x-coordinate of the intersection point. By either method, we find that the x-coordinate of the intersection point is $x \approx 1.620$, to the nearest thousandth. We conclude that the solution to the equation $2^x + 3^x = 9$ is approximately 1.620. ◆

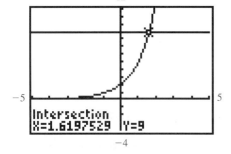

FIGURE 5.31

The graphs of $y = 2^x + 3^x$ and $y = 9$ intersect in one point.

PROBLEM 6 Solve the exponential equation in Example 6 by generating the graph of $f(x) = 2^x + 3^x - 9$ in the viewing rectangle of a graphing calculator and noting the x-intercept. You should obtain the same answer as in Example 6. ◆

◆ Application: Exponential Growth and Decay

Functions that represent an amount of a substance which changes by a fixed multiple over the same increment of time are of the form

$$\boxed{A(t) = A_0 e^{kt}}$$

where $A(t)$ is the amount present at time t, A_0 is the original amount present at time 0, and k is a constant related to the rate of growth or decay. If $k > 0$, then $A(t) = A_0 e^{kt}$ is called an **exponential growth function** and if $k < 0$, then $A(t) = A_0 e^{kt}$ is called an **exponential decay function**.

The compound interest formula

$$A = Pe^{rt}$$

is an example of an exponential growth function. In Section 5.1 we used this formula to find the amount A in a savings account when a certain principal P is invested at an interest rate r per year and interest is compounded continuously. Now that we have defined logarithms, we can also use this formula to find the time t or the interest rate r required for a given amount to reach a particular value.

EXAMPLE 7 An investor deposits $10,000 into a savings account paying 8% interest per year compounded continuously. Assuming the investor makes no subsequent deposit or withdrawal, find the time required for this investment to accumulate to $50,000.

Solution If $10,000 is deposited in a savings account at an interest rate of 8% per year and interest is compounded continuously, then the amount A in the account after t years is given by the formula

$$A = 10{,}000 e^{0.08t}$$

We want to find the time t when $A = \$50{,}000$. Thus, we replace A with 50,000 and solve for t as follows:

$$50{,}000 = 10{,}000 e^{0.08t}$$
$$5 = e^{0.08t} \qquad \text{Divide both sides by 10,000}$$
$$\ln 5 = 0.08t \qquad \text{Take the natural logarithm of both sides}$$
$$t = \frac{\ln 5}{0.08} \approx 20.1 \text{ years} \qquad \text{Divide by 0.08}$$

Thus, it takes approximately 20 years for the original investment to accumulate to $50,000. The graph in Figure 5.32 supports our work.

PROBLEM 7 At what interest rate, compounded continuously, must an amount of money be invested if the amount is to double in 5 years 6 months?

Under controlled laboratory conditions, a population P of living organisms increases exponentially as a function of time t, and the growth pattern is described by the **Malthusian model**

$$P(t) = P_0 e^{kt}$$

where P_0 is the initial population and k is a positive constant. The Malthusian model (named after the Englishman Thomas Malthus, 1766–1834) is also an exponential growth function.

FIGURE 5.32

The graph of $A = 10{,}000 e^{0.08t}$ for $t \geq 0$ shows that $A = \$50{,}000$ when $t \approx 20.1$ years.

EXAMPLE 8 Suppose 18 bacteria per microliter are present initially in a culture, and 20 minutes later 270 bacteria per microliter are present.

(a) Determine the Malthusian model that describes this growth pattern.

(b) Determine the time t (to the nearest minute) when 360 bacteria per microliter are present in the culture.

◆ **Solution**

(a) Since 18 bacteria are present initially, we know that $P_0 = 18$. Thus, starting with the Malthusian model $P(t) = P_0 e^{kt}$, we can write

$$P(t) = 18e^{kt}$$

We must now determine the constant k. Using the fact that if $t = 20$, then $P(t) = 270$, we have

$$270 = 18e^{20k}$$
$$15 = e^{20k} \qquad \text{Divide both sides by 18}$$
$$\ln 15 = 20k \qquad \text{Take the natural logarithm of both sides}$$
$$k = \frac{\ln 15}{20} \approx 0.1354 \qquad \text{Solve for } k$$

Replacing k with 0.1354 gives us the Malthusian model that describes this growth pattern. Thus, we have

$$P(t) = 18e^{0.1354t}$$

(b) Using the exponential function developed in part (a), we replace $P(t)$ with 360 and solve for t:

$$360 = 18e^{0.1354t}$$
$$20 = e^{0.1354t} \qquad \text{Divide both sides by 18}$$
$$\ln 20 = 0.1354t \qquad \text{Take the natural logarithm of both sides}$$
$$t = \frac{\ln 20}{0.1354} \approx 22 \text{ minutes} \qquad \text{Solve for } t$$

The graph in Figure 5.33 supports our work. ◆

PROBLEM 8 For the growth model described in Example 8, determine the time t (to the nearest minute) when 540 bacteria per microliter are present in the culture. ◆

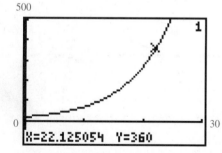

FIGURE 5.33

The graph of $P(t) = 18e^{0.1354t}$ for $t \geq 0$ shows that when 360 bacteria per microliter are present in the culture, $t \approx 22$ minutes.

Radioactive materials disintegrate exponentially over time, and their decay patterns can be described by exponential decay functions. To describe the decay of radioactive materials, the term *half-life* is often used. **Half-life** is defined as the time required for a given mass of a radioactive material to disintegrate to half its original mass.

EXAMPLE 9 Strontium-90, a waste product from nuclear reactors, has a half-life of 28 years. It is estimated that a certain quantity of this material will be safe to handle when its mass is $\frac{1}{1000}$ of its original amount. Determine the time required to store strontium-90 until it disintegrates to this level of safety.

◆ **Solution** If the half-life of strontium-90 is 28 years, then half of the original amount is present after 28 years. Thus, starting with the exponential decay function $A(t) = A_0 e^{kt}$, we replace $A(t)$ with $\frac{1}{2}A_0$ and t with 28, and then solve for k as follows:

$$\tfrac{1}{2} A_0 = A_0 e^{k(28)}$$

$$\tfrac{1}{2} = e^{k(28)} \qquad \text{Divide both sides by } A_0$$

$$\ln \tfrac{1}{2} = 28k \qquad \text{Take the natural logarithm of both sides}$$

$$k = \frac{\ln \tfrac{1}{2}}{28} \approx -0.02476 \qquad \text{Solve for } k$$

Thus, we have

$$A(t) = A_0 e^{-0.02476t}$$

To determine the time t required to store strontium-90 until its mass reaches $\frac{1}{1000}$ of its original amount, we replace $A(t)$ with $0.001 A_0$, and solve for t as follows:

$$0.001 A_0 = A_0 e^{-0.02476t}$$

$$0.001 = e^{-0.02476t} \qquad \text{Divide both sides by } A_0$$

$$\ln 0.001 = -0.02476t \qquad \text{Take the natural logarithm of both sides}$$

$$t = \frac{\ln 0.001}{-0.02476} \approx 279 \text{ years} \qquad \text{Solve for } t$$

PROBLEM 9 According to the decay function developed in Example 9, how long does it take 30 grams of strontium-90 to decay to 27 grams?

Exercises 5.4

Basic Skills

In Exercises 1–24, solve each exponential equation. Check graphically with a graphing calculator.

1. $9^x = 27$
2. $16^x = \tfrac{1}{8}$
3. $2^{x-1} = 16^x$
4. $5^{x+3} = 25^{x-2}$
5. $7^x = 35$
6. $3^x = 36$
7. $e^{3x} = \tfrac{2}{3}$
8. $10^{-2x} = 60$
9. $10^{3-2x} = 28$
10. $e^{2x-1} = 2$
11. $3^{2x-1} = 40$
12. $6^{x+5} = 75$
13. $4e^{-7x} = 15$
14. $12(10^{1-x}) = 60$
15. $3 - 2(10^{-x}) = -117$
16. $1 + 2e^x = 9$

17. $10^{2x-1} = 4^{-x}$
18. $5^{x-2} = 4^{2x+1}$
19. $e^{x/2} = 2^{x-1}$
20. $4^{-2x} = e^{x+1}$
21. $3 \cdot 2^{x-2} = 6^x$
22. $5 \cdot 3^{-x} = 4 \cdot 2^{x+2}$
23. $2^x - 6 \cdot 8^{3-x} = 0$
24. $\dfrac{10^{x-1}}{e^{2x}} - 4 = 0$

Solve each logarithmic equation in Exercises 25–44. Check graphically with a graphing calculator.

25. $\log x = -2$
26. $\ln x = 3$
27. $2 \log_4 x = 3$
28. $3 \log_{1/8} x = -4$
29. $\ln(x - e) = 1$
30. $-3 \log_8(x + 9) = 1$
31. $\log_2(2x^2 + 5x + 5) = 3$
32. $\ln(x^2 - 3x - 3) = 0$
33. $\log x + \log 5 = 1$
34. $\ln x = 1 + \ln 5$
35. $3 \ln(x + 1) - \ln 27 = 3$
36. $2 \log_6(2 - x) = 2 - \log_6 3$
37. $\log_4(x + 12) - \log_4(x - 3) = 2$
38. $\log_3 x + \log_3(x - 6) = 3$
39. $\log_5 x + \log_5(x + 2) = \log_5(x + 6)$
40. $\log 2 + \log(2x - 3) = \log 3x - \log 3$
41. $\log(x + 6) + 1 = 2 \log(3x - 2)$
42. $\log_2(x - 1) + 2 \log_2 x = 2 + \log_2 3x$
43. $\tfrac{1}{2} \ln(3 - 2x) - \ln x = 0$
44. $\tfrac{1}{2} \log(x + 3) + \log 2 = 1$

In Exercises 45–52, solve each literal equation for y. Assume a, b, and c are constants.

45. $\ln y + 2x = \ln c$
46. $2(x + \ln x) = \ln y - \ln c$
47. $2 \log(x + a) = 1 - \log y$
48. $\log y = \log(x + y) + 2$
49. $\ln(x + y) - \ln(x - y) = c$
50. $\ln(x + 2y) + 2 \ln x = \ln y + c$
51. $\dfrac{1}{a} \ln y - \dfrac{1}{a} \ln(a - by) = c$
52. $-\tfrac{1}{4} \ln(y + 2) + \tfrac{1}{4} \ln(y - 2) = x + c$

In Exercises 53–58, use factoring and the zero product property to help solve each equation. Check graphically with a graphing calculator.

53. $(\ln x)^2 = \ln x^2$
54. $2[\log(x + 1)]^2 - \log(x + 1)^3 = 5$
55. $e^{4x} - 3e^{2x} = 4$
56. $2^{3x} + 4 \cdot 2^{-3x} = 5$
57. $3 \cdot 10^x - 10^{-x} = 2$
58. $\tfrac{1}{2}(e^x - 9e^{-x}) = 4$

In Exercises 59–64, use a graphing calculator to find the approximate solution(s) of the equation. Round each answer to four significant digits.

59. $2x + \ln x = 5$
60. $e^{-x} - x^3 = 0$
61. $xe^x = 4$
62. $2^{x+1} + 5^x = 1$
63. $2^x = 3 - x^2$
64. $\ln(x + 1) = x^2 - 4x$

65. An investor deposits $1000 into a savings account paying 7% interest per year compounded continuously. Assuming the investor makes no subsequent deposit or withdrawal, find the time required (in years) for the investment to accumulate to the given amount:
    (a) $1500 (b) $6000 (c) $100,000

66. A certain amount of money is deposited into a savings account paying 10% interest per year compounded continuously. Assuming no subsequent deposit or withdrawal is made, find the time required for the amount to double.

67. When a child is born, his grandfather invests $10,000 for the child's college education. When the child is 18 years old, the balance in the account is $73,000. If interest is compounded continuously, find the interest rate (to the nearest percent) at which the investment is made. Assume no subsequent deposit or withdrawal is made and the interest rate does not change.

68. At what interest rate, compounded continuously, must money be invested if the amount is to quadruple in 18 years?

69. In 1982 foresters found 24 eagles living in a certain national park. In 1992 they counted 144 eagles living in this area.
    (a) Determine the Malthusian model that describes this growth pattern.
    (b) In what year is the eagle population expected to reach 300?

70. Approximately 10^4 bacteria per milliliter are present initially in a culture. Five hours later, approximately 10^7 bacteria per milliliter are present.
    (a) Determine the Malthusian model that describes this growth pattern.
    (b) Approximately how many minutes does it take for this population to double in size?

71. When a plant or animal dies, the amount of carbon-14 in the organism decreases exponentially according to the exponential decay function. Suppose that in 1986 an archaeologist discovers a human skull at an ancient burial site and determines that 60% of the original carbon-14 is still present. If the half-life of carbon-14 is 5600 years, approximate the ancient's year of death.

72. The radioactive substance iodine-131 decays from 30 grams to 25 grams in 50.5 hours.

(a) Determine the exponential decay function that describes this decay pattern.
(b) How many hours will it take for this radioactive substance to decay to 20 grams?
(c) Find the half-life (in days) of this radioactive substance.

73. Due to the presence of acid rain, the population P of fish in a pond in New Hampshire is decreasing according to the equation

$$\log_2 P = -\tfrac{1}{3}t + \log_2 P_0$$

where t is time in years after 1985 and P_0 is the original population in 1985.

(a) Solve the equation for P.
(b) In the year 1991, what percent of the original fish population remained?

74. A student infected with influenza returns to a college campus of 1000 students. If no student leaves the campus, then the time t (in days) it takes for N students to become infected with the virus is given by

$$t = \ln 999N - \ln(1000 - N).$$

(a) In how many days will 20 students be infected?
(b) Determine the number of infected students after 6 days.

Critical Thinking

75. In Section 5.1 we developed the compound interest formula

$$A = P\left(1 + \frac{r}{n}\right)^{nt}.$$

Solve this literal equation for t and express the answer in terms of natural logarithms.

76. Explain the fallacy in the following argument:

$$1 < 2$$

$\tfrac{1}{4} < \tfrac{1}{2}$ Divide both sides by 4

$\ln \tfrac{1}{4} < \ln \tfrac{1}{2}$ Take the natural logarithm of both sides

$\ln (\tfrac{1}{2})^2 < \ln \tfrac{1}{2}$ Rewrite $\tfrac{1}{4}$

$2 \ln \tfrac{1}{2} < \ln \tfrac{1}{2}$ Apply log property 3

$2 < 1$ Divide both sides by $\ln \tfrac{1}{2}$

77. If y varies directly as the mth power of x, then

$$y = kx^m$$

where k is the variation constant.

(a) Take the natural logarithm of both sides of this equation and show that

$$\ln y = m \ln x + \ln k.$$

(b) Construct the graph of $\ln y = m \ln x + \ln k$ in an XY-plane where $X = \ln x$ and $Y = \ln y$. Describe the graph.

78. Solve for x in terms of b.

(a) $\log_b (x - 3) = -1 + \log_b 5$
(b) $2 + \log_b (4x + 1) = \log_b (2 - x)$

Find the inverse of each function given in Exercises 79–86. Use a graphing calculator to verify each answer by showing that the graph of the function and the graph of the inverse are reflections of one another in the line $y = x$.

79. $h(x) = e^{x/2}$
80. $F(x) = 10^{2x/3}$
81. $G(x) = 1 - 10^{2x}$
82. $H(x) = 3e^{2x+1}$
83. $h(x) = -\ln \dfrac{x}{2}$
84. $F(x) = 2 \log \dfrac{3x}{2}$
85. $G(x) = 1 + 2 \log 3x$
86. $H(x) = 3 \ln(2x - 1)$

In Exercises 87–90, solve each equation for x.

87. $\dfrac{e^x + e^{-x}}{2} = 10$
88. $10^x + 10^{-x} = 6$
89. $2 \log_x 3 + \log_3 x = 3$
90. $2 \log_4 x - 3 \log_x 4 = 5$

91. When a cable or rope is suspended between two points at the same height and allowed to hang under its own weight, it forms a curve called a **catenary** (after the Latin word for chain). The power line shown in the figure is an example of a catenary, and its equation is given by

$$y = 25(e^{x/50} + e^{-x/50})$$

where x and y are measured in feet.

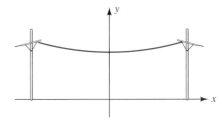

Exercise 91 continues

(a) Find the height of the power line at $x = 0$.

(b) Find x (to the nearest tenth of a foot) when $y = 75$ ft.

92. In a series circuit containing a capacitor C, a resistance R, and a battery source E, the instantaneous current i at any time t is given by

$$\ln i = -\frac{t}{RC} + \ln E - \ln R.$$

(a) Solve the equation for i.

(b) What happens to the current i as t increases without bound $(t \to +\infty)$?

93. In a particular electric circuit, the current i (in amperes) is a function of the time t (in seconds) after the switch is closed and is given by $i(t) = 20(1 - e^{-4.5t})$.

(a) Discuss the behavior of $i(t)$ as t increases without bound $(t \to +\infty)$.

(b) Find the time (in milliseconds) at which $i(t) = 6$ amperes.

94. When limited resources such as food supply and space are taken into account, the population P of a living organism as a function of time t is more accurately described by the **logistic law** than by the Malthusian model. The logistic law is given by the function

$$P(t) = \frac{cP_0}{P_0 + (c - P_0)e^{-kt}}$$

where P_0 is the initial population and c and k are positive constants. Suppose the population of wolves in a certain forest follows the logistic law with $P_0 = 200$, $c = 1500$, $k = 0.2$, and time t in years.

(a) Discuss the behavior of $P(t)$ as t increases without bound $(t \to +\infty)$.

(b) Find the time (in years) when $P(t) = 600$ wolves.

If a heated object of temperature T_0 is placed in a cooler medium that has a constant temperature of T_1, then the temperature T of the object at any time t is given by the formula

$$T = T_1 + (T_0 - T_1)e^{-kt}$$

where k is a positive constant. This formula is **Newton's law of cooling**, named after Sir Isaac Newton (1642–1727). Use this formula to answer Exercises 95 and 96.

95. A pizza baked at 400°F is removed from the oven and placed in a room with a constant temperature of 70°F. The pizza cools to 300°F in two minutes.

(a) State the formula that describes this cooling process.

(b) Find the temperature of the pizza after it cools for three minutes.

(c) How soon after it is removed from the oven does the pizza cool to 150°F?

96. A bottle of white zinfandel wine is stored in a room with a temperature of 70°F. The distributor recommends that the wine is best when served slightly chilled to 40°F. At 4:00 P.M. the bottle is placed in a refrigerator having a constant temperature of 35°F, and after 30 minutes the temperature of the wine is 50°F. At what time should the wine be removed from the refrigerator if it is to be served when chilled to 40°F?

Chapter 5 Review

◆ Questions for Writing or Group Discussion

1. If $b > 0$ but $b \neq 1$, how is b^r defined when r is a rational number? an irrational number?

2. Under what conditions is the *exponential function* $f(x) = b^x$ an increasing function? a decreasing function?

3. How is the graph of $f(x) = b^x$ related to the graph of $g(x) = b^{-x}$? Discuss the behavior of $f(x)$ and $g(x)$ as $x \to +\infty$.

4. Are the functions $f(x) = e^{x+1}$ and $g(x) = e \cdot e^x$ identical? Explain.

5. Why is an interest rate that is *compounded continuously* better than one that is compounded a fixed number of times per year?

6. What is the meaning of the *base b logarithm* $\log_b x$? How can a calculator be used to evaluate $\log_b x$? When is $\log_b x$ undefined?

7. Complete each *logarithmic identity*.
    (a) $\log_b 1 = $ _____
    (b) $\log_b b = $ _____
    (c) $\log_b b^u = $ _____
    (d) $b^{\log_b u} = $ _____

8. What is the base for *common logarithms*? for *natural logarithms*?

9. State the change of base formula for logarithms. How are the expressions $\log_b a$ and $\log_a b$ related?

10. List some characteristics of the graph of the *logarithmic function* $f(x) = \log_b x$.

11. How can we rewrite the logarithm of a product? of a quotient? of a quantity raised to a power?

12. Are the functions $f(x) = \ln(x+1)^2$ and $g(x) = 2\ln(x+1)$ identical? Explain.

13. How is the graph of $f(x) = \log_b x$ related to the graph of $g(x) = \log_b(1/x)$?

14. What is the domain and range of the *natural logarithmic function* $f(x) = \ln x$? What is the inverse function of $f(x) = \ln x$?

15. What *logarithmic scale* is used to describe the magnitude of an earthquake? the loudness of a sound?

16. Explain the general procedure for solving an *exponential equation*. Illustrate with an example.

17. Explain the general procedure for solving a *logarithmic equation*. Illustrate with an example.

18. Give an example of an *exponential growth function* and an *exponential decay function*.

19. Why is 1 excluded as a base for exponential functions?

20. What is the value of $(1 + 1/n)^n$ as n increases without bound?

Review Exercises

In Exercises 1–6, use the properties of real exponents to write each expression as a constant or as an expression in the form b^x or b^{-x} for $b > 1$.

1. 1^{3x-2}
2. $(3^{x+\pi})^0$
3. $e^{2x} \cdot e^{1-2x}$
4. $(2^{-3x})^2$
5. $\dfrac{4^{x+3}}{8^{x+2}}$
6. $(3^{-x} \cdot 9^x)^2$

In Exercises 7–24, find the value of each expression.

7. $\log_7 49$
8. $\log_2 32$
9. $\log \sqrt[3]{100}$
10. $\ln e^3$
11. $\log_{1/2} 8$
12. $\log_{1/3} 9$
13. $\log_8 2$
14. $\log_{25} \frac{1}{5}$
15. $\log_8 \frac{1}{4}$
16. $\log_{16} 64$
17. $\log_3 45$
18. $\log_6 30$
19. $\log_{1/2} 100$
20. $\log_{2/3} \frac{1}{2}$
21. $e^{\ln 4}$
22. $10^{\log 8}$
23. $3^{\log_7 7}$
24. $8^{\log_2 1}$

In Exercises 25–34, simplify each expression. Assume that each variable is restricted only to real numbers that allow the logarithmic expression to be defined.

25. $\log_7 7^{2x}$
26. $5^{\log_5(2-x)}$
27. $e^{\ln(x^2+1)}$
28. $\ln e^{2-3x}$
29. $\log_3 27^{1-x}$
30. $\log_4 2^{4x-4}$
31. $e^{2x+\ln x}$
32. $e^{-\ln(x-1)}$
33. $(\log_3 x)(\log_x 3)$
34. $\dfrac{1}{\log_x(x/2)} - \dfrac{1}{\log_2(x/2)}$

In Exercises 35–40, expand each expression to eliminate logarithms of products, quotients, and powers. Assume that each variable is restricted only to real numbers for which the properties of logarithms can be applied.

35. $\log_3[9x^2(x-1)]$
36. $\log_4[4x\sqrt{x+2}]$
37. $\log \dfrac{1}{x\sqrt{2x-3}}$
38. $\log \dfrac{1000}{xy^{1/3}}$
39. $\ln \dfrac{xe^{-x^2}}{e^x - 1}$
40. $\ln \dfrac{4e^{2x}}{\sqrt{e^x + 2}}$

In Exercises 41–46, condense each expression to a single logarithm with a coefficient of 1. Assume that each variable is restricted only to real numbers for which the properties of logarithms can be applied.

41. $\ln 12 - \ln 3 + \ln 2$
42. $-2 \log_2 6 + 3 \log_2 3 - \frac{1}{2} \log_2 9$
43. $3 \log_3(x+1) - \log_3(x^2 - 1) + \log_3(x-1)$
44. $2 \ln x + \frac{1}{2} \ln(x+1) + \ln 2$
45. $\log \dfrac{1}{\sqrt{x} - \sqrt{x-1}} - \log(\sqrt{x} + \sqrt{x-1})$
46. $\log(\sqrt{x+1} + 1) + \log(\sqrt{x+1} - 1)$

In Exercises 47–58, sketch the graph of each function. Label any x- and y-intercept(s) and horizontal or vertical asymptote.

47. $f(x) = 5^x$
48. $g(x) = 6^{-x}$
49. $h(x) = e^{x-2}$
50. $F(x) = 2 + e^x$
51. $G(x) = -(2 + e^{x+1})$
52. $H(x) = 3^x + 3^{-x}$
53. $h(x) = \log_{1/5} x$
54. $F(x) = \log_6(-x)$
55. $G(x) = -(2 + \log_4 x)$
56. $H(x) = \log_4(x+2)$
57. $f(x) = \ln \dfrac{1}{x+1}$
58. $g(x) = \ln \sqrt{1-x}$

In Exercises 59–80, solve each equation.

59. $10^{2x} = 80$
60. $e^{2x-1} = 9$
61. $16^x = \frac{1}{8}$
62. $3^{x+3} = 27^{1-x}$
63. $3^{2x} = 4^{1-x}$
64. $e^{2x} = 2^{x+1}$
65. $2e^{-x} = 3^{2x-1}$
66. $5 \cdot 2^{-x} = 3 \cdot 5^{x+2}$
67. $\log_5 x = 1$
68. $\ln x = 10$
69. $\ln |x+2| - 3 = 0$
70. $\log_3(x^2 - 3x + 5) = 2$
71. $\log_3(x-7) - \log_3(x+1) = 2$
72. $\log 3 - \log(2-x) = \log 2 - \log 2x$
73. $\frac{1}{2} \log(12-x) - \log x = 0$
74. $\log_2(3x-2) + 2 = 2 \log_2(x+2)$
75. $\ln y = c + 2 \ln |x|$, where c is a constant; solve for y
76. $3(x - \ln x) = \ln(2y - 3) + \ln c$, where c is a constant; solve for y
77. $x^2 e^{-x} - e^{-x} = 0$
78. $x \ln x - 3 \ln x = 0$
79. $\log_x 3 + 2 \log_3 x = 3$
80. $e^x + 1 = 2e^{-x}$

In Exercises 81–86, find the inverse of each function.

81. $f(x) = 8^x$
82. $g(x) = e^{-2x}$
83. $h(x) = 2e^{3x-2}$
84. $F(x) = 1 - 8^{1-x}$
85. $G(x) = \log_{1/2}(2x)$
86. $H(x) = -\ln(x-1)$

87. Suppose that $5000 is invested at an interest rate of 9% per year. Assuming the investor makes no subsequent deposit or withdrawal, determine the amount in the account after 10 years if the interest is compounded (a) semiannually, (b) monthly, or (c) continuously.

88. Suppose that $100,000 is invested at an interest rate of $7\frac{1}{2}$% per year. Assuming the investor makes no subsequent deposit or withdrawal, determine the amount in the account after 9 months if the interest is compounded (a) quarterly, (b) daily, or (c) continuously.

89. An investor deposits $500 into a savings account paying $6\frac{3}{4}$% interest per year compounded continuously. Assuming the investor makes no subsequent deposit or withdrawal, find the time it takes for the investment to accumulate to $800.

90. At what interest rate compounded continuously must money be invested if the amount is to triple in 12 years?

91. Initially, approximately 10^3 bacteria per milliliter are present in a culture. Three hours later approximately 10^5 bacteria per milliliter are present.
   (a) Determine the Malthusian model that describes this growth pattern.
   (b) Approximately how many minutes does it take for this population to double in size?

92. A radioactive substance disintegrates from 60 grams to 50 grams in four years.
   (a) Determine the exponential decay function that describes this disintegration.
   (b) How many years will it take for this radioactive substance to decay to 10 grams?
   (c) Find the half-life of this radioactive substance.

93. One of the strongest earthquakes ever recorded occurred in Japan in 1993 and measured 8.9 on the Richter scale. How many times more intense was this earthquake than one that measures 6.0 on the Richter scale?

94. Students taking their last French course are given a final exam, and the average score of the group is recorded. Each month thereafter the students are given an equivalent exam, and for each exam their average score is recorded. It is found that the average score S is a function of the time t (in months) and is given by

$$S(t) = 77 - 17 \ln(t + 1) \quad \text{for } 0 \leq t \leq 12.$$

(a) What is the average score on the original final exam?
(b) What is the average score on an exam given three months later?
(c) How much time has elapsed when the average score is 40?

95. When a hot metal object with a temperature of 400°F is placed in a room whose temperature is 70°F, the object cools at a rate such that its temperature T is a function of the time t (in minutes) that the object has been in the room and is given by Newton's cooling law

$$T(t) = 70 + (400 - 70)e^{-kt}$$

where k is a constant.

(a) Determine the value of the constant k if the temperature of the object after 10 minutes is 290°F.
(b) Find $T(60)$, the temperature of the object after 1 hour.
(c) How much time has elapsed when $T(t) = 150°F$?

96. In a certain forest, the population P of foxes after t years is described by the logistic law

$$P(t) = \frac{25000}{100 + 150e^{-0.1t}}.$$

(a) Find $P(10)$, the number of foxes in the forest after 10 years.
(b) Discuss the behavior of $P(t)$ as t increases without bound ($t \to +\infty$).
(c) Find the time in years when $P(t) = 210$ foxes.

The parabolic reflector of a searchlight has a diameter of 3 feet at the open end and a depth of 1 foot. How far from the vertex of the parabola should the filament of the bulb be located so that all the light rays emanating from the bulb are parallel to the axis of the parabola?

For the solution, see Example 5 in Section 6.2.

Conic Sections

6.1 The Circle
6.2 The Parabola
6.3 The Ellipse
6.4 The Hyperbola

6.1 The Circle

◆ Introductory Comments

Early Greek mathematicians noted that when a double cone is sliced with a plane, a special family of curves is formed. As illustrated in Figure 6.1, this family of curves has four main members: *circle, parabola, ellipse,* and *hyperbola.* Collectively, this family of curves is called the **conic sections**.

(a) Circle

(b) Parabola

(c) Ellipse

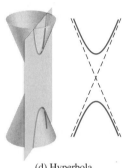
(d) Hyperbola

FIGURE 6.1
The conic sections

When the plane intersects the double cone, it is also possible to obtain a single point, one line, or a pair of lines. Do you see how? These special cases are called the **degenerate conic sections**.

An equation of the form

$$Ax^2 + Bxy + Cy^2 + Dx + Ey + F = 0$$

in which the coefficients A, B, and C are not all zero, is called a **general quadratic equation in two unknowns**. As we will see, if the graph of this type of equation exists, then it is either a conic section or a degenerate conic section.

In this chapter we study the geometric properties of each conic section and the distinguishing characteristics of their equations. We limit our discussion to general quadratic equations in two unknowns that contain no xy-term. By letting $B = 0$, we keep at least one of the coordinate axes parallel to an axis of symmetry of the conic section. If $B \neq 0$, then neither coordinate axis is parallel to an axis of symmetry of the conic section. For example, the graph of the general quadratic equation $xy = 1$ [or $y = 1/x$, the reciprocal function in Figure 2.42(f)] is a hyperbola whose axis of symmetry $y = x$ makes an angle of 45° with the coordinate axes.

◆ Determining the Equation of a Circle

We begin our discussion with the circle. Geometrically, we define a circle as follows.

Geometric Definition of a Circle

A **circle** is the set of all points in a plane that lie a fixed distance from a given point. We call the fixed distance the **radius** and the given point the **center** of the circle.

Shown in Figure 6.2 is a circle with radius r and center (h, k). The point (x, y) lies on this circle if and only if the distance from the center (h, k) to (x, y) is r. By using the distance formula (Section 2.1), we have

$$r = \sqrt{(x - h)^2 + (y - k)^2}.$$

If we square both sides of this equation and then interchange the sides, we obtain

$$(x - h)^2 + (y - k)^2 = r^2.$$

We refer to this last equation as the *algebraic definition of a circle*, or the *equation of a circle in standard form*.

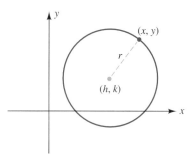

FIGURE 6.2

A circle with center (h, k) and radius r

Equation of a Circle in Standard Form

The **equation of a circle in standard form** with center (h, k) and radius r is

$$(x - h)^2 + (y - k)^2 = r^2.$$

We can determine the equation of a circle if we know its center (h, k) and the value of its radius r.

EXAMPLE 1 Determine the equation of a circle that has center $(2, -1)$ and passes through the point $(4, 0)$.

Solution   The circle with the given characteristics is shown in Figure 6.3. Since the center is $(2, -1)$, the equation of this circle must have the form

$$(x - 2)^2 + (y + 1)^2 = r^2$$

Now, the radius r of this circle is the distance from the center $(2, -1)$ to the point $(4, 0)$. By the distance formula, we have

$$r = \sqrt{(4 - 2)^2 + [0 - (-1)]^2} = \sqrt{5}$$

Hence, $r^2 = (\sqrt{5})^2 = 5$. Therefore, the equation of a circle that has center $(2, -1)$ and passes through the point $(4, 0)$ is

$$(x - 2)^2 + (y + 1)^2 = 5.$$

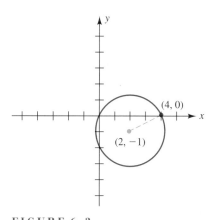

FIGURE 6.3

A circle that has center $(2, -1)$ and passes through the point $(4, 0)$

To generate the graph of $(x - h)^2 + (y - k)^2 = r^2$ in the viewing rectangle of a graphing calculator, we solve the equation for y:

$$(x - h)^2 + (y - k)^2 = r^2$$

$$(y - k)^2 = r^2 - (x - h)^2 \qquad \text{Subtract } (x - h)^2 \text{ from both sides}$$

$$y - k = \pm\sqrt{r^2 - (x - h)^2} \qquad \text{Take the square root of both sides}$$

$$y = k \pm \sqrt{r^2 - (x - h)^2} \qquad \text{Add } k \text{ to both sides}$$

We then graph both equations:

$$y = k + \sqrt{r^2 - (x - h)^2} \quad \text{and} \quad y = k - \sqrt{r^2 - (x - h)^2}$$

To obtain a true geometric perspective of the graph, we use a square set viewing rectangle.

PROBLEM 1 Generate the graph of the circle $(x - 2)^2 + (y + 1)^2 = 5$ in the viewing rectangle of a graphing calculator. Verify that this circle passes through the point $(4, 0)$ and that the coordinates of its center are $(2, -1)$. ◆

A **tangent** to a circle is a line that intersects the circle in one and only one point. Referring to Figure 6.4, we say that the line and the circle are *tangent* at point P. Point P is called the **point of tangency** of the line and circle. *Every tangent to a circle is perpendicular to a radius drawn to the point of tangency.*

EXAMPLE 2 Determine the equation of a circle that has center $(-2, 1)$ and is tangent to the line $x + 2y = 5$.

◆ **Solution** The circle with the given characteristics is shown in Figure 6.5. The slope of the line $x + 2y = 5$ (or $y = -\frac{1}{2}x + \frac{5}{2}$) is $-\frac{1}{2}$. Since perpendicular lines have negative reciprocal slopes, the slope of the radius drawn to the point of tangency is 2. We can find the equation of the radius drawn to the point of tangency by using the point-slope form of the equation of a straight line (Section 3.2):

$$y - y_1 = m(x - x_1)$$
$$y - 1 = 2(x + 2)$$
$$y = 2x + 5$$

To find the coordinates of the point of tangency P, we solve the equations $x + 2y = 5$ and $y = 2x + 5$ simultaneously by using substitution (Section 3.5):

$$x + 2y = 5$$

$$x + 2(2x + 5) = 5 \qquad \text{Substitute } y = 2x + 5$$

$$5x + 10 = 5 \qquad \text{Solve for } x$$

$$x = -1$$

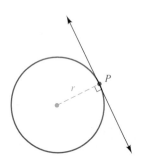

FIGURE 6.4
The line and the circle are *tangent* at point P.

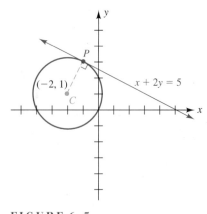

FIGURE 6.5
A circle that has center $(-2, 1)$ and is tangent to the line $x + 2y = 5$

Substituting $x = -1$ into the equation $y = 2x + 5$ gives us $y = 3$. Hence, the coordinates of the point of tangency P are $(-1, 3)$. Now, by the distance formula, the radius of the circle is

$$CP = \sqrt{(3-1)^2 + [-1-(-2)]^2} = \sqrt{5}$$

Therefore, the equation of a circle that has center $(-2, 1)$ and is tangent to the line $x + 2y = 5$ is

$$(x + 2)^2 + (y - 1)^2 = 5.$$

Figure 6.6 shows the graphs of $(x + 2)^2 + (y - 1)^2 = 5$ and $x + 2y = 5$ in the viewing rectangle of a graphing calculator. This display confirms our work. ◆

PROBLEM 2 Determine the x- and y-intercepts for the equation of the circle $(x + 2)^2 + (y - 1)^2 = 5$ obtained in Example 2. ◆

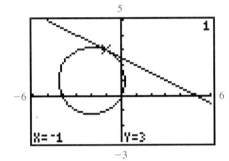

FIGURE 6.6
The circle $(x + 2)^2 + (y - 1)^2 = 5$ is tangent to the line $x + 2y = 5$ at $(-1, 3)$.

◆ General Form of a Circle

Squaring the expressions $(x - h)$ and $(y - k)$ in the equation

$$(x - h)^2 + (y - k)^2 = r^2$$

gives us

$$x^2 - 2hx + h^2 + y^2 - 2ky + k^2 = r^2$$
$$x^2 + y^2 - 2hx - 2ky + (h^2 + k^2 - r^2) = 0.$$

Replacing the constants $-2h$ with D, $-2k$ with E, and $(h^2 + k^2 - r^2)$ with F, we have

$$\boxed{x^2 + y^2 + Dx + Ey + F = 0}$$

This general quadratic equation in two unknowns is called the **equation of a circle in general form** and is characterized by the presence of x^2 and y^2 terms, each having a coefficient of 1.

If the equation of a circle is given in general form, then we can find the radius of the circle and the coordinates of its center by converting the equation to standard form and observing the values of h, k, and r. For example, the equation

$$x^2 + y^2 = 25$$

is of the form $x^2 + y^2 + Dx + Ey + F = 0$ with $D = E = 0$ and $F = -25$. By thinking of this equation as

$$(x - 0)^2 + (y - 0)^2 = 5^2$$

standard form of a circle

we know that its graph is a circle with center (0, 0) and radius 5. Figure 6.7 shows the graph of $x^2 + y^2 = 25$ (or $y = \pm\sqrt{25 - x^2}$) in the viewing rectangle of a graphing calculator. The graph confirms the center and radius.

In general, the graph of any equation of the form

$$x^2 + y^2 = r^2$$

with $r > 0$ is a circle with center at the origin and radius r.

To find the center and radius of a circle defined by the equation

$$x^2 + y^2 + Dx + Ey + F = 0 \quad \text{with } D \neq 0 \text{ or } E \neq 0,$$

we use the algebraic process of completing the square (Section 3.3) to write the equation in standard form. We can use a graphing calculator to check our work by solving the equation $x^2 + y^2 + Dx + Ey + F = 0$ for y and graphing *both* equations that result. To generate the graph of this equation when $E \neq 0$, we rewrite the equation as a quadratic equation in y:

$$y^2 + Ey + (x^2 + Dx + F) = 0$$

We apply the quadratic formula to this equation:

$$y = \frac{-E \pm \sqrt{E^2 - 4(x^2 + Dx + F)}}{2}$$

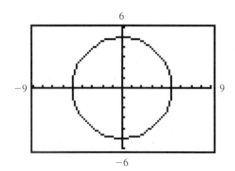

FIGURE 6.7
The graph of $x^2 + y^2 = 25$ is a circle with center (0, 0) and radius 5.

EXAMPLE 3 Find the radius and the coordinates of the center of the circle given by each equation.

(a) $x^2 + y^2 + 6x = 0$ **(b)** $4x^2 + 4y^2 - 8x + 20y + 13 = 0$

◆ **Solution**

(a) The equation $x^2 + y^2 + 6x = 0$ appears to be the equation of a circle in general form with $D = 6$, $E = 0$, and $F = 0$. Using the process of completing the square, we have

$(x^2 + 6x \quad) + y^2 = 0$ **Regroup**

$(x^2 + 6x + 9) + y^2 = 9$ **Complete the square by adding 9 to both sides**

$(x + 3)^2 + y^2 = 9$ **Factor the perfect square trinomial**

$[x - (-3)]^2 + (y - 0)^2 = 3^2$ **Write in standard form**

SECTION 6.1 ◆ The Circle

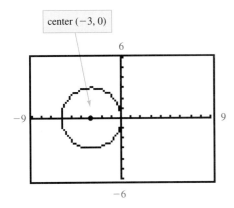

FIGURE 6.8
The graph of $x^2 + y^2 + 6x = 0$ is a circle with center $(-3, 0)$ and radius 3.

Thus, the center of the circle is $(-3, 0)$ and the radius is 3. Figure 6.8 shows the graph of $x^2 + y^2 + 6x = 0$ (or $y = \pm\sqrt{-x^2 - 6x}$) in the viewing rectangle of a graphing calculator. This graph confirms our work.

(b) Dividing both sides of the equation by 4, we obtain

$$x^2 + y^2 - 2x + 5y + \tfrac{13}{4} = 0.$$

This equation appears to be the equation of a circle in general form with $D = -2$, $E = 5$, and $F = \tfrac{13}{4}$. Using the process of completing the square, we have

$(x^2 - 2x \quad) + (y^2 + 5y + \quad) = -\tfrac{13}{4}$ Regroup

Add 1 to both sides.

$(x^2 - 2x + 1) + (y^2 + 5y + \tfrac{25}{4}) = -\tfrac{13}{4} + (1 + \tfrac{25}{4})$ Complete the squares

Add 25/4 to both sides.

$(x - 1)^2 + (y + \tfrac{5}{2})^2 = 4$ Factor

$(x - 1)^2 + [y - (-\tfrac{5}{2})]^2 = 2^2$ Write in standard form

Thus, the center of this circle is $(1, -\tfrac{5}{2})$ and the radius is 2.

To generate the graph of the equation $4x^2 + 4y^2 - 8x + 20y + 13 = 0$ in the viewing rectangle of a graphing calculator, we rewrite the equation as a quadratic equation in y,

$$4y^2 + 20y + (4x^2 - 8x + 13) = 0,$$

and apply the quadratic formula:

$$y = \frac{-(20) \pm \sqrt{(20)^2 - 4(4)(4x^2 - 8x + 13)}}{2(4)} = -\tfrac{5}{2} \pm \sqrt{3 + 2x - x^2}$$

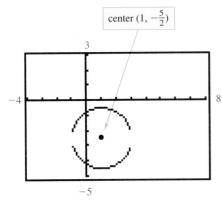

FIGURE 6.9
The graph of $4x^2 + 4y^2 - 8x + 20y + 13 = 0$ is a circle with center $(1, -\tfrac{5}{2})$ and radius 2.

Entering both of these equations into the calculator, we graph the circle shown in Figure 6.9. This graph confirms our work. ◆

CAUTION Not every equation of the form $x^2 + y^2 + Dx + Ey + F = 0$ is the equation of a circle. For example, if the equation is written in the standard form

$$(x - h)^2 + (y - k)^2 = N \quad \text{and} \quad N = 0,$$

then the graph of this equation is the single point (h, k). We refer to this single point as a *degenerate circle*. Also, if

$$(x - h)^2 + (y - k)^2 = N \quad \text{and} \quad N < 0,$$

then this equation has no graph, since no point (x, y) with real coordinates satisfies this equation.

PROBLEM 3 Is $x^2 + y^2 + 4x + 6y + 13 = 0$ the equation of a circle? Explain.

◆ Application: Layout of a Circular Curve

Two straight roadways are often connected by a circular curve that has a *central angle* θ (read "theta") and radius r, as shown in Figure 6.10. The point where roadway A is perpendicular to the radius of the circular curve is called the *point of curvature* (PC), and the point where roadway B is perpendicular to the radius of the circular curve is called the *point of tangency* (PT).

A surveyor lays out a circular curve by placing stakes at various points along the curve between PC and PT. We can use the Pythagorean theorem and the ideas discussed in this section to lay out a circular curve in which $\theta = 120°$, $90°$, or $60°$. The methods of trigonometry are needed to lay out circular curves with other central angles.

FIGURE 6.10
Two straight roadways connected by a circular curve

EXAMPLE 4 Two straight roadways are connected by a circular curve that has central angle $\theta = 90°$ and radius $r = 80$ ft, as shown in the Figure 6.11. A surveyor starts to lay out this circular curve by placing stakes at PC and PT and also at several other points along the curve, such as point A. Take the coordinate axes as shown in the figure.

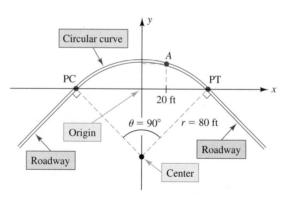

FIGURE 6.11

(a) Find the coordinates of PC and PT and the equation of the circular curve.

(b) Determine the coordinates of point A.

◆ **Solution**

(a) If $\theta = 90°$, then the triangle with vertices at the center, PT, and PC is an isosceles right triangle (45°–45°–90°). This implies that the triangle with vertices at the origin, center, and PT (or PC) is also an isosceles right triangle. Hence, the distance from the origin to the center is equal to the distance from the origin to PT. If we call this common distance a, then by the Pythagorean theorem we have

$$a^2 + a^2 = 80^2$$
$$2a^2 = 6400$$
$$a^2 = 3200$$
$$a = \sqrt{3200} = 40\sqrt{2} \approx 56.57$$

Hence, the coordinates of the center of the circular curve are $(0, -56.57)$, the coordinates of PT are $(56.57, 0)$ and, by symmetry, the coordinates of PC are $(-56.57, 0)$, with each coordinate measured to the nearest hundredth of a foot. The equation of the circular curve is given by

$$(x - h)^2 + (y - k)^2 = r^2$$
$$x^2 + (y + 56.57)^2 = 80^2, \text{ for } -56.57 < x < 56.57$$

(b) The x-coordinate of point A on the circular curve is 20 ft. We find the y-coordinate of this point by substitution:

$$x^2 + (y + 56.57)^2 = 80^2$$
$$20^2 + (y + 56.57)^2 = 80^2$$
$$(y + 56.57)^2 = 80^2 - 20^2$$
$$y + 56.57 = \sqrt{80^2 - 20^2}$$
$$y = \sqrt{80^2 - 20^2} - 56.57 \approx 20.89$$

So the coordinates of A are $(20, 20.89)$.

Figure 6.12 shows the graph of $x^2 + (y + 56.57)^2 = 80^2$ (or $y = -56.57 \pm \sqrt{80^2 - x^2}$) in the viewing rectangle of a graphing calculator between PC and PT. Note that $(20, 20.89)$ is a point on this graph, which supports our work.

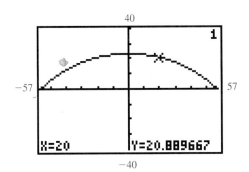

FIGURE 6.12
Graph of $x^2 + (y + 56.57)^2 = 80^2$ between PC and PT

PROBLEM 4 Referring to Example 4, show that the distance from the center of the circular curve to point A is 80 ft, the radius of the circle.

Exercises 6.1

 Basic Skills

In Exercises 1–16, determine the radius and the coordinates of the center of the circle given by each equation of a circle. Check graphically with a graphing calculator.

1. $x^2 + (y - 2)^2 = 16$
2. $(x + 5)^2 + (y + 2)^2 = 4$
3. $x^2 + y^2 = 9$
4. $x^2 + y^2 - 36 = 0$
5. $x^2 + y^2 - 2y = 3$
6. $x^2 + y^2 + 5x = 0$
7. $x^2 + y^2 - 6x + 8y + 9 = 0$
8. $x^2 + y^2 - x - 4y = 2$
9. $3x^2 + 3y^2 + 8x = 0$
10. $5x^2 + 5y^2 - 10y = 1$
11. $4x^2 + 4y^2 - 8x + 12y = 3$
12. $2x^2 + 2y^2 - x + 20y + 40 = 0$
13. $x^2 + y^2 + 10x - 2y + 26 = 0$
14. $4x^2 + 4y^2 + 20x - 16y + 41 = 0$
15. $3x^2 + 3y^2 - 4x - 8y + 24 = 0$
16. $x^2 + y^2 - 12x + 45 = 0$

Find the x- and y-intercepts of the circle whose equation is given in the indicated exercise.

17. Exercise 1
18. Exercise 2
19. Exercise 7
20. Exercise 8

In Exercises 21–32, determine the equation of a circle that has the given characteristics. Record each equation in general form.

21. Center at $(0, 0)$; radius 2
22. Center at $(-3, 0)$; radius 3
23. Center at $(2, -3)$; radius $\sqrt{2}$
24. Center at $(-1, 2)$; radius $\frac{2}{3}$
25. Center at the origin; passes through the point $(-3, 4)$
26. Center at $(-1, 3)$; passes through the point $(1, -1)$

27. Line segment from (2, 3) to (−2, 5) is the diameter
28. Line segment from (−1, −1) to (3, 1) is the radius
    [*Hint:* Two answers are possible.]
29. Center at (−2, 5); tangent to the *y*-axis
30. Center at (3, −1); tangent to the line $x - 2y = -5$
31. Center at (4, 2); tangent to the line $4x + 3y = -3$
32. Tangent to both the *x*- and *y*-axes; diameter of 10
    [*Hint:* Four answers are possible.]

In Exercises 33–36, use a graphing calculator to generate the graph of each circle. Then estimate its center and radius, rounding each answer to three significant digits. Check algebraically.

33. $32x^2 + 32y^2 + 343x - 876 = 0$
34. $x^2 + y^2 + 526x - 365y = 0$
35. $x^2 + y^2 - 7.22x - 4.84y + 12.34 = 0$
36. $4.1x^2 + 4.1y^2 - 24.1x - 16.3y + 25.2 = 0$

Two straight roadways are connected by a circular curve that has central angle θ and radius r, as shown in the sketch. A surveyor wants to lay out this circular curve by placing stakes at PC and PT and also at several other points along the curve, such as point A. Use this information for Exercises 37–40:

(a) *Find the coordinates of PC and PT and the equation of the circular curve for the indicated values of θ and r.*

(b) *Determine the coordinates of point A for the indicated value of x.*

Round all coordinates to the nearest hundredth of a foot.

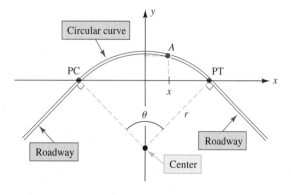

37. $\theta = 90°$, $r = 125$ ft, $x = 50$ ft
38. $\theta = 90°$, $r = 75$ ft, $x = 10$ ft

39. $\theta = 60°$, $r = 54$ ft, $x = 20$ ft
40. $\theta = 120°$, $r = 210$ ft, $x = 80$ ft
41. A castle entrance has the form of a Gothic arch, as shown in the figure.

    (a) Taking the axes as shown in the figure, find the equation of the circle that forms the left-hand side of the arch.

    (b) Use the equation in part (a) to find the height *h* of the arch, rounded to three significant digits.

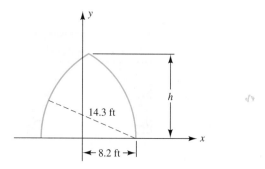

42. The radius of a Ferris wheel is 10.1 meters, as shown in the figure.

    (a) Taking the axes as shown, find the equation of the circle that forms the Ferris wheel.

    (b) Use the equation in part (a) to find the distance *x* from a chair to the vertical axis of the wheel when the height of the chair is 4.2 meters above the ground. Round the answer to three significant digits.

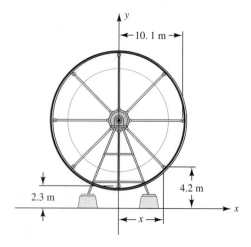

Critical Thinking

43. Two circles having the same center but different radii are called **concentric circles**.

    (a) If the ratio of the radii of two concentric circles is 2 : 1 and the equation of the circle with the larger radii is $x^2 + y^2 - 10x + 6y = 2$, determine the equation of the other circle.

    (b) Find the area *between* the two circles described in part (a).

44. Two circles have equations $x^2 + y^2 + 2x + 4y + 1 = 0$ and $x^2 + y^2 - 8y = 0$.

    (a) Find the distance between their centers.

    (b) Find the coordinates of the midpoint of the line segment joining their centers.

45. Find the points of intersection of the circle $x^2 + y^2 - 2x + 4y - 20 = 0$ and the line $y = x - 2$.

46. Show that the line $3x + 4y = -25$ is tangent to the circle $x^2 + y^2 = 25$, and find the coordinates of the point of tangency.

47. Find the equation of the tangent to the circle at the indicated point.

    (a) $x^2 + y^2 = 25$; at (4, 3)

    (b) $x^2 + y^2 - 6x + 4y = 4$; at (2, 2)

48. For the circle $x^2 + y^2 + Dx + Ey + F = 0$, use the method of completing the square to determine a formula in terms of D, E, and F for the radius of the circle.

49. Find the value(s) of h in the equation $(x - h)^2 + (y + 1)^2 = 25$ so that the circle passes through the point (4, 2).

50. Find the value of D so that the circle with the equation $x^2 + y^2 + Dx + 2y - 1 = 0$ has its center on the line $x + 3y = 2$.

51. Find the value of m for which the line $y = mx + 5$ is tangent to the circle $x^2 + y^2 = 9$.

52. Find the value of r^2 for which the circle $x^2 + y^2 = r^2$ is tangent to the line $x + y = 4$.

53. Show that $x_1 x + y_1 y = r^2$ is the equation of the tangent to $x^2 + y^2 = r^2$ at the point (x_1, y_1) on the circle.

54. Angle P is inscribed in a semicircle of radius r, as shown in the sketch.

    (a) Find the slopes of $\overline{AP}$ and $\overline{BP}$. Then show that the product of these slopes is -1.

    (b) Using the result of part (a), what conclusion can you make about any angle inscribed in a semicircle?

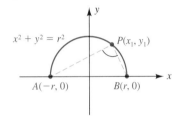

6.2 The Parabola

♦ Introductory Comments

Recall from Section 3.3 that the graph of the equation $y = x^2$ is a parabola with the vertical line $x = 0$ (the y-axis) as its *axis of symmetry*. The point at which the parabola intersects its axis of symmetry is called the *vertex* of the parabola (see Figure 6.13).

By the vertical stretch and compress rules (Section 2.5), we know that the graph of $y = ax^2$ is also a parabola, but with a *wider* opening than $y = x^2$ if $0 < a < 1$, or a *narrower* opening if $a > 1$, as shown in Figure 6.14. By the x-axis reflection rule (Section 2.5), the parabola opens *upward* if $a > 0$ or *downward* if $a < 0$ (see Figure 6.15). In summary, the value of a in the equation $y = ax^2$ determines the width and direction of the opening of the parabola.

FIGURE 6.13

The graph of $y = x^2$ is a parabola with vertex (0, 0) and axis of symmetry $x = 0$.

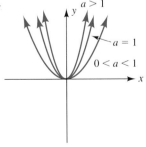

FIGURE 6.14

Graph of $y = ax^2$ for $0 < a < 1$, $a = 1$, and $a > 1$

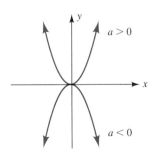

FIGURE 6.15

Graph of $y = ax^2$ for $a > 0$ and $a < 0$

354   CHAPTER 6 ◆ Conic Sections

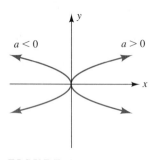

FIGURE 6.16
Graph of $x = ay^2$ for $a > 0$ and $a < 0$

Interchanging x and y in the equation $y = ax^2$ gives us $x = ay^2$, whose graph is also a parabola with its vertex at the origin. However, its axis of symmetry is the horizontal line $y = 0$ (the x-axis) instead of vertical line $x = 0$. If $a > 0$ in the equation $x = ay^2$, the parabola opens to the *right* and if $a < 0$, the parabola opens to the *left,* as shown in Figure 6.16. In this section, we study the geometric properties of a parabola and the distinguishing characteristics of its equation.

◆ Standard Form of a Parabola

If h and k are real numbers and we replace x with $(x - h)$ and y with $(y - k)$ in the equations $y = ax^2$ and $x = ay^2$, we obtain two additional equations:

$$y = ax^2 \qquad\qquad x = ay^2$$
$$y - k = a(x - h)^2 + k \qquad x - h = a(y - k)^2$$
$$y = a(x - h)^2 + k \qquad x = a(y - k)^2 + h$$

By the vertical and horizontal shift rules (Section 2.5), we can state that the graph of $y = a(x - h)^2 + k$ is the same as the graph of $y = ax^2$ with the vertex shifted to the point (h, k) and axis of symmetry the line $x = h$ (see Figure 6.17). The graph of $x = a(y - k)^2 + h$ is also a parabola with vertex (h, k), but with axis of symmetry $y = k$ (see Figure 6.18). We refer to $y = a(x - h)^2 + k$ and $x = a(y - k)^2 + h$ as the *algebraic definitions of a parabola,* or the *equations of a parabola in standard form.*

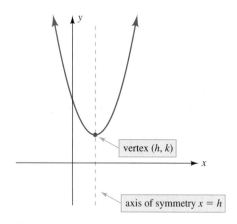

FIGURE 6.17
Graph of $y = a(x - h)^2 + k$ with $a > 0$, $h > 0$, and $k > 0$

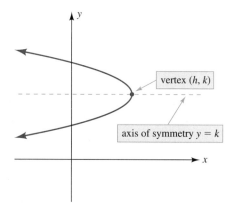

FIGURE 6.18
Graph of $x = a(y - k)^2 + h$ with $a < 0$, $h > 0$, and $k > 0$

◀ **Equations of a Parabola in Standard Form**

> The **equation of a parabola in standard form** with vertex (h, k) is
> $$y = a(x - h)^2 + k$$
> if the axis of symmetry of the parabola is *vertical,* or
> $$x = a(y - k)^2 + h$$
> if the axis of symmetry of the parabola is *horizontal.*
> In each case, a is a nonzero constant that determines the width and direction of the opening of the parabola.

SECTION 6.2 ◆ The Parabola

EXAMPLE 1 Determine the equation of a parabola in standard form with vertex (2, 3) and horizontal axis of symmetry if the parabola passes through the point (4, 1).

◆ **Solution** The parabola with the given characteristics is shown in Figure 6.19. Since the vertex is (2, 3) and the axis of symmetry is the *horizontal* line $y = 3$, the equation of this parabola must have the form

$$x = a(y - 3)^2 + 2.$$

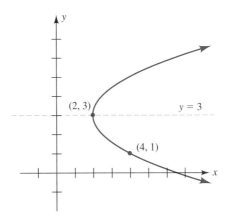

FIGURE 6.19

A parabola passing through (4, 1) with vertex (2, 3) and horizontal axis of symmetry

Since the parabola opens to the *right*, we know that $a > 0$. To determine the value of a, we use the fact that the parabola passes through the point (4, 1).

$$x = a(y - 3)^2 + 2$$

$$4 = a(1 - 3)^2 + 2 \qquad \text{Replace } x \text{ with 4 and } y \text{ with 1}$$

$$4 = 4a + 2 \qquad \text{Solve for } a$$

$$a = \tfrac{1}{2}$$

Thus, the required equation is $x = \tfrac{1}{2}(y - 3)^2 + 2$. ◆

To generate the graph of $x = a(y - k)^2 + h$ in the viewing rectangle of a graphing calculator, we solve the equation for y:

$$x = a(y - k)^2 + h$$

$$x - h = a(y - k)^2 \qquad \text{Subtract } h \text{ from both sides}$$

$$\frac{x - h}{a} = (y - k)^2 \qquad \text{Divide both sides by } a$$

$$y - k = \pm\sqrt{\frac{x - h}{a}} \qquad \text{Take the square root of both sides}$$

$$y = k \pm \sqrt{\frac{x - h}{a}} \qquad \text{Add } k \text{ to both sides}$$

We then enter both equations into the graphing calculator:

$$y = k + \sqrt{\frac{x-h}{a}} \quad \text{and} \quad y = k - \sqrt{\frac{x-h}{a}}$$

PROBLEM 1 Generate the graph of $x = \frac{1}{2}(y - 3)^2 + 2$ in the viewing rectangle of a graphing calculator. Verify that this parabola passes through the point (4, 1) and that the coordinates of its vertex are (2, 3). ◆

◆ General Form of a Parabola

If we expand $(x - h)^2$ in the equation $y = a(x - h)^2 + k$, distribute a, and collect all terms on one side of the equation, we obtain a general quadratic equation in two unknowns of the form

$$Ax^2 + Dx + Ey + F = 0$$

where A, D, E, and F are real numbers with $A \neq 0$ and $E \neq 0$. Similarly, if we expand $(y - k)^2$ in the equation $x = a(y - k)^2 + h$, distribute a, and collect all terms on one side of the equation, we obtain a general quadratic equation in two unknowns of the form

$$Cy^2 + Dx + Ey + F = 0$$

where C, D, E, and F are real numbers with $C \neq 0$ and $D \neq 0$. We refer to

$$\boxed{Ax^2 + Dx + Ey + F = 0} \quad \text{and} \quad \boxed{Cy^2 + Dx + Ey + F = 0}$$

as the **equations of a parabola in general form** with vertical and horizontal axes of symmetry, respectively. These general quadratic equations in two unknowns are characterized by the presence of either an x^2 term or y^2 term, but never both.

To generate the graph of the equation $Ax^2 + Dx + Ey + F = 0$ in the viewing rectangle of a graphing calculator, we simply solve the equation for y and enter the equation that results:

$$y = \frac{-Ax^2 - Dx - F}{E}$$

To generate the graph of the equation $Cy^2 + Dx + Ey + F = 0$ in the viewing rectangle of a graphing calculator, we solve the equation for y and enter *both* equations that result. If $E \neq 0$, we rewrite the equation as a quadratic equation in y, namely, $Cy^2 + Ey + (Dx + F) = 0$, and apply the quadratic formula:

$$y = \frac{-E \pm \sqrt{E^2 - 4C(Dx + F)}}{2C}$$

If the equation of a parabola is given in general form, we can find the coordinates of the vertex of the parabola by converting the equation to standard form and observing the values of h and k. To accomplish this task, it may be necessary to use the algebraic process of completing the square (Section 3.3).

EXAMPLE 2 Find the coordinates of the vertex and the direction of the opening of the parabola given by each equation.

(a) $2x^2 + 3y = 6$ (b) $y^2 - 2x - 4y + 10 = 0$

◀ Solution

(a) The x^2 term indicates that this is the equation of a parabola in general form with a vertical axis of symmetry. To determine the coordinates of the vertex, we solve for y and write the equation in standard form:

$$2x^2 + 3y = 6$$
$$3y = -2x^2 + 6 \qquad \text{Subtract } 2x^2 \text{ from both sides}$$
$$y = -\tfrac{2}{3}x^2 + 2 \qquad \text{Divide both sides by 3}$$
$$y = -\tfrac{2}{3}(x - 0)^2 + 2 \qquad \text{Write in standard form}$$

Thus, the vertex has coordinates $(0, 2)$. Since $a = -\tfrac{2}{3}$ (so $a < 0$), the parabola opens downward from its vertex $(0, 2)$. Figure 6.20 shows the graph of $2x^2 + 3y = 6$ (or $y = -\tfrac{2}{3}x^2 + 2$) in the viewing rectangle of a graphing calculator. This graph confirms our work.

(b) The y^2 term indicates that this is the equation of a parabola in general form with a horizontal axis of symmetry. To determine the coordinates of the vertex, we solve for x and then complete the square as follows:

$$y^2 - 2x - 4y + 10 = 0$$
$$x = \tfrac{1}{2}y^2 - 2y + 5 \qquad \text{Solve for } x$$
$$x = (\tfrac{1}{2}y^2 - 2y \quad) + 5 \qquad \text{Regroup}$$
$$x = \tfrac{1}{2}(y^2 - 4y \quad) + 5 \qquad \text{Factor out 1/2}$$

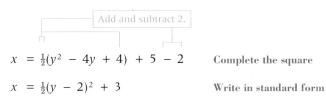
Add and subtract 2.

$$x = \tfrac{1}{2}(y^2 - 4y + 4) + 5 - 2 \qquad \text{Complete the square}$$
$$x = \tfrac{1}{2}(y - 2)^2 + 3 \qquad \text{Write in standard form}$$

Thus, the vertex has coordinates $(3, 2)$. Since $a = \tfrac{1}{2}$ (so $a > 0$), the parabola opens to the right from its vertex $(3, 2)$.

To generate the graph of the equation $y^2 - 2x - 4y + 10 = 0$ in the viewing rectangle of a graphing calculator, we rewrite the equation as a quadratic equation in y,

$$y^2 - 4y + (-2x + 10) = 0,$$

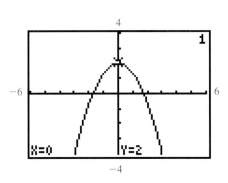

FIGURE 6.20

The graph of the equation $2x^2 + 3y = 6$ is a parabola. It opens downward from its vertex $(0, 2)$.

358 CHAPTER 6 ◆ Conic Sections

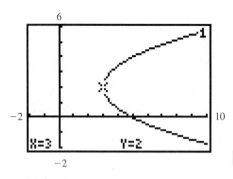

FIGURE 6.21
The graph of the equation
$y^2 - 2x - 4y + 10 = 0$ is a
parabola. It opens to the right from its
vertex (3, 2).

◆ **Geometric Definition
of a Parabola**

and apply the quadratic formula:

$$y = \frac{-(-4) \pm \sqrt{(-4)^2 - 4(1)(-2x + 10)}}{2(1)} = 2 \pm \sqrt{2x - 6}$$

Entering both of these equations into the calculator, we obtain the parabola that is shown in Figure 6.21. This graph confirms our work. ◆

PROBLEM 2 Find the *x*-intercepts for the parabolas defined by the equations in Example 2. Check graphically with a graphing calculator. ◆

◆ Geometric Definition of a Parabola

Geometrically, we define a parabola as follows.

> A **parabola** is the set of all points in a plane that are equidistant from a fixed line and a fixed point not on the line. The fixed point is called the **focus** of the parabola and the fixed line is called the **directrix** of the parabola.

We now show that our geometric definition of a parabola is consistent with our algebraic definition. Applying our geometric definition, a parabola is the set of all points *P* such that

$$PQ = PF$$

as shown in Figure 6.22. For convenience, let's place the parabola that is shown in Figure 6.22 in the *xy*-coordinate plane so that it opens upward with vertex at the origin and axis of symmetry $x = 0$ (see Figure 6.23). The vertex,

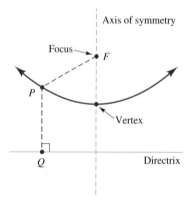

FIGURE 6.22
A parabola is the set of all points *P* such that $PQ = PF$.

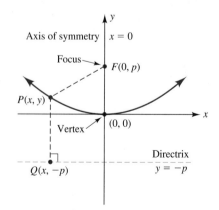

FIGURE 6.23
The vertex of a parabola is the halfway point between the focus and directrix.

like every other point on the parabola, is equidistant from the focus and the directrix. Hence, if the focus F has coordinates $(0, p)$, then the equation of the directrix must be $y = -p$. Referring to Figure 6.23, we have

$$PQ = PF$$

$$\sqrt{(x-x)^2 + (y+p)^2} = \sqrt{(x-0)^2 + (y-p)^2} \quad \text{Apply distance formula}$$

$$(x-x)^2 + (y+p)^2 = (x-0)^2 + (y-p)^2 \quad \text{Square both sides}$$

$$y^2 + 2py + p^2 = x^2 + y^2 - 2py + p^2 \quad \text{Expand}$$

$$4py = x^2 \quad \text{Solve for } y$$

$$y = \frac{1}{4p}x^2$$

From our algebraic definition of a parabola, we recognize

$$y = \frac{1}{4p}x^2 = \frac{1}{4p}(x-0)^2 + 0$$

as the equation of a parabola in standard form with vertex $(0, 0)$ and vertical axis of symmetry $x = 0$. In this equation, $a = 1/(4p)$ is the nonzero constant that determines the width and direction of the opening of the parabola. Hence, we have shown that our geometric and algebraic definitions of a parabola are consistent.

The equation of a parabola in standard form with a vertical axis of symmetry may be written as

$$\boxed{y = \frac{1}{4p}(x-h)^2 + k}$$

where (h, k) is the vertex and $|p|$ is the distance from the focus to the vertex. If $p > 0$, then the focus is *above* the vertex and the parabola opens upward, and if $p < 0$ then the focus is *below* the vertex and the parabola opens downward.

In a similar manner, we can show that if a parabola is placed in the xy-coordinate plane so that it opens to the right with vertex at the origin, then the geometric definition of a parabola yields

$$x = \frac{1}{4p}y^2.$$

Hence, we may rewrite the equation of a parabola in standard form with horizontal axis of symmetry as

$$\boxed{x = \frac{1}{4p}(y-k)^2 + h}$$

where (h, k) is the vertex and $|p|$ is the distance from the focus to the vertex. If $p > 0$, then the focus is to the *right* of the vertex, and the parabola opens toward the right, and if $p < 0$, then the focus is to the *left* of the vertex and the parabola opens toward the left.

EXAMPLE 3 Find the focus and directrix of each parabola.

(a) $y^2 + 12x = 0$ (b) $x^2 + 3x - 6y = 0$

◆ **Solution**

(a) Solving for x, we obtain

$$x = -\tfrac{1}{12}y^2 = -\tfrac{1}{12}(y - 0)^2 + 0,$$

which is the equation of a parabola in standard form with horizontal axis of symmetry, vertex at (0, 0), and

$$\frac{1}{4p} = -\frac{1}{12}$$
$$-4p = 12$$
$$p = -3.$$

Since $p < 0$, the focus is 3 units to the *left* of the vertex, and the directrix is 3 units to the right of the vertex. Hence, the coordinates of the focus are $(-3, 0)$ and the directrix is the vertical line $x = 3$, as shown in Figure 6.24.

(b) First, we write the equation in standard form by completing the square as follows:

$$x^2 + 3x - 6y = 0$$

$$y = \tfrac{1}{6}(x^2 + 3x \qquad) \qquad \text{Solve for } y$$

Add and subtract 3/8.

$$y = \tfrac{1}{6}(x^2 + 3x + \tfrac{9}{4}) - \tfrac{3}{8} \qquad \text{Complete the square}$$

$$y = \tfrac{1}{6}(x + \tfrac{3}{2})^2 - \tfrac{3}{8} \qquad \text{Factor}$$

We recognize this last equation as the equation of a parabola in standard form with vertical axis of symmetry, vertex at $(-\tfrac{3}{2}, -\tfrac{3}{8})$, and

$$\frac{1}{4p} = \frac{1}{6}$$
$$4p = 6$$
$$p = \tfrac{3}{2}$$

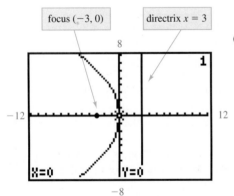

FIGURE 6.24
The focus of the parabola $y^2 + 12x = 0$ is $(-3, 0)$ and its directrix is $x = 3$.

SECTION 6.2 ◆ *The Parabola*   361

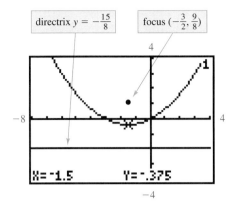

FIGURE 6.25
The focus of the parabola
$x^2 + 3x - 6y = 0$ is $\left(-\frac{3}{2}, \frac{9}{8}\right)$
and the directrix is $y = -\frac{15}{8}$.

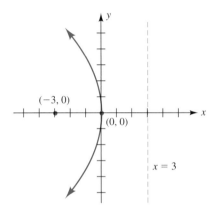

FIGURE 6.26
A parabola with a vertical directrix must have a horizontal axis of symmetry.

Since $p > 0$, the focus is $\frac{3}{2}$ units *above* the vertex, and the directrix is $\frac{3}{2}$ units below the vertex. Hence, the coordinates of the focus are $\left(-\frac{3}{2}, \frac{9}{8}\right)$ and the directrix is the horizontal line $y = -\frac{15}{8}$, as shown in Figure 6.25. ◆

PROBLEM 3 Find the focus and directrix of the parabola given by $3y^2 - 8x = 0$. ◆

If we know the focus and directrix of a parabola, then we can determine its equation.

EXAMPLE 4 Find the equation of a parabola with focus $F(-5, 1)$ and directrix $x = 2$.

◆ **Solution**   Since the directrix is the vertical line $x = 2$, the parabola must have a horizontal axis of symmetry that passes through the focus $F(-5, 1)$, as shown in Figure 6.26. Thus, the equation of this parabola has the form

$$x = \frac{1}{4p}(y - k)^2 + h,$$

where (h, k) is the vertex and $|p|$ is the distance from the focus to the vertex.

The vertex of a parabola is always located halfway between its focus and directrix. Hence, by the midpoint formula (Section 2.1), the coordinates of the vertex are

$$\left(\frac{-5 + 2}{2}, 1\right) = \left(-\frac{3}{2}, 1\right).$$

The distance from the focus to the vertex is

$$|p| = \left|-5 - \left(-\frac{3}{2}\right)\right| = \frac{7}{2}.$$

Since the focus is to the left of the vertex, the parabola opens to the left. Hence, we choose $p = -\frac{7}{2}$. Therefore, the equation of this parabola is

$$x = \frac{1}{4p}(y - k)^2 + h$$

$$x = \frac{1}{4\left(-\frac{7}{2}\right)}(y - 1)^2 + \left(-\frac{3}{2}\right)$$

$$x = -\frac{1}{14}(y - 1)^2 - \frac{3}{2}$$ ◆

PROBLEM 4 Referring to Example 4, express the equation of the parabola in general form with integer coefficients. ◆

◆ **Application: Reflection Property of a Parabola**

A law of physics states that the angle of incidence equals the angle of reflection; that is, the angle at which a light or sound wave hits a surface is equal to the angle at which it reflects from the surface, as shown in Figure 6.27. When this physical law is applied to a parabolic surface, an interesting phenomenon occurs: all light or sound waves that move parallel to the axis of the parabola bounce off the parabola and then pass through the focus of the parabola (see Figure 6.28). We refer to this phenomenon as the **reflection property of a parabola**. Engineers and scientists use this reflection property to design telescopes, radar antennas, satellite dishes, solar heating devices, microphones, and other communications equipment.

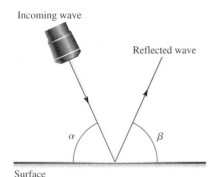

FIGURE 6.27
The angle of incidence α equals the angle of reflection β.

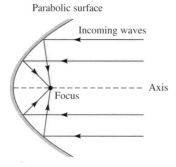

FIGURE 6.28
Light or sound waves entering parallel to the axis reflect to the focus.

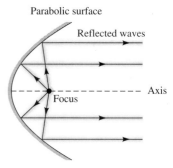

FIGURE 6.29
Light or sound waves emanating from the focus reflect parallel to the axis.

The reflection property of a parabola is also used in the design of searchlights and flashlights: The light rays emanating from a bulb, located at the focus of a parabolic surface, bounce off the surface and shine parallel to the axis of the parabola, thus keeping the light beam at maximum intensity (see Figure 6.29).

EXAMPLE 5 *Chapter Opening Problem*

The parabolic reflector of a searchlight has a diameter of 3 feet at the open end and a depth of 1 foot. How far from the vertex of the parabola should the filament of the bulb be located so that all the light rays emanating from the bulb are parallel to the axis of the parabola?

◆ **Solution**   We place the parabolic reflector in the xy-coordinate plane so that its vertex is at the origin and it opens to the right, as shown in Figure 6.30. Now, if the diameter of the opening of the parabolic reflector is 3 feet and the depth is 1 foot, then the coordinates of point Q in Figure 6.30 are $(1, \frac{3}{2})$. Hence,

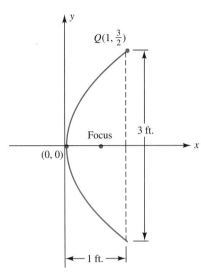

FIGURE 6.30
If the diameter of the parabolic reflector is 3 feet and the depth is 1 foot, then point Q has coordinates $(1, \frac{3}{2})$.

$$x = \frac{1}{4p} y^2$$

$$1 = \frac{1}{4p}\left(\frac{3}{2}\right)^2 \qquad \text{Substitute } x = 1 \text{ and } y = \tfrac{3}{2}$$

$$\frac{4}{9} = \frac{1}{4p} \qquad \text{Multiply both sides by } \tfrac{4}{9}$$

$$p = \frac{9}{16} \qquad \text{Solve for } p$$

Since p represents the distance from the vertex to the focus and light rays emanating from the focus are all parallel to the axis of the parabola, we conclude that the filament of the bulb should be placed $\frac{9}{16}$ foot (or $6\frac{3}{4}$ inches) from the vertex of the parabolic reflector.

PROBLEM 5 Taking the axes as shown in Figure 6.30, determine the equation of the parabola that may be rotated about the x-axis to generate this parabolic reflector.

Exercises 6.2

◆ Basic Skills

In Exercises 1–16, find the coordinates of the vertex and the direction of the opening of the parabola with the given equation. Check graphically with a graphing calculator.

1. $x = y^2$
2. $x = -3y^2$
3. $y = 2x^2 - 8$
4. $x = 1 - 4y^2$
5. $x = -(y - 1)^2 + 4$
6. $y = 2(x + 1)^2 - 2$
7. $y = 4x - x^2$
8. $4x = y^2 - 4$
9. $x = y^2 - 3y - 4$
10. $y = 6 - x - \frac{1}{3}x^2$
11. $y^2 - 2x - 4y + 2 = 0$
12. $y^2 + 3x + 6y + 3 = 0$
13. $x^2 + 8x - 2y + 16 = 0$
14. $x^2 - 3y - 6x + 15 = 0$
15. $2y^2 + 3x + 16y + 26 = 0$
16. $3y^2 + 4x - 6y + 3 = 0$

In Exercises 17–22, find the focus and directrix of the parabola whose equation is given in the indicated exercise.

17. Exercise 3
18. Exercise 6
19. Exercise 9
20. Exercise 10
21. Exercise 15
22. Exercise 16

In Exercises 23–32, determine the equation of the parabola with the given characteristics.

23. Vertex at the origin, passing through the point $(2, 4)$, and symmetric with respect to the x-axis
24. Vertex at the origin, passing through the point $(-1, 3)$, and symmetric with respect to the y-axis
25. Vertex at $(-1, -1)$, passing through the point $(0, 1)$, and symmetric with respect to the line $x = -1$
26. Vertex at $(2, 1)$, passing through the origin, and symmetric with respect to the line $y = 1$
27. Focus at $(0, 1)$ and directrix $x = 2$
28. Focus at $(-4, -2)$ and directrix $y = 2$
29. Vertex at $(-3, 2)$ and focus at $(-3, -3)$
30. Vertex at $(9, -12)$ and focus at $(1, -12)$

31. Vertex at (3, −1) and directrix $x = -1$
32. Vertex at (−8, −4) and directrix $y = 2$

In Exercises 33–36, use a graphing calculator to generate the graph of each parabola. Then estimate the coordinates of the vertex, rounding each coordinate to three significant digits.

33. $12.5x^2 - 18.6x + 2.4y = 10.7$
34. $1.24y^2 + 4.56x = 9.55$
35. $23.4y^2 - 45.8y - 12.6x = 0$
36. $y^2 - 2.34x - 5.74y = -6.87$

37. The cable that supports the weight of a suspension bridge approximates a parabola with vertex located halfway between the supports, as shown in the sketch. Determine the length x of the vertical cable that is 30 meters from the end of the support.

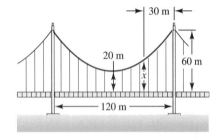

38. The main entrance into a new shopping mall is a parabolic arch with span 12 ft and height 18 ft, as shown in the sketch. Calculate the height h of a security camera that is located 4 feet from the center of the arch.

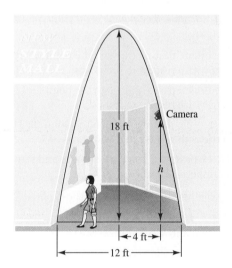

39. A parabolic satellite dish for receiving television signals has its receiver located at the focus, as shown in the sketch. Find the depth d of the satellite dish.

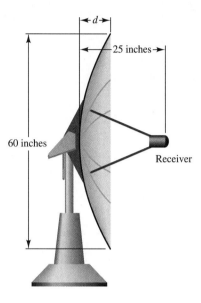

40. A solar collector for a domestic hot water heater is parabolic in shape, as shown in the sketch. At what distance d from the vertex should the water pipe be located in order to take advantage of the reflection property of a parabola?

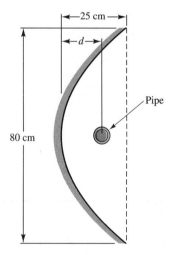

41. In an engineer's new design for an automobile headlight, a parabola is rotated about the x-axis, as shown in the sketch. Determine the equation of the parabola that is used to generate the shape of the headlight. Write the

equation in the form $x = ay^2$, with a rounded to three significant digits.

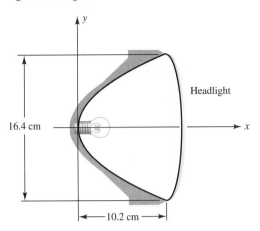

Headlight

42. A simply supported beam with a concentrated load at midspan deflects in the shape of a parabola, as shown in the sketch. Determine the equation of the parabola, which is called the *elastic curve*. Write the equation in the form $y = ax^2 + bx$ with a and b each rounded to three significant digits.

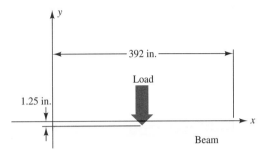

◆ Critical Thinking

43. If a parabola has two distinct x-intercepts a and b, what is the x-coordinate of its vertex?

44. Find all values of k such that the vertex of the graph of $y = x^2 + kx + 9$ lies on the x-axis.

45. If the graph of $x = ay^2 + by + c$ passes through the origin, what is the value of c?

46. Determine the value of a such that the parabola $x = ay^2 - 4y + 3$ has its axis of symmetry along the line $y = 5$.

47. What positive values of p make the opening of the parabola $y = \dfrac{1}{4p}x^2$ narrower than the opening of the parabola $y = x^2$? Wider than $y = x^2$?

48. The chord of a parabola that passes through its focus and is perpendicular to its axis of symmetry is called the *latus rectum* of the parabola. The latus rectum of the parabola $x = \dfrac{1}{4p}y^2$ is the line segment $\overline{AB}$, as shown in the sketch. What is the length of the latus rectum?

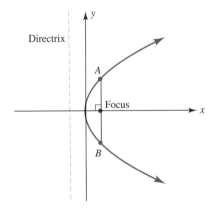

49. The equation of a parabola is $Ax^2 + Ey = 0$. Find the coordinates of its focus in terms of A and E.

50. What point on a parabola is always closest to its focus?

6.3 The Ellipse

◆ Introductory Comments

In this section, we study the geometric properties of the ellipse and the distinguishing characteristics of its equation. First, we consider the equation

$$\frac{x^2}{a^2} + \frac{y^2}{b^2} = 1$$

with $a > 0$ and $b > 0$. This is a general quadratic equation in two unknowns with $A = 1/a^2$, $C = 1/b^2$, $F = -1$, and $B = D = E = 0$. Using the tests for symmetry (Section 2.2), we find that the graph of this equation is symmetric with respect to the x-axis, the y-axis, and the origin. To find the x-intercepts of the graph we let $y = 0$ and obtain $x = \pm a$. Similarly, letting $x = 0$ gives us the y-intercepts $\pm b$. If we solve the original equation for y, we obtain

$$\frac{x^2}{a^2} + \frac{y^2}{b^2} = 1$$

$$\frac{y^2}{b^2} = 1 - \frac{x^2}{a^2} = \frac{a^2 - x^2}{a^2} \qquad \text{Subtract } x^2/a^2 \text{ from both sides and write as a single fraction}$$

$$y^2 = \frac{b^2}{a^2}(a^2 - x^2) \qquad \text{Multiply both sides by } b^2$$

$$y = \pm \frac{b}{a}\sqrt{a^2 - x^2} \qquad \text{Take the square root of both sides}$$

Now, if y is to be a real number, then $(a^2 - x^2)$ must be either positive or zero, that is

$$a^2 - x^2 \geq 0 \qquad \text{or} \qquad -a \leq x \leq a.$$

This means that the graph of $(x^2/a^2) + (y^2/b^2) = 1$ exists only between the x-intercepts $\pm a$. Also, from this form of the equation we observe that as x increases from 0 to a, y decreases from b to 0. Figure 6.31 shows the three possibilities for the graph of $(x^2/a^2) + (y^2/b^2) = 1$.

If $a = b$, as shown in Figure 6.31(a), then we have

$$\frac{x^2}{a^2} + \frac{y^2}{a^2} = 1 \qquad \text{or} \qquad x^2 + y^2 = a^2.$$

This is the equation of a circle with center at the origin and radius a (see Section 6.1). If a and b have different values, as shown in parts (b) and (c) of the figure, then the circle is flattened or stretched to form an egg-shaped curve called an **ellipse**.

In each of the graphs shown in part (b) or (c) of Figure 6.31, the horizontal and vertical line segments $\overline{AB}$ and $\overline{CD}$ are the **axes** of the ellipse. We refer to the longer line segment as the **major axis** and the shorter line segment as the **minor axis**. If $a > b$, as in Figure 6.31(b), then $\overline{AB}$ is the major axis and $\overline{CD}$ is the minor axis. However, if $a < b$, as in Figure 6.31(c), then $\overline{CD}$ is the major axis and $\overline{AB}$ is the minor axis. We refer to the endpoints of the major axis as the **vertices** of the ellipse and the midpoint of the major axis and minor axis as its **center**. Half of an axis of an ellipse is called a **semiaxis**. Note that the lengths of the semiaxes are a and b.

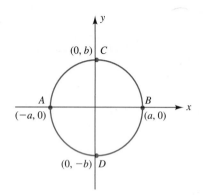

(a) If $a = b$, the graph is a circle with center at the origin.

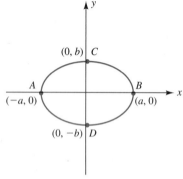

(b) If $a > b$, the graph is an ellipse, elongated horizontally, with center at the origin.

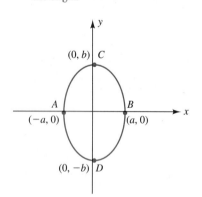

(c) If $a < b$, the graph is an ellipse, elongated vertically, with center at the origin.

FIGURE 6.31
The three possibilities for the graph of $\dfrac{x^2}{a^2} + \dfrac{y^2}{b^2} = 1$

◆ Standard Form of an Ellipse

If we replace x with $(x - h)$ and y with $(y - k)$ in the equation $(x^2/a^2) + (y^2/b^2) = 1$, we obtain

$$\frac{(x - h)^2}{a^2} + \frac{(y - k)^2}{b^2} = 1.$$

If a and b are positive real numbers such that $a = b$, we have

$$\frac{(x - h)^2}{a^2} + \frac{(y - k)^2}{a^2} = 1 \quad \text{or} \quad (x - h)^2 + (y - k)^2 = a^2$$

the equation of a circle in standard form with center (h, k) and radius a (see Section 6.1). If a and b have different positive values, we refer to $(x - h)^2/a^2 + (y - k)^2/b^2 = 1$ as the *algebraic definition of an ellipse*, or the *equation of an ellipse in standard form*.

◀ **Equation of an Ellipse in Standard Form**

> The **equation of an ellipse in standard form** with center (h, k), horizontal axis of length $2a$, and vertical axis of length $2b$ is
>
> $$\frac{(x - h)^2}{a^2} + \frac{(y - k)^2}{b^2} = 1.$$

We can determine the equation of an ellipse if we know its center (h, k) and the lengths of its semiaxes a and b.

EXAMPLE 1 Determine the equation of an ellipse with vertices $(3, 3)$ and $(-5, 3)$ and minor axis of length 4.

◀ **Solution** The ellipse with the given characteristics is shown in Figure 6.32. Since the vertices of an ellipse are the endpoints of its major axis, we know the major axis is horizontal. Thus, the minor axis is vertical. Now, if the minor axis is vertical and has a length of 4, we have

$$2b = 4 \quad \text{or} \quad b = 2.$$

The length of the major axis is the distance between the x-coordinates of the vertices, or $|3 - (-5)| = 8$. Thus,

$$2a = 8 \quad \text{or} \quad a = 4.$$

The midpoint of the major axis is the center of the ellipse. Using the midpoint formula (Section 2.1), we find the coordinates of the center:

$$(h, k) = \left(\frac{3 + (-5)}{2}, \frac{3 + 3}{2}\right) = (-1, 3).$$

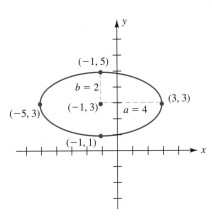

FIGURE 6.32
An ellipse with vertices $(3, 3)$ and $(-5, 3)$ and minor axis of length 4

Thus, the equation of the ellipse is

$$\frac{(x-h)^2}{a^2} + \frac{(y-k)^2}{b^2} = 1$$

$$\frac{[x-(-1)]^2}{4^2} + \frac{(y-3)^2}{2^2} = 1 \quad \text{Substitute}$$

$$\frac{(x+1)^2}{16} + \frac{(y-3)^2}{4} = 1 \quad \text{Simplify}$$

◆

To generate the graph of $(x-h)^2/a^2 + (y-k)^2/b^2 = 1$ in the viewing rectangle of a graphing calculator, we must solve the equation for y and enter both equations that result. To obtain a true geometric perspective of the graph, we use a square set viewing rectangle.

PROBLEM 1 Generate the graph of

$$\frac{(x+1)^2}{16} + \frac{(y-3)^2}{4} = 1$$

in the viewing rectangle of a graphing calculator. Verify that this ellipse has vertices $(3, 3)$ and $(-5, 3)$ and minor axis of length 4. ◆

◆ **General Form of an Ellipse**

If we take the equation of an ellipse in standard form and multiply both sides of the equation by a^2b^2, we obtain an equation of the form

$$A(x-h)^2 + C(y-k)^2 = N$$

with $A = b^2$, $C = a^2$, and $N = a^2b^2$. If we expand $(x-h)^2$ and $(y-k)^2$, distribute A and C, and collect all terms on one side of the equation, we obtain an equation of the form

$$\boxed{Ax^2 + Cy^2 + Dx + Ey + F = 0}$$

We refer to this general quadratic equation in two unknowns as the **equation of an ellipse in general form** with vertical and horizontal axes of symmetry. This equation is characterized by the presence of x^2 and y^2 terms that have *different coefficients* but *like signs*. The difference between the general form of the equation of an ellipse and that of a circle is that the coefficients of the x^2 and y^2 terms of an ellipse are different, whereas those of a circle are the same.

If the equation of an ellipse is given in general form, then we can find the coordinates of its center and vertices by converting the equation to standard form and observing the values of h, k, a, and b. We can use a graphing calculator to check our work by solving the equation

$$Ax^2 + Cy^2 + Dx + Ey + F = 0$$

SECTION 6.3 ◆ The Ellipse

for y and graphing both equations that result. To generate the graph of this equation when $E \neq 0$, we rewrite the equation as a quadratic equation in y,

$$Cy^2 + Ey + (Ax^2 + Dx + F) = 0,$$

and apply the quadratic formula:

$$y = \frac{-E \pm \sqrt{E^2 - 4C(Ax^2 + Dx + F)}}{2C}$$

EXAMPLE 2 Determine the coordinates of the center and vertices of the ellipse given by each equation.

(a) $4x^2 + y^2 = 9$ **(b)** $x^2 + 9y^2 + 8x - 36y + 43 = 0$

◆ Solution

(a) We begin by writing the equation in standard form:

$$4x^2 + y^2 = 9$$

$$\frac{4x^2}{9} + \frac{y^2}{9} = 1 \quad \text{Divide both sides by 9}$$

$$\frac{x^2}{\frac{9}{4}} + \frac{y^2}{9} = 1 \quad \text{Invert 4/9 and divide}$$

$$\frac{(x-0)^2}{\left(\frac{3}{2}\right)^2} + \frac{(y-0)^2}{3^2} = 1 \quad \text{Write in standard form}$$

This last equation tells us that the graph of $4x^2 + y^2 = 9$ is an ellipse with $a = \frac{3}{2}$, $b = 3$, and center at the origin. Since $b > a$, the ellipse is elongated vertically with a major axis of length $2b = 6$. Hence, the coordinates of the vertices are $(0, 3)$ and $(0, -3)$. Figure 6.33 shows the graph of $4x^2 + y^2 = 9$ (or $y = \pm\sqrt{9 - 4x^2}$) in the viewing rectangle of a graphing calculator, which confirms our work.

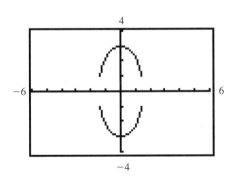

FIGURE 6.33
The graph of $4x^2 + y^2 = 9$ is an ellipse with center $(0, 0)$ and vertices $(0, 3)$ and $(0, -3)$.

(b) We begin by writing the equation in standard form. To do this, we use the process of completing the square (Section 3.3) and proceed as follows:

$$x^2 + 9y^2 + 8x - 36y + 43 = 0$$

$$(x^2 + 8x\ \ ) + (9y^2 - 36y\ \ ) = -43 \quad \text{Regroup}$$

$$(x^2 + 8x\ \ ) + 9(y^2 - 4y\ \ ) = -43 \quad \text{Factor out 9}$$

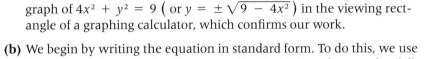

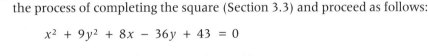

$$(x^2 + 8x + 16) + 9(y^2 - 4y + 4) = -43 + (16 + 36) \quad \text{Complete the squares}$$

$$(x + 4)^2 + 9(y - 2)^2 = 9 \quad \text{Factor}$$

$$\frac{(x + 4)^2}{9} + \frac{(y - 2)^2}{1} = 1 \quad \text{Divide both sides by 9}$$

$$\frac{[x - (-4)]^2}{3^2} + \frac{(y - 2)^2}{1^2} = 1 \quad \text{Write in standard form}$$

This last equation tells us that the graph of $x^2 + 9y^2 + 8x - 36y + 43 = 0$ is an ellipse with $a = 3$, $b = 1$, and center $(h, k) = (-4, 2)$. Since $a > b$, the ellipse is elongated horizontally with major axis of length $2a = 6$. Hence, the coordinates of the vertices are $(-7, 2)$ and $(-1, 2)$.

To generate the graph of the equation $x^2 + 9y^2 + 8x - 36y + 43 = 0$ in the viewing rectangle of a graphing calculator, we rewrite the equation as a quadratic equation in y,

$$9y^2 - 36y + (x^2 + 8x + 43) = 0,$$

and apply the quadratic formula:

$$y = \frac{-(-36) \pm \sqrt{(-36)^2 - 4(9)(x^2 + 8x + 43)}}{2(9)} = 2 \pm \frac{\sqrt{-7 - 8x - x^2}}{3}$$

Entering both of these equations into the calculator, we obtain the ellipse that is shown in Figure 6.34. This graph confirms our work.

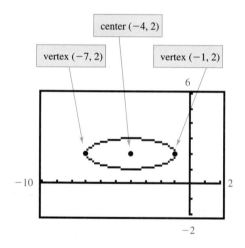

FIGURE 6.34
The graph of
$x^2 + 9y^2 + 8x - 36y + 43 = 0$ is an ellipse with center $(-4, 2)$ and vertices $(-7, 2)$ and $(-1, 2)$.

CAUTION Not every equation of the form

$$Ax^2 + Cy^2 + Dx + Ey + F = 0,$$

where the x^2 and y^2 terms have different coefficients but like signs, represents an ellipse. For example, if this equation is written in the form

$$A(x - h)^2 + C(y - k)^2 = N \quad \text{and} \quad N = 0,$$

then the graph of the equation is the single point (h, k). We refer to this single point as a *degenerate ellipse*. Also, if

$$A(x - h)^2 + C(y - k)^2 = N \quad \text{and} \quad N < 0,$$

then this equation has no graph, since no point (x, y) with real coordinates satisfies this equation.

PROBLEM 2 Is $3x^2 + 2y^2 - 6x + 8y + 15 = 0$ the equation of an ellipse? Explain.

SECTION 6.3 ◆ *The Ellipse*

◆ Geometric Definition of an Ellipse

Geometrically, we define an ellipse as follows.

◀ **Geometric Definition of an Ellipse**

> An **ellipse** is the set of all points in a plane, the sum of whose distances from two fixed points is a constant, which is equal to the length of the major axis of the ellipse. The two fixed points are called the **foci** (plural of focus).

We now show that our geometric definition of an ellipse is consistent with our algebraic definition. Applying our geometric definition, an ellipse is the set of all points P such that

$$PF_1 + PF_2 = \text{length of major axis}$$

as shown in Figure 6.35. For convenience, we can place the ellipse that is shown in Figure 6.35 in the xy-coordinate plane so that its center is at the origin, its major axis is horizontal with length $2a$, and its minor axis is vertical with length $2b$ (see Figure 6.36). We have labeled the coordinates of the foci $(-c, 0)$ and $(c, 0)$, with $a > c$.

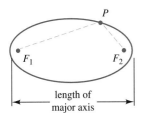

FIGURE 6.35
An ellipse is the set of all points P such that $PF_1 + PF_2 = $ length of major axis.

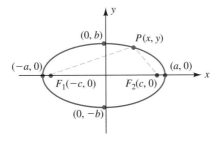

FIGURE 6.36
An ellipse in the xy-coordinate plane with center at the origin, major axis of length $2a$, minor axis of length $2b$, and foci with coordinates $(-c, 0)$ and $(c, 0)$

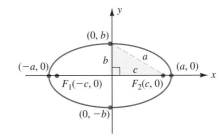

FIGURE 6.37
The constants a, b, and c are related by the Pythagorean theorem.

The constants a, b, and c have a special relationship. Referring to Figure 6.37, the point $(0, b)$ is on the ellipse and is equidistant from each of the foci. Since the sum of the distances from the foci to the point $(0, b)$ must be $2a$, the distance from each focus to $(0, b)$ is a. Thus, the shaded right triangle in Figure 6.37 has legs b and c and hypotenuse a. Hence, by the Pythagorean theorem, we have the relation

$$\boxed{c^2 = a^2 - b^2}$$

in an ellipse with horizontal semimajor axis of length a, vertical semiminor axis of length b, and distance c from the center to each focus. For an ellipse with vertical semimajor axis of length b and horizontal semiminor axis of length a, we have the relation

$$c^2 = b^2 - a^2$$

Now referring to Figure 6.36, we have

$PF_1 + PF_2 = 2a$	
$\sqrt{(x+c)^2 + y^2} + \sqrt{(x-c)^2 + y^2} = 2a$	Apply the distance formula
$\sqrt{(x+c)^2 + y^2} = 2a - \sqrt{(x-c)^2 + y^2}$	Isolate one of the radicals
$(x+c)^2 + y^2 = 4a^2 - 4a\sqrt{(x-c)^2 + y^2} + (x-c)^2 + y^2$	Square both sides
$4a\sqrt{(x-c)^2 + y^2} = 4a^2 + (x-c)^2 + y^2 - (x+c)^2 - y^2$	Isolate the remaining radical
$a\sqrt{(x-c)^2 + y^2} = a^2 - cx$	Simplify the right side then divide both sides by 4
$a^2[(x-c)^2 + y^2] = a^4 - 2a^2cx + c^2x^2$	Square both sides
$(a^2 - c^2)x^2 + a^2y^2 = a^2(a^2 - c^2)$	Collect like terms
$\dfrac{x^2}{a^2} + \dfrac{y^2}{a^2 - c^2} = 1$	Divide both sides by $a^2(a^2 - c^2)$
$\dfrac{x^2}{a^2} + \dfrac{y^2}{b^2} = 1$	Replace c^2 with $a^2 - b^2$

From our algebraic definition of an ellipse, we recognize this last equation as the equation of an ellipse in standard form with center (0, 0), horizontal axis of length $2a$, and vertical axis of length $2b$. We have shown that our geometric and algebraic definitions of an ellipse are consistent.

The **eccentricity** e of an ellipse is a measure of its "ovalness" and is defined as

$$e = \frac{\text{distance from center to focus}}{\text{distance from center to vertex}}$$

(Do not confuse the eccentricity e with base e in exponential and logarithmic expressions.) The eccentricity e of an ellipse is a number between 0 and 1. If e is close to 1, then the ellipse is very long and narrow and if e is close to 0, then the ellipse is nearly circular.

EXAMPLE 3 Find the foci and eccentricity of the ellipse defined by each equation.

(a) $16x^2 + 9y^2 = 144$ **(b)** $x^2 + 4y^2 - 6x + 5 = 0$

◆ Solution

(a) Dividing both sides by 144, we obtain

$$\frac{x^2}{9} + \frac{y^2}{16} = 1 \quad \text{or} \quad \frac{(x-0)^2}{3^2} + \frac{(y-0)^2}{4^2} = 1,$$

the equation of an ellipse in standard form with $a = 3$, $b = 4$, and center at the origin. Since $b > a$, the ellipse is elongated vertically with a semimajor axis of length 4 and semiminor axis of length 3. To find the coordinates of the foci, we apply the relation

$$c^2 = 4^2 - 3^2,$$

which implies $|c| = \sqrt{7} \approx 2.646$. Hence, the coordinates of the foci are

$$\left(0, \sqrt{7}\right) \approx (0, 2.646) \quad \text{and} \quad \left(0, -\sqrt{7}\right) \approx (0, -2.646)$$

as shown in Figure 6.38. The eccentricity e of this ellipse is

$$e = \frac{\text{distance from center to focus}}{\text{distance from center to vertex}} = \frac{\sqrt{7}}{4} \approx 0.66.$$

(b) First, we write the equation in standard form by completing the square:

$$x^2 + 4y^2 - 6x + 5 = 0$$

$$(x^2 - 6x \quad) + 4y^2 = -5 \qquad \text{Regroup}$$

Add 9 to both sides.

$$(x^2 - 6x + 9) + 4y^2 = -5 + 9 \qquad \text{Complete the square}$$

$$(x - 3)^2 + 4y^2 = 4 \qquad \text{Factor}$$

$$\frac{(x-3)^2}{2^2} + \frac{(y-0)^2}{1^2} = 1 \qquad \text{Divide both sides by 4 and write in standard form}$$

This is the equation of an ellipse in standard form with $a = 2$, $b = 1$, and center $(3, 0)$. Since $a > b$, the ellipse is elongated horizontally with a semimajor axis of length 2 and semiminor axis of length 1. To find the coordinates of the foci, we apply the relation

$$c^2 = 2^2 - 1^2,$$

which implies that $c = \sqrt{3} \approx 1.732$. Hence, the coordinates of the foci are

$$\left(3 - \sqrt{3}, 0\right) \approx (1.268, 0) \quad \text{and} \quad \left(3 + \sqrt{3}, 0\right) \approx (4.732, 0)$$

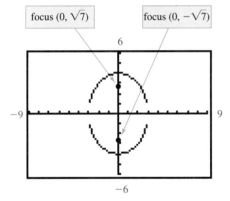

FIGURE 6.38
The foci of the ellipse
$16x^2 + 9y^2 = 144$ are $(0, \sqrt{7})$
and $(0, -\sqrt{7})$.

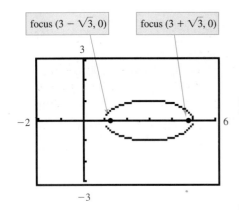

FIGURE 6.39
The foci of the ellipse
$x^2 + 4y^2 - 6x + 5 = 0$ are
$(3 - \sqrt{3}, 0)$ and $(3 + \sqrt{3}, 0)$.

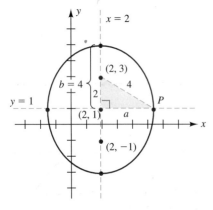

FIGURE 6.40
An ellipse with foci $(2, 3)$ and $(2, -1)$ has center $(2, 1)$.

as shown in Figure 6.39. The eccentricity e of this ellipse is

$$e = \frac{\text{distance from center to focus}}{\text{distance from center to vertex}} = \frac{\sqrt{3}}{2} \approx 0.87.$$

PROBLEM 3 Find the foci and eccentricity for the ellipse defined by $9x^2 + y^2 - 9 = 0$.

If we know the foci and eccentricity of an ellipse, then we can determine its equation.

EXAMPLE 4 Find the equation of an ellipse in standard form with foci $(2, 3)$ and $(2, -1)$ and eccentricity $\tfrac{1}{2}$.

◆ **Solution** The center of an ellipse is located halfway between its foci. Hence, the center of this ellipse is $(h, k) = (2, 1)$, as shown in Figure 6.40. Therefore, the equation of this ellipse must have the form

$$\frac{(x - 2)^2}{a^2} + \frac{(y - 1)^2}{b^2} = 1.$$

We now proceed to find the lengths of the semimajor axis b and the semiminor axis a. Since the distance from the center to a focus is $c = 2$ and the eccentricity of this ellipse is given to be $\tfrac{1}{2}$, we have

$$\frac{\text{distance from center to focus}}{\text{distance from center to vertex}} = \frac{2}{b} = \frac{1}{2},$$

which implies that $b = 4$. We find the value of a by applying the relation $c^2 = b^2 - a^2$:

$$2^2 = 4^2 - a^2 \quad \text{Substitute } c = 2 \text{ and } b = 4$$
$$a^2 = 4^2 - 2^2 \quad \text{Solve for } a$$
$$a^2 = 12$$
$$a = 2\sqrt{3}$$

Since $(h, k) = (2, 1)$, $a^2 = (2\sqrt{3})^2 = 12$, and $b^2 = (4)^2 = 16$, the equation of this ellipse in standard form is

$$\frac{(x - 2)^2}{12} + \frac{(y - 1)^2}{16} = 1.$$

PROBLEM 4 Referring to Example 4, express the equation of the ellipse in general form with integer coefficients.

◆ **Application: Reflection Property of an Ellipse**

Like the parabola, the ellipse has a *reflection property*. This property explains the "whispering gallery effect" in elliptically shaped dome rooms. If a person standing at a special point on one side of the room whispers, all the sound waves bounce off the sides of the dome and return to another special point on the other side of the room, allowing a person standing at that point to hear the whisper (see Figure 6.41). The two special points in the room where this phenomenon occurs are the *foci* of the ellipse. The whispering gallery effect may be observed in the National Statuary Hall in the Capitol building in Washington, D.C. (Exercises 41 and 42 are applied problems that use the reflection property of an ellipse.)

FIGURE 6.41
In an elliptically shaped dome room, the whisper of one person can be heard by another person on the other side of the room.

Exercises 6.3

Basic Skills

In Exercises 1–16, determine the coordinates of the center and vertices of the ellipse defined by each equation. Check graphically with a graphing calculator.

1. $\dfrac{x^2}{16} + \dfrac{y^2}{4} = 1$
2. $\dfrac{x^2}{4} + \dfrac{y^2}{9} = 1$
3. $25x^2 + 4y^2 = 100$
4. $16x^2 + 25y^2 = 400$
5. $x^2 + 16y^2 = 9$
6. $4x^2 + 9y^2 = 16$
7. $3x^2 + 4y^2 - 9 = 0$
8. $x^2 + 5y^2 - 10 = 0$
9. $\dfrac{(x-2)^2}{16} + \dfrac{(y+1)^2}{4} = 1$
10. $\dfrac{x^2}{9} + \dfrac{(y-3)^2}{25} = 1$
11. $9x^2 + 4(y+2)^2 = 36$
12. $16(x-4)^2 + (y-5)^2 = 16$
13. $25x^2 + 4y^2 - 50x + 24y + 45 = 0$
14. $9x^2 + 25y^2 - 36x + 50y = 39$
15. $4x^2 + y^2 + 40x - 4y + 103 = 0$
16. $2x^2 + y^2 + 8x + 8y + 23 = 0$

In Exercises 17–22, find the foci and eccentricity of the ellipse whose equation is given in the indicated exercise.

17. Exercise 3
18. Exercise 4
19. Exercise 9
20. Exercise 12
21. Exercise 13
22. Exercise 16

In Exercises 23–32, determine the equation of the ellipse with the given characteristics.

23. Center at the origin, minor axis of length 2, and horizontal major axis of length 5
24. Vertices $(0, \pm 3)$ and minor axis of length 2
25. Vertices $(2, -6)$ and $(2, 2)$, and minor axis of length 3
26. Vertices $(-1, 2)$ and $(5, 2)$, and endpoints of minor axis $(2, 4)$ and $(2, 0)$
27. Vertices $(-2, -3)$ and $(-2, 3)$, and one of the foci at $(-2, 1)$
28. Foci $(-4, -1)$ and $(2, -1)$, and one of the vertices at $(4, -1)$
29. Foci $(\pm 2, 0)$ and major axis of length 6

30. Foci (1, ±4) and minor axis of length 6
31. Foci (0, 0) and (3, 0), and eccentricity $\frac{1}{2}$
32. Foci (1, −3) and (1, 5), and eccentricity $\frac{2}{3}$

In Exercises 33–36, use a graphing calculator to generate the graph of each ellipse. Then estimate the coordinates of the vertices, rounding each coordinate to three significant digits. Check algebraically.

33. $23.4x^2 + 19.6y^2 = 87.3$
34. $8.67x^2 + 3.67y^2 = 9.87y$
35. $0.25x^2 + 0.14y^2 - 1.24y - 8.12 = 0$
36. $5.4x^2 + 3.4y^2 - 2.2x - 4.8y - 11.6 = 0$

37. A road passes through a tunnel whose cross-section is a semiellipse, which is 12 feet high at the center and 30 feet wide, as shown in the sketch. What is the tallest tractor-trailer rig that can fit through the tunnel if the trailer is 10 feet wide?

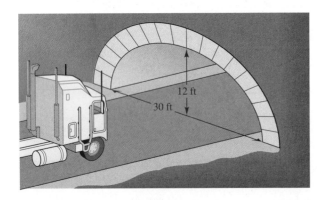

38. A window positioned above an 8-foot-wide glass door is a semiellipse, as shown in the sketch. What is the length of each brace?

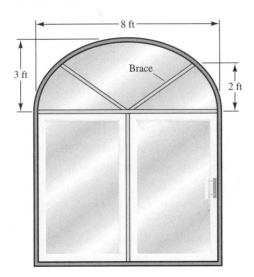

39. A stream 2.3 meters deep passes through an elliptical culvert, as shown in the sketch. What is the width w of the stream?

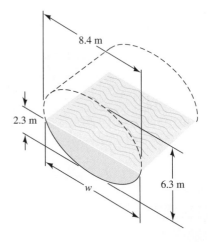

40. In astronomy, the average distance from the Earth to the Sun is called 1 astronomical unit (AU), and 1 AU ≈ 9.26×10^7 miles. It is known that Halley's comet travels around the Sun in an elliptical orbit with the Sun at one focus. The major axis of the ellipse is 36.18 AU and the minor axis is 9.12 AU.

  (a) What is the distance between the foci of the ellipse?
  (b) What is the eccentricity of Halley's comet?
  (c) What is the closest that Halley's comet comes to the sun?
  (d) What is the furthest that Halley's comet gets from the sun?

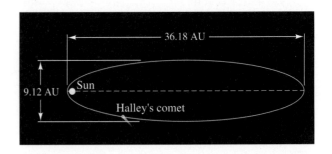

41. An architect's design for an elliptically shaped ceiling calls for rotating the upper portion of the ellipse $x^2 + 4y^2 = 400$, 180° about its major axis, as shown in the sketch.

(a) Find the length and width of this ceiling if x and y are measured in feet.

(b) By the reflection property of an ellipse, a whisper spoken at one focus can be heard at the other focus. Where are the whispering and listening positions located?

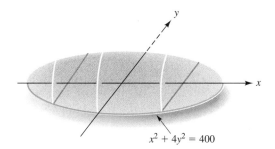

42. A *lithotripter* is a medical device that uses the reflection property of an ellipse. Ultrahigh-frequency (UHF) sound waves emanating from one focus of a lithotripter are reflected to the other focus of the device to break up a kidney stone. Suppose a lithotripter is formed by rotating the portion of an ellipse below its minor axis about its major axis, as shown in the sketch. How far from the kidney stone should the opening of the lithotripter be placed in order to break up the stone?

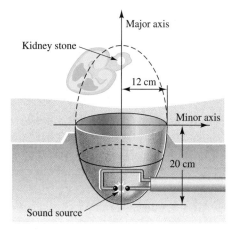

Critical Thinking

43. Suppose you are given two thumbtacks, a piece of string, and a pencil. Using this material, explain how to construct an ellipse.

44. The eccentricity of one ellipse is the constant k_1 and the eccentricity of another ellipse is the constant k_2 such that $k_1 > k_2$. Which ellipse is more circular? Explain.

45. Find the equation of the largest circle that can fit inside the ellipse

$$\frac{x^2}{a^2} + \frac{y^2}{b^2} = 1 \quad \text{if } a < b.$$

46. The equation of an ellipse with a horizontal major axis is $Ax^2 + Cy^2 = E$. Find the coordinates of its foci in terms of A, C, and E.

47. The chord of an ellipse that passes through a focus and is perpendicular to its major axis is called the *latus rectum* of the ellipse. For the ellipse $(x^2/a^2) + (y^2/b^2) = 1$ with $a > b$, the latus rectum is the line segment $\overline{AB}$ shown in the sketch. Show that the length of the latus rectum is $2b^2/a$.

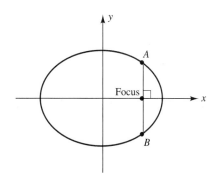

48. The sketch shows a square inscribed in the ellipse $9x^2 + 16y^2 = 144$. What is the length of the side of the square?

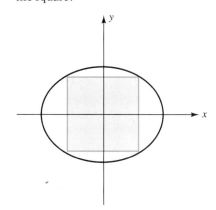

If a and b are positive real numbers, then solving $(x^2/a^2) + (y^2/b^2) = 1$ for y yields $y = \pm(b/a)\sqrt{a^2 - x^2}$. Taken separately, the equations

$$y = \frac{b}{a}\sqrt{a^2 - x^2} \quad \text{and} \quad y = -\frac{b}{a}\sqrt{a^2 - x^2}$$

define y as a function of x, and the graph of each function is a semi-ellipse (half an ellipse) whenever $a \neq b$. In Exercises 49 and 50, state the domain and range of each function, and sketch its graph.

49. $y = \frac{2}{3}\sqrt{9 - x^2}$
50. $y = -\sqrt{25 - 9x^2}$

6.4 The Hyperbola

◆ Introductory Comments

In this section, we study the geometric properties of the hyperbola and the distinguishing characteristics of its equation. First, we consider the equation

$$\frac{x^2}{a^2} - \frac{y^2}{b^2} = 1$$

with $a > 0$ and $b > 0$. This is a general quadratic equation in two unknowns with $A = 1/a^2$, $C = -1/b^2$, $F = -1$, and $B = D = E = 0$. Using the tests for symmetry (Section 2.2), we find that the graph of this equation is symmetric with respect to the x-axis, the y-axis, and the origin. To find the x-intercept of the graph we let $y = 0$ and obtain $x = \pm a$. However, if we let $x = 0$, we find that the graph has no y-intercept because the equation

$$-\frac{y^2}{b^2} = 1 \quad \text{or} \quad y^2 = -b^2$$

has no real solution. If we solve the original equation $\dfrac{x^2}{a^2} - \dfrac{y^2}{b^2} = 1$ for y, we obtain

$$y = \pm\frac{b}{a}\sqrt{x^2 - a^2}.$$

Now if y is to be a real number, then $x^2 - a^2$ must be either positive or zero, that is,

$$x^2 - a^2 \geq 0,$$

so $\quad x \geq a \quad \text{or} \quad x \leq -a.$

This means that the graph exists only to the right of the x-intercept a and to the left of the x-intercept $-a$. Intuitively, we can see that for very large values of x (either negative or positive),

$$y = \pm\frac{b}{a}\sqrt{x^2 - a^2} \approx \pm\frac{b}{a}\sqrt{x^2} = \pm\frac{b}{a}x.$$

In other words, the graph of $(x^2/a^2) - (y^2/b^2) = 1$ approaches the lines $y = \pm\dfrac{b}{a}x$ as $|x|$ increases without bound. We refer to the lines

$$y = \pm \frac{b}{a}x$$

as the **asymptotes** of the curve.

The diagonals of a rectangle with vertices (a, b), $(-a, b)$, $(-a, -b)$, and $(a, -b)$ have slopes $\pm b/a$. Extending these diagonals gives us the asymptotes of the curve, as shown in Figure 6.42. Using the x-intercepts and the asymptotes, we can sketch the graph of $(x^2/a^2) - (y^2/b^2) = 1$, which is the **hyperbola** shown in Figure 6.43. The two distinct curves that make up the graph are the **branches** of the hyperbola.

Figure 6.43 shows the characteristics of a hyperbola. The points $(-a, 0)$ and $(a, 0)$ are the **vertices** of the hyperbola, and the horizontal and vertical line segments $\overline{AB}$ and $\overline{CD}$ are called the **axes** of the hyperbola. Half of an axis of a hyperbola is called a **semiaxis**. Note that the lengths of the semiaxes are a and b. The line segment that connects the vertices of the hyperbola is called the **transverse axis** and the other axis of the hyperbola is the **conjugate axis**. In this figure the horizontal line segment $\overline{AB}$ is the transverse axis and the vertical line segment $\overline{CD}$ is the conjugate axis. The midpoint of both the transverse axis and the conjugate axis is the **center** of the hyperbola. The hyperbola in Figure 6.43 has its center at the origin.

The graph of the equation

$$\frac{y^2}{b^2} - \frac{x^2}{a^2} = 1$$

with $a > 0$ and $b > 0$ is also a hyperbola. However, the y-intercepts are now $\pm b$ and the graph has no x-intercept. Thus, the vertices of this hyperbola are $(0, b)$ and $(0, -b)$. The transverse axis is vertical with length $2b$ and the conjugate axis is horizontal with length $2a$. The asymptotes are still the two lines $y = \pm(b/a)x$, and the center of the hyperbola remains at the origin. The graph of $(y^2/b^2) - (x^2/a^2) = 1$ is shown in Figure 6.44.

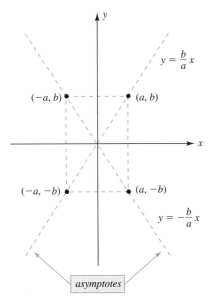

FIGURE 6.42

The asymptotes $y = \pm(b/a)x$ can be constructed from the diagonals of the rectangle with vertices (a, b), $(-a, b)$, $(-a, -b)$, and $(a, -b)$.

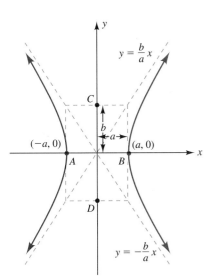

FIGURE 6.43

The graph of $\dfrac{x^2}{a^2} - \dfrac{y^2}{b^2} = 1$ is called a *hyperbola*. The x-intercepts are $\pm a$ and the asymptotes are $y = \pm(b/a)x$.

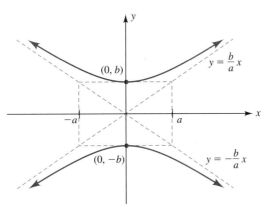

FIGURE 6.44

The graph of $\dfrac{y^2}{b^2} - \dfrac{x^2}{a^2} = 1$ is a hyperbola with y-intercepts $\pm b$ and asymptotes $y = \pm(b/a)x$.

◆ Standard Form of a Hyperbola

The graphs of the equations

$$\frac{x^2}{a^2} - \frac{y^2}{b^2} = 1 \quad \text{and} \quad \frac{y^2}{b^2} - \frac{x^2}{a^2} = 1$$

are the same as the graphs of

$$\frac{(x-h)^2}{a^2} - \frac{(y-k)^2}{b^2} = 1 \quad \text{and} \quad \frac{(y-k)^2}{b^2} - \frac{(x-h)^2}{a^2} = 1,$$

respectively, but the center of the hyperbola is now (h, k). We refer to these equations with $a > 0$ and $b > 0$ as the *algebraic definitions of a hyperbola,* or the *equations of a hyperbola in standard form.*

◆ Equations of a Hyperbola in Standard Form

> The **equation of a hyperbola in standard form** with center (h, k), *horizontal* transverse axis of length $2a$, and vertical conjugate axis of length $2b$ is
>
> $$\frac{(x-h)^2}{a^2} - \frac{(y-k)^2}{b^2} = 1.$$
>
> The **equation of a hyperbola in standard form** with center (h, k), *vertical* transverse axis of length $2b$, and horizontal conjugate axis of length $2a$ is
>
> $$\frac{(y-k)^2}{b^2} - \frac{(x-h)^2}{a^2} = 1.$$

We can determine the equation of a hyperbola if we know its center (h, k) and the lengths of its semiaxes a and b.

EXAMPLE 1 Determine the equation of a hyperbola if its asymptotes are $y = \frac{1}{2}x + 1$ and $y = -\frac{1}{2}x + 1$ and its horizontal transverse axis has length 6.

◆ **Solution** The hyperbola with the given characteristics is shown in Figure 6.45. The intersection point of the asymptotes is the center of the hyperbola. Hence, the center of this hyperbola is

$$(h, k) = (0, 1).$$

Since we know that the horizontal transverse axis has length 6, we have

$$2a = 6 \quad \text{or} \quad a = 3.$$

We can determine b by using the fact that the slopes of the asymptotes are $\pm b/a = \pm 1/2$. Substituting $a = 3$, we have

$$\pm \frac{b}{3} = \pm \frac{1}{2},$$

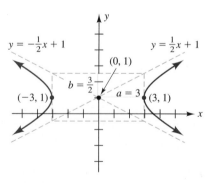

FIGURE 6.45
The hyperbola with asymptotes $y = \pm \frac{1}{2}x + 1$ and horizontal transverse axis of length 6

which implies that $b = \frac{3}{2}$. Therefore, the equation of the hyperbola is

$$\frac{(x - h)^2}{a^2} - \frac{(y - k)^2}{b^2} = 1$$

$$\frac{(x - 0)^2}{3^2} - \frac{(y - 1)^2}{(3/2)^2} = 1 \quad \text{Substitute}$$

$$\frac{x^2}{9} - \frac{4(y - 1)^2}{9} = 1 \quad \text{Simplify}$$

◆

To generate the graph of

$$\frac{(x - h)^2}{a^2} - \frac{(y - k)^2}{b^2} = 1 \quad \text{or} \quad \frac{(y - k)^2}{b^2} - \frac{(x - h)^2}{a^2} = 1$$

in the viewing rectangle of a graphing calculator, we must solve the equation for y and enter both equations that result.

PROBLEM 1 Generate the graph of $\dfrac{x^2}{9} - \dfrac{4(y - 1)^2}{9} = 1$ in the viewing rectangle of a graphing calculator. Verify that this hyperbola has horizontal transverse axis of length 6 and asymptotes $y = \pm\frac{1}{2}x + 1$. ◆

◆ **General Form of a Hyperbola**

If we take the equations of a hyperbola in standard form and multiply both sides of each of these equations by a^2b^2, we obtain equations of the forms

$$A(x - h)^2 - C(y - k)^2 = N \quad \text{and} \quad C(y - k)^2 - A(x - h)^2 = N$$

with $A = b^2$, $C = a^2$, and $N = a^2b^2$. Now, if we expand $(x - h)^2$ and $(y - k)^2$ in these equations, distribute A and C, and collect all terms on one side, we obtain general quadratic equations in two unknowns of the forms

$$\boxed{Ax^2 - Cy^2 + Dx + Ey + F = 0}$$

and

$$\boxed{Cy^2 - Ax^2 + Dx + Ey + F = 0}$$

We refer to these two equations as the **equations of a hyperbola in general form** with vertical and horizontal axes of symmetry, respectively. These general quadratic equations in two unknowns are characterized by the presence of x^2 and y^2 terms that have *different* signs.

If the equation of a hyperbola is given in general form, then we can find the coordinates of its center and vertices by converting the equation to standard form and observing the values of h, k, a, and b. We can use a graphing calculator to check our work by solving the equation in general form for y and graphing both equations that result.

EXAMPLE 2 Determine the coordinates of the center and vertices of the hyperbola given by each equation.

(a) $4x^2 - 5y^2 = 20$ (b) $4x^2 - 9y^2 + 8x + 36y + 4 = 0$

◆ **Solution**

(a) We begin by writing the equation in standard form:

$$4x^2 - 5y^2 = 20$$

$$\frac{x^2}{5} - \frac{y^2}{4} = 1 \qquad \text{Divide both sides by 20}$$

$$\frac{(x-0)^2}{(\sqrt{5})^2} - \frac{(y-0)^2}{2^2} = 1 \qquad \text{Write in standard form}$$

This last equation tells us that the graph of $4x^2 - 5y^2 = 20$ is a hyperbola with horizontal semitransverse axis of length $a = \sqrt{5}$, vertical semiconjugate axis of length $b = 2$, and center at the origin. Hence, the coordinates of the vertices are

$$(\sqrt{5}, 0) \approx (2.236, 0) \quad \text{and} \quad (-\sqrt{5}, 0) \approx (-2.236, 0).$$

Figure 6.46 shows the graph of $4x^2 - 5y^2 = 20$ $\left(\text{or } y = \pm 2\sqrt{\frac{x^2 - 5}{5}}\right)$ in the viewing rectangle of a graphing calculator. This graph confirms our work.

(b) We begin by writing the equation in standard form. To do this, we use the process of completing the square (Section 3.3):

$$4x^2 - 9y^2 + 8x + 36y + 4 = 0$$

$$(4x^2 + 8x \quad) + (-9y^2 + 36y \quad) = -4 \qquad \text{Regroup}$$

$$4(x^2 + 2x \quad) - 9(y^2 - 4y \quad) = -4 \qquad \text{Factor out 4 and } -9$$

Add 4 to both sides.

$$4(x^2 + 2x + 1) - 9(y^2 - 4y + 4) = -4 + (4 - 36) \qquad \text{Complete the squares}$$

Add −36 to both sides.

$$4(x + 1)^2 - 9(y - 2)^2 = -36 \qquad \text{Factor}$$

$$\frac{(y-2)^2}{4} - \frac{(x+1)^2}{9} = 1 \qquad \text{Divide both sides by } -36$$

$$\frac{(y-2)^2}{2^2} - \frac{(x+1)^2}{3^2} = 1 \qquad \text{Write in standard form}$$

The last equation tells us that the graph of $4x^2 - 9y^2 + 8x + 36y + 4 = 0$ is a hyperbola with vertical semitransverse axis of length $b = 2$, horizontal

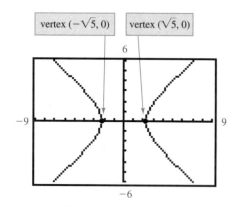

FIGURE 6.46
The graph of $4x^2 - 5y^2 = 20$ is a hyperbola with center at the origin and vertices $(\pm\sqrt{5}, 0)$.

semiconjugate axis of length $a = 3$, and center $(h, k) = (-1, 2)$. Hence the coordinates of the vertices are $(-1, 4)$ and $(-1, 0)$.

To generate the graph of the equation $4x^2 - 9y^2 + 8x + 36y + 4 = 0$ in the viewing rectangle of a graphing calculator, we rewrite the equation as a quadratic equation in y,

$$-9y^2 + 36y + (4x^2 + 8x + 4) = 0,$$

and apply the quadratic formula:

$$y = \frac{-36 \pm \sqrt{(36)^2 - 4(-9)(4x^2 + 8x + 4)}}{2(-9)}$$

$$= 2 \pm \frac{2\sqrt{x^2 + 2x + 10}}{3}$$

Entering both of these equations into the calculator, we obtain the hyperbola that is shown in Figure 6.47. This graph confirms our work. ◆

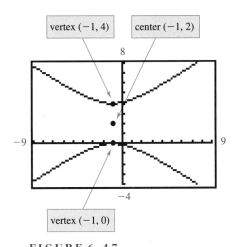

FIGURE 6.47
The graph of
$4x^2 - 9y^2 + 8x + 36y + 4 = 0$
is a hyperbola with center $(-1, 2)$ and vertices $(-1, 4)$ and $(-1, 0)$.

CAUTION Not every equation of the form

$$Ax^2 - Cy^2 + Dx + Ey + F = 0 \quad \text{or} \quad Cy^2 - Ax^2 + Dx + Ey + F = 0,$$

where the x^2 and y^2 terms have different signs, represents a hyperbola. For example, if $Ax^2 - Cy^2 + Dx + Ey + F = 0$ is written in the form

$$A(x - h)^2 - C(y - k)^2 = N \quad \text{and} \quad N = 0,$$

then the graph of the equation is a pair of lines that intersect at (h, k). We refer to these two lines as a *degenerate hyperbola*.

PROBLEM 2 Is $x^2 - 4y^2 - 6x + 8y + 5 = 0$ the equation of a hyperbola? Explain. ◆

◆ Geometric Definition of a Hyperbola

Geometrically, we define a hyperbola as follows.

◀ **Geometric Definition of a Hyperbola**

A **hyperbola** is the set of all points in a plane, the difference of whose distances from two fixed points is a constant, which is equal to the length of the transverse axis of the hyperbola. The two fixed points are called the **foci** (plural of focus).

For convenience, we place a hyperbola in the xy-coordinate plane so that its center is at the origin, its transverse axis is horizontal with length $2a$, and

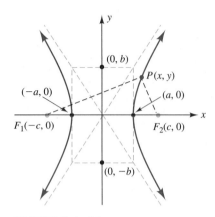

FIGURE 6.48
A hyperbola in the xy-coordinate plane with center at the origin, transverse axis of length $2a$, conjugate axis of length $2b$, and foci with coordinates $(-c, 0)$ and $(c, 0)$

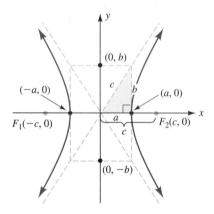

FIGURE 6.49
The constants a, b, and c are related by the Pythagorean theorem.

its conjugate axis is vertical with length $2b$ (see Figure 6.48). We have labeled the coordinates of the foci $(-c, 0)$ and $(c, 0)$, with $a < c$.

The relationship between the constants a, b, and c in any hyperbola is

$$c^2 = a^2 + b^2$$

Figure 6.49 illustrates this relationship. If c is the length of the line segment from the center to the focus, then c is also the length of the diagonal of the rectangle that is formed by a and b. Now, referring to Figure 6.48 and applying our geometric definition of a hyperbola, we have

$$PF_1 - PF_2 = 2a$$
$$\sqrt{(x + c)^2 + y^2} - \sqrt{(x - c)^2 + y^2} = 2a$$

Using the same algebraic procedure as we used for an ellipse in Section 6.3—isolating the radical, squaring both sides, and simplifying—and using the fact that $c^2 = a^2 + b^2$ in any hyperbola, we obtain the algebraic definition

$$\frac{x^2}{a^2} - \frac{y^2}{b^2} = 1.$$

Hence, we conclude that our algebraic definition of a hyperbola is consistent with our geometric definition.

The **eccentricity** e of a hyperbola is a measure of the "spread of its wings" and is defined as

$$e = \frac{\text{distance from center to focus}}{\text{distance from center to vertex}}$$

The eccentricity e of a hyperbola is always a number greater than 1. The larger the number, the wider the opening of each branch of the hyperbola.

EXAMPLE 3 Find the foci and eccentricity of the hyperbola defined by each equation.

(a) $16x^2 - 9y^2 = 144$ **(b)** $y^2 - 4x^2 - 6y + 5 = 0$

◆ Solution

(a) Dividing both sides by 144, we obtain

$$\frac{x^2}{9} - \frac{y^2}{16} = 1 \quad \text{or} \quad \frac{(x - 0)^2}{3^2} - \frac{(y - 0)^2}{4^2} = 1,$$

the equation of a hyperbola in standard form with center at the origin, horizontal semitransverse axis of length $a = 3$ and vertical semiconjugate axis of length $b = 4$. To find the coordinates of the foci, we apply the relation

$$c^2 = 3^2 + 4^2,$$

which implies that $|c| = 5$. Hence, the coordinates of the foci are $(-5, 0)$ and $(5, 0)$, as shown in Figure 6.50.

The eccentricity e of this hyperbola is

$$e = \frac{\text{distance from center to focus}}{\text{distance from center to vertex}} = \frac{5}{3} \approx 1.67.$$

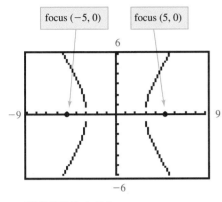

FIGURE 6.50

The foci of the hyperbola $16x^2 - 9y^2 = 144$ are $(-5, 0)$ and $(5, 0)$.

(b) First, we write the equation in standard form by completing the square:

$$y^2 - 4x^2 - 6y + 5 = 0$$

$$(y^2 - 6y \quad) - 4x^2 = -5 \qquad \text{Regroup}$$

Add 9 to both sides.

$$(y^2 - 6y + 9) - 4x^2 = -5 + 9 \qquad \text{Complete the square}$$

$$(y - 3)^2 - 4x^2 = 4 \qquad \text{Factor}$$

$$\frac{(y - 3)^2}{2^2} - \frac{(x - 0)^2}{1^2} = 1 \qquad \begin{array}{l}\text{Divide both sides by 4 and}\\\text{write in standard form}\end{array}$$

This is the equation of a hyperbola in standard form with center $(h, k) = (0, 3)$, vertical semitransverse axis of length $b = 2$, and horizontal semiconjugate axis of length $a = 1$. To find the coordinates of the foci, we apply the relation

$$c^2 = 1^2 + 2^2,$$

which implies that $|c| = \sqrt{5} \approx 2.236$. Therefore, the coordinates of the foci are

$$\left(0, 3 + \sqrt{5}\right) \approx (0, 5.236) \quad \text{and} \quad \left(0, 3 - \sqrt{5}\right) \approx (0, 0.764)$$

as shown in Figure 6.51.

The eccentricity e of this hyperbola is

$$e = \frac{\text{distance from center to focus}}{\text{distance from center to vertex}} = \frac{\sqrt{5}}{2} \approx 1.12.$$

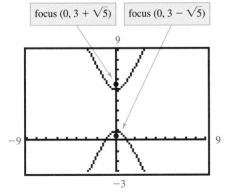

FIGURE 6.51

The foci of the hyperbola $y^2 - 4x^2 - 6y + 5 = 0$ are $(0, 3 + \sqrt{5})$ and $(0, 3 - \sqrt{5})$.

PROBLEM 3 Find the foci and eccentricity for the hyperbola defined by $9x^2 - y^2 = 9 = 0$.

If we know the foci and eccentricity of a hyperbola, then we can determine its equation.

EXAMPLE 4 Find the equation of the hyperbola in standard form with foci (2, 5) and (2, −3) and eccentricity 2.

Solution The center of a hyperbola is located halfway between its foci. Hence, the center of this hyperbola is $(h, k) = (2, 1)$, as shown in Figure 6.52. Therefore, since the transverse axis is vertical, the equation of this hyperbola must have the form

$$\frac{(y-1)^2}{b^2} - \frac{(x-2)^2}{a^2} = 1$$

We now proceed to find the lengths of the semitransverse axis b and the semiconjugate axis a. Since the distance from the center to a focus is $c = 4$ and the eccentricity of his hyperbola is given to be 2, we have

$$\frac{\text{distance from center to focus}}{\text{distance from center to vertex}} = \frac{4}{b} = 2,$$

which implies that $b = 2$. We find the value of a by applying the relation $c^2 = a^2 + b^2$:

$$4^2 = a^2 + 2^2 \quad \text{Substitute } c = 4 \text{ and } b = 2$$
$$a^2 = 4^2 - 2^2 \quad \text{Solve for } a$$
$$a^2 = 12$$
$$a = 2\sqrt{3}$$

Since $(h, k) = (2, 1)$, $a^2 = (2\sqrt{3})^2 = 12$, and $b^2 = (2)^2 = 4$, the equation of this hyperbola in standard form is

$$\frac{(y-1)^2}{4} - \frac{(x-2)^2}{12} = 1.$$ ◆

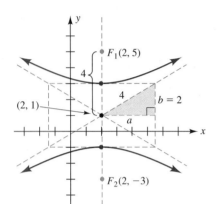

FIGURE 6.52
A hyperbola with foci (2, 5) and (2, −3) has center (2, 1).

PROBLEM 4 Referring to Example 4, express the equation of the hyperbola in general form with integer coefficients. ◆

◆ Application: Reflection Property of a Hyperbola

Like the parabola and ellipse, the hyperbola also has a *reflection property*. A ray of light directed at one focus of a hyperbolic mirror will be reflected so as to pass through its other focus. The reflection property of a hyperbola is used in the design of reflecting telescopes such as the Hale telescope on Mount Palomar, California. As illustrated in Figure 6.53, light from the moon hits the primary parabolic mirror in the telescope and is reflected back toward the parabola's focus F_1, which is also one of the two foci of the hyperbolic mirror. The light hits the hyperbolic mirror and is reflected toward the hyperbola's second focus F_2, which is also one of the two foci of an elliptical mirror. The light then hits the elliptical mirror and reflects to the ellipses's second focus F_3, which is the eyepiece of the telescope (see Exercises 37).

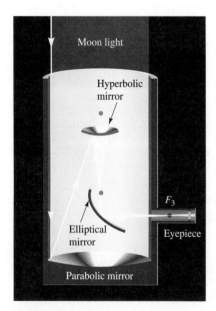

FIGURE 6.53
A reflecting telescope uses the reflection properties of the parabola, ellipse, and hyperbola to direct an image to the eyepiece.

Exercises 6.4

Basic Skills

In Exercises 1–16, determine the coordinates of the center and vertices of the hyperbola given by each equation. Check graphically with a graphing calculator.

1. $\dfrac{x^2}{4} - \dfrac{y^2}{16} = 1$
2. $\dfrac{y^2}{4} - \dfrac{x^2}{4} = 1$
3. $9y^2 - 4x^2 = 36$
4. $9x^2 - 16y^2 = 144$
5. $16x^2 - y^2 = 9$
6. $4y^2 - 16x^2 = 25$
7. $9x^2 - 25y^2 = -4$
8. $4y^2 - x^2 + 12 = 0$
9. $\dfrac{x^2}{9} - \dfrac{(y+2)^2}{4} = 1$
10. $\dfrac{(y-1)^2}{16} - (x+3)^2 = 1$
11. $(y-3)^2 - (x-1)^2 = 36$
12. $9(x+1)^2 - 25(y-4)^2 = 225$
13. $36x^2 - y^2 - 144x - 6y + 126 = 0$
14. $16y^2 - 9x^2 + 90x = 261$
15. $9y^2 - 3x^2 - 24x + 18y = 120$
16. $2x^2 - 4y^2 + 12x + 16y + 1 = 0$

In Exercises 17–22, find the foci and eccentricity of the hyperbola whose equation is given in the indicated exercise.

17. Exercise 3
18. Exercise 4
19. Exercise 9
20. Exercise 12
21. Exercise 13
22. Exercise 16

In Exercises 23–32, determine the equation of the hyperbola with the given characteristics.

23. Vertices $(0, \pm 4)$ and asymptotes $4y - x = 0$ and $4y + x = 0$
24. Vertices $(2, -2)$ and $(-2, -2)$, and endpoints of conjugate axis $(0, 0)$ and $(0, -4)$
25. Asymptotes $y = \pm x$ and horizontal transverse axis of length 4
26. Asymptotes $y = -x$ and $y = x + 2$, and vertical transverse axis of length 4
27. Vertices $(-2, -3)$ and $(-2, 3)$, and one of the foci at $(-2, 5)$
28. Foci $(-4, -1)$ and $(2, -1)$, and one of the vertices at $(1, -1)$
29. Foci $(\pm 2, 0)$ and transverse axis of length 2
30. Foci $(1, \pm 4)$ and conjugate axis of length 4
31. Foci $(0, 0)$ and $(3, 0)$, and eccentricity 3
32. Foci $(1, -3)$ and $(1, 5)$, and eccentricity $\tfrac{4}{3}$

In Exercises 33–36, use a graphing calculator to generate the graph of each hyperbola. Then estimate the coordinates of the vertices, rounding each coordinate to three significant digits. Check algebraically.

33. $356x^2 - 129y^2 = 926$
34. $1.25x^2 - 2.70y^2 + 8.55y = 2.35$
35. $0.35x^2 - 0.12y^2 - 2.34y - 7.89 = 0$
36. $3.4x^2 - 6.4y^2 - 1.4x - 4.8y - 18.6 = 0$

37. A Cassegrain telescope contains a primary parabolic mirror and a smaller hyperbolic mirror, as shown in the sketch. A ray of light from the moon strikes the parabolic mirror and is reflected toward the parabola's focus F_1, which is also a focus of the hyperbolic mirror. The ray of light is then reflected off the hyperbolic mirror toward the hyperbola's other focus F_2, which is the eyepiece of the telescope. If the equation of the hyperbolic mirror is $x^2/441 - y^2/400 = 1$ and x and y are measured in centimeters, determine the distance from the hyperbolic mirror to the eyepiece.

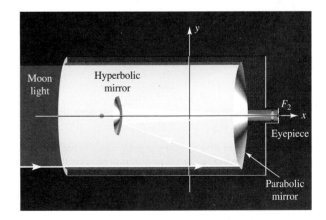

38. The figure shows the hyperbolic path of a UFO. This path moves closer toward Earth along the line $y = x/4$ and comes within 6000 miles of the Earth's surface before moving away. Assuming the radius of the Earth

is 4000 miles, determine the equation of the path of the UFO.

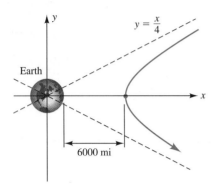

39. Allied tanks are located at fixed points A and B, as shown in the sketch. An enemy tank located at point P fires its gun and 1.3 seconds after the shot is heard at point B, the shot is heard at point A.

    (a) Assuming that the speed of sound is 1100 ft/s, find $PA - PB$ and explain why point P must lie on a branch of a hyperbola.

    (b) Determine the equation of the branch of the hyperbola described in part (a) if the coordinates (in feet) of points A and B are $(-1500, 0)$ and $(1500, 0)$, respectively.

    (c) An allied patrol unit at point $C(0, 0)$ calculates that the distance from them to the enemy tank is 950 ft. Determine the distances PA and PB from the allied tanks to the enemy tank.

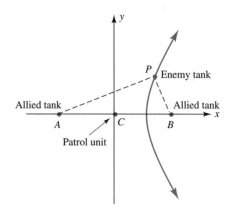

40. In an engineer's design for the horn on a megaphone, a hyperbola is rotated about the x-axis, as shown in the figure. If the distance from A to B is 10.26 cm, determine the radius of each end of the megaphone.

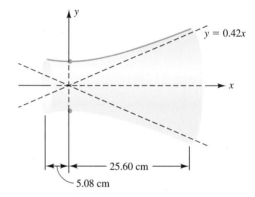

 Critical Thinking

41. Describe the general shape of a hyperbola whose eccentricity is (a) close to 1 or (b) very large.

42. What is the eccentricity of a hyperbola with asymptotes $y = \pm x$?

43. Suppose the transverse axis and conjugate axis of a hyperbola are both doubled. What effect does this have on the hyperbola's eccentricity and asymptotes?

44. The equation of a hyperbola with a horizontal transverse axis is $Ax^2 - Cy^2 = E$. Find the coordinates of its foci in terms of A, C, and E.

45. Describe the graph of each of the following degenerate conic sections.

    (a) $x^2 + y^2 - 6x - 8y + 25 = 0$
    (b) $9x^2 + 4y^2 + 54x - 16y + 97 = 0$
    (c) $4x^2 - y^2 - 16x - 4y + 12 = 0$
    (d) $y^2 - 9x^2 + 36x - 10y - 11 = 0$

46. When a supersonic jet breaks the sound barrier, it produces a conical shock wave, as shown in the sketch. Describe the intersection of the conical shock wave with the ground if the plane's path is parallel to the ground.

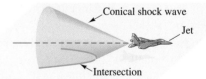

If a and b are positive real numbers, then solving

$$\frac{x^2}{a^2} - \frac{y^2}{b^2} = 1 \quad \text{and} \quad \frac{y^2}{b^2} - \frac{x^2}{a^2} = 1$$

for y yields

$$y = \pm \frac{b}{a}\sqrt{x^2 - a^2} \quad \text{and} \quad y = \pm \frac{b}{a}\sqrt{x^2 + a^2},$$

respectively. Taken separately, the equations

$$y = \frac{b}{a}\sqrt{x^2 - a^2} \qquad y = -\frac{b}{a}\sqrt{x^2 - a^2}$$

$$y = \frac{b}{a}\sqrt{x^2 + a^2} \qquad y = -\frac{b}{a}\sqrt{x^2 + a^2}$$

define y as a function of x, and the graph of each function is one-half (not necessarily one branch) of a hyperbola. In Exercises 47–50, state the domain and range of each function, and sketch its graph.

47. $y = \frac{1}{2}\sqrt{x^2 - 4}$
48. $y = -3\sqrt{x^2 - 25}$
49. $y = -\sqrt{x^2 + 9}$
50. $y = \frac{3}{4}\sqrt{4x^2 + 1}$

Chapter 6 Review

◆ Questions for Writing or Group Discussion

1. Name the four main *conic sections*. Show how each is formed by slicing a double cone with a plane.
2. What is meant by a *degenerate conic section*?
3. Give the *standard form* of the equation of each conic section if its axes of symmetry are parallel to the *xy*-axes.
4. Give the *general form* of the equation of each conic section if its axes of symmetry are parallel to the *xy*-axes.
5. State the geometric definition of a *circle*. Illustrate with a sketch.
6. What is the slope of the *tangent* to the circle $x^2 + y^2 = r^2$ at the point (x_1, y_1) on the circle?
7. State the geometric definition of a *parabola*. Illustrate with a sketch.
8. Under what conditions does the equation of a parabola define y as a function of x?
9. How can you tell from the equation of a parabola whether the curve opens upward, downward, to the left, or to the right?
10. State the geometric definition of an *ellipse*. Illustrate with a sketch.
11. What is meant by the *major axis* of an ellipse? How can you determine from the equation of an ellipse in standard form whether the major axis is horizontal or vertical?
12. Why is a *circle* a special case of an ellipse?
13. Explain the procedure for finding the coordinates of the *foci* of an ellipse from its equation. Illustrate with an example.
14. What is meant by the *eccentricity* of an ellipse? Describe the shape of an ellipse with eccentricity close to 1 or close to 0.
15. State the geometric definition of a *hyperbola*. Illustrate with a sketch.
16. What is meant by the *transverse axis* of a hyperbola? How can you determine from the equation of a hyperbola in standard form whether the transverse axis is horizontal or vertical?
17. What are the equations of the *asymptotes* of the hyperbola $(x^2/a^2) - (y^2/b^2) = 1$? How are they used as an aid in graphing a hyperbola?

18. What is meant by the *eccentricity* of a hyperbola? Describe the shape of a hyperbola with eccentricity close to 1 or very large.
19. Explain the procedure for finding the coordinates of the *foci* of a hyperbola from its equation. Illustrate with an example.
20. Discuss the *reflection properties* of a parabola, an ellipse, and a hyperbola.

Review Exercises

In Exercises 1–28, identify the graph of each equation as a circle, parabola, ellipse, hyperbola, or a degenerate conic section. If the graph is a circle, give its radius and the coordinates of its center. If the graph is a parabola, give the direction of its opening and the coordinates of its vertex. If the graph is an ellipse or a hyperbola, give the coordinates of its center and vertices.

1. $x^2 + y^2 = 16$
2. $9x^2 + 9y^2 = 1$
3. $y^2 - 4x = 0$
4. $2x^2 + y = 0$
5. $x^2 + 9y^2 = 16$
6. $4x^2 + 3y^2 = 9$
7. $x^2 - y^2 = 25$
8. $4y^2 - x^2 = 49$
9. $4(x + 2)^2 + 4(y + 1)^2 = 1$
10. $(x - 1)^2 + y^2 = 5$
11. $y = (x - 2)^2 - 3$
12. $x = -2(y + 1)^2 + 2$
13. $9x^2 + (y - 3)^2 = 9$
14. $4(x + 2)^2 + 9(y - 1)^2 = 36$
15. $9x^2 - 16(y - 1)^2 = 144$
16. $(x + 2)^2 - (y - 1)^2 = -4$
17. $x^2 + y^2 - 10x - 8y + 16 = 0$
18. $x^2 + y^2 - 6y - 16 = 0$
19. $y^2 + x - 2y - 15 = 0$
20. $x^2 + 2x - 2y + 5 = 0$
21. $4x^2 + y^2 - 8x + 4y - 8 = 0$
22. $x^2 + 16y^2 - 32y - 9 = 0$
23. $4y^2 - 9x^2 - 8y + 5 = 0$
24. $16x^2 - 9y^2 - 32x - 36y - 24 = 0$
25. $y^2 - 4x^2 + 8x - 4 = 0$
26. $x^2 + y^2 - 16y + 64 = 0$
27. $2x^2 + 3y^2 - 4x + 12y + 14 = 0$
28. $x^2 - y^2 - 4x - 2y + 3 = 0$

In Exercises 29–32, find the focus and directrix of the conic section whose equation is given in the indicated exercise.

29. Exercise 3
30. Exercise 12
31. Exercise 19
32. Exercise 20

In Exercises 33–36, find the foci and eccentricity of the conic section whose equation is given in the indicated exercise.

33. Exercise 7
34. Exercise 6
35. Exercise 21
36. Exercise 24

In Exercises 37–56, determine the equation of the conic section from the given information.

37. Circle with center (0, 0) and radius 7
38. Circle with center (0, −2) and passing through (1, 3)
39. Circle whose diameter is the line segment joining the points (0, 3) and (4, 6)
40. Circle with center (4, −2) and tangent to the *x*-axis
41. Circle with center (2, 3) and tangent to $4x - 3y = 24$
42. Circle with radius $\sqrt{5}$ and passing through (2, 1) and (2, 5)
43. Parabola with vertex (0, −4) and *x*-intercepts ±2
44. Parabola with vertex (2, 1) and *y*-intercepts −1 and 3
45. Parabola passing through the origin, vertex (−2, −3), and symmetric with respect to $y = -3$
46. Parabola passing through (−1, 0), vertex (2, 3), and symmetric with respect to $x = 2$
47. Parabola with focus (1, 0) and directrix $y = -1$
48. Parabola with vertex (3, −2) and focus (4, −2)
49. Ellipse with vertices (0, 0) and (6, 0), and minor axis of length 4
50. Ellipse with center (2, 5), minor axis of length 2, and horizontal major axis of length 8
51. Ellipse with foci (0, 0) and (4, 0), and eccentricity $\frac{1}{2}$
52. Ellipse with vertices (−1, −3) and (−1, 3), and one of the foci at (−1, 2)
53. Hyperbola with asymptotes $y = \pm \frac{2}{3}x$ and vertical transverse axis of length 6
54. Hyperbola with center (2, −1), horizontal transverse axis of length 10, and conjugate axis of length 4

55. Hyperbola with vertices $(-3, -1)$ and $(3, -1)$, and one of the foci at $(4, -1)$
56. Hyperbola with foci $(\pm 3, 0)$ and eccentricity 2
57. Discuss the position of the circle
    $(x - h)^2 + (y - k)^2 = r^2$ relative
    to the coordinate axes for the indicated condition.
    (a) $h = 0$ (b) $k = 0$
    (c) $h = r$ (d) $k = r$
58. A *secant* to a circle is a line that intersects the circle in two distinct points. Find the coordinates of the points of intersection of the secant $x + y = 4$ and the circle $x^2 + y^2 = 10$.
59. Two hyperbolas are *conjugate* to each other if they have the same asymptotes and if the transverse axis of one is equal to the conjugate axis of the other. Find the equation of the hyperbola that is conjugate to $3x^2 - 4y^2 = 12$.
60. Find the length of the side of a square inscribed in an ellipse with semimajor axis of length a and semiminor axis of length b.

For Exercises 61–64, refer to the hyperbola $(x^2/a^2) - (y^2/b^2) = 1$ with point P on the asymptote $y = -\dfrac{b}{a}x$ and foci at $(-c, 0)$ and $(c, 0)$, as shown in the figure.

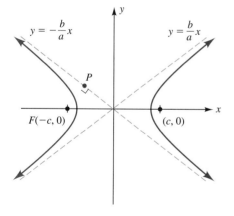

61. Find the slope of $\overline{FP}$.
62. Find the equation of the line that passes through F and P.
63. Find the coordinates of point P.
64. Find FP, the distance from F to P. What conclusion can you make about the distance between a focus and an asymptote of a hyperbola?
65. Two straight roadways are connected by a circular curve that has central angle $\theta = 90°$ and radius $r = 100$ ft, as shown in the sketch. A surveyor wants to lay out this circular curve by placing stakes at PC, PT, and at several other points along the curve, such as point A. Take the axes as shown in the figure.
    (a) Find the coordinates of PC and PT and the equation of the circular curve.
    (b) Determine the coordinates of point A. Round all coordinates to the nearest hundredth of a foot.

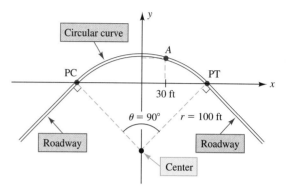

66. The arch of a bridge has the shape of the upper half of an ellipse. At the water level, the distance between the supports is 100 feet, and the greatest height above the water level is 20 feet, as shown in the sketch.
    (a) Taking the axes as shown in the sketch, find the equation of the ellipse that forms this arch.
    (b) Use a graphing calculator to generate the graph of the equation in part (a). Then estimate (to the nearest hundredth of a foot) the height of the arch above the surface of the water at points which are 10 feet apart along the surface of the water, beginning at $x = 10$ ft and ending at $x = 40$ ft.

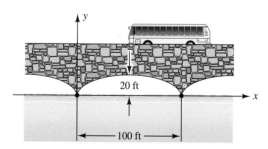

67. A parabolic satellite dish for receiving television signals has the dimensions shown in the sketch at the top of the next page. At what distance d (in inches) should the receiver be located in order to take advantage of the reflection property of a parabola?

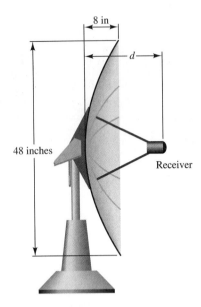

68. A simply supported beam with a concentrated load at midspan deflects in the shape of a parabola, as shown in the sketch. Determine the equation of the parabola, called the *elastic curve*. Write the equation in the form $y = ax^2 + bx$ with a and b each rounded to three significant digits.

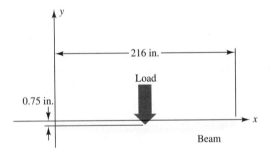

69. The equation of the path of a football kicked from the ground is $x^2 - 174x + 100y = 0$, where x and y are measured in feet and the x-axis is taken along the ground.

(a) Find the horizontal distance traveled by the ball.

(b) Find the greatest height reached by the ball.

70. The roof of a whispering gallery has the shape of an ellipse rotated about its major axis. The distance between the foci is 400 ft and the distance from a vertex to its corresponding focus is 75 ft. If a sound emitted at one focus strikes the roof, find the time before it is heard at the other focus. Assume that sound travels at 1100 ft/s.

71. Ultrahigh-frequency (UHF) sound waves emanating from one focus of a lithotripter are reflected to the other focus of the device to break up a kidney stone. Suppose a lithotripter is formed by rotating the portion of an ellipse below its minor axis about its major axis, as shown in the sketch. How far from the kidney stone (in cm) should the opening of the lithotripter be placed in order to break up the stone?

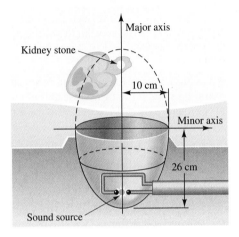

72. In an engineer's design for a cooling tower in a nuclear power plant, a hyperbola is rotated about the y-axis, as shown in the figure. If the distance from A to B is 12.3 meters, determine the radius of each end of the tower.

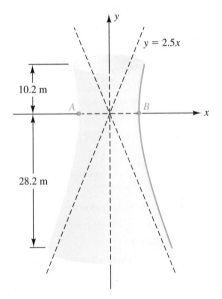

Cumulative Review Exercises for Chapters 4, 5, and 6

1. If the base is 4, find the logarithms of each number.
   (a) 64
   (b) $\frac{1}{16}$
   (c) 32
   (d) $\frac{1}{8}$

2. Find the zeros of the function f defined by
   $$f(x) = a + \frac{b}{\ln x}$$ if a and b are constants.

3. Identify the graph of each equation as a parabola, circle, ellipse, or hyperbola.
   (a) $x^2 + y^2 = 9$
   (b) $x^2 - y^2 = 9$
   (c) $x + y^2 = 9$
   (d) $2x^2 + y^2 = 9$

4. Discuss the position of the vertex of the parabola $y = a(x - h)^2 + k$ for each of the indicated conditions.
   (a) $h = 0$
   (b) $k = 0$
   (c) both $h = 0$ and $k = 0$

5. Factor completely over the integers:
   $x^4 + 2x^3 + x^2 - 8x - 20$

6. Find the remainder when $2x^4 - 3x^2 + 4x - 3$ is divided by $x - a$.

7. Solve for x:
   (a) $\log_x 3 - \log_4 16 - \log_3 \frac{1}{9} - \ln e = 2 \log 1$
   (b) $x^{\ln x} - e^2 x = 0$

8. For $a > 0$, show that $a^x = e^{x \ln a}$. Use this fact to write each of the following expressions as an exponential with base e.
   (a) 10^x
   (b) 3^{-x}
   (c) 5^{3x}
   (d) 2^{x+1}

9. Solve each equation for y.
   (a) $2 \ln x - \frac{1}{2} \ln y = 1$
   (b) $x + 2 = e^{x - \ln y}$
   (c) $2 \log_3(y + 1) = 2 - \log_3(y - 1)$

10. A function f is defined by $f(x) = 2 - e^{-x}$.
    (a) Show that f is one-to-one.
    (b) Find the inverse function f^{-1}, and state its domain.

11. A line is tangent to a circle at the point (3, 4). Find the equation of the line if the equation of the circle is $x^2 + y^2 = 25$.

12. Find the equations of the asymptotes for each of the following hyperbolas.
    (a) $4x^2 - y^2 = 16$
    (b) $y^2 - x^2 + 2x = 5$

13. Explain the behavior of the function f defined by
    $$f(x) = \frac{x}{x - 2}$$ as x approaches the given value.
    (a) $x \to 2^-$
    (b) $x \to 2^+$

14. Determine the oblique asymptote for the graph of the equation $x^2 - 2xy - y = 0$. Then graph the equation.

15. A sum of $10,000 is deposited in an account at a rate of 9% per year compounded continuously. Find the value of the account after eight years.

16. Given the functions $f(x) = \ln(x + 1)$ and $g(x) = e^{-x} - 1$, find each composite function.
    (a) $f \circ g$
    (b) $g \circ f$

17. Find the coordinates of the intersection points of the ellipses $x^2 + 2y^2 = 18$ and $9x^2 + 4y^2 - 24y = 0$. Round each coordinate to three significant digits.

18. Find the coordinates of the vertex and any x- and y-intercepts for each parabola.
    (a) $x^2 - y - 4x + 3 = 0$
    (b) $y^2 + 4y + x - 1 = 0$

19. Three hundred children are seated in an auditorium in a certain number of rows, with the same number of children seated in each row. Placing two more children in each row would result in five fewer rows being used. In how many rows are the children seated initially?

20. Solve the inequality $x^2 - 4 \leq \dfrac{1}{x^2}$ for x.

21. The circumference of a rear wheel on a dragster is 4 feet more than the circumference of a front wheel. A front wheel makes 88 more revolutions than a rear wheel in traveling the quarter-mile track. What is the circumference of each wheel?

22. Find the vertical and horizontal asymptotes for the graph of the function
    $$f(x) = \frac{x^2 - 2}{x^2 - 7x + 6}.$$
    Determine the value of x at the point where the graph of f crosses its horizontal asymptote.

23. The number N of golf balls (in dozens) sold by a certain company is a function of the amount x (in thousand of dollars) that is spent on advertising. This function is given by
    $$N(x) = 2500[1 + \ln(x + 1)].$$
    (a) If no money is spent on advertising, how many dozens of balls are sold?
    (b) If $500,000 is spent on advertising, approximately how many dozens of balls are sold?
    (c) How much money must be spent on advertising to sell 40,000 dozen balls?

24. A hot cup of coffee has an initial temperature of 200°F and is placed in a room with a temperature of 70°F. After t minutes, the temperature T of the coffee is $T = 70 + 130e^{-0.38t}$. How many *seconds* does it take for the temperature of the coffee to cool to 120°F?

25. Find the length of the line segment joining the intersection points of the curves $y = x + 1$ and $x^2 + y^2 = 25$.

26. Discuss the nature of the roots of the equation $x^n - 1 = 0$ if n is a positive integer and (a) n is even or (b) n is odd.

27. Determine the equation of a hyperbola in standard form if its asymptotes are $y = \pm 3x$ and its horizontal transverse axis has length 2.

28. Find the length of the side of a square inscribed in the circle $x^2 + y^2 = 2$.

29. The width of a cardboard box is twice its height, and the length of the box is 2 inches more than its width. Determine the dimensions of the box if its volume is 144 cubic inches.

30. The height of a tin can is 4 inches more than its radius. Determine to three significant digits the height of the can if its volume is 82 cubic inches.

31. Two circular stove pipes that are each 8 inches in diameter are cut at an angle of 45° to form an elbow, as shown in the sketch. Find the length of the major and minor axis of their elliptical intersection.

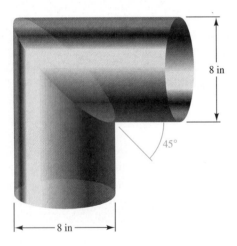

32. The planet Mercury travels in an elliptical orbit with the Sun at one focus, as shown in the sketch. The eccentricity of the orbit is 0.205, and the distance from the Sun to Mercury at perihelion (shortest distance from the Sun) is 2.85×10^7 miles.

    (a) Determine the distance from Mercury to the Sun at aphelion (greatest distance from the Sun).

    (b) Find the Cartesian equation of the orbit if the center of the ellipse is at the origin.

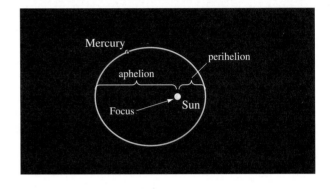

The chair lift at a ski area rises at an angle of 19.2° and attains a vertical rise of 1250 ft in 15 minutes. What is the speed (in mi/h) of the chair lift?

For the solution, see Example 4 in Section 7.7.

Introduction to Trigonometry

7.1 Angles and Their Measures

7.2 Defining the Trigonometric Functions

7.3 Evaluating the Trigonometric Functions

7.4 Properties and Graphs of the Sine and Cosine Functions

7.5 Properties and Graphs of the Other Trigonometric Functions

7.6 The Inverse Trigonometric Functions

7.7 Applications Involving Right Triangles and Harmonic Motion

7.1 Angles and Their Measures

◆ Introductory Comments

The word *trigonometry* is derived from the Greek language and means "triangle measurement." Early Greek mathematicians worked with the *trigonometry of angles*: They applied trigonometry to the fields of surveying, astronomy, and navigation in order to determine distances and angles in a triangle that could not be measured directly. After the advent of calculus in the 17th century, mathematicians developed the *trigonometry of real numbers* to describe cyclic patterns that vary with time, such as those observed in the weather, predator-prey relationships, vibrating springs, pendulums, and electrical circuits.

In this chapter, we discuss both approaches to trigonometry. We begin with a discussion of angles and various ways to measure them. Then, in Sections 7.2 and 7.3, we define and evaluate the trigonometric functions of angles and the trigonometric functions of real numbers.

◆ Angles

A *ray* with endpoint A that continues indefinitely through another point B is designated $\overrightarrow{AB}$. One such ray is shown in Figure 7.1. When we rotate this ray about its endpoint, we form an **angle** with *vertex A, initial side $\overrightarrow{AB}$* and *terminal side $\overrightarrow{AB'}$*. One method of naming this angle is to place a Greek letter, such as theta (θ), next to a curved arrow that designates the rotation (see Figure 7.2).

Angles formed by rotation are usually placed in a Cartesian coordinate plane using **standard position**—the angle has its vertex at the origin and its initial side along the positive *x*-axis, as shown in Figure 7.3.

By convention, an angle formed by *counterclockwise* rotation is said to be a **positive angle** and an angle formed by *clockwise* rotation is said to be a **negative angle**. Figure 7.4 illustrates a positive angle and a negative angle in standard position.

FIGURE 7.1
A ray with endpoint A, denoted $\overrightarrow{AB}$

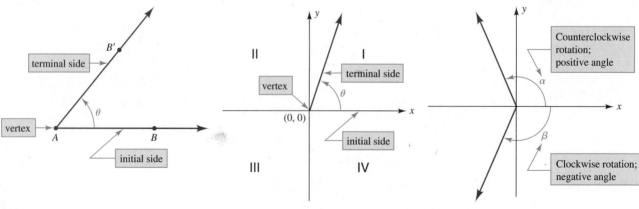

FIGURE 7.2
Angle θ, formed by rotating $\overrightarrow{AB}$ about its endpoint

FIGURE 7.3
Angle θ in standard position

FIGURE 7.4
Angle α (alpha) is positive and angle β (beta) is negative.

◆ Degree Measure

The *measure* of an angle is determined by the direction as well as the amount of the rotation from the initial side to the terminal side. One way to measure an angle is by using a unit of measure called a **degree** (°). One degree (1°) is formed when a ray is rotated $\frac{1}{360}$ of a full counterclockwise revolution about the origin of a Cartesian coordinate plane. Thus, one full counterclockwise revolution measures 360°. By convention, we write $\theta = 360°$ to indicate that the measure of angle θ is 360°. In Figure 7.5, several angles are shown in standard position. Each angle is measured in degrees.

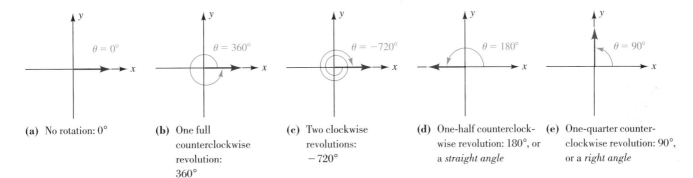

(a) No rotation: 0°
(b) One full counterclockwise revolution: 360°
(c) Two clockwise revolutions: −720°
(d) One-half counterclockwise revolution: 180°, or a *straight angle*
(e) One-quarter counterclockwise revolution: 90°, or a *right angle*

FIGURE 7.5
Angles measured in degrees in standard position

Although the angles shown in Figure 7.5 with measures 0°, 360°, and −720° differ in the amount and direction of their rotation, they have the same terminal side. Angles in standard position that have the same terminal side are called **coterminal angles**. We can find the measures of angles that are coterminal with another angle measured in degrees by adding and subtracting multiples of 360°.

EXAMPLE 1 Determine the degree measures of a positive angle and a negative angle that are coterminal with an angle having the given measure.

(a) 390°   (b) −420°

◆ Solution

(a) We find angles coterminal with 390° by adding and subtracting multiples of 360°. Thus, all angles of the form

$$(390 + 360n)° \quad \text{where } n \text{ is an integer}$$

are coterminal with 390°.

Letting $n = -1$, we obtain
$(390 - 360)° = 30°$
a *positive* angle coterminal with 390°

Letting $n = -2$, we obtain
$(390 - 720)° = -330°$
a *negative* angle coterminal with 390°

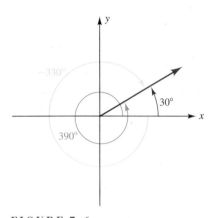

FIGURE 7.6
Angles of measure 30°, 390°, and −330° are coterminal.

Figure 7.6 shows angles of 30°, 390°, and −330° in standard position. Note that these angles have the same terminal side.

(b) We find angles coterminal with −420° by adding and subtracting multiples of 360°. Thus, all angles of the form

$$(-420 + 360n)° \quad \text{where } n \text{ is an integer}$$

are coterminal with −420°.

Letting $n = 1$, we obtain
$$(-420 + 360)° = -60°$$
<u>a *negative* angle coterminal with −420°</u>

Letting $n = 2$, we obtain
$$(-420 + 720)° = 300°$$
<u>a *positive* angle coterminal with −420°</u>

Figure 7.7 shows angles of −420°, 300°, and −60° in standard position. Note that these angles have the same terminal side. ◆

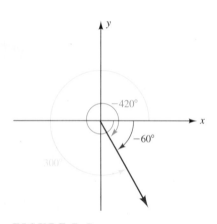

FIGURE 7.7
Angles of measure −420°, 300°, and −60° are coterminal.

PROBLEM 1 Find a positive angle and a negative angle that are coterminal with an angle of 600°. ◆

When greater precision is needed to measure an angle, we divide each degree into 60 equal parts, each called a **minute** (′). If still greater precision is needed to measure an angle, then each minute is divided into 60 equal parts, each called a **second** (″). We can use the following facts to make a conversion from degrees, minutes, seconds into decimal degrees, and vice versa.

$$1° = 60' \qquad 1' = 60'' \qquad 1° = 3600''$$

EXAMPLE 2 Convert each angle measure.

(a) 16°24′27″ to decimal degrees.

(b) 123.423° to degrees, minutes, seconds.

◆ **Solution**

(a) An angle written as 16°24′27″ is read "16 degrees, 24 minutes, 27 seconds." To convert to decimal degrees, we apply two conversion factors (Section 1.2):

$$16° \, 24' 27'' = 16° + 24'\left(\frac{1°}{60'}\right) + 27''\left(\frac{1°}{3600''}\right)$$

conversion factors

$$= 16° + 0.4° + 0.0075°$$
$$= 16.4075°$$

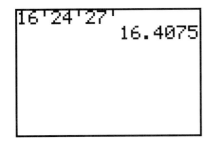

FIGURE 7.8

Typical display of converting 16°24′27″ to decimal degrees

Most graphing calculators have the capability of converting angle measures written in degrees, minutes, seconds into decimal degrees, and vice versa. Figure 7.8 shows a typical display of this conversion in the viewing rectangle of a graphing calculator.

(b) First, we convert the decimal part of the degrees (0.423°) to minutes:

$$0.423° = 0.423° \left(\frac{60'}{1°}\right) = 25.38'$$

conversion factor

Now, we convert the decimal part of the minutes (0.38′) to seconds:

$$0.38' = 0.38' \left(\frac{60''}{1'}\right) = 22.8''$$

conversion factor

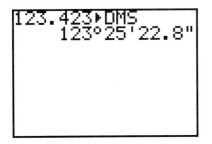

FIGURE 7.9

Typical display of converting 123.423° to degrees, minutes, seconds

Thus,

$$123.423° \approx 123°25'23'' \text{ (to the nearest second)}.$$

Figure 7.9 shows a typical display of this conversion in the viewing rectangle of a graphing calculator. ◆

PROBLEM 2 Convert the angle measure 48°15′30″ to decimal degrees (to the nearest thousandth). Use the angle feature of a graphing calculator to verify your answer. ◆

◆ Radian Measure

Another way to measure an angle is by using a unit of measure called a **radian**. An angle whose measure is one radian (1 rad) is formed when we rotate $\vec{AB}$ counterclockwise such that the *length of the arc BB′* equals the length of $\overline{AB}$, as shown in Figure 7.10.

Recall from elementary geometry that the circumference C of a circle is given by the formula $C = 2\pi r$ where r is the radius of the circle. Thus, for any circle, its radius r can be marked off 2π (≈ 6.28) times along its circumference, as shown in Figure 7.11. Thus, one full counterclockwise revolution corresponds to 2π rad (≈ 6.28 rad). By convention, if no unit of measure is specified with an angle, then the angle is measured in radians. Thus, if we write $\theta = 2\pi$, we mean that θ measures 2π rad. In Figure 7.12, several angles are shown in standard position and measured in radians.

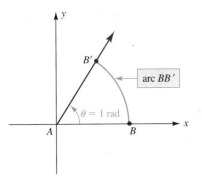

FIGURE 7.10

When the length of arc BB′ equals the length of $\overline{AB}$, an angle θ with radian measure 1 is formed.

400 CHAPTER 7 ◆ *Introduction to Trigonometry*

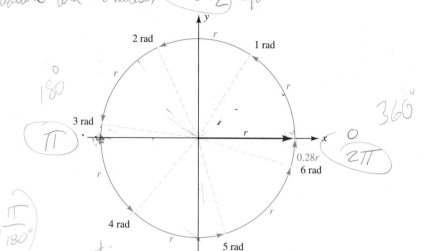

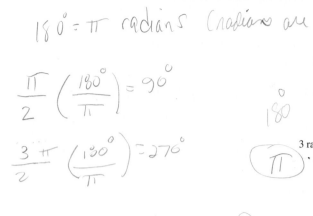

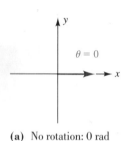

FIGURE 7.11
The radius of a circle can be marked off 2π (≈ 6.28) times along its circumference.

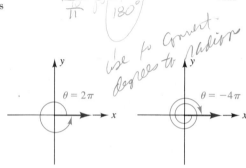

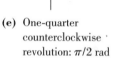

(a) No rotation: 0 rad **(b)** One full counterclockwise revolution: 2π rad **(c)** Two clockwise revolutions: -4π rad **(d)** One-half counterclockwise revolution: π rad **(e)** One-quarter counterclockwise revolution: $\pi/2$ rad

FIGURE 7.12
Angles measured in radians in standard position

One complete revolution measures 2π rad. One complete revolution also measures 360°. Thus,

$$2\pi \text{ rad} = 360° \quad \text{or} \quad \pi \text{ rad} = 180°.$$

From this relationship, we obtain

$$1 \text{ rad} = \frac{180°}{\pi} \approx 57.3° \quad \text{and} \quad 1° = \frac{\pi}{180} \text{ rad} \approx 0.01745 \text{ rad}$$

These equations give us the following conversions factors.

◆ **Conversion Factors: Radians to Degrees and Degrees to Radians**

To convert radians to degrees, multiply the radian measure by $\dfrac{180°}{\pi}$.

To convert degrees to radians, multiply the degree measure by $\dfrac{\pi}{180°}$.

SECTION 7.1 ♦ Angles and Their Measures

EXAMPLE 3 Convert each degree measure to radians and convert each radian measure to degrees.

(a) $\theta = 30°$ (b) $\theta = -240°$

(c) $\theta = \dfrac{9\pi}{4}$ (d) $\theta = 30$

♦ **Solution** The angular measures in parts (a) and (b) are in degrees. Thus, we convert to radian measure as follows:

(a) $30° = 30°\left(\dfrac{\pi}{180°}\right) = \dfrac{\pi}{6} \approx 0.5236$

(b) $-240° = -240°\left(\dfrac{\pi}{180°}\right) = -\dfrac{4\pi}{3} \approx -4.189$

Most graphing calculators have the capability of converting angle measures written in degrees into radians, and vice versa. Figure 7.13 shows a typical display of these conversions in the viewing rectangle of a graphing calculator. The conversions are done by selecting "radian" mode.

The angular measures in parts (c) and (d) are in radians, since no unit of measure is specified. We convert radians to degrees as follows:

(c) $\dfrac{9\pi}{4} = \dfrac{9\pi}{4}\left(\dfrac{180°}{\pi}\right) = 405°$

(d) $30 = 30\left(\dfrac{180°}{\pi}\right) = \dfrac{5400°}{\pi} \approx 1719°$

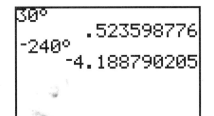

FIGURE 7.13
Typical display of converting 30° and −240° to radians

Figure 7.14 shows a typical display of these conversions in the viewing rectangle of a graphing calculator. The conversions are done by selecting "degree" mode. ♦

We can find the measures of angles that are coterminal with an angle measured in radians by adding and subtracting multiples of 2π. For example, all angles of the form

$$\dfrac{\pi}{6} + 2\pi n \quad \text{where } n \text{ is an integer}$$

are coterminal with $\pi/6$. Thus, the angles with measures

$$\dfrac{\pi}{6} + 2\pi = \dfrac{13\pi}{6}, \quad \dfrac{\pi}{6} + 4\pi = \dfrac{25\pi}{6}, \quad \dfrac{\pi}{6} - 2\pi = -\dfrac{11\pi}{6}$$

are coterminal with $\dfrac{\pi}{6}$.

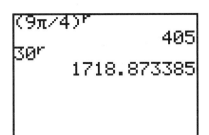

FIGURE 7.14
Typical display of converting $9\pi/4$ rad and 30 rad to degrees

PROBLEM 3 Determine the radian measure of the smallest positive angle that is coterminal with $\dfrac{9\pi}{4}$. ♦

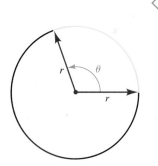

FIGURE 7.15
A circle of radius r with central angle θ that intercepts an arc of length s

◆ Application: Arc Length

Shown in Figure 7.15 is a circle of radius r with *central angle* θ that intercepts an arc of length s. For a central angle $\theta = 2\pi$, the arc length s is the circumference of the entire circle. Hence $s = 2\pi r$ (recall that circumference $= 2\pi r$). For a central angle of $\theta = \pi$, the arc length s is one-half the circumference of the circle. Hence $s = \pi r$. For a central angle of $\theta = \pi/2$, the arc length s is one-quarter the circumference of the circle. Hence $s = \pi r/2$. In each case, the ratio of the central angle θ to the arc length s is always $1/r$, that is,

$$\frac{\theta}{s} = \frac{1}{r}$$

$$s = \theta r \qquad \text{Solve for } s$$

◆ Arc Length Formula

If a central angle θ (in radians) intercepts an arc of length s on a circle with radius r, then

$$s = \theta r.$$

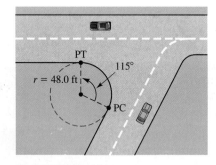

FIGURE 7.16
Intersecting highways

EXAMPLE 4 For the intersecting highways shown in Figure 7.16, determine the length of the circular arc from the point of tangency, PT, to the point of curvature, PC.

◆ **Solution** Before we apply the arc length formula, we first convert 115° to radians, as follows:

$$115° = 115° \left(\frac{\pi}{180°}\right) = \frac{23\pi}{36}$$

Thus, $\quad s = \theta r = \dfrac{23\pi}{36} \cdot 48.0 \text{ ft} \approx 96.3 \text{ ft}$

Hence, the arc length from PT to PC is approximately 96.3 ft. ◆

PROBLEM 4 Determine the length of the arc, as described in Example 4, if the central angle measures 125°. ◆

◆ Application: Angular Speed

Linear speed v is defined as the distance s traveled per unit of time t, that is,

$$v = \frac{s}{t}$$

SECTION 7.1 ♦ Angles and Their Measures

For example, if a boy walks a distance of 6 miles in 2 hours, then his linear speed is

$$v = \frac{s}{t} = \frac{6 \text{ mi}}{2 \text{ h}} = 3 \text{ mi/h}$$

Angular speed ω (denoted by the Greek letter *omega*) is defined as the amount of rotation θ per unit of time t, that is,

$$\omega = \frac{\theta}{t}$$

For example, if a wheel rotates 6 rad in 2 seconds, then its angular speed is

$$\omega = \frac{\theta}{t} = \frac{6 \text{ rad}}{2s} = 3 \text{ rad/s}$$

To develop a relationship between linear speed and angular speed, we proceed as follows:

$$s = \theta r$$
$$\frac{s}{t} = \frac{\theta}{t} r \quad \text{Divide both sides by } t$$
$$v = \omega r \quad \text{Substitute}$$

◆ Relationship between Linear Speed and Angular Speed

If v is the linear speed of an object, ω its angular speed in radians per unit of time, and r the radius of rotation, then

$$v = \omega r.$$

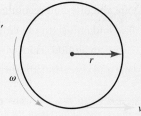

EXAMPLE 5 Each tire on an automobile has a radius of 1.25 ft. How many revolutions per minute (rpm) does a tire make when the automobile is traveling at a speed of 88.0 ft/s?

◆ Solution We first find the angular speed ω of a tire:

$$v = \omega r$$
$$88.0 \text{ ft/s} = \omega(1.25 \text{ ft})$$
$$\omega = 70.4 \text{ rad/s} \quad \text{Divide both sides by 1.25 ft}$$

Therefore, the angle through which a tire rotates in *one minute* is

$$\theta = \omega t = (70.4 \text{ rad/s})(60 \text{ s}) = 4224 \text{ rad}$$

Since every 2π rad equals one revolution (rev), we have

$$\theta = 4224 \text{ rad} \left(\frac{1 \text{ rev}}{2\pi \text{ rad}} \right) \approx 672 \text{ rev}$$

Thus, the tire is rotating at 672 rpm when the automobile travels at 88 ft/s. ◆

PROBLEM 5 Determine the rpm's of the tire described in Example 5 if the automobile is traveling at a speed of 45 mi/h. ◆

Exercises 7.1

Basic Skills

In Exercises 1–8, draw each angle in standard position, and give its degree measure and radian measure.

1. One-eighth of a revolution counterclockwise
2. Two-thirds of a revolution counterclockwise
3. Five-sixths of a revolution clockwise
4. One-fourth of a revolution clockwise
5. One and one-half revolutions counterclockwise
6. Two and one-sixth revolutions counterclockwise
7. Three revolutions clockwise
8. One and three-fourths revolutions clockwise

In Exercises 9–16, find the degree measure of the smallest positive angle that is coterminal with the given angular measure.

9. 580° 10. 695° 11. −1824° 12. 1200°
13. 1000° 14. −906° 15. 414°38′ 16. 628°22′15″

In Exercises 17–24, find the radian measure of the smallest positive angle that is coterminal with the given angular measure.

17. 7π 18. $\dfrac{5\pi}{2}$ 19. $-\dfrac{15\pi}{4}$ 20. $\dfrac{24\pi}{5}$
21. $\dfrac{13\pi}{3}$ 22. $\dfrac{11\pi}{4}$ 23. 8   24. −12

In Exercises 25–32, convert the degree measure to radian measure. Write the answer as a reduced fraction in terms of π.

25. 60°   26. 45°   27. 225°   28. −150°
29. 432°   30. 520°   31. −375°   32. 774°

In Exercises 33–40, convert each radian measure to degree measure.

33. $\dfrac{5\pi}{4}$ 34. $\dfrac{7\pi}{6}$ 35. $-\dfrac{\pi}{18}$ 36. $\dfrac{5\pi}{3}$
37. $\dfrac{13\pi}{8}$ 38. $-\dfrac{22\pi}{5}$ 39. 12π 40. $\dfrac{25\pi}{12}$

In Exercises 41–44, use the angle feature of a graphing calculator to convert each angular measure to (a) decimal degrees and (b) radians (to the nearest hundredth).

41. 22°36′   42. 178°54′
43. 215°28′48″   44. −6°2′49.2″

In Exercises 45–48, use the angle feature of a graphing calculator to convert each degree measure to (a) degrees, minutes, seconds and (b) radians (to the nearest hundredth).

45. 32.31°   46. −112.325°
47. 306.1225°   48. 258.0008°

In Exercises 49–52, use the angle feature of a graphing calculator to convert each radian measure to (a) decimal degrees (to the nearest ten-thousandth) and (b) degrees, minutes, and (nearest) seconds.

49. 1.5

50. −2.36

51. 10

52. 1.571

53. The radius of a circle is 12 cm. Find the length of an arc (in cm) intercepted on its circumference by the given central angle.

   (a) $\dfrac{\pi}{2}$ (b) $\dfrac{7\pi}{6}$ (c) 60°   (d) 135°

54. Find the diameter of a circle if the length of an arc intercepted on its circumference is 21π inches when the central angle is θ.

   (a) $\theta = \dfrac{\pi}{2}$ (b) $\theta = \dfrac{7\pi}{6}$

   (c) $\theta = 60°$ (d) $\theta = 135°$

55. On a carousel, the horses are placed 15 ft from the center pole. How many feet does a passenger travel when riding a horse if the carousel makes 20 revolutions before stopping?

56. Winnipeg, Canada, is approximately due north of Austin, Texas. Austin is at latitude 30° North and Winnipeg is at latitude 50° North, as shown in the sketch. If the radius of the earth is approximately 4000 miles, what is the distance between the two cities?

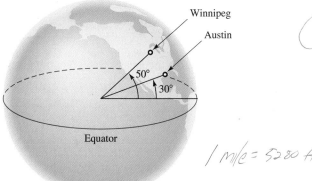

57. A freight train rounds a curve of radius 1320 ft at the speed of 30 mi/h. Through how many radians does a point on the train turn in 30 seconds?

58. A rotary-lawn-mower blade turns at 1800 revolutions per minute (rpm). What is the linear speed (in ft/s) of the tip of the blade if the diameter of the blade is 18 inches?

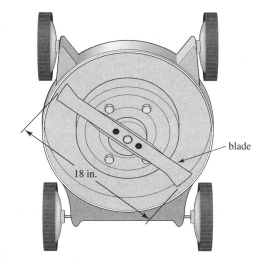

59. What is the angular speed (in rad/s) of each part of the clock?

   (a) the second hand   (b) the minute hand

   (c) the hour hand

60. A space shuttle orbits the moon once every 130 minutes and maintains a constant altitude of 70 miles above the surface of the moon. If the radius of the moon is approximately 1100 miles, what is the linear speed (in mi/h) of the satellite?

61. A stereo record has a diameter of 12 inches and rotates on a turntable at $33\tfrac{1}{3}$ revolutions per minute (rpm).

   (a) Find the angular speed of the record (in rad/min).

   (b) Find the linear speed of the record (in ft/s).

62. A gear in a piece of machinery makes 12 revolutions per minute (rpm). How many seconds does it take for the gear to turn through 144°?

63. A pendulum swinging through a central angle of 6°45′ sweeps an arc of length 10.0 cm. What is the length of the pendulum?

64. For the intersecting highways shown in the sketch at the top of the next page, find the central angle θ in degrees, minutes, seconds if the arc length from the point of tangency PT to the point of curvature PC is 162.35 ft.

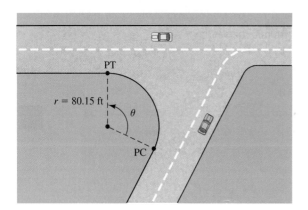

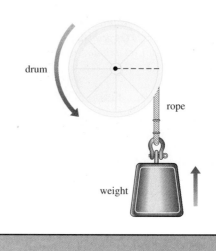

65. As the drum shown in the figure rotates counterclockwise, the rope is wound around the drum and the weight moves upward. After 1.5 seconds, how high is the weight above the ground if the angular speed of the drum is 2.25 rad/s and the radius of the drum is 25.6 cm?

66. Find the radius of the drum described in Exercise 65 if the weight is raised 30.2 cm when the drum rotates 85°.

Critical Thinking

67. Which is larger, an angle that measures 1 rad or an angle that measures 1°? Explain.

68. The wheels on a bicycle are turning at the rate of 30 revolutions per minute (rpm). Assuming the radius of each wheel is 1 foot, find the time necessary for the cyclist to travel a distance of $600\pi \approx 1885$ ft?

69. A particle moves along a circular path given by the equation $x^2 + y^2 = 144$, where x and y are measured in centimeters. How far has the particle moved after sweeping through a central angle of 150°?

70. A belt connects two pulleys with radii 8 cm and 12 cm, as shown in the sketch. If the larger pulley turns through an angle of 10 rad, find the angle through which the smaller pulley turns.

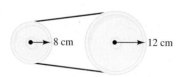

71. A *sector* of a circle is the shaded region bounded by the arc PQ and the two radii $\overline{OP}$ and $\overline{OQ}$, as shown in the figure.

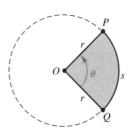

(a) Using the fact that the area A of a sector is directly proportional to the measure of its central angle θ (in radians), show that

$$A = \tfrac{1}{2} \theta r^2.$$

(b) Show that if s is the length of the arc from P to Q, then the area A of the sector is given by

$$A = \tfrac{1}{2} sr.$$

(c) Use the formulas in parts (a) and (b) to find the area of each sector that follows:

SECTION 7.2 ♦ *Defining the Trigonometric Functions* 407

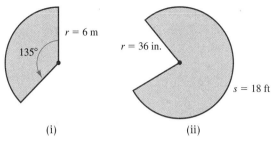

(i)   (ii)

72. Referring to Exercises 71, suppose the perimeter of a sector is 12 cm.

(a) Express the area A of the sector as a function of r.
(b) Generate the graph of the function defined in part (a) in the viewing rectangle of a graphing calculator.
(c) Find the value of the radius r that gives the maximum area of the sector. What is the maximum area?
(d) What is the measure of the central angle θ when the area of the sector is the largest possible?

7.2 Defining the Trigonometric Functions

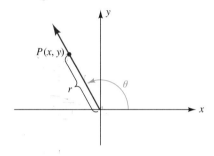

FIGURE 7.17
The values of x, y, and r determine the six trigonometric ratios for angle θ.

♦ Introductory Comments

Consider an angle θ in standard position with $P(x, y)$ a point other than $(0, 0)$ on the terminal side of θ, as shown in Figure 7.17. By the distance formula (Section 2.1), the distance r from the origin to the point $P(x, y)$ is always positive and is given by

$$r = \sqrt{x^2 + y^2}.$$

From the values of x, y, and r, we can form six ratios for angle θ. These six ratios are called the *trigonometric ratios* and are defined as *sine* (sin), *cosine* (cos), *tangent* (tan), *cosecant* (csc), *secant* (sec), and *cotangent* (cot).

◀ Trigonometric Ratios

If θ is an angle in standard position with $P(x, y)$ a point other than $(0, 0)$ on the terminal side of θ, then the six **trigonometric ratios** of angle θ are defined as follows:

1. $\sin \theta = \dfrac{y}{r}$ 4. $\csc \theta = \dfrac{r}{y}$ for $y \neq 0$

2. $\cos \theta = \dfrac{x}{r}$ 5. $\sec \theta = \dfrac{r}{x}$ for $x \neq 0$

3. $\tan \theta = \dfrac{y}{x}$ for $x \neq 0$ 6. $\cot \theta = \dfrac{x}{y}$ for $y \neq 0$

The value of each trigonometric ratio is determined by the angle θ, not by the particular point $P(x, y)$ that we choose on the terminal side of θ. To show this, consider an angle θ in standard position with points $P(x, y)$ and

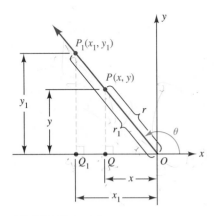

FIGURE 7.18
When the points $P(x, y)$ and $P_1(x_1, y_1)$ are on the terminal side of angle θ, the triangles OPQ and OP_1Q_1 are similar.

$P_1(x_1, y_1)$ on its terminal side, as shown in Figure 7.18. Since the triangles OPQ and OP_1Q_1 are similar, it follows from elementary geometry that their corresponding sides are proportional. Hence,

$$\sin \theta = \frac{y}{r} = \frac{y_1}{r_1}, \quad \cos \theta = \frac{x}{r} = \frac{x_1}{r_1}, \quad \tan \theta = \frac{y}{x} = \frac{y_1}{x_1}, \quad \text{and so on.}$$

Since the value of each ratio depends only on angle θ, we can say that the ratios are *functions* of angle θ. In turn, we call these ratios the **trigonometric functions** of angle θ. In this section, we work with the trigonometric functions of angles and discuss some of the fundamental relationships between these functions.

◆ Angle Domains of the Trigonometric Functions

The *domain* of each trigonometric function is the set of all angles θ such that the denominator of its corresponding ratio is not zero. If the terminal side of an angle θ is along the x-axis and $P(x, y)$ is a point on its terminal side, then $y = 0$ and, therefore,

$$\csc \theta = \frac{r}{y} \quad \text{and} \quad \cot \theta = \frac{x}{y} \quad \text{are undefined.}$$

If the terminal side of θ is along the y-axis and $P(x, y)$ is a point on its terminal side, then $x = 0$ and, therefore,

$$\tan \theta = \frac{y}{x} \quad \text{and} \quad \sec \theta = \frac{r}{x} \quad \text{are undefined.}$$

However, since r is always positive,

$$\sin \theta = \frac{y}{r} \quad \text{and} \quad \cos \theta = \frac{x}{r} \quad \text{are defined for all angles } \theta.$$

Table 7.1 summarizes our results.

TABLE 7.1
Angle domains of the trigonometric functions

Trigonometric function	Domain
sine or cosine	All angles
cosecant or cotangent	All angles except those whose terminal side is on the x-axis
tangent or secant	All angles except those whose terminal side is on the y-axis

Our definitions of the trigonometric functions as ratios allow us to find the following products:

SECTION 7.2 ♦ Defining the Trigonometric Functions

$$\sin \theta \csc \theta = \left(\frac{y}{r}\right)\left(\frac{r}{y}\right) = 1 \quad \text{for } y \neq 0$$

$$\cos \theta \sec \theta = \left(\frac{x}{r}\right)\left(\frac{r}{x}\right) = 1 \quad \text{for } x \neq 0$$

$$\tan \theta \cot \theta = \left(\frac{y}{x}\right)\left(\frac{x}{y}\right) = 1 \quad \text{for } x \neq 0 \text{ and } y \neq 0$$

When the product of two numbers equals 1, the two numbers are said to be *reciprocals* of each other. Thus, to find the six trigonometric functions of an angle θ whose terminal side passes through a given point, we begin by evaluating $\sin \theta$, $\cos \theta$, and $\tan \theta$. For $\csc \theta$, $\sec \theta$, and $\cot \theta$, we simply use the fact that these are reciprocals of $\sin \theta$, $\cos \theta$ and $\tan \theta$, respectively:

$$\csc \theta = \frac{1}{\sin \theta} \qquad \sec \theta = \frac{1}{\cos \theta} \qquad \cot \theta = \frac{1}{\tan \theta}$$

EXAMPLE 1 Given that $P(-1, 2)$ is a point on the terminal side of an angle θ in standard position, find the values of the six trigonometric functions of θ.

◆ **Solution** Referring to Figure 7.19, we have $x = -1$, $y = 2$, and

$$r = \sqrt{x^2 + y^2} = \sqrt{(-1)^2 + (2)^2} = \sqrt{5}.$$

Thus,

$$\sin \theta = \frac{y}{r} = \frac{2}{\sqrt{5}} \qquad \csc \theta = \frac{1}{\sin \theta} = \frac{\sqrt{5}}{2}$$

$$\cos \theta = \frac{x}{r} = \frac{-1}{\sqrt{5}} = -\frac{1}{\sqrt{5}} \qquad \sec \theta = \frac{1}{\cos \theta} = -\sqrt{5}$$

$$\tan \theta = \frac{y}{x} = \frac{2}{-1} = -2 \qquad \cot \theta = \frac{1}{\tan \theta} = -\frac{1}{2}$$

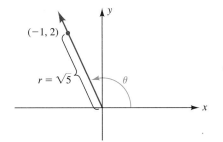

FIGURE 7.19
An angle θ in standard position with terminal side passing through the point $(-1, 2)$

PROBLEM 1 Given that $(-2, 4)$ is a point on the terminal side of an angle θ in standard position, find the values of the six trigonometric functions of θ.

When the terminal side of an angle θ lies in a particular quadrant, we say that angle θ is in that quadrant as well. As illustrated in the next example, if we know the value of one of the trigonometric functions of θ and the quadrant in which θ lies, then we can determine the values of the other five trigonometric functions of θ.

EXAMPLE 2 The value of one trigonometric function is given, along with the quadrant in which θ lies. Use the trigonometric ratios to find the values of the other five trigonometric functions of θ.

(a) $\tan \theta = -\frac{3}{4}$, θ in quadrant II

(b) $\cos \theta = 0.3$, θ in quadrant IV

◆ **Solution**

(a) In quadrant II, we have $x < 0$ and $y > 0$. Thus, for

$$\tan \theta = -\frac{3}{4} = \frac{y}{x},$$

we choose $P(x, y) = P(-4, 3)$ as a point on the terminal side of θ, as shown in Figure 7.20. Now, substituting $x = -4$ and $y = 3$, we find r:

$$r = \sqrt{x^2 + y^2} = \sqrt{(-4)^2 + (3)^2} = \sqrt{25} = 5$$

Using $x = -4$, $y = 3$ and $r = 5$, we find the values of the other five trigonometric functions of θ:

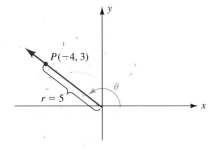

FIGURE 7.20
For $\tan \theta = -\frac{3}{4}$, θ in quadrant II, we have $x = -4$, $y = 3$, and $r = 5$.

$$\sin \theta = \frac{y}{r} = \frac{3}{5} \qquad \csc \theta = \frac{1}{\sin \theta} = \frac{5}{3}$$

$$\cos \theta = \frac{x}{r} = \frac{-4}{5} = -\frac{4}{5} \qquad \sec \theta = \frac{1}{\cos \theta} = -\frac{5}{4}$$

$$\cot \theta = \frac{1}{\tan \theta} = -\frac{4}{3}$$

(b) In quadrant IV, we have $x > 0$ and $y < 0$. Thus, for

$$\cos \theta = 0.3 = \frac{3}{10} = \frac{x}{r}$$

we choose a point on the terminal side of θ such that $x = 3$ and $r = 10$. To find y, we substitute $x = 3$ and $r = 10$ in the equation $r = \sqrt{x^2 + y^2}$:

$10 = \sqrt{(3)^2 + y^2}$ **Substitute for x and r**

$100 = 9 + y^2$ **Square both sides**

$y^2 = 91$ **Subtract 9 from both sides**

$y = \pm \sqrt{91}$ ──choose──→ $y = -\sqrt{91}$

Angle θ in quadrant IV implies $y < 0$.

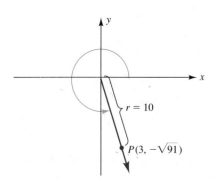

FIGURE 7.21
For $\cos \theta = 0.3$, θ in quadrant IV, we have $x = 3$, $y = -\sqrt{91}$, and $r = 10$.

Using $x = 3$, $y = -\sqrt{91}$, and $r = 10$, as shown in Figure 7.21, we find the values of the other five trigonometric functions of θ:

$$\sin \theta = \frac{y}{r} = \frac{-\sqrt{91}}{10} = -\frac{\sqrt{91}}{10} \qquad \csc \theta = \frac{1}{\sin \theta} = -\frac{10}{\sqrt{91}}$$

$$\tan \theta = \frac{y}{x} = \frac{-\sqrt{91}}{3} = -\frac{\sqrt{91}}{3} \qquad \cot \theta = \frac{1}{\tan \theta} = -\frac{3}{\sqrt{91}}$$

$$\sec \theta = \frac{1}{\cos \theta} = \frac{10}{3} \qquad \blacklozenge$$

As we observe in Example 2, the algebraic sign of a trigonometric function of θ depends entirely on the quadrant in which θ lies. To help remember the quadrants in which $\sin \theta > 0$, $\cos \theta > 0$, and $\tan \theta > 0$, we use the phrase "**A**ll **S**tudents **T**ake **C**ourses", as shown in Figure 7.22. Since the reciprocal of a positive number is positive, we have $\csc \theta > 0$ in quadrants I and II, $\sec \theta > 0$ in quadrants I and IV, and $\cot \theta > 0$ in quadrants I and III.

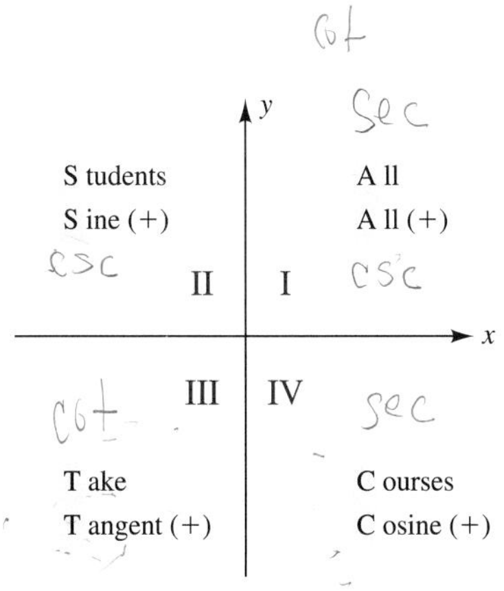

PROBLEM 2 Find the quadrant in which θ lies if $\tan \theta > 0$ and $\cos \theta < 0$. $\blacklozenge$

S tudents S ine (+)	A ll A ll (+)	
II	I	
III	IV	
T ake T angent (+)	C ourses C osine (+)	

FIGURE 7.22
Quadrants in which $\sin \theta$, $\cos \theta$, and $\tan \theta$ are positive

◆ The Fundamental Trigonometric Identities

Several relationships between the trigonometric functions are important for our future work with these functions. We have already discussed the following reciprocal relationships:

$$\sin \theta \csc \theta = 1 \qquad \cos \theta \sec \theta = 1 \qquad \tan \theta \cot \theta = 1$$

Now, consider the quotients $\dfrac{\sin \theta}{\cos \theta}$ and $\dfrac{\cos \theta}{\sin \theta}$. Using the trigonometric ratios, we have

$$\frac{\sin \theta}{\cos \theta} = \frac{y/r}{x/r} = \frac{y}{r} \cdot \frac{r}{x} = \frac{y}{x} = \tan \theta \quad \text{for } x \neq 0$$

and

$$\frac{\cos \theta}{\sin \theta} = \frac{x/r}{y/r} = \frac{x}{r} \cdot \frac{r}{y} = \frac{x}{y} = \cot \theta \quad \text{for } y \neq 0.$$

In summary, we can state two identities involving quotients:

$$\tan \theta = \frac{\sin \theta}{\cos \theta} \qquad \cot \theta = \frac{\cos \theta}{\sin \theta}$$

Next, we consider the expression $(\sin \theta)^2 + (\cos \theta)^2$. Using the trigonometric ratios, we have

$$(\sin \theta)^2 + (\cos \theta)^2 = \left(\frac{y}{r}\right)^2 + \left(\frac{x}{r}\right)^2 = \frac{x^2 + y^2}{r^2}.$$

However, since $r = \sqrt{x^2 + y^2}$, we have $r^2 = x^2 + y^2$. Hence,

$$(\sin \theta)^2 + (\cos \theta)^2 = \frac{x^2 + y^2}{r^2} = \frac{r^2}{r^2} = 1.$$

To avoid using parentheses, we usually write $(\sin \theta)^2$ as $\sin^2\theta$ and $(\cos \theta)^2$ as $\cos^2\theta$. This convention is applied to other positive powers of the trigonometric functions as well. In summary, we state an identity involving powers:

$$\boxed{\sin^2\theta + \cos^2\theta = 1}$$

CAUTION Do not confuse $\sin^2\theta$ with $\sin \theta^2$. Remember,

$$\sin^2\theta = (\sin \theta)(\sin \theta) = (\sin \theta)^2,$$

whereas
$$\sin \theta^2 = \sin(\theta \cdot \theta).$$

In general, $\sin^2\theta \neq \sin \theta^2$.

If we start with the relationship $(\sin \theta)^2 + (\cos \theta)^2 = 1$ and divide both sides by $(\cos \theta)^2$, provided that $\cos \theta \neq 0$, we obtain

$$\frac{(\sin \theta)^2}{(\cos \theta)^2} + \frac{(\cos \theta)^2}{(\cos \theta)^2} = \frac{1}{(\cos \theta)^2}$$

$$\left(\frac{\sin \theta}{\cos \theta}\right)^2 + 1 = \left(\frac{1}{\cos \theta}\right)^2$$

$$(\tan \theta)^2 + 1 = (\sec \theta)^2$$

In a similar manner, starting with the same relationship, $(\sin \theta)^2 + (\cos \theta)^2 = 1$, and dividing both sides by $(\sin \theta)^2$, provided that $\sin \theta \neq 0$, we obtain $1 + (\cot \theta)^2 = (\csc \theta)^2$. In summary, we state two more identities involving powers:

$$\boxed{1 + \tan^2\theta = \sec^2\theta \qquad 1 + \cot^2\theta = \csc^2\theta}$$

These basic relationships between the trigonometric functions, which lay the foundation for more advanced work in trigonometry, are often referred to as the **fundamental trigonometric identities**.

◀ **Fundamental Trigonometric Identities**

For θ in the domains of the trigonometric functions,

1. $(\sin \theta)(\csc \theta) = 1$
2. $(\cos \theta)(\sec \theta) = 1$
3. $(\tan \theta)(\cot \theta) = 1$
4. $\tan \theta = \dfrac{\sin \theta}{\cos \theta}$
5. $\cot \theta = \dfrac{\cos \theta}{\sin \theta}$
6. $\sin^2\theta + \cos^2\theta = 1$
7. $1 + \tan^2\theta = \sec^2\theta$
8. $1 + \cot^2\theta = \csc^2\theta$

EXAMPLE 3 Use the fundamental trigonometric identities to find the value of the indicated trigonometric function from the given information.

(a) Find $\tan \theta$ if $\sin \theta = \frac{12}{13}$ and θ is in quadrant I.

(b) Find $\sin \theta$ if $\cot \theta = \sqrt{3}$ and $\sin \theta < 0$.

◆ Solution

(a) From the trigonometric identity $\sin^2\theta + \cos^2\theta = 1$, we have

$$\cos^2\theta = 1 - \sin^2\theta$$
$$\cos \theta = \pm\sqrt{1 - \sin^2\theta}$$

However, since we are given that θ is in quadrant I, we know that all trigonometric functions of θ must be positive. Thus,

$$\cos \theta = \sqrt{1 - \sin^2\theta} = \sqrt{1 - \left(\tfrac{12}{13}\right)^2} = \sqrt{\tfrac{25}{169}} = \tfrac{5}{13}.$$

Now, using the trigonometric identity $\tan \theta = \sin \theta / \cos \theta$, we have

$$\tan \theta = \frac{\tfrac{12}{13}}{\tfrac{5}{13}} = \frac{12}{5}.$$

(b) For $\cot \theta = \sqrt{3} > 0$, we know that θ is in either the first or third quadrant. We know that in quadrant I, $\sin \theta > 0$ and in quadrant III, $\sin \theta < 0$. Since we want $\sin \theta < 0$, we conclude that θ is in quadrant III. Now, from the trigonometric identity $1 + \cot^2\theta = \csc^2\theta$, we obtain

$$\csc \theta = \pm\sqrt{1 + \cot^2\theta}$$

However, since θ is in quadrant III, we know that $\csc \theta < 0$. Thus,

$$\csc \theta = -\sqrt{1 + \cot^2\theta} = -\sqrt{1 + (\sqrt{3})^2} = -\sqrt{4} = -2.$$

From the identity $\sin \theta \csc \theta = 1$, we find

$$\sin \theta = \frac{1}{\csc \theta} = -\frac{1}{2}. \qquad ◆$$

PROBLEM 3 Given that $\tan \theta = -1$ and $\cos \theta < 0$, use the fundamental trigonometric identities to find $\cos \theta$. ◆

◆ Trigonometric Ratios for Right Triangles

Recall from elementary geometry that a **right triangle** is a triangle in which one of the interior angles is a *right angle.* Since the sum of the interior angles in a right triangle is 180°, and the right angle measures 90°, it follows that the other two angles must be positive angles that each measure less than 90°. We refer to a positive angle with measure less than 90° as an *acute angle.* Figure 7.23 shows a right triangle with acute angles A and B and right angle

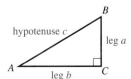

FIGURE 7.23

The sides of a right triangle are related by the Pythagorean theorem, $a^2 + b^2 = c^2$.

C. By convention, we call the side opposite the right angle the *hypotenuse* and assign its length the lowercase letter c. The sides opposite the angles A and B are called *legs*, and their lengths are denoted by lowercase letters a and b, respectively. The lengths of the sides of a right triangle are related by the **Pythagorean theorem**:

$$a^2 + b^2 = c^2$$

For the right triangle in Figure 7.23, we can place the vertex of angle A at the origin of a Cartesian coordinate plane and side b along the positive x-axis, as shown in Figure 7.24. We can then express the trigonometric functions of the acute angle A as ratios of the sides of the right triangle:

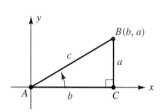

FIGURE 7.24
The right triangle in Figure 7.23 with acute angle A in standard position

$$\sin A = \frac{a}{c} = \frac{\text{side opposite } A}{\text{hypotenuse}} \qquad \csc A = \frac{c}{a} = \frac{\text{hypotenuse}}{\text{side opposite } A}$$

$$\cos A = \frac{b}{c} = \frac{\text{side adjacent to } A}{\text{hypotenuse}} \qquad \sec A = \frac{c}{b} = \frac{\text{hypotenuse}}{\text{side adjacent to } A}$$

$$\tan A = \frac{a}{b} = \frac{\text{side opposite } A}{\text{side adjacent to } A} \qquad \cot A = \frac{b}{a} = \frac{\text{side adjacent to } A}{\text{side opposite } A}$$

Similarly, by placing the vertex of angle B at the origin and side a along the positive x-axis, as shown in Figure 7.25, we can express the trigonometric functions of the acute angle B as the ratios of the sides of the right triangle:

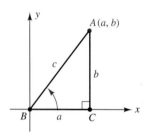

FIGURE 7.25
The right triangle in Figure 7.23 with acute angle B in standard position

$$\sin B = \frac{b}{c} = \frac{\text{side opposite } B}{\text{hypotenuse}} \qquad \csc B = \frac{c}{b} = \frac{\text{hypotenuse}}{\text{side opposite } B}$$

$$\cos B = \frac{a}{c} = \frac{\text{side adjacent to } B}{\text{hypotenuse}} \qquad \sec B = \frac{c}{a} = \frac{\text{hypotenuse}}{\text{side adjacent to } B}$$

$$\tan B = \frac{b}{a} = \frac{\text{side opposite } B}{\text{side adjacent to } B} \qquad \cot B = \frac{a}{b} = \frac{\text{side adjacent to } B}{\text{side opposite } B}$$

In summary, we state the **trigonometric ratios for right triangles**.

◀ **Trigonometric Ratios for Right Triangles**

If θ is an acute angle in a right triangle, then

$$\sin \theta = \frac{\text{opp}}{\text{hyp}} \qquad \cos \theta = \frac{\text{adj}}{\text{hyp}} \qquad \tan \theta = \frac{\text{opp}}{\text{adj}}$$

$$\csc \theta = \frac{\text{hyp}}{\text{opp}} \qquad \sec \theta = \frac{\text{hyp}}{\text{adj}} \qquad \cot \theta = \frac{\text{adj}}{\text{opp}}$$

where *opp* is the length of the side opposite θ, *adj* is the length of the side adjacent to θ, and *hyp* is the length of the hypotenuse of the right triangle.

SECTION 7.2 ♦ Defining the Trigonometric Functions

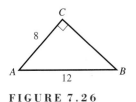

FIGURE 7.26

EXAMPLE 4 Given the right triangle in Figure 7.26, find the value of each trigonometric function.

(a) sin A **(b)** cos A **(c)** tan A

♦ **Solution** First, we apply the Pythagorean theorem to find the length of the leg opposite angle A. Letting a be the length of this leg, we find

$$a^2 = 12^2 - 8^2 = 80,$$

which implies

$$a = \sqrt{80} = 4\sqrt{5}.$$

Now, using the trigonometric ratios for right triangles, we obtain the values of the indicated trigonometric functions.

(a) $\sin A = \dfrac{\text{opp}}{\text{hyp}} = \dfrac{4\sqrt{5}}{12} = \dfrac{\sqrt{5}}{3}$ **(b)** $\cos A = \dfrac{\text{adj}}{\text{hyp}} = \dfrac{8}{12} = \dfrac{2}{3}$

(c) $\tan A = \dfrac{\text{opp}}{\text{adj}} = \dfrac{4\sqrt{5}}{8} = \dfrac{\sqrt{5}}{2}$ ♦

PROBLEM 4 Find the value of each trigonometric function for the right triangle in Figure 7.26.

(a) sin B **(b)** cos B **(c)** tan B ♦

Since the sum of the interior angles in a right triangle is 180° and the right angle measures 90°, it follows that the sum of the two acute angles is 180° − 90° = 90°. Any two acute angles whose sum is 90° are **complementary angles**. Now, referring to Figures 7.24 and 7.25, we observe the following facts concerning the trigonometric functions of complementary angles:

$\cos A = \dfrac{b}{c} = \sin B,$ $\cos B = \dfrac{a}{c} = \sin A$ ← The cosine of an acute angle is the sine of its complementary angle.

$\cot A = \dfrac{b}{a} = \tan B,$ $\cot B = \dfrac{a}{b} = \tan A$ ← The cotangent of an acute angle is the tangent of its complementary angle.

$\csc A = \dfrac{c}{a} = \sec B,$ $\csc B = \dfrac{c}{b} = \sec A$ ← The cosecant of an acute angle is the secant of its complementary angle.

Because of these relationships, we say that the *sine* and *cosine* are *cofunctions*, the *tangent* and *cotangent* are *cofunctions*, and the *secant* and *cosecant* are *cofunctions*. A trigonometric function of any acute angle θ is the same as the cofunction of its complementary angle $90° - \theta$, that is,

$$\sin \theta = \cos(90° - \theta) \qquad \cos \theta = \sin(90° - \theta)$$
$$\tan \theta = \cot(90° - \theta) \qquad \cot \theta = \tan(90° - \theta)$$
$$\sec \theta = \csc(90° - \theta) \qquad \csc \theta = \sec(90° - \theta)$$

EXAMPLE 5 Express each trigonometric function as the cofunction of its complementary angle.

(a) $\cos 42°$ (b) $\tan \dfrac{\pi}{8}$

◆ **Solution**

(a) $\cos 42° = \sin(90° - 42°) = \sin 48°$

(b) $\tan \dfrac{\pi}{8} = \cot\left(\dfrac{\pi}{2} - \dfrac{\pi}{8}\right) = \cot \dfrac{3\pi}{8}$

PROBLEM 5 Express $\sec 12°32'$ as the cofunction of its complementary angle.

Exercises 7.2

Basic Skills

In Exercises 1–6, a point P on the terminal side of θ is shown in the figure. Evaluate (if possible) each of the six trigonometric functions of θ.

In Exercises 7–14, find the quadrant in which θ lies if the given conditions are satisfied.

7. $\tan \theta > 0$ and $\sin \theta < 0$
8. $\sin \theta > 0$ and $\tan \theta < 0$
9. $\sec \theta < 0$ and $\tan \theta < 0$
10. $\cos \theta < 0$ and $\cot \theta > 0$
11. $\cot \theta > 0$ and $\csc \theta > 0$
12. $\sec \theta > 0$ and $\sin \theta < 0$
13. $\sin \theta < 0$ and $\cos \theta > 0$
14. $\cos \theta < 0$ and $\csc \theta < 0$

1.

2.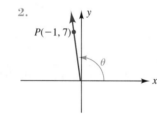

3.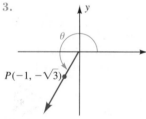

4.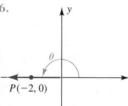

5.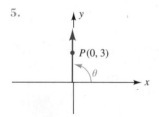

6.

In Exercises 15–24, the value of one of the trigonometric functions is given with a condition. Use this information to find the values of the other five trigonometric functions of θ.

15. $\sin \theta = \tfrac{3}{5}$, θ is in quadrant I
16. $\cos \theta = \tfrac{5}{13}$, θ is in quadrant I
17. $\tan \theta = -\tfrac{2}{3}$, θ is in quadrant II
18. $\sec \theta = -\tfrac{3}{2}$, θ is in quadrant III
19. $\cos \theta = \dfrac{\sqrt{3}}{2}$, $\sin \theta < 0$
20. $\sin \theta = \tfrac{1}{2}$, $\tan \theta < 0$
21. $\cot \theta = -1$, $\cos \theta > 0$
22. $\csc \theta = 2$, $\tan \theta > 0$
23. $\sec \theta = -1.2$, $\tan \theta > 0$
24. $\tan \theta = -0.8$, $\cos \theta > 0$

In Exercises 25–38, use the fundamental trigonometric identities to find the value of the indicated trigonometric function from the given information.

25. $\csc\theta = 3$;   find $\sin\theta$
26. $\sec\theta = -\frac{5}{2}$;   find $\cos\theta$
27. $\tan\theta = -0.2$;   find $\cot\theta$
28. $\sin\theta = -0.25$;   find $\csc\theta$
29. $\sin\theta = \dfrac{3}{\sqrt{10}}$, $\cos\theta = \dfrac{1}{\sqrt{10}}$;   find $\tan\theta$
30. $\sin\theta = \frac{4}{5}$, $\cot\theta = \frac{3}{4}$;   find $\cos\theta$
31. $\sec\theta = \frac{13}{12}$, $\tan\theta = \frac{5}{12}$;   find $\sin\theta$
32. $\sin\theta = \dfrac{3}{\sqrt{13}}$, $\tan\theta = \frac{3}{2}$;   find $\sec\theta$
33. $\cos\theta = \frac{3}{5}$, θ is in quadrant I;   find $\sin\theta$
34. $\sin\theta = -\frac{1}{2}$, θ is in quadrant III;   find $\cos\theta$
35. $\tan\theta = -\sqrt{2}$, $\sin\theta > 0$;   find $\sec\theta$
36. $\sec\theta = 3$, $\tan\theta < 0$;   find $\sin\theta$
37. $\csc\theta = -\sqrt{3}$, $\cot\theta > 0$;   find $\tan\theta$
38. $\tan\theta = -1$, $\sec\theta < 0$;   find $\sin\theta$

In Exercises 39 and 40, use the right triangle shown in the sketch to find the values of the indicated trigonometric function.

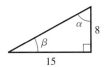

39. (a) $\sin\alpha$ (b) $\cos\alpha$ (c) $\cot\beta$
40. (a) $\tan\alpha$ (b) $\csc\beta$ (c) $\sec\beta$

In Exercises 41 and 42, use the right triangle shown in the figure to find the values of the indicated trigonometric function.

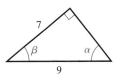

41. (a) $\tan\beta$ (b) $\csc\alpha$ (c) $\sec\alpha$
42. (a) $\sin\beta$ (b) $\cos\beta$ (c) $\cot\alpha$

In Exercises 43 and 44, use the right triangle shown in the sketch to find the values of the indicated trigonometric function.

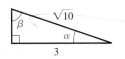

43. (a) $\sin\beta$ (b) $\cos\alpha$ (c) $\tan\beta$
44. (a) $\cot\alpha$ (b) $\sec\beta$ (c) $\csc\alpha$

In Exercises 45–50, express the trigonometric function as the cofunction of its complementary angle.

45. $\sin 30°$
46. $\cos 63°$
47. $\tan\dfrac{\pi}{4}$
48. $\sec\dfrac{5\pi}{12}$
49. $\cot 34°43'$
50. $\csc 56°17'45''$

Critical Thinking

In Exercises 51–54, sketch and label the sides of a right triangle corresponding to the given trigonometric function of the acute angle θ. (Note: Answers are not unique.)

51. $\sin\theta = \frac{1}{2}$
52. $\cos\theta = \frac{3}{4}$
53. $\tan\theta = 2$
54. $\csc\theta = 3$

55. For a given angle θ, is it possible to have $\sin\theta > 0$ and $\csc\theta < 0$? Explain.

56. For a given angle θ, is it possible to have $\tan\theta < 0$ and $\cot\theta > 0$? Explain.

57. Use the fundamental trigonometric identities to express $\cos\theta$, $\tan\theta$, $\csc\theta$, $\sec\theta$, and $\cot\theta$ in terms of $\sin\theta$ only.

58. In each of the eight fundamental trigonometric identities, replace each function with its cofunction. Are the resulting equations identities? Explain.

59. Use the properties of logarithms (Section 5.3) to rewrite each expression as a single logarithmic expression. Then apply the fundamental trigonometric identities to simplify your answer. Assume that θ is in quadrant I.
    (a) $\log(\sin\theta) + \log(\csc\theta)$
    (b) $\ln(\cos\theta) - \ln(\sin\theta)$
    (c) $2\ln(\sin\theta) - \ln(\tan\theta) - \ln(\cos\theta)$
    (d) $\frac{1}{2}\log(1 - \sin^2\theta) + \log(\sec\theta)$

60. The terminal side of an angle θ in standard position passes through the intersection point of the given curves. Find the trigonometric functions of θ, if they exist.
    (a) $2x - y = -10$ and $3x + y = -5$
    (b) $y = x^2 - 4x$ and $y = 4x - 16$

7.3 Evaluating the Trigonometric Functions

◆ Introductory Comments

To define the trigonometric functions of a real number t, we may imagine a real number line, called the t-axis, wrapped continuously around a **unit circle**—a circle whose equation is $x^2 + y^2 = 1$. The t-axis is tangent to the unit circle at the point $(1, 0)$ and has the same unit scale as the x-axis and y-axis, as shown in Figure 7.27. By placing the origin of the t-axis at the point $(1, 0)$ on the circle, we are able to wrap the positive end of the line around the circle in a counterclockwise direction and wrap the negative end of the line around the circle in a clockwise direction, with an angle of 0 rad corresponding to the number 0 on the t-axis

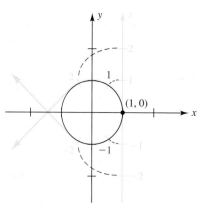

FIGURE 7.27

A real number line being wrapped around the unit circle, $x^2 + y^2 = 1$

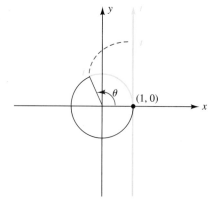

FIGURE 7.28

For each real number t, there corresponds a central angle θ.

Now for each real number t on the t-axis, there corresponds a central angle θ on the unit circle, as shown in Figure 7.28. Recall from Section 7.1 that if a central angle θ (in radians) intercepts an arc of length t on a circle with radius r, then

$$t = \theta r.$$

However, since the unit circle has radius 1 unit, we have

$$t = \theta.$$

Hence, we may find the values of the trigonometric functions for a real number t by considering the real number as the radian measure of the central angle θ.

◆ Trigonometric Functions of Real Numbers

If t is a real number, then

1. $\sin t = \sin \theta$ 4. $\csc t = \csc \theta$
2. $\cos t = \cos \theta$ 5. $\sec t = \sec \theta$
3. $\tan t = \tan \theta$ 6. $\cot t = \cot \theta$

where θ is an angle whose measure is t radians.

Note: As we stated in Section 7.1, if no unit of measure is specified with an angle, then the angle is assumed to be in radians. Thus, a trigonometric expression such as sin 2 means either

1. the sine of the real number 2, or
2. the sine of an angle with radian measure 2.

The context in which sin 2 appears determines which of these meanings is intended, but both meanings have the same numerical value.

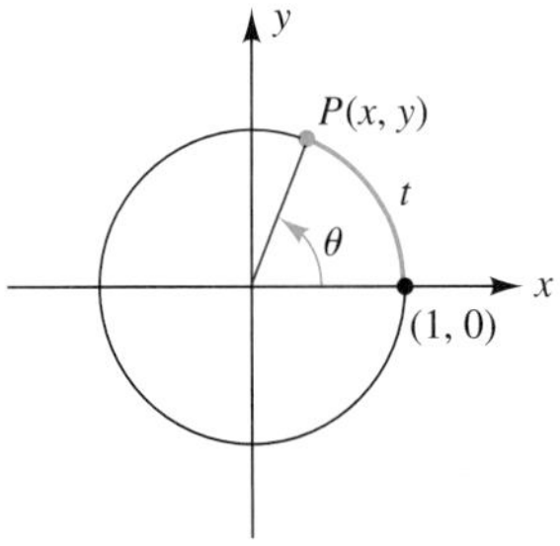

Suppose the arc length from $(1, 0)$ to a point $P(x, y)$ on the unit circle is t units, as shown in Figure 7.29. On the unit circle we have $r = 1$. Thus, it follows that angle θ in the figure has radian measure t, and the sine and cosine of the real number t can be defined as follows:

$$\sin t = \sin \theta = \frac{y}{r} = \frac{y}{1} = y \quad \text{and} \quad \cos t = \cos \theta = \frac{x}{r} = \frac{x}{1} = x$$

Hence, the coordinates of point P on the unit circle are

$$(x, y) = (\cos t, \sin t) = (\cos \theta, \sin \theta)$$

as illustrated in Figure 7.30. In this section, we use this fact to help evaluate the trigonometric functions of some special angles and real numbers.

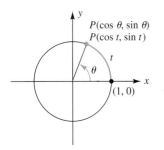

FIGURE 7.29
A unit circle whose arc length from $(1, 0)$ to $P(x, y)$ is t units

FIGURE 7.30
Renaming the coordinates of point P in Figure 7.29

◆ Quadrantal Numbers

Any value of t or θ, such as 0, $\pi/2$, π, or $3\pi/2$, that corresponds to a point of intersection of the unit circle with the x- or y-axis is called a **quadrantal number**. We can evaluate the trigonometric functions of a quadrantal number by using the coordinates of its corresponding point P on the unit circle.

EXAMPLE 1 Evaluate each trigonometric function, if it is defined.

(a) $\cos \pi$ (b) $\sin 90°$ (c) $\sec 0°$ (d) $\tan \dfrac{3\pi}{2}$

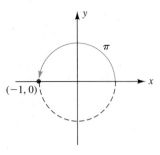

FIGURE 7.31
The point on the unit circle that corresponds to an arc length π is $(-1, 0)$.

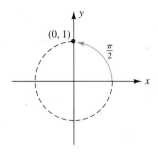

FIGURE 7.32
The point on the unit circle that corresponds to an arc length $\pi/2$ is $(0, 1)$.

◆ **Solution**

(a) The point on the unit circle that corresponds to a central angle π, or an arc length π, is $(-1, 0)$, as shown in Figure 7.31. The x-coordinate of this point is the value of $\cos \pi$. Hence,

$$\cos \pi = -1.$$

(b) The point on the unit circle that corresponds to a central angle 90° ($\pi/2$ rad), or an arc length $\pi/2$, is $(0, 1)$, as shown in Figure 7.32. The y-coordinate of this point is the value of $\sin 90°$. Hence,

$$\sin 90° = 1.$$

(c) The point on the unit circle that corresponds to a central angle 0° (0 rad), or an arc length 0, is $(1, 0)$, as shown in Figure 7.33. The x-coordinate of this point is the value of $\cos 0°$. Hence, $\cos 0° = 1$. Now, by one of our fundamental trigonometric identities (Section 7.2), we have

$$\sec 0° = \frac{1}{\cos 0°} = \frac{1}{1} = 1.$$

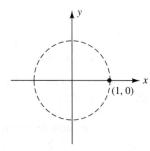

FIGURE 7.33
The point on the unit circle that corresponds to an arc length 0 is $(1, 0)$.

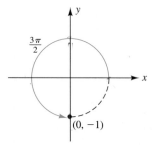

FIGURE 7.34
The point on the unit circle that corresponds to an arc length $3\pi/2$ is $(0, -1)$.

(d) The point on the unit circle that corresponds to a central angle $3\pi/2$, or an arc length $3\pi/2$, is $(0, -1)$, as shown in Figure 7.34. The x-coordinate of this point is the value of $\cos(3\pi/2)$, and the y-coordinate of this point is the value of $\sin(3\pi/2)$. By one of our fundamental trigonometric identities (Section 7.2), we have

$$\tan \frac{3\pi}{2} = \frac{\sin(3\pi/2)}{\cos(3\pi/2)}, \quad \text{which is undefined}$$

because division by zero is not allowed. ◆

Table 7.2 gives the values of the trigonometric functions for the quadrantal numbers, 0, $\pi/2$, π, and $3\pi/2$. By using the procedure shown in Example 1, we can verify each entry.

TABLE 7.2
Values of the trigonometric functions for the quadrantal numbers 0, $\pi/2$, π, and $3\pi/2$

θ	$\sin\theta$	$\cos\theta$	$\tan\theta$	$\csc\theta$	$\sec\theta$	$\cot\theta$
0° or 0	0	1	0	undefined	1	undefined
90° or $\pi/2$	1	0	undefined	1	undefined	0
180° or π	0	−1	0	undefined	−1	undefined
270° or $3\pi/2$	−1	0	undefined	−1	undefined	0

We can obtain values of the trigonometric functions for other quadrantal numbers from Table 7.2 by using the idea of coterminal angles (Section 7.1). For example, since 2π is coterminal with 0, we have

$$\sin 2\pi = \sin 0 = 0, \quad \cos 2\pi = \cos 0 = 1, \quad \tan 2\pi = \tan 0 = 0, \quad \text{and so on.}$$

PROBLEM 1 Evaluate each trigonometric function, if it is defined.

(a) $\cos 7\pi$ **(b)** $\tan 450°$ ◆

◆ Multiples of $\pi/6$, $\pi/4$, and $\pi/3$

By using the relationships between the sides of a 45°-45°-90° triangle and a 30°-60°-90° triangle, we can find the values of the trigonometric functions of 30°, 45°, and 60°. Recall from elementary geometry that an *isosceles right triangle* has legs of equal length, and its interior angles opposite the legs measure 45°. Referring to Figure 7.35, if the length of one leg is 1, then the length of the other leg is also 1 and, by the Pythagorean theorem, the length of the hypotenuse is $\sqrt{2}$.

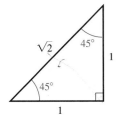

FIGURE 7.35
Relationship between the sides of a 45°-45°-90° triangle

Also recall from elementary geometry that in an *equilateral triangle* all sides have the same length and all interior angles measure 60°. If an altitude is drawn to one of the sides of an equilateral triangle, the vertex angle and base are both cut in half, and two identical 30°-60°-90° triangles are formed. The triangle in Figure 7.36 shows that if the length of the leg opposite the 30° angle is 1, then the length of the hypotenuse is 2 and, by the Pythagorean theorem, the length of the leg opposite the 60° angle is $\sqrt{3}$.

If we apply the trigonometric ratios for right triangles (Section 7.2) to the special right triangles in Figures 7.35 and 7.36, we can evaluate trigonometric functions of 30°, 45°, and 60°. Here are some examples:

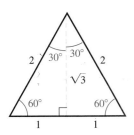

FIGURE 7.36
Relationship between the sides of a 30°-60°-90° triangle

$$\sin 30° = \frac{\text{opp}}{\text{hyp}} = \frac{1}{2} \qquad \cos 45° = \frac{\text{adj}}{\text{hyp}} = \frac{1}{\sqrt{2}} \qquad \tan 60° = \frac{\text{opp}}{\text{adj}} = \sqrt{3}$$

Table 7.3 gives a complete list of the values of the trigonometric functions of these special angles.

TABLE 7.3
Values of the trigonometric functions for $\pi/6$, $\pi/4$, and $\pi/3$

θ	$\sin\theta$	$\cos\theta$	$\tan\theta$	$\csc\theta$	$\sec\theta$	$\cot\theta$
30° or $\pi/6$	$\dfrac{1}{2}$	$\dfrac{\sqrt{3}}{2}$	$\dfrac{1}{\sqrt{3}}$	2	$\dfrac{2}{\sqrt{3}}$	$\sqrt{3}$
45° or $\pi/4$	$\dfrac{1}{\sqrt{2}}$	$\dfrac{1}{\sqrt{2}}$	1	$\sqrt{2}$	$\sqrt{2}$	1
60° or $\pi/3$	$\dfrac{\sqrt{3}}{2}$	$\dfrac{1}{2}$	$\sqrt{3}$	$\dfrac{2}{\sqrt{3}}$	2	$\dfrac{1}{\sqrt{3}}$

Figure 7.37 shows central angles or arc lengths $\pi/6$, $\pi/4$, and $\pi/3$ and the coordinates of their corresponding points ($\cos\theta$, $\sin\theta$) on the unit circle. As illustrated in the next example, we can evaluate trigonometric functions for multiples of $\pi/6$, $\pi/4$, and $\pi/3$ by using symmetry of this unit circle.

EXAMPLE 2 Use the unit circle in Figure 7.37 to evaluate each trigonometric function.

(a) $\cos\dfrac{7\pi}{4}$ (b) $\sin 150°$ (c) $\cot 225°$ (d) $\csc\dfrac{8\pi}{3}$

◆ **Solution**

(a) By symmetry, the point on the unit circle that corresponds to a central angle $7\pi/4$, or an arc length $7\pi/4$, is $(1/\sqrt{2}, -1/\sqrt{2})$, as shown in Figure 7.38. The x-coordinate of this point is the value of $\cos(7\pi/4)$. Hence,

$$\cos\frac{7\pi}{4} = \frac{1}{\sqrt{2}}.$$

(b) By symmetry, the point on the unit circle that corresponds to a central angle 150° ($5\pi/6$ rad), or an arc length $5\pi/6$, is $(-\sqrt{3}/2, \frac{1}{2})$, as shown in Figure 7.39. The y-coordinate of this point is the value of $\sin 150°$. Hence,

$$\sin 150° = \frac{1}{2}.$$

(c) By symmetry, the point on the unit circle that corresponds to a central angle 225° ($5\pi/4$ rad), or an arc length $5\pi/4$, is $(-1/\sqrt{2}, -1/\sqrt{2})$, as shown in Figure 7.40. The x-coordinate of this point is the value of $\cos 225°$, and the y-coordinate of this point is the value of $\sin 225°$.

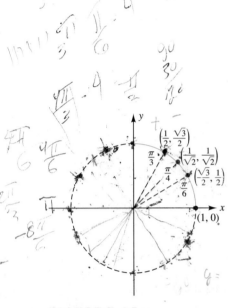

FIGURE 7.37
Arc lengths $\pi/6$, $\pi/4$, and $\pi/3$ and the coordinates of their points on the unit circle

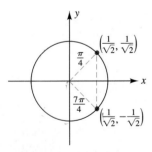

FIGURE 7.38
The point on the unit circle that corresponds to an arc length $7\pi/4$ is $(1/\sqrt{2}, -1/\sqrt{2})$.

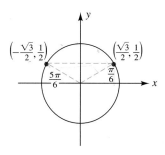

FIGURE 7.39
The point on the unit circle that corresponds to an arc length $5\pi/6$ is $\left(-\sqrt{3}/2, \tfrac{1}{2}\right)$.

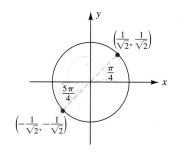

FIGURE 7.40
The point on the unit circle that corresponds to an arc length $5\pi/4$ is $\left(-1/\sqrt{2}, -1/\sqrt{2}\right)$.

By one of our fundamental trigonometric identities (Section 7.2), we have

$$\cot 225° = \frac{\cos 225°}{\sin 225°} = \frac{-1/\sqrt{2}}{-1/\sqrt{2}} = 1.$$

(d) Since

$$\frac{8\pi}{3} = 2\pi + \frac{2\pi}{3},$$

we know that $8\pi/3$ is coterminal with $2\pi/3$. By symmetry, the point on the unit circle that corresponds to a central angle $2\pi/3$, or an arc length $2\pi/3$, is $\left(-\tfrac{1}{2}, \sqrt{3}/2\right)$, as shown in Figure 7.41. The y-coordinate of this point is the value of $\sin(2\pi/3)$. Hence, $\sin(2\pi/3) = \sqrt{3}/2$. By one of our fundamental trigonometric identities (Section 7.2), we have

$$\csc \frac{2\pi}{3} = \frac{1}{\sin(2\pi/3)} = \frac{1}{\sqrt{3}/2} = \frac{2}{\sqrt{3}}.$$

Hence, $\csc \dfrac{8\pi}{3} = \dfrac{2}{\sqrt{3}}.$ ◆

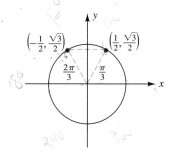

FIGURE 7.41
The point on the unit circle that corresponds to an arc length $2\pi/3$ is $\left(-\tfrac{1}{2}, \sqrt{3}/2\right)$.

PROBLEM 2 Use the unit circle in Figure 7.37 to evaluate each trigonometric function.

(a) $\tan \dfrac{3\pi}{4}$ **(b)** $\sec(-420°)$ ◆

◆ Evaluating Trigonometric Functions with a Calculator

As we illustrated in Examples 1 and 2, we can find the *exact values* of the trigonometric functions for some special angles and real numbers. For other angles and real numbers, however, we use the [SIN], [COS], or [TAN] keys on a calculator to help find the *approximate values* of the trigonometric functions. The following guidelines explain how to use a calculator to evaluate the trigonometric functions.

◆ Guideline for Calculator Usage

1. Set the calculator in degree mode to find the trigonometric functions of angles in degrees, or in radian mode to find the trigonometric functions of real numbers or angles in radians.
2. Press the appropriate key, [SIN], [COS], or [TAN], to evaluate the sine, cosine, or tangent of a given angle or real number.
3. For cosecant, secant, or cotangent, use the [SIN], [COS], or [TAN] key, respectively. Then press the reciprocal key, [1/x] or [x⁻¹], to obtain the value of the cosecant, secant, or cotangent of a given angle or real number.

EXAMPLE 3 Use a calculator to evaluate each trigonometric function. Round each answer to four significant digits.

(a) $\sin 2$ (b) $\tan\left(-\dfrac{3\pi}{8}\right)$ (c) $\sec 216°$ (d) $\csc 16°24'27''$

◆ Solution

(a) To find sin 2, we set the calculator in radian mode and use the [SIN] key. Rounding the display in Figure 7.42 to four significant digits, we obtain

$$\sin 2 \approx 0.9093.$$

(b) To find $\tan\left(-\dfrac{3\pi}{8}\right)$, we set the calculator in radian mode and use the [TAN] key. Rounding the display in Figure 7.42 to four significant digits, we obtain,

$$\tan\left(-\dfrac{3\pi}{8}\right) \approx -2.414.$$

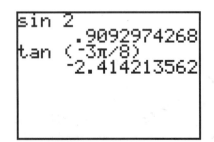

FIGURE 7.42
Typical display for evaluating sin 2 and tan(−3π/8) using radian mode

(c) To find sec 216°, we set the calculator in degree mode and use the [COS] key in conjunction with the reciprocal key, [1/x] or [x⁻¹]. Rounding the display in Figure 7.43 to four significant digits, we obtain

$$\sec 216° \approx -1.236.$$

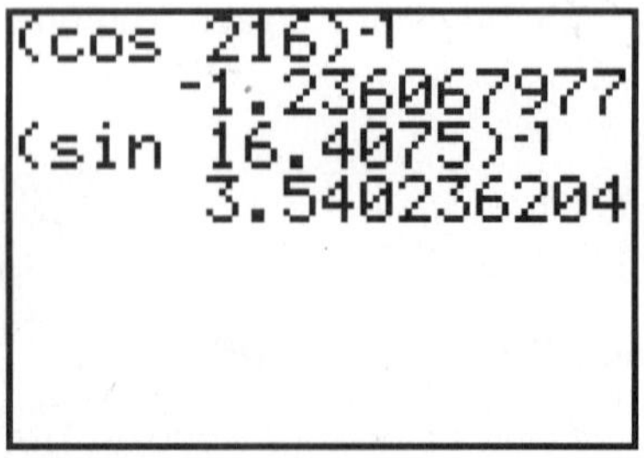

(d) To find csc 16°24'27'', we first convert to decimal degrees. From Example 2 in Section 7.1, we know that

$$\csc 16°24'27'' = \csc 16.4075°$$

Now, to find csc 16.4075°, we set the calculator in degree mode and use the [SIN] key in conjunction with the reciprocal key, [1/x] or [x⁻¹]. Rounding the display in Figure 7.43 to four significant digits, we obtain

$$\csc 16.4075° \approx 3.540.$$

FIGURE 7.43
Typical display for evaluating sec 216° and csc 16.4075° using degree mode

PROBLEM 3 Use a calculator to evaluate each trigonometric function. Round each answer to four significant digits.

(a) $\cos \dfrac{43\pi}{24}$ (b) $\cot 121°30'$

Exercises 7.3

Basic Skills

In Exercises 1–16, label a point on the unit circle that corresponds to the given central angle θ. Then use this point to find the exact values of each of the six trigonometric functions of θ that are defined. Do not use a calculator.

1. $\theta = 180°$
2. $\theta = 360°$
3. $\theta = 450°$
4. $\theta = 630°$
5. $\theta = 4\pi$
6. $\theta = \dfrac{9\pi}{2}$
7. $\theta = -\dfrac{\pi}{2}$
8. $\theta = -900°$
9. $\theta = 30°$
10. $\theta = \dfrac{\pi}{4}$
11. $\theta = \dfrac{13\pi}{3}$
12. $\theta = 780°$
13. $\theta = -315°$
14. $\theta = -1020°$
15. $\theta = \dfrac{9\pi}{4}$
16. $\theta = -\dfrac{11\pi}{6}$

In Exercises 17–40, use symmetry of the unit circle shown in the figure to find the exact value of each trigonometric function. Do not use a calculator.

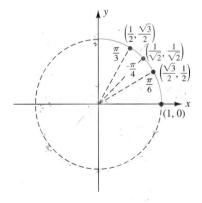

17. $\cos 120°$
18. $\tan 225°$
19. $\sin 210°$
20. $\sec 330°$
21. $\cot 570°$
22. $\sin 660°$
23. $\cot 1035°$
24. $\csc 960°$
25. $\csc \dfrac{7\pi}{4}$
26. $\cot \dfrac{5\pi}{6}$
27. $\sin \dfrac{2\pi}{3}$
28. $\cos \dfrac{5\pi}{4}$
29. $\sin \dfrac{11\pi}{4}$
30. $\cos \dfrac{11\pi}{3}$
31. $\sec \dfrac{19\pi}{6}$
32. $\cot \dfrac{31\pi}{6}$
33. $\tan(-210°)$
34. $\csc(-330°)$
35. $\cos(-870°)$
36. $\tan(-945°)$
37. $\sec\left(-\dfrac{5\pi}{3}\right)$
38. $\cot\left(-\dfrac{3\pi}{4}\right)$
39. $\csc\left(-\dfrac{41\pi}{6}\right)$
40. $\sin\left(-\dfrac{14\pi}{3}\right)$

In Exercises 41–56, assume the functions f, g, and h are defined as follows:

$$f(t) = 3 + \tan \dfrac{\pi t}{4}, \quad g(t) = -2 \sin 3\pi t, \quad \text{and}$$
$$h(t) = 3 \cos\left(\pi t - \dfrac{\pi}{3}\right)$$

Compute the functional value, if it is defined. Do not use a calculator.

41. $f(0)$
42. $f(1)$
43. $f(-3)$
44. $f(-8)$
45. $g\left(\tfrac{2}{3}\right)$
46. $g\left(\tfrac{3}{2}\right)$
47. $g\left(-\tfrac{1}{6}\right)$
48. $g\left(\tfrac{7}{18}\right)$
49. $h(0)$
50. $h(3)$
51. $h\left(\tfrac{1}{2}\right)$
52. $h\left(-\tfrac{2}{3}\right)$

53. $g(f(1))$
54. $f(h(\frac{1}{3}))$
55. $g(h(-2))$
56. $h(f(2))$

In Exercises 57–76, use a calculator to find the approximate value of each trigonometric function. Round each answer to four significant digits.

57. $\sin 25°$
58. $\cos 76°$
59. $\sin 195°$
60. $\tan 306°$
61. $\cos 617°$
62. $\csc 1224°$
63. $\sec(-216°)$
64. $\cot(-500°)$
65. $\sec \dfrac{\pi}{5}$
66. $\cot \dfrac{3\pi}{10}$
67. $\tan\left(-\dfrac{11\pi}{12}\right)$
68. $\cos \dfrac{17\pi}{10}$
69. $\tan 1$
70. $\cos 0.4$
71. $\csc 4.28$
72. $\sec 7.6$
73. $\sin 46°40'$
74. $\cot(-1674°30')$
75. $\sin 29°20'45''$
76. $\cos 600°00'30''$

77. The displacement d (in centimeters) of an oscillating spring from its equilibrium position is a function of time t (in seconds) and is given by

$$d(t) = 12 \cos \pi t.$$

Find the displacement at each value of t.
(a) 0 s (b) 5 s (c) $\frac{1}{2}$ s (d) $\frac{3}{4}$ s

78. The population P of mice in a certain field is a function of time t in years and is given by

$$P(t) = 350 + 200 \sin \dfrac{2\pi}{3} t.$$

(a) Find the initial mouse population.
(b) Find the mouse population after 3 years 9 months.

79. The voltage v (in volts) in an electric circuit is a function of time t (in seconds) and is given by

$$v(t) = 20 \sin 4t.$$

Find the voltage at each value of t.
(a) 1 s (b) 3 s (c) 4.5 s (d) 6.3 s

80. The expected daily high temperature T (in degrees Fahrenheit) for a certain city can be approximated by

$$T = 52 - 32 \cos\left[\dfrac{2\pi}{365}(t - 25)\right]$$

where t is the time (in days) with $t = 1$ corresponding to January 1. Find the expected daily high temperature on each day.
(a) January 1
(b) January 25
(c) July 4 ($t = 185$)
(d) September 1 ($t = 244$)

Critical Thinking

81. Choose various values of θ and evaluate $\sin \theta$ and $\sin(-\theta)$. How does the value of $\sin \theta$ compare to the value of $\sin(-\theta)$?

82. Choose various values of θ and evaluate $\cos \theta$ and $\cos(-\theta)$. How does the value of $\cos \theta$ compare to the value of $\cos(-\theta)$?

83. Only four of the eight fundamental trigonometric identities (Section 7.2) hold when θ is replaced by π. Which four do *not* hold? Explain why.

84. Only four of the eight fundamental trigonometric identities (Section 7.2) hold when θ is replaced by $\pi/2$. Which four do *not* hold? Explain why.

85. Evaluate each expression.
(a) $\log_6\left[\sin \dfrac{\pi}{2}\right]$
(b) $\log_2\left[\cos \dfrac{\pi}{3}\right]$
(c) $\ln[e^{\sin(\pi/6)}]$
(d) $e^{-2\ln[\cos(\pi/3)]}$

86. Compare the slope m of a line that passes through the origin and the point $P(x, y)$ to the value of $\tan \theta$, where θ is the positive angle in standard position whose terminal side passes through the point $P(x, y)$. What can you conclude about m and $\tan \theta$? Use your observation to find the slope of a line that is formed by the terminal side of each angle θ.
(a) $\theta = 60°$
(b) $\theta = 150°$
(c) $\theta = 225°$
(d) $\theta = 300°$

87. For a central angle θ there corresponds a point P on the unit circle with coordinates $(\cos \theta, \sin \theta)$, as shown in the figure.

SECTION 7.4 ◆ Sine and Cosine Functions

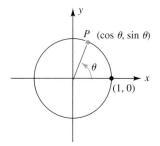

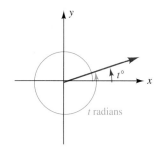

Complete the following table to show the changes in the values of cos θ, sin θ, and tan θ as we let θ increase from 0 to 2π.

θ	sin θ	cos θ	tan θ
0 to $\frac{\pi}{2}$			
$\frac{\pi}{2}$ to π			
π to $\frac{3\pi}{2}$			
$\frac{3\pi}{2}$ to 2π			

88. Usually, for a real number t, sin $t \neq$ sin $t°$. For instance, sin 1 ≈ 0.8415, but sin 1° ≈ 0.01752. However, if t radians and t degrees have the same terminal side, as shown in the figure, then sin t = sin $t°$. Find all real numbers t for which sin t = sin $t°$.

89. In order to use calculus to find the derivatives of the sine and cosine functions, it is necessary to evaluate the quotient (sin t)/t as t approaches zero. Use a calculator to complete the following table, and then state the value that (sin t)/t seems to approach as $t \to 0^+$.

t	1	0.1	0.01	0.001	0.0001
$\frac{\sin t}{t}$					

90. In order to use calculus to find the derivatives of the sine and cosine functions, it is necessary to evaluate the quotient (cos t − 1)/t as t approaches zero. Use a calculator to complete the following table, and then state the value that (cos t − 1)/t seems to approach as $t \to 0^+$.

t	1	0.1	0.01	0.001	0.0001
$\frac{\cos t - 1}{t}$					

7.4 Properties and Graphs of the Sine and Cosine Functions

◆ Introductory Comments

Recall from Section 7.3 that if the arc length from (1, 0) to a point P on the unit circle is t units, then the coordinates of point P are (cos t, sin t), as shown in Figure 7.44. We can obtain sketches of the graphs of the sine function

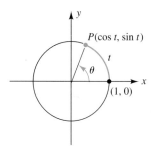

FIGURE 7.44
The coordinates of point P on the unit circle are (cos t, sin t).

$f(t) = \sin t$ and the cosine function $g(t) = \cos t$ in the interval $[0, 2\pi]$ by observing the changes in the y-coordinate $\sin t$ and the x-coordinate $\cos t$ as we let t increase from 0 to 2π. Table 7.4 summarizes these changes.

TABLE 7.4

Changes in $\sin t$ and $\cos t$ as t increases from 0 to 2π

t	y-coordinate $\sin t$	x-coordinate $\cos t$
0 to $\pi/2$	0 to 1	1 to 0
$\pi/2$ to π	1 to 0	0 to -1
π to $3\pi/2$	0 to -1	-1 to 0
$3\pi/2$ to 2π	-1 to 0	0 to 1

The graphs in Figure 7.45 and 7.46 illustrate these changes in $\sin t$ and $\cos t$ as t increases from 0 to 2π.

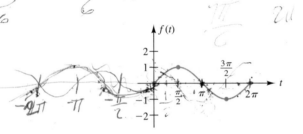

FIGURE 7.45
Graph of $f(t) = \sin t$ in the interval $[0, 2\pi]$

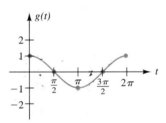

FIGURE 7.46
Graph of $g(t) = \cos t$ in the interval $[0, 2\pi]$

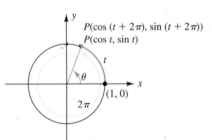

FIGURE 7.47
If t increases by a multiple of 2π, the coordinates of P remain unchanged.

The coordinates of point P in Figure 7.47 indicate that if t increases by a multiple of 2π (the circumference of the unit circle), the coordinates remain unchanged, that is

$$\sin(t + 2\pi n) = \sin t$$

and

$$\cos(t + 2\pi n) = \cos t$$

for any integer n. When a function repeats its value in a repetitive manner, as do $\sin t$ and $\cos t$, we say that the function is *periodic* and, in this case, has a *period* of 2π. In general, we have the following definition for a *periodic function*.

Periodic Function

A function F is said to be **periodic** if there exists a real number $c > 0$ such that

$$F(t + c) = F(t)$$

for all t in the domain of F. The least such positive number, if it exists, is called the **period** of F.

When we graph a periodic function over one period, we say that we have graphed *one cycle* of the function. The graphs in Figures 7.45 and 7.46 show one cycle of $f(t) = \sin t$ and $g(t) = \cos t$, respectively. Since the sine and cosine functions are periodic with period 2π, we can obtain the entire graphs of $f(t) = \sin t$ and $g(t) = \cos t$ simply by repeating the portion of the graph from 0 to 2π, as illustrated in Figures 7.48 and 7.49, respectively.

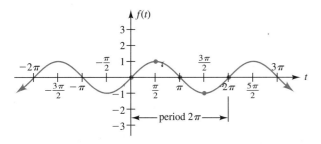

FIGURE 7.48
Graph of the sine function
$f(t) = \sin t$

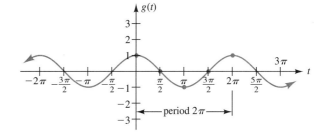

FIGURE 7.49
Graph of the cosine function
$g(t) = \cos t$

The following properties can be observed from the graphs of $f(t) = \sin t$ and $g(t) = \cos t$.

Properties of the Sine Function

Sine function: $f(t) = \sin t$

1. The *domain* is the set of all real numbers.
2. The *range* is the set of all real numbers in the closed interval $[-1, 1]$.
3. The *period* is 2π.
4. The *zeros* are 0, $\pm\pi$, $\pm 2\pi$, $\pm 3\pi$, ...
5. The function is *odd*, that is, $\sin(-t) = -\sin t$ for all t.

◆ Properties of the Cosine Function

Cosine function: $g(t) = \cos t$

1. The *domain* is the set of all real numbers.
2. The *range* is the set of all real numbers in the closed interval $[-1, 1]$.
3. The *period* is 2π.
4. The *zeros* are $\pm \dfrac{\pi}{2}, \pm \dfrac{3\pi}{2}, \pm \dfrac{5\pi}{2}, \ldots$
5. The function is *even*, that is, $\cos(-t) = \cos t$ for all t.

By setting a graphing calculator in radian mode and using the $\boxed{\text{SIN}}$ key and $\boxed{\text{COS}}$ key, we can generate the graphs of the sine and cosine functions. When using a graphing calculator, we choose x and y for the independent and dependent variables so that the graphs are displayed in the usual xy-coordinate plane. However, when entering $y = \sin x$ or $y = \cos x$, do not confuse the variables x and y with the coordinates of the point $P(x, y)$ on the unit circle.

In this section, we use the basic graphs of $y = \sin x$ and $y = \cos x$ in conjunction with the shift rules, axis reflection rules, and stretch and compress rules (Section 2.5) to discuss the graphs of the equations

$$y = a \sin(bx + c) \quad \text{and} \quad y = a \cos(bx + c).$$

◆ Graphs of $y = a \sin(bx + c)$ and $y = a \cos(bx + c)$

By the vertical stretch and compress rules (Section 2.5), we know that the graphs of $y = a \sin x$ and $y = a \cos x$, with $a > 0$, are similar to the graphs of $y = \sin x$ and $y = \cos x$, respectively, stretched or compressed vertically by a factor of a. When working with the sine and cosine functions, we refer to the absolute value of a as the *amplitude* of the graph.

◆ Amplitude

The **amplitude** of $y = a \sin x$ and $y = a \cos x$ is $|a|$ for $a \neq 0$.

Each of the graphs of $y = a \sin x$ and $y = a \cos x$ has period 2π, and each graph completes one cycle from $x = 0$ to $x = 2\pi$. Therefore, the graphs of $y = a \sin bx$, and $y = a \cos bx$ must each complete one cycle from

$$bx = 0 \quad \text{to} \quad bx = 2\pi$$
$$x = 0 \quad \text{to} \quad x = \frac{2\pi}{b}$$

SECTION 7.4 ♦ *Sine and Cosine Functions*

The period P of $y = a \sin bx$ and $y = a \cos bx$ is the length of the interval from $x = 0$ to $x = 2\pi/b$:

$$P = \left| \frac{2\pi}{b} - 0 \right| = \frac{2\pi}{|b|}$$

◀ **Period Formula for Sine and Cosine**

The period P of $y = a \sin bx$ or $y = a \cos bx$ is given by

$$P = \frac{2\pi}{|b|} \quad \text{for } b \neq 0.$$

Figures 7.50 and 7.51 show the graphs of one cycle of $y = a \sin bx$ and $y = a \cos bx$ for $a > 0$ and $b > 0$.

The graph of $y = a \sin bx$ has five key values in the interval $[0, 2\pi/b]$:

$$(0, 0) \quad \left(\frac{\pi}{b}, 0\right) \quad \left(\frac{2\pi}{b}, 0\right) \quad \left(\frac{\pi}{2b}, a\right) \quad \left(\frac{3\pi}{2b}, -a\right)$$

x-intercepts   maximum point   minimum point

The graph of $y = a \cos bx$ also has five key values in the interval $[0, 2\pi/b]$:

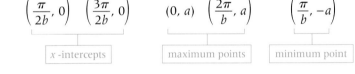

$$\left(\frac{\pi}{2b}, 0\right) \quad \left(\frac{3\pi}{2b}, 0\right) \quad (0, a) \quad \left(\frac{2\pi}{b}, a\right) \quad \left(\frac{\pi}{b}, -a\right)$$

x-intercepts   maximum points   minimum point

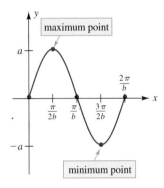

FIGURE 7.50

Graph of one cycle of $y = a \sin bx$ for $a > 0$ and $b > 0$

In each case, we have obtained the five key values by dividing the interval $[0, 2\pi/b]$ into four equal parts.

The equations $y = a \sin(bx + c)$ and $y = a \cos(bx + c)$ may be rewritten by factoring out b:

$$y = a \sin(bx + c) = a \sin b\left(x + \frac{c}{b}\right)$$

$$y = a \cos(bx + c) = a \cos b\left(x + \frac{c}{b}\right)$$

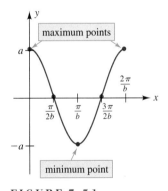

FIGURE 7.51

Graph of one cycle of $y = a \cos bx$ for $a > 0$ and $b > 0$

Now, by the horizontal shift rule (Section 2.5), we know that the graphs of these equations are the same as the graphs of $y = a \sin bx$ and $y = a \cos bx$, respectively, shifted horizontally $|c/b|$ units. If $c/b < 0$ the shift is to the right, or if $c/b > 0$ the shift is to the left. When working with the sine and cosine functions, we refer to the absolute value of c/b as the *phase shift* of the graph.

Phase Shift

The **phase shift** of $y = a\sin(bx + c)$ and $y = a\cos(bx + c)$ is

$\left|\dfrac{c}{b}\right|$ units to the right if $\dfrac{c}{b} < 0$ or $\left|\dfrac{c}{b}\right|$ units to the left if $\dfrac{c}{b} > 0$.

EXAMPLE 1 State the amplitude, period, and phase shift for the graph of each equation. Then list the five key values on one cycle of the graph.

(a) $y = 3\sin 2x$ **(b)** $y = \tfrac{1}{2}\cos\left(\dfrac{\pi}{4}x - \dfrac{\pi}{4}\right)$

Solution

(a) For $y = 3\sin 2x$, we have $a = 3$, $b = 2$, and $c = 0$. Therefore,

Amplitude: $|a| = |3| = 3$ Period: $\dfrac{2\pi}{|b|} = \dfrac{2\pi}{|2|} = \pi$

Phase shift: $\left|\dfrac{c}{b}\right| = \left|\dfrac{0}{2}\right| = 0$ (no shift)

Thus, this sine curve completes one cycle from $x = 0$ to $x = \pi$ and has a maximum y-value of 3. Dividing the interval $[0, \pi]$ into four equal parts gives us the five key values on this cycle:

$(0, 0)$ $\left(\dfrac{\pi}{2}, 0\right)$ $(\pi, 0)$ $\left(\dfrac{\pi}{4}, 3\right)$ $\left(\dfrac{3\pi}{4}, -3\right)$

x-intercepts   maximum point   minimum point

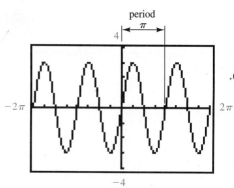

FIGURE 7.52
The graph of $y = 3\sin 2x$ has amplitude 3, period π, and no phase shift.

Figure 7.52 shows the graph of $y = 3\sin 2x$ in the viewing rectangle of a graphing calculator, and this graph supports our work. We have chosen $\pi/4$ as the distance between scale marks on the x-axis.

(b) For $y = \tfrac{1}{2}\cos\left(\dfrac{\pi}{4}x - \dfrac{\pi}{4}\right)$, we have $a = \tfrac{1}{2}$, $b = \pi/4$, and $c = -\pi/4$. Therefore,

Amplitude: $|a| = \left|\dfrac{1}{2}\right| = \dfrac{1}{2}$ Period: $\dfrac{2\pi}{|b|} = \dfrac{2\pi}{|\pi/4|} = 8$

Phase shift: $\left|\dfrac{c}{b}\right| = \left|\dfrac{-\pi/4}{\pi/4}\right| = |-1| = 1$ unit right, since $\dfrac{c}{b} < 0$.

Thus, this cosine curve completes one cycle from

$x = 1$ to $x = 1 + 8 = 9$,

Add the period.

SECTION 7.4 ◆ Sine and Cosine Functions

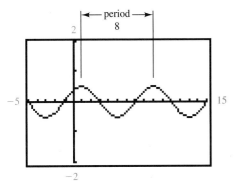

FIGURE 7.53

The graph of $y = \frac{1}{2} \cos\left(\frac{\pi}{4}x - \frac{\pi}{4}\right)$ has amplitude $\frac{1}{2}$, period 8, and phase shift 1 unit to the right.

and has a maximum value of $\frac{1}{2}$. Dividing the interval [1, 9] into four equal parts gives us the five key values on this cycle:

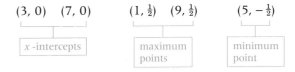

(3, 0) (7, 0)   (1, $\frac{1}{2}$) (9, $\frac{1}{2}$)   (5, $-\frac{1}{2}$)

x-intercepts   maximum points   minimum point

Figure 7.53 shows the graph of $y = \frac{1}{2} \cos\left(\frac{\pi}{4}x - \frac{\pi}{4}\right)$ in the viewing rectangle of a graphing calculator, and this graph supports our work. We have chosen 1 as the distance between scale marks on both the x-axis and y-axis. ◆

If b is negative in $y = a \sin(bx + c)$ or $y = a \cos(bx + c)$, then we may rewrite these equations by using the facts that, for any real number t,

$$\sin(-t) = -\sin t \quad \text{and} \quad \cos(-t) = \cos t.$$

For example, the equation $y = 3 \sin(-2x)$ may be rewritten as

$$y = 3 \sin(-2x) = -3 \sin 2x.$$

Now, by the x-axis reflection rule (Section 2.5), the graph of $y = -3 \sin 2x$ is the same as the graph of $y = 3 \sin 2x$, reflected about the x-axis.

PROBLEM 1 Compare the graph of $y = \frac{1}{2} \cos\left(\frac{\pi}{4} - \frac{\pi}{4}x\right)$ to the graph of $y = \frac{1}{2} \cos\left(\frac{\pi}{4}x - \frac{\pi}{4}\right)$. ◆

In the next example, we reverse the preceding procedure to determine the equation of a sine or cosine curve from its graph.

EXAMPLE 2 Write the equation of the graph shown in Figure 7.54 in the form

$$y = a \sin(bx + c).$$

Use the least nonnegative real number c with $a > 0$ and $b > 0$.

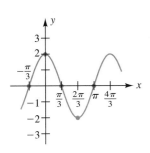

FIGURE 7.54

◆ **Solution**   From the graph, we note the following facts:

1. The amplitude is 2, so we have $a = 2$.

2. Since the sine curve completes one cycle from $x = -\pi/3$ to $x = \pi$, the period is $P = \pi - (-\pi/3) = 4\pi/3$. Now, using the period formula

$P = 2\pi/b$, we solve for b:

$$\frac{4\pi}{3} = \frac{2\pi}{b}$$

$$4\pi b = 6\pi \quad \text{Multiply both sides by } 3b.$$

$$b = \frac{3}{2} \quad \text{Divide both sides by } 4\pi$$

3. The phase shift c/b is $\pi/3$ units to the left. Thus,

$$\frac{c}{b} = \frac{\pi}{3}$$

$$\frac{c}{\frac{3}{2}} = \frac{\pi}{3} \quad \text{Replace } b \text{ with } \tfrac{3}{2}$$

$$c = \frac{\pi}{2} \quad \text{Multiply both sides by } \tfrac{3}{2}$$

Hence, the equation of the curve is $y = 2\sin\left(\frac{3}{2}x + \frac{\pi}{2}\right)$. Figure 7.55 shows the graph of $y = 2\sin\left(\frac{3}{2}x + \frac{\pi}{2}\right)$ in the viewing rectangle of a graphing calculator, and this graph verifies our work. We have chosen $\pi/3$ as the distance between scale marks on the x-axis. ◆

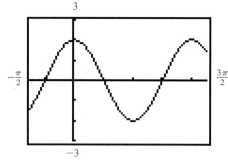

FIGURE 7.55

The graph of $y = 2\sin\left(\frac{3}{2}x + \frac{\pi}{2}\right)$ has the same characteristics as the graph in Figure 7.54.

PROBLEM 2   We may think of the curve shown in Figure 7.54 as a cosine curve that completes one cycle from $x = 0$ to $x = 4\pi/3$. Write the equation of this graph in the form $y = a\cos(bx + c)$ using the least nonnegative real number c with $a > 0$ and $b > 0$. ◆

◆ Application: Predator-Prey Relationships

Consider the relationship between mice (prey) and hawks (predators) in a certain field. If the number of mice is relatively large, the number of hawks in the area starts to increase, since the mice provide a plentiful food supply for the hawks. Now, as the number of hawks in the area increases, their food supply (the mice) begins to be depleted. This, in turn, causes the hawks to search elsewhere for food, which then allows the mice population to begin to increase again. This cyclic pattern repeats itself over and over again with both populations oscillating in a periodic manner about their average values, as shown in Figure 7.56. Hence, the population P at any time t in a **predator-prey relationship** may be described by the equation

$$P = d + a\sin(bt + c)$$

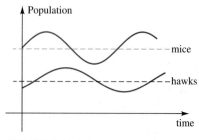

FIGURE 7.56

Mice and hawk populations in a predator-prey relationship oscillating in a periodic manner about their average values

where a, b, c, and d are constants. By the vertical shift rule (Section 2.5), the constant d shifts the graph of $y = a\sin(bt + c)$ vertically $|d|$ units.

SECTION 7.4 ♦ Sine and Cosine Functions

EXAMPLE 3 Rangers have determined that the population P of red-tailed hawks in a certain area of a state forest can be described by the equation

$$P = 70 + 50 \sin\left(\frac{\pi t}{6} - \frac{\pi}{3}\right)$$

where t is the time (in months) with $t = 1$ corresponding to January 1.

(a) Generate the graph of this equation in the viewing rectangle of a graphing calculator. Then determine the red-tailed hawk population on March 1 ($t = 3$).

(b) Determine the greatest number of red-tailed hawks supported in this area during one year, and state the month in which this maximum value occurs.

♦ Solution

(a) For $P = 70 + 50 \sin\left(\frac{\pi t}{6} - \frac{\pi}{3}\right)$, we have $a = 50$, $b = \pi/6$, $c = -\pi/3$, and $d = 70$. Therefore,

Amplitude: $|a| = |50| = 50$ *Period:* $\frac{2\pi}{|b|} = \frac{2\pi}{|\pi/6|} = 12$

Phase shift: $\left|\frac{c}{b}\right| = \left|\frac{-\pi/3}{\pi/6}\right| = |-2| = 2$ units right, since $\frac{c}{b} < 0$

Vertical shift: $d = 70$ units

This information suggests that an appropriate viewing rectangle for observing the graph of $P = 70 + 50 \sin\left(\frac{\pi t}{6} - \frac{\pi}{3}\right)$ is

```
tmin = 0
tmax = 16
tscl = 1
Pmin = 0
Pmax = 160
Pscl = 20
```

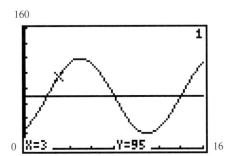

FIGURE 7.57
The graph of
$P = 70 + 50 \sin\left(\frac{\pi t}{6} - \frac{\pi}{3}\right)$
completes one cycle from $t = 2$ to $t = 14$.

Figure 7.57 shows the graph of $P = 70 + 50 \sin\left(\frac{\pi t}{6} - \frac{\pi}{3}\right)$ in this viewing rectangle. We also show the horizontal line $P = 70$, which represents the average value of the hawk population. As illustrated on this graph, this area supports 95 red-tailed hawks on March 1 ($t = 3$). We can verify this fact algebraically by substituting $t = 3$ in the equation $P = 70 + 50 \sin\left(\frac{\pi t}{6} - \frac{\pi}{3}\right)$:

$$P = 70 + 50 \sin\left(\frac{\pi(3)}{6} - \frac{\pi}{3}\right) = 70 + 50 \sin\frac{\pi}{6} = 95$$

Recall from Section 7.3 that $\sin(\pi/6) = \frac{1}{2}$.

(b) The graph of $P = 70 + 50 \sin\left(\frac{\pi t}{6} - \frac{\pi}{3}\right)$ completes one cycle from

$$t = 2 \quad \text{to} \quad t = 2 + 12 = 14,$$

Add the period.

as shown in Figure 7.57. Dividing the interval [2, 14] into four equal parts gives us the five key values on this cycle:

(2, 70)   (8, 70)   (14, 70)   (5, 120)   (11, 20)

intercepts along $P = 70$ maximum point   minimum point

Therefore, the greatest number of red-tailed hawks supported in this area is 120, and this occurs in the fifth month, May.

PROBLEM 3 Referring to Example 3, find the red-tailed hawk population on September 1 ($t = 9$). Use the graph in Figure 7.57 to confirm your answer.

Exercises 7.4

Basic Skills

Complete the following table of values for $f(x) = \sin x$ and $g(x) = \cos x$. Use the values in the table to plot points, and graph one cycle of the graphs of f and g.

1.

x	0	$\frac{\pi}{4}$	$\frac{\pi}{2}$	$\frac{3\pi}{4}$	π	$\frac{5\pi}{4}$	$\frac{3\pi}{2}$	$\frac{7\pi}{4}$	2π
$f(x) = \sin x$	0	$\frac{\sqrt{2}}{2}$	1	$\frac{\sqrt{2}}{2}$	0	$-\frac{\sqrt{2}}{2}$	-1	$-\frac{\sqrt{2}}{2}$	0
$g(x) = \cos x$	1	$\frac{\sqrt{2}}{2}$	0	$-\frac{\sqrt{2}}{2}$	-1	$-\frac{\sqrt{2}}{2}$	0	$\frac{\sqrt{2}}{2}$	1

2.

x	0	$\frac{\pi}{6}$	$\frac{\pi}{3}$	$\frac{\pi}{2}$	$\frac{2\pi}{3}$	$\frac{5\pi}{6}$	π	$\frac{7\pi}{6}$	$\frac{4\pi}{3}$	$\frac{3\pi}{2}$	$\frac{5\pi}{3}$	$\frac{11\pi}{6}$	2π
$f(x) = \sin x$													
$g(x) = \cos x$													

In Exercises 3–10, use the results of Exercises 1 and 2 and the fact that the sine and cosine functions are periodic with period 2π to find all values of x in the interval $[-4\pi, 4\pi]$ that satisfy the given equation.

3. $\sin x = 0$
4. $\cos x = 0$
5. $\cos x = -1$
6. $\sin x = 1$
7. $\cos x = \dfrac{1}{\sqrt{2}}$
8. $\sin x = -\dfrac{1}{\sqrt{2}}$
9. $\sin x = -\dfrac{1}{2}$
10. $\cos x = \dfrac{\sqrt{3}}{2}$

In Exercises 11–36, state the amplitude, period, and phase shift for the graph of each equation. Then show the five key values on one cycle of the graph. Use a graphing calculator to confirm each of your answers.

11. $y = 5 \sin x$
12. $y = \tfrac{2}{3} \cos x$
13. $y = 4 \cos 3x$
14. $y = 3 \sin 0.1x$
15. $y = -\tfrac{1}{2} \sin \tfrac{1}{6} x$
16. $y = -\cos 6x$
17. $y = \tfrac{4}{3} \cos(-\pi x)$
18. $y = \sqrt{3} \sin\left(-\dfrac{\pi}{3} x\right)$
19. $y = -\sin(-1.5x)$
20. $y = -\tfrac{5}{4} \cos \dfrac{\pi x}{5}$
21. $y = 1 + \cos \dfrac{3\pi x}{8}$
22. $y = \sin 2x - 3$
23. $y = \sin\left(x + \dfrac{\pi}{6}\right)$
24. $y = 2 \cos\left(x - \dfrac{\pi}{4}\right)$
25. $y = 4 \cos\left(\dfrac{x}{2} - \dfrac{3\pi}{8}\right)$
26. $y = \tfrac{1}{3} \sin\left(2x + \dfrac{\pi}{3}\right)$
27. $y = -\tfrac{1}{2} \sin\left(\pi x + \dfrac{3\pi}{4}\right)$
28. $y = -4 \cos \dfrac{\pi}{3}(x - 1)$
29. $y = \cos 2\left(\dfrac{\pi}{6} - x\right)$
30. $y = 3 \sin\left(\dfrac{2\pi}{3} - \pi x\right)$
31. $y = 1.5 \sin(0.75x + \pi)$
32. $y = 5 \cos 0.4(x - \pi)$
33. $y = 3 + 2 \cos 3\left(x - \dfrac{\pi}{12}\right)$
34. $y = 2 - 3 \sin\left(\dfrac{x}{3} + \dfrac{\pi}{9}\right)$
35. $y = \sin(6x + \pi) - 3$
36. $y = -\left[1 + \cos\left(3x - \dfrac{\pi}{2}\right)\right]$

In Exercises 37–42, generate the graph of each equation in the viewing rectangle of a graphing calculator. From the graph, approximate the period and phase shift, rounding each answer to three significant digits. Check each of your answers algebraically.

37. $y = 125 \sin 342x$
38. $y = 9.87 + 5.46 \cos 3.65x$
39. $y = 26.6 \cos(0.223x - 1.24)$
40. $y = 2.24 \sin(1.24 - 19.1x)$
41. $y = 5.46 + \sin(12.5x - 10.3)$
42. $y = 113 - 105 \cos[0.765(x - 3.95)]$

In Exercises 43–48, write the equation of the given graph in two forms:

  (i) $y = a \sin(bx + c)$ (ii) $y = a \cos(bx + c)$

In each form, use the least nonnegative real number c with $a > 0$ and $b > 0$.

43.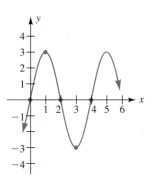

44.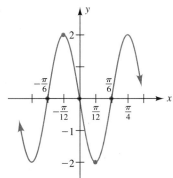

45.

46.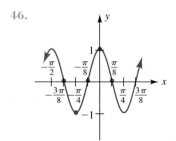

47.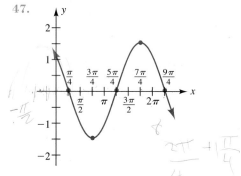

48.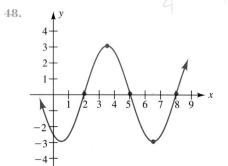

49. Ordinary household voltage v is given by

$$v = 170 \sin 377t$$

where t is time (in seconds) with $t \geq 0$.
   (a) Determine the amplitude and period for the graph of this equation.
   (b) The number of cycles that the voltage completes in one second is called its *frequency*. The frequency of household voltage, in cycles per second, or hertz (Hz), is the reciprocal of the period. Find the frequency for household voltage.

50. Rangers have determined that the population P of black bears in a state forest in Maine can be approximated by

$$P = 75 + 40 \sin \frac{\pi t}{4}$$

where t is the time (in years) with $t = 0$ corresponding to January 1, 1990.

   (a) Generate the graph of this equation in the viewing rectangle of a graphing calculator, then determine the black bear population on January 1, 1995.
   (b) Find the first year after the year 2000 when the population will reach its maximum value.

51. Over a period of years, the owner of a ski shop has found that the amount of monthly sales S (in dollars) can be approximated by

$$S = 4500 + 3200 \cos \frac{\pi}{6}(t - 2)$$

where t is the time (in months) with $t = 1$ corresponding to January.
   (a) Determine the month when sales are greatest, and state the amount of sales for that month.
   (b) Determine the month when sales are least, and state the amount of sales for that month.

52. Meteorologists project that the daily high temperature T (in degrees Fahrenheit) in Buffalo, New York, for 1998 may be found by using the equation

$$T = 42 - 55 \cos\left(\frac{\pi t}{183} + \frac{2\pi}{61}\right)$$

where t is the time (in days) with $t = 1$ corresponding to January 1.
   (a) Generate the graph of this equation in the viewing rectangle of a graphing calculator. Then determine the daily high temperature on February 1, 1998 ($t = 32$).
   (b) Determine the greatest temperature projected for Buffalo during 1998, and state the day on which it will occur.

53. The silver fox population P in a national forest in Canada is given by

$$P = 540 + 220 \sin 0.449t$$

where t is the time (in years) with $t = 0$ corresponding to January 1, 1990. What is the first year after 1990 when the fox population is expected to reach its maximum value? minimum value?

54. It is projected that the total amount of snowfall S (in inches) during the year in the Boston area is

$$S = 18.2 + 16.5 \cos[0.785(t + 1)],$$

where t is the time (in years) with $t = 0$ corresponding to January 1, 1990. What is the first year after 1998 when the snowfall is expected to reach its maximum value? minimum value?

55. The cross-section of an aquifer has the shape of a sine wave, as illustrated in the sketch.
    (a) Determine the depth d of the aquifer below level ground as a function of the distance x for $0 \leq x \leq 480$.
    (b) Find the depth of the aquifer below the ground when $x = 280$ ft.

56. A portion of a roller-coaster track is built in the shape of a cosine wave, as sketched in the figure.
    (a) Determine the height h of the roller-coaster track above level ground as a function of the distance x for $0 \leq x \leq 120$.
    (b) Find the height of the track above level ground when $x = 60$ ft.

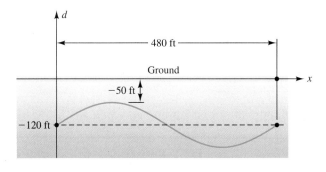

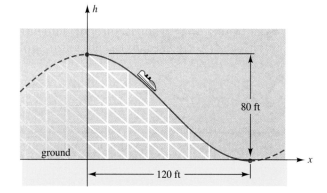

Critical Thinking

If the graph of $y = \sin x$ is shifted left $\pi/2$ units, we obtain the graph of $y = \cos x$. Based upon this observation, we conclude that

$$\sin\left(x + \frac{\pi}{2}\right) = \cos x$$

for all real numbers x. In Exercises 57–62, generate the graph of the function f in the viewing rectangle of a graphing calculator, and then compare the graph of f with the graphs of $y = \sin x$, $y = -\sin x$, $y = \cos x$, and $y = -\cos x$. Based upon your comparisons, state a trigonometric identity that relates $f(x)$ to either $\sin x$, $-\sin x$, $\cos x$, or $-\cos x$.

57. $f(x) = \sin\left(x - \dfrac{\pi}{2}\right)$

58. $f(x) = \cos\left(x + \dfrac{\pi}{2}\right)$

59. $f(x) = \cos\left(x + \dfrac{3\pi}{2}\right)$

60. $f(x) = \sin\left(x - \dfrac{3\pi}{2}\right)$

61. $f(x) = \sin(x - \pi)$

62. $f(x) = \cos(x + \pi)$

63. Find a function f of the form $f(x) = k + a \sin bx$ if its graph has a maximum point at $(2, 3)$ and a minimum point at $(6, -5)$.

64. Use the results of Exercises 1 and 2 and the fact that the sine and cosine functions are periodic with period 2π to find all values of x in the interval $[-4\pi, 4\pi]$ for which $\sin x = \cos x$.

65. Is it possible for a periodic function to be a one-to-one function? Explain.

66. Generate the graphs of the functions $f(x) = \sin|x|$ and $g(x) = |\sin x|$ in the viewing rectangle of a graphing calculator, and then determine whether each of these functions is periodic. If the function is periodic, state its period.

In calculus, it is shown that certain polynomial functions represent good approximations of the trigonometric functions. In Exercises 67 and 68, use a graphing calculator to generate the graph of the polynomial function P over the interval $[-\pi, \pi]$.

(a) What trigonometric function does P seem to approximate?
(b) Compare $P(1)$ to the value of the trigonometric function found in part (a) when $x = 1$. Round each answer to four significant digits.

67. $P(x) = x - \dfrac{x^3}{6} + \dfrac{x^5}{120} - \dfrac{x^7}{5040}$

68. $P(x) = 1 - \dfrac{x^2}{2} + \dfrac{x^4}{24} - \dfrac{x^6}{720}$

7.5 Properties and Graphs of the Other Trigonometric Functions

◆ Introductory Comments

By one of our fundamental trigonometric identities (Section 7.2), we have

$$\tan t = \frac{\sin t}{\cos t}$$

Hence, $\tan t = 0$ when $\sin t = 0$, that is, when

$$t = 0, \pm \pi, \pm 2\pi, \pm 3\pi, \text{ and so on.}$$

Since division by zero is undefined, $\tan t$ is undefined when $\cos t = 0$, that is, when

$$t = \pm \frac{\pi}{2}, \pm \frac{3\pi}{2}, \pm \frac{5\pi}{2}, \pm \frac{7\pi}{2}, \text{ and so on.}$$

If t increases by π units, as shown in Figure 7.58, then the coordinates of the new point are

$$(\cos(t + \pi), \sin(t + \pi)) = (-\cos t, -\sin t)$$

Therefore,

$$\tan(t + \pi) = \frac{-\sin t}{-\cos t} = \frac{\sin t}{\cos t} = \tan t$$

In general, for any integer n, we have

$$\boxed{\tan(t + \pi n) = \tan t}$$

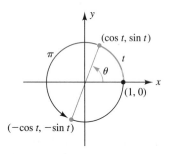

FIGURE 7.58
If t increases by π units, the coordinates of the point are $(-\cos t, -\sin t)$.

Since π is the smallest number c for which $\tan(t + c) = \tan t$, we conclude the tangent function is periodic with period π. Hence, it must complete one cycle between the undefined values of $t = -\pi/2$ and $t = \pi/2$.

Referring to Figure 7.58, as t approaches $\pi/2$ through values less than $\pi/2$, then $\sin t$ approaches 1 and $\cos t$ approaches 0. Thus, $(\sin t)/(\cos t)$ becomes very large. Hence, we say that $\tan t$ *increases without bound*. Symbolically, we write

$$\tan t \to \infty \quad \text{as} \quad t \to \frac{\pi}{2}^-.$$

Again, referring to Figure 7.58, as t approaches $-\pi/2$ through values greater than $-\pi/2$, then $\sin t$ approaches -1 and $\cos t$ approaches 0. Thus, (sin t)/(cos t) is negative and very large in absolute value. Hence, we say that tan t *decreases without bound*. Symbolically, we write

$$\tan t \to -\infty \quad \text{as} \quad t \to -\frac{\pi}{2}^+.$$

In summary, the lines $t = \pi/2$ and $t = -\pi/2$ are *vertical asymptotes* for the graph of the tangent function $f(t) = \tan t$. Figure 7.59 shows the appearance of the graph of $f(t) = \tan t$ in the interval $(-\pi/2, \pi/2)$. We can obtain the entire graph of $f(t) = \tan t$ by repeating the pattern in Figure 7.59 over successive intervals of length π, as illustrated in Figure 7.60.

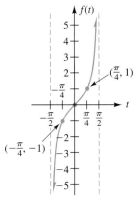

FIGURE 7.59

Graph of one cycle of $f(t) = \tan t$ in the interval $(-\pi/2, \pi/2)$

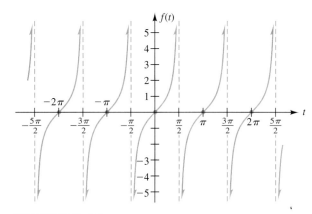

FIGURE 7.60

Graph of the tangent function $f(t) = \tan t$

The following properties can be observed from the graph of $f(t) = \tan t$.

◀ Properties of the Tangent Function

Tangent function: $f(t) = \tan t$

1. The *domain* is the set of all real numbers except

$$t = \pm \frac{\pi}{2}, \pm \frac{3\pi}{2}, \pm \frac{5\pi}{2}, \ldots.$$

2. The *range* is the set of all real numbers.
3. The *period* is π.
4. The *zeros* are $0, \pm \pi, \pm 2\pi, \pm 3\pi, \ldots$.
5. The function is *odd*, that is, $\tan(-t) = -\tan t$ for all t in the domain of t.

By setting a graphing calculator in radian mode and using the $\boxed{\text{TAN}}$ key, we can generate the graph of the tangent function. When using a graphing calculator, we choose x and y for the independent and dependent variables so that the graphs are displayed in the usual *xy*-coordinate plane. However, when entering $y = \tan x$, do not confuse the variables x and y with the coordinates of the point $P(x, y)$ on the unit circle.

Since

$$\csc t = \frac{1}{\sin t}, \quad \sec t = \frac{1}{\cos t}, \quad \text{and} \quad \cot t = \frac{1}{\tan t},$$

we can obtain the graphs of the cosecant function $F(t) = \csc t$, the secant function $G(t) = \sec t$, and the cotangent function $H(t) = \cot t$ by finding, whenever possible, the reciprocals of the *y*-coordinates on the graphs of $y = \sin t$, $y = \cos t$, and $y = \tan t$, respectively. All of these functions are shown in Figures 7.61, 7.62, and 7.63. Values of t for which $\sin t = 0$, $\cos t = 0$, and $\tan t = 0$ lead to vertical asymptotes on the graphs of $F(t) = \csc t$, $G(t) = \sec t$, and $H(t) = \cot t$, respectively. We can generate the graphs of the cosecant function, the secant function, and the cotangent function by setting a graphing calculator in radian mode and entering $y = 1/\sin x$, $y = 1/\cos x$, and $y = 1/\tan x$.

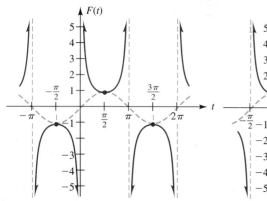

FIGURE 7.61
Graph of the cosecant function $F(t) = \csc t$

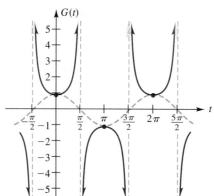

FIGURE 7.62
Graph of the secant function $G(t) = \sec t$

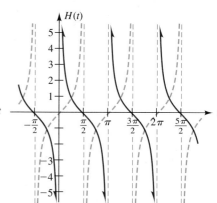

FIGURE 7.63
Graph of the cotangent function $H(t) = \cot t$

The following properties can be observed from the graphs of $F(t) = \csc t$, $G(t) = \sec t$, and $H(t) = \cot t$.

◀ Properties of the Cosecant Function

Cosecant function: $F(t) = \csc t$

1. The *domain* is the set of all real numbers except
$$t = 0, \pm\pi, \pm 2\pi, \pm 3\pi, \ldots.$$
2. The *range* is $(-\infty, -1] \cup [1, \infty)$.
3. The *period* is 2π.
4. The function has no *zeros*.
5. The function is *odd*, that is, $\csc(-t) = -\csc t$ for all t in the domain of F.

◆ **Properties of the Secant Function**

Secant function: $G(t) = \sec t$

1. The *domain* is the set of all real numbers except
$$t = \pm \frac{\pi}{2}, \pm \frac{3\pi}{2}, \pm \frac{5\pi}{2}, \ldots$$
2. The *range* is $(-\infty, -1] \cup [1, \infty)$.
3. The *period* is 2π.
4. The function has no *zeros*.
5. The function is *even*, that is, $\sec(-t) = \sec t$ for all t in the domain of G.

◆ **Properties of the Cotangent Function**

Cotangent function: $H(t) = \cot t$

1. The *domain* is the set of all real numbers except
$$t = 0, \pm\pi, \pm 2\pi, \pm 3\pi, \ldots$$
2. The *range* is the set of all real numbers.
3. The *period* is π.
4. The *zeros* are $\pm \frac{\pi}{2}, \pm \frac{3\pi}{2}, \pm \frac{5\pi}{2}, \ldots$
5. The function is *odd*, that is, $\cot(-t) = -\cot t$ for all t in the domain of H.

In this section, we use the basic graphs of $y = \tan x$, $y = \cot x$, $y = \csc x$, and $y = \sec x$ in conjunction with the shift rules, axis reflection rules, and stretch and compress rules (Section 2.5) to discuss the graphs of the following equations:

$$y = a \tan(bx + c) \qquad y = a \cot(bx + c)$$
$$y = a \csc(bx + c) \qquad y = a \sec(bx + c)$$

◆ **Graphs of $y = a \tan(bx + c)$ and $y = a \cot(bx + c)$**

By the vertical stretch and compress rules (Section 2.5), we know that the graphs of $y = a \tan x$ and $y = a \cot x$, with $a > 0$, are similar to the graphs of $y = \tan x$ and $y = \cot x$, respectively, stretched or compressed vertically by a factor of a.

The graph of $y = a \tan x$ has period π and completes one cycle from $x = -\pi/2$ to $x = \pi/2$. Therefore, the graph of $y = a \tan bx$ completes one cycle from

$$bx = -\frac{\pi}{2} \quad \text{to} \quad bx = \frac{\pi}{2}$$

$$x = -\frac{\pi}{2b} \quad \text{to} \quad x = \frac{\pi}{2b}$$

A pair of consecutive vertical asymptotes for the graph of $y = a \tan bx$.

Hence, the period P of $y = a \tan bx$ is the length of the interval from $x = -\pi/(2b)$ to $x = \pi/(2b)$:

$$P = \left| \frac{\pi}{2b} - \left(-\frac{\pi}{2b}\right) \right| = \frac{\pi}{|b|}$$

The graph of $y = a \cot x$ has period π and completes one cycle from $x = 0$ to $x = \pi$. Therefore, the graph of $y = a \cot bx$ completes one cycle from

$$bx = 0 \quad \text{to} \quad bx = \pi$$

$$x = 0 \quad \text{to} \quad x = \frac{\pi}{b}$$

A pair of consecutive vertical asymptotes for the graph of $y = a \cot bx$.

Hence, the period P of $y = a \cot bx$ is the length of the interval from $x = 0$ to $x = \pi/b$:

$$P = \left| \frac{\pi}{b} - 0 \right| = \frac{\pi}{|b|}$$

◀ **Period Formula for Tangent and Cotangent**

The period P of $y = a \tan bx$ or $y = a \cot bx$ is given by

$$P = \frac{\pi}{|b|} \quad \text{for } b \neq 0.$$

Figures 7.64 and 7.65 show the graphs of one cycle of $y = a \tan bx$ and $y = a \cot bx$ for $a > 0$ and $b > 0$.

SECTION 7.5 ♦ Other Trigonometric Functions

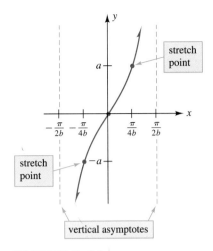

FIGURE 7.64
Graph of one cycle of $y = a \tan bx$ for $a > 0$ and $b > 0$

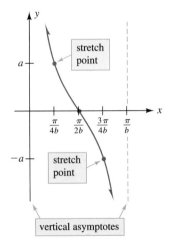

FIGURE 7.65
Graph of one cycle of $y = a \cot bx$ for $a > 0$ and $b > 0$.

The graph of $y = a \tan bx$ has five key values in the interval $[-\pi/(2b), \pi/(2b)]$, which we may obtain by dividing this interval into four equal parts:

$$x = -\frac{\pi}{2b} \quad x = \frac{\pi}{2b} \quad (0, 0) \quad \left(-\frac{\pi}{4b}, -a\right) \quad \left(\frac{\pi}{4b}, a\right)$$

vertical asymptotes — x-intercept — stretch (or compress) points

The graph of $y = a \cot bx$ has five key values in the interval $[0, \pi/b]$, which we may obtain by dividing this interval into four equal parts:

$$x = 0 \quad x = \frac{\pi}{b} \quad \left(\frac{\pi}{2b}, 0\right) \quad \left(\frac{\pi}{4b}, a\right) \quad \left(\frac{3\pi}{4b}, -a\right)$$

vertical asymptotes — x-intercept — stretch (or compress) points

The equations $y = a \tan(bx + c)$ and $y = a \cot(bx + c)$ may be rewritten by factoring out b:

$$y = a \tan(bx + c) = a \tan b\left(x + \frac{c}{b}\right)$$

$$y = a \cot(bx + c) = a \cot b\left(x + \frac{c}{b}\right)$$

Now, by the horizontal shift rule (Section 2.5), we know that the graphs of these equations are the same as the graphs of $y = a \tan bx$ and $y = a \cot bx$, respectively, shifted horizontally $|c/b|$ units. If $c/b < 0$ the shift is to the right, or if $c/b > 0$ the shift is to the left.

EXAMPLE 1 State the period for the graph of each equation. Then list the five key values on one cycle of the graph.

(a) $y = 3 \tan \dfrac{x}{2}$ **(b)** $y = \cot\left(2x - \dfrac{\pi}{2}\right)$

◆ **Solution**

(a) The graph of $y = \tan x$ completes one cycle from $x = -\pi/2$ to $x = \pi/2$. Therefore, the graph of $y = 3 \tan \dfrac{x}{2}$ completes one cycle from

$$\frac{x}{2} = -\frac{\pi}{2} \quad \text{to} \quad \frac{x}{2} = \frac{\pi}{2}$$
$$x = -\pi \quad \text{to} \quad x = \pi$$

and is stretched vertically by a factor of 3. The length of the interval from $x = -\pi$ to $x = \pi$ is 2π. Hence, the period is 2π, which we can verify by using the period formula for tangent:

$$P = \frac{\pi}{|b|} = \frac{\pi}{\left|\frac{1}{2}\right|} = 2\pi$$

Dividing the interval $[-\pi, \pi]$ into four equal parts gives us the five key values on this cycle:

$$\underbrace{x = -\pi \quad x = \pi}_{\text{vertical asymptotes}} \quad \underbrace{(0, 0)}_{x\text{-intercept}} \quad \underbrace{\left(-\frac{\pi}{2}, -3\right) \left(\frac{\pi}{2}, 3\right)}_{\text{stretch points}}$$

Figure 7.66 shows the graph of $y = 3 \tan \dfrac{x}{2}$ in the viewing rectangle of a graphing calculator, and this graph supports our work. We have chosen $\pi/2$ as the distance between scale marks on the x-axis.

(b) The graph of $y = \cot x$ completes one cycle from $x = 0$ to $x = \pi$. Therefore, the graph of $y = \cot\left(2x - \dfrac{\pi}{2}\right)$ completes one cycle from

$$2x - \frac{\pi}{2} = 0 \quad \text{to} \quad 2x - \frac{\pi}{2} = \pi$$

$$2x = \frac{\pi}{2} \quad \text{to} \quad 2x = \frac{3\pi}{2}$$

$$x = \frac{\pi}{4} \quad \text{to} \quad x = \frac{3\pi}{4}$$

The length of the interval from $\pi/4$ to $3\pi/4$ is $\pi/2$. Hence, the period is $\pi/2$, which we can verify by using the period formula for cotangent:

$$P = \frac{\pi}{|b|} = \frac{\pi}{|2|} = \frac{\pi}{2}$$

Dividing the interval $[\pi/4, 3\pi/4]$ into four equal parts gives us the five key values on this cycle:

$$\underbrace{x = \frac{\pi}{4} \quad x = \frac{3\pi}{4}}_{\text{vertical asymptotes}} \quad \underbrace{\left(\frac{\pi}{2}, 0\right),}_{x\text{-intercept}} \quad \underbrace{\left(\frac{3\pi}{8}, 1\right) \left(\frac{5\pi}{8}, -1\right)}_{\text{stretch points}}$$

Figure 7.67 shows the graph of $y = \cot\left(2x - \dfrac{\pi}{2}\right)$ in the viewing rectangle of a graphing calculator, and this graph supports our work. We have chosen $\pi/8$ as the distance between scale marks on the x-axis. ◆

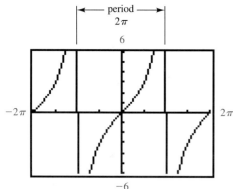

FIGURE 7.66

The graph of $y = 3 \tan \dfrac{x}{2}$ completes one cycle from $x = -\pi$ to $x = \pi$ and is stretched vertically by a factor of 3.

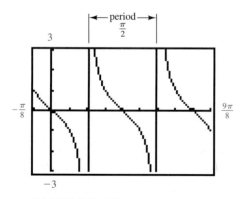

FIGURE 7.67

The graph of $y = \cot\left(2x - \dfrac{\pi}{2}\right)$ completes one cycle from $x = \pi/4$ to $x = 3\pi/4$.

SECTION 7.5 ♦ *Other Trigonometric Functions*

If b is negative in $y = a\tan(bx + c)$ or $y = a\cot(bx + c)$, then we may rewrite these equations by using the facts that, for any real number t,

$$\tan(-t) = -\tan t \quad \text{and} \quad \cot(-t) = -\cot t.$$

For example, the equation $y = 3\tan\left(-\dfrac{x}{2}\right)$ may be rewritten as

$$y = 3\tan\left(-\frac{x}{2}\right) = -3\tan\frac{x}{2}.$$

Now, by the x-axis reflection rule (Section 2.5), the graph of $y = -3\tan\dfrac{x}{2}$ is the same as the graph of $y = 3\tan\dfrac{x}{2}$, reflected about the x-axis.

PROBLEM 1 Compare the graph of $y = \cot\left(\dfrac{\pi}{2} - 2x\right)$ to the graph of $y = \cot\left(2x - \dfrac{\pi}{2}\right)$. ♦

♦ Graphs of $y = a\csc(bx + c)$ and $y = a\sec(bx + c)$

By the vertical stretch and compress rules (Section 2.5), we know that the graphs of $y = a\csc x$ and $y = a\sec x$, with $a > 0$, are similar to the graphs of $y = \csc x$ and $y = \sec x$, respectively, stretched or compressed vertically by a factor of a.

The graph of $y = a\csc x$ has period 2π and completes one cycle from $x = 0$ to $x = 2\pi$. Therefore, the graph of $y = a\csc bx$ completes one cycle from

$$bx = 0 \quad \text{to} \quad bx = 2\pi$$

$$x = 0 \quad \text{to} \quad x = \frac{2\pi}{b}$$

Hence, the period P of $y = a\csc bx$ is the length of the interval from $x = 0$ to $x = 2\pi/b$:

$$P = \left|\frac{2\pi}{b} - 0\right| = \frac{2\pi}{|b|}$$

The graph of $y = a\sec x$ has period 2π and completes one cycle from $x = -\pi/2$ to $x = 3\pi/2$. Therefore, the graph of $y = a\sec bx$ completes one cycle from

$$bx = -\frac{\pi}{2} \quad \text{to} \quad bx = \frac{3\pi}{2}$$

$$x = -\frac{\pi}{2b} \quad \text{to} \quad x = \frac{3\pi}{2b}$$

Hence, the period P of $y = a \sec bx$ is the length of the interval from $x = -\pi/(2b)$ to $x = 3\pi/(2b)$:

$$P = \left| \frac{3\pi}{2b} - \left(-\frac{\pi}{2b}\right) \right| = \frac{2\pi}{|b|}$$

◀ **Period Formula for Cosecant and Secant**

The period P of $y = a \csc bx$ or $y = a \sec bx$ is given by

$$P = \frac{2\pi}{|b|} \quad \text{for } b \neq 0.$$

Figures 7.68 and 7.69 show the graphs of one cycle of $y = a \csc bx$ and $y = a \sec bx$ with $a > 0$ and $b > 0$.

The graph of $y = a \csc bx$ has five key values in the interval $[0, 2\pi/b]$, which we may obtain by dividing this interval into four equal parts:

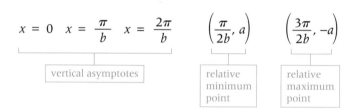

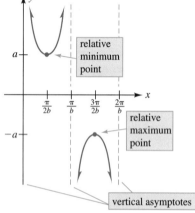

FIGURE 7.68

Graph of one cycle of $y = a \csc bx$ with $a > 0$ and $b > 0$

The graph of $y = a \sec bx$ has five key values in the interval $[-\pi/(2b), 3\pi/(2b)]$, which we may obtain by dividing this interval into four equal parts:

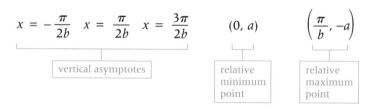

The equations $y = a \csc(bx + c)$ and $y = a \sec(bx + c)$ may be rewritten by factoring out b:

$$y = a \csc(bx + c) = a \csc b\left(x + \frac{c}{b}\right)$$

$$y = a \sec(bx + c) = a \sec b\left(x + \frac{c}{b}\right)$$

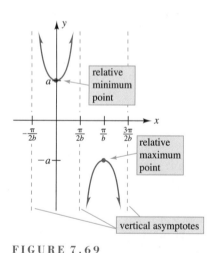

FIGURE 7.69

Graph of one cycle of $y = a \sec bx$ with $a > 0$ and $b > 0$

Now, by the horizontal shift rule (Section 2.5), we know that the graphs of these equations are the same as the graphs of $y = a \csc bx$ and $y = a \sec bx$, respectively, shifted horizontally $|c/b|$ units. If $c/b < 0$ the shift is to the right, or if $c/b > 0$ the shift is to the left.

EXAMPLE 2 State the period for the graph of each equation. Then list the five key values on one cycle of the graph.

(a) $y = 3 \csc 4x$ **(b)** $y = \sec\left(\pi x + \dfrac{\pi}{2}\right)$

◆ **Solution**

(a) The graph of $y = \csc x$ completes one cycle from $x = 0$ to $x = 2\pi$. Therefore, the graph of $y = 3 \csc 4x$ completes one cycle from

$$4x = 0 \quad \text{to} \quad 4x = 2\pi$$

$$x = 0 \quad \text{to} \quad x = \dfrac{\pi}{2}$$

and is stretched vertically by a factor of 3. The length of the interval from $x = 0$ to $x = \pi/2$ is $\pi/2$. Hence, the period is $\pi/2$, which we can verify by using the period formula for cosecant:

$$P = \dfrac{2\pi}{|b|} = \dfrac{2\pi}{|4|} = \dfrac{\pi}{2}$$

Dividing the interval $[0, \pi/2]$ into four equal parts gives us the five key values on this cycle:

$$x = 0 \quad x = \dfrac{\pi}{4} \quad x = \dfrac{\pi}{2} \qquad \left(\dfrac{\pi}{8}, 3\right) \qquad \left(\dfrac{3\pi}{8}, -3\right)$$

vertical asymptotes | relative minimum point | relative maximum point

Figure 7.70 shows the graph of $y = 3 \csc 4x$ in the viewing rectangle of a graphing calculator, and this graph supports our work. We have chosen $\pi/8$ as the distance between scale marks on the x-axis.

(b) The graph of $y = \sec x$ completes one cycle from $x = -\pi/2$ to $x = 3\pi/2$. Therefore, the graph of $y = \sec\left(\pi x + \dfrac{\pi}{2}\right)$ completes one cycle from

$$\pi x + \dfrac{\pi}{2} = -\dfrac{\pi}{2} \quad \text{to} \quad \pi x + \dfrac{\pi}{2} = \dfrac{3\pi}{2}$$

$$\pi x = -\pi \quad \text{to} \quad \pi x = \pi$$

$$x = -1 \quad \text{to} \quad x = 1$$

The length of the interval from $x = -1$ to $x = 1$ is 2. Hence, the period is 2, which we can verify by using the period formula for secant:

$$P = \dfrac{2\pi}{|b|} = \dfrac{2\pi}{|\pi|} = 2$$

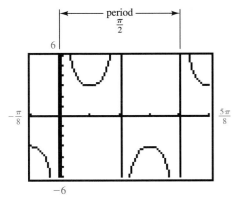

FIGURE 7.70

The graph of $y = 3 \csc 4x$ completes one cycle from $x = 0$ to $x = \pi/2$ and is stretched vertically by a factor of 3.

Dividing the interval $[-1, 1]$ into four equal parts gives us the following five key values on this cycle:

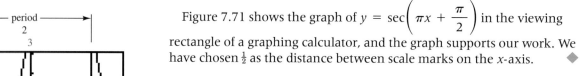

Figure 7.71 shows the graph of $y = \sec\left(\pi x + \dfrac{\pi}{2}\right)$ in the viewing rectangle of a graphing calculator, and the graph supports our work. We have chosen $\frac{1}{2}$ as the distance between scale marks on the x-axis. ◆

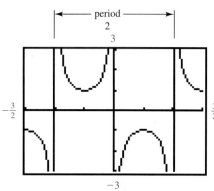

FIGURE 7.71

The graph of $y = \sec\left(\pi x + \dfrac{\pi}{2}\right)$ completes one cycle from $x = -1$ to $x = 1$.

If b is negative in $y = a \csc(bx + c)$ or $y = a \sec(bx + c)$, then we may rewrite these equations by using the facts that, for any real number t,

$$\csc(-t) = -\csc t \quad \text{and} \quad \sec(-t) = \sec t.$$

For example, the equation $y = 3 \csc(-4x)$ may be rewritten as follows:

$$y = 3 \csc(-4x) = -3 \csc 4x$$

Now, by the x-axis reflection rule (Section 2.5), the graph of $y = -3 \csc 4x$ is the same as the graph of $y = 3 \csc 4x$ reflected about the x-axis.

PROBLEM 2 Compare the graph of $y = \sec\left(-\pi x - \dfrac{\pi}{2}\right)$ to the graph of $y = \sec\left(\pi x + \dfrac{\pi}{2}\right)$. ◆

Exercises 7.5

◆ Basic Skills

In Exercises 1–4, complete the given table of values, and then plot points to obtain the graph of one cycle of the given function.

1. $f(x) = \tan x$

x	$-\dfrac{\pi}{2}$	$-\dfrac{\pi}{3}$	$-\dfrac{\pi}{4}$	$-\dfrac{\pi}{6}$	0	$\dfrac{\pi}{6}$	$\dfrac{\pi}{4}$	$\dfrac{\pi}{3}$	$\dfrac{\pi}{2}$
$f(x)$									

2. $f(x) = \cot x$

x	0	$\frac{\pi}{6}$	$\frac{\pi}{4}$	$\frac{\pi}{3}$	$\frac{\pi}{2}$	$\frac{2\pi}{3}$	$\frac{3\pi}{4}$	$\frac{5\pi}{6}$	π
$f(x)$									

3. $f(x) = \csc x$

x	0	$\frac{\pi}{4}$	$\frac{\pi}{2}$	$\frac{3\pi}{4}$	π	$\frac{5\pi}{4}$	$\frac{3\pi}{2}$	$\frac{7\pi}{4}$	2π
$f(x)$									

4. $f(x) = \sec x$

x	$-\frac{\pi}{2}$	$-\frac{\pi}{3}$	$-\frac{\pi}{6}$	0	$\frac{\pi}{6}$	$\frac{\pi}{3}$	$\frac{\pi}{2}$	$\frac{2\pi}{3}$	$\frac{5\pi}{6}$	π	$\frac{7\pi}{6}$	$\frac{4\pi}{3}$	$\frac{3\pi}{2}$
$f(x)$													

In Exercises 5–12, use the results of Exercises 1 and 2 and the fact that the tangent and cotangent functions are periodic with period π to find all values of x in the interval $[-4\pi, 4\pi]$ that satisfy the given equation.

5. $\tan x = 0$
6. $\cot x = 0$
7. $\cot x = -1$
8. $\tan x = 1$
9. $\tan x = \frac{1}{\sqrt{3}}$
10. $\cot x = -\frac{1}{\sqrt{3}}$
11. $\tan x = -\sqrt{3}$
12. $\cot x = \sqrt{3}$

In Exercises 13–20, use the results of Exercises 3 and 4 and the fact that the cosecant and secant functions are periodic with period 2π to find all values of x in the interval $[-4\pi, 4\pi]$ that satisfy the given equation.

13. $\csc x = 1$
14. $\sec x = -1$
15. $\sec x = -2$
16. $\csc x = \sqrt{2}$
17. $\sec x = \frac{2}{\sqrt{3}}$
18. $\csc x = -\sqrt{2}$
19. $\csc x = 0$
20. $\sec x = \frac{1}{2}$

In Exercises 21–48, state the period for the graph of each equation. Then show the five key values on one cycle of the graph. Use a graphing calculator to confirm each of your answers.

21. $y = \tan 3x$
22. $y = 2 \tan \frac{x}{3}$
23. $y = 2 \tan\left(-\frac{3}{4}x\right)$
24. $y = -\frac{1}{3} \tan 4x$
25. $y = \frac{1}{2} \cot 4x$
26. $y = 2 \cot 2x$
27. $y = -\cot \frac{2}{3} x$
28. $y = 3 \cot\left(-\frac{\pi x}{2}\right)$
29. $y = \csc \pi x$
30. $y = \csc \frac{2x}{5}$
31. $y = 2 \csc(-3x)$
32. $y = -0.5 \csc 1.2x$
33. $y = \sec 0.8x$
34. $y = \frac{2}{3} \sec 4x$
35. $y = -4 \sec \frac{3}{8} x$
36. $y = \frac{1}{2} \sec(-2x)$
37. $y = 3 + \tan x$
38. $y = 4 - \cot \frac{\pi}{12} x$
39. $y = 1 - \csc 2x$
40. $y = 2[1 + \sec(-2x)]$
41. $y = \tan(3x + \pi)$
42. $y = 3 \cot\left(2\pi x - \frac{\pi}{4}\right)$
43. $y = \frac{1}{3} \cot\left(\pi x - \frac{\pi}{6}\right)$
44. $y = \tan\left(2x + \frac{\pi}{8}\right)$
45. $y = \csc\left(\frac{\pi}{3} - x\right)$
46. $y = 2 \sec(4x + \pi)$
47. $y = \frac{3}{4} \sec \pi \left(x - \frac{1}{4}\right)$
48. $y = 2 + 3 \csc\left(x - \frac{\pi}{8}\right)$

In Exercises 49–54, generate the graph of each equation in the viewing rectangle of a graphing calculator. From the graph, ap-

proximate the period, rounding each answer to three significant digits. Check each of your answers algebraically.

49. $y = 1.50 \tan 4.33x$
50. $y = 0.46 \cot 3.65x$
51. $y = 2.67 \csc 0.87x$
52. $y = 2.12 \sec(-1.91x)$
53. $y = \cot(11.2 - 1.32x)$
54. $y = 1.2 \csc(1.09x - 0.52)$

In Exercises 55 and 56, write the equation of the given graph in the form $y = a \tan bx$ or $y = a \cot bx$ with $a > 0$ and $b > 0$.

55.
56.

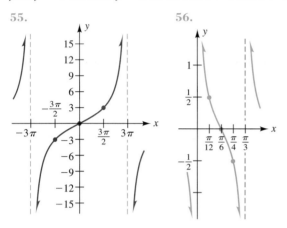

57.

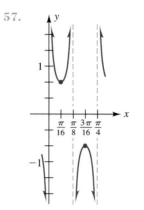

58.

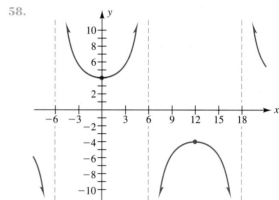

In Exercises 57 and 58, write the equation of the given graph in the form $y = a \csc bx$ or $y = a \sec bx$ with $a > 0$ and $b > 0$.

Critical Thinking

If the graph of $y = \tan x$ is shifted to the right $\pi/2$ units, we obtain the graph of $y = \cot x$, reflected about the x-axis. Based upon this observation, we conclude that

$$\tan\left(x - \frac{\pi}{2}\right) = -\cot x$$

for all real numbers x. In Exercises 59–62, generate the graph of the function f in the viewing rectangle of a graphing calculator, and then compare the graph of f with the graphs of $y = \tan x$, $y = -\tan x$, $y = \cot x$, and $y = -\cot x$. Based upon your comparisons, state a trigonometric identity that relates $f(x)$ to either $\tan x$, $-\tan x$, $\cot x$, or $-\cot x$.

59. $f(x) = \tan\left(x + \frac{\pi}{2}\right)$
60. $f(x) = \tan(x + \pi)$
61. $f(x) = \cot(\pi - x)$
62. $f(x) = \cot\left(x + \frac{3\pi}{2}\right)$

If the graph of $y = \sec x$ is shifted to the right $\pi/2$ units, we obtain the graph of $y = \csc x$. Based upon this observation, we conclude that

$$\sec\left(x - \frac{\pi}{2}\right) = \csc x$$

for all real numbers x. In Exercises 63–66, generate the graph of the function f in the viewing rectangle of a graphing calculator, and then compare the graph of f with the graphs of $y = \csc x$, $y = -\csc x$, $y = \sec x$, and $y = -\sec x$. Based upon your comparisons, state a trigonometric identity that relates $f(x)$ to either $\csc x$, $-\csc x$, $\sec x$, or $-\sec x$.

63. $f(x) = \csc\left(x + \frac{\pi}{2}\right)$
64. $f(x) = \csc(\pi - x)$
65. $f(x) = \sec\left(x - \frac{3\pi}{2}\right)$
66. $f(x) = \sec(x - \pi)$

67. Find a function f of the form $f(x) = k + a \csc bx$ if its graph has a relative minimum point at (2, 5) and a relative maximum point at (6, −1).

68. Generate the graphs of the functions $f(x) = \tan|x|$ and $g(x) = |\tan x|$ in the viewing rectangle of a graphing calculator, and then determine whether either of these functions is periodic. If the function is periodic, state its period.

69. Use the results of Exercises 1 and 2 and the fact that the tangent and cotangent functions are periodic with pe-

SECTION 7.6 ◆ *The Inverse Trigonometric Functions* 453

riod π to find all values of x in the interval $[-4\pi, 4\pi]$ for which $\tan x = \cot x$.

70. In calculus, it is shown that certain polynomial functions represent good approximations of the trigonometric functions. Use a graphing calculator to generate the graph of the polynomial function

$$P(x) = x + \frac{x^3}{3} + \frac{2x^5}{15} + \frac{17x^7}{315} + \frac{62x^9}{2835}$$

over the interval $(-\pi/2, \pi/2)$.

(a) What trigonometric function does P seem to approximate?

(b) Compare $P(1)$ to the value of that trigonometric function found in part (a) when $x = 1$. Round each answer to four significant digits.

7.6 The Inverse Trigonometric Functions

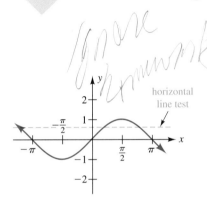

FIGURE 7.72

The graph of the sine function, $y = \sin x$, fails the horizontal line test. Thus, the sine function is not one-to-one and does not have an inverse.

◆ **Introductory Comments**

Recall that in order to have an inverse, a function must be one-to-one. By the horizontal line test (Section 2.6), the sine function $y = \sin x$ is not a one-to-one function because, as shown in Figure 7.72, at least one horizontal line intersects its graph in more than one point. Similarly, the cosine function $y = \cos x$ and the tangent function $y = \tan x$ are not one-to-one functions because each graph is intercepted by horizontal lines in more than one point. Thus, we conclude that neither the sine, cosine, nor tangent function has an inverse over its entire domain.

In this section, we restrict the domain of each of these trigonometric functions so that the new function which is formed is a one-to-one function that has the same range as the original function and, therefore, does have an inverse. We begin by restricting the domain of the sine function.

◆ **Restricted Sine Function and Its Inverse**

Consider the *restricted sine function*

$$y = \sin x \quad \text{for } -\pi/2 \leq x \leq \pi/2.$$

The graph of this function is shown in Figure 7.73. On the interval $[-\pi/2, \pi/2]$ the restricted sine function attains every value of the sine function once and only once. Thus, the restricted sine function is one-to-one and has an inverse. We refer to the inverse of the restricted sine function as the *inverse sine function*.

Using the method suggested in Section 2.6, we can attempt to find this inverse sine function by interchanging x and y in $y = \sin x$ for $-\pi/2 \leq x \leq \pi/2$, and then solving for y. If we proceed in this manner, we obtain

$$x = \sin y \quad \text{for } -\pi/2 \leq y \leq \pi/2.$$

However, we have no algebraic procedure enabling us to solve for y in this equation. By convention, we solve this equation for y by writing

$$y = \sin^{-1} x \quad \text{for } -\pi/2 \leq y \leq \pi/2.$$

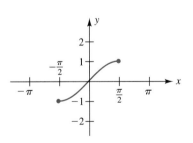

FIGURE 7.73

The restricted sine function $y = \sin x$, for $-\pi/2 \leq x \leq \pi/2$ is one-to-one and has an inverse.

Hence, $\sin^{-1} x$ denotes the real number (or angle in radians) between $-\pi/2$ and $\pi/2$ whose sine is x.

Inverse Sine Function

The **inverse sine function** is defined as

$$y = \sin^{-1} x \quad \text{if and only if} \quad x = \sin y \quad \text{for } -\pi/2 \leq y \leq \pi/2.$$

CAUTION In the notation $\sin^{-1} x$, the superscript -1 denotes an inverse, not an exponent, that is,

$$\sin^{-1} x \quad \text{does } not \text{ mean} \quad \frac{1}{\sin x}.$$

Note: The inverse sine function is also referred to as the **arcsine function**, and the notation arcsin x is used frequently instead of $\sin^{-1} x$. The arcsine notation comes from the fact that we can interpret the real number whose sine is x as the length of that arc on a unit circle whose sine is x.

Recall from Section 2.6 that the graphs of a function and its inverse function are symmetric with respect to the line $y = x$. Thus, the graph of the inverse sine function (or arcsine function) is the same as the graph of the restricted sine function (Figure 7.73) reflected in the line $y = x$. Figure 7.74 shows the graph of $y = \sin^{-1} x$. Note that the domain of $y = \sin^{-1} x$ is $[-1, 1]$ and the range is $[-\pi/2, \pi/2]$. Also note that the graph of $y = \sin^{-1} x$ is symmetric with respect to the origin, which indicates that the inverse sine function is an odd function. Thus, we have

$$\sin^{-1}(-x) = -\sin^{-1} x$$

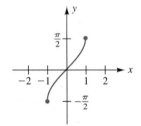

FIGURE 7.74
Graph of the inverse sine function, $y = \sin^{-1} x$. Its domain is $[-1, 1]$ and its range is $[-\pi/2, \pi/2]$.

for every real number x in the interval $[-1, 1]$.

EXAMPLE 1 Find the exact value of each expression, if it is defined.

(a) $\sin^{-1} \dfrac{1}{\sqrt{2}}$ **(b)** $\arcsin(-1)$ **(c)** $\sin^{-1} 2$

Solution

(a) Letting $y = \sin^{-1} \dfrac{1}{\sqrt{2}}$, we have

$$\sin y = \frac{1}{\sqrt{2}} \quad \text{for } -\pi/2 \leq y \leq \pi/2.$$

From our work with the unit circle in Section 7.3, we know that

$$\sin \frac{\pi}{4} = \frac{1}{\sqrt{2}}.$$

Since $\pi/4$ is in the interval $[-\pi/2, \pi/2]$, we conclude that

$$\sin^{-1}\frac{1}{\sqrt{2}} = \frac{\pi}{4}$$

(b) Letting $y = \arcsin(-1)$, we have

$$\sin y = -1 \quad \text{for } -\pi/2 \le y \le \pi/2.$$

From our work with the unit circle in Section 7.3, we know that $\sin(\pi/2) = 1$. Therefore, using the fact that $\sin^{-1}(-x) = -\sin^{-1}x$, we write

$$\arcsin(-1) = -\arcsin 1 = -\frac{\pi}{2}$$

(c) The domain of the inverse sine function $y = \sin^{-1}x$ is $[-1, 1]$. Thus, for $x > 1$ or $x < -1$, the inverse sine function is undefined. Hence,

$$\sin^{-1} 2 \quad \text{is undefined.} \qquad \blacklozenge$$

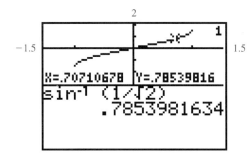

FIGURE 7.75
Graphical and numerical evaluation of $\sin^{-1}(1/\sqrt{2})$ using split-screen mode

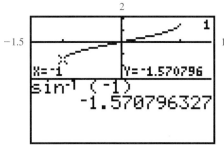

FIGURE 7.76
Graphical and numerical evaluation of $\arcsin(-1)$ using split-screen mode

We can use a graphing calculator to verify the results in Example 1. To evaluate the inverse sine function or to generate the graph of the inverse sine function in the viewing rectangle of a graphing calculator, we set the calculator in radian mode and use the inverse sine key, $\boxed{\text{SIN}^{-1}}$. (To access the inverse trigonometric functions on some calculators, we use the $\boxed{\text{2ND}}$ key or $\boxed{\text{INV}}$ key in conjunction with the trig key.) The split screens in Figures 7.75 and 7.76 verify the evaluations in Examples 1(a) and 1(b), both graphically and numerically. The evaluation in Example 1(c) gives the message "domain error" in the display of the calculator.

$\boxed{\text{PROBLEM 1}}$ Use a calculator to find the approximate value of each expression. Round each answer to four significant digits.

(a) $\arcsin 0.53$ **(b)** $\sin^{-1}(-0.766)$ $\qquad \blacklozenge$

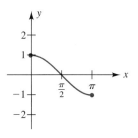

FIGURE 7.77
The restricted cosine function $y = \cos x$ for $0 \le x \le \pi$ is one-to-one and has an inverse.

◆ Restricted Cosine Function and Its Inverse

Like the sine function, the cosine function $y = \cos x$ is not one-to-one and does not have an inverse. However, consider the *restricted cosine function*

$$y = \cos x \quad \text{for } 0 \le x \le \pi.$$

The graph of this function is shown in Figure 7.77. On the interval $[0, \pi]$ the restricted cosine function attains every value of the cosine function once and only once. Thus, the restricted cosine function is one-to-one and has an inverse. We refer to the inverse of the restricted cosine function as the *inverse cosine function* and define it as follows.

Inverse Cosine Function

The **inverse cosine function** is defined as

$$y = \cos^{-1} x \quad \text{if and only if} \quad x = \cos y \quad \text{for } 0 \leq y \leq \pi.$$

Note: The inverse cosine function is also referred to as the **arccosine function**, and the notation arccos x is used frequently instead of $\cos^{-1} x$. Both $\cos^{-1} x$ and arccos x denote the real number (or angle in radians) between 0 and π whose cosine is x.

The graph of the inverse cosine function (or arccosine function) is the same as the graph of the restricted cosine function (Figure 7.77) reflected in the line $y = x$. Figure 7.78 shows the graph of $y = \cos^{-1} x$. Note that the domain of $y = \cos^{-1} x$ is $[-1, 1]$ and the range is $[0, \pi]$. Since the graph of $y = \cos^{-1} x$ is symmetric with respect to neither the origin nor the y-axis, the inverse cosine function is neither even nor odd. However, we do have the following relationship between positive and negative values of x in the interval $[-1, 1]$:

$$\cos^{-1}(-x) = \pi - \cos^{-1} x$$

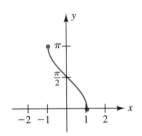

FIGURE 7.78
Graph of the inverse cosine function, $y = \cos^{-1} x$. Its domain is $[-1, 1]$ and its range is $[0, \pi]$.

In Exercise 73, we are asked to prove this relationship.

EXAMPLE 2 Find the exact value of each expression, if it is defined.

(a) $\cos^{-1} 1$ **(b)** $\cos^{-1}\left(-\frac{1}{2}\right)$ **(c)** $\arccos(-2)$

Solution

(a) Letting $y = \cos^{-1} 1$, we have

$$\cos y = 1 \quad \text{for } 0 \leq y \leq \pi.$$

From our work with the unit circle in Section 7.3, we have $\cos 0 = 1$. Since 0 is in the interval $[0, \pi]$, we conclude that

$$\cos^{-1} 1 = 0.$$

(b) Letting $y = \cos^{-1}\left(-\frac{1}{2}\right)$, we have

$$\cos y = -\tfrac{1}{2} \quad \text{for } 0 \leq y \leq \pi.$$

From our work with the unit circle in Section 7.3, we know that $\cos(\pi/3) = \frac{1}{2}$. Therefore, using the fact that $\cos^{-1}(-x) = \pi - \cos^{-1} x$, we write

$$\cos^{-1}\left(-\tfrac{1}{2}\right) = \pi - \cos^{-1} \tfrac{1}{2} = \pi - \frac{\pi}{3} = \frac{2\pi}{3}.$$

(c) The domain of the arccosine function $y = \arccos x$ is $[-1, 1]$. Thus, for $x > 1$ or $x < -1$, the arccosine function is undefined. Hence,

$$\arccos(-2) \quad \text{is undefined.}$$

We can use a graphing calculator to verify the results in Example 2. To evaluate the inverse cosine function or to generate the graph of the inverse cosine function in the viewing rectangle of a graphing calculator, we set the calculator in radian mode and use the inverse cosine key, $\boxed{\cos^{-1}}$. The split screens in Figures 7.79 and 7.80 verify the evaluations in Examples 2(a) and 2(b), both graphically and numerically. For the evaluation in Example 2(c), the calculator displays the message "domain error."

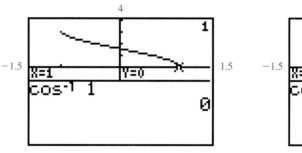

FIGURE 7.79
Graphical and numerical evaluation of $\cos^{-1} 1$ using split-screen mode

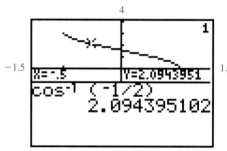

FIGURE 7.80
Graphical and numerical evaluation of $\cos^{-1}\left(-\frac{1}{2}\right)$ using split-screen mode

PROBLEM 2 Use a calculator to find the approximate value of each expression. Round each answer to four significant digits.

(a) $\arccos 0.809$ (b) $\cos^{-1}(-0.42)$

◆ Restricted Tangent Function and Its Inverse

Like the sine and cosine functions, the tangent function $y = \tan x$ is not one-to-one and does not have an inverse. However, consider the *restricted tangent function*

$$y = \tan x \quad \text{for} \quad -\pi/2 < x < \pi/2.$$

The graph of this function is shown in Figure 7.81. On the interval $(-\pi/2, \pi/2)$ the restricted tangent function attains every value of the tangent function once and only once. Thus, the restricted tangent function is one-to-one and has an inverse. We refer to the inverse of the restricted tangent function as the *inverse tangent function* and define it as follows.

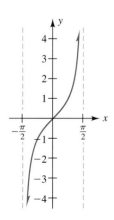

FIGURE 7.81
The restricted tangent function $y = \tan x$ for $-\pi/2 < x < \pi/2$ is one-to-one and has an inverse.

Inverse Tangent Function

The **inverse tangent function** is defined as

$$y = \tan^{-1} x \quad \text{if and only if} \quad x = \tan y \quad \text{for } -\pi/2 < y < \pi/2.$$

Note: The inverse tangent function is also referred to as the **arctangent function**, and the notation arctan x is used frequently instead of $\tan^{-1} x$. Both $\tan^{-1} x$ and arctan x denote the real number (or angle in radians) between $-\pi/2$ and $\pi/2$ whose tangent is x.

The graph of the inverse tangent function (or arc tangent function) is the same as the graph of the restricted tangent function (Figure 7.81) reflected in the line $y = x$. Figure 7.82 shows the graph of $y = \tan^{-1} x$. The domain of $y = \tan^{-1} x$ is $(-\infty, \infty)$ and its range is $(-\pi/2, \pi/2)$. Note that the graph of $y = \tan^{-1} x$ is symmetric with respect to the origin, which indicates that the inverse tangent function is an odd function. Thus, we have

$$\tan^{-1}(-x) = -\tan^{-1} x$$

for every real number x.

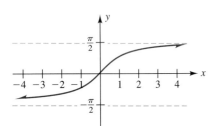

FIGURE 7.82

Graph of the inverse tangent function, $y = \tan^{-1} x$. The domain is $(-\infty, \infty)$ and the range is $(-\pi/2, \pi/2)$.

EXAMPLE 3 Find the exact value of each expression, if it is defined.

(a) arctan 0   **(b)** $\tan^{-1} 1$ **(b)** $\tan^{-1}(-\sqrt{3})$

Solution

(a) Letting $y = $ arctan 0, we have

$$\tan y = 0 \quad \text{for } -\pi/2 < y < \pi/2.$$

From our work with the unit circle in Section 7.3, we have tan 0 = 0. Since 0 is in the interval $(-\pi/2, \pi/2)$, we conclude that

$$\text{arctan } 0 = 0.$$

(b) Letting $y = \tan^{-1} 1$, we have

$$\tan y = 1 \quad \text{for } -\pi/2 < y < \pi/2.$$

From our work with the unit circle in Section 7.3, we know that $\tan (\pi/4) = 1$. Since $\pi/4$ is in the interval $(-\pi/2, \pi/2)$, we conclude that

$$\tan^{-1} 1 = \frac{\pi}{4}.$$

(c) Letting $y = \tan^{-1}(-\sqrt{3})$, we have

$$\tan y = -\sqrt{3} \quad \text{for } -\pi/2 < y < \pi/2.$$

SECTION 7.6 ◆ *The Inverse Trigonometric Functions* 459

From our work with the unit circle in Section 7.3, we have $\tan(\pi/3) = \sqrt{3}$. Therefore, using the fact that $\tan^{-1}(-x) = -\tan^{-1}x$, we write

$$\tan^{-1}(-\sqrt{3}) = -\tan^{-1}(\sqrt{3}) = -\frac{\pi}{3}.$$ ◆

We can use a graphing calculator to verify the results in Example 3. To evaluate the inverse tangent function or to generate the graph of the inverse tangent function in the viewing rectangle of a graphing calculator, we set the calculator in radian mode and use the inverse tangent key, TAN⁻¹. The split screens in Figures 7.83, 7.84, and 7.85 verify the evaluations in Example 3 both graphically and numerically.

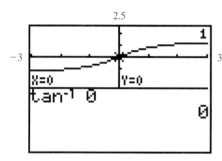

FIGURE 7.83

Graphical and numerical evaluation of arctan 0 using split-screen mode

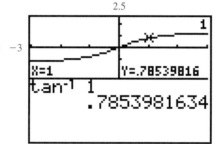

FIGURE 7.84

Graphical and numerical evaluation of $\tan^{-1}1$ using split-screen mode

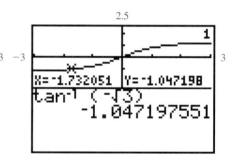

FIGURE 7.85

Graphical and numerical evaluation of $\tan^{-1}(-\sqrt{3})$ using split-screen mode

PROBLEM 3 Use a calculator to find the approximate value of each expression. Round each answer to four significant digits.

(a) arctan 0.977   **(b)** $\tan^{-1}(-1.6)$ ◆

Note: The inverse cosecant, inverse secant, and inverse cotangent functions are rarely used. For a brief discussion of these three inverse trigonometric functions, see Exercises 69–72.

◆ **Composing Trigonometric and Inverse Trigonometric Functions**

Recall from Section 2.6 that if f and f^{-1} are a pair of inverse functions, then the result of composing f with f^{-1} in either order is the identity function, that is,

$$f(f^{-1}(x)) = x \quad \text{for all } x \text{ in the domain of } f^{-1}$$

and

$$f^{-1}(f(x)) = x \quad \text{for all } x \text{ in the domain of } f.$$

If we apply this fact to the restricted trigonometric functions and their inverses, we obtain the following **trigonometric composition properties**.

Trigonometric Composition Properties

1. $\sin(\sin^{-1} x) = x$ for all x in $[-1, 1]$.
2. $\sin^{-1}(\sin x) = x$ for all x in $[-\pi/2, \pi/2]$.
3. $\cos(\cos^{-1} x) = x$ for all x in $[-1, 1]$.
4. $\cos^{-1}(\cos x) = x$ for all x in $[0, \pi]$.
5. $\tan(\tan^{-1} x) = x$ for all x.
6. $\tan^{-1}(\tan x) = x$ for all x in $(-\pi/2, \pi/2)$.

When we use these composition properties, it is important to adhere to the domain restrictions. As illustrated in the next example, it may be necessary to rewrite the composition in an equivalent form before applying these properties.

EXAMPLE 4 Find the exact value of each expression, if it is defined.

(a) $\tan[\tan^{-1}(-6)]$ (b) $\sin\left(\arcsin \frac{3}{2}\right)$

(c) $\cos^{-1}(\cos 3\pi)$ (d) $\arcsin\left(\sin \frac{5\pi}{4}\right)$

Solution

(a) The domain of the inverse tangent function is $(-\infty, \infty)$. Since -6 is in the domain, we apply trigonometric composition property 5, and write

$$\tan[\tan^{-1}(-6)] = -6.$$

(b) The domain of the arcsine function is $[-1, 1]$. Since $\frac{3}{2}$ is not in the domain, $\arcsin \frac{3}{2}$ is undefined. Hence,

$$\sin\left(\arcsin \tfrac{3}{2}\right) \quad \text{is undefined.}$$

(c) The domain of the restricted cosine function is $[0, \pi]$. Since 3π is not in the domain, we cannot apply trigonometric composition property 4, that is,

$$\cos^{-1}(\cos 3\pi) \neq 3\pi.$$

However, we do know that 3π is coterminal with π, and since π is in the domain of the restricted cosine function, we may rewrite the expression and then apply trigonometric composition property 4 as follows:

$$\cos^{-1}(\cos 3\pi) = \cos^{-1}(\cos \pi) = \pi.$$

(d) The domain of the restricted sine function is $[-\pi/2, \pi/2]$. Since $5\pi/4$ is not in the domain, we cannot apply trigonometric composition property 2, that is,

$$\arcsin\left(\sin \frac{5\pi}{4}\right) \neq \frac{5\pi}{4}.$$

However, by symmetry of the unit circle (see Figure 7.86), we do know that

$$\sin \frac{5\pi}{4} = \sin\left(-\frac{\pi}{4}\right)$$

and, since $-\pi/4$ is in the domain of the restricted sine function, we may rewrite the expression and then apply trigonometric composition property 2 as follows:

$$\arcsin\left(\sin \frac{5\pi}{4}\right) = \arcsin\left[\sin\left(-\frac{\pi}{4}\right)\right] = -\frac{\pi}{4}.$$

As shown in Figure 7.87, we can use a calculator to verify our work in Examples 4(a), 4(c), and 4(d). If we try to enter the expression in Example 4(b), the calculator displays the message "domain error." ◆

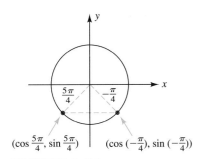

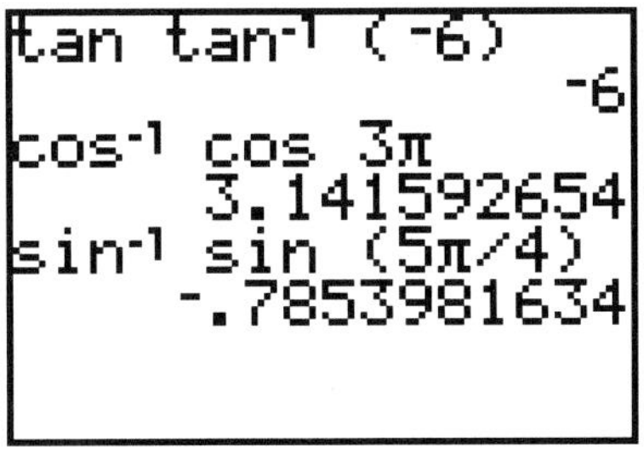

FIGURE 7.86
By symmetry of the unit circle, $\sin(5\pi/4) = \sin(-\pi/4)$.

PROBLEM 4 Find the exact value of each expression, if it is defined. Use a calculator to verify each answer.

(a) $\tan(\arctan 2)$ (b) $\sin^{-1}\left(\sin \frac{3\pi}{2}\right)$ ◆

We may use the trigonometric ratios for right triangles (Section 7.2) to evaluate other composite functions involving the trigonometric and inverse trigonometric functions. This procedure is illustrated in the next example.

EXAMPLE 5 Find the exact value of each expression. Use a calculator to verify each answer.

(a) $\sin\left(\arctan \frac{1}{2}\right)$ (b) $\sin\left[\arctan\left(-\frac{1}{2}\right)\right]$

◆ Solution

(a) Letting $\theta = \arctan \frac{1}{2}$, we have

$$\tan \theta = \frac{1}{2} \quad \text{for } -\pi/2 < \theta < \pi/2.$$

Since $\tan \theta > 0$, we may regard θ as an acute angle in a right triangle whose tangent is $\frac{1}{2}$. Now, by the trigonometric ratios for right triangles (Section 7.2), we have

$$\tan \theta = \frac{\text{opp}}{\text{adj}} = \frac{1}{2}$$

as shown in Figure 7.88. By the Pythagorean theorem, the length of the hypotenuse of this right triangle is $\sqrt{5}$ units. Thus,

$$\sin\left(\arctan \frac{1}{2}\right) = \sin \theta = \frac{\text{opp}}{\text{hyp}} = \frac{1}{\sqrt{5}}.$$

The display in Figure 7.89 supports our work.

FIGURE 7.87
Typical display of the evaluations in Examples 4(a), 4(c), and 4(d).

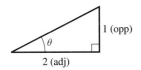

FIGURE 7.88
A right triangle with $\tan \theta = \frac{1}{2}$.

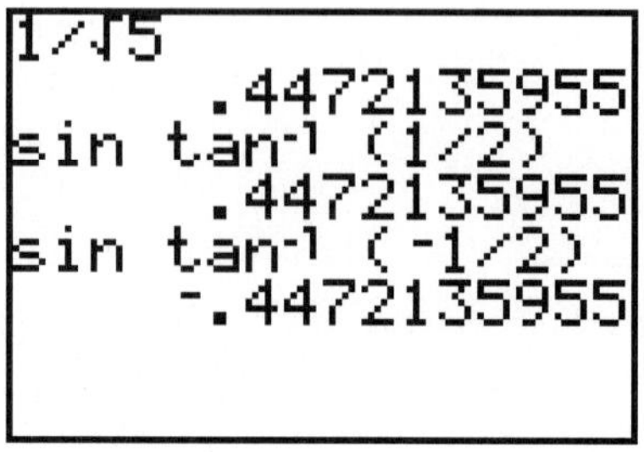

FIGURE 7.89
Typical display of the evaluations in Example 5

(b) To evaluate $\sin[\arctan(-\frac{1}{2})]$, we use the facts that

$$\arctan(-x) = -\arctan x \quad \text{and} \quad \sin(-x) = -\sin x$$

to obtain

$$\sin[\arctan(-\tfrac{1}{2})] = \sin(-\arctan \tfrac{1}{2}) = -\sin(\arctan \tfrac{1}{2}).$$

Using the result from part (a), we have

$$\sin[\arctan(-\tfrac{1}{2})] = -\sin(\arctan \tfrac{1}{2}) = -\frac{1}{\sqrt{5}}.$$

The display in Figure 7.89 supports our work. ◆

PROBLEM 5 Find the exact value of $\cos[\arctan(-\tfrac{1}{2})]$. ◆

In calculus, it is often necessary to express composite functions involving trigonometric and inverse trigonometric expressions in x as algebraic functions in x. We can use the right triangle approach illustrated in Example 5 to accomplish this task.

EXAMPLE 6 Express $\cos(\sin^{-1} 2x)$, for $0 \leq x \leq \tfrac{1}{2}$, as an algebraic expression in x.

◆ **Solution** Letting $\theta = \sin^{-1} 2x$, we have $\sin \theta = 2x$. Thus, the restricted domain, $0 \leq x \leq \tfrac{1}{2}$, implies $0 \leq \sin \theta \leq 1$. Hence, we may regard θ as an acute angle in a right triangle whose sine is $2x$. Now, by the trigonometric ratios for right triangles (Section 7.2), we have

$$\sin \theta = \frac{\text{opp}}{\text{hyp}} = \frac{2x}{1}$$

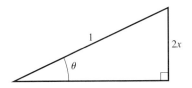

FIGURE 7.90
A right triangle with $\sin \theta = 2x/1$

as shown in Figure 7.90. By the Pythagorean theorem, the length of the other leg of this right triangle is $\sqrt{1-4x^2}$. Hence,

$$\cos(\sin^{-1} 2x) = \cos \theta = \sqrt{1-4x^2} \quad \text{for } 0 \le x \le \tfrac{1}{2}.$$

We can use a graphing calculator to confirm our work of expressing a composite function involving trigonometric and inverse trigonometric expressions as an algebraic function. Both forms of the function should yield the same graph in the viewing rectangle.

PROBLEM 6 Use a graphing calculator to verify the result of Example 6 by showing that the graph of $y = \cos(\sin^{-1} 2x)$ appears to be the same as the graph of $y = \sqrt{1-4x^2}$ in the interval $[0, \tfrac{1}{2}]$.

◆ Application: Baseball Diamond

We conclude this section with an applied problem that involves an inverse trigonometric function. We will discuss additional applications of the trigonometric functions and their inverses in Section 7.7.

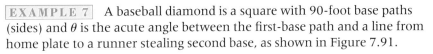

EXAMPLE 7 A baseball diamond is a square with 90-foot base paths (sides) and θ is the acute angle between the first-base path and a line from home plate to a runner stealing second base, as shown in Figure 7.91.

(a) Express angle θ as a function of the runner's distance d from home plate.

(b) A catcher cannot throw out a runner trying to steal second base if the runner's distance d from home plate is more than 100 ft when the catcher releases the ball. Use the function in part (a) to find the measure of angle θ when $d = 100$ ft.

◆ **Solution**

(a) The sketch of the baseball diamond in Figure 7.91 reveals a right triangle. In relation to angle θ, the first-base path is the adjacent side of the right triangle (with length 90 ft) and d is the hypotenuse. Thus, using the trigonometric ratios for right triangles, we can state that

$$\cos \theta = \frac{\text{adj}}{\text{hyp}} = \frac{90}{d}.$$

Since θ is an acute angle, θ is in the domain of the restricted cosine function. Taking the inverse cosine of both sides of this equation, we may write

$$\cos^{-1}(\cos \theta) = \cos^{-1} \frac{90}{d} \quad \text{or} \quad \theta = \cos^{-1} \frac{90}{d}$$

(b) When $d = 100$ ft, we have

$$\theta = \cos^{-1} \frac{90}{d} = \cos^{-1} \frac{90}{100} \approx 0.451 \text{ radian} \quad \text{or} \quad 25.8°.$$

FIGURE 7.91
The quantities d, θ, and 90 are parts of a right triangle.

The runner is a little less than half way to second base when $\theta = 25.8°$.

PROBLEM 7 Referring to Example 7, describe the domain and range of $\theta = \arccos \dfrac{90}{d}$.

Exercises 7.6

Basic Skills

In Exercises 1–16, find the exact value of each expression, if it is defined. Record each answer as a real number (or as an angle in radians). Use a calculator to confirm each of your answers.

1. $\sin^{-1} \frac{1}{2}$
2. $\cos^{-1} \frac{1}{2}$
3. $\arctan \sqrt{3}$
4. $\arcsin 1$
5. $\arccos \dfrac{1}{\sqrt{2}}$
6. $\cos^{-1} \dfrac{\sqrt{3}}{2}$
7. $\sin^{-1}\left(-\dfrac{\sqrt{3}}{2}\right)$
8. $\tan^{-1}\left(-\dfrac{1}{\sqrt{3}}\right)$
9. $\tan^{-1}(-1)$
10. $\arcsin\left(-\frac{1}{2}\right)$
11. $\cos^{-1}\left(-\dfrac{1}{\sqrt{2}}\right)$
12. $\cos^{-1}\left(-\dfrac{\sqrt{3}}{2}\right)$
13. $\cos^{-1} 0$
14. $\arcsin \sqrt{3}$
15. $\arccos(-5)$
16. $\cos^{-1}(-1)$

In Exercises 17–40, find the exact value of each expression, if it is defined. Use a calculator to confirm each of your answers.

17. $\sin\left(\sin^{-1} \frac{2}{3}\right)$
18. $\tan(\tan^{-1} 1.6)$
19. $\cos[\arccos(-12)]$
20. $\cos\left[\arccos\left(-\frac{3}{5}\right)\right]$
21. $\sin^{-1}\left(\sin \dfrac{\pi}{8}\right)$
22. $\arccos(\cos 2)$
23. $\tan^{-1}[\tan(-1)]$
24. $\arctan\left(\tan \dfrac{\pi}{2}\right)$
25. $\arccos(\cos 5\pi)$
26. $\arctan\left(\tan \dfrac{5\pi}{3}\right)$
27. $\sin^{-1}\left(\sin \dfrac{3\pi}{4}\right)$
28. $\cos^{-1}\left[\cos\left(-\dfrac{\pi}{2}\right)\right]$
29. $\sin^{-1}\left(\sin \dfrac{9\pi}{10}\right)$
30. $\cos^{-1}\left(\cos \dfrac{9\pi}{5}\right)$
31. $\tan^{-1}\left(\tan \dfrac{5\pi}{8}\right)$
32. $\cos^{-1}\left(\cos \dfrac{11\pi}{9}\right)$
33. $\sin(\arctan 3)$
34. $\sin\left(\arccos \frac{12}{13}\right)$
35. $\cos\left(\sin^{-1} \frac{2}{3}\right)$
36. $\cos(\tan^{-1} 0.3)$
37. $\csc\left[\tan^{-1}\left(-\frac{2}{3}\right)\right]$
38. $\sec\left[\sin^{-1}\left(-\dfrac{\sqrt{7}}{4}\right)\right]$
39. $\cot\left[\arccos\left(-\frac{1}{2}\right)\right]$
40. $\tan\left[\sin^{-1}(-\sqrt{2})\right]$

In Exercises 41–56, use a calculator to find the approximate value of each expression, if it exists. Record each answer as a real number (or as an angle in radians) rounded to four significant digits.

41. $\arcsin 0.28$
42. $\arccos 0.92$
43. $\tan^{-1} 8$
44. $\sin^{-1}(-0.6)$
45. $\cos^{-1}(-0.135)$
46. $\arctan 2.876$
47. $\arcsin 0.9501$
48. $\cos^{-1}(-0.045)$
49. $\arctan\left(-\frac{5}{12}\right)$
50. $\cos^{-1} \dfrac{\sqrt{5}}{3}$
51. $\arcsin\left(-\frac{4}{3}\right)$
52. $\cos^{-1} \pi$
53. $\sin(\sin^{-1} 0.887)$
54. $\arccos(\cos 3.156)$
55. $\arctan[\tan(-7)]$
56. $\cos[\sin^{-1}(-0.6781)]$

In Exercises 57–64, use a right triangle to write each composite trigonometric expression as an algebraic expression in x.

57. $\sin(\cos^{-1} x)$ for $0 \le x \le 1$
58. $\cos(\sin^{-1} 3x)$ for $0 \le x \le \frac{1}{3}$
59. $\tan\left(\arcsin \dfrac{1}{x}\right)$ for $1 < x < \infty$
60. $\sec\left(\arctan \dfrac{x}{2}\right)$ for $0 \le x < \infty$
61. $\sec[\cos^{-1}(x-1)]$ for $1 < x \le 2$
62. $\cot[\tan^{-1}(x+2)]$ for $-2 < x < \infty$
63. $\cos\left(\arctan \dfrac{1}{x-1}\right)$ for $1 < x < \infty$

64. $\csc\left(\arccos\dfrac{x}{\sqrt{x^2+4}}\right)$ for $0 \le x < \infty$

65. An observer is 550 ft from the launchpad when a rocket is fired vertically upward, as shown in the sketch.

   (a) Express angle θ as a function of the height h of the rocket.

   (b) Determine the domain and range of the function described in part (a).

   (c) Use the answer from part (a) to find the exact value of θ (in radians) when $h = 550$ ft.

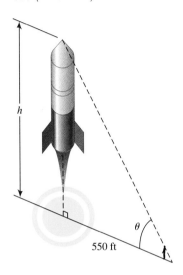

66. A 32-foot ladder is leaned against the side of a building with the foot of the ladder and the base of the building on level ground.

   (a) Express the acute angle θ between the ladder and the ground as a function of the distance x from the foot of the ladder to the base of the building.

   (b) Determine the domain and range of the function described in part (a).

   (c) Use the answer from part (a) to find the exact value of θ (in radians) when $x = 16$ ft.

67. A flight of stairs is constructed, as shown in the sketch.

   (a) Express the acute angle θ as a function of the tread length t.

   (b) Determine the domain and range of the function defined in part (a).

   (c) Use the answer from part (a) to find the approximate value of θ (in degrees) if $t = 12$ inches.

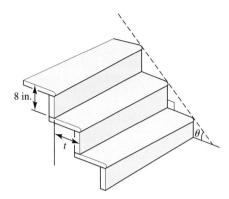

68. A pendulum hangs on a 1-meter string that is attached to the ceiling of a room. When the pendulum is pulled to one side, an angle θ is formed between the new position and the original, vertical position, as shown in the sketch.

   (a) Express angle θ as a function of the height h of the pendulum above its lowest position.

   (b) Determine the domain and range of the function defined in part (a).

   (c) Use the answer from part (a) to find the approximate value of θ (in radians) when $h = 0.46$ meter.

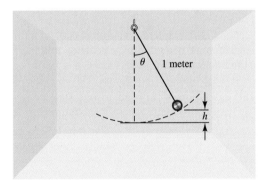

 Critical Thinking

69. The inverse of the *restricted cosecant function*,

$$y = \csc x \quad \text{for} \quad -\pi/2 \le x \le \pi/2 \text{ with } x \ne 0,$$

   is called the *inverse cosecant function* and is defined as

   $$y = \csc^{-1} x$$

   if and only if $\csc x = y$ for $-\pi/2 \le y \le \pi/2$ with $y \ne 0$.

   (a) Sketch the graph of the restricted cosecant function, and show that it is one-to-one.

   (b) Sketch the graph of the inverse cosecant function, and state its domain and range.

70. The inverse of the *restricted secant function*,

$$y = \sec x \quad \text{for } 0 \leq x \leq \pi \text{ with } x \neq \pi/2,$$

is called the *inverse secant function* and is defined as

$$y = \sec^{-1} x$$

if and only if $\sec x = y$ for $0 \leq y \leq \pi$ with $y \neq \pi/2$.

(a) Sketch the graph of the restricted secant function, and show that it is one-to-one.

(b) Sketch the graph of the inverse secant function, and state its domain and range.

71. The inverse of the *restricted cotangent function*,

$$y = \cot x \quad \text{for } 0 < x < \pi,$$

is called the *inverse cotangent function* and is defined as

$$y = \cot^{-1} x$$

if and only if $\cot x = y$ for $0 < y < \pi$.

(a) Sketch the graph of the restricted cotangent function, and show that it is one-to-one.

(b) Sketch the graph of the inverse cotangent function, and state its domain and range.

72. Use the definitions in Exercises 69–71 to find the exact value of each expression.

(a) $\csc^{-1} 2$
(b) $\cot^{-1}(-1)$
(c) $\sec(\sec^{-1} 2.63)$
(d) $\csc^{-1}\left(\csc \dfrac{5\pi}{3}\right)$

73. Generate the graph of $f(t) = \cos(\pi - t)$ in the viewing rectangle of a graphing calculator, and then compare it to the graphs of $y = \sin t$, $y = -\sin t$, $y = \cos t$, and $y = -\cos t$.

(a) Based upon your comparisons, state a trigonometric identity that relates $f(t)$ to either $y = \sin t$, $y = -\sin t$, $y = \cos t$, or $y = -\cos t$.

(b) Let $t = \cos^{-1} x$ in the identity found in part (a), and show that

$$\cos^{-1}(-x) = \pi - \cos^{-1} x,$$

provided that x is in the interval $[-1, 1]$.

74. Generate the graph of $f(t) = \cos\left(\dfrac{\pi}{2} - t\right)$ in the viewing rectangle of a graphing calculator and compare it to the graphs of $y = \sin t$, $y = -\sin t$, $y = \cos t$, and $y = -\cos t$.

(a) Based upon your comparisons, state a trigonometric identity that relates $f(t)$ to either $y = \sin t$, $y = -\sin t$, $y = \cos t$, or $y = -\cos t$.

(b) Let $t = \sin^{-1} x$ in the identity from part (a), and show that

$$\sin^{-1} x + \cos^{-1} x = \dfrac{\pi}{2},$$

provided that x is in the interval $[-1, 1]$.

75. Find a value of x for which $\tan^{-1}(\tan x) \neq x$. Does this contradict the inverse function concept, which states that $f^{-1}(f(x)) = x$? Explain.

76. From the fundamental trigonometric identities (Section 7.2), we know that

$$\sin^2 x + \cos^2 x = 1 \quad \text{and} \quad \tan x = \dfrac{\sin x}{\cos x}.$$

Do these identities hold true when we replace $\sin x$, $\cos x$, and $\tan x$ with their corresponding inverse trigonometric functions? Use numerical examples to support your conclusion.

77. Explain the difference between the notations $\sin^{-1} x$ and $(\sin x)^{-1}$.

78. Given that $\csc x = k$ for $k > 1$, express x in terms of the inverse sine function.

In Exercises 79–82, solve each equation for x.

79. $\sin^{-1}(3x - 2) = \dfrac{\pi}{6}$

80. $\cos^{-1}(x^2 - 5x + 3) = \pi$

81. $4 \arctan(2x^2 - 3x) = -\pi$

82. $2 \arcsin(x - 3) = 3\pi$

83. Use the shift, axis reflection, and stretch and compress rules (Section 2.5) to sketch the graph of each function. Use a graphing calculator to verify each sketch.

(a) $f(x) = \tan^{-1}(x - 2)$

(b) $f(x) = \tfrac{1}{2} \arccos x - \dfrac{\pi}{2}$

84. Use a graphing calculator to generate the graph of the polynomial function

$$P(x) = x - \dfrac{x^3}{3} + \dfrac{x^5}{5} - \dfrac{x^7}{7} + \dfrac{x^9}{9}$$

over the interval $[-1, 1]$.

(a) What inverse trigonometric function does P seem to approximate?

(b) Compare $P(0.5)$ to the value of that inverse trigonometric function when $x = 0.5$. Round each answer to four significant digits.

7.7 Applications Involving Right Triangles and Harmonic Motion

◆ Introductory Comments

In this section, we discuss applied problems involving right triangles and *harmonic motion*, or periodic motion about an equilibrium position. For the right-triangle problems, we use the trigonometric functions of angles, and for the harmonic motion problems, we use the trigonometric functions of real numbers.

We begin with problems requiring that we find the unknown parts of a right triangle when given either one side and one acute angle or two sides of the right triangle. The procedure is called *solving a right triangle*. The Pythagorean theorem and the trigonometric ratios for right triangles (Section 7.2) are used to solve a right triangle. For our convenience, these ratios are listed once again.

◀ **Trigonometric Ratios for Right Triangles**

If θ is an acute angle in a right triangle, then

$$\sin \theta = \frac{\text{opp}}{\text{hyp}} \qquad \cos \theta = \frac{\text{adj}}{\text{hyp}} \qquad \tan \theta = \frac{\text{opp}}{\text{adj}}$$

$$\csc \theta = \frac{\text{hyp}}{\text{opp}} \qquad \sec \theta = \frac{\text{hyp}}{\text{adj}} \qquad \cot \theta = \frac{\text{adj}}{\text{opp}}$$

where *opp* is the length of the side opposite θ, *adj* is the length of the side adjacent to θ, and *hyp* is the length of the hypotenuse of the right triangle.

◆ Solving Right Triangles

It is best to use the sine, cosine, or tangent function to solve a right triangle, since each of these functions can be evaluated directly by a key on a calculator. In the first example, we solve a right triangle in which one of the sides and one of the acute angles are known.

EXAMPLE 1 Solve the right triangle shown in Figure 7.92. Round x and y to three significant digits.

◆ **Solution** To solve the right triangle means to find the missing parts—in this case, angle α and sides x and y. Since the acute angles in a right triangle are complementary, we know that

$$\alpha = 90° - 33° = 57°.$$

In relation to the given angle of 33°, x is the adjacent side, y is the opposite side, and 18 is the hypotenuse. Thus, using the trigonometric ratios for right triangles,

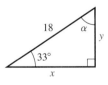

FIGURE 7.92

$$\cos\theta = \frac{\text{adj}}{\text{hyp}} \quad \text{and} \quad \sin\theta = \frac{\text{opp}}{\text{hyp}},$$

we have $\quad \cos 33° = \dfrac{x}{18} \qquad\qquad \sin 33° = \dfrac{y}{18}$

$$x = 18 \cos 33° \qquad\qquad y = 18 \sin 33°$$

To evaluate x and y, we use the sine key, $\boxed{\text{SIN}}$, and cosine key, $\boxed{\text{COS}}$, on a calculator set in degree mode:

$$x = 18 \cos 33° \approx 15.1 \quad \text{and} \quad y = 18 \sin 33° \approx 9.80$$

It is good practice to check the values of x and y by applying the Pythagorean theorem. Use a calculator to show that $18^2 \approx 15.1^2 + 9.80^2$. ◆

PROBLEM 1 Solve the right triangle shown in Figure 7.93. Round x and z to three significant digits.

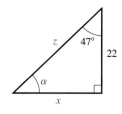

FIGURE 7.93 ◆

In the next example, we solve a right triangle in which two of the sides are known. We apply the inverse trigonometric functions to help solve problems of this type.

EXAMPLE 2 Solve the right triangle shown in Figure 7.94. Round α and β to the nearest tenth of a degree.

◆ Solution   By the Pythagorean theorem, we have

$$z^2 = 5^2 + 12^2 = 169$$

Now, taking the square root of both sides of this equation, we obtain

$$z = \sqrt{169} = 13$$

In relation to angle α, 12 is the opposite side and 5 is the adjacent side. Thus, by the trigonometric ratios for right triangles,

$$\tan \alpha = \frac{\text{opp}}{\text{adj}} = \frac{12}{5}$$

Since α is an acute angle, α is in the domain of the restricted tangent function. Taking the inverse tangent of both sides of this equation, we may write

$$\tan^{-1}(\tan \alpha) = \tan^{-1} \tfrac{12}{5} \quad \text{or} \quad \alpha = \tan^{-1} \tfrac{12}{5}$$

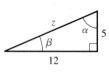

FIGURE 7.94

SECTION 7.7 ♦ Right Triangles and Harmonic Motion

Now, using the inverse tangent key, $\boxed{\text{TAN}^{-1}}$, on a calculator set in degree mode, we obtain

$$\alpha = \tan^{-1} \tfrac{12}{5} \approx 67.4°$$

Since the acute angles in a right triangle are complementary, we find

$$\beta = 90° - \alpha \approx 22.6°.$$

♦

PROBLEM 2 Solve the right triangle shown in Figure 7.95. Round α and β to the nearest tenth of a degree and round x to three significant digits.

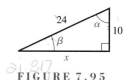

FIGURE 7.95 ♦

♦ Applied Problems Involving Right Triangles

In the next three examples, we solve some applied problems by finding the missing parts of right triangles.

EXAMPLE 3 To sharpen the cutting teeth of a saw blade, as shown in Figure 7.96, we must find the measure of angle β. Find β to the nearest tenth of a degree.

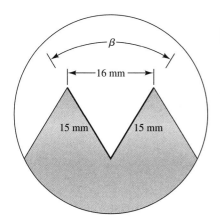

FIGURE 7.96

♦ **Solution** As Figure 7.96 shows, the gap between the teeth forms an isosceles triangle. Recall from elementary geometry that if we construct an altitude from the vertex angle to the base of an isosceles triangle, both the vertex angle β and the base (16 mm) are cut in half, as shown in Figure 7.97. Working with either right triangle in Figure 7.97 we find that, in relation to angle $\beta/2$, 8 is the opposite side and 15 is the hypotenuse. Thus, by the trigonometric ratios for right triangles,

$$\sin \frac{\beta}{2} = \frac{\text{opp}}{\text{hyp}} = \frac{8}{15}.$$

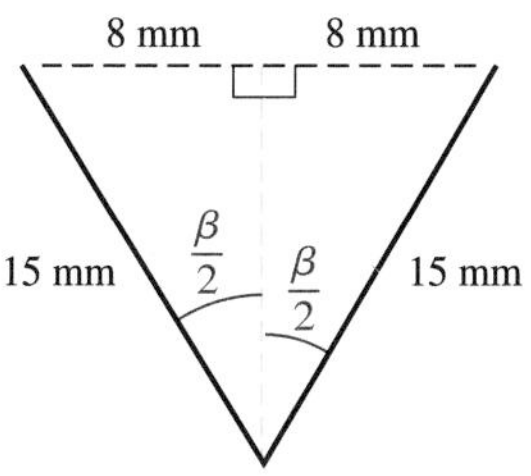

Since $\beta/2$ is an acute angle, $\beta/2$ is in the domain of the restricted sine function. Taking the inverse sine of both sides of this equation, we write

$$\frac{\beta}{2} = \sin^{-1} \frac{8}{15} \quad \text{or} \quad \beta = 2 \sin^{-1} \frac{8}{15}.$$

Using the inverse sine key, $\boxed{\text{SIN}^{-1}}$, on a calculator set in the degree mode, we obtain

$$\beta = 2 \sin^{-1} \tfrac{8}{15} \approx 64.4°.$$

♦

FIGURE 7.97
The altitude to the base of an isosceles triangle cuts the vertex angle and base in half.

PROBLEM 3 For the cutting teeth of a saw blade, as shown in Figure 7.98, find the distance x. Round to three significant digits.

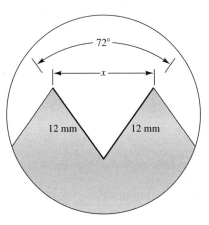

FIGURE 7.98

EXAMPLE 4 *Chapter Opening Problem*
The chair lift at a ski area rises at an angle of 19.2° and attains a vertical rise of 1250 ft in 15 minutes. What is the speed (in mi/h) of the chair lift?

◆ **Solution** Letting

s = the distance (in feet) the chair lift travels,

we set up a sketch of the given information, as shown in Figure 7.99. In relation to the given angle of 19.2°, 1250 ft is the opposite side and s is the hypotenuse. Thus, using the fact that $\sin \theta = $ opp/hyp, we have

$$\sin 19.2° = \frac{1250}{s}$$

$s \sin 19.2° = 1250$ Multiply both sides by s

$s = \dfrac{1250}{\sin 19.2°}$ ft   Divide both sides by $\sin 19.2°$

Recall from Section 7.1 that linear speed v is defined as the distance s traveled per unit of time t. Thus, the speed of the chair lift is

$$v = \frac{s}{t} = \frac{(1250/\sin 19.2°)\text{ft}}{15 \text{ min}} \cdot \frac{60 \text{ min}}{1 \text{ hr}} \cdot \frac{1 \text{ mi}}{5280 \text{ ft}} \approx 2.9 \text{ mi/h}$$

conversion factors

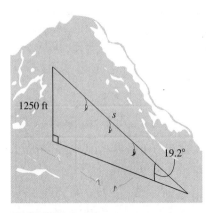

FIGURE 7.99

A right triangle in which the side opposite the given acute angle is known and the hypotenuse is to be determined.

PROBLEM 4 If the chair lift described in Example 4 rises at an angle of 15.5° and has a speed of 3 mi/h, find the time (in minutes) required to attain a vertical rise of 1800 ft.

A *transit* is an instrument that can be used to measure an angle from the horizontal to a point that is either above or below the horizontal. If the point sighted is above the horizontal, as shown in Figure 7.100(a), the angle is called an **angle of elevation**. If the point sighted is below the horizontal, as shown in Figure 7.100(b), then the angle is called an **angle of depression**.

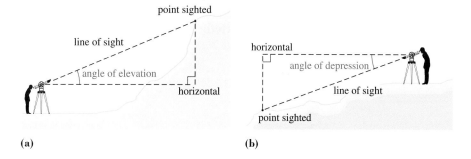

FIGURE 7.100
An angle of elevation and angle of depression

(a)   (b)

EXAMPLE 5 The horizontal distance from a transit to a flagpole is 32.6 meters. The angle of elevation from the transit to the top of the flagpole is 38°27′ and the angle of depression from the transit to the bottom of the flagpole is 6°30′. Find the height of the flagpole.

◆ **Solution**   Letting

$$h = \text{the height (in meters) of the flagpole,}$$

we set up a sketch of the given information, as shown in Figure 7.101. In relation to the angle of elevation (38°27′), a is the opposite side and 32.6 is the adjacent side. In relation to the angle of depression (6°30′), b is the opposite side and 32.6 is the adjacent side. Thus, using the fact that $\tan \theta = \text{opp}/\text{adj}$, we have

$$\tan 38°27' = \frac{a}{32.6} \qquad \text{and} \qquad \tan 6°30' = \frac{b}{32.6}$$
$$a = 32.6 \tan 38°27' \qquad\qquad b = 32.6 \tan 6°30'$$

Thus, the height h of the flagpole is

$$h = a + b$$
$$= 32.6 \tan 38°27' + 32.6 \tan 6°30'$$
$$\approx 29.6 \text{ m.} \qquad \blacklozenge$$

FIGURE 7.101
The height h of the flagpole is the sum of the legs a and b.

PROBLEM 5 If the horizontal distance from a transit to a flagpole is 42.8 meters and the angles of elevation to the bottom and the top of the flagpole are 5°36′ and 22°45′, respectively, find the height of the pole. ◆

◆ **Simple Harmonic Motion**

Thus far we have looked at applied problems involving the trigonometric functions of angles. We now turn our attention to applied problems involving the trigonometric functions of real numbers. Specifically, we look at periodic motion that is symmetric about an equilibrium position. We refer to motion of this type as **simple harmonic motion**.

Consider a spring with an attached weight, as shown in Figure 7.102. If the weight is pulled down and then released, the weight will oscillate back and forth through the equilibrium position. If we ignore frictional forces, this oscillating motion will repeat itself over and over again.

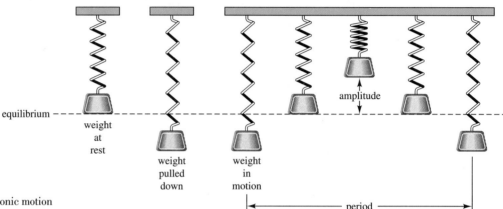

FIGURE 7.102
An illustration of simple harmonic motion

In Figure 7.102, the dashed line indicating the movement of the weight suggests that we may describe the displacement of the weight from its equilibrium position by a sine or cosine function. Suppose we consider the maximum displacement of the weight from its equilibrium position as the amplitude. Also, suppose we consider the time required for the weight to travel from its maximum displacement below the equilibrium position to its maximum displacement above the equilibrium position and back again to its maximum displacement below the equilibrium position as the period. Now, we can express the displacement d of the weight from its equilibrium position as a function of time t by writing

$$d = a \sin bt \quad \text{or} \quad d = a \cos bt$$

where a is the amplitude and b is related to the period P by the formula $P = 2\pi/b$.

Since the period P is the time required for one complete oscillation, the reciprocal of P must represent the number of oscillations per unit of time. We refer to $1/P$ as the **frequency** f. Hence, if $P = 2\pi/b$, then

$$b = \frac{2\pi}{P} = 2\pi \cdot \frac{1}{P} = 2\pi f.$$

We obtain the models for simple harmonic motion by replacing b with $2\pi f$ in the equations $d = a \sin bt$ and $d = a \cos bt$:

$$d = a \sin 2\pi ft \quad \text{or} \quad d = a \cos 2\pi ft$$

In these equations

> d is the displacement from the equilibrium position after t units of time,
>
> $|a|$ is the maximum displacement, and
>
> f is the frequency, the number of oscillations per unit of time, $f = 1/P$.

Which model we use depends on the displacement d of the weight at $t = 0$. If $d = 0$ at $t = 0$ then we use $d = a \sin 2\pi ft$, and if $d = a$ at $t = 0$, then we use $d = a \cos 2\pi ft$.

EXAMPLE 6 A weight attached to a spring is pulled down 15 cm from its equilibrium position and is then released at $t = 0$ (seconds). The weight completes one cycle in $\frac{1}{4}$ second.

(a) Write an equation for this simple harmonic motion.

(b) State the amplitude and period for the graph of the equation in part (a). Then list the five key points on one cycle of the graph.

(c) Find the time at which the weight, moving downward for the second time, passes through the equilibrium position.

◆ Solution

(a) We shall assume that displacements below the equilibrium position are negative and displacements above the equilibrium position are positive. Since $d = -15$ cm at $t = 0$, we begin with the model

$$d = -15 \cos 2\pi ft.$$

Moreover, since the weight completes one cycle in $\frac{1}{4}$ second, we have

$$f = \frac{1}{P} = \frac{1}{\frac{1}{4}} = 4.$$

Hence, the equation that describes the motion is

$$d = -15 \cos 2\pi(4)t = -15 \cos 8\pi t \quad \text{for } t \geq 0.$$

(b) For $d = -15 \cos 8\pi t$, we have

Amplitude: $|a| = |-15| = 15$ Period: $\dfrac{2\pi}{|b|} = \dfrac{2\pi}{|8\pi|} = \dfrac{1}{4}$

Thus, this cosine curve completes one cycle from $t = 0$ to $t = \frac{1}{4}$ second and has a maximum displacement of 15 cm. Dividing the interval $[0, \frac{1}{4}]$ into four equal parts gives us the five key values on this cycle:

474   CHAPTER 7 ♦ Introduction to Trigonometry

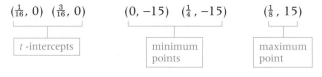

Figure 7.103 shows the graph of $d = -15 \cos 8\pi t$ in the viewing rectangle of a graphing calculator, and this graph supports our work. We have chosen $\frac{1}{16}$ as the distance between scale marks on the horizontal axis and 5 as the distance between scale marks on the vertical axis.

(c) Referring to Figure 7.103, the weight moves upward on the interval $(0, \frac{1}{8})$ passing through the equilibrium position at $t = \frac{1}{16}$ second. It then moves downward on the interval $(\frac{1}{8}, \frac{1}{4})$, passing through the equilibrium position again at $t = \frac{3}{16}$ second. Because of the periodic nature of this function, the weight moves downward for the second time on the interval

$$\left(\tfrac{1}{8} + \tfrac{1}{4}, \tfrac{1}{4} + \tfrac{1}{4}\right) = \left(\tfrac{3}{8}, \tfrac{1}{2}\right)$$

and passes through the equilibrium position at

$$t = \tfrac{3}{16} + \tfrac{1}{4} = \tfrac{7}{16} \text{ second.} \qquad \blacklozenge$$

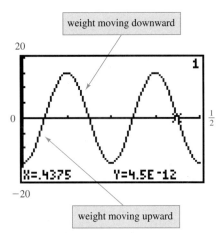

FIGURE 7.103
The graph of $d = -15 \cos 8\pi t$ for $t \geq 0$ illustrates the motion of the spring.

PROBLEM 6 For the weight described in Example 6, find its displacement at $t = 2$ seconds. ♦

Exercises 7.7

◆ **Basic Skills**

In Exercises 1–6, solve each right triangle. Round x and y to three significant digits.

1.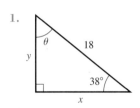

2.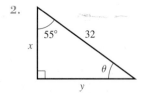

5. & 6.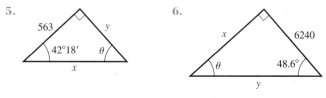

In Exercises 7–12, solve each right triangle. Round α and β to the nearest tenth of a degree, and round x to three significant digits.

3.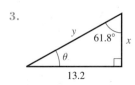

4.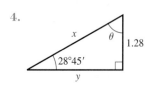

7.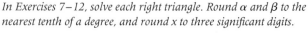

8.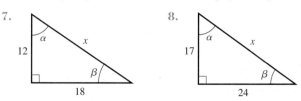

SECTION 7.7 ◆ Exercises

9. 10.
11. 12.

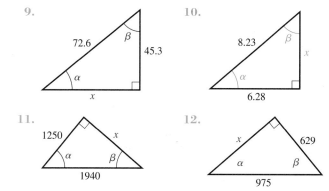

Solve each problem in Exercises 13–32. Round angles to the nearest tenth of a degree and all other measurements to three significant digits.

13. Find the vertical height of a kite if 350 ft of string has been let out and the angle from the ground to the kite string is 75°. Assume the string is taut.

14. An inclined ramp leading into a parking garage is 170 m long and rises 18 m. What is the angle of incline of the ramp?

15. When a helicopter is 1.2 km above one end of an island, the angle of depression to the other end of the island is 22.7°. What is the length of the island?

16. As illustrated in the sketch, the length of the shadow of a telephone pole is 62 ft when the angle of elevation of the sun is 32°. What is the height of the pole?

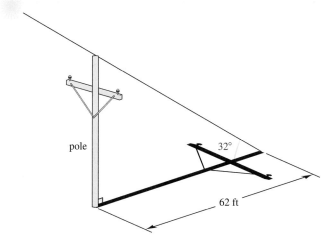

17. A guy wire from the top of an antenna is anchored 43.9 m from the base of the antenna and makes an angle of 72.6° with the ground, as shown in the sketch. What is the length of the guy wire?

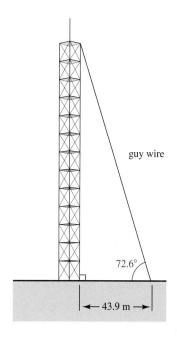

18. A security camera is used by a bank to monitor a teller's counter. The camera is mounted on a wall at a height of 12 ft. If the line-of-sight distance from the camera to the counter is 20 ft and the counter is 4 ft high, what is the angle of depression of the camera?

19. A radiologist uses a gamma ray to treat a tumor that is 3.2 cm beneath a patient's skin. The ray is directed into the skin at an angle of 48°. How far does the ray travel through the patient's body before striking the tumor?

20. The equal sides of an isosceles triangle are 36 cm and the nonequal side is 42 cm. Determine the measure of the interior angles of the triangle.

21. For the drill bit shown in the sketch, find angle θ.

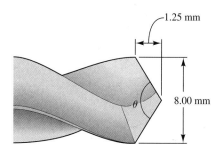

22. Three drill holes are located in an aluminum plate, as shown in the sketch at the top of the next page. Find the offset distances x and y.

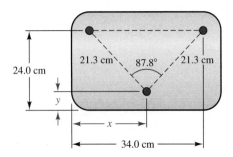

23. For the trapezoidal utility knife blade shown in the sketch, find the distance x.

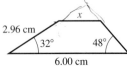

24. For the trapezoidal piece of land shown in the sketch, find angles α and β.

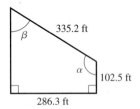

25. For the flat-headed machine screw shown in the sketch, find angles α and β.

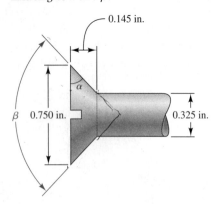

26. Each side of a hexagonal nut is 6 mm, as shown in the sketch. Find the distances x and y.

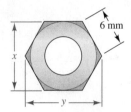

27. For the five intersecting highways sketched in the figure, find the distance between points B and C.

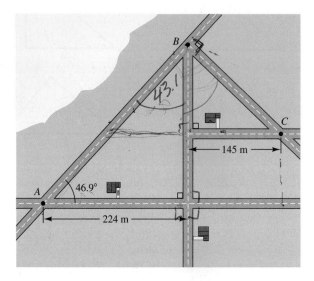

28. For the four-panel Pratt truss shown in the sketch, find the length of each lettered member.

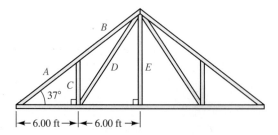

29. An airplane climbs at an angle of 6.5° at a constant speed of 475 mi/h. How many seconds does it take to ascend from an altitude of 10,500 ft to 12,800 ft?

30. A missile is traveling at a speed of 925 m/s and at an angle of 42.8° from the horizontal. What is the missile's vertical displacement after one minute?

31. An observer in a lighthouse determines that the angles of depression to two sailboats directly in line with the lighthouse are 3°36′ and 5°45′. If the observer is 125 ft above sea level, find the distance between the two ships.

32. The horizontal distance from a transit to a smokestack is 126 ft. The angle of elevation to the top of the smokestack is 65°33′ and the angle of depression to the bottom of the smokestack is 3°21′. Find the height of the smokestack.

The equations in Exercises 33–36 describe the simple harmonic motion of an object. Find

(a) the maximum displacement from the equilibrium position,
(b) the frequency, and
(c) the first time ($t > 0$) that the object passes through the equilibrium position.

33. $d = 8 \cos 6\pi t$
34. $d = \frac{1}{2} \cos 30t$
35. $d = \frac{2}{3} \sin \frac{\pi t}{10}$
36. $d = 24 \sin 120\pi t$

37. The wake made by a large ship sets a floating bottle in simple harmonic motion, as shown in the sketch. In 3 seconds the bottle rides from the crest of one wave to the crest of the next wave.
    (a) Find the equation that describes the displacement d of the bottle from its equilibrium position if $d = 0$ (inches) at $t = 0$ (seconds).
    (b) Find the displacement of the bottle at $t = 4\frac{3}{4}$ seconds.

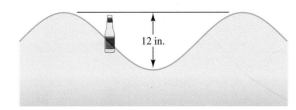

38. A guitar string is lifted up 2 mm from the fret board and then released at $t = 0$ (seconds). Suppose the note that is struck has a frequency of 440 vibrations per second. Write the equation of simple harmonic motion of a point on the string.

39. A spring with an attached weight is pushed up 20 cm from its equilibrium position and then released at $t = 0$ (seconds). The weight completes one cycle in 0.4 second.
    (a) Write the equation of simple harmonic motion, and generate its graph in the viewing rectangle of a graphing calculator.
    (b) Find the displacement of the weight when $t = 6.7$ seconds.
    (c) Find the time at which the spring, moving upward for the third time, passes through the equilibrium position.

40. A spring with an attached weight is pulled down 8.5 cm from its equilibrium position and is then released at $t = 0$ s. The weight completes one cycle in $\pi/32$ second.
    (a) Write the equation of simple harmonic motion, and generate its graph in the viewing rectangle of a graphing calculator.
    (b) Find the displacement of the weight when $t = 1$ second.
    (c) Find the time at which the spring, moving downward for the second time, passes through the equilibrium position.

Critical Thinking

41. For the cross-belted pulley system shown in the sketch, find the distance x and the angle θ.

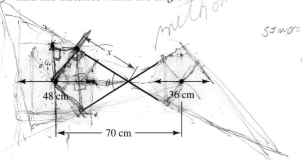

42. A cylindrical oil tank is 36 inches in diameter and 50 inches in length, as shown in the sketch. How many gallons of oil are stored in the tank if the depth of the oil is 22 inches? [*Hint:* One gallon of oil occupies a space equivalent to 231 cubic inches.]

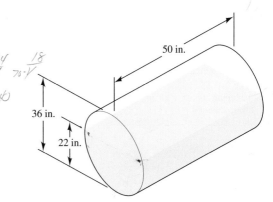

43. A piece of sheet metal 24 inches wide is bent along its center line to form a V-shaped gutter. Does the gutter have a greater water-carrying capacity when the angle of the V is 75° or 120°?

44. A painter has two ladders, one of which is twice as long as the other. Each ladder rests on the floor of a room and reaches the same height on the wall. If the shorter ladder makes an angle of 80° with the floor, find the angle between the longer ladder and the floor.

45. A regular octagon is inscribed in a circle. If the radius of the circle is 18 cm, find the length of a side of the octagon.

46. Find the angle between the diagonal of a cube and the diagonal of a face of the cube.

47. Another example of simple harmonic motion is the pendulum on a grandfather clock swinging back and forth through its equilibrium position, as shown in the sketch. In physics, it is shown that the frequency f of a pendulum is given by

$$f = \frac{1}{2\pi} \sqrt{\frac{g}{l}}$$

where g is the gravitational constant 32.2 feet per second squared (ft/s²), and l is the length of the pendulum (in feet). Suppose the length of the pendulum in a grandfather clock is 3.26 feet and the angle θ between the equilibrium position and the position of maximum displacement is 4.00°. Find the equation that describes the displacement d of the pendulum from its equilibrium position if $d = 0$ (feet) at $t = 0$ (seconds).

48. Referring to Exercise 47, how many seconds tick off the clock each time the pendulum passes through the equilibrium position?

49. A rain gutter is to be made from a piece of sheet metal 3 feet wide by turning up one-foot strips to make equal angles θ, as shown in the sketch.

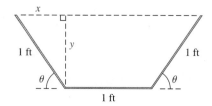

(a) Express the distances x and y as trigonometric functions of θ.

(b) Express the trapezoidal cross-sectional area A as a function of θ.

(c) Use a graphing calculator to generate the graph of the function described in part (b). Then use this graph to estimate the angle θ (in radians) that yields the maximum cross-sectional area and, hence, the greatest water-carrying capacity. Round the answer to four significant digits and convert this angle to the nearest degree.

(d) State the greatest cross-sectional area that can be formed.

50. A length of rigid copper pipe, when held horizontally, touches the walls of a 6-ft corridor and a 10-ft corridor at points A, B, and C, as shown in the sketch.

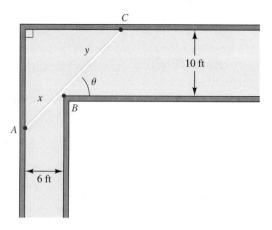

(a) Express the lengths x and y as trigonometric functions of θ.

(b) Express the total length L of the rigid copper pipe as a function of θ.

(c) Use a graphing calculator to generate the graph of the function in part (b). Then use this graph to estimate the shortest length of pipe that, when held horizontally, touches the walls of the corridors at points A, B, and C. Round the answer to three significant digits.

(d) The length found in part (c) actually represents the longest pipe that can be held horizontally and moved from one corridor to the other without bending the pipe or denting the walls. If the height of the ceiling in the corridors is 12 ft and the pipe can be tilted upward, determine the longest pipe that can be moved from one corridor to the other without bending the pipe. Round the answer to three significant digits.

Models for simple harmonic motion are unrealistic because they ignore frictional forces and assume that the oscillating motion will repeat itself over and over again. Of course, this is not the case unless the motion occurs in a vacuum. In Exercise 51 and 52, use a graphing calculator to generate the graph of each function. Do these functions seem to describe more accurately the oscillating motion of a spring with an attached weight? What effect does the coefficient e^{-kt} have on the amplitude? Explain.

51. $y = 3e^{-t} \sin 5t$ for $t \geq 0$

52. $y = -5e^{-2t} \cos 3\pi t$ for $t \geq 0$

Chapter 7 Review

◆ Questions for Writing or Group Discussion

1. When is an angle in *standard position*?
2. How are *negative angles* developed?
3. State the procedure for converting *radian* measure to *degree* measure. Illustrate with an example.
4. State the procedure for finding the measures of angles that are *coterminal*. Use an angle measured in radians as an example.
5. Discuss the procedure for converting angles measured in *degrees, minutes, seconds* to decimal degrees.
6. What is the measure of a *central angle* that intercepts an arc of length a on the unit circle?
7. Discuss the algebraic signs of the trigonometric functions of angle θ when θ lies in each of the four quadrants.
8. List the eight *fundamental trigonometric identities*. Using $\theta = \pi/6$, verify each identity.
9. Explain the difference between $\sin^2 \theta$ and $\sin \theta^2$.
10. For a given angle θ, is it possible to have both $\cos \theta > 0$ and $\sec \theta < 0$? Explain.
11. Given that θ is an acute angle, express the trigonometric functions of θ in terms of their *cofunctions*.
12. What is a *quadrantal number*? Explain the procedure for evaluating the trigonometric functions of a quadrantal number without using a calculator.
13. Explain the procedure of using symmetry of the *unit circle* to evaluate the trigonometric functions that are multiples of $\pi/6$, $\pi/4$, and $\pi/3$. Illustrate with examples.

14. Discuss the procedure for finding the trigonometric functions of a real number by using a calculator.
15. State the real number domain of each of the trigonometric functions.
16. State the range of each trigonometric function.
17. Which trigonometric functions are even functions and which are odd functions?
18. Explain how to obtain the graph of $y = \sin x$ from the graph of $y = \cos x$.
19. Explain how to obtain the graph of $y = \cot x$ from the graph of $y = \tan x$.
20. Is each trigonometric function a *periodic function*? Give the *period* of each trigonometric function that is periodic.
21. Discuss the value that $f(t)$ approaches as $t \to \dfrac{\pi}{2}^-$:
    (a) $f(t) = \sin t$ (b) $f(t) = \cos t$
    (c) $f(t) = \tan t$ (d) $f(t) = \sec t$
22. Discuss *amplitude, period,* and *phase shift* in relation to the graph of $y = a \sin(bx + c)$.
23. Why must the domain of the sine, cosine, and tangent functions be *restricted* in order to define inverses for these functions?
24. Discuss the domain and range of the *inverse sine function,* the *inverse cosine function,* and the *inverse tangent function.* List some properties of these inverse functions.
25. State the conditions under which each of these relationships is true:
    (a) $\sin(\arcsin x) = x$ (b) $\arccos(\cos x) = x$
26. Which pair of trigonometric functions are reciprocals as well as cofunctions of each other?
27. What is meant by *solving a right triangle*? Discuss the procedure for solving a right triangle in which an acute angle and the hypotenuse are known.
28. What is meant by *simple harmonic motion*?
29. Try using a calculator to evaluate $\tan 90°$. Explain the result.
30. If $-2\pi \le x \le 2\pi$, how many different values of x are there for which $\sin x = \frac{1}{2}$? for which $\sin x = 2$? Explain.

Review Exercises

In Exercises 1–6, an angle is described in terms of revolutions.
(a) *Draw the angle in standard position.*
(b) *Give the degree measure and radian measure of the angle.*
(c) *Find the degree measure and radian measure of the smallest positive angle that is coterminal with the given angle.*

1. one and two-thirds revolutions counterclockwise
2. one and one-ninth revolutions counterclockwise
3. two and two-fifths revolutions counterclockwise
4. one and five-sixths revolutions counterclockwise
5. one and one-twelfth revolutions clockwise
6. three and three-eighths revolutions clockwise

In Exercises 7–10, convert the degree measure to radian measure. Write the answer as a reduced fraction in terms of π.

7. $200°$ 8. $75°$ 9. $780°$ 10. $-405°$

In Exercises 11–14, convert the radian measure to degree measure.

11. $\dfrac{11\pi}{6}$ 12. $\dfrac{5\pi}{18}$ 13. -8π 14. $\dfrac{31\pi}{10}$

15. Convert 123°42′18″ to (a) decimal degrees and (b) radians (to the nearest thousandth).
16. Convert 57.975° to (a) degrees, minutes, seconds and (b) radians (to the nearest thousandth).
17. Convert 2.65 radians to (a) decimal degrees (to the nearest thousandth) and (b) degrees, minutes, seconds.
18. Find the radian measure of the smallest positive angle that is coterminal with 456°36′.

In Exercises 19–22, a point on the terminal side of θ is shown. Evaluate each of the six trigonometric functions of θ.

19.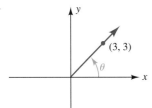

20.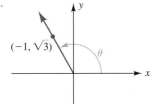

21.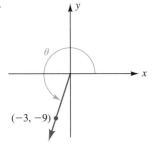

22.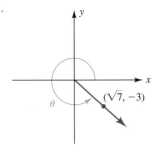

In Exercises 23–30, find the exact value of the indicated trigonometric function from the given information.

23. $\tan \theta = \frac{1}{2}$; find $\cot \theta$
24. $\sec \theta = -4$; find $\cos \theta$
25. $\sin \theta = \frac{4}{5}$, θ in quadrant II; find $\cos \theta$
26. $\cos \theta = -\frac{2}{3}$, $\tan \theta > 0$; find $\sin \theta$
27. $\tan \theta = \sqrt{2}$, $\sin \theta < 0$; find $\sec \theta$
28. $\csc \theta = 2$, $\tan \theta < 0$; find $\cot \theta$
29. $\sin \theta = -\frac{12}{13}$, $\cos \theta > 0$; find $\tan \theta$
30. $\cot \theta = -\sqrt{3}$, θ in quadrant IV; find $\cos \theta$

In Exercises 31–34, use the right triangle in the sketch to find the exact value of the indicated trigonometric function.

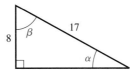

31. $\sin \alpha$
32. $\cos \alpha$
33. $\tan \beta$
34. $\sec \beta$

In Exercises 35–50, find the exact value of each expression, if it is defined. Do not use a calculator.

35. $\sin 210°$
36. $\cos 225°$
37. $\tan 420°$
38. $\cot(-540°)$
39. $\sec \dfrac{7\pi}{4}$
40. $\csc \dfrac{2\pi}{3}$
41. $\cos(-3\pi)$
42. $\sin \dfrac{13\pi}{6}$
43. $\cos^{-1} \tfrac{1}{2}$
44. $\arctan(-1)$
45. $\arcsin \tfrac{3}{2}$
46. $\sin^{-1}\left(-\dfrac{1}{\sqrt{2}}\right)$
47. $\arctan\left(\tan \dfrac{\pi}{3}\right)$
48. $\sin^{-1}\left(\sin \dfrac{5\pi}{6}\right)$
49. $\sin\left(\cos^{-1} \tfrac{1}{3}\right)$
50. $\cos(\arctan 4)$

In Exercises 51–62, use a calculator to find the approximate value of each expression. Round each answer to four significant digits.

51. $\cos 72°$
52. $\sin 305°$
53. $\cot 132°27′$
54. $\tan(-24°22′36″)$
55. $\csc 1.24$
56. $\sec 12$
57. $\sin\left(-\dfrac{7\pi}{8}\right)$
58. $\cos \dfrac{17\pi}{12}$
59. $\tan^{-1} 2$
60. $\sin^{-1}\left(-\tfrac{2}{3}\right)$
61. $\arccos(-0.4561)$
62. $\arctan 1.254$

In Exercises 63–66, compute the functional value rounding each answer to four significant digits. Assume that

$f(t) = \sin t \quad g(t) = 2 \cos 3t \quad h(t) = \tan(t - 1)$

63. $f(2)$
64. $g\left(-\frac{5}{3}\right)$
65. $h(4)$
66. $g(h(2))$

Given the functions f and g in Exercises 67–70, define the composite function $f \circ g$ by using an algebraic expression.

67. $f(x) = \sin x, g(x) = \cos^{-1} 4x$
68. $f(x) = \cos x, g(x) = \tan^{-1} \frac{x}{2}$
69. $f(x) = \tan x, g(x) = \arcsin \frac{1}{\sqrt{x}}$
70. $f(x) = \csc x, g(x) = \arccos(x - 2)$

In Exercises 71–78, give the amplitude, period, and phase shift for the graph of each equation. Then show the five key values on one cycle of the graph. Use a graphing calculator to confirm each of your answers.

71. $y = 3 \cos 4x$
72. $y = 2 \sin \frac{2x}{5}$
73. $y = \frac{5}{4} \sin(-3x)$
74. $y = 2 \cos(-0.3x)$
75. $y = \cos\left(x - \frac{\pi}{3}\right)$
76. $y = \frac{1}{3} \sin\left(\pi x + \frac{\pi}{12}\right)$
77. $y = 2 - 3 \sin\left(2x - \frac{\pi}{4}\right)$
78. $y = 1 + \cos\left(\frac{x}{2} + \pi\right)$

In Exercises 79–86, give the period for the graph of each equation. Then show the five key values on one cycle of the graph. Use a graphing calculator to confirm each of your answers.

79. $y = 2 \tan \frac{\pi x}{2}$
80. $y = \sec 2x$
81. $y = -\csc \frac{3x}{4}$
82. $y = -3 \cot \pi x$
83. $y = 0.5 \cot(2x + 0.125\pi)$
84. $y = -2 \tan\left(3x - \frac{2\pi}{3}\right)$
85. $y = 3 + \sec \frac{\pi x}{6}$
86. $y = 1 - \csc 3x$

87. Solve the right triangle in the sketch. Round x and y to three significant digits.

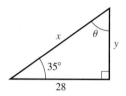

88. Solve the right triangle in the sketch. Round α and β to the nearest tenth of a degree, and round x to three significant digits.

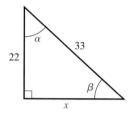

89. Montreal, Canada, is approximately due north of Santiago, Chile. Montreal is at latitude 45° North and Santiago is at latitude 33° South, as shown in the figure. If the radius of the earth is approximately 4000 miles, find the distance between the two cities.

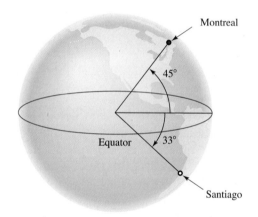

90. What is the linear speed (in miles per hour) of the Earth in its orbit around the Sun? Assume that the Earth's orbit is circular with a radius of 9.3×10^7 miles.

91. Express the length L of the *chord* in the sketch as a function of the central angle θ. Then complete the following table.

θ	1	$\frac{\pi}{12}$		
L			5	8

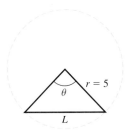

92. An observer is located 360 ft from the liftoff point of a hot-air balloon, as sketched in the figure. If the balloon rises vertically from its liftoff position at the rate of 3 ft/s, express angle θ as a function of the time t (in seconds) after liftoff.

93. For the piece of land shown in the sketch, find angles α and β. Round each answer to the nearest tenth of a degree.

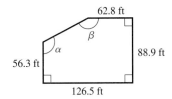

94. Three drill holes are located in an aluminum plate, as shown in the sketch.
(a) Express θ as a function of the distance x.
(b) Determine the domain and range of the function described in part (a).
(c) Use the answer from part (a) to find θ when $x = 18\sqrt{3}$ cm.

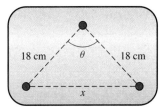

95. Four straight roads intersect, as shown in the sketch.
(a) Express α as a function of the distance x.
(b) Express β as a function of the distance x.
(c) Express θ as a function of the distance x.

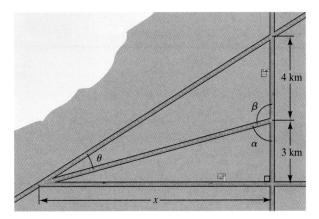

96. A tractor-trailer rig maintains a speed of 30 mi/h while climbing a hill that rises at an angle of 8.6°. How many minutes are required for the rig to attain a vertical rise of 2450 ft?

97. An air-traffic controller in a tower notes that two small planes on the ground are directly in line with the tower. The angles of depression to the planes are 9°15′ and 12°30′. If the controller is 150 ft above the ground, find the distance between the two planes.

98. The wolf population P in a national forest in Canada is given by

$$P = 340 + 180 \cos 0.785t$$

where t is the time (in years) with $t = 0$ corresponding to January 1, 1990. What is the first year after 1990 when the wolf population is expected to reach its maximum value? minimum value?

99. The wake from a small boat sets a fishing bob in simple harmonic motion, as shown in the sketch at the top of the next page. In 2 seconds the bob rides from the crest of one wave to the crest of the next wave.

(a) Find the equation that describes the displacement d of the bob from its equilibrium position if $d = 0$ (inches) at $t = 0$ (seconds).

(b) Find the displacement of the bob at $t = 3\frac{3}{4}$ seconds.

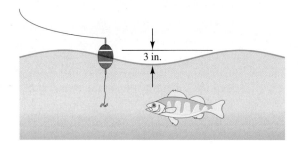

100. The periodic motion of a pendulum is an example of simple harmonic motion. The displacement d of the pendulum from its vertical position after t seconds can be described by the equation

$$d = a \cos \sqrt{\frac{g}{L}}\, t$$

where $|a|$ is the maximum displacement from the vertical position, g is the gravitational constant 980 cm/s², and L is the length of the pendulum (in centimeters). Suppose a pendulum is 80 cm long and the maximum displacement, 7.0 cm, occurs at $t = 0$ (seconds).

(a) Write the equation for the simple harmonic motion.

(b) Use a graphing calculator to generate the graph of the motion described in part (a).

(c) Find the time at which the pendulum passes through the vertical position for the third time.

The number of hours N of daylight on a particular day d of the year in the Boston area may be approximated by

$$N = 12 + 3 \sin\left[\frac{2\pi}{365}(d - 80)\right]$$

Determine the days of the year that have exactly 13 hours 30 minutes of daylight.

For the solution, see Example 6 in Section 8.2.

Analytic Trigonometry

- 8.1 Algebraic Manipulations of Trigonometric Expressions
- 8.2 Trigonometric Equations
- 8.3 Sum and Difference Formulas
- 8.4 Multiple-Angle Formulas
- 8.5 Product-to-Sum Formulas and Sum-to-Product Formulas

8.1 Algebraic Manipulations of Trigonometric Expressions

◆ Introductory Comments

In this chapter we study *analytic trigonometry,* a branch of mathematics in which algebraic procedures are applied to trigonometry. We apply the algebraic concepts from Chapter 1 to the trigonometric functions from Chapter 7 in order to

1. write trigonometric expressions in simpler form,
2. verify trigonometric identities,
3. solve trigonometric equations, and
4. develop several useful trigonometric formulas.

We begin by discussing the basic algebraic operations of addition, subtraction, multiplication, and division with trigonometric expressions.

◆ Basic Operations with Trigonometric Expressions

Because trigonometric expressions, such as $\sin x$, $\cos x$, $\tan x$, $\csc x$, $\sec x$, and $\cot x$, represent real numbers whenever the variable x is a real number (or an angle) for which the trigonometric expression is defined, we can manipulate trigonometric expressions in the same manner as we manipulate algebraic expressions. Hence, to expand

$$(\sin x + \tan x)^2,$$

we think of $\sin x$ and $\tan x$ as real numbers A and B. Now, since

$$(A + B)^2 = A^2 + 2AB + B^2,$$

we have

$$(\sin x + \tan x)^2 = \sin^2 x + 2 \sin x \tan x + \tan^2 x$$

Note: In this example, we have followed convention and omitted parentheses when squaring $\sin x$ and $\tan x$, that is, we prefer $\sin^2 x$ to $(\sin x)^2$ and $\tan^2 x$ to $(\tan x)^2$. We also omit parentheses when writing the product of two or more trigonometric expressions, that is, we prefer $\sin x \tan x$ to $(\sin x)(\tan x)$.

EXAMPLE 1 Perform the indicated operations.

(a) $(-2 \sin^3 x)^4$

(b) $3 \cos^2 t (2 \cos^3 t - 5 \cos^4 2t)$

(c) $(2 \tan \theta - 1)(\tan \theta + 3)$

(d) $\dfrac{5}{2 \sec^2 x} - \dfrac{1}{3 \sec x}$

SECTION 8.1 ♦ Manipulations of Trigonometric Expressions

♦ **Solution**

(a) $(-2\sin^3 x)^4 = (-2)^4(\sin^3 x)^4$ Apply $(ab)^n = a^n b^n$

$\qquad\qquad\qquad = 16\sin^{12} x$ Apply $(a^m)^n = a^{mn}$

(b) $3\cos^2 t(2\cos^3 t - 5\cos^4 2t)$

$\qquad = 3\cos^2 t(2\cos^3 t) - 3\cos^2 t(5\cos^4 2t)$ Apply distributive property

$\qquad = 6\cos^5 t - 15\cos^2 t\cos^4 2t$ Apply $a^m a^n = a^{m+n}$

Since t is different from $2t$, we cannot add these exponents.

(c) $(2\tan\theta - 1)(\tan\theta + 3)$

$\qquad = 2\tan^2\theta + 6\tan\theta - \tan\theta - 3$ Multiply like binomials

$\qquad = 2\tan^2\theta + 5\tan\theta - 3$ Combine like terms

(d) $\dfrac{5}{2\sec^2 x} - \dfrac{1}{3\sec x} = \dfrac{3 \cdot 5}{3 \cdot 2\sec^2 x} - \dfrac{2\sec x \cdot 1}{2\sec x \cdot 3\sec x}$ Change to equivalent fractions with the same LCD

The LCD is $6\sec^2 x$.

$\qquad\qquad\qquad\qquad = \dfrac{15 - 2\sec x}{6\sec^2 x}$ Subtract fractions ♦

PROBLEM 1   Perform the indicated operation:
$\dfrac{4}{\csc\theta - 1} + \dfrac{3}{1 - \csc\theta}$ ♦

♦ **Factoring Trigonometric Expressions**

As we illustrate in the next example, the techniques of factoring, which we discussed in Section 1.6, can also be applied to trigonometric expressions.

EXAMPLE 2   Factor each expression completely over the set of integers.

(a) $3\cos x\sin^2 x - 6\cos x$ (b) $\sin^2\theta - \cos^2\theta$

(c) $6\tan^2 3x - \tan 3x - 2$ (d) $\sec^3 t + 1$

♦ **Solution**

(a) The two terms of the expression $3\cos x\sin^2 x - 6\cos x$ contain a common factor, $3\cos x$, which we can factor out:

$3\cos x\sin^2 x - 6\cos x = 3\cos x(\sin^2 x - 2).$

(b) The expression $\sin^2\theta - \cos^2\theta$ is the difference of squares. Thus,

$$\sin^2\theta - \cos^2\theta = (\sin\theta + \cos\theta)(\sin\theta - \cos\theta).$$

(c) The expression $6\tan^2 3x - \tan 3x - 2$ is quadratic in form and factors as a trinomial square:

$$6\tan^2 3x - \tan 3x - 2 = (3\tan 3x - 2)(2\tan 3x + 1).$$

(d) The expression $\sec^3 t + 1$ is the sum of cubes. Thus,

$$\sec^3 t + 1 = (\sec t + 1)(\sec^2 t - \sec t + 1).$$

PROBLEM 2 Factor the expression $\csc^2 x - 3\csc x - 10$ completely over the set of integers.

◆ Simplifying Trigonometric Expressions

The eight fundamental trigonometric identities given in Section 7.2 are frequently used to help simplify trigonometric expressions. For our convenience, these identities are listed here.

◆ Fundamental Trigonometric Identities

1. $\sin x \csc x = 1$
2. $\cos x \sec x = 1$
3. $\tan x \cot x = 1$
4. $\tan x = \dfrac{\sin x}{\cos x}$
5. $\cot x = \dfrac{\cos x}{\sin x}$
6. $\sin^2 x + \cos^2 x = 1$
7. $1 + \tan^2 x = \sec^2 x$
8. $1 + \cot^2 x = \csc^2 x$

EXAMPLE 3 Use algebraic manipulations and the fundamental trigonometric identities to simplify each expression to a single trigonometric function or a constant.

(a) $\cos x \tan^2 x + \cos x$ **(b)** $\dfrac{\cos^2 x}{1 - \sin x} - \sin x$

◆ **Solution**

(a) $\cos x \tan^2 x + \cos x$

$= \cos x (\tan^2 x + 1)$ Factor

$= \cos x \sec^2 x$ Apply trig identity 7

$= (\cos x \sec x) \sec x$ Rewrite

$= (1) \sec x$ Apply trig identity 2

$= \sec x$ Simplify

SECTION 8.1 ♦ Manipulations of Trigonometric Expressions

(b) $\dfrac{\cos^2 x}{1 - \sin x} - \sin x$

$= \dfrac{\cos^2 x - \sin x (1 - \sin x)}{1 - \sin x}$ Subtract fractions

$= \dfrac{\cos^2 x - \sin x + \sin^2 x}{1 - \sin x}$ Apply the distributive property

$= \dfrac{1 - \sin x}{1 - \sin x}$ Apply trig identity 6

$= 1$ Reduce to lowest terms provided that $1 - \sin x \neq 0$ ◆

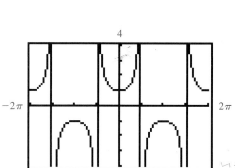

FIGURE 8.1
The graph of $y = \cos x \tan^2 x + \cos x$ is the same as the graph of $y = \sec x$.

We can use a graphing calculator to verify our work in Example 3. If $f(x)$ is the original expression and $g(x)$ is the simplified expression, then the graph of $y = f(x)$ should be identical to the graph of $y = g(x)$. Figure 8.1 shows the graph of $y = \cos x \tan^2 x + \cos x$ in the viewing rectangle of a graphing calculator. The graph of this equation seems to be the same as the graph of $y = \sec x$, which confirms our answer to Example 3(a). Figure 8.2 shows the graph of $y = \dfrac{\cos^2 x}{1 - \sin x} - \sin x$ in the viewing rectangle of a graphing calculator. The graph of this equation seems to be the same as the graph of $y = 1$, which confirms our answer to Example 3(b).

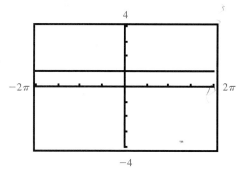

FIGURE 8.2
The graph of $y = \dfrac{\cos^2 x}{1 - \sin x} - \sin x$ is the same as the graph of $y = 1$, provided that $1 - \sin x \neq 0$.

PROBLEM 3 Express $\sin^2\theta \cot^2\theta + \sin^2\theta$ as a single trigonometric function or constant. Use a graphing calculator to confirm your answer. ◆

The procedures we used to simplify the trigonometric expressions in Example 3 are *not* unique. For instance, another method of simplifying $\cos x \tan^2 x + \cos x$ is to rewrite the expression in terms of sines and cosines and then proceed as follows.

$\cos x \tan^2 x + \cos x = \cos x \left(\dfrac{\sin x}{\cos x} \right)^2 + \cos x$ Apply trig identity 4

$= \cos x \dfrac{\sin^2 x}{\cos^2 x} + \cos x$ Apply $(a/b)^n = a^n/b^n$

$= \dfrac{\sin^2 x}{\cos x} + \cos x$ Multiply fractions

$= \dfrac{\sin^2 x + \cos^2 x}{\cos x}$ Add fractions

$= \dfrac{1}{\cos x}$ Apply trig identity 6

$= \sec x$ Apply trig identity 2

When simplifying a trigonometric expression the procedure of rewriting the expression in terms of sines and cosines often works well.

EXAMPLE 4 Express $\dfrac{\cot x - 1}{1 - \tan x}$ as a single trigonometric function or a constant.

Solution We begin by writing the expression in terms of sines and cosines:

$$\dfrac{\cot x - 1}{1 - \tan x} = \dfrac{\dfrac{\cos x}{\sin x} - 1}{1 - \dfrac{\sin x}{\cos x}} \qquad \text{Apply trig identities 5 and 4}$$

$$= \dfrac{\dfrac{\cos x - \sin x}{\sin x}}{\dfrac{\cos x - \sin x}{\cos x}} \qquad \text{Perform the indicated subtractions in the numerator and denominator}$$

$$= \dfrac{\cos x - \sin x}{\sin x} \cdot \dfrac{\cos x}{\cos x - \sin x} \qquad \text{Rewrite as a product}$$

$$= \dfrac{\cos x}{\sin x} \qquad \text{Multiply and reduce, provided that } \cos x - \sin x \neq 0$$

$$= \cot x \qquad \text{Apply trig identity 5}$$

The graph in Figure 8.3 supports our work. ◆

PROBLEM 4 Express $\sin x + \cos x \cot x$ as a single trigonometric function or constant. ◆

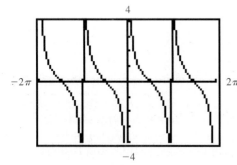

FIGURE 8.3

The graph of $y = \dfrac{\cot x - 1}{1 - \tan x}$ is the same as the graph of $y = \cot x$, provided that $\cos x - \sin x \neq 0$.

◆ Verifying Trigonometric Identities

A **trigonometric equation** is a statement declaring that two trigonometric expressions are equal. A trigonometric equation that is true for all permissible values of the variables that it involves is called a **trigonometric identity.** To verify that a trigonometric equation is an identity, we can use the fundamental trigonometric identities in conjunction with algebraic manipulations to transform one side of the equation into the other. Although no general method works for verifying every trigonometric identity, we offer the following suggestion.

◆ **Suggestion for Verifying a Trigonometric Identity**

> Reduce the more complicated side of the equation to the simpler side. If no simplification is obvious, express every trigonometric expression on the more complicated side of the equation in terms of sines and cosines, and then reduce this expression to the simpler side.

CAUTION When verifying a trigonometric identity, we are trying to show that the equation is true for all permissible values of the variables that it involves. Thus, it is incorrect to assume the equation is already true, and then apply the rules for

SECTION 8.1 ♦ Manipulations of Trigonometric Expressions

generating equivalent equations, such as adding an expression or multiplying an expression to both sides.

EXAMPLE 5 A trigonometric equation of the form $f(x) = g(x)$ is given. Use a graphing calculator to generate the graphs of $y = f(x)$ and $y = g(x)$. Compare the graphs, and then state whether the given equation seems to be a trigonometric identity. If the equation seems to be an identity, verify this fact algebraically.

(a) $\sec x - \sin x \cot x = \tan x \sin x$

(b) $2 \sec^2 x - 1 = \dfrac{1 + \sin^2 x}{\cos^2 x}$

◀ Solution

(a) As illustrated in Figure 8.4, the graphs of $y = \sec x - \sin x \cot x$ and $y = \tan x \sin x$ appear to be identical. Hence, the trigonometric equation

$$\sec x - \sin x \cot x = \tan x \sin x$$

seems to be an identity. To verify this fact algebraically, we reduce the more complicated, left-hand side of the equation to the simpler, right-hand side:

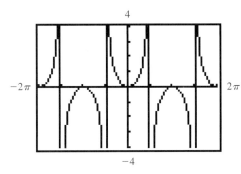

FIGURE 8.4
The graphs of $y = \sec x - \sin x \cot x$ and $y = \tan x \sin x$ appear to be identical.

$\sec x - \sin x \cot x = \tan x \sin x$	
$\dfrac{1}{\cos x} - \sin x \dfrac{\cos x}{\sin x}$	Apply trig identities 2 and 5
$\dfrac{1}{\cos x} - \cos x$	Multiply fractions
$\dfrac{1 - \cos^2 x}{\cos x}$	Subtract fractions
$\dfrac{\sin^2 x}{\cos x}$	Apply trig identity 6
$\dfrac{\sin x}{\cos x} \sin x$	Rewrite
$\tan x \sin x$	Apply trig identity 4

Since we have reduced the left-hand side of this equation to the right-hand side, we conclude that

$$\sec x - \sin x \cot x = \tan x \sin x \quad \text{is a trigonometric identity.}$$

(b) As illustrated in Figure 8.5, the graphs of $y = 2 \sec^2 x - 1$ and $y = \dfrac{1 + \sin^2 x}{\cos^2 x}$ appear to be identical. Hence, the trigonometric equation

$$2 \sec^2 x - 1 = \dfrac{1 + \sin^2 x}{\cos^2 x}$$

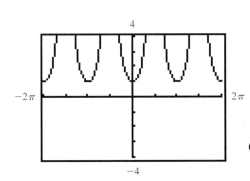

FIGURE 8.5
The graphs of $y = 2 \sec^2 x - 1$ and $y = \dfrac{1 + \sin^2 x}{\cos^2 x}$ appear to be identical.

seems to be an identity. To verify this fact algebraically, we reduce the more complicated, right-hand side of the equation to the simpler, left-hand side:

$$2\sec^2 x - 1 = \frac{1 + \sin^2 x}{\cos^2 x}$$

$$\frac{1}{\cos^2 x} + \frac{\sin^2 x}{\cos^2 x} \qquad \text{Separate into two fractions}$$

$$\sec^2 x + \tan^2 x \qquad \text{Apply trig identities 2 and 4}$$

$$\sec^2 x + (\sec^2 x - 1) \qquad \text{Apply trig identity 7}$$

$$2\sec^2 x - 1 \qquad \text{Combine like terms}$$

Since we have reduced the right-hand side of this equation to the left-hand side, we conclude that

$$2\sec^2 x - 1 = \frac{1 + \sin^2 x}{\cos^2 x} \quad \text{is a trigonometric identity.} \quad \blacklozenge$$

PROBLEM 5 Repeat the directions in Example 5 for $\tan\theta - \cot\theta = \sec\theta\csc\theta$. $\quad \blacklozenge$

As illustrated in the next example, we can sometimes verify a trigonometric identity by using the fundamental property of fractions (Section 1.7) or by using the techniques of factoring.

EXAMPLE 6 Show graphically and verify algebraically that each equation is a trigonometric identity.

(a) $\dfrac{\sin\theta}{1 + \cos\theta} = \dfrac{1 - \cos\theta}{\sin\theta}$

(b) $\dfrac{\cos^3 x + \sin^3 x}{\cos x + \sin x} = 1 - \sin x \cos x$

$\blacklozenge$ **Solution**

(a) As illustrated in Figure 8.6, the graphs of $y = \dfrac{\sin\theta}{1 + \cos\theta}$ and $y = \dfrac{1 - \cos\theta}{\sin\theta}$ appear to be identical. Hence, the trigonometric equation

$$\frac{\sin\theta}{1 + \cos\theta} = \frac{1 - \cos\theta}{\sin\theta}$$

seems to be an identity. To verify this fact algebraically, we proceed as follows:

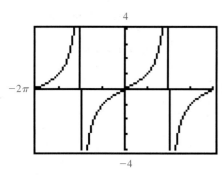

FIGURE 8.6

The graphs of $y = \dfrac{\sin\theta}{1 + \cos\theta}$ and $y = \dfrac{1 - \cos\theta}{\sin\theta}$ appear to be identical.

SECTION 8.1 ♦ *Manipulations of Trigonometric Expressions* 493

$$\frac{\sin\theta}{1+\cos\theta} = \frac{1-\cos\theta}{\sin\theta}$$

$\dfrac{\sin\theta}{1+\cos\theta} \cdot \dfrac{1-\cos\theta}{1-\cos\theta}$	Apply the fundamental property of fractions
$\dfrac{\sin\theta\,(1-\cos\theta)}{1-\cos^2\theta}$	Multiply
$\dfrac{\sin\theta\,(1-\cos\theta)}{\sin^2\theta}$	Apply trig identity 6
$\dfrac{1-\cos\theta}{\sin\theta}$	Reduce, provided that $\sin\theta \neq 0$

Since we have reduced the left-hand side of this equation to the right-hand side, we conclude that

$$\frac{\sin\theta}{1+\cos\theta} = \frac{1-\cos\theta}{\sin\theta} \quad \text{is a trigonometric identity.}$$

(b) As illustrated in Figure 8.7, the graphs of $y = \dfrac{\cos^3 x + \sin^3 x}{\cos x + \sin x}$ and $y = 1 - \sin x \cos x$ appear to be identical. Hence, the trigonometric equation

$$\frac{\cos^3 x + \sin^3 x}{\cos x + \sin x} = 1 - \sin x \cos x$$

seems to be an identity. To verify this fact algebraically, we note that the numerator of the fraction on the left-hand side is the sum of cubes:

$$\frac{\cos^3 x + \sin^3 x}{\cos x + \sin x} = 1 - \cos x \sin x$$

$\dfrac{(\cos x + \sin x)(\cos^2 x - \cos x \sin x + \sin^2 x)}{\cos x + \sin x}$	Factor
$\cos^2 x - \cos x \sin x + \sin^2 x$	Reduce, provided that $\cos x + \sin x \neq 0$
$1 - \cos x \sin x$	Apply trig identity 6

Since we have reduced the left-hand side of this equation to the right-hand side, we conclude that

$$\frac{\cos^3 x + \sin^3 x}{\cos x + \sin x} = 1 - \sin x \cos x \quad \text{is a trigonometric identity,}$$

provided $\sin x + \cos x \neq 0$. ◆

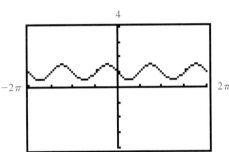

FIGURE 8.7
The graphs of $y = \dfrac{\cos^3 x + \sin^3 x}{\cos x + \sin x}$ and $y = 1 - \sin x \cos x$ appear to be identical.

PROBLEM 6 Show graphically and verify algebraically that

$$\sec^4 x - \tan^4 x = \frac{1 + \sin^2 x}{\cos^2 x} \quad \text{is a trigonometric identity.} \quad ◆$$

◆ Trigonometric Substitution

In calculus, we must sometimes rewrite algebraic expressions by making a *trigonometric substitution*. This technique is illustrated in the next example.

EXAMPLE 7 Make the trigonometric substitution $x = 3 \tan \theta$ with $0 < \theta < \pi/2$, and then express $\sqrt{x^2 + 9}$ as a function of θ in simplified form.

◆ Solution Replacing x with $3 \tan \theta$, we have

$$\sqrt{x^2 + 9} = \sqrt{(3 \tan \theta)^2 + 9}$$
$$= \sqrt{9 \tan^2 \theta + 9} \qquad \text{Apply } (ab)^n = a^n b^n$$
$$= \sqrt{9(\tan^2 \theta + 1)} \qquad \text{Factor}$$
$$= \sqrt{9 \sec^2 \theta} \qquad \text{Apply trig identity 7}$$
$$= 3 \sec \theta \qquad \text{since } \sec \theta > 0 \text{ for } 0 < \theta < \pi/2 \qquad \blacklozenge$$

We can check the results of the trigonometric substitution in Example 7 by using the trigonometric ratios for right triangles (Section 7.2). Since $x = 3 \tan \theta$, we have

$$\tan \theta = \frac{x}{3} = \frac{\text{opp}}{\text{adj}}$$

as shown in Figure 8.8. By the Pythagorean theorem, the hypotenuse is $\sqrt{x^2 + 9}$. Thus,

$$\sec \theta = \frac{\text{hyp}}{\text{adj}} = \frac{\sqrt{x^2 + 9}}{3}.$$

Hence, $\sqrt{x^2 + 9} = 3 \sec \theta$.

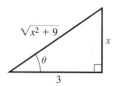

FIGURE 8.8
A right triangle with $\tan \theta = x/3$

PROBLEM 7 Make the trigonometric substitution $x = 2 \sin \theta$ with $0 < \theta < \pi/2$, and then express $\sqrt{4 - x^2}$ as a function of θ in simplified form. ◆

Exercises 8.1

◆ Basic Skills

In Exercises 1–10, perform the indicated operations.

1. $(3 \cos^2 y \sin y)^3$
2. $(2 \tan^2 x \sin x)(-3 \tan^3 x \sin 3x)$
3. $\dfrac{12 \cos^2 x \cos^{-4} 2x}{3 \cos^{-3} 2x \cos x}$
4. $2 \cos^{-3} y (\cos^3 y - 3 \cos^4 y)$
5. $(3 \tan \theta + 2)^2$
6. $(3 \sin \theta - 2)(\sin \theta + 1)$
7. $\dfrac{8 \tan x \cos 2x}{5 \sin x} \cdot \dfrac{15 \sin^2 x}{\cos 2x}$

8. $\dfrac{3}{2 \sin x} - \dfrac{4}{\sin^2 x}$

9. $\dfrac{1}{1 - \cos x} + \dfrac{1}{\cos x}$

10. $\dfrac{1 + \dfrac{1}{\tan 3y}}{\dfrac{1}{\tan^2 3y}}$

In Exercises 11–16, factor completely over the set of integers.

11. $3 \tan \theta - 15 \tan^3 \theta$
12. $12 \cos x \cot^2 x - 15 \cos^2 x \cot x$
13. $\cos^2 \theta - \sin^2 \theta$
14. $4 \tan^2 x - 1$
15. $\sin^2 x - 2 \sin x - 3$
16. $2 \csc^2 x - 3 \csc x - 5$

In Exercises 17–20, reduce each fraction to lowest terms.

17. $\dfrac{\sin x + 1}{\sin^2 x - 1}$

18. $\dfrac{\tan^3 x - \tan^2 x - 12 \tan x}{3 + \tan x}$

19. $\dfrac{\sin^4 x - \cos^4 x}{\sin x + \cos x}$

20. $\dfrac{8 - \cot^3 y}{2 - \cot y}$

In Exercises 21–36, simplify each expression to a single trigonometric function or a constant. Use a graphing calculator to confirm each of your answers.

21. $\cos x \csc x$
22. $\sin \theta \sec \theta$
23. $\dfrac{\sec x}{\csc x}$
24. $\dfrac{\cot x}{\cos x}$
25. $\csc^2 x \tan^2 x - \tan^2 x$
26. $\sin^2 x + \cot^2 x \sin^2 x$
27. $\cos^2 y(1 - \sec^2 y) + \sin^2 y$
28. $(\csc \theta - \cot \theta)(\csc \theta + \cot \theta)$
29. $(\sin \theta + \cos \theta)^2 + (\sin \theta - \cos \theta)^2$
30. $\dfrac{\sin x \cot x + \cos x}{2 \cot x}$
31. $\dfrac{\tan^2 x + 1}{\tan x + \cot x}$
32. $\dfrac{\sin y + \tan y}{1 + \sec y}$
33. $\dfrac{1 + \csc x}{\sec x} - \cos x$
34. $\dfrac{\sin x}{\sec x + 1} + \dfrac{\sin x}{\sec x - 1}$
35. $(\sec \theta + \tan \theta)^4 (\sec \theta - \tan \theta)^4$
36. $\left(\dfrac{1 + \sin x}{2 \cos^2 x}\right)^3 (1 - \sin x)^3$

In Exercises 37–60, a trigonometric equation of the form $f(x) = g(x)$ is given. Use a graphing calculator to generate the graphs of $y = f(x)$ and $y = g(x)$. Compare the graphs, and then state whether the given equation seems to be a trigonometric identity. If the equation seems to be an identity, verify this fact algebraically.

37. $\sin \theta (\csc \theta - \sin \theta) = \cos^2 \theta$
38. $\cos^2 x (\sec^2 x + \csc^2 x) = \csc^2 x$
39. $(\sin x + \cos x)^2 = 1 + 2 \sin x \cos x$
40. $(\tan \theta + 1)^2 = \sec^2 \theta + 2 \tan \theta$
41. $\cos^2 x - \sin^2 x = 2 \cos^2 x - 1$
42. $\cos^2 x - \sin^2 x = 1 + 2 \sin^2 x$
43. $\dfrac{1 - \cos^2 y}{\cos y} = \sin y \tan y$
44. $\dfrac{\sec^2 y}{\sec^2 y - 1} = \csc^2 y$
45. $\dfrac{1 + \csc x}{\sec x} = \cos x - \cot x$
46. $\dfrac{\sin^2 x + \sin x + \cos^2 x}{\sin x} = 1 + \csc x$
47. $(1 - \cos \theta)(1 + \sec \theta) = \sin \theta \tan \theta$
48. $(\sin x + \cos x)(\csc x - \sec x) = \cot x - \tan x$
49. $\cot^2 y - \cos^2 y = \cot^2 y \cos^2 y$
50. $\sec^2 \theta + \csc^2 \theta = \sec^2 \theta \csc^2 \theta$
51. $\dfrac{\cos x + \sin x}{\sec x + \csc x} = \sin 2x$
52. $\dfrac{\cos x + \sin x}{\cos x - \sin x} = \dfrac{1 + \tan x}{1 - \tan x}$
53. $\dfrac{\cos x}{1 - \sin x} = \dfrac{1 + \sin x}{\cos x}$
54. $\dfrac{1 + \sin x}{1 - \sin x} = (\tan x - \sec x)^2$
55. $\dfrac{1 - \csc^2 x}{1 + \csc x} = \dfrac{\sin x - 1}{\sin x}$
56. $\dfrac{\tan^3 x - 1}{\tan x - 1} = \sec^2 x + \tan x$
57. $\dfrac{\cos^4 y - \sin^4 y}{1 - \tan^4 y} = \cos^4 y$
58. $\dfrac{\sec^4 y - 1}{\tan^2 y} = 2 + \tan^2 y$
59. $2 \tan \theta + \sec^2 \theta = \sin^2 \theta (\sec \theta + \csc \theta)^2$
60. $\cot \theta - \tan \theta = 2 \cos \theta \csc \theta - \sec \theta \csc \theta$

In Exercises 61–68, make the indicated trigonometric substitution and simplify the given expression. Assume that $0 < \theta < \pi/2$.

61. For $\sqrt{1 - x^2}$, let $x = \sin \theta$.

62. For $\sqrt{x^2 - 25}$, let $x = 5 \sec \theta$.

63. For $\dfrac{\sqrt{x^2 + 16}}{x}$, let $x = 4 \tan \theta$.

64. For $\dfrac{x}{\sqrt{4 - x^2}}$, let $x = 2 \sin \theta$.

65. For $\dfrac{x^3}{(x^2 - 9)^{3/2}}$, let $x = 3 \sec \theta$.

66. For $\dfrac{x^2}{x^2 + 16}$, let $x = 4 \tan \theta$.

67. For $\dfrac{x}{\sqrt{4 - (x + 1)^2}}$, let $x + 1 = 2 \sin \theta$.

68. For $\sqrt{1 + e^{2x}}$, let $e^x = \tan \theta$.

◆ Critical Thinking

In Exercises 69–74, show that each equation is an identity.

69. $\log(1 - \cos x) + \log(1 + \cos x) = 2 \log |\sin x|$
70. $2 \log |\sec x| - \log |\tan x| = \log |\sec x \csc x|$
71. $-\ln |\csc x - \cot x| = \ln |\csc x + \cot x|$
72. $-\ln |\sec x - \tan x| = \ln |\sec x + \tan x|$
73. $\dfrac{e^{\csc^2 x}}{e^{\cos^2 x} e^{\cot^2 x}} = e^{\sin^2 x}$
74. $e^{x - 2\ln|\cos x|} = e^x \sec^2 x$
75. Simplify the left-hand side of the equation
    $(x \sin x - y \cos x)^2 + (x \cos x + y \sin x)^2 = 25$.
    Then describe the graph of this equation.

76. One of the fundamental trigonometric identities is $1 + \tan^2 x = \sec^2 x$. Does it follow that
    $$\sqrt{1 + \tan^2 x} = \sec x$$
    is also a trigonometric identity? Explain.

77. Is $\sin(x + y) = \sin x + \sin y$ a trigonometric identity? Select some values of x and y, evaluate each side of the equation for these values, and then state your conclusion.

78. Is $\cos(x - y) = \cos x - \cos y$ a trigonometric identity? Select some values of x and y, evaluate each side of the equation for these values, and then state your conclusion.

8.2 Trigonometric Equations

◆ Introductory Comments

A trigonometric equation that is true for some values of the variables it involves, but is false for others, or is never true for any real number, is called a **conditional trigonometric equation.** In this section, we discuss procedures for solving conditional trigonometric equations. To *solve* a conditional trigonometric equation means to find all values of the variables for which the equation is true. Unless specifically stated otherwise, the solutions are expressed as real numbers or, equivalently, as angles in radians. We begin with some simple types of equations and then expand to types that are solved by factoring or by applying the quadratic formula.

◆ Simple Trigonometric Equations

The rules and procedures for solving algebraic equations apply to conditional trigonometric equations as well. For instance, to solve the equation $2 \sin x - 1 = 0$ for x, we begin by generating an equivalent equation:

$$2 \sin x - 1 = 0$$
$$2 \sin x = 1 \quad \text{Add 1 to both sides}$$
$$\sin x = \tfrac{1}{2} \quad \text{Divide both sides by 2}$$

SECTION 8.2 ◆ Trigonometric Equations

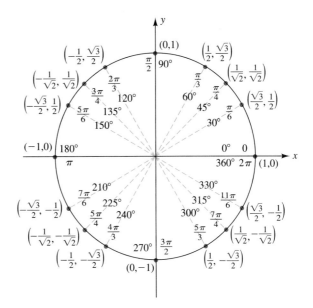

FIGURE 8.9
Special values on the unit circle

As illustrated on the unit circle in Figure 8.9, we select

$$\frac{\pi}{6} \quad \text{and} \quad \frac{5\pi}{6}$$

as the only values of θ in the interval $[0, 2\pi)$ for which $\sin \theta = \frac{1}{2}$. Since the sine function is periodic with period 2π, we add multiples of 2π to $\pi/6$ and $5\pi/6$ to obtain all real solutions of this equation. Thus, the expressions

$$\frac{\pi}{6} + 2\pi n \quad \text{and} \quad \frac{5\pi}{6} + 2\pi n \quad \text{where } n \text{ is any integer}$$

describe the *general solution* of the equation $2 \sin x - 1 = 0$.

To obtain *particular solutions* of a trigonometric equation in some given interval, we substitute appropriate integers for n in the general solution. For example, to find the particular solutions of the equation $2 \sin x - 1 = 0$ in the interval $[-2\pi, 4\pi]$. We choose integers n for

$$\frac{\pi}{6} + 2\pi n \quad \text{and} \quad \frac{5\pi}{6} + 2\pi n$$

that yield solutions in this interval. If we let $n = -1, 0,$ and 1 in the general solution, we obtain the particular solutions:

$$\underbrace{-\frac{11\pi}{6} \quad -\frac{7\pi}{6} \quad \frac{\pi}{6} \quad \frac{5\pi}{6} \quad \frac{13\pi}{6} \quad \frac{17\pi}{6}}_{\text{particular solutions of } 2 \sin x - 1 = 0 \text{ in the interval } [-2\pi, 4\pi]}$$

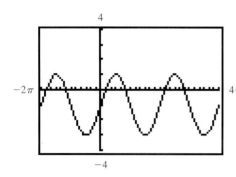

FIGURE 8.10
Graph of $y = 2 \sin x - 1$ in the interval $[-2\pi, 4\pi]$ with x-scale $\pi/6$

We can use a graphing calculator to verify the solutions of a trigonometric equation of the form $f(x) = 0$ by noting the points at which the graph of $y = f(x)$ intersects the x-axis. Figure 8.10 shows the graph of $y = 2 \sin x - 1$ in the interval $[-2\pi, 4\pi]$. In this interval the graph intersects the x-axis at

$$-\frac{11\pi}{6}, \quad -\frac{7\pi}{6}, \quad \frac{\pi}{6}, \quad \frac{5\pi}{6}, \quad \frac{13\pi}{6}, \quad \text{and} \quad \frac{17\pi}{6}.$$

These facts confirm our work.

We can solve any trigonometric equation containing $\cos x$, $\csc x$, or $\sec x$ by the preceding method. For equations containing $\tan x$ or $\cot x$, the procedure is slightly different, because the period of the tangent and cotangent functions is π. (The sine, cosine, cosecant, and secant functions have period 2π.) For equations containing $\tan x$ or $\cot x$, we find all solutions in the interval $[0, \pi)$, and then add multiples of π to these solutions to obtain the general solution of the equation. We now summarize the steps for solving a trigonometric equation containing either $\sin x$, $\cos x$, $\csc x$, $\sec x$, $\tan x$, or $\cot x$.

◆ Procedure for Solving a Trigonometric Equation

An equation containing $\sin x$, $\cos x$, $\csc x$, or $\sec x$:

1. Solve the equation for the trigonometric function.
2. Find all x in the interval $[0, 2\pi)$ that satisfy the equation in step 1.
3. Add $2\pi n$, where n is an integer, to each solution in step 2.

Equations containing $\tan x$ or $\cot x$:

1. Solve the equation for the trigonometric function.
2. Find all x in the interval $[0, \pi)$ that satisfy the equation in step 1.
3. Add πn, where n is an integer, to each solution in step 2.

EXAMPLE 1 Find the general solution of each trigonometric equation.

(a) $\sqrt{2} \cos x + 1 = 0$ (b) $\tan^2 x - 3 = 0$

◆ Solution

(a) We begin by solving for $\cos x$:

$$\sqrt{2} \cos x + 1 = 0$$
$$\sqrt{2} \cos x = -1 \quad \text{Subtract 1 from both sides}$$
$$\cos x = -\frac{1}{\sqrt{2}} \quad \text{Divide both sides by } \sqrt{2}$$

Now, from the unit circle in Figure 8.9, we select

$$\frac{3\pi}{4} \quad \text{and} \quad \frac{5\pi}{4}$$

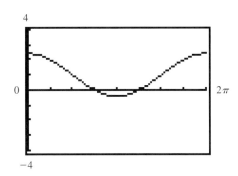

FIGURE 8.11
Graph of $y = \sqrt{2} \cos x + 1$ in the interval $[0, 2\pi)$ with x-scale $\pi/4$

as the only values of θ in the interval $[0, 2\pi)$ for which $\cos \theta = -1\sqrt{2}$. Figure 8.11 shows the graph of $y = \sqrt{2} \cos x + 1$ in the interval $[0, 2\pi)$. In this interval the graph intersects the x-axis at $3\pi/4$ and $5\pi/4$, which confirms our work.

Since the cosine function is periodic and its period is 2π, we add multiples of 2π to $3\pi/4$ and $5\pi/4$ to obtain all real solutions. Thus, the expressions

$$\frac{3\pi}{4} + 2\pi n \quad \text{and} \quad \frac{5\pi}{4} + 2\pi n \quad \text{where } n \text{ is an integer}$$

describe the general solution of the equation $\sqrt{2} \cos x + 1 = 0$.

(b) We begin by solving for $\tan x$:

$$\tan^2 x - 3 = 0$$

$$\tan^2 x = 3 \qquad \text{Add 3 to both sides}$$

$$\tan x = \pm\sqrt{3} \qquad \text{Take the square root of both sides}$$

Now, from the unit circle in Figure 8.9, we select

$$\frac{\pi}{3} \quad \text{and} \quad \frac{2\pi}{3}$$

as the only values of θ in the interval $[0, \pi)$ for which $\tan \theta = \pm\sqrt{3}$. Figure 8.12 shows the graph of $y = \tan^2 x - 3$ in the interval $[0, \pi)$. In this interval the graph intersects the x-axis at $\pi/3$ and $2\pi/3$, which confirms our work.

Since the tangent function is periodic and its period is π, we add multiples of π to $\pi/3$ and $2\pi/3$ to obtain all real solutions. Thus, the expressions

$$\frac{\pi}{3} + \pi n \quad \text{and} \quad \frac{2\pi}{3} + \pi n \quad \text{where } n \text{ is an integer}$$

describe the general solution of the equation $\tan^2 x - 3 = 0$. ◆

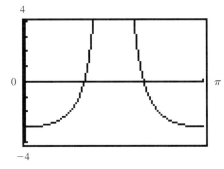

FIGURE 8.12
Graph of $y = \tan^2 x - 3$ in the interval $[0, \pi)$ with x-scale $\pi/3$

PROBLEM 1 Find the general solution of the trigonometric equation $\sqrt{3} \csc x - 2 = 0$.

◆

The method we outlined in Example 1 for solving a trigonometric equation may be extended to trigonometric equations of the forms

$$\sin(bx + c) = k \qquad \cos(bx + c) = k \qquad \tan(bx + c) = k$$
$$\csc(bx + c) = k \qquad \sec(bx + c) = k \qquad \cot(bx + c) = k$$

where b, c, and k are constants. We begin by finding all values of $(bx + c)$ that satisfy the equation, and then solve for x by using ordinary algebraic methods. This procedure is illustrated in the next example.

EXAMPLE 2 Find the general solution of each trigonometric equation. Then find the particular solutions in the interval $[0, 2\pi)$.

(a) $2 \cot 3x - 2 = 0$ (b) $4 + \sin\left(2x - \dfrac{\pi}{3}\right) = 5$

◀ **Solution**

(a) We begin by solving for $\cot 3x$ as follows:

$$2 \cot 3x - 2 = 0$$

$$2 \cot 3x = 2 \quad \text{Add 2 to both sides}$$

$$\cot 3x = 1 \quad \text{Divide both sides by 2}$$

From the unit circle in Figure 8.9, we know that $\cot(\pi/4) = 1$ in the interval $[0, \pi)$. Hence,

$$3x = \dfrac{\pi}{4} \quad \text{and, consequently,} \quad 3x = \dfrac{\pi}{4} + \pi n.$$

Dividing both sides of this equation by 3 gives us the general solution:

$$x = \dfrac{\pi}{12} + \dfrac{\pi}{3}n, \quad \text{where } n \text{ is an integer}$$

To find the particular solutions in the interval $[0, 2\pi)$, we let $n = 0, 1, 2, 3, 4, 5$ in the general solution and obtain

$$\dfrac{\pi}{12}, \dfrac{5\pi}{12}, \dfrac{3\pi}{4}, \dfrac{13\pi}{12}, \dfrac{17\pi}{12}, \text{ and } \dfrac{7\pi}{4}.$$

Figure 8.13 shows the graph of $y = 2 \cot 3x - 2$ in the interval $[0, 2\pi)$. The six x-intercepts confirm our work.

(b) We begin by solving for $\sin\left(2x - \dfrac{\pi}{3}\right)$ as follows:

$$4 + \sin\left(2x - \dfrac{\pi}{3}\right) = 5$$

$$\sin\left(2x - \dfrac{\pi}{3}\right) = 1 \quad \text{Subtract 4 from both sides}$$

From the unit circle in Figure 8.9, we know that $\sin(\pi/2) = 1$ in the interval $[0, 2\pi)$. Hence,

$$2x - \dfrac{\pi}{3} = \dfrac{\pi}{2} \quad \text{and, consequently,} \quad 2x - \dfrac{\pi}{3} = \dfrac{\pi}{2} + 2\pi n.$$

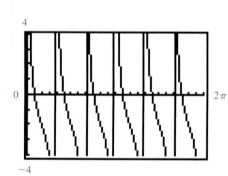

FIGURE 8.13
Graph of $y = 2 \cot 3x - 2$ in the interval $[0, 2\pi)$ with x-scale $\pi/12$

Solving this equation for x gives us the general solution:

$$2x - \frac{\pi}{3} = \frac{\pi}{2} + 2\pi n$$

$$2x = \frac{5\pi}{6} + 2\pi n \quad \text{Add } \pi/3 \text{ to both sides}$$

$$x = \frac{5\pi}{12} + \pi n \quad \text{Divide both sides by 2}$$

To find the particular solutions in the interval $[0, 2\pi)$, we let $n = 0$ and $n = 1$ in the general solution and obtain

$$\frac{5\pi}{12} \quad \text{and} \quad \frac{17\pi}{12}$$

Figure 8.14 shows the graph of $y = \sin\left(2x - \frac{\pi}{3}\right) - 1$ in the interval $[0, 2\pi)$. The two x-intercepts confirm our work. ◆

PROBLEM 2 Find the general solution of the trigonometric equation $2 \cos 2x + 1 = 0$. Then find the particular solutions in the interval $[0, 2\pi)$. ◆

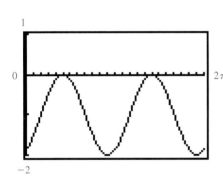

FIGURE 8.14

Graph of $y = \sin\left(2x - \frac{\pi}{3}\right) - 1$ in the interval $[0, 2\pi)$ with x-scale $\pi/12$

◆ Techniques of Solving Trigonometric Equations

Some trigonometric equations can be solved by factoring. This method relies on the zero product property (Section 2.2): We try to write the equation in such a form that a product of trigonometric expressions is zero. The procedure is illustrated in the next example.

EXAMPLE 3 Find the general solution of each trigonometric equation.

(a) $\tan x \sin x - \sin x = 0$

(b) $\sin^2 x + \sin x - 2 = 0$

◆ **Solution**

(a) Using common-term factoring and the zero product property, we have

$$\tan x \sin x - \sin x = 0$$
$$\sin x (\tan x - 1) = 0$$
$$\sin x = 0 \quad \text{or} \quad \tan x - 1 = 0$$
$$x = 0, \pi \qquad \qquad \tan x = 1$$
$$\underbrace{}_{\text{solutions in the interval }[0, 2\pi)} \qquad x = \frac{\pi}{4}$$
$$\underbrace{}_{\text{solution in the interval }[0, \pi)}$$

Although

$$0 + 2\pi n, \quad \pi + 2\pi n, \quad \text{and} \quad \frac{\pi}{4} + \pi n \quad \text{where } n \text{ is an integer}$$

describe all solutions of the original equation, the expressions $0 + 2\pi n$ and $\pi + 2\pi n$ can be combined and written more compactly as πn. Thus, we say that the expressions

$$\pi n \quad \text{and} \quad \frac{\pi}{4} + \pi n \quad \text{where } n \text{ is an integer}$$

describe the general solution of the equation $\tan x \sin x - \sin x = 0$. The graph in Figure 8.15 supports our work.

(b) We factor the left-hand side of the equation as a trinomial and apply the zero product property:

$$\sin^2 x + \sin x - 2 = 0$$
$$(\sin x - 1)(\sin x + 2) = 0$$
$$\sin x = 1 \quad \text{or} \quad \sin x = -2$$
$$x = \frac{\pi}{2} \qquad \qquad \text{no solution}$$

solution in the interval $[0, 2\pi)$

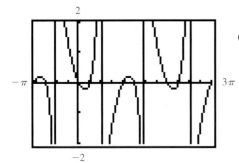

FIGURE 8.15
Graph of $y = \tan x \sin x - \sin x$ in the interval $[-\pi, 3\pi]$ with x-scale $\pi/4$

Thus, the expression

$$\frac{\pi}{2} + 2\pi n, \quad \text{where } n \text{ is an integer}$$

describes the general solution of the equation $\sin^2 x + \sin x - 2 = 0$. The graph in Figure 8.16 supports our work. ◆

PROBLEM 3 Find the general solution of the trigonometric equation $\sin x \cos x - \cos x = 0$. ◆

To solve some trigonometric equations, we must apply the fundamental trigonometric identities in order to write the equation as a product that equals zero. As with algebraic equations, when squaring both sides of a trigonometric equation or multiplying both sides by a variable expression, we must check for *extraneous roots*, which don't satisfy the original equation.

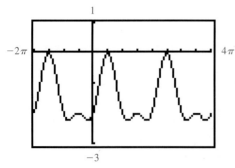

FIGURE 8.16
Graph of $y = \sin^2 x + \sin x - 2$ in the interval $[-2\pi, 4\pi]$ with x-scale $\pi/2$

EXAMPLE 4 Find the general solution of each trigonometric equation.

(a) $2 \sin x + \cot x = \csc x$ **(b)** $\sin x + \cos x = 1$

◆ **Solution**

(a) We begin by applying the fundamental trigonometric identities:

$$2 \sin x + \cot x = \csc x$$
$$2 \sin x + \frac{\cos x}{\sin x} = \frac{1}{\sin x} \quad \text{Apply trig identities 5 and 1}$$
$$2 \sin^2 x + \cos x = 1 \quad \text{Multiply both sides by } \sin x, \text{ provided that } \sin x \neq 0$$
$$2(1 - \cos^2 x) + \cos x = 1 \quad \text{Apply trig identity 6}$$
$$2 \cos^2 x - \cos x - 1 = 0 \quad \text{Collect like terms}$$

The equation is now written in terms of a single trigonometric function, cosine, so we can use factoring and the zero product property to solve this equation:

$$2 \cos^2 x - \cos x - 1 = 0$$
$$(2 \cos x + 1)(\cos x - 1) = 0$$
$$\cos x = -\tfrac{1}{2} \quad \text{or} \quad \cos x = 1$$
$$x = \frac{2\pi}{3}, \frac{4\pi}{3} \qquad x = 0$$

possible solutions in the interval $[0, 2\pi)$

To obtain these apparent solutions, we multiplied both sides of an equation by a variable expression, namely, $\sin x$. Thus, we must check for extraneous roots. Since $\sin 0 = 0$, we must discard $x = 0$ as a root. Thus, on the interval $[0, 2\pi)$, the only solutions are $2\pi/3$ and $4\pi/3$. Hence, the expressions

$$\frac{2\pi}{3} + 2\pi n \quad \text{and} \quad \frac{4\pi}{3} + 2\pi n \quad \text{where } n \text{ is an integer}$$

describe the general solution of the equation $2 \sin x + \cot x = \csc x$. The graph in Figure 8.17 supports our work. If we trace along the curve, we note holes in the graph at $x = 0, \pm 2\pi, \pm 4\pi$, and so on—these values correspond to the extraneous solution $0 + 2\pi n$.

(b) Squaring both sides of this equation leads to the expressions $\sin^2 x$ and $\cos^2 x$, which are related by the trigonometric identity $\sin^2 x + \cos^2 x = 1$. Thus,

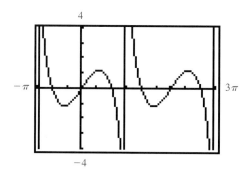

FIGURE 8.17
Graph of $y = 2 \sin x + \cot x - \csc x$ in the interval $[-\pi, 3\pi]$ with x-scale $\pi/3$

$$\sin x + \cos x = 1$$
$$\sin^2 x + 2 \sin x \cos x + \cos^2 x = 1 \quad \text{Square both sides}$$
$$1 + 2 \sin x \cos x = 1 \quad \text{Apply } \sin^2 x + \cos^2 x = 1$$
$$2 \sin x \cos x = 0 \quad \text{Subtract 1 from both sides}$$
$$\sin x \cos x = 0 \quad \text{Divide both sides by 2}$$
$$\sin x = 0 \quad \text{or} \quad \cos x = 0$$
$$x = 0, \pi \quad\quad x = \frac{\pi}{2}, \frac{3\pi}{2}$$

possible solutions in the interval $[0, 2\pi)$

To obtain these apparent solutions, we squared both sides of an equation. Thus, we must check for extraneous roots. Of these four possible solutions, we find that $x = \pi$ and $x = 3\pi/2$ are extraneous. Thus, in the interval $[0, 2\pi)$, the only solutions are 0 and $\pi/2$. Hence, the expressions

$$2\pi n \quad \text{and} \quad \frac{\pi}{2} + 2\pi n \quad \text{where } n \text{ is an integer}$$

describe the general solution of the equation $\sin x + \cos x = 1$. The graph in Figure 8.18 supports our work.

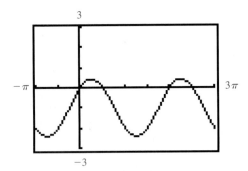

FIGURE 8.18

Graph of $y = \sin x + \cos x - 1$ in the interval $[-\pi, 3\pi]$ with x-scale $\pi/2$

PROBLEM 4 Find the general solution of the trigonometric equation $\sin x - \cos x = 1$.

◆ Formulas for Solving Trigonometric Equations

In the preceding discussion, we chose trigonometric equations in which the solutions unfolded from special values, such as $\pi/6$, $\pi/4$, $\pi/3$, $\pi/2$, and so on. For equations in which these special values do not occur, we use the inverse trigonometric functions and symmetry of the unit circle to obtain the following formulas.

Formulas for Solving Trigonometric Equations

Equation	Formula for general solution	Display of two solutions on the unit circle	
$\sin x = k$	$x = \begin{cases} \sin^{-1}k + 2\pi n \\ \pi - \sin^{-1}k + 2\pi n \end{cases}$	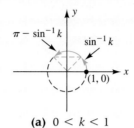 **(a)** $0 < k < 1$	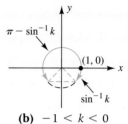 **(b)** $-1 < k < 0$
$\cos x = k$	$x = \begin{cases} \cos^{-1}k + 2\pi n \\ -\cos^{-1}k + 2\pi n \end{cases}$	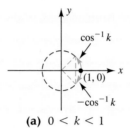 **(a)** $0 < k < 1$	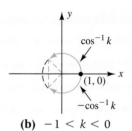 **(b)** $-1 < k < 0$
$\tan x = k$	$x = \tan^{-1}k + \pi n$	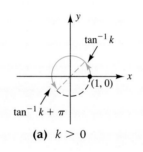 **(a)** $k > 0$	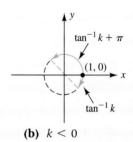 **(b)** $k < 0$

We can also apply these formulas to the equations $\csc x = k$, $\sec x = k$, and $\cot x = k$ by rewriting them as $\sin x = 1/k$, $\cos x = 1/k$, and $\tan x = 1/k$, respectively.

EXAMPLE 5 Use the formulas for solving a trigonometric equation to approximate to four significant digits the particular solutions in the interval $[0, 2\pi)$.

(a) $5 \sin x + 2 = 0$ **(b)** $\cos^2 x - 2 \cos x = 2$

◀ Solution

(a) Solving for $\sin x$, we have

$$\sin x = -\tfrac{2}{5}$$

By the formula for sin $x = k$, the general solution is

$$x = \begin{cases} \sin^{-1}(-\frac{2}{5}) + 2\pi n \\ \pi - \sin^{-1}(-\frac{2}{5}) + 2\pi n \end{cases}$$

To obtain the solutions in the interval $[0, 2\pi)$, we let $n = 0$ in the solution $\pi - \sin^{-1}(-\frac{2}{5}) + 2\pi n$ and let $n = 1$ in $\sin^{-1}(-\frac{2}{5}) + 2\pi n$. Using the inverse sine key on a calculator set in radian mode, we evaluate these expressions and find that, rounded to four significant digits

$$x \approx 3.553, 5.872.$$

Figure 8.19 shows the graph of $y = 5 \sin x + 2$ in the interval $[0, 2\pi)$, which confirms our work.

(b) Writing this quadratic type equation in standard form, we obtain

$$\cos^2 x - 2 \cos x - 2 = 0.$$

Since this equation is not factorable over the set of integers, we apply the quadratic formula (Section 3.4) with $a = 1$, $b = -2$, and $c = -2$, and solve for $\cos x$:

$$\cos x = \frac{2 \pm \sqrt{(-2)^2 - 4(1)(-2)}}{2(1)} = \frac{2 \pm \sqrt{12}}{2} = 1 \pm \sqrt{3}.$$

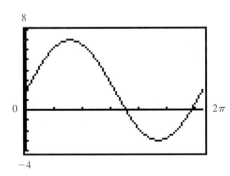

FIGURE 8.19

Graph of $y = 5 \sin x + 2$ in the interval $[0, 2\pi)$ with x-scale 1

Remember that the range of the cosine function is $[-1, 1]$. Hence, the equation

$$\cos x = 1 + \sqrt{3} \approx 2.732$$

has no solution. By the formula for $\cos x = k$, the general solution of the equation $\cos x = 1 - \sqrt{3} \approx -0.73205$ is

$$x = \begin{cases} \cos^{-1}(1 - \sqrt{3}) + 2\pi n \\ -\cos^{-1}(1 - \sqrt{3}) + 2\pi n \end{cases}$$

To obtain the solutions in the interval $[0, 2\pi)$, we let $n = 0$ in the solution $\cos^{-1}(1 - \sqrt{3}) + 2\pi n$ and let $n = 1$ in the solution $-\cos^{-1}(1 - \sqrt{3}) + 2\pi n$. Using the inverse cosine key on a calculator set in the radian mode, we evaluate these expressions and find that, rounded to four significant digits

$$x \approx 2.392, 3.891.$$

Figure 8.20 shows the graph of $y = \cos^2 x - 2 \cos x - 2$ in the interval $[0, 2\pi)$, which confirms our work. ◆

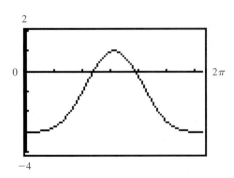

FIGURE 8.20

Graph of $y = \cos^2 x - 2 \cos x - 2$ in the interval $[0, 2\pi)$ with x-scale 1

PROBLEM 5 Repeat the directions in Example 5 for $\tan^2 x - 4 = 0$. ◆

♦ Application: Daylight Hours in the Northern Hemisphere

At two times during the year, the sun crosses the plane of the Earth's equator, making night and day equal in length all over the Earth. These times are called the *vernal equinox* (about March 21, the 80th day of the year) and the *autumnal equinox* (about September 21). At any given latitude in the Northern Hemisphere, the number of hours of daylight oscillates in a periodic fashion about the equilibrium position of 12 hours, reaching its maximum value at the *summer solstice* (about June 21) and its minimum value at the *winter solstice* (about December 21), as shown in Figure 8.21. Hence, in the Northern Hemisphere, the number of hours N of daylight for a given latitude on a particular day d of the year may be approximated by the equation

$$N = 12 + k \sin\left[\frac{2\pi}{365}(d - 80)\right]$$

where k is a constant that depends on the given latitude.

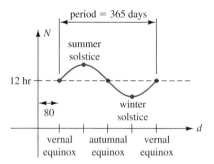

FIGURE 8.21

Daylight hours at a given latitude in the Northern Hemisphere

EXAMPLE 6 *Chapter Opening Problem*

The number of hours N of daylight on a particular day d of the year in the Boston area may be approximated by

$$N = 12 + 3 \sin\left[\frac{2\pi}{365}(d - 80)\right].$$

Determine the days of the year that have exactly 13 hours 30 minutes of daylight.

◀ **Solution** To determine the days of the year that have exactly 13 hours 30 minutes of daylight, we must solve the equation

$$13\tfrac{1}{2} = 12 + 3 \sin\left[\frac{2\pi}{365}(d - 80)\right]$$

for d.

Graphical approach: We generate the graphs of

$$y = 13.5 \quad \text{and} \quad y = 12 + 3 \sin\left[\frac{2\pi}{365}(d - 80)\right]$$

in the same viewing rectangle of a graphing calculator. The points of intersection of the two graphs represent the days of the year that have 13 hours 30 minutes of daylight. As shown in Figure 8.22, 13 hours 30 minutes of daylight occurs on day 110 (April 20th) and on day 232 (August 20th).

FIGURE 8.22
The graph of $y = 13.5$ and
$y = 12 + 3 \sin\left[\dfrac{2\pi}{365}(d - 80)\right]$
intersect at $d \approx 110, 232$.

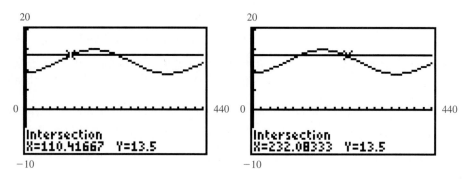

Algebraic approach: We subtract 12 from both sides of the original equation and then divide both sides by 3 to obtain

$$\sin\left[\frac{2\pi}{365}(d - 80)\right] = \frac{1}{2}$$

Now, from the unit circle in Figure 8.9, we select

$$\frac{\pi}{6} \quad \text{and} \quad \frac{5\pi}{6}$$

as the only values of θ in the interval $[0, 2\pi)$ for which $\sin \theta = 1/2$. Hence,

$$\frac{2\pi}{365}(d - 80) = \frac{\pi}{6} \quad \text{and} \quad \frac{2\pi}{365}(d - 80) = \frac{5\pi}{6}$$

$$d - 80 = \frac{365}{12} \qquad\qquad d - 80 = \frac{1825}{12} \quad \text{Divide both sides by } 2\pi/365$$

$$d \approx 110 \qquad\qquad\qquad d \approx 232 \qquad \text{Add 80 to both sides}$$

Thus, 13 hours 30 minutes of daylight occurs on day 110 (April 20th) and on day 232 (August 20th), which agrees with our graphical approach. ◆

PROBLEM 6 Use the equation from Example 6 to determine the days of the year that have exactly 10 hours 30 minutes of daylight in the Boston area. ◆

Exercises 8.2

◆ Basic Skills

In Exercises 1–16, find the general solution of each equation, if it exists. Use a graphing calculator to check each answer.

1. $\sin x = 1$
2. $\cos x = -1$
3. $\tan x = 0$
4. $\sin x = -\frac{1}{2}$
5. $\cos x = \dfrac{\sqrt{3}}{2}$
6. $\tan x = -1$
7. $\csc x = \sqrt{2}$
8. $\sec x = -2$
9. $\sqrt{3} \cot x = 1$
10. $2 \sin x = -\sqrt{3}$

11. $2 \cos x - 1 = 0$
12. $\sqrt{3} \tan x + 3 = 0$
13. $\sin^2 x = \frac{1}{4}$
14. $\tan^2 x - 1 = 0$
15. $4 \cos^2 x + 3 = 0$
16. $2 \sin^2 x - 1 = 0$

In Exercises 17–32, find
(a) the general solution of each equation, if it exists, and
(b) all solutions that exist in the interval $[0, 2\pi)$.
Use a graphing calculator to check each answer.

17. $\sin 2x = 1$
18. $\cot 3x = 1$
19. $\csc \frac{x}{2} + 1 = 0$
20. $2 \cos 3x + 1 = 0$
21. $\sqrt{2} \cos 4x + 1 = 0$
22. $\tan \frac{2x}{3} - \sqrt{3} = 0$
23. $4 \cos^2 2x - 1 = 0$
24. $2 \sin^2 5x + 2 = 0$
25. $\tan\left(2x - \frac{\pi}{6}\right) = 1$
26. $\sin\left(3x + \frac{\pi}{4}\right) = 0$
27. $4 \sec(x + \pi) = 8$
28. $\sqrt{3} \tan\left(4x - \frac{\pi}{3}\right) = 1$
29. $2 \sin 3\left(x + \frac{\pi}{4}\right) + 1 = 0$
30. $\sqrt{3} - \cot 2(x - \pi) = 0$
31. $\tan^2\left(\frac{x - \pi}{2}\right) = 3$
32. $4 \cos^2\left(2x + \frac{\pi}{6}\right) - 3 = 0$

In Exercises 33–60, find the general solution of each equation, if it exists. Use a graphing calculator to check each answer.

33. $(\sin x + 1)(2 \sin x - 1) = 0$
34. $(\cos x - 1)(\cos x + 1) = 0$
35. $(\cot x - 1)(2 \sin^2 x - 1) = 0$
36. $(2 \cos^2 x - 1)(\csc x - 1) = 0$
37. $\cos^2 x - \cos x = 0$
38. $2 \sin^2 x + \sin x = 0$
39. $\tan x - 2 \tan x \cos x = 0$
40. $\tan x \csc x - \csc x = 0$
41. $\cos x \csc^2 x - 2 \cos x = 0$
42. $\sin x \cot^2 x = 3 \sin x$
43. $2 \cos^2 x - \cos x = 1$
44. $\csc^2 x - 3 \csc x + 2 = 0$
45. $2 \sec^3 x + \sec^2 x - 8 \sec x - 4 = 0$
46. $2 \sin^3 x - \sin^2 x + 6 \sin x = 3$
47. $2 \cos^2 x - \sin x - 1 = 0$
48. $2 \sin^2 x - 3 \cos x - 3 = 0$
49. $\tan^2 x + 3 \sec x = -3$
50. $\csc^2 x - \cot x = 1$
51. $4 \tan^2 x = 3 \sec^2 x$
52. $2(1 - 2 \cos^2 x) = 1 - 2 \sin^2 x$
53. $\sec x + \tan x = \cos x$
54. $2 \sin x - 3 \cot x = 3 \csc x$
55. $\cot x - \cos x = 0$
56. $\csc x = \sin x - \cot x$
57. $\sin x + 1 = \cos x$
58. $\cot x + 1 = \csc x$
59. $\tan x = \sec x + 1$
60. $\cot x = 3 \tan x$

In Exercises 61–76, use the formulas for solving a trigonometric equation to approximate to four significant digits all solutions in the interval $[0, 2\pi)$. Use a graphing calculator to check each answer.

61. $\sin x = \frac{2}{3}$
62. $\cos x = -0.25$
63. $\cot x + 5 = 0$
64. $2 \csc x - 7 = 0$
65. $\sec^2 x - 9 = 0$
66. $4 \tan^2 x - 3 = 0$
67. $5 \sin 3x - 4 = 0$
68. $\cos(2x - 1) = 0.7$
69. $(\cos x + 3)(4 \sin x - 1) = 0$
70. $(2 \sec x - 1)(\sec x + 4) = 0$
71. $\tan^2 x - \tan x = 6$
72. $2 \sin^2 x - 3 \sin x - 1 = 0$
73. $4 \cos^2 x - 7 \sin x + 2 = 0$
74. $2 \tan^2 x - 5 \sec x - 1 = 0$
75. $2 \sin x - \cos x = 0$
76. $\cot x = \csc x + 2$

In Exercises 77–80, an equation of the form $f(x) = 0$ is given. Use a graphing calculator to generate the graph of the function $y = f(x)$ in the interval $[0, 2\pi)$. Then use this graph to approximate the solutions of the given equation in the interval $[0, 2\pi)$. Round each answer to three significant digits.

77. $2 \cos^2 x - \cos x - 2 = 0$
78. $3 \sin^2 2x - 2 \cos 2x - 1 = 0$
79. $2 \tan^3 2x - \tan 2x - 1 = 0$
80. $\sin^4 x - \tan x + 2 = 0$

81. The displacement d (in centimeters) of an oscillating spring from its equilibrium position is a function of time t (in seconds) and is given by

$$d = 12 \cos 2t$$

Determine the times in the interval $[0, 2\pi)$ at which $d = -6$ cm.

82. The voltage v (in volts) of an electrical circuit is a function of time t (in seconds) and is given by

$$v = 20 \sin 4\pi t.$$

Determine the times in the interval $[0, 1]$ for which $v = 10$ volts.

83. Rangers have determined that the deer population P in a state forest in Vermont can be approximated by the equation

$$P = 186 + 84\sqrt{2} \sin \frac{\pi}{8} t$$

where t is the time (in years) with $t = 0$ corresponding to January 1, 1980. Determine the years between 1980 and 2000 when this deer population is 102.

84. The number of hours N of daylight on a particular day d in the Boston area may be approximated by the equation

$$N = 12 + 3 \sin\left[\frac{2\pi}{365}(d - 80)\right].$$

Determine the day of the year that has exactly 9 hours of daylight in the Boston area.

85. The number of hours N of daylight on a particular day d in Fairbanks may be approximated by the equation

$$N = 12 + 11 \sin\left[\frac{2\pi}{365}(d - 80)\right].$$

Determine the days of the year that have exactly 20 hours of daylight.

86. The voltage v (in volts) of an electric circuit is a function of time t (in milliseconds) and is given by

$$v = 12.5 \sin(2.4t - 4.5).$$

Approximate to three significant digits the times in the interval $[0, 10]$ for which $v = 9.2$ volts.

◀ Critical Thinking

87. Suppose that x_0 is a solution of the trigonometric equation $\sin bx = k$. State some other solutions of this equation.

88. Does the trigonometric equation $\tan x = k$ have an infinite number of solutions for all real numbers k? Does $\sin x = k$? Does $\cos x = k$? Explain each answer.

89. The graphs of $y = \sin x$ and $y = k$ for $-1 < k < 1$ intersect twice in the interval $[0, 2\pi)$. Hence, the equation $\sin x = k$ must have two solutions in the interval $[0, 2\pi)$. How many solutions does the equation $\sin nx = k$, where n is an integer and $-1 < k < 1$, have in the interval $[0, 2\pi)$? Explain.

90. A *chord* of a circle is a line segment whose endpoints lie on the circle. The chord $\overline{AB}$ is shown in the sketch.
    (a) Show that the length of the chord $\overline{AB}$ (denoted AB), with central angle θ, is given by

    $$AB = 2r \sin \frac{\theta}{2}.$$

    (b) Find θ (in degrees) if $r = 6$ cm and $AB = 6\sqrt{3}$ cm.

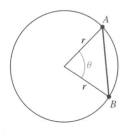

91. Find the general solution of each equation.
    (a) $\ln(2 - \cos^2 x) = 0$
    (b) $\log_2(1 - \sin x) = 1 + 2 \log_2(\sin x)$

92. Find the general solution of each equation.
    (a) $(16^{\cos x})^{\tan x} = 4$
    (b) $3^{\sec x} \cdot 9^{-\sin^2 x} = 9^{\cos^2 x}$

93. Given that $\sin^2 x = 1$, find the fallacy in the following argument:

    $\sin^2 x = 1$

    $\dfrac{\sin^2 x}{\cos^2 x} = \dfrac{1}{\cos^2 x}$ Divide both sides by $\cos^2 x$

    $\tan^2 x = \sec^2 x$ Apply trig identities

    $\tan^2 x + 1 = \sec^2 x + 1$ Add 1 to both sides

    $\sec^2 x = \sec^2 x + 1$ Apply trig identity

    $0 = 1$ Subtract $\sec^2 x$ from both sides

94. Show algebraically how to find the points of intersection of the two given curves. Check graphically.
    (a) $y = 5 \sin x + 1$ and $y = \sin x - 1$
    (b) $y = \cos 2x - 1$ and $y = \sin 2x$

8.3 Sum and Difference Formulas

◆ Introductory Comments

In this section, we derive formulas for the trigonometric functions of $u \pm v$, where u and v are real numbers or are angles measured in degrees or radians. We refer to these formulas as the *sum and difference formulas* and we use them to help prove other trigonometric identities and to solve some trigonometric equations. The sum and difference formulas are especially useful for working with trigonometric functions in calculus.

◆ The Sum and Difference Formulas for Cosine

We begin by developing a formula for $\cos(u - v)$. Consider a unit circle in which u and v are angles in standard position, as shown in Figure 8.23. Note that angles u and v intersect the unit circle at the points $A(\cos u, \sin u)$ and $B(\cos v, \sin v)$, respectively, and that the angle between $\overline{OA}$ and $\overline{OB}$ is $u - v$. Figure 8.24 shows angle $u - v$ with a line segment joining points A and B. In Figure 8.25 we redraw angle $u - v$ in standard position and relabel the points on the unit circle $A'(\cos(u - v), \sin(u - v))$ and $B'(1, 0)$, respectively.

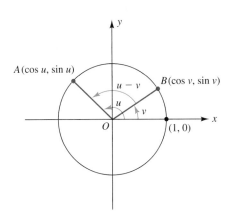

FIGURE 8.23

A unit circle with angles u and v in standard position

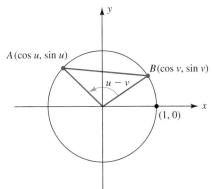

FIGURE 8.24

Angle $u - v$ with a line segment joining points A and B.

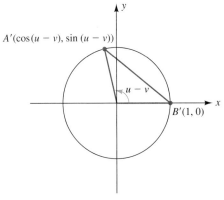

FIGURE 8.25

Angle $u - v$ redrawn in standard position

Now, line segments AB and $A'B'$, in Figure 8.24 and 8.25, must have the same length. By the distance formula,

$$\begin{aligned} AB &= \sqrt{(\cos u - \cos v)^2 + (\sin u - \sin v)^2} \\ &= \sqrt{(\cos^2 u - 2\cos u \cos v + \cos^2 v) + (\sin^2 u - 2\sin u \sin v + \sin^2 v)} \\ &= \sqrt{2 - 2\cos u \cos v - 2\sin u \sin v} \quad \text{Apply } \sin^2\theta + \cos^2\theta = 1 \end{aligned}$$

and

$$A'B' = \sqrt{[\cos(u-v) - 1]^2 + [\sin(u-v) - 0]^2}$$
$$= \sqrt{[\cos^2(u-v) - 2\cos(u-v) + 1] + \sin^2(u-v)}$$
$$= \sqrt{2 - 2\cos(u-v)} \quad \text{Apply } \sin^2\theta + \cos^2\theta = 1$$

Equating these lengths, we obtain

$$\sqrt{2 - 2\cos(u-v)} = \sqrt{2 - 2\cos u \cos v - 2\sin u \sin v}$$
$$2 - 2\cos(u-v) = 2 - 2\cos u \cos v - 2\sin u \sin v \quad \text{Square both sides}$$
$$-2\cos(u-v) = -2\cos u \cos v - 2\sin u \sin v \quad \text{Subtract 2 from both sides}$$
$$\cos(u-v) = \cos u \cos v + \sin u \sin v \quad \text{Divide both sides by } -2$$

We refer to this last equation as the *difference formula for cosine.* Although we have developed this formula for positive angles u and v, the formula is also valid for negative angles and for real numbers.

Recall from Section 7.4 that the sine function is odd and the cosine function is even, that is,

$$\sin(-v) = -\sin v \quad \text{and} \quad \cos(-v) = \cos v.$$

Using these facts, we can obtain a *sum formula for cosine* as follows:

$$\cos(u+v) = \cos(u - (-v)) \quad \text{Rewrite as a difference}$$
$$= \cos u \cos(-v) + \sin u \sin(-v) \quad \text{Apply the difference formula for cosine}$$
$$= \cos u \cos v - \sin u \sin v \quad \text{Simplify}$$

◀ **Sum and Difference Formulas for Cosine**

1. $\cos(u+v) = \cos u \cos v - \sin u \sin v$
2. $\cos(u-v) = \cos u \cos v + \sin u \sin v$

EXAMPLE 1 Use the sum and difference formulas for cosine to simplify each expression.

(a) $\cos\left(\dfrac{\pi}{2} - x\right)$

(b) $\cos 3x \cos x - \sin 3x \sin x$

◀ **Solution**

(a) Applying the difference formula for cosine with $u = \pi/2$ and $v = x$, we have

$$\cos\left(\dfrac{\pi}{2} - x\right) = \cos\dfrac{\pi}{2}\cos x + \sin\dfrac{\pi}{2}\sin x$$
$$= (0)\cos x + (1)\sin x \quad \text{Evaluate}$$
$$= \sin x \quad \text{Simplify}$$

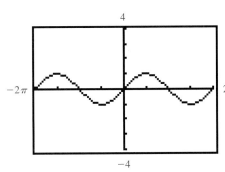

FIGURE 8.26
The graph of $y = \cos\left(\dfrac{\pi}{2} - x\right)$ is the same as the graph of $y = \sin x$.

Figure 8.26 shows the graph of $y = \cos\left(\dfrac{\pi}{2} - x\right)$ in the viewing rectangle of a graphing calculator. The graph of this equation is the same as the graph of $y = \sin x$; this fact confirms our work.

(b) Applying the sum formula for cosine, from right to left, with $u = 3x$ and $v = x$, we have

$$\cos 3x \cos x - \sin 3x \sin x = \cos(3x + x) = \cos 4x$$

Figure 8.27 shows the graph of $y = \cos 3x \cos x - \sin 3x \sin x$ in the viewing rectangle of a graphing calculator. The graph of this equation is the same as the graph of $y = \cos 4x$; this fact confirms our work. ◆

PROBLEM 1 Use the sum or difference formula for cosine to simplify $\cos 3x \cos 2x + \sin 3x \sin 2x$. ◆

◆ The Sum and Difference Formulas for Sine

Recall from Section 7.2 that the sine and cosine are cofunctions and that a trigonometric function of any acute angle θ is the same as the cofunction of its complementary angle $90° - \theta$. In Example 1(a), we have shown that this relationship holds for any arbitrary angle or real number, that is,

$$\sin x = \cos\left(\dfrac{\pi}{2} - x\right) \quad \text{for all } x.$$

Now, if we replace x with $\left(\dfrac{\pi}{2} - u\right)$, we obtain

$$\sin\left(\dfrac{\pi}{2} - u\right) = \cos\left[\dfrac{\pi}{2} - \left(\dfrac{\pi}{2} - u\right)\right] = \cos u.$$

Hence, we can also state that

$$\cos x = \sin\left(\dfrac{\pi}{2} - x\right) \quad \text{for all } x.$$

By using these cofunction relationships, we can derive a formula for $\sin(u + v)$:

$$\sin(u + v) = \cos\left[\dfrac{\pi}{2} - (u + v)\right] \quad \text{Use cofunction relationship}$$

$$= \cos\left[\left(\dfrac{\pi}{2} - u\right) - v\right] \quad \text{Rewrite}$$

$$= \cos\left(\dfrac{\pi}{2} - u\right)\cos v + \sin\left(\dfrac{\pi}{2} - u\right)\sin v \quad \text{Apply the difference formula for cosine}$$

$$= \sin u \cos v + \cos u \sin v \quad \text{Use cofunction relationship}$$

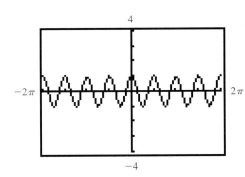

FIGURE 8.27
The graph of
$y = \cos 3x \cos x - \sin 3x \sin x$ is the same as the graph of $y = \cos 4x$.

We refer to

$$\sin(u + v) = \sin u \cos v + \cos u \sin v$$

as the *sum formula for sine*.

We can obtain the *difference formula for sine* as follows:

$$\sin(u - v) = \sin(u + (-v)) \qquad \text{Rewrite as a sum}$$
$$= \sin u \cos(-v) + \cos u \sin(-v) \qquad \text{Apply the sum formula for sine}$$
$$= \sin u \cos v - \cos u \sin v \qquad \text{Simplify}$$

◆ **Sum and Difference Formulas for Sine**

1. $\sin(u + v) = \sin u \cos v + \cos u \sin v$
2. $\sin(u - v) = \sin u \cos v - \cos u \sin v$

EXAMPLE 2 Use the sum and difference formulas for sine to simplify each expression.

(a) $\sin\left(x + \dfrac{3\pi}{2}\right)$ **(b)** $\sin\left(x + \dfrac{\pi}{6}\right)\cos x - \cos\left(x + \dfrac{\pi}{6}\right)\sin x$

◆ **Solution**

(a) Applying the sum formula for sine with $u = x$ and $v = 3\pi/2$, we have

$$\sin\left(x + \dfrac{3\pi}{2}\right) = \sin x \cos \dfrac{3\pi}{2} + \cos x \sin \dfrac{3\pi}{2}$$
$$= \sin x \,(0) + \cos x \,(-1) \qquad \text{Evaluate}$$
$$= -\cos x \qquad \text{Simplify}$$

Figure 8.28 shows the graph of $y = \sin\left(x + \dfrac{3\pi}{2}\right)$ the viewing rectangle of a graphing calculator. The graph of this equation is the same as the graph of $y = -\cos x$, and this fact confirms our work.

(b) Applying the difference formula for sine, from right to left, with $u = x + \dfrac{\pi}{6}$ and $v = x$, we have

$$\sin\left(x + \dfrac{\pi}{6}\right)\cos x - \cos\left(x + \dfrac{\pi}{6}\right)\sin x = \sin\left[\left(x + \dfrac{\pi}{6}\right) - x\right]$$
$$= \sin \dfrac{\pi}{6} \qquad \text{Simplify}$$
$$= \dfrac{1}{2} \qquad \text{Evaluate}$$

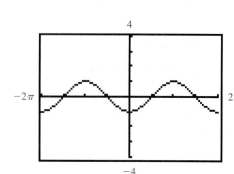

FIGURE 8.28

The graph of $y = \sin\left(x + \dfrac{3\pi}{2}\right)$ is the same as the graph of $y = -\cos x$.

The graph of $y = \sin\left(x + \dfrac{\pi}{6}\right)\cos x - \cos\left(x + \dfrac{\pi}{6}\right)\sin x$ is shown in Figure 8.29. The graph of this equation is the same as the graph of $y = \tfrac{1}{2}$, thus confirming our work. ◆

PROBLEM 2 As an alternate method for simplifying the expression in Example 2(b), expand $\sin\left(x + \dfrac{\pi}{6}\right)$ and $\cos\left(x + \dfrac{\pi}{6}\right)$ by using the sum formulas for sine and cosine, respectively. Then perform the indicated operations. ◆

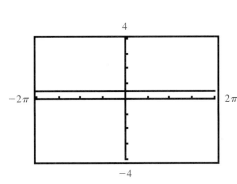

FIGURE 8.29
The graph of $y = \sin\left(x + \dfrac{\pi}{6}\right)\cos x - \cos\left(x + \dfrac{\pi}{6}\right)\sin x$ is the same as the graph of $y = \tfrac{1}{2}$.

◆ The Sum and Difference Formulas for Tangent

We now derive a *sum formula for tangent*:

$$\tan(u + v) = \dfrac{\sin(u + v)}{\cos(u + v)} \qquad \text{Apply } \tan x = \dfrac{\sin x}{\cos x}$$

$$= \dfrac{\sin u \cos v + \cos u \sin v}{\cos u \cos v - \sin u \sin v} \qquad \text{Apply sum formulas for sine and cosine}$$

$$= \dfrac{\dfrac{\sin u \cos v + \cos u \sin v}{\cos u \cos v}}{\dfrac{\cos u \cos v - \sin u \sin v}{\cos u \cos v}} \qquad \text{Divide both numerator and denominator by } \cos u \cos v, \text{ provided that } \cos u \cos v \neq 0$$

$$= \dfrac{\dfrac{\sin u}{\cos u} + \dfrac{\sin v}{\cos v}}{1 - \dfrac{\sin u \sin v}{\cos u \cos v}} \qquad \text{Split and reduce fractions}$$

$$= \dfrac{\tan u + \tan v}{1 - \tan u \tan v} \qquad \text{Apply } \dfrac{\sin x}{\cos x} = \tan x$$

Recall that the tangent function is an odd function, that is,

$$\tan(-v) = -\tan v.$$

We use this fact to obtain a *difference formula for tangent*:

$$\tan(u - v) = \tan(u + (-v)) \qquad \text{Rewrite as a sum}$$

$$= \dfrac{\tan u + \tan(-v)}{1 - \tan u \tan(-v)} \qquad \text{Apply the sum formula for tangent}$$

$$= \dfrac{\tan u - \tan v}{1 + \tan u \tan v} \qquad \text{Simplify}$$

◆ Sum and Difference Formulas for Tangent

1. $\tan(u + v) = \dfrac{\tan u + \tan v}{1 - \tan u \tan v}$

2. $\tan(u - v) = \dfrac{\tan u - \tan v}{1 + \tan u \tan v}$

EXAMPLE 3 Use the sum and difference formulas for tangent to simplify each expression.

(a) $\tan(\pi + x)$ (b) $\dfrac{\tan 87° - \tan 27°}{1 + \tan 87° \tan 27°}$

◆ **Solution**

(a) Applying the sum formula for tangent with $u = \pi$ and $v = x$, we have

$$\tan(\pi + x) = \dfrac{\tan \pi + \tan x}{1 - \tan \pi \tan x}$$

$$= \dfrac{0 + \tan x}{1 - (0) \tan x} \qquad \text{Evaluate}$$

$$= \tan x \qquad \text{Simplify}$$

Figure 8.30 shows the graph of $y = \tan(\pi + x)$ in the viewing rectangle of a graphing calculator. The graph of this equation is the same as the graph of $y = \tan x$, and this fact confirms our work.

(b) Applying the difference formula for tangent, from right to left, with $u = 87°$ and $v = 27°$, we have

$$\dfrac{\tan 87° - \tan 27°}{1 + \tan 87° \tan 27°} = \tan(87° - 27°) = \tan 60° = \sqrt{3}. \quad \blacklozenge$$

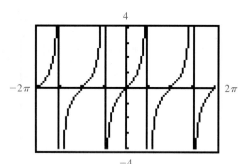

FIGURE 8.30
The graph of $y = \tan(\pi + x)$ is the same as the graph of $y = \tan x$.

PROBLEM 3 Use a calculator to find the approximate value of the expression

$$\dfrac{\tan 87° - \tan 27°}{1 + \tan 87° \tan 27°}.$$

Compare your answer with the exact value in Example 3(b). ◆

◆ Applying the Sum and Difference Formulas

The next four examples illustrate a variety of uses of the sum and difference formulas. We begin by showing how to evaluate a trigonometric function whose angle is the sum (or difference) of the special angles 30°, 45°, and 60°.

EXAMPLE 4 Find the exact value of cos 75°.

◆ **Solution**   We can think of 75° as the sum of the special angles 30° and 45°. Hence,

$$\cos 75° = \cos(30° + 45°) \quad \text{Rewrite with special angles}$$

$$= \cos 30° \cos 45° - \sin 30° \sin 45° \quad \text{Apply the sum formula for cosine}$$

$$= \frac{\sqrt{3}}{2} \cdot \frac{1}{\sqrt{2}} - \frac{1}{2} \cdot \frac{1}{\sqrt{2}} \quad \text{Evaluate}$$

$$= \frac{\sqrt{3} - 1}{2\sqrt{2}} \quad \text{or} \quad \frac{\sqrt{6} - \sqrt{2}}{4} \quad \text{Simplify}$$

PROBLEM 4 Use a calculator to find the approximate value of cos 75°. Compare your answer with the exact value found in Example 4.

If a trigonometric function of u and a trigonometric function of v are given and the quadrants containing u and v are known, then we can find the values of the trigonometric functions of $u \pm v$. This procedure is shown in the next example.

EXAMPLE 5 Given that u and v are second-quadrant angles with $\sin u = \frac{3}{5}$ and $\tan v = -\frac{1}{4}$, find the exact value of $\tan(u + v)$.

Solution We can use the sum formula for tangent to evaluate $\tan(u + v)$, provided that we know the values of $\tan u$ and $\tan v$. We are given that $\tan v = -\frac{1}{4}$. From the trigonometric identity $\sin^2 u + \cos^2 u = 1$, we have

$$\cos u = \pm\sqrt{1 - \sin^2 u}.$$

However, since it is given that u is a second-quadrant angle, we know that $\cos u$ must be negative. Thus,

$$\cos u = -\sqrt{1 - \sin^2 u} = -\sqrt{1 - (\tfrac{3}{5})^2} = -\sqrt{\tfrac{16}{25}} = -\tfrac{4}{5}.$$

Therefore,

$$\tan u = \frac{\sin u}{\cos u} = \frac{\tfrac{3}{5}}{-\tfrac{4}{5}} = -\frac{3}{4}$$

Hence,

$$\tan(u + v) = \frac{\tan u + \tan v}{1 - \tan u \tan v} = \frac{(-\tfrac{3}{4}) + (-\tfrac{1}{4})}{1 - (-\tfrac{3}{4})(-\tfrac{1}{4})} = -\frac{16}{13}.$$

PROBLEM 5 Use the information in Example 5 to find the exact value of $\sin(u - v)$.

As illustrated in the next example, sum and difference formulas can be useful in verifying certain trigonometric identities.

EXAMPLE 6 Show graphically and verify algebraically that the trigonometric equation is an identity:

$$\frac{\cos x}{\sec 3x} + \frac{\sin x}{\csc 3x} = \cos 2x$$

Solution In Figure 8.31, the graphs of $y = \dfrac{\cos x}{\sec 3x} + \dfrac{\sin x}{\csc 3x}$ and $y = \cos 2x$ appear to be identical. Hence, the trigonometric equation

$$\frac{\cos x}{\sec 3x} + \frac{\sin x}{\csc 3x} = \cos 2x$$

seems to be an identity. To verify this fact algebraically, we proceed as follows:

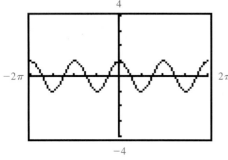

FIGURE 8.31

The graphs of $y = \dfrac{\cos x}{\sec 3x} + \dfrac{\sin x}{\csc 3x}$ and $y = \cos 2x$ appear to be identical.

$$\frac{\cos x}{\sec 3x} + \frac{\sin x}{\csc 3x} = \cos 2x$$

$\dfrac{\cos x}{1/\cos 3x} + \dfrac{\sin x}{1/\sin 3x}$	Apply fundamental trig identities
$\cos 3x \cos x + \sin 3x \sin x$	Divide
$\cos(3x - x)$	Apply difference formula for cosine
$\cos 2x$	Simplify

Since we have reduced the left-hand side of this equation to the right-hand side, we conclude that

$$\frac{\cos x}{\sec 3x} + \frac{\sin x}{\csc 3x} = \cos 2x \quad \text{is a trigonometric identity.} \quad \blacklozenge$$

PROBLEM 6 Show that the following trigonometric equation is an identity:

$$\sin 3x \cot x + \cos 3x = \sin 4x \csc x \quad \blacklozenge$$

Finally, we illustrate the use of the sum and difference formulas to help solve a trigonometric equation.

EXAMPLE 7 Find the general solution of the equation $\sin 3x \cos x = 1 + \cos 3x \sin x$.

Solution To solve this equation, we proceed as follows:

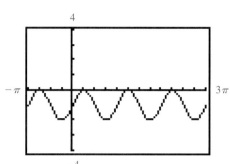

FIGURE 8.32

Graph of
$y = \sin 3x \cos x - \cos 3x \sin x - 1$
in the interval $[-\pi, 3\pi]$ with x-scale $\pi/4$.

$$\sin 3x \cos x = 1 + \cos 3x \sin x$$

$$\sin 3x \cos x - \cos 3x \sin x = 1 \quad \text{Subtract } \cos 3x \sin x \text{ from both sides}$$

$$\sin(3x - x) = 1 \quad \text{Apply the difference formula for sine}$$

$$\sin 2x = 1 \quad \text{Simplify}$$

Solving for x, we obtain

$$2x = \frac{\pi}{2} + 2\pi n$$

$$x = \frac{\pi}{4} + \pi n \quad \text{where } n \text{ is an integer.}$$

The graph in Figure 8.32 supports our work. ◆

PROBLEM 7 Find the general solution of the equation $\sin 3x \cos x = 1 - \cos 3x \sin x$. ◆

◆ Application: The Function $f(t) = A \sin bt + B \cos bt$

In physics and many branches of engineering, we may encounter functions of the form

$$f(t) = A \sin bt + B \cos bt$$

where A, B, and b are real numbers and t is time (in seconds). When working with this type of function, we usually replace the expression $A \sin bt + B \cos bt$ with

$$a \sin(bt + c)$$

where a is the distance from the origin to a point P having coordinates (A, B) and c is the measure of the angle in standard position with terminal side $\overline{OP}$, as shown in Figure 8.33. To show that these two expressions are equivalent, refer to the figure and observe that

$$\cos c = \frac{A}{a} \quad \text{and} \quad \sin c = \frac{B}{a}$$

$$A = a \cos c \qquad B = a \sin c$$

FIGURE 8.33

The point $P(A, B)$ determines distance a and angle c.

Thus,

$$A \sin bt + B \cos bt = (a \cos c) \sin bt + (a \sin c) \cos bt$$

$$= a \sin bt \cos c + a \cos bt \sin c \quad \text{Rewrite}$$

$$= a(\sin bt \cos c + \cos bt \sin c) \quad \text{Factor}$$

$$= a \sin (bt + c) \quad \text{Apply the sum formula for sine}$$

When a function f is written in the form

$$f(t) = a \sin(bt + c)$$

we can determine the amplitude, period, and phase shift of f quite easily. Recall from Section 7.4 that

Amplitude: $|a|$ Period: $\dfrac{2\pi}{|b|}$ Phase shift: $\left|\dfrac{c}{b}\right|$

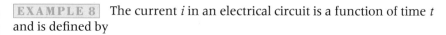

EXAMPLE 8 The current i in an electrical circuit is a function of time t and is defined by

$$i(t) = \sqrt{3} \sin 4t + \cos 4t$$

where i is measured in amperes (amp) and t is given in seconds.

(a) Express this function in the form $i(t) = a \sin(bt + c)$.

(b) Determine the amplitude, period, and phase shift of i.

(c) Use both forms of $i(t)$ to find $i(\pi/24)$, the current in the circuit at $\pi/24$ seconds.

◆ **Solution**

(a) For $i(t) = \sqrt{3} \sin 4t + \cos 4t$, we have $A = \sqrt{3}$ and $B = 1$. Plotting the point $P(\sqrt{3}, 1)$, as shown in Figure 8.34, we find that

$$a = \sqrt{A^2 + B^2} = \sqrt{(\sqrt{3})^2 + (1)^2} = \sqrt{4} = 2$$

and

$$\tan c = \frac{B}{A} = \frac{1}{\sqrt{3}}.$$

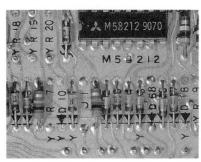

FIGURE 8.34

The values of a and c can be determined from the point $P(\sqrt{3}, 1)$.

Since $0 < c \leq \pi/2$, we have

$$c = \tan^{-1} \frac{1}{\sqrt{3}} = \frac{\pi}{6}.$$

Hence, $\quad i(t) = \sqrt{3} \sin 4t + \cos 4t = 2 \sin\left(4t + \dfrac{\pi}{6}\right)$.

(b) For $i(t) = 2 \sin\left(4t + \dfrac{\pi}{6}\right)$, we have $a = 2$, $b = 4$, and $c = \pi/6$. Thus,

Amplitude: $|a| = 2$ Period: $\dfrac{2\pi}{|b|} = \dfrac{2\pi}{4} = \dfrac{\pi}{2}$

Phase shift: $\left|\dfrac{c}{b}\right| = \dfrac{\pi/6}{4} = \dfrac{\pi}{24}$ unit to the left

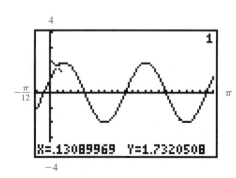

FIGURE 8.35

Graph of $i(t) = \sqrt{3} \sin 4t + \cos 4t$ in the interval $[-\pi/12, \pi]$ with t-scale $\pi/24$

The graph of $i(t) = \sqrt{3} \sin 4t + \cos 4t = 2 \sin\left(4t + \dfrac{\pi}{6}\right)$ is shown in Figure 8.35, and this graph supports our work.

(c) Using $i(t) = \sqrt{3} \sin 4t + \cos 4t$, we find

$$i\left(\frac{\pi}{24}\right) = \sqrt{3} \sin \frac{\pi}{6} + \cos \frac{\pi}{6} = \sqrt{3}\left(\frac{1}{2}\right) + \frac{\sqrt{3}}{2} = \sqrt{3} \text{ amp.}$$

Using $i(t) = 2 \sin\left(4t + \frac{\pi}{6}\right)$, we find

$$i\left(\frac{\pi}{24}\right) = 2 \sin\left(\frac{\pi}{6} + \frac{\pi}{6}\right) = 2 \sin\left(\frac{\pi}{3}\right) = 2\left(\frac{\sqrt{3}}{2}\right) = \sqrt{3} \text{ amp}$$

As we expected, both forms of $i(t)$ yield the same answer. The decimal form of this answer is shown in Figure 8.35. ◆

PROBLEM 8 For the electrical current described in Example 8, find all times t in the interval $[0, \pi/2)$ for which $i(t) = 1$ amp. Use the graph in Figure 8.35 to check your answers. ◆

Exercises 8.3

◆ Basic Skills

In Exercises 1–16, use the sum and difference formulas to simplify each expression. Use a graphing calculator to confirm each answer.

1. $\cos\left(\dfrac{\pi}{2} + x\right)$
2. $\cos(\pi - x)$
3. $\sin(\pi - x)$
4. $\sin\left(\dfrac{3\pi}{2} + x\right)$
5. $\tan(2\pi + x)$
6. $\tan(\pi - x)$
7. $\sin\left(\dfrac{\pi}{6} - x\right) + \cos\left(x - \dfrac{2\pi}{3}\right)$
8. $\sin\left(\dfrac{\pi}{4} + x\right) - \sin\left(\dfrac{\pi}{4} - x\right)$
9. $\sin 35° \cos 55° + \cos 35° \sin 55°$
10. $\cos 230° \cos 50° + \sin 230° \sin 50°$
11. $\cos 3x \cos 5x - \sin 3x \sin 5x$
12. $\sin 5x \cos 6x - \cos 5x \sin 6x$
13. $\sin\left(\dfrac{\pi}{4} - x\right) \cos \dfrac{\pi}{4} + \cos\left(\dfrac{\pi}{4} - x\right) \sin \dfrac{\pi}{4}$
14. $\cos\left(\dfrac{\pi}{6} - x\right) \cos x - \sin\left(\dfrac{\pi}{6} - x\right) \sin x$
15. $\dfrac{\tan 205° + \tan 20°}{1 - \tan 205° \tan 20°}$
16. $\dfrac{\tan\left(\dfrac{\pi}{3} + x\right) - \tan x}{1 + \tan\left(\dfrac{\pi}{3} + x\right) \tan x}$

For the trigonometric functions in Exercises 17–22, express each angle as the sum or difference of angles with measure 30°, 45°, or 60°. Then find the exact value of the trigonometric function by applying a sum or difference formula. Use a calculator to check each value.

17. $\sin 75°$
18. $\sin 15°$
19. $\cos 15°$
20. $\cos 105°$
21. $\tan 105°$
22. $\tan 15°$

In Exercises 23–30, find the exact value of the indicated trigonometric function from the given information.

23. $\tan x = 4$; find $\tan\left(x + \dfrac{\pi}{4}\right)$.
24. $\tan x = -2$; find $\tan\left(x - \dfrac{\pi}{4}\right)$.
25. $\sin x = \frac{4}{5}$, x in quadrant I; find $\sin\left(x - \dfrac{\pi}{6}\right)$.

26. $\cos x = \frac{12}{13}$, x in quadrant I; find $\cos\left(x + \frac{2\pi}{3}\right)$.

27. $\sin x = \frac{3}{5}$, $\cos y = -\frac{5}{13}$, x and y in quadrant II; find $\cos(x + y)$.

28. $\cos x = \frac{1}{3}$, $\tan y = -\sqrt{2}$, x and y in quadrant IV; find $\tan(x - y)$.

29. $\cot x = \frac{3}{4}$, $\tan y = 3$, x in quadrant I and y in quadrant III; find $\sin(x + y)$.

30. $\sec x = 4$, $\sec y = 2$, x and y in quadrant I; find $\cos(x - y)$.

In Exercises 31–38, a trigonometric equation of the form $f(x) = g(x)$ is given. Use a graphing calculator to generate the graphs of $y = f(x)$ and $y = g(x)$. Compare the graphs, and then state whether the given equation seems to be a trigonometric identity. If the equation seems to be an identity, verify this fact algebraically.

31. $\sin 2x \cot x - \cos 2x = 1$

32. $\cos 2x - \sin 2x \tan 3x = \cos 5x \sec 3x$

33. $\dfrac{\cos x}{\csc 4x} + \dfrac{\cos 4x}{\csc x} = \sin 5x$

34. $\dfrac{\sin x}{\sec 2x} - \dfrac{\cos x}{\csc 2x} = -\sin x$

35. $\dfrac{1}{\cot 2x} + \dfrac{1}{\tan 4x} = \sec 4x$

36. $\tan 6x - \tan 3x = \dfrac{\tan 3x}{\cos 6x}$

37. $\dfrac{1 + \tan x}{\tan\left(x + \dfrac{\pi}{4}\right)} = 1 - \tan x$

38. $\dfrac{\sin(2x + \pi)}{\cos 2x \cos \pi} = \tan x$

In Exercises 39–42, show that each trigonometric equation is an identity.

39. $\sin(x + y)\sin(x - y) = \sin^2 x - \sin^2 y$

40. $\cos(x + y)\cos(x - y) = \cos^2 x - \sin^2 y$

41. $\sec(x + y) = \dfrac{\sec x \sec y}{1 - \tan x \tan y}$

42. $\csc(x - y) = \dfrac{\csc x \csc y}{\cot y - \cot x}$

In Exercises 43–54, find the general solution of each equation, if it exists. Use a graphing calculator to check each answer.

43. $\cos 3x \cos x + \sin 3x \sin x = 0$

44. $\cos 2x \cos x = 1 + \sin 2x \sin x$

45. $\sqrt{2} \sin 3x \cos 2x = 1 + \sqrt{2} \cos 3x \sin 2x$

46. $2 \sin x \cos 2x + 2 \cos x \sin 2x = 1$

47. $\dfrac{\tan 2x + \tan x}{1 - \tan 2x \tan x} = 1$

48. $1 + \tan x \tan 3x = \tan x - \tan 3x$

49. $\tan 2x \cos x = \sin x$

50. $\cot 3x + \tan x = 0$

51. $\cos x + \sin\left(\dfrac{\pi}{2} - x\right) = 1$

52. $2 \tan x + \tan(\pi - x) = \sqrt{3}$

53. $\cos\left(x + \dfrac{\pi}{4}\right) + \cos\left(x - \dfrac{\pi}{4}\right) = 1$

54. $\sin\left(x + \dfrac{\pi}{6}\right) - \sin\left(x - \dfrac{\pi}{6}\right) = \dfrac{1}{2}$

In Exercises 55–58,
(a) express each function f in the form $f(x) = a \sin(bx + c)$, and
(b) determine the amplitude, period, and phase shift of f.

55. $f(x) = 3 \sin 2x + 3 \cos 2x$

56. $f(x) = 2 \sin 3x - 2 \cos 3x$

57. $f(x) = -\sqrt{3} \sin 5x + \cos 5x$

58. $f(x) = -\sin 4x - \sqrt{3} \cos 4x$

In Exercises 59 and 60, express each function f in the form $f(x) = a \sin(bx + c)$, rounding a and c each to three significant digits. Check each answer by generating the graphs of both forms of the function in the viewing rectangle of a graphing calculator.

59. $f(x) = 12.6 \sin 5.2x - 10.3 \cos 5.2x$

60. $f(x) = -1.35 \sin 3.7x + 2.88 \cos 3.7x$

61. The current i (in amperes) in an electrical circuit is a function of time t (in seconds) and is given by

$$i(t) = 5 \sin 6\pi t + 12 \cos 6\pi t.$$

(a) Determine a and c such that $i(t) = a \sin(6\pi t + c)$.

(b) Find the times t between 0 and 1 second for which $i = 8.5$ amp.

62. The motion of a weight attached to an oscillating spring is given by

$$d = 24 \sin 8t - 7 \cos 8t,$$

where d is the displacement (in centimeters) from its equilibrium position at time t (in seconds).

(a) Determine a and c such that $d = a \sin(8t + c)$.

(b) Find the times t between 0 and 1 second for which $d = -9.5$ cm.

Critical Thinking

63. In general, $\sin(u + v) \neq \sin u + \sin v$.
    (a) Use $u = \pi/3$ and $v = \pi/6$ to illustrate this fact.
    (b) Do any real numbers u and v exist for which $\sin(u + v) = \sin u + \sin v$? If so, give an example.

64. In general, $\tan(u - v) \neq \tan u - \tan v$.
    (a) Use $u = \pi/3$ and $v = \pi/6$ to illustrate this fact.
    (b) Do any real numbers u and v exist for which $\tan(u - v) = \tan u - \tan v$? If so, give an example.

65. Use the sum and difference formulas to find the exact value of each expression.
    (a) $\sin(\arctan 2 + \arctan 3)$
    (b) $\cos(\cos^{-1} \tfrac{1}{4} - \tan^{-1} \tfrac{1}{2})$

66. State the period of each function.
    (a) $f(x) = \cos 5x \cos 3x - \sin 5x \sin 3x$
    (b) $f(x) = \sin 5x \cos 3x - \cos 5x \sin 3x$

67. Suppose A, B, and C are the interior angles of an acute triangle with $\sin A = a$ and $\sin B = b$. Express $\sin C$ in terms of a and b. [*Hint:* The sum of the interior angles of a triangle is $180°$.]

68. Given the right triangle in the sketch, express each function in terms of a, b, and c.
    (a) $\tan \alpha$
    (b) $\tan \beta$
    (c) $\tan \theta$ [*Hint:* Use the fact that $\theta = \beta - \alpha$ and apply the difference formula for tangent.]

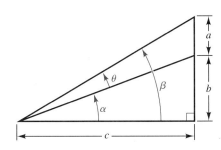

69. For functions such as $f(t) = A \sin bt + B \cos bt$, we may also replace the expression $A \sin bt + B \cos bt$ with $a \cos(bt - c)$, where a is the distance from the origin to a point P having coordinates (B, A) and c is the measure of the angle in standard position with terminal side $\overline{OP}$, as shown in the sketch. Show that these two expressions are equivalent.

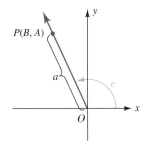

70. Given that $f(x) = \sin x$, show that the difference quotient is equivalent to the right-hand side of this equation:

$$\frac{f(x + h) - f(x)}{h} = \sin x \left(\frac{\cos h - 1}{h} \right) + \cos x \left(\frac{\sin h}{h} \right)$$

In Exercises 71–76, assume that α is a fourth-quadrant angle and β is a second-quadrant angle with $\cos \alpha = 0.5299$ and $\sin \beta = 0.5592$. Use a calculator to evaluate to three significant digits each trigonometric expression by each of the following methods:

(a) *Determine the values of α and β and then directly evaluate the given expression.*

(b) *Apply the appropriate sum or difference formula.*

71. $\sin(\alpha + \beta)$
72. $\sin(\alpha - \beta)$
73. $\cos(\alpha - \beta)$
74. $\cos(\alpha + \beta)$
75. $\tan(\alpha + \beta)$
76. $\tan(\alpha - \beta)$

8.4 Multiple-Angle Formulas

◆ Introductory Comments

The sum and difference formulas (Section 8.3) may be used to develop several other trigonometric formulas. In this section, we derive formulas for the trigonometric functions of ku, where k is a constant and u is a real number or an angle measured in degrees or radians. We refer to these formulas as the *multiple-angle formulas,* and we will use them to help prove other trigonometric identities and to solve some trigonometric equations. The multiple-angle formulas are especially useful for working with trigonometric functions in calculus.

◆ Double-Angle Formulas

The sum formula for sine, given in Section 8.3, states that

$$\sin(u + v) = \sin u \cos v + \cos u \sin v.$$

If we replace v with u in this formula, then we obtain

$$\sin(u + u) = \sin u \cos u + \cos u \sin u$$
$$\sin 2u = 2 \sin u \cos u \quad \text{Simplify both sides}$$

We refer to this last equation as the *double-angle formula for sine.* In a similar manner, we can use the sum formulas for cosine and tangent to derive *double-angle formulas for cosine and tangent.*

◆ Double-Angle Formulas

1. $\sin 2u = 2 \sin u \cos u$
2. $\cos 2u = \cos^2 u - \sin^2 u$
3. $\tan 2u = \dfrac{2 \tan u}{1 - \tan^2 u}$

EXAMPLE 1 Use a double-angle formula to simplify each expression.

(a) $6 \sin 5x \cos 5x$ (b) $\cos^2 \dfrac{x}{2} - \sin^2 \dfrac{x}{2}$

◆ Solution

(a) Applying the double-angle formula for sine, from right to left, with $u = 5x$, we have

$$6 \sin 5x \cos 5x = 3(2 \sin 5x \cos 5x) \quad \text{Rewrite}$$
$$= 3 \sin 2(5x) \quad \text{Apply double-angle formula}$$
$$= 3 \sin 10x \quad \text{Simplify}$$

Figure 8.36 shows the graph of $y = 6 \sin 5x \cos 5x$ in the viewing rectangle of a graphing calculator. The graph of this equation is the same as the graph of $y = 3 \sin 10 x$, thus confirming our work.

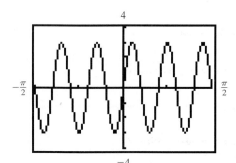

FIGURE 8.36
The graph of $y = 6 \sin 5x \cos 5x$ is the same as the graph of $y = 3 \sin 10x$.

(b) Applying the double-angle formula for cosine, from right to left, with $u = x/2$, we have

$$\cos^2 \frac{x}{2} - \sin^2 \frac{x}{2} = \cos 2\left(\frac{x}{2}\right) = \cos x$$

Figure 8.37 shows the graph of $y = \cos^2 \frac{x}{2} - \sin^2 \frac{x}{2}$ in the viewing rectangle of a graphing calculator. This graph is the same as the graph of $y = \cos x$, which confirms our work. ◆

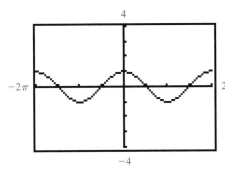

FIGURE 8.37
The graph of $y = \cos^2 \frac{x}{2} - \sin^2 \frac{x}{2}$ is the same as the graph of $y = \cos x$.

PROBLEM 1 Use a double-angle formula to simplify the expression $\dfrac{2 \tan 3x}{1 - \tan^2 3x}$.

◆ **Applying the Double-Angle Formulas**

Double-angle formulas have a variety of uses. In the next example, we use these formulas to help verify a trigonometric identity.

EXAMPLE 2 Show graphically and verify algebraically that the trigonometric equation is an identity:

$$\csc 2x - \cot 2x = \tan x$$

◆ **Solution** As illustrated in Figure 8.38, the graphs of $y = \csc 2x - \cot 2x$ and $y = \tan x$ appear to be identical. Hence, the trigonometric equation

$$\csc 2x - \cot 2x = \tan x$$

seems to be an identity. To verify this fact algebraically, we proceed as follows:

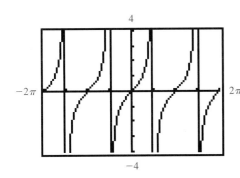

FIGURE 8.38
The graphs of $y = \csc 2x - \cot 2x$ and $y = \tan x$ appear to be identical.

$\csc 2x - \cot 2x = \tan x$	
$\dfrac{1}{\sin 2x} - \dfrac{\cos 2x}{\sin 2x}$	Apply fundamental trig identities
$\dfrac{1 - \cos 2x}{\sin 2x}$	Subtract fractions
$\dfrac{1 - (\cos^2 x - \sin^2 x)}{2 \sin x \cos x}$	Apply double-angle formulas
$\dfrac{(1 - \cos^2 x) + \sin^2 x}{2 \sin x \cos x}$	Rewrite
$\dfrac{2 \sin^2 x}{2 \sin x \cos x}$	Replace $1 - \cos^2 x$ with $\sin^2 x$ and simplify
$\dfrac{\sin x}{\cos x}$	Reduce, provided that $\sin x \neq 0$
$\tan x$	Apply fundamental trig identity

Since we have reduced the left-hand side of this equation to the right-hand side, we conclude that

$$\csc 2x - \cot 2x = \tan x \quad \text{is a trigonometric identity.} \qquad \blacklozenge$$

PROBLEM 2 Show graphically and verify algebraically that $2 \cos x \csc 2x = \csc x$ is a trigonometric identity. $\qquad \blacklozenge$

To help solve certain trigonometric equations in which the angles have the ratio 2 to 1, we apply the double-angle formulas to obtain trigonometric functions with the same angle.

EXAMPLE 3 Find the general solution of the trigonometric equation $\cos 2x = 2 \sin^2 x$.

◆ **Solution** To solve this equation, we must first express the trigonometric functions with the same angle. If we apply the double-angle formula for cosine, we obtain an equation in which all angles are x. Thus, we proceed as follows:

$$\cos 2x = 2 \sin^2 x$$

$$\cos^2 x - \sin^2 x = 2 \sin^2 x \qquad \textbf{Apply double-angle formula for cosine}$$

$$(1 - \sin^2 x) - \sin^2 x = 2 \sin^2 x \qquad \textbf{Replace } \cos^2 x \textbf{ with } 1 - \sin^2 x$$

$$1 = 4 \sin^2 x \qquad \textbf{Add } 2 \sin^2 x \textbf{ to both sides}$$

$$\sin^2 x = \tfrac{1}{4} \qquad \textbf{Divide both sides by 4}$$

$$\sin x = \pm \tfrac{1}{2} \qquad \textbf{Solve for } \sin x$$

Consequently, for $\sin x = \tfrac{1}{2}$, we have

$$x = \frac{\pi}{6} + 2\pi n \quad \text{and} \quad x = \frac{5\pi}{6} + 2\pi n$$

and for $\sin x = -\tfrac{1}{2}$, we have

$$x = \frac{7\pi}{6} + 2\pi n \quad \text{and} \quad x = \frac{11\pi}{6} + 2\pi n$$

The general form of the solutions may be written more compactly as

$$x = \frac{\pi}{6} + \pi n \quad \text{and} \quad x = \frac{5\pi}{6} + \pi n \quad \text{where } n \text{ is an integer.}$$

The graph in Figure 8.39 supports our work. $\qquad \blacklozenge$

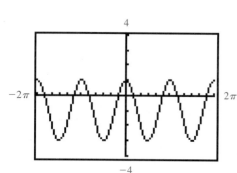

FIGURE 8.39

Graph of $y = \cos 2x - 2 \sin^2 x$ in the interval $[-2\pi, 2\pi]$ with x-scale $\pi/6$

PROBLEM 3 Find the general solution of the trigonometric equation $\cos x + \sin 2x = 0$.

By using double-angle formulas in conjunction with the sum formulas from Section 8.3, we can develop other multiple-angle formulas. In the next example, we develop a triple-angle formula for sine.

EXAMPLE 4 Derive a formula for $\sin 3x$ in terms of $\sin x$.

◀ **Solution** Writing $3x$ as $2x + x$, we proceed as follows:

$\sin 3x = \sin(2x + x)$

$= \sin 2x \cos x + \cos 2x \sin x$	Apply sum formula for sine
$= (2 \sin x \cos x) \cos x + (\cos^2 x - \sin^2 x) \sin x$	Apply double-angle formulas
$= 2 \sin x \cos^2 x + \cos^2 x \sin x - \sin^3 x$	Multiply
$= 2 \sin x (1 - \sin^2 x) + (1 - \sin^2 x) \sin x - \sin^3 x$	Replace $\cos^2 x$ with $1 - \sin^2 x$
$= 2 \sin x - 2 \sin^3 x + \sin x - \sin^3 x - \sin^3 x$	Multiply
$= 3 \sin x - 4 \sin^3 x$	Collect like terms

Hence, the *triple angle formula for sine* is

$$\sin 3x = 3 \sin x - 4 \sin^3 x.$$

The graph in Figure 8.40 supports our work.

PROBLEM 4 Show that the triple-angle formula for sine developed in Example 4 is true when x is replaced by $\pi/6$.

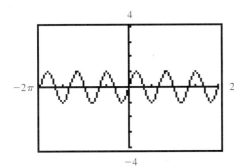

FIGURE 8.40
The graphs of $y = 3 \sin x - 4 \sin^3 x$ and $y = \sin 3x$ appear to be identical.

◆ Power Reduction Formulas

We now develop two alternative forms for the double-angle formula for cosine:

$\cos 2u = \cos^2 u - \sin^2 u$ $\cos 2u = \cos^2 u - \sin^2 u$
$\qquad = (1 - \sin^2 u) - \sin^2 u$ $\qquad = \cos^2 u - (1 - \cos^2 u)$
$\qquad = 1 - 2 \sin^2 u$ $\qquad = 2 \cos^2 u - 1$

Hence,

$$\cos 2u = 1 - 2 \sin^2 u \qquad \cos 2u = 2 \cos^2 u - 1$$

From these two equations, we can derive *power reduction formulas* for sine, cosine, and tangent. Solving the equation $\cos 2u = 1 - 2\sin^2 u$ for $\sin^2 u$ and the equation $\cos 2u = 2\cos^2 u - 1$ for $\cos^2 u$, we obtain

$$\sin^2 u = \tfrac{1}{2}(1 - \cos 2u) \quad \text{and} \quad \cos^2 u = \tfrac{1}{2}(1 + \cos 2u).$$

Since $\tan^2 u = \dfrac{\sin^2 u}{\cos^2 u}$, we also have

$$\tan^2 u = \frac{1 - \cos 2u}{1 + \cos 2u}.$$

◆ Power Reduction Formulas

1. $\sin^2 u = \tfrac{1}{2}(1 - \cos 2u)$
2. $\cos^2 u = \tfrac{1}{2}(1 + \cos 2u)$
3. $\tan^2 u = \dfrac{1 - \cos 2u}{1 + \cos 2u}$

Power reduction formulas are used in calculus to express even powers of sine and cosine as first powers of the cosine function. The procedure is illustrated in the next example.

EXAMPLE 5 Express $\sin^4 x$ in terms of first powers of the cosine function.

◆ Solution We proceed as follows:

$\sin^4 x = \sin^2 x \sin^2 x$ **Rewrite with exponents of 2**

$= \tfrac{1}{2}(1 - \cos 2x)\tfrac{1}{2}(1 - \cos 2x)$ **Apply power reduction formula for sine**

$= \tfrac{1}{4}(1 - 2\cos 2x + \cos^2 2x)$ **Multiply**

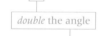
double the angle

$= \tfrac{1}{4}[1 - 2\cos 2x + \tfrac{1}{2}(1 + \cos 4x)]$ **Apply power reduction formula for cosine**

$= \tfrac{1}{4}(1 - 2\cos 2x + \tfrac{1}{2} + \tfrac{1}{2}\cos 4x)$ **Distribute $\tfrac{1}{2}$**

$= \tfrac{1}{4}(\tfrac{3}{2} - 2\cos 2x + \tfrac{1}{2}\cos 4x)$ **Combine like terms**

$= \tfrac{1}{8}(3 - 4\cos 2x + \cos 4x)$ **Factor out $\tfrac{1}{2}$**

The graph in Figure 8.41 supports our work. ◆

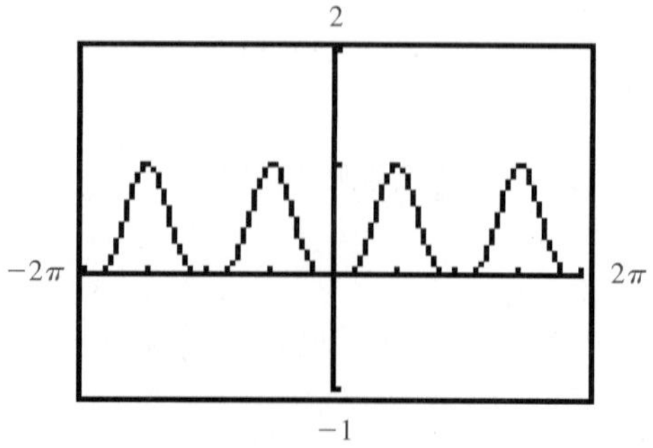

FIGURE 8.41
The graphs of $y = \sin^4 x$ and $y = \tfrac{1}{8}(3 - 4\cos 2x + \cos 4x)$ appear to be identical.

SECTION 8.4 ◆ *Multiple-Angle Formulas*

PROBLEM 5 Express $\cos^4 x$ in terms of first powers of the cosine function. ◆

◆ Half-Angle Formulas

Alternate forms of the power reduction formulas are obtained by replacing u with $v/2$ and then taking the square root of both sides of the equation. We refer to the equations that are formed by this procedure as *half-angle formulas*.

◀ **Half-Angle Formulas**

1. $\sin \dfrac{v}{2} = \pm \sqrt{\dfrac{1 - \cos v}{2}}$

2. $\cos \dfrac{v}{2} = \pm \sqrt{\dfrac{1 + \cos v}{2}}$

3. $\tan \dfrac{v}{2} = \pm \sqrt{\dfrac{1 - \cos v}{1 + \cos v}}$

Note: We select the positive or negative square root in a half-angle formula depending on the sign of $\sin \dfrac{v}{2}$, $\cos \dfrac{v}{2}$, or $\tan \dfrac{v}{2}$ in the quadrant containing $v/2$.

EXAMPLE 6 Find the exact value of $\sin \dfrac{\pi}{12}$.

◀ **Solution** We can think of $\pi/12$ as a first-quadrant angle that is half of $\pi/6$. Since the sine of a first-quadrant angle is positive, we use the positive square root in the half-angle formula for sine and proceed as follows:

$$\sin \dfrac{\pi}{12} = \sin \dfrac{\pi/6}{2} = \sqrt{\dfrac{1 - \cos(\pi/6)}{2}}$$

$$= \sqrt{\dfrac{1 - \sqrt{3}/2}{2}} \qquad \text{Evaluate } \cos(\pi/6)$$

$$= \dfrac{\sqrt{2 - \sqrt{3}}}{2} \qquad \text{Simplify} \qquad ◆$$

PROBLEM 6 Use a calculator to find the approximate value of $\sin \dfrac{\pi}{12}$. Compare your answer with the exact value found in Example 6. ◆

EXAMPLE 7 Given that θ is a fourth-quadrant angle with $\dfrac{3\pi}{2} < \theta < 2\pi$ and $\cos \theta = 0.62$, find the exact value of $\cos \dfrac{\theta}{2}$.

Solution If θ is a fourth-quadrant angle with

$$\frac{3\pi}{2} < \theta < 2\pi$$

then $\quad\dfrac{3\pi}{4} < \dfrac{\theta}{2} < \pi \quad$ **Divide each member by 2**

Therefore, $\theta/2$ is a second-quadrant angle. Since the cosine of a second-quadrant angle is negative, we use the negative square root in the half-angle formula for cosine and proceed as follows:

$$\cos\frac{\theta}{2} = -\sqrt{\frac{1+\cos\theta}{2}} = -\sqrt{\frac{1+0.62}{2}} = -\sqrt{0.81} = -0.9. \quad\blacklozenge$$

PROBLEM 7 Given that θ is a third-quadrant angle with $\pi < \theta < \dfrac{3\pi}{2}$ and $\cos\theta = -0.28$, find the exact value of $\sin\dfrac{\theta}{2}$. $\quad\blacklozenge$

Alternate forms of the half-angle formula for tangent may be obtained as follows:

$$\tan\frac{v}{2} = \frac{\sin(v/2)}{\cos(v/2)}$$

$$= \frac{2\sin(v/2)\cos(v/2)}{2\cos^2(v/2)} \quad \textbf{Multiply numerator and denominator by } 2\cos(v/2)$$

$$= \frac{\sin v}{2\left(\dfrac{1+\cos v}{2}\right)} \quad \textbf{Apply double angle and power reduction formulas}$$

$$= \frac{\sin v}{1+\cos v} \quad \textbf{Simplify}$$

$$= \frac{1-\cos v}{\sin v} \quad \textbf{From Example 6(a) in Section 8.1}$$

We now state these half-angle formulas for tangent:

$$\tan\frac{v}{2} = \frac{\sin v}{1+\cos v} \qquad \tan\frac{v}{2} = \frac{1-\cos v}{\sin v}$$

The advantage of using either one of these alternate forms is that it contains no radical and thus has no ambiguity of sign.

EXAMPLE 8 Show graphically and verify algebraically that the trigonometric equation is an identity:

$$\tan\frac{x}{2} + \cot\frac{x}{2} = 2\csc x$$

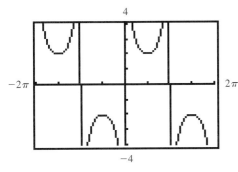

FIGURE 8.42

The graphs of $y = \tan \dfrac{x}{2} + \cot \dfrac{x}{2}$ and $y = 2 \csc x$ appear to be identical.

◆ **Solution** As illustrated in Figure 8.42, the graphs of $y = \tan \dfrac{x}{2} + \cot \dfrac{x}{2}$ and $y = 2 \csc x$ appear to be identical. Hence, the trigonometric equation

$$\tan \frac{x}{2} + \cot \frac{x}{2} = 2 \csc x$$

seems to be an identity. To verify this fact algebraically, we proceed as follows:

$\tan \dfrac{x}{2} + \cot \dfrac{x}{2} = 2 \csc x$	
$\tan \dfrac{x}{2} + \dfrac{1}{\tan \dfrac{x}{2}}$	Apply fundamental trig identity
$\dfrac{\sin x}{1 + \cos x} + \dfrac{1 + \cos x}{\sin x}$	Apply half-angle formula
$\dfrac{\sin^2 x + 1 + 2 \cos x + \cos^2 x}{(1 + \cos x) \sin x}$	Add fractions
$\dfrac{2 + 2 \cos x}{(1 + \cos x) \sin x}$	Replace $\sin^2 x + \cos^2 x$ with 1
$\dfrac{2(1 + \cos x)}{(1 + \cos x) \sin x}$	Factor
$\dfrac{2}{\sin x}$	Reduce, provided that $1 + \cos x \neq 0$
$2 \csc x$	Apply fundamental trig identity

Since we have reduced the left-hand side of this equation to the right-hand side, we conclude that

$$\tan \frac{x}{2} + \cot \frac{x}{2} = 2 \csc x \quad \text{is a trigonometric identity} \quad ◆$$

PROBLEM 8 Rework Example 8 by applying the formula $\tan \dfrac{v}{2} = \dfrac{1 - \cos v}{\sin v}$. ◆

◆ Application: Design of an A-Frame Structure

We conclude this section with an applied problem concerning an A-frame structure (a structure in the form of an inverted V).

EXAMPLE 9 An architect is to design an A-frame house with rafters of length 20 feet.

532 CHAPTER 8 ◆ *Analytic Trigonometry*

(a) Express the cross-sectional area A of the house as a trigonometric function of the angle θ between the rafters.

(b) Determine the largest possible cross-sectional area.

◆ **Solution**

(a) The cross-sectional area A of the house is the area of an isosceles triangle. Recall from elementary geometry that if we construct an altitude to the base of an isosceles triangle, the vertex angle θ and the base are both cut in half, as shown in Figure 8.43. Working with either right triangle in Figure 8.43, we find x and h:

$$\sin \frac{\theta}{2} = \frac{x}{20} \qquad \text{and} \qquad \cos \frac{\theta}{2} = \frac{h}{20}$$

$$x = 20 \sin \frac{\theta}{2} \qquad\qquad\qquad h = 20 \cos \frac{\theta}{2}$$

Thus, using the fact that the area of a triangle is half the product of its base and height, we have

$$A = xh = \left(20 \sin \frac{\theta}{2}\right)\left(20 \cos \frac{\theta}{2}\right)$$

$$= 400 \sin \frac{\theta}{2} \cos \frac{\theta}{2}$$

$$= 200\left(2 \sin \frac{\theta}{2} \cos \frac{\theta}{2}\right) \qquad \textbf{Rewrite}$$

$$= 200 \sin \theta \qquad \textbf{Apply double-angle formula}$$

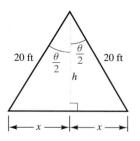

FIGURE 8.43
The altitude to the base of an isosceles triangle cuts the vertex angle and base in half.

(b) Note that angle θ must be between 0° and 180°. Since $\sin \theta$ reaches a maximum value of 1 when $\theta = 90°$, we conclude that the largest cross-sectional area is 200 square feet, and this occurs when $\theta = 90°$. ◆

PROBLEM 9 For the A-frame house described in Example 9, find an acute angle θ (to the nearest tenth of a degree) associated with a cross-sectional area of 170 square feet. ◆

Exercises 8.4

◆ **Basic Skills**

In Exercises 1–16, use double-angle formulas to simplify each expression. Use a graphing calculator to check each answer.

1. $10 \sin 3x \cos 3x$

2. $2 \sin \frac{x}{2} \cos \frac{x}{2}$

3. $\cos^2 2x - \sin^2 2x$

4. $\dfrac{6 \tan 2x}{1 - \tan^2 2x}$

5. $\sec 6x (\sin^2 3x - \cos^2 3x)$

6. $2 \sin 4x (\cos^2 2x - \sin^2 2x)$

7. $\frac{1}{2} \cot 3x (1 - \tan^2 3x)$

8. $\frac{1}{2} \sec 4x \csc 4x$

9. $1 - 2 \sin^2 \frac{x}{2}$

10. $2 \cos^2 5x - 1$

11. $\dfrac{\sin 2x}{2 \sin x}$

12. $2 \cos x - \sin 2x \csc x$

13. $(\sin x + \cos x)^2 - \sin 2x$

14. $\cos^4 2x - \sin^4 2x$

15. $\dfrac{2}{1 + \cos 6x}$

16. $\dfrac{2}{\cot 3x - \tan 3x}$

In Exercises 17–28, use double-angle formulas to help show that each trigonometric equation is an identity.

17. $2 \cos x \csc 2x = \csc x$

18. $\sec^2 x \cos 2x + 2 \tan^2 x = \sec^2 x$

19. $2 \sin 2x - 4 \sin 2x \sin^2 x = \sin 4x$

20. $2 \tan x - \sin 2x = 2 \tan x \sin^2 x$

21. $\dfrac{\tan 2x}{2 \tan x} = \dfrac{1}{2 - \sec^2 x}$

22. $\dfrac{2 \tan 3x}{1 + \tan^2 3x} = \sin 6x$

23. $\dfrac{\sin 6x}{1 + \cos 6x} = \tan 3x$

24. $\dfrac{1 + \cos 2x}{1 - \cos 2x} = \cot^2 x$

25. $\dfrac{2 \cos x}{\csc x - 2 \sin x} = \tan 2x$

26. $\dfrac{2 \cos x - \sec x}{2 \sin x} = \cot 2x$

27. $\dfrac{1 + \cot x}{\cot x - 1} = \dfrac{1 + \sin 2x}{\cos 2x}$

28. $2 \sin x + \sin 2x = \dfrac{2 \sin^3 x}{1 - \cos x}$

In Exercises 29–38, use double-angle formulas to help find the general solution of each equation. Use a graphing calculator to check each answer.

29. $2 \sin x = \sin 2x$

30. $\sin 2x - \cos x = 0$

31. $\sin x + \cos 2x = 1$

32. $\cos 2x + 3 \sin x = 2$

33. $2 \sin^2 2x = 2 + \cos 4x$

34. $2 \cos^2 3x + \cos 6x = 1$

35. $\tan 2x - 2 \sin x = 0$

36. $\sin 2x = \cot x$

37. $2 \tan 2x + \tan x = 0$

38. $\tan 4x - \tan 2x = 0$

In Exercises 39–42, derive a formula for the indicated trigonometric expression. Use a graphing calculator to verify each answer.

39. $\cos 3x$ in terms of $\cos x$

40. $\cos 4x$ in terms of $\cos x$

41. $\sin 5x$ in terms of $\sin x$

42. $\cos 5x$ in terms of $\cos x$

In Exercises 43–46, use power reduction formulas to express the even powers of sine and cosine in terms of first powers of the cosine function.

43. $\cos^2 3x$

44. $\sin^2 x \cos^2 x$

45. $\sin^4 x \cos^2 x$

46. $\sin^6 x$

In Exercises 47–50, find the exact value of each trigonometric function by applying a half-angle formula.

47. $\cos 15°$

48. $\tan 15°$

49. $\sin 22.5°$

50. $\cos 22.5°$

In Exercises 51–54, find the exact value of the indicated trigonometric function from the given information. Assume that $0 \le x < 2\pi$.

51. Given $\sin x = \frac{3}{5}$ and x in quadrant I, find
    (a) $\sin 2x$ (b) $\cos 2x$ (c) $\sin \frac{x}{2}$ (d) $\cos \frac{x}{2}$

52. Given $\cos x = -\frac{12}{13}$ and x in quadrant III, find
    (a) $\cos 2x$ (b) $\tan 2x$ (c) $\cos \frac{x}{2}$ (d) $\tan \frac{x}{2}$

53. Given $\tan x = -3$ and x in quadrant IV, find
    (a) $\tan 2x$ (b) $\sec 2x$ (c) $\sin \frac{x}{2}$ (d) $\cot \frac{x}{2}$

54. Given $\csc x = 1.25$ and x in quadrant II, find
    (a) $\sin 2x$ (b) $\cot 2x$ (c) $\cos \frac{x}{2}$ (d) $\csc \frac{x}{2}$

In Exercises 55–60, use half-angle formulas to help show that each trigonometric equation is an identity.

55. $2 \cos^2 \dfrac{x}{2} - \cos x = 1$

56. $2 \sin^2 \dfrac{x}{2} \sec x + 1 = \sec x$

57. $\csc^2 \dfrac{x}{2} - 1 = \dfrac{1 + \cos x}{1 - \cos x}$

58. $\tan \dfrac{x}{2} = \csc x - \cot x$

59. $\dfrac{2 \tan \dfrac{x}{2}}{1 + \tan^2 \dfrac{x}{2}} = \sin x$

60. $\cot \dfrac{x}{2} - \tan \dfrac{x}{2} = 2 \cot x$

In Exercises 61–66, use half-angle formulas to help find the general solution of each equation. Use a graphing calculator to check each answer.

61. $\sin^2 \dfrac{x}{2} - \cos x + 1 = 0$

62. $2 \cos^2 \dfrac{x}{2} = 2 - \cos x$

63. $\sin \dfrac{x}{2} - \sin x = 0$

64. $\cos x - \cos \dfrac{x}{2} + 1 = 0$

65. $\tan \dfrac{x}{2} + \sin 2x = \csc x$

66. $\tan \dfrac{x}{2} + 1 = \cos x$

A function and its graph are shown in Exercises 67 and 68. Approximate the x-intercepts of the graph, rounding each answer to three significant digits.

67.

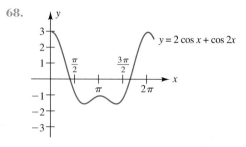

$y = \cos x - 2 \sin 2x$

68.

$y = 2 \cos x + \cos 2x$

In Exercises 69–72, an equation of the form $f(x) = 0$ is given. Use a graphing calculator to estimate the solutions of each equation in the interval $(0, 2\pi)$. Round each answer to three significant digits. Then check each answer algebraically and explain any discrepancy.

69. $2 \cos^2 x - \cos 2x - 2 \sin 2x = 0$

70. $2 \sin^2 2x + 2 \cos 4x - 1 = 0$

71. $\tan 2x + 3 \cot x = 0$

72. $3 \tan 2x - \tan 4x = 0$

In Exercises 73–76, an equation of the form $f(x) = g(x)$ is stated. Use a graphing calculator to generate the graphs of $y = f(x)$ and $y = g(x)$. Compare the graphs, and then state whether the equation seems to be a trigonometric identity.

73. $\sin 4x = 8 \cos^3 x \sin x - 4 \cos x \sin x$

74. $\cos 6x = 32 \cos^6 x - 48 \cos^4 x + 18 \cos^2 x - 1$

75. $\sin 6x = 32 \cos^5 x \sin x - 32 \cos^3 x \sin x + 6 \cos x \sin x$

76. $\tan 3x = \dfrac{3 \tan x - \tan^3 x}{1 - 3 \tan^2 x}$

77. The cross-section of a water trough has the shape of an isosceles triangle with dimensions as shown in the sketch.

   (a) Express the volume V of the trough as a trigonometric function of the angle θ between its sides.

   (b) Determine the largest possible volume of the trough.

   (c) Find θ (to the nearest tenth of a degree) if the volume is 30 cubic feet.

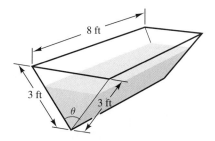

78. A rectangle is inscribed in a semicircle of radius 1 meter, as shown in the sketch.

   (a) Express the area A of the rectangle as a trigonometric function of angle θ.

   (b) What value of θ yields the rectangle with the maximum area? What is the maximum area?

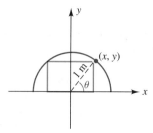

79. The height of one building is three times the height of another building. When standing midway between the two buildings, an observer notes that the angle of ele-

vation to the top of the taller building is twice the angle of elevation to the top of the other building. Assuming the buildings are on level ground, find these angles of elevation.

80. An amateur radio tower is attached to the roof of a house. The house is 15 ft high and the tower is 25 ft high, as shown in the sketch. Using a transit, a surveyor observes that the angle of elevation to the top of the roof is the same as the angle from the top of the roof to the top of the tower. How far away from the house is the transit located?

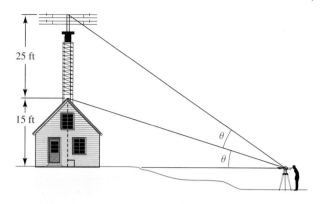

Critical Thinking

81. Find the amplitude and period of each function.
    (a) $y = 6 \sin 4x \cos 4x$ (b) $y = \sin x \cos x$

82. In general, $\sin 2u \neq 2 \sin u$.
    (a) Use $u = \pi/6$ to illustrate this fact.
    (b) Do any real numbers u exist for which $\sin 2u = 2 \sin u$? If so, give an example.

83. Use double-angle formulas to find the exact value of each expression.
    (a) $\sin[2(\arctan 3)]$ (b) $\cos[2(\cos^{-1} \frac{1}{4})]$

84. Use half-angle formulas to find the exact value of each expression.
    (a) $\tan[\frac{1}{2}(\arcsin \frac{2}{3})]$ (b) $\sin[\frac{1}{2}(\tan^{-1} 2)]$

85. A *rhombus* is a parallelogram with all sides of equal length. The diagonals of a rhombus are perpendicular to one another and bisect the interior angles, as shown in the sketch. Use this fact to show that the area A of a rhombus is

    $$A = a^2 \sin \theta$$

where a is the length of one side of the rhombus and θ is the measure of one of its interior angles.

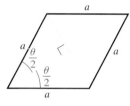

86. Refer to the right triangle in the sketch.
    (a) Use the double-angle formula for sine to find $\sin 2\alpha$ and $\sin 2\beta$. What do you observe? Explain.
    (b) Use the double-angle formula for cosine to find $\cos 2\alpha$ and $\cos 2\beta$. What do you observe? Explain.

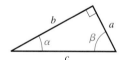

8.5 Product-to-Sum Formulas and Sum-to-Product Formulas

◆ Introductory Comments

In this section, we discuss trigonometric formulas that allow us to write the *product* of certain trigonometric functions as a *sum*, and the *sum* of certain trigonometric functions as a *product*. We refer to these formulas as the *product-to-sum formulas* and *sum-to-product formulas*, and we often use them to help prove other trigonometric identities and to solve some trigonometric equations. We begin by deriving the *product-to-sum formulas*.

◆ Deriving the Product-to-Sum Formulas

Recall (from Section 8.3) sum and difference formulas for cosine and sine:

1. $\cos(u + v) = \cos u \cos v - \sin u \sin v$
2. $\cos(u - v) = \cos u \cos v + \sin u \sin v$
3. $\sin(u + v) = \sin u \cos v + \cos u \sin v$
4. $\sin(u - v) = \sin u \cos v - \cos u \sin v$

If we add the respective members of the second equation to the first, we obtain

$$\cos(u + v) = \cos u \cos v - \sin u \sin v$$
$$\underline{\cos(u - v) = \cos u \cos v + \sin u \sin v}$$
$$\cos(u + v) + \cos(u - v) = 2 \cos u \cos v$$

Now, multiplying both sides of this equation by $\frac{1}{2}$ gives us

$$\cos u \cos v = \tfrac{1}{2}[\cos(u + v) + \cos(u - v)].$$

Note that this formula expresses the *product* $\cos u \cos v$ as a *sum*. We refer to this formula as a *product-to-sum formula*.

In a similar manner, by subtracting the respective members of the first equation from the second, adding the respective members of the fourth equation to the third, and subtracting the respective members of the fourth equation from the third, we obtain three other product-to-sum formulas.

◆ Product-to-Sum Formulas

1. $\cos u \cos v = \tfrac{1}{2}[\cos(u + v) + \cos(u - v)]$
2. $\sin u \sin v = \tfrac{1}{2}[\cos(u - v) - \cos(u + v)]$
3. $\sin u \cos v = \tfrac{1}{2}[\sin(u + v) + \sin(u - v)]$
4. $\cos u \sin v = \tfrac{1}{2}[\sin(u + v) - \sin(u - v)]$

EXAMPLE 1 Express the product as a sum or difference.

(a) $2 \sin 5x \sin x$ (b) $\sin 3x \cos 2x$

(a) Using product-to-sum formula 2 with $u = 5x$ and $v = x$, we have

$$2 \sin 5x \sin x = 2\{\tfrac{1}{2}[\cos(5x - x) - \cos(5x + x)]\}$$
$$= \cos 4x - \cos 6x.$$

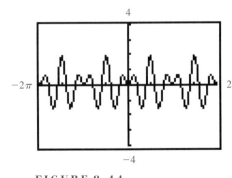

FIGURE 8.44

The graphs of $y = 2 \sin 5x \sin x$ and $y = \cos 4x - \cos 6x$ are identical.

As illustrated in Figure 8.44, the graphs of $y = 2 \sin 5x \sin x$ and $y = \cos 4x - \cos 6x$ appear to be identical, which confirms our work.

(b) Using product-to-sum formula 3 with $u = 3x$ and $v = 2x$, we have

$$\sin 3x \cos 2x = \tfrac{1}{2}[\sin(3x + 2x) + \sin(3x - 2x)]$$
$$= \tfrac{1}{2} \sin 5x + \tfrac{1}{2} \sin x$$

As illustrated in Figure 8.45, the graphs of $y = \sin 3x \cos 2x$ and $y = \tfrac{1}{2} \sin 5x + \tfrac{1}{2} \sin x$ appear to be identical; this fact confirms our work. ◆

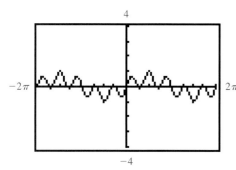

FIGURE 8.45
The graphs of $y = \sin 3x \cos 2x$ and $y = \tfrac{1}{2} \sin 5x + \tfrac{1}{2} \sin x$ are identical.

PROBLEM 1 Rework Example 1(b) by using product-to-sum formula 4 with $u = 2x$ and $v = 3x$. Explain why the answer is the same as that obtained in the example. ◆

◆ **Deriving the Sum-to-Product Formulas**

If we let

$$u + v = a \quad \text{and} \quad u - v = b,$$

then

$$(u + v) + (u - v) = a + b \quad \text{and} \quad (u + v) - (u - v) = a - b$$
$$2u = a + b \qquad\qquad 2v = a - b$$
$$u = \frac{a + b}{2} \qquad\qquad v = \frac{a - b}{2}$$

Making the substitutions

$$u + v = a, \quad u - v = b, \quad u = \frac{a + b}{2}, \quad \text{and} \quad v = \frac{a - b}{2}$$

into product-to-sum formula 1, we obtain

$$\cos \frac{a + b}{2} \cos \frac{a - b}{2} = \tfrac{1}{2}[\cos a + \cos b].$$

Now, multiplying both sides of this equation by 2 gives us

$$\cos a + \cos b = 2 \cos \frac{a + b}{2} \cos \frac{a - b}{2}.$$

This formula expresses the *sum* $\cos a + \cos b$ as a *product*, and we refer to this formula as a *sum-to-product formula*. Similarly, by making these substitutions into the other product-to-sum formulas, we obtain three other sum-to-product formulas.

Sum-to-Product Formulas

1. $\cos a + \cos b = 2 \cos \dfrac{a+b}{2} \cos \dfrac{a-b}{2}$

2. $\cos a - \cos b = -2 \sin \dfrac{a+b}{2} \sin \dfrac{a-b}{2}$

3. $\sin a + \sin b = 2 \sin \dfrac{a+b}{2} \cos \dfrac{a-b}{2}$

4. $\sin a - \sin b = 2 \cos \dfrac{a+b}{2} \sin \dfrac{a-b}{2}$

EXAMPLE 2 Express the sum or difference as a product.

(a) $\cos 8x + \cos 2x$ **(b)** $\sin x - \sin 4x$

(b) Using sum-to-product formula 1 with $a = 8x$ and $b = 2x$, we have

$$\cos 8x + \cos 2x = 2 \cos \frac{8x + 2x}{2} \cos \frac{8x - 2x}{2}$$

$$= 2 \cos 5x \cos 3x$$

As illustrated in Figure 8.46, the graphs of $y = \cos 8x + \cos 2x$ and $y = 2 \cos 5x \cos 3x$ appear to be identical, which confirms our work.

(b) Using sum-to-product formula 4 with $a = x$ and $b = 4x$, we have

$$\sin x - \sin 4x = 2 \cos \frac{x + 4x}{2} \sin \frac{x - 4x}{2}$$

$$= 2 \cos \frac{5x}{2} \sin\left(-\frac{3x}{2}\right)$$

$$= -2 \cos \frac{5x}{2} \sin \frac{3x}{2} \qquad \text{Use } \sin(-u) = -\sin u$$

As illustrated in Figure 8.47, the graphs of $y = \sin x - \sin 4x$ and $y = -2 \cos \dfrac{5x}{2} \sin \dfrac{3x}{2}$ appear to be identical, thus confirming our work. ◆

PROBLEM 2 Express $\sin 5x + \sin 7x$ as a product.

◆ Applying the Formulas

The next four examples illustrate a variety of uses of the product-to-sum formulas and sum-to-product formulas. In the next example, we apply a product-to-sum formula to find the exact value of the product of two trigonometric functions.

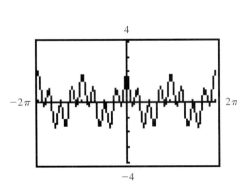

FIGURE 8.46
The graphs of $y = \cos 8x + \cos 2x$ and $y = 2 \cos 5x \cos 3x$ are identical.

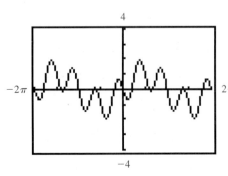

FIGURE 8.47
The graphs of $y = \sin x - \sin 4x$ and $y = -2 \cos \dfrac{5x}{2} \sin \dfrac{3x}{2}$ are identical.

SECTION 8.5 ◆ Product-to-Sum and Sum-to-Product Formulas

EXAMPLE 3 Use a product-to-sum formula to find the exact value of $\cos 75° \cos 15°$.

◆ **Solution** Using product-to-sum formula 1 with $u = 75°$ and $v = 15°$, we have

$$\cos 75° \cos 15° = \tfrac{1}{2}[\cos(75° + 15°) + \cos(75° - 15°)]$$

$$= \tfrac{1}{2}[\cos 90° + \cos 60°]$$

$$= \tfrac{1}{2}[0 + \tfrac{1}{2}] \qquad \text{Evaluate}$$

$$= \tfrac{1}{4} \qquad \text{Simplify} \qquad ◆$$

PROBLEM 3 Use a sum-to-product formula to find the exact value of $\cos 75° - \cos 15°$. ◆

The product-to-sum formulas and sum-to-product formulas are also helpful in verifying certain trigonometric identities.

EXAMPLE 4 Show graphically and verify algebraically that the trigonometric equation is an identity:

$$2 \cos^2 x \sin x - 2 \sin^3 x = \sin 3x - \sin x$$

◆ **Solution** As illustrated in Figure 8.48, the graphs of $y = 2 \cos^2 x \sin x - 2 \sin^3 x$ and $y = \sin 3x - \sin x$ appear to be identical. Hence, the trigonometric equation

$$2 \cos^2 x \sin x - 2 \sin^3 x = \sin 3x - \sin x$$

seems to be an identity. To verify this fact algebraically, we proceed as follows:

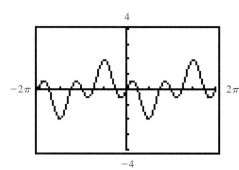

FIGURE 8.48
The graphs of
$y = 2 \cos^2 x \sin x - 2 \sin^3 x$
and $y = \sin 3x - \sin x$ appear
to be identical.

$2 \cos^2 x \sin x - 2 \sin^3 x = \sin 3x - \sin x$	
$2 \sin x (\cos^2 x - \sin^2 x)$	Factor
$2 \sin x \cos 2x$	Apply double-angle formula
$2\{\tfrac{1}{2}[\sin(x + 2x) + \sin(x - 2x)]\}$	Apply product-to-sum formula
$\sin 3x + \sin(-x)$	Simplify
$\sin 3x - \sin x$	Use $\sin(-u) = -\sin u$

Since we have reduced the left-hand side of this equation to the right-hand side, we conclude that

$$2 \cos^2 x \sin x - 2 \sin^3 x = \sin 3x - \sin x \text{ is a trigonometric identity.}$$
◆

PROBLEM 4 Rework Example 4 by working with the right-hand side of the equation.

We can find the solutions to certain trigonometric equations with the aid of product-to-sum and sum-to-product formulas.

EXAMPLE 5 Find the general solution of the trigonometric equation. $\sin 2x + \sin 4x = \sin 3x$.

◆ **Solution** To solve this equation, we proceed as follows:

$$\sin 2x + \sin 4x = \sin 3x$$
$$2 \sin 3x \cos(-x) = \sin 3x \quad \text{Apply sum-to-product formula}$$
$$2 \sin 3x \cos x = \sin 3x \quad \text{Apply } \cos(-u) = \cos u$$
$$2 \sin 3x \cos x - \sin 3x = 0 \quad \text{Subtract } \sin 3x \text{ from both sides}$$
$$\sin 3x (2 \cos x - 1) = 0 \quad \text{Factor}$$
$$\sin 3x = 0 \quad \text{or} \quad 2 \cos x - 1 = 0 \quad \text{Apply zero product property}$$
$$3x = \pi n \qquad \cos x = \tfrac{1}{2}$$
$$x = \frac{\pi n}{3} \qquad x = \frac{\pi}{3} + 2\pi n, \frac{5\pi}{3} + 2\pi n$$

Thus, the expression

$$\frac{\pi n}{3} \quad \text{where } n \text{ is an integer}$$

describes the general solution of this equation. The graph in Figure 8.49 supports our work. ◆

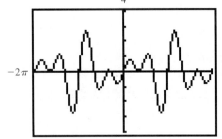

FIGURE 8.49
Graph of
$y = \sin 2x + \sin 4x - \sin 3x$ in the interval $[-2\pi, 2\pi]$ with x-scale $\pi/3$.

PROBLEM 5 Referring to Example 5, find the six solutions in the interval $[0, 2\pi)$. ◆

We may use sum-to-product formulas to help find the x-intercepts of the graph of an equation that involves the sum of sine or cosine functions with different angles.

EXAMPLE 6 Find the x-intercepts of the graph of the given equation in the interval $[0, 2\pi)$:

$$y = \cos 3x + \cos x$$

◆ **Solution** We find the x-intercepts by letting $y = 0$ and solving for x:

$$\cos 3x + \cos x = 0$$

$$2 \cos 2x \cos x = 0 \qquad \text{Apply sum-to-product formula}$$

$$2 \cos 2x = 0 \quad \text{or} \quad \cos x = 0 \qquad \text{Apply zero product property}$$

$$2x = \frac{\pi}{2} + \pi n \qquad x = \frac{\pi}{2} + \pi n$$

$$x = \frac{\pi}{4} + \frac{\pi}{2} n$$

Thus, in the interval $[0, 2\pi)$, the x-intercepts are

$$\frac{\pi}{4}, \frac{3\pi}{4}, \frac{5\pi}{4}, \frac{7\pi}{4}, \quad \text{and} \quad \frac{\pi}{2}, \frac{3\pi}{2}.$$

The graph in Figure 8.50 supports our work. ◆

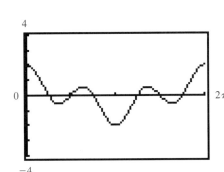

FIGURE 8.50
Graph of $y = \cos 3x + \cos x$ in the interval $[0, 2\pi)$ with x-scale $\pi/4$

PROBLEM 6 In calculus, it is shown that the solutions of the equation $3 \sin 3x + \sin x = 0$ yield the x-coordinates of the relative maximum and minimum points for the graph of $y = \cos 3x + \cos x$ (see Figure 8.50). Use the triple-angle formula for sine (see Example 4 in Section 8.4) to help estimate the relative maximum and minimum points in the interval $[0, 2\pi)$. ◆

◆ Summary of Trigonometric Identities

We conclude this chapter by listing many of the important trigonometric identities that we have discussed in Chapters 7 and 8.

Fundamental Trigonometric Identities

1. $\sin x \csc x = 1$	**2.** $\cos x \sec x = 1$	**3.** $\tan x \cot x = 1$
4. $\tan x = \dfrac{\sin x}{\cos x}$	**5.** $\cot x = \dfrac{\cos x}{\sin x}$	**6.** $\sin^2 x + \cos^2 x = 1$
7. $1 + \tan^2 x = \sec^2 x$	**8.** $1 + \cot^2 x = \csc^2 x$	

Even-Odd Identities

1. $\sin(-x) = -\sin x$	**2.** $\cos(-x) = \cos x$	**3.** $\tan(-x) = -\tan x$
4. $\csc(-x) = -\csc x$	**5.** $\sec(-x) = \sec x$	**6.** $\cot(-x) = -\cot x$

Cofunction Identities

1. $\sin\left(\dfrac{\pi}{2} - x\right) = \cos x$	**2.** $\cos\left(\dfrac{\pi}{2} - x\right) = \sin x$	**3.** $\tan\left(\dfrac{\pi}{2} - x\right) = \cot x$
4. $\cot\left(\dfrac{\pi}{2} - x\right) = \tan x$	**5.** $\sec\left(\dfrac{\pi}{2} - x\right) = \csc x$	**6.** $\csc\left(\dfrac{\pi}{2} - x\right) = \sec x$

Identities continued

Sum and Difference Formulas

1. $\cos(u + v) = \cos u \cos v - \sin u \sin v$
2. $\cos(u - v) = \cos u \cos v + \sin u \sin v$
3. $\sin(u + v) = \sin u \cos v + \cos u \sin v$
4. $\sin(u - v) = \sin u \cos v - \cos u \sin v$
5. $\tan(u + v) = \dfrac{\tan u + \tan v}{1 - \tan u \tan v}$
6. $\tan(u - v) = \dfrac{\tan u - \tan v}{1 + \tan u \tan v}$

Double-Angle Formulas

1. $\sin 2u = 2 \sin u \cos u$
2. $\cos 2u = \cos^2 u - \sin^2 u$
   $= 2 \cos^2 u - 1$
   $= 1 - 2 \sin^2 u$
3. $\tan 2u = \dfrac{2 \tan u}{1 - \tan^2 u}$

Power Reduction Formulas

1. $\sin^2 u = \dfrac{1 - \cos 2u}{2}$
2. $\cos^2 u = \dfrac{1 + \cos 2u}{2}$
3. $\tan^2 u = \dfrac{1 - \cos 2u}{1 + \cos 2u}$

Half-Angle Formulas

1. $\sin \dfrac{v}{2} = \pm \sqrt{\dfrac{1 - \cos v}{2}}$
2. $\cos \dfrac{v}{2} = \pm \sqrt{\dfrac{1 + \cos v}{2}}$
3. $\tan \dfrac{v}{2} = \pm \sqrt{\dfrac{1 - \cos v}{1 + \cos v}}$

$= \dfrac{\sin v}{1 + \cos v}$

$= \dfrac{1 - \cos v}{\sin v}$

Note: Selecting the positive or negative square root in a half-angle formula depends on the quadrant in which $v/2$ lies.

Product-to-Sum Formulas

1. $\cos u \cos v = \tfrac{1}{2}[\cos(u + v) + \cos(u - v)]$
2. $\sin u \sin v = \tfrac{1}{2}[\cos(u - v) - \cos(u + v)]$
3. $\sin u \cos v = \tfrac{1}{2}[\sin(u + v) + \sin(u - v)]$
4. $\cos u \sin v = \tfrac{1}{2}[\sin(u + v) - \sin(u - v)]$

Sum-to-Product Formulas

1. $\cos a + \cos b = 2 \cos \dfrac{a + b}{2} \cos \dfrac{a - b}{2}$
2. $\cos a - \cos b = -2 \sin \dfrac{a + b}{2} \sin \dfrac{a - b}{2}$
3. $\sin a + \sin b = 2 \sin \dfrac{a + b}{2} \cos \dfrac{a - b}{2}$
4. $\sin a - \sin b = 2 \cos \dfrac{a + b}{2} \sin \dfrac{a - b}{2}$

Exercises 8.5

Basic Skills

In Exercises 1–8, express each product as a sum or difference with positive multiples of x. Use a graphing calculator to confirm each answer.

1. $2 \sin 3x \sin x$
2. $2 \sin 5x \sin 4x$
3. $2 \cos 2x \cos 6x$
4. $2 \cos 4x \cos x$

5. $\cos 3x \sin 4x$

6. $\cos \dfrac{x}{2} \sin \dfrac{3x}{2}$

7. $6 \sin 5x \cos(-2x)$

8. $7 \sin(-3x) \cos 7x$

In Exercises 9–16, express each sum or difference as a product with positive multiples of x. Use a graphing calculator to confirm each answer.

9. $\cos 4x + \cos 2x$

10. $\cos x + \cos 3x$

11. $\sin 5x + \sin(-3x)$

12. $\sin 4x + \sin 5x$

13. $\cos \dfrac{2x}{3} - \cos \dfrac{4x}{3}$

14. $\cos(-2x) - \cos 7x$

15. $\tfrac{1}{2}(\sin 3x - \sin 6x)$

16. $5 \sin 9x - 5 \sin 4x$

In Exercises 17–24, use the product-to-sum or sum-to-product formulas to find the exact value of each trigonometric expression.

17. $\sin 75° \sin 15°$

18. $\cos 75° \sin 15°$

19. $\cos 15° \sin 105°$

20. $\cos 75° \cos 105°$

21. $\cos 195° + \cos 105°$

22. $\sin 165° + \sin 105°$

23. $\sin 165° - \sin 75°$

24. $\cos 15° - \cos 195°$

In Exercises 25–36, show that each trigonometric equation is an identity.

25. $2 \cos 4x \sin x + \sin 3x = \sin 5x$

26. $2 \sin 2x \cos x - \sin x = \sin 3x$

27. $\tfrac{1}{2} \sec x (\sin x + \sin 3x) = \sin 2x$

28. $\cos 3x + 3 \cos x = 4\cos^3 x$

29. $2 \sin x \cos 3x + \sin 2x = \sin 4x$

30. $4 \sin 6x \cos^2 2x - 2 \sin 6x = \sin 10x + \sin 2x$

31. $\dfrac{\sin 7x + \sin 3x}{\cos 3x + \cos 7x} = \tan 5x$

32. $\dfrac{\sin 3x - \sin 5x}{\cos 5x - \cos 3x} = \cot 4x$

33. $\dfrac{\cos x - \cos 3x}{\sin^2 x} = 4 \cos x$

34. $\dfrac{\sin 3x + \sin x}{\sin x \cos x} = 4 \cos x$

35. $\sin 12x + \sin 4x - \sin 8x = 4 \cos 6x \cos 4x \sin 2x$

36. $\cos 10x - \cos 6x + \cos 4x = 1 - 4 \sin 5x \sin 2x \cos 3x$

In Exercises 37–48, find the general solution of each trigonometric equation. Use a graphing calculator to check each answer.

37. $\sin 3x + \sin x = 0$

38. $\cos 3x + \cos 7x = 0$

39. $\cos 6x = \cos 2x$

40. $\sin x = \sin 5x$

41. $\cos 2x - \cos 4x - 2 \sin 3x = 0$

42. $\sin x - \cos 2x = \sin 3x$

43. $\sin 2x + \sin 4x = \sin 3x$

44. $\cos 5x + \cos x = 2 \cos 2x$

45. $\dfrac{\cos 5x - \cos x}{2 \sin 3x} = 1$

46. $\dfrac{\cos 4x}{\sin 5x - \sin 3x} = -1$

47. $\dfrac{\sin 2x}{\cos 3x} + 2 \sin x = 0$

48. $\dfrac{\cos 3x + \cos x}{2 \cos^3 x - 2 \cos x} = 2$

In Exercises 49–54, use the sum-to-product formulas to find the x-intercepts of the graph of f in the interval $[0, 2\pi)$. Use a graphing calculator to confirm each answer.

49. $f(x) = \cos 2x + \cos x$

50. $f(x) = \cos x - \cos 2x$

51. $f(x) = \sin 4x - \sin 2x$

52. $f(x) = \sin 3x + \sin 2x$

53. $f(x) = \cos 4x - \cos 2x$

54. $f(x) = \sin 3x - \sin x$

55. In calculus, it is shown that the solutions of the equation $2 \sin 2x + \sin x = 0$ yield the x-coordinates of the relative maximum and minimum points for the graph of $f(x) = \cos 2x + \cos x$ (see Exercise 49). Find the coordinates of the relative maximum and minimum points in the interval $[0, 2\pi)$. Use a graphing calculator to confirm each answer.

56. In calculus, it is shown that the solutions of the equation $2 \sin 2x - \sin x = 0$ yield the x-coordinates of the relative maximum and minimum points for the graph of $f(x) = \cos x - \cos 2x$ (see Exercise 50). Find the coordinates of the relative maximum and minimum points in the interval $[0, 2\pi)$. Use a graphing calculator to confirm each answer.

Critical Thinking

57. Express $\sin 190° + \sin 70°$ as the sine of an acute angle.

58. Express $2 \cos 130° \cos 140°$ as the cosine of an acute angle.

59. Given that x and y are supplementary angles, use the sum-to-product formulas to evaluate each expression. What do you observe? Explain.

(a) $\cos x + \cos y$ (b) $\sin x - \sin y$

60. Given that x and y are supplementary angles, use the sum-to-product formulas to write each expression as trigonometric functions of x only. What do you observe? Explain.

(a) $\cos x - \cos y$ (b) $\sin x + \sin y$

61. The graph of $y = \cos 3x + \cos x$ (Figure 8.50) completes one cycle from 0 to 2π. Hence, the period of $y = \cos 3x + \cos x$ is 2π. Is the period of $y = \cos nx + \cos x$, where n is an integer, always 2π? Explain.

62. Given that $f(x) = \cos x$, use a sum-to-product formula to show that the difference quotient is equivalent to the right-hand side of this equation:

$$\frac{f(x+h) - f(x)}{h} = -\frac{\sin(h/2)}{h/2} \sin\left(x + \frac{h}{2}\right)$$

Chapter 8 Review

◆ **Questions for Writing or Group Discussion**

1. What is a *trigonometric identity*? Discuss the procedure for determining whether a trigonometric equation is an identity.

2. Explain why the equation $\cos x = \sqrt{1 - \sin^2 x}$ is *not* a trigonometric identity. Give another example of a trigonometric equation that is true for several values of the variable but is not an identify.

3. Discuss the procedure for solving a *conditional trigonometric equation*.

4. What is meant by the *general solution* of a trigonometric equation? How can *particular solutions* be developed from the general solution?

5. Discuss the procedure for using a calculator to solve a trigonometric equation. Illustrate with an example.

6. When is it necessary to check a trigonometric equation for extraneous roots?

7. Complete each *sum and difference formula*:

(a) $\sin(x + y) =$ (b) $\cos(x + y) =$ (c) $\tan(x + y) =$
(d) $\sin(x - y) =$ (e) $\cos(x - y) =$ (f) $\tan(x - y) =$

8. Discuss $\tan(x + y)$ if $\sin x = 0$ and if $\cos x = 0$.

9. Complete each *double-angle formula* and discuss any restriction on x:

(a) $\sin 2x =$ (b) $\cos 2x =$ (c) $\tan 2x =$

10. Use the double-angle formula for cosine to derive the *power reduction formulas*.

11. Complete each *half-angle formula* and discuss any restriction on x:

(a) $\sin \frac{x}{2} =$ (b) $\cos \frac{x}{2} =$ (c) $\tan \frac{x}{2} =$

12. How is it possible to determine whether to select the positive or negative square root in a half-angle formula?

13. Complete each *product-to-sum formula*:

(a) $\cos x \cos y =$ (b) $\sin x \sin y =$ (c) $\sin x \cos y =$

14. Discuss how the *sum-to-product formulas* can be used to find the *x*-intercepts of the graph of $y = \cos ax + \cos bx$, where *a* and *b* are positive integers.

15. Explain the procedure for writing $A \sin bt + B \cos bt$ in the form $a \sin(bt + c)$.

16. For the function *f* defined by $f(t) = A \sin bt + B \cos bt$, explain the advantage of replacing the expression $A \sin bt + B \cos bt$ with $a \sin(bt + c)$.

Review Exercises

In Exercises 1–14, write each expression as a single trigonometric function or a constant. Use a graphing calculator to confirm each answer.

1. $\tan 2x \cos 2x$
2. $\dfrac{\csc x}{\sec x}$
3. $\sec^2 \theta - \tan^2 \theta$
4. $\sqrt{1 - \sin^2 x}$
5. $\cos 5x \cos 4x - \sin 5x \sin 4x$
6. $\sin(\theta - y) \cos y + \cos(\theta - y) \sin y$
7. $\sin\left(\dfrac{\pi}{2} + x\right)$
8. $\tan(\pi + x)$
9. $2 \sin 3x \cos 3x$
10. $\cos^2 3y - \sin^2 3y$
11. $\dfrac{2 \tan 4x}{1 - \tan^2 4x}$
12. $\dfrac{\tan 5x - \tan 3x}{1 + \tan 5x \tan 3x}$
13. $\sqrt{\dfrac{1 - \cos 10x}{2}}$
14. $\dfrac{\sin 4x}{1 + \cos 4x}$

In Exercises 15–18, express each product as a sum and each sum as a product. Use a graphing calculator to confirm each answer.

15. $2 \cos x \sin 5x$
16. $\cos 4y \cos 3y$
17. $\cos 3x - \cos 2x$
18. $\sin 8x + \sin 4x$

In Exercises 19–22, find the exact value of each trigonometric expression.

19. $\sin \dfrac{\pi}{8}$
20. $\tan 75°$
21. $\sin 75° \cos 15°$
22. $\cos \dfrac{5\pi}{12} + \cos \dfrac{7\pi}{12}$

In Exercises 23–26, find the exact value of the indicated trigonometric function from the given information.

23. Given $\sin x = \tfrac{3}{5}$ for $0 < x < \pi/2$, find
    (a) $\sin\left(x + \dfrac{\pi}{4}\right)$
    (b) $\cos\left(x - \dfrac{\pi}{6}\right)$
    (c) $\sin 2x$
    (d) $\cos \dfrac{x}{2}$

24. Given $\tan x = -2$ for $\pi/2 < x < \pi$, find
    (a) $\tan\left(x - \dfrac{\pi}{4}\right)$
    (b) $\cos\left(x + \dfrac{\pi}{3}\right)$
    (c) $\cos 2x$
    (d) $\tan \dfrac{x}{2}$

25. Given $\sin x = -\tfrac{2}{3}$ and $\cos y = -\tfrac{12}{13}$, with *x* and *y* in quadrant III, find
    (a) $\cos(x - y)$
    (b) $\sin(x + y)$
    (c) $\tan 2y$
    (d) $\sin \dfrac{x}{2}$, $\pi < x < \dfrac{3\pi}{2}$

26. Given $\sec x = 3$ and $\csc y = -2$, with *x* and *y* in quadrant IV, find
    (a) $\sin(x - y)$
    (b) $\tan(x + y)$
    (c) $\sin 2y$
    (d) $\cos \dfrac{x}{2}$, $\dfrac{3\pi}{2} < x < 2\pi$

In Exercises 27 and 28, make the indicated trigonometric substitution and simplify the given expression. Assume that $0 < \theta < \pi/2$.

27. For $\sqrt{x^2 + 9}$, let $x = 3 \tan \theta$.
28. For $\dfrac{\sqrt{16 - x^2}}{x}$, let $x = 4 \sin \theta$.

In Exercises 29 and 30, use the power reduction formulas to express the even powers of the sine and cosine in terms of first powers of the cosine function. Then use a product-to-sum formula to eliminate products of trigonometric functions. Use a graphing calculator to confirm each answer.

29. $\sin^2 x \cos^4 x$
30. $\cos^6 x$

In Exercises 31–60, a trigonometric equation of the form $f(x) = g(x)$ is given. Use a graphing calculator to generate the graphs of $y = f(x)$ and $y = g(x)$. Compare the graphs, and then state whether the given equation seems to be a trigonometric identity. If the equation seems to be an identity, verify this fact algebraically.

31. $1 - 2 \sin^2 x = 2 \cos^2 x - 1$
32. $\tan x \csc x \cos x = 1$

33. $\dfrac{1 + \tan^2 x}{\csc^2 x} = \tan^2 x$

34. $\dfrac{\cos x}{\csc^2 x - 1} = \sin x \tan x$

35. $\csc x - \cot x = \dfrac{\sin x}{1 - \cos x}$

36. $\cot x + \tan x = \cot x \sec^2 x$

37. $\dfrac{1 + \sin x}{1 - \sin x} = \dfrac{\csc x + 1}{\csc x - 1}$

38. $\dfrac{1 + \sec x}{\csc x} = \sin x \tan x$

39. $\dfrac{1}{\cos x + 1} - \dfrac{1}{\cos x - 1} = 2 \csc^2 x$

40. $\dfrac{\sec^3 x + \cos^3 x}{\sec x + \cos x} = \sec^2 x + \sin^2 x$

41. $\dfrac{\cos^2 x}{1 + \sin x} = 1 - \cos x$

42. $1 + \tan x = \dfrac{2 \tan x + \sec^2 x}{1 + \tan x}$

43. $\dfrac{1 + \sin^2 x - \cos^2 x}{\sin 2x} = \tan x$

44. $\dfrac{1 + \tan^2 x}{1 - \tan^2 x} = \sec 2x$

45. $2 \csc 4x = \csc 2x \sec 2x$

46. $\tan x \sin 2x = 1 - \sin 2x$

47. $\dfrac{2 \sin x \cos x}{\cos^2 x - \sin^2 x} = \dfrac{2 \tan x}{1 - \tan^2 x}$

48. $\cot 2x + \csc 2x = \cot x$

49. $\dfrac{\sin 2x}{\sin x} - \dfrac{\cos 2x}{\cos x} = \sec x$

50. $\dfrac{\csc x - \sec x}{\csc x + \sec x} = \dfrac{\cos 2x}{1 + \sin 2x}$

51. $\tan\left(x + \dfrac{\pi}{4}\right) = \dfrac{1 + \tan x}{1 - \tan x}$

52. $\sin(x + \pi) + \sin(x - \pi) = 2 \sin x$

53. $\dfrac{\sin 3x}{\sec x} - \dfrac{\cos 3x}{\csc x} = \sin 2x$

54. $\tan 3x + \cot 6x = \csc 6x$

55. $2 \sin^2 \dfrac{x}{2} + \cos x = 1$

56. $\tan \dfrac{x}{2} = \cos x - \cot x$

57. $2 \sin 7x \sin 2x + \cos 9x = \cos 4x$

58. $\dfrac{\cos 2x + \cos 6x}{\cos^2 x - \sin^2 x} = 2 \cos 4x$

59. $\dfrac{\sin 7x + \sin 3x}{\sin 10x} = \dfrac{\sec 5x}{\sec 2x}$

60. $\dfrac{2 \sin 3x \cos 2x - \sin x}{\cos x - 2 \sin 3x \sin 2x} = \tan x$

In Exercises 61–86, find (a) the general solution of each equation, and (b) the solutions in the interval $[0, 2\pi)$. Use a graphing calculator to check each of your answers.

61. $2 \cos x + 1 = 0$
62. $\sqrt{3} \tan x + 1 = 0$
63. $4 \sin^2 x - 3 = 0$
64. $\sec^2 x - 2 = 0$
65. $\cot 2x + 1 = 0$
66. $\sin 3x - 1 = 0$
67. $\sin\left(3x - \dfrac{\pi}{6}\right) = 1$
68. $\tan\left(2x + \dfrac{\pi}{4}\right) = \sqrt{3}$
69. $2 \tan x \sin x - \tan x = 0$
70. $\sin x + \sin x \sin 3x = 0$
71. $\sec^2 x - \sec x = 2$
72. $2 \cos^2 x - 5 \cos x + 3 = 0$
73. $1 + \cos^2 x + \sin x = 0$
74. $\sec^2 x + \tan x = 1$
75. $\csc x = 1 - \cot x$
76. $\sin x + \cos x + 1 = 0$
77. $\sin 2x + \sin x = 0$
78. $\sin^2 x + \cos 2x = 0$
79. $\sin 3x \cos x - \cos 3x \sin x = 1$
80. $\tan 5x \sin 3x + \cos 3x = 0$
81. $\cos^2 \dfrac{x}{2} + \cos x = 2$
82. $2 \tan \dfrac{x}{2} = \csc x$
83. $\sin 5x + \sin 3x = 0$
84. $\cos 3x - \cos x = \sin x$
85. $\cos 4x + 2 \cos^2 2x = 2$
86. $\sin 2x = \sqrt{2} \sin\left(\dfrac{3\pi}{2} - x\right)$

In Exercises 87–96, use the formulas for solving a trigonometric equation (Section 8.2) to approximate to four significant digits all the solutions in the interval $[0, 2\pi)$. Use a graphing calculator to check each answer.

87. $5 \sin x + 2 = 0$
88. $2 \sec x - 5 = 0$
89. $\tan^2 x - 4 = 0$
90. $9 \sin^2 x - 4 = 0$
91. $\cot^2 x - 5 \cot x + 6 = 0$
92. $3 \sin^2 x + 4 \sin x - 4 = 0$
93. $3 \sin x + \sin^2 x = \cos^2 x$
94. $\tan^2 x + 2 = 3 \sec x$

95. $9 \cos^2 x - 9 \cos 2x = 1$

96. $6 \sin\left(\dfrac{\pi}{2} + x\right) + 4 \cos 2x = 5$

97. Express each function f in the form $f(x) = a \sin(bx + c)$. Then state the amplitude, period, and phase shift of f.
    (a) $f(x) = \sin 3x + \sqrt{3} \cos 3x$
    (b) $f(x) = 2 \sin 4\pi x - 2 \cos 4\pi x$

98. Determine the x-intercepts for the graph of each function. Use a graphing calculator to check the answer.
    (a) $y = \sin 2x + \sin x$
    (b) $y = \sin \dfrac{\pi x}{2} + \cos \pi x$

99. The voltage v (in volts) in an electrical circuit is a function of time t (in seconds) and is given by
    $$v(t) = 24 \sin 8\pi t + 32 \cos 8\pi t.$$
    (a) Determine a and c such that $v(t) = a \sin(8\pi t + c)$.
    (b) Find the approximate times t between 0 and 1 second for which $v = 10$ volts.

100. A mansarded roof has a cross-section in the shape of an isosceles trapezoid with dimensions as shown in the sketch.
    (a) Show that the cross-sectional area A of the mansarded roof is given by $A = 16 \sin \theta + 8 \sin 2\theta$.
    (b) Using calculus, it can be shown that the value of θ for maximum cross-sectional area is a solution of the equation $16 \cos \theta + 16 \cos 2\theta = 0$. Find the maximum cross-sectional area of this roof.

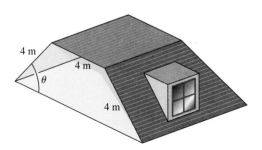

101. In physics it is shown that if a golf ball is driven from a tee with an initial velocity of v ft/s at an angle θ from the horizontal, as shown in the sketch, then the distance d (in feet) of the drive is given by
    $$d = \dfrac{1}{32} v^2 \sin \theta \cos \theta.$$
    If $v = 240$ ft/s, find values of θ (to the nearest tenth of a degree) that result in a 250-yard drive.

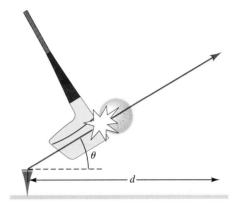

102. The number of hours N of daylight on a particular day d in Quebec, Canada, may be approximated by the equation
    $$N = 12 + 4 \sin\left[\dfrac{2\pi}{365}(d - 80)\right].$$
    Determine the days of the year that have exactly 15 hours of daylight.

A football is kicked with an initial velocity of 80 feet per second (ft/s) at an angle of 60° with the horizontal. Assuming that the ball is caught at the same height as it is kicked, determine (a) the parametric equations that describe the path of the ball and (b) the distance (in yards) downfield that the football is caught.

For the solution, see Example 6 in Section 9.7.

Additional Topics in Trigonometry

- 9.1 Law of Sines
- 9.2 Law of Cosines
- 9.3 Vectors
- 9.4 Trigonometric Form of Complex Numbers
- 9.5 Powers and Roots of Complex Numbers
- 9.6 Polar Coordinate System
- 9.7 Parametric Equations

9.1 Law of Sines

◆ Introductory Comments

In Section 7.7 we discussed the procedure for solving a right triangle. We now turn our attention to methods for solving an **oblique triangle**—a triangle that contains no right angle. In order to solve an oblique triangle, we need to be given three parts of the triangle:

1. two angles and any one side
   [AAS (angle-angle-side) or ASA (angle-side-angle)],
2. two sides and an angle opposite one of them (SSA),
3. two sides and their included angle (SAS), or
4. three sides (SSS).

In this section we derive a formula, called the *law of sines*, which enables us to solve an oblique triangle for cases 1 and 2. In Section 9.2, we derive a formula, called the *law of cosines*, which enables us to solve an oblique triangle for cases 3 and 4.

To derive the law of sines for the oblique triangle in Figure 9.1, we use a as the side opposite angle A, b as the side opposite angle B, and c as the side opposite angle C. By drawing altitude h from angle C to side c, as illustrated in Figure 9.2, we separate the oblique triangle into two right triangles.

Using the right triangle on the left in Figure 9.2, we have

$$\sin A = \frac{h}{b} \quad \text{or} \quad h = b \sin A.$$

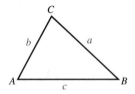

FIGURE 9.1
An oblique triangle

Using the right triangle on the right in the figure, we have

$$\sin B = \frac{h}{a} \quad \text{or} \quad h = a \sin B.$$

Equating these values of h gives us

$$a \sin B = b \sin A$$

$$\frac{a}{\sin A} = \frac{b}{\sin B} \qquad \text{Divide both sides by } \sin A \sin B$$

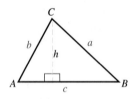

FIGURE 9.2
An oblique triangle with altitude h drawn to side c

Similarly, by drawing the altitude from angle B to side b, we can also show that

$$\frac{a}{\sin A} = \frac{c}{\sin C}.$$

In summary, we conclude that *in a given triangle the ratio of the length of a side to the sine of the angle opposite that side has the same constant value*. We refer to this fact as the **law of sines**.

Law of Sines

If A, B, and C are the angles of a triangle and a, b, and c are, respectively, the sides opposite these angles, then

$$\frac{a}{\sin A} = \frac{b}{\sin B} = \frac{c}{\sin C}.$$

Note: We have developed the law of sines for an oblique triangle with angle C acute ($C < 90°$). The law is also valid when C is obtuse ($90° < C < 180°$) or when C is a right angle ($C = 90°$).

We can use the law of sines to find the missing parts of an oblique triangle whenever one of the ratios

$$\frac{a}{\sin A}, \quad \frac{b}{\sin B}, \quad \text{or} \quad \frac{c}{\sin C}$$

is known and an additional side or angle is given. Hence, the law of sines is particularly useful when we are given either of two cases:

1. two angles and any one side (AAS or ASA)
2. two sides and an angle opposite one of them (SSA)

◆ Applying the Law of Sines

In the first example, we use the law of sines to solve an oblique triangle in which two angles and the side between them are known (case 1, ASA).

EXAMPLE 1 Solve the oblique triangle in Figure 9.3. Round x and y to three significant digits.

◆ **Solution**   To solve the oblique triangle means to find the missing parts, namely, angle θ and sides x and y. Since the sum of the interior angles in a triangle is $180°$, we have

$$\theta = 180° - (62° + 48°) = 70°.$$

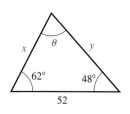

FIGURE 9.3

Now we apply the law of sines by matching each side with the sine of the angle opposite that side:

$$\frac{52}{\sin 70°} = \frac{x}{\sin 48°} = \frac{y}{\sin 62°}$$

Hence,

$$\frac{52}{\sin 70°} = \frac{x}{\sin 48°} \quad \text{and} \quad \frac{52}{\sin 70°} = \frac{y}{\sin 62°}$$

$$x = \frac{52 \sin 48°}{\sin 70°} \qquad\qquad y = \frac{52 \sin 62°}{\sin 70°}$$

$$x \approx 41.1 \qquad\qquad\qquad y \approx 48.9$$

PROBLEM 1 In any triangle, the longest side is opposite the largest angle and the shortest side is opposite the smallest angle. It is a good practice to use this geometric fact as a rough check of the solutions of an oblique triangle. Show that this fact holds for the oblique triangle in Figure 9.3.

In the next example, we solve an oblique triangle in which two sides and the angle opposite the larger one are known (case 2, SSA). To avoid *rounding errors* (errors that develop from using approximate values), we carry a few extra significant digits through the calculating process and then round the final answers to the desired accuracy. (For a discussion of rounding errors, see Appendix.)

EXAMPLE 2 Solve the oblique triangle shown in Figure 9.4. Round α and β to the nearest tenth of a degree and round x to three significant digits.

Solution To solve this oblique triangle, we must find angles α and β and side x. We apply the law of sines by matching each side with the sine of the angle opposite that side:

$$\frac{22}{\sin 32°} = \frac{18}{\sin \alpha} = \frac{x}{\sin \beta}$$

Hence,

$$\frac{22}{\sin 32°} = \frac{18}{\sin \alpha}$$

$$\sin \alpha = \frac{18 \sin 32°}{22}$$

Since the side opposite α is less than the side opposite 32° (that is, $18 < 22$), we know that $\alpha < 32°$. Using the inverse sine function, in conjunction with a calculator set in degree mode, we obtain

$$\alpha = \sin^{-1}\left(\frac{18 \sin 32°}{22}\right) \approx 25.694°.$$

Since the sum of the interior angles in a triangle is 180°, we have

$$\beta \approx 180° - (32° + 25.694°) = 122.306°.$$

Finally, returning to the law of sines with $\beta \approx 122.306°$, we find x as follows:

$$\frac{22}{\sin 32°} \approx \frac{x}{\sin 122.306°}$$

$$x \approx \frac{22 \sin 122.306°}{\sin 32°} \approx 35.089.$$

Rounding to the desired accuracy, we have $\alpha \approx 25.7°$, $\beta \approx 122.3°$, and $x \approx 35.1$.

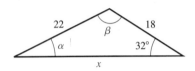

FIGURE 9.4

SECTION 9.1 ◆ *Law of Sines*   553

CAUTION Referring to Example 2, we cannot apply the Pythagorean theorem to find side x. The Pythagorean theorem holds for right triangles only and cannot be applied to an oblique triangle.

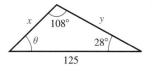

FIGURE 9.5

PROBLEM 2 Solve the oblique triangle shown in Figure 9.5. Round x and y to three significant digits. ◆

◆ The Ambiguous Case

For the oblique triangle in Example 2, we were given two sides and the angle opposite the *larger* of these sides. This condition determines a unique triangle. However, if the given angle is acute and the side opposite this angle is *smaller* than the other side, then, as illustrated in Figure 9.6, we can construct *two* oblique triangles (provided the data allows us to construct any triangle). Because of this ambiguity, we refer to this situation as the **ambiguous case**. *The ambiguous case occurs only when the given angle is acute and the side opposite this angle is smaller than the other given side.*

FIGURE 9.6
Two constructions are possible whenever $\theta < 90°$ and $a < b$.

EXAMPLE 3 Solve the oblique triangles shown in Figure 9.7. Round α and β to the nearest tenth of a degree and round x to three significant digits.

◀ Solution   To solve these oblique triangles, we must find angles α and β and side x. Applying the law of sines by matching each side with the sine of the angle opposite that side, we have

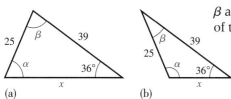

FIGURE 9.7

$$\frac{25}{\sin 36°} = \frac{39}{\sin \alpha} = \frac{x}{\sin \beta}.$$

Hence,

$$\frac{25}{\sin 36°} = \frac{39}{\sin \alpha}$$

$$\sin \alpha = \frac{39 \sin 36°}{25}$$

Since the sine function is positive in both quadrants I and II, we cannot tell from the sine of angle α whether angle α is acute or obtuse. If angle α is acute, then we can use the inverse sine function, in conjunction with a calculator set in degree mode, to obtain

$$\alpha = \sin^{-1}\left(\frac{39 \sin 36°}{25}\right) \approx 66.483°.$$

However, if α is obtuse, then by symmetry of the unit circle we have

$$\alpha \approx 180° - 66.483° = 113.517°,$$

as illustrated in Figure 9.8.

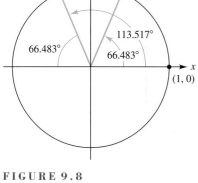

FIGURE 9.8
In an oblique triangle, $\sin \alpha = \dfrac{39 \sin 36°}{25}$ implies that $\alpha = 66.483°$ or $113.517°$.

(a) Using $\alpha = 66.483°$ and the fact that the sum of the interior angles in a triangle is 180°, we have

$$\beta \approx 180° - (36° + 66.483°) = 77.517°.$$

Finally, returning to the law of sines with $\beta \approx 77.517°$, we find x as follows:

$$\frac{25}{\sin 36°} \approx \frac{x}{\sin 77.517°}$$

$$x \approx \frac{25 \sin 77.517°}{\sin 36°} \approx 41.527$$

Rounding to the desired accuracy, we have $\alpha \approx 66.5°$, $\beta \approx 77.5°$, and $x \approx 41.5$.

(b) Using $\alpha \approx 113.517°$ and the fact that the sum of the interior angles in a triangle is 180°, we have

$$\beta \approx 180° - (36° + 113.517°) = 30.483°.$$

Finally, returning to the law of sines with $\beta \approx 30.483°$, we find x as follows:

$$\frac{25}{\sin 36°} \approx \frac{x}{\sin 30.483°}$$

$$x \approx \frac{25 \sin 30.483°}{\sin 36°} \approx 21.576$$

Rounding to the desired accuracy, we have $\alpha \approx 113.5°$, $\beta \approx 30.5°$, and $x \approx 21.6$. ◆

PROBLEM 3 Given that β is obtuse, solve the oblique triangle in Figure 9.9. Round α and β to the nearest tenth of a degree and round x to three significant digits. ◆

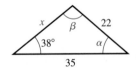

FIGURE 9.9

In order to solve an oblique triangle, we must be given sufficient data for the construction of a triangle. For example, consider the triangle in Figure 9.10. Using the law of sines to solve this oblique triangle for angle α, we obtain

$$\frac{12}{\sin 32°} = \frac{24}{\sin \alpha}$$

$$\sin \alpha = \frac{24 \sin 32°}{12} \approx 1.0598$$

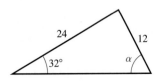

FIGURE 9.10
A triangle with erroneous given information

SECTION 9.1 ◆ *Law of Sines* 555

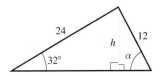

FIGURE 9.11
For this triangle to exist, we must have $h < 12$

However, 1.0598 is not in the range of the sine function. Hence, this last equation has no solution. To illustrate what is happening, we draw the altitude h as shown in Figure 9.11. For this triangle to exist, we must have $h < 12$. Using the right triangle on the left, we find

$$\sin 32° = \frac{h}{24}$$
$$h = 24 \sin 32° \approx 12.7$$

Since $h > 12$, we conclude that the data supplied is erroneous and no such triangle exists.

The following summarizes the possibilities when we are given two sides and an acute angle opposite one of them.

◆ **Constructions for SSA**

Given sides a and b and an acute angle α opposite side a, four possibilities exist for construction of a triangle.

1. If $a \geq b$, one triangle exists.

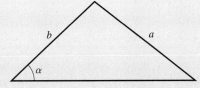

2. If $a < b$, two triangles exist whenever $a > h$.

3. If $a < b$, no triangle exists whenever $a < h$.

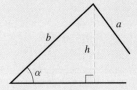

4. If $a < b$, one right triangle exists whenever $a = h$.

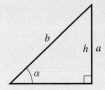

◆ **Application: Surveying**

The next example illustrates the procedure a surveyor might use to determine a distance that is inaccessible by direct measurement.

EXAMPLE 4 From the surveyor's notes shown in Figure 9.12, determine the width of the river.

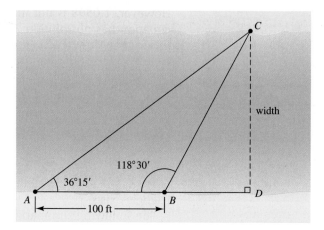

FIGURE 9.12

◆ **Solution**   Since the sum of the interior angles in a triangle is 180°, we have

$$\text{angle } ACB = 180° - (36°15' + 118°30') = 25°15'.$$

Working with oblique triangle ABC, we can determine the length AC from the law of sines:

$$\frac{AC}{\sin 118°30'} = \frac{100}{\sin 25°15'}$$

$$AC = \frac{100 \sin 118°30'}{\sin 25°15'}$$

Now, we can obtain the width of the river CD from right triangle ADC:

$$\sin 36°15' = \frac{\text{opp}}{\text{hyp}} = \frac{CD}{AC}$$

$CD = AC \sin 36°15'$ **Multiply both sides by AC**

$CD = \dfrac{100 \sin 118°30'}{\sin 25°15'} \sin 36°15'$ **Substitute for AC**

$CD \approx 121.8$ ft   **Evaluate**

Thus the width of the river is approximately 121.8 feet. ◆

PROBLEM 4   Find the width of the river CD in Figure 9.12 by using the law of sines on right triangle ADC. You should obtain the same value for CD as in Example 4. ◆

Exercises 9.1

Basic Skills

In Exercises 1–6, solve each oblique triangle. Round x and y to three significant digits.

1.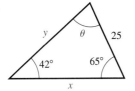

2.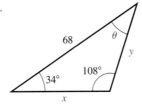

3.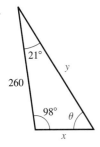

4.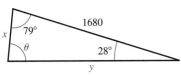

5.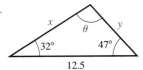

6.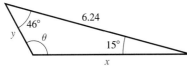

In Exercises 7–12, solve each oblique triangle. Round α and β to the nearest tenth of a degree and round x to three significant digits.

7.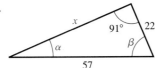

8.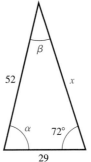

9.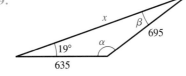

10.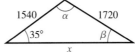

11.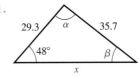

12.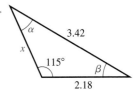

In Exercises 13–24, three parts of an oblique triangle are given. With the understanding that side a is opposite angle α, side b is opposite angle β, and side c is opposite angle θ, solve the oblique triangle with these three given parts. Round angles to the nearest tenth of a degree and sides to three significant digits. If the data allows for the construction of two triangles, find both solutions. If the data does not allow for the construction of a triangle, state that no such triangle exists.

13. $c = 55.4, \quad \alpha = 57°, \quad \beta = 86°$
14. $c = 2.74, \quad \alpha = 123°, \quad \theta = 28°$
15. $b = 35, \quad c = 22, \quad \beta = 62°$
16. $a = 0.189, \quad c = 0.297, \quad \theta = 59°$
17. $a = 680, \quad c = 990, \quad \alpha = 38°$
18. $b = 3.91, \quad c = 9.82, \quad \beta = 6°$
19. $b = 10.4, \quad c = 9.75, \quad \theta = 67°$

20. $a = 3.84$, $b = 4.67$, $\alpha = 52°$
21. $a = 280$, $\alpha = 126°$, $\beta = 58°$
22. $a = 151$, $b = 126$, $\beta = 108°$
23. $a = 11$, $b = 16$, $\alpha = 56°$
24. $a = 3.64$, $c = 1.94$, $\theta = 42°$

In Exercises 25–40, solve each applied problem. Round angles to the nearest tenth of a degree and all other measurements to three significant digits.

25. Two angles of a triangle are $62°15'$ and $68°30'$. If the shortest side is 34.8 m, find the length of the longest side.

26. Two angles of a triangle are $42°20'$ and $58°10'$. If the longest side is 234 ft, find the length of the shortest side.

27. From the surveyor's notes given in the sketch, find the length x of the pond.

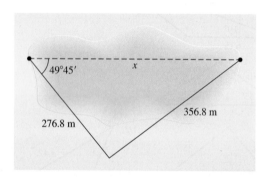

28. For the pulley system shown in the figure, find the distance x between the pulleys.

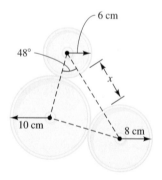

29. A slide at a playground is 22.5 ft long and is inclined $46.8°$ from the horizontal. The access ladder that reaches to the top of the slide is 18.2 ft long, as shown in the sketch. What acute angle θ does the ladder make with the horizontal?

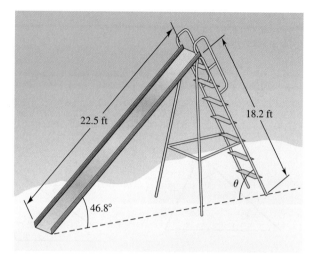

30. A piston and rod assembly has the dimensions shown in the sketch. Find two values of angle β when $\alpha = 18°$.

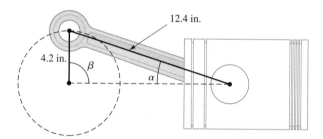

31. Find the amount of fencing needed to enclose the triangular piece of land shown in the figure given that angle C is (a) acute or (b) obtuse.

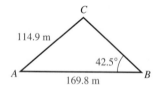

32. Three drill holes are located in a metal plate, as shown in the sketch. Determine the distance x given that angle θ is (a) acute or (b) obtuse.

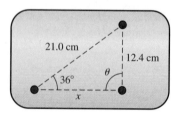

33. The angle of elevation from one edge of a ravine to the top of a tree on the opposite edge is $32.3°$. From a point

100 ft back from the edge of the ravine, the angle of elevation to the top of the same tree is 22.9°, as sketched in the figure. Find (a) the width of the ravine and (b) the height of the tree.

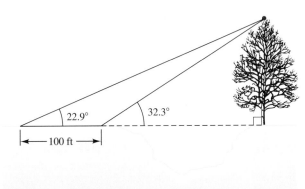

34. For the roof truss shown in the sketch, find the following measurements.
    (a) α (b) β (c) BD
    (d) θ (e) BC (f) CD

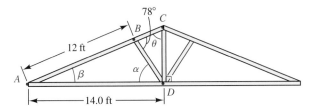

35. The length of a property line from point A to point B is impossible to measure directly. To find the distance AB indirectly, a surveyor selects a point C and finds the distances $BC = 865$ ft, $AC = 520$ ft, and angle $CAB = 38°20'$. What is the length of property line AB?

36. A person in a hot-air balloon 2500 ft above the ground notes that the angle of depression of an object on the ground is 32.6°. After ascending vertically for 15 minutes, the person finds that the angle of depression to the same object is 61.5°. What is the rate of ascent (in mi/h) of the balloon?

37. A flagpole is set vertically on a hill that is inclined at an angle of 7° to the horizontal. Find the height of the pole if the angle of elevation of the sun is 55° and the pole casts a shadow of length 26 ft directly (a) down the hill or (b) up the hill.

38. When the angle of elevation of the sun is 66°20′, a leaning telephone pole on level ground casts a shadow of length 10.6 ft. Find the length of the pole if it leans 12°30′ from the vertical (a) toward the sun or (b) away from sun.

39. The angles of elevation to the top of a tower are measured from two points on level ground that are 100 meters apart and are found to be 46.7° and 68.3°. Find the height of the tower if the two points are in the same vertical plane with the tower and are (a) on the same side of the tower or (b) on opposite sides of the tower.

40. The angles of depression to point A at the floor of a canyon are measured from two level points B and C on opposite sides of the canyon at the rim and are found to be 59.5° and 71.2°, respectively. Find the depth of the canyon if the three points are in the same vertical plane and the distance from B to C is 750 ft.

Critical Thinking

41. Suppose the interior angles of a triangle are A, B, and C, and the sides opposite these angles are a, b, and c, respectively. Furthermore, suppose that $b < a < c$. What does this imply about the measures of angles A, B, and C? Explain.

42. Given the oblique triangle in the sketch, find the fallacy in the following argument:

$$\frac{20}{\sin 30°} = \frac{20\sqrt{3}}{\sin \alpha} \quad \text{Apply law of sines}$$

$$\sin \alpha = \sqrt{3} \sin 30° \quad \text{Solve for } \sin \alpha$$

$$\sin \alpha = \frac{\sqrt{3}}{2} \quad \text{Evaluate } \sin 30° \text{ and multiply}$$

$$\alpha = \sin^{-1}\left(\frac{\sqrt{3}}{2}\right) = 60° \quad \text{Solve for } \alpha$$

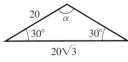

43. Apply the law of sines to the right triangle in the sketch and solve for $\sin \theta$. What familiar formula does this represent?

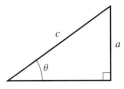

44. Apply the law of sines to an oblique triangle in which one angle is twice as large as another, as shown in the

sketch. Then apply a double-angle formula and express $\cos\theta$ in terms of a and b.

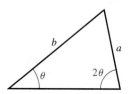

45. For the triangles in the sketch, use the law of sines to help show that

$$h = \frac{a \sin(\alpha - \beta)}{\cos\alpha \cos\beta}.$$

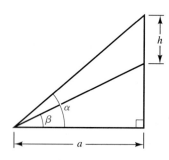

46. For the triangles in the figure, use the law of sines to help show that

$$h = \frac{a \sin\alpha \sin\beta}{\sin(\alpha - \beta)}.$$

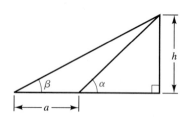

Suppose the interior angles of a triangle are A, B, and C, and the sides opposite these angles are a, b, and c, respectively. Mollweide's formula (after the German mathematician Karl Mollweide, 1774–1825) states that these six parts of a triangle are related by the formula

$$\frac{a-b}{c} = \frac{\sin\frac{A-B}{2}}{\cos\frac{C}{2}}.$$

We can use this formula to check the solutions of a triangle by comparing the ratio on the left-hand side to the ratio on the right-hand side. In Exercises 47–50, use a calculator to compare these ratios and state whether the given information appears to be the solution of a triangle.

47. $A = 34.4°$, $B = 115.7°$, $C = 29.9°$
$a = 3.61$, $b = 5.76$, $c = 3.19$

48. $A = 52.0°$, $B = 65.8°$, $C = 62.2°$
$a = 75.2$, $b = 87.0$, $c = 84.4$

49. $A = 19.7°$, $B = 53.7°$, $C = 106.6°$
$a = 2.508$, $b = 5.001$, $c = 7.134$

50. $A = 35.3°$, $B = 124.8°$, $C = 19.9°$
$a = 7005$, $b = 10{,}367$, $c = 5468$

In physics, a formula that mirrors the law of sines provides a means of calculating the direction taken by a refracted beam of light when it passes from one medium to a second medium. Referred to as Snell's law, this formula is given by

$$\frac{v_1}{\sin\alpha} = \frac{v_2}{\sin\beta}$$

where v_1 is the speed of light in the first medium, v_2 is the speed of light in the second medium, and angles α and β are the corresponding angles of refraction, as shown in the sketch. Use this formula in conjunction with a calculator to answer the questions in Exercises 51 and 52.

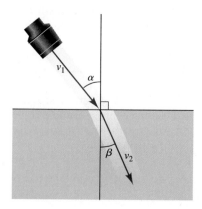

51. Given that the speed of light in air is $v_1 = 2.998 \times 10^{10}$ cm/s, $\alpha = 32.83°$, and $\beta = 21.45°$, find v_2, the speed of light in the second medium. Record your answer in scientific notation.

52. The speed of light in air is $v_1 = 2.998 \times 10^{10}$ cm/s and the speed of light in water is $v_2 = 2.254 \times 10^{10}$ cm/s. Given that $\alpha = 41.34°$, find angle β.

9.2 Law of Cosines

◆ Introductory Comments

As we stated in Section 9.1, in order to solve an oblique triangle, we must be given three parts of the triangle:

1. two angles and any one side (AAS or ASA),
2. two sides and an angle opposite one of them (SSA),
3. two sides and their included angle (SAS), or
4. three sides (SSS).

Cases 1 and 2 are solved by the law of sines (Section 9.1). Cases 3 and 4 cannot be solved by the law of sines because none of the ratios $a/\sin A$, $b/\sin B$, or $c/\sin C$ is known. For these cases we develop a new formula called the *law of cosines*.

Consider the oblique triangle in Figure 9.13. We use a as the side opposite angle A, b as the side opposite angle B, and c as the side opposite angle C. By drawing altitude h from angle C to side c, as illustrated in Figure 9.14, we separate the oblique triangle into two right triangles and separate side c into two parts, x and $c - x$.

Referring to Figure 9.14 and using the right triangle on the left, we have

$$\cos A = \frac{x}{b} \quad \text{or} \quad x = b \cos A.$$

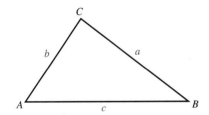

FIGURE 9.13
An oblique triangle

Also, by the Pythagorean theorem,

$$b^2 = h^2 + x^2 \quad \text{or} \quad h^2 = b^2 - x^2.$$

Referring to Figure 9.14 and using the right triangle on the right, we have

$$a^2 = h^2 + (c - x)^2 \quad \text{or} \quad h^2 = a^2 - (c - x)^2.$$

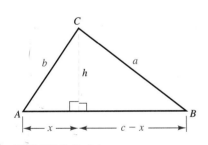

FIGURE 9.14
An oblique triangle with altitude h drawn to side c

Equating these values of h^2, we obtain

$$a^2 - (c - x)^2 = b^2 - x^2$$

$$a^2 = b^2 - x^2 + (c - x)^2 \qquad \text{Add } (c - x)^2 \text{ to both sides}$$

$$a^2 = b^2 - x^2 + c^2 - 2cx + x^2 \qquad \text{Expand}$$

$$a^2 = b^2 + c^2 - 2cx \qquad \text{Simplify}$$

$$a^2 = b^2 + c^2 - 2bc \cos A \qquad \text{Replace } x \text{ with } b \cos A$$

Similarly, by drawing the altitudes from angle B to side b and from angle A to side a, we can show that

$$b^2 = a^2 + c^2 - 2ac \cos B \quad \text{and} \quad c^2 = a^2 + b^2 - 2ab \cos C.$$

In summary, we conclude that *the square of the length of any side of a triangle is the sum of the squares of the lengths of the other two sides minus twice the product of the lengths of those two sides and the cosine of the angle between them.* We refer to this fact as the **law of cosines**.

◆ Law of Cosines

If A, B, and C are the angles of a triangle and a, b, and c are, respectively, the sides opposite these angles, then

$$a^2 = b^2 + c^2 - 2bc \cos A,$$
$$b^2 = a^2 + c^2 - 2ac \cos B,$$

and
$$c^2 = a^2 + b^2 - 2ab \cos C.$$

Note: We have developed the law of cosines for an oblique triangle with acute angles. The law is also valid when one of the angles is obtuse or when one of the angles is a right angle. Observe that if $C = 90°$, as shown in Figure 9.15, then we obtain

$$c^2 = a^2 + b^2 - 2ab \cos 90°$$
$$c^2 = a^2 + b^2 - 2ab(0)$$
$$c^2 = a^2 + b^2$$

Pythagorean theorem

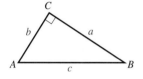

FIGURE 9.15
When $C = 90°$, the law of cosines yields the Pythagorean theorem, $c^2 = a^2 + b^2$.

◆ Applying the Law of Cosines

In the following example we use the law of cosines to solve an oblique triangle in which two sides and the included angle are given (case 3, SAS). When using the law of cosines, we can avoid rounding errors by carrying a few extra significant digits through the calculating process and then rounding the final answers to the desired accuracy.

EXAMPLE 1 Solve the oblique triangle shown in Fig. 9.16. Round α and β to the nearest tenth of a degree and round x to three significant digits.

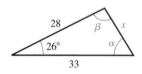

FIGURE 9.16

◆ **Solution** In this oblique triangle, we are given two sides and the angle between them (SAS). To solve the oblique triangle, we must find angles α and β and side x. To find side x, we apply the law of cosines:

side opposite 26°   sides adjacent to 26°

$$x^2 = 28^2 + 33^2 - 2(28)(33) \cos 26°$$
$$x = \sqrt{28^2 + 33^2 - 2(28)(33) \cos 26°} \approx 14.561$$

Since we now know the length of side x and the angle opposite that side (26°), we can use the law of sines and write

SECTION 9.2 ◆ *Law of Cosines* 563

$$\frac{14.561}{\sin 26°} \approx \frac{28}{\sin \alpha} = \frac{33}{\sin \beta}.$$

Hence, $\quad \sin \alpha \approx \dfrac{28 \sin 26°}{14.561} \quad$ and $\quad \sin \beta \approx \dfrac{33 \sin 26°}{14.561}.$

Remember that *we cannot tell from the sine of an angle whether the angle is acute or obtuse.* Thus, in determining the values of α and β, a good strategy is to find the smaller of these angles first (since the smaller angle *must* be acute). Since $28 < 33$, we begin by finding angle α (the angle opposite 28). Using the inverse sine function, in conjunction with a calculator set in degree mode, we have

$$\alpha \approx \sin^{-1}\left(\frac{28 \sin 26°}{14.561}\right) \approx 57.454°.$$

Now, since the sum of the interior angles of a triangle must be 180°, we conclude that

$$\beta \approx 180° - (26° + 57.454°) = 96.546°.$$

Rounding to the desired accuracy, we have $\alpha \approx 57.5°$, $\beta \approx 96.5°$, and $x \approx 14.6$. ◆

PROBLEM 1 From the second part of the law of sines in Example 1, we have

$$\sin \beta = \frac{33 \sin 26°}{14.561}$$

Show how to determine β from this equation. ◆

Since the cosine function is positive in quadrant I and negative in quadrant II, *we can determine from the cosine of the angle whether that angle is acute or obtuse,* that is,

$\quad \cos \theta > 0$ implies θ acute $\quad$ and $\quad \cos \theta < 0$ implies θ obtuse.

Thus, to solve an oblique triangle in which three sides are given (case 4, SSS), we first apply the law of cosines to find the angle opposite the *largest* side. We can then find the other two angles of the triangle—which must be acute—by using the law of sines. This procedure is illustrated in the next example.

EXAMPLE 2 Solve the oblique triangle shown in Figure 9.17. Round α, β, and θ to the nearest tenth of a degree.

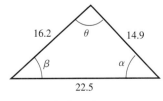

FIGURE 9.17

Solution To solve this oblique triangle, we must find angles α, β, and θ. Since θ is opposite the largest side, we apply the law of cosines to write

$$22.5^2 = \underbrace{16.2^2 + 14.9^2}_{\text{side opposite }\theta\ \ \text{sides adjacent to }\theta} - 2(16.2)(14.9)\cos\theta$$

(side opposite θ | sides adjacent to θ)

We now solve for $\cos\theta$:

$$22.5^2 = 16.2^2 + 14.9^2 - 2(16.2)(14.9)\cos\theta$$

$$22.5^2 - 16.2^2 - 14.9^2 = -2(16.2)(14.9)\cos\theta \qquad \text{Subtract } 16.2^2 \text{ and } 14.9^2 \text{ from both sides}$$

$$\cos\theta = \frac{22.5^2 - 16.2^2 - 14.9^2}{-2(16.2)(14.9)} \qquad \text{Divide both sides by } -2(16.2)(14.9)$$

If $\cos\theta > 0$, then θ is acute, and if $\cos\theta < 0$, then θ is obtuse. Using the inverse cosine function, in conjunction with a calculator set in degree mode, we find that θ is obtuse:

$$\theta = \cos^{-1}\left(\frac{22.5^2 - 16.2^2 - 14.9^2}{-2(16.2)(14.9)}\right) \approx 92.588°$$

Since we now know angle θ and the length of the side opposite θ ($= 22.5$), we can use the law of sines to write

$$\frac{22.5}{\sin 92.588°} \approx \frac{16.2}{\sin\alpha} = \frac{14.9}{\sin\beta}.$$

Hence,

$$\sin\alpha \approx \frac{16.2\ \sin 92.588°}{22.5} \quad \text{and} \quad \sin\beta \approx \frac{14.9\ \sin 92.588°}{22.5}.$$

Since both α and β must be acute angles (why?), we use the inverse sine function, in conjunction with a calculator set in degree mode, to obtain

$$\alpha \approx \sin^{-1}\left(\frac{16.2\ \sin 92.588°}{22.5}\right) \approx 45.994°$$

and

$$\beta \approx \sin^{-1}\left(\frac{14.9\ \sin 92.588°}{22.5}\right) \approx 41.418°.$$

Rounding to the desired accuracy, we have $\alpha \approx 46.0°$, $\beta \approx 41.4°$, and $\theta \approx 92.6°$. Note that the sum of these angles is 180.0°. ◆

In Example 2, we switched from the law of cosines to the simpler law of sines once we had found an additional piece of the triangle. As an alternative approach, we could have continued with the law of cosines, writing

$$16.2^2 = 22.5^2 + 14.9^2 - 2(22.5)(14.9) \cos \alpha$$

and

$$14.9^2 = 16.2^2 + 22.5^2 - 2(16.2)(22.5) \cos \beta$$

side opposite β sides adjacent to β

PROBLEM 2 Solve the equations

$$16.2^2 = 22.5^2 + 14.9^2 - 2(22.5)(14.9) \cos \alpha$$

and

$$14.9^2 = 16.2^2 + 22.5^2 - 2(16.2)(22.5) \cos \beta$$

for α and β, respectively. The results should agree with those in Example 2. ◆

As we discussed in Section 9.1, in order to solve an oblique triangle, we must be given sufficient data for the construction of a triangle. For example, consider the triangle in Figure 9.18. To find angle α, we apply the law of cosines and write

$$32^2 = 18^2 + 13^2 - 2(18)(13) \cos \alpha.$$

Hence,

$$\cos \alpha = \frac{32^2 - 18^2 - 13^2}{-2(18)(13)} \approx -1.1346.$$

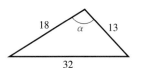

FIGURE 9.18
A triangle with erroneous given information

However, -1.1346 is not in the range of the cosine function. Hence, this last equation has no solution. In order to construct a triangle, *the sum of the lengths of any two sides must be greater than the length of the third side.* For the triangle in Figure 9.18, this is not the case:

$$18 + 13 < 32$$

Therefore, we conclude the data supplied is erroneous, and no such triangle exists.

◆ Area Formulas for Triangles

Consider the oblique triangle in Figure 9.19 in which two sides, a and b, and the angle between them, θ, are known. From elementary geometry we know that the area A of this triangle is half the product of base b and height h, that is,

$$A = \frac{bh}{2}.$$

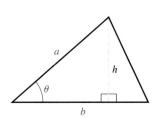

FIGURE 9.19
An oblique triangle in which two sides, a and b, and the angle between them, θ, are given

For the right triangle on the left in Figure 9.19, we have

$$\sin \theta = \frac{h}{a} \quad \text{or} \quad h = a \sin \theta.$$

Replacing h with $a \sin \theta$ in the basic area formula $A = bh/2$ gives us

$$A = \frac{ab \sin \theta}{2}.$$

◆ Area of a Triangle

> The area A of an oblique triangle in which two sides, a and b, and the angle between them, θ, are known is given by
>
> $$A = \frac{ab \sin \theta}{2}.$$

Note: We have developed this area formula for an oblique triangle in which θ is acute. The formula is also valid when θ is obtuse or when θ is a right angle.

EXAMPLE 3 Find the area of the triangle in Figure 9.17 in Example 2.

◆ **Solution** In Example 2, we showed that the angle between 16.2 and 14.9 is $\theta \approx 92.588°$. Thus, the area of this triangle is

$$A = \frac{ab \sin \theta}{2} = \frac{(16.2)(14.9) \sin 92.588°}{2} \approx 121 \text{ square units.} \quad ◆$$

PROBLEM 3 Find the area of the triangle in Figure 9.17 by using

(a) the sides 16.2 and 22.5 and the angle between them.

(b) the sides 22.5 and 14.9 and the angle between them.

Do you obtain the same answer as we did in Example 3? ◆

Around 100 A.D., Hero of Alexandria (Heron) derived a formula for the area of a triangle in which all three sides are known. The formula is now called **Hero's formula**. The proof of this formula is outlined in Exercise 52 of this section.

◆ Hero's Formula

> If a, b, and c are the sides of a triangle, then the area A of the triangle is given by
>
> $$A = \sqrt{s(s - a)(s - b)(s - c)}$$
>
> where s is half the perimeter (semiperimeter) of the triangle, that is, $s = \frac{1}{2}(a + b + c)$.

SECTION 9.2 ◆ Law of Cosines

EXAMPLE 4 Use Hero's formula to find the area of the triangle in Figure 9.17.

◆ **Solution** Since the sides of the triangle in Figure 9.17 are 16.2, 14.9, and 22.5, the semiperimeter s is

$$s = \tfrac{1}{2}(16.2 + 14.9 + 22.5) = 26.8.$$

Thus, by Hero's formula, the area A of this triangle is

$$\begin{aligned} A &= \sqrt{s(s-a)(s-b)(s-c)} \\ &= \sqrt{26.8(26.8 - 16.2)(26.8 - 14.9)(26.8 - 22.5)} \\ &= \sqrt{26.8(10.6)(11.9)(4.3)} \approx 121 \text{ square units} \end{aligned}$$

This area agrees with the answer we obtained in Example 3. ◆

PROBLEM 4 Use Hero's formula to find the area of a right triangle with sides 3, 4, and 5. Check the answer by using the basic area formula $A = bh/2$. ◆

◆ Application: Playing Golf

The law of cosines is useful for several applications. Our next example illustrates the procedure a golfer could use to determine the distance to the hole after the initial drive.

EXAMPLE 5 On a golf course, the distance from a tee to a hole is 355 yards. Suppose a golfer drives the ball 190 yd down the right side of the fairway at an angle of 20° from the center of the fairway. How far is the ball from the hole?

◆ **Solution** A sketch of the information is shown in Figure 9.20. Note that we are given two sides of a triangle (355 yd and 190 yd) and the angle between them (20°). Thus, to find the distance x to the hole, we apply the law of cosines:

$$x^2 = 355^2 + 190^2 - 2(355)(190)\cos 20°$$
$$x = \sqrt{355^2 + 190^2 - 2(355)(190)\cos 20°} \approx 188.04 \text{ yd}$$

Hence, the ball lies approximately 188 yards from the hole. ◆

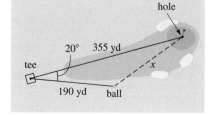

FIGURE 9.20
An oblique triangle in which two sides and the included angle are known

PROBLEM 5 Referring to Example 5, suppose that on the golfer's second shot the ball heads directly toward the hole but falls 40 yards short. How far is the ball from the tee after the second shot? ◆

Exercises 9.2

Basic Skills

In Exercises 1–10,
(a) solve each oblique triangle, and round α, β, and θ to the nearest tenth of a degree and x to three significant digits.
(b) determine the area of each triangle, and round the answer to three significant digits.

1.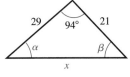

2.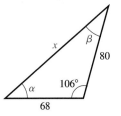

3.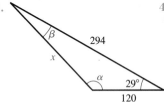

4.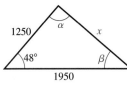

5.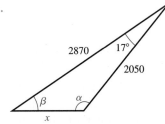

6.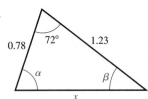

7.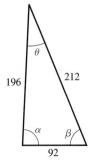

8.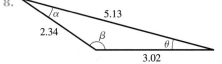

9.

10.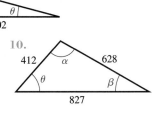

In Exercises 11–24, three parts of an oblique triangle are given. With the understanding that side a is opposite angle α, side b is opposite angle β, and side c is opposite angle θ, solve the oblique triangle with these three given parts. Round angles to the nearest tenth of a degree and sides to three significant digits. If the data does not allow for the construction of a triangle, state that no such triangle exists.

11. $a = 19$, $b = 21$, $\theta = 42°$
12. $a = 470$, $c = 622$, $\beta = 18°$
13. $b = 3.26$, $c = 5.11$, $\alpha = 112°$
14. $a = 49.2$, $b = 8.3$, $\theta = 142°$
15. $a = 325$, $c = 625$, $\beta = 48°$
16. $b = 3.26$, $c = 5.11$, $\alpha = 112°$
17. $a = 24$, $b = 19$, $c = 10$
18. $a = 0.56$, $b = 1.22$, $c = 0.87$
19. $a = 971$, $b = 1120$, $c = 793$
20. $a = 4200$, $b = 2100$, $c = 5300$
21. $a = 3.67$, $b = 2.27$, $c = 1.76$
22. $a = 12.5$, $b = 11.3$, $c = 19.1$
23. $a = 17$, $b = 32$, $c = 10$
24. $a = 1.72$, $b = 0.98$, $c = 0.62$

In Exercises 25–40, solve each applied problem. Round angles to the nearest tenth of a degree and all other measurements to three significant digits.

25. The sides of a triangle are 7 cm, 10 cm, and 12 cm. Find the measure of the largest angle.

26. The sides of a triangle are 7 ft, 8 ft, and 10 ft. Find the measure of the smallest angle.

27. Determine the length *x* of the brace for the traffic light shown in the figure.

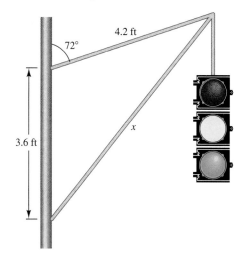

28. Two straight roads intersect with an angle of 64.5° between them. From the intersection point, it is 22.6 miles to a town at point *A* on one road and 15.8 miles to a town at *B* on the other road, as shown in the sketch. How far apart are the two towns?

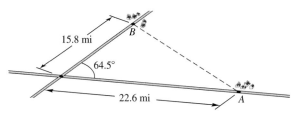

29. The distance between two points *A* and *B*, which are on opposite sides of a building, cannot be measured directly. To find the distance *AB*, a surveyor selects a point *C* and finds the distances *AC* and *BC* and angle *ACB* to be 132.4 ft, 156.9 ft, and 72°33′, respectively. What is the distance *AB*?

30. A triangular building lot has frontage of 237.5 ft on one street and frontage of 345.6 ft on another street. If the third side of the lot is 302.6 ft, find (a) the acute angle between the streets and (b) the area of the lot.

31. Two cyclists leave from the same point at the same time and travel along straight courses, which form an angle of 68° with each other. Both cyclists travel at a constant rate of speed—one at 18 mi/h and the other at 24 mi/h. What is the distance between the cyclists after 10 minutes of travel?

32. A baseball diamond is a square with 90-ft base paths. The pitcher's mound is 60 ft from home plate and lies directly between home plate and second base. How far is the pitcher's mound from first base?

33. Five drill holes are spaced equally around a circle whose radius is 8 cm, as shown in the figure. Find the distance between the centers of (a) holes *A* and *B* and (b) holes *A* and *D*.

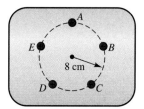

34. For the roof truss shown in the sketch, find the following measurements.
    (a) *BC* (b) α (c) β
    (d) θ   (e) *BD* (f) *CD*

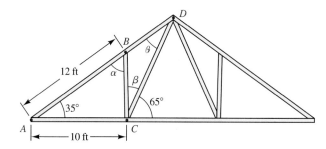

35. Two sides of a parallelogram are 62 m and 86 m, and the acute angle between the sides is 40°.
    (a) Find the length of each diagonal of the parallelogram.
    (b) Find the area of the parallelogram.

36. The diagonals of a parallelogram are 8 cm and 12 cm, and the obtuse angle between them is 130°.
    (a) Find the length of each side of the parallelogram.
    (b) Find the area of the parallelogram.

37. A 120-ft amateur radio tower is set vertically on a hill that is inclined at an angle of 9° to the horizontal. Find the length of a guy wire that is attached to the top of the tower and is anchored 85 ft from the tower (a) directly down the hill and (b) directly up the hill.

38. Four airports are located at points *A*, *B*, *C*, and *D* so that *B* is 220 mi directly east of *A*, *C* is 360 mi directly north of *B*, and *D* is 430 mi directly northwest of *C*. Find the distance between airports (a) at points *B* and *D* and (b) at points *A* and *D*.

39. From the surveyor's notes shown in the sketch that follows, find the length of the pond from *A* to *B*.

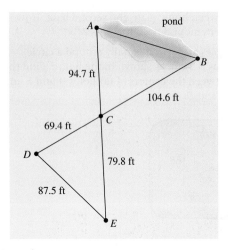

In Exercises 41 and 42, determine the number of acres in each tract of land. [*Hint: 43,560 square feet = 1 acre.*]

41.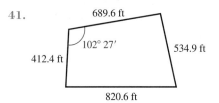

42.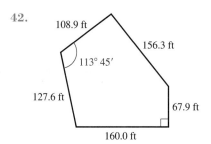

40. From the surveyor's notes shown in the sketch, find the distance from D to E.

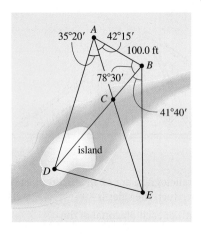

Critical Thinking

43. If a triangle does exist, is it always possible to solve the triangle when given any three of its six parts? Explain.

44. Suppose the sides of a triangle are a, b, and c, and the angle θ opposite side c is nearly 180°. Apply the law of cosines to this oblique triangle, and then give a geometric interpretation of the result.

45. The points $A(2, 3)$, $B(-2, 6)$, and $C(3, -6)$ are the vertices of an oblique triangle in the coordinate plane.
    (a) What is the measure of the largest interior angle of the triangle?
    (b) What is the area of the triangle?

46. The points $A(5, 0)$, $B(0, 0)$, and $C(x, y)$ are the vertices of an oblique triangle in the coordinate plane with point C in quadrant I, angle $ABC = 60°$, and side $BC = 8$ units. Use the law of cosines to find AC. Then use the distance formula (Section 2.1) to find the coordinates of point C.

For Exercises 47–50, refer to the oblique triangle shown in the figure.

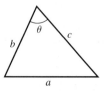

47. Find θ if $a^2 = b^2 + c^2 + bc$.
48. Find $\cos \theta$ if $b = c = 2a$.
49. Find b and c if $\theta = 120°$, $a = 3\sqrt{7}$, and $b = 2c$.
50. Find b and c if $\theta = 60°$, $a = 2\sqrt{7}$, and $b - c = 2$.

51. In a parallelogram, suppose the adjacent sides have lengths a and b and the diagonals have lengths x and y. Express the sum of the squares of the diagonals in terms of a and b.

52. The area A of the triangle in the figure is given by
$$A = \tfrac{1}{2}ab \sin \theta.$$

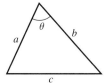

To prove Hero's formula, complete the following steps:

(a) Square both sides of the equation $A = \tfrac{1}{2}ab \sin \theta$ and show that
$$A^2 = \tfrac{1}{4}a^2b^2(1 + \cos \theta)(1 - \cos \theta).$$

(b) Solve the law of cosines, $c^2 = a^2 + b^2 - 2ab \cos \theta$, for $\cos \theta$. Then substitute for $\cos \theta$ in part (a) and show that
$$A^2 = \tfrac{1}{16}[(a + b)^2 - c^2][c^2 - (a - b)^2].$$

(c) Factor the right-hand side of the equation in part (b) and show that
$$A^2 = \tfrac{1}{16} 2s(2s - 2c)(2s - 2b)(2s - 2a)$$
where s is the semiperimeter of the triangle.

(d) Take the square root of each side of the equation in part (c) to obtain
$$A = \sqrt{s(s - a)(s - b)(s - c)}$$
which is Hero's formula.

53. In triangle ABC, line segment $\overline{CD}$ bisects angle C and separates side c into two parts, x and y, as shown in the sketch.

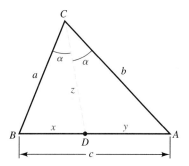

Given that $a = 4.27$, $b = 6.12$, and $c = 8.38$, solve the triangle as follows:

(a) Use the law of cosines to find angles A, B, and ACB, rounding each answer to the nearest hundredth of a degree.

(b) Use the results of part (a) and the law of sines to find the distances x, y, and z, rounding each answer to three significant digits.

(c) Find and compare the ratios $\dfrac{x}{y}$ and $\dfrac{a}{b}$, rounding each answer to three significant digits. What do you observe?

(d) Prove the observation you noted in part (c).

(e) Compute $\sqrt{ab - xy}$, rounding the answer to three significant digits. Compare this value to the value of z from part (b). What do you observe?

(f) Prove the observation you noted in part (e).

54. Using the observations you gathered in Exercise 53, find the distances x, y, and z without using the law of sines or cosines. Round each answer to three significant digits.

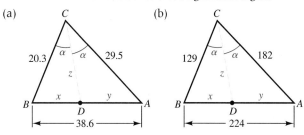

9.3 Vectors

◆ Introductory Comments

Many quantities cannot be described by *magnitude* alone. For example, if we state that a 100-lb force acts at the end of a diving board, we have not fully described the action taking place on the board. We need to know in what *direction* the force acts. As illustrated in Figure 9.21, a 100-lb force that

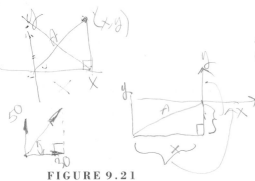

FIGURE 9.21
The action that takes place on a diving board is influenced by both the magnitude and the direction of the force that is applied.

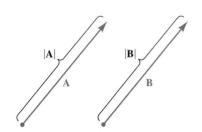

FIGURE 9.22
Vector **A** with magnitude |**A**| and vector **B** with magnitude |**B**|

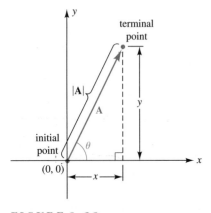

FIGURE 9.23
Vector **A** in standard position

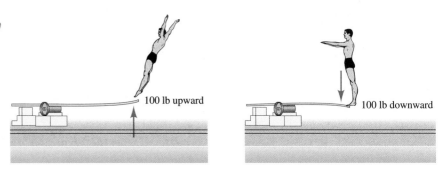

acts upward has a different effect on the board than a 100-lb force that acts downward.

Quantities that are described by both magnitude and direction are called **vector quantities**. Examples of vector quantities are force, velocity, electrical current, displacement, and so on. In this section, we introduce vector operations and show some applications of vectors.

◆ Vectors in Standard Position

In mathematics, we represent vectors with arrows. The length of the arrow (drawn to scale) is the *magnitude* of the vector, and the arrow's orientation in the plane tells us its *direction*. In this text, we use a boldface letter to denote a vector and absolute value bars around the boldface letter to denote its magnitude. Figure 9.22 illustrates vector **A** with magnitude |**A**| and vector **B** with magnitude |**B**|. These vectors have the same length (same magnitude) and are parallel (same direction). We say that vectors with the same magnitude and same direction are **equal**. Hence, we may write

$$\mathbf{A} = \mathbf{B}.$$

The arrowhead of a vector represents its **terminal point**, and the opposite end of the vector represents its **initial point**. In order to have some sense of direction, we usually place a vector in the coordinate plane with its initial point at the origin (0, 0). The terminal point of the vector then corresponds to some ordered pair (x, y) in the coordinate plane, as illustrated in Figure 9.23. This is called **standard position** of a vector, and the coordinates x and y are called the **components** of the vector in standard position. The notation

$$\mathbf{A} = \langle x, y \rangle$$

is called the *component form* of the vector. We refer to the vector with both initial point and terminal point at the origin as the **zero vector** and denote it by $\mathbf{0} = \langle 0, 0 \rangle$.

Associated with vector $\mathbf{A} = \langle x, y \rangle$ with $\mathbf{A} \neq \mathbf{0}$ is a **direction angle** θ, as shown in Figure 9.23. Usually, angle θ is measured counterclockwise from the positive x-axis with $0° \leq \theta < 360°$. From our discussion of trigonometry in Chapter 7, we know that x, y, $|\mathbf{A}|$, and θ are related as follows.

◆ **Relationships between x, y, $|\mathbf{A}|$, and θ**

If $|\mathbf{A}|$ is the magnitude and θ the direction angle of the nonzero vector $\mathbf{A} = \langle x, y \rangle$, then

1. $|\mathbf{A}| = \sqrt{x^2 + y^2}$

2. $\tan \theta = \dfrac{y}{x}$, provided that $x \neq 0$

3. $\cos \theta = \dfrac{x}{|\mathbf{A}|}$ or $x = |\mathbf{A}| \cos \theta$

4. $\sin \theta = \dfrac{y}{|\mathbf{A}|}$ or $y = |\mathbf{A}| \sin \theta$

Our first example illustrates the procedure for translating vector $\mathbf{A}$ into standard position and expressing it in component form $\mathbf{A} = \langle x, y \rangle$. If the component form of a vector is known, then we may find the magnitude and direction angle of the vector with the equations

$$|\mathbf{A}| = \sqrt{x^2 + y^2} \quad \text{and} \quad \tan \theta = \frac{y}{x}.$$

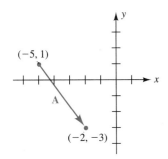

FIGURE 9.24

EXAMPLE 1 The vector $\mathbf{A}$ is shown in Figure 9.24.

(a) Translate $\mathbf{A}$ into standard position.

(b) Express $\mathbf{A}$ in component form.

(c) Find the magnitude of $\mathbf{A}$.

(d) Find the direction angle θ (with $0° \leq \theta < 360°$) for $\mathbf{A}$.

◆ **Solution**

(a) To be placed in standard position, vector $\mathbf{A}$ must have its initial point at the origin. Therefore, we must adjust its initial point $(-5, 1)$ by moving it five units right and one unit down. To preserve the magnitude and direction of this vector, we must make a similar adjustment to its terminal point. Moving the terminal point $(-2, -3)$ five units right and one unit down, we obtain $(3, -4)$, as illustrated in Figure 9.25.

(b) In Figure 9.25 we observe that the x-component of vector $\mathbf{A}$ is 3 and its y-component is -4. Hence, the component form of vector $\mathbf{A}$ is given by

$$\mathbf{A} = \langle 3, -4 \rangle.$$

(c) The magnitude of $\mathbf{A} = \langle 3, -4 \rangle$ is

$$|\mathbf{A}| = \sqrt{x^2 + y^2} = \sqrt{(3)^2 + (-4)^2} = 5.$$

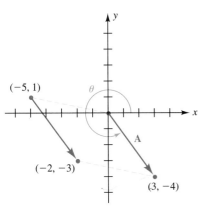

FIGURE 9.25

Vector $\mathbf{A}$ translated into standard position

(d) The direction angle θ for $\mathbf{A} = \langle 3, -4 \rangle$ must satisfy the equation

$$\tan \theta = \frac{y}{x} = \frac{-4}{3}.$$

Note that x is positive and y is negative. Hence, we know that the direction angle θ is a fourth-quadrant angle. Using the inverse tangent key on a calculator set in degree mode, we find that

$$\theta = \tan^{-1}\left(\frac{-4}{3}\right) \approx -53.1°.$$

Thus, we conclude that

$$\theta \approx 360° - 53.1° = 306.9°.$$

PROBLEM 1 Rework Example 1 if vector $\mathbf{A}$ has initial point $(-2, -3)$ and terminal point $(-5, 1)$.

If the magnitude and direction angle of a vector are known, then we may find the x- and y-components of the vector with the equations

$$x = |\mathbf{A}| \cos \theta \quad \text{and} \quad y = |\mathbf{A}| \sin \theta.$$

EXAMPLE 2 Given that vector $\mathbf{A}$ has a direction angle 158° and a magnitude 4, find its x- and y-components. Round each answer to three significant digits.

◆ **Solution** Vector $\mathbf{A}$ is shown in Figure 9.26. Its x-component is given by

$$x = |\mathbf{A}| \cos \theta = 4 \cos 158° \approx -3.71$$

and its y-component is given by

$$y = |\mathbf{A}| \sin \theta = 4 \sin 158° \approx 1.50.$$

Hence, $\mathbf{A} \approx \langle -3.71, 1.50 \rangle.$

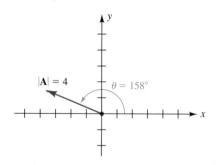

FIGURE 9.26
A vector with magnitude 4 and direction angle 158°

PROBLEM 2 Find the x- and y-components of vector $\mathbf{A}$ given that its direction angle is 212° and its magnitude is 8.

◆ **Vector Operations**

We discuss two fundamental vector operations in this text:

1. scalar multiplication and 2. vector addition.

We begin with *scalar multiplication*. The **product** of *scalar k* (a real number) and vector $\mathbf{A} = \langle x, y \rangle$ is the vector $k\mathbf{A}$ whose components are k times the corresponding components of vector $\mathbf{A}$, as shown in Figure 9.27.

SECTION 9.3 ◆ Vectors

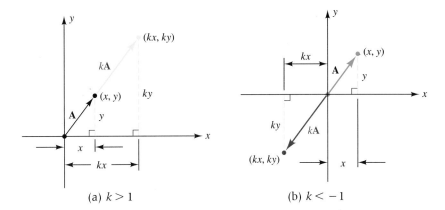

FIGURE 9.27
The components of $k\mathbf{A}$ are k times the corresponding components of $\mathbf{A}$.

(a) $k > 1$ (b) $k < -1$

◆ **Scalar Multiplication**

The **scalar multiplication** of real number k times vector $\mathbf{A} = \langle x, y \rangle$ is the vector $k\mathbf{A}$ defined by

$$k\mathbf{A} = \langle kx, ky \rangle.$$

In Figure 9.27, we note that the right triangles formed with the x-axis are *similar* (have corresponding angles equal and corresponding sides proportional). Hence, if k is positive, vector $\mathbf{A}$ and vector $k\mathbf{A}$ have the same direction angle (same direction), and if k is negative, vector $\mathbf{A}$ and vector $k\mathbf{A}$ have direction angles that differ by 180° (opposite directions). Also, by similar triangles, the magnitude of vector $k\mathbf{A}$ is $|k|$ times as large as the magnitude of vector $\mathbf{A}$, that is,

$$|k\mathbf{A}| = |k|\,|\mathbf{A}|$$

EXAMPLE 3 Given that $\mathbf{A} = \langle -3, 4 \rangle$, find each vector.

(a) $3\mathbf{A}$ **(b)** $-\tfrac{2}{3}\mathbf{A}$

◆ **Solution**

(a) Using scalar multiplication, we have

$$3\mathbf{A} = 3\langle -3, 4 \rangle = \langle 3(-3), 3(4) \rangle = \langle -9, 12 \rangle.$$

Vectors $\mathbf{A}$ and $3\mathbf{A}$ are sketched in Figure 9.28.

(b) Using scalar multiplication, we have

$$-\tfrac{2}{3}\mathbf{A} = -\tfrac{2}{3}\langle -3, 4 \rangle = \langle -\tfrac{2}{3}(-3), -\tfrac{2}{3}(4) \rangle = \langle 2, -\tfrac{8}{3} \rangle.$$

This vector is also sketched in Figure 9.28. ◆

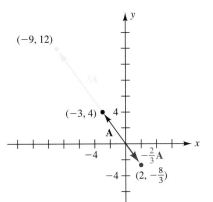

FIGURE 9.28
Relationship between vectors $\mathbf{A}$, $3\mathbf{A}$, and $-\tfrac{2}{3}\mathbf{A}$

The **negative** of $\mathbf{A} = \langle x, y \rangle$ is the vector $-\mathbf{A}$ defined by

$$-\mathbf{A} = -1\,\mathbf{A} = -1\langle x, y \rangle = \langle -x, -y \rangle.$$

PROBLEM 3 Given vector $\mathbf{A} = \langle -3, 4 \rangle$, find $-\mathbf{A}$.

The **sum** of two vectors $\mathbf{A} = \langle x_1, y_1 \rangle$ and $\mathbf{B} = \langle x_2, y_2 \rangle$ is the vector $\mathbf{A} + \mathbf{B}$ whose components are the sum of the corresponding components of vectors $\mathbf{A}$ and $\mathbf{B}$, as shown in Figure 9.29.

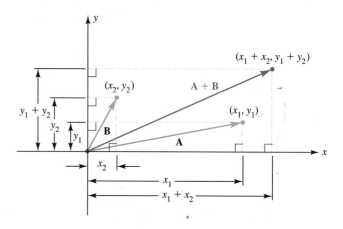

FIGURE 9.29

The components of $\mathbf{A} + \mathbf{B}$ are the sum of the corresponding components of $\mathbf{A}$ and $\mathbf{B}$.

Vector Addition

The **vector addition** of $\mathbf{A} = \langle x_1, y_1 \rangle$ and $\mathbf{B} = \langle x_2, y_2 \rangle$ is the vector $\mathbf{A} + \mathbf{B}$ defined as

$$\mathbf{A} + \mathbf{B} = \langle x_1 + x_2, y_1 + y_2 \rangle.$$

Note: In applied problems, we usually refer to the vector sum $\mathbf{A} + \mathbf{B}$ as the **resultant** of vectors $\mathbf{A}$ and $\mathbf{B}$.

Graphically, we may find the sum of vectors $\mathbf{A} = \langle x_1, y_1 \rangle$ and $\mathbf{B} = \langle x_2, y_2 \rangle$ by translating vector $\mathbf{A}$ such that its initial point coincides with the terminal point of vector $\mathbf{B}$, as shown in Figure 9.30 (or, alternatively, by translating vector $\mathbf{B}$ such that its initial point coincides with the terminal point of vector $\mathbf{A}$). Note in Figure 9.30 that the vector $\mathbf{A} + \mathbf{B}$ is the *diagonal of a parallelogram* having $\mathbf{A}$ and $\mathbf{B}$ as adjacent sides. We refer to this observation as the **parallelogram law** for vectors.

EXAMPLE 4 Given that $\mathbf{A} = \langle -2, 3 \rangle$ and $\mathbf{B} = \langle -4, -4 \rangle$, find each vector sum.

(a) $\mathbf{A} + \mathbf{B}$ (b) $-2\mathbf{A} + \frac{1}{4}\mathbf{B}$

Solution

(a) Using vector addition, we have

$$\begin{aligned}\mathbf{A} + \mathbf{B} &= \langle -2, 3 \rangle + \langle -4, -4 \rangle \\ &= \langle -2 + (-4), 3 + (-4) \rangle \\ &= \langle -6, -1 \rangle.\end{aligned}$$

A sketch of this vector is shown in Figure 9.31.

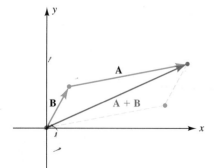

FIGURE 9.30

Vector $\mathbf{A} + \mathbf{B}$ is the diagonal of a parallelogram with adjacent sides $\mathbf{A}$ and $\mathbf{B}$.

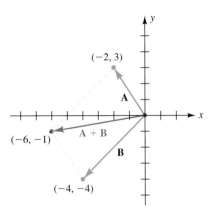

FIGURE 9.31

A + **B** is the diagonal of the parallelogram with sides **A** and **B**.

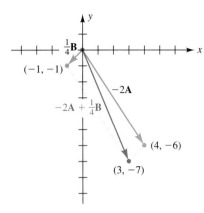

FIGURE 9.32

$-2\mathbf{A} + \frac{1}{4}\mathbf{B}$ is the diagonal of the parallelogram with sides $-2\mathbf{A}$ and $\frac{1}{4}\mathbf{B}$.

(b) Using scalar multiplication first, then vector addition, we have

$$-2\mathbf{A} + \tfrac{1}{4}\mathbf{B} = -2\langle -2, 3\rangle + \tfrac{1}{4}\langle -4, -4\rangle$$
$$= \langle 4, -6\rangle + \langle -1, -1\rangle$$
$$= \langle 3, -7\rangle$$

This vector is sketched in Figure 9.32. ◆

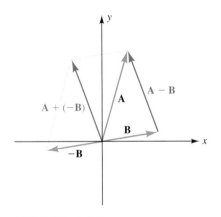

FIGURE 9.33

A − **B** may be described by the vector from the terminal point of **B** to the terminal point of **A**.

PROBLEM 4 Given vectors **A** = ⟨−2, 3⟩ and **B** = ⟨−4, −4⟩, find 2 **A** + 3 **B**. ◆

To subtract vector **B** = ⟨x_2, y_2⟩ from vector **A** = ⟨x_1, y_1⟩, we use vector addition to add the *negative* of vector **B** to vector **A**. Hence, we have the following rule for **vector subtraction**:

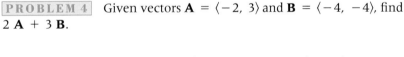

Since **A** + (−**B**) is the diagonal of a parallelogram formed by sides **A** and −**B**, we may describe **A** − **B** as the vector from the terminal point of **B** to the terminal point of **A**, as shown in Figure 9.33.

EXAMPLE 5 Referring to vectors **A** = ⟨−2, 3⟩ and **B** = ⟨−4, −4⟩, find **B** − **A**.

◆ **Solution** Using vector subtraction, we have

$$\mathbf{B} - \mathbf{A} = \langle -4, -4\rangle - \langle -2, 3\rangle$$
$$= \langle -4 - (-2), -4 - 3\rangle$$
$$= \langle -2, -7\rangle.$$

FIGURE 9.34

B − **A** may be described by the vector from the terminal point of **A** to the terminal point of **B**.

This vector subtraction is illustrated graphically in Figure 9.34. ◆

PROBLEM 5 Referring to the vectors **A** and **B** in Example 4, find **A** − **B**. ◆

Many of the properties of scalar multiplication and vector addition are similar to the properties of real numbers that we stated in Section 1.1. Next, we list nine fundamental properties of scalar multiplication and vector addition.

◆ **Fundamental Properties of Scalar Multiplication and Vector Addition**

If **A**, **B**, and **C** are vectors, **0** is the zero vector, and c and d are scalars, then

1. **A** + **B** = **B** + **A**   Commutative property of vector addition
2. (**A** + **B**) + **C** = **A** + (**B** + **C**)   Associative property of vector addition
3. **A** + **0** = **A**   Identity property of vector addition
4. **A** + (−**A**) = **0**   Inverse property of vector addition
5. (cd)**A** = $c(d$**A**$)$ Associative property of scalar multiplication
6. 1**A** = **A**   Identity property of scalar multiplication
7. 0**A** = **0**   Multiplicative property of zero
8. c(**A** + **B**) = c**A** + c**B**   Distributive property of a scalar over vector addition
9. $(c + d)$**A** = c**A** + d**A**   Distributive property of a vector over scalar addition

◆ **Unit Vectors**

Any vector whose magnitude is 1 is called a **unit vector**. Two special unit vectors, designated by **i** and **j**, are defined as

$$\mathbf{i} = \langle 1, 0 \rangle \quad \text{and} \quad \mathbf{j} = \langle 0, 1 \rangle.$$

We can use the vectors **i** and **j** to obtain an alternative way of denoting vector **A** = $\langle x, y \rangle$ as follows:

$$\mathbf{A} = \langle x, y \rangle = x \langle 1, 0 \rangle + y \langle 0, 1 \rangle$$
$$= x\mathbf{i} + y\mathbf{j}$$

Hence, **A** = $\langle x, y \rangle$ and **A** = $x\mathbf{i} + y\mathbf{j}$ are simply two different ways of stating that **A** is a vector with components x and y. We can use the unit vectors to describe any vector, such as

$$\langle 2, 3 \rangle = 2\mathbf{i} + 3\mathbf{j} \quad \text{and} \quad \langle 1, -6 \rangle = \mathbf{i} - 6\mathbf{j}$$

or even vectors that have at least one component equal to zero, such as

$$\langle -4, 0\rangle = -4\mathbf{i} + 0\mathbf{j} = -4\mathbf{i} \quad \text{and} \quad \langle 0, 0\rangle = 0\mathbf{i} + 0\mathbf{j} = \mathbf{0}.$$

We can also define scalar multiplication, vector addition, and vector subtraction in terms of the unit vectors $\mathbf{i}$ and $\mathbf{j}$:

1. $k(x\mathbf{i} + y\mathbf{j}) = (kx)\mathbf{i} + (ky)\mathbf{j}$
2. $(x_1\mathbf{i} + y_1\mathbf{j}) + (x_2\mathbf{i} + y_2\mathbf{j}) = (x_1 + x_2)\mathbf{i} + (y_1 + y_2)\mathbf{j}$
3. $(x_1\mathbf{i} + y_1\mathbf{j}) - (x_2\mathbf{i} + y_2\mathbf{j}) = (x_1 - x_2)\mathbf{i} + (y_1 - y_2)\mathbf{j}$

EXAMPLE 6 Given that $\mathbf{A} = 2\mathbf{i} - 3\mathbf{j}$ and $\mathbf{B} = \mathbf{i} - 6\mathbf{j}$, find $2\mathbf{A} - \mathbf{B}$.

◆ **Solution** Using scalar multiplication first, then vector subtraction, we obtain

$$\begin{aligned}
2\mathbf{A} - \mathbf{B} &= 2(2\mathbf{i} - 3\mathbf{j}) - (\mathbf{i} - 6\mathbf{j}) \\
&= (4\mathbf{i} - 6\mathbf{j}) - (\mathbf{i} - 6\mathbf{j}) \\
&= (4 - 1)\mathbf{i} + [-6 - (-6)]\mathbf{j} \\
&= 3\mathbf{i} + 0\mathbf{j} \quad \text{or} \quad \langle 3, 0\rangle.
\end{aligned}$$

◆

PROBLEM 6 Given the vectors $\mathbf{A}$ and $\mathbf{B}$ in Example 6, find $-\mathbf{A} + 2\mathbf{B}$. ◆

◆ **Application: Forces Acting on an Object**

Many applied problems in physics and mechanics deal with a *force* or a *group of forces* acting on a particular object. To analyze such problems, we let a vector represent each force and form a *vector diagram* in the coordinate plane.

EXAMPLE 7 A dogsled is pulled with a force of 100 lb at an angle of 30° with the horizontal.

(a) Find the force that pulls the sled horizontally.

(b) Find the force that lifts the sled vertically.

◆ **Solution** We begin by letting $\mathbf{F}$ be a vector that represents this force and drawing a vector diagram in the coordinate plane, as shown in Figure 9.35. Note that the magnitude of $\mathbf{F}$ is 100 and the direction angle for $\mathbf{F}$ is 30°.

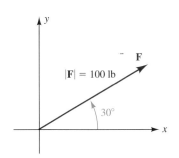

FIGURE 9.35

A vector diagram showing a force of 100 lb applied at an angle of 30°

(a) The force that pulls the sled horizontally is the *x*-component of the vector **F**. Using the fact that $x = |\mathbf{F}|\cos\theta$, we have

$$x = 100 \cos 30° \approx 86.6 \text{ lb}.$$

Hence, the force tending to pull the sled horizontally is 86.6 lb.

(b) The force that lifts the sled vertically is the *y*-component of the vector **F**. Using the fact that $y = |\mathbf{F}|\sin\theta$, we have

$$y = 100 \sin 30° = 50 \text{ lb}.$$

Hence, the force tending to lift the sled vertically is 50 lb. ◆

PROBLEM 7 If the sled in Example 7 weighs 75 lb, determine the magnitude of the force that must be applied at an angle of 30° with the horizontal in order to just lift the sled from the surface. ◆

A **force system** is a group of forces acting on a particular object. For our discussion, a force system is *concurrent* and *coplanar* (that is, the lines of action of all forces pass through a common point and lie in the same plane). A single force that can replace a force system and has the same physical effect upon the object as the system it replaces is called the **resultant force**. To determine the resultant force, we find the vector sum of all the forces in the system.

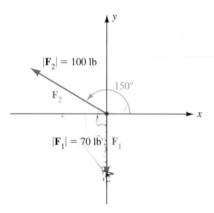

FIGURE 9.36

EXAMPLE 8 For the force system in Figure 9.36, find the magnitude and direction angle of the resultant force.

◆ **Solution** The resultant force **F** is the vector sum of $\mathbf{F}_1$ and $\mathbf{F}_2$. To find this sum, we express $\mathbf{F}_1$ and $\mathbf{F}_2$ in component form and then apply the rule for vector addition:

$$\begin{aligned}\mathbf{F} &= \mathbf{F}_1 + \mathbf{F}_2 \\ &= \langle 70 \cos 270°, 70 \sin 270°\rangle + \langle 100 \cos 150°, 100 \sin 150°\rangle \\ &= \langle 70 \cos 270° + 100 \cos 150°, 70 \sin 270° + 100 \sin 150°\rangle \\ &\approx \langle -86.6, -20\rangle\end{aligned}$$

The resultant force **F** is shown in Figure 9.37. Note that it is the diagonal of the parallelogram with sides $\mathbf{F}_1$ and $\mathbf{F}_2$. The magnitude of **F** is

$$|\mathbf{F}| \approx \sqrt{(-86.6)^2 + (-20)^2} \approx 89 \text{ lb}.$$

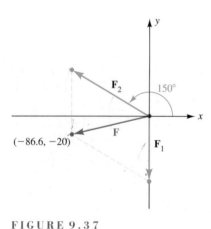

FIGURE 9.37

The resultant force **F** is the diagonal of the parallelogram with sides $\mathbf{F}_1$ and $\mathbf{F}_2$.

Note that both the *x*- and *y*-components of **F** are negative. Hence, we know that the direction angle θ for **F** is between 180° and 270° and satisfies the equation

$$\tan \theta = \frac{-20}{-86.6} = \frac{20}{86.6}$$

Hence, $\theta \approx 180° + \tan^{-1}\left(\dfrac{20}{86.6}\right) \approx 193°.$ ◆

If an object is in **equilibrium** (at rest), *the vector sum of all forces acting on the object must equal zero.* For equilibrium to occur in the force system of Example 8, we must add an additional force of the same magnitude as the resultant force **F** but in the opposite direction.

PROBLEM 8 Referring to the force system of Example 8, find an additional force $\mathbf{F}_3$ that produces equilibrium. ◆

◆ **Application: Displacement to a Landing Strip**

The **displacement** between two points is the length and direction of the straight line between the two points. Since displacement is described by both length (magnitude) and direction, it is considered a vector quantity. When displacement is discussed in navigation problems, the direction is usually given with respect to a compass reading, called a **bearing**. A bearing is expressed in degrees east (E) or west (W) of the north (N) or south (S) direction. Referring to Figure 9.38, the direction from point *A* to point *B* is given by the bearing S-30°-E, and the direction from point *A* to point *C* is given by the bearing N-75°-W. The direction North (N) or South (S) is always given as the first reference of a bearing.

FIGURE 9.38

Bearings are measured from the North or South axis.

EXAMPLE 9 The displacement from an airport to an old landing strip is known to be 200 miles S-85°-W. A small aircraft located 260 miles N-88°-W of the airport has engine trouble and needs to reach the old landing strip. What is the airplane's displacement to the landing strip?

◆ **Solution** We begin by drawing the given displacements as vectors $\mathbf{D}_1$ and $\mathbf{D}_2$ (see the axes on the right in Figure 9.39).

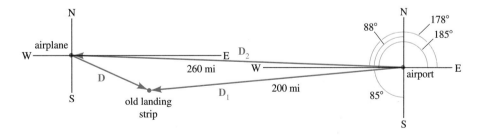

FIGURE 9.39

Graphical display of displacements $\mathbf{D}_1$, $\mathbf{D}_2$, and $\mathbf{D}$

Note in Figure 9.39 that the airplane's displacement $\mathbf{D}$ to the old landing strip is the vector from the terminal point of $\mathbf{D}_2$ to the terminal point of $\mathbf{D}_1$. Hence,

$$\mathbf{D} = \mathbf{D}_1 - \mathbf{D}_2.$$

To find this difference, we express $\mathbf{D}_1$ and $\mathbf{D}_2$ in component form and then apply the rule for vector subtraction:

$$\begin{aligned}\mathbf{D} &= \mathbf{D}_1 - \mathbf{D}_2 \\ &= \langle 200\cos 185°, 200\sin 185°\rangle - \langle 260\cos 178°, 260\sin 178°\rangle \\ &= \langle 200\cos 185° - 260\cos 178°, 200\sin 185° - 260\sin 178°\rangle \\ &\approx \langle 60.6, -26.5\rangle\end{aligned}$$

Hence, the magnitude of $\mathbf{D}$ is

$$|\mathbf{D}| = \sqrt{(60.6)^2 + (-26.5)^2} \approx 66 \text{ mi.}$$

The direction angle θ for $\mathbf{D} = \langle 60.6, -26.5\rangle$ is a fourth-quadrant angle. Hence,

$$\theta \approx \tan^{-1}\left(\frac{-26.5}{60.6}\right) \approx -23.6°.$$

Measuring from the South axis, we obtain $90° - 23.6° = 66.4°$. Therefore, the airplane should fly 66 miles S-66.4°-E in order to reach the old landing strip. ◆

An alternative approach to solving Example 9 is to use the law of cosines and law of sines. Observe that the vectors in Figure 9.39 form an oblique triangle with sides 200 miles and 260 miles. The angle between these sides is $180° - (88° + 85°) = 7°$. Since we know two sides and the angle between them (SAS), we can apply the law of cosines to find $|\mathbf{D}|$, and the law of sines to find the angle θ between $\mathbf{D}$ and $\mathbf{D}_2$. Note that the angle between $\mathbf{D}_2$ and the West axis on the right in Figure 9.39 is 2°. Thus, the angle between $\mathbf{D}_2$ and the East axis on the left is also 2°. Hence, the bearing is S-$[90° - (\theta + 2°)]$-E.

PROBLEM 9 Referring to Figure 9.39, use the law of cosines and law of sines to find the airplane's displacement to the landing strip. ◆

Exercises 9.3

Basic Skills

In Exercises 1–8, translate the vector into standard position and express it in component form.

1.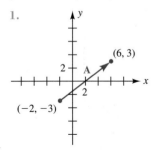

2.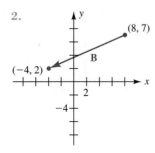

3.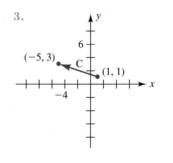

4.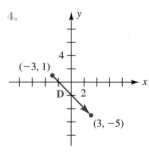

SECTION 9.3 ◆ Exercises

5.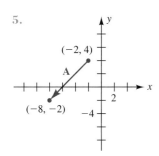
6.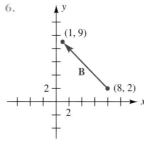
7.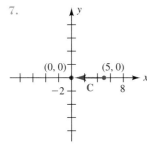
8.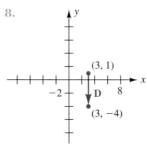

9. Determine the magnitude of vectors **A** and **C** in Exercises 1 and 3.
10. Determine the magnitude of vectors **B** and **D** in Exercises 2 and 4.
11. Determine the direction angle θ of vectors **A** and **C** in Exercises 5 and 7.
12. Determine the direction angle θ of vectors **B** and **D** in Exercises 6 and 8.

In Exercises 13–20, determine the magnitude and direction angle θ of the given vector.

13. $\mathbf{A} = \langle 5, -12 \rangle$
14. $\mathbf{B} = \langle 15, 8 \rangle$
15. $\mathbf{C} = \langle -6, -9 \rangle$
16. $\mathbf{D} = \langle -25, 35 \rangle$
17. $\mathbf{A} = \langle \frac{1}{2}, \frac{1}{4} \rangle$
18. $\mathbf{B} = \langle -\frac{2}{9}, \frac{4}{3} \rangle$
19. $\mathbf{C} = \langle -1, \sqrt{7} \rangle$
20. $\mathbf{D} = \langle -\sqrt{5}, -2 \rangle$

In Exercises 21–24, determine the magnitude and direction angle θ of each vector. Round the magnitude to three significant digits and round the direction angle to the nearest tenth of a degree.

21. $\mathbf{A} = \langle -1.25, 0.76 \rangle$
22. $\mathbf{B} = \langle 56.5, -82.7 \rangle$
23. $\mathbf{C} = \langle 0.356, 1.98 \rangle$
24. $\mathbf{D} = \langle -123.7, -987.2 \rangle$

In Exercises 25–32, determine the x- and y-components of a vector whose magnitude and direction angle θ are given.

25. $|\mathbf{A}| = 62.0, \theta = 45°$
26. $|\mathbf{B}| = 58, \theta = 30°$
27. $|\mathbf{C}| = 84, \theta = 210°$
28. $|\mathbf{F}| = 6\sqrt{2}, \theta = 315°$
29. $|\mathbf{A}| = \sqrt{2}, \theta = 225°$
30. $|\mathbf{B}| = 18, \theta = 120°$
31. $|\mathbf{C}| = \frac{9}{2}, \theta = 270°$
32. $|\mathbf{F}| = 7, \theta = 180°$

In Exercises 33–36, determine the x- and y-components of a vector whose magnitude and direction angle θ are given. Round the x- and y-components to three significant digits.

33. $|\mathbf{A}| = 5.61, \theta = 125.2°$
34. $|\mathbf{B}| = 11.2, \theta = 232.4°$
35. $|\mathbf{C}| = 79.6, \theta = 342.6°$
36. $|\mathbf{D}| = 0.758, \theta = 21.9°$

*In Exercises 37–52, perform the indicated vector operations given the vectors **A**, **B**, **C**, and **D**:*

$$\mathbf{A} = \langle 2, 5 \rangle \qquad \mathbf{B} = \langle -1, 6 \rangle$$
$$\mathbf{C} = -8\mathbf{i} - 2\mathbf{j} \qquad \mathbf{D} = 6\mathbf{i} - 3\mathbf{j}$$

Express each answer in component form.

37. $2\mathbf{A}$
38. $-\mathbf{B}$
39. $-\frac{1}{2}\mathbf{C}$
40. $\frac{8}{3}\mathbf{D}$
41. $\mathbf{A} + \mathbf{B}$
42. $\mathbf{C} + \mathbf{D}$
43. $\mathbf{A} - \mathbf{C}$
44. $\mathbf{D} - \mathbf{B}$
45. $4\mathbf{A} - 3\mathbf{C}$
46. $-\mathbf{A} + 3\mathbf{B}$
47. $\frac{1}{3}\mathbf{D} + \frac{1}{4}\mathbf{C}$
48. $\frac{9}{2}\mathbf{C} - 9\mathbf{B}$
49. $\mathbf{A} + (\mathbf{C} + \mathbf{D})$
50. $2\mathbf{A} - (\mathbf{B} - \frac{3}{2}\mathbf{C})$
51. $|\mathbf{A}|(\mathbf{D} + 4\mathbf{C})$
52. $|\mathbf{A} + \mathbf{B}|(\mathbf{C} + \mathbf{D})$

In Exercises 53–58, illustrate graphically the indicated operation.

53. the scalar multiplication in Exercise 39
54. the scalar multiplication in Exercise 40
55. the vector sum in Exercise 41
56. the vector sum in Exercise 42
57. the vector difference in Exercise 43
58. the vector difference in Exercise 44

For the force systems in Exercises 59–64, find the magnitude and direction angle θ of the resultant force. Round the magnitude to three significant digits and round θ to the nearest tenth of a degree.

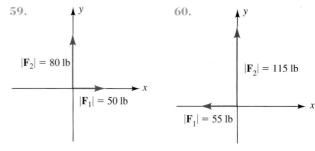

59. 60.

584 CHAPTER 9 ◆ *Additional Topics in Trigonometry*

61.
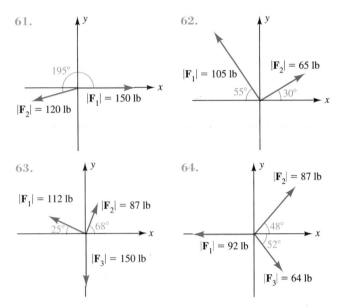

62.

63.

64.

65. A rocket lifts off at an angle of 45° with a constant velocity of 1200 mi/h, as shown in the sketch.
   (a) At what rate does the rocket rise vertically?
   (b) At what rate does the rocket move horizontally?

66. A 200-lb force is exerted horizontally to a crate resting on an inclined plane, as shown in the figure. If the angle of inclination of the plane is 30°, find the force that tends to push the crate up the plane.

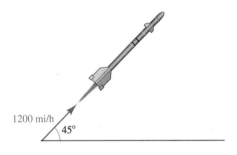

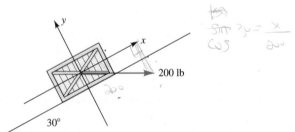

67. For the force system in Exercise 59, find an additional force $\mathbf{F}_3$ that produces equilibrium.

68. For the force system in Exercise 62, find an additional force $\mathbf{F}_3$ that produces equilibrium.

69. A man rows at a rate of 5.2 mi/h in still water. He wants to row directly across a river, so he keeps the boat moving perpendicular to the flow of the current. The river is 190 ft wide and flows at the rate of 2.3 mi/h, as shown in the accompanying sketch.
   (a) Find the magnitude and direction angle θ of the resultant velocity of the boat.
   (b) Find the landing distance downstream from the boat's launching point.
   (c) In what direction should the rower head the boat in order to land across the river, directly opposite the starting point?

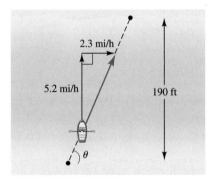

70. Two tugboats are towing an oil barge along a canal. One tugboat exerts a force of 5500 lb on its cable and the other tugboat exerts a force of 4200 lb on its cable, as shown in the figure. The angle between the cables is 17°.
   (a) Find the magnitude of the resultant force.
   (b) Find the angle between the resultant force and the cable with the force of 5500 lb.

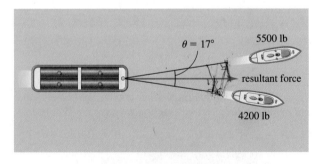

71. A small aircraft leaves airport A and flies 132 miles S-12°-E. It then changes course and flies 214 miles S-69°-E to land at airport B. What is the displacement from airport A to airport B?

72. A sailboat leaves island A and sails 82 miles N-31°-E. It then changes course and sails 107 miles N-78°-E to dock at island B. Find the displacement from island B to island A.

73. The displacement from a harbor on the mainland to a small island is known to be 62.2 miles N-43°30'-E. A sailboat, located 85.3 miles N-37°15'-E of the harbor, learns of an impending tropical storm and heads toward the island for safety. What is the sailboat's displacement to the island?

74. The displacement from an observation tower to a fire is known to be 12.7 miles N-8°30'-E. A group of firemen, located 15.2 miles N-12°20'-W of the tower, must reach the fire. What is the displacement from the firemen to the fire?

Critical Thinking

75. Express the magnitude of $\mathbf{A} + \mathbf{B}$ in terms of $|\mathbf{A}|$ and $|\mathbf{B}|$, given that vectors $\mathbf{A}$ and $\mathbf{B}$ have (a) the same direction or (b) opposite direction.

76. Suppose that $\mathbf{A}$ and $\mathbf{B}$ are two nonzero vectors such that $\mathbf{A} + \mathbf{B} = \mathbf{0}$. What is the relationship between the magnitude and direction of these two vectors?

77. Suppose that $\mathbf{A} = \langle x, y \rangle$ with $\mathbf{A} \neq \mathbf{0}$. Express the vector $\frac{1}{|\mathbf{A}|}\mathbf{A}$ in component form and then find the magnitude of this vector. What type of vector is $\frac{1}{|\mathbf{A}|}\mathbf{A}$?

78. Use the results of Exercise 77 to find a unit vector that has the same direction as each vector given.
(a) $\mathbf{A} = \langle -3, 4 \rangle$ (b) $\mathbf{B} = \langle 1, -3 \rangle$
(c) $\mathbf{C} = 5\mathbf{i} + 12\mathbf{j}$ (d) $\mathbf{D} = -8\mathbf{i} - 4\mathbf{j}$

79. Find the scalars c and d such that the given scalar multiplication and vector addition is true.
(a) $c\langle 1, 3 \rangle + d\langle -2, 6 \rangle = \langle 7, -3 \rangle$
(b) $c\langle 2, -5 \rangle + d\langle -1, -1 \rangle = \langle 2, 9 \rangle$

80. The *dot product* of two vectors is a way of multiplying one vector by another. The dot product of two vectors $\mathbf{A} = \langle x_1, y_1 \rangle$ and $\mathbf{B} = \langle x_2, y_2 \rangle$ is denoted $\mathbf{A} \cdot \mathbf{B}$, and

$$\mathbf{A} \cdot \mathbf{B} = x_1 x_2 + y_1 y_2.$$

Note that $\mathbf{A} \cdot \mathbf{B}$ is a *real number*, not a vector. The real number is found by multiplying corresponding components of vectors $\mathbf{A}$ and $\mathbf{B}$ and then finding their sum. Find $\mathbf{A} \cdot \mathbf{B}$ for the given vectors.
(a) $\mathbf{A} = \langle -3, 5 \rangle$ and $\mathbf{B} = \langle 4, 2 \rangle$
(b) $\mathbf{A} = 4\mathbf{i} + 8\mathbf{j}$ and $\mathbf{B} = -2\mathbf{i} + \mathbf{j}$

81. Use the definition of the dot product from Exercise 80 and vectors $\mathbf{A} = \langle x_1, y_1 \rangle$, $\mathbf{B} = \langle x_2, y_2 \rangle$, and $\mathbf{C} = \langle x_3, y_3 \rangle$ to show that each statement is true.
(a) $\mathbf{A} \cdot \mathbf{B} = \mathbf{B} \cdot \mathbf{A}$
(b) $\mathbf{A} \cdot (\mathbf{B} + \mathbf{C}) = \mathbf{A} \cdot \mathbf{B} + \mathbf{A} \cdot \mathbf{C}$
(c) $\mathbf{A} \cdot \mathbf{A} = |\mathbf{A}|^2$

82. The figure shows vectors $\mathbf{A}$, $\mathbf{B}$, and $\mathbf{A} - \mathbf{B}$, with angle θ ($0 < \theta \leq 180°$) between vectors $\mathbf{A}$ and $\mathbf{B}$. By the law of cosines (Section 9.2),

$$|\mathbf{A} - \mathbf{B}|^2 = |\mathbf{A}|^2 + |\mathbf{B}|^2 - 2|\mathbf{A}||\mathbf{B}| \cos \theta.$$

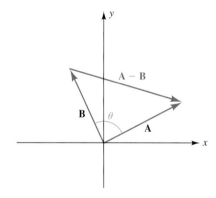

(a) Use the results of Exercise 81 to show that
$$|\mathbf{A} - \mathbf{B}|^2 = (\mathbf{A} - \mathbf{B}) \cdot (\mathbf{A} - \mathbf{B})$$
$$= |\mathbf{A}|^2 + |\mathbf{B}|^2 - 2\mathbf{A} \cdot \mathbf{B}.$$

(b) Substitute $|\mathbf{A}|^2 + |\mathbf{B}|^2 - 2\mathbf{A} \cdot \mathbf{B}$ for $|\mathbf{A} - \mathbf{B}|^2$ in the law of cosines formula, and then state an alternate formula for $\mathbf{A} \cdot \mathbf{B}$ that involves the angle θ.

(c) What is the value of $\mathbf{A} \cdot \mathbf{B}$ if the angle θ between two vectors $\mathbf{A}$ and $\mathbf{B}$ is 90°? 180°?

9.4 Trigonometric Form of Complex Numbers

◆ Introductory Comments

In Section 1.4 we introduced the complex number $a + bi$, where a and b are real numbers and $i = \sqrt{-1}$. We refer to $a + bi$ as a **complex number in standard form**. In this section we introduce the *trigonometric form* of a complex number. Expressing complex numbers in trigonometric form makes the operations of multiplication and division easier to perform. In order to express a complex number in trigonometric form, we must first discuss the graphical representation of a complex number.

For a graphical representation of a complex number, we assign $a + bi$ the ordered pair (a, b) in the coordinate plane. When the coordinate plane is used in this manner, we call it the **complex plane**. In the complex plane, the horizontal axis is referred to as the **real axis** and the vertical axis as the **imaginary axis**. In Figure 9.40, we plot the complex numbers

$$3 + 4i, \quad 1 - i, \quad 2i, \quad \text{and} \quad -3$$

in the complex plane.

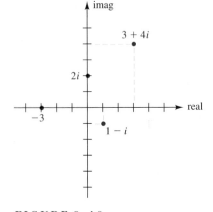

FIGURE 9.40

Four complex numbers plotted in the complex plane

◆ Changing between Trigonometric and Standard Form

In the complex plane, the distance r from the origin $(0, 0)$ to the complex number $a + bi$ is called the **modulus** (or **absolute value**) of the complex number. The angle θ, measured counterclockwise from the positive real axis as shown in Figure 9.41, is called an **argument** of the complex number. Normally, angle θ is measured in radians with $0 \leq \theta < 2\pi$. However, any angle coterminal with θ may also be considered an argument of the complex number. From our discussion of trigonometry in Chapter 7, we know that a, b, r, and θ are related as follows.

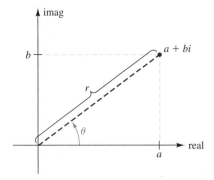

FIGURE 9.41

Complex number $a + bi$ with modulus r and argument θ

◀ **Relationships between a, b, r, and θ**

If r is the modulus and θ is an argument of a nonzero complex number $a + bi$, then

1. $r = \sqrt{a^2 + b^2}$

2. $\tan \theta = \dfrac{b}{a}$, provided that $a \neq 0$

3. $\cos \theta = \dfrac{a}{r}$ or $a = r \cos \theta$

4. $\sin \theta = \dfrac{b}{r}$ or $b = r \sin \theta$

SECTION 9.4 ♦ Trigonometric Form of Complex Numbers

If we substitute $r \cos \theta$ for a and $r \sin \theta$ for b in the complex number $a + bi$, we obtain

$$a + bi = (r \cos \theta) + (r \sin \theta) i = r(\cos \theta + i \sin \theta).$$

We refer to

$$\boxed{r(\cos \theta + i \sin \theta)}$$

as the **trigonometric form** of a complex number. To change from standard form to trigonometric form, we apply the equations

$$r = \sqrt{a^2 + b^2} \quad \text{and} \quad \tan \theta = \frac{b}{a}.$$

EXAMPLE 1 Express each complex number in trigonometric form with $0 \leq \theta < 2\pi$.

(a) $3 + 3i$ **(b)** $4 - 2i$

◆ Solution

(a) For $3 + 3i$, we have $a = 3$ and $b = 3$. Thus, the modulus is

$$r = \sqrt{a^2 + b^2} = \sqrt{(3)^2 + (3)^2} = \sqrt{18} = 3\sqrt{2}.$$

Since $3 + 3i$ lies in quadrant I, the argument θ must be a first-quadrant angle that satisfies the equation

$$\tan \theta = \frac{b}{a} = \frac{3}{3} = 1.$$

Hence,

$$\theta = \tan^{-1} 1 = \frac{\pi}{4}$$

as shown in Figure 9.42. Thus,

$$3 + 3i = 3\sqrt{2} \left(\cos \frac{\pi}{4} + i \sin \frac{\pi}{4} \right).$$

    standard form   trigonometric form

(b) For $4 - 2i$, we have $a = 4$ and $b = -2$. Thus, the modulus is

$$r = \sqrt{a^2 + b^2} = \sqrt{(4)^2 + (-2)^2} = \sqrt{20} = 2\sqrt{5}.$$

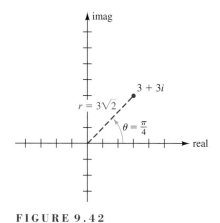

FIGURE 9.42
For the complex number $3 + 3i$, the modulus is $3\sqrt{2}$ and the argument is $\pi/4$.

Since $4 - 2i$ lies in quadrant IV, the argument θ must be a fourth-quadrant angle that satisfies the equation

$$\tan\theta = \frac{b}{a} = \frac{-2}{4} = -\frac{1}{2}.$$

Using the inverse tangent key on a calculator set in radian mode, we find

$$\theta = 2\pi + \tan^{-1}\left(-\tfrac{1}{2}\right) \approx 5.82$$

as shown in Figure 9.43. Thus,

$$4 - 2i \approx 2\sqrt{5}\,(\cos 5.82 + i\sin 5.82).$$

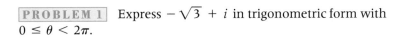

PROBLEM 1 Express $-\sqrt{3} + i$ in trigonometric form with $0 \leq \theta < 2\pi$.

To check the results of Example 1, we reverse the procedure and change trigonometric form to standard form.

EXAMPLE 2 Change the complex number $3\sqrt{2}\left(\cos\dfrac{\pi}{4} + i\sin\dfrac{\pi}{4}\right)$ to standard form.

◆ **Solution** To express this complex number in standard form, we proceed as follows:

$$3\sqrt{2}\left(\cos\frac{\pi}{4} + i\sin\frac{\pi}{4}\right) = 3\sqrt{2}\left[\left(\frac{1}{\sqrt{2}}\right) + i\left(\frac{1}{\sqrt{2}}\right)\right] \quad \text{Evaluate the trigonometric functions}$$

$$= 3\sqrt{2}\left(\frac{1}{\sqrt{2}}\right) + 3\sqrt{2}\left(\frac{1}{\sqrt{2}}\right)i \quad \text{Multiply}$$

$$= 3 + 3i \quad \text{Simpliify}$$

This result agrees with Example 1(a).

PROBLEM 2 Using a calculator set in radian mode, check the result of Example 1(b) by changing $2\sqrt{5}\,(\cos 5.82 + i\sin 5.82)$ to standard form.

◆ Multiplying and Dividing Complex Numbers

In Section 1.4 we found the sum, difference, product, and quotient of complex numbers expressed in standard form. You may wish to review these operations before reading further.

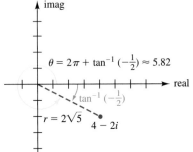

FIGURE 9.43
For the complex number $4 - 2i$, the modulus is $2\sqrt{5}$ and the argument is $2\pi + \tan^{-1}\left(-\tfrac{1}{2}\right) \approx 5.82$.

SECTION 9.4 ◆ *Trigonometric Form of Complex Numbers*

Consider the complex numbers z_1 and z_2 defined as

$$z_1 = r_1(\cos\theta_1 + i\sin\theta_1) \quad \text{and} \quad z_2 = r_2(\cos\theta_2 + i\sin\theta_2).$$

The product of z_1 and z_2 is

$$\begin{aligned}
z_1 z_2 &= [r_1(\cos\theta_1 + i\sin\theta_1)][r_2(\cos\theta_2 + i\sin\theta_2)] \\
&= r_1 r_2 (\cos\theta_1 + i\sin\theta_1)(\cos\theta_2 + i\sin\theta_2) &&\text{Rewrite the product} \\
&= r_1 r_2 [\cos\theta_1\cos\theta_2 + i\cos\theta_1\sin\theta_2 + i\sin\theta_1\cos\theta_2 - \sin\theta_1\sin\theta_2] &&\text{Multiply} \\
&= r_1 r_2 [(\cos\theta_1\cos\theta_2 - \sin\theta_1\sin\theta_2) + i(\sin\theta_1\cos\theta_2 + \cos\theta_1\sin\theta_2)] &&\text{Rearrange terms}
\end{aligned}$$

Apply sum formula for cosine (Section 8.3). Apply sum formula for sine (Section 8.3).

$$= r_1 r_2 [\cos(\theta_1 + \theta_2) + i\sin(\theta_1 + \theta_2)]$$

product of moduli of z_1 and z_2; sum of arguments of z_1 and z_2.

Observe that *the modulus of the product $z_1 z_2$ is the product of the moduli of z_1 and z_2, and an argument of $z_1 z_2$ is the sum of the arguments of z_1 and z_2.*

The quotient of z_1 and z_2 is

$$\begin{aligned}
\frac{z_1}{z_2} &= \frac{r_1(\cos\theta_1 + i\sin\theta_1)}{r_2(\cos\theta_2 + i\sin\theta_2)} \\
&= \frac{r_1(\cos\theta_1 + i\sin\theta_1)}{r_2(\cos\theta_2 + i\sin\theta_2)} \cdot \frac{(\cos\theta_2 - i\sin\theta_2)}{(\cos\theta_2 - i\sin\theta_2)} &&\text{Introduce the conjugate} \\
&= \frac{r_1}{r_2}\left[\frac{\cos\theta_1\cos\theta_2 - i\cos\theta_1\sin\theta_2 + i\sin\theta_1\cos\theta_2 + \sin\theta_1\sin\theta_2}{\cos^2\theta_2 + \sin^2\theta_2}\right] &&\text{Multiply} \\
&= \frac{r_1}{r_2}[(\cos\theta_1\cos\theta_2 + \sin\theta_1\sin\theta_2) + i(\sin\theta_1\cos\theta_2 - \cos\theta_1\sin\theta_2)] &&\text{Simplify and rewrite}
\end{aligned}$$

Apply difference formula for cosine (Section 8.3). Apply difference formula for sine (Section 8.3).

$$= \frac{r_1}{r_2}[\cos(\theta_1 - \theta_2) + i\sin(\theta_1 - \theta_2)]$$

quotient of moduli of z_1 and z_2; difference of arguments of z_1 and z_2.

Observe that *the modulus of the quotient z_1/z_2 is the quotient of the moduli of z_1 and z_2, and an argument of z_1/z_2 is the difference of the arguments of z_1 and z_2.*

Rules for Multiplying and Dividing Complex Numbers in Trigonometric Form

If $z_1 = r_1(\cos\theta_1 + i\sin\theta_1)$ and $z_2 = r_2(\cos\theta_2 + i\sin\theta_2)$, then

1. $z_1 z_2 = r_1 r_2 [\cos(\theta_1 + \theta_2) + i\sin(\theta_1 + \theta_2)]$, and

2. $\dfrac{z_1}{z_2} = \dfrac{r_1}{r_2}[\cos(\theta_1 - \theta_2) + i\sin(\theta_1 - \theta_2)]$, provided that $z_2 \neq 0$.

EXAMPLE 3 Given that $z_1 = 12\left(\cos\dfrac{\pi}{3} + i\sin\dfrac{\pi}{3}\right)$ and $z_2 = 3\left(\cos\dfrac{7\pi}{6} + i\sin\dfrac{7\pi}{6}\right)$, perform the indicated operations and record the results in standard form.

(a) $z_1 z_2$ (b) $\dfrac{z_1}{z_2}$

Solution

(a)
$$z_1 z_2 = \left[12\left(\cos\dfrac{\pi}{3} + i\sin\dfrac{\pi}{3}\right)\right]\left[3\left(\cos\dfrac{7\pi}{6} + i\sin\dfrac{7\pi}{6}\right)\right]$$
$$= 12 \cdot 3\left[\cos\left(\dfrac{\pi}{3} + \dfrac{7\pi}{6}\right) + i\sin\left(\dfrac{\pi}{3} + \dfrac{7\pi}{6}\right)\right] \quad \text{Multiply moduli and add arguments}$$
$$= 36\left[\cos\dfrac{3\pi}{2} + i\sin\dfrac{3\pi}{2}\right] \quad \text{Simplify}$$
$$= 36[0 + i(-1)] = -36i \quad \text{Change to standard form}$$

(b)
$$\dfrac{z_1}{z_2} = \dfrac{12\left(\cos\dfrac{\pi}{3} + i\sin\dfrac{\pi}{3}\right)}{3\left(\cos\dfrac{7\pi}{6} + i\sin\dfrac{7\pi}{6}\right)}$$
$$= 4\left[\cos\left(\dfrac{\pi}{3} - \dfrac{7\pi}{6}\right) + i\sin\left(\dfrac{\pi}{3} - \dfrac{7\pi}{6}\right)\right] \quad \text{Divide moduli and subtract arguments}$$
$$= 4\left[\cos\left(-\dfrac{5\pi}{6}\right) + i\sin\left(-\dfrac{5\pi}{6}\right)\right] \quad \text{Simplify}$$
$$= 4\left[-\dfrac{\sqrt{3}}{2} + i\left(-\dfrac{1}{2}\right)\right] = -2\sqrt{3} - 2i \quad \text{Change to standard form}$$

We can check the results of Example 3 by first changing z_1 and z_2 to standard form and then performing the operations of multiplication and division by using the methods discussed in Section 1.4:

$$\begin{aligned} z_1 &= 12\left(\cos\frac{\pi}{3} + i\sin\frac{\pi}{3}\right) \\ &= 12\left[\left(\frac{1}{2}\right) + i\left(\frac{\sqrt{3}}{2}\right)\right] \\ &= 6 + 6\sqrt{3}\,i \end{aligned} \qquad \text{and} \qquad \begin{aligned} z_2 &= 3\left(\cos\frac{7\pi}{6} + i\sin\frac{7\pi}{6}\right) \\ &= 3\left[\left(-\frac{\sqrt{3}}{2}\right) + i\left(-\frac{1}{2}\right)\right] \\ &= -\frac{3\sqrt{3}}{2} - \frac{3}{2}i \end{aligned}$$

standard forms of z_1 and z_2

Multiplying the standard forms of z_1 and z_2, we obtain

$$\begin{aligned} z_1 z_2 &= (6 + 6\sqrt{3}\,i)\left(-\frac{3\sqrt{3}}{2} - \frac{3}{2}i\right) \\ &= -9\sqrt{3} - 9i - 27i + 9\sqrt{3} = -36i. \end{aligned}$$

This agrees with the answer in Example 3(a).

PROBLEM 3 Check the result of Example 3(b) by dividing $z_1 = 6 + 6\sqrt{3}\,i$ by $z_2 = -\dfrac{3\sqrt{3}}{2} - \dfrac{3}{2}i.$ ◆

◆ Application: Alternating Current (AC) Circuits

An electrical circuit in which the voltage source is sinusoidal (a sine wave) is called an *alternating current* (AC) *circuit.* The **impedance** of an AC circuit is the opposition to the flow of electrons in the circuit. The unit of measure for impedance is the ohm (Ω).

Impedance (Z) consists of three components: **resistance** (R), **inductive reactance** (X_L), and **capacitive reactance** (X_C). In a purely resistive circuit, the voltage and current are *in phase,* that is, they both reach their peak values at the same time. In a purely inductive circuit, the voltage *leads* the current by 90°, that is, the voltage reaches its peak $\frac{1}{4}$ cycle sooner than the current. In a purely capacitive circuit, the voltage *lags* the current by 90°, that is, the current reaches its peak $\frac{1}{4}$ cycle sooner than the voltage.

The phase relationships of the three components make the complex plane ideally suited for describing the impedance Z of an AC circuit. Resistance R is represented by the positive real axis, inductive reactance X_L is represented by the positive imaginary axis, and capacitive reactance X_C is represented by the negative imaginary axis, as shown in Figure 9.44. Mathematically, we have

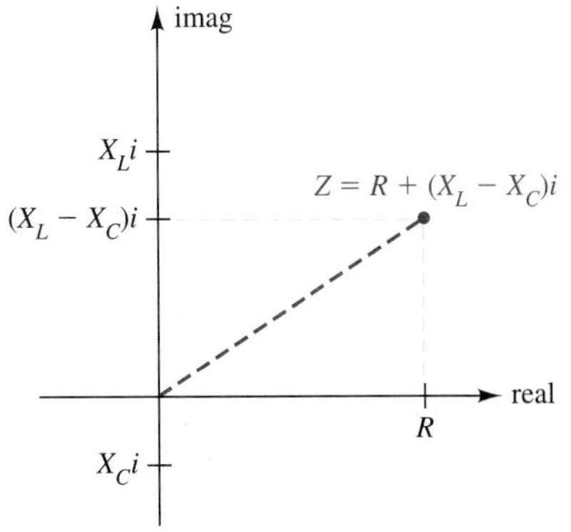

FIGURE 9.44
Representation of impedance Z and its components R, X_L, and X_C with $X_L > X_C$

$$\boxed{Z = R + (X_L - X_C)i}$$

EXAMPLE 4 Given an AC circuit in which $R = 6\,\Omega$, $X_L = 5\,\Omega$, and $X_C = 8\,\Omega$, find (a) the magnitude of the impedance and (b) the phase angle between the voltage and current.

◆ Solution

(a) The impedance Z is given by

$$Z = R + (X_L + X_C)i$$
$$= 6 + (5 - 8)i$$
$$= 6 - 3i$$

Figure 9.45 shows Z plotted in the complex plane. The magnitude (or modulus) of Z, written $|Z|$, is

$$|Z| = \sqrt{R^2 + (X_L - X_C)^2}$$
$$= \sqrt{6^2 + (-3)^2} = \sqrt{45} = 3\sqrt{5}\,\Omega \approx 6.71\,\Omega.$$

(b) The phase angle is the angle θ measured from R to Z, as shown in Figure 9.45. Using the fact that

$$\tan\theta = \frac{X_L - X_C}{R} = \frac{-3}{6} = -\frac{1}{2},$$

we have

$$\theta = \tan^{-1}\left(-\tfrac{1}{2}\right) \approx -26.6°.$$

The negative value indicates that the voltage lags the current by 26.6°. ◆

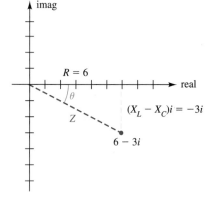

FIGURE 9.45
Representation of the impedance $Z = 6 - 3i$ in the complex plane

PROBLEM 4 Rework Example 4 given that $R = 6\,\Omega$, $X_L = 5\,\Omega$, and $X_C = 3\,\Omega$. ◆

Exercises 9.4

◆ Basic Skills

For each complex number in Exercises 1–20,
(a) plot the number in the complex plane,
(b) find its modulus,
(c) find its argument, and
(d) express the complex number in trigonometric form with $0 \le \theta < 2\pi$.

1. $1 + i$
2. $-4 + 4i$
3. $\tfrac{5}{2} - \tfrac{5}{2}i$
4. $-\sqrt{5} - \sqrt{5}i$
5. $-1 - \sqrt{3}i$
6. $\sqrt{3} + i$
7. $-2\sqrt{3} - 2i$
8. $7 - 7\sqrt{3}i$
9. $-4i$
10. $\sqrt{2}i$
11. π
12. $-\tfrac{8}{3}$
13. $3 + 4i$
14. $-12 + 5i$
15. $7 - i$
16. $-2 - 3i$
17. $-\sqrt{7} + 3i$
18. $5 + \sqrt{11}i$
19. $-\tfrac{1}{2} - \tfrac{2}{3}i$
20. $\tfrac{5}{4} - \sqrt{2}i$

In Exercises 21–24, express each complex number in trigonometric form with $0 \le \theta < 2\pi$. Round the modulus and argument to three significant digits.

21. $2.65 + 7.91i$
22. $-23 + 18i$
23. $0.65 - 1.23i$
24. $-567.1 - 129.1i$

In Exercises 25–34, express each complex number in standard form.

25. $12\left(\cos\dfrac{\pi}{6} + i\sin\dfrac{\pi}{6}\right)$

26. $5\left(\cos\dfrac{5\pi}{6} + i\sin\dfrac{5\pi}{6}\right)$

27. $\dfrac{1}{2}\left(\cos\dfrac{2\pi}{3} + i\sin\dfrac{2\pi}{3}\right)$

28. $4\left(\cos\dfrac{5\pi}{3} + i\sin\dfrac{5\pi}{3}\right)$

29. $\sqrt{2}\left(\cos\dfrac{5\pi}{4} + i\sin\dfrac{5\pi}{4}\right)$

30. $6\left(\cos\dfrac{\pi}{4} + i\sin\dfrac{\pi}{4}\right)$

31. $9\left(\cos\dfrac{\pi}{2} + i\sin\dfrac{\pi}{2}\right)$

32. $3\,(\cos 270° + i\sin 270°)$

33. $\tfrac{2}{3}\,(\cos 180° + i\sin 180°)$

34. $8\,(\cos 8\pi + i\sin 8\pi)$

In Exercises 35–38, express each complex number in the standard form $a + bi$ with a and b rounded to three significant digits.

35. $3\sqrt{5}\,(\cos 1.107 + i\sin 1.107)$

36. $\sqrt{34}\,(\cos 3.682 + i\sin 3.682)$

37. $14\,(\cos 2 + i\sin 2)$

38. $33\,(\cos 175° + i\sin 175°)$

In Exercises 39–50, perform the indicated operations and record the results (a) in trigonometric form with $0 \le \theta < 2\pi$ and (b) in standard form.

39. $\left[2\left(\cos\dfrac{\pi}{3} + i\sin\dfrac{\pi}{3}\right)\right]\left[9\left(\cos\dfrac{7\pi}{6} + i\sin\dfrac{7\pi}{6}\right)\right]$

40. $\left[4\left(\cos\dfrac{3\pi}{4} + i\sin\dfrac{3\pi}{4}\right)\right]\left[5\left(\cos\dfrac{\pi}{4} + i\sin\dfrac{\pi}{4}\right)\right]$

41. $\left[\dfrac{2}{3}\left(\cos\dfrac{\pi}{2} + i\sin\dfrac{\pi}{2}\right)\right]\left[6\left(\cos\dfrac{3\pi}{4} + i\sin\dfrac{3\pi}{4}\right)\right]$

42. $\left(\cos\dfrac{11\pi}{9} + i\sin\dfrac{11\pi}{9}\right)\left(\cos\dfrac{17\pi}{18} + i\sin\dfrac{17\pi}{18}\right)$

43. $\left[5\left(\cos\dfrac{17\pi}{12} + i\sin\dfrac{17\pi}{12}\right)\right]\left[\sqrt{2} + \sqrt{2}\,i\right]$

44. $\left[\sqrt{3}\,(\cos 150° + i\sin 150°)\right]\,[\cos 240° + i\sin 240°]$

45. $\dfrac{8\left(\cos\dfrac{4\pi}{3} + i\sin\dfrac{4\pi}{3}\right)}{2\left(\cos\dfrac{5\pi}{6} + i\sin\dfrac{5\pi}{6}\right)}$

46. $\dfrac{12\left(\cos\dfrac{5\pi}{4} + i\sin\dfrac{5\pi}{4}\right)}{16\left(\cos\dfrac{\pi}{4} + i\sin\dfrac{\pi}{4}\right)}$

47. $\dfrac{16\left(\cos\dfrac{19\pi}{20} + i\sin\dfrac{19\pi}{20}\right)}{\dfrac{8}{3}\left(\cos\dfrac{\pi}{5} + i\sin\dfrac{\pi}{5}\right)}$

48. $\dfrac{3\left(\cos\dfrac{\pi}{4} + i\sin\dfrac{\pi}{4}\right)}{15\left(\cos\dfrac{5\pi}{12} + i\sin\dfrac{5\pi}{12}\right)}$

49. $\dfrac{2 - 2\sqrt{3}\,i}{4\left(\cos\dfrac{\pi}{6} + i\sin\dfrac{\pi}{6}\right)}$

50. $\dfrac{\sqrt{2}\left(\cos\dfrac{\pi}{2} + i\sin\dfrac{\pi}{2}\right)}{-1 - i}$

In Exercises 51–56, perform the indicated operations and record the answer in the standard form $a + bi$ with a and b rounded to three significant digits.

51. $[1.67\,(\cos 2.36 + i\sin 2.36)] \times [2.45\,(\cos 1.61 + i\sin 1.61)]$

52. $[1 - 3i]\,[9\,(\cos 3 + i\sin 3)]$

53. $\dfrac{7\,(\cos 2 + i\sin 2)}{4\,(\cos 6 + i\sin 6)}$

54. $\dfrac{\cos 42.1° + i\sin 42.1°}{\cos 73.8° + i\sin 73.8°}$

55. $\dfrac{(-2.34 + 1.66i)\,(\cos 1.89 + i\sin 1.89)}{\cos 1.34 + i\sin 1.34}$

56. $\dfrac{(0.56 + 1.02i)\,(\cos 2.46 + i\sin 2.46)}{1.25i}$

In Exercises 57–60, the values of R, X_L, and X_C in an AC circuit are given. Find (a) the magnitude of the impedance Z and (b) the phase angle θ (in degrees) between the voltage and current.

57. $R = 5\,\Omega$, $X_L = 7\,\Omega$, $X_C = 2\,\Omega$
58. $R = 9\,\Omega$, $X_L = 16\,\Omega$, $X_C = 25\,\Omega$
59. $R = 4\sqrt{3}\,\Omega$, $X_L = 4\,\Omega$, $X_C = 8\,\Omega$
60. $R = 2\,\Omega$, $X_L = 4\sqrt{3}\,\Omega$, $X_C = 2\sqrt{3}\,\Omega$

In an AC circuit, the relationship between the voltage V (in volts), current I (in amps), and impedance Z (in ohms) is given by $V = IZ$. Use this formula to answer Exercises 61 and 62.

61. Find the magnitude and phase angle of the voltage V in an AC circuit in which $R = 4.2\ \Omega$, $X_L = 10.1\ \Omega$, $X_C = 13.2\ \Omega$, and $I = 8.3\ (\cos 1.25 + i \sin 1.25)$ amps.

62. Find the magnitude and phase angle of the current I in an AC circuit in which $R = 5.2\ \Omega$, $X_L = 15.3\ \Omega$, $X_C = 3.9\ \Omega$, and $V = 65.1\ (\cos 1.38 + i \sin 1.38)$ volts.

◆ Critical Thinking

63. If the product of two complex numbers in trigonometric form equals 1, then the complex numbers are said to be *multiplicative inverses* (or *reciprocals*) of each other. Find the multiplicative inverse of $r\ (\cos \theta + i \sin \theta)$.

64. If the product of two complex numbers in trigonometric form equals r^2, then the complex numbers are said to be *complex conjugates* of each other. Find the complex conjugate of $r\ (\cos \theta + i \sin \theta)$.

65. Find the product $(1 + i\sqrt{3})(1 + i)$ by
 (i) using the technique discussed in Section 1.4, that is, by multiplying as you would two binomials, and
 (ii) changing each factor to trigonometric form, and then using the technique discussed in this section.

Equate the real and imaginary parts of your two answers to find the exact values of each trigonometric function:
 (a) $\cos 105°$ (b) $\sin 105°$

66. Find the product $(\cos \theta + i \sin \theta)(\cos \theta + i \sin \theta)$ by
 (i) multiplying as you would two binomials.
 (ii) using the technique discussed in this section.

Equate the real and imaginary parts of your two answers. What two familiar trigonometric identities evolve?

67. Another form of a complex number, which is used frequently in calculus, is called the **exponential form**. It is written $re^{i\theta}$, where $e \approx 2.718$ (Section 5.1), and r and θ have the same meaning as in the trigonometric form, with θ expressed in radians. The exponential form of a complex number is defined as

$$re^{i\theta} = r\ (\cos \theta + i \sin \theta).$$

Write each complex number in the standard form $a + bi$.
 (a) $2e^{i\pi}$ (b) $e^{i(\pi/2)}$
 (c) $\sqrt{2}\ e^{i(\pi/4)}$ (d) $e^{-i(\pi/6)}$

68. Using the definition of the exponential form of a complex number in Exercise 67, show that each statement is an identity.

(a) $\dfrac{e^{i\theta} + e^{-i\theta}}{2} = \cos \theta$

(b) $\dfrac{e^{i\theta} - e^{-i\theta}}{2i} = \sin \theta$

These identities are usually referred to as *Euler's formula* (named after the mathematician Leonhard Euler, 1707–1783).

9.5 Powers and Roots of Complex Numbers

◆ Introductory Comments

In this section, we discuss the procedures for raising a complex number to an integer power and for finding the nth roots of a complex number. Consider the complex number

$$z = r\ (\cos \theta + i \sin \theta).$$

To raise z to a positive integer power, we apply the rule for multiplying complex numbers in trigonometric form (Section 9.4):

Second power:

$$z^2 = z \cdot z = r\ (\cos \theta + i \sin \theta)\ r\ (\cos \theta + i \sin \theta)$$

$$= r^2\ (\cos 2\theta + i \sin 2\theta) \quad \text{Multiply moduli and add arguments}$$

Third power:

$$z^3 = z \cdot z^2 = r(\cos\theta + i\sin\theta)\, r^2(\cos 2\theta + i\sin 2\theta)$$
$$= r^3(\cos 3\theta + i\sin 3\theta) \qquad \text{Multiply moduli and add arguments}$$

Fourth power:

$$z^4 = z \cdot z^3 = r(\cos\theta + i\sin\theta)\, r^3(\cos 3\theta + i\sin 3\theta)$$
$$= r^4(\cos 4\theta + i\sin 4\theta) \qquad \text{Multiply moduli and add arguments}$$

Do you see the pattern? If $z = r(\cos\theta + i\sin\theta)$ and n is a positive integer, then the modulus of z^n is r^n and an argument of z^n is $n\theta$. This important fact concerning powers of complex numbers is attributed to mathematician Abraham De Moivre and appropriately called *De Moivre's theorem*.

◆ **De Moivre's Theorem**

If $z = r(\cos\theta + i\sin\theta)$ and n is a positive integer, then

$$z^n = r^n(\cos n\theta + i\sin n\theta).$$

◆ **Positive Integer Powers of Complex Numbers**

In the first example, we use De Moivre's theorem to raise a complex number to a positive integer power.

EXAMPLE 1 Find the indicated power of the given complex number and express the answer in the standard form $a + bi$.

(a) $\left[2\left(\cos\dfrac{5\pi}{12} + i\sin\dfrac{5\pi}{12}\right)\right]^4$ **(b)** $(-1 + i)^{12}$

◆ **Solution**

(a) By De Moivre's theorem, we have

$$\left[2\left(\cos\dfrac{5\pi}{12} + i\sin\dfrac{5\pi}{12}\right)\right]^4 = 2^4\left[\cos\left(4\cdot\dfrac{5\pi}{12}\right) + i\sin\left(4\cdot\dfrac{5\pi}{12}\right)\right]$$
$$= 16\left(\cos\dfrac{5\pi}{3} + i\sin\dfrac{5\pi}{3}\right)$$
$$= 16\left[\dfrac{1}{2} + i\left(-\dfrac{\sqrt{3}}{2}\right)\right]$$
$$= 8 - 8\sqrt{3}\, i$$

(b) The complex number must be in trigonometric form before we can apply De Moivre's theorem. For $-1 + i$ the modulus is

$$r = \sqrt{a^2 + b^2} = \sqrt{(-1)^2 + (1)^2} = \sqrt{2}.$$

Since $-1 + i$ lies in quadrant II, the argument θ must be a second-quadrant angle that satisfies the equation

$$\tan \theta = \frac{b}{a} = \frac{1}{-1} = -1.$$

Thus, $\theta = 3\pi/4$ and

$$-1 + i = \sqrt{2}\left(\cos \frac{3\pi}{4} + i \sin \frac{3\pi}{4}\right).$$

Now, applying De Moivre's theorem, we have

$$\begin{aligned}(-1 + i)^{12} &= \left[\sqrt{2}\left(\cos \frac{3\pi}{4} + i \sin \frac{3\pi}{4}\right)\right]^{12} \\ &= (\sqrt{2})^{12}\left[\cos\left(12 \cdot \frac{3\pi}{4}\right) + i \sin\left(12 \cdot \frac{3\pi}{4}\right)\right] \\ &= 64(\cos 9\pi + i \sin 9\pi) \\ &= -64 + 0i \quad \text{or} \quad -64 \end{aligned}$$

PROBLEM 1 Express the power $(\sqrt{3} + i)^9$ in standard form $a + bi$.

◆ Zero and Negative Integer Exponents

How should we define z^0 or z^{-n} for n a positive integer? Certainly, we would like De Moivre's theorem to hold for zero and negative integer exponents. If $z = r(\cos \theta + i \sin \theta)$ and De Moivre's theorem is valid for the zero exponent, then

$$\begin{aligned} z^0 &= r^0 [\cos(0 \cdot \theta) + i \sin(0 \cdot \theta)] \\ &= 1(\cos 0 + i \sin 0) &&\text{Simplify} \\ &= 1[(1) + i(0)] &&\text{Evaluate trigonometric functions} \\ &= 1 &&\text{Multiply} \end{aligned}$$

If $z = r(\cos \theta + i \sin \theta)$ and De Moivre's theorem is valid for a negative integer exponent, then

$$z^{-n} = r^{-n}[\cos(-n\theta) + i \sin(-n\theta)].$$

Since the cosine function is an even function and the sine function is an odd function, we have

$$\cos(-n\theta) = \cos n\theta \quad \text{and} \quad \sin(-n\theta) = -\sin n\theta.$$

Thus

$$z^{-n} = r^{-n}(\cos n\theta - i\sin n\theta)$$

$$= \frac{\cos n\theta - i\sin n\theta}{r^n} \cdot \frac{\cos n\theta + i\sin n\theta}{\cos n\theta + i\sin n\theta} \quad \text{Introduce the conjugate of the numerator}$$

$$= \frac{\cos^2 n\theta + \sin^2 n\theta}{r^n(\cos n\theta + i\sin n\theta)} \quad \text{Multiply}$$

$$= \frac{1}{r^n(\cos n\theta + i\sin n\theta)} \quad \text{Apply trig identity}$$

$$= \frac{1}{z^n} \quad \text{Sustitute}$$

Hence, we make the following definitions, which are consistent with the definitions of the zero and negative integer exponents of real numbers.

◆ **Definitions of z^0 and z^{-n}**

For all complex numbers z, with $z \neq 0$, and any integer n,

$$z^0 = 1 \quad \text{and} \quad z^{-n} = \frac{1}{z^n}.$$

With this understanding of the zero and negative integer exponents, we extend De Moivre's theorem to include any integral power.

◆ **Extension of De Moivre's Theorem**

If $z = r(\cos\theta + i\sin\theta)$ and k is any integer, then

$$z^k = r^k(\cos k\theta + i\sin k\theta).$$

EXAMPLE 2 Find the indicated power of the given complex number.

(a) $\left[2\left(\cos\frac{5\pi}{12} + i\sin\frac{5\pi}{12}\right)\right]^0$

(b) $(-1+i)^{-12}$

◆ Solution

(a) By definition $z^0 = 1$, provided that $z \neq 0$. Therefore, we conclude that

$$\left[2\left(\cos\frac{5\pi}{12} + i\sin\frac{5\pi}{12}\right)\right]^0 = 1.$$

Also, by De Moivre's theorem, we have

$$\left[2\left(\cos\frac{5\pi}{12} + i\sin\frac{5\pi}{12}\right)\right]^0 = 2^0\left[\cos\left(0 \cdot \frac{5\pi}{12}\right) + i\sin\left(0 \cdot \frac{5\pi}{12}\right)\right]$$
$$= 1\,(\cos 0 + i\sin 0)$$
$$= 1[1 + i(0)]$$
$$= 1$$

(b) In Example 1(b), we used De Moivre's theorem to show that

$$(-1 + i)^{12} = \left[\sqrt{2}\left(\cos\frac{3\pi}{4} + i\sin\frac{3\pi}{4}\right)\right]^{12} = -64.$$

Since $z^{-n} = \dfrac{1}{z^n}$, provided that $z \neq 0$, we have

$$(-1 + i)^{-12} = \frac{1}{(-1 + i)^{12}} = -\frac{1}{64}.$$

Also, by De Moivre's theorem, we have

$$(-1 + i)^{-12} = \left[\sqrt{2}\left(\cos\frac{3\pi}{4} + i\sin\frac{3\pi}{4}\right)\right]^{-12}$$
$$= (\sqrt{2})^{-12}\left[\cos\left(-12 \cdot \frac{3\pi}{4}\right) + i\sin\left(-12 \cdot \frac{3\pi}{4}\right)\right]$$
$$= \frac{1}{(\sqrt{2})^{12}}\,[\cos(-9\pi) + i\sin(-9\pi)]$$
$$= \frac{1}{64}\,[(-1) + i(0)] = -\frac{1}{64}. \quad \blacklozenge$$

PROBLEM 2 Find the power of the complex number: $(\sqrt{3} + i)^{-9}$

◆ **nth Root of a Complex Number**

We define the *n*th root of a complex number as follows.

◆ **nth Root of z**

> If *z* and *w* are complex numbers, *n* is a positive integer, and $w^n = z$, then *w* is an **nth root of z**.

Now, if *w* is an *n*th root of *z*, and

$$w = s\,(\cos\alpha + i\sin\alpha) \quad \text{and} \quad z = r\,(\cos\theta + i\sin\theta),$$

then by the definition of an nth root and De Moivre's theorem, we have

$$w^n = z$$
$$[s(\cos \alpha + i \sin \alpha)]^n = r(\cos \theta + i \sin \theta)$$
$$s^n(\cos n\alpha + i \sin n\alpha) = r(\cos \theta + i \sin \theta)$$

Since two equal complex numbers must have the same modulus, it follows that

$$s^n = r \quad \text{or} \quad s = r^{1/n}.$$

Furthermore, since the sine and cosine functions have a period 2π, the arguments of equal complex numbers must either be equal or differ by an integral multiple of 2π. Thus,

$$n\alpha = \theta + 2\pi k \quad \text{or} \quad \alpha = \frac{\theta + 2\pi k}{n}$$

where k is an integer. For $k = 0, 1, 2, \ldots, n - 1$, we obtain n distinct values of α, which in turn identify n distinct complex roots of z. For $k = n, n + 1, n + 2, \ldots$ as well as for $k = -1, -2, -3, \ldots$, we obtain repetitions of these n distinct complex roots. For instance, if $k = n$, then

$$\alpha = \frac{\theta + 2\pi k}{n} = \frac{\theta + 2\pi n}{n} = \frac{\theta}{n} + 2\pi.$$

This last argument is coterminal with the angle we obtain when $k = 0$. In summary, we state the **nth root formula** for complex numbers.

◆ **nth Root Formula**

If $z = r(\cos \theta + i \sin \theta)$, with $z \neq 0$, and n is a positive integer, then z has exactly n distinct nth roots $w_0, w_1, w_2, \ldots, w_{n-1}$ given by

$$w_k = r^{1/n}\left[\cos\left(\frac{\theta + 2\pi k}{n}\right) + i \sin\left(\frac{\theta + 2\pi k}{n}\right)\right]$$

for $k = 0, 1, 2, \ldots, n - 1$.

EXAMPLE 3 Find the indicated roots of the given complex number.

(a) Square roots of $2\left(\cos \dfrac{5\pi}{12} + i \sin \dfrac{5\pi}{12}\right)$

(b) Cube roots of $-1 + i$

◆ Solution

(a) By the nth root formula, $2\left(\cos\dfrac{5\pi}{12} + i\sin\dfrac{5\pi}{12}\right)$ has two distinct square roots, and they are given by

$$w_k = 2^{1/2}\left\{\cos\left[\dfrac{(5\pi/12) + 2\pi k}{2}\right] + i\sin\left[\dfrac{(5\pi/12) + 2\pi k}{2}\right]\right\}$$

for $k = 0, 1$. Hence, the two square roots are as follows:

$$w_0 = 2^{1/2}\left\{\cos\left[\dfrac{(5\pi/12) + 2\pi(0)}{2}\right] + i\sin\left[\dfrac{(5\pi/12) + 2\pi(0)}{2}\right]\right\}$$

$$= 2^{1/2}\left(\cos\dfrac{5\pi}{24} + i\sin\dfrac{5\pi}{24}\right) \approx 1.122 + 0.861\,i$$

$$w_1 = 2^{1/2}\left\{\cos\left[\dfrac{(5\pi/12) + 2\pi(1)}{2}\right] + i\sin\left[\dfrac{(5\pi/12) + 2\pi(1)}{2}\right]\right\}$$

$$= 2^{1/2}\left(\cos\dfrac{29\pi}{24} + i\sin\dfrac{29\pi}{24}\right) \approx -1.122 - 0.861\,i$$

(b) In Example 1(b), we showed that

$$-1 + i = \sqrt{2}\left(\cos\dfrac{3\pi}{4} + i\sin\dfrac{3\pi}{4}\right).$$

By the nth root formula, this complex number has three distinct cube roots, and they are given by

$$w_k = (\sqrt{2})^{1/3}\left\{\cos\left[\dfrac{(3\pi/4) + 2\pi k}{3}\right] + i\sin\left[\dfrac{(3\pi/4) + 2\pi k}{3}\right]\right\}$$

for $k = 0, 1, 2$. Hence, the three cube roots are as follows:

$$w_0 = (\sqrt{2})^{1/3}\left\{\cos\left[\dfrac{(3\pi/4) + 2\pi(0)}{3}\right] + i\sin\left[\dfrac{(3\pi/4) + 2\pi(0)}{3}\right]\right\}$$

$$= 2^{1/6}\left(\cos\dfrac{\pi}{4} + i\sin\dfrac{\pi}{4}\right) \approx 0.794 + 0.794i$$

$$w_1 = (\sqrt{2})^{1/3}\left\{\cos\left[\dfrac{(3\pi/4) + 2\pi(1)}{3}\right] + i\sin\left[\dfrac{(3\pi/4) + 2\pi(1)}{3}\right]\right\}$$

$$= 2^{1/6}\left(\cos\dfrac{11\pi}{12} + i\sin\dfrac{11\pi}{12}\right) \approx -1.084 + 0.291i$$

$$w_2 = (\sqrt{2})^{1/3}\left\{\cos\left[\dfrac{(3\pi/4) + 2\pi(2)}{3}\right] + i\sin\left[\dfrac{(3\pi/4) + 2\pi(2)}{3}\right]\right\}$$

$$= 2^{1/6}\left(\cos\dfrac{19\pi}{12} + i\sin\dfrac{19\pi}{12}\right) \approx 0.291 - 1.084\,i$$

SECTION 9.5 ♦ Powers and Roots of Complex Numbers

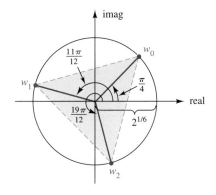

FIGURE 9.46
The cube roots of $-1 + i$ represent the vertices of an equilateral triangle in the complex plane.

Figure 9.46 shows the three cube roots of $-1 + i$ in the complex plane. Note that all roots are spaced equally from each other on a circle with center at the origin and radius $2^{1/6}$. Hence, we can think of w_0, w_1, and w_2 as the vertices of an equilateral triangle. In general, the nth roots of a complex number always represent the vertices of a *regular n-sided polygon* in the complex plane whenever $n \geq 3$. It is a good practice to use this fact to check your solutions whenever you find the roots of a complex number.

PROBLEM 3 Find the four fourth roots of $\sqrt{3} + i$. Show that these roots represent the vertices of a square in the complex plane. ♦

♦ Applying the *n*th Root Formula to Polynomial Equations

In Section 4.3, we used the fundamental theorem of algebra to show that every polynomial function P of degree $n \geq 1$ has exactly n zeros. Consequently, every nth-degree polynomial equation has n solutions in the complex number system. For example, consider the polynomial equation

$$x^4 + 64 = 0.$$

Since this is a fourth-degree equation, it must have four solutions in the complex number system. One method of finding these four solutions is to use factoring and the zero product property:

$$x^4 + 64 = 0$$

$$(x^4 + 16x^2 + 64) - 16x^2 = 0 \quad \text{Add and subtract } 16x^2 \text{ and group terms as shown}$$

$$(x^2 + 8)^2 - 16x^2 = 0 \quad \text{Factor the perfect square trinomial}$$

$$[(x^2 + 8) - 4x][(x^2 + 8) + 4x] = 0 \quad \text{Factor the difference of squares}$$

$$(x^2 - 4x + 8)(x^2 + 4x + 8) = 0 \quad \text{Rewrite}$$

$$x^2 - 4x + 8 = 0 \quad \text{or} \quad x^2 + 4x + 8 = 0 \quad \text{Apply the zero product property}$$

$$x = 2 \pm 2i \quad \text{or} \quad x = -2 \pm 2i \quad \text{Apply the quadratic formula}$$

Hence, the four solutions of the equation $x^4 + 64 = 0$ are $2 + 2i$, $2 - 2i$, $-2 + 2i$, and $-2 - 2i$. As illustrated in the next example, we may also use the nth root formula to find the solutions of the equation $x^4 + 64 = 0$.

EXAMPLE 4 Use the nth root formula to find the solutions of the equation $x^4 + 64 = 0$.

◆ **Solution**   Subtracting 64 from both sides of the equation, we obtain

$$x^4 = -64.$$

The solutions of this equation are the four fourth roots of -64. By the nth root formula, the four fourth roots of

$$-64 = -64 + 0i = 64(\cos \pi + i \sin \pi)$$

are given by

$$w_k = 64^{1/4}\left[\cos\left(\frac{\pi + 2\pi k}{4}\right) + i \sin\left(\frac{\pi + 2\pi k}{4}\right)\right]$$

for $k = 0, 1, 2, 3$. Hence, the four fourth roots are as follows:

$$w_0 = 64^{1/4}\left\{\cos\left[\frac{\pi + 2\pi(0)}{4}\right] + i \sin\left[\frac{\pi + 2\pi(0)}{4}\right]\right\}$$
$$= 2\sqrt{2}\left(\cos\frac{\pi}{4} + i \sin\frac{\pi}{4}\right) = 2 + 2i$$

$$w_1 = 64^{1/4}\left\{\cos\left[\frac{\pi + 2\pi(1)}{4}\right] + i \sin\left[\frac{\pi + 2\pi(1)}{4}\right]\right\}$$
$$= 2\sqrt{2}\left(\cos\frac{3\pi}{4} + i \sin\frac{3\pi}{4}\right) = -2 + 2i$$

$$w_2 = 64^{1/4}\left\{\cos\left[\frac{\pi + 2\pi(2)}{4}\right] + i \sin\left[\frac{\pi + 2\pi(2)}{4}\right]\right\}$$
$$= 2\sqrt{2}\left(\cos\frac{5\pi}{4} + i \sin\frac{5\pi}{4}\right) = -2 - 2i$$

$$w_3 = 64^{1/4}\left\{\cos\left[\frac{\pi + 2\pi(3)}{4}\right] + i \sin\left[\frac{\pi + 2\pi(3)}{4}\right]\right\}$$
$$= 2\sqrt{2}\left(\cos\frac{7\pi}{4} + i \sin\frac{7\pi}{4}\right) = 2 - 2i$$

Thus, the four solutions of the equation $x^4 + 64 = 0$ are $2 \pm 2i$ and $-2 \pm 2i$. As shown in Figure 9.47, these roots form the vertices of a square in the complex plane. ◆

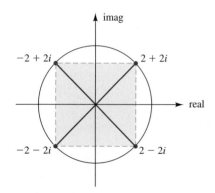

FIGURE 9.47
The solutions of $x^4 + 64 = 0$ represent the vertices of a square in the complex plane.

PROBLEM 4   Solve the equation $x^3 + 1 = 0$ by using the nth root formula. ◆

Exercises 9.5

Basic Skills

In Exercises 1–20, find the indicated power of each complex number. Record the results (a) in trigonometric form with $0 \leq \theta < 2\pi$ and (b) in standard form.

1. $\left[4\left(\cos\dfrac{\pi}{6} + i\sin\dfrac{\pi}{6}\right)\right]^3$
2. $\left[\sqrt{3}\left(\cos\dfrac{\pi}{4} + i\sin\dfrac{\pi}{4}\right)\right]^4$
3. $\left[\sqrt{2}\left(\cos\dfrac{15\pi}{8} + i\sin\dfrac{15\pi}{8}\right)\right]^6$
4. $\left[\dfrac{1}{2}\left(\cos\dfrac{11\pi}{12} + i\sin\dfrac{11\pi}{12}\right)\right]^3$
5. $[4\sqrt{5}(\cos 2.137 + i\sin 2.137)]^0$
6. $[9(\cos 240° + i\sin 240°)]^2$
7. $\left[\dfrac{2}{3}\left(\cos\dfrac{11\pi}{9} + i\sin\dfrac{11\pi}{9}\right)\right]^{-3}$
8. $\left(\cos\dfrac{5\pi}{3} + i\sin\dfrac{5\pi}{3}\right)^{-5}$
9. $(\cos 210° + i\sin 210°)^{-7}$
10. $\left[5\left(\cos\dfrac{17\pi}{18} + i\sin\dfrac{17\pi}{18}\right)\right]^{-3}$
11. $(1 + i)^8$
12. $(-2 + 2i)^4$
13. $(-1 - \sqrt{3}\,i)^5$
14. $(\sqrt{3} - i)^{10}$
15. $(0 - \sqrt{2}\,i)^{14}$
16. $(0 + 3i)^6$
17. $(3 - 3i)^{-2}$
18. $(1 - \sqrt{3}\,i)^{-4}$
19. $(-2\sqrt{3} - 2i)^{-3}$
20. $(9 - 3i)^0$

In Exercises 21–28, find the indicated power of each complex number. Record the answer in standard form, $a + bi$, with a and b rounded to three significant digits.

21. $[1.2(\cos 5 + i\sin 5)]^5$
22. $[3.3(\cos 70° + i\sin 70°)]^6$
23. $[2(\cos 1 + i\sin 1)]^{-3}$
24. $[2.3(\cos 1.24 + i\sin 1.24)]^{-4}$
25. $(1 - 2i)^3$
26. $(1.56 + 2.23i)^5$
27. $(-12.2 - 15.7i)^{-4}$
28. $(-14 + 8i)^{-3}$

In Exercises 29–42, find the indicated roots of each complex number. Record the results (a) in trigonometric form with $0 \leq \theta < 2\pi$ and (b) in standard form.

29. Square roots of $49\left(\cos\dfrac{5\pi}{3} + i\sin\dfrac{5\pi}{3}\right)$
30. Square roots of $18\left(\cos\dfrac{2\pi}{3} + i\sin\dfrac{2\pi}{3}\right)$
31. Cube roots of $8\left(\cos\dfrac{\pi}{2} + i\sin\dfrac{\pi}{2}\right)$
32. Fourth roots of $16\left(\cos\dfrac{4\pi}{3} + i\sin\dfrac{4\pi}{3}\right)$
33. Fourth roots of $25(\cos 240° + i\sin 240°)$
34. Cube roots of $64(\cos 180° + i\sin 180°)$
35. Square roots of i
36. Square roots of -4
37. Cube roots of 1
38. Cube roots of $-8i$
39. Fourth roots of -16
40. Sixth roots of 1
41. Square roots of $-4 + 4\sqrt{3}\,i$
42. Square roots of $9 - 9\sqrt{3}\,i$

In Exercises 43–46, plot the indicated solutions in the complex plane. Describe the regular polygon for which these points are vertices.

43. Solutions to Exercise 31
44. Solutions to Exercise 32
45. Solutions to Exercise 39
46. Solutions to Exercise 40

In Exercises 47–54, find the indicated roots of each complex number. Record each answer in standard form, $a + bi$, with a and b rounded to three significant digits.

47. Cube roots of $18\left(\cos\dfrac{3\pi}{4} + i\sin\dfrac{3\pi}{4}\right)$
48. Fourth roots of $54(\cos 200° + i\sin 200°)$
49. Square roots of $5 - 12i$
50. Square roots of $-15 - 8i$
51. Sixth roots of i
52. Fifth roots of -32

53. Fourth roots of $6.2 + 3.4i$

54. Cube roots of $-12.4 + 17.3i$

In Exercises 55–62, use the nth root formula to find all the solutions of the given equation. Write each answer in standard form, $a + bi$.

55. $x^3 + 8 = 0$
56. $x^3 - 1 = 0$
57. $x^4 - 81 = 0$
58. $x^6 + 729 = 0$
59. $x^3 - i = 0$
60. $x^3 + 64i = 0$
61. $x^2 - 1 = \sqrt{3}\, i$
62. $x^4 + 1 = \sqrt{3}\, i$

◆ Critical Thinking

63. What is the relationship between the solutions of the equation $x^8 = 1$ and a regular octagon? Explain.

64. What is the relationship between the zeros of the function $f(x) = x^{10} + 1$ and a regular decagon? Explain.

65. One of the cube roots of a complex number is

$$5\left(\cos\frac{3\pi}{8} + i \sin\frac{3\pi}{8}\right).$$

Determine the other two cube roots. [*Hint:* Consider the graphical representation of the roots in the complex plane.]

66. One of the sixth roots of a complex number is $2\sqrt{3} + 2i$. Determine the other five sixth roots. [*Hint:* Consider the graphical representation of the roots in the complex plane.]

67. In Exercise 67 of Section 9.4, we defined the *exponential form* of a complex number as $re^{i\theta} = r(\cos\theta + i\sin\theta)$. For $r = 1$, we have

$$e^{i\theta} = \cos\theta + i\sin\theta.$$

Show that the usual laws of exponents continue to apply to imaginary exponents, that is, show that each of the following statements is true.

(a) $e^{i\theta_1} \cdot e^{i\theta_2} = e^{i(\theta_1 + \theta_2)}$
(b) $\dfrac{e^{i\theta_1}}{e^{i\theta_2}} = e^{i(\theta_1 - \theta_2)}$
(c) $(e^{i\theta})^n = e^{i(n\theta)}$

68. Many objects in nature have a spiral pattern similar to the nautilus seashell shown in the sketch. Show that the nonnegative integer powers of $z = 1 + i$, when plotted in the same complex plane, form a spiral pattern similar to that of this seashell.

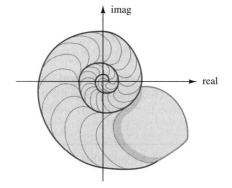

9.6 Polar Coordinate System

◆ Introductory Comments

Throughout this text, we have plotted points and graphed equations in the Cartesian coordinate system. In this section, we describe another coordinate system that is used to locate points in a plane and to graph equations—the *polar coordinate system*. Many equations, especially those representing conic sections, are easier to work with in the polar coordinate system than in the Cartesian coordinate system.

The **polar coordinate system** consists of a fixed point O, called the **pole**, and the **polar axis**, a reference ray with endpoint at the pole. It is customary to let the pole coincide with the origin in the Cartesian coordinate system and the polar axis coincide with the positive x-axis. A point P in the polar

SECTION 9.6 ◆ *Polar Coordinate System*

coordinate system is assigned an ordered pair $P(r, \theta)$, where r and θ are called the **polar coordinates** of point P. If $r > 0$, then r is the distance from the pole to point P and θ is the measure of the angle formed by ray OP and the polar axis (see Figure 9.48). If $r < 0$, then the absolute value of r is the distance from the pole to point P and $\theta + 180°$ (or $\theta + \pi$ if θ is in radians) is the measure of the angle formed by ray OP and the polar axis (see Figure 9.49).

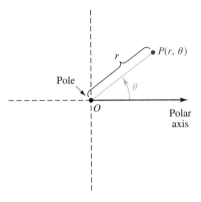

FIGURE 9.48
Plot of the point $P(r, \theta)$ when $r > 0$

◆ **Plotting Points in the Polar Coordinate System**

In the first example, we plot some points in the polar coordinate system.

EXAMPLE 1 Plot the points with the given polar coordinates.

(a) $P(3, 60°)$ (b) $Q(-3, 60°)$

(c) $R\left(-4, -\dfrac{\pi}{4}\right)$ (d) $S(0, \pi)$

◆ Solution

(a) For $P(3, 60°)$, we have $r = 3$ and $\theta = 60°$. Hence, the distance from the pole O to point P is 3 units. Since $r > 0$, the measure of the angle formed by ray OP and the polar axis is $60°$, as shown in Figure 9.50.

(b) For $Q(-3, 60°)$, we have $r = -3$ and $\theta = 60°$. Hence, the distance from the pole O to point Q is $|-3| = 3$ units. Since $r < 0$, the measure of the angle formed by ray OQ and the polar axis is $60° + 180° = 240°$, as shown in Figure 9.50.

(c) For $R\left(-4, -\dfrac{\pi}{4}\right)$, we have $r = -4$ and $\theta = -\dfrac{\pi}{4}$, or $\theta = -45°$. Hence, the distance from the pole O to point R is $|-4| = 4$ units. Since $r < 0$, the measure of the angle formed by ray OR and the polar axis is

$$-\frac{\pi}{4} + \pi = \frac{3\pi}{4} \quad \text{or} \quad 135°,$$

as shown in Figure 9.50.

(d) For $S(0, \pi)$, we have $r = 0$ and $\theta = \pi$, or $180°$. Since $r = 0$, the point S must be located at the pole O, as shown in Figure 9.50. In fact, all points of the form $(0, \theta)$ are located at the pole. ◆

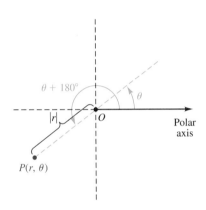

FIGURE 9.49
Plot of the point $P(r, \theta)$ when $r < 0$

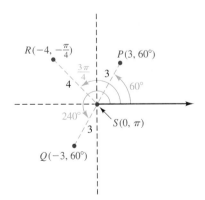

FIGURE 9.50
Points P, Q, and R plotted in the polar coordinate system

For each point P in the Cartesian plane there corresponds a unique ordered pair of real numbers (x, y), and for each ordered pair of real numbers (x, y) there corresponds a unique point P in the Cartesian plane. As we observed in Example 1(d), this one-to-one correspondence between ordered pairs and points in the polar coordinate system does not hold true for a point at the pole. Even a point not at the pole has infinitely many polar coordinate representations. For example, the point $P(r, \theta)$, with θ in radians, may also be represented by

$$P(r, \theta + 2\pi n) \quad \text{or} \quad P(-r, (\theta + \pi) + 2\pi n)$$

for any integer n.

PROBLEM 1 Find two other polar coordinate representations for $P(3, \pi)$, one with a positive value of r and the other with a negative value of r.

◆ Relationship between (r, θ) and (x, y)

From our discussion of trigonometry in Chapter 7, we note the following four relationships between a point P with polar coordinates (r, θ) and the same point P with Cartesian coordinates (x, y).

◀ **Relationship between (r, θ) and (x, y)**

If point P has coordinates (x, y) in the Cartesian coordinate system and coordinates (r, θ) in the polar coordinate system, then

1. $x = r \cos \theta$
2. $y = r \sin \theta$
3. $r^2 = x^2 + y^2$
4. $\tan \theta = \dfrac{y}{x}$, provided that $x \neq 0$

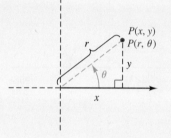

EXAMPLE 2 Convert the given coordinates to the indicated coordinates.

(a) $(-2, 135°)$ to (x, y)

(b) $\left(-2, 2\sqrt{3}\right)$ to (r, θ) with $r > 0$ and $0 \leq \theta < 2\pi$

◆ **Solution**

(a) We can convert the polar coordinates $(-2, 135°)$ to Cartesian coordinates (x, y) by using the relationships $x = r \cos \theta$ and $y = r \sin \theta$:

$$\begin{aligned}
x &= r \cos \theta & y &= r \sin \theta \\
&= -2 \cos 135° & &= -2 \sin 135° \\
&= -2\left(-\dfrac{1}{\sqrt{2}}\right) & &= -2\left(\dfrac{1}{\sqrt{2}}\right) \\
&= \sqrt{2} \approx 1.414 & &= -\sqrt{2} \approx -1.414
\end{aligned}$$

Hence, a point with polar coordinates $(-2, 135°)$ has Cartesian coordinates $\left(\sqrt{2}, -\sqrt{2}\right)$, as illustrated in Figure 9.51.

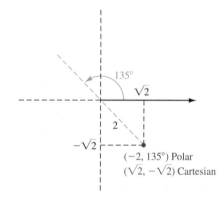

FIGURE 9.51

A point with polar coordinates $(-2, 135°)$ has Cartesian coordinates $\left(\sqrt{2}, -\sqrt{2}\right)$.

(b) We can convert the Cartesian coordinates $(-2, 2\sqrt{3})$ to polar coordinates (r, θ) by using the relationships $r^2 = x^2 + y^2$ and $\tan\theta = \dfrac{y}{x}$:

$$r^2 = x^2 + y^2$$
$$= (-2)^2 + (2\sqrt{3})^2$$
$$= 4 + 12$$
$$= 16$$

$$\tan\theta = \frac{y}{x}$$
$$= \frac{2\sqrt{3}}{-2}$$
$$= -\sqrt{3}$$

which implies that $r = 4$, since we want $r > 0$.

which implies that $\theta = 2\pi/3$, since θ is in quadrant II.

Thus, a point with Cartesian coordinates $(-2, 2\sqrt{3})$ has polar coordinates $(4, 2\pi/3)$, as illustrated in Figure 9.52.

PROBLEM 2 Convert $(-2, 2\sqrt{3})$ to (r, θ) with $r < 0$ and $0 \leq \theta < 2\pi$.

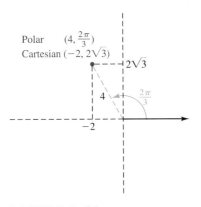

FIGURE 9.52

A point with Cartesian coordinates $(-2, 2\sqrt{3})$ has polar coordinates $(4, 2\pi/3)$.

◆ Sketching the Graph of a Polar Equation

We refer to any equation that is written in terms of r and θ as a **polar equation**. The **graph** of a polar equation is the set of all ordered pairs (r, θ) in the polar plane that satisfy the given equation. The point-plotting method (introduced in Section 2.2) for sketching the graph of an elementary Cartesian equation in x and y may also be used to help sketch the graph of a simple polar equation.

EXAMPLE 3 Sketch the graph of each polar equation.

(a) $r = 3$ **(b)** $\theta = \dfrac{\pi}{4}$

◀ Solution

(a) We can think of this equation as

$$r + 0 \cdot \theta = 3.$$

Regardless of what value we choose for θ, the quantity $0 \cdot \theta$ is always zero. This means that the ordered pairs satisfying the equation $r = 3$ can have any θ-value, provided the r-value is 3. The following table of values shows some points that satisfy the equation $r = 3$:

θ	0	$\dfrac{\pi}{4}$	$\dfrac{\pi}{2}$	$\dfrac{3\pi}{4}$	π	$\dfrac{5\pi}{4}$	$\dfrac{3\pi}{2}$	$\dfrac{7\pi}{4}$
r	3	3	3	3	3	3	3	3

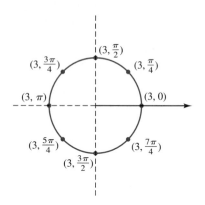

FIGURE 9.53

The graph of $r = 3$ is a circle with center at the pole and radius 3.

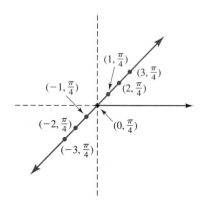

FIGURE 9.54

The graph of $\theta = \pi/4$ is a line that passes through the pole and makes an angle of $\pi/4$ with the polar axis.

After plotting these points in the polar coordinate system, we conclude that the graph of $r = 3$ is a circle with radius 3, as shown in Figure 9.53.

(b) We can think of the equation $\theta = \pi/4$ as

$$0 \cdot r + \theta = \frac{\pi}{4}.$$

Regardless of what value we choose for r, the quantity $0 \cdot r$ is always zero. This means that the ordered pairs satisfying the equation $\theta = \pi/4$ can have any r-value, provided the θ-value is $\pi/4$. The following table of values shows some points that satisfy the equation $\theta = \pi/4$:

r	0	1	2	3	-1	-2	-3
θ	$\dfrac{\pi}{4}$	$\dfrac{\pi}{4}$	$\dfrac{\pi}{4}$	$\dfrac{\pi}{4}$	$\dfrac{\pi}{4}$	$\dfrac{\pi}{4}$	$\dfrac{\pi}{4}$

After plotting these points in the polar coordinate system, we conclude that the graph of $\theta = \pi/4$ is a line that passes through the pole and makes an angle of $\pi/4$ (45°) with the polar axis, as shown in Figure 9.54. ◆

We can verify the graphs in Example 3 by converting the polar equations to equivalent Cartesian equations in x and y. To convert the polar equation $r = 3$ in Example 3(a), we square both sides and obtain $r^2 = 9$. Now, replacing r^2 with $x^2 + y^2$ gives us

$$x^2 + y^2 = 9,$$

which we recognize as the equation of a circle with center at the origin and radius 3. In general, the graph of

$$\boxed{r = a}$$

with a a nonzero constant, is a circle with center at the pole and radius $|a|$. For the polar equation $\theta = \pi/4$ in Example 3(b), we can write

$$\tan \theta = \tan \frac{\pi}{4} = 1.$$

Now, replacing $\tan \theta$ with y/x gives us

$$\frac{y}{x} = 1 \quad \text{or} \quad y = x,$$

which we recognize as the equation of a line that passes through the origin and separates quadrant I and quadrant III in half. In general, the graph of

$$\theta = k$$

with k a constant, is a line that passes through the pole and makes an angle of k radians with the polar axis.

PROBLEM 3 Describe the graph of each polar equation.

(a) $r = 5$ (b) $\theta = \dfrac{2\pi}{3}$

EXAMPLE 4 Sketch the graph of each polar equation and give its equivalent Cartesian equation.

(a) $r = \dfrac{3}{\sin \theta}$ (b) $r = -6 \sin \theta$

◆ Solution

(a) The following table of values shows some points that satisfy the equation $r = 3/\sin \theta$:

θ	0	$\dfrac{\pi}{6}$	$\dfrac{\pi}{2}$	$\dfrac{5\pi}{6}$	π	$\dfrac{7\pi}{6}$	$\dfrac{3\pi}{2}$	$\dfrac{11\pi}{6}$
r	undefined	6	3	6	undefined	-6	-3	-6

By plotting these points in the polar coordinate system, we see that the graph of $r = 3/\sin \theta$ is a line parallel to the polar axis and 3 units above the pole, as shown in Figure 9.55. We can confirm that the graph is a horizontal line by converting the equation to an equivalent Cartesian equation:

$$r = \dfrac{3}{\sin \theta}$$

$r \sin \theta = 3$ **Multiply both sides by $\sin \theta$**

$y = 3$ **Replace $r \sin \theta$ with y**

This last equation is a horizontal line 3 units above the x-axis.

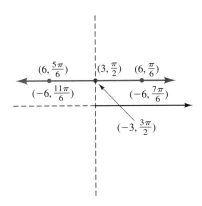

FIGURE 9.55

The graph of $r = 3/\sin \theta$ is a horizontal line 3 units above the polar axis.

(b) The following table of values shows some points that satisfy the equation $r = -6 \sin \theta$:

θ	0	$\dfrac{\pi}{6}$	$\dfrac{\pi}{3}$	$\dfrac{\pi}{2}$	$\dfrac{2\pi}{3}$	$\dfrac{5\pi}{6}$	π
r	0	-3	$-3\sqrt{3}$	-6	$-3\sqrt{3}$	-3	0

By plotting these points in the polar coordinate system, we see that the graph of $r = -6\sin\theta$ is a circle with center $(-3, \pi/2)$, as shown in Figure 9.56. Our table of values contains only values of θ between 0 and π. Values of θ between π and 2π simply trace the circle again.

We can confirm that this graph is a circle by converting the equation to an equivalent Cartesian equation:

$$r = -6\sin\theta$$

$$r^2 = -6r\sin\theta \quad \text{Multiply both sides by } r$$

$$x^2 + y^2 = -6y \quad \begin{array}{l}\text{Replace } r^2 \text{ with } x^2 + y^2 \\ \text{and } r\sin\theta \text{ with } y\end{array}$$

$$x^2 + (y^2 + 6y +) = 0 \quad \text{Add } 6y \text{ to both sides}$$

$$x^2 + (y^2 + 6y + 9) = 9 \quad \begin{array}{l}\text{Complete the square by} \\ \text{adding 9 to both sides}\end{array}$$

$$(x - 0)^2 + (y + 3)^2 = 3^2 \quad \text{Write in standard form}$$

We recognize this last equation as the equation of a circle in standard form with center $(0, -3)$ and radius 3. ◆

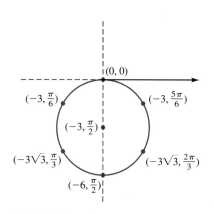

FIGURE 9.56
The graph of $r = -6\sin\theta$ is a circle with center $(-3, \pi/2)$ and radius 3.

PROBLEM 4 Sketch the graph of each polar equation and state its equivalent Cartesian equation.

(a) $r = \dfrac{3}{\cos\theta}$ **(b)** $r = -6\cos\theta$ ◆

Based on the results of Example 4 and Problem 4, we state the following conclusions. (See Exercises 65 and 66 for the proofs of these statements.)

1. The graph of

$$r = \dfrac{a}{\sin\theta}$$

with a a nonzero constant, is a horizontal line a units above the pole if $a > 0$, or $|a|$ units below the pole if $a < 0$.

2. The graph of

$$r = \dfrac{a}{\cos\theta}$$

SECTION 9.6 ♦ Polar Coordinate System

with a a nonzero constant, is a vertical line a units to the right of the pole if $a > 0$, or $|a|$ units to the left of the pole if $a < 0$.

3. The graph of

$$r = a \sin \theta$$

with a a nonzero constant, is a circle with radius $|a|/2$ and center $(a/2, 90°)$.

4. The graph of

$$r = a \cos \theta$$

with a a nonzero constant, is a circle with radius $|a|/2$ and center $(a/2, 0°)$.

♦ Polar Equations and the Graphing Calculator

Most graphing calculators are capable of generating graphs of polar equations. For example, to generate the line and circle defined by the polar equations in Example 4, we begin by selecting polar mode and radian mode on the calculator. Next, we press the WINDOW key or RANGE key and enter maximum and minimum values of θ, x, and y. To obtain a true geometric perspective of the graphs in Figures 9.55 and 9.56, we choose the θ-values

θ-min = 0 θ-max = 2π θ-step = 0.1

(smallest θ-value to be evaluated) (largest θ-value to be evaluated) (increment between θ-values)

with a square set viewing rectangle such as

$[-9, 9]$ by $[-6, 6]$

(x-min, x-max, y-min, y-max)

We now enter the polar equations $r = 3/\sin \theta$ and $r = -6 \sin \theta$ and activate the graphing feature of the calculator. The result is shown in Figure 9.57.

EXAMPLE 5 Given the polar equation $r = \sqrt{2} \sin 2\theta$, find values of θ in the interval $[0, 2\pi)$ at which $|r|$ attains its maximum value. Use a graphing calculator to verify your answer.

♦ **Solution** Since the range of the sine function is $[-1, 1]$, we know that values of r lie in the interval $[-\sqrt{2}, \sqrt{2}]$. Hence, the maximum

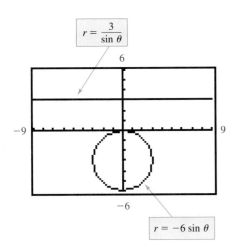

FIGURE 9.57
Graphs of $r = 3/\sin \theta$ and $r = -6 \sin \theta$ shown in the viewing rectangle of a graphing calculator

value of $|r|$ is $\sqrt{2}$ and this occurs when

$$2\theta = \frac{\pi}{2} + 2\pi n \quad \text{or} \quad 2\theta = \frac{3\pi}{2} + 2\pi n$$

$$\theta = \frac{\pi}{4} + \pi n \qquad\qquad \theta = \frac{3\pi}{4} + \pi n$$

Thus, the values of θ in the interval $[0, 2\pi)$ at which $|r|$ attains its maximum value are

$$\frac{\pi}{4}, \quad \frac{3\pi}{4}, \quad \frac{5\pi}{4}, \quad \text{and} \quad \frac{7\pi}{4}.$$

First selecting polar mode and radian mode on a graphing calculator, we then generate the graph of $r = \sqrt{2} \sin 2\theta$ as shown in Figure 9.58. The graph of this equation is called a *four-leafed rose* and each loop of the graph is called a *petal*. At $\theta = \pi/4$ (≈ 0.78539816), we obtain the Cartesian coordinates $(1, 1)$, which implies $r = \sqrt{2}$. By symmetry of the graph, we see that $\theta = 3\pi/4, 5\pi/4, 7\pi/4$ also give $|r| = \sqrt{2}$, which confirms our work. ◆

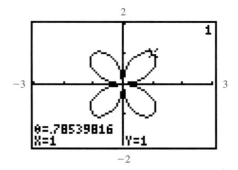

FIGURE 9.58
The graph of $r = \sqrt{2} \sin 2\theta$ is a four-leafed rose with maximum value $|r| = \sqrt{2}$ at $\theta = \pi/4, 3\pi/4, 5\pi/4,$ and $7\pi/4$.

CAUTION The vertical line test for functions (Section 2.3) is not valid for graphs of equations in the polar coordinate system. The equation $r = \sqrt{2} \sin 2\theta$ defines r as a function of θ even though its graph appears to fail the vertical line test (see Figure 9.58). For each value of θ that we select, we obtain one and only one value of r. To emphasize this functional relationship, we write

$$r = f(\theta) = \sqrt{2} \sin 2\theta.$$

PROBLEM 5 Given the polar equation $r = \sqrt{2} \cos 3\theta$, find values of θ in the interval $[0, \pi)$ at which $|r|$ attains its maximum value. Use a graphing calculator to verify your answer. ◆

We conclude this section by listing some basic polar equations and their graphs. We can verify each graph by using a graphing calculator or by plotting points.

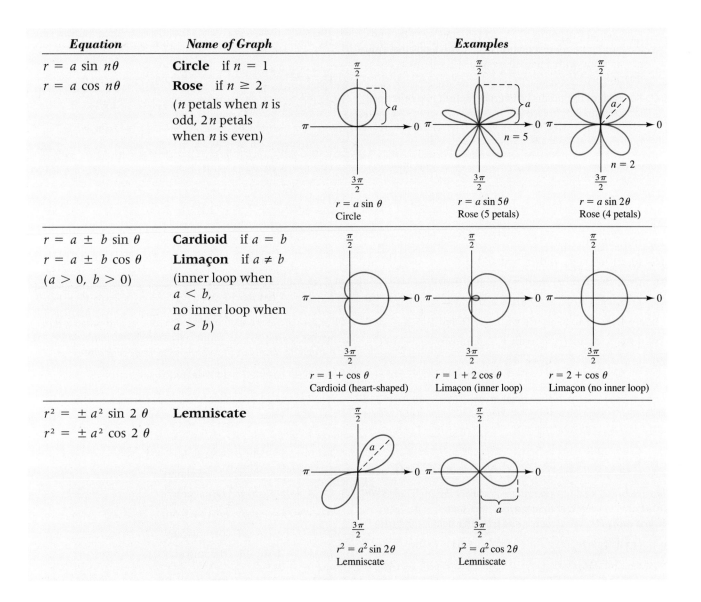

Equation	Name of Graph	Examples
$r = a \sin n\theta$ $r = a \cos n\theta$	**Circle** if $n = 1$ **Rose** if $n \geq 2$ (n petals when n is odd, $2n$ petals when n is even)	$r = a \sin \theta$ Circle $r = a \sin 5\theta$ Rose (5 petals) $r = a \sin 2\theta$ Rose (4 petals)
$r = a \pm b \sin \theta$ $r = a \pm b \cos \theta$ ($a > 0, b > 0$)	**Cardioid** if $a = b$ **Limaçon** if $a \neq b$ (inner loop when $a < b$, no inner loop when $a > b$)	$r = 1 + \cos \theta$ Cardioid (heart-shaped) $r = 1 + 2 \cos \theta$ Limaçon (inner loop) $r = 2 + \cos \theta$ Limaçon (no inner loop)
$r^2 = \pm a^2 \sin 2\theta$ $r^2 = \pm a^2 \cos 2\theta$	**Lemniscate**	$r^2 = a^2 \sin 2\theta$ Lemniscate $r^2 = a^2 \cos 2\theta$ Lemniscate

Exercises 9.6

Basic Skills

In Exercises 1–8, plot the points with the given polar coordinates.

1. $A(1, 180°)$, $B(2, 60°)$, $C(3, 225°)$, $D(0, 210°)$
2. $A(3, \pi/2)$, $B(1, 5\pi/6)$, $C(2, 2\pi/3)$, $D(4, 7\pi/4)$
3. $A(2, -\pi/2)$, $B(1, -\pi/6)$, $C(4, -3\pi/4)$, $D(0, -3)$
4. $A(2, -90°)$, $B(1, -30°)$, $C(4, -135°)$, $D(0, -330°)$
5. $A(-3, 0°)$, $B(-2, 45°)$, $C(-5, 150°)$, $D(-4, 300°)$
6. $A(-1, \pi/2)$, $B(-2, \pi/6)$, $C(-4, 3\pi/4)$, $D(-6, 0)$
7. $A(-4, -3\pi/2)$, $B(-1, -\pi/4)$, $C(-5, -7\pi/6)$, $D(-2, -5\pi/3)$
8. $A(-1, -270°)$, $B(-4, -60°)$, $C(-3, -315°)$, $D(-5, -240°)$

9. For each of the points in Exercise 5, find two other polar coordinate representations, one with a positive value of r and one with a negative value of r. (Answers are not unique.)

10. For each of the points in Exercise 8, find two other polar coordinate representations, one with a positive value of r and one with a negative value of r. (Answers are not unique.)

In Exercises 11–20, convert the given coordinates to the indicated coordinates.

11. $(1, 180°)$ to (x, y)
12. $(-2, \pi/2)$ to (x, y)
13. $(-4\sqrt{2}, 7\pi/4)$ to (x, y)
14. $(6, -30°)$ to (x, y)
15. $(0, 2)$ to (r, θ) with $r > 0$ and $0 \le \theta < 360°$
16. $(-3, 0)$ to (r, θ) with $r > 0$ and $0 \le \theta < 2\pi$
17. $(-5, -5)$ to (r, θ) with $r > 0$ and $0 \le \theta < 2\pi$
18. $(-4, 4\sqrt{3})$ to (r, θ) with $r > 0$ and $0 \le \theta < 360°$
19. $(3\sqrt{3}, 3)$ to (r, θ) with $r < 0$ and $0 \le \theta < 360°$
20. $(2\sqrt{2}, -2\sqrt{2})$ to (r, θ) with $r < 0$ and $-2\pi < \theta \le 0$

In Exercises 21–24, use a calculator to convert the given polar coordinates to Cartesian coordinates. Round each of the coordinates x and y to three significant digits.

21. $(2.45, 32.8°)$
22. $(343, -128.2°)$
23. $(-0.943, 1.23)$
24. $(-5.56, -2.44)$

In Exercises 25–28, use a calculator to convert the given Cartesian coordinates to polar coordinates with $r > 0$ and $0 \le \theta < 2\pi$. Round each of the coordinates r and θ to three significant digits.

25. $(12.4, 15.3)$
26. $(1.04, -4.78)$
27. $(-126, 236)$
28. $(-0.976, -0.561)$

In Exercises 29–48, express each polar equation as an equivalent Cartesian equation, then sketch its graph. Use a graphing calculator in polar mode to verify each answer.

29. $r = 1$
30. $r = 4$
31. $\theta = \dfrac{\pi}{3}$
32. $\theta = \dfrac{5\pi}{6}$
33. $r = \dfrac{-2}{\sin \theta}$
34. $r = 4 \csc \theta$
35. $r = \dfrac{4}{\cos \theta}$
36. $r = -\sec \theta$
37. $r = 10 \cos \theta$
38. $r = -8 \cos \theta$
39. $r = -5 \sin \theta$
40. $r = 3 \sin \theta$
41. $r = 4 \sin \theta + 2 \cos \theta$
42. $r = 6 \sin \theta - 2 \cos \theta$
43. $r = 8 \cos \theta - \sin \theta$
44. $r = -2 \sin \theta - 3 \cos \theta$
45. $r = \dfrac{2}{\cos \theta + \sin \theta}$
46. $r = \dfrac{12}{4 \sin \theta - 3 \cos \theta}$
47. $r = \dfrac{-15}{3 \cos \theta + 5 \sin \theta}$
48. $r = \dfrac{6}{2 \cos \theta - \sin \theta}$

For each of the equations in Exercises 49–58, find values of θ in one trace of the curve at which $|r|$ attains its maximum value. Use a graphing calculator in polar mode to verify each answer.

49. $r = 2 \cos 2\theta$ (four-leafed rose)
50. $r = 2 \cos 3\theta$ (three-leafed rose)
51. $r = 3 \sin 5\theta$ (five-leafed rose)
52. $r = -4 \sin 6\theta$ (twelve-leafed rose)
53. $r = 1 - \sin \theta$ (cardioid)
54. $r = 2 + 2 \cos \theta$ (cardioid)
55. $r = 2 + \sin \theta$ (limaçon)
56. $r = 1 - 3 \cos \theta$ (limaçon)
57. $r^2 = 4 \cos 2\theta$ (lemniscate)
58. $r^2 = 4 \sin 2\theta$ (lemniscate)

The polar equations in Exercise 59 and 60 are usually refered to as Archimedean spirals (named after the Greek mathematician Archimedes, 287?–212 B.C.). Choose an appropriate viewing rectangle on a graphing calculator that shows three loops of each spiral.

59. $r = \dfrac{\theta}{\pi}$ for $\theta \ge 0$
60. $r\theta = \pi$ for $\theta > 0$

Critical Thinking

61. (a) State a polar equation for the x-axis in the Cartesian coordinate system.

    (b) State a polar equation for the y-axis in the Cartesian coordinate system.

62. Find the polar form of the general linear equation $Ax + By = C$. Express the answer in the form $r = f(\theta)$.

63. Find the equivalent polar equation of each Cartesian equation. Express each answer in the form $r = f(\theta)$.
    (a) $(x - 1)^2 + (y - 1)^2 = 2$
    (b) $x^2 - 4y - 4 = 0$
    (c) $4x^2 + 3y^2 + 2y = 1$
    (d) $x^2 - 3y^2 + 4y = 1$

64. Describe the symmetry of the graph of the polar equation $r = f(\theta)$ for each condition.
    (a) $f(-\theta) = f(\theta)$ (b) $f(\theta + \pi) = f(\theta)$

65. Given that a is a nonzero real number, show that
    (a) the graph of $r = \dfrac{a}{\sin \theta}$ is a horizontal line a units above the pole if $a > 0$, or $|a|$ units below the pole if $a < 0$.
    (b) the graph of $r = \dfrac{a}{\cos \theta}$ is a vertical line a units to the right of the pole if $a > 0$, or $|a|$ units to the left of the pole if $a < 0$.

66. Given that a is a nonzero real number, show that
    (a) the graph of $r = a \sin \theta$ is a circle with radius $|a|/2$ and center $(a/2, 90°)$.
    (b) the graph of $r = a \cos \theta$ is a circle with radius $|a|/2$ and center $(a/2, 0°)$.

67. Unlike the intersection of two Cartesian equations, two polar equations that are graphed in the same plane may have intersection points whose coordinates do not satisfy both equations.
    (a) Determine the point of intersection of the polar equations $r = 4 \sin \theta$ and $r = 4 \cos \theta$, with $0 \le \theta \le \pi$, by solving the equations simultaneously.
    (b) Generate the graphs of the equations $r = 4 \sin \theta$ and $r = 4 \cos \theta$, with $0 \le \theta \le \pi$, in the same viewing rectangle of a graphing calculator. Note that these graphs have two points of intersection—only one was found by solving the equations simultaneously. Do the polar coordinates of this other intersection point satisfy both equations? Explain.

68. Generate the graphs of $r = 2 \sin 2\theta$ and $r = 1$ in the same viewing rectangle of a graphing calculator. In how many distinct points do the graphs of these equations intersect? Which of these point(s) satisfy both equations?

69. Suppose a circle has center with polar coordinates (a, α) and radius a, as shown in the sketch. Use the law of cosines (Section 9.2) to find the polar equation of this circle. Express the equation in the form $r = f(\theta)$.

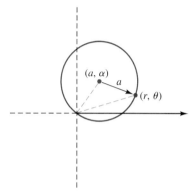

70. Show that the distance d between two points $P(a, \alpha)$ and $Q(b, \beta)$ in the polar plane is given by
$$d = \sqrt{a^2 + b^2 - 2ab \cos(\alpha - \beta)}.$$
Then use this *distance formula* to find the distance between the given points.
    (a) $(3, 2\pi/3)$ and $(4, \pi/6)$
    (b) $(-2, \pi/2)$ and $(4, -5\pi/6)$

9.7 Parametric Equations

◆ Introductory Comments

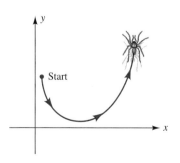

FIGURE 9.59
Path of a spider over a certain time interval

Over a certain time interval, a spider crawls along the path that is shown in Figure 9.59. To describe this situation mathematically, we might find a Cartesian equation in x and y (or a polar equation in r and θ) for the path of the spider. However, the Cartesian (or polar) equation does not tell us the position of the spider at a particular time t, nor does it tell us the direction in which the spider travels along this path. Since the x- and y-coordinates of the spider vary with time t, both x and y are functions of t:

$$x = f(t) \quad \text{and} \quad y = g(t)$$

where t ranges over some interval. We refer to the equations $x = f(t)$ and $y = g(t)$ as the **parametric equations** that describe the path of the spider and the auxilliary variable *t*, on which both *x* and *y* depend, as the **parameter**.

Parametric equations not only specify a set of points (x, y) that describe the curve, but they also describe the location on the curve at time *t* and the direction in which the curve is traced. In this section, we discuss the parametric equations of several curves. We begin by defining the parametric equations of a line.

◆ Parametric Equations of a Line

The set of equations

$$x = x_1 + t \quad \text{and} \quad y = y_1 + mt$$

are **parametric equations of a line** with *t* as the parameter. The line passes through the point (x_1, y_1) in the Cartesian plane and has slope *m*. To show that this is true, we eliminate the parameter *t* by solving for *t* in the first equation, $t = x - x_1$, and substituting this expression in the other equation:

$$y = y_1 + mt$$

$$y = y_1 + m(x - x_1) \quad \text{Substitute } x - x_1 \text{ for } t$$

$$y - y_1 = m(x - x_1) \quad \text{Subtract } y_1 \text{ from both sides}$$

We recognize the equation $y - y_1 = m(x - x_1)$ as the point-slope form of a line that passes through the point (x_1, y_1) and has slope *m* (see Section 3.2).

EXAMPLE 1 Identify and sketch the curve defined by the parametric equations

$$x = -1 + t \quad \text{and} \quad y = 3 + \tfrac{2}{3}t \quad \text{for } 0 \le t \le 6.$$

Indicate with arrows the direction of travel along the curve.

◆ **Solution** We recognize this set of equations as parametric equations of a line. The line passes through the point $(-1, 3)$ and has slope $\tfrac{2}{3}$. To determine the portion of the line traversed in the specified time interval and the direction of travel along that portion, we select values of *t* in the interval [0, 6] and find the corresponding values of *x* and *y*:

t	0 (start)	3	6 (end)
$x = -1 + t$	-1	2	5
$y = 3 + \tfrac{2}{3}t$	3	5	7

SECTION 9.7 ♦ *Parametric Equations*

We plot these points (x, y) in the Cartesian plane and then observe that the graph starts at the point $(-1, 3)$ and travels along a linear path until it reaches the point $(5, 7)$ as illustrated in Figure 9.60.

PROBLEM 1 Show that the parametric equations

$$x = 5 - t \quad \text{and} \quad y = 7 - \tfrac{2}{3}t \quad \text{for } 0 \leq t \leq 6$$

define the line segment shown in Figure 9.60, but with opposite direction.

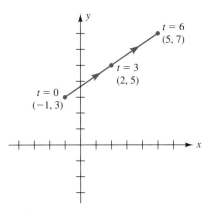

FIGURE 9.60
The parametric equations $x = -1 + t$ and $y = 3 + \tfrac{2}{3}t$, for $0 \leq t \leq 6$, define a line segment that passes through the point $(-1, 3)$ and has slope $\tfrac{2}{3}$.

◆ Parametric Equations of a Parabola

The set of equations

$$x = 2pt \quad \text{and} \quad y = pt^2$$

are **parametric equations of a parabola** with t as the parameter. The parabola has a *vertical axis* of symmetry, vertex $(0, 0)$, and focus $(0, p)$. To show that this is true, we eliminate the parameter t by solving for t in the first equation, $t = x/(2p)$, and substituting this expression in the other equation:

$$\begin{aligned} y &= pt^2 \\ &= p\left(\frac{x}{2p}\right)^2 \quad &\text{Substitute } \frac{x}{2p} \text{ for } t \\ &= \frac{1}{4p}x^2 \quad &\text{Square the quotient and simplify} \end{aligned}$$

We recognize the equation $y = \dfrac{1}{4p}x^2$ as a parabola with vertical axis of symmetry, vertex at the origin, and focus $(0, p)$ (see Section 6.2). In a similar manner, we can show that the set of equations

$$x = pt^2 \quad \text{and} \quad y = 2pt$$

are **parametric equations of a parabola** with *horizontal axis* of symmetry, vertex $(0, 0)$, and focus $(p, 0)$.

EXAMPLE 2 Identify and sketch the curve defined by the parametric equations

$$x = -t \quad \text{and} \quad y = -\tfrac{1}{2}t^2 \quad \text{for } t \geq 0.$$

Indicate with arrows the direction of travel along the curve.

◆ **Solution** We recognize this set of equations as parametric equations of a parabola. The parabola has vertical axis of symmetry, vertex (0, 0), and focus $(0, -\frac{1}{2})$. To determine the portion of the parabola traversed and the direction of travel along that portion, we select values of t in the interval $[0, \infty)$ and find the corresponding values of x and y:

t	0 (start)	2	4	$+\infty$ (approach)
$x = -t$	0	-2	-4	$-\infty$
$y = -\frac{1}{2}t^2$	0	-2	-8	$-\infty$

We plot these points (x, y) in the Cartesian plane and then observe that the graph starts at the point (0, 0) and travels downward to the left along a parabolic path as $t \to \infty$ (see Figure 9.61). ◆

PROBLEM 2 Identify and sketch the curve defined by the parametric equations

$$x = t^2 \quad \text{and} \quad y = 2t \quad \text{for } -2 \leq t \leq 2.$$

Indicate with arrows the direction of travel along the curve. ◆

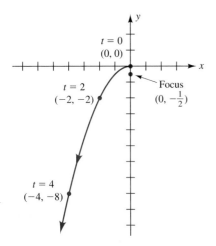

FIGURE 9.61
The parametric equations $x = -t$ and $y = -\frac{1}{2}t^2$, for $t \geq 0$, define the left-hand side of a parabola with vertical axis of symmetry, vertex (0, 0), and focus $(0, -\frac{1}{2})$.

◆ Parametric Equations of an Ellipse or Circle

The set of equations

$$x = a \cos \theta \quad \text{and} \quad y = b \sin \theta$$

with $a > 0$, $b > 0$, and $0 \leq \theta < 2\pi$ are **parametric equations of an ellipse** (or a **circle** if $a = b$) with θ as the parameter. The center of the ellipse is the origin, the horizontal axis has length $2a$, and the vertical axis has length $2b$. To show that this is true, we eliminate the parameter θ by squaring both x/a and y/b and then adding these expressions together as follows:

$$\left(\frac{x}{a}\right)^2 + \left(\frac{y}{b}\right)^2 = (\cos \theta)^2 + (\sin \theta)^2 \qquad \text{Substitute } \frac{x}{a} = \cos \theta, \frac{y}{b} = \sin \theta$$

$$\frac{x^2}{a^2} + \frac{y^2}{b^2} = \cos^2\theta + \sin^2\theta \qquad \text{Square the expressions}$$

$$\frac{x^2}{a^2} + \frac{y^2}{b^2} = 1 \qquad \text{Apply } \sin^2\theta + \cos^2\theta = 1$$

We recognize the equation $(x^2/a^2) + (y^2/b^2) = 1$ as the equation of an ellipse with center at the origin, horizontal axis of length $2a$, and vertical axis

SECTION 9.7 ◆ Parametric Equations

of length $2b$ (see Section 6.3). Of course, if $a = b$ we have the equation of a circle with center at the origin and radius a.

EXAMPLE 3 Identify and sketch the curve defined by the parametric equations

$$x = 2 \cos \pi t \quad \text{and} \quad y = 4 \sin \pi t \quad \text{for } 0 \leq t \leq 2.$$

Indicate with arrows the direction of travel along the curve.

◆ **Solution** We recognize this set of equations as parametric equations of an ellipse with center at the origin, horizontal axis of length $2(2) = 4$, and vertical axis of length $2(4) = 8$. To determine the portion of the ellipse traversed and the direction of travel along that portion, we select values of t in the interval $[0, 2]$ and find the corresponding values of x and y:

t	0 (start)	$\frac{1}{2}$	1	$\frac{3}{2}$	2 (end)
$x = 2 \cos \pi t$	2	0	-2	0	2
$y = 4 \sin \pi t$	0	4	0	-4	0

We plot these points (x, y) in the Cartesian plane and then observe that the curve starts at the point $(2, 0)$ and travels counterclockwise along an elliptical path until it returns to the point at which it started, $(2, 0)$, as illustrated in Figure 9.62. ◆

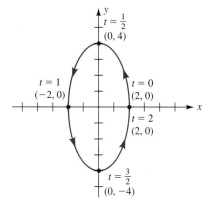

FIGURE 9.62

The parametric equations $x = 2 \cos \pi t$ and $y = 4 \sin \pi t$, for $0 \leq t \leq 2$, define an ellipse with center at the origin, horizontal axis of length 4, and vertical axis of length 8.

The parametric equations

$$x = a \sin \theta \quad \text{and} \quad y = b \cos \theta$$

with $a > 0$, $b > 0$, and $0 \leq \theta < 2\pi$, also represent an ellipse with center at the origin, horizontal axis of length $2a$ and vertical axis of length $2b$. However, in this case, the direction of travel along the curve is clockwise, starting at $(0, b)$.

PROBLEM 3 Identify and sketch the curve defined by the parametric equations

$$x = 2 \sin \pi t \quad \text{and} \quad y = 4 \cos \pi t \quad \text{for } 0 \leq t \leq 2.$$

Indicate with arrows the direction of travel along the curve. ◆

◆ Parametric Equations of a Hyperbola

The set of equations

$$x = a \sec \theta \quad \text{and} \quad y = b \tan \theta$$

with $a > 0$, $b > 0$, and $0 \leq \theta < 2\pi$ ($\theta \neq \pi/2, 3\pi/2$) are **parametric equations of a hyperbola** with θ as the parameter. The center of the hyperbola is the origin, the transverse axis is horizontal with length $2a$, and the conjugate axis is vertical with length $2b$. To show that this is true, we eliminate the parameter θ by squaring both x/a and y/b and then subtracting these expressions as follows:

$$\left(\frac{x}{a}\right)^2 - \left(\frac{y}{b}\right)^2 = (\sec \theta)^2 - (\tan \theta)^2 \quad \text{Substitute } \frac{x}{a} = \sec \theta, \frac{y}{b} = \tan \theta$$

$$\frac{x^2}{a^2} - \frac{y^2}{b^2} = \sec^2 \theta - \tan^2 \theta \quad \text{Square the expressions}$$

$$\frac{x^2}{a^2} - \frac{y^2}{b^2} = 1 \quad \text{Apply } \sec^2 \theta = 1 + \tan^2 \theta$$

We recognize the equation $(x^2/a^2) - (y^2/b^2) = 1$ as the equation of a hyperbola with center at the origin, horizontal transverse axis of length $2a$, and vertical conjugate axis of length $2b$ (see Section 6.4).

In a similar manner, we can show that the set of equations

$$x = a \tan \theta \quad \text{and} \quad y = b \sec \theta$$

are **parametric equations of a hyperbola** with center at the origin, vertical transverse axis of length $2b$, and horizontal conjugate axis of length $2a$.

EXAMPLE 4 Identify and sketch the curve defined by the parametric equations

$$x = 2 \sec \frac{\pi}{3} t \quad \text{and} \quad y = 4 \tan \frac{\pi}{3} t \quad \text{for } 0 \leq t < \tfrac{3}{2}.$$

Indicate with arrows the direction of travel along the curve.

◆ **Solution** We recognize this set of equations as parametric equations of a hyperbola with center at the origin, horizontal transverse axis of length $2(2) = 4$, and vertical conjugate axis of length $2(4) = 8$. To determine the portion of the hyperbola traversed and the direction of travel along that portion, we select values of t in the interval $[0, \tfrac{3}{2})$ and find the corresponding values of x and y:

SECTION 9.7 ◆ Parametric Equations

t	0 (start)	$\frac{1}{2}$	1	$\frac{3}{2}$ (approach)
$x = 2 \sec \frac{\pi}{3} t$	2	$\frac{4}{\sqrt{3}} \approx 2.3$	4	$+\infty$
$y = 4 \tan \frac{\pi}{3} t$	0	$\frac{4}{\sqrt{3}} \approx 2.3$	$4\sqrt{3} \approx 6.9$	$+\infty$

We plot these points (x, y) in the Cartesian plane and then observe that the curve starts at the point (2, 0) and travels upward to the right along a hyperbolic path as $t \to \frac{3}{2}$ (see in Figure 9.63). ◆

PROBLEM 4 Identify and sketch the curve defined by the parametric equations

$$x = 2 \tan \frac{\pi}{3} t \quad \text{and} \quad y = 4 \sec \frac{\pi}{3} t \quad \text{for } 0 \le t < \tfrac{3}{2}.$$

Indicate with arrows the direction of travel along the curve. ◆

FIGURE 9.63

The parametric equations $x = 2 \sec \frac{\pi}{3} t$ and $y = 4 \tan \frac{\pi}{3} t$, for $0 \le t < \tfrac{3}{2}$, define the upper right branch of a hyperbola with center at the origin, horizontal transverse axis of length 4, and vertical conjugate axis of length 8.

◆ Parametric Equations and the Graphing Calculator

Most graphing calculators are capable of generating the curve defined by a set of parametric equations. For example, to generate the ellipse defined by the parametric equations in Example 3,

$$x = 2 \cos \pi t \quad \text{and} \quad y = 4 \sin \pi t \quad \text{for } 0 \le t \le 2$$

we begin by selecting parametric mode and radian mode on the calculator. Next, we press the [WINDOW] key or [RANGE] key and enter maximum and minimum values of t, x, and y. Since we want $0 \le t \le 2$, we choose

t-min = 0	t-max = 2	t-step = 0.1
smallest t-value to be evaluated	largest t-value to be evaluated	increment between t-values

To obtain a true geometric perspective of the ellipse, we choose a square set viewing rectangle. We now enter the parametric equations $x = 2 \cos \pi t$ and $y = 4 \sin \pi t$ to generate the graph in Figure 9.64. Note that the calculator generates the ellipse in the same direction of travel as we observed in Example 3, that is, the curve starts at the point (2, 0) and travels counterclockwise until it returns to the point (2, 0).

We can also generate the graph of a polar equation $r = f(\theta)$ by using the parametric feature on a graphing calculator. Recall from Section 9.6 these relationships between polar and Cartesian coordinates:

$$x = r \cos \theta \quad \text{and} \quad y = r \sin \theta$$

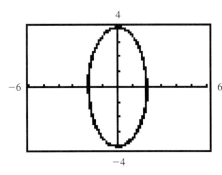

FIGURE 9.64

Graph of the parametric equations $x = 2 \cos \pi t$ and $y = 4 \sin \pi t$, for $0 \le t \le 2$, as shown on a graphing calculator

If we replace r with $f(\theta)$, we obtain the parametric equations

$$x = f(\theta) \cos \theta \quad \text{and} \quad y = f(\theta) \sin \theta$$

with parameter θ. Now replacing θ with t gives us the polar equation $r = f(\theta)$ in parametric form with parameter t.

◆ Polar Equation in Parametric Form

> The graph of the polar equation $r = f(\theta)$ is the curve defined by the parametric equations
>
> $$x = f(t) \cos t \quad \text{and} \quad y = f(t) \sin t$$
>
> with t as the parameter.

EXAMPLE 5 Express the polar equation $r = f(\theta) = \sqrt{2} \sin 2\theta$ as a set of parametric equations. Then generate the graph of these parametric equations in the viewing rectangle of a graphing calculator.

◆ **Solution** The polar equation $r = f(\theta) = \sqrt{2} \sin 2\theta$ may be described by the parametric equations

$$x = f(t) \cos t \quad \text{and} \quad y = f(t) \sin t$$
$$x = \sqrt{2} \sin 2t \cos t \quad \quad y = \sqrt{2} \sin 2t \sin t$$

The graph of these parametric equations is shown in Figure 9.65. We have chosen the t-values

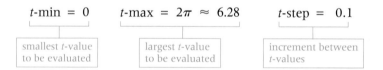

and a square set viewing rectangle. Note that the graph of the parametric equations $x = \sqrt{2} \sin 2t \cos t$ and $y = \sqrt{2} \sin 2t \sin t$ is identical to the graph of the polar equation $r = \sqrt{2} \sin 2\theta$ (see Figure 9.58 in Section 9.6). ◆

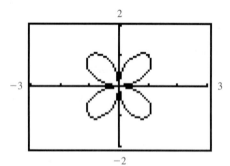

FIGURE 9.65

Graph of the parametric equations
$x = \sqrt{2} \sin 2t \cos t$ and
$y = \sqrt{2} \sin 2t \sin t$, as shown on a graphing calculator.

PROBLEM 5 Rework Example 5 for the polar equation $r = f(\theta) = 2 \sin 3\theta$. ◆

◆ Application: Projectile Motion

Nearly four hundred years ago, Galileo discovered the laws of motion for freely falling objects. He discovered than when a projectile is fired straight upward from the ground with an initial velocity v (in feet per second), its distance d (in feet) above the ground after time t (in seconds) is given by

$$d = vt - 16t^2.$$

We can use parametric equations to extend this idea to cases in which the initial velocity v is at an angle θ other than 90° from the horizontal. As shown in Figure 9.66, we may resolve the initial velocity v into its horizontal component v_x and its vertical component v_y, where

$$v_x = v \cos \theta \quad \text{and} \quad v_y = v \sin \theta.$$

The motion in the horizontal direction is one of constant velocity v_x. Hence, the position x of the projectile in the horizontal direction at any time t is given by

$$x = v_x t = (v \cos \theta)t.$$

The motion in the vertical direction is the same as a projectile fired straight upward with initial velocity v_y. Thus, the position y of the projectile in the vertical direction at any time t is given by

$$y = v_y t - 16t^2 = (v \sin \theta)t - 16t^2.$$

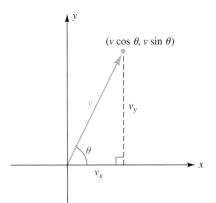

FIGURE 9.66
Initial velocity v resolved into its horizontal component v_x and vertical component v_y

◆ **Parametric Equations for Projectile Motion**

If a projectile is fired from the ground with an initial velocity v at an angle θ with the horizontal, then the position (x, y) of the projectile t seconds after it is fired is given by the parametric equations

$$x = (v \cos \theta)t \quad \text{and} \quad y = (v \sin \theta)t - 16t^2$$

where x and y are measured in feet.

EXAMPLE 6 *Chapter Opening Problem*

A football is kicked with an initial velocity of 80 feet per second (ft/s) at an angle of 60° with the horizontal. Assuming that the ball is caught at the same height as it is kicked, determine

(a) the parametric equations that describe the path of the ball, and

(b) the distance (in yards) downfield that the football is caught.

◆ Solution

(a) If $v = 80$ ft/s and $\theta = 60°$, then the parametric equations that describe the path of the ball are

$$x = (v \cos \theta)t \quad \text{and} \quad y = (v \sin \theta)t - 16t^2$$
$$= (80 \cos 60°)t \qquad\qquad = (80 \sin 60°)t - 16t^2$$
$$= 40t \qquad\qquad\qquad\quad = 40\sqrt{3}\,t - 16t^2$$

(b) The football is caught when $y = 0$, which implies that

$$y = 40\sqrt{3}\,t - 16t^2 = 0$$
$$t(40\sqrt{3} - 16t) = 0$$
$$t = 0 \quad \text{or} \quad 40\sqrt{3} - 16t = 0$$
$$t = \frac{5\sqrt{3}}{2} \approx 4.33$$

In football jargon, $t \approx 4.33$ seconds is referred to as the "hang time" of the kick—the time the ball is in the air. Hence, the horizontal distance downfield at which the football is caught is given by

$$x = 40t = 40\left(\frac{5\sqrt{3}}{2}\right) \approx 173 \text{ ft} \quad \text{or} \quad 58 \text{ yd}$$

Figure 9.67 shows the path of the ball as graphed in the viewing rectangle of a graphing calculator. This graph confirms our work. ◆

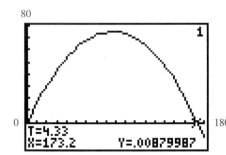

FIGURE 9.67
Graph of the parametric equations $x = 40t$, and $y = 40\sqrt{3}\,t - 16t^2$, for $0 \leq t \leq 4.33$

PROBLEM 6 Identify the graph of the familiar Cartesian equation that results if we eliminate the parameter t in the parametric equations developed in Example 6. Use this Cartesian equation to determine how high the ball rises in its path of motion. Use a graphing calculator to check your answer by generating the graph of the parametric equations in the viewing rectangle and tracing to the maximum point. ◆

Exercises 9.7

◆ Basic Skills

In Exercises 1–20, identify and sketch the curve defined by each set of parametric equations, and indicate with arrows the direction of travel along the curve as t increases in value. Use a graphing calculator set in parametric mode to confirm each sketch.

1. $x = t$ and $y = 2t$ for $0 \leq t \leq 4$
2. $x = t$ and $y = -\frac{1}{2}t$ for $0 \leq t \leq 8$
3. $x = -2 + t$ and $y = 1 - 3t$ for $t \geq 0$
4. $x = 3 + t$ and $y = -1 + 4t$ for $t \leq 0$
5. $x = -2t$ and $y = -t^2$ for $t \leq 0$
6. $x = 2t^2$ and $y = 4t$ for $t \geq 0$
7. $x = 3t^2$ and $y = 6t$ for $-2 \leq t \leq 2$
8. $x = -\frac{2}{3}t$ and $y = -\frac{1}{3}t^2$ for $0 \leq t \leq 6$
9. $x = \cos \pi t$ and $y = \sin \pi t$ for $0 \leq t \leq 2$
10. $x = 2\cos\frac{2\pi}{3}t$ and $y = 2\sin\frac{2\pi}{3}t$ for $0 \leq t \leq 3$
11. $x = 3\sin\frac{\pi}{3}t$ and $y = 3\cos\frac{\pi}{3}t$ for $0 \leq t \leq 3$
12. $x = 5\sin \pi t$ and $y = 5\cos \pi t$ for $0 \leq t \leq \frac{1}{2}$
13. $x = \cos\frac{\pi}{2}t$ and $y = 2\sin\frac{\pi}{2}t$ for $0 \leq t \leq 4$
14. $x = 4\cos 3\pi t$ and $y = 3\sin 3\pi t$ for $0 \leq t \leq \frac{2}{3}$
15. $x = 3\sin 2\pi t$ and $y = 2\cos 2\pi t$ for $0 \leq t \leq \frac{1}{4}$
16. $x = \sin\frac{3\pi}{4}t$ and $y = 3\cos\frac{3\pi}{4}t$ for $0 \leq t \leq 2$
17. $x = \sec \pi t$ and $y = \tan \pi t$ for $\frac{1}{2} < t < \frac{3}{2}$
18. $x = 2\tan\frac{\pi}{2}t$ and $y = 3\sec\frac{\pi}{2}t$ for $1 < t < 3$

19. $x = 3 \tan \frac{3\pi}{4} t$ and $y = 2 \sec \frac{3\pi}{4} t$
for $0 \leq t \leq 2$, provided $t \neq \frac{2}{3}$

20. $x = 4 \sec 4\pi t$ and $y = 2 \tan 4\pi t$
for $0 \leq t \leq \frac{1}{4}$, provided $t \neq \frac{1}{8}$

In Exercises 21–30, find a set of parametric equations that describes a curve with the given characteristics. (Answers are not unique).

21. Line with slope -4 and passing through $(-1, 4)$
22. Line that passes through $(-3, -5)$ and $(0, 1)$
23. Parabola with vertex $(0, 0)$ and focus $(0, 2)$
24. Parabola with vertex $(0, 0)$ and directrix $x = 1$
25. Circle with center $(0, 0)$ and radius 4
26. Circle with center $(0, 0)$ and radius 3
27. Ellipse with center $(0, 0)$, horizontal axis of length 6, and vertical axis of length 4
28. Ellipse with center $(0, 0)$, horizontal axis of length 10, and eccentricity $\frac{3}{5}$
29. Hyperbola with center $(0, 0)$, horizontal transverse axis of length 4, and vertical conjugate axis of length 4
30. Hyperbola with center $(0, 0)$, vertical transverse axis of length 8, and eccentricity $\frac{5}{4}$

In Exercises 31–40,
(a) *find the Cartesian equation in x and y for the given parametric equations and*
(b) *use the vertical and horizontal shift rules and the x-axis reflection rule (Section 2.5) to sketch the curve defined by the parametric equations. Use a graphing calculator set in parametric mode to confirm each sketch.*

31. $x = t$ and $y = |t| - 1$ for $-9 \leq t \leq 9$
32. $x = t$ and $y = 2 - t^2$ for $-3 \leq t \leq 3$
33. $x = t - 2$ and $y = \frac{1}{t}$
for $-8 \leq t \leq 8$ provided $t \neq 0$
34. $x = t + 4$ and $y = |t|$ for $-5 \leq t \leq 5$
35. $x = t + 1$ and $y = -t^3$ for $-3 \leq t \leq 3$
36. $x = t - 3$ and $y = -\sqrt{t}$ for $t \geq 0$
37. $x = t + 3$ and $y = \sqrt{t} - 1$ for $t \geq 0$
38. $x = t - 1$ and $y = t^3 + 1$ for $-3 \leq t \leq 3$
39. $x = t - 5$ and $y = 4 - t^2$ for $-10 \leq t \leq 10$
40. $x = t + 2$ and $y = 1 - \sqrt[3]{t}$ for $-27 \leq t \leq 27$

In Exercises 41–50, express each polar equation as a set of parametric equations. Then generate the graph of these parametric equations in the viewing rectangle of a graphing calculator.

41. $r = 2 \cos 2\theta$ (four-leafed rose)
42. $r = 2 \cos 3\theta$ (three-leafed rose)
43. $r = 3 \sin 5\theta$ (five-leafed rose)
44. $r = -4 \sin 6\theta$ (twelve-leafed rose)
45. $r = 1 - \sin \theta$ (cardioid)
46. $r = 2 + 2 \cos \theta$ (cardioid)
47. $r = 2 + \sin \theta$ (limaçon)
48. $r = 1 - 3 \cos \theta$ (limaçon)
49. $r^2 = 4 \cos 2\theta$ (lemniscate)
50. $r^2 = 4 \sin 2\theta$ (lemniscate)

51. A batter hits a baseball with an initial velocity of 90 feet per second (ft/s) at an angle of 45° with the horizontal, as shown in the sketch. Assuming that the ball is caught at the same height as it is hit, determine
   (a) the parametric equations that describe the path of the ball,
   (b) the distance d from the place where the ball is hit to the place where it is caught,
   (c) the Cartesian equation in x and y that descibes the path of the ball, and
   (d) the maximum height h the ball rises in its path of motion.

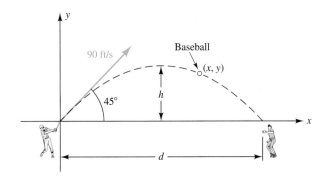

52. A rocket is fired from the ground with an initial velocity of 1200 feet per second (ft/s) at an angle of 30° with the horizontal, as shown in the sketch. Assuming that the ground is level, determine
   (a) the parametric equations that describe the path of the rocket,
   (b) the distance d from the firing point to the landing point,
   (c) the Cartesian equation in x and y that descibes the path of the rocket, and

Exercise 52 continues

(d) the maximum height h the rocket rises in its path of motion.

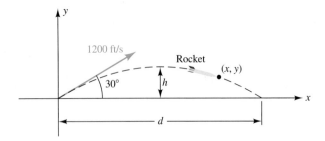

53. When a certain rocket is fired from the ground with an initial velocity v at an angle θ, its path is described by the curve shown in the sketch. Its position t seconds after it is fired is given by the parametric equations

$$x = 544t \quad \text{and} \quad y = 748t - 16t^2$$

with x and y measured in feet. Find the range R (the horizontal distance from the firing point to the landing point) and maximum height h to which the rocket rises.

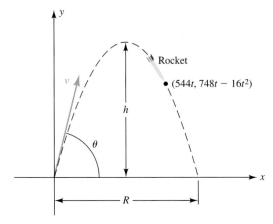

54. For the rocket described in Exercise 53, find the initial velocity v and the angle θ at which the rocket is fired.

Critical Thinking

55. Parametric equations of the form

$$v_1 = a \sin \omega_1 t \quad \text{and} \quad v_2 = b \cos \omega_2 t \quad \text{for } t \geq 0$$

where a, b, ω_1, and ω_2 are constants, occur frequently in electrical applications. The variables v_1 and v_2 represent voltage. The graph of these parametric equations is called a **Lissajous figure** (after the French physicist Jules Lissajous, 1822–1880). Use a graphing calculator to determine the type of Lissajous figure that results for the given ratio of ω_1 to ω_2.

(a) 1 to 1   (b) 2 to 1   (c) 1 to 3   (d) 2 to 3

56. Show that the set of equations

$$x = x_1 + (x_2 - x_1)t \quad \text{and} \quad y = y_1 + (y_2 - y_1)t$$

are parametric equations of a line through the points $P(x_1, y_1)$ and $Q(x_2, y_2)$. Use this set of equations to write parametric equations of a line through the points $(2, 3)$ and $(-4, 5)$ with the given direction of travel along the line.

(a) from $(2, 3)$ to $(-4, 5)$ (b) from $(-4, 5)$ to $(2, 3)$

57. Show that the set of equations

$$x = h + a \cos \theta \quad \text{and} \quad y = k + b \sin \theta$$

for $0 \leq \theta < 2\pi$ are parametric equations of an ellipse (or a circle if $a = b$) with center at (h, k), horizontal axis of length $2a$, and vertical axis of length $2b$.

58. Show that the set of equations

$$x = h + a \sec \theta \quad \text{and} \quad y = k + b \tan \theta$$

for $-\pi/2 < \theta < \pi/2$ and $\pi/2 < \theta < 3\pi/2$, are parametric equations of a hyperbola with center at (h, k), horizontal transverse axis of length $2a$, and vertical conjugate axis of length $2b$.

In Exercises 59–62, use the formulas in Exercises 57 and 58 to find a set of parametric equations that describes the conic section with the given characteristics. Use a graphing calculator set in parametric mode to check each answer.

59. Ellipse with center $(2, 1)$, horizontal axis of length 4, and vertical axis of length 2

60. Ellipse with center $(-2, 3)$, horizontal axis of length 3, and vertical axis of length 5

61. Hyperbola with center $(-1, -2,)$, horizontal transverse axis of length 2, and vertical conjugate axis of length 2

62. Circle with center $(2, -4)$ and radius 6

63. Suppose that $x = f(t)$ and $y = g(t)$, for $0 \leq t \leq 2$, are parametric equations of a curve C. Describe the graph of the given set of parametric equations in relation to the graph of C.

(a) $x = f(2t)$ and $y = g(2t)$ for $0 \leq t \leq 1$
(b) $x = f(2 - t)$ and $y = g(2 - t)$ for $0 \leq t \leq 2$
(c) $x = f(t) + 2$ and $y = g(t) - 1$ for $0 \leq t \leq 2$
(d) $x = g(t)$ and $y = f(t)$ for $0 \leq t \leq 2$

64. If a projectile is fired from the ground with an initial velocity v at an angle θ with the horizontal, then the position (x, y) of the projectile t seconds after it is fired is given by the parametric equations

$$x = (v \cos \theta)t \quad \text{and} \quad y = (v \sin \theta)t - 16t^2$$

where x and y are measured in feet.

(a) Show that the path of the projectile is always a parabola.

(b) Show that the range R of the projectile (the horizontal distance from the firing point to the landing point) is given by

$$R = \frac{v^2}{32} \sin 2\theta.$$

(c) Determine the angle θ that makes the range R as large as possible.

Chapter 9 Review

◆ **Questions for Writing or Group Discussion**

1. What is an *oblique triangle*?
2. State the four cases in which it is possible to solve an oblique triangle. In which of these cases is the *law of sines* applied first? the *law of cosines* applied first?
3. Is it possible to solve an oblique triangle in which only the three angles are given? Explain.
4. Discuss the conditions when the *ambiguous case* may occur in an oblique triangle.
5. Give some examples of oblique triangles with erroneous data that have no solution.
6. When solving an oblique triangle in which all three sides are known, why is it a good strategy to find the angle opposite the largest side first?
7. What is a *vector quantity*? Give some examples of vector quantities.
8. When is a vector in *standard position*? How is a vector in standard position specified?
9. What is meant by the *x*- and *y*-components of a vector?
10. Explain the procedure for finding the *magnitude* and *direction angle* of a vector in component form.
11. Illustrate graphically the *scalar multiplication* of real number k times vector **A** when (a) $k > 1$, (b) $0 < k < 1$, (c) $k < -1$, and (d) $-1 < k < 0$.
12. Discuss the *parallelogram law* for vectors.
13. Describe graphically the vectors $\mathbf{A} - \mathbf{B}$ and $\mathbf{B} - \mathbf{A}$.
14. What is meant by the *resultant force* of a *force system*? Explain the procedure for determining the resultant force.
15. When is a force system in *equilibrium*?
16. In the *complex plane*, what names are assigned to the horizontal and vertical axes?
17. Explain the procedure for finding the *modulus* and *argument* of a complex number in standard form.

18. What is the *trigonometric form* of a complex number? Explain the procedure for multiplying and dividing complex numbers in trigonometric form.
19. State *De Moivre's theorem* and illustrate its use with an example.
20. Discuss the geometrical significance of the *n*th roots of a complex number when plotted in the complex plane.
21. How is the *polar coordinate system* formed?
22. Is there a one-to-one correspondence between ordered pairs and points in the polar coordinate system? Explain.
23. Explain the procedure for changing *polar coordinates* to Cartesian coordinates, and vice versa.
24. Given that *a* is a constant, describe the graph of each *polar equation*:

    (a) $r = a$ (b) $\theta = a$ (c) $r = \dfrac{a}{\sin \theta}$ (d) $r = a \cos \theta$

25. Discuss the procedure for writing a polar equation in *parametric form*. Illustrate with an example.
26. What are the parametric equations for *projectile motion*? Discuss the path of the projectile.

Review Exercises

In Exercises 1–10, (a) solve each oblique triangle and (b) find the area of the triangle.

1.

2.

3.

4.

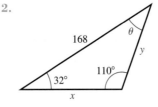

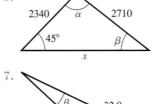

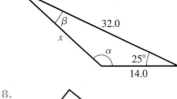

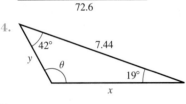

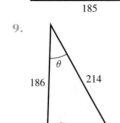

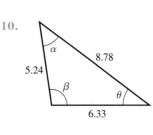

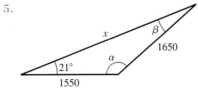

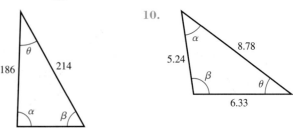

In Exercises 11–22, three parts of an oblique triangle are given. With the understanding that side a is opposite angle α, side b is opposite angle β, and side c is opposite angle θ, solve the oblique triangle with the three given parts. Round angles to the nearest tenth of a degree and sides to three significant digits. If the data allows for the construction of two triangles, find both solutions. If the data does not allow for the construction of any triangle, state that no such triangle exists.

11. $b = 60.4$, $c = 55.4$, $\beta = 78°$
12. $c = 3.71$, $\alpha = 121°$, $\theta = 33°$
13. $a = 19$, $b = 35$, $c = 22$
14. $a = 0.123$, $b = 0.231$, $\theta = 59°$
15. $a = 610$, $c = 950$, $\alpha = 38°$
16. $b = 5.65$, $c = 4.85$, $\beta = 106°$
17. $b = 12.9$, $c = 9.32$, $\theta = 65°$
18. $a = 456$, $b = 575$, $\alpha = 52°$
19. $a = 2.80$, $b = 6.94$, $\theta = 48°$
20. $a = 152$, $b = 355$, $c = 170$
21. $a = 11$, $b = 16$, $c = 5$
22. $a = 3.64$, $c = 1.94$, $\theta = 42°$

23. Use Hero's formula to find the area A of an equilateral triangle with side a.

24. Given that θ is the angle between two adjacent sides a and b of a parallelogram, express the area of the parallelogram in terms of θ, a, and b.

For the vectors in Exercises 25–28, (a) translate the vector into standard position, (b) express the vector in component form, (c) determine its magnitude, and (d) determine its direction angle.

25.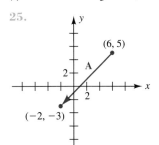

26.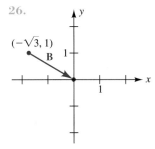

27.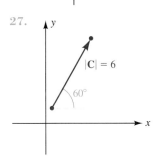

28.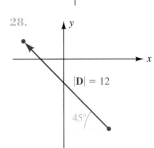

29. Determine the magnitude and direction angle θ of each vector.
   (a) $\mathbf{A} = \langle 12, -5 \rangle$ (b) $\mathbf{C} = \langle \sqrt{7}, 3 \rangle$

30. Determine the x- and y-components of a vector whose magnitude and direction angle θ are given.
   (a) $|\mathbf{A}| = 72.4$, $\theta = 132.7°$
   (b) $|\mathbf{C}| = 643$, $\theta = 250°$

In Exercises 31–38, perform the indicated vector operations given the vectors $\mathbf{A}$, $\mathbf{B}$, $\mathbf{C}$, and $\mathbf{D}$:

$$\mathbf{A} = \langle -3, 4 \rangle \qquad \mathbf{B} = \langle -2, -6 \rangle$$
$$\mathbf{C} = 3\mathbf{i} - 7\mathbf{j} \qquad \mathbf{D} = 3\mathbf{i} + \mathbf{j}$$

Express each answer in component form.

31. $-\mathbf{C}$
32. $\tfrac{1}{2}\mathbf{B}$
33. $\mathbf{A} + \mathbf{C}$
34. $\mathbf{D} - \mathbf{B}$
35. $2\mathbf{A} - 5\mathbf{C}$
36. $-\mathbf{A} + 3(\mathbf{B} - 2\mathbf{C})$
37. $|\mathbf{A}|(\mathbf{D} + 4\mathbf{C})$
38. $|\mathbf{A} + \mathbf{C}|(\mathbf{C} + \mathbf{D})$

39. Illustrate graphically the vector sum in Exercise 33.

40. Illustrate graphically the vector difference in Exercise 34.

In Exercises 41–44, (a) plot each complex number in the complex plane, (b) find its modulus, (c) find its argument, and (d) express the complex number in trigonometric form with $0 \le \theta < 2\pi$.

41. $6 - 6i$
42. $-\sqrt{2} + \sqrt{2}\,i$
43. $-5\sqrt{3} - 5i$
44. $-\tfrac{5}{4}i$

In Exercises 45–54, perform the indicated operations and record the results (a) in trigonometric form with $0 \le \theta < 2\pi$ and (b) in standard form.

45. $\left[3\left(\cos\dfrac{\pi}{3} + i\sin\dfrac{\pi}{3}\right)\right]\left[6\left(\cos\dfrac{\pi}{4} + i\sin\dfrac{\pi}{4}\right)\right]$

46. $\left[5\left(\cos\dfrac{3\pi}{4} + i\sin\dfrac{3\pi}{4}\right)\right]$
    $\times \left[12\left(\cos\dfrac{17\pi}{12} + i\sin\dfrac{17\pi}{12}\right)\right]$

47. $\dfrac{\cos 81° + i\sin 81°}{\cos 141° + i\sin 141°}$

48. $\dfrac{12\,[\cos(4\pi/3) + i\sin(4\pi/3)]}{4 + 4i}$

49. $\left[2\left(\cos\dfrac{17\pi}{12} + i\sin\dfrac{17\pi}{12}\right)\right]^{8}$

50. $\left[\tfrac{1}{3}(\cos 36° + i\sin 36°)\right]^{-5}$

51. $(1 + i)^{-10}$

52. $(\sqrt{2} - \sqrt{2}\,i)(\sqrt{3} + i)^6$

53. $\dfrac{\{\sqrt{2}\,[\cos(\pi/2) + i\sin(\pi/2)]\}^4}{(-1 - i)^5}$

54. $\dfrac{(8 + 6i)^0}{(1 - \sqrt{3}\,i)^3(1 + i)^{-4}}$

In Exercises 55–60, find the indicated roots of each complex number. Record the results (a) in trigonometric form with $0 \leq \theta < 2\pi$ and (b) in standard form.

55. Square roots of $\dfrac{1}{4}\left(\cos\dfrac{2\pi}{3} + i\sin\dfrac{2\pi}{3}\right)$

56. Cube roots of $27\left(\cos\dfrac{3\pi}{4} + i\sin\dfrac{3\pi}{4}\right)$

57. Fourth roots of 1

58. Fifth roots of i

59. Cube roots of $4\sqrt{2} - 4\sqrt{2}\,i$

60. Fourth roots of $-2 - 2\sqrt{3}\,i$

61. One of the fourth roots of a complex number is $\sqrt{3} - i$. Determine the other three fourth roots.

62. Find all the solutions of the given equation.
    (a) $x^3 - 27 = 0$ (b) $x^5 + 1 = -i$

63. Convert the given polar coordinates to Cartesian coordinates.
    (a) $(-2, 45°)$ (b) $\left(6, -\dfrac{\pi}{3}\right)$

64. Convert the given Cartesian coordinates to polar coordinates.
    (a) $(-2, 2)$ with $r < 0$ and $0° \leq \theta < 360°$
    (b) $(-\sqrt{3}, 1)$ with $r > 0$ and $-2\pi < \theta \leq 0$

In Exercises 65–72, sketch the graph of each polar equation and state its equivalent Cartesian equation.

65. $r = 2$

66. $\theta = \pi$

67. $r = \dfrac{1}{\sin\theta}$

68. $r = -3\sec\theta$

69. $r = 5\cos\theta$

70. $r = -2\sin\theta$

71. $r = 6\sin\theta - 8\cos\theta$

72. $r = \dfrac{12}{3\cos\theta - 4\sin\theta}$

In Exercises 73–82, discuss the curve defined by each set of parametric equations.

73. $x = 2 + t$ and $y = 3 - 2t$ for $0 \leq t \leq 6$

74. $x = 3t$ and $y = 1 + t$ for $t \geq 0$

75. $x = -4t$ and $y = -2t^2$ for $t \geq 0$

76. $x = \tfrac{1}{2}t^2$ and $y = t$ for $-2 \leq t \leq 2$

77. $x = 2\sin\pi t$ and $y = 2\cos\pi t$ for $0 \leq t \leq 1$

78. $x = \sin\dfrac{\pi}{3}t$ and $y = \cos\dfrac{\pi}{3}t$ for $0 \leq t \leq 6$

79. $x = 3\sin\dfrac{\pi}{2}t$ and $y = 4\cos\dfrac{\pi}{2}t$ for $0 \leq t \leq 4$

80. $x = \sin\pi t$ and $y = 2\cos\pi t$ for $0 \leq t \leq \tfrac{1}{2}$

81. $x = 2\sec 2\pi t$ and $y = 3\tan 2\pi t$ for $0 \leq t < \dfrac{\pi}{4}$

82. $x = \tan\dfrac{3\pi}{4}t$ and $y = \sec\dfrac{3\pi}{4}t$ for $0 \leq t \leq \tfrac{4}{3}$, provided $t \neq \tfrac{2}{3}$

83. Because of a large oak tree directly on line, a surveyor is unable to measure directly the distance between two points A and B. To determine this distance indirectly, he selects a third point C, which is 75.6 ft from A and 93.6 ft from B. With his transit, he then finds angle BCA to be $72°30'$. Determine, to the nearest tenth of a foot, the distance from A to B.

84. Two passenger trains depart from the same terminal at the same time and travel along straight railroads that make an angle of $65°$ with each other. One of the trains travels at 32 mi/h and the other at 40 mi/h. How far apart are the trains at the end of 2 hours 15 minutes?

85. In a baseball game, a right fielder positions himself 290 ft from home plate along a line that is $18°$ from the foul line, as shown in the sketch. Given that a baseball diamond is a square with 90-ft base paths, find the distance from the right fielder to (a) first base, (b) second base, and (c) third base.

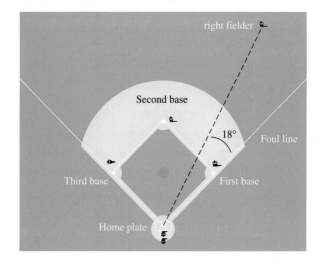

86. A carpenter plans to build a dormer addition to the roof of a Cape Cod cottage, as shown in the figure. Find (a) the length x, (b) angle α, and (c) angle β.

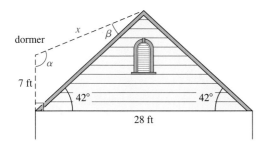

87. As a ship sails into a harbor along a straight course, the captain notices that a buoy off the port side makes an angle of 12° with the path of the ship. Sailing an additional 1200 meters, the captain now observes that the buoy makes an angle of 32° with the path of the ship, as shown in the sketch. What is the closest the ship comes to the buoy?

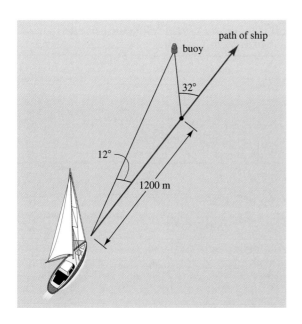

88. The building codes in the city of Boston specify that a single-family dwelling cannot be placed on a lot whose area is less than 15,000 sq ft. If the angle at the street corner of a triangular lot is 73°27′ and the frontage of the lot on one street is 165.6 ft, as shown in the sketch, determine the minimum frontage x on the other street to make the lot the minimum size for a buildable lot.

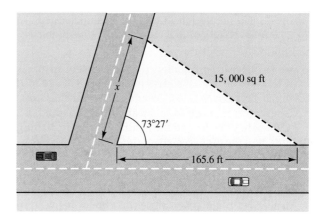

89. A 45-lb force is applied to the handle of a child's wagon at an angle of 34° with the horizontal. Find (a) the force that pulls the wagon horizontally and (b) the force that lifts the wagon vertically.

90. For the force system shown in the sketch, find the magnitude and direction angle of the resultant force.

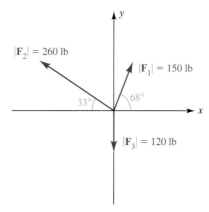

91. For the force system in Exercise 90, find the magnitude and direction angle of an additional force $\mathbf{F}_4$ that produces equilibrium.

92. Two cranes lift a large fishing boat from the water. One crane exerts a force of 18,500 lb on its cable and the other crane exerts a force of 14,200 lb on its cable. If the angle between the cables is 24°, find the magnitude of the resultant force.

93. A sailboat leaves island A and sails 75 miles S-31°-W. It then changes course and sails 112 miles S-18°-E to dock at island B. What is the displacement from island B to island A?

94. The displacement from a Coast Guard station to a disabled ship is known to be 22.7 miles N-9°40′-W. Find

the displacement from a Coast Guard ship, located 25.2 miles N-12°20′-E of the station, to the disabled ship.

95. An aircraft carrier is stationed directly east of an airbase. A jet takes off from the carrier at the speed of 480 mi/h in the direction N-78°15′-W and, at the same instant, a helicopter takes off from the airbase at the speed of 150 mi/h. If the jet and the helicopter meet in 30 minutes, find the displacement of the helicopter from the base.

96. Given an AC circuit in which $R = 12 \, \Omega$, $X_L = 3 \, \Omega$, and $X_C = 12 \, \Omega$, find (a) the magnitude of the impedance and (b) the phase angle (in degrees) between the voltage and current.

97. Two straight roadways are connected by a circular curve with central angle $\theta = 98°$ and radius $r = 100$ ft. A surveyor wants to lay out this circular curve by placing stakes at PC and PT and also at several other points along the curve, such as point A. Assume the curve is placed in a Cartesian plane, as shown in the sketch.

   (a) Find the coordinates of PC and PT and the equation of the circular curve.

   (b) Determine the coordinates of point A, rounded to the nearest hundredth of a foot.

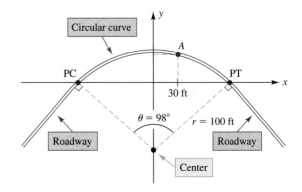

98. A golf ball is hit from the tee with an initial velocity of 150 ft/s at an angle of 35° with the horizontal. Assume that the ball is hit straight down the fairway (no hook, no slice) and that the ground is level.

   (a) Determine the parametric equations that describe the path of the ball.

   (b) Find the distance (in yards) from the tee to the point at which the ball strikes the ground.

   (c) State the Cartesian equation in x and y that describes the path of the ball.

   (d) Find the maximum height the ball rises in its path of motion.

Cumulative Review Exercises for Chapters 7, 8, and 9

1. Use Hero's formula to find the area A of an isosceles triangle with equal sides a and nonequal side b.

2. Use the law of cosines to show that the sum of the squares of the two diagonals of a parallelogram equals the sum of the squares of the four sides of the parallelogram.

3. Find the exact value of each expression.
   (a) $\sin 210° + \cos(-30°) + \tan 405°$
   (b) $\sin \frac{2\pi}{3} \cot \frac{17\pi}{6} - \cos \frac{3\pi}{4}$
   (c) $\arctan(-\sqrt{3}) + \arccos\left(\cos \frac{5\pi}{4}\right)$
   (d) $\tan\left[90° - \sin^{-1} \frac{12}{13}\right]$

4. Find the radian measure of acute angles x and y if $\tan(x + 3y) = 1$ and $\cos(x - 2y) = 1$.

5. Simplify each expression to a constant.
   (a) $\sin^2\theta + \sin\theta \cot\theta \cos\theta$
   (b) $\sin\left(x - \frac{\pi}{6}\right) - \cos\left(\frac{2\pi}{3} - x\right)$
   (c) $\cot^2\theta - \frac{2 \cot\theta}{\tan 2\theta}$
   (d) $\frac{\cos^2(x/2)}{\cos x + \sin^2(x/2)}$

6. If θ is a fourth-quadrant angle with $\cos\theta = \sqrt{\frac{2}{3}}$, find the exact value of each expression.
   (a) $\sin\theta$ (b) $\tan\theta$
   (c) $\sin(\theta + \pi)$ (d) $\cos 2\theta$

7. Show that each trigonometric equation is an identity.
   (a) $(\tan x + \sec x)^2 = \frac{1 + \sin x}{1 - \sin x}$
   (b) $\sec^2 x \sin^2 x + (\sin x + \cos x)^2 - \sec^2 x = \sin 2x$
   (c) $1 + \tan x \tan \frac{x}{2} = \sec x$
   (d) $\frac{1 - \cos 6x}{1 + \cos 6x} = \tan^2 3x$
   (e) $\sqrt{\frac{\csc x - 1}{\csc x + 1}} = \frac{1 - \sin x}{|\cos x|}$
   (f) $\frac{\sin 4x - \sin 2x}{\cos 4x + \cos 2x} = \tan x$

8. Sketch the graph of each function.
   (a) $f(x) = 2 \tan \frac{3x}{2}$ (b) $g(x) = \sin(-5\pi x)$
   (c) $F(x) = \sec\left(x - \frac{\pi}{3}\right)$
   (d) $G(x) = 2 + 3 \cos(2x + \pi)$
   (e) $y = \frac{\pi}{2} - \arccos x$
   (f) $y = \sin^2 x$
   (*Hint:* Rewrite using a power reduction formula.)

9. From a point P, tangent lines are drawn to a circle of circumference 48π. If the angle between the tangent lines is $50°$, determine (a) the distance from point P to the center of the circle and (b) the length of the intercepted arc.

10. State which of the trigonometric functions are even functions and which are odd functions.

11. For each equation, find the solutions in the interval $[0, 2\pi)$.
    (a) $\tan 3x = \sqrt{3}$
    (b) $2 \sin x - 3 \csc x + 5 = 0$
    (c) $3 \cos x - \cos 2x = 1$
    (d) $\sin 3x + \sin x + \cos x = 0$
    (e) $\cot 4x + \cot 2x = 0$ (f) $(25^{\sin x})^{\cot x} = \frac{1}{5}$

12. Two sides of a triangle are 63.0 m and 86.0 m and the area of the triangle is 2360 sq m. Determine two possible values for the third side of the triangle.

13. Determine the distance from A to B along the centerline of the highway shown in the sketch.

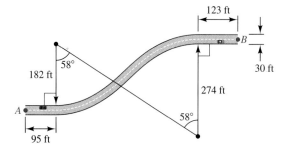

14. The latitude of Boston is $42°20'$. Assuming the Earth is a sphere with radius 3960 mi, determine the distance from Boston (a) to the equator and (b) to the North Pole.

15. Expand $(\cos\theta + i \sin\theta)^3$ by using De Moivre's theorem and also by using the cube of a binomial formula (Section 1.5). Equate the two results and obtain a formula for $\sin 3\theta$ in terms of $\sin\theta$ and a formula for $\cos 3\theta$ in terms of $\cos\theta$.

16. Given that α and β are the acute angles of a right triangle, a is the side opposite angle α, b is the side opposite angle β, and c is the hypotenuse, express
    (a) a in terms of b and α
    (b) c in terms of b and β
    (c) α in terms of b and c
    (d) β in terms of a and b

17. A building lot has the shape of a quadrilateral. The lengths of two adjacent sides of the lot are 250 ft and 354 ft, and the angle between these sides is 69°45'. The other two sides are, respectively, perpendicular to these adjacent sides. Find (a) the length of the other two sides and (b) the area of the building lot.

18. Sketch an angle of radian measure 8 in standard position. What is the smallest positive angle that is coterminal with 8 radians?

19. Given that $\overline{AD}$ bisects angle A, as shown in the sketch, find AB.

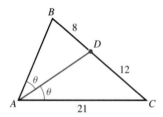

20. Given that $\overline{DE}$ is perpendicular to $\overline{AB}$, as shown in the sketch, find AE.

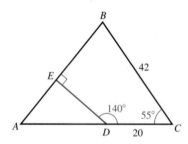

21. Perform the indicated operations and record the results in standard form.
    (a) $(1 - i)^3(1 + i)^5$
    (b) $\dfrac{(-1 + i)^{12}}{(1 + \sqrt{3}\,i)^3(-\sqrt{3} - i)^6}$

22. Find all the roots of the equation $x^5 = 243$, and plot the roots in the complex plane.

23. Simplify each expression.
    (a) $e^{-\ln(\sin x)}$ for $0 < x \leq \pi/2$
    (b) $\ln |\cot x| + \ln |\sin x| - \ln |\cos x|$

24. Given that θ is an obtuse angle, express $\csc \theta$, $\cos \theta$, and $\tan \theta$ in terms of $\sin \theta$.

25. A surveyor finds that the angle of elevation to the top of a radio transmission tower is 32°. Moving directly toward the tower a distance of 150 ft, she finds that the angle of elevation of the tower is now 43°. Assuming the ground is inclined at an angle of 12°, as shown in the sketch, find the height of the tower.

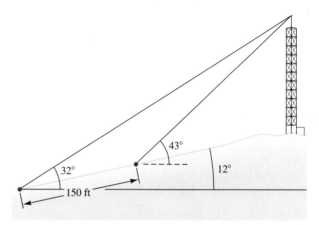

26. The airspeed of a plane heading in the direction N-62°-E is 375 mi/h. The wind, blowing directly from the south, causes the plane to drift to the north 9°, as shown in the sketch. Determine the velocity of the wind and the ground speed of the plane (its speed relative to the ground).

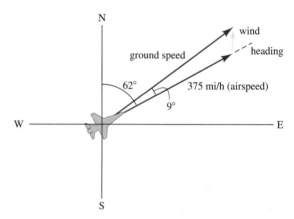

27. Given that $\mathbf{A} = \langle 2, 2 \rangle$ and $\mathbf{B} = \langle -3, 6 \rangle$, find the magnitude and direction angle of each vector difference.
    (a) $\mathbf{A} - \mathbf{B}$ (b) $\mathbf{B} - \mathbf{A}$

28. Rangers have determined that the population P of deer in a state forest in Massachusetts can be approximated by the equation

$$P = 54 + 29 \sin \frac{\pi}{8} t$$

Exercise 28 continues

where t is the time in years with $t = 0$ corresponding to 1990. Determine the expected number of deer in the forest in the year 2002.

29. Solve each triangle.

    (a)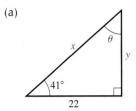

    (b)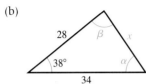

30. The structure in the sketch is in equilibrium.

    (a) Draw a vector diagram showing the weight W supported by the cables.

    (b) Find the tension (stretching force) in the left and right cables.

    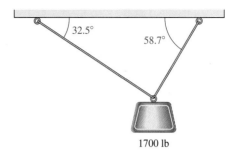

31. Find the polar equation that has the same graph as the given Cartesian equation. Express the answer in the form $r = f(\theta)$.

    (a) $2x - 3y = 5$ (b) $y^2 - 6x - 9 = 0$

32. Find the Cartesian equation in x and y for each set of parametric equations. Express each answer in the form $y = f(x)$.

    (a) $x = e^{t-1}$ and $y = 3t$

    (b) $x = \ln 2t$ and $y = \sin 2t$

The Centrum civic center in Worcester, Massachusetts, sold 12,000 tickets to a recent rock concert. The ticket prices were $12, $18, and $24, and the total income from ticket sales was $201,000. How many tickets of each type were sold if the number of $12 tickets sold was twice the number of $24 tickets sold?

For the solution, see Example 5 in Section 10.1.

Linear Systems and Matrices

10.1 Systems of Linear Equations

10.2 Matrices and Their Properties

10.3 Determinants and Inverses of Matrices

10.4 Systems of Linear Inequalities and Linear Programming

10.1 Systems of Linear Equations

◆ Introductory Comments

In Section 3.5, we graphed the linear equations

$$A_1 x + B_1 y = C_1 \quad \text{and} \quad A_2 x + B_2 y = C_2$$

on the same coordinate plane and determined their points of intersection by using either the substitution method or the elimination method. Figure 10.1 illustrates the three possibilities for the graphs of these two linear equations.

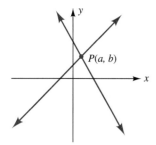

(a) The lines intersect at one point $P(a, b)$.

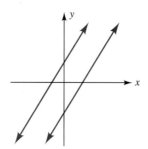

(b) The lines are parallel and do not intersect.

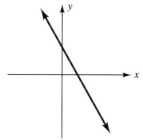

(c) The lines coincide and intersect at an infinite number of points.

FIGURE 10.1
The three possibilities for the intersection of two lines graphed on the same coordinate plane

We say that the pair of linear equations

$$A_1 x + B_1 y = C_1$$
$$A_2 x + B_2 y = C_2$$

represent a **system** of two linear equations in two variables x and y. This system has three possible types of solutions, as illustrated in Figure 10.1:

1. *Exactly one solution*, namely, the ordered pair (a, b). The lines intersect at one point $P(a, b)$, as shown in Figure 10.1(a).
2. *No solution*, an **inconsistent system**. The lines are parallel and do not intersect, as shown in Figure 10.1(b).
3. *Infinitely many solutions*, a **dependent system**. The lines coincide and intersect at an infinite number of points, as shown in Figure 10.1(c).

In this section, we discuss some methods of solving n linear equations in n variables with $n \geq 3$. A system of n linear equations in n variables $x_1, x_2, x_3, \ldots, x_n$ is called an $n \times n$ (read "n by n") linear system:

SECTION 10.1 ♦ Systems of Linear Equations

$$
\begin{array}{l}
a_{11}x_1 + a_{12}x_2 + a_{13}x_3 + \cdots + a_{1n}x_n = k_1 \quad \leftarrow \text{first equation} \\
a_{21}x_1 + a_{22}x_2 + a_{23}x_3 + \cdots + a_{2n}x_n = k_2 \quad \leftarrow \text{second equation} \\
a_{31}x_1 + a_{32}x_2 + a_{33}x_3 + \cdots + a_{3n}x_n = k_3 \quad \leftarrow \text{third equation} \\
\quad \vdots \qquad\qquad \vdots \qquad\qquad \vdots \qquad\qquad\qquad \vdots \qquad\qquad \vdots \\
a_{n1}x_1 + a_{n2}x_2 + a_{n3}x_3 + \cdots + a_{nn}x_n = k_n \quad \leftarrow n\text{th equation}
\end{array}
$$

To describe the coefficients of each variable we use a *double subscript* notation in which each of the numbers in the subscript has a definite meaning. The first number corresponds to the equation and the second number corresponds to the variable. Thus, a_{32} represents the coefficient in the third equation attached to the second variable (x_2), whereas a_{23} represents the coefficient in the second equation attached to the third variable (x_3).

Although an $n \times n$ linear system with $n \geq 3$ has no graphical representation in the coordinate plane, this system still has three possible types of solutions:

1. Exactly one solution, namely, $x_1 = c_1, x_2 = c_2, \ldots, x_n = c_n$, which is written as an ordered n-tuple $(c_1, c_2, \ldots, c_n)$
2. No solution, an inconsistent system
3. Infinitely many solutions, a dependent system

In more advanced math courses, it is shown that the graph of three linear equations in three variables consists of three planes in space. Figure 10.2 shows some ways in which three planes may intersect to obtain exactly one solution, no solution, or infinitely many solutions.

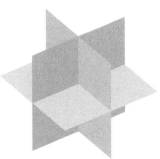
(a) The planes intersect in exactly one point, and the system has exactly one solution.

(b) The planes have no point in common, and the system has no solution.

(c) The planes intersect in a line, and the system has infinitely many solutions.

FIGURE 10.2
Some graphical illustrations of the three types of solutions for a 3×3 linear system

♦ The Elimination Method

We can use the *elimination method* (from Section 3.5) to help solve a 3×3 linear system. The key step in this method is to obtain, for one of the variables, coefficients that are equal in absolute value so that by adding or subtracting the two equations, this variable is eliminated. (See Example 2 in Section 3.5.) For a 3 × 3 linear system, this method of solution involves eliminating one of the variables between a pair of equations, and then eliminating the same variable between a different pair of equations. This procedure yields two linear equations in two variables, which we can solve by using the substitution method or the elimination method. Although the order in which we eliminate the variables from a system of equations is immaterial, we can often reduce the amount of work needed by using some insight and making intelligent choices.

The elimination method relies on generating **equivalent systems**—systems that have the same solution. We can generate equivalent systems by performing any of the following manipulations.

◀ **Generating Equivalent Systems**

Performing any of the following manipulations on a system of equations generates an equivalent system:

1. Interchange the position of two equations.
2. Multiply both sides of an equation by a nonzero constant.
3. Add a multiple of one equation to another equation.

EXAMPLE 1 Use the elimination method to solve the given system of linear equations.

$$\begin{aligned} 2x - y + 4z &= -3 \\ x + 2y - 3z &= 1 \\ 3x + y - 3z &= 6 \end{aligned}$$

◆ **Solution** If we add the first and third equations, we eliminate y and obtain an equation in x and z:

$$\begin{aligned} 2x - y + 4z &= -3 \\ 3x + y - 3z &= 6 \\ \hline 5x + z &= 3 \end{aligned}$$

Next, we choose another pair of equations and eliminate the same variable. Choosing the first and second equations, we eliminate y and obtain another equation in x and z:

$$\begin{aligned} 2x - y + 4z = -3 & \xrightarrow{\text{Multiply both sides by 2}} & 4x - 2y + 8z &= -6 \\ x + 2y - 3z = 1 & \xrightarrow{\text{No change}} & x + 2y - 3z &= 1 \\ & & \hline 5x + 5z &= -5 \end{aligned}$$

We now have two linear equations in the same two variables, x and z. If we subtract one of these equations from the other, we eliminate x:

SECTION 10.1 ♦ Systems of Linear Equations

$$\begin{aligned} 5x + z &= 3 \\ 5x + 5z &= -5 \\ \hline -4z &= 8 \end{aligned}$$

We find z, x, and y by solving for z, and then substituting back through the equations:

$-4z = 8$: $\quad\quad\quad\quad\quad\quad\quad\quad\quad\quad\quad\quad$ implies $\quad z = -2$

$5x + z = 3$: $\quad\quad\quad 5x + (-2) = 3 \quad$ implies $\quad x = 1$

$3x + y - 3z = 6$: $\quad 3(1) + y - 3(-2) = 6 \quad$ implies $\quad y = -3$

Thus, the solution of the original system of equations is $x = 1$, $y = -3$, and $z = -2$, which we write as the *ordered triple* $(1, -3, -2)$. We can check this solution by verifying that each of the three original equations is satisfied when we substitute $x = 1$, $y = -3$, and $z = -2$. ♦

In Example 1 we generated the equivalent system

$$3x + y - 3z = 6$$
$$5x + z = 3$$
$$-4z = 8$$

to obtain the solution of the original system of equation. This system is not a unique choice. For the first equation in this equivalent system we can use either of the other original equations, and for the second equation we can use the other equation in x and z, namely, $5x + 5z = -5$.

PROBLEM 1 For the system of equations given in Example 1, show that the solution $(1, -3, -2)$ is also given by the equivalent system

$$2x - y + 4z = -3$$
$$5x + 5z = -5$$
$$-4z = 8$$
♦

In the next example, we solve a dependent system and show how to express its infinitely many solutions as an ordered triple.

EXAMPLE 2 Use the elimination method to solve the given system of linear equations.

$$x - 2y - 3z = 2$$
$$x - 4y - 13z = 14$$
$$-3x + 5y + 4z = 0$$

Solution If we subtract the second equation from the first equation, we eliminate x and obtain an equation in y and z:

$$\begin{array}{r} x - 2y - 3z = 2 \\ x - 4y - 13z = 14 \\ \hline 2y + 10z = -12 \end{array}$$

Next, we choose a different pair of equations and eliminate the same variable. Choosing the first and third equations, we eliminate x and obtain another equation in y and z:

$$\begin{array}{ll} x - 2y - 3z = 2 & \xrightarrow{\text{Multiply both sides by 3}} \quad 3x - 6y - 9z = 6 \\ -3x + 5y + 4z = 0 & \xrightarrow{\text{No change}} \quad \underline{-3x + 5y + 4z = 0} \\ & \qquad\qquad\qquad\qquad\quad -y - 5z = 6 \end{array}$$

We now have two linear equations in two unknowns, y and z, which we can try to solve by the elimination method:

$$\begin{array}{ll} 2y + 10z = -12 & \xrightarrow{\text{No change}} \quad 2y + 10z = -12 \\ -y - 5z = 6 & \xrightarrow{\text{Multiply both sides by 2}} \quad \underline{-2y - 10z = 12} \\ & \qquad\qquad\qquad\qquad\quad 0 = 0 \end{array}$$

Recall from Section 3.5 that if the variables drop out and the resulting statement is true ($0 = 0$), then the lines coincide and the system has infinitely many solutions. Hence, this is a dependent system.

We can express the infinitely many solutions as an ordered triple. First, we solve the equation $-y - 5z = 6$ (or $2y + 10z = -12$) for either y or z. Solving for y is easier, and we obtain

$$y = -5z - 6.$$

Now, we substitute $y = -5z - 6$ into one of the original equations, say $x - 2y - 3z = 2$, to express x in terms of z:

$$\begin{aligned} x - 2y - 3z &= 2 \\ x - 2(-5z - 6) - 3z &= 2 \\ x &= -7z - 10 \end{aligned}$$

If we let $z = a$, where a is an arbitrary real number, then y and x can also be expressed in terms of a:

$$y = -5a - 6 \quad \text{and} \quad x = -7a - 10$$

Hence, the original system of equations has an infinite number of solutions of the form

$$(-7a - 10, -5a - 6, a) \quad \text{where } a \text{ is any real number.}$$

We can obtain a *particular solution* of this system by letting a take on any real value. For example, if we let $a = 0$, then we obtain the particular solution $(-10, -6, 0)$.

◆

PROBLEM 2 Use the elimination method to show that the given system of linear equations is an inconsistent system.

$$x - 2y - 3z = -3$$
$$x - 4y - 13z = 14$$
$$-3x + 5y + 4z = 1$$

◆

The elimination method may be extended to an $n \times n$ system of linear equations with $n \geq 4$. For example, with four variables, we eliminate one of the variables between three different pairs of equations. The result is three equations in the remaining three variables, which we can solve by using the procedure illustrated in the previous two examples.

EXAMPLE 3 Use the elimination method to solve the given system of linear equations.

$$x + 2y + t = 3$$
$$y + 2z = 3$$
$$x - 2z = 0$$
$$3y - 4z + t = 2$$

◆ **Solution** If we subtract the first and third equations, we eliminate x and obtain an equation in y, z, and t:

$$\begin{array}{r} x + 2y + t = 3 \\ x - 2z = 0 \\ \hline 2y + 2z + t = 3 \end{array}$$

Joining this equation with the second and fourth equations gives us a 3×3 linear system:

$$2y + 2z + t = 3$$
$$y + 2z = 3$$
$$3y - 4z + t = 2$$

If we subtract the first and third equations of this 3×3 system, we eliminate t and obtain an equation in y and z:

$$\begin{array}{r} 2y + 2z + t = 3 \\ 3y - 4z + t = 2 \\ \hline -y + 6z = 1 \end{array}$$

We now have two linear equations in the same two variables, y and z. If we add these equations, we eliminate y:

$$\begin{aligned} y + 2z &= 3 \\ -y + 6z &= 1 \\ \hline 8z &= 4 \end{aligned}$$

We find z, y, t, and x by solving for z and then substituting back through the equations:

$$\begin{aligned} 8z = 4: & & & & \text{implies} & & z = \tfrac{1}{2} \\ y + 2z = 3: & & y + 2\left(\tfrac{1}{2}\right) = 3 & & \text{implies} & & y = 2 \\ 2y + 2z + t = 3: & & 2(2) + 2\left(\tfrac{1}{2}\right) + t = 3 & & \text{implies} & & t = -2 \\ x - 2z = 0: & & x - 2\left(\tfrac{1}{2}\right) = 0 & & \text{implies} & & x = 1 \end{aligned}$$

Thus, the solution of the original system of equations is $x = 1$, $y = 2$, $z = \tfrac{1}{2}$, and $t = -2$, which we write as the *ordered quadruple* $\left(1, 2, \tfrac{1}{2}, -2\right)$. You should check this solution. ◆

PROBLEM 3 Rework Example 3 by first eliminating the variable t. ◆

◆ Nonsquare Systems

Thus far we have worked with linear systems in which the number of equations equals the number of variables. Such a system is called a **square system**. If the number of equations does not equal the number of variables, the system is called a **nonsquare system**. A nonsquare system may not have a unique solution.

EXAMPLE 4 Use the elimination method to solve the given system of linear equations.

$$\begin{aligned} 3x + 3y - 2z &= 3 \\ 2x + y - z &= 2 \end{aligned}$$

◆ Solution Choosing to eliminate z first, we proceed as follows:

$$\begin{array}{lcl} 3x + 3y - 2z = 3 & \xrightarrow{\text{No change}} & 3x + 3y - 2z = 3 \\ 2x + y - z = 2 & \xrightarrow{\text{Multiply both sides by } -2} & -4x - 2y + 2z = -4 \\ & & \hline \\ & & -x + y = -1 \end{array}$$

Since this system does not contain a third equation, we cannot form two equations in x and y. Therefore, we solve the equation $-x + y = -1$ for y in term of x (or for x in terms of y). Solving for y, we obtain

$$y = x - 1.$$

Now, substituting $y = x - 1$ into one of the original equations, we find z in terms of x:

$$2x + y - z = 2: \quad 2x + (x - 1) - z = 2 \quad \text{implies} \quad z = 3x - 3$$

If we let $x = a$, where a is any arbitrary real number, then y and z can also be expressed in terms of a:

$$y = a - 1 \quad \text{and} \quad z = 3a - 3$$

Hence, the original system of equations has an infinite number of solutions of the form

$$(a, a - 1, 3a - 3) \quad \text{where } a \text{ is any real number.} \quad \blacklozenge$$

PROBLEM 4 Solve the system in Example 4 by solving the equation $-x + y = -1$ for x in terms of y (instead of solving for y in terms of x). Then find z in terms of y. Show that the solution of this system of equations can also be written as the ordered triple $(b + 1, b, 3b)$ where b is any real number. $\blacklozenge$

◆ Application: Solving Word Problems

In Section 2.4, we solved word problems by assigning one of the unknowns the variable x and expressing the other unknowns in terms of x. Using this method, it was possible to develop one equation in one unknown and to solve this equation for x. Many word problems, however, are set up more naturally by assigning a different variable to each of the unknowns and then solving a system of equations. The next example illustrates this procedure.

EXAMPLE 5 *Chapter Opening Problem*
The Centrum civic center in Worcester, Massachusetts, sold 12,000 tickets to a recent rock concert. The ticket prices were $12, $18, and $24, and the total income from ticket sales was $201,000. How many tickets of each type were sold if the number of $12 tickets sold was twice the number of $24 tickets sold?

◆ Solution Let

$$x = \text{the number of \$12 tickets sold}$$
$$y = \text{the number of \$18 tickets sold}$$

and
$$z = \text{the number of \$24 tickets sold.}$$

Since 12,000 tickets were sold, we have

$$x + y + z = 12{,}000.$$

Since the ticket prices were $12, $18, and $24, and the total income from ticket sales was $201,000, we have

$$12x + 18y + 24z = 201{,}000.$$

Now, since the Centrum sold twice as many $12 tickets as $24 tickets, we have

$$x = 2z, \quad \text{or} \quad x - 2z = 0.$$

Hence, we can write a system of linear equations to describe this problem:

$$\begin{aligned} x + y + z &= 12{,}000 \\ 12x + 18y + 24z &= 201{,}000 \\ x \qquad\quad - 2z &= 0 \end{aligned}$$

In Problem 5, you are asked to use the elimination method to show that the solution of this system of linear equations is

$$(5000,\ 4500,\ 2500).$$

In summary, the Centrum sold 5000 tickets at $12, 4500 tickets at $18, and 2500 tickets at $24 each. ◆

PROBLEM 5 Use the elimination method to verify that the solution of the system of linear equations in Example 5 is (5000, 4500, 2500). ◆

Exercises 10.1

Basic Skills

In Exercises 1–24, use the elimination method to solve each system of linear equations.

1. $\begin{aligned} x \qquad\ \ + z &= 11 \\ y + z &= 5 \\ 4x + 2y \qquad &= 48 \end{aligned}$

2. $\begin{aligned} x + y + z &= 19 \\ 2x - y \qquad &= 3 \\ y - z &= 10 \end{aligned}$

3. $\begin{aligned} 2x + 3y \qquad &= 2 \\ x \qquad - 2z &= 7 \\ 2y - z &= 3 \end{aligned}$

4. $\begin{aligned} 3x + y - 2z &= -2 \\ y - 3z &= -1 \\ 2x + 3y + 5z &= 3 \end{aligned}$

5. $\begin{aligned} x + 3y - z &= 5 \\ 2x + 3y \qquad &= 4 \\ -x - y + 3z &= -1 \end{aligned}$

6. $\begin{aligned} x - 2y - 5z &= -1 \\ -3x + y - 2z &= 0 \\ 2x \qquad + 3z &= 1 \end{aligned}$

7. $\begin{aligned} 4x - 3y - z &= 0 \\ x - 3y + 2z &= 7 \\ 3x + 9y - z &= -2 \end{aligned}$

8. $\begin{aligned} 5x - 2y + z &= -2 \\ 3x - y - 3z &= -2 \\ x + 2y - z &= 4 \end{aligned}$

9. $\begin{aligned} 3x - 2y + z &= 3 \\ 4x + y \qquad &= 1 \\ 11y - 4z &= -9 \end{aligned}$

10. $\begin{aligned} 2x + 3y \qquad &= 2 \\ 8x \qquad - 4z &= 3 \\ 3y + z &= -1 \end{aligned}$

11. $\begin{aligned} 2x + 3y + z &= 7 \\ 5x - y - z &= 2 \\ 4x - y + 3z &= 9 \end{aligned}$

12. $\begin{aligned} 4x - y + 3z &= 3 \\ 3x - 4y + 2z &= 4 \\ 2x + 3y - z &= 1 \end{aligned}$

13. $\begin{aligned} x - 2y + z &= 1 \\ 3x - y + z &= 6 \\ -2x + y - 4z &= 1 \end{aligned}$

14. $5x - 3y + 4z = -1$
    $3x + 8y - z = -4$
    $2x - 11y + 4z = 2$

15. $6x - 4y - 5z = 4$
    $3x - 2y + z = 9$
    $\frac{3}{2}x - \frac{2}{3}y + 2z = \frac{41}{6}$

16. $0.2x + 0.3y + 2z = 3$
    $3x - 2y + 4z = 6$
    $0.5x - y + 3z = -0.5$

17. $x + y + z = 1$
    $x + z + t = 2$
    $x + y + t = 6$
    $y + z + t = 3$

18. $3x + 2y + z = 1$
    $x + 3t = -4$
    $y - 2z - t = -2$
    $2x + y - z = -3$

19. $x - 2y + t = 1$
    $2y - z + 3t = 12$
    $x - 2z = 2$
    $2y + t = 5$

20. $x + y - 2t = 7$
    $2x - z + t = -4$
    $3x + y + z = 5$
    $2y - z + 3t = -2$

21. $x_1 + x_2 = 2$
    $x_2 - x_3 + x_4 = 5$
    $x_1 + x_3 + x_5 = -1$
    $x_2 + x_4 = 0$
    $x_4 + x_5 = 2$

22. $x_1 + x_2 + x_3 + x_4 = 12$
    $x_1 + x_2 + x_3 + x_5 = 14$
    $x_1 + x_2 + x_4 + x_5 = 16$
    $x_1 + x_3 + x_4 + x_5 = 18$
    $x_2 + x_3 + x_4 + x_5 = 20$

23. $x_1 + 2x_2 + x_5 = 5$
    $x_2 - x_3 + x_4 = 1$
    $x_1 + 3x_3 - x_5 = -2$
    $3x_2 - x_4 + 2x_5 = 8$
    $x_3 + 2x_4 - x_5 = -5$

24. $2x_1 + x_2 - x_3 + x_5 = -1$
    $x_1 - 2x_2 + x_4 = 5$
    $3x_2 - x_3 + 2x_5 = -5$
    $3x_1 + 2x_3 - x_4 - x_5 = 7$
    $4x_2 + 3x_3 + x_4 - x_5 = -1$

In Exercises 25–32, use the elimination method to solve each non-square system of linear equations.

25. $2x + 2y + z = 4$
    $4y - z = 0$

26. $3x - y + 4z = 8$
    $y + 2z = 1$

27. $x + 2y - z = 1$
    $3x + 2y - z = 5$

28. $x - 3y + 2z = -2$
    $2x - 4y + 5z = 1$

29. $3x - 2y + 4t = 4$
    $y - 2z + t = 1$
    $x - 2y + 8t = 0$

30. $2x - 3y + z = 1$
    $3x - y - t = 0$
    $y + 3z - 2t = 3$

31. $x + y = 7$
    $2x - y = 8$
    $x - 3y = -1$

32. $2x - 5y = 8$
    $3x + y = 2$
    $4x + 3y = 0$

In Exercises 33–44, use a system of linear equations to solve each applied problem.

33. A car rental agency charges a dollars per day to rent a car plus b dollars per mile. Find a and b if the one-day rental charge is $45 for 100 miles or $55 for 150 miles.

34. An investor buys two types of stocks—one costs $40 per share and the other costs $24 per share. Altogether, the investor purchased 55 shares of stock for a total cost of $1720. How many shares of each stock were purchased?

35. A total of $10,000 is invested, part of the money at 10% and the rest at 12%. If the total annual interest is $1090, how much money is invested at each rate?

36. It takes a ferryboat 4 hours to travel 20 miles upstream and 3 hours to make the return trip downstream. What is the speed of the boat and the speed of the river current?

37. The sum of the three interior angles of any triangle is 180°. Suppose the largest angle is six times as large as the smallest angle and is equal to twice the sum of the two smaller angles. Find the three angles.

38. A contractor buys three parcels of land for $200,000. Parcel A costs three times as much as parcel B and $20,000 less than parcels B and C together. Find the cost of each parcel of land.

39. A woman invests $50,000 at the simple interest rates of 8%, 9%, and $9\frac{1}{2}$% per year. Altogether these investments earn $4550 per year. How much does she have invested at each rate if she has the same amount invested at 9% and $9\frac{1}{2}$%?

40. A merchant mixes three types of tea, costing $2.40, $3.00, and $3.20 per pound, in order to obtain 50 pounds of a blend that costs $2.92 per pound. How many pounds of each type does the merchant use if the blend contains 10 pounds more of the $3.20 type than of the $2.40 type?

41. The greens fee for a round of golf is $15 for a child, $20 for an adult, and $18 for a senior citizen. On a certain

day, 120 rounds of golf were played and the receipts totaled $2220. If twice as many seniors played as children, find the number of adults who played golf that day.

42. Three circular pulleys are tangent to each other, as shown in the sketch. If the center-to-center distances AB, AC, and BC are 18 cm, 22 cm, and 16 cm, respectively, find the radius of each pulley.

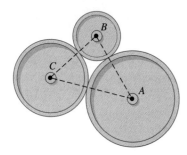

43. A photography studio develops four sizes of prints (in inches); 2×3, 3×5, 5×8, and 8×10. After taking senior-class pictures, the studio offers the following four packages to the students.

Package	Number of prints			
	2×3	3×5	5×8	8×10
A	4	2	1	—
B	—	2	2	1
C	8	—	4	2
D	12	4	2	2

If package A sells for $12, package B for $24, package C for $46, and package D for $47, determine the cost per print for each size print.

44. The fueling facility at an airport uses three pumps to load each jet with 1500 gallons of fuel before takeoff. When all three pumps are used, each plane can be loaded with fuel in 1 hour 15 minutes. However, on a certain day, the first pump malfunctions and works at only half its normal rate. Fueling now requires 1 hour 40 minutes for each plane. When the malfunctioning pump is closed completely for repair, the fueling takes 2 hours 30 minutes for each plane. What is the normal pumping rate (in gallons per minute) for the malfunctioning pump?

 Critical Thinking

45. A system of linear equations has the following equations:

$$x + 3y - 2z = B$$
$$-2x + y + Az = 8$$
$$3x - 5y - z = 12$$

Find values of A and B for which the system is (a) inconsistent or (b) dependent.

46. Find a system of linear equations whose solution is the ordered triple $(-1, 2, 5)$. (The answer is not unique.)

47. Use a system of linear equations to find the values of A, B, and C such that

$$A(x - 1)(x + 2) + B(x - 4)(x + 2) + C(x - 4)(x - 1)$$

is equivalent to $x^2 + 8x$.

48. The graph of the circle $x^2 + y^2 + Dx + Ey + F = 0$ passes through the points $(0, 3)$, $(-2, 0)$, and $(4, -4)$. Use a system of linear equations to determine the values for D, E, and F.

49. The graph of the quadratic function $f(x) = ax^2 + bx + c$ is sketched in the figure. Use a system of linear equations to determine the values for a, b, and c.

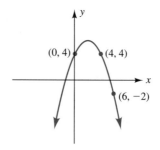

50. The graph of the cubic function

$$f(x) = ax^3 + bx^2 + cx + d$$

is shown in the sketch. Use a system of linear equations to determine the values for a, b, c, and d.

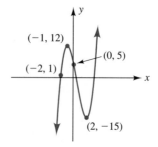

A system of linear equations is said to be **homogeneous** if all the constant terms are zero. A homogeneous system of equations always has the trivial solution $(0, 0, 0, \ldots, 0)$. However, some homogeneous systems also have nontrivial solutions. In Exercises 51 and 52, find the solution to each homogeneous system.

51. $\quad x + 3y - 2z = 0$
    $\quad 3x - 2y + 5z = 0$
    $\quad -2x - y - z = 0$

52. $\quad 3x - 5y + 5z = 0$
    $\quad x - 4y - 3z = 0$
    $\quad -4x + 15y + 10z = 0$

In Exercises 53 and 54, develop a system of linear equation for each applied problem, then solve. [Hint: Each system is dependent and has more than one solution.]

53. A group of 20 skiers spent $260 to ski at Deer Mountain and $180 to ski at Mt. Cabott. The cost of lift tickets at these ski areas are shown in the following table:

	Deer Mountain	Mt. Cabott
Adult	$18	$24
Student	$15	$15
Child	$10	Free

How many adults, students, and children are in this group of 20 skiers?

54. The following table shows the weight composition of four alloys. How many pounds of each alloy must be melted together to form 100 pounds of an alloy that is 20% copper, 46% gold, 21% tin, and 13% zinc?

Alloy	Copper	Gold	Tin	Zinc
A	20%	50%	20%	10%
B	—	60%	40%	—
C	40%	40%	—	20%
D	—	20%	50%	30%

10.2 Matrices and Their Properties

◆ Introductory Comments

A **matrix** is a rectangular array of numbers enclosed by a pair of brackets. Each particular number in the matrix is referred to as an **element** of the matrix. We use a capital letter, such as A, B, or C, to represent a matrix. Consider matrix A:

$$A = \begin{bmatrix} a_{11} & a_{12} & a_{13} & \cdots & a_{1j} & \cdots & a_{1n} \\ a_{21} & a_{22} & a_{23} & \cdots & a_{2j} & \cdots & a_{2n} \\ a_{31} & a_{32} & a_{33} & \cdots & a_{3j} & \cdots & a_{3n} \\ \vdots & \vdots & \vdots & & \vdots & & \vdots \\ a_{i1} & a_{i2} & a_{i3} & \cdots & a_{ij} & \cdots & a_{in} \\ \vdots & \vdots & \vdots & & \vdots & & \vdots \\ a_{m1} & a_{m2} & a_{m3} & \cdots & a_{mj} & \cdots & a_{mn} \end{bmatrix}$$

(jth column; ith row)

We use double subscript notation for each element. Each of the two numbers in the subscript has a definite meaning: the first number corresponds to the **row** and the second number to the **column** in which the element stands. Thus, a_{ij} represents the element in the ith row and jth column. We say that matrix A has m rows and n columns and is of **order** $m \times n$ (read "m by n").

Here are some examples of matrices (plural of matrix):

$$B = \begin{bmatrix} 2 & 0 & 6 \\ 5 & -1 & 3 \end{bmatrix} \qquad C = \begin{bmatrix} 0 & 1 \\ 8 & 4 \\ -2 & 3 \end{bmatrix}$$

Matrix B has two rows and three columns and is of order 2×3. Matrix C has three rows and two columns and is of order 3×2.

For expressing the rules of matrix algebra, we usually write matrix A in the compact form

$$A = [\underbrace{a_{ij}}_{\text{the matrix with elements } a_{ij}}]$$

Two matrices are **equal** if and only if their corresponding elements are equal. More formally, we state the following definition of equality of matrices.

◆ Equality of Matrices

> Two matrices $A = [a_{ij}]$ and $B = [b_{ij}]$ are **equal** if they both are of order $m \times n$ and
>
> $$a_{ij} = b_{ij}$$
>
> for all $i = 1, 2, 3, \ldots, m$ and $j = 1, 2, 3, \ldots, n$.

For example,

$$\begin{bmatrix} a_{11} & a_{12} \\ a_{21} & a_{22} \end{bmatrix} = \begin{bmatrix} -2 & -6 \\ -5 & 7 \end{bmatrix}$$

if and only if $a_{11} = -2$, $a_{12} = -6$, $a_{21} = -5$, and $a_{22} = 7$.

◆ Using a Matrix to Solve a System of Linear Equations

We can use a matrix to help solve a system of linear equations. Consider the following nonsquare system of m linear equations in n variables x_1, x_2, $x_3, \ldots, x_n$:

$$\begin{aligned} a_{11}x_1 + a_{12}x_2 + a_{13}x_3 + \cdots + a_{1n}x_n &= k_1 &&\text{first equation} \\ a_{21}x_1 + a_{22}x_2 + a_{23}x_3 + \cdots + a_{2n}x_n &= k_2 &&\text{second equation} \\ a_{31}x_1 + a_{32}x_2 + a_{33}x_3 + \cdots + a_{3n}x_n &= k_3 &&\text{third equation} \\ &\vdots \\ a_{m1}x_1 + a_{m2}x_2 + a_{m3}x_3 + \cdots + a_{mn}x_n &= k_m &&m\text{th equation} \end{aligned}$$

SECTION 10.2 ♦ Matrices and Their Properties

For every such $m \times n$ linear system, there corresponds an enlarged or **augmented matrix** of order $m \times (n + 1)$ that is written as follows:

$$\begin{bmatrix} a_{11} & a_{12} & a_{13} & \cdots & a_{1n} & k_1 \\ a_{21} & a_{22} & a_{23} & \cdots & a_{2n} & k_2 \\ a_{31} & a_{32} & a_{33} & \cdots & a_{3n} & k_3 \\ \vdots & \vdots & \vdots & & \vdots & \vdots \\ a_{m1} & a_{m2} & a_{m3} & \cdots & a_{mn} & k_m \end{bmatrix}$$

$\underbrace{\qquad\qquad\qquad\qquad}_{\text{coefficients of the variable terms in the linear system}}$ $\underbrace{\quad}_{\text{constant terms}}$

For example, the system of linear equations

$$\begin{aligned} x + 3y - 2z &= 1 \\ -2x - y &= 0 \\ 3x - z &= -2 \end{aligned}$$

corresponds to the augmented matrix

$$\begin{bmatrix} 1 & 3 & -2 & 1 \\ -2 & -1 & 0 & 0 \\ 3 & 0 & -1 & -2 \end{bmatrix}$$

We use a vertical dashed line in the augmented matrix to separate the coefficients of the variable terms in the linear system from the constant terms in the system.

An augmented matrix is in **echelon form** if the elements $a_{ij} = 0$ when $i > j$; that is, the element in the first column in every row after the first is 0; the element in the second column in every row after the second is 0; the element in the third column in every row after the third is 0; and so on. For example, the augmented matrix

$$\begin{bmatrix} 1 & 2 & 5 & -5 \\ 0 & 2 & -1 & 3 \\ 0 & 0 & 2 & 4 \end{bmatrix}$$

is in echelon form and corresponds to the system of linear equations

$$\begin{aligned} x + 2y + 5z &= -5 \\ 2y - z &= 3 \\ 2z &= 4 \end{aligned}$$

In the previous section, we discussed three manipulations for generating equivalent systems of equations:

1. Interchange the position of two equations.
2. Multiply both sides of an equation by a nonzero constant.
3. Add a multiple of one equation to another.

Each of these manipulations corresponds to an **elementary row operation** that can be performed on a matrix. If one matrix is obtained from another by a sequence of elementary row operations, we say that the two matrices are **row equivalent**.

◆ Generating Row-Equivalent Matrices

> Performing any of the following elementary row operations on a matrix generates a row-equivalent matrix:
>
> 1. Interchange the position of two rows.
> 2. Multiply all elements in a row by a nonzero constant.
> 3. Add a multiple of one row to another row.

If two matrices are row equivalent, we write the symbol $\sim$ between the two matrices.

CAUTION When working with matrices, do not confuse *equality* (written as $=$) with *equivalency* (written as $\sim$). For example, when interchanging rows, we write

$$\begin{bmatrix} 2 & 5 \\ -3 & 4 \end{bmatrix} \sim \begin{bmatrix} -3 & 4 \\ 2 & 5 \end{bmatrix}$$

However, since corresponding elements of these matrices are *not* the same, we write

$$\begin{bmatrix} 2 & 5 \\ -3 & 4 \end{bmatrix} \neq \begin{bmatrix} -3 & 4 \\ 2 & 5 \end{bmatrix}$$

In this text, we use the following notation to describe an operation that generates a row-equivalent matrix:

$R_i \leftrightarrow R_j$ Interchange the ith row and jth row of a matrix.

$cR_i \to R_i$ Replace the ith row with a multiple of c times the ith row.

$cR_i + R_j \to R_j$ Replace the jth row with the sum of c times the ith row and the jth row.

In the next example, we solve a system of linear equations by using elementary row operations on an augmented matrix. The scheme is to write the augmented matrix in echelon form, write the corresponding system of equations, and solve for the variables. We refer to this procedure of solving a system of linear equations as the **Gaussian method** (named after German mathematician Carl Friedrich Gauss).

SECTION 10.2 ♦ Matrices and Their Properties

The Gaussian method is the procedure used by most computer programs (and some graphing calculators) to solve systems of linear equations. This method often requires lots of arithmetic with awkward fractions—something that computers do quickly and accurately, but humans do not. To avoid unnecessary fractional computations, we usually begin by generating a row-equivalent matrix in which the element in the first row, first column is 1.

EXAMPLE 1 Use the Gaussian method to solve each system of linear equations.

(a) $13x + 3y = 55$
$12x + 7y = 55$

(b) $2x - y + 4z = -3$
$x + 2y - 3z = 1$
$3x + y - 3z = 6$

♦ Solution

(a) Beginning with the augmented matrix that corresponds to this system of linear equations, we use elementary row operations to write the augmented matrix in echelon form:

$$\begin{bmatrix} 13 & 3 & | & 55 \\ 12 & 7 & | & 55 \end{bmatrix} \sim \begin{bmatrix} 1 & -4 & | & 0 \\ 12 & 7 & | & 55 \end{bmatrix} \quad -R_2 + R_1 \to R_1$$

$$\sim \begin{bmatrix} 1 & -4 & | & 0 \\ 0 & 55 & | & 55 \end{bmatrix} \quad -12R_1 + R_2 \to R_2$$

This last matrix is in echelon form and corresponds to the following system of linear equations:

$$x - 4y = 0$$
$$55y = 55$$

The equation $55y = 55$ implies that $y = 1$. Substituting this value of y back into the other equation, $x - 4y = 0$, we find that $x = 4$. Thus, the solution of the original system is $(4, 1)$. (You should check this solution.)

(b) Beginning with the augmented matrix that corresponds to this system of linear equations, we use elementary row operations to write the augmented matrix in echelon form:

$$\begin{bmatrix} 2 & -1 & 4 & | & -3 \\ 1 & 2 & -3 & | & 1 \\ 3 & 1 & -3 & | & 6 \end{bmatrix} \sim \begin{bmatrix} 1 & 2 & -3 & | & 1 \\ 2 & -1 & 4 & | & -3 \\ 3 & 1 & -3 & | & 6 \end{bmatrix} \quad R_1 \leftrightarrow R_2$$

$$\sim \begin{bmatrix} 1 & 2 & -3 & | & 1 \\ 0 & -5 & 10 & | & -5 \\ 0 & -5 & 6 & | & 3 \end{bmatrix} \quad \begin{array}{l} -2R_1 + R_2 \to R_2 \\ -3R_1 + R_3 \to R_3 \end{array}$$

$$\sim \begin{bmatrix} 1 & 2 & -3 & | & 1 \\ 0 & -5 & 10 & | & -5 \\ 0 & 0 & -4 & | & 8 \end{bmatrix} \quad -R_2 + R_3 \to R_3$$

This last matrix is in echelon form and corresponds to the following system of linear equations:

$$\begin{aligned} x + 2y - 3z &= 1 \\ -5y + 10z &= -5 \\ -4z &= 8 \end{aligned}$$

The equation $-4z = 8$ implies that $z = -2$. Substituting this value of z back into the other equation, we find that $y = -3$ and $x = 1$. Thus, the solution of the original system is $(1, -3, -2)$. ◆

Remember, a system of linear equations may have infinitely many solutions or no solution. If we perform row operations on an augmented matrix and obtain a row of zeros except for the last entry in the row, then we conclude that the system is inconsistent (it has no solution).

PROBLEM 1 Use the Gaussian method to show that the given system of linear equations is inconsistent.

$$\begin{aligned} 3x - 9y + 6z &= 1 \\ 2x - y - z &= 3 \\ x - 2y + z &= 1 \end{aligned}$$

◆

◆ Matrix Addition and Scalar Multiples

As we observed in Example 1, matrices can be used to help solve a system of linear equations. Their use, however, far exceeds this single purpose. When English mathematician Arthur Cayley (1821–1895) introduced matrices in 1858, he defined the operations of addition and multiplication with matrices. In doing so, he created a new type of mathematics that has served as a model for various applications, from economics to quantum mechanics.

To add two $m \times n$ matrices, we simply add the elements in corresponding positions. More formally, we state the following rule for matrix addition.

◆ **Matrix Addition**

If $A = [a_{ij}]$ and $B = [b_{ij}]$ are matrices of order $m \times n$, then their **sum** $A + B$ is a matrix of order $m \times n$ defined by

$$A + B = [a_{ij} + b_{ij}].$$

EXAMPLE 2 Matrices A, B, and C are defined as follows.

$$A = \begin{bmatrix} 6 & 5 & -2 \\ 4 & 0 & -1 \end{bmatrix} \quad B = \begin{bmatrix} 3 & -5 & 2 \\ 0 & 1 & -3 \end{bmatrix}$$

$$C = \begin{bmatrix} 2 & 3 \\ -1 & 1 \\ 0 & -2 \end{bmatrix}$$

SECTION 10.2 ♦ Matrices and Their Properties

Determine each sum.

(a) $A + B$ **(b)** $C + C$

◆ Solution

(a) $A + B = \begin{bmatrix} 6+3 & 5+(-5) & -2+2 \\ 4+0 & 0+1 & -1+(-3) \end{bmatrix} = \begin{bmatrix} 9 & 0 & 0 \\ 4 & 1 & -4 \end{bmatrix}$

(b) $C + C = \begin{bmatrix} 2+2 & 3+3 \\ -1+(-1) & 1+1 \\ 0+0 & -2+(-2) \end{bmatrix} = \begin{bmatrix} 4 & 6 \\ -2 & 2 \\ 0 & -4 \end{bmatrix}$ ◆

CAUTION   We cannot add two matrices of different orders. For the matrices A and C in Example 2, we say that the sum $A + C$ is undefined.

When we add matrix C to itself, as we did in Example 2(b), we obtain a matrix of the same order as C in which each element is twice the corresponding element of C. We refer to the matrix $2C$ as the one obtained by multiplying each element of C by 2. Hence, if C is a matrix, then

$$2C = C + C.$$

When working with matrices, we refer to the number 2 as a **scalar**. In this text, *a scalar is a real number.* To multiply a matrix by a scalar k, we simply multiply each element of the matrix by k.

◆ **Scalar Multiple of a Matrix**

> If $A = [a_{ij}]$ is a matrix of order $m \times n$ and k is a scalar, then the **scalar multiple** of A is of order $m \times n$ and is defined by
>
> $$kA = [ka_{ij}].$$

PROBLEM 2   Given the matrices A and B in Example 2, find $-2A + 3B$.   ◆

The matrix $-A$ (read "the opposite of A") represents the scalar multiple $-1A$. Thus, $-A$ is a matrix of the same order as A in which each element is the *opposite* of the corresponding element of A. Therefore, for every matrix A we have

$$A + (-A) = \mathbf{0},$$

where **0** represents the **zero matrix**, a matrix of the same order as A in which each element is zero. We define matrix subtraction in terms of matrix addition:

$$A - B = A + (-B)$$

EXAMPLE 3 Given the matrices

$$A = \begin{bmatrix} 2 & 5 \\ -3 & 0 \\ 1 & 4 \end{bmatrix} \quad \text{and} \quad B = \begin{bmatrix} 4 & 6 \\ -2 & -5 \\ 0 & 3 \end{bmatrix}$$

determine $2A - B$.

Solution By the definition of matrix subtraction, we have

$$2A - B = 2A + (-B) = \begin{bmatrix} 4 & 10 \\ -6 & 0 \\ 2 & 8 \end{bmatrix} + \begin{bmatrix} -4 & -6 \\ 2 & 5 \\ 0 & -3 \end{bmatrix}$$

$$= \begin{bmatrix} 0 & 4 \\ -4 & 5 \\ 2 & 5 \end{bmatrix}$$

PROBLEM 3 Given the matrices A and B in Example 3, find $A - 2B$.

◆ Matrix Multiplication

The product of two matrices A and B, denoted AB, is found by multiplying each row of matrix A by each column of matrix B. If the product AB is to exist, then the number of columns in matrix A must equal the number of rows in matrix B. The product AB has exactly as many rows as A and exactly as many columns as B:

Matrix A: $m \times n$ Matrix B: $n \times p$

These must be equal for the product AB to be defined.

The order of matrix AB is $m \times p$.

If A and B are matrices such that

$$A = \begin{bmatrix} a_{11} & a_{12} & a_{13} & \cdots & a_{1n} \\ a_{21} & a_{22} & a_{23} & \cdots & a_{2n} \\ \vdots & \vdots & \vdots & & \vdots \\ a_{i1} & a_{i2} & a_{i3} & \cdots & a_{in} \\ \vdots & \vdots & \vdots & & \vdots \\ a_{m1} & a_{m2} & a_{m3} & \cdots & a_{mn} \end{bmatrix} \quad \text{and} \quad B = \begin{bmatrix} b_{11} & b_{12} & \cdots & b_{1j} & \cdots & b_{1p} \\ b_{21} & b_{22} & \cdots & b_{2j} & \cdots & b_{2p} \\ b_{31} & b_{32} & \cdots & b_{3j} & \cdots & b_{3p} \\ \vdots & \vdots & & \vdots & & \vdots \\ b_{n1} & b_{n2} & \cdots & b_{nj} & \cdots & b_{np} \end{bmatrix}$$

SECTION 10.2 ♦ Matrices and Their Properties

then the ith row, jth column element of the matrix AB is

$$a_{i1}b_{1j} + a_{i2}b_{2j} + a_{i3}b_{3j} + \cdots + a_{in}b_{nj}.$$

We now formalize the definition of matrix multiplication.

◀ **Matrix Multiplication**

If $A = [a_{ij}]$ is a matrix of order $m \times n$ and $B = [b_{ij}]$ is a matrix of order $n \times p$, then the **product** AB is a matrix of order $m \times p$ defined by

$$AB = [c_{ij}] \quad \text{where} \quad c_{ij} = a_{i1}b_{1j} + a_{i2}b_{2j} + a_{i3}b_{3j} + \cdots + a_{in}b_{nj}.$$

EXAMPLE 4 Matrices A, B, C, and D are defined as follows.

$$A = \begin{bmatrix} 2 & 3 & -1 \end{bmatrix} \quad B = \begin{bmatrix} 4 \\ 1 \\ -3 \end{bmatrix} \quad C = \begin{bmatrix} 2 & -1 \\ 3 & 4 \\ 0 & -2 \end{bmatrix}$$

$$D = \begin{bmatrix} -3 & 2 \\ 1 & 4 \end{bmatrix}$$

Determine each product, if it is defined.

(a) AB **(b)** BA **(c)** CD **(d)** DC

◆ **Solution**

(a) The product AB has exactly as many rows as A and exactly as many columns as B. Thus, the product AB is a 1×1 matrix. Using matrix multiplication, we find

$$AB = \begin{bmatrix} 2 & 3 & -1 \end{bmatrix} \begin{bmatrix} 4 \\ 1 \\ -3 \end{bmatrix} = [(2)(4) + (3)(1) + (-1)(-3)] = [14]$$

(b) The product BA has exactly as many rows as B and exactly as many columns as A. Thus, the product BA is a 3×3 matrix. Using matrix multiplication, we find

$$BA = \begin{bmatrix} 4 \\ 1 \\ -3 \end{bmatrix} \begin{bmatrix} 2 & 3 & -1 \end{bmatrix} = \begin{bmatrix} (4)(2) & (4)(3) & (4)(-1) \\ (1)(2) & (1)(3) & (1)(-1) \\ (-3)(2) & (-3)(3) & (-3)(-1) \end{bmatrix}$$

$$= \begin{bmatrix} 8 & 12 & -4 \\ 2 & 3 & -1 \\ -6 & -9 & 3 \end{bmatrix}$$

(c) Since the number of columns of matrix C is the same as the number of rows of matrix D, we know that the product CD is defined. The product

CD has the same number of rows as C and the same number of columns as D, that is, CD is of order 3×2. Using matrix multiplication, we find

$$CD = \begin{bmatrix} 2 & -1 \\ 3 & 4 \\ 0 & -2 \end{bmatrix} \begin{bmatrix} -3 & 2 \\ 1 & 4 \end{bmatrix} = \begin{bmatrix} (2)(-3) + (-1)(1) & (2)(2) + (-1)(4) \\ (3)(-3) + (4)(1) & (3)(2) + (4)(4) \\ (0)(-3) + (-2)(1) & (0)(2) + (-2)(4) \end{bmatrix}$$

$$= \begin{bmatrix} -7 & 0 \\ -5 & 22 \\ -2 & -8 \end{bmatrix}$$

(d) Note that the number of columns of matrix D is *not* the same as the number of rows of matrix C. Hence, the product DC is not defined. ◆

PROBLEM 4 For the matrices A and C defined in Example 4, find the product AC. ◆

CAUTION Unlike multiplication of real numbers, matrix multiplication is *not* commutative. Observe from Example 4 that $AB \neq BA$ and $CD \neq DC$.

Although matrix multiplication is not commutative, many of the properties of matrix addition and matrix multiplication are similar to the properties of real numbers:

◆ Properties of Matrix Addition and Matrix Multiplication

For matrices A, B, and C of the appropriate orders so that the operations are defined,

Commutative Property of Addition	$A + B = B + A$
Associative Property of Addition	$(A + B) + C = A + (B + C)$
Associative Property of Multiplication	$A(BC) = (AB)C$
Distributive Property (left)	$A(B + C) = AB + AC$
Distributive Property (right)	$(B + C)A = BA + CA$

◆ Matrices and the Graphing Calculator

We can use the matrix features of a graphing calculator to perform operations with matrices. First, we use the MATRIX key to enter the order and elements of the matrices. We now return to the home (primary) screen on the calculator and perform the indicated operations. Figure 10.3 shows typical displays for finding the product in Example 4(c). For a more detailed explanation about the matrix operations that can be performed on a graphing calculator, consult the manual that came with your graphing calculator or the *Graphing Technology Laboratory Manual* that accompanies this text. After reading this information, use a graphing calculator to verify the products in Example 4(a) and 4(b).

SECTION 10.2 ◆ *Matrices and Their Properties*   659

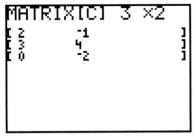

(a) Enter matrix C.

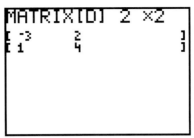

(b) Enter matrix D.

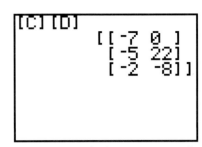

(c) Return to the home screen and perform the indicated operation.

FIGURE 10.3
Typical displays for finding the product in Example 4(c)

 ◆ **Application: Inventory and Cost**

We conclude this section with an application of matrix multiplication.

EXAMPLE 5   The cost (in dollars) of each size of a certain model of brass bed is given by matrix C, and the current inventory of this model of brass bed at two furniture stores in the greater Boston area is given by matrix B. Find CB and state what it represents.

$$C = \begin{matrix} \text{Twin} & \text{Double} & \text{Queen} & \text{King} \\ [300 & 350 & 400 & 600] \end{matrix} \text{ Cost} \qquad B = \begin{matrix} & \text{Brookline} & \text{Cambridge} \\ \begin{bmatrix} 2 & 5 \\ 6 & 0 \\ 0 & 2 \\ 2 & 3 \end{bmatrix} & \begin{matrix} \text{Twin} \\ \text{Double} \\ \text{Queen} \\ \text{King} \end{matrix} \end{matrix}$$

◆ **Solution**

$$CB = [300 \quad 350 \quad 400 \quad 600] \begin{bmatrix} 2 & 5 \\ 6 & 0 \\ 0 & 2 \\ 2 & 3 \end{bmatrix} = [3900 \quad 4100]$$

$300(2) + 350(6) + 400(0) + 600(2)$
$300(5) + 350(0) + 400(2) + 600(3)$

The 1×2 matrix

$$CB = \begin{matrix} \text{Brookline} & \text{Cambridge} \\ [3900 & 4100] \end{matrix} \text{ Total cost}$$

represents the total cost (in dollars) of the inventory of this type of bed at each store.   ◆

PROBLEM 5   For the matrix B given in Example 5, find the matrix $[1 \quad 1 \quad 1 \quad 1]B$, and state what it represents.   ◆

Exercises 10.2

Basic Skills

In Exercises 1–16, use the Gaussian method to solve each system of linear equations.

1. $2x - 3y = 1$
   $x + 4y = 6$

2. $4x - 3y = 0$
   $2x - y = -10$

3. $5x - 20y = -10$
   $-2x + 8y = 4$

4. $3x - 12y = 6$
   $4x - 16y = -8$

5. $x - 3y = 7$
   $2x + z = 5$
   $4y + 2z = -2$

6. $4x - 2z = -1$
   $2x + 2y = 1$
   $x - 3y + z = -7$

7. $2x - y - 2z = 0$
   $3x + 3y + z = 6$
   $y + 3z = -5$

8. $5x - y + z = 3$
   $2x + 3z = -5$
   $2x - 3y - 4z = 4$

9. $2x + 3y - 2z = 3$
   $3x + 5y + z = 2$
   $x + 2y - 5z = 1$

10. $3x - 4y + 19z = 6$
    $x - y + 4z = 2$
    $5x - 2y - z = 10$

11. $4x - 3y - z = 5$
    $3x + 2y - 3z = 2$
    $-2x + y - z = 0$

12. $2x - y + z = 7$
    $4x - 3y - 3z = 11$
    $5x + y - z = 14$

13. $x + 2y + t = 3$
    $y + 2z = 3$
    $x - 2z = 0$
    $3y - 4z + t = 2$

14. $2x - y + t = 9$
    $3y + 2z = -6$
    $y - 2z + 4t = -14$
    $3x - z + t = -4$

15. $x + 3y + 2t = 3$
    $3x + 2z - t = 3$
    $3y - z + 2t = 1$
    $x + 2y - 3z = -4$

16. $5x - y + 2z = -3$
    $2x - 5y - t = -3$
    $4y - 3z + 4t = -3$
    $-x + 4z - 2t = -3$

In Exercises 17–56, perform the indicated matrix operations given that A, B, C, D, E, F, G, and H are defined as follows. If an operation is not defined, state the reason. Use the matrix features of a graphing calculator to verify each answer.

$$A = \begin{bmatrix} 1 & 2 & 0 \\ -2 & 3 & -4 \end{bmatrix} \quad B = \begin{bmatrix} 2 & -3 & -5 \\ -1 & 4 & 4 \end{bmatrix}$$

$$C = \begin{bmatrix} 2 & -1 \\ 3 & 2 \\ -1 & 4 \end{bmatrix} \quad D = \begin{bmatrix} 6 & 0 \\ -3 & 9 \\ 0 & 3 \end{bmatrix}$$

$$E = \begin{bmatrix} 2 & -4 & -5 \end{bmatrix} \quad F = \begin{bmatrix} 1 \\ -3 \\ 2 \end{bmatrix}$$

$$G = \begin{bmatrix} -1 & 1 \\ 1 & -1 \end{bmatrix} \quad H = \begin{bmatrix} 5 & 5 \\ -2 & -2 \end{bmatrix}$$

17. $2A$
18. $-5G$
19. $-\frac{1}{3}D$
20. $\frac{3}{2}E$
21. $A + B$
22. $C + D$
23. $E + F$
24. $F + F$
25. $3G + H$
26. $2C + 5G$
27. $-4A + 3B$
28. $-\frac{1}{3}D + 2C$
29. $H - B$
30. $D - C$
31. $3G - 2H$
32. $\frac{1}{2}A - \frac{3}{2}B$
33. $-2(3E)$
34. $3A - 2A$
35. $\frac{1}{2}F - \frac{3}{2}F$
36. $2A + 2(B - A)$
37. AC
38. CB
39. GB
40. ED
41. DH
42. HD
43. EF
44. FE
45. HG
46. GH
47. $C(A + B)$
48. $CA + CB$
49. $(A - B)D$
50. $AD - BD$
51. $(EC)A$
52. $E(CA)$

53. $(BD)G$ 54. $B(DG)$ 57. $37.5CB$ 58. $[1 \; 1 \; 1](37.5CB)$

55. $F(2E)$ 56. $2(FE)$ 59. $37.5(B - A)$ 60. $37.5C(B - A)$

An engineering firm employs three types of union workers: technicians, engineers, and senior engineers. The company and the union have agreed to a new contract, which states that the hourly wages (in dollars) increase from the figures given in matrix A to those given in matrix B.

Old hourly wage
$$A = \begin{bmatrix} 8.35 \\ 13.45 \\ 22.15 \end{bmatrix} \begin{matrix} \text{Technician} \\ \text{Engineer} \\ \text{Senior engineer} \end{matrix}$$

New hourly wage
$$B = \begin{bmatrix} 9.65 \\ 16.75 \\ 28.25 \end{bmatrix} \begin{matrix} \text{Technician} \\ \text{Engineer} \\ \text{Senior engineer} \end{matrix}$$

and that each person works 37.5 hours per week. The company maintains offices in Boston, Atlanta, and Seattle, and the number of technicians, engineers, and senior engineers at each facility is given in matrix C:

$$\begin{matrix} & \text{Technician} & \text{Engineer} & \text{Senior engineer} \\ C = & \begin{bmatrix} 81 & 67 & 19 \\ 57 & 42 & 16 \\ 122 & 88 & 31 \end{bmatrix} & & \begin{matrix} \text{Boston} \\ \text{Atlanta} \\ \text{Seattle} \end{matrix} \end{matrix}$$

In Exercises 57–60, find the indicated matrix, and describe what it represents.

A car rental agency rents three types of cars: subcompact, midsize, and large cars. The daily rental rates (in dollars) charged by the agency are given in matrix C.

$$\begin{matrix} & \text{Subcompact} & \text{Mid-size} & \text{Large} \\ C = & [20 & 32 & 40] & \text{Cost} \end{matrix}$$

Within a certain city, the agency has three terminals. The number of subcompact, midsize, and large cars rented at each terminal on Monday and Tuesday are given in matrices M and T, respectively.

$$\begin{matrix} & \text{Terminal 1} & \text{Terminal 2} & \text{Terminal 3} \\ M = & \begin{bmatrix} 12 & 14 & 18 \\ 8 & 14 & 10 \\ 2 & 5 & 6 \end{bmatrix} & & \begin{matrix} \text{Subcompact} \\ \text{Midsize} \\ \text{Large} \end{matrix} \end{matrix}$$

$$\begin{matrix} & \text{Terminal 1} & \text{Terminal 2} & \text{Terminal 3} \\ T = & \begin{bmatrix} 10 & 8 & 9 \\ 2 & 4 & 10 \\ 0 & 1 & 6 \end{bmatrix} & & \begin{matrix} \text{Subcompact} \\ \text{Midsize} \\ \text{Large} \end{matrix} \end{matrix}$$

In Exercises 61–64, find the indicated matrix, and state what it represents.

61. $(M + T)\begin{bmatrix} 1 \\ 1 \\ 1 \end{bmatrix}$ 62. $[1 \; 1 \; 1](M + T)$

63. $C(M + T)$ 64. $C(M + T)\begin{bmatrix} 1 \\ 1 \\ 1 \end{bmatrix}$

Critical Thinking

65. The zero product property (Section 2.2) states that if a and b are real numbers and $ab = 0$, then either $a = 0$ or $b = 0$. This property does not hold for matrices: Find a pair of 2×2 matrices A and B for which $AB = 0$, but $A \neq 0$ and $B \neq 0$.

66. Determine the values of x and y such that the following product is true.

$$\begin{bmatrix} 2 & -3 \\ x & y \end{bmatrix} \begin{bmatrix} 4 & 3 \\ -6 & 5 \end{bmatrix} = \begin{bmatrix} 26 & -9 \\ 24 & -1 \end{bmatrix}$$

67. If A is an $m \times m$ matrix, then we define the *square* of matrix A by writing $A^2 = AA$.

(a) Given

$$A = \begin{bmatrix} 1 & -2 & 3 \\ 0 & 3 & -2 \\ 3 & 1 & 4 \end{bmatrix}$$

$$B = \begin{bmatrix} 2 & 0 & -1 \\ 1 & 3 & 4 \\ 0 & -1 & -1 \end{bmatrix}$$

perform each matrix operation:

(i) A^2 (ii) B^2
(iii) $A^2 - B^2$ (iv) $(A + B)(A - B)$

Exercise 67 continues

(b) Compare the answers to (iii) and (iv) of part (a). Does $A^2 - B^2 = (A + B)(A - B)$?

68. Suppose matrices A, X, and B are defined as follows.

$$A = \begin{bmatrix} 1 & -1 & 0 \\ 2 & -1 & 1 \\ 0 & 4 & 5 \end{bmatrix}$$

$$X = \begin{bmatrix} x \\ y \\ z \end{bmatrix} \qquad B = \begin{bmatrix} 2 \\ 9 \\ 23 \end{bmatrix}$$

(a) Express the product AX as a 3×1 matrix.

(b) Use the result of part (a) and equality of matrices to determine x, y, and z if $AX = B$.

In Exercises 69 and 70, use matrices A, B, A^T, and B^T defined as follows:

$$A = \begin{bmatrix} a_{11} & a_{12} \\ a_{21} & a_{22} \end{bmatrix} \qquad B = \begin{bmatrix} b_{11} & b_{12} \\ b_{21} & b_{22} \end{bmatrix}$$

$$A^T = \begin{bmatrix} a_{11} & a_{21} \\ a_{12} & a_{22} \end{bmatrix} \quad \text{and} \quad B^T = \begin{bmatrix} b_{11} & b_{21} \\ b_{12} & b_{22} \end{bmatrix}$$

Show that the given statement is true. (The matrices A^T and B^T are called the transposes of matrices A and B, respectively. The transpose of a matrix is formed by switching its rows and columns.)

69. $(AB)^T = B^T A^T$

70. $(A + B)^T = A^T + B^T$

71. Cayley defined matrix multiplication to help simplify double systems of equation such as

$$u = a_{11}r + a_{12}s \qquad x = b_{11}u + b_{12}v$$
$$v = a_{21}r + a_{22}s \qquad y = b_{21}u + b_{22}v$$

(a) Use substitution to find x and y in terms of r and s.

(b) Express the product

$$\begin{bmatrix} b_{11} & b_{12} \\ b_{21} & b_{22} \end{bmatrix} \begin{bmatrix} a_{11} & a_{12} \\ a_{21} & a_{22} \end{bmatrix} \begin{bmatrix} r \\ s \end{bmatrix}$$

as a 2×1 matrix. What do the elements in this 2×1 matrix represent?

72. Use an appropriate substitution to change the given system of equations into a linear system. Apply the Gaussian method to the linear system of equations, and then state the solution of the original system.

(a) $\dfrac{2}{x} + \dfrac{3}{y} = 15$

$\dfrac{4}{x} - \dfrac{5}{y} = -14$

(b) $e^x - e^y + 2e^z = 0$
$2e^x - e^y + 3e^z = 3$
$5e^x - 2e^y - 3e^z = -1$

10.3 Determinants and Inverses of Matrices

◆ **Introductory Comments**

A matrix of order $n \times n$ is called a **square matrix**. In a square matrix, the elements $a_{11}, a_{22}, a_{33}, \ldots, a_{nn}$ are called the **main diagonal** elements. Associated with every square matrix A is a real number called the *determinant of A*, which we denote by writing $|A|$. The 2×2 matrix $A = \begin{bmatrix} a_{11} & a_{12} \\ a_{21} & a_{22} \end{bmatrix}$ has a **second-order determinant**

$$|A| = \begin{vmatrix} a_{11} & a_{12} \\ a_{21} & a_{22} \end{vmatrix}$$

whose value is found by taking the product of the main diagonal elements and subtracting the product of the elements on the other diagonal.

Determinant of a 2 × 2 Matrix

The **determinant** of the 2 × 2 matrix $A = \begin{bmatrix} a_{11} & a_{12} \\ a_{21} & a_{22} \end{bmatrix}$ is

$$|A| = \begin{vmatrix} a_{11} & a_{12} \\ a_{21} & a_{22} \end{vmatrix} = a_{11}a_{22} - a_{21}a_{12}$$

Here are some illustrations of evaluating a second-order determinant:

$$\begin{vmatrix} 8 & -4 \\ 3 & -2 \end{vmatrix} = (8)(-2) - (3)(-4) = -16 + 12 = -4$$

$$\begin{vmatrix} 6 & -1 \\ 0 & 7 \end{vmatrix} = (6)(7) - (0)(-1) = 42 - 0 = 42$$

The determinant of any $n \times n$ matrix in echelon form is the product of the elements along its main diagonal.

Determinant of a Matrix in Echelon Form

If A is an $n \times n$ matrix in echelon form, then $|A|$ is the product of the elements along its main diagonal.

Here are some illustrations of evaluating a determinant in echelon form:

$$\begin{vmatrix} 3 & -1 & 4 \\ 0 & 4 & -2 \\ 0 & 0 & -5 \end{vmatrix} = (3)(4)(-5) = -60$$

$$\begin{vmatrix} 2 & -3 & 1 & 0 \\ 0 & 2 & 4 & -7 \\ 0 & 0 & -3 & 1 \\ 0 & 0 & 0 & -1 \end{vmatrix} = (2)(2)(-3)(-1) = 12$$

In this section, we discuss some methods that may be used to evaluate the determinant of an $n \times n$ matrix. We then introduce the inverse of a matrix and show how it may be used to help solve a system of linear equations. We begin by looking at the effects of elementary row operations on a determinant.

◆ Evaluating the Determinant of a Matrix

Consider the 2 × 2 matrix $A = \begin{bmatrix} a_{11} & a_{12} \\ a_{21} & a_{22} \end{bmatrix}$ and the effects of elementary row operations on its determinant

$$|A| = \begin{vmatrix} a_{11} & a_{12} \\ a_{21} & a_{22} \end{vmatrix} = a_{11}a_{22} - a_{21}a_{12}$$

1. If the rows of matrix A are interchanged to form the row-equivalent matrix

$$B = \begin{bmatrix} a_{21} & a_{22} \\ a_{11} & a_{12} \end{bmatrix},$$

then

$$\begin{aligned} |B| &= a_{21}a_{12} - a_{11}a_{22} \\ &= -(a_{11}a_{22} - a_{21}a_{12}) \\ &= -|A|. \end{aligned}$$

2. If a row of matrix A is replaced by c times that row to form the row-equivalent matrix

$$B = \begin{bmatrix} ca_{11} & ca_{12} \\ a_{21} & a_{22} \end{bmatrix},$$

then

$$\begin{aligned} |B| &= ca_{11}a_{22} - ca_{21}a_{12} \\ &= c(a_{11}a_{22} - a_{21}a_{12}) \\ &= c|A|. \end{aligned}$$

3. If a row of matrix A is replaced by the sum of that row and c times another row to form the row equivalent matrix

$$B = \begin{bmatrix} a_{11} & a_{12} \\ ca_{11} + a_{21} & ca_{12} + a_{22} \end{bmatrix},$$

then

$$\begin{aligned} |B| &= a_{11}(ca_{12} + a_{22}) - a_{12}(ca_{11} + a_{21}) \\ &= a_{11}a_{22} - a_{21}a_{12} \\ &= |A|. \end{aligned}$$

We now generalize these observations for any $n \times n$ matrix A.

◆ Effects of Row Operations on a Determinant

Suppose A is an $n \times n$ matrix.

1. If two rows of A are interchanged to form the row-equivalent matrix B, then $|B| = -|A|$.
2. If a row of A is replaced by c times that row to form the row-equivalent matrix B, then $|B| = c|A|$.
3. If a row of A is replaced by the sum of that row and c times another row to form the row-equivalent matrix B, then $|B| = |A|$.

SECTION 10.3 ♦ Determinants and Inverses of Matrices

Operation 2 has the effect of factoring out a common factor from a row. For example,

$$\begin{vmatrix} 8 & -4 \\ 3 & -2 \end{vmatrix} = 4 \begin{vmatrix} 2 & -1 \\ 3 & -2 \end{vmatrix} = 4[-4 - (-3)] = -4.$$

Factor 4 from row 1.

We can evaluate the determinant of any $n \times n$ matrix by using elementary row operations. The procedure is illustrated in the next example.

EXAMPLE 1 Use elementary row operations to evaluate the determinant of matrix A.

$$A = \begin{bmatrix} 2 & -2 & -3 \\ 0 & 6 & 3 \\ 1 & -2 & -5 \end{bmatrix}$$

◆ **Solution** We write the determinant in echelon form, and record the effects of the row operations on the determinant:

$$|A| = \begin{vmatrix} 2 & -2 & -3 \\ 0 & 6 & 3 \\ 1 & -2 & -5 \end{vmatrix} = - \begin{vmatrix} 1 & -2 & -5 \\ 0 & 6 & 3 \\ 2 & -2 & -3 \end{vmatrix} \quad R_1 \leftrightarrow R_3$$

determinant changes sign

$$= - \begin{vmatrix} 1 & -2 & -5 \\ 0 & 6 & 3 \\ 0 & 2 & 7 \end{vmatrix} \quad -2R_1 + R_3 \to R_3$$

$$= -3 \begin{vmatrix} 1 & -2 & -5 \\ 0 & 2 & 1 \\ 0 & 2 & 7 \end{vmatrix} \quad \tfrac{1}{3} R_2 \to R_2$$

Factor out 3 from R_2.

$$= -3 \begin{vmatrix} 1 & -2 & -5 \\ 0 & 2 & 1 \\ 0 & 0 & 6 \end{vmatrix} \quad -R_2 + R_3 \to R_3$$

Now, multiplying the elements along the main diagonal, we have

$$|A| = -3[(1)(2)(6)] = -36 \qquad ♦$$

We can use elementary row operations to show that any third-order determinant can be expressed in terms of second-order determinants (see Exercise 49):

$$\begin{vmatrix} a_{11} & a_{12} & a_{13} \\ a_{21} & a_{22} & a_{23} \\ a_{31} & a_{32} & a_{33} \end{vmatrix} = a_{11} \begin{vmatrix} a_{22} & a_{23} \\ a_{32} & a_{33} \end{vmatrix} - a_{12} \begin{vmatrix} a_{21} & a_{23} \\ a_{31} & a_{33} \end{vmatrix} + a_{13} \begin{vmatrix} a_{21} & a_{22} \\ a_{31} & a_{32} \end{vmatrix}$$

This formula is called the **expansion of a third-order determinant by its first row**. Using this formula on matrix A in Example 1, we have

$$\begin{vmatrix} 2 & -2 & -3 \\ 0 & 6 & 3 \\ 1 & -2 & -5 \end{vmatrix} = 2 \begin{vmatrix} 6 & 3 \\ -2 & -5 \end{vmatrix} - (-2) \begin{vmatrix} 0 & 3 \\ 1 & -5 \end{vmatrix} + (-3) \begin{vmatrix} 0 & 6 \\ 1 & -2 \end{vmatrix}$$
$$= 2[-30 - (-6)] + 2[0 - 3] - 3[0 - 6]$$
$$= -36,$$

which agrees with our result in Example 1.

We can also evaluate the determinant of an $n \times n$ matrix by selecting the appropriate line from the matrix menu on a graphing calculator. Figure 10.4 shows typical displays for evaluating the determinant of matrix A in Example 1.

When we perform elementary row operations on a determinant, it is important to watch for a pair of rows in which corresponding elements are proportional. If we add a certain multiple of one of these rows to the other, we generate a row in which every element is zero. *If every element in a row of matrix A is zero, then* $|A| = 0$.

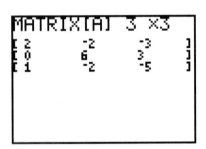

(a) Enter matrix A

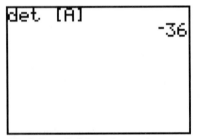

(b) Select "det" from the matrix menu

FIGURE 10.4

Typical displays for evaluating the determinant in Example 1

PROBLEM 1 Explain why the determinant of the given matrix A is zero. Then enter this matrix into a graphing calculator and select the "det" command from the matrix menu to confirm this answer.

$$A = \begin{bmatrix} 1 & -3 & -2 & 6 \\ 2 & 4 & 2 & -1 \\ 0 & 2 & 5 & -6 \\ 6 & 12 & 6 & -3 \end{bmatrix}$$

◆ **Inverse of a Matrix**

A square matrix with the digit 1 along its main diagonal and zeros elsewhere is called the $n \times n$ **identity matrix** and is denoted by

$$I_n = \begin{bmatrix} 1 & 0 & 0 & \ldots & 0 \\ 0 & 1 & 0 & \ldots & 0 \\ 0 & 0 & 1 & \ldots & 0 \\ \vdots & \vdots & \vdots & & \vdots \\ 0 & 0 & 0 & \ldots & 1 \end{bmatrix}$$

Anytime we multiply an $n \times n$ matrix by the identity matrix I_n, we obtain the original $n \times n$ matrix from which we started. We refer to this fact as the **identity property of matrix multiplication**.

◆ Identity Property of Matrix Multiplication

If A is a square matrix of order $n \times n$ and I_n is the $n \times n$ identity matrix, then

$$AI_n = I_nA = A.$$

We say that two square $n \times n$ matrices A and B are **inverses** of each other if their product is the identity matrix I_n, that is, if

$$AB = BA = I_n.$$

For example, the square matrices

$$A = \begin{bmatrix} 2 & -5 \\ -1 & 3 \end{bmatrix} \quad \text{and} \quad B = \begin{bmatrix} 3 & 5 \\ 1 & 2 \end{bmatrix}$$

are inverses of each other, since

$$AB = \begin{bmatrix} 2 & -5 \\ -1 & 3 \end{bmatrix}\begin{bmatrix} 3 & 5 \\ 1 & 2 \end{bmatrix} = \begin{bmatrix} 6 + (-5) & 10 + (-10) \\ -3 + 3 & -5 + 6 \end{bmatrix} = \begin{bmatrix} 1 & 0 \\ 0 & 1 \end{bmatrix}$$

and

$$BA = \begin{bmatrix} 3 & 5 \\ 1 & 2 \end{bmatrix}\begin{bmatrix} 2 & -5 \\ -1 & 3 \end{bmatrix} = \begin{bmatrix} 6 + (-5) & -15 + 15 \\ 2 + (-2) & -5 + 6 \end{bmatrix} = \begin{bmatrix} 1 & 0 \\ 0 & 1 \end{bmatrix}$$

(identity matrix of order 2×2)

CAUTION Two nonsquare matrices A and B cannot be inverses of each other, since their products AB and BA (if they exist) are always of different orders and, therefore, are not equal. In general, a nonsquare matrix cannot have an inverse.

We usually denote the inverse of matrix A by using the notation A^{-1} (read "A inverse"). We now state the **inverse property of matrix multiplication**.

◆ Inverse Property of Matrix Multiplication

If A is a square $n \times n$ matrix and A^{-1} is its inverse, then

$$AA^{-1} = A^{-1}A = I_n$$

where I_n is the identity matrix of order $n \times n$.

CAUTION If a is a real number, then $a^{-1} = 1/a$; however, if A is a matrix, the notation A^{-1} does *not* mean $1/A$. In fact, matrix division is *not* defined.

An $n \times n$ matrix A has an inverse if and only if the determinant of A is not zero. If $|A| \neq 0$, then we say that matrix A is **invertible**. If $|A| = 0$, then we say that matrix A is **singular**.

◆ **Procedure for Finding A^{-1}**

> If A is an $n \times n$ matrix and $|A| \neq 0$, then its inverse A^{-1} may be found as follows.
>
> 1. Write the $n \times 2n$ matrix $[A \mid I_n]$ consisting of matrix A on the left-hand side and the identity matrix I_n on the right-hand side of a dashed line.
>
> 2. Use elementary row operations on the matrix $[A \mid I_n]$ to obtain the matrix $[I_n \mid B]$. The matrix B is the inverse matrix of A, that is, $B = A^{-1}$.

EXAMPLE 2 Find the inverse of the given matrix, if it exists.

(a) $A = \begin{bmatrix} -3 & -1 \\ 2 & 2 \end{bmatrix}$ (b) $A = \begin{bmatrix} 1 & -1 & 0 \\ 2 & -1 & 1 \\ 0 & 4 & 5 \end{bmatrix}$

◆ **Solution**

(a) The determinant of matrix A is $(-6) - (-2) = -4 \neq 0$. Thus, we know that A is invertible.

Beginning with the 2×4 matrix $[A \mid I_2]$, we perform elementary row operations until we obtain the matrix $[I_2 \mid B]$.

$\begin{bmatrix} -3 & -1 & \mid & 1 & 0 \\ 2 & 2 & \mid & 0 & 1 \end{bmatrix} \sim \begin{bmatrix} 1 & 3 & \mid & 1 & 2 \\ 2 & 2 & \mid & 0 & 1 \end{bmatrix}$ $2R_2 + R_1 \to R_1$

$\sim \begin{bmatrix} 1 & 3 & \mid & 1 & 2 \\ 0 & -4 & \mid & -2 & -3 \end{bmatrix}$ $-2R_1 + R_2 \to R_2$

$\sim \begin{bmatrix} 1 & 3 & \mid & 1 & 2 \\ 0 & 1 & \mid & \frac{1}{2} & \frac{3}{4} \end{bmatrix}$ $-\frac{1}{4}R_2 \to R_2$

$\sim \begin{bmatrix} 1 & 0 & \mid & -\frac{1}{2} & -\frac{1}{4} \\ 0 & 1 & \mid & \frac{1}{2} & \frac{3}{4} \end{bmatrix}$ $-3R_2 + R_1 \to R_1$

Therefore, the inverse of matrix A is

$$A^{-1} = \begin{bmatrix} -\frac{1}{2} & -\frac{1}{4} \\ \frac{1}{2} & \frac{3}{4} \end{bmatrix} = \frac{1}{4} \begin{bmatrix} -2 & -1 \\ 2 & 3 \end{bmatrix}$$

> To avoid fractional elements, factor the scalar $\frac{1}{4}$ from each element of the matrix.

(b) Using an algebraic procedure for evaluating a third-order determinant, or selecting "det" from the matrix menu on a graphing calculator, we find that $|A| = 1 \neq 0$. Thus, we know that A is invertible.

Beginning with the 3×6 matrix $[A \mid I_3]$, we perform elementary row operations until we obtain the matrix $[I_3 \mid B]$.

$$\begin{bmatrix} 1 & -1 & 0 & 1 & 0 & 0 \\ 2 & -1 & 1 & 0 & 1 & 0 \\ 0 & 4 & 5 & 0 & 0 & 1 \end{bmatrix} \sim \begin{bmatrix} 1 & -1 & 0 & 1 & 0 & 0 \\ 0 & 1 & 1 & -2 & 1 & 0 \\ 0 & 4 & 5 & 0 & 0 & 1 \end{bmatrix} \quad -2R_1 + R_2 \to R_2$$

$$\sim \begin{bmatrix} 1 & -1 & 0 & 1 & 0 & 0 \\ 0 & 1 & 1 & -2 & 1 & 0 \\ 0 & 0 & 1 & 8 & -4 & 1 \end{bmatrix} \quad -4R_2 + R_3 \to R_3$$

$$\sim \begin{bmatrix} 1 & -1 & 0 & 1 & 0 & 0 \\ 0 & 1 & 0 & -10 & 5 & -1 \\ 0 & 0 & 1 & 8 & -4 & 1 \end{bmatrix} \quad -R_3 + R_2 \to R_2$$

$$\sim \begin{bmatrix} 1 & 0 & 0 & -9 & 5 & -1 \\ 0 & 1 & 0 & -10 & 5 & -1 \\ 0 & 0 & 1 & 8 & -4 & 1 \end{bmatrix} \quad R_2 + R_1 \to R_1$$

Therefore, the inverse matrix of A is

$$A^{-1} = \begin{bmatrix} -9 & 5 & -1 \\ -10 & 5 & -1 \\ 8 & -4 & 1 \end{bmatrix}$$

◆

We can use the matrix features of a graphing calculator to find the inverse of a square matrix. First, we use the MATRIX key and enter the elements of the matrix whose inverse we wish to find. We now return to the home (primary) screen on the calculator and find the inverse matrix by using the inverse function key. Figure 10.5 shows typical displays for finding the inverse matrix in Example 2(a). For a more detailed explanation about finding the inverse of a matrix by using a graphing calculator, consult the manual that came with your calculator or the *Graphing Technology Laboratory Manual* that accompanies this text. After reading this information, use a graphing calculator to verify the inverse in Example 2(b).

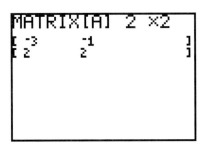

(a) Enter the elements of matrix A.

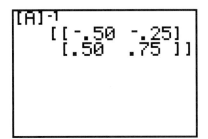

(b) Return to the home screen to find the inverse matrix.

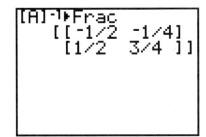

(c) Express the inverse matrix with fractional elements (optional).

FIGURE 10.5

Typical displays for finding the inverse in Example 2(a)

PROBLEM 2 Show that the given matrix A is singular.

$$A = \begin{bmatrix} 1 & 2 & -3 \\ 1 & 0 & 2 \\ 2 & 2 & -1 \end{bmatrix}$$

Try to find A^{-1} by using the matrix features on a graphing calculator. What error message appears? ◆

◆ Using Inverses to Solve a System of Linear Equations

We can use the inverse of a matrix to help solve a system of linear equations. Consider a system of n linear equations in n unknowns:

$$\begin{aligned} a_{11}x_1 + a_{12}x_2 + a_{13}x_3 + \cdots + a_{1n}x_n &= k_1 \\ a_{21}x_1 + a_{22}x_2 + a_{23}x_3 + \cdots + a_{2n}x_n &= k_2 \\ a_{31}x_1 + a_{32}x_2 + a_{33}x_3 + \cdots + a_{3n}x_n &= k_3 \\ &\vdots \\ a_{n1}x_1 + a_{n2}x_2 + a_{n3}x_3 + \cdots + a_{nn}x_n &= k_n \end{aligned}$$

We can represent this $n \times n$ linear system of equations by the matrix equation

$$AX = B$$

where

$$A = \begin{bmatrix} a_{11} & a_{12} & a_{13} & \cdots & a_{1n} \\ a_{21} & a_{22} & a_{23} & \cdots & a_{2n} \\ a_{31} & a_{32} & a_{33} & \cdots & a_{3n} \\ \vdots & \vdots & \vdots & & \vdots \\ a_{n1} & a_{n2} & a_{n3} & \cdots & a_{nn} \end{bmatrix} \quad X = \begin{bmatrix} x_1 \\ x_2 \\ x_3 \\ \vdots \\ x_n \end{bmatrix} \quad B = \begin{bmatrix} k_1 \\ k_2 \\ k_3 \\ \vdots \\ k_n \end{bmatrix}$$

We refer to matrix A as the **coefficient matrix**, matrix X as the **variable matrix**, and matrix B as the **constant matrix**. If the coefficient matrix A is invertible, then we can solve the equation $AX = B$ for X as follows:

$$\begin{aligned} AX &= B \\ A^{-1}(AX) &= A^{-1}B & &\text{Multiply both sides by } A^{-1} \\ (A^{-1}A)X &= A^{-1}B & &\text{Apply associative property of matrix multiplication} \\ I_n X &= A^{-1}B & &\text{Apply inverse property of matrix multiplication} \\ X &= A^{-1}B & &\text{Apply identity property of matrix multiplication} \end{aligned}$$

The product $A^{-1}B$ represents the solution to the $n \times n$ linear system of equations. We refer to this procedure of solving a system of linear equations as the **inverse method**.

SECTION 10.3 ♦ *Determinants and Inverses of Matrices* 671

CAUTION Since matrix division is undefined, we cannot solve the matrix equation $AX = B$ for X by dividing both sides by A. Also, since matrix multiplication is not commutative, $X \neq BA^{-1}$.

EXAMPLE 3 Use the inverse method to solve the given system of linear equations.

$$\begin{aligned} x - y &= 2 \\ 2x - y + z &= 9 \\ 4y + 5z &= 23 \end{aligned}$$

♦ **Solution** If we let

$$A = \begin{bmatrix} 1 & -1 & 0 \\ 2 & -1 & 1 \\ 0 & 4 & 5 \end{bmatrix} \quad X = \begin{bmatrix} x \\ y \\ z \end{bmatrix} \quad B = \begin{bmatrix} 2 \\ 9 \\ 23 \end{bmatrix}$$

$\underbrace{\qquad}_{\text{coefficient matrix}} \quad \underbrace{\qquad}_{\text{variable matrix}} \quad \underbrace{\qquad}_{\text{constant matrix}}$

then the matrix equation

$$AX = B$$

$$\begin{bmatrix} 1 & -1 & 0 \\ 2 & -1 & 1 \\ 0 & 4 & 5 \end{bmatrix} \begin{bmatrix} x \\ y \\ z \end{bmatrix} = \begin{bmatrix} 2 \\ 9 \\ 23 \end{bmatrix}$$

is equivalent to the given system of three equations in three unknowns. From Example 2(b), we know that the coefficient matrix A is invertible and that

$$A^{-1} = \begin{bmatrix} -9 & 5 & -1 \\ -10 & 5 & -1 \\ 8 & -4 & 1 \end{bmatrix}$$

Thus, this system of equations has a unique solution given by

$$\begin{aligned} X = A^{-1}B &= \begin{bmatrix} -9 & 5 & -1 \\ -10 & 5 & -1 \\ 8 & -4 & 1 \end{bmatrix} \begin{bmatrix} 2 \\ 9 \\ 23 \end{bmatrix} \\ &= \begin{bmatrix} -18 + 45 + (-23) \\ -20 + 45 + (-23) \\ 16 + (-36) + 23 \end{bmatrix} \\ &= \begin{bmatrix} 4 \\ 2 \\ 3 \end{bmatrix} \end{aligned}$$

The solution of this system of linear equations is the ordered triple (4, 2, 3). You can check the solution by substituting $x = 4$, $y = 2$, and $z = 3$ into each of the original equations in the system. ◆

This inverse method is one of the most efficient procedures we have for solving a system of linear equations with a graphing calculator. Figure 10.6 shows typical displays for solving the linear system in Example 3.

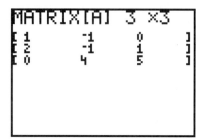 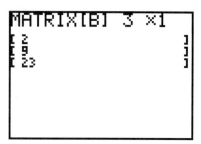 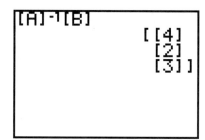

(a) Enter the dimension and elements of the coefficient matrix A.

(b) Enter the dimension and elements of the constant matrix B.

(c) Return to the home screen, and compute the product $A^{-1}B$.

FIGURE 10.6
Typical displays for solving the linear system in Example 3

If the coefficient matrix is singular, then the system of linear equations does not have a unique solution.

PROBLEM 3 Show that the given system of linear equations does not have a unique solution. Is this system inconsistent or dependent?

$$x + 2y - 3z = -3$$
$$x + 3z = 4$$
$$2x + 2y - z = 1$$

◆

Exercises 10.3

◆ Basic Skills

In Exercises 1–12, evaluate the determinant of each matrix by inspection.

1. $\begin{bmatrix} 1 & 4 \\ 6 & 2 \end{bmatrix}$

2. $\begin{bmatrix} 9 & 0 \\ 1 & 2 \end{bmatrix}$

3. $\begin{bmatrix} 8 & -4 \\ -1 & 0 \end{bmatrix}$

4. $\begin{bmatrix} 5 & -2 \\ 3 & -4 \end{bmatrix}$

5. $\begin{bmatrix} -10 & -12 \\ 4 & -17 \end{bmatrix}$

6. $\begin{bmatrix} 12 & 6 \\ -15 & -18 \end{bmatrix}$

7. $\begin{bmatrix} 2 & 3 & -1 \\ 0 & 3 & 9 \\ 0 & 0 & -2 \end{bmatrix}$

8. $\begin{bmatrix} -3 & 9 & 0 \\ 0 & -4 & 5 \\ 0 & 0 & 4 \end{bmatrix}$

9. $\begin{bmatrix} 1 & -2 & 3 & 0 \\ 0 & 2 & 8 & 0 \\ 0 & 0 & -8 & 12 \\ 0 & 0 & 0 & -13 \end{bmatrix}$

10. $\begin{bmatrix} -4 & 0 & -8 & 4 \\ 0 & -4 & 5 & 10 \\ 0 & 0 & 13 & 0 \\ 0 & 0 & 0 & -5 \end{bmatrix}$

11. $\begin{bmatrix} 4 & -2 & 10 & 8 \\ 0 & 0 & 0 & 0 \\ -2 & 4 & 12 & 9 \\ 2 & 9 & 4 & -7 \end{bmatrix}$

12. $\begin{bmatrix} 4 & -2 & 6 & 14 \\ 1 & 2 & 3 & 1 \\ 2 & -1 & 3 & 7 \\ 1 & 3 & -4 & 9 \end{bmatrix}$

In Exercises 13–20, evaluate the determinant of each matrix by using the following two methods.
 (i) Elementary row operations.
(ii) The formula for the expansion of a third-order determinant by its first row.

13. $\begin{bmatrix} 1 & 0 & 2 \\ 2 & -3 & 4 \\ -3 & 1 & 9 \end{bmatrix}$

14. $\begin{bmatrix} 1 & 2 & -3 \\ 3 & 0 & 3 \\ -4 & 1 & -2 \end{bmatrix}$

15. $\begin{bmatrix} 4 & 0 & -3 \\ 3 & -7 & 2 \\ 1 & -5 & 0 \end{bmatrix}$

16. $\begin{bmatrix} 2 & -3 & 5 \\ -1 & 7 & 3 \\ 6 & -9 & 15 \end{bmatrix}$

17. $\begin{bmatrix} 3 & 2 & 4 \\ 0 & 1 & 3 \\ 0 & -2 & 2 \end{bmatrix}$

18. $\begin{bmatrix} 3 & -3 & 2 \\ 2 & 0 & 1 \\ 2 & 0 & -4 \end{bmatrix}$

19. $\begin{bmatrix} 2 & 6 & 3 \\ 0 & 4 & 4 \\ 2 & -1 & 1 \end{bmatrix}$

20. $\begin{bmatrix} 8 & 2 & -6 \\ 1 & 4 & 8 \\ -6 & 3 & 6 \end{bmatrix}$

In Exercises 21–24, evaluate the determinant of each matrix by using elementary row operations. Check each answer by entering the matrix into a graphing calculator and selecting the "det" command from the matrix menu.

21. $\begin{bmatrix} 1 & -2 & 0 & 4 \\ 3 & 0 & 8 & -2 \\ 1 & 0 & -4 & 5 \\ 2 & -3 & -4 & 1 \end{bmatrix}$

22. $\begin{bmatrix} 2 & -4 & 5 & 9 \\ 1 & -3 & 4 & 7 \\ 2 & -3 & 0 & 1 \\ 0 & -1 & 2 & 6 \end{bmatrix}$

23. $\begin{bmatrix} 2 & 3 & 0 & 4 \\ 0 & 1 & 5 & 2 \\ 2 & -1 & 3 & 4 \\ -4 & 2 & -2 & 0 \end{bmatrix}$

24. $\begin{bmatrix} 3 & 2 & -1 & 4 \\ 2 & -2 & 4 & 0 \\ -2 & 3 & 6 & 1 \\ 1 & 4 & -3 & 0 \end{bmatrix}$

In Exercises 25–38, find the inverse of matrix A. If A does not have an inverse, state that it is singular. Check each answer by using the matrix features of a graphing calculator.

25. $A = \begin{bmatrix} 3 & 4 \\ -2 & -3 \end{bmatrix}$

26. $A = \begin{bmatrix} 6 & 7 \\ 5 & 6 \end{bmatrix}$

27. $A = \begin{bmatrix} 1 & 4 \\ 2 & 3 \end{bmatrix}$

28. $A = \begin{bmatrix} 4 & 2 \\ -1 & 2 \end{bmatrix}$

29. $A = \begin{bmatrix} 2 & 3 \\ -6 & -9 \end{bmatrix}$

30. $A = \begin{bmatrix} 5 & -2 \\ -10 & 4 \end{bmatrix}$

31. $A = \begin{bmatrix} 1 & 2 & -3 \\ 0 & 2 & -1 \end{bmatrix}$

32. $A = \begin{bmatrix} 3 & 5 \\ 0 & -2 \\ -2 & 3 \end{bmatrix}$

33. $A = \begin{bmatrix} 1 & 2 & -3 \\ 2 & 3 & -4 \\ -1 & 0 & 3 \end{bmatrix}$

34. $A = \begin{bmatrix} 1 & 0 & 2 \\ 0 & 1 & 4 \\ -1 & 0 & 2 \end{bmatrix}$

35. $A = \begin{bmatrix} 2 & 0 & 5 \\ 1 & -1 & 4 \\ -3 & 2 & -8 \end{bmatrix}$

36. $A = \begin{bmatrix} -3 & 9 & 0 \\ 4 & -20 & 1 \\ 1 & -5 & 0 \end{bmatrix}$

37. $A = \begin{bmatrix} -2 & -1 & -8 \\ 3 & 2 & 10 \\ 4 & 4 & 6 \end{bmatrix}$

38. $A = \begin{bmatrix} 3 & -8 & 12 \\ -2 & 6 & -8 \\ 5 & -8 & 12 \end{bmatrix}$

Exercises 39–44, use the inverse method to solve each system of linear equations. Use the matrix features of a graphing calculator to confirm your work.

39. $\begin{aligned} x - y + z &= -3 \\ 2x + y &= 1 \\ y - 3z &= 7 \end{aligned}$

40. $\begin{aligned} x + 3y &= 4 \\ y - 2z &= 3 \\ x - y + 3z &= -3 \end{aligned}$

41. $\begin{aligned} x - 3y + z &= -3 \\ 2x \quad - 3z &= 7 \\ -x + y + 2z &= -4 \end{aligned}$

42. $\begin{aligned} 3x - y - 2z &= 0 \\ 2x + 8y &= -1 \\ x - y + 10z &= 27 \end{aligned}$

43. $\begin{aligned} 2x - y + z &= 4 \\ 3x + 2y - 3z &= 2 \\ 4x + 5y - 6z &= -3 \end{aligned}$

44. $\begin{aligned} -2x - y + 4z &= -7 \\ 3x + 2y - z &= 4 \\ 4x - 3y + 2z &= -1 \end{aligned}$

45. Two cables support an 800-lb weight, as shown in the figure. The tensions T_1 and T_2 (in pounds) in each cable may be found by solving the following system of linear equations:

$$0.4226 T_1 + 0.7431 T_2 = 800$$
$$0.9063 T_1 - 0.6691 T_2 = 0$$

Use the inverse method in conjunction with the matrix features of a graphing calculator to find T_1 and T_2. Round each answer to three significant digits.

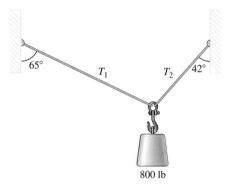

46. For the electrical circuit shown in the figure, the branch currents i_1, i_2, and i_3 (in amperes) may be found by solving the following system of linear equations:

$$i_1 - i_2 - i_3 = 0$$
$$68 i_1 \quad + 48 i_3 = 12.5$$
$$32 i_2 - 48 i_3 = -3.5$$

Use the inverse method in conjunction with the matrix features of a graphing calculator to find i_1, i_2, and i_3. Round each answer to three significant digits.

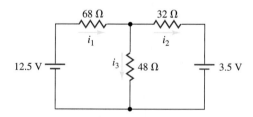

47. When an object is propelled vertically *upward*, its distance s (in feet) above the ground after t second(s) is given by $s = -16t^2 + v_0 t + s_0$, where v_0 is the initial

velocity of the object and s_0 is the initial distance above the ground. Use the inverse method in conjunction with the matrix features of a graphing calculator to find v_0 and s_0 if $s = 98.3$ ft when $t = 1.2$ s and $s = 123.2$ ft when $t = 2.1$ s. Round each answer to three significant digits.

48. The forces F_1, F_2, and F_3 (in pounds), which act on the beam shown in the sketch, may be found by solving the following system of linear equations:

$$0.7547F_1 \quad\quad - 0.4848F_3 = 0$$
$$0.6561F_1 - F_2 + 0.8746F_3 = 200$$
$$\quad\quad -18.0F_2 + 21.0F_3 = 2800$$

Use the inverse method in conjunction with the matrix features of a graphing calculator to find F_1, F_2, and F_3. Round each answer to three significant digits.

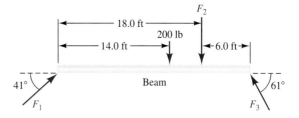

Critical Thinking

49. Use elementary row operations to show that

$$\begin{vmatrix} a_{11} & a_{12} & a_{13} \\ a_{21} & a_{22} & a_{23} \\ a_{31} & a_{32} & a_{33} \end{vmatrix} = $$

$$a_{11}\begin{vmatrix} a_{22} & a_{23} \\ a_{32} & a_{33} \end{vmatrix} - a_{12}\begin{vmatrix} a_{21} & a_{23} \\ a_{31} & a_{33} \end{vmatrix} + a_{13}\begin{vmatrix} a_{21} & a_{22} \\ a_{31} & a_{32} \end{vmatrix}$$

50. Find the value(s) of x for which the given determinant is equal to zero.

(a) $\begin{vmatrix} x & -2 \\ -3 & 18 \end{vmatrix}$ (b) $\begin{vmatrix} x & 0 & 3 \\ 4 & -2 & x \\ 6 & 1 & -1 \end{vmatrix}$

51. The triangle shaded in the figure has vertices $(0, 0)$, (x_1, y_1), and (x_2, y_2).
    (a) Find the area of this triangle by subtracting the area of the three right triangles from the area of the rectangle.
    (b) Evaluate
    $$\tfrac{1}{2}\begin{vmatrix} x_1 & y_1 \\ x_2 & y_2 \end{vmatrix},$$
    and compare your answer with part (a). What does this determinant represent?

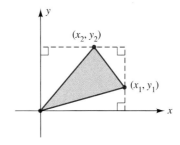

52. Use the determinant in Exercise 51(b) to find the area of the triangle with the given vertices.
    (a) $A(0, 0)$, $B(6, 2)$, $C(4, 4)$
    (b) $A(0, 0)$, $B(3\sqrt{2}, 3)$, $C(-\sqrt{2}, 3)$

53. Suppose A, B, and C are $n \times n$ matrices and $AB = AC$. Is it possible to "cancel" matrix A from both sides of this equation and conclude that $B = C$? Explain.

54. Describe the inverse of an $n \times n$ matrix having nonzero elements $a_{11}, a_{22}, a_{33}, \ldots, a_{nn}$ along its main diagonal and zeros elsewhere.

55. Given the 2 × 2 matrix $A = \begin{bmatrix} a & b \\ c & d \end{bmatrix}$, where $a, b, c,$ and d are real numbers, find a formula for A^{-1} and list any restriction(s) on the formula.

56. Use the formula from Exercise 55 to find the inverse of each of the following matrices.

(a) $\begin{bmatrix} 2 & 1 \\ 3 & 7 \end{bmatrix}$ (b) $\begin{bmatrix} 4 & -5 \\ -9 & 4 \end{bmatrix}$

(c) $\begin{bmatrix} 3 & 2 \\ -4 & -3 \end{bmatrix}$ (d) $\begin{bmatrix} 15 & 12 \\ -9 & -8 \end{bmatrix}$

10.4 Systems of Linear Inequalities and Linear Programming

◆ Introductory Comments

Each of the following statements is a **linear inequality** in two variables x and y:

$$Ax + By > C \qquad Ax + By < C$$
$$Ax + By \geq C \qquad Ax + By \leq C$$

If a linear inequality in two variables is true when x is replaced with x_1 and y with y_1, then the ordered pair (x_1, y_1) is a **solution** of the linear inequality. For example, the ordered pair $(2, -1)$ is a solution of the linear inequality $3x - 2y > 7$ because

$$3(2) - 2(-1) > 7$$
$$8 > 7 \quad \text{is true.}$$

However, the ordered pair $(0, 1)$ is *not* a solution of this linear inequality because

$$3(0) - 2(1) > 7$$
$$-2 > 7 \quad \text{is false.}$$

Linear inequalities occur when solving certain types of decision-making problems, called *linear programming problems*. Since these types of problems involve working with the graphs of linear inequalities, we begin with a discussion of graphing linear inequalities.

◆ Graphing Linear Inequalities

One way to illustrate *all* the ordered pairs that are solutions of a linear inequality is to draw its graph. We begin by graphing the equation $Ax + By = C$ using a dashed line if the inequality symbol is $<$ or $>$, and a solid line if the inequality symbol is $\leq$ or $\geq$. This dashed or solid line divides the xy-coordinate plane into two half-planes. We then select any ordered pair form either half-plane as a *test point* in order to determine whether it is a solution of the inequality. If the ordered pair we select is a solution, then every ordered pair in that half-plane is also a solution. Shading this half-plane gives us the *graph of the linear inequality*. The procedure is illustrated in the next example.

EXAMPLE 1 Graph the solution set of each linear inequality.

(a) $3x + y > 0$ **(b)** $3x - 2y \geq 6$

SECTION 10.4 ♦ Linear Inequalities and Linear Programming

◆ **Solution**

(a) Since the inequality symbol is >, the graph of the corresponding equation $3x + y = 0$ is sketched as a dashed line, as shown in Figure 10.7. Selecting $(1, 0)$ as an arbitrary test point for this inequality, we find that

$$3(1) + (0) > 0$$
$$3 > 0 \quad \text{is true}$$

Shading the half-plane that contains the point $(1, 0)$ gives us the graph of $3x + y > 0$ (see Figure 10.7). Every ordered pair in the shaded region, but not on the dashed line, is a solution of the inequality $3x + y > 0$.

(b) Since the inequality symbol is ≥, the graph of the corresponding equation $3x - 2y = 6$ is shown as a solid line in Figure 10.8. Selecting $(0, 0)$ as an arbitrary test point for this inequality, we find that

$$3(0) - 2(0) \geq 6$$
$$0 \geq 6 \quad \text{is false}$$

Shading the half-plane that does *not* contain the point $(0, 0)$ gives us the graph of $3x - 2y \geq 6$ (see Figure 10.8). Every ordered pair in the shaded region or on the solid line is a solution of the inequality $3x - 2y \geq 6$. ◆

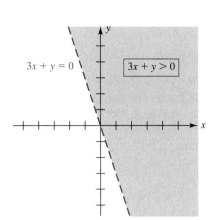

FIGURE 10.7
Graph of the solution set for the inequality $3x + y > 0$

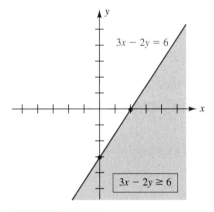

FIGURE 10.8
Graph of the solution set for the inequality $3x - 2y \geq 6$

Each of the following statements is a linear inequality in two variables x and y written in slope-intercept form:

$$y < ax + b \qquad y > ax + b$$
$$y \leq ax + b \qquad y \geq ax + b$$

The graph of the solution set of a linear inequality in slope-intercept form is the half-plane *above* the line $y = ax + b$ if the inequality symbol is > or ≥, or the half-plane *below* the line $y = ax + b$ if the inequality symbol is < or ≤. Thus, an alternate method of graphing the solution set of the inequalities in Example 1 is to write the inequality in slope-intercept form as follows:

$$\begin{array}{ll} 3x + y > 0 & 3x - 2y \geq 6 \\ y > -3x & -2y \geq -3x + 6 \\ & y \leq \tfrac{3}{2}x - 3 \end{array}$$

Reverse the inequality when multipying or dividing both sides by a negative number.

The graph of the solution set for $y > -3x$ is the half-plane *above* the line $y = -3x$, as shown in Figure 10.7. The graph of the solution set for $y \leq \tfrac{3}{2}x - 3$ is the half-plane *below* the line $y = \tfrac{3}{2}x - 3$, as shown in Figure 10.8.

PROBLEM 1 Graph the solution set for each linear inequality.

(a) $y < -2$ (b) $y \leq -2x + 5$ ◆

◆ Graphing a System of Linear Inequalities

When we have several linear inequalities in two unknowns x and y to be solved simultaneously, we say that we have a **system of linear inequalities**. If an ordered pair (x_1, y_1) satisfies *every* inequality in the system, then (x_1, y_1) is a solution of the system. *All* ordered pairs that satisfy the system can be illustrated graphically by shading the region in the xy-coordinate plane that is *common* to the graphs of all the inequalities in the system.

EXAMPLE 2 Graph the solution set of each system of linear inequalities.

(a) $3x + y > 0$
$3x - 2y \geq 6$

(b) $y \geq x - 3$
$x + 2y \leq 12$
$x \geq 0$
$y \geq 0$

◆ Solution

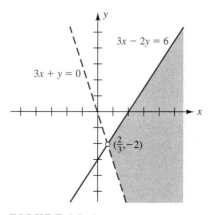

FIGURE 10.9

Graph of the solution set for the system

$3x + y > 0$
$3x - 2y \geq 6$

(a) The shaded regions in Figures 10.7 and 10.8 represent the solution sets of the individual inequalities $3x + y > 0$ and $3x - 2y \geq 6$, respectively. Now, the solution set of this system of linear inequalities consists of all points common to both these shaded regions, as illustrated in Figure 10.9. We find the intersection point of the two lines, $\left(\frac{2}{3}, -2\right)$, by solving the equations $3x + y = 0$ and $3x - 2y = 6$ *simultaneously*. Every ordered pair within the shaded region of Figure 10.9 or along the portion of the solid line $3x - 2y = 6$ that touches this region, *except* the point $\left(\frac{2}{3}, -2\right)$, is a solution of this system of inequalities.

(b) We begin by graphing each of the four individual inequalities on the same coordinate plane. The solution set of this system of linear inequalities is the region that is common to all these inequalities. The shaded region in Figure 10.10 shows the solution set. Each of the points where the sides of the shaded region intersect is called a *vertex*, and we find these points (the *vertices* of the graph) as follows:

1. Solving $y = x - 3$ and $y = 0$ simultaneously yields the vertex $(3, 0)$.

2. Solving $y = x - 3$ and $x + 2y = 12$ simultaneously yields the vertex $(6, 3)$.

3. Solving $x + 2y = 12$ and $x = 0$ simultaneously yields the vertex $(0, 6)$.

4. Solving $x = 0$ and $y = 0$ simultaneously yields the vertex $(0, 0)$.

Every ordered pair in the shaded region of Figure 10.10 or along the portions of the solid lines that touch this region is a solution of this system. ◆

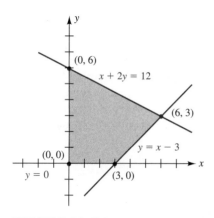

FIGURE 10.10

Graph of the solution set for the system

$y \geq x - 3$
$x + 2y \leq 12$
$x \geq 0$
$y \geq 0$

Most graphing calculators have the capability of generating the graph of a system of linear inequalities. To graph the system in Example 2(a), we first generate the graphs of the corresponding equations $3x + y = 0$ (or $y = -3x$) and $3x - 2y = 6$ (or $y = \frac{3}{2}x - 3$) in the viewing rectangle [see Figure 10.11(a)]. Next, we select "shade" from the [DRAW] menu to shade the area above $y = -3x$ and below $y = \frac{3}{2}x - 3$ [see Figure 10.11(b)]. Finally, using the built-in program for finding the intersection points of two graphs (or the [TRACE] key and [ZOOM] key), we find the vertex of the region [see Figure 10.11(c)].

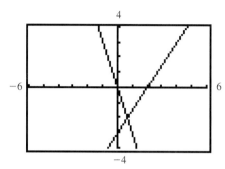

(a) Generate the graphs of the corresponding equations.

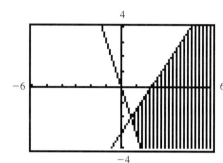

(b) Shade the region above $y = -3x$ and below $y = \frac{3}{2}x - 3$.

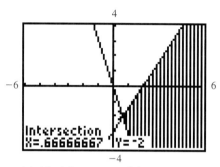

(c) Find the vertex of the region.

FIGURE 10.11

Typical displays for graphing the system of inequalities in Example 2(a)

PROBLEM 2 Use a graphing calculator to generate the graph of the solution set of the system of inequalities given in Example 2(b). Verify the vertices of the shaded region. ◆

◆ An Introduction to Linear Programming

In Figure 10.10, every line segment joining any two points in the shaded region is contained in that region. A region with this characteristic is said to be **convex**. In **linear programming** (a branch of mathematics), we maximize or minimize a linear function f in two variables x and y over a convex region. The function f, defined by

$$f(x, y) = Ax + By,$$

is called the **objective function** with slope $-A/B$. The convex region over which we maximize or minimize the objective function is formed from a system of linear inequalities called the **constraints**.

Note: The linear programming problems that we discuss in this section represent an introduction to this branch of mathematics. More advanced linear programming problems use more than two unknowns and require sophisticated matrices and computer programming to find maximum and minimum values.

Suppose we wish to find the maximum and minimum values of an objective function f, defined by

$$f(x, y) = 2x + 3y,$$

over the convex region in Figure 10.10. Any line of this form has slope $-\frac{2}{3}$. Figure 10.12 displays the convex region for the constraints given in Example 2(b). It also shows several lines with slope $-\frac{2}{3}$ that pass through some points of the convex region. From the information shown in the figure, we can state the following facts about values of this objective function:

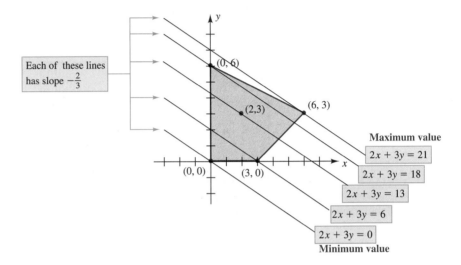

FIGURE 10.12

The maximum value of $f(x, y) = 2x + 3y$ over the given convex region is 21, and this occurs at the vertex (6, 3). The minimum value of $f(x, y) = 2x + 3y$ over the given convex region is 0, and this occurs at the vertex (0, 0).

1. The value of $f(x, y)$ is constant along each line and this functional value may be determined by using any point on the line. Note that $f(0, 0) = 0$, $f(3, 0) = 6$, $f(2, 3) = 13$, $f(0, 6) = 18$, and $f(6, 3) = 21$.
2. The value of $f(x, y)$ increases as lines with slope $-\frac{2}{3}$ move from left to right through the convex region.
3. The maximum value of $f(x, y)$ over the convex region is 21, and this occurs at a vertex of the region, namely, (6, 3).
4. The minimum value of $f(x, y)$ over the convex region is 0, and this also occurs at a vertex of the region, namely, (0, 0).

In summary, we state the following **fundamental principle of linear programming**.

◆ **Fundamental Principle of Linear Programming**

The maximum or minimum value (if it exists) of an objective function f, defined by

$$f(x, y) = Ax + By,$$

always occurs at a vertex of the convex region that is formed from a system of linear inequalities, called constraints.

EXAMPLE 3 Find the maximum and minimum values (if they exist) of the objective function f defined by $f(x, y) = 2x + 5y$, subject to the following set of constraints:

$$y \geq 7 - 2x$$
$$x + 3y \geq 7$$
$$x + y \geq 5$$
$$x \geq 0$$
$$y \geq 0$$

◆ **Solution** We begin by graphing the system of linear inequalities given as constraints (see Figure 10.13). Note that the constraints form an *unbounded* convex region—at least one side of this region is not bounded by a line of constraint.

We find the vertices of this convex region as follows:

1. Solving $y = 7 - 2x$ and $x = 0$ simultaneously yields the vertex $(0, 7)$.
2. Solving $y = 7 - 2x$ and $x + y = 5$ simultaneously yields the vertex $(2, 3)$.
3. Solving $x + y = 5$ and $x + 3y = 7$ simultaneously yields the vertex $(4, 1)$.
4. Solving $x + 3y = 7$ and $y = 0$ simultaneously yields the vertex $(7, 0)$.

According to the fundamental principle of linear programming, the maximum and minimum values of f (if they exist) occur at vertices of this convex region. Thus, we evaluate $f(x, y)$ at each vertex, as shown in the following table.

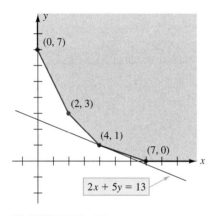

FIGURE 10.13
The minimum value of $f(x, y) = 2x + 5y$ over this unbounded convex region is 13, and it occurs when $x = 4$ and $y = 1$. The function f has no maximum value.

Vertex	Value of $f(x, y) = 2x + 5y$
(0, 7)	$f(0, 7) = 2(0) + 5(7) = 35$
(2, 3)	$f(2, 3) = 2(2) + 5(3) = 19$
(4, 1)	$f(4, 1) = 2(4) + 5(1) = 13$ ← minimum value
(7, 0)	$f(7, 0) = 2(7) + 5(0) = 14$

Using this table and Figure 10.13, we observe that the minimum value of f is $f(4, 1) = 13$.

We cannot conclude from the table that the maximum value of f is $f(0, 7) = 35$. Since values of $f(x, y)$ increase without bound as lines with slope $-\frac{2}{5}$ move from left to right through this unbounded convex region, we must conclude that a maximum value of f does not exist. ◆

PROBLEM 3 Find the maximum and minimum values (if they exist) of the objective function f, defined by $f(x, y) = 2x - 5y$, subject to the set of constraints given in Example 3. ◆

◆ **Application: Maximizing a Profit**

Linear programming has many applications in business and economics. We conclude this section with an example showing how we can maximize a profit.

EXAMPLE 4 A ski company manufactures a slalom ski and a racing ski. The profit on each pair of slalom skis is $30 and on each pair of racing skis is $50. The company can produce at most 60 pairs of slalom skis per day and at most 40 pairs of racing skis per day. Production of a pair of slalom skis requires 2 hours of labor, and production of a pair of racing skis requires 3 hours of labor. If the maximum number of hours available for the production of skis is 150 hours per day, determine the number of each type of ski that should be manufactured to maximize the profit. (Assume that all skis can be sold.)

◆ **Solution**   We begin by letting

$$x = \text{number of slalom skis produced}$$

and

$$y = \text{number of racing skis produced}.$$

If P represents the profit, then the objective function P, defined by

$$P(x, y) = 30x + 50y,$$

is the function we wish to maximize, subject to the following constraints:

$$x \geq 0$$
$$y \geq 0$$
$$x \leq 60$$
$$y \leq 40$$
$$2x + 3y \leq 150$$

The graph of the constraints is shown in Figure 10.14. According to the fundamental principle of linear programming, the maximum value of P must occur at a vertex of this convex region. Thus, we evaluate $P(x, y)$ at each vertex, as shown in the following table.

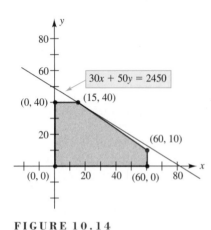

FIGURE 10.14
The maximum value of $P(x, y) = 30x + 50y$ over this convex region is $P(15, 40) = 2450$.

Vertex	Value of $P(x, y) = 30x + 50y$
(0, 0)	$P(0, 0) = 30(0) + 50(0) = 0$
(0, 40)	$P(0, 40) = 30(0) + 50(40) = 2000$
(15, 40)	$P(15, 40) = 30(15) + 50(40) = 2450$ ← maximum value
(60, 10)	$P(60, 10) = 30(60) + 50(10) = 2300$
(60, 0)	$P(60, 0) = 30(60) + 50(0) = 1800$

From the table, we observe that the maximum value of P is P(15, 40) = $2450 (see Figure 10.14). Thus, to maximize profit the company should manufacture 15 pairs of slalom skis and 40 pairs of racing skis per day.

PROBLEM 4 Suppose the ski manufacturer in Example 4 makes some changes in production methods. The company's profit is still $30 on each pair of slalom skis but is now $45 on each pair of racing skis. Show that we can find more than one way to maximize the profit subject to these same constraints.

Exercises 10.4

Basic Skills

In Exercises 1–12, graph the solution set of each linear inequality.

1. $3x + 2y \geq 12$
2. $2x - 5y < 10$
3. $2x - 7y < 10$
4. $6x + 4y > 18$
5. $x > 3$
6. $y \leq -2$
7. $2y - 5 > 0$
8. $3x + 4 > 0$
9. $y \leq 2x - 3$
10. $y \geq 4 - 3x$
11. $y < \dfrac{2 - 3x}{4}$
12. $y \geq \dfrac{2x + 8}{3}$

In Exercises 13–30, graph the solution set of each system of linear inequalities. Label the vertices of the shaded region.

13. $x \geq 2$
    $y \leq 1$
14. $x + y < 4$
    $x > 3$
15. $x + 2y < 7$
    $x - y < 1$
16. $x - y \leq 2$
    $y < x + 3$
17. $x + 2y < 6$
    $x \geq 0$
    $y \geq 0$
18. $y < x$
    $x < 4$
    $y > -2$
19. $5x - 2y > -6$
    $x - 2y \leq 2$
    $x + 2y \leq 6$
20. $3x + 2y \geq -6$
    $11x - 2y < 34$
    $x - 4y \geq -16$
21. $y > x$
    $x + 2y \geq 6$
    $2x + y \leq 15$
22. $3y \geq x$
    $x - 3y \geq -12$
    $x \leq 6$
23. $x + 3y \leq 15$
    $2y \geq x$
    $y > 1$
    $x \geq 0$
24. $2y \leq x + 9$
    $y \leq -3x + 15$
    $x > -3$
    $y \geq 0$
25. $3x + 2y \geq 6$
    $5x + 4y \leq 20$
    $x \geq 0$
    $y \geq 0$
26. $x + y \geq 10$
    $x + y \leq 15$
    $y \leq 20$
    $x \leq 25$
27. $y \geq 2x - 8$
    $x + y \leq 7$
    $x \leq 4$
    $x \geq 0$
    $y \geq 0$
28. $x + 2y \geq 3$
    $x - 2y \geq -3$
    $3x + 2y \leq 23$
    $y \geq x - 6$
    $x \geq 1$
29. $2x + y \geq 4$
    $x + 2y \geq 5$
    $x + 4y \geq 7$
    $x \geq 0$
    $y \geq 0$
30. $x + 4y \geq 14$
    $x + 2y \geq 10$
    $x + y \geq 7$
    $x \geq 2$
    $y \geq 0$

In Exercises 31–44, find the maximum and minimum values (if they exist) of the objective function subject to the given constraints.

Objective function	Constraints
31. $f(x, y) = 3x + y$	$x + y \leq 6$ $x \leq 3$ $x \geq 0$ $y \geq 0$
32. $f(x, y) = 2x - 3y$	$x + y \leq 6$ $x \leq 3$ $x \geq 0$ $y \geq 0$
33. $f(x, y) = 4x - 5y$	$y \leq x + 1$ $y \geq x - 2$ $y \leq 7$ $y \geq 2$

	Objective function	Constraints
34.	$f(x, y) = 3x + 2y$	$y \leq x + 1$
		$y \geq x - 2$
		$y \leq 7$
		$y \geq 2$
35.	$f(x, y) = 2x - y$	$2x + 3y \leq 16$
		$x - 4y \leq -3$
		$y \leq 3x - 2$
36.	$f(x, y) = x + 5y$	$2x + 3y \leq 16$
		$x - 4y \leq -3$
		$y \leq 3x - 2$
37.	$f(x, y) = x + y$	$x + 2y \leq 10$
		$y \leq 6 - x$
		$x \leq 4$
		$x \geq 0$
		$y \geq 0$
38.	$f(x, y) = 5x + 2y$	$x + 2y \leq 10$
		$y \leq 6 - x$
		$x \leq 4$
		$x \geq 0$
		$y \geq 0$
39.	$f(x, y) = 5x + 2y$	$y \leq 27 - 3x$
		$3x - 2y \leq 18$
		$3x + 4y \geq 18$
		$3x - 5y \geq -9$
40.	$f(x, y) = 3x + y$	$y \leq 27 - 3x$
		$3x - 2y \leq 18$
		$3x + 4y \geq 18$
		$3x - 5y \geq -9$
41.	$f(x, y) = 3x + 4y$	See Exercise 29
42.	$f(x, y) = 2x - y$	See Exercise 29
43.	$f(x, y) = 7x - 2y$	See Exercise 30
44.	$f(x, y) = 2x + 7y$	See Exercise 30

In Exercises 45–52, use linear programming to solve each problem.

45. A doctor is going to purchase bookcases for his office. One type of bookcase costs $100, holds 30 cubic feet of books, and takes up 6 square feet of floor space. Another type costs $200, holds 48 cubic feet of books, and takes up 9 square feet of floor space. The doctor wants to spend no more than $1200 for the bookcases and use no more than 66 square feet of office floor space. How many of each type bookcase should he purchase to *maximize* storage capacity?

46. A sand and gravel company wishes to purchase six-wheel and ten-wheel dump trucks for delivery vehicles. Each six-wheel dump truck sells for $20,000, has an average monthly fuel cost of $100, and has a carrying capacity of 12 cubic yards. Each ten-wheel dump truck sells for $30,000, has an average monthly fuel cost of $300, and has a carrying capacity of 17 cubic yards. Suppose the company wishes to spend no more than $240,000 for the trucks and no more than $1500 per month for fuel. How many trucks of each type should the company purchase to *maximize* their delivery capacity?

47. A company manufactures oak tables and chairs. Each table requires 2 hours 15 minutes to assemble, 1 hour to finish, and 30 minutes to pack for shipping. Each chair requires 2 hours to assemble, 2 hours to finish, and 10 minutes to pack for shipping. The maximum number of labor hours available each day for assembling is 90 hours, for finishing is 80 hours, and for packing is 15 hours. If the manufacturer makes a profit of $80 on each table and $30 on each chair, determine the number of each that should be produced each day to *maximize* profit. (Assume that all tables and chairs can be sold.)

48. A kennel raises Doberman pinschers and German shepherds. The kennel can raise no more than 40 dogs and wishes to have no more than 24 Doberman pinschers. The cost of raising a Doberman pinscher is $50, the cost of raising a German shepherd is $30, and the kennel can invest no more than $1500 for this purpose. If the kennel makes a profit of $150 per pinscher and $100 per shepherd, how many dogs of each type should be raised in order to *maximize* profit? (Assume that all dogs can be sold.)

49. A grain company stocks two types of wild-bird seed: sunflower and cracked corn. The company buys the sunflower seed for $10 per bag and the cracked corn for $5 per bag. They sell the sunflower seed for $16 per bag and the cracked corn for $10 per bag. From past experience, the company knows that they will sell at least twice as many bags of sunflower seed as bags of cracked corn. If the company does not wish to order more than $600 worth of seed, how many bags of each type should be ordered to *maximize* profit? (Assume that all bags of seed can be sold.)

50. A woman inherits $48,000 from her grandfather's estate, subject to the following conditions: Part or all of the money must be invested either in treasury bills that pay 8¢ on each dollar or in bonds that pay 10¢ on each dollar. At least $10,000 must be invested in treasury bills and at most $30,000 in bonds. The amount invested in treasury bills must be no more than twice the amount invested in bonds. How much should she invest in each to *maximize* her investment?

51. A company makes two brands of cereal P and Q, both of which are enriched with vitamins A and B. Cereal P cost 10¢ per ounce and cereal Q cost 15¢ per ounce. The number of milligrams of each vitamin that is con-

tained in each ounce of cereal is given in the following table.

	Cereal P	Cereal Q
Vitamin A	1.2	2.3
Vitamin B	3.3	1.4

Suppose a woman needs at least 8.0 mg of vitamin A and at least 9.0 mg of vitamin B to satisfy her daily requirement. How many ounces of each type of cereal should she consume if she wishes to *minimize* her food cost? Round each answer to the nearest tenth of an ounce.

52. A farmer has 200 acres of land available and wishes to plant at least 40 acres of corn and beans. The cost of the corn seed is $115 per acre and the cost of the bean seed is $185 per acre. Past experience has shown that the cost of maintaining and harvesting the corn crop is $260 per acre, and the cost of maintaining and harvesting the bean crop is $145 per acre. The farmer wants to spend no more than $12,500 for seed and no more than $16,500 on maintenance and harvest. If the expected profits from the corn crop and bean crop are $550 per acre and $520 per acre, respectively, how many acres of each crop should be planted to *maximize* profit? Round each answer to the nearest tenth of an acre.

Critical Thinking

53. Given the convex region shown in the figure, find the slope m of an objective function $f(x, y) = Ax + By$ that has a maximum value at the given point(s).

(a) at both points P and Q
(b) at both points Q and R
(c) at point P only
(d) at point Q only
(e) at point R only

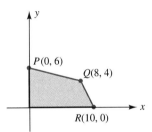

54. Given the convex region shown in the figure, find the slope m of an objective function $f(x, y) = Ax + By$ that has a minimum value at the given point(s).

(a) at both points P and Q
(b) at both points Q and R
(c) at point P only
(d) at point Q only
(e) at point R only

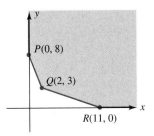

55. A landscape gardening company has two brands of lawn fertilizer. The nutrient contents in kilograms per bag is shown in the following table:

	Brand A	Brand B
Nitrogen (N)	30	20
Phosphoric acid (P_2O_5)	2	4
Potash (K_2O)	1	4

It has been determined that a certain lawn needs at least 120 kg of nitrogen, at least 16 kg of phosphoric acid, and at least 12 kg of potash. If brand A costs $22 per bag and brand B costs $18 per bag, how many bags of each brand should be used to *minimize* the cost? What is the minimum cost?

56. A computer company has two distribution centers, one in Boston and one in Providence. The company receives an order to ship 40 computers to college A, and 60 computers to college B. The cost of shipping each computer is shown in the following table:

	College A	College B
From **Boston**	$9	$8
From **Providence**	$8	$10

Suppose that the distribution center in Boston has 80 computers available for shipping and the distribution center in Providence has 30 computers available. How should the order be filled in order to *minimize* the total shipping cost? What is the minimum cost?

Chapter 10 Review

♦ **Questions for Writing or Group Discussion**

1. State the three possible types of solutions for a system of n linear equations in n variables.
2. Explain how the *elimination method* may be used to solve a 3×3 linear system.
3. Can a *nonsquare system* of linear equations have a unique solution? Explain.
4. What is meant by the *order* of a matrix? Give an example of a matrix with order 4×3.
5. What is an *augmented matrix*? How is it formed?
6. List three *elementary row operations* that can be performed on a matrix to generate a row-equivalent matrix. Illustrate each operation with an example.
7. Give an example of two matrices that are equivalent but not equal.
8. Is it possible to find the *sum of two matrices* having different orders? Explain.
9. Explain the procedure for finding the *product of two matrices*. Under what conditions does the product exist?
10. If matrix A has order 3×2 and matrix B has order 2×5, what is the order of AB?
11. Is matrix multiplication commutative? associative? Illustrate with examples.
12. Suppose A and B are $n \times n$ matrices and $AB = BA = I_n$, where I_n is the identity matrix of order $n \times n$. What is the relationship between A and B?
13. Explain the procedure for finding the *inverse* of a square matrix. Does every square matrix have an inverse?
14. What is the value of $|A|$ if matrix A is *singular*? Explain.
15. When solving the matrix equation $AX = B$ for X, what restrictions must be placed on A and B for the solution to exist?
16. What is the procedure for finding the *determinant* of a 2×2 matrix? of a 3×3 matrix?
17. What are the effects of elementary row operations on the determinant of an $n \times n$ matrix?
18. List some conditions for which the determinant of an $n \times n$ matrix is zero.
19. Explain the procedure for graphing a *system of linear inequalities*. Illustrate with an example.
20. What is *linear programming*? Where does the maximum or minimum value of the objective function always occur?

Review Exercises

In Exercises 1–18 solve each system of linear equations by using any convenient method.

1. $\begin{aligned} x - y &= 5 \\ 2x + 3y - z &= -4 \\ y + 3z &= -6 \end{aligned}$

2. $\begin{aligned} 2x + y &= 1 \\ x + z &= -4 \\ 3y - 2z &= 11 \end{aligned}$

3. $\begin{aligned} 3x - 2y + z &= -3 \\ 2x + y - 3z &= 2 \\ 4y - 2z &= 4 \end{aligned}$

4. $\begin{aligned} 5x - 2y + 3z &= 11 \\ 2x - y + 4z &= 5 \\ 4x - 3y + 2z &= 9 \end{aligned}$

5. $\begin{aligned} 2x + 6y - 5z &= 1 \\ 3x + 4y - 5z &= -1 \\ x + 8y - 5z &= 3 \end{aligned}$

6. $\begin{aligned} -4x - 3y + z &= 2 \\ 3x - y + 4z &= -3 \\ 4x + 3y - z &= 0 \end{aligned}$

7. $\begin{aligned} \tfrac{1}{2}x - 3y + 5z &= 18 \\ 2x - \tfrac{3}{8}y - z &= 15 \\ 3x + 5y + 6z &= 4 \end{aligned}$

8. $\begin{aligned} 0.2x - 1.6y - 3.2z &= -12 \\ 1.4x + 2.3y - 0.7z &= 104 \\ 9.8x - 2.4y + 1.1z &= 113 \end{aligned}$

9. $\begin{aligned} 13x - 12y - 5z &= 6 \\ 4x - 11y + 15z &= 28 \\ 6x - 10y &= 9 \end{aligned}$

10. $\begin{aligned} 3x + 10y + 4z &= -8 \\ 8x + 5y - 9z &= 16 \\ 5x - 7y - 11z &= 16 \end{aligned}$

11. $\begin{aligned} x - 2y + 3z &= -9 \\ 2x + y - z &= 13 \\ 3x + 4y - z &= 23 \end{aligned}$

12. $\begin{aligned} 2x + 3y + z &= -3 \\ 5x - 2y + 3z &= 13 \\ 3x - y - 4z &= -10 \end{aligned}$

13. $\begin{aligned} x - 3y + z &= 6 \\ 4x - 2y - z &= -1 \end{aligned}$

14. $\begin{aligned} 2x + y - 3z &= 4 \\ 3x + 2y - z &= 7 \end{aligned}$

15. $\begin{aligned} x + y + t &= -1 \\ 2y - t &= 2 \\ x - z + 3t &= 11 \\ 4y + 3z &= 4 \end{aligned}$

16. $\begin{aligned} 2x - y + 3z - t &= 1 \\ x + 3y + t &= -5 \\ 2y + 4z + 5t &= 1 \\ 4x - z + 2t &= 2 \end{aligned}$

17. $\begin{aligned} 3x - y + z &= 0 \\ 4y - 4z + 3t &= 12 \\ 2x - 2y - t &= 0 \\ 5x - 3y - 2z + 5t &= 12 \end{aligned}$

18. $\begin{aligned} x_1 + 2x_2 - x_5 &= -7 \\ x_2 - x_3 + x_4 &= -6 \\ x_1 + 2x_3 + x_5 &= 11 \\ 3x_2 - 2x_4 &= -4 \\ 4x_3 - x_4 - 2x_5 &= 5 \end{aligned}$

In Exercises 19–34, perform the indicated matrix operations given that A, B, C, D, E, F, G, and H are defined as follows.

$$A = \begin{bmatrix} 4 & 5 \\ 3 & 4 \end{bmatrix} \qquad B = \begin{bmatrix} -2 & 3 \\ 1 & 0 \end{bmatrix}$$

$$C = \begin{bmatrix} 1 & 0 & -3 \\ -2 & 4 & 5 \end{bmatrix}$$

$$D = \begin{bmatrix} 4 & 2 & -1 \\ 0 & -3 & -6 \end{bmatrix}$$

$$E = \begin{bmatrix} 2 & -3 \\ -1 & 5 \\ 3 & 2 \end{bmatrix} \qquad F = \begin{bmatrix} 3 & 5 \\ -2 & 7 \\ 1 & 0 \end{bmatrix}$$

$$G = \begin{bmatrix} 1 & 4 & -2 \end{bmatrix} \qquad H = \begin{bmatrix} -6 \\ 3 \\ -9 \end{bmatrix}$$

19. $-3G$
20. $\tfrac{1}{3}H$
21. $A + B$
22. $F - E$
23. $\tfrac{1}{2}A - \tfrac{3}{2}A$
24. $\tfrac{1}{3}(-3D)$
25. $2C - 3D$
26. $3(A + 2B)$
27. AB
28. CE
29. GH
30. HG
31. $(CF)A$
32. $-2(DH)$
33. $G(2E + F)$
34. $\tfrac{1}{2}(A^2 + B)$

In Exercises 35–40, find the inverse of each matrix, if it exists.

35. $\begin{bmatrix} 4 & 3 \\ 2 & 2 \end{bmatrix}$

36. $\begin{bmatrix} -6 & 3 \\ 5 & -2 \end{bmatrix}$

37. $\begin{bmatrix} 1 & 4 & 0 \\ 2 & 6 & -1 \\ 3 & 2 & 5 \end{bmatrix}$

38. $\begin{bmatrix} 2 & -1 & 3 \\ 3 & -5 & 8 \\ 4 & 6 & -2 \end{bmatrix}$

39. $\begin{bmatrix} 2 & 8 & -3 & 7 \\ 4 & 1 & -6 & 0 \\ 0 & 3 & 0 & -3 \\ -6 & 9 & 9 & -7 \end{bmatrix}$

40. $\begin{bmatrix} 4 & -9 & -12 & -3 \\ 0 & 4 & 9 & -4 \\ -3 & 7 & 10 & 2 \\ 2 & -2 & 0 & 7 \end{bmatrix}$

In Exercises 41–48, evaluate the determinant of each matrix.

41. $\begin{bmatrix} 2 & -3 \\ 5 & -2 \end{bmatrix}$
42. $\begin{bmatrix} 12 & 15 \\ -7 & 9 \end{bmatrix}$

43. $\begin{bmatrix} 1 & -2 & 0 \\ 2 & -3 & 1 \\ 6 & 0 & -2 \end{bmatrix}$

44. $\begin{bmatrix} 3 & -2 & 4 \\ 8 & 2 & -3 \\ -1 & 7 & 3 \end{bmatrix}$

45. $\begin{bmatrix} 3 & -4 & 7 \\ 0 & -2 & 3 \\ 0 & 0 & 5 \end{bmatrix}$

46. $\begin{bmatrix} 1 & 9 & 6 & -3 \\ 0 & -4 & 5 & 1 \\ 0 & 0 & -4 & 2 \\ 0 & 0 & 0 & -2 \end{bmatrix}$

47. $\begin{bmatrix} 2 & -3 & 0 & 1 \\ 1 & -4 & 2 & -5 \\ 0 & -1 & 0 & 2 \\ 2 & 6 & -4 & 0 \end{bmatrix}$

48. $\begin{bmatrix} 1 & 3 & -2 & 4 \\ 0 & 6 & 2 & -4 \\ 2 & 3 & -5 & 10 \\ 3 & 1 & 0 & 6 \end{bmatrix}$

In Exercises 49–52, graph the solution set of each system of linear inequalities. Label the vertices of the shaded region.

49. $\begin{array}{l} y < x + 2 \\ 3x + 2y < 6 \\ x \leq 6 \end{array}$
50. $\begin{array}{l} 3x + 4y \geq 12 \\ 6x + 5y \leq 30 \\ x \geq 0 \\ y \geq 0 \end{array}$

51. $\begin{array}{l} x \leq 18 - 3y \\ 3x + 2y \leq 19 \\ 2x + y \leq 12 \\ x \geq 0 \\ y \geq 0 \end{array}$
52. $\begin{array}{l} 4x + y \geq 8 \\ y \geq 5 - x \\ x + 5y \geq 13 \\ x \geq 0 \\ y \geq 0 \end{array}$

In Exercises 53 and 54, find the maximum and minimum values (if they exist) of the objective function subject to the given constraints.

53. Objective function: $f(x, y) = 2x + 3y$
    Constraints: See Exercise 51

54. Objective function: $f(x, y) = 2x + y$
    Constraints: See Exercise 52

55. Find k such that the determinant is equal to the given value.

   (a) $\begin{vmatrix} k & 3 \\ 5 & 7 \end{vmatrix} = 6$
   (b) $\begin{vmatrix} k & 2 \\ 5 & k \end{vmatrix} = 15$

   (c) $\begin{vmatrix} 0 & k & 0 \\ -1 & 2 & 4 \\ 2 & 5 & k \end{vmatrix} = -12$

   (d) $\begin{vmatrix} 2 & k & 3 \\ 0 & 3 & k \\ 5 & 1 & 1 \end{vmatrix} = 0$

56. The rows and columns of $A = \begin{bmatrix} a_{11} & a_{12} \\ a_{21} & a_{22} \end{bmatrix}$ are interchanged to form the matrix A^T (called the *transpose* of A). What is the relationship between $|A|$ and $|A^T|$?

57. Recall from Section 3.2 that the equation of a nonvertical line passing through two distinct points (x_1, y_1) and (x_2, y_2) is given by

    $$y - y_1 = m(x - x_1)$$

    where m is the slope of the line and

    $$m = \frac{y_2 - y_1}{x_2 - x_1}.$$

    (a) Replace m in the equation $y - y_1 = m(x - x_1)$ with $\frac{y_2 - y_1}{x_2 - x_1}$ and solve for y. Express the answer as a single fraction with parentheses removed from the numerator.

    (b) Solve the determinant equation

    $$\begin{vmatrix} 1 & x & y \\ 1 & x_1 & y_1 \\ 1 & x_2 & y_2 \end{vmatrix} = 0$$

    for y and express the answer as a single fraction. Compare your answer to part (a). What does this determinant equation represent?

58. Use the determinant equation in Exercise 57(b) to find the equation of the line passing through the given points A and B. Write your answer in slope-intercept form, $y = mx + b$.

    (a) $A(-2, 5)$, $B(2, -3)$
    (b) $A(\sqrt{2}, -\sqrt{3})$, $B(-\sqrt{3}, \sqrt{2})$

In Exercises 59–64, use a system of linear equations to solve each problem.

59. A refinery has a "regular" gasoline that sells for 83¢ per gallon and a "super" gasoline that sells for 99¢ per gal-

lon. How many gallons of each type gasoline should be mixed together to form 12,000 gallons of a gasoline that sells for 86¢ per gallon?

60. A man has invested $10,000 in two bank accounts. One account has an annual yield of 8% and the other, an annual yield of 9%. The interest he earns from these accounts is taxable at the rate of 30%. After taxes, he earns $602 from these accounts for the year. How much is invested in each account?

61. Three types of tickets are available for baseball games: box seats for $16, grandstand seats for $10, and bleacher seats for $5. The seating capacity at the ball park is 32,500, and the park has 2500 more grandstand seats than bleacher seats. When a game is sold out, the revenue from ticket sales is $343,500. How many seats of each type are in the ball park?

62. At the baseball game, Kristen buys 2 hot dogs, 1 bag of peanuts, and 2 sodas for $8.50. Caryn buys 1 hot dog, 2 bags of peanuts, 1 ice cream bar, and 1 soda for $7.00. David buys 4 hot dogs and 4 sodas for $16.00. Regina buys 1 hot dog, 2 ice cream bars, and 1 bag of peanuts for $7.00. What is the price of each item: a hot dog? a bag of peanuts? a soda? an ice cream bar?

63. A hiking trail, connecting campsite A to campsite B, has a pond $2\frac{1}{2}$ miles from campsite A. Starting from campsite A, the trail goes downhill for 1 mile, is level for 6 miles, and goes uphill for 2 miles. Suppose a woman can hike from campsite A to campsite B in 3 hours 15 minutes; from campsite B to campsite A in 3 hours; and from campsite A to the pond and back again to campsite A in 1 hour 45 minutes. Find her rates of walking (in miles per hour) downhill, on level ground, and uphill.

64. Find the equation of the parabola $x = ay^2 + by + c$ that passes through the points $(k_1, 1)$, $(k_2, -2)$, and $(k_3, 0)$, given the following information.
    (a) $k_1 = 4$, $k_2 = 1$, and $k_3 = -1$
    (b) $k_1 = -2$, $k_2 = 1$, and $k_3 = -3$
    (c) $k_1 = 3$, $k_2 = 8$, and $k_3 = 4$
    (d) $k_1 = 2$, $k_2 = 5$, and $k_3 = 0$

An appliance store has a two-day sale for three styles of microwave ovens. The sale prices (in dollars) are given in matrix C:

	Model		
	400-watt	500-watt	600-watt
$C = $	[120	160	220]	Price

Within the Boston area, the store has three outlets. The number of 400-watt, 500-watt, and 600-watt models of microwave ovens sold at each outlet during the first and second days of the sale are given in matrices A and B, respectively.

$$A = \begin{bmatrix} 8 & 10 & 16 \\ 12 & 8 & 9 \\ 6 & 5 & 6 \end{bmatrix} \begin{matrix} \text{400-watt} \\ \text{500-watt} \\ \text{600-watt} \end{matrix}$$

with columns labeled Outlet 1, Outlet 2, Outlet 3.

$$B = \begin{bmatrix} 5 & 12 & 8 \\ 7 & 3 & 6 \\ 8 & 3 & 6 \end{bmatrix} \begin{matrix} \text{400-watt} \\ \text{500-watt} \\ \text{600-watt} \end{matrix}$$

with columns labeled Outlet 1, Outlet 2, Outlet 3.

In Exercises 65–68, find the indicated matrix, and state what it represents.

65. $A \begin{bmatrix} 1 \\ 1 \\ 1 \end{bmatrix}$

66. $\begin{bmatrix} 1 & 1 & 1 \end{bmatrix}(A + B)$

67. CB

68. $C(A + B)\begin{bmatrix} 1 \\ 1 \\ 1 \end{bmatrix}$

69. Two cables support a 750-lb weight, as shown in the figure. The tensions T_1 and T_2 (in pounds) on each cable may be found by solving the following system of linear equations:

$$0.5150T_1 + 0.7880T_2 = 750$$
$$0.8572T_1 - 0.6157T_2 = 0$$

Use the inverse method in conjunction with the matrix features of a graphing calculator to find T_1 and T_2. Round each answer to three significant digits.

70. The *Wheatstone bridge circuit*, shown in the sketch, is used in electrical measurement applications. The currents i_1, i_2, and i_m (in amperes) may be found by solving the following system of linear equations:

$$(10.2 + 31.5)i_1 \qquad\qquad - \qquad 31.5i_m = 125$$
$$\qquad\qquad (20.6 + 41.2)i_2 + \qquad 41.2i_m = 125$$
$$10.2i_1 + \qquad 41.2i_2 + (50.0 + 41.2)i_m = 125$$

Exercise 70 continues

Use the inverse method in conjunction with the matrix features of a graphing calculator to find i_1, i_2, and i_m. Round each answer to three significant digits.

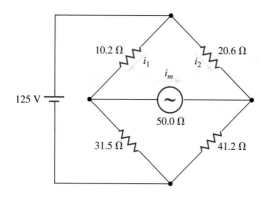

In Exercises 71 and 72, use linear programming to solve each problem.

71. A college football stadium has 27,500 seats and, of these, 7500 are midfield seats. Tickets for a midfield seat cost $20 each, and tickets for all other seats cost $12 each. To make a profit, the college must make at least $300,000 from ticket sales. What is the minimum number of tickets that the college must sell to make a profit?

72. A manufacturer produces two models of log-homes kits: a one-bedroom model and a two-bedroom model. The profit on each one-bedroom kit is $4500 and on each two-bedroom kit, $6000. The company can produce at most 4 one-bedroom kits per week and at most 3 two-bedroom kits per week. To produce a one-bedroom kit requires 200 hours of labor, and to produce a two-bedroom kit requires 300 hours of labor. If the maximum number of hours available for the production of log-home kits is 1100 hours per week, determine the number of each model that should be manufactured to maximize the profit. (Assume all log-home kits can be sold.)

A father decides to deposit $150 each month in a savings account for his daughter's college education. The savings account earns interest at the rate of 9% per year compounded monthly. How much will be in the account at the end of eight years if no money is in the account today?

For the solution, see Example 6 in Section 11.4.

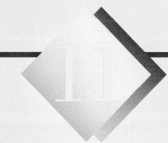

Sequences and Series

11.1   An Introduction to Sequences and Series

11.2   The Sum of a Series and Mathematical Induction

11.3   Arithmetic Sequences and Series

11.4   Geometric Sequences and Series

11.5   Infinite Geometric Series

11.6   Factorials and Their Applications

11.1 An Introduction to Sequences and Series

◆ Introductory Comments

The phrase "a sequence of events" means the following of one event after another in a given order of succession. In mathematics, we refer to a **sequence** as an ordered list of numbers, and each number in the list is called an **element** of the sequence. When writing a sequence, we use commas to separate the elements and an ellipsis (three dots, . . .) to show an omission of elements that are obviously understood. For example, we write the sequence that represents the first eleven odd, positive integers as

$$1, 3, 5, 7, 9, 11, 13, 15, 17, 19, 21$$

or, more compactly, as

$$1, 3, 5, 7, \ldots, 21$$

We refer to this sequence as a *finite sequence* because the sequence stops after a finite number of elements. In a sequence such as

$$1, 3, 5, 7, \ldots$$

the ellipsis at the end indicates that the sequence continues forever. We refer to this type of sequence as an *infinite sequence*.

The indicated sum of the elements in a sequence is called a **series**. Associated with the finite sequence $1, 3, 5, 7, \ldots, 21$ is the *finite series*

$$1 + 3 + 5 + 7 + \cdots + 21$$

and associated with the infinite sequence $1, 3, 5, 7, \ldots$ is the *infinite series*

$$1 + 3 + 5 + 7 + \cdots$$

Each number in a series is called a **term** of the series. The terms in a series are identical to the elements in its corresponding sequence.

In this chapter, we study sequences, series, and their applications. We begin by using functional notation to define a sequence. If we think of a sequence in terms of functional notation, then we can use a graphing calculator to support our work with sequences.

◆ Using Functional Notation to Define a Sequence

We define a *finite sequence function* as a function whose domain is the set of positive integers from 1 to k and an *infinite sequence function* as a function whose domain is the set of all positive integers. In either case, the elements in the range of the function form the sequence.

SECTION 11.1 ♦ An Introduction to Sequences and Series

♦ Definitions of Finite and Infinite Sequences

A function u, whose domain is the set of positive integers from 1 to k, is a *finite sequence function*. The elements in the range of u, taken in the order

$$u(1), u(2), u(3), \ldots, u(k)$$

form a **finite sequence**.

A function v, whose domain is the set of *all* positive integers, is an *infinite sequence function*. The elements in the range of v, taken in the order

$$v(1), v(2), v(3), \ldots$$

form an **infinite sequence**.

EXAMPLE 1 Find the sequence represented by each sequence function.

(a) $u(n) = 2n - 1$ for domain {1, 2, 3, 4}

(b) $v(n) = (-1)^{n+1} \, 2^n$ for domain {1, 2, 3, 4, ...}

♦ Solution

(a) The function u is a finite sequence function. The elements in the range of u are as follows:

$$u(1) = 2(1) - 1 = 1 \qquad u(2) = 2(2) - 1 = 3$$
$$u(3) = 2(3) - 1 = 5 \qquad u(4) = 2(4) - 1 = 7$$

The elements in the range of u, written in the order

$$1, 3, 5, 7,$$

form the finite sequence. This sequence is the first four odd, positive integers.

(b) The function v is an infinite sequence function. The first four elements in the range of v are as follows:

$$v(1) = (-1)^{1+1} \, 2^1 = 2 \qquad v(2) = (-1)^{1+2} \, 2^2 = -4$$
$$v(3) = (-1)^{1+3} \, 2^3 = 8 \qquad v(4) = (-1)^{1+4} \, 2^4 = -16$$

The elements in the range of v, written in the order

$$2, -4, 8, -16, \ldots$$

ending with an ellipsis, form the infinite sequence. Note that the signs of the elements in this sequence alternate between positive and negative values. We refer to a sequence with this characteristic as an *alternating sequence*. ◆

The graph of a sequence function u is formed by plotting the points $(n, u(n))$ in the coordinate plane. Since a sequence function has only integers in its domain, we do *not* connect these points. The graph of the sequence function u, defined in Example 1(a), is shown in Figure 11.1(a).

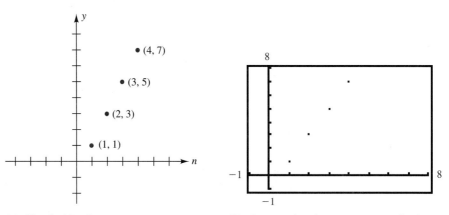

FIGURE 11.1

Graph of the sequence function $u(n) = 2n - 1$ for domain $\{1, 2, 3, 4\}$

(a) Sketched by plotting points

(b) Generated in the viewing rectangle of a graphing calculator

We can generate the graph of a sequence function on most graphing calculators. First, we select "sequence" mode and "dot" mode. We then press the $\boxed{y=}$ key and enter the sequence function we wish to graph. The $\boxed{\text{WINDOW}}$ key (or $\boxed{\text{RANGE}}$ key) is used to select the size of the viewing rectangle as well as other "window" values, such as the value of n at which calculation begins (n-start), the value of $u(n)$ when $n = n$-start [$u(n)$-start], and the values of n at which plotting begins and ends (n-min and n-max, respectively). Figure 11.1(b) shows the graph of $u(n) = 2n - 1$ for domain $\{1, 2, 3, 4\}$ in the viewing rectangle of a graphing calculator. For this sequence we select these "window" values:

$n\text{-start} = 1 \qquad u(n)\text{-start} = 1 \qquad n\text{-min} = 1 \qquad n\text{-max} = 4$

PROBLEM 1 Find the sequence represented by the sequence function

$$u(n) = (-1)^n \, 2^{n-1} \quad \text{for domain } \{1, 2, 3, 4\}$$

Then sketch the graph of this function. ◆

◆ General Element of a Sequence

It is customary to denote the first element of a sequence as a_1, the second element as a_2, the third element as a_3, and so on. The nth element of a sequence, denoted as a_n, is called the **general element** of the sequence. Thus, for a sequence function u, we have

$$u(1) = a_1 \quad \text{First element}$$
$$u(2) = a_2 \quad \text{Second element}$$
$$u(3) = a_3 \quad \text{Third element}$$
$$\vdots$$
$$u(n) = a_n \quad \text{General element}$$
$$\vdots$$

The general element of a sequence may be given **explicitly** in terms of n, or it may be expressed in terms of preceding elements. When it is expressed in terms of preceding elements, we say that the sequence is defined **recursively**.

We can generate a sequence in the viewing rectangle of most graphing calculators by selecting the variable u_n (which corresponds to the general element a_n) from the "y-vars" menu. First, we enter the general element and store it in u_n. Next, we select the appropriate "window" values for n-start and u_n-start. Now, we can generate the sequence defined by this general element by entering u_n (n-start, n-stop). Figure 11.2 shows a typical display of the sequence defined in Example 1(a). For a more detailed explanation about the sequence features of a graphing calculator, consult the manual that came with your calculator or the *Graphing Technology Laboratory Manual* that accompanies this text. After reading this information, use a graphing calculator to generate the first four elements of the infinite sequence in Example 1(b).

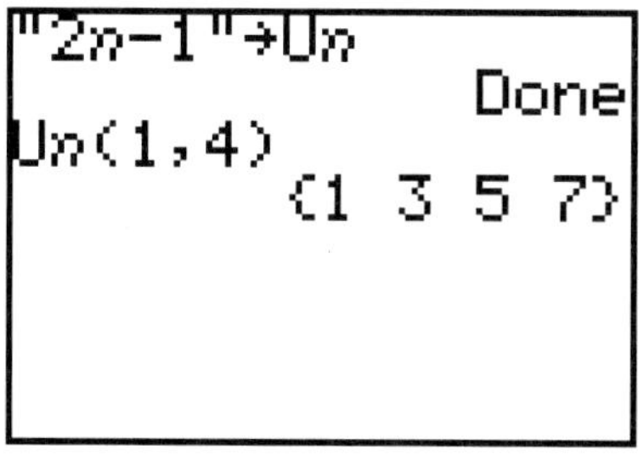

EXAMPLE 2 Find the first four elements of a sequence whose general element a_n is given.

(a) $a_n = \dfrac{1 + n}{n}$ **(b)** $a_n = \begin{cases} 1 & \text{if } n = 1 \\ na_{n-1} & \text{if } n \geq 2 \end{cases}$

◆ Solution

(a) This general element is defined explicitly in terms of n. For $n = 1, 2, 3, 4$, we have

$$a_1 = \frac{1+1}{2} = 2 \qquad a_2 = \frac{1+2}{2} = \frac{3}{2}$$

$$a_3 = \frac{1+3}{3} = \frac{4}{3} \qquad a_4 = \frac{1+4}{4} = \frac{5}{4}$$

Thus, the first four elements of this sequence are

$$2, \tfrac{3}{2}, \tfrac{4}{3}, \tfrac{5}{4}$$

The display in Figure 11.3 confirms our work. We have chosen n-start $= 1$ and u_n-start $= 2$ as "window" values.

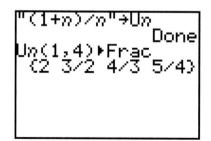

FIGURE 11.2

Typical display of the sequence defined by $u(n) = 2n - 1$, with domain $\{1, 2, 3, 4\}$

FIGURE 11.3

Typical display of the first four elements of the sequence with general element $a_n = \dfrac{1 + n}{n}$

(b) Note that when $n \geq 2$, the general element a_n is defined in terms of the preceding element a_{n-1}. Hence, this is a recursively defined sequence. For $n = 1, 2, 3, 4$, we have

$$a_1 = 1 \qquad a_2 = 2a_1 = 2 \cdot 1 = 2$$
$$a_3 = 3a_2 = 3 \cdot 2 = 6 \qquad a_4 = 4a_3 = 4 \cdot 6 = 24$$

Thus, the first four elements of this sequence are

$$1, 2, 6, 24$$

We can confirm our answer by using the sequence features on a graphing calculator. For recursively defined functions, we define u_n by using the variable u_{n-1} on the calculator. Figure 11.4 shows a typical display of this sequence. We have chosen n-start $= 1$ and u_n-start $= 1$ as "window" values. ◆

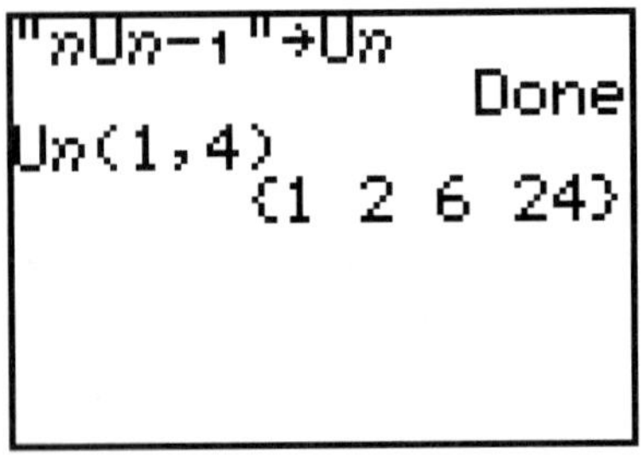

Referring to Example 2(b), the numbers in the sequence 1, 2, 6, 24 are called the *factorials* of 1, 2, 3, and 4, respectively. A **factorial** of a given positive integer is the product of that integer and all lesser positive integers. In Section 11.6, we discuss factorials and their applications in more detail.

FIGURE 11.4

Typical display of the first four elements of the sequence with the general element

$$a_n = \begin{cases} 1 & \text{if } n = 1 \\ na_{n-1} & \text{if } n \geq 2 \end{cases}$$

PROBLEM 2 Use the general element in Example 2(b) to find the factorials of 5 and 6. ◆

Given the first few elements of a sequence, we cannot *uniquely* determine the general element. For example, consider the finite sequence

$$2, 4, 6$$

Note that 2, 4, and 6 are even numbers or multiples of 2. Thus, one possibility for the general element is simply

$$a_n = 2n.$$

However, another possibility for the general element is

$$a_n = \begin{cases} 2n & \text{if } n \text{ is odd} \\ 2a_{n-1} & \text{if } n \text{ is even} \end{cases}$$

In the next example, we use **pattern recognition** to determine the general element of a given sequence. Keep in mind that the answers we give are not unique. They do, however, represent the most obvious choices.

EXAMPLE 3 Use pattern recognition to find the general element of each sequence.

(a) 1, 3, 5, 7, ..., 21   **(b)** 1, 3, 9, 27, 81, 243

(c) $\frac{2}{3}, \frac{3}{4}, \frac{4}{5}, \frac{5}{6}, \ldots$ **(d)** $-x, x^5, -x^9, x^{13}, \ldots$

Solution

(a) We recognize the elements of the finite sequence 1, 3, 5, 7, ..., 21 as odd, positive integers. Since these odd numbers are one less than the even numbers, which have the general element $2n$, we may write the general element for these odd, positive integers as

$$a_n = 2n - 1.$$

(b) The elements in the finite sequence 1, 3, 9, 27, 81, 243 are the powers of three. Since we must obtain 1 when $n = 1$, we can write the general element as

$$a_n = 3^{n-1}.$$

(c) For the infinite sequence $\frac{2}{3}, \frac{3}{4}, \frac{4}{5}, \frac{5}{6}, \ldots$, each denominator is one more than its corresponding numerator. Since we must obtain $\frac{2}{3}$ when $n = 1$, we can write the general element as

$$a_n = \frac{n+1}{n+2}.$$

(d) In the infinite sequence $-x, x^5, -x^9, x^{13}, \ldots$, each exponent after the first is four more than the preceding exponent. This suggests an exponent containing $4n$. However, since we must obtain an exponent of 1 when $n = 1$, we write these exponents as $4n - 3$ for $n = 1, 2, 3, \ldots$. Also, since the *odd-numbered* elements are negative, we must include a factor of $(-1)^n$. Thus, we may write the general element for this sequence as

$$a_n = (-1)^n \, x^{4n-3}.$$

◆

PROBLEM 3 Verify that each general element found for each part of Example 3 does give the first four elements of its sequence. ◆

Note: It is not possible to find a formula that defines the general element of every sequence. For the sequence of prime numbers

$$2, 3, 5, 7, 11, 13, 17, 19, 23, 29, \ldots, a_n, \ldots$$

where a_n is the nth prime number, we cannot write a formula that defines a_n.

◆ Sigma Form of a Series

In a sequence, we refer to a_n as the general element, but in a series we refer to a_n as the **general term** of the series.

The finite series

$$a_1 + a_2 + a_3 + \cdots + a_n$$

and the infinite series

$$a_1 + a_2 + a_3 + \cdots + a_n + \cdots$$

are written here in **expanded form**. These series may also be written more compactly by using **sigma form**. The Greek letter Σ (read "sigma") is used as the "summation symbol" in sigma form.

Sigma Form

The finite series

$$a_1 + a_2 + a_3 + \cdots + a_n$$

may be written in sigma form as

$$\sum_{i=1}^{n} a_i$$

The infinite series

$$a_1 + a_2 + a_3 + \cdots + a_n + \cdots$$

may be written in sigma form as

$$\sum_{i=1}^{\infty} a_i$$

The letter i in the sigma form is called the **index of summation** or the **summation variable**. We often refer to the summation variable as a *dummy variable* because any other letter can be used without changing the series. The letters i, j, and k are commonly used as summation variables.

To change a series from sigma form to expanded form, we replace the summation variable (index) with successive integers, starting with the integer written below Σ and ending with the integer written above Σ. If the series is infinite, ∞ is written above Σ to indicate that the summation has no ending value.

EXAMPLE 4 Write each series in expanded form.

(a) $\displaystyle\sum_{i=1}^{4} (5i - 2)$ **(b)** $\displaystyle\sum_{j=1}^{\infty} (-1)^j j^2$

Solution

(a) To express $\displaystyle\sum_{i=1}^{4} (5i - 2)$ [read "the sum of $(5i - 2)$ from $i = 1$ to $i = 4$"] in expanded form, we let the index i take on successive integers from $i = 1$ to $i = 4$. Thus,

$$\sum_{i=1}^{4} (5i - 2) = [5(1) - 2] + [5(2) - 2] + [5(3) - 2] + [5(4) - 2]$$
$$= 3 + 8 + 13 + 18$$

expanded form

SECTION 11.1 ♦ An Introduction to Sequences and Series

Since the terms in a series are identical to the corresponding elements in its associated sequence, we can confirm our answer by using the sequence features of a graphing calculator, as shown in Figure 11.5.

(b) To express $\sum_{j=1}^{\infty} (-1)^j j^2$ in expanded form, we let the index j take on *all* successive integers starting with $j = 1$. To do this, we write the first few terms of the series, followed by an ellipsis, then the nth term, followed by an ellipsis:

$$\sum_{j=1}^{\infty} (-1)^j j^2 = (-1)^1\, 1^2 + (-1)^2\, 2^2 + (-1)^3\, 3^2 + \cdots + (-1)^n n^2 + \cdots$$
$$= \underbrace{-1 + 4 - 9 + \cdots + (-1)^n n^2 + \cdots}_{\text{expanded form}}$$

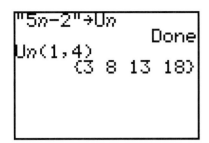

FIGURE 11.5

Typical display of the terms in the series $\sum_{i=1}^{4} (5i - 2)$

The display in Figure 11.6 shows the first three terms of this infinite series and confirms our work. ♦

If a finite series contains many terms and we wish to write the series in expanded form, then we usually write the first few terms followed by an ellipsis, the nth term followed by an ellipsis, and the last term. For example,

$$\sum_{i=1}^{40} (5i - 2) = 3 + 8 + 13 + \cdots + (5n - 2) + \cdots + 198$$

We find the last term by evaluating $[5(40) - 2]$.

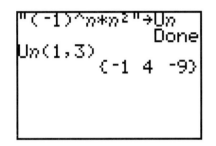

FIGURE 11.6

Typical display of the first three terms in the series $\sum_{j=1}^{\infty} (-1)^j j^2$

PROBLEM 4 Write the series $\sum_{k=1}^{27} \dfrac{k}{k+1}$ in expanded form. ♦

If the general term of a series in expanded form is known, then we can write the series in sigma form. The letter we use for the summation variable is immaterial—only the general term and the beginning and ending values of the summation variable are significant.

EXAMPLE 5 Use pattern recognition to find the general term of each series. Then write each series in sigma form.

(a) $1 + 3 + 5 + 7 + 9$ **(b)** $2 + 5 + 8 + 11 + \cdots + 83$
(c) $\frac{1}{2} - \frac{1}{4} + \frac{1}{8} - \frac{1}{16} + \cdots$ **(d)** $2x^2 + 3x^3 + 4x^4 + 5x^5 + \cdots$

♦ **Solution**

(a) We recognize the terms in the series $1 + 3 + 5 + 7 + 9$ as the first five odd, positive integers. Thus, from Example 3(a), we know that the general term of this finite series is

$$a_n = 2n - 1.$$

In our definition of sigma form for a finite series, the summation variable i begins at a value of 1 and ends at a value of n, where n is the number of terms in the finite series. Since this series contains 5 terms, we write

$$1 + 3 + 5 + 7 + 9 = \underbrace{\sum_{i=1}^{5} (2i - 1)}_{\text{sigma form}}$$

(b) In the finite series $2 + 5 + 8 + 11 + \cdots + 83$, each term after the first is three more than the preceding term. This suggests a general term containing $3n$. However, since we must obtain 2 when $n = 1$, we write the general term of this series as

$$a_n = 3n - 1.$$

Now we must find the upper value of the summation variable. To do this, we set the general term equal to the last term in the series and solve for the number of terms n, as follows:

$$3n - 1 = 83$$
$$3n = 84$$
$$n = 28$$

Thus, using i as the summation variable, we write

$$2 + 5 + 8 + 11 + \cdots + 83 = \underbrace{\sum_{i=1}^{28} (3i - 1)}_{\text{sigma form}}$$

(c) The terms of the infinite series $\frac{1}{2} - \frac{1}{4} + \frac{1}{8} - \frac{1}{16} + \cdots$ are powers of one-half with alternating signs. Powers of one-half can be written as $\left(\frac{1}{2}\right)^n$ for $n = 1, 2, 3, \ldots$. Since the even-numbered terms are negative, we must also include the factor $(-1)^{n+1}$. Thus, the general term for this series is

$$a_n = (-1)^{n+1} \left(\frac{1}{2}\right)^n \quad \text{or} \quad \frac{(-1)^{n+1}}{2^n}.$$

The upper value of the summation variable for an infinite series is always ∞. Thus, using i as the summation variable, we have

$$\frac{1}{2} - \frac{1}{4} + \frac{1}{8} - \frac{1}{16} + \cdots = \underbrace{\sum_{i=1}^{\infty} \frac{(-1)^{i+1}}{2^i}}_{\text{sigma form}}$$

(d) The coefficients of x in the infinite series $2x^2 + 3x^3 + 4x^4 + 5x^5 + \cdots$ are consecutive integers. Since we must obtain 2 when $n = 1$, we can write these coefficients as $n + 1$ for $n = 1, 2, 3, \ldots$. In each term, the exponent of x is the same as its coefficient. Thus, the general term for this series is

$$a_n = (n + 1)x^{n+1}$$

The upper value of the summation variable for an infinite series is always ∞. Thus, using i as the summation variable, we have

$$2x^2 + 3x^3 + 4x^4 + 5x^5 + \cdots = \underbrace{\sum_{i=1}^{\infty} (i + 1) x^{i+1}}_{\text{sigma form}}$$

Our definition of sigma form uses $i = 1$ as the lower value of the summation variable. We can use other lower values to describe a series. Consider the series in Example 5(a):

$$1 + 3 + 5 + 7 + 9 = \sum_{i=1}^{5} (2i - 1)$$

If we write the upper value as $i = 5$, instead of 5, and then perform a *shift of index* by letting $i = k + 1$, we obtain

$$\sum_{i=1}^{i=5} (2i - 1) = \sum_{k+1=1}^{k+1=5} [2(k + 1) - 1] = \sum_{k=0}^{k=4} (2k + 1).$$

Observe that

$$\sum_{k=0}^{4} (2k + 1) = 1 + 3 + 5 + 7 + 9 = \sum_{i=1}^{5} (2i - 1).$$

Note: Unless otherwise instructed, we shall use $i = 1$ as the lower value of the summation variable when changing from expanded form to sigma form.

PROBLEM 5 Perform a shift of index on the sigma form $\sum_{i=1}^{\infty} (i + 1)x^{i+1}$ by letting $i = k - 1$.

Exercises 11.1

Basic Skills

In Exercises 1–6, find the sequence represented by each sequence function. Then sketch the graph of each function.

1. $u(n) = 6 - 3n$ for domain $\{1, 2, 3, 4, 5\}$

2. $u(n) = n^2 - 1$ for domain $\{1, 2, 3, 4, 5, 6\}$

3. $u(n) = (-1)^{n+1}n$ for domain $\{1, 2, 3, 4\}$

4. $u(n) = \dfrac{n}{n + 1}$ for domain $\{1, 2, 3, 4, 5, 6, 7\}$

5. $v(n) = \dfrac{2^{2n-1}}{n^2}$ for domain $\{1, 2, 3, 4, \ldots\}$

6. $v(n) = \dfrac{(-1)^n}{2^n - 1}$ for domain $\{1, 2, 3, 4, \ldots\}$

In Exercises 7–20, find the first six elements of a sequence with the given general element. When possible, use the sequence features of a graphing calculator to confirm your answer.

7. $a_n = 5n - 4$
8. $a_n = 4 - 3n$
9. $a_n = 3(-2)^n$
10. $a_n = 9\left(\dfrac{2}{3}\right)^{n-1}$
11. $a_n = 2 + \dfrac{1}{n}$
12. $a_n = \dfrac{n}{2^n}$
13. $a_n = 1 + (-1)^{n+1}$
14. $a_n = \dfrac{(-1)^n}{3n - 4}$
15. $a_n = \begin{cases} n & \text{if } n \text{ is odd} \\ 2n - 1 & \text{if } n \text{ is even} \end{cases}$
16. $a_n = \begin{cases} 0 & \text{if } n \text{ is odd} \\ n^2 - 1 & \text{if } n \text{ is even} \end{cases}$
17. $a_n = \begin{cases} 3 & \text{if } n = 1 \\ 2a_{n-1} & \text{if } n \geq 2 \end{cases}$
18. $a_n = \begin{cases} -1 & \text{if } n = 1 \\ (a_{n-1})^{n-1} & \text{if } n \geq 2 \end{cases}$
19. $a_n = \begin{cases} 1 & \text{if } n = 1 \\ 2 & \text{if } n = 2 \\ a_{n-1} - a_{n-2} & \text{if } n \geq 3 \end{cases}$
20. $a_n = \begin{cases} -5 & \text{if } n = 1 \\ 2 & \text{if } n = 2 \\ a_{n-1}a_{n-2} & \text{if } n \geq 3 \end{cases}$

In Exercises 21–32, use pattern recognition to find the general element of each sequence. (Answers are not unique.)

21. $1, 4, 9, 16, 25, \ldots$
22. $1, 8, 27, 81, 243, \ldots$
23. $2, -4, 6, -8, 10, -12$
24. $-1, 3, -5, 7, -9, \ldots, 27$
25. $3, 4, 5, 6, 7, \ldots, 29$
26. $1, 4, 7, 10, 13, \ldots$
27. $\tfrac{1}{2}, \tfrac{3}{4}, \tfrac{5}{6}, \tfrac{7}{8}, \tfrac{9}{10}, \ldots$
28. $\tfrac{1}{2}, \tfrac{3}{5}, \tfrac{5}{8}, \tfrac{7}{11}, \tfrac{9}{14}, \tfrac{11}{17}, \tfrac{13}{20}$
29. $2, 0.2, 0.02, 0.002, 0.0002, \ldots$
30. $0.31, 0.0031, 0.000031, 0.00000031, \ldots$
31. $-x, \dfrac{x^2}{3}, -\dfrac{x^3}{5}, \dfrac{x^4}{7}, -\dfrac{x^5}{9}, \ldots$
32. $x, -\dfrac{x^3}{5}, \dfrac{x^5}{9}, -\dfrac{x^7}{13}, \dfrac{x^9}{17}, \ldots$

In Exercises 33–42, write each series in expanded form. When possible, use the sequence features of a graphing calculator to confirm your answer.

33. $\sum\limits_{i=1}^{5} 2i$
34. $\sum\limits_{i=1}^{4} (7i - 3)$
35. $\sum\limits_{n=1}^{10} (-1)^n n^3$
36. $\sum\limits_{j=1}^{7} (-1)^{j+1} (j^2 - 1)$
37. $\sum\limits_{k=1}^{\infty} k^k$
38. $\sum\limits_{k=1}^{\infty} \dfrac{1}{k}$
39. $\sum\limits_{i=0}^{50} \dfrac{2i - 3}{2}$
40. $\sum\limits_{i=2}^{25} \dfrac{i}{i^2 + 1}$
41. $\sum\limits_{i=1}^{\infty} \dfrac{(-1)^i (2x^{i-1})}{i}$
42. $\sum\limits_{i=1}^{\infty} \dfrac{(-1)^{i+1} x^i}{3i + 1}$

In Exercises 43–54, write the series in sigma form using $i = 1$ as the lower value of the summation variable.

43. $2 + 4 + 6 + 8 + 10 + 12 + 14$
44. $1 - 3 + 5 - 7 + 9 - 11 + 13 - 15$
45. $2 + 7 + 12 + 17 + \cdots + 147$
46. $3 + 8 + 13 + 18 + \cdots + 98$
47. $1 - 1 + 1 - 1 + 1 - 1 + \cdots$
48. $-1 + 4 - 9 + 16 - 25 + \cdots$
49. $-1 + \tfrac{1}{2} - \tfrac{1}{4} + \tfrac{1}{8} - \tfrac{1}{16} + \cdots + \tfrac{1}{512}$
50. $2 + \tfrac{3}{2} + \tfrac{4}{3} + \tfrac{5}{4} + \cdots + \tfrac{54}{53}$
51. $2 + \tfrac{4}{3} + \tfrac{8}{9} + \tfrac{16}{27} + \tfrac{32}{81} + \cdots$
52. $1 + \tfrac{4}{3} + \tfrac{16}{5} + \tfrac{64}{7} + \tfrac{256}{9} + \cdots$
53. $3x + 4x^3 + 5x^5 + 6x^7 + 7x^9$
54. $\dfrac{1}{x^4} - \dfrac{2}{x^{10}} + \dfrac{4}{x^{16}} - \dfrac{8}{x^{22}} + \cdots$

In Exercises 55–60, perform the indicated shift of index.

55. Rewrite $\sum\limits_{i=1}^{4} (2i - 5)$ by letting $i = k + 1$.

56. Rewrite $\sum\limits_{i=0}^{50} (-1)^{i+1} (i + 1)$ by letting $i = k - 1$.

57. Rewrite $\sum\limits_{i=1}^{\infty} \dfrac{(-1)^i (2x^{i-1})}{i}$ using $k = 0$ as the lower value of the summation variable.

58. Rewrite $\sum\limits_{i=1}^{\infty} \dfrac{(-1)^{i+1} x^i}{3i + 1}$ using $k = 2$ as the lower value of the summation variable.

59. Rewrite $\sum\limits_{i=2}^{25} \dfrac{i}{i^2 + 1}$ using $k = 0$ as the lower value of the summation variable.

60. Rewrite $\sum\limits_{i=0}^{10} \dfrac{(-1)^{i+1} 3^{i+2}}{2i + 4}$ using $k = 2$ as the lower value of the summation variable.

Critical Thinking

61. One possibility for the general element of a sequence that begins with 1, 3, 5, is $a_n = 2n - 1$. Find another possibility for the general element of a sequence that begins with 1, 3, 5.

62. How does the graph of the sequence function $u(n) = 2n - 3$ differ from the graph of the linear function $f(x) = 2x - 3$?

63. Find a finite sequence function u indicated by each graph, and state its domain.

(a)
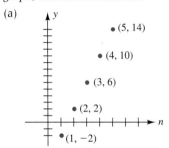

(b)
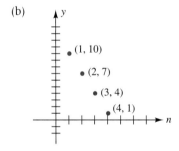

64. Give a recursive definition for the general element of each infinite sequence.
(a) 6, 10, 14, 18, 22, . . .   (b) 8, 4, 2, 1, $\tfrac{1}{2}$, . . .
(c) 1, 4, 10, 22, 46, 94, . . .   (d) 1, 2, 5, 26, 677, . . .

65. The general element of the *Fibonacci sequence* [named after Italian mathematician Leonardo Fibonacci (1180?–1250?)] is given by

$$a_n = \begin{cases} 1 & \text{if } n = 1 \\ 1 & \text{if } n = 2 \\ a_{n-1} + a_{n-2} & \text{if } n \geq 3 \end{cases}$$

(a) Find the first eight elements of Fibonacci sequence.

(b) The elements in a Fibonacci sequence occur frequently in the geometry of nature. Explain how the branching of the tree shown in the figure forms a Fibonacci sequence.

(c) Investigate the branchings of trees in your area. Do they follow a Fibonacci sequence?

66. Each element in the sequence with general element

$$a_n = \begin{cases} \dfrac{N}{2} & \text{if } n = 1 \\ \dfrac{1}{2}\left(a_{n-1} + \dfrac{N}{a_{n-1}}\right) & \text{if } n \geq 2 \end{cases}$$

gives a better approximation of $\sqrt{N}$ than does the element that precedes it. For each of the given square roots, find the fifth element, a_5, rounding your answer to five significant digits. Compare a_5 to the value you obtain by using the $\boxed{\sqrt{}}$ key on your calculator.

(a) $\sqrt{2}$ (b) $\sqrt{3}$
(c) $\sqrt{22}$ (d) $\sqrt{85}$

11.2 The Sum of a Series and Mathematical Induction

◆ Introductory Comments

To every finite series, a number can be assigned as its *sum*. For example, we find the sum of the finite series

$$\sum_{i=1}^{4}(2i - 1)$$

by adding the terms in the series, as follows:

$$\sum_{i=1}^{4} (2i - 1) = 1 + 3 + 5 + 7 = 16 \quad \text{— sum of the first four odd integers}$$

Some, but not all, infinite series can be assigned a sum. For example, in Section 11.5, we show that

$$\sum_{i=1}^{\infty} (-1)^{i+1} \left(\frac{1}{2}\right)^{i} = \frac{1}{2} - \frac{1}{4} + \frac{1}{8} - \frac{1}{16} + \cdots = \frac{1}{3} \quad \text{— sum of the infinite series}$$

However, other infinite series such as

$$\sum_{i=1}^{\infty} i = 1 + 2 + 3 + 4 + \cdots$$

do not appear to sum to any finite number. The sums of infinite series are usually studied in calculus. In this section, we limit our discussion to methods of finding the sums of finite series.

◆ Finding the Sum of a Series by Direct Computation

If the number of terms in the series is small, then we can find the sum of the series by direct computation and adding the terms in the series.

EXAMPLE 1 Find the sum of each series.

(a) $\sum_{i=1}^{8} 7$ (b) $\sum_{i=1}^{6} \frac{i}{2^i}$

◆ Solution

(a) $\sum_{i=1}^{8} 7 = \underbrace{7 + 7 + 7 + 7 + 7 + 7 + 7 + 7}_{8 \text{ terms}} = 8(7) = 56$

(b) $\sum_{i=1}^{6} \frac{i}{2^i} = \frac{1}{2} + \frac{2}{4} + \frac{3}{8} + \frac{4}{16} + \frac{5}{32} + \frac{6}{64}$

$= \frac{1(32) + 2(16) + 3(8) + 4(4) + 5(2) + 6}{64}$

$= \frac{120}{64} = \frac{15}{8}$

To avoid this awkward fractional computation, we can use the sequence features of a graphing calculator in conjunction with its sum feature. We usually access the sum feature of a graphing calculator from the "List" menu. Most graphing calculators have the capability of finding the sum

SECTION 11.2 ♦ The Sum of a Series and Mathematical Induction

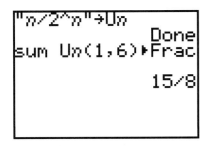

FIGURE 11.7
Typical display of the sum of the series in Example 1(b)

of up to 99 numbers in a list. Figure 11.7 shows a typical display of the sum of this series and confirms our work. ♦

The series given in Example 1(a) has the form $\sum_{i=1}^{n} c$ where c is constant. For any series of this form, we can find the sum by the following formula:

$$\sum_{i=1}^{n} c = \underbrace{c + c + c + \cdots + c}_{n \text{ terms}} = nc$$

Formulas like this one enable us to quickly find the sum of a series containing a large number of terms.

PROBLEM 1 Find the sum of the series $\sum_{i=1}^{1000} 4$. ♦

♦ Proof by Mathematical Induction

The direct computation procedure for finding the sum of a series may be tedious if the series contains a large number of terms. To find the sum of a series with a large number of terms, it is advantageous for us to develop and apply a formula. In mathematics, we often use pattern recognition to help develop formulas. For example, suppose we wish to find a formula for the sum of the first n odd, positive integers. We can represent this sum by the series

$$\sum_{i=1}^{n} (2i - 1) = 1 + 3 + 5 + \cdots + (2n - 1).$$

We begin by calculating the sum of the odd, positive integers for the first few values of n and observe any pattern that develops. Table 11.1 shows the sum of the first n odd, positive integers from $n = 1$ to $n = 6$.

The sums in Table 11.1 form the sequence

$$1, 4, 9, 16, 25, 36.$$

Do you detect the pattern? From this sequence, we are led to believe that the sum of the first n odd, positive integers is n^2 or, in series notation,

$$\sum_{i=1}^{n} (2i - 1) = 1 + 3 + 5 + \cdots + (2n - 1) = n^2.$$

However, *choosing a formula based on a few observations does not guarantee the validity of the formula for all integers n*. What we need is a "chain reaction" procedure that will guarantee that once the formula is valid for a particular integer n, then the formula is valid for the next integer, and the next, and the next, and so on, indefinitely.

Consider the row of dominoes placed on end, as shown in Figure 11.8.

TABLE 11.1

Sum of the first n odd, positive integers for $n = 1$ to $n = 6$

n	Sum of the first n odd positive integers
1	$1 = 1$
2	$1 + 3 = 4$
3	$1 + 3 + 5 = 9$
4	$1 + 3 + 5 + 7 = 16$
5	$1 + 3 + 5 + 7 + 9 = 25$
6	$1 + 3 + 5 + 7 + 9 + 11 = 36$

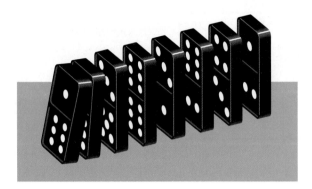

FIGURE 11.8
A row of dominoes placed on end

If we know that

1. the first domino is knocked over, and
2. if any domino in the row is knocked over, then the next one in line is also knocked over,

then we can conclude that *all* dominoes in the row will be knocked over. By analogy with the domino problem, we can state that a *mathematical statement,* such as

$$\sum_{i=1}^{n}(2i-1) = 1 + 3 + 5 + \cdots + (2n-1) = n^2$$

is true for all positive integers n provided that

1. the statement is true when $n = 1$, and
2. if the statement is true for $n = k$, where k is an arbitrary positive integer, then the statement is true for the next integer, namely, $n = k + 1$.

We refer to this chain reaction procedure as *proof by mathematical induction.*

◆ Proof by Mathematical Induction

The following two-step procedure, called proof by mathematical induction, may be used to show that a mathematical statement is true for all positive integers n:

Step 1 Show that the statement is true when $n = 1$.

Step 2 Assume that the statement is true for an arbitrary positive integer k and show, by using this assumption, that the statement is also true when $n = k + 1$.

EXAMPLE 2 Use mathematical induction to prove that

$$\sum_{i=1}^{n}(2i-1) = 1 + 3 + 5 + \cdots + (2n-1) = n^2$$

for all positive integers n.

Solution

Step 1 We must first show that the statement is true for $n = 1$:

$$1 = 1^2 \quad ?$$
$$1 = 1 \quad \text{is } true.$$

Step 2 Assuming that the statement is true for $n = k$, we must show that, on the basis of that assumption, the statement is true for $n = k + 1$.

Assume: $\quad 1 + 3 + 5 + \cdots + (2k - 1) = k^2$

Prove: $\quad 1 + 3 + 5 + \cdots + (2k - 1) + [2(k + 1) - 1] = (k + 1)^2$

Proof:

$1 + 3 + 5 + \cdots + (2k - 1) + [2(k + 1) - 1] = \{1 + 3 + 5 + \cdots + (2k - 1)\} + [2(k + 1) - 1]$

Replace these terms with the *assumed* statement.

$$= k^2 \qquad\qquad + [(2k + 2) - 1]$$

$$= k^2 + 2k + 1 \qquad \textbf{Combine like terms}$$

$$= (k + 1)^2 \qquad \textbf{Factor}$$

Thus, the truth of the statement when $n = k$ implies the truth of the statement when $n = k + 1$. Hence, by mathematical induction, we can state that

$$\sum_{i=1}^{n} (2i - 1) = 1 + 3 + 5 + 7 + \cdots + (2n - 1) = n^2$$

for all positive integers n. ◆

PROBLEM 2 Find the sum of the series $\sum_{i=1}^{50} (2i - 1)$ by using the formula we proved in Example 2. Verify the answer by using the sequence and sum features of a graphing calculator. ◆

EXAMPLE 3 Use mathematical induction to prove that the sum of the first n positive integers is $\dfrac{n(n + 1)}{2}$ by showing that

$$\sum_{i=1}^{n} i = 1 + 2 + 3 + \cdots + n = \frac{n(n + 1)}{2}$$

for all positive integers n.

◆ Solution

Step 1 We must first show that the statement is true for $n = 1$:

$$1 = \frac{1(1 + 1)}{2} \quad ?$$

$$1 = 1 \qquad \text{is } true.$$

Step 2 Assuming that the statement is true for $n = k$, we must show that, on the basis of that assumption, the statement is true for $n = k + 1$.

Assume: $\quad 1 + 2 + 3 + 4 + \cdots + k = \dfrac{k(k + 1)}{2}$

Prove: $\quad 1 + 2 + 3 + 4 + \cdots + k + (k + 1) = \dfrac{(k + 1)[(k + 1) + 1]}{2}$

Proof:

$1 + 2 + 3 + 4 + \cdots + k + (k + 1) = [\underbrace{1 + 2 + 3 + 4 + \cdots + k}] + (k + 1)$

Replace these terms with the *assumed* statement.

$= \dfrac{k(k + 1)}{2} + (k + 1)$

$= \dfrac{k(k + 1) + 2(k + 1)}{2} \qquad$ **Write as a single fraction**

$= \dfrac{(k + 1)(k + 2)}{2} \qquad$ **Factor**

$= \dfrac{(k + 1)[(k + 1) + 1]}{2} \qquad$ **Rewrite**

Thus, the truth of the statement when $n = k$ implies the truth of the statement when $n = k + 1$. Hence, by mathematical induction, we can state that

$$\sum_{i=1}^{n} i = 1 + 2 + 3 + \cdots + n = \frac{n(n + 1)}{2}$$

for all positive integers n. ◆

The formula for the sum of positive integers (from Example 3), along with formulas for the *sums of powers of positive integers,* are used in calculus to help find the area under a curve. Here we list the formulas for sums of powers up to and including 3. Mathematical induction may be used to prove formulas 2 and 3 as well (see Exercises 13 and 14).

Formulas for the Sums of Powers of Positive Integers

1. $\sum_{i=1}^{n} i = 1 + 2 + 3 + \cdots + n = \dfrac{n(n+1)}{2}$

2. $\sum_{i=1}^{n} i^2 = 1^2 + 2^2 + 3^2 + \cdots + n^2 = \dfrac{n(n+1)(2n+1)}{6}$

3. $\sum_{i=1}^{n} i^3 = 1^3 + 2^3 + 3^3 + \cdots + n^3 = \left[\dfrac{n(n+1)}{2}\right]^2$

PROBLEM 3 Find the sum of each series by using the formulas for the sums of powers of positive integers. Verify each answer by using the sequence and sum features of a graphing calculator.

(a) $\sum_{i=1}^{19} i$ (b) $\sum_{i=1}^{19} i^2$ (c) $\sum_{i=1}^{19} i^3$

◆ Summation Properties

We now state three fundamental **summation properties**. Each property can be proved by mathematical induction or by writing the property in expanded form and then applying the properties of real numbers (see Exercises 62–64).

Summation Properties

1. $\sum_{i=1}^{n} c a_i = c \sum_{i=1}^{n} a_i$

2. $\sum_{i=1}^{n} (a_i + b_i) = \sum_{i=1}^{n} a_i + \sum_{i=1}^{n} b_i$

3. $\sum_{i=1}^{n} (a_i - b_i) = \sum_{i=1}^{n} a_i - \sum_{i=1}^{n} b_i$

We use these summation properties in conjunction with the formulas for the sums of powers of integers to find the sums of several other series. For example, suppose we wish to find a formula for the sum of the first n even, positive integers. We can represent this sum by the series

$$\sum_{i=1}^{n} 2i = 2 + 4 + 6 + \cdots + 2n.$$

Now, applying summation property 1, we have

$$\sum_{i=1}^{n} 2i = 2 \sum_{i=1}^{n} i.$$

However, from Example 3, we know that

$$\sum_{i=1}^{n} i = \frac{n(n+1)}{2}.$$

Hence,

$$\sum_{i=1}^{n} 2i = 2 \sum_{i=1}^{n} i = 2 \frac{n(n+1)}{2} = n(n+1).$$

Thus, a formula for the sum of the first n even, positive integers is

$$\sum_{i=1}^{n} 2i = 2 + 4 + 6 + \cdots + 2n = n(n+1).$$

EXAMPLE 4 Use the summation properties and the formulas for the sums of powers of integers to find the sum of the following series:

$$\sum_{i=1}^{9} (i^2 + 3i - 4)$$

◆ **Solution** Using the summation properties, we write

$$\sum_{i=1}^{9} (i^2 + 3i - 4) = \sum_{i=1}^{9} i^2 + 3 \sum_{i=1}^{9} i - \sum_{i=1}^{9} 4$$

Now, using the fact that

$$\sum_{i=1}^{n} i^2 = \frac{n(n+1)(2n+1)}{6}, \quad \sum_{i=1}^{n} i = \frac{n(n+1)}{2}, \quad \text{and} \quad \sum_{i=1}^{n} c = nc,$$

we have

$$\sum_{i=1}^{9} (i^2 + 3i - 4) = \sum_{i=1}^{9} i^2 + 3 \sum_{i=1}^{9} i - \sum_{i=1}^{9} 4$$

$$= \frac{9(9+1)[2(9)+1]}{6} + 3 \left[\frac{9(9+1)}{2} \right] - 9(4)$$

$$= 285 + 135 - 36$$

$$= 384$$

The display in Figure 11.9 confirms our work. ◆

PROBLEM 4 Use the methods of Example 4 to find the sum of the series $\sum_{i=1}^{20} (5i - 2)$. Verify the answer by using the sequence and sum features of a graphing calculator. ◆

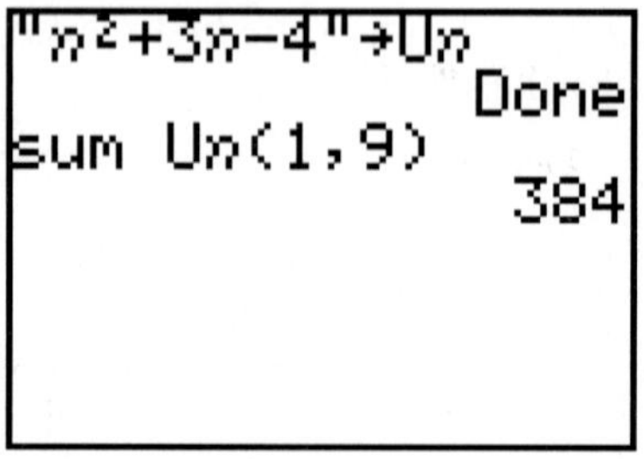

FIGURE 11.9

Typical display of the sum of the series in Example 4

◆ Other Applications of Mathematical Induction

Some mathematical statements are *not* true for the first $(j - 1)$ positive integers, but *are* true for integers greater than or equal to j. To prove such statements for $n \geq j$, we can *extend* mathematical induction by showing that

1. the statement is true when $n = j$, and
2. the truth of the statement $n = k$ for $k \geq j$ implies the truth of the statement $n = k + 1$.

EXAMPLE 5 Use mathematical induction to prove that the sum of the interior angles of an n-sided convex polygon is $(n - 2)180°$ for all positive integers $n \geq 3$.

◆ Solution

Step 1 We must first show that the statement is true for $n = 3$:

$$(3 - 2)180° = 180° \quad \text{sum of the interior angles of a triangle (a 3-sided polygon).}$$

Hence, the statement is true for $n = 3$.

Step 2 Assuming that the statement is true for $n = k$ for $k \geq 3$, we must show that, on the basis of that assumption, the statement is true for $n = k + 1$.

Assume: $(k - 2)180°$ is the sum of the interior angles of a k-sided convex polygon.

Prove: $[(k + 1) - 2]180°$ is the sum of the interior angles of a $(k + 1)$-sided convex polygon.

Proof: By attaching a triangle to a k-sided convex polygon, as shown in Figure 11.10, we form a $(k + 1)$-sided convex polygon. Hence, the sum of the interior angles of the $(k + 1)$-sided polygon is

$$\underbrace{(k - 2)180°}_{\text{sum of the interior angles of the } k\text{-sided convex polygon}} + \underbrace{180°}_{\text{sum of the interior angles of the triangle}}$$

$$[(k - 2) + 1]180° \quad \text{Factor out } 180°$$

$$[(k + 1) - 2]180° \quad \text{Rewrite}$$

Thus, the truth of the statement when $n = k$ for $k \geq 3$ implies the truth of the statement when $n = k + 1$. Therefore, by mathematical induction, we can state that the sum of the interior angles of an n-sided convex polygon is $(n - 2)180°$ for all positive integers $n \geq 3$. ◆

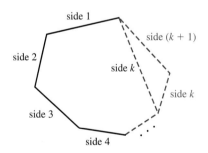

FIGURE 11.10
Attaching a triangle to a k-sided convex polygon forms a $(k + 1)$-sided convex polygon.

PROBLEM 5 Use the result of Example 5 to determine the sum of the interior angles of a 12-sided convex polygon. ◆

Exercises 11.2

Basic Skills

In Exercises 1–10, use direct computation to find the sum of each series. When possible, use the sequence and sum features of a graphing calculator to confirm each answer.

1. $\sum_{i=1}^{10} 15$
2. $\sum_{i=1}^{24} \frac{5}{6}$
3. $\sum_{i=1}^{100} \pi$
4. $\sum_{i=1}^{250} \sqrt{2}$
5. $\sum_{i=1}^{8} (2 - i)$
6. $\sum_{i=1}^{5} 3(-2)^i$
7. $\sum_{i=1}^{7} (2^i - 1)$
8. $\sum_{i=1}^{6} (i^2 + 1)$
9. $\sum_{i=1}^{5} \frac{(-1)^{i+1} 2^i}{i}$
10. $\sum_{i=1}^{4} \frac{(-1)^i i^2}{i + 1}$

In Exercises 11–28, prove each statement by using mathematical induction.

11. $2 + 6 + 10 + 14 + \cdots + (4n - 2) = 2n^2$
12. $1 + 4 + 7 + 10 + \cdots + (3n - 2) = \frac{n(3n - 1)}{2}$
13. $1^2 + 2^2 + 3^2 + \cdots + n^2 = \frac{n(n + 1)(2n + 1)}{6}$
14. $1^3 + 2^3 + 3^3 + \cdots + n^3 = \left[\frac{n(n + 1)}{2} \right]^2$
15. $1^2 + 3^2 + 5^2 + 7^2 + \cdots + (2n - 1)^2$
    $= \frac{n(2n - 1)(2n + 1)}{3}$
16. $2^2 + 4^2 + 6^2 + 8^2 + \cdots + (2n)^2$
    $= \frac{2n(n + 1)(2n + 1)}{3}$
17. $2 \cdot 4 + 4 \cdot 6 + 6 \cdot 8 + 8 \cdot 10 + \cdots + 2n(2n + 2)$
    $= \frac{4n(n + 1)(n + 2)}{3}$
18. $1 \cdot 3 + 2 \cdot 5 + 3 \cdot 7 + 4 \cdot 9 + \cdots + n(2n + 1)$
    $= \frac{n(n + 1)(4n + 5)}{6}$
19. $\frac{1}{1 \cdot 2} + \frac{1}{2 \cdot 3} + \frac{1}{3 \cdot 4} + \frac{1}{4 \cdot 5} + \cdots + \frac{1}{n(n + 1)}$
    $= \frac{n}{n + 1}$
20. $\frac{1}{1 \cdot 3} + \frac{1}{3 \cdot 5} + \frac{1}{5 \cdot 7} + \cdots + \frac{1}{(2n - 1)(2n + 1)}$
    $= \frac{n}{2n + 1}$
21. $2 + 2^2 + 2^3 + 2^4 + \cdots + 2^n = 2(2^n - 1)$
22. $3 + 3^2 + 3^3 + 3^4 + \cdots + 3^n = \frac{3}{2}(3^n - 1)$
23. $4 + 4^2 + 4^3 + 4^4 + \cdots + 4^n = \frac{4}{3}(4^n - 1)$
24. $5 + 5^2 + 5^3 + 5^4 + \cdots + 5^n = \frac{5}{4}(5^n - 1)$
25. $1 + 2 \cdot 2 + 3 \cdot 2^2 + 4 \cdot 2^3 + \cdots + n \cdot 2^{n-1}$
    $= 1 + (n - 1)2^n$
26. $1 + 2 \cdot 3 + 3 \cdot 3^2 + 4 \cdot 3^3 + \cdots + n \cdot 3^{n-1}$
    $= \frac{1 + (2n - 1)3^n}{4}$
27. $\frac{1}{2} + \frac{1}{4} + \frac{1}{8} + \frac{1}{16} + \cdots + \frac{1}{2^n} = 1 - \frac{1}{2^n}$
28. $\frac{3}{2} + \frac{5}{2^2} + \frac{7}{2^3} + \frac{9}{2^4} + \cdots + \frac{2n + 1}{2^n}$
    $= 5 - \frac{2n + 5}{2^n}$

In Exercises 29–46, use the formulas of Exercises 11–28 to find the sum of the given series. When possible, use the sequence and sum features of a graphing calculator to confirm your answer.

29. $\sum_{i=1}^{20} (4i - 2)$
30. $\sum_{i=1}^{26} (3i - 2)$
31. $\sum_{i=1}^{36} i^2$
32. $\sum_{i=1}^{24} i^3$
33. $\sum_{i=1}^{25} (2i - 1)^2$
34. $\sum_{i=1}^{18} (2i)^2$
35. $\sum_{i=1}^{15} 2i(2i + 2)$
36. $\sum_{i=1}^{18} i(2i + 1)$
37. $\sum_{i=1}^{499} \frac{1}{i(i + 1)}$
38. $\sum_{i=1}^{100} \frac{1}{(2i - 1)(2i + 1)}$
39. $\sum_{i=1}^{10} 2^i$
40. $\sum_{i=1}^{8} 3^i$
41. $\sum_{i=1}^{6} 4^i$
42. $\sum_{i=1}^{5} 5^i$

43. $\sum_{i=1}^{12} i \cdot 2^{i-1}$
44. $\sum_{i=1}^{9} i \cdot 3^{i-1}$
45. $\sum_{i=1}^{8} \frac{1}{2^i}$
46. $\sum_{i=1}^{10} \frac{2i+1}{2^i}$

In Exercises 47–54, use the summation properties and the formulas for the sums of powers of integers to find the sum of the given series. When possible, use the sequence and sum features of a graphing calculator to confirm your answer.

47. $\sum_{i=1}^{50} 3i$
48. $\sum_{i=1}^{36} 5i^2$
49. $\sum_{i=1}^{24} (2i^2 - 5)$
50. $\sum_{i=1}^{10} (4 - i^3)$
51. $\sum_{i=1}^{20} i^2(i-3)$
52. $\sum_{i=1}^{30} i(i+4)$
53. $\sum_{i=1}^{18} (2i^2 - i + 1)$
54. $\sum_{i=1}^{12} (3i^3 - 2i - 6)$

In Exercises 55–58, use the sequence and sum features of a graphing calculator to find the sum of each series.

55. $\sum_{i=1}^{84} (9.75 - 1.02i)$
56. $\sum_{i=1}^{75} 0.04(5 + 0.08i)$
57. $\sum_{i=1}^{42} (3.6i^2 - 5.6i + 8.9)$
58. $\sum_{i=1}^{25} (2.78i^3 - 4.32i)$

The series in Exercises 59 and 60 are used in calculus to approximate the area of the shaded region that is shown in the figure. Use the sequence and sum features of a graphing calculator to find the shaded area to the nearest tenth of a square unit.

59. $\sum_{i=1}^{50} 0.04[4 - (0.04i)^2]$

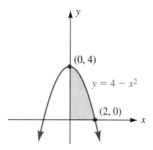

60. $\sum_{i=1}^{50} 0.04[8 - (0.04i)^3]$

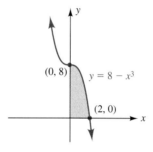

Critical Thinking

61. Find the sum of each series, and express the answer as a single logarithm.

(a) $\sum_{i=1}^{5} \log i$ (b) $\sum_{i=1}^{5} \log \frac{i+1}{i}$

In Exercises 62–64, prove each summation property by writing it in expanded form and then applying the properties of real numbers.

62. $\sum_{i=1}^{n} ca_i = c \sum_{i=1}^{n} a_i$

63. $\sum_{i=1}^{n} (a_i + b_i) = \sum_{i=1}^{n} a_i + \sum_{i=1}^{n} b_i$

64. $\sum_{i=1}^{n} (a_i - b_i) = \sum_{i=1}^{n} a_i - \sum_{i=1}^{n} b_i$

In Exercises 65–67,

(a) use pattern recognition to state the general term of the series.
(b) find the sum of the first 20 terms of the series by using the summation properties and the formulas for the sums of powers of integers.

65. $1 \cdot 2 + 3 \cdot 4 + 5 \cdot 6 + 7 \cdot 8 + \cdots$

66. $1 \cdot 3 + 3 \cdot 5 + 5 \cdot 7 + 7 \cdot 9 + \cdots$

67. $1 \cdot 2 \cdot 3 + 4 \cdot 5 \cdot 6 + 7 \cdot 8 \cdot 9 + 10 \cdot 11 \cdot 12 + \cdots$

68. On the basis of the results from Exercises 21–24, make a conjecture about the sum of the series

$$\sum_{i=1}^{n} r^i \quad \text{for } r \neq 1.$$

Then prove it by using mathematical induction.

In Exercises 69–76, use mathematical induction to prove the given statement.

69. A regular convex polygon with n sides ($n \geq 4$) has $\frac{1}{2}n(n-3)$ diagonals.

70. If n people are in a room ($n \geq 2$) and every person shakes hands with every other person, then $\frac{1}{2}n(n-1)$ handshakes take place.

71. $x - 1$ is a factor of $x^n - 1$ for all positive integers n. [*Hint:* For step 2, add and subtract x.]

72. $a^n - b^n$ is divisible by $a - b$ for all positive integers n. [*Hint:* For step 2, add and subtract ab^k.]

73. The product of any two consecutive, positive integers n and $n + 1$ is an even number.

74. The sum of cubes of any three consecutive, positive integers n, $n + 1$, and $n + 2$ is a multiple of 9.

75. $2^n > 2n$ for all positive integers $n \geq 3$.

76. $2^n > n^2$ for all positive integers $n \geq 5$.

11.3 Arithmetic Sequences and Series

◆ Introductory Comments

Consider the sequence

$$3, 7, 11, 15, 19, 23, \ldots$$

Each element after the first is obtained by adding 4 to the preceding element. A sequence in which every element after the first is obtained by *adding* a fixed number d to the preceding element is called an *arithmetic sequence*.

◆ Arithmetic Sequence

> The sequence $a_1, a_2, a_3, \ldots, a_n, \ldots$ is called an **arithmetic sequence** if there exists a fixed number d such that, for all positive integers k,
>
> $$a_{k+1} = a_k + d.$$

It follows from the definition of an arithmetic sequence that the difference between any two successive elements must be the constant d, that is,

$$a_{k+1} - a_k = d.$$

We refer to the constant d as the **common difference** of the sequence.

If the first element and the common difference of an arithmetic sequence are known, then we can generate the sequence. For example, if the first element is $a_1 = -5$ and the common difference is $d = 3$, then the first five elements of this arithmetic sequence are

$$-5, -2, 1, 4, 7$$

In this section we study arithmetic sequences and series and their applications.

◆ The General Element of an Arithmetic Sequence

Our definition of an arithmetic sequence allows us to define the general element a_n in terms of the element that precedes it, namely, a_{n-1}. Thus, we can define the general element a_n by the recursive formula

$$a_n = a_{n-1} + d$$

where d is a fixed number (the common difference). We now develop a formula that defines the general element a_n *explicitly*. If the first element in an arithmetic sequence is a_1 and the common difference is d, then the first six elements of the sequence are as follows:

First element:	a_1
Second element:	$a_2 = a_1 + d$
Third element:	$a_3 = a_2 + d = (a_1 + d) + d = a_1 + 2d$
Fourth element:	$a_4 = a_3 + d = (a_1 + 2d) + d = a_1 + 3d$
Fifth element:	$a_5 = a_4 + d = (a_1 + 3d) + d = a_1 + 4d$
Sixth element:	$a_6 = a_5 + d = (a_1 + 4d) + d = a_1 + 5d$

Do you see the pattern? Each element is the sum of a_1 and a multiple of d, where the coefficient of d is one less than the number of the element. Thus, it appears that the general element of an arithmetic sequence may be written as

$$a_n = a_1 + (n - 1)d.$$

We now verify this formula by using mathematical induction (see Section 11.2).

Step 1 We must first show that the statement is true for $n = 1$.

$$a_1 = a_1 + (1 - 1)d \quad ?$$

$$a_1 = a_1 \quad \text{is } true.$$

Step 2 Assuming that the statement is true for $n = k$, we must show that, on the basis of that assumption, the statement is true for $n = k + 1$.

Assume: $\quad a_k = a_1 + (k - 1)d$

Prove: $\quad a_{k+1} = a_1 + [(k + 1) - 1]d$

Proof: Beginning with the definition of an arithmetic sequence, we have

$$a_{k+1} = a_k + d$$

Rewrite using the *assumed* statement.

$$= [a_1 + (k-1)d] + d$$
$$= a_1 + [(k-1)d + d] \quad \text{Apply associative property}$$
$$= a_1 + [(k-1) + 1]d \quad \text{Factor}$$
$$= a_1 + [(k+1) - 1]d \quad \text{Apply commutative and associative properties}$$

Thus, the truth of the statement when $n = k$ implies the truth of the statement when $n = k + 1$. Hence, by mathematical induction, we have the following formula.

◆ General Element of an Arithmetic Sequence

> The **general element** of an arithmetic sequence with first term a_1 and common difference d is given by
> $$a_n = a_1 + (n-1)d.$$

EXAMPLE 1 Find the general element of the arithmetic sequence $23, 17, 11, 5, \ldots$.

◆ Solution The first element of this sequence is $a_1 = 23$. We can find the common difference d by selecting any two successive elements and computing the difference $a_{k+1} - a_k$. Choosing the third and fourth elements, we find

$$d = a_4 - a_3 = 5 - 11 = -6.$$

Thus, the general element of this arithmetic sequence is

$$a_n = a_1 + (n-1)d$$
$$= 23 + (n-1)(-6)$$
$$= 29 - 6n$$

We can use the sequence features of a graphing calculator to confirm our answer. Figure 11.11 shows a typical display of the first four elements of a sequence with general element $a_n = 29 - 6n$. ◆

PROBLEM 1 Find the general element of the arithmetic sequence $3, 7, 11, 15, \ldots$. ◆

Once the general element of an arithmetic sequence is known, we can find any element in the sequence. For example, to find the 25th element (a_{25}) of

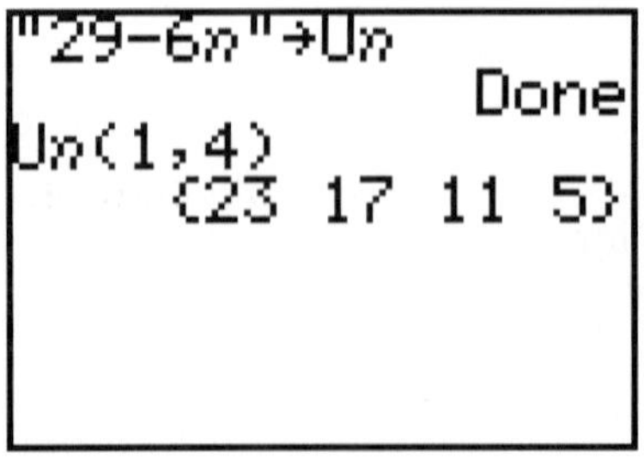

FIGURE 11.11
The first four elements of a sequence with general element $a_n = 29 - 6n$ are 23, 17, 11, 5.

the arithmetic sequence given in Example 1, we simply replace n with 25 in the general element $a_n = 29 - 6n$ as follows:

$$a_{25} = 29 - 6(25) = -121.$$

We can also use the sequence features of a graphing calculator to find the 25th element, as shown in Figure 11.12.

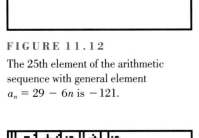

FIGURE 11.12
The 25th element of the arithmetic sequence with general element $a_n = 29 - 6n$ is -121.

EXAMPLE 2 Find the 30th element a_{30} of an arithmetic sequence whose first element a_1 is 3 and whose common difference d is 4.

◆ Solution For this arithmetic sequence, $a_1 = 3$ and $d = 4$. Thus, the general element of the arithmetic sequence is

$$a_n = 3 + (n - 1)4 = -1 + 4n$$

Therefore, the 30th element is

$$a_{30} = -1 + 4(30) = 119.$$

The display in Figure 11.13 confirms our work. ◆

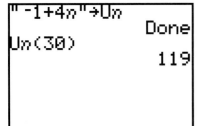

FIGURE 11.13
The 30th element of the arithmetic sequence with general element $a_n = -1 + 4n$ is 119.

PROBLEM 2 Find the first element a_1 of an arithmetic sequence whose 60th element a_{60} is -98 and whose common difference d is -4. ◆

EXAMPLE 3 Find the common difference d and the first element a_1 of an arithmetic sequence whose 5th element a_5 is 20 and whose 11th element a_{11} is 62.

◆ Solution Using the formula $a_n = a_1 + (n - 1)d$, we have

$$a_5 = a_1 + (5 - 1)d \quad \text{and} \quad a_{11} = a_1 + (11 - 1)d$$
$$20 = a_1 + 4d \qquad\qquad\qquad 62 = a_1 + 10d$$

To find d and a_1, we solve these equations simultaneously as follows:

$$\begin{array}{rl} a_1 + 4d = 20 & \xrightarrow{\text{Multiply by } -1} \quad -a_1 - 4d = -20 \\ a_1 + 10d = 62 & \xrightarrow{\text{No change}} \quad \underline{a_1 + 10d = 62} \\ & \qquad\qquad\qquad\qquad 6d = 42 \quad \text{Add and solve for } d \\ & \qquad\qquad\qquad\qquadd = 7 \end{array}$$

Substituting $d = 7$ into the equation $a_1 + 4d = 20$, we find

$$a_1 + 4(7) = 20$$
$$a_1 = -8$$

Thus, the common difference is 7 and the first element is -8. ◆

PROBLEM 3 Find the general element of the arithmetic sequence described in Example 3. Then use the general element to verify that $a_5 = 20$ and $a_{11} = 62$. ◆

◆ Arithmetic Series

The indicated sum of the elements of an arithmetic sequence is called an **arithmetic series**. The arithmetic series associated with the finite arithmetic sequence

$$a_1, a_1 + d, a_1 + 2d, a_1 + 3d, \ldots, a_1 + (n-1)d$$

can be written in expanded form as

$$a_1 + (a_1 + d) + (a_1 + 2d) + (a_1 + 3d) + \cdots + [a_1 + (n-1)d]$$

or in sigma form as

$$\sum_{i=1}^{n} [a_1 + (i-1)d].$$

We now develop a formula that we can use to find the sum of a finite arithmetic series. If we represent the sum of these n terms by S_n, then

$$S_n = a_1 + (a_1 + d) + (a_1 + 2d) + (a_1 + 3d) + \cdots + [a_1 + (n-1)d].$$

Alternatively, we can find the sum by starting with the nth term a_n and subtracting the common difference d each time to get to the next term as follows:

$$S_n = a_n + (a_n - d) + (a_n - 2d) + (a_n - 3d) + \cdots + [a_n - (n-1)d].$$

If we add these two versions of S_n, term by term, the multiples of d drop out and we obtain a formula for the sum of an arithmetic series:

$$\begin{aligned}
S_n &= a_1 + (a_1 + d) + (a_1 + 2d) + (a_1 + 3d) + \cdots + [a_1 + (n-1)d] \\
+\ S_n &= a_n + (a_n - d) + (a_n - 2d) + (a_n - 3d) + \cdots + [a_n - (n-1)d] \\
\hline
2S_n &= (a_1 + a_n) + (a_1 + a_n) + (a_1 + a_n) + (a_1 + a_n) + \cdots + (a_1 + a_n)
\end{aligned}$$

n terms

$$2S_n = n(a_1 + a_n)$$

$$S_n = n\left(\frac{a_1 + a_n}{2}\right)$$

number of terms | average value of the first and last terms

SECTION 11.3 ◆ *Arithmetic Sequences and Series* 719

◆ **Sum of a Finite Arithmetic Series**

The **sum S_n of a finite arithmetic series** with n terms is given by

$$S_n = n\left(\frac{a_1 + a_n}{2}\right)$$

where a_1 is the first term and a_n is the last term in the series.

EXAMPLE 4   Find the sum of each arithmetic series.

(a) $\sum_{i=1}^{40} (4i - 3)$ (b) $3 + 9 + 15 + \cdots + 273$

◆ Solution

(a) Since the summation variable i starts at 1 and ends at 40, the number of terms in this series is $n = 40$. Writing this series in expanded form, we have

$$\sum_{i=1}^{40} (4i - 3) = 1 + 5 + 9 + \cdots + 157$$

This is an arithmetic series with common difference $d = 4$, first term $a_1 = 1$, and last term $a_{40} = 157$. The average value of the first and last terms is

$$\frac{a_1 + a_{40}}{2} = \frac{1 + 157}{2} = \frac{158}{2} = 79$$

Thus, the sum of the arithmetic series $\sum_{i=1}^{40} (4i - 3)$ is

$$S_{40} = 40(79) = 3160$$

where 40 is the number of terms and 79 is the average value of the first and last terms.

The display in Figure 11.14 confirms our work.

(b) The first term of the arithmetic series $3 + 9 + 15 + \cdots + 273$ is $a_1 = 3$ and the common difference is $d = 6$. Hence, the general term of this series is

$$a_n = 3 + (n - 1)6 = 6n - 3$$

To find the number of terms n in this series, we let $a_n = 273$ and solve for n as follows:

$$a_n = 6n - 3$$
$$273 = 6n - 3$$
$$276 = 6n$$
$$n = 46$$

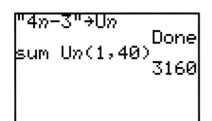

FIGURE 11.14
The sum of the first 40 terms of a series with general term $a_n = 4n - 3$ is 3160.

The average value of the first and last terms is

$$\frac{a_1 + a_{46}}{2} = \frac{3 + 273}{2} = \frac{276}{2} = 138$$

Thus, the sum of the arithmetic series $3 + 9 + 15 + \cdots + 273$ is

$$S_{46} = \underbrace{46}_{\text{number of terms}}\underbrace{(138)}_{\text{average value of the first and last terms}} = 6348$$

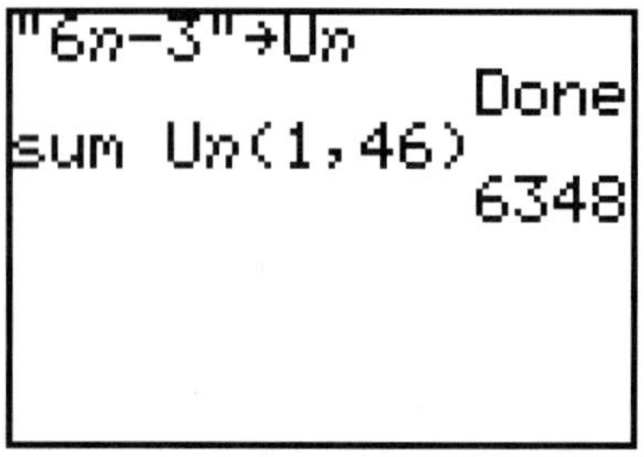

The display in Figure 11.15 supports our work.

If the summation variable i starts at a number other than 1, we must be certain to determine the correct number of terms in the series. For instance, the arithmetic series $\sum_{i=0}^{40}(4i - 3)$ has 41 terms, not 40.

PROBLEM 4 Find the sum of the arithmetic series $\sum_{i=0}^{40}(4i - 3)$.

If we replace a_n with $a_1 + (n - 1)d$ in the formula $S_n = n\left(\dfrac{a_1 + a_n}{2}\right)$, we obtain an alternative formula for the sum of a finite arithmetic series:

$$S_n = n\left(\frac{a_1 + [a_1 + (n - 1)d]}{2}\right) = \frac{n}{2}[2a_1 + (n - 1)d]$$

FIGURE 11.15
The sum of the first 46 terms of a series with general term $a_n = 6n - 3$ is 6348.

◆ **Sum of a Finite Arithmetic Series (Alternative Formula)**

> The **sum S_n of a finite arithmetic series** with n terms is given by
>
> $$S_n = \frac{n}{2}[2a_1 + (n - 1)d]$$
>
> where d is the common difference and a_1 is the first term in the series.

EXAMPLE 5 Find the sum of the first 50 terms of an arithmetic series whose first term is 24 and whose second term is 20.

◆ **Solution** The first term is $a_1 = 24$ and the second term is $a_2 = 20$. Thus, the common difference is

$$d = a_2 - a_1 = 20 - 24 = -4$$

and the general term of this arithmetic series is

$$a_n = 24 + (n - 1)(-4) = 28 - 4n$$

SECTION 11.3 ♦ Arithmetic Sequences and Series

Now, using the formula $S_n = \dfrac{n}{2}[2a_1 + (n - 1)d]$, we find the sum of the first 50 terms:

$$S_{50} = \frac{50}{2}[2(24) + (50 - 1)(-4)]$$
$$= 25[48 + (49)(-4)]$$
$$= -3700$$

FIGURE 11.16
The sum of the first 50 terms of a series with general term $a_n = 28 - 4n$ is -3700.

Hence, the sum of the first 50 terms of an arithmetic series with first term 24, and second term 20 is -3700.

The display in Figure 11.16 supports our work. ♦

PROBLEM 5 The first term a_1 of an arithmetic series is 8 and the sum of the first 10 terms is 215. Find the common difference d. ♦

♦ Application: Installment Plan

Arithmetic sequences and series are used in many types of applied problems. We conclude this section with an application concerning the payment of a debt by using an *installment plan* (a system for paying a debt by making payments at successive fixed times).

EXAMPLE 6 A stereo system is purchased for $1200 on an installment plan. The plan consists of 24 monthly payments in the amount of $50 plus $1\frac{1}{2}\%$ interest on the unpaid balance for that month.

(a) Find the amount of the first payment, the second payment, and the third payment.

(b) Find the general element of the sequence of payments.

(c) Find the total amount paid for the stereo system.

◆ Solution

(a) The first monthly payment a_1 is $50 plus $1\frac{1}{2}\%$ of $1200, that is,

$$a_1 = \$50 + 0.015(\$1200) = \$68.00$$

The unpaid balance after the first payment is $1200 − $50 = $1150. Thus, the second payment a_2 is $50 + $1\frac{1}{2}\%$ of $1150, or

$$a_2 = \$50 + 0.015(\$1150) = \$67.25$$

The unpaid balance after the second payment is $1150 − $50 = $1100. Thus, the third payment a_3 is $50 + $1\frac{1}{2}\%$ of $1100, or

$$a_3 = \$50 + 0.015(\$1100) = \$66.50$$

(b) The first three payments form the arithmetic sequence

$$\$68.00, \quad \$67.25, \quad \$66.50$$

The common difference d of this sequence is

$$d = 1\tfrac{1}{2}\% \text{ of } (-\$50) = -\$0.75.$$

Since $a_1 = \$68.00$ and $d = -\$0.75$, the general element (in dollars) is

$$a_n = 68.00 + (n - 1)(-0.75) = 68.75 - 0.75n$$

(c) Since the plan consists of 24 monthly payments, the total amount paid (in dollars) for the stereo system is represented by the arithmetic series

$$\sum_{i=1}^{24} (68.75 - 0.75i)$$

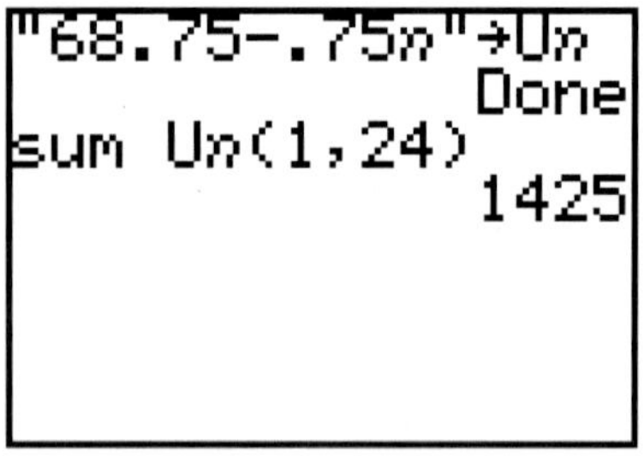

Using the formula $S_n = \dfrac{n}{2}[2a_1 + (n - 1)d]$ or the sequence and sum features of a graphing calculator (see Figure 11.17), we find that the total amount paid for the stereo system with this installment plan is $\$1,425$.

FIGURE 11.17
The sum of the first 24 terms of a series with general term $a_n = 68.75 - 0.75n$ is 1425.

PROBLEM 6 Find the amount of the 24th payment for the installment plan described in Example 6. Then use the formula $S_n = n\left(\dfrac{a_1 + a_n}{2}\right)$ to show that the total amount paid for this stereo system is $\$1425$.

Exercises 11.3

Basic Skills

In Exercises 1–6, write the first five elements of an arithmetic sequence whose first element a_1 and common difference d are given.

1. $a_1 = 3, \ d = 4$
2. $a_1 = -28, \ d = 12$
3. $a_1 = 2, \ d = -\tfrac{2}{3}$
4. $a_1 = -\tfrac{1}{2}, \ d = -\tfrac{3}{4}$
5. $a_1 = -1, \ d = \pi$
6. $a_1 = \sqrt{2}, \ d = -\sqrt{2}$

In Exercises 7–16, find the general element of the given arithmetic sequence. When possible, use the sequence features of a graphing calculator to verify your answer.

7. $3, 9, 15, 21, \ldots$
8. $1, 8, 15, 22, \ldots$
9. $11, 4, -3, -10, \ldots$
10. $-2, -7, -12, -17, \ldots$
11. $-\tfrac{2}{3}, 1, \tfrac{8}{3}, \tfrac{13}{3}, \ldots$
12. $\tfrac{3}{4}, -2, -\tfrac{19}{4}, -\tfrac{15}{2}, \ldots$
13. $\dfrac{\pi}{6}, -\dfrac{\pi}{6}, -\dfrac{\pi}{2}, -\dfrac{5\pi}{6}, \ldots$
14. $\dfrac{\sqrt{3}}{2}, 2\sqrt{3}, \dfrac{7\sqrt{3}}{2}, 5\sqrt{3}, \ldots$
15. $x - 4, \ x - 2, \ x, \ x + 2, \ldots$
16. $x, \ y, \ 2m - x, \ 3y - 2x, \ldots$

In Exercises 17–26, information is given about an arithmetic sequence $a_1, a_2, a_3, \ldots, a_n$ with common difference d. Find the indicated unknown.

17. $a_1 = 4, \ d = 7; \ a_{28} = \ ?$

18. $a_1 = 12$, $d = -9$; $a_{32} = ?$
19. $a_{82} = 17$, $d = 4$; $a_1 = ?$
20. $a_{31} = -5$, $d = -\frac{2}{5}$; $a_1 = ?$
21. $a_2 = 9$, $a_{14} = 3$; $d = ?$
22. $a_{13} = 46$, $a_4 = 100$; $a_1 = ?$
23. $a_6 = 2.25$, $a_{16} = 6.25$; $a_{80} = ?$
24. $a_5 = 72$, $a_{21} = -8$; $a_{62} = ?$
25. $a_{19} - a_5 = -36$; $d = ?$
26. $a_5 - a_{24} = 9$; $d = ?$

In Exercises 27–36, find the sum of the arithmetic series. When possible, use the sequence and sum features of a graphing calculator to verify your answer.

27. $\sum_{i=1}^{35} 4i$
28. $\sum_{i=1}^{18} (2i - 5)$
29. $\sum_{i=1}^{58} (7 - 5i)$
30. $\sum_{i=1}^{100} \frac{3i + 2}{2}$
31. $\sum_{i=0}^{25} \left(\frac{i}{2} - 1\right)$
32. $\sum_{i=3}^{30} (0.6i + 0.8)$
33. $8 + 11 + 14 + 17 + \cdots + 170$
34. $-5 + 1 + 7 + 13 + \cdots + 283$
35. $2.4 + 1.8 + 1.2 + 0.6 + \cdots + (-18)$
36. $\frac{3}{2} + \frac{9}{4} + 3 + \frac{15}{4} + \cdots + 78$

In Exercises 37–46, information is given about an arithmetic series $a_1 + a_2 + a_3 + \cdots + a_n$ with sum S_n and common difference d. Find the indicated unknown.

37. $a_1 = 12$, $d = 5$; $S_{31} = ?$
38. $a_1 = \frac{5}{4}$, $d = -\frac{1}{2}$; $S_{10} = ?$
39. $S_{32} = 448$, $a_1 = 2$; $a_{32} = ?$
40. $S_{28} = -98$, $a_{28} = -181$; $a_1 = ?$
41. $S_{54} = 11{,}583$, $a_1 = 2.5$; $d = ?$
42. $S_{18} = -45$, $d = -9$; $a_1 = ?$
43. $a_1 = -9$, $a_{40} = 147$; $S_{62} = ?$
44. $a_5 = 15$, $a_{19} = 43$; $S_{101} = ?$
45. $S_{24} = 432$, $a_1 = -28$; $a_{33} = ?$
46. $S_{65} = -1560$, $d = -3$; $a_{25} = ?$

In Exercises 47 and 48, use the sequence features of a graphing calculator to find the 79th element of the given arithmetic sequence.

47. $-8.32, -6.36, -4.40, -2.44, \ldots$
48. $837.6, 736.4, 635.2, 534.0, \ldots$

In Exercises 49–52, use the sequence and sum features of a graphing calculator to find the sum of each arithmetic series.

49. $\sum_{i=1}^{35} (2.78i - 19.4)$
50. $\sum_{i=2}^{54} (437.3 - 17.6i)$
51. $0.0824 + 0.0044 + (-0.0736) + \cdots + (-1.5556)$
52. $1176.2 + 1989.1 + 2802.0 + \cdots + 22{,}311.6$
53. Find the sum of the first 50 even, positive integers.
54. Find the sum of the first 80 odd, positive integers.
55. Find the number of multiples of three between 100 and 599.
56. Find the sum of the integers between 1 and 1000 whose last digit is 9.
57. While making out his will, a man decides to leave $300,000 to his six children in such a way that each child receives $10,000 less than the next oldest child. How much is willed to the oldest child?
58. To drill an artesian well, a company charges $5 per foot until it reaches bedrock and, for each foot thereafter, 10¢ more than the preceding foot. How much does it cost to drill a well 350 feet deep if bedrock is reached at 150 feet?
59. An auditorium has 32 seats in the first row, and each row thereafter has 4 more seats than the preceding row. If the auditorium has 25 rows, find the number of seats
    (a) in the last row.   (b) in the auditorium.
60. An object falling freely from a position of rest travels 16 ft during the first second, 48 ft during the second second, 80 ft during the third second, and so on.
    (a) Find the distance fallen during the tenth second.
    (b) Find the total distance fallen after 10 seconds.
61. A sailboat costing $36,000 is purchased under an agreement to pay $600 per month for 60 months plus 1.25% interest on the unpaid balance for that month.
    (a) Find the amount of the first payment, the second payment, and the third payment.
    (b) Find the general element of the sequence of payments.
    (c) Find the total amount paid for the sailboat.
62. A woman buys a new automobile and agrees to pay $10,929.60 in 24 monthly installments that form an arithmetic sequence. After making the 12th payment, one-third of the debt remains unpaid.
    (a) Find the amount of the first payment.
    (b) Find the amount of the 24th payment.

Critical Thinking

63. Is the sequence $\ln a, \ln a^2, \ln a^3, \ln a^4, \ldots$ an arithmetic sequence? If it is an arithmetic sequence, state the common difference.

64. Write the first five elements of the sequence defined recursively by

$$a_n = \begin{cases} 4 & \text{if } n = 1 \\ \dfrac{2a_{n-1} - 1}{2} & \text{if } n \geq 2 \end{cases}$$

Is this an arithmetic sequence? Express a_n explicitly in terms of n.

65. Find x if $10^{-2}, x, 10^{-4}$ are the first three elements of an arithmetic sequence.

66. Find x if $5x + 3, x + 4, 2x - 5$ are the first three elements of an arithmetic sequence.

67. If the numbers $a_1, a_2, a_3, \ldots, a_{n-1}, a_n$ form an arithmetic sequence, then $a_2, a_3, \ldots, a_{n-1}$ are called the $(n - 2)$ *arithmetic means* between a_1 and a_n. Insert four arithmetic means between 19 and 32.

68. The sum of the first four terms of an arithmetic series is 80. The fourth term is ten more than twice the second term. Find the first term and the common difference for this series.

69. A person borrows $20,000 from a loan shark and agrees to repay $34,800 by making payments of $300 the first month and, for each month thereafter, $100 more than the preceding month.
    (a) How many payments are required to repay this loan?
    (b) How much is the last payment?

70. A contractor has agreed to build an in-ground swimming pool in 20 days. If he does not complete the project on time, he must forfeit $150 for the first day beyond the given time allotment and, for each additional day thereafter, he must forfeit $20 more than the preceding day. How many days are required to put in the swimming pool if the contractor forfeits $1760?

11.4 Geometric Sequences and Series

◆ Introductory Comments

Consider the sequence

$$2, 6, 18, 54, 162, 486, \ldots$$

Each element after the first is obtained by multiplying the preceding element by 3. A sequence in which every element after the first is obtained by *multiplying* the preceding element by a fixed, nonzero number r is called a *geometric sequence*.

◆ Geometric Sequence

> The sequence $a_1, a_2, a_3, \ldots, a_n, \ldots$ is called a **geometric sequence** if there exists a fixed, nonzero number r such that for all positive integers k
>
> $$a_{k+1} = a_k r.$$

It follows from the definition of a geometric sequence that the ratio of any two successive elements must be the constant r, that is,

SECTION 11.4 ♦ Geometric Sequences and Series

$$\frac{a_{k+1}}{a_k} = r$$

We refer to the constant r as the **common ratio** of the sequence.

If the first element and the common ratio of a geometric sequence are known, then we can generate the sequence. For example, if the first element is $a_1 = 8$, and the common ratio is $-\frac{1}{2}$, then the first five elements of this geometric sequence are

$$8, -4, 2, -1, \tfrac{1}{2}$$

In this section we study geometric sequences and series and their applications.

♦ The General Element of a Geometric Sequence

Our definition of a geometric sequence allows us to define the general element a_n in terms of the element that precedes it, namely, a_{n-1}. That is, we can define the general element a_n by the recursive formula

$$a_n = a_{n-1}r$$

where r is a fixed number (the common ratio). We now proceed to find a formula that defines the general element a_n *explicitly*. If the first element in a geometric sequence is a_1 and the common ratio is r, then the first six elements of the sequence are as follows:

First element: $\quad a_1$

Second element: $\quad a_2 = a_1 r$

Third element: $\quad a_3 = a_2 r = (a_1 r)r = a_1 r^2$

Fourth element: $\quad a_4 = a_3 r = (a_1 r^2)r = a_1 r^3$

Fifth element: $\quad a_5 = a_4 r = (a_1 r^3)r = a_1 r^4$

Sixth element: $\quad a_6 = a_5 r = (a_1 r^4)r = a_1 r^5$

Do you see the pattern? Each element is the product of a_1 and a power of r, where the exponent of r is one less than the number of the element. Thus, it appears that the general element of a geometric sequence can be written as

$$a_n = a_1 r^{n-1}$$

We now verify this formula by using mathematical induction (see Section 11.2).

Step 1 We must first show that the statement is true for $n = 1$.

$$a_1 = a_1 r^{1-1} \quad ?$$

$$a_1 = a_1 r^0 \quad ?$$

$$a_1 = a_1 \text{ is } true$$

Step 2 Assuming that the statement is true for $n = k$, we must show that, on the basis of that assumption, the statement is true for $n = k + 1$.

Assume: $$a_k = a_1 r^{k-1}$$

Prove: $$a_{k+1} = a_1 r^{(k+1)-1}$$

Proof: Beginning with the definition of a geometric sequence, we have

$$a_{k+1} = a_k \, r$$

Rewrite using the *assumed* statement.

$$= [a_1 \, r^{k-1}] \, r$$
$$= a_1 \, r^{(k-1)+1} \quad \text{Add exponents}$$
$$= a_1 \, r^{(k+1)-1} \quad \text{Apply commutative and associative properties}$$

Thus, the truth of the statement when $n = k$ implies the truth of the statement when $n = k + 1$. Hence, by mathematical induction, we have the following formula.

◆ General Element of a Geometric Sequence

The **general element** of a geometric sequence with first element a_1 and common ratio r is given by

$$a_n = a_1 r^{n-1}.$$

EXAMPLE 1 Find the general element of the geometric sequence $2, -3, \frac{9}{2}, -\frac{27}{4}, \ldots$.

◆ **Solution** The first element of this sequence is $a_1 = 2$. We can find the common ratio r by selecting any two successive elements and computing the ratio $\frac{a_{k+1}}{a_k}$. Choosing the first and second elements, we find

$$r = \frac{a_2}{a_1} = \frac{-3}{2} = -\frac{3}{2}$$

Thus, the general element of this geometric sequence is

$$a_n = a_1 r^{n-1}$$
$$= 2\left(-\tfrac{3}{2}\right)^{n-1}$$

We can use the sequence features of a graphing calculator to confirm our answer. Figure 11.18 shows a typical display of the first four elements of a sequence with general element $a_n = 2\left(-\tfrac{3}{2}\right)^{n-1}$. ◆

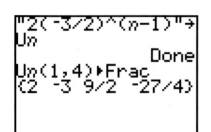

FIGURE 11.18
The first four elements of a sequence with general element $a_n = 2\left(-\tfrac{3}{2}\right)^{n-1}$ are $2, -3, \tfrac{9}{2}, -\tfrac{27}{4}$.

PROBLEM 1 Find the general element of the geometric sequence 5, 15, 45, 135, ...

Once the general element of a geometric sequence is known, we can find any element in the sequence. For example, to find the tenth element a_{10} of the geometric sequence in Example 1, we simply replace n with 10 in the general element $a_n = 2\left(-\frac{3}{2}\right)^{n-1}$, as follows:

$$a_{10} = 2\left(-\frac{3}{2}\right)^{10-1} = \frac{(-3)^9}{2^8} = -\frac{19{,}683}{256}$$

We can also use the sequence features of a graphing calculator to find the tenth element, as shown in Figure 11.19.

EXAMPLE 2 Find the 13th element a_{13} of a geometric sequence whose first element a_1 is 3 and whose common ratio r is 4.

◆ **Solution** For this geometric sequence, $a_1 = 3$ and $r = 4$. Thus, the general element of the geometric sequence is

$$a_n = 3(4)^{n-1}.$$

Therefore, the 13th element is

$$a_{13} = 3(4)^{13-1} = 50{,}331{,}648.$$

The display in Figure 11.20 confirms our work.

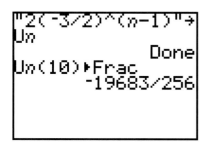

FIGURE 11.19
The tenth element of the geometric sequence with general element $a_n = 2\left(-\frac{3}{2}\right)^{n-1}$ is $-\frac{19683}{256}$.

PROBLEM 2 Find the first element a_1 of a geometric sequence whose ninth element a_9 is $\frac{1}{81}$ and whose common ratio r is $\frac{1}{3}$.

EXAMPLE 3 Find the common ratio r and the first element a_1 of a geometric sequence if the third element a_3 is 24 and the sixth element a_6 is 3.

◆ **Solution** Using the formula $a_n = a_1 r^{n-1}$, we have

$$a_3 = a_1 r^{3-1} \quad \text{and} \quad a_6 = a_1 r^{6-1}$$
$$24 = a_1 r^2 \qquad\qquad\quad 3 = a_1 r^5$$

To find a_1 and r, we solve the equations simultaneously by using substitution. Solving for a_1 in the equation $24 = a_1 r^2$, we obtain $a_1 = 24/r^2$. Now, substituting $24/r^2$ for a_1 in the other equation, $3 = a_1 r^5$, we obtain

$$3 = \left(\frac{24}{r^2}\right) r^5$$
$$3 = 24 r^3$$
$$\tfrac{1}{8} = r^3$$
$$r = \tfrac{1}{2}$$

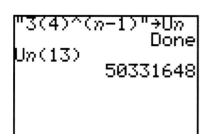

FIGURE 11.20
The 13th element of the geometric sequence with general element $a_n = 3(4)^{n-1}$ is 50,331,648.

Substituting $r = \frac{1}{2}$ into the equation $3 = a_1 r^5$, we find

$$3 = a_1 \left(\tfrac{1}{2}\right)^5$$

$$3 = \frac{a_1}{32}$$

$$a_1 = 96$$

Thus, the common ratio is $\frac{1}{2}$ and the first element is 96.   ◆

PROBLEM 3 Find the general element of the geometric sequence described in Example 3. Then use the general element to verify that $a_3 = 24$ and $a_6 = 3$.   ◆

◆ Geometric Series

The indicated sum of the elements of a geometric sequence is called a **geometric series**. The geometric series associated with the finite geometric sequence

$$a_1, a_1 r, a_1 r^2, a_1 r^3, \ldots, a_1 r^{n-1}$$

can be written in expanded form as

$$a_1 + a_1 r + a_1 r^2 + a_1 r^3 + \cdots + a_1 r^{n-1}$$

or in sigma form as

$$\sum_{i=1}^{n} a_1 r^{i-1}.$$

We now develop a formula that can be used to find the sum of a finite geometric series. If we represent the sum of these n terms by S_n, then

$$S_n = a_1 + a_1 r + a_1 r^2 + a_1 r^3 + \cdots + a_1 r^{n-2} + a_1 r^{n-1}.$$

Multiplying both sides of this equation by $-r$, we obtain

$$-rS_n = -a_1 r - a_1 r^2 - a_1 r^3 - a_1 r^4 - \cdots - a_1 r^{n-1} - a_1 r^n.$$

If we add these two equations, term by term, all terms on the right-hand side drop out except for a_1 and $-a_1 r^n$, and we obtain a formula for the sum of a finite geometric series:

$$\begin{aligned}
S_n &= a_1 + a_1 r + a_1 r^2 + a_1 r^3 + \cdots + a_1 r^{n-2} + a_1 r^{n-1} \\
-rS_n &= - a_1 r - a_1 r^2 - a_1 r^3 - a_1 r^4 - \cdots - a_1 r^{n-1} - a_1 r^n \\
\hline
S_n - rS_n &= a_1 \phantom{+ a_1 r + a_1 r^2 + a_1 r^3 + \cdots + a_1 r^{n-2}} - a_1 r^n
\end{aligned}$$

$$S_n(1 - r) = a_1(1 - r^n) \qquad \text{Factor both sides}$$

$$S_n = \frac{a_1(1 - r^n)}{1 - r} \quad \text{for } r \neq 1 \qquad \text{Divide both sides by } (1 - r)$$

Sum of a Finite Geometric Series

The **sum S_n of a finite geometric series** with a common ratio r, provided that $r \neq 1$, is given by

$$S_n = \frac{a_1(1 - r^n)}{1 - r}$$

where a_1 is the first term and n is the number of terms in the series.

Note: If the common ratio of a finite geometric series is $r = 1$, then the sum of the series $\sum_{i=1}^{n} a_1 r^{i-1}$ is

$$\underbrace{a_1 + a_1 + a_1 + \cdots + a_1}_{n \text{ terms}} = na_1.$$

EXAMPLE 4 Find the sum of each finite geometric series.

(a) $\sum_{i=1}^{10} 5(-2)^{i-1}$ (b) $\sum_{i=0}^{8} \left(\frac{1}{2}\right)^i$

Solution

(a) Since the summation variable i starts at 1 and ends at 10, the number of terms in this series is $n = 10$. Writing this series in expanded form, we have

$$\sum_{i=1}^{10} 5(-2)^{i-1} = 5 - 10 + 20 - 40 + \cdots - 2560.$$

This is a geometric series with common ratio $r = -2$ and first term $a_1 = 5$. Thus, the sum of this geometric series is

$$S_{10} = \frac{a_1(1 - r^{10})}{1 - r}$$
$$= \frac{5[1 - (-2)^{10}]}{1 - (-2)} = \frac{5(-1023)}{3} = -1705$$

The display in Figure 11.21 confirms our work.

(b) For this series the summation variable i ranges from 0 to 8, so the number of terms in this series is $n = 9$. Writing this series in expanded form, we have

$$\sum_{i=0}^{8} \left(\frac{1}{2}\right)^i = 1 + \frac{1}{2} + \frac{1}{4} + \cdots + \frac{1}{256}$$

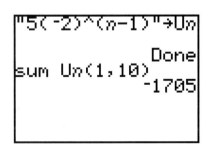

FIGURE 11.21
The sum of the first ten terms of a series with general term $a_n = 5(-2)^{n-1}$ is -1705.

This is a geometric series with common ratio $r = \frac{1}{2}$ and first term $a_1 = 1$. Thus, the sum of this geometric series is

$$S_9 = \frac{a_1(1-r^9)}{1-r}$$

$$= \frac{1[1-(\frac{1}{2})^9]}{1-\frac{1}{2}} = \frac{\frac{511}{512}}{\frac{1}{2}} = \frac{511}{256}$$

The display in Figure 11.22 confirms our work.

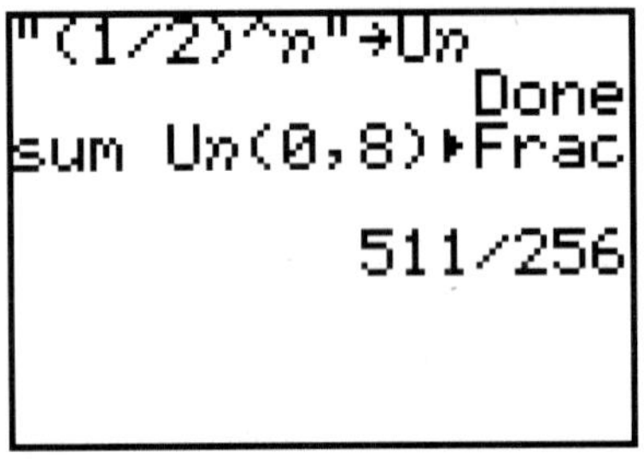

FIGURE 11.22
The sum of the first nine terms of a series with general term $a_n = (\frac{1}{2})^n$ is $\frac{511}{256}$.

PROBLEM 4 Find the first term a_1 of a geometric series if the sum of the first 12 terms is 8190 and the common ratio r is 2.

The formula for the sum of a finite geometric series can be rewritten to obtain an alternative formula:

$$S_n = \frac{a_1(1-r^n)}{1-r} = \frac{a_1 - a_1 r^n}{1-r} = \frac{a_1 - r(a_1 r^{n-1})}{1-r} \quad \text{Rewrite the numerator}$$

$$= \frac{a_1 - ra_n}{1-r} \quad \text{Replace } a_1 r^{n-1} \text{ with } a_n$$

◆ **Sum of a Finite Geometric Series (Alternative Formula)**

The **sum S_n of a finite geometric series** with a common ratio r, provided that $r \neq 1$, is given by

$$S_n = \frac{a_1 - ra_n}{1-r}$$

where a_1 is the first term and a_n is the last term in the series.

EXAMPLE 5 Find the sum of the geometric series:
$\frac{2}{3} + 2 + 6 + \cdots + 1458$

◆ **Solution** The common ratio for this geometric series is $r = 3$. The first term is $a_1 = \frac{2}{3}$ and the last term is $a_n = 1458$. Using the alternative formula for the sum of a finite geometric series, we have

$$S_n = \frac{a_1 - ra_n}{1-r}$$

$$= \frac{\frac{2}{3} - 3(1458)}{1-3}$$

$$= \frac{\frac{2}{3}}{-2} - \frac{3(1458)}{-2}$$

$$= -\frac{1}{3} + 2187$$

$$= \frac{6560}{3}$$

SECTION 11.4 ♦ Geometric Sequences and Series

We can find the number of terms n in the series

$$\tfrac{2}{3} + 2 + 6 + \cdots + 1458$$

by using the formula $a_n = a_1 r^{n-1}$ with $a_1 = \tfrac{2}{3}$, $a_n = 1458$, and $r = 3$. The equation becomes

$$1458 = \tfrac{2}{3}(3)^{n-1} \quad \text{or} \quad 3^{n-1} = 2187.$$

To solve this exponential equation (Section 5.4) for n, we begin by taking the natural logarithm of both sides as follows:

$$\ln 3^{n-1} = \ln 2187$$
$$(n - 1) \ln 3 = \ln 2187 \qquad \text{Apply log property 3}$$
$$n = 1 + \frac{\ln 2187}{\ln 3} = 8 \qquad \text{Solve for } n$$

Hence, the geometric series $\tfrac{2}{3} + 2 + 6 + \cdots + 1458$ has eight terms and may be written in sigma form as follows:

$$\sum_{i=1}^{8} \frac{2}{3}(3)^{n-1}$$

The display in Figure 11.23 confirms that the sum of the series in Example 5 is $\tfrac{6560}{3}$.

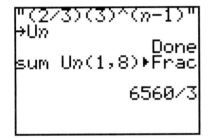

FIGURE 11.23
The sum of the first eight terms of the series with general term $a_n = \tfrac{2}{3}(3)^{n-1}$ is $\tfrac{6560}{3}$.

PROBLEM 5 Find the number of terms in the geometric series
$3 + 12 + 48 + \cdots + 201{,}326{,}592$

♦ Application: Simple Annuities

Recall from Section 5.1 that if a certain principal P is deposited in a savings account at an interest rate r per year and interest is compounded n times per year, then the amount A in the savings account after t years is given by the formula

$$A = P\left(1 + \frac{r}{n}\right)^{nt}.$$

In this formula, r/n is the *interest rate per time interval*, which we denote by i, and nt is the *total number of compounding intervals*, which we denote by N. Thus, the compound interest formula is often written as

$$A = P(1 + i)^N.$$

A **simple annuity** is a sequence of equal periodic payments of R dollars that are made over N equal time intervals at an interest rate per time interval of i. By applying the formula $A = P(1 + i)^N$, we can find the amount accumulated immediately after each periodic payment. Beginning with the Nth payment, we have the following information:

Payment number	Amount accumulated on each payment	Comment
N	R	Earns no interest
$N-1$	$R(1+i)$	Earns interest on one time interval
$N-2$	$R(1+i)^2$	Earns interest on two time intervals
$N-3$	$R(1+i)^3$	Earns interest on three time intervals
$\vdots$	$\vdots$	$\vdots$
1 or $N-(N-1)$	$R(1+i)^{N-1}$	Earns interest on $N-1$ time intervals

The sequence

$$R, R(1+i), R(1+i)^2, R(1+i)^3, \ldots, R(1+i)^{N-1}$$

is a geometric sequence with first term $a_1 = R$, common ratio $r = 1 + i$, and general element

$$\boxed{a_n = R(1+i)^{n-1}}$$

To find the total amount S of an annuity immediately after the Nth payment, we simply sum the amounts accumulated on each payment:

$$S = R + R(1+i) + R(1+i)^2 + R(1+i)^3 + \cdots + R(1+i)^{N-1}$$

Using the formula for the sum of a finite geometric series, we have

$$S = \frac{a_1(1-r^n)}{1-r} = \frac{R[1-(1+i)^N]}{1-(1+i)} = \frac{R[(1+i)^N - 1]}{i}.$$

◆ **Amount of an Annuity**

> If equal period payments of R dollars are made over N equal time intervals at an interest rate per time interval of i, then the total **amount S of an annuity** immediately after the Nth payment is
>
> $$S = \frac{R[(1+i)^N - 1]}{i}.$$

Saving plans, rent payments, and life-insurance payments are examples of an annuity. We conclude this section with an example concerning a savings plan.

EXAMPLE 6 *Chapter Opening Problem*

A father decides to deposit $150 each month in a savings account for his daughter's college education. The savings account earns interest at the rate of 9% per year compounded monthly. How much money will be in the account at the end of eight years if no money is in the account today?

Solution For this annuity we have equal monthly payments of $150 made over 12(8), or 96, months at an interest rate *per month* of $0.09 \div 12$, or 0.0075. Using the formula for the amount of an annuity with $R = 150$, $N = 96$, and $i = 0.0075$, we obtain

$$S = \frac{R[(1 + i)^N - 1]}{i}$$

$$= \frac{150[(1 + 0.0075)^{96} - 1]}{0.0075}$$

$$\approx 20{,}978.42$$

Thus, after eight years, the account will contain $20,978.42.
The general element of the sequence of payments is

$$a_n = 150(1 + 0.0075)^{n-1}.$$

Using this general element with the sequence and sum features of a graphing calculator, we can verify the amount in the account after eight years, as shown in Figure 11.24.

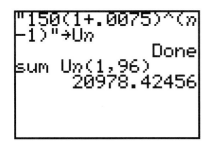

FIGURE 11.24
The sum of the first 96 terms with general element $a_n = 150(1 + 0.0075)^{n-1}$ is approximately 20978.42.

PROBLEM 6 Suppose the father's savings account described in Example 6 already contains $5000. How much will be in the account after eight years?

Exercises 11.4

Basic Skills

In Exercises 1–6, write the first five elements of a geometric sequence whose first element a_1 and common ratio r are given.

1. $a_1 = 3$, $r = 4$
2. $a_1 = -2$, $r = 5$
3. $a_1 = \frac{1}{4}$, $r = -\frac{2}{3}$
4. $a_1 = -16$, $r = -\frac{3}{4}$
5. $a_1 = -1$, $r = \pi$
6. $a_1 = 1 + \sqrt{2}$, $r = 1 - \sqrt{2}$

In Exercises 7–16, find the general element of the given geometric sequence. When possible, use the sequence features of a graphing calculator to verify your answer.

7. 1, 3, 9, 27, ...
8. 1, −8, 64, −512, ...
9. −64, 16, −4, 1, ...
10. 24, 36, 54, 81, ...
11. $\frac{1}{4}, \frac{1}{6}, \frac{1}{9}, \frac{2}{27}, \ldots$
12. $-\frac{3}{16}, \frac{1}{4}, -\frac{1}{3}, \frac{4}{9}, \ldots$
13. $-\sqrt{2}, \sqrt{6}, -3\sqrt{2}, 3\sqrt{6}, \ldots$
14. $-3\sqrt{3}, -3, -\sqrt{3}, -1, \ldots$
15. $3, 6x^2, 12x^4, 24x^6, \ldots$
16. $2x^3, 10x^2y^2, 50xy^4, 250y^6, \ldots$

In Exercises 17–26, information is given about a geometric sequence $a_1, a_2, a_3, \ldots, a_n$ with common ratio r. Find the indicated unknown.

17. $a_1 = 4$, $r = 3$; $a_9 = ?$
18. $a_1 = 12$, $r = -2$; $a_{12} = ?$
19. $a_6 = 36$, $r = -6$; $a_1 = ?$
20. $a_7 = -5$, $r = -\frac{5}{2}$; $a_1 = ?$
21. $a_2 = -4$, $a_6 = -\frac{81}{4}$; $r = ?$
22. $a_9 = \frac{3}{32}$, $a_4 = 3$; $a_1 = ?$
23. $a_5 = \frac{8}{27}$, $a_8 = -1$; $a_{13} = ?$

24. $a_5 = 0.5$, $a_9 = 312.5$; $a_2 = ?$

25. $\dfrac{a_4}{a_8} = 16$; $r = ?$

26. $\dfrac{a_5}{a_8} = -\dfrac{1}{27}$; $r = ?$

In Exercises 27–36, find the sum of the geometric series. When possible, use the sequence and sum features of a graphing calculator to verify your answer.

27. $\sum_{i=1}^{6} 2(-3)^{i-1}$

28. $\sum_{i=1}^{9} -3(2)^{i-1}$

29. $\sum_{i=1}^{5} \dfrac{5}{3^{i+1}}$

30. $\sum_{i=0}^{10} (-2)^i$

31. $\sum_{i=0}^{8} 42(0.1)^i$

32. $\sum_{i=2}^{7} \dfrac{4^i}{3^{i-1}}$

33. $1 + 4 + 16 + 64 + \cdots + 1{,}048{,}576$

34. $-243 + 162 - 108 + 72 - \cdots + \dfrac{128}{9}$

35. $192 - 96 + 48 - 24 + \cdots - \dfrac{3}{8}$

36. $2 + \sqrt{2} + 1 + \dfrac{\sqrt{2}}{2} + \cdots + \dfrac{1}{8}$

In Exercises 37–46, information about a geometric series $a_1 + a_2 + a_3 + \cdots + a_n$ with sum S_n and common ratio r is given. Find the indicated unknown.

37. $a_1 = 16$, $r = -\tfrac{1}{2}$; $S_{11} = ?$

38. $a_1 = -\tfrac{5}{4}$, $r = 2$; $S_{10} = ?$

39. $S_6 = 728$, $r = 3$; $a_1 = ?$

40. $S_3 = 93$, $a_1 = 3$; $r = ?$

41. $a_3 = \tfrac{3}{4}$, $r = \tfrac{2}{3}$; $S_6 = ?$

42. $a_6 = 2048$, $r = -4$; $S_{10} = ?$

43. $a_1 = -64$, $a_4 = 27$; $S_7 = ?$

44. $a_3 = 24$, $a_6 = 3$; $S_{11} = ?$

45. $S_3 = 57$, $a_1 = 12$; $a_6 = ?$

46. $S_7 = 57\tfrac{7}{8}$, $r = -\tfrac{3}{2}$; $a_4 = ?$

In Exercises 47 and 48, use the sequence features of a graphing calculator to find the 12th element of the given geometric sequence. Round your answer to three significant digits.

47. -8.36, -2.926, -1.0241, -0.358435, ...

48. 6.4, -11.52, 20.736, -37.3248, ...

In Exercises 49 and 50, use the sequence and sum features of a graphing calculator to find the sum of each geometric series. Round your answer to three significant digits.

49. $\sum_{i=1}^{18} 2.28(0.32)^{i-1}$

50. $\sum_{i=1}^{21} 13.3(-1.25)^{i-1}$

51. Find the number of elements in the geometric sequence $0.375, 1.5, 6, \ldots, 6{,}291{,}456$.

52. The first three terms of a geometric series are 6, 12, 24, and the sum of the series is 3066. How many terms are in the series?

53. The population of a city that now has 125,400 inhabitants is increasing by $6\tfrac{1}{2}\%$ each year. What will the population be six years from now?

54. A car radiator currently contains 3 gallons of the original antifreeze solution. One gallon of the original antifreeze is removed and replaced with one gallon of new antifreeze. One year later, one gallon of this mixture is removed and again replaced with one gallon of new antifreeze. If this process of removing a gallon and replacing a gallon is repeated each year for three more years, how many gallons of the original antifreeze will remain in the radiator?

55. An automobile that costs $16,000 when new depreciates 50% in value each year. What is it worth at the end of five years?

56. A union contract states that the workers shall receive a 6% pay increase per year for the next n years. Suppose that at the beginning of the first year of the contract, each worker is paid $28,540 per year. Express a worker's annual pay P for each succeeding year as a function of n.

57. Suppose you have a choice between two scholarships. One pays $10,000 per year for four years. The other pays $10 the first month, $20 the second month, $40 the third month, and so on, doubling the amount each month for one year only. Which offer should you accept?

58. How many ancestors (parents, grandparents, great-grandparents, and so on) do you have if you trace ten generations?

In Exercises 59 and 60, find the amount accumulated under each annuity. Round your answer to the nearest cent.

59. $100 per month for 120 months, at the rate of 8% per year compounded monthly.

60. $2500 per year for 20 years, at the rate of 9% per year compounded yearly.

61. Suppose your parents had deposited $10,000 in an account on your first birthday and then deposited $1000 on each birthday thereafter. Furthermore, suppose that the account had earned interest at the rate of 8.75% compounded yearly. How much money would be in the account on your 18th birthday?

62. A woman decides to deposit $500 each month in an account that earns interest at the rate of 8.25% per year compounded monthly. How much will she have in the account at the end of 5 years if the account presently contains $22,420?

Critical Thinking

63. Suppose $2x, 5x, 8x, 11x, \ldots$ is an arithmetic sequence.
    (a) What type of sequence is $e^{2x}, e^{5x}, e^{8x}, e^{11x}, \ldots$?
    (b) State the general element of the sequence in part (a).

64. Suppose $2x, 4x, 8x, 16x, \ldots$ is a geometric sequence.
    (a) What type of sequence is $\ln 2x, \ln 4x, \ln 8x, \ln 16x, \ldots$?
    (b) State the general element of the sequence in part (a).

65. Find x if $10^{-2}, x, 10^{-4}$ are the first three elements of a geometric sequence.

66. Find x if $x - 4, 5 - 2x, 4x - 1$ are the first three elements of a geometric sequence.

67. If the numbers $a_1, a_2, a_3, \ldots, a_{n-1}, a_n$ form a geometric sequence, then $a_2, a_3, \ldots, a_{n-1}$ are called the $(n - 2)$ *geometric means* between a_1 and a_n. Insert three geometric means between 16 and 81.

68. The sum of the first three terms of a geometric series is 63, and the third term is 45 more than the first. What is the first term of the series?

69. The product of the first three elements of a geometric sequence is -125 and their sum is 21. What is the common ratio of this sequence?

70. A ball is dropped from a height of 12 meters and rebounds after hitting the ground each time one-third the height from which it last fell. Find the *total distance* the ball has traveled when it hits the ground for the sixth time.

11.5 Infinite Geometric Series

◆ Introductory Comments

The series associated with the infinite geometric sequence

$$a_1, \; a_1 r, \; a_1 r^2, \; a_1 r^3, \; \ldots, \; a_1 r^{n-1}, \; \ldots$$

is called an **infinite geometric series**. We can write this series in expanded form as

$$a_1 + a_1 r + a_1 r^2 + a_1 r^3 + \cdots + a_1 r^{n-1} + \cdots$$

or in sigma form as

$$\sum_{i=1}^{\infty} a_1 r^{i-1}$$

To understand what is meant by the sum of an infinite geometric series, we first consider some finite sums:

$$S_1 = a_1$$
$$S_2 = a_1 + a_1 r$$
$$S_3 = a_1 + a_1 r + a_1 r^2$$
$$S_4 = a_1 + a_1 r + a_1 r^2 + a_1 r^3$$
$$\vdots$$
$$S_n = a_1 + a_1 r + a_1 r^2 + a_1 r^3 + \cdots + a_1 r^{n-1} = \frac{a_1(1 - r^n)}{1 - r} \quad \text{provided } r \neq 1$$

We refer to the real number S_n as the **nth partial sum** of the infinite series. For the infinite geometric series

$$\sum_{i=1}^{\infty} \left(\frac{1}{2}\right)^{n-1} = 1 + \tfrac{1}{2} + \tfrac{1}{4} + \tfrac{1}{8} + \cdots$$

the first four partial sums are

$S_1 = 1$ $\qquad S_2 = 1 + \tfrac{1}{2} = \tfrac{3}{2}$

$S_3 = 1 + \tfrac{1}{2} + \tfrac{1}{4} = \tfrac{7}{4}$ $\qquad S_4 = 1 + \tfrac{1}{2} + \tfrac{1}{4} + \tfrac{1}{8} = \tfrac{15}{8}$

The display in Figure 11.25 shows the sums S_{10} and S_{20}. Note how these sums appear to be approaching a value of 2.

The nth partial sum of the series $\sum_{i=1}^{\infty} \left(\tfrac{1}{2}\right)^{n-1}$ is

$$S_n = \frac{a_1(1 - r^n)}{1 - r} = \frac{1\left[1 - \left(\tfrac{1}{2}\right)^n\right]}{1 - \tfrac{1}{2}} = 2\left[1 - \left(\tfrac{1}{2}\right)^n\right].$$

The values given in Table 11.2 indicate that as the number of terms n gets larger and larger, the value of $\left(\tfrac{1}{2}\right)^n$ gets smaller and smaller. Hence, we conclude that as the number of terms n increases without bound ($n \to \infty$), the nth partial sum approaches 2:

$$S_n = 2\left[1 - \left(\tfrac{1}{2}\right)^n\right] \approx 2[1 - 0] = 2$$

Since the nth partial sum approaches 2 as $n \to \infty$, we write

$$\sum_{i=1}^{\infty} \left(\frac{1}{2}\right)^{n-1} = 2$$

We can find the sum of some, but not of every infinite geometric series. As we will show, the value of the common ratio r determines whether an infinite geometric series can be assigned a sum.

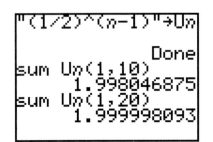

FIGURE 11.25

The 10th and 20th partial sums of the infinite geometric series $\sum_{i=1}^{\infty} \left(\tfrac{1}{2}\right)^{n-1}$.

TABLE 11.2

Values of $\left(\tfrac{1}{2}\right)^n$ as n gets larger and larger

n	1	2	3	4	5
$\left(\tfrac{1}{2}\right)^n$	$\tfrac{1}{2}$	$\tfrac{1}{4}$	$\tfrac{1}{8}$	$\tfrac{1}{16}$	$\tfrac{1}{32}$

◆ Sum of an Infinite Geometric Series

If the nth partial sum of a geometric series approaches a finite number S as $n \to \infty$, then the infinite series is said to **converge** to the sum S, and we write

$$\sum_{i=1}^{\infty} a_1 r^{i-1} = \lim_{n \to \infty} S_n = S$$

"the limit of S_n as n approaches ∞"

If the nth partial sum approaches positive infinity, negative infinity, or no specific value as $n \to \infty$, then the infinite geometric series cannot be assigned a sum and is said to **diverge**.

Let's examine what happens to the nth partial sum of an infinite geometric series as $n \to \infty$. We must consider four cases for the value of the common ratio r.

Case 1 If an infinite geometric series has a common ratio r such that $|r| < 1$, then $|r^n|$ approaches 0 ($|r^n| \to 0$) as $n \to \infty$. Hence,

$$\sum_{i=1}^{\infty} a_1 r^{i-1} = \lim_{n \to \infty} S_n = \lim_{n \to \infty} \frac{a_1(1 - r^n)}{1 - r}$$

$$= \frac{a_1(1 - 0)}{1 - r}$$

$$= \frac{a_1}{1 - r} \quad \text{if } |r| < 1$$

Case 2 If an infinite geometric series has a common ratio r such that $|r| > 1$, then $|r^n| \to \infty$ as $n \to \infty$. Hence, $\lim_{n \to \infty} S_n$ does not exist and the series diverges.

Case 3 If an infinite geometric series has common ratio $r = 1$, then the sequence of partial sums is

$$a_1, 2a_1, 3a_1, \ldots, na_1, \ldots$$

Hence, $\lim_{n \to \infty} S_n$ does not exist and the series diverges.

Case 4 If an infinite geometric series has common ratio $r = -1$, then the sequence of partial sums is

$$a_1, 0, a_1, 0, a_1, 0, \ldots$$

Hence, $\lim_{n \to \infty} S_n$ does not exist and the series diverges.

In summary, it is the value of the common ratio r that determines whether an infinite geometric series converges or diverges.

◀ Sum of an Infinite Geometric Series

The infinite geometric series

$$\sum_{i=1}^{\infty} a_1 r^{i-1} = a_1 + a_1 r + a_1 r^2 + a_1 r^3 + \cdots + a_1 r^{n-1} + \cdots$$

converges if $|r| < 1$ and has the finite **sum**

$$S = \frac{a_1}{1 - r}$$

or *diverges* if $|r| \geq 1$ and has no finite sum.

EXAMPLE 1 Determine whether the infinite geometric series converges or diverges. If the series converges, find its sum.

(a) $\displaystyle\sum_{n=0}^{\infty} \left(\frac{3}{2}\right)^n$ (b) $\displaystyle\sum_{i=1}^{\infty} (-1)^{i+1} \left(\frac{1}{2}\right)^i$

Solution

(a) Writing this series in expanded form, we have

$$\sum_{n=0}^{\infty} \left(\frac{3}{2}\right)^n = 1 + \frac{3}{2} + \frac{9}{4} + \frac{27}{8} + \cdots$$

This is an infinite geometric series with first term $a_1 = 1$ and common ratio $r = \frac{3}{2}$. Since

$$|r| = \left|\frac{3}{2}\right| = \frac{3}{2} \geq 1,$$

the series diverges and has no finite sum.

(b) Writing this series in expanded form, we have

$$\sum_{i=1}^{\infty} (-1)^{i+1} \left(\frac{1}{2}\right)^i = \frac{1}{2} - \frac{1}{4} + \frac{1}{8} - \frac{1}{16} + \cdots$$

This is an infinite geometric series with first term $a_1 = \frac{1}{2}$ and common ratio $r = -\frac{1}{2}$. Since

$$|r| = \left|-\frac{1}{2}\right| = \frac{1}{2} < 1,$$

the infinite geometric series converges and the sum S of the series is

$$S = \frac{a_1}{1 - r} = \frac{\frac{1}{2}}{1 - \left(-\frac{1}{2}\right)} = \frac{\frac{1}{2}}{\frac{3}{2}} = \frac{1}{3}.$$

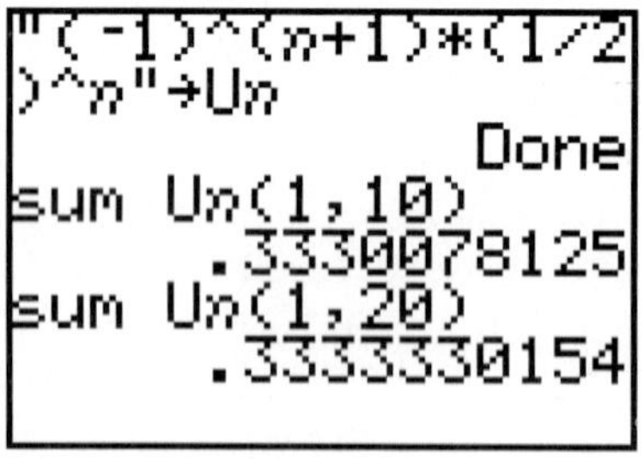

The display in Figure 11.26 shows the sums S_{10} and S_{20}. These sums appear to be approaching a value of $\frac{1}{3}$. ◆

PROBLEM 1 Determine whether the following infinite geometric series converges or diverges.

$$1 + 0.1 + 0.01 + 0.001 + \cdots$$

If the series converges, find its sum. ◆

FIGURE 11.26
The 10th and 20th partial sums of the infinite geometric series $\sum_{i=1}^{\infty} (-1)^{i+1} \left(\frac{1}{2}\right)^i$

◆ Repeating Decimal Numbers

In Section 1.1 we stated that all decimal numbers that either terminate or repeat the same block of digits are rational numbers. We can express the terminating decimal 0.25 in *fractional form* by writing $\frac{25}{100}$ or in *reduced fractional form* by writing $\frac{1}{4}$. To find the reduced fractional form of a repeating decimal such as

$$0.\overline{12} = 0.121212\ldots$$

The block of digits under the bar repeats indefinitely.

SECTION 11.5 ♦ Infinite Geometric Series

we can use the formula for the sum of an infinite geometric series. The procedure is outlined in the next example.

EXAMPLE 2 Find the reduced fractional form of each repeating decimal number.

(a) $0.\overline{12} = 0.121212\ldots$ (b) $6.3\overline{450} = 6.3450450450\ldots$

♦ **Solution**

(a) The decimal number $0.\overline{12}$ can be written as the infinite series

$$0.12 + 0.0012 + 0.000012 + \cdots$$

This is an infinite geometric series with first term $a_1 = 0.12$ and common ratio $r = 0.01$. Since $r < 1$, the series converges to the sum

$$S = \frac{a_1}{1 - r} = \frac{0.12}{1 - 0.01} = \frac{0.12}{0.99} = \frac{4}{33}.$$

Hence, the reduced fractional form of the repeating decimal number $0.\overline{12}$ is $\frac{4}{33}$.

(b) The decimal number $6.3\overline{450}$ can be written as

$$6.3 + [0.0450 + 0.0000450 + 0.0000000450 + \cdots]$$

The expression within brackets is an infinite geometric series with first term $a_1 = 0.0450$ and common ratio $r = 0.001$. Since $r < 1$, the series converges to the sum

$$S = \frac{a_1}{1 - r} = \frac{0.0450}{1 - 0.001} = \frac{45}{999} = \frac{5}{111}.$$

Hence, the reduced fractional form of the repeating decimal number $6.3\overline{450}$ is

$$6.3 + \frac{5}{111} = \frac{63}{10} + \frac{5}{111} = \frac{7043}{1110}$$

Most graphing calculators have the capability of converting a repeating decimal number to reduced fractional form, as shown in Figure 11.27. ♦

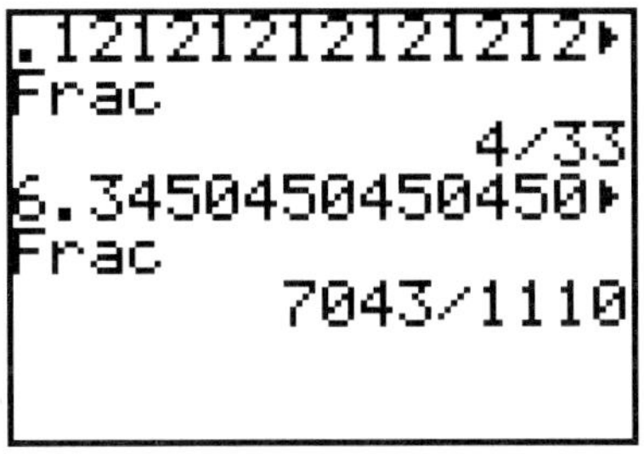

PROBLEM 2 Find the reduced fractional form of the repeating decimal number $0.88888\ldots$. Use a graphing calculator to verify the answer. ♦

FIGURE 11.27

Converting the repeating decimal numbers in Example 2 to reduced fractional form.

♦ **Interval of Convergence**

In some infinite series, each term contains a variable. Such series occur in calculus and its applications. For example, consider the infinite series

$$1 + (cx) + (cx)^2 + (cx)^3 + \cdots$$

where x is a variable and c a constant such that $c \neq 0$. Obviously, if $x = 0$, this series converges to 1. If $x \neq 0$, we can think of this series as an infinite geometric series with first term $a_1 = 1$ and common ratio $r = cx$. Depending on the real number that we substitute for x, the infinite geometric series either converges if $|cx| < 1$, or diverges if $|cx| \geq 1$. For an infinite geometric series with variable terms, the subset of real numbers for which the series converges is called its **interval of convergence**.

EXAMPLE 3 Find the interval of convergence and the sum of the infinite series

$$1 + 2x + 4x^2 + 8x^3 + \cdots$$

Solution If $x = 0$, the series converges to 1. If $x \neq 0$, we can think of this series as an infinite geometric series with first term $a_1 = 1$ and common ratio $r = 2x$. The infinite geometric series converges whenever

$$|2x| < 1$$
$$-1 < 2x < 1$$
$$-\tfrac{1}{2} < x < \tfrac{1}{2}$$

Thus, the interval of convergence for this infinite series is $\left(-\tfrac{1}{2}, \tfrac{1}{2}\right)$.

We can find the sum S of this infinite series by using the formula for the sum of an infinite geometric series. Thus,

$$S = \frac{a_1}{1 - r} = \frac{1}{1 - 2x} \quad \text{provided } -\tfrac{1}{2} < x < \tfrac{1}{2}.$$

PROBLEM 3 Use polynomial long division (Section 4.2) to show that

$$\frac{1}{1 - 2x} = 1 + 2x + 4x^2 + 8x^3 + \cdots$$

◆ Application: Hammering a Nail

Infinite geometric series occur in many types of applied problems. We conclude this section with an application.

EXAMPLE 4 A nail $2\tfrac{1}{2}$ inches long is being hammered into a board. The first impact drives the nail $\tfrac{3}{4}$ inch into the board, and each additional impact drives the nail two-thirds the distance of the preceding impact. If the nail is hammered indefinitely, will its head ever be flush with the board?

Solution The first impact drives the nail $\tfrac{3}{4}$ inch into the board. The second impact drives the nail two-thirds of $\tfrac{3}{4}$ inch, or $\tfrac{1}{2}$ inch into the board. The third impact drives the nail two-thirds of $\tfrac{1}{2}$ inch, or $\tfrac{1}{3}$ inch into the board, and so on.

The total distance that the nail is driven into the board can be represented by the infinite geometric series

$$\tfrac{3}{4} + \tfrac{1}{2} + \tfrac{1}{3} + \cdots$$

where $a_1 = \tfrac{3}{4}$ and $r = \tfrac{2}{3}$. Since $|r| < 1$, the sum S of this infinite geometric series is

$$S = \frac{a_1}{1-r} = \frac{\tfrac{3}{4}}{1-\tfrac{2}{3}} = \frac{\tfrac{3}{4}}{\tfrac{1}{3}} = \frac{9}{4} \text{ or } 2\tfrac{1}{4}.$$

Thus, after infinitely many hits, the nail is driven into the board $2\tfrac{1}{4}$ inches. Since the nail is $2\tfrac{1}{2}$ inches long, $\tfrac{1}{4}$ inch of the nail is not hammered into the board. Hence, the head of the nail will never be flush with the board under this condition. ◆

PROBLEM 4 If an initial hammer impact drives a $2\tfrac{1}{2}$-inch nail into the board $\tfrac{7}{8}$ inch, and each successive impact drives the nail two-thirds the distance of the preceding impact, will the head of the nail ever be flush with the board?

Exercises 11.5

◆ Basic Skills

In Exercises 1–8, find the first four partial sums and the nth partial sum for each infinite geometric series.

1. $\displaystyle\sum_{i=1}^{\infty} \left(\frac{2}{3}\right)^{i-1}$
2. $\displaystyle\sum_{i=1}^{\infty} \left(-\frac{3}{4}\right)^{i}$
3. $\displaystyle\sum_{i=0}^{\infty} 2(-3)^{i}$
4. $\displaystyle\sum_{i=2}^{\infty} 3(0.1)^{i-2}$
5. $-243 + 162 - 108 + 72 - \cdots$
6. $1 + 4 + 16 + 64 + \cdots$
7. $\tfrac{9}{8} + \tfrac{3}{4} + \tfrac{1}{2} + \tfrac{1}{3} + \cdots$
8. $15 - 5 + \tfrac{5}{3} - \tfrac{5}{9} + \cdots$

In Exercises 9–20, determine whether the infinite geometric series converges or diverges. If the series converges, find its sum.

9. $\displaystyle\sum_{i=1}^{\infty} 4(3)^{i-1}$
10. $\displaystyle\sum_{i=0}^{\infty} 2\left(-\frac{3}{5}\right)^{i}$
11. $\displaystyle\sum_{i=2}^{\infty} (-0.2)^{i}$
12. $\displaystyle\sum_{i=1}^{\infty} 8\left(\frac{3}{2}\right)^{i-1}$
13. $\displaystyle\sum_{i=1}^{\infty} (-1)^{i+1}\left(\frac{5}{6}\right)^{i}$
14. $\displaystyle\sum_{i=1}^{\infty} \frac{2}{3^{i+1}}$

15. $1 - 2 + 4 - 8 + \cdots$
16. $3.375 + 2.25 + 1.5 + 1 + \cdots$
17. $-6 + 3 - \tfrac{3}{2} + \tfrac{3}{4} - \cdots$
18. $1 - \dfrac{1}{\pi} + \dfrac{1}{\pi^2} - \dfrac{1}{\pi^3} + \cdots$
19. $5\sqrt{10} + 5\sqrt{2} + \sqrt{10} + \sqrt{2} + \cdots$
20. $-\sqrt{3} - \sqrt{6} - 2\sqrt{3} - 2\sqrt{6} - \cdots$

In Exercises 21–28, information is given about an infinite geometric series $a_1 + a_2 + a_3 + \cdots$ with sum S and common ratio r. Find the indicated unknown.

21. $S = -36, \quad a_1 = -12; \quad r = ?$
22. $S = 24, \quad r = -\tfrac{1}{2}; \quad a_1 = ?$
23. $a_3 = \tfrac{3}{4}, \quad r = \tfrac{2}{3}; \quad S = ?$
24. $a_6 = 1.28, \quad r = 0.4; \quad S = ?$
25. $a_1 = -64, \quad a_4 = 27; \quad S = ?$
26. $a_3 = 24, \quad a_6 = 3; \quad S = ?$
27. $S = 18, \quad a_2 = 4; \quad a_1 = ?$
28. $S = 32, \quad a_2 = 6; \quad r = ?$

In Exercises 29–38, find the reduced fractional form of each repeating decimal number.

29. $0.33333\ldots$
30. $0.77777\ldots$
31. $0.090909\ldots$
32. $0.272727\ldots$
33. $2.\overline{108}$
34. $34.\overline{303}$
35. $15.020312031\ldots$
36. $2.00010101\ldots$
37. $0.8\overline{015}$
38. $5.25\overline{144}$

 In Exercises 39–46, find the interval of convergence and the sum S of each infinite series.

39. $x - \dfrac{x^2}{5} + \dfrac{x^3}{25} - \dfrac{x^4}{125} + \cdots$

40. $3 + 9x + 27x^2 + 81x^3 + \cdots$

41. $1 + 4x^2 + 16x^4 + 64x^6 + \cdots$

42. $x - 8x^4 + 64x^7 - 512x^{10} + \cdots$

43. $1 - (x - 2) + (x - 2)^2 - \cdots$

44. $2(x + 1) + 4(x + 1)^2 + 8(x + 1)^3 - \cdots$

45. $2x + 1 + \dfrac{1}{2x} + \dfrac{1}{4x^2} + \cdots$

46. $1 - \dfrac{2}{x} + \dfrac{4}{x^2} - \dfrac{8}{x^3} + \cdots$

47. A golf ball is dropped from a height of 18 meters and rebounds each time to two-thirds of the height from which it last fell, as shown by the arrows in the figure.

   (a) Find the distance the ball travels *downward* before coming to rest.
   (b) Find the distance the ball travels *upward* before coming to rest.
   (c) Find the *total distance* the ball travels before coming to rest.

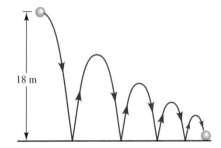

18 m

48. A lawn-mower blade makes 450 revolutions during the first second after the power is shut off and one-sixteenth as many revolutions during each succeeding second. How many revolutions does the blade make before coming to rest?

49. A man runs the 26-mile Boston Marathon at the rate of 10 miles the first hour, 6 miles the second hour, 3.6 miles the third hour, and so on. Does he ever finish the race?

50. A 24-foot steel pile is driven vertically into the soil to form part of a foundation. The first impact drives the pile 5 feet into the soil, and each additional impact drives the pile four-fifths the distance of the preceding impact. Will the top of the pile ever be flush with the ground?

51. A pendulum swings through an arc 22.54 cm long. On each succeeding swing, the pendulum travels an arc that is 0.84 the length of the preceding swing. How far does the pendulum travel before coming to rest?

52. A person with cancer is given 485.0 mg of an experimental drug the first week of treatment, and each succeeding week the patient is given 85% of the previous week's dosage. How many milligrams of this drug is the patient given? Round your answer to the nearest tenth of a milligram.

Critical Thinking

53. Determine the value of x if $\sum_{i=1}^{\infty} (3x)^{i-1} = \dfrac{2}{3}$.

54. Find the sum of each infinite series.

   (a) $\sum_{i=1}^{\infty} 2\left[\left(\dfrac{2}{3}\right)^{i-1} + \left(\dfrac{3}{4}\right)^i\right]$

   (b) $\sum_{i=0}^{\infty} \left(\dfrac{1}{2^i} - \dfrac{1}{3^i} + \dfrac{1}{4^i}\right)$

55. Find the sum of each infinite series, if it exists.

   (a) $1 - 1 + 1 - 1 + 1 - 1 + \cdots$
   (b) $(1 - 1) + (1 - 1) + (1 - 1) + \cdots$
   (c) $1 + (-1 + 1) + (-1 + 1) + \cdots$

   Does the associative property of real numbers seem to apply to infinite series?

56. Is it possible to have an infinite geometric series with first term 7 and sum 3? Explain.

57. Find the common ratio of an infinite geometric series in which each term is twice the sum of all the terms that follow it.

58. Find an infinite geometric series in which the first term is 4 and each term is 5 times the sum of all the terms that follow it.

59. A square has sides of length 24 cm. Inside this square, a second square is constructed by joining the midpoints

of the sides of the first square. Inside the second square, a third square is constructed by joining the midpoints of the sides of the second square, and so on indefinitely, as shown in the sketch.

(a) What is the sum of the areas of all the squares?
(b) What is the sum of the perimeters of all the squares?

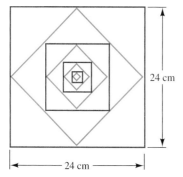

60. In a right triangle ABC, angle A is 30°, angle B is 60°, angle C is 90°, and the length of $\overline{AB}$ is 32 inches. Suppose $\overline{CD}$ is drawn perpendicular to $\overline{AB}$, $\overline{DE}$ is drawn perpendicular to $\overline{AC}$, $\overline{EF}$ is drawn perpendicular to $\overline{AB}$, $\overline{FG}$ is drawn perpendicular to $\overline{AC}$, and so on indefinitely, as shown in the sketch.

(a) What is the sum of all the perpendiculars to side $\overline{AC}$ if $\overline{BC}$ is considered the first of these perpendiculars?
(b) What is the sum of all the perpendiculars to side $\overline{AB}$?

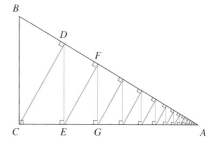

11.6 Factorials and Their Applications

◆ Introductory Comments

The sequence defined recursively by

$$a_n = \begin{cases} 1 & \text{if } n = 1 \\ na_{n-1} & \text{if } n \geq 2 \end{cases}$$

has the following elements:

$a_1 = 1$
$a_2 = 2a_1 \quad = 2 \cdot 1$
$a_3 = 3a_2 \quad = 3 \cdot 2 \cdot 1$
$a_4 = 4a_3 \quad = 4 \cdot 3 \cdot 2 \cdot 1$
$a_5 = 5a_4 \quad = 5 \cdot 4 \cdot 3 \cdot 2 \cdot 1$
$\vdots \qquad \vdots \qquad \vdots$
$a_n = na_{n-1} = n \cdot (n - 1) \cdot (n - 2) \cdot (n - 3) \cdot \ldots \cdot 2 \cdot 1$

The nth element a_n represents the product of all consecutive, positive integers from 1 up to and including n. This special product is called ***n* factorial** and is denoted by **$n!$**.

◆ n factorial

If n is a positive integer, then n factorial ($n!$) is given by

$$n! = n \cdot (n-1) \cdot (n-2) \cdot (n-3) \cdot \ldots \cdot 2 \cdot 1.$$

We usually extend this definition to include **zero factorial**, $0!$. As we will observe in this section, it is convenient to have

$$0! = 1$$

To simplify many factorial expressions, we apply the fact that if n and $n-1$ are consecutive positive integers, then

$$n \cdot (n-1)! = n!$$

Consider these illustrations of simplified factorial expressions:

$$8 \cdot 7! = 8!$$

$$15 \cdot 14 \cdot 13 \cdot 12 \cdot 11! = 15!$$

$$\frac{10!}{7!} = \frac{10 \cdot 9 \cdot 8 \cdot 7!}{7!} = 10 \cdot 9 \cdot 8 = 720$$

In this section, we discuss factorial and their applications to permutations, combinations, and the binomial theorem.

◆ Permutations

An arrangement of a set of n distinct objects in a given order is called a **permutation**. Any two objects a and b have two permutations:

$$ab \quad \text{and} \quad ba$$

We have two choices for the first position—either a or b. Once we have made a choice for this position, we have one choice for the last position. Note that the product

$$2 \cdot 1 \quad \text{or} \quad 2!$$

gives us the number of permutations of two distinct objects.

The three objects a, b, and c have six permutations:

$$abc, \ acb, \ bac, \ bca, \ cab, \ cba$$

We have three choices for the first position—either a, b, or c. Once we have made a choice for this position, we have two choices for the middle position,

SECTION 11.6 ◆ *Factorials and Their Applications* 745

and once we have made a choice for the middle position, we have one choice for the last position. The product

$$3 \cdot 2 \cdot 1 \quad \text{or} \quad 3!$$

gives us the number of permutations of three distinct objects. In general, the number of permutations of n distinct objects is

$$n \cdot (n - 1) \cdot (n - 2) \cdot \ldots \cdot 2 \cdot 1 \quad \text{or} \quad n!$$

◆ **Permutations of n Objects**

> The number of permutations of n distinct objects is $n!$

Suppose we wish to determine the number of permutations of n distinct objects taken r at a time, where $r \leq n$. Consider the following r boxes:

1st 2d 3d rth
☐   ☐   ☐ ... ☐

The 1st box can be filled in n ways.

the 2d box in $(n - 1)$ ways.

the 3d box in $(n - 2)$ ways.

⋮

the rth box in $[n - (r - 1)]$ ways.

Thus, the number of permutations of n distinct objects taken r at a time, where $r \leq n$, is

$$n \cdot (n - 1) \cdot (n - 2) \cdot \ldots \cdot [n - (r - 1)].$$

We can rewrite $[n - (r - 1)]$ as $[(n - r) + 1]$. Now, the next smaller integer after $[(n - r) + 1]$ is $(n - r)$. Thus, multiplying $n \cdot (n - 1) \cdot (n - 2) \cdot \ldots \cdot [n - (r - 1)]$ by $(n - r)!$ gives us $n!$. Hence,

$$n \cdot (n - 1) \cdot (n - 2) \cdot \ldots \cdot [n - (r - 1)] \cdot \frac{(n - r)!}{(n - r)!} = \frac{n!}{(n - r)!}.$$

◆ **Permutations of n Objects Taken r at a Time**

> The number of permutations of n distinct objects taken r at a time, where $r \leq n$, is denoted by $_nP_r$ and
>
> $$_nP_r = \frac{n!}{(n - r)!}.$$

Note: If $r = n$, we obtain

$$_nP_n = \frac{n!}{(n-n)!} = \frac{n!}{0!} = \frac{n!}{1} = n!$$

Recall that $n!$ is the number of permutations of n distinct objects.

EXAMPLE 1 At a track meet, eight contestants are entered in the 100-yard dash. Find the number of possible orders of finish for

(a) all eight contestants   **(b)** the first three positions.

◆ **Solution**

(a) The number of permutations of eight distinct objects taken 8 at a time is $_8P_8$. Thus, the number of possible orders of finish for the race is

$$_8P_8 = 8! \quad \text{or} \quad 40{,}320 \text{ orders.}$$

(b) The number of permutations of eight distinct objects taken 3 at a time is $_8P_3$. Thus, the number of possible orders of finish for the first three positions is

$$_8P_3 = \frac{8!}{(8-3)!} = \frac{8!}{5!} \quad \text{or} \quad 336 \text{ orders.}$$

Most graphing calculators have a probability menu for finding the number of permutations of n items taken r at a time. Figure 11.28 shows a typical display of evaluating the permutations in this Example. ◆

PROBLEM 1 For the race described in Example 1, find the number of possible orders of finish for the first four positions. Use the permutation feature of a graphing calculator to confirm the answer. ◆

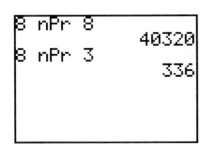

FIGURE 11.28

Evaluating the permutations in Example 1

Suppose we wish to find the number of permutations that can be made from the letters of the word PAPA. If we distinguish between the P's by putting one in boldface type (**P**), and between the A's by putting one in boldface type (**A**), then we have 4! or 24 permutations:

PAPA   P**A**PA   PA**P**A   P**AP**A
PPAA   **P**PAA   P**P**AA   **PP**AA
PAAP   **P**AAP   P**A**AP   **PA**AP
APPA   **A**PPA   A**P**PA   **AP**PA
AAPP   **A**APP   A**A**PP   **AA**PP
APAP   **A**PAP   AP**A**P   APA**P**.

For each of these six groups, we have 2! ways to arrange the P's and 2! ways to arrange the A's.

However, if we write all the letters in the same style of type, only six of these permutations are *distinguishable:*

PAPA   PPAA   PAAP   APPA   AAPP   APAP.

Thus, for four letters, two of which are P's and two of which are A's, we have

SECTION 11.6 ♦ Factorials and Their Applications

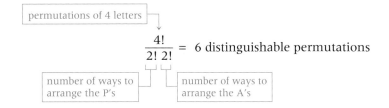

$$\frac{4!}{2!\,2!} = 6 \text{ distinguishable permutations}$$

In general, we have the following rule for distinguishable permutations.

◆ **Distinguishable Permutations**

If n objects are made up of n_1 of one kind, n_2 of a second kind, ..., n_k of a kth kind, such that $n_1 + n_2 + \cdots + n_k = n$, then the number of distinguishable permutations of these n objects is given by

$$\frac{n!}{n_1!n_2!\cdots n_k!}$$

EXAMPLE 2   How many orders of wins and losses are possible for a college football team to end the season with six wins and four losses?

◆ **Solution**   The team plays ten games; six of which are wins (W), and four of which are losses (L). The number of distinguishable permutations for WWWWWWLLLL is

$$\frac{10!}{6!\,4!} = 210$$

Thus, a football team can end the season with six wins and four losses in 210 ways.   ◆

PROBLEM 2   How many orders of wins, losses, and ties are possible for a team to end the season with three wins, five losses, and two ties?   ◆

◆ Combinations

The selection of r objects from a set of n distinct objects *without regard to order* is called a **combination** of n objects taken r at a time. We denote the number of such combinations by $_nC_r$.

We can choose two objects from a set of four distinct objects, a, b, c, and d, in $_4P_2$, or 12 permutations:

ab ba
ac ca
ad da   Once 2 objects are chosen, we
bc cb   have 2! ways of arranging them.
bd db
cd dc

Now, if we disregard order, then each of these groups is counted only once. Hence, the combinations of four objects taken two at a time are

$$ab \quad ac \quad ad \quad bc \quad bd \quad cd$$

The number of such combinations is

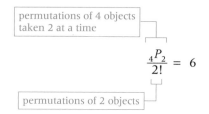

In general, we have the following rule for combinations.

◆ Combinations of n Objects Taken r at a Time

The number of combinations of n distinct objects taken r at a time, with $r \leq n$, is denoted by $_nC_r$, and

$$_nC_r = \frac{_nP_r}{r!} = \frac{n!}{r!(n-r)!}.$$

CAUTION Do not confuse permutations and combinations. Remember, in a permutation, order is considered. In a combination, order does not matter. For example, a signal made up of three flags that may be obtained from eight differently colored flags is a permutation, because a rearrangement of the three flags makes a different signal (order matters). However, a committee of three members that may be obtained from eight people is a combination, because a rearrangement of the three members is the same committee (the order of the member selection does not change the committee).

EXAMPLE 3 In a certain poker game, five cards are dealt to a player from an ordinary deck of 52 cards.

(a) How many different possible poker hands can be dealt?

(b) How many poker hands consist of all hearts?

(c) How many poker hands consist of three aces?

◆ **Solution** The order in which the five cards are dealt does not matter. Hence, we are finding combinations. A standard deck of 52 cards has four suits (hearts, diamonds, clubs, spades), and each suit contains thirteen cards (2, 3, 4, . . . , 10, jack, queen, king, ace).

(a) The number of possible poker hands that can be dealt is the number of combinations of 52 objects taken 5 at a time:

$$_{52}C_5 = \frac{52!}{5!(52-5)!} = \frac{52!}{5!\,47!} = 2{,}598{,}960$$

Thus, 2,598,960 poker hands can be dealt.

SECTION 11.6 ♦ Factorials and Their Applications

(b) The number of poker hands consisting of all hearts is the number of combinations of 13 objects taken 5 at a time:

$$_{13}C_5 = \frac{13!}{5!(13-5)!} = \frac{13!}{5!\,8!} = 1287 \text{ combinations}$$

Thus, a hand of all hearts can be dealt in 1287 ways.

(c) The number of ways of being dealt three of the four aces in the deck is

$$_4C_3 = \frac{4!}{3!\,1!} = 4 \text{ combinations.}$$

The number of ways of being dealt the remaining two cards from the 48 non-ace cards is

$$_{48}C_2 = \frac{48!}{2!\,46!} = 1128$$

Thus, a five-card poker hand consisting of three aces can be dealt in

$$_4C_3 \cdot {}_{48}C_2 = 4 \cdot 1128 = 4512 \text{ combinations.}$$

Most graphing calculators have the capability of finding the number of combinations of n items taken r at a time. Figure 11.29 shows a typical display of evaluating the combinations in this Example. ♦

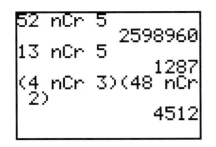

FIGURE 11.29
Evaluating the combinations in Example 3

PROBLEM 3 Find the number of five-card poker hands that consist of two queens and three kings. Use the combination feature of a graphing calculator to confirm your answer. ♦

♦ Binomial Coefficients and the Binomial Theorem

We usually denote the factorial expression for $_nC_r$ by writing the symbol

$$\binom{n}{r}$$

and refer to this as a *binomial coefficient*.

♦ **Binomial Coefficient**

> The symbol $\binom{n}{r}$ is called a **binomial coefficient** and is defined by
>
> $$\binom{n}{r} = \frac{n!}{r!(n-r)!}$$
>
> where r is a nonnegative integer such that $r \leq n$.

From the following illustrations of evaluating binomial coefficients,

$$\binom{5}{0} = \frac{5!}{0!(5-0)!} = \frac{5!}{0! \cdot 5!} = \frac{1}{0!} = \frac{1}{1} = 1$$

$$\binom{5}{1} = \frac{5!}{1!(5-1)!} = \frac{5!}{1! \cdot 4!} = \frac{5 \cdot 4!}{1! \cdot 4!} = \frac{5}{1} = 5$$

$$\binom{5}{2} = \frac{5!}{2!(5-2)!} = \frac{5!}{2! \cdot 3!} = \frac{5 \cdot 4 \cdot 3!}{2! \cdot 3!} = \frac{20}{2} = 10$$

$$\binom{5}{3} = \frac{5!}{3!(5-3)!} = \frac{5!}{3! \cdot 2!} = \frac{5 \cdot 4 \cdot 3!}{3! \cdot 2!} = \frac{20}{2} = 10$$

$$\binom{5}{4} = \frac{5!}{4!(5-4)!} = \frac{5!}{4! \cdot 1!} = \frac{5 \cdot 4!}{4! \cdot 1!} = \frac{5}{1} = 5$$

$$\binom{5}{5} = \frac{5!}{5!(5-5)!} = \frac{5!}{5! \cdot 0!} = \frac{1}{0!} = \frac{1}{1} = 1$$

we can make two general observations about binomial coefficients:

1. We observe the following equalities:

$$\binom{5}{0} = \binom{5}{5} \qquad \binom{5}{1} = \binom{5}{4} \qquad \binom{5}{2} = \binom{5}{3}$$

In general, if r and n are nonnegative integers such that $r \leq n$, then

$$\binom{n}{r} = \binom{n}{n-r}.$$

2. The binomial coefficients $\binom{5}{0}$ and $\binom{5}{5}$ both equal 1. In general, for any nonnegative integer n,

$$\binom{n}{0} = \binom{n}{n} = 1.$$

These are two *fundamental properties of binomial coefficients*.

◆ **Fundamental Properties of Binomial Coefficients**

If r and n are nonnegative integers such that $r \leq n$, then

1. $\binom{n}{r} = \binom{n}{n-r}$ **2.** $\binom{n}{0} = \binom{n}{n} = 1$

In Section 1.5 we developed formulas for the square and cube of a binomial. Recall that

SECTION 11.6 ◆ Factorials and Their Applications

$$(A + B)^2 = A^2 + 2AB + B^2$$
and
$$(A + B)^3 = A^3 + 3A^2B + 3AB^2 + B^3.$$

Note that the coefficients of the terms in the expansion of $(A + B)^2$ are equivalent to the binomial coefficients

$$\binom{2}{0} = 1, \quad \binom{2}{1} = 2, \quad \text{and} \quad \binom{2}{2} = 1$$

and the coefficients of the terms in the expansion of $(A + B)^3$ are equivalent to the binomial coefficients

$$\binom{3}{0} = 1, \quad \binom{3}{1} = 3, \quad \binom{3}{2} = 3, \quad \text{and} \quad \binom{3}{3} = 1.$$

Let's see if we can use pattern recognition to derive a formula for the expansion of $(A + B)^n$ where n is any positive integer.

Observation	Conjecture
1. The expansion of $(A + B)^2$ contains 3 terms. The first term is A^2 and the last term is B^2. The expansion of $(A + B)^3$ contains 4 terms. The first term is A^3 and the last term is B^3.	**1.** The expansion of $(A + B)^n$ should contain $(n + 1)$ terms. The first term should be A^n and the last term should be B^n.
2. In the expansion of $(A + B)^2$, the exponents of A decrease by 1 in each succeeding term after the first, and the sum of the exponents on A and B together in any term is 2. In the expansion of $(A + B)^3$, the exponents on A decrease by 1 in each succeeding term after the first, and the sum of the exponents on A and B together in any term is 3.	**2.** In the expansion of $(A + B)^n$, the exponents on A should decrease by 1 in each succeeding term after the first, and the sum of the exponents on A and B together in any term should be n.
3. The coefficients of the terms in the expansion of $(A + B)^2$ are equivalent to $$\binom{2}{0}, \binom{2}{1}, \binom{2}{2}$$ The coefficients of the terms in the expansion of $(A + B)^3$ are equivalent to $$\binom{3}{0}, \binom{3}{1}, \binom{3}{2}, \binom{3}{3}$$	**3.** The coefficients of the terms in the expansion of $(A + B)^n$ should be equivalent to $$\binom{n}{0}, \binom{n}{1}, \binom{n}{2}, \binom{n}{3}, \ldots, \binom{n}{n}$$

From our preceding conjectures, we write

$$(A + B)^n = \binom{n}{0} A^n + \binom{n}{1} A^{n-1}B + \binom{n}{2} A^{n-2}B^2 + \binom{n}{3} A^{n-3}B^3 + \binom{n}{4} A^{n-4}B^4 + \cdots + \binom{n}{n} B^n$$

We can use mathematical induction (Section 11.2) to prove this formula. The proof, however, is quite cumbersome and we shall simply accept its validity. We refer to this formula as the **binomial theorem** and state it more compactly, using sigma form with lower value $i = 0$:

◆ Binomial Theorem

If n is a positive integer, then

$$(A + B)^n = \sum_{i=0}^{n} \binom{n}{i} A^{n-i} B^i$$

EXAMPLE 4 Expand $(x - 2y)^5$ by using the binomial theorem.

◆ **Solution** For $(x - 2y)^5$, we have $A = x$, $B = -2y$, and $n = 5$. Thus, by the binomial theorem, we obtain

$$(x - 2y)^5 = \sum_{i=0}^{5} \binom{5}{i} (x)^{5-i} (-2y)^i = \binom{5}{0} x^5 + \binom{5}{1} x^4(-2y)$$

$$+ \binom{5}{2} x^3(-2y)^2 + \binom{5}{3} x^2(-2y)^3 + \binom{5}{4} x(-2y)^4 + \binom{5}{5} (-2y)^5$$

Now, from our previous discussion, we know that

$$\binom{5}{0} = \binom{5}{5} = 1, \quad \binom{5}{1} = \binom{5}{4} = 5, \quad \text{and} \quad \binom{5}{2} = \binom{5}{3} = 10.$$

Hence,

$$(x - 2y)^5 = 1x^5 + 5x^4(-2y) + 10x^3(4y^2) + 10x^2(-8y^3) + 10x(16y^4) + 1(-32y^5)$$
$$= x^5 - 10x^4y + 40x^3y^2 - 80x^2y^3 + 160xy^4 - 32y^5 \quad ◆$$

The **kth term** ($1 < k < n + 1$) in the expansion of $(A + B)^n$ can be written as follows.

◆ kth Term of $(A + B)^n$

$$\binom{n}{k - 1} A^{n-(k-1)} B^{k-1}$$

PROBLEM 4 Find the fifth term in the expansion of $(x + y)^9$. ◆

Exercises 11.6

Basic Skills

In Exercises 1–6, simplify each factorial expression.

1. (a) $10 \cdot 9 \cdot 8!$ (b) $k(k-1)(k-2)!$
2. (a) $\dfrac{15!}{14!}$ (b) $\dfrac{(k+1)!}{k!}$
3. (a) $\dfrac{21!}{21}$ (b) $\dfrac{k!}{k}$
4. (a) $\dfrac{12!}{11 \cdot 9!}$ (b) $\dfrac{(k+1)!}{k \cdot (k-2)!}$
5. (a) $\dfrac{1}{10!} + \dfrac{1}{9!}$ (b) $\dfrac{1}{k!} + \dfrac{1}{(k-1)!}$
6. (a) $\dfrac{1}{7! \cdot 11!} + \dfrac{1}{6! \cdot 12!}$
   (b) $\dfrac{1}{k! \cdot (r-1)!} + \dfrac{1}{(k-1)! \cdot r!}$

In Exercises 7–20, use the formulas for permutations and combinations to help answer each question. Use the permutation and combination features of a graphing calculator to verify each answer.

7. A family of five has bought seats to a baseball game.
   (a) In how many different ways can they be seated in the five seats?
   (b) In how many ways can they be seated if the father must sit in the middle seat?

8. Signals are sent from a battleship by placing four flags on a vertical pole. How many different signals can be formed by selecting the four flags from
   (a) four differently colored flags?
   (b) two identical red and two identical white flags?

9. How many distinct permutations can be made from the word ALGEBRA?

10. How many distinct permutations starting with M can be made from the word MISSISSIPPI?

11. How many ways can five apple trees, three pear trees, and two peach trees be planted along a driveway if it is not possible to distinguish between trees of the same type?

12. In how many ways is it possible for seven coin tosses to land as four heads and three tails?

13. From nine equally qualified students, how many ways may three of them be selected for scholarships?

14. A hockey lineup consists of two forwards, two defensemen, one center, and one goalie. How many different lineups can a coach form from a roster that has six forwards, six defensemen, three centers, and three goalies?

15. From a collection of records consisting of three hard-rock records, three soft-rock records, and two jazz records, a disk jockey selects five records at random to play.
    (a) How many different selections are possible?
    (b) How many selections are possible with two soft-rock records?
    (c) How many selections are possible with two hard-rock records and two soft-rock records if a certain jazz record must be played.

16. The Greek alphabet contains 24 letters. How many fraternity names can be formed using
    (a) three different Greek letters?
    (b) three Greek letters?

17. A college baseball team has 15 players. How many different ways can the nine starting positions be filled if each player can play any position?

18. In how many ways can a kindergarten teacher assign seats to her 12 students if the classroom has 15 chairs?

19. How many ways can 14 professors be assigned to teach nine sections of college algebra if no professor is assigned more than one section?

20. Suppose 54 Republicans and 46 Democrats sit in the United States Senate. How many ways can a committee of 4 Republicans and 3 Democrats be chosen?

In Exercises 21–34, evaluate each binomial coefficient expression.

21. $\dbinom{42}{1}$ 22. $\dbinom{75}{74}$
23. $\dbinom{19}{19}$ 24. $\dbinom{25}{25}$
25. $\dbinom{31}{0}$ 26. $\dbinom{15}{0}$
27. $\dbinom{6}{2}$ 28. $\dbinom{8}{3}$

29. $\binom{9}{6}$

30. $\binom{10}{7}$

31. $\binom{9}{7}\binom{7}{3}$

32. $\binom{11}{8}\binom{10}{1}\binom{8}{8}$

33. $\binom{12}{5} + \binom{12}{4} - \binom{13}{5}$

34. $\binom{9}{4} - \binom{8}{4} - \binom{8}{3}$

In Exercises 35–44, use the binomial theorem to expand each expression.

35. $(x + y)^5$
36. $(m - n)^7$
37. $(2x - 1)^6$
38. $(1 + 3y)^4$
39. $(3a + 2b)^4$
40. $(5x - 2y)^5$
41. $(2x^2 - 1)^7$
42. $(1 + m^3)^8$
43. $(\sqrt{x} + 2)^8$
44. $(2x^{1/3} - 3y)^6$

In Exercises 45–52, find the indicated term of the expansion.

45. Fourth term of $(x + y)^{12}$
46. Third term of $(a - b)^{16}$
47. Third term of $(3n - 2)^7$
48. Fourth term of $(2p + 1)^9$
49. Fifth term of $(x + 3y^2)^8$
50. Fifth term of $(x - 2y^{-1/2})^8$
51. Seventh term of $(\sqrt{t} - 2)^{10}$
52. Next to last term of $(9x^2 - y^5)^{12}$

Critical Thinking

In Exercises 53–58, solve for n.

53. $_nP_2 = 182$
54. $_nP_3 = 1320$
55. $_nC_2 = 36$
56. $_nC_3 = 120$
57. $_nC_3 = {_nP_2}$
58. $_nC_3 - {_{n-1}P_2} = 0$

59. Show that $_nC_r = {_nC_{n-r}}$.

60. Show that $_nC_r + {_nC_{r+1}} = {_{n+1}C_{r+1}}$.

61. In how many different ways can k distinguishable trees be planted in a circle if $k = 2$? 3? 4? 5? . . . n? [*Hint:* Two circular permutations are different only if corresponding objects are preceded or followed by different objects. Hence, simply rotating the objects clockwise or counterclockwise one or more positions does not make a different circular permutation.]

In Exercises 62 and 63, show that the given statement is true for all positive integers n.

62. $\binom{n}{0} + \binom{n}{1} + \binom{n}{2} + \cdots + \binom{n}{n} = 2^n$

63. $\binom{n}{0} - \binom{n}{1} + \binom{n}{2} - \cdots + (-1)^n \binom{n}{n} = 0$

64. Use the binomial theorem to find the indicated power of the complex number.
   (a) $(2 + 3i)^4$ (b) $(2 - i)^6$

65. Use the binomial theorem to expand each trinomial.
   (a) $(x^2 + x + 1)^5$ (b) $(x^2 - x - 1)^5$

 66. For each function f, defined as indicated, find the difference quotient $\dfrac{f(x + h) - f(x)}{h}$.

(a) $f(x) = x^4$
(b) $f(x) = x^n$ for n a positive integer

67. In the 17th century, French mathematician Blaise Pascal discovered a triangular pattern for the coefficients of the terms in the expansion of $(A + B)^n$. The first seven rows of *Pascal's triangle* are given:

Row														Coefficients in the expansion of:
1							1							$(A + B)^0$
2						1		1						$(A + B)^1$
3					1		2		1					$(A + B)^2$
4				1		3		3		1				$(A + B)^3$
5			1		4		6		4		1			$(A + B)^4$
6		1		5		10		10		5		1		$(A + B)^5$
7	1		6		15		20		15		6		1	$(A + B)^6$

Each row in the triangle starts and ends with a coefficient of 1. The other coefficients in each row can be obtained from the coefficients in the preceding row. Do you see the pattern? Use your observation to determine the coefficients for each row:

(a) row 8 (b) row 9 (c) row 10

68. Find the sum of the numbers in each row of Pascal's triangle (see Exercise 67). Do you observe a pattern to these sums? Guess a formula for the sum of the numbers in the nth row of Pascal's triangle, and verify your answer.

69. In calculus, it is shown that the expansion of $(1 + x)^n$, where n is a negative integer or a rational number and $|x| < 1$, is given by the *infinite binomial series*

$$1 + nx + \frac{n(n-1)}{2!}x^2 + \frac{n(n-1)(n-2)}{3!}x^3 + \cdots$$

Write the first four terms in the expansion of the given expression.

(a) $(1 + x)^{-1}$ provided $|x| < 1$
(b) $(1 - y)^{-2}$ provided $|y| < 1$
(c) $(1 - z)^{-1/2}$ provided $|z| < 1$
(d) $(1 + x)^{1/3}$ provided $|x| < 1$

70. Use mathematical induction to prove that $n! > 2^n$ for all integers $n \geq 4$.

71. In calculus, it is shown that the *power series* for e^x is given by

$$e^x = 1 + x + \frac{x^2}{2!} + \frac{x^3}{3!} + \frac{x^4}{4!} + \cdots$$

Use the first ten terms of this series to estimate the value of e. Compare this value of e to the value you obtain by using the $\boxed{e^x}$ key on your calculator with $x = 1$.

72. *Stirling's formula* states that a good approximation to $n!$, especially for large values of n, is given by

$$\sqrt{2\pi n}\left(\frac{n}{e}\right)^n.$$

(a) Use Stirling's formula and a calculator to estimate 5!, 10!, and 15!.
(b) Use the $\boxed{n!}$ key on your calculator to find the values of 5!, 10!, and 15!.
(c) Compute the error in the approximations of 5!, 10!, and 15! when using Stirling's formula.

Chapter 11 Review

◆ Questions for Writing or Group Discussion

1. What is the domain of an *infinite sequence function*?
2. Explain the similarities and differences between a *sequence* and a *series*.
3. Explain what is meant by defining a sequence *recursively* and *explicitly*.
4. Explain the procedure for finding the fifth element of a sequence whose *general element* is known.
5. How can the process of *mathematical induction* be used to show that a baby can climb a flight of stairs?
6. What is the upper value of the *summation variable* for an infinite series?
7. State the procedure for changing a finite series from *expanded form* to *sigma form*. Illustrate with an arithmetic series.
8. List three *summation properties*. Give an example of each that illustrates its use.
9. Explain the procedure for finding the *common ratio* of a geometric sequence and the *common difference* of an arithmetic sequence.
10. Does there exist a sequence that is both an *arithmetic sequence* and a *geometric sequence*? If so, give an example.
11. Do the opposites of the elements in an arithmetic sequence also form an arithmetic sequence? Explain.
12. Do the reciprocals of the elements in a geometric sequence also form a geometric sequence? Explain.
13. What is a *simple annuity*?
14. What can we conclude about an infinite geometric series whose n*th partial sum* approaches a finite value as n increases without bound ($n \to \infty$)?

15. Under what conditions does an infinite geometric series *converge* or *diverge*?
16. Is the sequence log 2, log 4, log 8, log 16, ... arithmetic, geometric, or neither? Explain.
17. Give an example of an *infinite geometric series* that
    (a) converges and has a sum of 3.   (b) diverges.
18. Explain the procedure for changing a *repeating decimal* to reduced fractional form.
19. What is meant by the *interval of convergence* of an infinite geometric series with variable terms?
20. State another *binomial coefficient* that is equal to $\binom{n}{k}$.
21. What is the difference between a *permutation* and a *combination*? Illustrate with examples.
22. State the *binomial theorem*. Illustrate with an example how it is used to expand a binomial.

Review Exercises

In Exercises 1 and 2, list the first six elements of each sequence function, then sketch its graph.

1. $u(n) = \dfrac{n+1}{n}$
2. $v(n) = (-1)^{n+1}(2n)$

In Exercises 3–8, find the first six elements of a sequence with the given general element.

3. $a_n = 3n - 7$
4. $a_n = 6\left(-\tfrac{1}{2}\right)^{n-1}$
5. $a_n = (-1)^{n+1} n!$
6. $a_n = \dfrac{n}{n+2}$
7. $a_n = \begin{cases} n - 1 & \text{if } n \text{ is odd} \\ (a_{n-1})^2 & \text{if } n \text{ is even} \end{cases}$
8. $a_n = \begin{cases} 3 & \text{if } n = 1 \\ 5 & \text{if } n = 2 \\ a_{n-1} + a_{n-2} & \text{if } n \geq 3 \end{cases}$

In Exercises 9–22,

(a) *determine whether each sequence appears to be arithmetic, geometric, or neither.*
(b) *use pattern recognition to find the general element of each sequence.*

9. $-1, 3, -5, 7, -9, 11, -13, 15, \ldots$
10. $2, -4, 6, -8, 10, -12, \ldots$
11. $3, 9, 15, 21, 27, 33, 39, \ldots$
12. $49, 38, 27, 16, 5, -6, -17, -28, \ldots$
13. $768, 192, 48, 12, 3, \ldots$
14. $5, 15, 45, 135, 405, 1215, \ldots$
15. $\tfrac{1}{2}, -\tfrac{3}{4}, \tfrac{5}{6}, -\tfrac{7}{8}, \tfrac{9}{10}, \ldots$
16. $3, \tfrac{4}{3}, 1, \tfrac{6}{7}, \tfrac{7}{9}, \ldots$
17. $-4, 2, -1, \tfrac{1}{2}, -\tfrac{1}{4}, \tfrac{1}{8}, \ldots$
18. $2500, 25, 0.25, 0.0025, 0.000025, \ldots$
19. $\tfrac{13}{3}, 3, \tfrac{5}{3}, \tfrac{1}{3}, -1, \ldots$
20. $-\tfrac{3}{2}, -\tfrac{1}{2}, \tfrac{1}{2}, \tfrac{3}{2}, \tfrac{5}{2}, \tfrac{7}{2}, \ldots$
21. $0, 3, 8, 15, 24, 35, 48, 63, 80, \ldots$
22. $1, 1, 2, 6, 24, 120, 720, \ldots$

In Exercises 23–34,

(a) *determine whether each series is arithmetic, geometric, or neither.*
(b) *find the sum of each series.*

23. $\displaystyle\sum_{i=1}^{35} 2i$
24. $\displaystyle\sum_{i=1}^{20} (7 - 3i)$
25. $\displaystyle\sum_{i=2}^{10} 8\left(-\tfrac{3}{2}\right)^{i-1}$
26. $\displaystyle\sum_{i=0}^{12} (-3)^i$
27. $\displaystyle\sum_{i=1}^{18} 3i^2$
28. $\displaystyle\sum_{i=1}^{24} i(i - 3)$
29. $\displaystyle\sum_{i=1}^{\infty} 5\left(\tfrac{1}{2}\right)^{i-1}$
30. $\displaystyle\sum_{i=2}^{\infty} \dfrac{1}{(-2)^i}$
31. $\displaystyle\sum_{i=0}^{30} \left(\dfrac{i}{3} - 2\right)$
32. $\displaystyle\sum_{i=0}^{32} \dfrac{5i + 4}{6}$
33. $\displaystyle\sum_{i=1}^{16} (2i^3 - 3i + 5)$
34. $\displaystyle\sum_{i=1}^{6} \dfrac{(-1)^{i+1} 2^i}{i}$

In Exercises 35–40, prove each statement using mathematical induction.

35. $2 + 8 + 14 + 20 + \cdots + (6n - 4) = 3n^2 - n$

36. $6 + 11 + 16 + 21 + \cdots + (5n + 1) = \dfrac{n(5n + 7)}{2}$

37. $1 \cdot 2 + 2 \cdot 3 + 3 \cdot 4 + 4 \cdot 5 + \cdots + n(n + 1)$
$= \dfrac{n(n + 1)(n + 2)}{3}$

38. $1 \cdot 2 \cdot 3 + 2 \cdot 3 \cdot 4 + 3 \cdot 4 \cdot 5 + 4 \cdot 5 \cdot 6 + \cdots$
$+ n(n + 1)(n + 2) = \dfrac{n(n + 1)(n + 2)(n + 3)}{4}$

39. $1^3 + 3^3 + 5^3 + 7^3 + \cdots + (2n - 1)^3 = n^2(2n^2 - 1)$

40. $\dfrac{1}{2} + \dfrac{2}{2^2} + \dfrac{3}{2^3} + \dfrac{4}{2^4} + \cdots + \dfrac{n}{2^n} = 2 - \dfrac{n + 2}{2^n}$

In Exercises 41–44, use the results of Exercises 35–40 to find the sum of the given series.

41. $\sum_{i=1}^{28} i(i + 1)$

42. $\sum_{i=1}^{36} i(i + 1)(i + 2)$

43. $\sum_{i=1}^{15} (2i - 1)^3$

44. $\sum_{i=1}^{12} \dfrac{i}{2^i}$

In Exercises 45 and 46, write each arithmetic series in sigma form using (a) $i = 1$ and (b) $i = 0$ as the lower value of the summation variable.

45. $5 + 9 + 13 + \cdots + 1001$

46. $88 + 76 + 64 + 52 + \cdots$

In Exercises 47 and 48, write each geometric series in sigma form using $i = 1$ as the lower value of the summation variable.

47. $243 - 162 + 108 - 72 + \cdots$

48. $2 + 6 + 18 + 54 + \cdots + 9{,}565{,}938$

In Exercises 49 and 50,

(a) *find the first four partial sums and the nth partial sum for each infinite geometric series.*

(b) *find the sum of the series, if it exists.*

49. $384 - 96 + 24 - 6 + \cdots$

50. $\sum_{i=1}^{\infty} \left(\dfrac{5}{4}\right)^{i-1}$

51. What is the 26th element of an arithmetic sequence whose first element is 92 and whose common difference is -7?

52. What is the general element of an arithmetic sequence whose 6th element is 34 and whose 12th element is 52?

53. What is the sum of the first 60 terms of an arithmetic series whose first term is 9 and whose second term is 24?

54. The first term of an arithmetic series is -19, and the sum of the first 30 terms is 300. What is the 15th term of the series?

55. What is the first element of a geometric sequence whose 11th element is $\tfrac{3}{32}$ and whose common ratio is $\tfrac{1}{2}$?

56. What is the general element of a geometric sequence whose second element is 12 and whose eighth element is 8748?

57. Find x so that $x - 3, 2x - 3, 6x + 1$ are the first three terms of a geometric series.

58. The sum of the first three terms of a geometric series is 26, and the third term is 16 more than the first term. What is the general term of the series?

59. What is the sum of an infinite geometric series whose first term is $\tfrac{27}{16}$ and whose fifth term is $\tfrac{1}{3}$?

60. What is the common ratio of an infinite geometric series in which each term is three times the sum of all the terms that follow it?

61. Find the reduced fractional form represented by each repeating decimal:
    (a) $3.030303\ldots$ (b) $2.7\overline{306}$

62. Find the interval of convergence and the sum of each infinite series:
    (a) $9 + 6x + 4x^2 + \cdots$
    (b) $3 - 9(x - 1) + 27(x - 1)^2 - \cdots$

63. A Christmas tree farm has 20 trees planted in the first row, and each row thereafter has 6 more trees than the preceding row. If the farm has 116 rows of trees, how many trees are planted altogether?

64. A car costing $12,000 is purchased under an agreement to pay $250 per month plus $\tfrac{3}{4}$% interest on the unpaid balance for that month. Find the total amount paid for the car.

65. To save for a down payment on a home, a married couple deposits $300 each month in an account that earns interest at the rate of 8% per year compounded monthly. Determine how much money will be in the account at the end of 3 years given the indicated account balance.
    (a) The account is empty today.
    (b) The account has $6000 today.

66. The enrollment at a certain college is presently 5411 students, but the enrollment is decreasing by 4% each year. Determine the enrollment five years from now.

67. A boy pedals his bicycle up a 105-ft hill at the rate of 36 ft the first second, 24 ft the second second, 16 ft the third second, and so on. Does he ever reach the top of the hill?

68. A tennis ball is dropped from a height of 16 feet, and rebounds each time one-quarter the height from which it last fell. Find the total distance the ball travels before coming to rest.

In Exercise 69–74, evaluate each binomial coefficient expression.

69. $\binom{16}{1}$

70. $\binom{22}{0}$

71. $\binom{13}{11}$

72. $\binom{20}{4}$

73. $\binom{15}{5} + \binom{15}{4} - \binom{16}{5}$

74. $\dfrac{\binom{7}{2}\binom{3}{1}}{\binom{10}{3}}$

In Exercises 75 and 76, use the binomial theorem to expand each expression.

75. $(3x - 2)^6$

76. $(2 + a^{1/2})^8$

77. Find the fourth term of $(a^{-1} + b)^{10}$.

78. Find the tenth term of $(4x^2 - 1)^{12}$.

79. How many ways can six sedans, three coupes, and four station wagons be lined up in a row at a dealership, if one does not distinguish between vehicles of the same type?

80. A college athletic director must hire 3 assistant football coaches from 12 equally qualified candidates. How many ways can he select the 3 coaches?

81. In the game of bridge, 13 cards are dealt to each of 4 players from a deck of 52 cards.
    (a) How many different bridge hands are possible?
    (b) How many bridge hands consist of 4 aces?
    (c) How many bridge hands consist of 8 spades?

82. How many ways can 10 people be assigned to play the four infield positions of a baseball team if each person can play any infield position?

Cumulative Review Exercises for Chapters 10 and 11

1. Find $f(1) + f(2) + f(3) + \cdots + f(20)$ for the given function f.
   (a) $f(x) = 2x$ (b) $f(x) = 2^x$

2. The sum of a finite arithmetic series with n terms is n^2. Find the sum of the first and last terms of the series.

3. (a) Use mathematical induction to show that
$$\sum_{i=1}^{n} i^4 = 1^4 + 2^4 + 3^4 + \cdots + n^4 = \frac{n(n+1)(2n+1)(3n^2+3n-1)}{30}.$$
   (b) Find the sum of the series $1^4 + 2^4 + 3^4 + \cdots + 20^4$.

4. (a) Use mathematical induction to show that
$$\sum_{i=1}^{n} i^5 = 1^5 + 2^5 + 3^5 + \cdots + n^5 = \frac{n^2(n+1)^2(2n^2+2n-1)}{12}.$$
   (b) Find the sum of the series $1^5 + 2^5 + 3^5 + \cdots + 20^5$.

5. The nth partial sum of an infinite series is $S_n = 4[1 - (2/3)^n]$. What is the sum of the series?

6. How many consecutive multiples of 6, beginning with 6, must be taken so that their sum is 7650?

7. Solve each system of linear equations.
   (a) $\begin{aligned} 2x + y + z &= 6 \\ x - 2y + 2z &= 10 \\ 3x - y - z &= 4 \end{aligned}$
   (b) $\begin{aligned} x - y \phantom{{}+2z} - 2w &= 2 \\ 2y + 3z + 2w &= -8 \\ x \phantom{{}-y} + 2z - 4w &= 1 \\ x + 4y + z \phantom{{}- 4w} &= 6 \end{aligned}$

8. Given that the sequence a^2, b^2, c^2 is an arithmetic sequence, show that
$$\frac{1}{b+c}, \frac{1}{a+c}, \frac{1}{a+b}$$
is also an arithmetic sequence.

9. Given the matrices
$$A = \begin{bmatrix} 7 & -2 & 6 \\ 2 & 3 & 5 \\ 5 & -1 & 4 \end{bmatrix} \quad X = \begin{bmatrix} x \\ y \\ z \end{bmatrix} \quad B = \begin{bmatrix} 5 \\ 2 \\ 5 \end{bmatrix}$$
find A^{-1}, and solve the matrix equation $AX = B$ for X.

10. Find x and y if $x + 3y$, $2x - y$, $x - 8y$, $8x + 6$ are the first four elements in an arithmetic sequence.

11. Find x and y if $4, x, y$ are the first three elements in an arithmetic sequence and $x, y, 18$ are the first three elements in a geometric sequence.

12. Solve for x.
    (a) $\begin{vmatrix} x & 2 \\ 1 & 3 \end{vmatrix} = 7$
    (b) $\begin{vmatrix} 2 & x & 3 \\ -2 & 4 & x \\ 5 & -3 & -2 \end{vmatrix} = -51$

13. Determine the nature of the sequence
$$\log a_1, \log a_2, \log a_3, \log a_4, \ldots$$
if $a_1, a_2, a_3, a_4, \ldots$ is a geometric sequence with positive elements and common ratio r.

14. If $a_1, a_2, a_3, a_4, \ldots$ is an arithmetic sequence with common difference d, what type of sequence is $10^{a_1}, 10^{a_2}, 10^{a_3}, 10^{a_4}, \ldots$?

15. Perform the indicated operations, given that matrices A and B are defined as follows.
$$A = \begin{bmatrix} 2 & 6 & -1 & 0 \\ 5 & 5 & -4 & 2 \\ 0 & 8 & 2 & -3 \\ 6 & 8 & 2 & -3 \end{bmatrix}$$
$$B = \begin{bmatrix} 5 & -1 & 4 & -7 \\ 3 & -1 & 0 & 3 \\ 9 & 2 & 4 & 0 \\ -6 & -3 & 7 & 2 \end{bmatrix}$$
   (a) $A + B$ (b) $3B - 2A$
   (c) AB (d) BA

16. Find the interval of convergence and the sum of the infinite geometric series.
$$\frac{1}{x+1} + \frac{x}{(x+1)^2} + \frac{x^2}{(x+1)^3} + \cdots$$

17. The product of the first seven terms in a geometric series is 0.0002187. What is the fourth term of the series?

18. Find the sum of each infinite geometric series.
    (a) $2, 2 - \sqrt{2}, \dfrac{\sqrt{2}-1}{\sqrt{2}+1}, \ldots$
    (b) $\sum_{i=1}^{\infty} \left[\left(\tfrac{1}{3}\right)^{i-1} + \left(\tfrac{1}{2}\right)^i \right]$

19. Evaluate each series.

(a) $\sum_{k=0}^{6} \binom{6}{k}$

(b) $\sum_{k=0}^{6} \left[\binom{6}{k} + \binom{6}{6-k} \right]$

(c) $\sum_{k=0}^{6} \left[\binom{6}{k} - \binom{6}{6-k} \right]$

(d) $\sum_{k=0}^{6} \left[\binom{6}{k} \binom{6}{6-k} \right]$

20. Find the middle term in the expansion of $(x - y)^{20}$.

21. The sides of a right triangle form an arithmetic sequence with common difference 6. Find the lengths of the sides of the triangle.

22. Fifty thousand dollars is invested—part at 5%, part at 6%, and part at 7%—yielding an annual interest of $3080. The income from the 6% investment yields $60 more than the income from the 7% investment. How much money is invested at each percentage rate?

23. Suppose 12 points lie in a plane, but no two points lie on the same straight line.

    (a) How many different lines are determined by the twelve points?

    (b) How many different triangles are determined by the twelve points?

24. It is estimated that a new car costing $12,000 depreciates in value by 25% each year. What is its approximate value after seven years?

25. The sum of the first four terms of a series is $\frac{25}{12}$. The first term is twice the second term, and the second term is twice the fourth term. Six times the third term is one less than twice the sum of the first two terms. Determine the first four terms of the series, and then use pattern recognition to describe the general term.

26. An equilateral triangle has sides of length 12 cm. Inside this triangle, a second equilateral triangle is constructed by joining the midpoints of the sides of the first triangle. Inside the second triangle, a third equilateral triangle is constructed by joining the midpoints of the sides of the second triangle, and so on, indefinitely.

    (a) What is the sum of the perimeters of all the equilateral triangles?

    (b) What is the sum of the areas of all the equilateral triangles?

27. Given that

$$A = \begin{bmatrix} 3 & 4 \\ 2 & 3 \end{bmatrix} \quad B = \begin{bmatrix} -6 & 3 \\ 1 & 2 \end{bmatrix}$$

$$C = \begin{bmatrix} 2 & 3 \\ -1 & 4 \end{bmatrix}$$

solve the matrix equation $2AX + B = C$ for the matrix X.

28. Use appropriate substitutions to change the given system of equations to a linear system. Solve the linear system and then state the solution of the original system.

$$\frac{2}{x} + \frac{3}{y} + \frac{1}{z} = 4$$

$$\frac{3}{x} - \frac{5}{y} + \frac{2}{z} = -5$$

$$\frac{4}{x} - \frac{6}{y} + \frac{3}{z} = -7$$

29. A cross-country ski trail connects condominium A to condominium B, with a warming hut located 4 miles from A. Starting from condominium A, the trail goes uphill for 3 miles, level for 4 miles, and then downhill for 6 miles. Suppose a man can ski from A to B in 2 hours; from B to A in 2 hours 45 minutes; and from A to the warming hut and back again to A in 1 hour 30 minutes. What are this skier's rates of skiing uphill, on level ground, and downhill (in miles per hour, mi/h).

30. When two resistors R_a and R_b are connected in parallel, as shown in the figure, the total resistance R_t between points A and B may be found by using the equation

$$\frac{1}{R_t} = \frac{1}{R_a} + \frac{1}{R_b}.$$

Find the values of three resistors R_1, R_2, and R_3 if the total resistance is 60 ohms (Ω) when R_1 and R_2 are connected in parallel, 100 Ω when R_2 and R_3 are connected in parallel, and 75 Ω when R_1 and R_3 are connected in parallel.

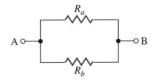

31. How long does it take for an annuity to amount to a million dollars if $600 is invested each month into an account that earns interest at the rate of 9% per year compounded monthly?

32. A company manufactures different types of prefinished oak flooring and packs the flooring in 30-square-foot cartons for shipping. The company receives an order for 65 cartons of a parquet flooring from customer A and an order for 90 cartons of the same parquet flooring

from customer B. The company has 100 cartons of this product at warehouse W_1 and 70 cartons at warehouse W_2. The shipping costs from the warehouses to the customers are given in the following table.

(a) How should the order be filled in order to minimize the shipping cost?
(b) What is the minimum shipping cost?

Warehouse	Customer	Shipping cost per carton
W_1	A	$1.25
W_2	A	$1.40
W_1	B	$1.30
W_2	B	$1.50

APPENDIX ◆ Significant Digits

Consider measuring the width of the wooden block in Figure A.1 with a ruler marked in intervals of 0.1-inch. The width of the block appears to be 1.8 inches. However, does the end of the block fall exactly in the middle of the marking for 1.8 inches, or slightly to the left or right of this marking? If we use a powerful magnifying glass, we might attempt to answer this question, but we could never determine the *exact* width of the block. For this reason, we say that every number found by a measuring process is an *approximate number*.

The number 8 in the tenths position of the number "1.8 inches" does have some *significance*, since the width seems to be closer to 1.8 inches than to either 1.7 inches or 1.9 inches. If a digit contributes to our knowledge of how good an approximation is, it is called a **significant digit**. Throughout this text, you are asked to round answers to a certain number of significant digits. The following may be used as a guide for this purpose.

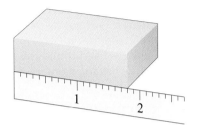

FIGURE A.1
Measuring the width of a block with a ruler marked in 0.1-inch intervals.

◆ Significant Digits

1. All nonzero digits are significant.

   Examples: 475 has 3 significant digits.
   12.827 has 5 significant digits.

2. Zeros between nonzero digits are significant.

   Examples: 6506 has 4 significant digits.
   42.0072 has 6 significant digits.

3. Zeros appearing at the end of a decimal number are significant.

   Examples: 7.00 has 3 significant digits.
   76.40 has 4 significant digits.

4. Zeros at the beginning of a decimal number are *not* significant and serve only to locate the decimal point correctly.

   Examples: 0.002 has 1 significant digit.
   0.023 has 2 significant digits.

5. Zeros at the end of an integer are *not* significant unless a tilde (˜) is placed above one of the zeros. The tilde is placed over the last zero that is significant.

   Examples: 22,00̃0 has 2 significant digits.
   22,0̃00 has 4 significant digits.

If a number is written in scientific notation as

$$k \times 10^n, \quad \text{where } 1 \leq |k| < 10 \text{ and } n \text{ is an integer,}$$

then the number of significant digits in that number is the same as the number of significant digits in *k*.

Examples:
$6.2 \times 10^3 = 6200$ has 2 significant digits.
$6.20 \times 10^3 = 62\tilde{0}0$ has 3 significant digits.
$6.2 \times 10^{-3} = 0.0062$ has 2 significant digits.
$6.20 \times 10^{-3} = 0.00620$ has 3 significant digits.

To avoid making a *rounding error* when working with approximate numbers, we carry along a few extra significant digits through the calculating process and then round the final answer to the desired accuracy. For example, suppose we wish to find the area of the gable end of the house shown in Figure A.2 and round this answer to three significant digits. First, we find the area of the rectangular part of the gable, without rounding the answer, as follows:

$$A = lw = (24.26 \text{ ft})(9.78 \text{ ft}) \approx 237.2628 \text{ sq ft.}$$

Next, we find the area of the triangular part of the gable, without rounding the answer, as follows:

$$A = \tfrac{1}{2}bh = \tfrac{1}{2}(24.26 \text{ ft})(8.52 \text{ ft}) \approx 103.3476 \text{ sq ft.}$$

Finally, we add these areas and then round to the desired accuracy of three significant digits. Hence, the area of the gable end of the house is

$$237.2628 \text{ sq ft} + 103.3476 \text{ sq ft} = 340.6104 \text{ sq ft} \approx 341 \text{ sq ft.}$$

round to 3 significant digits

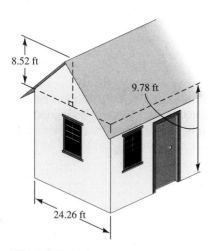

FIGURE A.2
Dimensions of the gable end of a house

If we round the areas of the rectangular and triangular parts to three significant digits and then add these areas, we accumulate a *rounding error*. Note that

$$237 \text{ sq ft} + 103 \text{ sq ft} = 34\tilde{0} \text{ sq ft.}$$

This answer does not have the desired accuracy. When working with approximate numbers, we usually use a calculator to perform the basic operations. It is best to let the calculator store all the digits and then round the final display to the desired accuracy. By using this procedure, you will be sure to avoid rounding errors.

Some exercises in this text that deal with approximate numbers may not specify a required number of significant digits in the final answer. For exercises of this nature, look at the data given in the problem and determine which approximate number has the *fewest* number of significant digits. It is common practice to round the final answer to the same number of significant digits as the approximate number with the fewest number of significant digits. For example, note in Figure A.2 that both 8.52 ft and 9.78 ft have the fewest number of significant digits, namely, three. Hence, it is acceptable to round the area of the gable to three significant digits as well.

Solutions to Problems and Answers to Odd-Numbered Exercises

Chapter 1

SECTION 1.1

Problems

1. (a) $a \leq 8$ (b) $-4 \leq b < 0$
2. $|c - (-1)| \geq 5$ or $|c + 1| \geq 5$
3. (a) number line with solid dot at 0, open circle at 3
   (b) number line with solid dot at 4, open circle at 9
4. (a) number line from -2 (open) to 2 (solid), with -1 marked
   (b) number line with arrow left from -3, and segment 3 to 5

Exercises

1. $x < 0$ 3. $a \leq 7$ 5. $2 < p \leq 10$ 7. $0 < c < 8$
9. $-2 \leq t \leq 0$ 11. 15 13. 1.9 15. $\frac{29}{24}$ 17. $9\frac{1}{6}$
19. $\pi - \sqrt{2}$ 21. $|a - 7| \geq 3$ 23. $|x + 4| < 2$
25. $|a + 1| > 4$ 27. $|1 - y| < |y|$
29. $|b + 2| > |b - 5|$ 31. $[-4, -2]$ 33. $(2, 6)$
35. $[-1, 4)$ 37. $(-\infty, -1]$ 39. $(10, \infty)$
41. $(-\infty, 0) \cup (0, \infty)$ 43. $(-\infty, -7] \cup [-4, \infty)$
45. $[0, 2) \cup (2, \infty)$ 47. $(-\infty, 1) \cup (1, 6) \cup (6, \infty)$
49. $(-\infty, 0) \cup [2, 3) \cup (3, \infty)$
51. number line
53. number line
55. number line
57. number line
59. number line
61. $x - \pi$ 63. 1 65. $-5 < x < 5$
67. The interval notation $[a, a]$ represents the single point a on the real number line. The interval notation (a, a) is meaningless.
69. $(-\infty, -1) \cup (0, 1)$ 71. $\frac{157}{50}, \pi, \frac{22}{7}, 3.145, \sqrt{10}, 3.2$

SECTION 1.2

Problems

1. $\dfrac{4^{-2}x^{-3}y^4}{4x^{-4}y^{-3}} = \dfrac{x^4 y^3 y^4}{4 \cdot 4^2 x^3} = \dfrac{xy^7}{4^3}$ or $\dfrac{xy^7}{64}$

2. $\dfrac{9.3 \times 10^7 \text{ mi}}{1.86 \times 10^5 \text{ mi/s}} = 5 \times 10^2 \text{ s}$
   $= 500 \text{ s} \cdot \dfrac{1 \text{ min}}{60 \text{ s}} \approx 8.3 \text{ min}$

3. $P = \dfrac{(6 \times 10^5)^2}{3 \times 10^1} = \dfrac{36 \times 10^{10}}{3 \times 10^1}$
   $= 12 \times 10^9 \text{ W} = 12{,}000 \times 10^6 \text{ W} = 12{,}000 \text{ MW}$

Exercises

1. 36 3. -16 5. $\frac{1}{64}$ 7. $-\frac{1}{64}$ 9. $\frac{81}{16}$ 11. $\frac{1}{81}$ 13. $\frac{2}{9}$
15. $-\frac{1}{4}$ 17. $\frac{7}{12}$ 19. $\frac{6}{5}$ 21. $-\dfrac{8}{xy^3}$ 23. $\dfrac{10}{(1 - 2x)^6}$
25. $-50x^4 y^{13}$ 27. $\dfrac{16}{y^4}$ 29. $2x^3 y^4$ 31. $\dfrac{1}{x + 3}$
33. $(x - 2)^8$ 35. $(2y + 3)^2$ 37. $\dfrac{1}{18mn^2}$
39. $8p^3(q - r)^7$ 41. 0.0000000069

43. $-175{,}000{,}000{,}000$ 45. 0.000392
47. 5.43×10^4 49. 1.3×10^{-7}
51. 2.40×10^5 53. 2×10^3 55. 3.1×10^{-18}
57. 3.125×10^{48} 59. 2.4×10^{-6} 61. 1.119
63. $42{,}3\tilde{0}0$ 65. 1.07×10^{-23} 67. 1.99×10^{-30}
69. 12 yr   71. $6.7\ \mu A$ 73. $33{,}100$ sq m   75. 1
77. Since $a/0$ is undefined, the expression $1 = (a/0)^0$ is meaningless.
79. (a) 643.70 (b) Approximately $231{,}732$

73. $\sqrt[3]{5xy^2}$ 75. $\dfrac{x\sqrt{2x}}{y}$ 77. $|x+1|$
79. $6\sqrt{2}$ 81. $\dfrac{\sqrt[3]{12x^2y}}{3xy}$ 83. $\dfrac{42a}{5\sqrt{6a}}$
85. $\dfrac{x+2}{\sqrt[3]{3x(x+2)}}$ 87. $\sqrt[6]{675}$ 89. $\sqrt[6]{x^3 y^4}$
91. (a) $2 - 2\sqrt[3]{2} + 2\sqrt[3]{5}$
    (b) $a^2b\sqrt{c} + b^2c\sqrt{a} + ac^2\sqrt{b}$
93. $\dfrac{\pi\sqrt{2L}}{4}$ 95. Approximately 8.00 ft

SECTION 1.3

Problems

1. (a) $\sqrt[5]{(x^2+2)^2} = (x^2+2)^{2/5}$
   (b) $(2x^2)^{3/7} = \sqrt[7]{(2x^2)^3} = \sqrt[7]{8x^6}$
2. (a) $[(-2)^4 x^4]^{1/4} = [(-2)^4]^{1/4}(x^4)^{1/4} = |-2|x = 2x$
   (b) $[(-2)^4]^{1/4}(x^4)^{1/4} = |-2||x| = 2|x|$
3. (a) $\sqrt[3]{32(x+y)^3} = \sqrt[3]{8(x+y)^3 \cdot 4} = 2(x+y)\sqrt[3]{4}$
   (b) $\sqrt[4]{\sqrt[3]{x^2 y^6}} = \sqrt[12]{x^2 y^6} = \sqrt[12]{(xy^3)^2} = \sqrt[6]{xy^3}$
4. $\dfrac{\sqrt[3]{5x^2}}{\sqrt[3]{2y}} = \dfrac{\sqrt[3]{5x^2}}{\sqrt[3]{2y}} \cdot \dfrac{\sqrt[3]{4y^2}}{\sqrt[3]{4y^2}} = \dfrac{\sqrt[3]{20x^2 y^2}}{2y}$

Exercises

1. 2   3. -2 5. 343   7. 25   9. $\frac{1}{1000}$
11. $-\frac{1}{27}$ 13. $\frac{1}{4}$ 15. $\frac{2}{3}$ 17. -1 19. 8
21. 2.290   23. 0.001123   25. 16.58   27. 0.6553
29. $(2a)^{1/2}$ 31. $(y^2-3)^{4/7}$ 33. $x(x^2+y^2)^{-1/2}$
35. $x^{3/2}$ 37. $\sqrt{x}$ 39. $\sqrt[4]{27m^6}$ or $m\sqrt[4]{27m^2}$
41. $\sqrt[5]{(x+y)^4}$ 43. $\dfrac{2}{\sqrt[3]{x^2}}$ 45. $4x$ 47. $\dfrac{4}{x^2}$
49. $(x+y)^{17/6}$ 51. $\dfrac{y^2}{16x^2}$ 53. $\dfrac{x^{4/3}}{(x^2+4)^{1/4}}$ 55. $\dfrac{y^2}{2}$
57. $\dfrac{x}{(x^2-1)^{1/2}}$ 59. $2|x| + 2x$ 61. $2\sqrt{6}$
63. $x+1$ 65. $x^4\sqrt{x}$ 67. $2xy\sqrt[3]{2x}$
69. $7(a+b)\sqrt{a^2+b^2}$ 71. $\dfrac{3xy\sqrt[3]{2y}}{5z}$

SECTION 1.4

Problems

1. $(3 + \sqrt{-25}) - (3 - \sqrt{-36})$
   $= (3 + 5i) - (3 - 6i) = 11i$
2. $(1+3i)(2+5i) = 2 + 11i + 15i^2 = -13 + 11i$
3. $\dfrac{1+\sqrt{-9}}{1-\sqrt{-9}} = \dfrac{(1+3i)(1+3i)}{(1-3i)(1+3i)}$
   $= \dfrac{1 + 6i + 9i^2}{1 - 9i^2}$
   $= \dfrac{-8+6i}{10} = -\dfrac{4}{5} + \dfrac{3}{5}i$
4. $(2 - \sqrt{-9})^2 = (2-3i)^2 = (2-3i)(2-3i)$
   $= 4 - 12i + 9i^2 = -5 - 12i$

Exercises

1. $3 + 8i$ 3. $15 + 3i$ 5. $24 + 16i$ 7. $39 + 33i$
9. 6   11. $2 - 4i$ 13. $\frac{21}{29} + \frac{9}{29}i$ 15. $-\frac{3}{5} - \frac{4}{5}i$
17. $9i$ 19. -18 21. $3i\sqrt{5}$ 23. $-2 + 33i$
25. 3   27. $30 + 19i$ 29. $-\dfrac{\sqrt{6}}{5} + \dfrac{2}{5}i$ 31. -5
33. $-2i$ 35. $-27i$ 37. $108i$ 39. $5 - 12i$
41. $-\frac{9}{1681} - \frac{40}{1681}i$ 43. 0
45. Since $(1 + 0i)(a + bi) = a + bi$, the complex number $1 + 0i$ is the multiplicative identity of $a + bi$.
47. (a) 0   (b) 0   49. (a) 0   (b) 0   51. $\pm 3i$
53. (a) $4.24i$ (b) $22.5i$ (c) $12.4i$ (d) $0.936i$
55. 1.07 A

SECTION 1.5

◆ Problems

1. $(5x^2 - 3x) + (5x - x^2) - (2x + 4x^2)$
   $= 5x^2 - 3x + 5x - x^2 - 2x - 4x^2$
   $= 0$

2. $m^3 - m^2 + 3m + 4$
   $ m^2 + 2$
   $\overline{m^5 - m^4 + 3m^3 + 4m^2 }$
   $ 2m^3 - 2m^2 + 6m + 8$
   $\overline{m^5 - m^4 + 5m^3 + 2m^2 + 6m + 8}$

3. (a) $(2x - 3)(2x + 3) = (2x)^2 - (3)^2 = 4x^2 - 9$
   (b) $(3x - 5y)^2 = (3x)^2 + 2(3x)(-5y) + (-5y)^2$
   $= 9x^2 - 30xy + 25y^2$
   (c) $(x^4 + 4)^3 = (x^4)^3 + 3(x^4)^2(4) + 3(x^4)(4)^2 + (4)^3$
   $= x^{12} + 12x^8 + 48x^4 + 64$

4. $x^2 - [y(x + 2y) - (x + y)(2y - x)]$
   $= x^2 - [xy + 2y^2 - (2y^2 + xy - x^2)]$
   $= x^2 - [xy + 2y^2 - 2y^2 - xy + x^2]$
   $= x^2 - x^2 = 0$

5. $(x^{-4} + 4)^3 = (x^{-4})^3 + 3(x^{-4})^2(4) + 3(x^{-4})(4)^2 + (4)^3$
   $= x^{-12} + 12x^{-8} + 48x^{-4} + 64$

◆ Exercises

1. $-x^2$ 3. $-5x^2y$ 5. $-2x^{-1}$ 7. $10\sqrt{3y}$

9. $\sqrt{10x}$ 11. polynomial of degree 6

13. not a polynomial, because y has a negative exponent

15. polynomial of degree 2 17. polynomial of degree 8

19. not a polynomial, because variables appear in the radicand

21. $7m^2 - 5m - 9$ 23. $2xy - x^2$

25. $-32m^5n + 8m^4 - 4m^3 + 4m^2$

27. $3x^3 + 7x^2 - 27xy - 63y$ 29. $12x^2 - xy - y^2$

31. $9x^2 - y^2$ 33. $2n^3 + n^2 - 12n + 4$

35. $x^3 + 27$ 37. $8x^3 - 1$ 39. $x^3\sqrt{3} - 9$

41. $m^2 + 12m + 36$ 43. $4a^2 - 12ab^2 + 9b^4$

45. $9x^6 + 30x^3 + 25$ 47. $x^3 + 9x^2 + 27x + 27$

49. $64x^3 - 144x^2y^2 + 108xy^4 - 27y^6$

51. $-4n^2 - 9n + 4$ 53. m^3

55. $32x^3 - 32x^2 + 8x$

57. $a^4 - 32a^2 + 256$ 59. $162x^4 - 162x^3 + 51x^2$

61. $4x^{-2}$ 63. $x - 3y$ 65. $4x - 28y\sqrt{x} + 49y^2$

67. $x^{1/3} - x^{2/3} - 8$

69. degree of $P + Q$ is m; degree of $P - Q$ is m; degree of PQ is $m + n$; and degree of P^2 is $2m$

71. $5x^2 + 12x - 10$ 73. $10\pi r - 25\pi$

75. (a) 19 (b) -5 77. $1.06x$

79. (a) $(6.32 \times 10^{-3})x^2 - (3.46 \times 10^{-4})x^3 + (4.72 \times 10^{-6})x^4$
   (b) 0.53 inch

SECTION 1.6

◆ Problems

1. $(9 - 3y) + (2xy - 6x) = 3(3 - y) - 2x(3 - y)$
   $= (3 - y)(3 - 2x)$

2. $x^2 + 10x + 25 = (x + 5)(x + 5) = (x + 5)^2$

3. (a) $(x + y)^2 - 25y^4 = [(x + y) + 5y^2][(x + y) - 5y^2]$
   $= (x + y + 5y^2)(x + y - 5y^2)$
   (b) $64 - x^3 = (4 - x)(16 + 4x + x^2)$

◆ Exercises

1. $ab(3a - 1)$ 3. $3xy(2x^2 - 3x + y)$

5. $(x - 1)(5 - x)$ 7. $7m(m + 3)^3$

9. $(a + b)(c + d)$ 11. $(x - 3)(x^2 + 2)$

13. $(2 - x)(3 - y)$ 15. $(5 - x^2)(3 - 2y)$

17. $(a + 5)(a - 2)$ 19. $(x - 9y)(x - 2y)$

21. $(12 + x)(2 + x)$ 23. $(x^2 + 6)(x^2 - 3)$

25. $(2x + 3)(x + 1)$ 27. $(3a - 4b)(a + 5b)$

29. $(4x + 1)(x - 3)$ 31. $(5 - 2t)(3 - 2t)$

33. $(4x - 5y)(3x - 2y)$ 35. $(2n^2 + 5)(n^2 + 1)$

37. $(x^3 - 2y^2)(6x^3 - 5y^2)$ 39. $(y^6 + 2)(8y^6 - 3)$

41. $(x + 9)(x - 9)$ 43. $(5x + y)(5x - y)$

45. $(2x^2 - 5x + 10)(2x^2 + 5x + 10)$

47. $(t + 2)(t^2 - 2t + 4)$

49. $(3x + 2y^2)(9x^2 - 6xy^2 + 4y^4)$

51. $(2 - 5n)(4 + 10n + 25n^2)$

53. $(t + 4)^2$ 55. $(2a^2 - 3b^2)^2$

57. $4xy(x - 3)(x - 4)$ 59. $2x(3x + 2)(x - 2)$

61. $5a(a + 3)(a - 3)$ 63. $2xy(2x + 1)(4x^2 - 2x + 1)$

65. $2(n + 2)(n - 2)(n^2 + 4)$

67. $(2x + 3)(2x - 3)(x + 1)(x - 1)$

69. $3a(a^3 + 3)(a - 1)(a^2 + a + 1)$

71. $2y(x + 1)(2x + 3)(2x - 3)$

73. $(x + 1)(x - 1)(x^2 + x + 1)(x^2 - x + 1)$

75. $(x + y - 2)(x - y - 2)$

77. $(2x + 2y + 3)(2x - 2y + 3)$

79. $(x + \sqrt{10})(x - \sqrt{10})$

81. $(a + b\sqrt[3]{5})(a^2 - ab\sqrt[3]{5} + b^2\sqrt[3]{25})$

83. $(x + 5i)(x - 5i)$

85. $(t + \sqrt{6})(t - \sqrt{6})(t + i\sqrt{6})(t - i\sqrt{6})$

87. $2\pi rt(h + r)$ 89. $(3.2 + A)(3.2 - A)(1 + A^2)$

SECTION 1.7

◆ **Problems**

1. $\dfrac{16 - x^2}{2x^3 - 5x^2 - 12x} = \dfrac{-(4 + x)(4 - x)}{x(x - 4)(2x + 3)}$
$= \dfrac{-(4 + x)}{x(2x + 3)}$

2. $\dfrac{x^2}{2x^2 - 5x + 2} \div \dfrac{4x}{4 - x^2}$
$= \dfrac{x^2}{(2x - 1)(x - 2)} \cdot \dfrac{(2 + x)(2 - x)}{4x}$
$= \dfrac{x(2 + x)}{4(1 - 2x)}$

3. $\dfrac{2x^2 - 9}{2x^2 + 11x + 5} - \dfrac{x^2 + 16}{2x^2 + 11x + 5}$
$= \dfrac{(2x^2 - 9) - (x^2 + 16)}{2x^2 + 11x + 5}$
$= \dfrac{x^2 - 25}{(2x + 1)(x + 5)} = \dfrac{x - 5}{2x + 1}$

4. $\dfrac{x + 1}{x^2 - 2x + 1} - \dfrac{1}{x - 1} = \dfrac{x + 1}{(x - 1)^2} - \dfrac{x - 1}{(x - 1)^2}$
$= \dfrac{(x + 1) - (x - 1)}{(x - 1)^2} = \dfrac{2}{(x - 1)^2}$

5. $\dfrac{\sqrt{(x + h) + 1} - \sqrt{x + 1}}{h} \cdot \dfrac{\sqrt{(x + h) + 1} + \sqrt{x + 1}}{\sqrt{(x + h) + 1} + \sqrt{x + 1}}$
$= \dfrac{[(x + h) + 1] - (x + 1)}{h(\sqrt{(x + h) + 1} + \sqrt{x + 1})}$
$= \dfrac{1}{\sqrt{(x + h) + 1} + \sqrt{x + 1}}$

6. $\dfrac{\dfrac{3}{x + h} - \dfrac{3}{x}}{h} = \dfrac{\dfrac{3x - 3(x + h)}{x(x + h)}}{h}$
$= \dfrac{-3h}{x(x + h)} \cdot \dfrac{1}{h} = \dfrac{-3}{x(x + h)}$

7. $x[-2(1 - 3x)^{-3}(-3)] + (1 - 3x)^{-2}$
$= \dfrac{6x}{(1 - 3x)^3} + \dfrac{1}{(1 - 3x)^2}$
$= \dfrac{6x + (1 - 3x)}{(1 - 3x)^3}$
$= \dfrac{3x + 1}{(1 - 3x)^3}$

◆ **Exercises**

1. $\dfrac{1}{1 - 2x}$ 3. $-\dfrac{1}{3}$ 5. $\dfrac{2x + 1}{x - 1}$ 7. $-\dfrac{n + 1}{n + 6}$

9. $\dfrac{1}{x + 3}$ 11. $\dfrac{3 + x + y}{3 - x + y}$ 13. $\dfrac{3(x + 1)}{x - 1}$

15. $\dfrac{2 + n}{1 - n}$ 17. $\dfrac{2x}{x + y}$ 19. $-\dfrac{3}{3b + a}$ 21. $n + 2$

23. $\dfrac{x}{3}$ 25. $\dfrac{m}{m^2 - 9}$ 27. $\dfrac{x}{(x - 1)^3}$ 29. $\dfrac{4 - 2x^2}{x(x + 2)^2}$

31. $\dfrac{-2}{(1 + a)(1 - a)}$ 33. $\dfrac{1}{a - 1}$ 35. $\dfrac{x^2 + y^2}{xy}$

37. $-\dfrac{a + b + c}{bc}$ 39. $\dfrac{-1}{3 + \sqrt{t + 3}}$

41. $\dfrac{2}{\sqrt{2(x + h)} + \sqrt{2x}}$

43. $\dfrac{2x + h}{\sqrt{(x + h)^2 + 1} + \sqrt{x^2 + 1}}$ 45. $\dfrac{a}{1 - a}$

47. $\dfrac{-1}{x(x + h)}$ 49. $\dfrac{-(2x + h)}{x^2(x + h)^2}$ 51. $(8x + 6)(x + 3)^2$

53. $\dfrac{2 + 6t}{(2 - 3t)^4}$ 55. $\dfrac{5 - 16n}{5(1 - 4n)^{6/5}}$ 57. $\dfrac{3 - 5t}{(3 - 4t)^{3/4}}$

59. $\dfrac{5t^2 - 1}{t^{4/3}}$

61. The expressions are *not* equal when $x = 1$ because $x^2 - 1 = 0$, and division by 0 is undefined.

63. (a) $y + 7 - \dfrac{5}{y}$ (b) $\dfrac{14}{t + 7} - t - 7$

65. (a) $x + \sqrt{x^2 - 1}$ (b) $\sqrt{a^2 + b^2} - b$

67. $\dfrac{2E}{2R + r}$ 69. $\dfrac{120 - 60\sqrt{n} - 4n + n^2}{4 - n}$

SECTION 1.8

◆ Problems

1. Adding -3 to both sides, we obtain $|4t - 3| = -1$. However, since the absolute value of a real number is always *nonnegative*, we conclude that this equation has no solution.

2. Multiplying both sides by 8, we obtain
$$4(y + 1) - 5(y - 1) > 8$$
$$4y + 4 - 5y + 5 > 8$$
$$-y > -1$$
$$y < 1$$
Hence, the solution set is $(-\infty, 1)$.

3. $ax + b = cx + d$
$b - d = cx - ax$
$b - d = x(c - a)$
$x = \dfrac{b - d}{c - a} \cdot \dfrac{-1}{-1} = \dfrac{d - b}{a - c}$

4. $A = \dfrac{1}{2}h(b_1 + b_2)$
$\dfrac{2A}{h} = b_1 + b_2$
$b_1 = \dfrac{2A}{h} - b_2$

◆ Exercises

1. -7 3. -4 5. -4 7. $-\tfrac{3}{2}$ 9. 5
11. 2   13. $\tfrac{2}{5}$ 15. $\tfrac{7}{6}$ 17. ± 4 19. $1, \tfrac{1}{5}$
21. $-\tfrac{1}{3}, \tfrac{19}{3}$ 23. $\pm\sqrt{3}$ 25. ± 4 27. $\pm\sqrt{2}$
29. $-6, 8$ 31. $(-\infty, 3)$ 33. $(-\infty, 2]$ 35. $(-7, \infty)$
37. $\left(-\tfrac{3}{2}, \infty\right)$ 39. $(-\infty, 10]$ 41. $(-3, 3)$
43. $\left(-\infty, -\tfrac{3}{2}\right) \cup \left(\tfrac{3}{2}, \infty\right)$ 45. $(-\infty, 1) \cup (4, \infty)$

47. $\left[\tfrac{5}{2}, \tfrac{11}{2}\right]$ 49. $\left(-\infty, -\tfrac{14}{3}\right] \cup \left[\tfrac{10}{3}, \infty\right)$

51. $x = \dfrac{bc}{a}$, $a \neq 0$, $b \neq 0$

53. $x = \dfrac{c}{a - b}$, $a \neq b$

55. $x = \dfrac{ak - a}{b - 3k}$, $k \neq 0$, $b \neq 3k$

57. $x = \dfrac{-b \pm 1}{a}$, $a \neq 0$

59. $x \leq \dfrac{-b}{a - c}$ 61. $x \geq \dfrac{abc - ay}{b}$ 63. $15°C$

65. 5 cm   67. 9π ft   69. (a) $E = IR$ (b) $R = \dfrac{E}{I}$

71. (a) $L_0 = \dfrac{L}{1 + \mu\Delta t}$ (b) $\Delta t = \dfrac{L - L_0}{L_0\mu}$

73. (a) $r = \dfrac{A - P}{Pt}$ (b) $P = \dfrac{A}{1 + rt}$

75. (a) $t = \dfrac{mv - mv_0}{F}$ (b) $m = \dfrac{Ft}{v - v_0}$

77. (a) $R_1 = \dfrac{P - I^2 R_2}{I^2}$ (b) $I = \pm\sqrt{\dfrac{P}{R_1 + R_2}}$

79. $A - B = 0$ is an equivalent equation; $A/B = 1$ and $A^2 = B^2$ are *not* equivalent equations.

81. $\left(-\infty, -\tfrac{8}{3}\right)$

83. If $x > 5$, then $5 - x < 0$. Hence, when dividing both sides by $5 - x$, we must reverse the direction of the inequality.

85. $48.6°F$

CHAPTER 1 REVIEW EXERCISES

1. $a \geq 7$ 3. $-10 < c < 0$ 5. $|a - 3| \leq 2$
7. 15   9. $[3, 7]$ 11. $(-1, \infty)$
13. $(-\infty, -6] \cup [-1, \infty)$ 15. $[0, 4) \cup (4, \infty)$
17. $-50x^6 y^{13}$ 19. $\dfrac{1}{(2x + 1)^4}$ 21. $x - 1$
23. $x + 1$ 25. $\dfrac{y^6}{8x^{3/2}}$ 27. 3×10^{-24} 29. $(x + 2)^{2/3}$

31. $\dfrac{1}{\sqrt[6]{x+4}}$ 33. $4(x+y)\sqrt{2(x+y)}$

35. $2y\sqrt[4]{4x^3}$ 37. $3|xy|$ 39. $22i\sqrt{5}$

41. -64 43. $12 - i$ 45. $47 + i$

47. $2a^2 + a$ 49. $2\sqrt{10x}$

51. $2x^6 + 4x^5 - 3x^4 - 13x^3 - 2x^2 + 9x + 3$

53. $30x^4 - 13x^2y - 10y^2$ 55. $4t^2 - 25$

57. $x^3 - 64$ 59. $9x^2 - 42xy + 49y^2$

61. $8x^3 - 60x^2 + 150x - 125$

63. $x^6 - 27x^4 + 243x^2 - 729$ 65. $-3y^2$

67. $x^2 + 3$ 69. $3y(y-2)(y+2)$

71. $(m-3)(m^2+1)$ 73. $(3x+4y)(2x-3y)$

75. $(2x^2 - 3)(x^2 - 5)$ 77. $(x+5)(x-3)$

79. $(5x + 3y^2)(5x - 3y^2)$ 81. $(2x-1)(4x^2 + 2x + 1)$

83. $(t+2)^3$ 85. $2xy(4x-3)(x+1)$

87. $2xy(1+2x)(1+3x)(1-3x)$ 89. $\dfrac{n+9}{3n-2}$

91. $\dfrac{1}{x}$ 93. $-6x$ 95. $\dfrac{21y^2 - 10x^3}{24xy}$

97. $\dfrac{-x(2x+3)}{3y(x+3)}$ 99. $\dfrac{-3}{x+3}$ 101. $\dfrac{12}{x(x+4)^2}$

103. $\dfrac{2x}{(x+2y)(x-2y)^2}$ 105. a

107. $\dfrac{x-5}{\sqrt[4]{3x(x-5)}}$ 109. $2x + \sqrt{4x^2 - 1}$ 111. 2

113. $-\tfrac{1}{2}$ 115. $-4, \tfrac{20}{3}$ 117. ± 4 119. $L = \dfrac{Rd^2}{\mu + \mu t}$

121. $(-\infty, -3)$ 123. $[2, \infty)$

125. $(-\infty, -1) \cup \left(\tfrac{13}{3}, \infty\right)$ 127. 40 m

129. $103{,}000$ sq ft

131. (a) $r = \sqrt{\dfrac{3V}{\pi h}}$ (b) 6 inches

Chapter 2

SECTION 2.1

◆ Problems

1.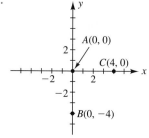

2. $PQ = \sqrt{[1 - (-3)]^2 + (-2 - 1)^2} = \sqrt{25} = 5$

3. $\left(\dfrac{-4+5}{2}, \dfrac{-3+(-1)}{2}\right) = \left(\dfrac{1}{2}, -2\right)$

4. $AB = \sqrt{[2-(-2)]^2 + (5-3)^2} = \sqrt{20} = 2\sqrt{5}$
   $AC = \sqrt{(4-2)^2 + (1-5)^2} = \sqrt{20} = 2\sqrt{5}$
   $BC = \sqrt{[4-(-2)]^2 + (1-3)^2} = \sqrt{40} = 2\sqrt{10}$

   Since $AB = AC$, we conclude that triangle ABC is isosceles. Also, since $(AB)^2 + (AC)^2 = (BC)^2$, we conclude that triangle ABC is a right triangle with hypotense $\overline{BC}$.

   Area $= \tfrac{1}{2}(2\sqrt{5})(2\sqrt{5}) = 10$ square units.

◆ Exercises

1. III   3. positive; negative   5. x-axis

7. $A(2,1), B(-4,4), C(-3,-1), D(4,-2), E(3,0), F(0,-3)$

9. (a) 5 (b) $\left(\tfrac{5}{2}, 4\right)$

11. (a) $\sqrt{34}$ (b) $\left(-\tfrac{1}{2}, \tfrac{11}{2}\right)$

13. (a) $\sqrt{113}$ (b) $\left(-1, -\tfrac{3}{2}\right)$

15. (a) $\tfrac{5}{12}$ (b) $\left(-\tfrac{5}{8}, \tfrac{5}{6}\right)$

17. (a) $3\sqrt{2}$ (b) $\left(-2, \dfrac{\sqrt{2}}{2}\right)$

19. $d = \sqrt{x^2 + y^2}$ 21. $3\sqrt{13} + 3\sqrt{5}$

23. $(AB)^2 = (AC)^2 + (BC)^2$, where $AB = 5\sqrt{2}$, $AC = 2\sqrt{10}$, $BC = \sqrt{10}$; area $= 10$ square units

25. $AB = AC = BC = 2\sqrt{2}$; area $= 2\sqrt{3}$ square units

27. $\left(-1, \tfrac{3}{2}\right), (2, -3), \left(5, -\tfrac{15}{2}\right)$

Chapter 2

29. $2\sqrt{17}$, $5\sqrt{2}$, $\sqrt{74}$ **31.** $(0, \frac{7}{9})$ **33.** $\left(\dfrac{a}{2}, \dfrac{b}{2}\right)$

35. $\dfrac{\sqrt{a^2 + b^2}}{2}$ **37.** $(a + b, c)$ **39.** $\left(\dfrac{a + b}{2}, \dfrac{c}{2}\right)$

41. $\left(\frac{14}{3}, 2\right)$ and $\left(\frac{10}{3}, 0\right)$

43. radius ≈ 4.33 units, area ≈ 59.0 square units

SECTION 2.2

◆ **Problems**

1. 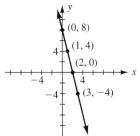   2.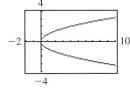

3. The graph of $y^2 + x = 4$ is symmetric with respect to the x-axis, since replacing y with $-y$ yields an equivalent equation:

$$(-y)^2 + x = 4$$
$$y^2 + x = 4$$

To find the x-intercept, let $y = 0$:

$$(0)^2 + x = 4$$
$$x = 4$$

To find the y-intercepts, let $x = 0$:

$$y^2 + (0) = 4$$
$$y = \pm 2$$

4. The graph of $y = x^3 - 3x$ is symmetric with respect to the origin, since replacing x with $-x$ and y with $-y$ yields an equivalent equation:

$$-y = (-x)^3 - 3(-x)$$
$$-y = -x^3 + 3x$$
$$y = x^3 - 3x$$

The graph of $x^2 + 3y^2 - 4x - 12 = 0$ is symmetric with respect to the x-axis, since replacing y with $-y$ yields an equivalent equation:

$$x^2 + 3(-y)^2 - 4x - 12 = 0$$
$$x^2 + 3y^2 - 4x - 12 = 0$$

5.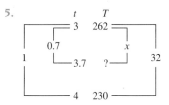

Setting up the direct proportion,

$$\frac{1}{0.7} = \frac{32}{x},$$

we find $x = 22.4$. Therefore, for $t = 3.7$ min, $T \approx 262 - 22.4 = 239.6°$F

◆ **Exercises**

1. 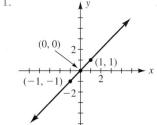   3.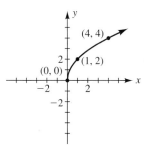

5. 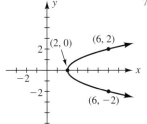   7.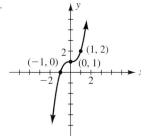

9. 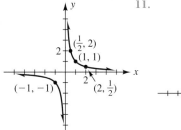   11.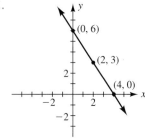

13. Symmetric with respect to the y-axis; x-intercepts ± 3; y-intercept $\frac{9}{4}$

15. Symmetric with respect to the origin; no x-intercept; no y-intercept

17. Symmetric with respect to the y-axis; x-intercepts ± 3; y-intercept 3

19. Symmetric with respect to the origin; no x-intercept; no y-intercept

21. Symmetric with respect to the x-axis; x-intercept 2; no y-intercept

23. None of these types of symmetry; x-intercept $\frac{5}{3}$; y-intercept 2

25. Symmetric with respect to the y-axis; x-intercepts ± 6; y-intercept 12

27. None of these types of symmetry; x-intercepts 0, 2; y-intercept 0

29. Symmetric with respect to the x-axis; x-intercepts 8, -2; y-intercepts ± 4

31. Symmetric with respect to the origin; x-intercepts $\pm\sqrt{2}$; no y-intercept

33. None of these types of symmetry; x-intercepts $-\frac{3}{2}$, $-\frac{15}{2}$; y-intercept 1

35. Symmetric with respect to the origin; x-intercepts 0, ± 1; y-intercept 0

37. Symmetric with respect to the x-axis; x-intercept -2; y-intercepts ± 2

39. Symmetric with respect to the y-axis; no x-intercept; y-intercept 1

41. (a)   (b) 59.8°F

43. (a)   (b) 211.2 ft

45. The graph of $y = |x| + c$, with $c > 0$, is the same as the graph of $y = |x|$, but shifted vertically upward c units.

47. The graph of $y = \sqrt{x - c}$, with $c > 0$, is the same as the graph of $y = \sqrt{x}$, but shifted horizontally to the right c units.

49. (a)

C (in °C)	3.9	23.3	−19.4	12	31	−6
F (in °F)	39	74	−3	53.6	87.8	21.2

(b)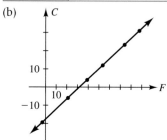

(c) F-intercept 32, the Fahrenheit temperature when the Celsius temperature is 0. C-intercept $-\frac{160}{9}$, the Celsius temperature when the Fahrenheit temperature is 0.

(d) $-40°F = -40°C$. Replace C with F in the equation $5F - 9C = 160$ and solve for F.

SECTION 2.3

Problems

1. Solving the equation for y, we obtain $y = \pm\sqrt{4 - 2x}$. Since there are two outputs for each input value $x < 2$, we conclude that the equation $2x + y^2 = 4$ does *not* define y as a function of x. Also, note that the graph of this equation fails the vertical line test.

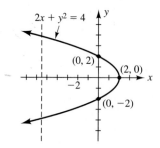

2. $f(\pi) = \pi^2 - 4\pi + 3 \approx 0.303$

3. Step 1: $g(x + h) = 4 - 3(x + h)$
$= 4 - 3x - 3h$
Step 2: $g(x + h) - g(x)$
$= (4 - 3x - 3h) - (4 - 3x)$
$= -3h$
Step 3: $\dfrac{g(x + h) - g(x)}{h} = \dfrac{-3h}{h} = -3$

4. $h(1) = (1) - 1 = 0$

5. The radicand $2 - x$ must be positive. Thus,
$$2 - x > 0$$
$$-x > -2$$
$$x < 2$$
Hence, the domain is $(-\infty, 2)$.

◆ **Exercises**

1. Defines y as a function of x
3. Defines y as a function of x
5. Defines y as a function of x
7. Does not define y as a function of x
9. Defines y as a function of x
11. Defines y as a function of x
13. Defines y as a function of x 15. 13   17. 5
19. $9 - 2\sqrt{2}$ 21. $|2ab + 3|$ 23. $t - 4$
25. $n^2 - n + 1$ 27. $\sqrt{x}$ 29. $\sqrt{1 + x^2} - 4$
31. $4x - 2\sqrt{x - 2} - 7$ 33. $4x - 34\sqrt{x} + 73$
35. (a) 9   (b) -11 (c) 4   (d) 0
37. (a) 7   (b) 3   (c) 5   (d) $\begin{cases} x + 1 \text{ if } x > -1 \\ 0 \text{ if } x = -1 \\ -(x + 1) \text{ if } x < -1 \end{cases}$
39. 2   41. $2x + h$ 43. $2x + 2 + h$ 45. $\dfrac{-1}{x(x + h)}$
47. $(-\infty, \infty)$ 49. $(-\infty, \infty)$ 51. $(-\infty, 4]$
53. $[-4, 4]$ 55. $(-\infty, -2) \cup (-2, \infty)$
57. $(-\infty, -2) \cup (-2, 2) \cup (2, \infty)$
59. $(-\infty, -5) \cup (-5, 2) \cup (2, \infty)$
61. $\left[\frac{1}{2}, 4\right) \cup (4, \infty)$
63. $r(25) = 7$ meters, which represents the radius of the oil spill after 25 minutes
65. $C(12.4) = \$19.48$, which represents the cost of a taxicab fare when the cab is driven 12.4 miles
67. No; several values of y correspond to each input value of x.
69. (a) Yes; for each element in set X there corresponds exactly one element in set Y.
(b) No; the element 2 in set X is not assigned a value.
(c) Yes; for each element in set X there corresponds exactly one element in set Y.
(d) No; the element 2 in set X corresponds to two elements in set Y.
71. (a) $A = 4$ (b) $A = 134$
73. The population is one-fourth as large.

SECTION 2.4

◆ **Problems**

1. (a) $F(0) = -1 + \sqrt{15} \approx 2.87$
(b) Since 6 is not an element in the domain of F, the functional value $F(6)$ is undefined.

2. The function g has domain $(-\infty, -2] \cup [2, \infty)$ and range $[0, \infty)$. The function h has domain $(-\infty, \infty)$ and range $(-\infty, \infty)$.

3. $$\sqrt{2x - 1} + 2 = x$$
$$\sqrt{2x - 1} = x - 2$$
$$2x - 1 = x^2 - 4x + 4$$
$$x^2 - 6x + 5 = 0$$
$$(x - 1)(x - 5) = 0$$
$$x = 1, 5$$
However, 1 is an extraneous root. Thus, the only solution is 5.

4. $$\dfrac{x + 1}{2} - \dfrac{5(x - 1)}{8} \geq 1$$
$$4(x + 1) - 5(x - 1) \geq 8$$
$$-x + 9 \geq 8$$
$$-x \geq -1$$
$$x \leq 1$$
Hence, the solution set is $(-\infty, 1]$, which agrees with the graphical result in Example 4(b).

5. $$(x + 6)^2 = x^2 + 18^2$$
$$x^2 + 12x + 36 = x^2 + 324$$
$$12x = 288$$
$$x = 24$$

◆ **Exercises**

1. (a) $(-\infty, \infty)$ (b) $[-2, \infty)$ (c) ± 2 (d) Even
3. (a) $(-\infty, \infty)$ (b) $[-1, \infty)$ (c) $0, -2$
   (d) Neither
5. (a) $[-\sqrt{5}, \sqrt{5}]$ (b) $[0, \sqrt{5}]$ (c) $\pm\sqrt{5}$
   (d) Even
7. (a) $(-\infty, \infty)$ (b) $[-2, 0)$ (c) None   (d) Even

9. (a) $[-2, 2]$ (b) $[-22, 22]$ (c) $0, \pm\sqrt[4]{5}$
   (d) Odd

11. Odd   13. Even   15. Odd   17. Even

19. Neither   21. $\frac{5}{3}$ 23. $8, -5$ 25. $3, \frac{1}{2}$ 27. $\frac{3}{2}$

29. ± 2 31. 5 33. $-\frac{1}{2}, \frac{3}{2}$ 35. $17, -25$ 37. $\frac{7}{9}$

39. -3 41. $0, -2$ 43. $(-\infty, 2]$ 45. $(-\infty, 10]$

47. $(-\infty, 1) \cup (4, \infty)$

49. $(-\infty, -3] \cup [-\sqrt{3}, \sqrt{3}] \cup [3, \infty)$ 51. 13 inches

53. 9,728 box seats

55. 441 sq ft for the square room; 459 sq ft for the rectangular room

57. 170 ft   59. 1, 3, 5 or 3, 5, 7   61. (2, 3)

63. It has no real solution

65. One y-intercept; infinitely many x-intercepts

67. No, it would fail the vertical line test.

69. $6\frac{1}{2}$ years

SECTION 2.5

◆ Problems

1. The graph of $h(x) = x^3 - 1$ is the same as the graph of $f(x) = x^3$, but shifted vertically *downward* 1 unit:

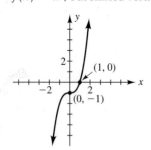

2. The graph of $h(x) = (x - 1)^3$ is the same as the graph of $f(x) = x^3$, but shifted horizontally *to the right* 1 unit:

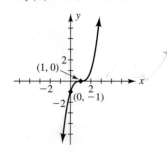

3. The graph of $F(x) = (x + 1)^2 - 4$ is obtained by shifting the graph of $f(x) = x^2$ to the left 1 unit and downward 4 units:

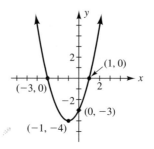

4. The graph of $F(x) = -|x + 2|$ is obtained by shifting the graph of $f(x) = |x|$ to the left 2 units and then reflecting this graph about the x-axis:

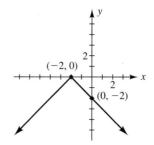

5. The graph of $F(x) = 4x - 1$ is formed by stretching the graph of $f(x) = x$ by a factor of 4 and then shifting this graph downward 1 unit:

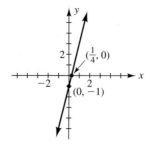

6.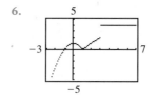

Exercises

1.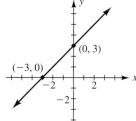
3.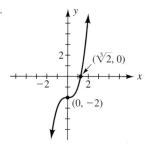
5.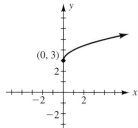
7.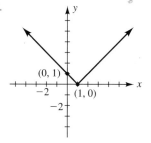
9.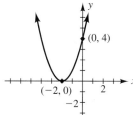
11.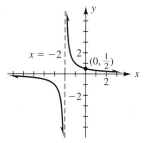
13.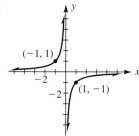
15.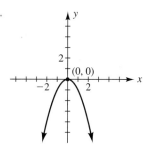
17.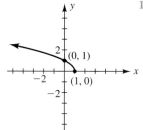
19.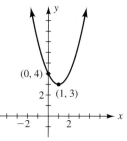
21.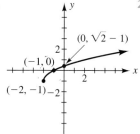
23.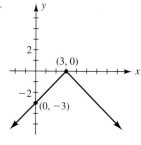
25.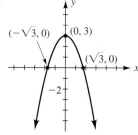
27.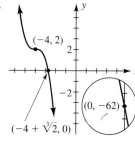
29.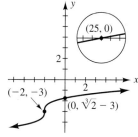
31.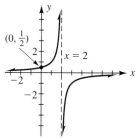
33.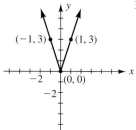
35.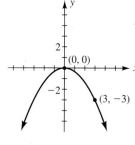
37.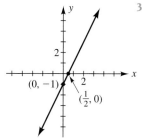
39.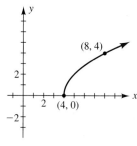

41.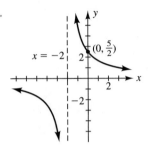

43. *Increasing functions:* Identity, cubing, square root, cube root; *decreasing function:* Reciprocal

45. Increasing   47. Decreasing   49. Neither

51. Increasing

53.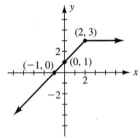

Increasing: $(-\infty, 3)$
Constant: $(3, \infty)$

55.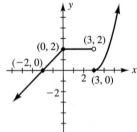

Increasing: $(-\infty, 0) \cup (3, \infty)$
Constant: $(0, 3)$

57.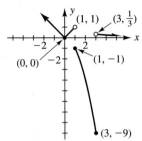

Increasing: $(0, 1)$
Decreasing: $(-\infty, 0) \cup (1, 3) \cup (3, \infty)$

59.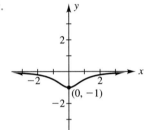

61.

63. $y = 1 + \sqrt{4 - x^2}$

65. $y = -2 + \sqrt{4 - (x + 1)^2}$

67. Vertical shift rule

69. x- and y-axis reflection rules

71.

To graph:	Draw the graph of $y = f(x)$ and then
$y = f(x) + c$, for $c > 0$	shift vertically upward c units
$y = f(x) - c$, for $c > 0$	shift vertically downward c units
$y = f(x + c)$, for $c > 0$	shift horizontally to the left c units
$y = f(x - c)$, for $c > 0$	shift horizontally to the right c units
$y = -f(x)$	reflect about the x-axis
$y = f(-x)$	reflect about the y-axis
$y = cf(x)$, for $c > 1$	stretch vertically by a factor of c
$y = cf(x)$, for $0 < c < 1$	compress vertically by a factor of $1/c$

73. (a) 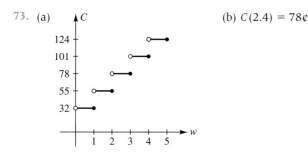 (b) $C(2.4) = 78¢$

SECTION 2.6

◆ Problems

1. $(f \circ g)(0) = f(g(0)) = f(0) = -2$

2. (a) $f(9) = 3$. Hence, $(g \circ f)(9) = g(f(9))$
   $= g(3)$
   $= 3^2 - 4 = 5$
   (b) $(g \circ f)(9) = 9 - 4 = 5$

3. $H(a) = H(b)$ implies
$$a^2 - 1 = b^2 - 1$$
$$a^2 = b^2$$
$$a = b \quad \text{for } a \geq 0, b \geq 0$$

Also, note that the graph of $H(x) = x^2 - 1$ with $x \geq 0$ passes the horizontal line test.

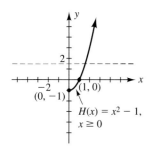

4. As shown in the figure, the functions f and g are one-to-one functions and their domains and ranges are the set of all real numbers. The graphs of f and g appear to be reflections of one another in the line $y = x$, which indicates that these functions might be inverses of each other.

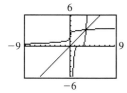

Note that
$$f(g(x)) = f(\sqrt[5]{x} + 2) = [(\sqrt[5]{x} + 2) - 2]^5$$
$$= (\sqrt[5]{x})^5 = x$$
and $g(f(x)) = g[(x - 2)^5] = \sqrt[5]{(x - 2)^5} + 2$
$$= (x - 2) + 2 = x.$$

Hence, we conclude that f and g are inverses of each other.

5. Replacing $f(x)$ with y, we solve for x as follows:
$$y = \frac{1}{x + 2}$$
$$x + 2 = \frac{1}{y}$$
$$x = \frac{1}{y} - 2$$

Interchanging x and y, we obtain $y = \frac{1}{x} - 2$.

Hence, the inverse of f is $f^{-1}(x) = \frac{1}{x} - 2$.

◆ Exercises

1. 0 3. Undefined 5. 2 7. 1

9. $(f \circ g)(x) = 2x^2 + 3; (-\infty, \infty)$

11. $(G \circ f)(x) = \frac{1}{2x + 1}; (-\infty, -\frac{1}{2}) \cup (-\frac{1}{2}, \infty)$

13. $(F \circ h)(x) = \sqrt{x - 4}; [4, \infty)$

15. $(G \circ H)(x) = \frac{x - 1}{2}; (-\infty, 1) \cup (1, \infty)$

17. $(H \circ f)(x) = \frac{1}{x}; (-\infty, 0) \cup (0, \infty)$

19. $(f \circ f)(x) = 4x + 3; (-\infty, \infty)$ 21. $\frac{1}{2}$

23. (a) One-to-one

25. (a) Not one-to-one (b) $[0, \infty)$ is one restriction.

27. (a) One-to-one

29. (a) Not one-to-one (b) $[-2, \infty)$ is one restriction.

31. (a) Not one-to-one (b) $[2, \infty)$ is one restriction.

33. f and g are inverses. 35. f and g are not inverses.

37. f and g are inverses. 39. f and g are inverses.

41. f and g are inverses. 43. f and g are inverses.

45. (a) $f^{-1}(x) = \dfrac{x}{2}$ (b)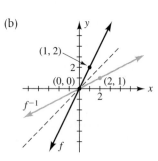

47. (a) $f^{-1}(x) = 3 - x$ (b)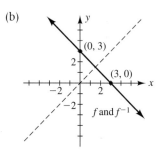

49. (a) $f^{-1}(x) = \sqrt[3]{2 - x}$ (b)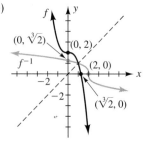

51. (a) $f^{-1}(x) = x^3 + 4$ (b)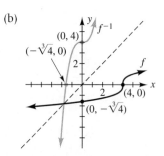

53. (a) $f^{-1}(x) = \dfrac{1}{x} - 1$ (b)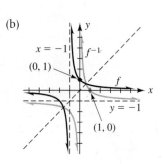

55. (a) $f^{-1}(x) = (x - 1)^2 + 4,\ x \geq 1$ (b)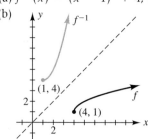

57. (a) $f^{-1}(x) = \sqrt{x} + 1$ (b)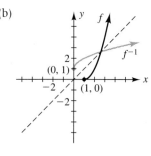

59. (a) $f^{-1}(x) = \sqrt{9 - x^2},\ 0 \leq x \leq 3$ (b)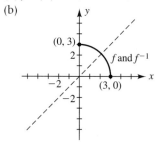

61. (a) $f^{-1}(x) = -\sqrt{x - 3}$ (b)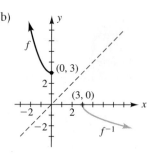

63. 4 **65.** 3 **67.** 2

69.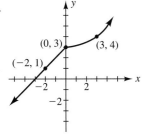

x	-2	0	3
$f^{-1}(x)$	1	3	4

71.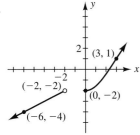

x	-2	0	3
$f^{-1}(x)$	undefined	-2	1

73. $(f \circ g)^{-1}(x) = (g^{-1} \circ f^{-1})(x) = \dfrac{3x + 1}{3}$

75. $(-\infty, 0) \cup (0, \infty)$ 77. $\left(-\infty, \tfrac{2}{3}\right) \cup \left(\tfrac{2}{3}, \infty\right)$

79. $f(x) = \tfrac{1}{3}x - 2$

81. (a) $(V \circ r)t = 36\pi(t + 1)^3$
(b) Volume of the balloon as a function of time after the inflation process begins
(c) 2304π cubic inches   (d) 9 seconds

SECTION 2.7

Problems

1. If $x = 50$ ft and $y = 100 - x$, then $y = 50$ ft. Maximum area occurs when the rectangle is a square.

2. From the similar triangles, we have $h = 3r$. Hence,
$$V = \tfrac{1}{3}\pi r^2 h = \tfrac{1}{3}\pi r^2 (3r) = \pi r^3$$

3. $T = \dfrac{168{,}000}{40} = \$4200$

4. $t = \dfrac{240}{50} = 4.8$ h

5. Since the electrical resistance R of a wire varies directly as its length L and inversely as the square of its radius r, we write $R = \dfrac{kL}{r^2}$. Doubling both L and r, we obtain

$$R = \dfrac{k(2L)}{(2r)^2} = \dfrac{1}{2} \cdot \dfrac{kL}{r^2}.$$

Thus, the resistance becomes one-half as large.

Exercises

1. $d = \dfrac{C}{\pi}$ 3. $C = 40 + \dfrac{n}{5}$

5. (a) $s = \dfrac{d}{\sqrt{2}}$ (b) $A = \dfrac{d^2}{2}$ (c) $P = 2\sqrt{2}\,d$

7. $W = 125{,}000 + 4000t$

9. (a) $V = \tfrac{4}{3}\pi h^3$ (b) $V = \tfrac{1}{6}\pi r^3$

11. (a) $d = \sqrt{2 - 2x}$ (b) $[-1, 1]$

13. (a) $A = \dfrac{x\sqrt{x}}{2}$ (b) $P = x + \sqrt{x} + \sqrt{x^2 + x}$

15. $S = 2x^2 + \dfrac{256}{x}$

17. (a) $V = 9h^2$
(b) Domain $[0, 4]$
(c)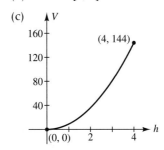

19. (a) $d = 30\sqrt{9 + t^2}$
(b) Domain $[0, 3]$; range $\left[90, 90\sqrt{2}\right]$

21. (a) $A = \dfrac{x\sqrt{262.44 - x^2}}{2}$
(b) Approximately 64.6 sq in.

23. (a) $d = \dfrac{F}{50}$

(b)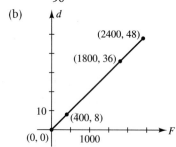

(c) 36 cm

25. (a) $W = 125l$

(b)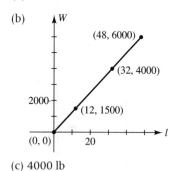

(c) 4000 lb

27. (a) $N = \dfrac{10{,}000}{d}$

(b)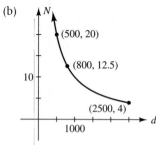

(c) 12.5 calls

29. (a) $W = \dfrac{300}{f}$

(b)

(c) 10 m

31. (a) $F = \dfrac{160{,}000}{d^2}$

(b)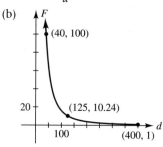

(c) 10.24 N

33. (a) $T \approx 2.007\sqrt{l}$

(b)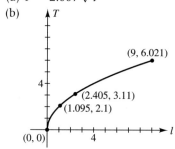

(c) Approximately 3.11 s

35. The force is tripled.

37. (a) $A = 1000x - 2x^2$ (b) (0, 500)
(c) 250 ft by 500 ft
(d) The ratio of x to y is 1 to 2.

39. (a) $A = 30r - 2r^2 - \dfrac{\pi r^2}{2}$ (b) $r \approx 4.20$ ft
(c) $y = 4.20$ ft; the ratio of r to y is 1 to 1.

41. $d = \begin{cases} 30t & \text{if } t \le 3 \\ 30\sqrt{9 + (t-3)^2} & \text{if } 3 < t \le 6 \end{cases}$

43. 25 seconds

CHAPTER 2 REVIEW EXERCISES

1. (a) 5 (b) $\left(2, \dfrac{9}{2}\right)$

3. (a) $2\sqrt{5}$ (b) (0, 2)

5. (a) $\sqrt{97}$ (b) $\left(\dfrac{3}{2}, -3\right)$

7. $AB = BC = CD = DA = 2\sqrt{2}$; perimeter is $8\sqrt{2}$; area is 8

9. $\dfrac{a}{2}$; half as long as OA

11. $\dfrac{\sqrt{(a-b)^2 + c^2}}{2}$; half as long as AB

13. (a) *x*-intercept 0; *y*-intercept 0
    (b) Symmetric with respect to origin
    (c)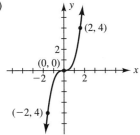
    (d) Defines *y* as a function of *x*

15. (a) *x*-intercept 0; *y*-intercept 0
    (b) Symmetric with respect to *x*-axis
    (c)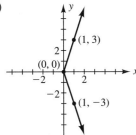
    (d) Does not define *y* as a function of *x*

17. (a) No axis intercepts
    (b) Symmetric with respect to *y*-axis
    (c)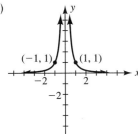
    (d) Defines *y* as a function of *x*

19. (a) *x*-intercept 3; *y*-intercept 2
    (b) No symmetry
    (c)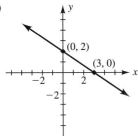
    (d) Defines *y* as a function of *x*

21. (a) *x*-intercepts ± 3; *y*-intercept -3
    (b) Symmetric with respect to *y*-axis
    (c)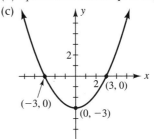
    (d) Defines *y* as a function of *x*

23. (a) *x*-intercepts ± 2; *y*-intercepts ± 1
    (b) Symmetric with respect to *x*-axis, *y*-axis and origin
    (c)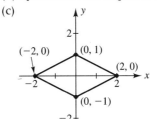
    (d) Does not define *y* as a function of *x*

25. 1   27. 5   29. $2x^2 - 13x + 21$ 31. -5

33. 5   35. $18x^4 + 45x^2 + 28$ 37. $\dfrac{3x+1}{x-1}$

39. $\dfrac{x-2}{3}$ 41. 2   43. $4x - 1 + 2h$

45. (a) $(-\infty, \infty)$
    (b) -3
    (c) Neither
    (d)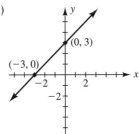
    (e) Increasing $(-\infty, \infty)$
    (f) One-to-one
    (g) $(-\infty, \infty)$

47. (a) $(-\infty, \infty)$
    (b) 0

(c) Odd
(d)
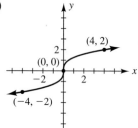

(e) Increasing $(-\infty, \infty)$ (f) One-to-one   (g) $(-\infty, \infty)$

49. (a) $(-\infty, \infty)$ (b) ± 3 (c) Even
(d)
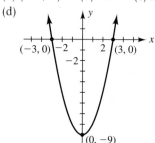

(e) Decreasing on $(-\infty, 0)$; increasing on $(0, \infty)$
(f) Not one-to-one   (g) $[-9, \infty)$

51. (a) $(-\infty, \infty)$ (b) -3 (c) Neither
(d)
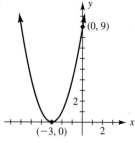

(e) Decreasing on $(-\infty, -3)$; increasing on $(-3, \infty)$
(f) Not one-to-one   (g) $[0, \infty)$

53. (a) $(-\infty, \infty)$ (b) -2 (c) Neither
(d)
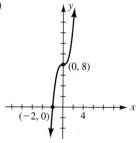

(e) Increasing $(-\infty, \infty)$ (f) One-to-one   (g) $(-\infty, \infty)$

55. (a) $(-\infty, \infty)$ (b) 2 and 4   (c) Neither
(d)
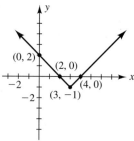

(e) Decreasing on $(-\infty, 3)$; increasing on $(3, \infty)$
(f) Not one-to-one   (g) $[-1, \infty)$

57. (a) $(-\infty, \infty)$ (b) 0 and 6   (c) Neither
(e) Increasing on $(-\infty, 3)$; decreasing on $(3, \infty)$
(f) Not one-to-one   (g) $(-\infty, 9]$

59. (a) $[-10, 10]$ (b) ± 10 (c) Even
(e) Increasing on $(-10, 0)$; decreasing on $(0, 10)$
(f) Not one-to-one   (g) $[0, 10]$

61. (a) $(-\infty, \infty)$ (b) 0 and $\pm \sqrt{2}$ (c) Even
(e) Increasing on $(-\infty, -1) \cup (0, 1)$; decreasing on $(-1, 0) \cup (1, \infty)$ (f) Not one-to-one
(g) $(-\infty, 1]$

63. (a) $(-\infty, 3]$ (b) None   (c) Neither
(d)

(e) Decreasing on $(-\infty, 0)$; constant on $(0, 3)$
(f) Not one-to-one   (g) $[1, \infty)$

65. 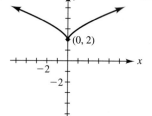   67.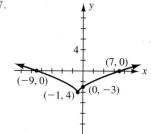

69.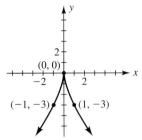

71. (a) $f^{-1}(x) = -\sqrt{x}$ (b)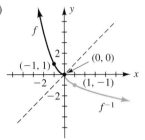

73. (a) $f^{-1}(x) = \sqrt[3]{3 - x}$ (b)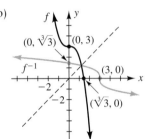

75. (a) $f^{-1}(x) = x^2 + 2, \; x \geq 0$
(b)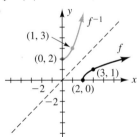

77. (a) $f^{-1}(x) = \begin{cases} \sqrt[3]{x} & \text{if } x < 1 \\ 2x - 1 & \text{if } x \geq 1 \end{cases}$
(b)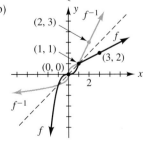

79. $(-\infty, 3]$ **81.** 3 **83.** 4

85.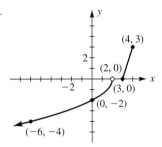

87. $-\frac{1}{2}$ **89.** $\frac{1}{9}$ **91.** $\left(-\infty, \frac{7}{4}\right]$

93. $8400 at 8%; $4100 at 9%

95. (a) $63.43 (b) 425.6 mi

97. $C = \begin{cases} 0.015n & \text{if } n \leq 1000 \\ 15 + 0.02(n - 1000) & \text{if } n > 1000 \end{cases}$

99. (a) $A = 3x - x^3$ (b) $(0, \sqrt{3})$ (c) $P(1, 2)$

101. (a) $T = \dfrac{V}{25}$

(b)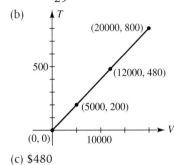

(c) $480

103. (a) $G = \dfrac{n(2000 - n)}{19{,}200}$

(b) Approximately 39 field mice per acre

Chapter 3

SECTION 3.1

Problems

1. (a) $m = \dfrac{2 - 4}{-1 - 3} = \dfrac{-2}{-4} = \dfrac{1}{2}$

(b) $m = \dfrac{-3 - 4}{2 - (-2)} = \dfrac{-7}{4} = -\dfrac{7}{4}$

(c) $m = \dfrac{3 - 3}{-2 - 3} = \dfrac{0}{-5} = 0$

Relabeling the points does not change the value of the slope.

2.
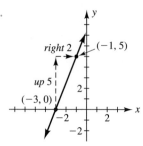

3. slope is 3; y-intercept is $-\frac{1}{2}$

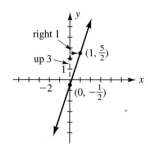

4. *x-intercept:* $3x + 8(0) = 12$
$3x = 12$
$x = 4$

   y-intercept: $3(0) + 8y = 12$
$8y = 12$
$y = \frac{3}{2}$

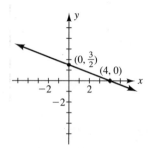

5. $0.33x - 4656.50 = 15000$
$0.33x = 19656.50$
$x = \$59{,}565.15$

◆ Exercises

1. 2   3. $-\frac{2}{3}$ 5. Undefined   7. -4 9. $\dfrac{3-b}{4}$

11. -1

13.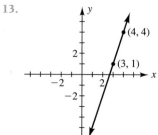

15.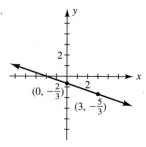

17.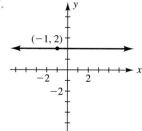

19. $m = 3$, $b = 0$

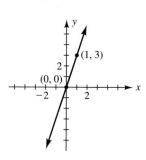

21. $m = -\frac{1}{3}$, $b = 0$ 23. $m = 2$, $b = 3$

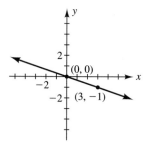
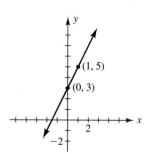

25. $m = -4$, $b = 1$ 27. $m = \frac{1}{2}$, $b = -1$

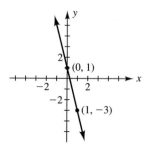
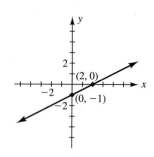

29. $m = -\frac{3}{4}$, $b = \frac{1}{4}$

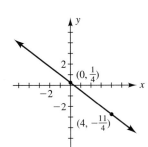

31.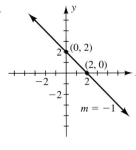

33.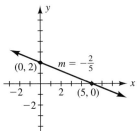

35.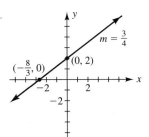

37.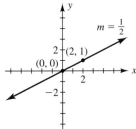

39.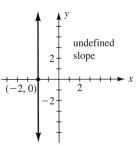

41.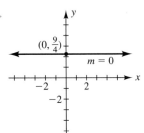

43.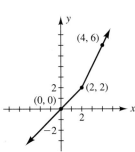

45.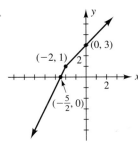

47.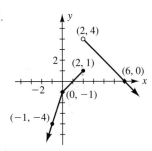

49. x-intercept 3.43; y-intercept 6.00

51. x-intercept -4.07; y-intercept 3.17

53. x-intercept 5.35; y-intercept -3.94

55. x-intercept -12.0; y-intercept 1.88

57. (a) $m = \frac{9}{5}$, $b = 32$
(b)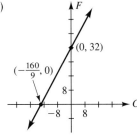

59. (a) $m = -3$, $b = 44$
(b)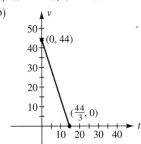

(c) $t = \frac{44}{3}$ s is the time when car comes to rest.

61. (a) $C = \begin{cases} 3 & \text{if } x \leq 30 \\ 0.06x + 1.2 & \text{if } 30 < x \leq 100 \\ 0.04x + 3.2 & \text{if } 100 < x \leq 300 \\ 0.01x + 12.2 & \text{if } x > 300 \end{cases}$

(b)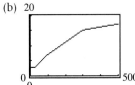

(c) $12.80 (d) 272 kilowatt-hours

63. $(-5, 5)$ and $(3, -1)$ **65.** (a) 1 (b) 9 (c) -5

67. $f^{-1}(x) = \dfrac{x - b}{m}$

69. $\dfrac{x}{C/A} - \dfrac{y}{C/B} = 1$; C/A is the x-intercept of the line and C/B is the y-intercept of the line.

SECTION 3.2

◆ Problems

1. $y - 0 = \frac{1}{2}[x - (-3)]$, or $y = \frac{1}{2}x + \frac{3}{2}$
2. $y - 1 = 1(x - 2)$
   $y - 1 = x - 2$
   $y = x - 1$
3. $y - 0 = \frac{2}{3}[x - (-3)]$
   $y = \frac{2}{3}x + 2$
4. Slope of line through (3, 0) and (2, −3):
   $$m = \frac{0 - (-3)}{3 - 2} = 3$$
   Equation of the perpendicular line through (3, −4):
   $$y - (-4) = -\frac{1}{3}(x - 3)$$
   $$y = -\frac{1}{3}x - 3$$
5. The V-intercept ($11,000) represents the value of the car when new.

◆ Exercises

1. $y = \frac{3}{4}x + 3$ 3. $y = 3x + 5$ 5. $y = -\frac{7}{3}x + \frac{28}{3}$
7. $y = -3x$ 9. $y = -3x + 11$ 11. $y = -\frac{7}{15}x + \frac{2}{5}$
13. $y = -\frac{23}{18}x + \frac{1}{6}$ 15. $y = \frac{2}{3}x - 2$
17. $y = -24x + 6$ 19. $y = \frac{3}{2}$ 21. $x - 3 = 0$
23. $y = \frac{1}{5}x - 2$ 25. $y = -x + 5$
27. $y = -\frac{1}{2}x - 1$ 29. $y = 3x + 8$
31. $y = \frac{2}{3}x - \frac{13}{3}$ 33. $y = \frac{1}{4}x + 3$ 35. $y = \frac{3}{4}x + \frac{3}{4}$
37. $y = 2x$
39. Both lines have slope 1.72 and, thus, are parallel.
41. (a) $r = \frac{5}{2}T - 100$ (b) 0 chirps/min
43. (a) $V = 60,000 - 9000x$ (b) $24,000
45. (a) $V = 15,000x + 50,000$ (b) $230,000
47. (a) $w = -0.040625T + 4.57185$
    (b) Approximately 3.15 mm
    (c) Approximately 112.5 °F
49. (a) $y = -\frac{1}{3}x - \frac{16}{3}$ or $x + 3y + 16 = 0$
    (b) $y = \frac{4}{3}x - \frac{11}{6}$ or $8x - 6y - 11 = 0$
51. $f(x) = \frac{4}{3}x + \frac{13}{3}$ 53. $g(x) = 3x - 7$
55. $\frac{12}{5}$ 57. (a) −2 (b) 0 (c) −6 (d) 4
59. $y = -6x + 4$ 61. $y = -\frac{A}{B}x + \left(\frac{A}{B}x_1 + y_1\right)$

SECTION 3.3

◆ Problems

1. Vertex: (3, −1)
   x-intercepts: $0 = (x - 3)^2 - 1$
   $1 = (x - 3)^2$
   $\pm 1 = x - 3$
   $x = 4$ or $x = 2$
   y-intercept: $f(0) = (0 - 3)^2 - 1 = 8$

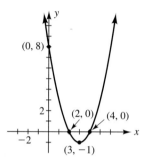

2. The basic form is $y = a(x - 4)^2 + 1$. To find a, we substitute $x = 2$ and $y = 3$:
   $$3 = a(2 - 4)^2 + 1$$
   $$3 = 4a + 1$$
   $$a = \frac{1}{2}$$
   Hence, the equation is $y = \frac{1}{2}(x - 4)^2 + 1$.

3. To find the x-intercepts, we solve the equation $f(x) = 0$:
   $$x^2 - 3x - 4 = 0$$
   $$(x - 4)(x + 1) = 0$$
   $x - 4 = 0$ or $x + 1 = 0$
   $x = 4$ $x = -1$
   Hence, the x-intercepts are 4 and −1.

4. Since $a = -3 < 0$, the parabola opens downward with a maximum value at
   $$x = -\frac{0}{2(-3)} = 0.$$
   Hence, the maximum value is $y = 3 - 3(0)^2 = 3$, and this occurs when $x = 0$.

5. If fencing all four sides, then $100 = 2l + 2w$, or $l = 50 - w$. Hence,
   $$A = lw = (50 - w)(w) = 50w - w^2.$$
   The graph of this equation is a parabola that opens downward with a maximum value at
   $$w = -\frac{50}{2(-1)} = 25 \text{ ft.}$$
   Hence,
   $$l = 50 - 25 = 25 \text{ ft.}$$

Thus, the largest area that can be enclosed is

$$(25 \text{ ft})(25 \text{ ft}) = 625 \text{ sq ft.}$$

6. We let $s = 4$ and solve for t:

$$4 = -16t^2 + 96t + 4$$
$$16t^2 - 96t = 0$$
$$16t(t - 6) = 0$$
$$t = 0 \quad \text{or} \quad t = 6$$

Hence, the ball is in flight 6 seconds.

Exercises

1.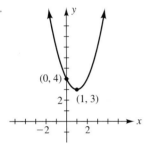

3.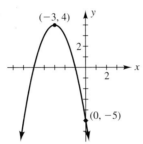

5.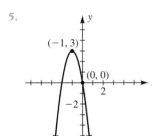

7.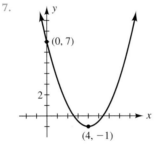

9.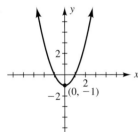

11.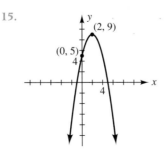

13.

15.

17.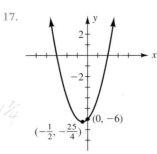

19.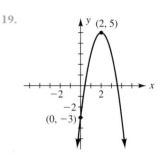

21.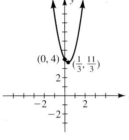

23.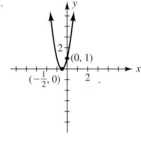

25. $y = \frac{1}{4}x^2$ 27. $y = -\frac{1}{2}x^2 + 2$

29. $y = -\frac{2}{9}x^2 + \frac{4}{3}x$ 31. $y = \frac{2}{3}x^2 + \frac{4}{3}x + 2$

33. Maximum value of 16   35. Minimum value of $\frac{3}{4}$

37. Maximum value of 9   39. (0.744, 3.78)

41. (2.63, 1.51)

43. (a) $-16t^2 + 64t + 80$ (b) 2 s
    (c) 144 ft   (d) 5 s

45. (a) 50 pairs   (b) $2300   47. 62 and 62

49. 11 in. by 11 in.   51. $33\frac{1}{3}$ ft by 100 ft

53. Approximately 343 watts

55. (a) $12   (b) $3600   57. $\dfrac{a + b}{2}$

59. $y = \frac{1}{2}(x - 1)^2 - 2$ 61. $c = 0$

63. No effect; the axis of symmetry of $y = x^2 + 4x + c$ is $x = -2$ for *all* values of c.

SECTION 3.4

Problems

1. $x^2 - x - 12 = 3(x + 2)(x - 2)$
   $x^2 - x - 12 = 3x^2 - 12$
   $2x^2 + x = 0$ (standard form)
   $x(2x + 1) = 0$
   $x = 0 \quad \text{or} \quad x = -\frac{1}{2}$

2. Check $1 + 2i$: $(1 + 2i)^2 + 5 \stackrel{?}{=} 2(1 + 2i)$
   $(1 + 4i + 4i^2) + 5 \stackrel{?}{=} 2 + 4i$
   $(-3 + 4i) + 5 \stackrel{?}{=} 2 + 4i$
   $2 + 4i = 2 + 4i$

   Check $1 - 2i$: $(1 - 2i)^2 + 5 \stackrel{?}{=} 2(1 - 2i)$
   $(1 - 4i + 4i^2) + 5 \stackrel{?}{=} 2 - 4i$
   $(-3 - 4i) + 5 \stackrel{?}{=} 2 - 4i$
   $2 - 4i = 2 - 4i$

3. $x^2 < 16x$
   $x^2 - 16x < 0$
   $x(x - 16) < 0$

   Boundary numbers: 0, 16
   Algebraic signs: ++++ ---- ++++

   Solution set: $(0, 16)$

4. Let $x =$ uniform width (in feet) of the additional parking area. Then
   $(220 + x)(100 + x) = 2(220)(100)$
   $x^2 + 320x - 22{,}000 = 0.$
   So, $x = \dfrac{-320 \pm \sqrt{(320)^2 - 4(1)(-22{,}000)}}{2(1)}$
   $x = \dfrac{-320 \pm \sqrt{190{,}400}}{2}$
   $x \approx 58.2$ ft

5.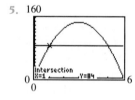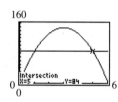

Exercises

1. $-4, 2$ 3. $0, 2$ 5. ± 6 7. $-1, \frac{5}{2}$
9. $-6, -3$ 11. $-4, 2$
13. $\dfrac{11 \pm 5\sqrt{5}}{2}$ 15. $\frac{3}{4}, -\frac{4}{5}$ 17. $\dfrac{2 \pm i}{5}$
19. $\dfrac{-5 \pm 3\sqrt{5}}{2}$ 21. $\dfrac{\sqrt{7} \pm 5i}{4}$ 23. $\pm \dfrac{\sqrt{5}}{2}$
25. $\pm \dfrac{\sqrt{15}}{2}$ 27. $\pm \frac{7}{2}i$ 29. $\dfrac{-3 \pm 4\sqrt{2}}{4}$
31. $5 \pm 3i\sqrt{2}$ 33. $-6 \pm 4\sqrt{3}$ 35. $6, -2$
37. $\dfrac{3 \pm 3\sqrt{5}}{2}$ 39. $\frac{7}{6}, -\frac{3}{5}$ 41. $(-3, 3)$

43. $(-\infty, 0] \cup [12, \infty)$ 45. $[-1, 9]$
47. $(-\infty, -\frac{7}{2}) \cup (\frac{1}{2}, \infty)$
49. $\left(\dfrac{11 - 5\sqrt{5}}{2}, \dfrac{11 + 5\sqrt{5}}{2} \right)$
51. $[-4, -\frac{2}{3}]$ 53. $x \approx -0.991, 3.34$
55. $x \approx -0.927, 1.12$ 57. $(-\infty, -1.82) \cup (1.82, \infty)$
59. $[-1.27, 1.70]$ 61. $18, 14$
63. Base is 9 in., height is 12 in.
65. Width is 5 ft, length is 7 ft
67. Approximately 3.57 ft 69. $700
71. Between 2 m and 11 m
73. $0, 2, 4$ or $2, 4, 6$ or $4, 6, 8$ 75. When $p > \$3$
77. $5\,s < t < 15\,s$ 79. $0, -\dfrac{b}{a}$ 81. $-a \pm b$
83. $a, -(a + 2b)$ 85. $k = 64$ 87. $k = \pm 2\sqrt{6}$
89. $k = 8$ 91. $(-4, -2) \cup (3, 5)$ 93. $[3, 8]$
95. $(-2, 4)$ 97. $k \leq -2\sqrt{6}$ or $k \geq 2\sqrt{6}$
99. $k \leq 0$ or $k \geq 4$

SECTION 3.5

Problems

1. $y = 3 - 5x$: $-\frac{31}{13} \stackrel{?}{=} 3 - 5\left(\frac{14}{13}\right)$
   $-\frac{31}{13} \stackrel{?}{=} \frac{39}{13} - \frac{70}{13}$
   $-\frac{31}{13} = -\frac{31}{13}$

   $y = \frac{3}{2}x - 4$: $-\frac{31}{13} \stackrel{?}{=} \frac{3}{2}\left(\frac{14}{13}\right) - 4$
   $-\frac{31}{13} \stackrel{?}{=} \frac{21}{13} - \frac{52}{13}$
   $-\frac{31}{13} = -\frac{31}{13}$

2. (a) $2x - y = 3$
   $2x - (4x + 5) = 3$
   $-2x = 8$
   $x = -4$

   Hence, $y = 4(-4) + 5 = -11$. The lines intersect at the point $(-4, -11)$.

   (b) $2x - y = 3$
   $\underline{-4x + y = 5}$
   $-2x\ \ \ \ \ = 8$
   $x = -4$

   Hence, $y = 4(-4) + 5 = -11$. The lines intersect at the point $(-4, -11)$.

3. Substituting $x^2 - 2x + 3$ for y, we obtain

$$4x - (x^2 - 2x + 3) = 6$$
$$x^2 - 6x + 9 = 0$$
$$(x - 3)^2 = 0$$
$$x = 3$$

Hence, $y = (3)^2 - 2(3) + 3 = 6$. The line is tangent to the parabola at the intersection point $(3, 6)$.

4. Substituting x^2 for y, we obtain

$$x^2 = 4x - x^2$$
$$2x^2 - 4x = 0$$
$$2x(x - 2) = 0$$
$$x = 0, 2$$

Hence, $y = 0^2 = 0$ and $y = 2^2 = 4$. The parabolas intersect at $(0, 0)$ and $(2, 4)$.

5.
$$x^6 + 4x^2 - 16 = 0$$
$$(x^6 - 8) + 4x^2 - 8 = 0$$
$$(x^2 - 2)(x^4 + 2x^2 + 4) + 4(x^2 - 2) = 0$$
$$(x^2 - 2)(x^4 + 2x^2 + 8) = 0$$

$x^2 - 2 = 0$ or $x^4 + 2x^2 + 8 = 0$

$x = \pm\sqrt{2}$ No real solution

Hence, the exact coordinates of the intersection points are $(-\sqrt{2}, -2\sqrt{2})$ and $(\sqrt{2}, 2\sqrt{2})$.

6.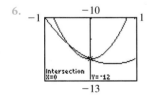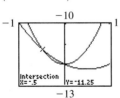

◆ **Exercises**

1. $\left(2, \frac{2}{5}\right)$ 3. $\left(-\frac{1}{2}, -1\right)$ 5. $\left(-\frac{1}{2}, \frac{11}{2}\right)$

7. $(-3, -23)$ 9. No intersection point   11. $\left(\frac{7}{3}, \frac{5}{6}\right)$

13. $(-3, -6), (1, 2)$ 15. $(2, 5), \left(-\frac{3}{2}, \frac{3}{2}\right)$

17. $\left(\frac{1}{2}, -\frac{7}{4}\right), (5, 5)$ 19. $(-2, -3), \left(\frac{1}{4}, \frac{21}{8}\right)$

21. $\left(\dfrac{-11 + \sqrt{41}}{4}, \dfrac{-11 + \sqrt{41}}{8}\right)$,

$\left(\dfrac{-11 - \sqrt{41}}{4}, \dfrac{-11 - \sqrt{41}}{8}\right)$

23. $\left(5 + \sqrt{13}, \dfrac{19 + 5\sqrt{13}}{2}\right)$,

$\left(5 - \sqrt{13}, \dfrac{19 - 5\sqrt{13}}{2}\right)$

25. $(0, 3), (3, 0)$ 27. $(4, 2\sqrt{5}), (4, -2\sqrt{5})$

29. $(3, 4), \left(-\frac{24}{5}, \frac{7}{5}\right)$ 31. $\left(-\frac{1}{2}, -2\right), (1, 1)$

33. $(4, 0), (8, 2)$ 35. No intersection point

37. $(1, 3\sqrt{3}), (1, -3\sqrt{3}), (-1, 3\sqrt{3}), (-1, -3\sqrt{3})$

39. $(0, 1), \left(-\frac{8}{5}, -\frac{3}{5}\right)$ 41. $(2, \sqrt{2})$

43. $(0, 2), (0, -2)$ 45. $\left(\frac{5}{4}, -\frac{3}{4}\right)$

47. $(9, 3), (9, -3)$ 49. $(1, 2), \left(\frac{2}{5}, -\frac{8}{5}\right)$

51. $(-1.879, -6.638), (0.347, 0.042), (1.532, 3.596)$

53. $(-1.195, 3.817), (2.930, 2.723)$

55. $(2.000, 1.000), (0.400, -0.600)$

57. $(-1.328, 6.220), (0.816, 0.886)$

59. -4 61. 2 63. $-1, \frac{5}{2}$ 65. $-3, -6$

67. $-4, 2$ 69. No real solution

71. Approximately $0.854, -5.854$

73. No real solution   75. $(5, -1), (2, 2), (-1, -7)$

77. $(1, 5), (9, 1), (-1, 1)$ 79. $4\sqrt{5}$

81. $P\left(-1, \frac{9}{2}\right), Q(4, 2), R\left(4, -\frac{7}{4}\right), S\left(-1, -\frac{1}{2}\right)$

83. $P(2, 1), Q(4, 5), R(4, -5)$

85. $P(-2, 5), Q(1, 2), R(-2, -3), S\left(\frac{7}{2}, -3\right)$

CHAPTER 3 REVIEW EXERCISES

1.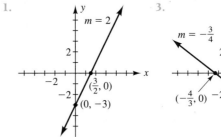

3.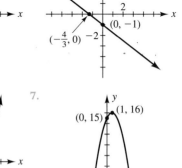

5.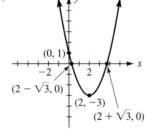

7.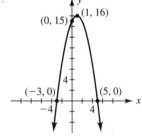

9.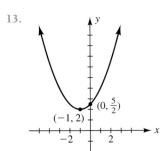

11.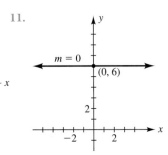

13.

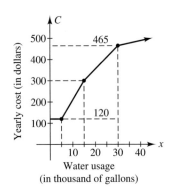

15. $y = \frac{3}{4}x - 2$ 17. $y = 3x + 10$
19. $y = -\frac{3}{2}x - 2$ 21. $y = x^2 - 4$
23. $y = \frac{3}{4}(x + 2)^2 - 3$ 25. $8, -2$ 27. $\pm 4i$
29. $2 \pm \sqrt{14}$ 31. $1 \pm 2i$ 33. $\sqrt{6}, -\frac{\sqrt{6}}{2}$
35. $[-3, \frac{2}{3}]$ 37. $(-\infty, -1 - \sqrt{5}) \cup (-1 + \sqrt{5}, \infty)$
39. $(-\infty, -\frac{2}{3}) \cup (2, \infty)$ 41. -2 43. $-1, \frac{1}{9}$
45. Approximately $0.764, 5.236$ 47. $(\frac{1}{2}, -3)$
49. $(4, 6), (1, 3)$ 51. $(\sqrt{11}, 5), (-\sqrt{11}, 5), (0, -6)$
53. $(1, \sqrt{2}), (1, -\sqrt{2}), (-1, \sqrt{2}), (-1, -\sqrt{2})$
55. $(6, \frac{3}{2}), (-\frac{10}{3}, -\frac{5}{6})$ 57. $(1.104, 1.344)$
59. (a) $x = \dfrac{C_1 B_2 - C_2 B_1}{A_1 B_2 - A_2 B_1}$ (b) $y = \dfrac{A_1 C_2 - A_2 C_1}{A_1 B_2 - A_2 B_1}$
61. (a) $y = \dfrac{B}{A}x$ (b) $\left(\dfrac{-AC}{A^2 + B^2}, \dfrac{-BC}{A^2 + B^2}\right)$
63. $(-2, 2), (\frac{4}{3}, \frac{16}{3}), (-\frac{1}{3}, \frac{7}{6}), (3, \frac{9}{2})$
65. Minimum value of $\frac{11}{3}$ 67. $a = -2$
69. (a) $V = -400x + 2400$ (b) 6 years
71. $C(x) = \begin{cases} 120 & \text{if } x \le 5000 \\ 0.018x + 30 & \text{if } 5000 < x \le 15{,}000 \\ 0.011x + 135 & \text{if } 15{,}000 < x \le 30{,}000 \\ 0.003x + 375 & \text{if } x > 30{,}000 \end{cases}$

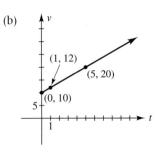

73. 200 ft along tar road by 125 ft along gravel road
75. 30 m   77. 1.5 in.   79. $\pm 10\sqrt{26}$

Cumulative Review: Chapters 1, 2, 3

1. (a) $(x + 1)(x - 4)(2x + 3)$
   (b) $(x - 2a)(x + 2a - 4y)$
3. 8
5. (a) $k = 0$ (b) $k = 9$ (c) $k < 9$ (d) $k > 9$
7. $2i$ 9. $A = \dfrac{x^2}{18}$
11. (a) $\dfrac{1}{b - c}$ (b) $\dfrac{x^2 - x + 1}{x^2}$
13. (a) $v = 2t + 10$ (b)

The v-intercept represents the initial velocity.

15. $f(0)$ 17. At least $7\frac{1}{2}$ quarts   19. 70 square units
21. (a) Symmetric with respect to the y-axis
    (b) Symmetric with respect to the origin
    (c) Symmetric with respect to the x-axis
    (d) Symmetric with respect to the x-axis, y-axis, and origin
23. (a) $y = 1$ (b) $y = x - 2$
    (c) $3x + 2y = 11$ (d) $y = -3x + 10$

25. (a) 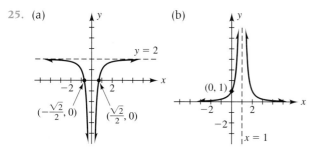 (b)

27. $m = \frac{5}{3}$

29. All squares from 9 in. by 9 in. to 11 in. by 11 in.

31. (a) $A = 6000 + 100x - 20x^2$ (b) in $2\frac{1}{2}$ weeks

Chapter 4

SECTION 4.1

Problems

1. *x-intercept*: $1 - (x + 2)^6 = 0$
$$(x + 2)^6 = 1$$
$$x + 2 = \pm 1$$
$$x = -2 \pm 1$$
$$x = -1, -3$$
y-intercept: $P(0) = 1 - (0 + 2)^6 = -63$

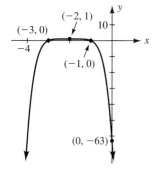

2.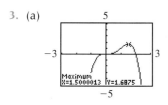

3. (a)

(b)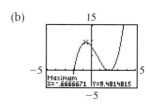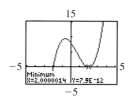

4. M is an odd function:
$$M(-x) = 100(-x) - (-x)^3$$
$$= -100x + x^3$$
$$= -M(x)$$

Exercises

1. 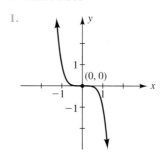 3.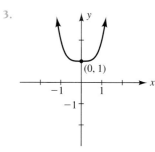

5. 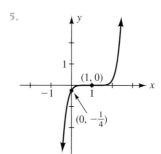 7.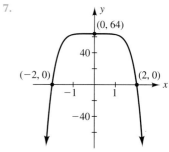

9. 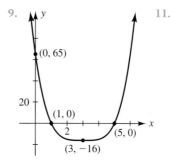 11.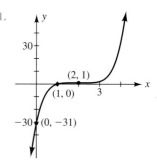

13. (a) Down to the left as $x \to -\infty$; up to the right as $x \to \infty$
    (b) Two or none

15. (a) Down to the left as $x \to -\infty$; down to the right as $x \to \infty$
    (b) Three or one

17. (a) Up to the left as $x \to -\infty$; down to the right as $x \to \infty$
    (b) Six, four, two, or none

19. (a) Up to the left as $x \to -\infty$; up to the right as $x \to \infty$
    (b) Five, three, or one

21. Maximum $(0.239, -1.66)$; minimum $(2.09, -8.05)$

23. Maxima $(-0.951, 11.0)$, $(0.824, 7.48)$; minimum $(0.128, 5.87)$

25. Maximum $(0.734, -0.605)$; minimum $(-0.734, -5.40)$

27. Minimum $(0.000, 1.000)$

29.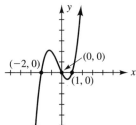

31.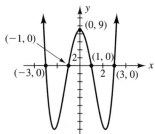

33.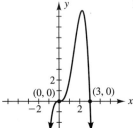

35.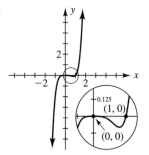

37.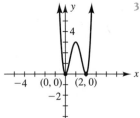

39.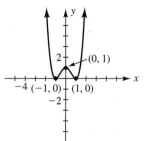

41.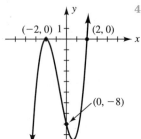

43.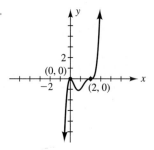

45.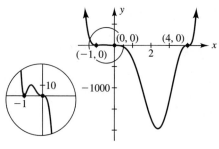

47. 13 years

49. (a) Domain $(0, 6)$
    (b)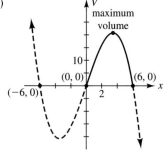
    (c) $x \approx 3.464$ ft

51. (a) $V = 180x^2 - 4x^3$; domain $(0, 45)$
    (b)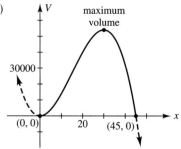
    (c) $x \approx 30$ inches

53. (a) $V = 64x - 28x^2 + 3x^3$
    (b)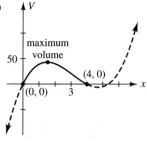
    (c) $x \approx 1.51$ cm

55. (a)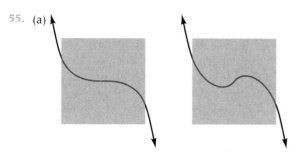

(b) $y = -x^3 - 2x$; $y = -x^3 + 2x$

57. (a)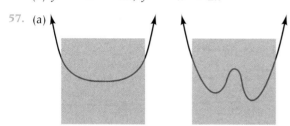

(b) $y = x^4 + 2x^2$; $y = x^4 - 2x^2$

59. (a)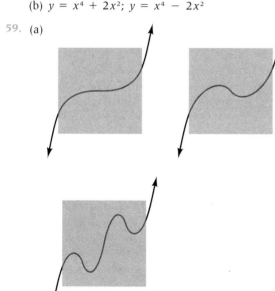

(b) $y = x^5 + 3x^3$; $y = x^5 - 3x^3$;
$y = x^5 - 3x^3 + 2x$

61. (a)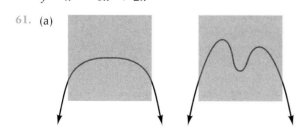

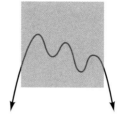

(b) $y = -x^6 - 2x^4$; $y = -x^6 + 2x^4$;
$y = -x^6 + 2x^4 - x^2$

63. (a) $(1, 0)$, $(\sqrt{3}, 3\sqrt{3} - 3)$, $(-\sqrt{3}, -3\sqrt{3} - 3)$
(b) $(3, -12)$, $(\sqrt{5}, \sqrt{5} - 15)$,
$(-\sqrt{5}, -\sqrt{5} - 15)$

65.

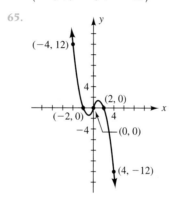

67. No; its graph is never symmetric with respect to the y-axis.

SECTION 4.2

Problems

1. $(x - 2)(2x^2 - x + 1) + (-2)$
$= (2x^3 - x^2 + x - 4x^2 + 2x - 2) + (-2)$
$= 2x^3 - 5x^2 + 3x - 4$

2.
$$3x^2 + 0x - 2 \overline{\smash{\big)}\begin{array}{r} 2x - 1 \\ 6x^3 - 3x^2 + 0x + 0 \end{array}}$$
$$\underline{6x^3 + 0x^2 - 4x}$$
$$-3x^2 + 4x$$
$$\underline{-3x^2 + 0x + 2}$$
$$4x - 2$$

Hence, $\dfrac{6x^3 - 3x^2}{3x^2 - 2} = 2x - 1 + \dfrac{4x - 2}{3x^2 - 2}$.

3.
$$\begin{array}{r|rrrr} 1) & 1 & -5 & 7 & -3 \\ & & 1 & -4 & 3 \\ \hline & 1 & -4 & 3 & 0 \end{array}$$

Hence, $\dfrac{x^3 - 5x^2 + 7x - 3}{x - 1} = x^2 - 4x + 3$.

4. $(x + 3)(2x^3 - 6x^2 + 3x) + 1$
   $= (2x^4 - 6x^3 + 3x^2 + 6x^3 - 18x^2 + 9x) + 1$
   $= 2x^4 - 15x^2 + 9x + 1$

5.
$$3\overline{)\begin{array}{ccccc} 1 & -3 & -4 & 13 & -8 \\ & 3 & 0 & -12 & 3 \\ \hline 1 & 0 & -4 & 1 & \boxed{-5} \end{array}} \text{ remainder}$$

6.
$$\tfrac{1}{2}\overline{)\begin{array}{ccccc} 4 & -6 & 0 & 4 & -3 \\ & 2 & -2 & -1 & \tfrac{3}{2} \\ \hline 4 & -4 & -2 & 3 & \boxed{-\tfrac{3}{2}} \end{array}} \text{ remainder}$$

Hence, $P(\tfrac{1}{2}) = -\tfrac{3}{2}$.

7.
$$1 - i\overline{)\begin{array}{cccc} 3 & -8 & 10 & -4 \\ & 3 - 3i & -8 + 2i & 4 \\ \hline 3 & -5 - 3i & 2 + 2i & 0 \end{array}}$$

Since the remainder is 0, we conclude that $1 - i$ is a zero of P.

◆ Exercises

1. $3x^2 + x + 2 + \dfrac{-1}{x - 1}$

3. $3x^2 - 4x - 1 + \dfrac{-4x + 6}{x^2 - 2x + 3}$

5. $2x^2 + x + 2 + \dfrac{-1}{2x - 1}$

7. $x^2 + 1 + \dfrac{-3x^2 + 4}{2x^3 + 1}$

9. $x^2 - x + 1 + \dfrac{-1}{x + 1}$

11. $-3x^3 - x - 1$ 13. $-3 + \dfrac{6}{2x + 1}$

15. $2x + 2 + \dfrac{6x - 2}{(2x - 1)^2}$

17. $2x - 1 + \dfrac{3}{x - 1}$

19. $x^2 + 3x - 8 + \dfrac{20}{x + 2}$

21. $2x^3 + x^2 + x + 4$

23. $4x^3 + x^2 - 6x + 4 + \dfrac{-24}{x + 6}$

25. $-x^4 - 2x^3 - 4x^2 - 8x - 16 + \dfrac{-29}{x - 2}$

27. $3x^2 - 6x + 15$

29. $2x^3 - 4x^2 + 10x - 44 + 118/(x + \tfrac{5}{2})$

31. $14.5x^2 + 2.5x - 45.5$

33. $275x^3 + 495x^2 + 765x + 750$

35. $P(1) = 8$ 37. $F(-2) = 7$ 39. $P(4) = 25$

41. $Q(\tfrac{1}{2}) = 0$ 43. $f(-\tfrac{1}{3}) = 0$

45. $h(1 - i) = 7 - 4i$ 47. $P(\sqrt{2}) = 12$

49. $P(1.7) = -5.0582$ 51. No   53. Yes   55. Yes

57. Yes   59. No   61. Yes   63. Yes   65. Yes

67. 455 m   69. $P(x) = x^4 + 7x^3 + 7x^2 + 11x + 18$

71. $2x^2 - 3x + 2$ 73. $8x^3 + x^2 + 2x - 2$

75. $-x^3 - 2x + 3$ 77. $k = 16$ 79. $k = 2$

SECTION 4.3

◆ Problems

1.
$$2i\overline{)\begin{array}{cccc} 2 & -3 & 8 & -12 \\ & 0 + 4i & -8 - 6i & 12 \\ \hline 2 & -3 + 4i & -6i & 0 \end{array}}$$

Since $P(2i) = 0$, we conclude that $x - 2i$ is a factor of $P(x) = 2x^3 - 3x^2 + 8x - 12$.

2. $P(x) = a(x - 0)^3(x + 3)$
   $8 = a(1 - 0)^3(1 + 3)$
   $8 = 4a$
   $a = 2$

Hence, the polynomial function P with these characteristics is

$$P(x) = 2(x - 0)^3(x + 3) = 2x^4 + 6x^3$$

3. The possible rational zeros of $P(x) = 3x^3 + 22x^2 + 25x + 6$ are

$$\dfrac{\pm 1}{\pm 1}, \dfrac{\pm 2}{\pm 1}, \dfrac{\pm 3}{\pm 1}, \dfrac{\pm 6}{\pm 1}, \text{ and } \dfrac{\pm 1}{\pm 3}, \dfrac{\pm 2}{\pm 3}, \dfrac{\pm 3}{\pm 3}, \dfrac{\pm 6}{\pm 3}$$

These simplify to $\pm 1, \pm 2, \pm 3, \pm 6, \pm \tfrac{1}{3}, \pm \tfrac{2}{3}$. Testing -1 using synthetic division, we find

$$-1\overline{)\begin{array}{cccc} 3 & 22 & 25 & 6 \\ & -3 & -19 & -6 \\ \hline 3 & 19 & 6 & 0 \end{array}}$$

Hence, $P(x) = (x + 1)(3x^2 + 19x + 6)$
   $= (x + 1)(x + 6)(3x + 1)$.
By the factor theorem, the zeros of P are -1, -6, and $-\tfrac{1}{3}$.

4. $(x - 3)^2[x - (1 + 2i)][x - (1 - 2i)]$
   $= (x^2 - 6x + 9)[x^2 - (1 + 2i)x - (1 - 2i)x$
   $\quad + (1 + 2i)(1 - 2i)]$
   $= (x^2 - 6x + 9)(x^2 - 2x + 5)$
   $= x^4 - 8x^3 + 26x^2 - 48x + 45$

5.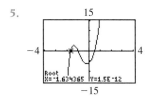

   Hence, the irrational zero in the interval $(-2, -1)$ is approximately -1.634.

6.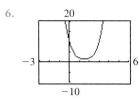

   Since the graph does not intersect the x-axis, we conclude that this function has no real zero.

◆ Exercises

1. No   3. No   5. Yes   7. Yes   9. No   11. Yes

13. $P(x) = 2x^3 - 4x^2 - 2x + 4$

15. $P(x) = -4x^4 + 16x^3 - 12x^2 - 16x + 16$

17. $P(x) = x^4 - 1$ 19. $P(x) = -5x^5 + 10x^3$

21. (a) $P(x) = (x - 1)(x - 1)(x + 3)$
    (b) 1 of multiplicity two, -3

23. (a) $P(x) = (2x + 1)(3x - 4)(x + 1)$
    (b) $-\frac{1}{2}, \frac{4}{3}, -1$

25. (a) $P(x) = (x - 2)^2(x + i)(x - i)$
    (b) 2 of multiplicity two, $\pm i$

27. (a) $P(x) = (2x - 3)(x - 3)(x - 2)^2$
    (b) $\frac{3}{2}$, 3, 2 of multiplicity two

29. (a) $P(x) = (x + 2i)(x - 2i)(x - 4)(x + 1)$
    (b) $\pm 2i, 4, -1$

31. (a) $P(x) = [x - (3 - i)][x - (3 + i)]$
    $\quad \times [x - (-1 + \sqrt{6})][x - (-1 - \sqrt{6})]$
    (b) $3 \pm i, -1 \pm \sqrt{6}$

33. (a) 1, 2, 3   (b) $P(x) = (x - 1)(x - 2)(x - 3)$

35. (a) $4, -2, \frac{1}{2}$ (b) $f(x) = (x - 4)(x + 2)(2x - 1)$

37. (a) $-\frac{1}{4}$ (b) $H(x) = (4x + 1)(x^2 + 2x + 6)$

39. (a) $\frac{1}{2}, \frac{2}{3}, -\frac{3}{2}$ (b) $P(x) = (2x - 1)(3x - 2)(2x + 3)$

41. (a) $-2, \frac{3}{2}$ (b) $f(x) = (x + 2)(2x - 3)(x^2 + 3)$

43. (a) $\frac{8}{3}$ (b) $g(x) = (3x - 8)(x^3 - 2x^2 - 1)$

45. (a) -1 of multiplicity two, $-2, -\frac{1}{2}$
    (b) $P(x) = (x + 1)^2(x + 2)(2x + 1)$

47. (a) ± 2
    (b) $f(x) = (x - 2)(x + 2)(x^3 + x^2 - 2x + 4)$

49. (a) 1 of multiplicity two, -2 of multiplicity two, $-\frac{1}{2}$
    (b) $F(x) = (x - 1)^2(x + 2)^2(2x + 1)$

51. For $f(x) = x^n - a^n$, $f(-a) = 0$ when n is even, but $f(-a) = -2a^n$ when n is odd. Hence, $x + a$ is a factor of $x^n - a^n$ only when n is even.

53. $k = -7$ 55. $k = 4, -2$ 57. $\frac{1}{3}$ 59. $\frac{2}{3}, -2$

61. 2   63. $f(x) = -2x^3 + 4x^2 + 10x - 12$

65. $f(x) = 6x^4 + 6x^3 - 18x^2 - 30x - 12$

67. Approximately 5 h 55 min after high tide.

SECTION 4.4

◆ Problems

1. $(x^2 + 4)(x^2 - 3x - 4) = 0$
   $x^2 + 4 = 0$ or $x^2 - 3x - 4 = 0$
   $x^2 = -4$ $(x + 1)(x - 4) = 0$
   $x = \pm 2i$ $x = -1, 4$

   Hence, the four solutions are $\pm 2i, -1$, and 4.

2. Writing the equation in standard form, we obtain $x^3 - x^2 - 10x - 8 = 0$. The possible rational roots are $\pm 1, \pm 2, \pm 4$, and ± 8. The graph in the figure indicates that $-2, -1$, and 4 may be rational roots.

   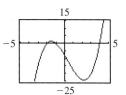

   Test -1: $\quad -1 \overline{\smash{)}\,1 \quad -1 \quad -10 \quad -8}$
   $\qquad\qquad\qquad\quad\; -1 \quad\;\; 2 \quad\;\;\; 8$
   $\qquad\qquad\quad \overline{\;\;1 \quad -2 \quad\; -8 \quad\;\; \circled{0}} \;\text{remainder}$

   Since the remainder is zero, -1 is a root. Thus, $x + 1$ is a factor:

$$x^3 - x^2 - 10x - 8 = 0$$
$$(x + 1)(x^2 - 2x - 8) = 0$$
$$(x + 1)(x + 2)(x - 4) = 0$$
$$x = -1, -2, 4$$

Hence, the three solutions are -1, -2, and 4.

3. Writing the equation in standard form, we obtain $x^4 + 2x^3 + x^2 - 2x - 8 = 0$. The possible rational roots are ± 1, ± 2, ± 4, and ± 8. The graph in the figure indicates that -2 may be a rational root:

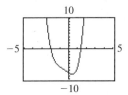

Test -2:
$$\begin{array}{r|rrrrr}
-2) & 1 & 2 & 1 & -2 & -8 \\
 & & -2 & 0 & -2 & 8 \\
\hline
 & 1 & 0 & 1 & -4 & \boxed{0} \text{ remainder}
\end{array}$$

The root in the interval $(1, 2)$ must be irrational. Tracing to this x-intercept, we find that this root is approximately 1.379. Hence, the two real roots of this equation are -2 and 1.379 (nearest thousandth).

4.
$$x^3 + 6 \geq 7x$$
$$x^3 - 7x + 6 \geq 0$$
$$(x - 1)(x^2 + x - 6) \geq 0$$
$$(x - 1)(x - 2)(x + 3) \geq 0$$

Boundary numbers: $1, 2, -3$
Algebraic signs: $---+++++++++++--+++$

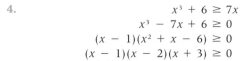

Solution set: $[-3, 1] \cup [2, \infty)$

5. To find the radius x when the volume is 4000 cubic feet, we solve the equation

$$4000 = \pi x^2 (60 - x) + \frac{2}{3}\pi x^3$$
$$\pi x^3 - 180\pi x^2 + 12000 = 0$$

The graph in the figure indicates that $y = \pi x^3 - 180\pi x^2 + 12000$ intersects the x-axis only once in the interval $(0, 60)$. Hence, the equation $\pi x^3 - 180\pi x^2 + 12000 = 0$ has only one real root. As shown in this figure, the radius $x \approx 4.6675$ ft.

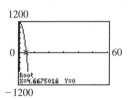

6. Using the result of the synthetic division in Example 6, we factor completely over the set of integers:

$$2x^3 - 23x^2 + 60x - 45$$
$$(x - \tfrac{3}{2})(2x^2 - 20x + 30)$$
$$(2x - 3)(x^2 - 10x + 15)$$

The boundary number in the interval $(1, 2)$ is the positive root of the equation $x^2 - 10x + 15 = 0$. To find the exact value of this boundary number, we apply the quadratic formula:

$$x = \frac{-(-10) \pm \sqrt{(-10)^2 - 4(1)(15)}}{2(1)}$$
$$= \frac{10 \pm \sqrt{40}}{2} = \frac{10 \pm 2\sqrt{10}}{2}$$
$$= 5 \pm \sqrt{10}$$

Hence, the exact value of the boundary number in the interval $(1, 2)$ is $5 - \sqrt{10}$.

◆ Exercises

1. $0, \pm 1$ 3. 0 of multiplicity two, $\dfrac{3 \pm \sqrt{17}}{2}$

5. 0 of multiplicity three, $\dfrac{3 \pm \sqrt{17}}{4}$ 7. $\tfrac{3}{2}, \pm\sqrt{2}\,i$

9. $0, -2, -5$ of multiplicity two   11. $1, -\tfrac{8}{3}, \tfrac{3}{2}$

13. $\tfrac{1}{3}$ of multiplicity two, $\pm i$ 15. $0, -3, \dfrac{3 \pm 3i\sqrt{3}}{2}$

17. $\pm 2, -1 \pm 2i$ 19. $-\tfrac{1}{2}, \tfrac{2}{3}, 3, 4$ 21. 1.38

23. -1.55 25. $-1.88, 0.78, 4.11$

27. $2.09, -1.15, -0.28$ 29. $0.69, 1.78$ 31. $-1, 1.13$

33. $-\tfrac{3}{2}, -0.20$ 35. $-1, 2, 1.22$

37. $\pm 2, -3.26, -0.28, 0.54$ 39. $1, -1.25$

41. $(-\infty, 2 - 2\sqrt{3}] \cup [0, 2 + 2\sqrt{3}]$

43. $(-\infty, -3) \cup (-\sqrt{2}, 0) \cup (\sqrt{2}, \infty)$

45. $(-\sqrt{5}, \sqrt{5})$ 47. $[-2, \tfrac{1}{2}] \cup [\tfrac{2}{3}, \infty)$

49. $[-1, \tfrac{5}{2}]$ 51. $(-\infty, 1) \cup (1, 3)$ 53. $[\tfrac{1}{3}, \tfrac{1}{2}]$

55. $(-\infty, -\sqrt{3}] \cup [1, \sqrt{3}] \cup [5, \infty)$

57. Approximately 5.1 years   59. 8 units or more

61. Approximately 4.44 in. by 4.44 in. by 2.44 in.

63. 4 ft by 3 ft by $1\tfrac{1}{2}$ ft   65. 5 inches

67. Positive and less than 0.177 cm, or greater than $1\frac{1}{2}$ cm but less than 2 cm

69. $1.3788,\ -0.6894 \pm 1.5575i$

71. (a) $x^3 - 12x + 16 = 0$ (b) 3 of multiplicity two, -3

73. $(2, 8),\ \left(\frac{1}{2}, \frac{1}{8}\right),\ \left(-\frac{5}{2}, -\frac{125}{8}\right)$

75. One negative real root and two imaginary roots

77. One positive real root, one negative real root, and two imaginary roots

SECTION 4.5

Problems

1. (a) The zero of the denominator $D(x) = x - 1$ is 1. Hence, the domain of F is $(-\infty, 1) \cup (1, \infty)$.
   (b) The zeros of the denominator $D(x) = x^2 - 2x - 8 = (x + 2)(x - 4)$ are -2 and 4. Thus, the domain of f is $(-\infty, -2) \cup (-2, 4) \cup (4, \infty)$.

2. For $F(x) = 3$, we have
$$\frac{3x - 2}{x - 1} = 3$$
$$3x - 2 = 3(x - 1)$$
$$3x - 2 = 3x - 3$$
$$-2 = -3,$$
which is a contradiction.

3. We solve the equation $g(x) = 2x$:
$$\frac{2x^3}{x^2 + 3} = 2x$$
$$2x^3 = 2x(x^2 + 3)$$
$$2x^3 = 2x^3 + 6x$$
$$6x = 0$$
$$x = 0$$
Hence, the rational function g intersects its oblique asymptote at $(0, 0)$.

4. From the graph in the figure, we conclude that the coordinates of the relative maximum point are approximately $(-0.354, -0.369)$.

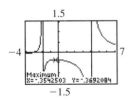

5. The graph is shown in the figure. We cannot evaluate the function at $x = 1$. This suggests that a hole occurs in the graph at $(1, 3)$.

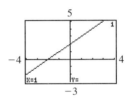

6. To determine the number of kits that are produced if the average cost per kit is $16,640, we solve the equation
$$16{,}640 = \frac{141{,}120 + 12{,}000x + 20x^2}{x}$$
$$16{,}640x = 141{,}120 + 12{,}000x + 20x^2$$
$$20x^2 - 4640x + 141{,}120 = 0$$
$$x^2 - 232x + 7056 = 0$$
So, $x = \dfrac{-(-232) \pm \sqrt{(-232)^2 - 4(1)(7056)}}{2(1)}$
$= 36$ or 196 kits.

Exercises

1. (a) $x = -2$
   (b) As $x \to -2^-$, $f(x) \to \infty$; as $x \to -2^+$, $f(x) \to -\infty$.

3. (a) None

5. (a) $x = \frac{1}{2},\ x = -3$
   (b) As $x \to \frac{1}{2}^-$, $f(x) \to \infty$; as $x \to \frac{1}{2}^+$, $f(x) \to -\infty$. As $x \to -3^-$, $f(x) \to -\infty$; as $x \to -3^+$, $f(x) \to \infty$.

7. (a) $x = 2$
   (b) As $x \to 2^-$, $f(x) \to -\infty$; as $x \to 2^+$, $f(x) \to \infty$.

9. (a) $x = 2,\ x = 4,\ x = -1$
   (b) As $x \to 2^-$, $f(x) \to -\infty$; as $x \to 2^+$, $f(x) \to \infty$. As $x \to 4^-$, $f(x) \to \infty$; as $x \to 4^+$, $f(x) \to -\infty$. As $x \to -1^-$, $f(x) \to -\infty$; as $x \to -1^+$, $f(x) \to \infty$.

11. (a) $y = 1$
    (b) As $x \to \infty$, $f(x) \to 1^+$; as $x \to -\infty$, $f(x) \to 1^-$.
    (c) Does not cross.

13. (a) $y = \frac{1}{2}$
    (b) As $x \to \infty$, $f(x) \to \frac{1}{2}^-$; as $x \to -\infty$, $f(x) \to \frac{1}{2}^+$.
    (c) Crosses at $\left(\frac{13}{7}, \frac{1}{2}\right)$

15. None exist

17. (a) $y = -3$
    (b) As $x \to \infty$, $f(x) \to -3^-$; as $x \to -\infty$, $f(x) \to -3^+$.
    (c) Crosses at $\left(-\frac{1}{10}, -3\right)$

19. (a) $y = 0$
(b) As $x \to \infty$, $f(x) \to 0^+$; as $x \to -\infty$, $f(x) \to 0^+$.
(c) Crosses at $\left(\frac{1}{3}, 0\right)$

21.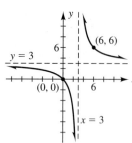

23.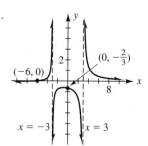

25.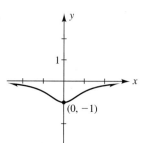

27.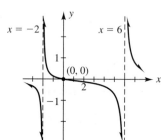

29.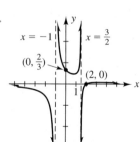

31.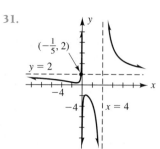

33.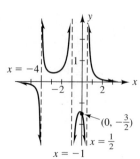

35.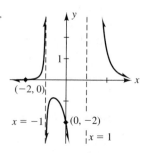

37. (a) $y = x$ (b) Does not cross
(c)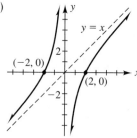

39. (a) $y = x + 3$ (b) Does not cross
(c)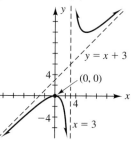

41. (a) $y = x - 6$ (b) Does not cross
(c)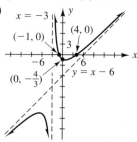

43. (a) $y = x + 3$ (b) Crosses at $\left(-\frac{20}{13}, \frac{19}{13}\right)$
(c)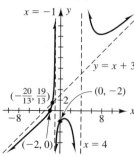

45.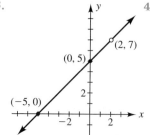

47.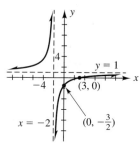

49.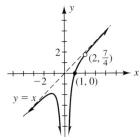

51. (b) $x = 18$ in. by $y = 9$ in.

53. (a) $C(x) = \dfrac{81{,}000 + (400x + 0.1x^2)}{x}$

    (b) Approximately 900 stoves

55. No; the procedure of dividing numerator and denominator by the highest power of x that appears in the rational function yields at most one horizontal asymptote.

57. One example is $f(x) = \dfrac{x^2 - 6x + 8}{x^3}$; its graph crosses the horizontal asymptote $y = 0$ at $(2, 0)$ and at $(4, 0)$.

59. No. If $P(1) = 0$, then the graph of f has no vertical asymptote but, instead, has a hole at $x = 1$.

61. (a)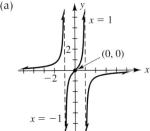

    (b)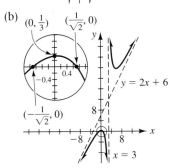

63. (a) $C(x) \to \infty$, a prohibitive cost

SECTION 4.6

Problems

1. The LCD is $2(m + 2)$:

   $$f(m) = \dfrac{3}{2} - \dfrac{1}{m + 2} - \dfrac{m}{2m + 4}$$
   $$= \dfrac{3(m + 2) - 1(2) - m}{2(m + 2)} = \dfrac{2m + 4}{2m + 4}$$

   Reducing to lowest terms, gives us a new function F:

   $$F(m) = \dfrac{2m + 4}{2m + 4} = 1 \quad \text{for } m \ne -2.$$

2.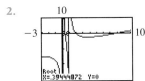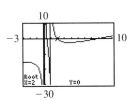

   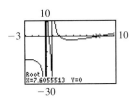

3. $$\dfrac{x^2 + 2x - 4}{x + 2} \ge 1$$
    $$\dfrac{x^2 + 2x - 4}{x + 2} - 1 \ge 0$$
    $$\dfrac{(x^2 + 2x - 4) - (x + 2)}{x + 2} \ge 0$$
    $$\dfrac{(x + 3)(x - 2)}{x + 2} \ge 0$$

    Boundary numbers: $-3, -2, 2$
    Algebraic signs: $- - + + - - - - + + + +$

    Solution set: $[-3, -2) \cup [2, \infty)$

4. Let x = crew's speed (in mi/h) in still water.
   Then $x + 3$ = crew's speed downstream
   and $x - 3$ = crew's speed upstream

   Since it takes the same time to row 4 mi downstream as it does to row 3 mi upstream, we write the following equation:

   $$\dfrac{4}{x + 3} = \dfrac{3}{x - 3}$$
   $$4(x - 3) = 3(x + 3)$$
   $$x = 21 \text{ mi/h}$$

   Hence, the team will qualify for the NCAA trials.

5. Let x = amount of time (in hours) to flood the bog using both spillways.

Then $\dfrac{1}{48} + \dfrac{1}{36} = \dfrac{1}{x}$
$3x + 4x = 144$
$x = 20\frac{4}{7}$ hours

◆ Exercises

1. 8   3. $-\frac{1}{5}$ 5. 1   7. No solution   9. 5, -4

11. 2, $-\frac{3}{7}$ 13. 4, 2   15. 2   17. 4   19. $\pm\dfrac{\sqrt{5}}{2}$

21. -3 23. $1 \pm \sqrt{3}$ 25. 2, $\pm\dfrac{\sqrt{3}}{3}$

27. 0, -1, $\dfrac{3 \pm \sqrt{321}}{6}$ 29. $\left(0, \frac{1}{2}\right)$ 31. $(-3, 1)$

33. $\left[-\frac{7}{5}, -\frac{1}{2}\right)$ 35. $(-\infty, -4] \cup [1, 2)$

37. $\left(-3 - \sqrt{7}, -5\right) \cup \left(-3 - \sqrt{7}, 5\right)$

39. $(-8, -2) \cup (0, 2)$ 41. $(-3, 2)$ 43. $(-3, 4]$

45. $(-2, 2)$ 47. $\left[-\frac{5}{2}, -2\right) \cup \left(-1, -\frac{1}{3}\right] \cup (2, 3]$

49. 4 and 12   51. 6 mi/h

53. Approximately 10.4 knots, 12.6 knots   55. $37\frac{1}{2}$ hours

57. Large pump, 40 min; small pump, 60 min

59. 12 students   61. $x = \dfrac{2mn}{n - m}$ 63. $x = -\dfrac{k}{5}$

65. 0, $\dfrac{a + b}{b}$ 67. $a, -\dfrac{1}{a}$ 69. $(-\infty, -3] \cup [-1, \infty)$

71. $\left(-\infty, -\frac{3}{2}\right] \cup (1, \infty)$

73. We cannot multiply both sides by $x - 1$ because we don't know whether $x - 1$ is positive or negative. The correct solution set is $(-1, 0)$.

75. $0\,\Omega < R_2 \le 36.2\,\Omega$

CHAPTER 4 REVIEW EXERCISES

1. $3x + 2 + \dfrac{-6x + 2}{x^2 + 1}$

3. $2x^3 - 2x^2 + 4x - 5 + \dfrac{9}{x + 2}$

5. $6x^3 + 9x + 3 + \dfrac{-3}{x - \frac{1}{3}}$

7. $x^3 - x - 1 + \dfrac{x^2 + 2x + 2}{x^3 + x + 1}$

9. 6   11. 0   13. Yes   15. No   17. Yes

19. (a) $P(x) = (x - 2)(x - 4)(x + 1)$ (b) 2, 4, -1

21. (a) $P(x) = (3x - 4)\left[x - \left(1 + \sqrt{6}\right)\right]$
$\times \left[x - \left(1 - \sqrt{6}\right)\right]$
(b) $\frac{4}{3}, 1 + \sqrt{6}, 1 - \sqrt{6}$

23. (a) $P(x) = (2x - 1)(x + 3)(x + 2i)(x - 2i)$
(b) $\frac{1}{2}, -3, \pm 2i$

25. (a) $P(x) = (x - 1)(x + 1)(x - 2)(2x - 1)(2x + 3)$
(b) $\pm 1, 2, \frac{1}{2}, -\frac{3}{2}$

27. (a) $P(x) = (x + i)(x - i)[x - (2 + 2i)]$
$\times [x - (2 - 2i)]$
(b) $\pm i, 2 \pm 2i$

29. (a) $P(x) = [x - (2 + i)][x - (2 - i)]$
$\times (2x - 1)(x + 2i)(x - 2i)$
(b) $2 \pm i, \frac{1}{2}, \pm 2i$

31. 1.67   33. $\frac{5}{2}, -1.15$ 35. $-5.98, -0.92$

37. 2 of multiplicity two, 4.17   39. $\frac{1}{6}$ 41. 7

43. $3 \pm \sqrt{5}$ 45. $\left[-1 - \sqrt{5}, 0\right] \cup \left[-1 + \sqrt{5}, \infty\right)$

47. $(2, \infty)$ 49. $(-\infty, -2) \cup [4, \infty)$

51. $(-\infty, -4) \cup \left(-\frac{2}{3}, 0\right) \cup (2, \infty)$

53.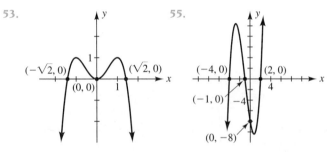

55.

57.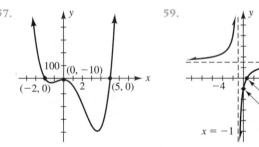

59.

61.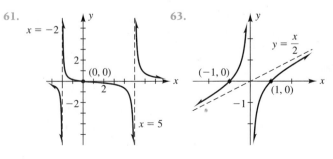

63.

65.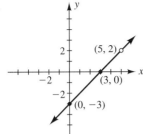

67. (a) $P(x) = \frac{1}{2}(x^3 - 2x^2 - 5x + 6)$
    (b) $P(x) = 2(x^3 - 2x^2 - 5x + 6)$

69.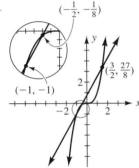

71. $y = \frac{2}{3}$; crosses at $\left(\frac{5}{3}, \frac{2}{3}\right)$ 73. $2\frac{1}{2}$ ft   75. 4 cm

77. (a) $V = 4x^3 - 70x^2 + 300x$
    (b) 5 cm or approximately 1.10 cm
    (c) About 2.83 cm

79. (b) 5 m by 5 m by 4 m   81. 24 mi/h

Chapter 5

SECTION 5.1

Problems

1. The domain and range of H may be determined from its graph (see Figure 5.4): domain $(-\infty, \infty)$; range $(-\infty, 3)$.

2.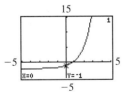

3. $\dfrac{125^{x-2}}{25^{2x-3}} = \dfrac{(5^3)^{x-2}}{(5^2)^{2x-3}} = \dfrac{5^{3x-6}}{5^{4x-6}} = 5^{-x} = \left(\dfrac{1}{5}\right)^x$

4. As illustrated in the figure, in about 8.75 years the amount in the account will be $2000.

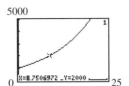

5. Continuous compounding yields $5.87 more than quarterly compounding over the five-year period.

Exercises

1. (a) $(0, 1)$ (b) Largest $y = 2^x$; smallest $y = 3^x$
   (c) Largest $y = 3^x$; smallest $y = 2^x$

3.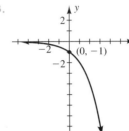

5.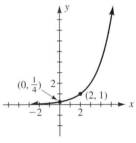

7.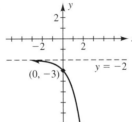

9.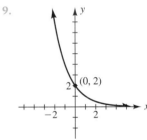

11.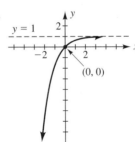

13.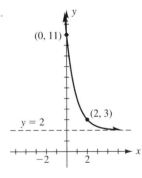

15.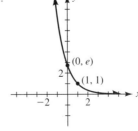

17. 1 **19.** 36^{-x} **21.** 1 **23.** e^x **25.** 9^x **27.** $\frac{1}{3}$ **29.** 5

31. (a) $A = 5000(1.09)^t$ (b) $7057.91

33. (a) $A = 10{,}000(1.02125)^{4t}$ (b) $27,437.17

35. (a) $A = 2000e^{0.06t}$ (b) $3432.01

37. (a) $A = 120{,}000e^{0.08t}$ (b) $124,897.29

39. (a) $k = 15$ (b) 6.74 psi

41. (a) 18 bacteria (b) 133 bacteria

43. (a) $12,000 (b) $4,466
(c) The machine's scrap value is $2000.

45. (a) $4^x - 4^{-x}$ (b) 4 **47.** 2

49. 2 **51.** ± 1 **53.** $a = \frac{1}{4}$

55.

x	0	1	2	3	4	-1	-2	-3	-4
$f(x)$	1	-2	4	-8	16	$-\frac{1}{2}$	$\frac{1}{4}$	$-\frac{1}{8}$	$\frac{1}{16}$

The function f is not defined when $x = 1/n$, where n is an even integer. Thus, we cannot connect these points to form the graph.

57. (ii) 8.0% compounded monthly

59. (b)

h	1	0.1	0.01	0.001	0.0001
$\dfrac{e^h - 1}{h}$	1.7183	1.0517	1.0050	1.0005	1.0001

(c) $\dfrac{e^h - 1}{h} \to 1$ as $h \to 0^+$.

SECTION 5.2

Problems

1. (a) $\log 1 = \log_{10} 1 = 0$ (b) $\ln e = \log_e e = 1$

2.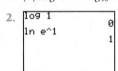

3. The graph of $g(x) = \log_{1/4} x = \dfrac{\ln x}{\ln \frac{1}{4}}$:

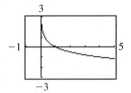

4. The graph of $F(x) = \log_4(x + 2) = \dfrac{\ln(x + 2)}{\ln 4}$:

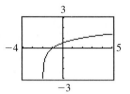

The graph of $G(x) = 1 + \log_4(-x) = 1 + \dfrac{\ln(-x)}{\ln 4}$:

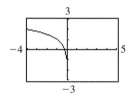

5. $\log_3 9^{2x-3} = \log_3 (3^2)^{2x-3} = \log_3 3^{4x-6} = 4x - 6$
As shown in the figure, the graph of $y = \log_3 9^{2x-3}$ is identical to the graph of $y = 4x - 6$:

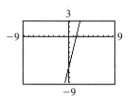

6. $R(5) = 100 - 29 \log[27(5) + 1]$
$= 100 - 29 \log 136 \approx 38.1\%$

Exercises

1. 2 **3.** 3 **5.** 1 **7.** -4 **9.** -2

11. 2 **13.** 0 **15.** 2

17. Undefined **19.** $\frac{3}{2}$ **21.** $-\frac{4}{3}$ **23.** 3

25. 5 **27.** 10 **29.** 0 **31.** 1.447 **33.** 2.079

35. 3.585 **37.** 2.757 **39.** -0.3691 **41.** -8.697

43. (a) (1, 0) (b) Largest $y = \log x$; smallest $y = \log_2 x$
(c) Largest $y = \log_2 x$; smallest $y = \log x$

45.   **47.**

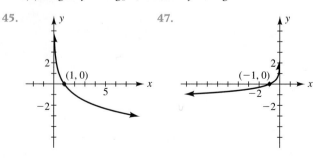

49.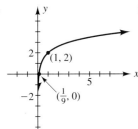

51.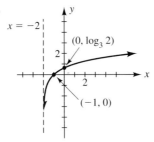

53.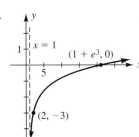

55.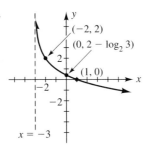

57.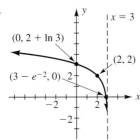

59. $x - 3$ **61.** $x + 2$ **63.** $x^2 + 4$ **65.** $x^2 + 2x$

67. 1   **69.** $3x + 6$ **71.** xe^x **73.** x^2

75. pH ≈ 6.4

77. (a) 100%   (b) 69%
(c) About 4 days
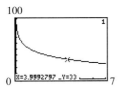

79. (a) 1000   (b) 5615
(c) About $19,000
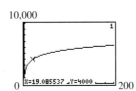

81. (a) 3   (b) 4

83. (a) $(0, e) \cup (e, \infty)$ (b) $(0, 10^{10}]$

85. (a) $f^{-1}(x) = \log_6 x$

(b)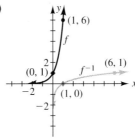

87. (a) $f^{-1}(x) = 8^x$

(b)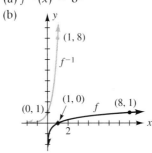

89. (a) It appears that $\ln(xy) = \ln x + \ln y$.
(b) It appears that $\ln \dfrac{x}{y} = \ln x - \ln y$.
(c) It appears that $\ln x^y = y \ln x$.

91. 2.01

93. (a) $(0, \infty)$ (b) $x = 1$ (c) $f(x) \to 0$
(d) $f(x) \to -\infty$
(e) $x = e$

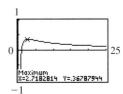

SECTION 5.3

◆ **Problems**

1. For example, $(\log_2 8)^2 \neq 2 \log_2 8$. Note that $(\log_2 8)^2 = 3^2 = 9$, whereas $2 \log_2 8 = 2 \cdot 3 = 6$.

2. The graph of $H(x) = \log_4 \dfrac{1}{x - 1} = \dfrac{\ln[1/(x - 1)]}{\ln 4}$:

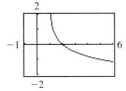

3.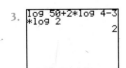

4. (a) [calculator display: ln 2/ln 12+ln 6/ln 12 = 1] (b) [calculator display: ln 16/ln 2 · 3·ln 27/ln 2 = 12]

5. Let I_a = the intensity of the 1933 earthquake
   and I_b = the intensity of the 1989 earthquake.

   Then
   $$8.9 = \log I_a - \log I_0$$
   $$7.1 = \log I_b - \log I_0$$
   $$\overline{1.8 = \log I_a - \log I_b}$$

   Hence, $\dfrac{I_a}{I_b} = 10^{1.8} \approx 63$ times as intense.

◆ Exercises

1. $\log_2(8+8) = \log_2 16 = 4$ and
   $\log_2 8 + \log_2 8 = 3 + 3 = 6$

3. $\dfrac{\log_2 8}{\log_2 2} = \dfrac{3}{1} = 3$ and $\log_2 \dfrac{8}{2} = \log_2 4 = 2$

5. 10   7. $3 + \log_2 x + \log_2(x+2)$

9. $2 + \ln x - \ln 10$ 11. $\tfrac{2}{3} \log_3 x - 1$

13. $2 \log_b x - \tfrac{2}{3} \log_b(x+1)$

15. $-(2 + \tfrac{3}{2} \log_3 x + \tfrac{1}{3} \log_3 y)$

17. $2x^2 - 2 \ln(e^x + 1)$

19. $\ln 24$ 21. 2   23. -2

25. $\log_5 \dfrac{x^2 + 2x - 3}{x^2}$

27. $\ln \dfrac{x-2}{x+2}$ 29. $\ln \dfrac{9\sqrt{x^2-9}}{x^2}$

31.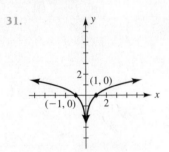

33.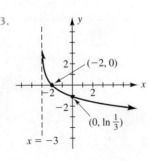

35.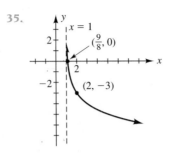

37.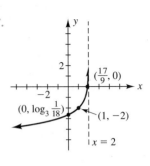

39. 1   41. 4   43. 2   45. 1

47. About 32 times more intense

49. About 1.32×10^{10} times brighter

51. 10^9 times more intense

53. No; the domain of f is $(-\infty, 0) \cup (0, \infty)$, whereas the domain of g is $(0, \infty)$.

55. (a) $\tfrac{1}{2}$ (b) $\tfrac{1}{3}$ (c) 1   (d) $\tfrac{1}{6}$
    (e) $\tfrac{1}{5}$ (f) $-\tfrac{1}{5}$ (g) -1 (h) 1

57. $\log(6.2 \times 10^k) = \log 6.2 + \log 10^k = \log 6.2 + k$

59. (a) 1   (b) $\ln e = 1$

SECTION 5.4

◆ Problems

1. $$5^{x-1} = 325$$
   $$(x-1) \ln 5 = \ln 325$$
   $$x - 1 = \dfrac{\ln 325}{\ln 5}$$
   $$x = 1 + \dfrac{\ln 325}{\ln 5} \approx 4.594$$

2. $$e^{x-1} = 3^x$$
   $$(x-1) \ln e = x \ln 3$$
   $$x - 1 = x \ln 3$$
   $$x - x \ln 3 = 1$$
   $$x(1 - \ln 3) = 1$$
   $$x = \dfrac{1}{1 - \ln 3} \approx -10.14$$

3. $$2 \ln x - \ln 9 = 4$$
   $$\ln \dfrac{x^2}{9} = 4$$
   $$e^4 = \dfrac{x^2}{9}$$
   $$x^2 = 9e^4$$
   $$x = 3e^2$$

4. $x + \ln y = \ln c$
    $\ln y - \ln c = -x$
$$\ln \frac{y}{c} = -x$$
$$e^{-x} = \frac{y}{c}$$
$$y = ce^{-x}$$

5. $\log \sqrt{x} = \sqrt{\log x}$
    $\tfrac{1}{2} \log x = \sqrt{\log x}$
    $\tfrac{1}{4} (\log x)^2 = \log x$
    $(\log x)^2 - 4 \log x = 0$
    $\log x (\log x - 4) = 0$
    $\log x = 0$ or $\log x - 4 = 0$
        $x = 1$ $x = 10^4$

6.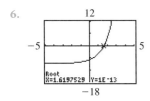

7. $2 = e^{r(5.5)}$
    $r(5.5) = \ln 2$
$$r = \frac{\ln 2}{5.5} \approx 0.126 \quad \text{or} \quad 12.6\%$$

8. $540 = 18e^{0.1354t}$
    $30 = e^{0.1354t}$
    $0.1354t = \ln 30$
$$t = \frac{\ln 30}{0.1354} \approx 25 \text{ minutes}$$

9. $27 = 30e^{-0.02476t}$
    $0.9 = e^{-0.02476t}$
    $-0.02476t = \ln 0.9$
$$t = \frac{\ln 0.9}{-0.02476} \approx 4.26 \text{ years}$$

◆ Exercises

1. $\tfrac{3}{2}$ 3. $-\tfrac{1}{3}$ 5. $\dfrac{\ln 35}{\ln 7} \approx 1.827$

7. $\dfrac{\ln \tfrac{2}{3}}{3} \approx -0.135$

9. $\dfrac{3 - \log 28}{2} \approx 0.7764$ 11. $\dfrac{\ln 120}{\ln 9} \approx 2.179$

13. $-\dfrac{\ln \tfrac{15}{4}}{7} \approx -0.1888$ 15. $-\log 60 \approx -1.778$

17. $\dfrac{1}{2 + \log 4} \approx 0.384$ 19. $\dfrac{\ln 4}{\ln 4 - 1} \approx 3.589$

21. $\dfrac{\ln \tfrac{3}{4}}{\ln 3} \approx -0.2619$ 23. $\dfrac{\ln 3072}{\ln 16} \approx 2.896$ 25. $\tfrac{1}{100}$

27. 8   29. $2e$ 31. $\tfrac{1}{2}, -3$ 33. 2   35. $3e - 1$

37. 4   39. 2   41. 4   43. 1   45. $y = ce^{-2x}$

47. $y = \dfrac{10}{(x + a)^2}$ 49. $y = \dfrac{kx - x}{1 + k}$ for $k = e^c$

51. $y = \dfrac{ka}{1 + kb}$ for $k = e^{ac}$ 53. $1, e^2$ 55. $\dfrac{\ln 4}{2}$

57. 0   59. 2.123   61. 1.202   63. $-1.637, 1.000$

65. (a) 5.8 yr   (b) 25.6 yr   (c) 65.8 yr

67. 11%

69. (a) $A(t) = 24e^{0.1792t}$ (b) 1996

71. Approximately 2140 B.C.

73. (a) $P = P_0 2^{-(1/3)t}$ (b) 25%

75. $t = \dfrac{\ln(A/P)}{n \ln[1 + (r/n)]}$ 77. (b) A line

79. (a) $h^{-1}(x) = 2 \ln x$
    (b)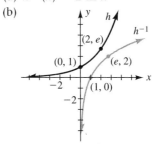

81. (a) $G^{-1}(x) = \tfrac{1}{2} \log(1 - x)$
    (b)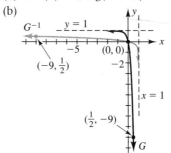

83. (a) $h^{-1}(x) = 2e^{-x}$
    (b)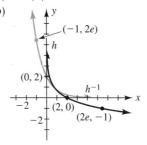

85. (a) $G^{-1}(x) = \frac{1}{3}(10^{(x-1)/2})$
    (b)
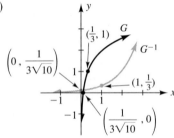

87. $\ln(10 \pm 3\sqrt{11}) \approx \pm 2.993$ 89. 3, 9

91. (a) 50 ft   (b) $50 \ln\left(\frac{3 \pm \sqrt{5}}{2}\right) \approx \pm 48.1$ ft

93. (a) $i(t) \to 20$ amp   (b) Approximately 79 ms

95. (a) $T = 70 + 330e^{-0.1805t}$ (b) 262°F
    (c) About 7.85 minutes

CHAPTER 5 REVIEW EXERCISES

1. 1   3. e 5. 2^{-x} 7. 2   9. $\frac{2}{3}$ 11. -3

13. $\frac{1}{3}$ 15. $-\frac{2}{3}$ 17. Approximately 3.465

19. Approximately -6.644 21. 4   23. 3   25. $2x$

27. $x^2 + 1$ 29. $3 - 3x$ 31. xe^{2x} 33. 1

35. $2 + 2\log_3 x + \log_3(x - 1)$

37. $-\log x - \frac{1}{2}\log(2x - 3)$

39. $-x^2 + \ln x - \ln(e^x - 1)$

41. $\ln 8$ 43. $\log_3(x + 1)^2$ 45. 0

47. 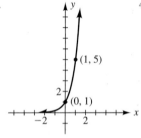   49.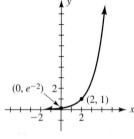

51. 53.
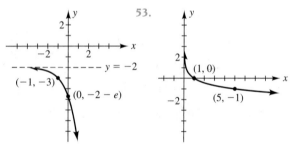

55. 57.
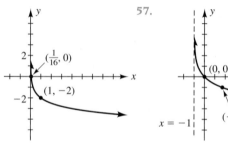

59. $\frac{\log 80}{2} \approx 0.9515$ 61. $-\frac{3}{4}$ 63. $\frac{\ln 4}{\ln 36} \approx 0.3869$

65. $\frac{\ln 6}{1 + \ln 9} \approx 0.5604$ 67. 5   69. $-2 \pm e^3$

71. No solution   73. 3   75. $y = kx^2$ for $k = e^c$

77. ± 1 79. 3, $\sqrt{3}$ 81. $f^{-1}(x) = \log_8 x$

83. $h^{-1}(x) = \frac{2 + \ln(x/2)}{3}$ 85. $G^{-1}(x) = 2^{-(x+1)}$

87. (a) $12,058.57   (b) $12,256.79   (c) $12,298.02

89. Approximately 7 years

91. (a) $A(t) = 10^3 e^{1.535t}$ for t in hours
    (b) Approximately 27 minutes

93. About 794 times more intense

95. (a) $k \approx 0.0405$ (b) Approximately 99°F
    (c) About 35 minutes

Chapter 6

SECTION 6.1

Problems

1. The graph of the equation $(x - 2)^2 + (y + 1)^2 = 5$ (or $y = -1 \pm \sqrt{5 - (x - 2)^2}$) is shown in the figure using a square set viewing rectangle. The circle passes through the point (4, 0) and the coordinates of its center appear to be (2, −1).

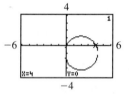

2. x-intercepts: $(x + 2)^2 + (0 - 1)^2 = 5$
$(x + 2)^2 = 4$
$x + 2 = \pm 2$
$x = -2 \pm 2$
$x = 0, -4$

y-intercepts:
$$(0 + 2)^2 + (y - 1)^2 = 5$$
$$(y - 1)^2 = 1$$
$$y - 1 = \pm 1$$
$$y = 1 \pm 1$$
$$y = 0, 2$$

3. Completing the square, we have $(x + 2)^2 + (y + 3)^2 = 0$. Since $r^2 = 0$, the graph of this equation is the single point $(-2, -3)$, so it is not the graph of a circle.

4. The distance from the center $C(0, -56.57)$ to point $A(20, 20.89)$ is 80 ft:
$$CA = \sqrt{(20 - 0)^2 + [20.89 - (-56.57)]^2}$$
$$= \sqrt{6400.0516} \approx 80 \text{ ft}$$

◆ Exercises

1. Radius 4; center $(0, 2)$ 3. Radius 3; center $(0, 0)$
5. Radius 2; center $(0, 1)$ 7. Radius 4; center $(3, -4)$
9. Radius $\frac{4}{3}$; center $\left(-\frac{4}{3}, 0\right)$
11. Radius 2; center $\left(1, -\frac{3}{2}\right)$
13. The equation defines a degenerate circle, and its graph is the single point $(-5, 1)$.
15. The equation does not define a circle and has no graph.
17. *x*-intercepts $\pm 2\sqrt{3}$; *y*-intercepts $-2, 6$
19. *x*-intercept 3; *y*-intercepts $-4 \pm \sqrt{7}$
21. $x^2 + y^2 = 4$ 23. $x^2 + y^2 - 4x + 6y + 11 = 0$
25. $x^2 + y^2 = 25$ 27. $x^2 + y^2 - 8y + 11 = 0$
29. $x^2 + y^2 + 4x - 10y + 25 = 0$
31. $x^2 + y^2 - 8x - 4y - 5 = 0$
33. Center $(-5.36, 0.00)$; radius 7.49
35. Center $(3.61, 2.42)$; radius 2.56
37. (a) $PC(-88.39, 0)$; $PT(88.39, 0)$; $x^2 + (y + 88.39)^2 = 125^2$
    (b) $A(50, 26.17)$
39. (a) $PC(-27.00, 0)$; $PT(27.00, 0)$; $x^2 + (y + 46.77)^2 = 54^2$
    (b) $A(20, 3.39)$
41. (a) $(x - 8.2)^2 + y^2 = 14.3^2$ (b) 12.4 ft
43. (a) $x^2 + y^2 - 10x + 6y + 25 = 0$ (b) 27π
45. $(4, 2)$ and $(-3, -5)$
47. (a) $4x + 3y = 25$ (b) $x - 4y = -6$
49. $h = 0$ or 8   51. $m = \pm \frac{4}{3}$

SECTION 6.2

◆ Problems

1. The graph of the equation $x = \frac{1}{2}(y - 3)^2 + 2$ (or $y = 3 \pm \sqrt{2x - 4}$) is shown in the figure. The parabola passes through the point $(4, 1)$ and the coordinates of its vertex appear to be $(2, 3)$.

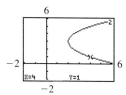

2. *x*-intercepts of $2x^2 + 3y = 6$:
$$2x^2 + 3(0) = 6$$
$$2x^2 = 6$$
$$x^2 = 3$$
$$x = \pm\sqrt{3}$$

x-intercept of $y^2 - 2x - 4y + 10 = 0$:
$$(0)^2 - 2x - 4(0) + 10 = 0$$
$$-2x + 10 = 0$$
$$-2x = -10$$
$$x = 5$$

3. Solving the equation $3y^2 - 8x = 0$ for x, we obtain $x = \frac{3}{8}y^2$. Hence, $1/(4p) = \frac{3}{8}$, which implies that $p = \frac{2}{3}$. Thus, the coordinates of the focus are $\left(\frac{2}{3}, 0\right)$ and the directrix is the vertical line $x = -\frac{2}{3}$.

4. We can change to general form as follows:
$$x = -\frac{1}{14}(y - 1)^2 - \frac{3}{2}$$
$$-14x = (y - 1)^2 + 21$$
$$-14x = (y^2 - 2y + 1) + 21$$
$$y^2 + 14x - 2y + 22 = 0$$

5. Replacing p with $\frac{9}{16}$, we obtain
$$x = \frac{1}{4p}y^2 = \frac{1}{4\left(\frac{9}{16}\right)}y^2 = \frac{4}{9}y^2$$

Hence, the equation of the parabola is $x = \frac{4}{9}y^2$.

◆ Exercises

1. Vertex $(0, 0)$; opens to the right
3. Vertex $(0, -8)$; opens upward
5. Vertex $(4, 1)$; opens to the left
7. Vertex $(2, 4)$; opens downward
9. Vertex $\left(-\frac{25}{4}, \frac{3}{2}\right)$; opens to the right
11. Vertex $(-1, 2)$; opens to the right
13. Vertex $(-4, 0)$; opens upward

15. Vertex $(2, -4)$; opens to the left

17. Focus $\left(0, -\frac{63}{8}\right)$; directrix $y = -\frac{65}{8}$

19. Focus $\left(-6, \frac{3}{2}\right)$; directrix $x = -\frac{13}{2}$

21. Focus $\left(\frac{13}{8}, -4\right)$; directrix $x = \frac{19}{8}$

23. $x = \frac{1}{8}y^2$ 25. $y = 2(x + 1)^2 - 1$

27. $x = -\frac{1}{4}(y - 1)^2 + 1$

29. $y = -\frac{1}{20}(x + 3)^2 + 2$

31. $x = \frac{1}{16}(y + 1)^2 + 3$

33. $(0.744, 7.34)$ 35. $(-1.78, 0.979)$

37. 30 meters   39. $d = 9$ inches   41. $x = 0.152y^2$

43. $\dfrac{a + b}{2}$ 45. $c = 0$

47. Narrower for $0 < p < \frac{1}{4}$; wider for $p > \frac{1}{4}$

49. $\left(0, -\dfrac{E}{4A}\right)$

SECTION 6.3

◆ **Problems**

1. The graph of the equation $\dfrac{(x + 1)^2}{16} + \dfrac{(y - 3)^2}{4} = 1$
   (or $y = 3 \pm \frac{1}{2}\sqrt{16 - (x + 1)^2}$) is shown in the figure. The ellipse appears to have minor axis of length 4 and vertices at $(3, 3)$ and $(-5, 3)$.

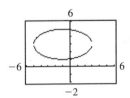

2. Completing the square, we have
   $3(x - 1)^2 + 2(y + 2)^2 = -4$. Since $-4 < 0$, this equation does not define an ellipse. No point (x, y) with real coordinates satisfies this equation.

3. Writing the equation in standard form, we obtain
   $$\dfrac{x^2}{1^2} + \dfrac{y^2}{3^2} = 1,$$
   which we recognize as an ellipse with $a = 1$, $b = 3$, and center at the origin. To locate the foci, we apply $c^2 = 3^2 - 1^2$, which implies that $c = \pm 2\sqrt{2}$. Hence, the coordinates of the foci are $F_1(0, 2\sqrt{2})$ and $F_2(0, -2\sqrt{2})$. The eccentricity is $e = 2\sqrt{2}/3 \approx 0.94$.

4. To change to general form, we proceed as follows:
   $$\dfrac{(x - 2)^2}{12} + \dfrac{(y - 1)^2}{16} = 1$$
   $$4(x - 2)^2 + 3(y - 1)^2 = 48$$
   $$4(x^2 - 4x + 4) + 3(y^2 - 2y + 1) = 48$$
   $$4x^2 + 3y^2 - 16x - 6y - 29 = 0$$

◆ **Exercises**

1. Center $(0, 0)$; vertices $(-4, 0)$, $(4, 0)$
3. Center $(0, 0)$; vertices $(0, 5)$, $(0, -5)$
5. Center $(0, 0)$; vertices $(-3, 0)$, $(3, 0)$
7. Center $(0, 0)$; vertices $\left(-\sqrt{3}, 0\right)$, $\left(\sqrt{3}, 0\right)$
9. Center $(2, -1)$; vertices $(-2, -1)$, $(6, -1)$
11. Center $(0, -2)$; vertices $(0, 1)$, $(0, -5)$
13. Center $(1, -3)$; vertices $(1, -1)$, $(1, -5)$
15. Center $(-5, 2)$; vertices $(-5, 3)$, $(-5, 1)$
17. Foci $\left(0, \sqrt{21}\right)$, $\left(0, -\sqrt{21}\right)$; eccentricity $\sqrt{21}/5 \approx 0.92$
19. Foci $\left(2 + 2\sqrt{3}, -1\right)$, $\left(2 - 2\sqrt{3}, -1\right)$; eccentricity $\sqrt{3}/2 \approx 0.87$
21. Foci $\left(1, \dfrac{-15 + 2\sqrt{21}}{5}\right)$, $\left(1, \dfrac{-15 - 2\sqrt{21}}{5}\right)$; eccentricity $\dfrac{\sqrt{21}}{5} \approx 0.92$
23. $\dfrac{4x^2}{25} + y^2 = 1$ 25. $\dfrac{4(x - 2)^2}{9} + \dfrac{(y + 2)^2}{16} = 1$
27. $\dfrac{(x + 2)^2}{8} + \dfrac{y^2}{9} = 1$ 29. $\dfrac{x^2}{9} + \dfrac{y^2}{5} = 1$
31. $\dfrac{(x - \frac{3}{2})^2}{9} + \dfrac{4y^2}{27} = 1$ 33. $(0, 2.11)$, $(0, -2.11)$
35. $(0, 13.2)$, $(0, -4.38)$ 37. $8\sqrt{2} \approx 11.3$ ft
39. $w \approx 8.1$ m
41. (a) 40 ft by 20 ft
    (b) Along the major axis, $10\sqrt{3} \approx 17.3$ ft on either side of the center
43. Tack down the ends of the string with the thumbtacks and insert the pencil point into the loop of the string. Now, keeping the string taut, move the pencil to trace out an ellipse.

45. $x^2 + y^2 = a^2$

49. Domain $[-3, 3]$; range $[0, 2]$

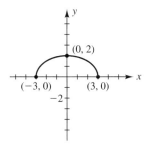

SECTION 6.4

Problems

1. The graph of the equation $\dfrac{x^2}{9} - \dfrac{4(y-1)^2}{9} = 1$

   (or $y = 1 \pm \tfrac{1}{2}\sqrt{x^2 - 9}$) and its asymptotes $y = \pm\tfrac{1}{2}x + 1$ are shown in the figure. The hyperbola appears to have transverse axis of length 6 and vertices $(-3, 1)$ and $(3, 1)$.

2. Completing the square, we obtain $(x - 3)^2 - 4(y - 1)^2 = 0$. Since the right-hand side of this equation is zero, we conclude this equation does not define a hyperbola. The graph of this equation is the pair of intersecting lines $y = \tfrac{1}{2}x - \tfrac{1}{2}$ and $y = -\tfrac{1}{2}x + \tfrac{5}{2}$.

3. Writing the equation in standard form, we obtain
$$\frac{x^2}{1^2} - \frac{y^2}{3^2} = 1,$$
which we recognize as a hyperbola with $a = 1$, $b = 3$, and center at the origin. To locate the foci, we apply $c^2 = 3^2 + 1^2$, which implies that $c = \pm\sqrt{10}$. Hence, the coordinates of the foci are $F_1(\sqrt{10}, 0)$ and $F_2(-\sqrt{10}, 0)$. The eccentricity is $e = \sqrt{10}/1 \approx 3.2$.

4. To change to general form, we proceed as follows:
$$\frac{(y-1)^2}{4} - \frac{(x-2)^2}{12} = 1$$
$$3(y-1)^2 - (x-2)^2 = 12$$
$$3(y^2 - 2y + 1) - (x^2 - 4x + 4) = 12$$
$$3y^2 - x^2 + 4x - 6y - 13 = 0$$

Exercises

1. Center $(0, 0)$; vertices $(-2, 0)$, $(2, 0)$

3. Center $(0, 0)$; vertices $(0, 2)$, $(0, -2)$

5. Center $(0, 0)$; vertices $\left(-\tfrac{3}{4}, 0\right)$, $\left(\tfrac{3}{4}, 0\right)$

7. Center $(0, 0)$; vertices $\left(0, \tfrac{2}{5}\right)$, $\left(0, -\tfrac{2}{5}\right)$

9. Center $(0, -2)$; vertices $(-3, -2)$, $(3, -2)$

11. Center $(1, 3)$; vertices $(1, 9)$, $(1, -3)$

13. Center $(2, -3)$; vertices $\left(\tfrac{3}{2}, -3\right)$, $\left(\tfrac{5}{2}, -3\right)$

15. Center $(-4, -1)$; vertices $(-4, 2)$, $(-4, -4)$

17. Foci $\left(0, \sqrt{13}\right)$, $\left(0, -\sqrt{13}\right)$; eccentricity $\sqrt{13}/2 \approx 1.8$

19. Foci $\left(\sqrt{13}, -2\right)$, $\left(-\sqrt{13}, -2\right)$; eccentricity $\sqrt{13}/3 \approx 1.2$

21. Foci $\left(\dfrac{4+\sqrt{37}}{2}, -3\right)$, $\left(\dfrac{4-\sqrt{37}}{2}, -3\right)$; eccentricity $\sqrt{37} \approx 6.1$

23. $\dfrac{y^2}{16} - \dfrac{x^2}{256} = 1$ 25. $\dfrac{x^2}{4} - \dfrac{y^2}{4} = 1$

27. $\dfrac{y^2}{9} - \dfrac{(x+2)^2}{16} = 1$ 29. $\dfrac{x^2}{1} - \dfrac{y^2}{3} = 1$

31. $4\left(x - \tfrac{3}{2}\right)^2 - \dfrac{y^2}{2} = 1$ 33. $(1.61, 0)$, $(-1.61, 0)$

35. $(0, -15.2)$, $(0, -4.34)$ 37. 50 cm

39. (a) $PA - PB = 1430$ ft; P must lie on a branch of a hyperbola because the difference in distances from two fixed points A and B is constant (1430 ft).

    (b) $\dfrac{x^2}{511{,}225} - \dfrac{y^2}{1{,}738{,}755} = 1$

    (c) $PA = 2340$ ft, $PB = 910$ ft

41. (a) The opening of each branch of the hyperbola is very narrow.
    (b) The opening of each branch of the hyperbola is very wide.

43. No effect

45. (a) Degenerate circle; a single point $(3, 4)$
    (b) Degenerate ellipse; a single point $(-3, 2)$
    (c) Degenerate hyperbola; a pair of intersecting lines, $y = 2x - 6$ and $y = -2x + 2$
    (d) Degenerate hyperbola; a pair of intersecting lines, $y = 3x - 1$ and $y = -3x + 11$

47. Domain $(-\infty, -2] \cup [2, \infty)$; range $[0, \infty)$

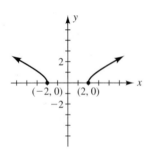

49. Domain $(-\infty, \infty)$; range $(-\infty, -3]$

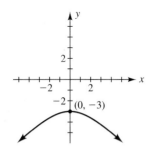

CHAPTER 6 REVIEW EXERCISES

1. Circle; radius 4; center (0, 0)
3. Parabola; vertex (0, 0); opens to the right
5. Ellipse; center (0, 0); vertices (−4, 0), (4, 0)
7. Hyperbola; center (0, 0); vertices (−5, 0), (5, 0)
9. Circle; radius $\frac{1}{2}$; center (−2, −1)
11. Parabola; vertex (2, −3); opens upward
13. Ellipse; center (0, 3); vertices (0, 6), (0, 0)
15. Hyperbola; center (0, 1); vertices (−4, 1), (4, 1)
17. Circle; radius 5, center (5, 4)
19. Parabola; vertex (16, 1); opens to the left
21. Ellipse; center (1, −2); vertices (1, 2), (1, −6)
23. Hyperbola; center (0, 1); vertices $\left(-\frac{1}{3}, 1\right), \left(\frac{1}{3}, 1\right)$
25. Degenerate hyperbola; a pair of intersecting lines $y = 2x - 2$ and $y = -2x + 2$
27. Degenerate ellipse; a single point (1, −2)
29. Focus (1, 0); directrix $x = -1$
31. Focus $\left(\frac{63}{4}, 1\right)$; directrix $x = \frac{65}{4}$
33. Foci $\left(\pm 5\sqrt{2}, 0\right)$; eccentricity $\sqrt{2} \approx 1.4$
35. Foci $\left(1, -2 \pm 2\sqrt{3}\right)$; eccentricity $\sqrt{3}/2 \approx 0.87$
37. $x^2 + y^2 = 49$ 39. $(x - 2)^2 + \left(y - \frac{9}{2}\right)^2 = \frac{25}{4}$
41. $(x - 2)^2 + (y - 3)^2 = 25$ 43. $y = x^2 - 4$
45. $x = \frac{2}{9}(y + 3)^2 - 2$ 47. $y = \frac{1}{2}(x - 1)^2 - \frac{1}{2}$
49. $\dfrac{(x - 3)^2}{9} + \dfrac{y^2}{4} = 1$ 51. $\dfrac{(x - 2)^2}{16} + \dfrac{y^2}{12} = 1$
53. $\dfrac{y^2}{9} - \dfrac{4x^2}{81} = 1$ 55. $\dfrac{x^2}{9} - \dfrac{(y + 1)^2}{7} = 1$
57. (a) Center on the y-axis   (b) Center on the x-axis
    (c) Tangent to the y-axis   (d) Tangent to the x-axis
59. $3y^2 - 4x^2 = 12$ 61. a/b
63. $\left(\dfrac{-a^2}{c}, \dfrac{ab}{c}\right)$
65. (a) PC(−70.71, 0); PT(70.71 0); $x^2 + (y + 70.71)^2 = 10{,}000$
    (b) $A(30, 24.68)$
67. $d = 18$ in.   69. (a) 174 ft   (b) 75.69 ft
71. 24 cm

Cumulative Review: Chapters 4, 5, 6

1. (a) 3   (b) −2   (c) $\frac{5}{2}$ (d) $-\frac{3}{2}$
3. (a) Circle   (b) Hyperbola   (c) Parabola   (d) Ellipse
5. $(x + 2)(x - 2)(x^2 + 2x + 5)$
7. (a) 3   (b) e^2, e^{-1}
9. (a) $y = e^{-2}x^4$ (b) $y = \dfrac{e^x}{x + 2}$ (c) $y = 2$
11. $3x + 4y = 25$ 13. (a) $f(x) \to -\infty$ (b) $f(x) \to \infty$
15. $20{,}544.33   17. (−1.99, 2.65), (1.99, 2.65)
19. 30 rows   21. Rear wheel 10 ft; front wheel 6 ft
23. (a) 2500 dozen   (b) 35,306 dozen
    (c) $3,269,016.00
25. $7\sqrt{2}$ 27. $9x^2 - y^2 = 9$
29. Height 3 in.; width 6 in.; length 8 in.
31. Length of major axis $8\sqrt{2} \approx 11.3$ in.; length of minor axis 8 inches

Chapter 7

SECTION 7.1

Problems

1. Angles coterminal with 600° have the form $(600 + 360n)°$, where n is an integer. Some positive angles coterminal with 600° are 240°, 960°, and 1320°. Some negative angles coterminal with 600° are $-120°$, $-480°$, and $-840°$.

2. $48°15'30'' = 48° + 15'\left(\dfrac{1°}{60'}\right) + 30''\left(\dfrac{1°}{3600''}\right)$
   $\approx 48.258°$.

3. Angles coterminal with $9\pi/4$ have the form $9\pi/4 + 2\pi n$, where n is an integer. By letting $n = -1$, we find the smallest positive coterminal angle to be $\pi/4$.

4. Converting to radians, we have
$$125° = 125°\left(\dfrac{\pi}{180°}\right) = \dfrac{25\pi}{36}.$$
Thus,
$$s = \theta r = \left(\dfrac{25\pi}{36}\right)(48.0) \approx 104.7 \text{ ft}.$$

5. We first convert 45 mi/h to ft/s:
$$45 \text{ mi/h} = 45 \dfrac{\text{mi}}{\text{h}}\left(\dfrac{5280 \text{ ft}}{1 \text{ mi}}\right)\left(\dfrac{1 \text{ h}}{3600 \text{ s}}\right) = 66 \text{ ft/s}.$$
Thus, the angular speed ω is
$$\omega = \dfrac{66 \text{ ft/s}}{1.25 \text{ ft}} = 52.8 \text{ rad/s}.$$
Hence,
$$\theta = \left(52.8 \dfrac{\text{rad}}{\text{s}}\right)\left(\dfrac{60 \text{ s}}{1 \text{ min}}\right)\left(\dfrac{1 \text{ rev}}{2\pi \text{ rad}}\right) \approx 504 \text{ rpm}.$$

Exercises

1.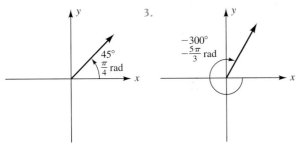
3.

5.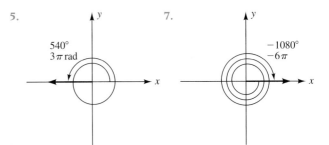

7.

9. 220° 11. 336° 13. 280° 15. 54°38' 17. π

19. $\dfrac{\pi}{4}$ 21. $\dfrac{\pi}{3}$ 23. $8 - 2\pi$ 25. $\dfrac{\pi}{3}$ 27. $\dfrac{5\pi}{4}$

29. $\dfrac{12\pi}{5}$ 31. $-\dfrac{25\pi}{12}$ 33. 225° 35. $-10°$

37. $292\frac{1}{2}°$ 39. 2160°

41. (a) 22.6° (b) 0.39 43. (a) 215.48° (b) 3.76

45. (a) 32°18'36'' (b) 0.56

47. (a) 306°07'21'' (b) 5.34

49. (a) 85.9437° (b) 85°56'37''

51. (a) 572.9578° (b) 572°57'28''

53. (a) 6π cm (b) 14π cm (c) 4π cm (d) 9π cm

55. $600\pi \approx 1885$ ft 57. 1 rad

59. (a) $\dfrac{\pi}{30}$ rad/s (b) $\dfrac{\pi}{1800}$ rad/s (c) $\dfrac{\pi}{21,600}$ rad/s

61. (a) $\dfrac{200\pi}{3}$ rad/min (b) $\dfrac{5\pi}{9}$ ft/s

63. About 84.9 cm 65. 86.4 cm

67. Since 1 rad $\approx 57.3°$, we conclude that 1 radian is larger than 1°.

69. $10\pi \approx 31.4$ cm

71. (c) sector (i) $\dfrac{27\pi}{2}$ sq m; sector (ii) 27 sq ft

SECTION 7.2

Problems

1. If $x = -2$ and $y = 4$, then
   $r = \sqrt{(-2)^2 + (4)^2} = \sqrt{20} = 2\sqrt{5}$. Hence,

$$\sin\theta = \frac{y}{r} = \frac{4}{2\sqrt{5}} = \frac{2}{\sqrt{5}}$$

$$\cos\theta = \frac{x}{r} = \frac{-2}{2\sqrt{5}} = -\frac{1}{\sqrt{5}}$$

$$\tan\theta = \frac{y}{x} = \frac{4}{-2} = -2$$

$$\csc\theta = \frac{1}{\sin\theta} = \frac{\sqrt{5}}{2}$$

$$\sec\theta = \frac{1}{\cos\theta} = -\sqrt{5}$$

$$\cot\theta = \frac{1}{\tan\theta} = -\frac{1}{2}$$

2. If $\tan\theta > 0$ and $\cos\theta < 0$, then θ is in quadrant III.

3. From the identity $1 + \tan^2\theta = \sec^2\theta$, we have $\sec\theta = \pm\sqrt{1 + \tan^2\theta}$. Since $\tan\theta = -1 < 0$ and $\cos\theta < 0$, we know that θ is in quadrant II, and hence $\sec\theta < 0$. Thus,

$$\sec\theta = -\sqrt{1 + (-1)^2} = -\sqrt{2}$$

and $\cos\theta = \dfrac{1}{\sec\theta} = -\dfrac{1}{\sqrt{2}}$.

4. (a) $\sin B = \dfrac{\text{side opposite } B}{\text{hypotenuse}} = \dfrac{8}{12} = \dfrac{2}{3}$

 (b) $\cos B = \dfrac{\text{side adjacent to } B}{\text{hypotenuse}} = \dfrac{4\sqrt{5}}{12} = \dfrac{\sqrt{5}}{3}$

 (c) $\tan B = \dfrac{\text{side opposite } B}{\text{side adjacent to } B} = \dfrac{8}{4\sqrt{5}} = \dfrac{2}{\sqrt{5}}$

5. $\sec 12°32' = \csc(90° - 12°32')$
$= \csc(89°60' - 12°32')$
$= \csc 77°28'$.

◆ Exercises

1. $\sin\theta = \frac{4}{5}$, $\cos\theta = \frac{3}{5}$, $\tan\theta = \frac{4}{3}$, $\csc\theta = \frac{5}{4}$, $\sec\theta = \frac{5}{3}$, $\cot\theta = \frac{3}{4}$

3. $\sin\theta = -\dfrac{\sqrt{3}}{2}$, $\cos\theta = -\frac{1}{2}$, $\tan\theta = \sqrt{3}$, $\csc\theta = -\dfrac{2}{\sqrt{3}}$, $\sec\theta = -2$, $\cot\theta = \dfrac{1}{\sqrt{3}}$

5. $\sin\theta = 1$, $\cos\theta = 0$, $\tan\theta$ is undefined, $\csc\theta = 1$, $\sec\theta$ is undefined, $\cot\theta = 0$

7. Quadrant III   9. Quadrant II   11. Quadrant I

13. Quadrant IV

15. $\csc\theta = \frac{5}{3}$, $\cos\theta = \frac{4}{5}$, $\sec\theta = \frac{5}{4}$, $\tan\theta = \frac{3}{4}$, $\cot\theta = \frac{4}{3}$

17. $\cot\theta = -\frac{3}{2}$, $\sin\theta = \dfrac{2}{\sqrt{13}}$, $\csc\theta = \dfrac{\sqrt{13}}{2}$, $\cos\theta = -\dfrac{3}{\sqrt{13}}$, $\sec\theta = -\dfrac{\sqrt{13}}{3}$

19. $\sec\theta = \dfrac{2}{\sqrt{3}}$, $\sin\theta = -\frac{1}{2}$, $\csc\theta = -2$, $\tan\theta = -\dfrac{1}{\sqrt{3}}$, $\cot\theta = -\sqrt{3}$

21. $\tan\theta = -1$, $\sin\theta = -\dfrac{1}{\sqrt{2}}$, $\csc\theta = -\sqrt{2}$, $\cos\theta = \dfrac{1}{\sqrt{2}}$, $\sec\theta = \sqrt{2}$

23. $\cos\theta = -\frac{5}{6}$, $\sin\theta = -\dfrac{\sqrt{11}}{6}$, $\csc\theta = -\dfrac{6}{\sqrt{11}}$, $\tan\theta = \dfrac{\sqrt{11}}{5}$, $\cot\theta = \dfrac{5}{\sqrt{11}}$

25. $\sin\theta = \frac{5}{13}$ 27. $\cot\theta = -5$ 29. $\tan\theta = 3$

31. $\sin\theta = \frac{5}{13}$ 33. $\sin\theta = \frac{4}{5}$ 35. $\sec\theta = -\sqrt{3}$

37. $\tan\theta = -\dfrac{1}{\sqrt{2}}$

39. (a) $\sin\alpha = \frac{15}{17}$ (b) $\cos\alpha = \frac{8}{17}$ (c) $\cot\beta = \frac{15}{8}$

41. (a) $\tan\beta = \dfrac{4\sqrt{2}}{7}$
 (b) $\csc\alpha = \frac{9}{7}$
 (c) $\sec\alpha = \dfrac{9}{4\sqrt{2}}$

43. (a) $\sin\beta = \dfrac{3}{\sqrt{10}}$
 (b) $\cos\alpha = \dfrac{3}{\sqrt{10}}$
 (c) $\tan\beta = 3$

45. $\cos 60°$ 47. $\cot\dfrac{\pi}{4}$ 49. $\tan 55°17'$

51.   53.

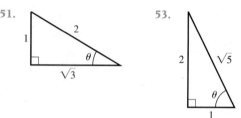

55. No; reciprocals can't have different algebraic signs.

57. $\cos\theta = \pm\sqrt{1-\sin^2\theta}$, $\tan\theta = \dfrac{\pm\sin\theta}{\sqrt{1-\sin^2\theta}}$,

$\csc\theta = \dfrac{1}{\sin\theta}$, $\sec\theta = \dfrac{\pm 1}{\sqrt{1-\sin^2\theta}}$,

$\cot\theta = \dfrac{\pm\sqrt{1-\sin^2\theta}}{\sin\theta}$

59. (a) 0   (b) $\ln(\cot\theta)$ (c) $\ln(\sin\theta)$ (d) 0

SECTION 7.3

◆ Problems

1. (a) Since 7π radians is coterminal with π radians and $\cos\pi = -1$, we conclude that $\cos 7\pi = -1$.
   (b) Since 450° is coterminal with 90° and tan 90° is undefined, we conclude that tan 450° is undefined.

2. (a) By symmetry, the point on the unit circle that corresponds to an arc length $3\pi/4$ is $(-1/\sqrt{2},\, 1/\sqrt{2})$, as shown in the sketch:

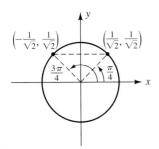

   Thus, $\tan\dfrac{3\pi}{4} = -\tan\dfrac{\pi}{4} = -1$.

   (b) By symmetry, the point on the unit circle that corresponds to a central angle of $-420°$ (which is coterminal with 300°) is $(\tfrac{1}{2}, -\sqrt{3}/2)$, as shown in the sketch:

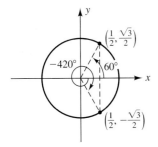

   Thus, $\sec(-420°) = \sec 300° = \sec 60° = 2$.

3. (a) Setting the calculator in radian mode, we obtain $\cos\dfrac{43\pi}{24} \approx 0.7934$, as shown in the figure:

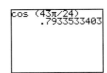

   (b) Setting the calculator in degree mode, we obtain $\cot 121°30' = \cot 121.5° \approx -0.6128$, as shown in the figure:

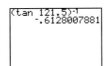

◆ Exercises

1.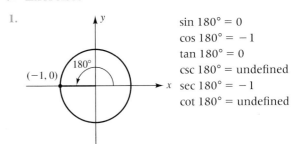

   sin 180° = 0
   cos 180° = −1
   tan 180° = 0
   csc 180° = undefined
   sec 180° = −1
   cot 180° = undefined

3.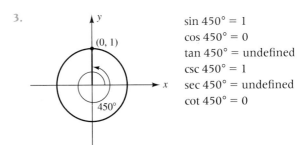

   sin 450° = 1
   cos 450° = 0
   tan 450° = undefined
   csc 450° = 1
   sec 450° = undefined
   cot 450° = 0

5.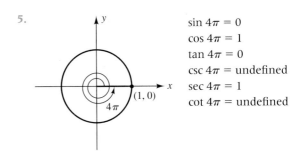

   $\sin 4\pi = 0$
   $\cos 4\pi = 1$
   $\tan 4\pi = 0$
   $\csc 4\pi$ = undefined
   $\sec 4\pi = 1$
   $\cot 4\pi$ = undefined

7.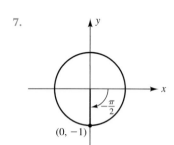

$\sin(-\pi/2) = -1$
$\cos(-\pi/2) = 0$
$\tan(-\pi/2) =$ undefined
$\csc(-\pi/2) = -1$
$\sec(-\pi/2) =$ undefined
$\cot(-\pi/2) = 0$

9.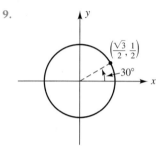

$\sin 30° = \frac{1}{2}$
$\cos 30° = \sqrt{3}/2$
$\tan 30° = 1/\sqrt{3}$
$\csc 30° = 2$
$\sec 30° = 2/\sqrt{3}$
$\cot 30° = \sqrt{3}$

11.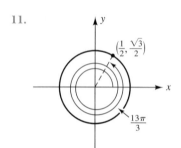

$\sin(13\pi/3) = \sqrt{3}/2$
$\cos(13\pi/3) = \frac{1}{2}$
$\tan(13\pi/3) = \sqrt{3}$
$\csc(13\pi/3) = 2/\sqrt{3}$
$\sec(13\pi/3) = 2$
$\cot(13\pi/3) = 1/\sqrt{3}$

13.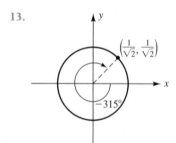

$\sin(-315°) = 1/\sqrt{2}$
$\cos(-315°) = 1/\sqrt{2}$
$\tan(-315°) = 1$
$\csc(-315°) = \sqrt{2}$
$\sec(-315°) = \sqrt{2}$
$\cot(-315°) = 1$

15.

$\sin(9\pi/4) = 1/\sqrt{2}$
$\cos(9\pi/4) = 1/\sqrt{2}$
$\tan(9\pi/4) = 1$
$\csc(9\pi/4) = \sqrt{2}$
$\sec(9\pi/4) = \sqrt{2}$
$\cot(9\pi/4) = 1$

17. $-\frac{1}{2}$ 19. $-\frac{1}{2}$ 21. $\sqrt{3}$ 23. -1

25. $-\sqrt{2}$ 27. $\frac{\sqrt{3}}{2}$ 29. $\frac{1}{\sqrt{2}}$ 31. $-\frac{2}{\sqrt{3}}$

33. $-\frac{1}{\sqrt{3}}$ 35. $-\frac{\sqrt{3}}{2}$ 37. 2   39. -2 41. 3

43. 4   45. 0   47. 2   49. $\frac{3}{2}$ 51. $\frac{3\sqrt{3}}{2}$ 53. 0

55. -2 57. 0.4226   59. -0.2588 61. -0.2250

63. -1.236 65. 1.236   67. 0.2679   69. 1.557

71. -1.101 73. 0.7274   75. 0.4901

77. (a) 12 cm   (b) -12 cm   (c) 0 cm   (d) $-6\sqrt{2}$ cm

79. (a) About -15.1 volts   (b) About -10.7 volts
    (c) About -15.0 volts   (d) About 1.34 volts

81. It appears that $\sin(-\theta) = -\sin\theta$.

83. Since $\csc\pi$ and $\cot\pi$ are undefined,

$$(\sin\pi)(\csc\pi) \neq 1, \ (\tan\pi)(\cot\pi) \neq 1,$$
$$\cot\pi \neq \frac{\cos\pi}{\sin\pi}, \text{ and } 1 + \cot^2\pi \neq \csc^2\pi$$

85. (a) 0   (b) -1 (c) $\frac{1}{2}$ (d) 4

87.

θ	$\sin\theta$	$\cos\theta$	$\tan\theta$
0 to $\pi/2$	0 to 1	1 to 0	0 to ∞
$\pi/2$ to π	1 to 0	0 to -1	$-\infty$ to 0
π to $3\pi/2$	0 to -1	-1 to 0	0 to ∞
$3\pi/2$ to 2π	-1 to 0	0 to 1	$-\infty$ to 0

89. $\frac{\sin t}{t}$ seems to approach 1 as $t \to 0^+$.

SECTION 7.4

◆ Problems

1. Since the cosine function is an even function, we have

$$y = \frac{1}{2}\cos\left(\frac{\pi}{4} - \frac{\pi}{4}x\right) = \frac{1}{2}\cos\left[-\left(\frac{\pi}{4}x - \frac{\pi}{4}\right)\right]$$
$$= \frac{1}{2}\cos\left(\frac{\pi}{4}x - \frac{\pi}{4}\right)$$

Hence, the graph of $y = \frac{1}{2}\cos\left(\frac{\pi}{4} - \frac{\pi}{4}x\right)$ is exactly the same as the graph of $y = \frac{1}{2}\cos\left(\frac{\pi}{4}x - \frac{\pi}{4}\right)$.

2. We can think of the graph in Figure 7.54 as a cosine curve with amplitude 2, period $4\pi/3$, and no phase shift. Hence, using $a = 2$, $b = \frac{3}{2}$, and $c = 0$, we have the equation $y = 2\cos\frac{3}{2}x$.

3. The population P on September 1 ($t = 9$) is

$$P = 70 + 50\sin\left(\frac{9\pi}{6} - \frac{\pi}{3}\right)$$
$$= 70 + 50\sin\frac{7\pi}{6}$$
$$= 70 + 50\left(-\frac{1}{2}\right) = 45 \text{ hawks}$$

◆ Exercises

1.

x	0	$\frac{\pi}{4}$	$\frac{\pi}{2}$	$\frac{3\pi}{4}$	π	$\frac{5\pi}{4}$	$\frac{3\pi}{2}$	$\frac{7\pi}{4}$	2π
$f(x) = \sin x$	0	$\frac{1}{\sqrt{2}}$	1	$\frac{1}{\sqrt{2}}$	0	$-\frac{1}{\sqrt{2}}$	-1	$-\frac{1}{\sqrt{2}}$	0
$g(x) = \cos x$	1	$\frac{1}{\sqrt{2}}$	0	$-\frac{1}{\sqrt{2}}$	-1	$-\frac{1}{\sqrt{2}}$	0	$\frac{1}{\sqrt{2}}$	1

Note: See Figures 7.45 and 7.46 for the graphs of f and g.

3. $0, \pm\pi, \pm 2\pi, \pm 3\pi, \pm 4\pi$ 5. $\pm\pi, \pm 3\pi$

7. $\pm\frac{\pi}{4}, \pm\frac{7\pi}{4}, \pm\frac{9\pi}{4}, \pm\frac{15\pi}{4}$

9. $\frac{7\pi}{6}, \frac{11\pi}{6}, \frac{19\pi}{6}, \frac{23\pi}{6}, -\frac{\pi}{6}, -\frac{5\pi}{6}, -\frac{13\pi}{6}, -\frac{17\pi}{6}$

11.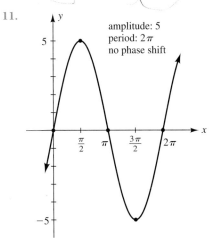
amplitude: 5
period: 2π
no phase shift

13.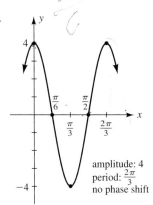
amplitude: 4
period: $\frac{2\pi}{3}$
no phase shift

15.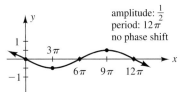
amplitude: $\frac{1}{2}$
period: 12π
no phase shift

17.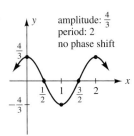
amplitude: $\frac{4}{3}$
period: 2
no phase shift

19.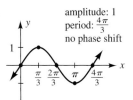
amplitude: 1
period: $\frac{4\pi}{3}$
no phase shift

21.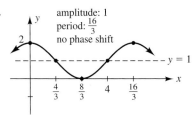
amplitude: 1
period: $\frac{16}{3}$
no phase shift

23.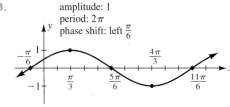
amplitude: 1
period: 2π
phase shift: left $\frac{\pi}{6}$

25. 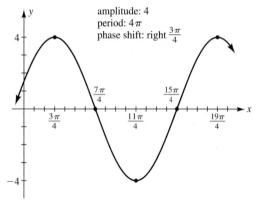 amplitude: 4
period: 4π
phase shift: right $\frac{3\pi}{4}$

27. 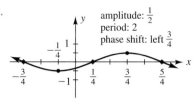 amplitude: $\frac{1}{2}$
period: 2
phase shift: left $\frac{3}{4}$

29. 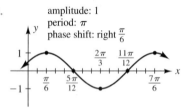 amplitude: 1
period: π
phase shift: right $\frac{\pi}{6}$

31. 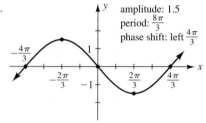 amplitude: 1.5
period: $\frac{8\pi}{3}$
phase shift: left $\frac{4\pi}{3}$

33. 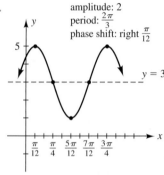 amplitude: 2
period: $\frac{2\pi}{3}$
phase shift: right $\frac{\pi}{12}$

35. 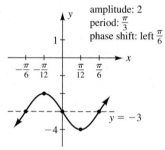 amplitude: 2
period: $\frac{\pi}{3}$
phase shift: left $\frac{\pi}{6}$

37. Period 0.0184, no phase shift

39. Period 28.2, phase shift right 5.56 units

41. Period 0.503, phase shift left 0.824 units

43. (i) $y = 3 \sin \frac{\pi}{2} x$ (ii) $y = 3 \cos\left(\frac{\pi}{2}x + \frac{3\pi}{2}\right)$

45. (i) $y = 2 \sin\left(x + \frac{\pi}{2}\right)$ (ii) $y = 2 \cos x$

47. (i) $y = \frac{3}{2} \sin\left(x + \frac{3\pi}{4}\right)$ (ii) $y = \frac{3}{2} \cos\left(x + \frac{\pi}{4}\right)$

49. (a) Amplitude: 170
Period: $\frac{1}{60}$

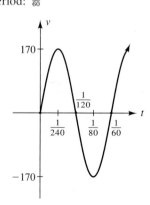

(b) 60 Hz

51. (a) February, $7700   (b) August, $1300

53. Greatest during the year 1993; least during the year 2000

55. (a) $d = -120 + 70 \sin \frac{\pi x}{240}$
(b) 155 ft below ground

57. $\sin\left(x - \frac{\pi}{2}\right) = -\cos x$

59. $\cos\left(x + \frac{3\pi}{2}\right) = \sin x$

61. $\sin(x - \pi) = -\sin x$ **63.** $f(x) = -1 + 4\sin\dfrac{\pi x}{4}$

65. If c is the period of a periodic function f, then $f(t + c) = f(t)$ for all t in the domain of f. Since the two inputs $t + c$ and t give the same output value, we conclude that f is *not* one-to-one.

67. (a) $f(x) = \sin x$ (b) $P(1) \approx 0.841$, $f(1) \approx 0.841$

SECTION 7.5

◀ Problems

1. Since the cotangent function is an odd function, we have

$$y = \cot\left(\dfrac{\pi}{2} - 2x\right) = \cot\left[-\left(2x - \dfrac{\pi}{2}\right)\right]$$
$$= -\cot\left(2x - \dfrac{\pi}{2}\right)$$

Hence, by the x-axis reflection rule (Section 2.5), the graph of $y = \cot\left(\dfrac{\pi}{2} - 2x\right)$ is the same as the graph of $y = \cot\left(2x - \dfrac{\pi}{2}\right)$ reflected about the x-axis.

2. Since the secant function is an even function, we have

$$y = \sec\left(-\pi x - \dfrac{\pi}{2}\right) = \sec\left[-\left(\pi x + \dfrac{\pi}{2}\right)\right]$$
$$= \sec\left(\pi x + \dfrac{\pi}{2}\right)$$

Hence, the graph of $y = \sec\left(-\pi x - \dfrac{\pi}{2}\right)$ is exactly the same as the graph of $y = \sec\left(\pi x + \dfrac{\pi}{2}\right)$.

◀ Exercises

1.

x	$-\dfrac{\pi}{2}$	$-\dfrac{\pi}{3}$	$-\dfrac{\pi}{4}$	$-\dfrac{\pi}{6}$	0	$\dfrac{\pi}{6}$	$\dfrac{\pi}{4}$	$\dfrac{\pi}{3}$	$\dfrac{\pi}{2}$
$f(x) = \tan x$	undef.	$-\sqrt{3}$	-1	$-\dfrac{1}{\sqrt{3}}$	0	$\dfrac{1}{\sqrt{3}}$	1	$\sqrt{3}$	undef.

Note: See Figure 7.59 for the graph of $f(x) = \tan x$.

3.

x	0	$\dfrac{\pi}{4}$	$\dfrac{\pi}{2}$	$\dfrac{3\pi}{4}$	π	$\dfrac{5\pi}{4}$	$\dfrac{3\pi}{2}$	$\dfrac{7\pi}{4}$	2π
$f(x) = \csc x$	undef.	$\sqrt{2}$	1	$\sqrt{2}$	undef.	$-\sqrt{2}$	-1	$-\sqrt{2}$	undef.

Note: See Figure 7.61 for the graph of $f(x) = \csc x$.

5. $0, \pm\pi, \pm 2\pi, \pm 3\pi, \pm 4\pi$

7. $\dfrac{3\pi}{4}, \dfrac{7\pi}{4}, \dfrac{11\pi}{4}, \dfrac{15\pi}{4}, -\dfrac{\pi}{4}, -\dfrac{5\pi}{4}, -\dfrac{9\pi}{4}, -\dfrac{13\pi}{4}$

9. $\dfrac{\pi}{6}, \dfrac{7\pi}{6}, \dfrac{13\pi}{6}, \dfrac{19\pi}{6}, -\dfrac{5\pi}{6}, -\dfrac{11\pi}{6}, -\dfrac{17\pi}{6}, -\dfrac{23\pi}{6}$

11. $\dfrac{2\pi}{3}, \dfrac{5\pi}{3}, \dfrac{8\pi}{3}, \dfrac{11\pi}{3}, -\dfrac{\pi}{3}, -\dfrac{4\pi}{3}, -\dfrac{7\pi}{3}, -\dfrac{10\pi}{3}$

13. $\dfrac{\pi}{2}, \dfrac{5\pi}{2}, -\dfrac{3\pi}{2}, -\dfrac{7\pi}{2}$

15. $\pm\dfrac{2\pi}{3}, \pm\dfrac{4\pi}{3}, \pm\dfrac{8\pi}{3}, \pm\dfrac{10\pi}{3}$

17. $\pm\dfrac{\pi}{6}, \pm\dfrac{11\pi}{6}, \pm\dfrac{13\pi}{6}, \pm\dfrac{23\pi}{6}$ **19.** No solution

21. 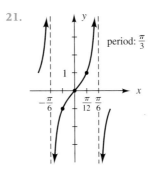 period: $\dfrac{\pi}{3}$

23. 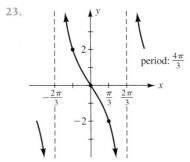 period: $\dfrac{4\pi}{3}$

25. 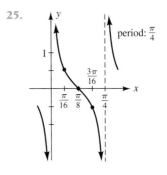 period: $\dfrac{\pi}{4}$

27. 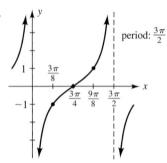 period: $\frac{3\pi}{2}$

29. 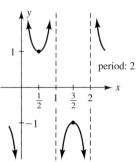 period: 2

31. 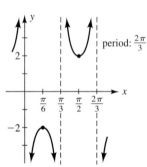 period: $\frac{2\pi}{3}$

33. 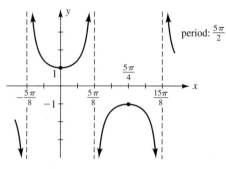 period: $\frac{5\pi}{2}$

35. 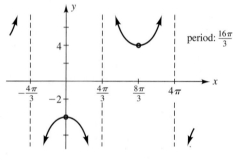 period: $\frac{16\pi}{3}$

37. 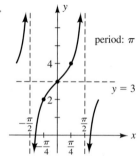 period: π

39. 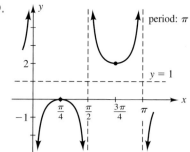 period: π

41. 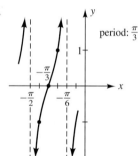 period: $\frac{\pi}{3}$

43. 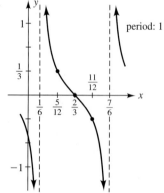 period: 1

45.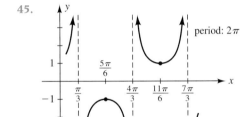
period: 2π

47.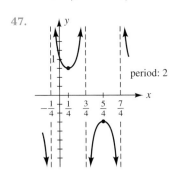
period: 2

49. 0.726   51. 7.22   53. 2.38   55. $y = 3 \tan \frac{1}{6} x$

57. (a) $y = \frac{2}{3} \csc 8x$ 59. $\tan\left(x + \frac{\pi}{2}\right) = -\cot x$

61. $\cot(\pi - x) = -\cot x$ 63. $\csc\left(x + \frac{\pi}{2}\right) = \sec x$

65. $\sec\left(x - \frac{3\pi}{2}\right) = -\csc x$ 67. $f(x) = 2 + 3 \csc \frac{\pi x}{4}$

69. $\pm \frac{\pi}{4}, \pm \frac{3\pi}{4}, \pm \frac{5\pi}{4}, \pm \frac{7\pi}{4}, \pm \frac{9\pi}{4}, \pm \frac{11\pi}{4}, \pm \frac{13\pi}{4}, \pm \frac{15\pi}{4}$

SECTION 7.6

◆ Problems

1. (a) Setting a calculator in radian mode, we find arcsin $0.53 \approx 0.5586$.
   (b) Setting a calculator in radian mode, we find $\sin^{-1}(-0.766) \approx -0.8726$.

2. (a) Setting a calculator in radian mode, we find arccos $0.809 \approx 0.6283$.
   (b) Setting a calculator in radian mode, we find $\cos^{-1}(-0.42) \approx 2.004$.

3. (a) Setting a calculator in radian mode, we find arctan $0.977 \approx 0.7738$.
   (b) Setting a calculator in radian mode, we find $\tan^{-1}(-1.6) \approx -1.012$.

4. (a) Since 2 is in the domain of the inverse tangent function, we have $\tan(\arctan 2) = 2$.

(b) Since $3\pi/2$ is coterminal with $-\pi/2$ and $-\pi/2$ is in the domain of the restricted sine function, we have
$$\sin^{-1}\left(\sin \frac{3\pi}{2}\right) = \sin^{-1}\left[\sin\left(-\frac{\pi}{2}\right)\right] = -\frac{\pi}{2}.$$

5. We begin by rewriting the expression:
$$\cos\left[\arctan\left(-\tfrac{1}{2}\right)\right] = \cos\left[-\arctan \tfrac{1}{2}\right] = \cos\left[\arctan \tfrac{1}{2}\right].$$

If we let $\theta = \arctan \tfrac{1}{2}$, then $\tan \theta = \tfrac{1}{2} = \dfrac{\text{opp}}{\text{adj}}$, and by the Pythagorean theorem, the hypotenuse is $\sqrt{5}$, as shown in the figure:

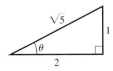

From the right triangle shown, we have
$$\cos\left[\arctan \frac{1}{2}\right] = \cos \theta = \frac{2}{\sqrt{5}}.$$

6.

7. Domain $[90, 90\sqrt{2}]$; range $[0, \pi/4]$

◆ Exercises

1. $\dfrac{\pi}{6}$ 3. $\dfrac{\pi}{3}$ 5. $\dfrac{\pi}{4}$ 7. $-\dfrac{\pi}{3}$ 9. $-\dfrac{\pi}{4}$

11. $\dfrac{3\pi}{4}$ 13. $\dfrac{\pi}{2}$

15. Undefined   17. $\tfrac{2}{3}$ 19. Undefined

21. $\dfrac{\pi}{8}$ 23. -1 25. π 27. $\dfrac{\pi}{4}$ 29. $\dfrac{\pi}{10}$

31. $-\dfrac{3\pi}{8}$ 33. $\dfrac{3}{\sqrt{10}}$ 35. $\dfrac{\sqrt{5}}{3}$ 37. $-\dfrac{\sqrt{13}}{2}$

39. $-\dfrac{1}{\sqrt{3}}$ 41. 0.2838   43. 1.446   45. 1.706

47. 1.254   49. -0.3948 51. Undefined   53. 0.887

55. -0.7168 57. $\sqrt{1 - x^2}$ 59. $\dfrac{1}{\sqrt{x^2 - 1}}$

61. $\dfrac{1}{x - 1}$ 63. $\dfrac{x - 1}{\sqrt{x^2 - 2x + 2}}$

65. (a) $\theta = \arctan \dfrac{h}{550}$

(b) Domain $[0, \infty)$; range $\left[0, \dfrac{\pi}{2}\right)$ (c) $\dfrac{\pi}{4}$

67. (a) $\theta = \arctan \dfrac{8}{t}$

(b) Domain $(0, \infty)$; range $\left(0, \dfrac{\pi}{2}\right)$

(c) About $33.7°$

69. (a) 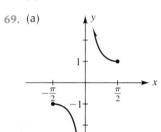 (b)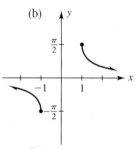

Domain $(-\infty, -1] \cup [1, \infty)$
Range $[-\pi/2, 0) \cup (0, \pi/2]$

71. (a) 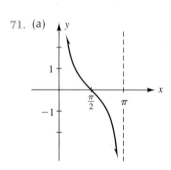 (b)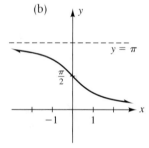

Domain $(-\infty, \infty)$
Range $(0, \pi)$

73. (a) $\cos(\pi - t) = -\cos t$

75. One example is $\tan^{-1}(\tan \pi) = 0 \neq \pi$. This does not contradict the inverse function concept, since π is not in the domain of the restricted tangent function.

77. The notation $\sin^{-1} x$ denotes the inverse sine, and $(\sin x)^{-1}$ denotes the reciprocal of the sine, which is the cosecant.

79. $\tfrac{5}{6}$ 81. $1, \tfrac{1}{2}$

83. (a) (b)

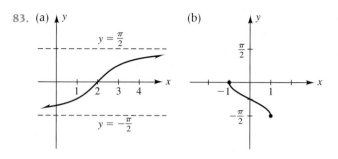

SECTION 7.7

◆ Problems

1. $\alpha = 90° - 47° = 43°$

$\tan 47° = \dfrac{x}{22}$; $x = 22 \tan 47° \approx 23.6$

$\cos 47° = \dfrac{22}{z}$; $z = \dfrac{22}{\cos 47°} \approx 32.3$

2. $x = \sqrt{24^2 - 10^2} \approx 21.8$

$\cos \alpha = \tfrac{10}{24}$; $\alpha = \arccos \tfrac{10}{24} \approx 65.4°$

$\sin \beta = \tfrac{10}{24}$; $\beta = \arcsin \tfrac{10}{24} \approx 24.6°$

3.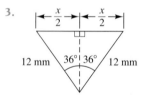

Using either of the right triangles in the figure, we have $\sin 36° = \dfrac{x/2}{12}$. Therefore,

$\dfrac{x}{2} = 12 \sin 36°$

$x = 24 \sin 36° \approx 14.1$ mm.

4.

From the right triangle in the figure, we have $\sin 15.5° = \dfrac{1800}{s}$, which implies that

$s = \dfrac{1800}{\sin 15.5°} \approx 6735.6$ ft. Hence,

$t = \dfrac{s}{v} = \dfrac{6735.6 \text{ ft}}{3 \dfrac{\text{mi}}{\text{h}}} \cdot \dfrac{1 \text{ mi}}{5280 \text{ ft}}$

≈ 0.425 h or 25.5 min.

5.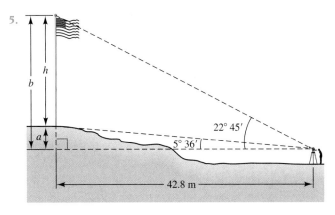

From the right triangles in the figure, we find a and b:

$$\tan 5°36' = \frac{a}{42.8}; \quad a = 42.8 \tan 5°36' \approx 4.197 \text{ m}$$

$$\tan 22°45' = \frac{b}{42.8}; \quad b = 42.8 \tan 22°45' \approx 17.948 \text{ m}.$$

Hence, the height h of the pole is $h = b - a \approx 13.8$ m.

6. Substituting $t = 2$ into the equation $d = -15 \cos 8\pi t$, we find

$$d = -15 \cos 16\pi = -15(1) = -15 \text{ cm}.$$

Exercises

1. $\theta = 52°$, $x \approx 14.2$, $y \approx 11.1$

3. $\theta = 28.2°$, $x \approx 7.08$, $y \approx 15.0$

5. $\theta = 47°42'$, $x \approx 761$, $y \approx 512$

7. $\alpha \approx 56.3°$, $\beta \approx 33.7°$, $x \approx 21.6$

9. $\alpha \approx 38.6°$, $\beta \approx 51.4°$, $x \approx 56.7$

11. $\alpha \approx 49.9°$, $\beta \approx 40.1°$, $x \approx 1480$

13. 338 ft   15. 2.87 km   17. 147 m   19. 4.31 cm

21. 145.3°   23. 2.08 cm   25. $\alpha = 34.3°$, $\beta = 111.4°$

27. 199 m   29. 29.2 s   31. 745 ft

33. (a) 8   (b) 3   (c) $\frac{1}{12}$ 35. (a) $\frac{2}{3}$ (b) $\frac{1}{20}$ (c) 10

37. (a) $d = 6 \sin \frac{2\pi}{3} t$
   (b) 3 inches below the equilibrium position

39. (a) $d = 20 \cos 5\pi t$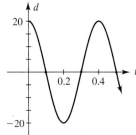

(b) 0 cm   (c) 1.1 s

41. $x = 32$ cm, $\theta \approx 73.7°$ 43. 75°

45. About 13.8 cm   47. $d \approx 0.227 \sin 3.14t$

49. (a) $x = \cos \theta$, $y = \sin \theta$ (b) $A = \sin \theta \cos \theta + \sin \theta$
   (c) $\theta \approx 1.047$ rad $\approx 60°$ (d) About 1.30 sq ft

51. Since $e^{-kt} \to 0$ as $t \to \infty$, the coefficient e^{-kt} causes a decrease in amplitude of successive oscillations.

CHAPTER 7 REVIEW EXERCISES

1. (a) 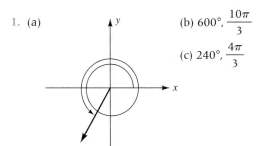   (b) 600°, $\frac{10\pi}{3}$
   (c) 240°, $\frac{4\pi}{3}$

3. (a) 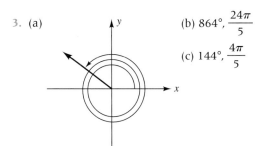   (b) 864°, $\frac{24\pi}{5}$
   (c) 144°, $\frac{4\pi}{5}$

5. (a) 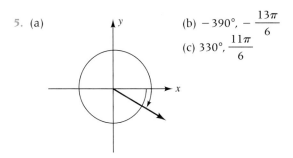   (b) $-390°$, $-\frac{13\pi}{6}$
   (c) 330°, $\frac{11\pi}{6}$

7. $\dfrac{10\pi}{9}$ 9. $\dfrac{13\pi}{3}$ 11. $330°$ 13. $-1440°$

15. (a) 123.705   (b) 2.159

17. (a) $151.834°$ (b) $151°50'02''$

19. $\sin\theta = \dfrac{1}{\sqrt{2}}$, $\cos\theta = \dfrac{1}{\sqrt{2}}$, $\tan\theta = 1$, $\csc\theta = \sqrt{2}$, $\sec\theta = \sqrt{2}$, $\cot\theta = 1$

21. $\sin\theta = -\dfrac{3}{\sqrt{10}}$, $\cos\theta = -\dfrac{1}{\sqrt{10}}$, $\tan\theta = 3$, $\csc\theta = -\dfrac{\sqrt{10}}{3}$, $\sec\theta = -\sqrt{10}$, $\cot\theta = \tfrac{1}{3}$

23. 2   25. $-\tfrac{3}{5}$ 27. $-\sqrt{3}$ 29. $-\tfrac{12}{5}$ 31. $\tfrac{8}{17}$

33. $\tfrac{15}{8}$ 35. $-\tfrac{1}{2}$ 37. $\sqrt{3}$ 39. $\sqrt{2}$ 41. -1

43. $\dfrac{\pi}{3}$ 45. Undefined   47. $\dfrac{\pi}{3}$ 49. $\dfrac{2\sqrt{2}}{3}$

51. 0.3090   53. -0.9147 55. 1.057   57. -0.3827

59. 1.107   61. 2.044   63. $\sin 2 \approx 0.9093$

65. $\tan 3 \approx -0.1425$

67. $(f \circ g)(x) = \sqrt{1 - 16x^2}$ for $-\tfrac{1}{4} \le x \le \tfrac{1}{4}$

69. $(f \circ g)(x) = \dfrac{1}{\sqrt{x-1}}$ for $x > 1$

71.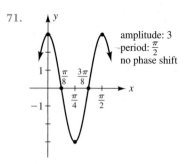
amplitude: 3
period: $\dfrac{\pi}{2}$
no phase shift

73.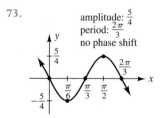
amplitude: $\tfrac{5}{4}$
period: $\dfrac{2\pi}{3}$
no phase shift

75.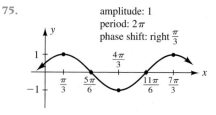
amplitude: 1
period: 2π
phase shift: right $\dfrac{\pi}{3}$

77.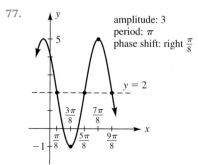
amplitude: 3
period: π
phase shift: right $\dfrac{\pi}{8}$

79.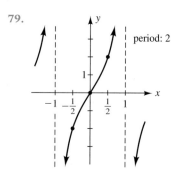
period: 2

81.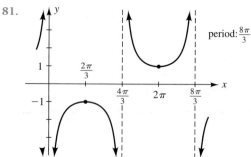
period: $\dfrac{8\pi}{3}$

83.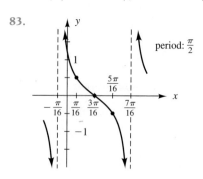
period: $\dfrac{\pi}{2}$

85.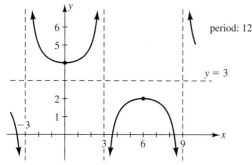
period: 12

87. $\theta = 55°$, $x \approx 34.2$, $y \approx 19.6$ 89. About 5400 mi

91. $L = 10 \sin \dfrac{\theta}{2}$,

θ	1	$\dfrac{\pi}{12}$	$\dfrac{\pi}{3}$	1.85
L	4.79	1.31	5	8

93. $\alpha \approx 117.1°$, $\beta \approx 152.9°$

95. (a) $\alpha = \arctan \dfrac{x}{3}$ (b) $\beta = \pi - \arctan \dfrac{x}{3}$

   (c) $\theta = \arctan \dfrac{x}{3} + \arctan \dfrac{7}{x} - \dfrac{\pi}{2}$

97. About 244 ft

99. (a) $d = \tfrac{3}{2} \sin \pi t$

   (b) $\dfrac{3}{2\sqrt{2}} \approx 1.06$ inches below the equilibrium position

Chapter 8

SECTION 8.1

Problems

1. $\dfrac{4}{\csc \theta - 1} + \dfrac{3}{1 - \csc \theta} = \dfrac{4}{\csc \theta - 1} + \dfrac{-3}{\csc \theta - 1}$
$= \dfrac{1}{\csc \theta - 1}$

2. Factoring as a trinomial square, we have
$\csc^2 x - 3 \csc x - 10 = (\csc x - 5)(\csc x + 2)$

3. $\sin^2\theta \cot^2\theta + \sin^2\theta = \sin^2\theta (\cot^2\theta + 1)$
$= \sin^2\theta \csc^2\theta$
$= (\sin \theta \csc \theta)^2$
$= 1^2$
$= 1$

4. $\sin x + \cos x \cot x = \sin x + \cos x \dfrac{\cos x}{\sin x}$
$= \sin x + \dfrac{\cos^2 x}{\sin x}$
$= \dfrac{\sin^2 x + \cos^2 x}{\sin x}$
$= \dfrac{1}{\sin x}$
$= \csc x$

5. As shown in the following figures, the graphs of
$$y = \tan \theta - \cot \theta \quad \text{and} \quad y = \sec \theta \csc \theta$$
are not identical. Hence, we conclude that
$$\tan \theta - \cot \theta = \sec \theta \csc \theta$$
is not a trigonometric identity.

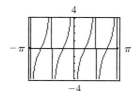 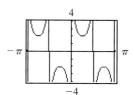

6.
$$\sec^4 x - \tan^4 x = \dfrac{1 + \sin^2 x}{\cos^2 x}$$

$(\sec^2 x - \tan^2 x)(\sec^2 x + \tan^2 x)$
$1(\sec^2 x + \tan^2 x)$
$\dfrac{1}{\cos^2 x} + \dfrac{\sin^2 x}{\cos^2 x}$
$\dfrac{1 + \sin^2 x}{\cos^2 x}$

7. $\sqrt{4 - x^2} = \sqrt{4 - (2 \sin \theta)^2}$
$= \sqrt{4 - 4\sin^2\theta}$
$= \sqrt{4(1 - \sin^2\theta)}$
$= \sqrt{4\cos^2\theta}$
$= 2 \cos \theta$ for $0 < \theta < \pi/2$.

Exercises

1. $27 \cos^6 y \sin^3 y$ 3. $\dfrac{4 \cos x}{\cos 2x}$

5. $9 \tan^2\theta + 12 \tan \theta + 4$ 7. $24 \tan x \sin x$

9. $\dfrac{1}{\cos x(1 - \cos x)}$ 11. $3 \tan \theta (1 - 5 \tan^2\theta)$

13. $(\cos\theta + \sin\theta)(\cos\theta - \sin\theta)$

15. $(\sin x - 3)(\sin x + 1)$ 17. $\dfrac{1}{\sin x - 1}$

19. $\sin x - \cos x$ 21. $\cot x$ 23. $\tan x$ 25. 1

27. 0   29. 2   31. $\tan x$ 33. $\cot x$ 35. 1

37. Identity   39. Identity   41. Identity   43. Identity

45. Not an identity   47. Identity   49. Identity

51. Not an identity   53. Identity   55. Identity

57. Identity   59. Identity   61. $\cos\theta$ 63. $\csc\theta$

65. $\csc^3\theta$ 67. $\tan\theta - \tfrac{1}{2}\sec\theta$

75. $x^2 + y^2 = 25$, a circle with center $(0, 0)$ and radius 5

77. No; for example, $\sin(30° + 60°) = \sin 90° = 1$, whereas
$\sin 30° + \sin 60° = \dfrac{1}{2} + \dfrac{\sqrt{3}}{2} = \dfrac{1 + \sqrt{3}}{2} \approx 1.37$.

SECTION 8.2

◆ Problems

1. $\sqrt{3}\csc x - 2 = 0$
$$\csc x = \dfrac{2}{\sqrt{3}}$$
$$x = \dfrac{\pi}{3}, \dfrac{2\pi}{3} \text{ in the interval } [0, 2\pi)$$

Hence, $\dfrac{\pi}{3} + 2\pi n$ and $\dfrac{2\pi}{3} + 2\pi n$ describe the general solution of the equation.

2. $2\cos 2x + 1 = 0$
$$\cos 2x = -\tfrac{1}{2}$$
$$2x = \dfrac{2\pi}{3} + 2\pi n \quad \text{and} \quad 2x = \dfrac{4\pi}{3} + 2\pi n$$

Hence, $\dfrac{\pi}{3} + \pi n$ and $\dfrac{2\pi}{3} + \pi n$ describe the general solution of the equation.
The particular solutions in the interval $[0, 2\pi)$ are $\dfrac{\pi}{3}, \dfrac{2\pi}{3}, \dfrac{4\pi}{3}, \dfrac{5\pi}{3}$.

3. $\sin x \cos x - \cos x = 0$
$$\cos x (\sin x - 1) = 0$$
$$\cos x = 0 \quad \text{or} \quad \sin x - 1 = 0$$
$$\sin x = 1$$
$$x = \dfrac{\pi}{2}, \dfrac{3\pi}{2} \qquad x = \dfrac{\pi}{2}$$

in the interval $[0, 2\pi)$. Thus, $\dfrac{\pi}{2} + \pi n$ describes the general solution of the equation.

4. Squaring both sides of $\sin x - \cos x = 1$, we obtain
$$\sin^2 x - 2\sin x \cos x + \cos^2 x = 1$$
$$1 - 2\sin x \cos x = 1$$
$$2\sin x \cos x = 0$$
$$\sin x = 0 \quad \text{or} \quad \cos x = 0$$
$$x = 0, \pi \qquad x = \dfrac{\pi}{2}, \dfrac{3\pi}{2}$$

as possible solutions in the interval $[0, 2\pi)$. However, 0 and $3\pi/2$ are extraneous roots. Hence, $\pi + 2\pi n$ and $\dfrac{\pi}{2} + 2\pi n$ describe the general solution of the equation.

5. For $\tan^2 x - 4 = 0$, we obtain $\tan x = \pm 2$. Thus,
$$x = \tan^{-1} 2 + \pi n \quad \text{and} \quad x = \tan^{-1}(-2) + \pi n$$
$$\approx 1.107 + \pi n \qquad \approx -1.107 + \pi n$$

Hence, in the interval $[0, 2\pi)$ the solutions are 1.107, 2.034, 4.249, and 5.176.

6. Replacing N with $10\tfrac{1}{2}$, we obtain
$$10\tfrac{1}{2} = 12 + 3\sin\left[\dfrac{2\pi}{365}(d - 80)\right]$$
$$-\tfrac{3}{2} = 3\sin\left[\dfrac{2\pi}{365}(d - 80)\right]$$
$$-\tfrac{1}{2} = \sin\left[\dfrac{2\pi}{365}(d - 80)\right].$$

Hence,
$$\dfrac{2\pi}{365}(d - 80) = \dfrac{7\pi}{6}$$
$$d = \dfrac{7\pi}{6} \cdot \dfrac{365}{2\pi} + 80 \approx 293$$

and $\dfrac{2\pi}{365}(d - 80) = \dfrac{11\pi}{6}$
$$d = \dfrac{11\pi}{6} \cdot \dfrac{365}{2\pi} + 80 \approx 415$$

Thus, 10 hours 30 minutes of daylight occurs on day 293 (October 20) and day $415 - 365 = 50$ (February 19).

◆ Exercises

1. $\dfrac{\pi}{2} + 2\pi n$ 3. πn 5. $\dfrac{\pi}{6} + 2\pi n, \dfrac{11\pi}{6} + 2\pi n$

7. $\dfrac{\pi}{4} + 2\pi n, \dfrac{3\pi}{4} + 2\pi n$ 9. $\dfrac{\pi}{3} + \pi n$

11. $\dfrac{\pi}{3} + 2\pi n, \dfrac{5\pi}{3} + 2\pi n$ 13. $\dfrac{\pi}{6} + \pi n, \dfrac{5\pi}{6} + \pi n$

15. No solution 17. (a) $\frac{\pi}{4} + \pi n$ (b) $\frac{\pi}{4}, \frac{5\pi}{4}$

19. (a) $3\pi + 4\pi n$ (b) None

21. (a) $\frac{3\pi}{16} + \frac{\pi}{2}n, \frac{5\pi}{16} + \frac{\pi}{2}n$
    (b) $\frac{3\pi}{16}, \frac{5\pi}{16}, \frac{11\pi}{16}, \frac{13\pi}{16}, \frac{19\pi}{16}, \frac{21\pi}{16}, \frac{27\pi}{16}, \frac{29\pi}{16}$

23. (a) $\frac{\pi}{6} + \frac{\pi}{2}n, \frac{\pi}{3} + \frac{\pi}{2}n$
    (b) $\frac{\pi}{6}, \frac{\pi}{3}, \frac{2\pi}{3}, \frac{5\pi}{6}, \frac{7\pi}{6}, \frac{4\pi}{3}, \frac{5\pi}{3}, \frac{11\pi}{6}$

25. (a) $\frac{5\pi}{24} + \frac{\pi}{2}n$ (b) $\frac{5\pi}{24}, \frac{17\pi}{24}, \frac{29\pi}{24}, \frac{41\pi}{24}$

27. (a) $\pm\frac{2\pi}{3} + 2\pi n$ (b) $\frac{2\pi}{3}, \frac{4\pi}{3}$

29. (a) $\frac{5\pi}{36} + \frac{2\pi}{3}n, \frac{13\pi}{36} + \frac{2\pi}{3}n$
    (b) $\frac{5\pi}{36}, \frac{13\pi}{36}, \frac{29\pi}{36}, \frac{37\pi}{36}, \frac{53\pi}{36}, \frac{61\pi}{36}$

31. (a) $\frac{5\pi}{3} + 2\pi n, \frac{7\pi}{3} + 2\pi n$ (b) $\frac{\pi}{3}, \frac{5\pi}{3}$

33. $\frac{3\pi}{2} + 2\pi n, \frac{\pi}{6} + 2\pi n, \frac{5\pi}{6} + 2\pi n$ 35. $\frac{\pi}{4} + \frac{\pi}{2}n$

37. $\frac{\pi}{2} + \pi n, 2\pi n$ 39. $\pi n, \frac{\pi}{3} + 2\pi n, \frac{5\pi}{3} + 2\pi n$

41. $\frac{\pi}{2} + \pi n, \frac{\pi}{4} + \frac{\pi}{2}n$ 43. $2\pi n, \frac{2\pi}{3} + 2\pi n, \frac{4\pi}{3} + 2\pi n$

45. $\frac{\pi}{3} + \pi n, \frac{2\pi}{3} + \pi n$

47. $\frac{\pi}{6} + 2\pi n, \frac{5\pi}{6} + 2\pi n, \frac{3\pi}{2} + 2\pi n$

49. $\pi + 2\pi n, \frac{2\pi}{3} + 2\pi n, \frac{4\pi}{3} + 2\pi n$

51. $\frac{\pi}{3} + \pi n, \frac{2\pi}{3} + \pi n$

53. πn 55. $\frac{\pi}{2} + \pi n$ 57. $2\pi n, \frac{3\pi}{2} + 2\pi n$

59. $\pi + 2\pi n$ 61. 0.7297, 2.412

63. 2.944, 6.086 65. 1.231, 1.911, 4.373, 5.052

67. 0.3091, 0.7381, 2.403, 2.832, 4.498, 4.927

69. 0.2527, 2.889 71. 1.249, 2.034, 4.391, 5.176

73. 0.6818, 2.460 75. 0.4636, 3.605 77. 2.47, 3.82

79. 0.393, 1.96, 3.53, 5.11

81. In seconds: $\frac{\pi}{3}, \frac{2\pi}{3}, \frac{4\pi}{3}, \frac{5\pi}{3}$ 83. 1990, 1994

85. Day 127 (May 7) and day 215 (August 3)

87. $x_0 + \frac{2\pi}{b}n$

89. Since the graph of $y = \sin nx$ completes n cycles in the interval $[0, 2\pi)$, the equation $\sin nx = k$ for $-1 < k < 1$ must have $2n$ solutions.

91. (a) πn (b) $\frac{\pi}{6} + 2\pi n, \frac{5\pi}{6} + 2\pi n$

93. Since $\sin^2 x + \cos^2 x = 1$, $\sin^2 x = 1$ implies that $\cos^2 x = 0$. Hence, dividing both sides by $\cos^2 x$ does not generate an equivalent equation.

SECTION 8.3

Problems

1. $\cos 3x \cos 2x + \sin 3x \sin 2x = \cos(3x - 2x) = \cos x$

2. $\sin\left(x + \frac{\pi}{6}\right) \cos x - \cos\left(x + \frac{\pi}{6}\right) \sin x$
   $= \left(\sin x \cos \frac{\pi}{6} + \cos x \sin \frac{\pi}{6}\right) \cos x$
   $\quad - \left(\cos x \cos \frac{\pi}{6} - \sin x \sin \frac{\pi}{6}\right) \sin x$
   $= \cos^2 x \sin \frac{\pi}{6} + \sin^2 x \sin \frac{\pi}{6}$
   $= \sin \frac{\pi}{6} (\cos^2 x + \sin^2 x)$
   $= \sin \frac{\pi}{6} = \frac{1}{2}$

3. Setting a calculator in degree mode, we obtain
   $$\frac{\tan 87° - \tan 27°}{1 + \tan 87° \tan 27°} \approx 1.732 \quad \text{and} \quad \sqrt{3} \approx 1.732$$
   as shown in the figure:

4. Setting a calculator in degree mode, we obtain
$\cos 75° \approx 0.2588$ and $(\sqrt{6} - \sqrt{2})/4 \approx 0.2588$
as shown in the figure:

```
cos 75
        .2588190451
(√6-√2)/4
        .2588190451
```

5. From Example 5, we have $\sin u = \frac{3}{5}$ and $\cos u = -\frac{4}{5}$. Also, if $\tan v = -\frac{1}{4}$ with v in quadrant II, then

$$\sec v = -\sqrt{1 + \tan^2 v}$$
$$= -\sqrt{1 + (-\frac{1}{4})^2} = -\frac{\sqrt{17}}{4}$$

Hence, $\cos v = -\dfrac{4}{\sqrt{17}}$ and $\sin v = \dfrac{1}{\sqrt{17}}$. Thus,

$\sin(u - v) = \sin u \cos v - \cos u \sin v$
$= \left(\dfrac{3}{5}\right)\left(-\dfrac{4}{\sqrt{17}}\right) - \left(-\dfrac{4}{5}\right)\left(\dfrac{1}{\sqrt{17}}\right)$
$= -\dfrac{8}{5\sqrt{17}}$

6. $\sin 3x \cot x + \cos 3x = \sin 4x \csc x$
    $\sin 3x \cdot \dfrac{\cos x}{\sin x} + \cos 3x$
    $\dfrac{\sin 3x \cos x + \cos 3x \sin x}{\sin x}$
    $\dfrac{\sin 4x}{\sin x}$
    $\sin 4x \csc x$

7. $\sin 3x \cos x = 1 - \cos 3x \sin x$
    $\sin 3x \cos x + \cos 3x \sin x = 1$
    $\sin 4x = 1$
    $4x = \dfrac{\pi}{2} + 2\pi n$
    $x = \dfrac{\pi}{8} + \dfrac{\pi}{2}n$

8. If $i(t) = 1$ amp, then
$1 = 2\sin\left(4t + \dfrac{\pi}{6}\right)$ or $\dfrac{1}{2} = \sin\left(4t + \dfrac{\pi}{6}\right)$

which implies that

$4t + \dfrac{\pi}{6} = \dfrac{\pi}{6} + 2\pi n$ and $4t + \dfrac{\pi}{6} = \dfrac{5\pi}{6} + 2\pi n$
$t = \dfrac{\pi}{2}n$ $\qquad t = \dfrac{\pi}{6} + \dfrac{\pi}{2}n$

Thus, the only times in the interval $[0, \pi/2)$ for which $i(t) = 1$ amp is when $t = 0$ or $\pi/6$ second.

◆ **Exercises**

1. $-\sin x$ 3. $\sin x$ 5. $\tan x$ 7. 0 9. 1
11. $\cos 8x$ 13. $\cos x$ 15. 1
17. $\dfrac{1 + \sqrt{3}}{2\sqrt{2}}$ or $\dfrac{\sqrt{2} + \sqrt{6}}{4}$
19. $\dfrac{1 + \sqrt{3}}{2\sqrt{2}}$ or $\dfrac{\sqrt{2} + \sqrt{6}}{4}$ 21. $\dfrac{1 + \sqrt{3}}{1 - \sqrt{3}}$ 23. $-\dfrac{5}{3}$
25. $\dfrac{4\sqrt{3} - 3}{10}$ 27. $-\dfrac{16}{65}$ 29. $-\dfrac{13}{5\sqrt{10}}$
31. Identity 33. Identity 35. Not an identity
37. Identity 43. $\dfrac{\pi}{4} + \dfrac{\pi}{2}n$ 45. $\dfrac{\pi}{4} + 2\pi n, \dfrac{3\pi}{4} + 2\pi n$
47. $\dfrac{\pi}{12} + \dfrac{\pi}{3}n$ 49. πn 51. $\dfrac{\pi}{3} + 2\pi n, \dfrac{5\pi}{3} + 2\pi n$
53. $\dfrac{\pi}{4} + 2\pi n, \dfrac{7\pi}{4} + 2\pi n$

55. (a) $f(x) = 3\sqrt{2}\sin\left(2x + \dfrac{\pi}{4}\right)$
    (b) Amplitude $3\sqrt{2}$, period π, phase shift $\pi/8$ unit left

57. (a) $f(x) = 2\sin\left(5x + \dfrac{5\pi}{6}\right)$
    (b) Amplitude 2, period $\dfrac{2\pi}{5}$, phase shift $\pi/6$ unit left

59. $f(x) = 16.3\sin(5.2x + 5.60)$

61. (a) $a = 13, c = 1.176$
    (b) In seconds: approximately
    0.67, 0.309, 0.400, 0.642, 0.733, 0.975

63. (a) $\sin\left(\dfrac{\pi}{6} + \dfrac{\pi}{3}\right) = \sin\dfrac{\pi}{2} = 1$,
    but $\sin\dfrac{\pi}{6} + \sin\dfrac{\pi}{3} = \dfrac{1}{2} + \dfrac{\sqrt{3}}{2} = \dfrac{1 + \sqrt{3}}{2} \approx 1.366$
    (b) Yes; for example, let $u = \pi/6$ and $v = 11\pi/6$.
    Then $\sin\left(\dfrac{\pi}{6} + \dfrac{11\pi}{6}\right) = \sin 2\pi = 0$,
    and $\sin\dfrac{\pi}{6} + \sin\dfrac{11\pi}{6} = \dfrac{1}{2} + \left(-\dfrac{1}{2}\right) = 0$

65. (a) $\dfrac{1}{\sqrt{2}}$ (b) $\dfrac{2 + \sqrt{15}}{4\sqrt{5}}$ or $\dfrac{2\sqrt{5} + 5\sqrt{3}}{20}$

67. $\sin C = a\sqrt{1 - b^2} + b\sqrt{1 - a^2}$

71. (a) 0.999 (b) 0.999

73. (a) -0.914 (b) -0.914

75. (a) 28.6 (b) 28.6

SECTION 8.4

◆ Problems

1. $\dfrac{2\tan 3x}{1-\tan^2 3x} = \tan 2(3x) = \tan 6x$

2. $\quad 2\cos x \csc 2x = \csc x$

$\quad 2\cos x \cdot \dfrac{1}{\sin 2x}$

$\quad 2\cos x \cdot \dfrac{1}{2\sin x \cos x}$

$\quad \dfrac{1}{\sin x}$

$\quad \csc x$

3. $\cos x + \sin 2x = 0$
   $\cos x + 2\sin x \cos x = 0$
   $\cos x(1 + 2\sin x) = 0$
   $\cos x = 0 \quad \text{or} \quad 1 + 2\sin x = 0$
   $\sin x = -\tfrac{1}{2}$

   Hence, $x = \dfrac{\pi}{2} + \pi n, \dfrac{7\pi}{6} + 2\pi n, \dfrac{11\pi}{6} + 2\pi n$.

4. $\sin 3\left(\dfrac{\pi}{6}\right) \stackrel{?}{=} 3\sin\dfrac{\pi}{6} - 4\sin^3\dfrac{\pi}{6}$
   $\sin\dfrac{\pi}{2} \stackrel{?}{=} 3\left(\tfrac{1}{2}\right) - 4\left(\tfrac{1}{2}\right)^3$
   $1 \stackrel{?}{=} \tfrac{3}{2} - \tfrac{1}{2}$
   $1 = 1$

5. $\cos^4 x = \cos^2 x \cos^2 x$
   $= \tfrac{1}{2}(1 + \cos 2x) \cdot \tfrac{1}{2}(1 + \cos 2x)$
   $= \tfrac{1}{4}(1 + 2\cos 2x + \cos^2 2x)$
   $= \tfrac{1}{4}\left(1 + 2\cos 2x + \tfrac{1}{2}(1 + \cos 4x)\right)$
   $= \tfrac{1}{4}\left(\tfrac{3}{2} + 2\cos 2x + \tfrac{1}{2}\cos 4x\right)$
   $= \tfrac{1}{8}(3 + 4\cos 2x + \cos 4x)$

6. Setting a calculator in radian mode, we obtain
   $\sin\dfrac{\pi}{12} \approx 0.2588 \quad \text{and} \quad \dfrac{\sqrt{2-\sqrt{3}}}{2} \approx 0.2588$

   as shown in the following figure:

   ```
   sin (π/12)
              .2588190451
   √(2-√3)/2
              .2588190451
   ```

7. If θ is in quadrant III, then
   $$\pi < \theta < \dfrac{3\pi}{2} \quad \text{or} \quad \dfrac{\pi}{2} < \dfrac{\theta}{2} < \dfrac{3\pi}{4}.$$
   Hence, $\sin\dfrac{\theta}{2}$ is *positive* and
   $$\sin\dfrac{\theta}{2} = \sqrt{\dfrac{1-\cos\theta}{2}} = \sqrt{\dfrac{1-(-0.28)}{2}}$$
   $$= \sqrt{0.64}$$
   $$= 0.8$$

8. $\tan\dfrac{x}{2} + \cot\dfrac{x}{2} = 2\csc x$

   $\dfrac{1-\cos x}{\sin x} + \dfrac{\sin x}{1-\cos x}$

   $\dfrac{(1 - 2\cos x + \cos^2 x) + \sin^2 x}{\sin x(1-\cos x)}$

   $\dfrac{2(1-\cos x)}{\sin x(1-\cos x)}$

   $2\csc x$

9. If $A = 170$ sq ft, then $170 = 200\sin\theta$, or $\sin\theta = \tfrac{170}{200}$. Hence, $\theta = \arcsin\tfrac{170}{200} \approx 58.2°$.

◆ Exercises

1. $5\sin 6x$ 3. $\cos 4x$ 5. -1 7. $\cot 6x$ 9. $\cos x$

11. $\cos x$ 13. 1   15. $\sec^2 3x$ 29. πn

31. $\pi n, \dfrac{\pi}{6} + 2\pi n, \dfrac{5\pi}{6} + 2\pi n$ 33. $\dfrac{\pi}{6} + \dfrac{\pi}{2}n, \dfrac{\pi}{3} + \dfrac{\pi}{2}n$

35. $\pi n, \dfrac{2\pi}{3} + 2\pi n, \dfrac{4\pi}{3} + 2\pi n$

37. $\pi n, \dfrac{\pi}{3} + \pi n, \dfrac{2\pi}{3} + \pi n$

39. $\cos 3x = 4\cos^3 x - 3\cos x$

41. $\sin 5x = 5\sin x - 20\sin^3 x + 16\sin^5 x$

43. $\tfrac{1}{2}(1 + \cos 6x)$

45. $\tfrac{1}{16}(1 - \cos 2x - \cos 4x + \cos 4x \cos 2x)$

47. $\tfrac{1}{2}\sqrt{2+\sqrt{3}}$ 49. $\tfrac{1}{2}\sqrt{2-\sqrt{2}}$

51. (a) $\tfrac{24}{25}$ (b) $\tfrac{7}{25}$ (c) $\dfrac{1}{\sqrt{10}}$ (d) $\dfrac{3}{\sqrt{10}}$

53. (a) $\tfrac{3}{4}$ (b) $-\tfrac{5}{4}$ (c) $\tfrac{1}{10}\sqrt{50 - 5\sqrt{10}}$
    (d) $-\tfrac{1}{3}\sqrt{11 + 2\sqrt{10}}$

61. $2\pi n$ 63. $2\pi n, \frac{2\pi}{3} + 2\pi n$ 65. $\frac{\pi}{2} + \pi n, \frac{\pi}{4} + \frac{\pi}{2}n$

67. 0.253, 1.57, 2.89, 4.71 69. 0.262, 1.31, 3.40, 4.45

71. 1.05, 1.57, 2.09, 4.19, 4.71, 5.24 (*Note:* The double-angle formula does not generate the solutions $\pi/2$ and $3\pi/2$, because $\tan(\pi/2)$ and $\tan(3\pi/2)$ are undefined.)

73. Identity 75. Identity

77. (a) $V = 36 \sin \theta$ (b) 36 cu ft (c) $\theta \approx 56.4°$

79. 30° and 60°

81. (a) Amplitude 3, period $\frac{\pi}{4}$

   (b) Amplitude $\frac{1}{2}$, period π

83. (a) $\frac{3}{5}$ (b) $-\frac{7}{8}$

SECTION 8.5

Problems

1. $\sin 3x \cos 2x = \cos 2x \sin 3x$
   $= \frac{1}{2}[\sin(2x + 3x) - \sin(2x - 3x)]$
   $= \frac{1}{2}[\sin 5x - \sin(-x)]$.
   However, since the sine function is an odd function and $\sin(-x) = -\sin x$, we have
   $\sin 3x \cos 2x = \frac{1}{2}(\sin 5x + \sin x)$
   $= \frac{1}{2} \sin 5x + \frac{1}{2} \sin x$.

2. $\sin 5x + \sin 7x = 2 \sin \frac{5x + 7x}{2} \cos \frac{5x - 7x}{2}$
   $= 2 \sin 6x \cos(-x)$.
   However, since the cosine function is an even function and $\cos(-x) = \cos x$, we have
   $\sin 5x + \sin 7x = 2 \sin 6x \cos x$.

3. $\cos 75° - \cos 15° = -2 \sin \frac{75° + 15°}{2} \sin \frac{75° - 15°}{2}$
   $= -2 \sin 45° \sin 30°$
   $= -2 \cdot \frac{1}{\sqrt{2}} \cdot \frac{1}{2} = -\frac{1}{\sqrt{2}}$.

4. $2 \cos^2 x \sin x - 2 \sin^3 x = \sin 3x - \sin x$

   | $2 \cos \frac{3x + x}{2} \sin \frac{3x - x}{2}$
   | $2 \cos 2x \sin x$
   | $2(\cos^2 x - \sin^2 x) \sin x$
   | $2 \cos^2 x \sin x - 2 \sin^3 x$

5. To find the six solutions in the interval $[0, 2\pi)$, we let $n = 0, 1, 2, 3, 4, 5$ in the general solution $\pi n/3$, and obtain

   $0, \frac{\pi}{3}, \frac{2\pi}{3}, \pi, \frac{4\pi}{3}, \frac{5\pi}{3}$.

6. $3 \sin 3x + \sin x = 0$
   $3(3 \sin x - 4 \sin^3 x) + \sin x = 0$
   $10 \sin x - 12 \sin^3 x = 0$
   $2 \sin x(5 - 6 \sin^2 x) = 0$

   $2 \sin x = 0$ or $5 - 6 \sin^2 x = 0$
   $\sin x = 0$ $\sin x = \pm\sqrt{\frac{5}{6}}$

   Hence, $x = \pi n$ and $x \approx 1.15 + \pi n, 1.99 + \pi n$. Thus, in the interval $[0, 2\pi)$ we have

   Relative maxima: $(0, 2), (1.99, 0.544), (4.29, 0.544)$
   Relative minima: $(1.15, -0.544), (\pi, -2), (5.13, -0.544)$

Exercises

1. $\cos 2x - \cos 4x$ 3. $\cos 8x + \cos 4x$
5. $\frac{1}{2} \sin 7x + \frac{1}{2} \sin x$ 7. $3 \sin 3x + 3 \sin 7x$
9. $2 \cos 3x \cos x$ 11. $2 \sin x \cos 4x$
13. $2 \sin x \sin \frac{x}{3}$ 15. $-\cos \frac{9x}{2} \sin \frac{3x}{2}$ 17. $\frac{1}{4}$
19. $\frac{\sqrt{3} + 2}{4}$ 21. $-\frac{\sqrt{6}}{2}$ 23. $-\frac{1}{\sqrt{2}}$ 37. $\frac{\pi}{2}n$
39. $\frac{\pi}{4}n$ 41. $\frac{\pi}{3}n, \frac{\pi}{2} + 2\pi n$ 43. $\frac{\pi}{3}n$ 45. $\frac{3\pi}{4} + \pi n$
47. $\frac{\pi}{4} + \frac{\pi}{2}n, \pi n$ 49. $\frac{\pi}{3}, \pi, \frac{5\pi}{3}$
51. $0, \frac{\pi}{6}, \frac{\pi}{2}, \frac{5\pi}{6}, \pi, \frac{7\pi}{6}, \frac{3\pi}{2}, \frac{11\pi}{6}$
53. $0, \frac{\pi}{3}, \frac{2\pi}{3}, \pi, \frac{4\pi}{3}, \frac{5\pi}{3}$
55. Relative maxima: $(0, 2), (\pi, 0)$
    Relative minima: $(\arccos(-\frac{1}{4}), -\frac{9}{8})$,
    $(2\pi - \arccos(-\frac{1}{4}), -\frac{9}{8})$
57. $\sin 50°$
59. (a) 0; for supplementary angles x and y, $\cos x = -\cos y$.
    (b) 0; for supplementary angles x and y, $\sin x = \sin y$.
61. Yes, because the least common multiple for their periods, $2\pi/n$ and 2π, is 2π.

CHAPTER 8 REVIEW EXERCISES

1. $\sin 2x$ 3. 1 5. $\cos 9x$ 7. $\cos x$
9. $\sin 6x$ 11. $\tan 8x$ 13. $|\sin 5x|$

15. $\sin 6x + \sin 4x$ 17. $-2\sin\dfrac{5x}{2}\sin\dfrac{x}{2}$

19. $\dfrac{\sqrt{2-\sqrt{2}}}{2}$ 21. $\dfrac{2+\sqrt{3}}{4}$

23. (a) $\dfrac{7}{5\sqrt{2}}$ (b) $\dfrac{4\sqrt{3}+3}{10}$ (c) $\dfrac{24}{25}$ (d) $\dfrac{3}{\sqrt{10}}$

25. (a) $\dfrac{12\sqrt{5}+10}{39}$ (b) $\dfrac{5\sqrt{5}+24}{39}$ (c) $\dfrac{120}{119}$

(d) $\dfrac{\sqrt{18+6\sqrt{5}}}{6}$

27. $3\sec\theta$

29. $\tfrac{1}{16}(1+\cos 2x - \cos 4x - \cos 4x \cos 2x)$
or, applying a product-to-sum formula,
$\tfrac{1}{32}(2+\cos 2x - 2\cos 4x - \cos 6x)$

31. Identity   33. Identity   35. Not an identity

37. Identity   39. Identity   41. Not an identity

43. Identity   45. Identity   47. Identity   49. Identity

51. Identity   53. Identity   55. Identity

57. Not an identity   59. Identity

61. (a) $\dfrac{2\pi}{3}+2\pi n, \dfrac{4\pi}{3}+2\pi n$ (b) $\dfrac{2\pi}{3}, \dfrac{4\pi}{3}$

63. (a) $\dfrac{\pi}{3}+\pi n, \dfrac{2\pi}{3}+\pi n$ (b) $\dfrac{\pi}{3}, \dfrac{2\pi}{3}, \dfrac{4\pi}{3}, \dfrac{5\pi}{3}$

65. (a) $\dfrac{3\pi}{8}+\dfrac{\pi}{2}n$ (b) $\dfrac{3\pi}{8}, \dfrac{7\pi}{8}, \dfrac{11\pi}{8}, \dfrac{15\pi}{8}$

67. (a) $\dfrac{2\pi}{9}+\dfrac{2\pi}{3}n$ (b) $\dfrac{2\pi}{9}, \dfrac{8\pi}{9}, \dfrac{14\pi}{9}$

69. (a) $\pi n, \dfrac{\pi}{6}+2\pi n, \dfrac{5\pi}{6}+2\pi n$ (b) $0, \dfrac{\pi}{6}, \dfrac{5\pi}{6}, \pi$

71. (a) $\pi+2\pi n, \dfrac{\pi}{3}+2\pi n, \dfrac{5\pi}{3}+2\pi n$ (b) $\dfrac{\pi}{3}, \pi, \dfrac{5\pi}{3}$

73. (a) $\dfrac{3\pi}{2}+2\pi n$ (b) $\dfrac{3\pi}{2}$ 75. (a) $\dfrac{\pi}{2}+2\pi n$ (b) $\dfrac{\pi}{2}$

77. (a) $\pi n, \dfrac{2\pi}{3}+2\pi n, \dfrac{4\pi}{3}+2\pi n$ (b) $0, \pi, \dfrac{2\pi}{3}, \dfrac{4\pi}{3}$

79. (a) $\dfrac{\pi}{4}+\pi n$ (b) $\dfrac{\pi}{4}, \dfrac{5\pi}{4}$ 81. (a) $2\pi n$ (b) 0

83. (a) $\dfrac{\pi}{4}n$ (b) $\dfrac{\pi}{4}, \dfrac{\pi}{2}, \dfrac{3\pi}{4}, \pi, \dfrac{5\pi}{4}, \dfrac{3\pi}{2}, \dfrac{7\pi}{4}$

85. (a) $\dfrac{\pi}{12}+\dfrac{\pi}{2}n, \dfrac{5\pi}{12}+\dfrac{\pi}{2}n$

(b) $\dfrac{\pi}{12}, \dfrac{5\pi}{12}, \dfrac{7\pi}{12}, \dfrac{11\pi}{12}, \dfrac{13\pi}{12}, \dfrac{17\pi}{12}, \dfrac{19\pi}{12}, \dfrac{23\pi}{12}$

87. 3.553, 5.872   89. 1.107, 2.034, 4.249, 5.176

91. 0.3218, 0.4636, 3.463, 3.605   93. 0.2846, 2.857

95. 0.3398, 2.802, 3.481, 5.943

97. (a) $f(x)=2\sin\left(3x+\dfrac{\pi}{3}\right)$; amplitude 2, period $\dfrac{2\pi}{3}$,
phase shift $\dfrac{\pi}{9}$ unit left

(b) $f(x)=2\sqrt{2}\sin\left(4\pi x+\dfrac{7\pi}{4}\right)$; amplitude $2\sqrt{2}$,
period $\tfrac{1}{2}$, phase shift $\tfrac{7}{16}$ units left

99. (a) $a=40, c\approx 0.9273$
(b) In seconds: 0.078, 0.223, 0.328, 0.473, 0.578, 0.723, 0.828, 0.973

101. 28.2°, 61.8°

Chapter 9

SECTION 9.1

Problems

1. The longest side (52) is opposite the largest angle ($\theta=70°$) and the shortest side ($x=41.1$) is opposite the smallest angle (48°).

2. $\theta = 180° - (108° + 28°) = 44°$ Hence, by the law of sines

$$\dfrac{125}{\sin 108°} = \dfrac{x}{\sin 28°} = \dfrac{y}{\sin 44°}$$

$\dfrac{125}{\sin 108°} = \dfrac{x}{\sin 28°}$ and $\dfrac{125}{\sin 108°} = \dfrac{y}{\sin 44°}$

$x = \dfrac{125\sin 28°}{\sin 108°}$ $y = \dfrac{125\sin 44°}{\sin 108°}$

≈ 61.7 ≈ 91.3

3. By the law of sines, we have

$$\dfrac{22}{\sin 38°} = \dfrac{35}{\sin\beta} \text{ or } \sin\beta = \dfrac{35\sin 38°}{22}$$

Since β is an obtuse angle, we have

$$\beta = 180° - \sin^{-1}\left(\dfrac{35\sin 38°}{22}\right) \approx 101.6324°.$$

Using the fact that the sum of the interior angles in a triangle is 180°, we find

$\alpha \approx 180° - (38° + 101.6324°) = 40.3676°$. Applying the law of sines again gives us x:

$$\frac{22}{\sin 38°} = \frac{x}{\sin 40.3676°}$$

$$x = \frac{22 \sin 40.3676°}{\sin 38°} \approx 23.1444.$$

Rounding α and β to the nearest tenth of a degree and x to three significant digits, we write $\alpha \approx 40.4°$, $\beta \approx 101.6°$, and $x \approx 23.1$.

4. By the law of sines, $\dfrac{CD}{\sin 36°15'} = \dfrac{AC}{\sin 90°}$.

However, since $\sin 90° = 1$, we have

$CD = AC \sin 36°15'$
$= \left(\dfrac{100 \sin 118°30'}{\sin 25°15'}\right) \sin 36°15' \approx 121.8$ ft

◆ Exercises

1. $\theta = 73°$, $x \approx 35.7$, $y \approx 33.9$
3. $\theta = 61°$, $x \approx 107$, $y \approx 294$
5. $\theta = 101°$, $x \approx 9.31$, $y \approx 6.75$
7. $\alpha \approx 22.7°$, $\beta \approx 66.3°$, $x \approx 52.2$
9. $\alpha \approx 143.7°$, $\beta \approx 17.3°$, $x \approx 1260$
11. $\alpha \approx 94.4°$, $\beta \approx 37.6°$, $x \approx 47.9$
13. $\theta = 37°$, $a \approx 77.2$, $b \approx 91.8$
15. $\alpha \approx 84.3°$, $\theta \approx 33.7°$, $a \approx 39.4$
17. $\theta \approx 63.7°$, $\beta \approx 78.3°$, $b \approx 1080$ or $\theta \approx 116.3°$, $\beta \approx 25.7°$, $b \approx 479$
19. $\alpha \approx 33.9°$, $\beta \approx 79.1°$, $a \approx 5.91$ or $\alpha \approx 12.1°$, $\beta \approx 100.9°$, $a \approx 2.22$
21. No such triangle exists   23. No such triangle exists
25. 42.7 m   27. 466 m   29. 64.3°
31. (a) 416 m   (b) 403 m   33. (a) 201 ft   (b) 127 ft
35. 1210 ft   37. (a) 33.7 ft   (b) 40.0 ft
39. (a) 184 m   (b) 74.6 m
41. $B < A < C$; the smallest angle is opposite the shortest side and the largest angle is opposite the longest side.
43. $\sin \theta = \dfrac{a}{c} = \dfrac{\text{opp}}{\text{hyp}}$; right triangle definition of sine
47. Appears to be a solution of a triangle
49. Does not appear to be a solution of a triangle
52. $v_2 \approx 2.02 \times 10^{10}$ cm/s

SECTION 9.2

◆ Problems

1. Since the side opposite β is the largest side of the triangle, we cannot tell whether β is acute or obtuse. Thus,

$$\beta = \sin^{-1}\left(\frac{33 \sin 26°}{14.561}\right) \approx 83.5°$$

or

$$\beta = 180° - \sin^{-1}\left(\frac{33 \sin 26°}{14.561}\right) \approx 96.5°$$

If we choose $\beta \approx 83.5°$, then $\alpha \approx 180° - (26° + 83.5°) = 70.5°$. However, this would contradict the law of sines, since

$$\sin 70.5° \neq \frac{28 \sin 26°}{14.561}.$$

Therefore, we conclude that $\beta \approx 96.5°$.

2. Solving the equation
$16.2^2 = 22.5^2 + 14.9^2 - 2(22.5)(14.9) \cos \alpha$ for $\cos \alpha$, we obtain

$$\cos \alpha = \frac{16.2^2 - 22.5^2 - 14.9^2}{-2(22.5)(14.9)}$$

which implies that

$$\alpha = \cos^{-1}\left(\frac{16.2^2 - 22.5^2 - 14.9^2}{-2(22.5)(14.9)}\right) \approx 46.0°.$$

Solving the equation
$14.9^2 = 16.2^2 + 22.5^2 - 2(16.2)(22.5) \cos \beta$ for $\cos \beta$, we obtain

$$\cos \beta = \frac{14.9^2 - 16.2^2 - 22.5^2}{-2(16.2)(22.5)}$$

which implies that

$$\beta = \cos^{-1}\left(\frac{14.9^2 - 16.2^2 - 22.5^2}{-2(16.2)(22.5)}\right) \approx 41.4°.$$

3. (a) $A = \dfrac{ab \sin \theta}{2} = \dfrac{(16.2)(22.5) \sin 41.418°}{2}$
≈ 121 square units

   (b) $A = \dfrac{ab \sin \theta}{2} = \dfrac{(22.5)(14.9) \sin 45.994°}{2}$
≈ 121 square units

4. For a triangle with sides $a = 3$, $b = 4$, and $c = 5$, the semi-perimeter $s = \frac{1}{2}(3 + 4 + 5) = 6$. Thus, by Hero's formula,

the area of the triangle is

$$A = \sqrt{s(s-a)(s-b)(s-c)}$$
$$= \sqrt{6(6-3)(6-4)(6-5)}$$
$$= \sqrt{36}$$
$$= 6 \text{ square units.}$$

Since this triangle is a right triangle with base $b = 4$ and height $h = c = 3$, we have

$$A = \frac{bh}{2} = \frac{(4)(3)}{2} = 6 \text{ square units.}$$

This area agrees with the answer we obtained from Hero's formula.

5. The information is shown in the sketch.

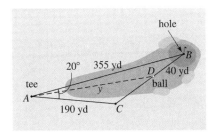

Using the fact that $BC \approx 188.04$ yd (see Example 5), we can apply the law of sines to triangle ABC in order to determine angle B:

$$\frac{188.04}{\sin 20°} = \frac{190}{\sin B} \quad \text{or} \quad \sin B = \frac{190 \sin 20°}{188.04}$$

Hence, $B = \sin^{-1}\left(\frac{190 \sin 20°}{188.04}\right) \approx 20.218°$.

Now, by applying the law of cosines to triangle ABD, we can find the distance y from the tee to the ball:

$$y^2 = 40^2 + 355^2 - 2(40)(355) \cos 20.218°$$
$$y = \sqrt{40^2 + 355^2 - 2(40)(355) \cos 20.218°} \approx 318 \text{ yd}$$

◆ Exercises

1. (a) $x \approx 37.0, \alpha \approx 34.5°, \beta = 51.5°$
   (b) 304 square units

3. (a) $x \approx 198, \alpha \approx 133.9°, \beta = 17.1°$
   (b) 8550 square units

5. (a) $x \approx 1090, \alpha \approx 129.6°, \beta = 33.4°$
   (b) 860,000 square units

7. (a) $\alpha \approx 86.9°, \beta \approx 67.4°, \theta \approx 25.7°$
   (b) 9000 square units

9. (a) $\alpha \approx 37.6°, \beta \approx 96.3°, \theta \approx 46.1°$
   (b) 221 square units

11. $c \approx 14.5, \alpha \approx 61.6°, \beta \approx 76.4°$

13. $a \approx 7.02, \beta \approx 25.5°, \theta \approx 42.5°$

15. $b \approx 474, \alpha \approx 30.7°, \theta \approx 101.3°$

17. $\alpha \approx 107.6°, \beta \approx 49.0°, \theta \approx 23.4°$

19. $\alpha \approx 58.0°, \beta \approx 78.1°, \theta \approx 43.9°$

21. $\alpha \approx 130.8°, \beta \approx 27.9°, \theta \approx 21.3°$

23. No such triangle exists   25. 88.0°   27. 6.32 ft

29. 172 ft   31. 4.00 mi   33. (a) 9.40 cm   (b) 15.2 cm

35. (a) 55.4 m and 139 m   (b) 3430 sq m

37. (a) 158 ft   (b) 136 ft   39. 117 ft   41. 8.07 acres

43. No, not if we are given only three angles

45. (a) About 133.2°   (b) About 16.5 square units

47. $\theta = 120°$ 49. $b = 6, c = 3$

51. $x^2 + y^2 = 2a^2 + 2b^2$

53. (a) $A \approx 29.30°, B \approx 44.55°, ACB \approx 106.15°$
    (b) $x \approx 3.44, y \approx 4.94, z \approx 3.02$
    (c) $\frac{x}{y} \approx 0.696, \frac{a}{b} \approx 0.698$; it appears that $\frac{x}{y} = \frac{a}{b}$.
    (e) $\sqrt{ab - xy} \approx 3.02$; it appears that $\sqrt{ab - xy} = z$.

SECTION 9.3

◆ Problems

1. (a) We place vector **A** in standard position by moving its initial point $(-2, -3)$ and terminal point $(-5, 1)$ two units right and three units up, as shown in the sketch.

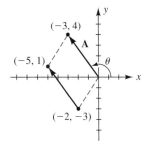

(b) Since the x-component is -3 and the y-component is 4 when placed in standard position, we write
$\mathbf{A} = \langle -3, 4 \rangle$.

(c) $|\mathbf{A}| = \sqrt{(-3)^2 + (4)^2} = \sqrt{25} = 5$

(d) Since θ is in quadrant II, we have

$\theta = 180° - \tan^{-1}\left|\dfrac{4}{-3}\right| \approx 126.9°$.

2. $x = |\mathbf{A}|\cos\theta = 8\cos 212° \approx -6.78$ and
$y = |\mathbf{A}|\sin\theta = 8\sin 212° \approx -4.24$. Hence,
$\mathbf{A} \approx \langle -6.78, -4.24\rangle$.

3. If $\mathbf{A} = \langle -3, 4\rangle$, then $-\mathbf{A} = \langle -(-3), -(4)\rangle = \langle 3, -4\rangle$.

4. $2\mathbf{A} + 3\mathbf{B} = 2\langle -2, 3\rangle + 3\langle -4, -4\rangle$
$= \langle -4, 6\rangle + \langle -12, -12\rangle$
$= \langle -4 + (-12), 6 + (-12)\rangle$
$= \langle -16, -6\rangle$

5. $\mathbf{A} - \mathbf{B} = \langle -2, 3\rangle - \langle -4, -4\rangle$
$= \langle -2 - (-4), 3 - (-4)\rangle$
$= \langle 2, 7\rangle$

6. $-\mathbf{A} + 2\mathbf{B} = -(2\mathbf{i} - 3\mathbf{j}) + 2(\mathbf{i} - 6\mathbf{j})$
$= (-2\mathbf{i} + 3\mathbf{j}) + (2\mathbf{i} - 12\mathbf{j})$
$= (-2 + 2)\mathbf{i} + (3 - 12)\mathbf{j}$
$= 0\mathbf{i} - 9\mathbf{j}$ or $\langle 0, -9\rangle$

7. In order to just lift the sled from the surface, the vertical component of a force $\mathbf{F}$, applied at an angle of 30°, must be 75 lb. Hence,

$75 = |\mathbf{F}|\sin 30°$ or $|\mathbf{F}| = \dfrac{75}{\sin 30°} = 150$ lb.

8. The force $\mathbf{F}_3$ that produces equilibrium must have the same magnitude as the resultant force $\mathbf{F}$ but opposite direction. Hence, $|\mathbf{F}_3| \approx 89$ lb with direction angle $\theta \approx 193° - 180° = 13°$.

9. Using the information shown in the sketch, we can find the magnitude of the displacement $\mathbf{D}$ by using the law of cosines:

$|\mathbf{D}|^2 = 200^2 + 260^2 - 2(200)(260)\cos 7°$

$|\mathbf{D}| = \sqrt{200^2 + 260^2 - 2(200)(260)\cos 7°} \approx 66.1453$.

Rounding to two significant digits, we obtain $|\mathbf{D}| \approx 66$ miles. Now by the law of sines,

$\dfrac{200}{\sin\theta} = \dfrac{66.1453}{\sin 7°}$ or $\sin\theta = \dfrac{200\sin 7°}{66.1453}$

which implies

$\theta = \sin^{-1}\left(\dfrac{200\sin 7°}{66.1453}\right) \approx 21.6°$.

Measured from the South axis, we obtain $90° - (21.6° + 2°) = 66.4°$. Thus, the bearing is S-66.4°-E.

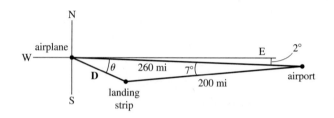

◆ Exercises

1. 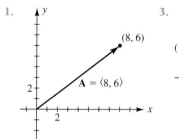   3.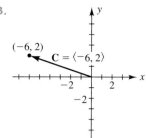

5. 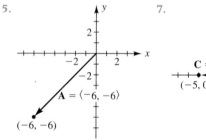   7.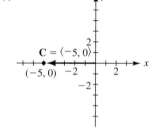

9. $|\mathbf{A}| = 10$, $|\mathbf{C}| = 2\sqrt{10} \approx 6.32$

11. For $\mathbf{A}$, $\theta = 225°$; for $\mathbf{C}$, $\theta = 180°$

13. $|\mathbf{A}| = 13$, $\theta \approx 292.6°$

15. $|\mathbf{C}| = \sqrt{117} \approx 10.8$, $\theta \approx 236.3°$

17. $|\mathbf{A}| = \dfrac{\sqrt{5}}{4} \approx 0.559$, $\theta \approx 26.6°$

19. $|\mathbf{C}| = 2\sqrt{2} \approx 2.83$, $\theta \approx 110.7°$

21. $|\mathbf{A}| \approx 1.46$, $\theta \approx 148.7°$ 23. $|\mathbf{C}| \approx 2.01$, $\theta \approx 79.8°$

25. $x = 310\sqrt{2} \approx 438$, $y = 310\sqrt{2} \approx 438$

27. $x = -42\sqrt{3} \approx -72.7$, $y = -42$

29. $x = -1$, $y = -1$ 31. $x = 0$, $y = -\dfrac{9}{2}$

33. $x \approx -3.23$, $y \approx 4.58$ 35. $x \approx 76.0$, $y \approx -23.8$

37. $\langle 4, 10\rangle$ 39. $\langle 4, 1\rangle$ 41. $\langle 1, 11\rangle$ 43. $\langle 10, 7\rangle$

45. $\langle 32, 26\rangle$ 47. $\langle 0, -\dfrac{3}{2}\rangle$ 49. $\langle 0, 0\rangle$

51. $\langle -26\sqrt{29}, -11\sqrt{29} \rangle$

53.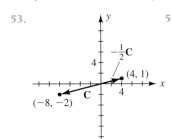

55.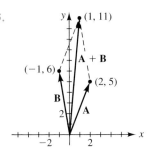

57.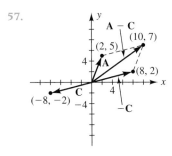

59. Magnitude 94.3 lb, direction angle $\theta \approx 58.0°$

61. Magnitude 46.1 lb, direction angle $\theta \approx 317.7°$

63. Magnitude 72.3 lb, direction angle $\theta \approx 197.7°$

65. (a) $600\sqrt{2} \approx 848.5$ mi/h
(b) $600\sqrt{2} \approx 848.5$ mi/h

67. $\mathbf{F}_3 = \langle -50, -80 \rangle$; $|\mathbf{F}_3| = 94.3$ lb, $\theta = 238°$

69. (a) Magnitude 5.7 mi/h, direction angle $\theta \approx 66°$
(b) About 84 ft   (c) $\theta \approx 114°$

71. About 307 mi, S-48°-E

73. About 24.4 mi, S-21°9′-W

75. (a) $|\mathbf{A}| + |\mathbf{B}|$ (b) $||\mathbf{A}| - |\mathbf{B}||$

77. $\left\langle \dfrac{x}{\sqrt{x^2+y^2}}, \dfrac{y}{\sqrt{x^2+y^2}} \right\rangle$; a unit vector, since its magnitude is 1

79. (a) $c = 3$, $d = -2$ (b) $c = -1$, $d = -4$

SECTION 9.4

◆ **Problems**

1. The modulus r of the complex number $-\sqrt{3} + i$ is

$$r = \sqrt{(-\sqrt{3})^2 + (1)^2} = 2.$$

Since $-\sqrt{3} + i$ lies in quadrant II, θ must be a second-quadrant angle that satisfies the equation

$$\tan \theta = \frac{1}{-\sqrt{3}}.$$

Therefore, $\theta = 5\pi/6$, and we conclude that

$$-\sqrt{3} + i = 2\left(\cos \frac{5\pi}{6} + i \sin \frac{5\pi}{6} \right).$$

2. Using a calculator set in radian mode, we find that $2\sqrt{5}\cos 5.82 \approx 4.00$ and $2\sqrt{5}\sin 5.82 \approx -2.00$. Hence, $2\sqrt{5}(\cos 5.82 + i\sin 5.82) \approx 4.00 - 2.00i$.

3. $\dfrac{6 + 6\sqrt{3}\,i}{-\dfrac{3\sqrt{3}}{2} - \dfrac{3}{2}i} = \dfrac{6 + 6\sqrt{3}\,i}{-\dfrac{3\sqrt{3}}{2} - \dfrac{3}{2}i} \cdot \dfrac{-\dfrac{3\sqrt{3}}{2} + \dfrac{3}{2}i}{-\dfrac{3\sqrt{3}}{2} + \dfrac{3}{2}i}$

$= \dfrac{-9\sqrt{3} + 9i - 27i - 9\sqrt{3}}{\dfrac{27}{4} + \dfrac{9}{4}}$

$= \dfrac{-18\sqrt{3} - 18i}{9} = -2\sqrt{3} - 2i$

4. For $R = 6\,\Omega$, $X_L = 5\,\Omega$, and $X_C = 3\,\Omega$, we have $Z = 6 + (5 - 3)i = 6 + 2i$.

(a) $|\mathbf{Z}| = \sqrt{6^2 + 2^2} = \sqrt{40} = 2\sqrt{10} \approx 6.32\,\Omega$.
(b) Since $6 + 2i$ lies in quadrant I, θ must be a first-quadrant angle that satisfies the equation $\tan \theta = \tfrac{2}{6} = \tfrac{1}{3}$. Hence, $\theta = \arctan \tfrac{1}{3} \approx 18.4°$. Therefore, we conclude that the voltage leads the current by 18.4°.

◆ **Exercises**

1. (a)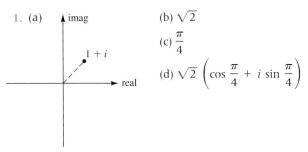
(b) $\sqrt{2}$
(c) $\dfrac{\pi}{4}$
(d) $\sqrt{2}\left(\cos \dfrac{\pi}{4} + i \sin \dfrac{\pi}{4} \right)$

3. (a)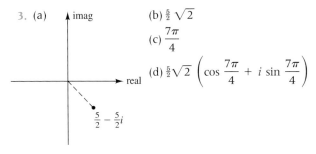
(b) $\tfrac{5}{2}\sqrt{2}$
(c) $\dfrac{7\pi}{4}$
(d) $\tfrac{5}{2}\sqrt{2}\left(\cos \dfrac{7\pi}{4} + i \sin \dfrac{7\pi}{4} \right)$

834   SOLUTIONS AND ANSWERS ◆

5. (a)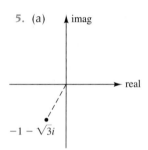
(b) 2
(c) $\dfrac{4\pi}{3}$
(d) $2\left(\cos\dfrac{4\pi}{3} + i\sin\dfrac{4\pi}{3}\right)$

15. (a)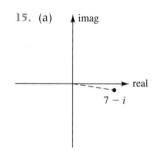
(b) $5\sqrt{2}$
(c) $2\pi - \arctan\frac{1}{7} \approx 6.14$
(d) $5\sqrt{2}\,(\cos 6.14 + i\sin 6.14)$

7. (a)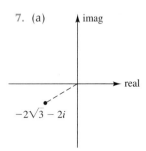
(b) 4
(c) $\dfrac{7\pi}{6}$
(d) $4\left(\cos\dfrac{7\pi}{6} + i\sin\dfrac{7\pi}{6}\right)$

17. (a)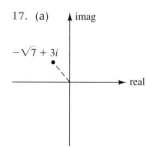
(b) 4
(c) $\pi - \arctan\dfrac{3}{\sqrt{7}} \approx 2.29$
(d) $4\,(\cos 2.29 + i\sin 2.29)$

9. (a)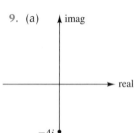
(b) 4
(c) $\dfrac{3\pi}{2}$
(d) $4\left(\cos\dfrac{3\pi}{2} + i\sin\dfrac{3\pi}{2}\right)$

19. (a)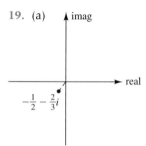
(b) $\frac{5}{6}$
(c) $\pi + \arctan\frac{4}{3} \approx 4.07$
(d) $\frac{5}{6}\,(\cos 4.07 + i\sin 4.07)$

11. (a)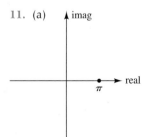
(b) π
(c) 0
(d) $\pi\,(\cos 0 + i\sin 0)$

21. $8.34\,(\cos 1.25 + i\sin 1.25)$

23. $1.39\,(\cos 5.20 + i\sin 5.20)$

25. $6\sqrt{3} + 6i$

27. $-\dfrac{1}{4} + \dfrac{\sqrt{3}}{4}i$ 29. $-1 - i$ 31. $9i$ 33. $-\frac{2}{3}$

35. $3.00 + 6.00i$ 37. $-5.83 + 12.7i$

13. (a)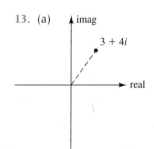
(b) 5
(c) $\arctan\frac{4}{3} \approx 0.927$
(d) $5\,(\cos 0.927 + i\sin 0.927)$

39. (a) $18\left(\cos\dfrac{3\pi}{2} + i\sin\dfrac{3\pi}{2}\right)$ (b) $-18i$

41. (a) $4\left(\cos\dfrac{5\pi}{4} + i\sin\dfrac{5\pi}{4}\right)$ (b) $-2\sqrt{2} - 2\sqrt{2}\,i$

43. (a) $10\left(\cos\dfrac{5\pi}{3} + i\sin\dfrac{5\pi}{3}\right)$ (b) $5 - 5\sqrt{3}\,i$

45. (a) $4\left(\cos\dfrac{\pi}{2} + i\sin\dfrac{\pi}{2}\right)$ (b) $4i$

47. (a) $6\left(\cos\dfrac{3\pi}{4} + i\sin\dfrac{3\pi}{4}\right)$ (b) $-3\sqrt{2} + 3\sqrt{2}\,i$

49. (a) $\cos \frac{3\pi}{2} + i \sin \frac{3\pi}{2}$ (b) $-i$

51. $-2.77 - 3.01 i$

53. $-1.14 + 1.32 i$ 55. $-2.86 + 0.192 i$

57. (a) $5\sqrt{2}\,\Omega$ (b) $45°$ 59. (a) $8\,\Omega$ (b) $-30°$

61. Magnitude ≈ 43.3 volts, phase angle $\approx 35.2°$

63. $\frac{1}{r}[\cos(-\theta) + i \sin(-\theta)]$ or $\frac{1}{r}(\cos \theta - i \sin \theta)$

65. (a) $\frac{\sqrt{2} - \sqrt{6}}{4}$ (b) $\frac{\sqrt{2} + \sqrt{6}}{4}$

67. (a) -2 (b) i (c) $1 + i$ (d) $\frac{\sqrt{3}}{2} - \frac{1}{2}i$

SECTION 9.5

◆ Problems

1. $(\sqrt{3} + i)^9 = \left[2\left(\cos \frac{\pi}{6} + i \sin \frac{\pi}{6}\right)\right]^9$
$= 2^9 \left(\cos \frac{3\pi}{2} + i \sin \frac{3\pi}{2}\right)$
$= -512\, i$

2. $(\sqrt{3} + i)^{-9} = \left[2\left(\cos \frac{\pi}{6} + i \sin \frac{\pi}{6}\right)\right]^{-9}$
$= 2^{-9}\left[\cos\left(-\frac{3\pi}{2}\right) + i \sin\left(-\frac{3\pi}{2}\right)\right]$
$= \frac{1}{512}\, i$

3. The four fourth roots of
$\sqrt{3} + i = 2\left(\cos \frac{\pi}{6} + i \sin \frac{\pi}{6}\right)$ are
$w_0 = 2^{1/4}\left[\cos\left(\frac{(\pi/6) + 2\pi(0)}{4}\right) + i \sin\left(\frac{(\pi/6) + 2\pi(0)}{4}\right)\right]$
$= 2^{1/4}\left(\cos \frac{\pi}{24} + i \sin \frac{\pi}{24}\right) \approx 1.179 + 0.155\, i$
$w_1 = 2^{1/4}\left[\cos\left(\frac{(\pi/6) + 2\pi(1)}{4}\right) + i \sin\left(\frac{(\pi/6) + 2\pi(1)}{4}\right)\right]$
$= 2^{1/4}\left(\cos \frac{13\pi}{24} + i \sin \frac{13\pi}{24}\right) \approx -0.155 + 1.179i$
$w_2 = 2^{1/4}\left[\cos\left(\frac{(\pi/6) + 2\pi(2)}{4}\right) + i \sin\left(\frac{(\pi/6) + 2\pi(2)}{4}\right)\right]$
$= 2^{1/4}\left(\cos \frac{25\pi}{24} + i \sin \frac{25\pi}{24}\right) \approx -1.179 - 0.155\, i$
$w_3 = 2^{1/4}\left[\cos\left(\frac{(\pi/6) + 2\pi(3)}{4}\right) + i \sin\left(\frac{(\pi/6) + 2\pi(3)}{4}\right)\right]$
$= 2^{1/4}\left(\cos \frac{37\pi}{24} + i \sin \frac{37\pi}{24}\right) \approx 0.155 - 1.179i$

As shown in the sketch, these roots form the vertices of a square in the complex plane.

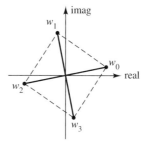

4. The solutions of the equation $x^3 + 1 = 0$, or $x^3 = -1$, are the three cube roots of -1. By the nth root formula, the three cube roots of $-1 + 0i = 1\,(\cos \pi + i \sin \pi)$ are as follows:

$w_0 = 1^{1/3}\left[\cos\left(\frac{\pi + 2\pi(0)}{3}\right) + i \sin\left(\frac{\pi + 2\pi(0)}{3}\right)\right]$
$= 1^{1/3}\left(\cos \frac{\pi}{3} + i \sin \frac{\pi}{3}\right) = \frac{1}{2} + \frac{\sqrt{3}}{2}\, i$

$w_1 = 1^{1/3}\left[\cos\left(\frac{\pi + 2\pi(1)}{3}\right) + i \sin\left(\frac{\pi + 2\pi(1)}{3}\right)\right]$
$= 1^{1/3}(\cos \pi + i \sin \pi) = -1$

$w_2 = 1^{1/3}\left[\cos\left(\frac{\pi + 2\pi(2)}{3}\right) + i \sin\left(\frac{\pi + 2\pi(2)}{3}\right)\right]$
$= 1^{1/3}\left(\cos \frac{5\pi}{3} + i \sin \frac{5\pi}{3}\right) = \frac{1}{2} - \frac{\sqrt{3}}{2}\, i$

◆ Exercises

1. (a) $64\left(\cos \frac{\pi}{2} + i \sin \frac{\pi}{2}\right)$ (b) $64\, i$

3. (a) $8\left(\cos \frac{5\pi}{4} + i \sin \frac{5\pi}{4}\right)$ (b) $-4\sqrt{2} - 4\sqrt{2}\, i$

5. (a) $\cos 0 + i \sin 0$ (b) 1

7. (a) $\frac{27}{8}\left(\cos \frac{\pi}{3} + i \sin \frac{\pi}{3}\right)$ (b) $\frac{27}{16} + \frac{27\sqrt{3}}{16}\, i$

9. (a) $\cos \frac{11\pi}{6} + i \sin \frac{11\pi}{6}$ (b) $\frac{\sqrt{3}}{2} - \frac{1}{2}\, i$

11. (a) $16\,(\cos 0 + i \sin 0)$ (b) 16

13. (a) $32\left(\cos \frac{2\pi}{3} + i \sin \frac{2\pi}{3}\right)$ (b) $-16 + 16\sqrt{3}\, i$

15. (a) $128\,(\cos \pi + i \sin \pi)$ (b) -128

17. (a) $\frac{1}{18}\left(\cos \frac{\pi}{2} + i \sin \frac{\pi}{2}\right)$ (b) $\frac{1}{18}\, i$

19. (a) $\dfrac{1}{64}\left(\cos \dfrac{\pi}{2} + i \sin \dfrac{\pi}{2}\right)$ (b) $\tfrac{1}{64} i$

21. $2.47 - 0.329 i$ 23. $-0.124 - 0.0176 i$

25. $-11.0 + 2.00 i$

27. $(-5.62 \times 10^{-6}) + (3.06 \times 10^{-6}) i$

29. (a) $7\left(\cos \dfrac{5\pi}{6} + i \sin \dfrac{5\pi}{6}\right)$, $7\left(\cos \dfrac{11\pi}{6} + i \sin \dfrac{11\pi}{6}\right)$

(b) $-\dfrac{7\sqrt{3}}{2} + \dfrac{7}{2} i, \dfrac{7\sqrt{3}}{2} - \dfrac{7}{2} i$

31. (a) $2\left(\cos \dfrac{\pi}{6} + i \sin \dfrac{\pi}{6}\right), 2\left(\cos \dfrac{5\pi}{6} + i \sin \dfrac{5\pi}{6}\right)$, $2\left(\cos \dfrac{3\pi}{2} + i \sin \dfrac{3\pi}{2}\right)$

(b) $\sqrt{3} + i, -\sqrt{3} + i, -2i$

33. (a) $\sqrt{5}\left(\cos \dfrac{\pi}{3} + i \sin \dfrac{\pi}{3}\right)$, $\sqrt{5}\left(\cos \dfrac{5\pi}{6} + i \sin \dfrac{5\pi}{6}\right)$, $\sqrt{5}\left(\cos \dfrac{4\pi}{3} + i \sin \dfrac{4\pi}{3}\right)$, $\sqrt{5}\left(\cos \dfrac{11\pi}{6} + i \sin \dfrac{11\pi}{6}\right)$

(b) $\dfrac{\sqrt{5}}{2} + \dfrac{\sqrt{15}}{2} i, -\dfrac{\sqrt{15}}{2} + \dfrac{\sqrt{5}}{2} i,$ $-\dfrac{\sqrt{5}}{2} - \dfrac{\sqrt{15}}{2} i, \dfrac{\sqrt{15}}{2} - \dfrac{\sqrt{5}}{2} i$

35. (a) $\cos \dfrac{\pi}{4} + i \sin \dfrac{\pi}{4}, \cos \dfrac{5\pi}{4} + i \sin \dfrac{5\pi}{4}$

(b) $\dfrac{1}{\sqrt{2}} + \dfrac{1}{\sqrt{2}} i, -\dfrac{1}{\sqrt{2}} - \dfrac{1}{\sqrt{2}} i$

37. (a) $\cos 0 + i \sin 0, \cos \dfrac{2\pi}{3} + i \sin \dfrac{2\pi}{3}$, $\cos \dfrac{4\pi}{3} + i \sin \dfrac{4\pi}{3}$

(b) $1, -\dfrac{1}{2} + \dfrac{\sqrt{3}}{2} i, -\dfrac{1}{2} - \dfrac{\sqrt{3}}{2} i$

39. (a) $2\left(\cos \dfrac{\pi}{4} + i \sin \dfrac{\pi}{4}\right), 2\left(\cos \dfrac{3\pi}{4} + i \sin \dfrac{3\pi}{4}\right)$, $2\left(\cos \dfrac{5\pi}{4} + i \sin \dfrac{5\pi}{4}\right), 2\left(\cos \dfrac{7\pi}{4} + i \sin \dfrac{7\pi}{4}\right)$

(b) $\sqrt{2} + \sqrt{2} i, -\sqrt{2} + \sqrt{2} i, -\sqrt{2} - \sqrt{2} i,$ $\sqrt{2} - \sqrt{2} i$

41. (a) $2\sqrt{2}\left(\cos \dfrac{\pi}{3} + i \sin \dfrac{\pi}{3}\right),$ $2\sqrt{2}\left(\cos \dfrac{4\pi}{3} + i \sin \dfrac{4\pi}{3}\right)$

(b) $\sqrt{2} + \sqrt{6} i, -\sqrt{2} - \sqrt{6} i$

43. Equilateral triangle

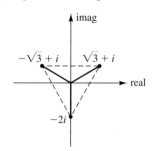

45. Square

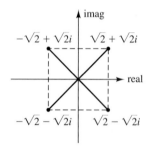

47. $1.85 + 1.85 i, -2.53 + 0.678 i, 0.678 - 2.53 i$

49. $-3.00 + 2.00 i, 3.00 - 2.00 i$

51. $0.966 + 0.259 i, 0.259 + 0.966 i, -0.707 + 0.707 i,$ $-0.966 - 0.259 i, -0.259 - 0.966 i, 0.707 - 0.707 i$

53. $1.62 + 0.204 i, -0.204 + 1.62 i, -1.62 - 0.204 i,$ $0.204 - 1.62 i$

55. $-2, 1 + \sqrt{3} i, 1 - \sqrt{3} i$ 57. $3, -3, 3 i, -3 i$

59. $-i, \dfrac{\sqrt{3}}{2} + \dfrac{1}{2} i, -\dfrac{\sqrt{3}}{2} + \dfrac{1}{2} i$

61. $\dfrac{\sqrt{6}}{2} + \dfrac{\sqrt{2}}{2} i, -\dfrac{\sqrt{6}}{2} - \dfrac{\sqrt{2}}{2} i$

63. The eight solutions of $x^8 = 1$, when plotted in the complex plane, form the vertices of a regular octagon.

65. $5\left(\cos\dfrac{25\pi}{24} + i\sin\dfrac{25\pi}{24}\right)$,
    $5\left(\cos\dfrac{41\pi}{24} + i\sin\dfrac{41\pi}{24}\right)$

SECTION 9.6

Problems

1. The point $P(3, \pi)$ may be represented by $P(3, \pi + 2\pi n)$ for any integer n. Letting $n = 1$, we have $P(3, 3\pi)$ as another name for point P. Also, the point $P(3, \pi)$ may be represented by $P(-3, (\pi + \pi) + 2\pi n)$ for any integer n. Letting $n = -1$, we have $P(-3, 0)$ as another name for point P.

2. From Example 2(b), the point $(-2, 2\sqrt{3})$ has polar coordinates $(4, 2\pi/3)$. This point may also be represented by $(-4, (2\pi/3 + \pi) + 2\pi n)$ for any integer n. Since we want $0 \le \theta < 2\pi$, we choose $n = 0$ and obtain $(-4, 5\pi/3)$.

3. (a) The polar equation $r = 5$ represents a circle with center at the pole and radius 5, as shown in the sketch:

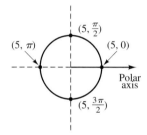

   (b) The polar equation $\theta = \dfrac{2\pi}{3}$ represents a line that passes through the pole and makes an angle of $2\pi/3$ rad with the polar axis, as shown in the sketch:

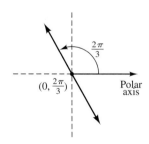

4. (a) Multiplying both sides of the equation $r = \dfrac{3}{\cos\theta}$ by $\cos\theta$ gives us $r\cos\theta = 3$. Now, replacing $r\cos\theta$ with x, we obtain the equivalent Cartesian equation $x = 3$, which is the vertical line shown in the sketch:

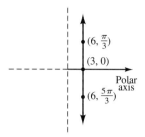

   (b) Multiplying both sides of the equation $r = -6\cos\theta$ by r gives us $r^2 = -6r\cos\theta$. Now, replacing r^2 with $x^2 + y^2$ and $r\cos\theta$ with x, we obtain the equivalent Cartesian equation $x^2 + y^2 + 6x = 0$, or $(x + 3)^2 + (y - 0)^2 = 3^2$. This is the graph of a circle with center $(-3, 0)$, as shown in the sketch:

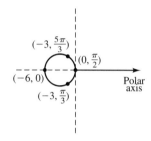

5. The maximum value of $|r|$ is $\sqrt{2}$ and this occurs when

$$3\theta = 0 + 2\pi n \quad \text{or} \quad 3\theta = \pi + 2\pi n$$
$$\theta = 0 + \dfrac{2\pi}{3}n \quad\quad \theta = \dfrac{\pi}{3} + \dfrac{2\pi}{3}n$$

Thus, the values of θ in the interval $(0, \pi]$ at which $|r|$ attains its maximum value are $\dfrac{\pi}{3}$, $\dfrac{2\pi}{3}$, and π as shown in the figure:

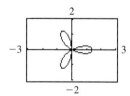

Exercises

1. 3.

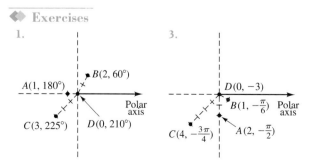

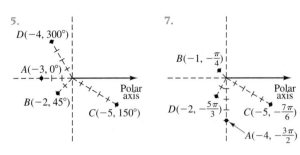

9. $A(3, 180°), A(-3, 360°); B(2, 225°), B(-2, 405°);$
$C(5, 330°), C(-5, 510°); D(4, 120°), D(-4, -60°)$

11. $(-1, 0)$ 13. $(-4, 4)$ 15. $(2, 90°)$

17. $(5\sqrt{2}, 5\pi/4)$ 19. $(-6, 210°)$ 21. $(2.06, 1.33)$

23. $(-0.315, -0.889)$ 25. $(19.7, 0.890)$

27. $(268, 2.06)$

29. $x^2 + y^2 = 1$ 31. $y = \sqrt{3}\,x$

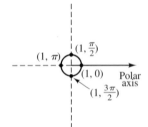 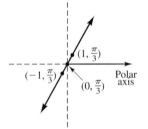

33. $y = -2$ 35. $x = 4$

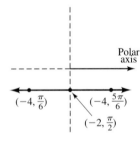 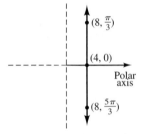

37. $(x - 5)^2 + y^2 = 25$ 39. $x^2 + \left(y + \frac{5}{2}\right)^2 = \frac{25}{4}$

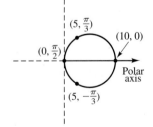 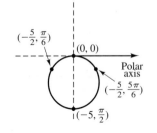

41. $(x - 1)^2 + (y - 2)^2 = 5$ 43. $(x - 4)^2 + \left(y + \frac{1}{2}\right)^2 = \frac{65}{4}$

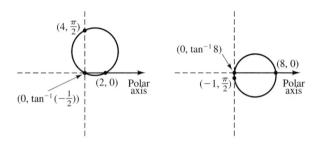

45. $x + y = 2$

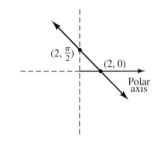

47. $3x + 5y = -15$

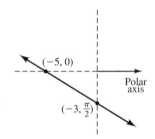

49. $\theta = 0, \dfrac{\pi}{2}, \pi, \dfrac{3\pi}{2}$ 51. $\theta = \dfrac{\pi}{10}, \dfrac{3\pi}{10}, \dfrac{\pi}{2}, \dfrac{7\pi}{10}, \dfrac{9\pi}{10}$

53. $\theta = \dfrac{3\pi}{2}$ 55. $\theta = \dfrac{\pi}{2}$ 57. $\theta = 0, \pi$

59.

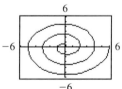

61. (a) $\theta = 0$ (b) $\theta = \pi/2$

63. (a) $r = 2\cos\theta + 2\sin\theta$ (b) $r = \dfrac{2}{1 - \sin\theta}$

(c) $r = \dfrac{-1}{2 - \sin\theta}$ (d) $r = \dfrac{-1}{1 - 2\sin\theta}$

67. (a) $(2\sqrt{2}, \pi/4)$

(b) No; although the graphs of both equations pass through the pole, the coordinates at the pole are different. For $r = 4 \sin \theta$, the coordinates at the pole are $(0, 0)$, but for $r = 4 \cos \theta$, the coordinates at the pole are $(0, \pi/2)$.

69. $r = 2a \cos(\alpha - \theta)$

SECTION 9.7

Problems

1. The equation $x = 5 - t$ implies that $t = 5 - x$. We eliminate the parameter t to obtain

$$y = 7 - \tfrac{2}{3}t = 7 - \tfrac{2}{3}(5 - x) = \tfrac{11}{3} + \tfrac{2}{3}x$$

which we recognize as a line with slope $\tfrac{2}{3}$. At $t = 0$ we have $(5, 7)$, and at $t = 6$ we have $(-1, 3)$. Hence, we conclude that the parametric equations

$$x = 5 - t \quad \text{and} \quad y = 7 - \tfrac{2}{3}t \quad \text{for } 0 \le t \le 6$$

define the line segment shown in Figure 9.60, but with opposite direction.

2. The equation $y = 2t$ implies that $t = y/2$. We eliminate the parameter t to obtain

$$x = t^2 = \left(\frac{y}{2}\right)^2 = \tfrac{1}{4}y^2$$

which we recognize as a parabola with horizontal axis of symmetry, vertex $(0, 0)$, and focus $(1, 0)$. The portion of the parabola traversed and the direction of travel along that portion is shown in the sketch:

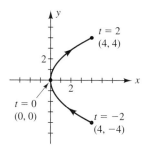

3. The equations $x = 2 \sin \pi t$ and $y = 4 \cos \pi t$ imply that $x/2 = \sin \pi t$ and $y/4 = \cos \pi t$. Now,

$$\left(\frac{x}{2}\right)^2 + \left(\frac{y}{4}\right)^2 = \sin^2 \pi t + \cos^2 \pi t = 1$$

or

$$\frac{x^2}{4} + \frac{y^2}{16} = 1$$

which we recognize as an ellipse with center at the origin, horizontal axis of length 4, and vertical axis of length 8. The direction of travel along the ellipse is shown in the sketch.

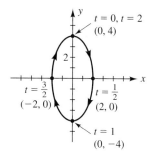

4. For $x = 2 \tan \dfrac{\pi}{3} t$ and $y = 4 \sec \dfrac{\pi}{3} t$, we have

$$\left(\frac{y}{4}\right)^2 - \left(\frac{x}{2}\right)^2 = \sec^2\left(\frac{\pi}{3}t\right) - \tan^2\left(\frac{\pi}{3}t\right) = 1$$

or

$$\frac{y^2}{16} - \frac{x^2}{4} = 1$$

which we recognize as the equation of a hyperbola with center at the origin, vertical transverse axis of length 8, and horizontal conjugate axis of length 4. The portion of the hyperbola traversed and the direction of travel along that portion is shown in the sketch:

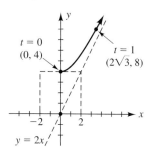

5. The parametric equations that describe the three-leafed rose $r = f(\theta) = 2 \sin 3\theta$ are $x = 2 \sin 3t \cos t$ and $y = 2 \sin 3t \sin t$. First selecting parametric mode and radian mode on a graphing calculator, we enter these equations to obtain the graph shown in the figure:

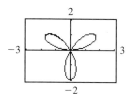

840 **SOLUTIONS AND ANSWERS** ◆

6. The equation $x = 40t$ implies that $t = x/40$. We eliminate the parameter t to obtain

$$y = 40\sqrt{3}\,t - 16t^2 = 40\sqrt{3}\left(\frac{x}{40}\right) - 16\left(\frac{x}{40}\right)^2$$
$$= \sqrt{3}\,x - \tfrac{1}{100}x^2$$
$$= -\tfrac{1}{100}\left(x - 50\sqrt{3}\right)^2 + 75$$

which we recognize as a parabola with vertex $\left(50\sqrt{3},\, 75\right)$. Hence, the maximum height that the ball rises in its parabolic path is 75 ft.

◆ Exercises

1. Line with slope 2, passing through (0, 0)

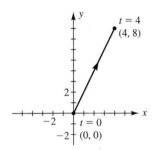

3. Line with slope -3, passing through $(-2, 1)$

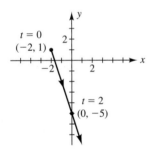

5. Parabola with vertex (0, 0) and focus (0, -1)

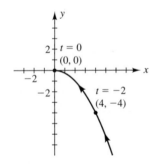

7. Parabola with vertex (0, 0) and focus (3, 0)

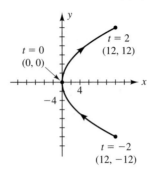

9. Circle with center (0, 0) and radius 1

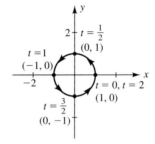

11. Circle with center (0, 0) and radius 3

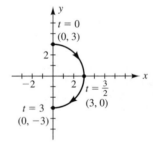

13. Ellipse with center (0, 0), horizontal minor axis 2, and vertical major axis 4

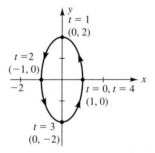

15. Ellipse with center (0, 0), horizontal major axis 6, and vertical minor axis 4

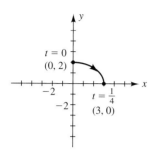

17. Hyperbola with center (0, 0), horizontal transverse axis 2, and vertical conjugate axis 2

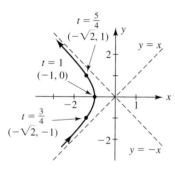

19. Hyperbola with center (0, 0), vertical transverse axis 4, and horizontal conjugate axis 6

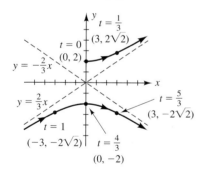

21. $x = -1 + t$ and $y = 4 - 4t$

23. $x = 4t$ and $y = 2t^2$

25. $x = 4\cos t$ and $y = 4\sin t$

27. $x = 3\cos t$ and $y = 2\sin t$

29. $x = 2\sec t$ and $y = 2\tan t$

31. (a) $y = |x| - 1$ (b)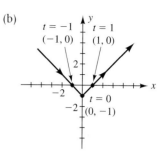

33. (a) $y = \dfrac{1}{x+2}$ (b)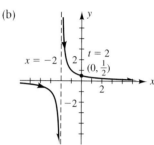

35. (a) $y = -(x-1)^3$ (b)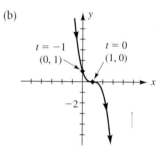

37. (a) $y = \sqrt{x-3} - 1$ (b)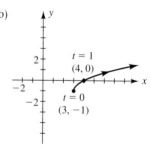

39. (a) $y = 4 - (x+5)^2$ (b)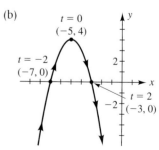

41. $x = 2 \cos 2t \cos t$ and $y = 2 \cos 2t \sin t$

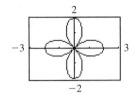

43. $x = 3 \sin 5t \cos t$ and $y = 3 \sin 5t \sin t$

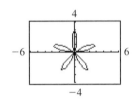

45. $x = (1 - \sin t) \cos t$ and $y = (1 - \sin t) \sin t$

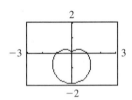

47. $x = (2 + \sin t) \cos t$ and $y = (2 + \sin t) \sin t$

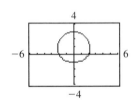

49. $x = 2 \sqrt{\cos 2t} \cos t$ and $y = 2 \sqrt{\cos 2t} \sin t$

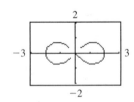

51. (a) $x = 45 \sqrt{2}\, t$ and $y = 45 \sqrt{2}\, t - 16t^2$
    (b) $253 \tfrac{1}{8}$ ft   (c) $y = x - \tfrac{8}{2025}x^2$ (d) $63 \tfrac{9}{32}$ ft

53. $R \approx 25{,}400$ ft, $h \approx 8740$ ft

55. (a)

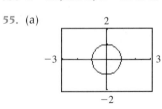

(b)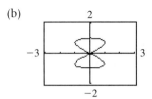

(c)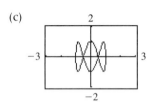

(d)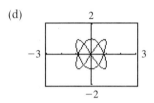

59. $x = 2 + 2 \cos t$ and $y = 1 + \sin t$

61. $x = -1 + \sec t$ and $y = -2 + \tan t$

63. (a) Same as C
    (b) Same as C, but with opposite direction
    (c) Same as C, but shifted 2 units right and 1 unit down
    (d) Same as C but reflected about the line $y = x$

CHAPTER 9 REVIEW EXERCISES

1. (a) $x \approx 544$, $y \approx 488$, $\theta = 80°$
   (b) About 81,600 square units

3. (a) $x \approx 56.2$, $y \approx 35.0$, $\theta = 103°$
   (b) About 958 square units

5. (a) $\alpha \approx 139.3°$, $\beta \approx 19.7°$, $x \approx 3000$
   (b) About 833,000 square units

7. (a) $\alpha \approx 138.0°$, $\beta \approx 17.0°$, $x \approx 20.2$
   (b) About 94.7 square units

9. (a) $\alpha \approx 88.2°$, $\beta \approx 60.3°$, $\theta \approx 31.5°$
   (b) About 10,400 square units

11. $a \approx 38.2$, $\alpha \approx 38.2°$, $\theta \approx 63.8°$

13. $\alpha \approx 28.9°$, $\beta \approx 117.0°$, $\theta \approx 34.1°$

15. $b \approx 922$, $\theta \approx 73.5°$, $\beta \approx 68.5°$ or $b \approx 575$, $\theta \approx 106.5°$, $\beta \approx 35.5°$

17. No such triangle exists

19. $\alpha \approx 22.3°, \beta \approx 109.7°, c \approx 5.48$

21. No such triangle exists.   23. $\dfrac{a^2}{4}\sqrt{3}$

25. (a) 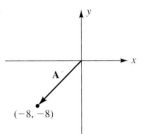   (b) $\langle -8, -8 \rangle$
   (c) $8\sqrt{2}$
   (d) $225°$

27. (a) 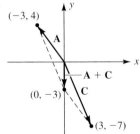   (b) $\langle 3, 3\sqrt{3} \rangle$
   (c) 6
   (d) $60°$

29. (a) $|\mathbf{A}| = 13, \theta \approx 337.4°$ (b) $|\mathbf{C}| = 4, \theta \approx 48.6°$

31. $\langle -3, 7 \rangle$ 33. $\langle 0, -3 \rangle$ 35. $\langle -21, 43 \rangle$

37. $\langle 75, -135 \rangle$

39.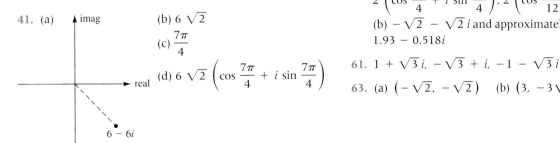

41. (a) (b) $6\sqrt{2}$
   (c) $\dfrac{7\pi}{4}$
   (d) $6\sqrt{2}\left(\cos\dfrac{7\pi}{4} + i\sin\dfrac{7\pi}{4}\right)$

43. (a) (b) 10
   (c) $\dfrac{7\pi}{6}$
   (d) $10\left(\cos\dfrac{7\pi}{6} + i\sin\dfrac{7\pi}{6}\right)$

45. (a) $18\left(\cos\dfrac{7\pi}{12} + i\sin\dfrac{7\pi}{12}\right)$
   (b) Approximately $-4.66 + 17.4\,i$

47. (a) $\cos\dfrac{5\pi}{3} + i\sin\dfrac{5\pi}{3}$ (b) $\dfrac{1}{2} - \dfrac{\sqrt{3}}{2}i$

49. (a) $256\left(\cos\dfrac{4\pi}{3} + i\sin\dfrac{4\pi}{3}\right)$
   (b) $-128 - 128\sqrt{3}\,i$

51. (a) $\dfrac{1}{32}\left(\cos\dfrac{3\pi}{2} + i\sin\dfrac{3\pi}{2}\right)$ (b) $-\dfrac{1}{32}i$

53. (a) $\dfrac{1}{\sqrt{2}}\left(\cos\dfrac{7\pi}{4} + i\sin\dfrac{7\pi}{4}\right)$ (b) $\dfrac{1}{2} - \dfrac{1}{2}i$

55. (a) $\dfrac{1}{2}\left(\cos\dfrac{\pi}{3} + i\sin\dfrac{\pi}{3}\right), \dfrac{1}{2}\left(\cos\dfrac{4\pi}{3} + i\sin\dfrac{4\pi}{3}\right)$
   (b) $\dfrac{1}{4} + \dfrac{\sqrt{3}}{4}i, -\dfrac{1}{4} - \dfrac{\sqrt{3}}{4}i$

57. (a) $\cos 0 + i\sin 0, \cos\dfrac{\pi}{2} + i\sin\dfrac{\pi}{2}, \cos\pi + i\sin\pi,$
   $\cos\dfrac{3\pi}{2} + i\sin\dfrac{3\pi}{2}$
   (b) $1, -1, i, -i$

59. (a) $2\left(\cos\dfrac{7\pi}{12} + i\sin\dfrac{7\pi}{12}\right),$
   $2\left(\cos\dfrac{5\pi}{4} + i\sin\dfrac{5\pi}{4}\right), 2\left(\cos\dfrac{23\pi}{12} + i\sin\dfrac{23\pi}{12}\right)$
   (b) $-\sqrt{2} - \sqrt{2}\,i$ and approximately $-0.518 + 1.93i,$
   $1.93 - 0.518i$

61. $1 + \sqrt{3}\,i, -\sqrt{3} + i, -1 - \sqrt{3}\,i$

63. (a) $\left(-\sqrt{2}, -\sqrt{2}\right)$ (b) $\left(3, -3\sqrt{3}\right)$

65. $x^2 + y^2 = 4$

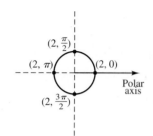

67. $y = 1$

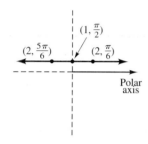

69. $\left(x - \frac{5}{2}\right)^2 + y^2 = \frac{25}{4}$

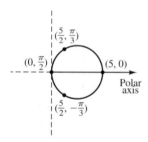

71. $(x + 4)^2 + (y - 3)^2 = 25$

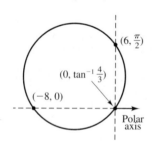

73. Line with slope -2, passing through $(2, 3)$

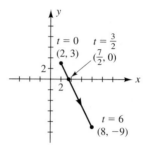

75. Parabola with vertex $(0, 0)$ and focus $(0, -2)$

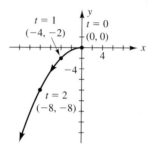

77. Circle with center $(0, 0)$ and radius 2

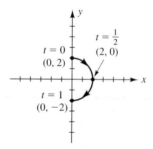

79. Ellipse with center $(0, 0)$, vertical major axis 8, and horizontal minor axis 6

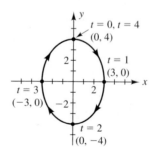

81. Hyperbola with center (0, 0), horizontal transverse axis 4, and vertical conjugate axis 6

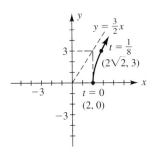

83. 101.1 ft

85. (a) About 206 ft   (b) About 186 ft   (c) About 276 ft

87. About 390 m

89. (a) About 37.3 lb   (b) About 25.2 lb

91. $\mathbf{F}_4 \approx \langle 161.86, -160.68 \rangle$; $|\mathbf{F}_4| \approx 228$ lb, $\theta \approx 315.2°$

93. Approximately 171 mi, N-1.3°-E

95. Approximately 75 mi, N-49°20'-E

97. (a) PC$(-75.47, 0)$, PT$(75.47, 0)$,
    $x^2 + (y + 65.6)^2 = 100^2$
    (b) $A(30.00, 29.79)$

Cumulative Review: Chapters 7, 8, 9

1. $A = \dfrac{b\sqrt{4a^2 - b^2}}{4}$

3. (a) $\dfrac{1 + \sqrt{3}}{2}$ (b) $\dfrac{-3 + \sqrt{2}}{2}$ (c) $\dfrac{5\pi}{12}$ (d) $\dfrac{5}{12}$

5. (a) 1   (b) 0   (c) 1   (d) 1

9. (a) About 56.8   (b) $\dfrac{52\pi}{3} \approx 54.5$

11. (a) $\dfrac{\pi}{9}, \dfrac{4\pi}{9}, \dfrac{7\pi}{9}, \dfrac{10\pi}{9}, \dfrac{13\pi}{9}, \dfrac{16\pi}{9}$ (b) $\dfrac{\pi}{6}, \dfrac{5\pi}{6}$
    (c) $\dfrac{\pi}{2}, \dfrac{3\pi}{2}$ (d) $\dfrac{\pi}{2}, \dfrac{7\pi}{12}, \dfrac{11\pi}{12}, \dfrac{3\pi}{2}, \dfrac{19\pi}{12}, \dfrac{23\pi}{12}$
    (e) $\dfrac{\pi}{6}, \dfrac{\pi}{3}, \dfrac{2\pi}{3}, \dfrac{5\pi}{6}, \dfrac{7\pi}{6}, \dfrac{4\pi}{3}, \dfrac{5\pi}{3}, \dfrac{11\pi}{6}$ (f) $\dfrac{2\pi}{3}, \dfrac{4\pi}{3}$

13. About 710 ft

15. $\sin 3\theta = 3 \sin \theta - 4 \sin^3 \theta$, $\cos 3\theta = 4 \cos^3 \theta - 3 \cos \theta$

17. (a) About 136 ft and 285 ft
    (b) Approximately 59,700 sq ft

19. 14   21. (a) $16i$ (b) $-\dfrac{1}{8}$

23. (a) $\csc x$ (b) 0   25. About 142 ft

27. (a) $|\mathbf{A} - \mathbf{B}| = \sqrt{41} \approx 6.40$, $\theta \approx 321.3°$
    (b) $|\mathbf{B} - \mathbf{A}| = \sqrt{41} \approx 6.40$, $\theta \approx 141.3°$

29. (a) $x \approx 29.2$, $y \approx 19.1$, $\theta = 49°$
    (b) $\alpha \approx 55.3°$, $\beta \approx 86.7°$, $x \approx 21.0$

31. (a) $r = \dfrac{5}{2\cos\theta - 3\sin\theta}$ (b) $r = \dfrac{3}{1 - \cos\theta}$

Chapter 10

SECTION 10.1

Problems

1. $\phantom{z = -2 \text{ and } 5x + 5z = -5}\ -4z = 8\quad$ implies $\quad z = -2$
   $z = -2$ and $5x + 5z = -5\quad$ implies $\quad x = 1$
   $z = -2$, $x = 1$, $2x - y + 4z = -3\quad$ implies $\quad y = -3$

2. Subtracting the second equation from the first, we eliminate x:

   $$\begin{array}{r} x - 2y - 3z = -3 \\ x - 4y - 13z = 14 \\ \hline 2y + 10z = -17 \end{array}$$

   Adding 3 times the first equation to the third equation, we again eliminate x:

   $$\begin{array}{r} 3x - 6y - 9z = -9 \\ -3x + 5y + 4z = 1 \\ \hline -y - 5z = -8 \end{array}$$

   We now have two linear equations in two unknowns y and z, which we can try to solve in the usual manner.

   $$\begin{array}{rcl} 2y + 10z = -17 & \xrightarrow{\text{No change}} & 2y + 10z = -17 \\ -y - 5z = -8 & \xrightarrow{\text{Multiply by 2}} & -2y - 10z = -16 \\ \hline & & 0 = -33 \end{array}$$

   Recall from Section 3.5 that if the variables drop out and the resulting statement is false ($0 = -33$), then the system has no solution. Hence, this is an inconsistent system.

3. Subtracting the fourth equation from the first equation, we eliminate t:

   $$\begin{array}{r} x + 2y + t = 3 \\ 3y - 4z + t = 2 \\ \hline x - y + 4z = 1 \end{array}$$

Adding this equation to the second equation in the system ($y + 2z = 3$), we eliminate y:

$$\begin{aligned} x - y + 4z &= 1 \\ y + 2z &= 3 \\ \hline x + 6z &= 4 \end{aligned}$$

Subtracting the third equation in the system ($x - 2z = 0$) from this equation, we eliminate x:

$$\begin{aligned} x + 6z &= 4 \\ x - 2z &= 0 \\ \hline 8z &= 4 \quad \text{implies} \quad z = \tfrac{1}{2} \end{aligned}$$

Finally, we find x, y, and t by substituting back through the system.

$z = \tfrac{1}{2}$ and $x - 2z = 0$ implies $x = 1$
$z = \tfrac{1}{2}$ and $y + 2z = 3$ implies $y = 2$
$x = 1$, $y = 2$, and $x + 2y + t = 3$ implies $t = -2$

4. Solving $-x + y = -1$ for x, we obtain $x = y + 1$. Using this equation, we express z in terms of y:

$$2x + y - z = 2 \quad \text{implies}$$
$$2(y + 1) + y - z = 2 \quad \text{or} \quad z = 3y$$

Now, if we let $y = b$, where b is any real number, we obtain

$$x = b + 1 \quad \text{and} \quad z = 3b$$

Hence, $(b + 1, b, 3b)$ describes the solution of this system.

5. Adding -18 times the first equation to the second equation, we eliminate y:

$$\begin{aligned} -18x - 18y - 18z &= -216{,}000 \\ 12x + 18y + 24z &= 201{,}000 \\ \hline -6x + 6z &= -15{,}000 \end{aligned}$$

Adding $-6x + 6z = -15{,}000$ to 3 times the third equation, we eliminate z:

$$\begin{aligned} -6x + 6z &= -15{,}000 \\ 3x - 6z &= 0 \\ \hline -3x &= -15{,}000 \quad \text{implies } x = 5000 \end{aligned}$$

Finally, we find z and y by substituting back through the system: $z = 2500$, $y = 4500$.

Exercises

1. $(10, 4, 1)$ 3. $(1, 0, -3)$ 5. $(-1, 2, 0)$

7. $\left(\tfrac{2}{3}, -\tfrac{1}{9}, 3\right)$

9. Dependent system; infinite number of solutions of the form $\left(\dfrac{5 - a}{11}, \dfrac{4a - 9}{11}, a\right)$, where a is any real number

11. $(1, 1, 2)$ 13. $\left(\tfrac{5}{2}, 0, -\tfrac{3}{2}\right)$ 15. $(1, -2, 2)$

17. $(1, 2, -2, 3)$ 19. $(0, 1, -1, 3)$

21. Dependent system; infinite number of solutions of the form $(4 - a, a - 2, -5, 2 - a, a)$, where a is any real number

23. $(1, 2, -1, -2, 0)$

25. $(2 - 3a, a, 4a)$, where a is any real number

27. $(2, a, 2a + 1)$, where a is any real number

29. $(2a + 2, 5a + 1, 3a, a)$, where a is any real number

31. $(5, 2)$ 33. $a = \$25$, $b = \$0.20$

35. $\$5500$ at 10%, $\$4500$ at 12% 37. $120°, 40°, 20°$

39. $\$6000$ at 8%, $\$22{,}000$ at 9%, $\$22{,}000$ at $9\tfrac{1}{2}$%

41. 20 children, 60 adults, 40 seniors

43. $\$0.75$ for 2×3 print, $\$2.00$ for 3×5 print, $\$5.00$ for 5×8 print, $\$10.00$ for 8×10 print

45. (a) $A = \tfrac{3}{2}$, B is any real number except -28
    (b) $A = \tfrac{3}{2}$, $B = -28$

47. $A = \tfrac{8}{3}$, $B = -1$, $C = -\tfrac{2}{3}$

49. $a = -\tfrac{1}{2}$, $b = 2$, $c = 4$

51. $(-a, a, a)$, where a is any real number

53. No adults, 12 students, 8 children or 5 adults, 4 students, 11 children

SECTION 10.2

Problems

1. $\begin{bmatrix} 3 & -9 & 6 & | & 1 \\ 2 & -1 & -1 & | & 3 \\ 1 & -2 & 1 & | & 1 \end{bmatrix}$

$\sim \begin{bmatrix} 1 & -2 & 1 & | & 1 \\ 2 & -1 & -1 & | & 3 \\ 3 & -9 & 6 & | & 1 \end{bmatrix} \quad R_1 \leftrightarrow R_3$

$\sim \begin{bmatrix} 1 & -2 & 1 & | & 1 \\ 0 & 3 & -3 & | & 1 \\ 0 & -3 & 3 & | & -2 \end{bmatrix} \quad \begin{array}{l} -2R_1 + R_2 \to R_2 \\ -3R_1 + R_3 \to R_3 \end{array}$

$\sim \begin{bmatrix} 1 & -2 & 1 & | & 1 \\ 0 & 3 & -3 & | & 1 \\ 0 & 0 & 0 & | & -1 \end{bmatrix} \quad R_2 + R_3 \to R_3$

Since all elements in the last row except the last entry are zero, we conclude the system is inconsistent and has no solution.

2. $-2A + 3B$

$$= -2 \begin{bmatrix} 6 & 5 & -2 \\ 4 & 0 & -1 \end{bmatrix} + 3 \begin{bmatrix} 3 & -5 & 2 \\ 0 & 1 & -3 \end{bmatrix}$$

$$= \begin{bmatrix} -12 & -10 & 4 \\ -8 & 0 & 2 \end{bmatrix} + \begin{bmatrix} 9 & -15 & 6 \\ 0 & 3 & -9 \end{bmatrix}$$

$$= \begin{bmatrix} -3 & -25 & 10 \\ -8 & 3 & -7 \end{bmatrix}$$

3. $A - 2B = A + (-2B)$

$$= \begin{bmatrix} 2 & 5 \\ -3 & 0 \\ 1 & 4 \end{bmatrix} + (-2) \begin{bmatrix} 4 & 6 \\ -2 & -5 \\ 0 & 3 \end{bmatrix}$$

$$= \begin{bmatrix} 2 & 5 \\ -3 & 0 \\ 1 & 4 \end{bmatrix} + \begin{bmatrix} -8 & -12 \\ 4 & 10 \\ 0 & -6 \end{bmatrix}$$

$$= \begin{bmatrix} -6 & -7 \\ 1 & 10 \\ 1 & -2 \end{bmatrix}$$

4. $AC = \begin{bmatrix} 2 & 3 & -1 \end{bmatrix} \begin{bmatrix} 2 & -1 \\ 3 & 4 \\ 0 & -2 \end{bmatrix}$

$= \begin{bmatrix} 4 + 9 + 0 & -2 + 12 + 2 \end{bmatrix}$

$= \begin{bmatrix} 13 & 12 \end{bmatrix}$

5. $\begin{bmatrix} 1 & 1 & 1 & 1 \end{bmatrix} \begin{bmatrix} 2 & 5 \\ 6 & 0 \\ 0 & 2 \\ 2 & 3 \end{bmatrix}$

$= \begin{bmatrix} 2+6+0+2 & 5+0+2+3 \end{bmatrix}$

$= \begin{bmatrix} 10 & 10 \end{bmatrix}$

The 1×2 matrix $\begin{bmatrix} 10 & 10 \end{bmatrix}$ represents the total inventory of this type bed at each outlet.

◆ Exercises

1. $(2, 1)$

3. Dependent system; infinite number of solutions of the form $(4a - 2, a)$, where a is any real number

5. $(1, -2, 3)$ 7. $(-1, 4, -3)$ 9. $\left(\frac{23}{4}, -3, -\frac{1}{4}\right)$

11. $\left(\frac{1}{10}, -\frac{11}{10}, -\frac{13}{10}\right)$ 13. $\left(1, 2, \frac{1}{2}, -2\right)$

15. $\left(\frac{2}{3}, -\frac{1}{3}, \frac{4}{3}, \frac{5}{3}\right)$ 17. $\begin{bmatrix} -2 & 4 & 0 \\ -4 & 6 & -8 \end{bmatrix}$

19. $\begin{bmatrix} -2 & 0 \\ 1 & -3 \\ 0 & -1 \end{bmatrix}$ 21. $\begin{bmatrix} 3 & -1 & -5 \\ -3 & 7 & 0 \end{bmatrix}$

23. Undefined; matrices of different orders cannot be added.

25. $\begin{bmatrix} 2 & 8 \\ 1 & -5 \end{bmatrix}$ 27. $\begin{bmatrix} 2 & -17 & -15 \\ 5 & 0 & 28 \end{bmatrix}$

29. Undefined; matrices of different orders cannot be subtracted.

31. $\begin{bmatrix} -13 & -7 \\ 7 & 1 \end{bmatrix}$ 33. $\begin{bmatrix} -12 & 24 & 30 \end{bmatrix}$

35. $\begin{bmatrix} -1 \\ 3 \\ -2 \end{bmatrix}$ 37. $\begin{bmatrix} 8 & 3 \\ 9 & -8 \end{bmatrix}$

39. $\begin{bmatrix} -3 & 7 & 9 \\ 3 & -7 & -9 \end{bmatrix}$ 41. $\begin{bmatrix} 30 & 30 \\ -33 & -33 \\ -6 & -6 \end{bmatrix}$

43. $\begin{bmatrix} 4 \end{bmatrix}$ 45. $\begin{bmatrix} 0 & 0 \\ 0 & 0 \end{bmatrix}$

47. $\begin{bmatrix} 9 & -9 & -10 \\ 3 & 11 & -15 \\ -15 & 29 & 5 \end{bmatrix}$ 49. $\begin{bmatrix} -21 & 60 \\ -3 & -33 \end{bmatrix}$

51. $\begin{bmatrix} 57 & -96 & 120 \end{bmatrix}$ 53. $\begin{bmatrix} -63 & 63 \\ 66 & -66 \end{bmatrix}$

55. $\begin{bmatrix} 4 & -8 & -10 \\ -12 & 24 & 30 \\ 8 & -16 & -20 \end{bmatrix}$

57. $\begin{bmatrix} 91{,}524.375 \\ 63{,}958.125 \\ 132{,}264.375 \end{bmatrix}$, new weekly payroll (in dollars) at each facility

59. $\begin{bmatrix} 48.75 \\ 123.75 \\ 228.75 \end{bmatrix}$, weekly pay increase (in dollars) for each technician, engineer, and senior engineer

61. $\begin{bmatrix} 71 \\ 48 \\ 20 \end{bmatrix}$, total number of subcompact, midsize, and large cars that were rented on Monday and Tuesday

63. $\begin{bmatrix} 840 & 1256 & 1660 \end{bmatrix}$, total revenue (in dollars) from cars that were rented at each terminal on Monday and Tuesday

65. One example is $A = \begin{bmatrix} a & a \\ b & b \end{bmatrix}$,

$$B = \begin{bmatrix} -1 & 1 \\ 1 & -1 \end{bmatrix},$$

where a and b are any real numbers.

67. (a) (i) $\begin{bmatrix} 10 & -5 & 19 \\ -6 & 7 & -14 \\ 15 & 1 & 23 \end{bmatrix}$

(ii) $\begin{bmatrix} 4 & 1 & -1 \\ 5 & 5 & 7 \\ -1 & -2 & -3 \end{bmatrix}$

(iii) $\begin{bmatrix} 6 & -6 & 20 \\ -11 & 2 & -21 \\ 16 & 3 & 26 \end{bmatrix}$

(iv) $\begin{bmatrix} 5 & -2 & 34 \\ -1 & 2 & -22 \\ 6 & 0 & 27 \end{bmatrix}$

(b) From parts (iii) and (iv), we see that
$A^2 - B^2 \neq (A + B)(A - B)$.

71. (a) $x = (a_{11}b_{11} + a_{21}b_{12})r + (a_{12}b_{11} + a_{22}b_{12})s$
$y = (a_{11}b_{21} + a_{21}b_{22})r + (a_{12}b_{21} + a_{22}b_{22})s$

(b) $\begin{bmatrix} (a_{11}b_{11} + a_{21}b_{12})r + (a_{12}b_{11} + a_{22}b_{12})s \\ (a_{11}b_{21} + a_{21}b_{22})r + (a_{12}b_{21} + a_{22}b_{22})s \end{bmatrix}$

The elements represent the values of x and y.

SECTION 10.3

◆ Problems

1. Corresponding elements in rows 2 and 4 are proportional. If we add $-3R_2$ to R_4, we obtain a row in which every element is zero. If every element in a row of a matrix is zero, then the determinant of the matrix is 0.

2. $\begin{bmatrix} 1 & 2 & -3 & | & 1 & 0 & 0 \\ 1 & 0 & 2 & | & 0 & 1 & 0 \\ 2 & 2 & -1 & | & 0 & 0 & 1 \end{bmatrix}$

$\sim \begin{bmatrix} 1 & 2 & -3 & | & 1 & 0 & 0 \\ 0 & -2 & 5 & | & -1 & 1 & 0 \\ 0 & -2 & 5 & | & -2 & 0 & 1 \end{bmatrix}$

$\sim \begin{bmatrix} 1 & 2 & -3 & | & 1 & 0 & 0 \\ 0 & -2 & 5 & | & -1 & 1 & 0 \\ 0 & 0 & 0 & | & -1 & -1 & 1 \end{bmatrix}$

Since we obtain a row of zeros on the A portion of this matrix, we conclude that A is singular.

3. The coefficient matrix for this system of equations is the same as matrix A defined in Problem 2. Since matrix A is singular (see Problem 2), A^{-1} does not exist and we cannot solve the matrix equation

$$\begin{bmatrix} 1 & 2 & -3 \\ 1 & 0 & 2 \\ 2 & 2 & -1 \end{bmatrix} \begin{bmatrix} x \\ y \\ z \end{bmatrix} = \begin{bmatrix} -3 \\ 4 \\ 1 \end{bmatrix}$$

uniquely for x, y, and z. The system of equations in Problem 3 is a dependent system with infinitely many solutions of the form $\left(4 - 2a, \dfrac{5a - 7}{2}, a\right)$.

◆ Exercises

1. -22 3. -4 5. 218 7. -12 9. 208

11. 0 13. -45 15. 64 17. 24 19. 40

21. 368 23. 304 25. $\begin{bmatrix} 3 & 4 \\ -2 & -3 \end{bmatrix}$

27. $\dfrac{1}{5}\begin{bmatrix} -3 & 4 \\ 2 & -1 \end{bmatrix}$ 29. Singular   31. Singular

33. $\dfrac{1}{4}\begin{bmatrix} -9 & 6 & -1 \\ 2 & 0 & 2 \\ -3 & 2 & 1 \end{bmatrix}$

35. $\dfrac{1}{5}\begin{bmatrix} 0 & -10 & -5 \\ 4 & 1 & 3 \\ 1 & 4 & 2 \end{bmatrix}$

37. $\begin{bmatrix} -14 & -13 & 3 \\ 11 & 10 & -2 \\ 2 & 2 & -\frac{1}{2} \end{bmatrix}$

39. $(0, 1, -2)$ 41. $\left(\frac{1}{2}, \frac{1}{2}, -2\right)$ 43. $(1, -5, -3)$

45. $T_1 = 560$ lb, $T_2 = 758$ lb

47. $v_0 = 80.5$ ft/s, $s_0 = 24.8$ ft

51. (a) $\dfrac{x_1 y_2 - x_2 y_1}{2}$

(b) The area of a triangle with vertices $(0, 0)$, (x_1, y_1), and (x_2, y_2)

53. Matrix A can be cancelled only if it is invertible.

55. $A^{-1} = \dfrac{1}{ad - bc}\begin{bmatrix} d & -b \\ -c & a \end{bmatrix}$ for $ad - bc \neq 0$

SECTION 10.4

◆ Problems

1. (a) 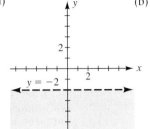 (b)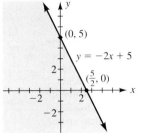

2.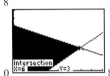

3. The objective function has no maximum or minimum value when lines with slope $\frac{2}{5}$ pass through the unbounded region shown in Figure 7.13.

4. The profit P is now given by $P = 30x + 45y$.

Vertex	Values of $P = 30x + 45y$
(0, 0)	0
(0, 40)	1800
(15, 40)	2250 ← maximum value
(60, 10)	2250 ←
(60, 0)	1800

Hence, to maximize profits, the company should manufacture either 15 pairs of slalom skis and 40 pairs of racing skis, or 60 pairs of slalom skis and 10 pairs of racing skis.

◆ Exercises

1.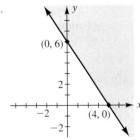

3.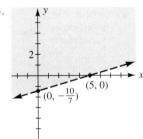

5.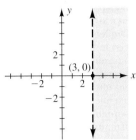

7.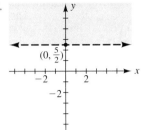

9.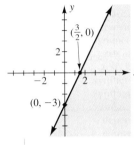

11.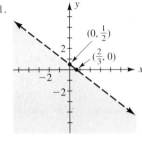

13.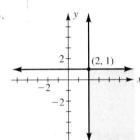

15.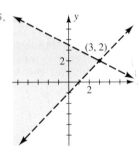

17.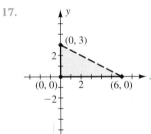

19.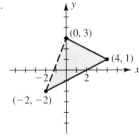

21.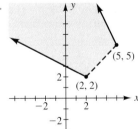

23.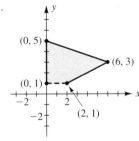

25. 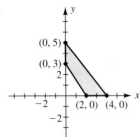 27.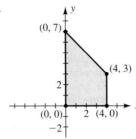

15. $(-11, 4, -4, 6)$ 17. $\left(\frac{3}{4}, \frac{1}{4}, -2, 1\right)$

19. $[-3 \quad -12 \quad 6]$ 21. $\begin{bmatrix} 2 & 8 \\ 4 & 4 \end{bmatrix}$

23. $\begin{bmatrix} -4 & -5 \\ -3 & -4 \end{bmatrix}$ 25. $\begin{bmatrix} -10 & -6 & -3 \\ -4 & 17 & 28 \end{bmatrix}$

27. $\begin{bmatrix} -3 & 12 \\ -2 & 9 \end{bmatrix}$ 29. $[24]$ 31. $\begin{bmatrix} 15 & 20 \\ 18 & 27 \end{bmatrix}$

33. $[-23 \quad 59]$ 35. $\begin{bmatrix} 1 & -\frac{3}{2} \\ -1 & 2 \end{bmatrix}$

29.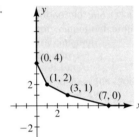

37. $\dfrac{1}{20} \begin{bmatrix} -32 & 20 & 4 \\ 13 & -5 & -1 \\ 14 & -10 & 2 \end{bmatrix}$ 39. Singular

41. 11 43. -14 45. -30 47. -16

49. 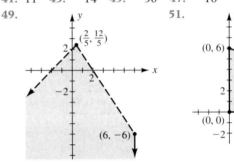 51.

31. Maximum value 12; minimum value 0

33. Maximum value 6; minimum value -11

35. Maximum value 8; minimum value 0

37. Maximum value 6; minimum value 0

39. Maximum value 47; minimum value 16

41. No maximum value; minimum value 11

43. No maximum value; no minimum value

45. Eight $100-cases, two $200-cases

47. 24 tables, 18 chairs

49. 24 bags of cracked corn, 48 bags of sunflower seeds

51. 1.6 oz. of cereal P, 2.6 oz. of cereal Q

53. (a) $m = -\frac{1}{4}$ (b) $m = -2$ (c) $-\frac{1}{4} < m \leq 0$
    (d) $-2 < m < -\frac{1}{4}$ (e) $m < -2$ or $m > 0$

55. 2 bags of brand A and 3 bags of brand B; minimum cost $98

53. Minimum value 0; maximum value 21

55. (a) $k = 3$ (b) $k = \pm 5$ (c) $k = 6, -2$
    (d) $k = 3, -\frac{13}{5}$

57. (a) $y = \dfrac{x_2 y_1 - x_1 y_2 + xy_2 - xy_1}{x_2 - x_1}$

    (b) The equation of a nonvertical line passing through two distinct points (x_1, y_1), and (x_2, y_2)

59. 9750 gallons of regular; 2250 gallons of super

61. 11,000 box seats; 12,000 grandstand seats; 9500 bleacher seats

63. 4 mi/h downhill; 3 mi/h on level ground; 2 mi/h uphill

65. $\begin{bmatrix} 34 \\ 29 \\ 17 \end{bmatrix}$, total number of 400-watt, 500-watt, and 600-watt microwave ovens sold during first day of sale

67. $[3480 \quad 2580 \quad 3240]$, revenue (in dollars) from sales of microwave ovens at each outlet on second day of sale

69. $T_1 = 465$ lb, $T_2 = 648$ lb 71. 20,000 seats

CHAPTER 10 REVIEW EXERCISES

1. $(2, -3, -1)$ 3. $\left(-\frac{1}{3}, \frac{2}{3}, -\frac{2}{3}\right)$

5. Dependent system; infinite number of solutions of the form $(2a - 2, a, 2a - 1)$, where a is any real number

7. $\left(7, -4, \frac{1}{2}\right)$ 9. $\left(\frac{1}{4}, -\frac{3}{4}, \frac{5}{4}\right)$ 11. $(4, 2, -3)$

13. $(a + 1, a, 2a + 5)$, where a is any real number

Chapter 11

SECTION 11.1

Problems

1. $u(1) = (-1)^1 2^0 = -1$, $\quad u(2) = (-1)^2 2^1 = 2$,
   $u(3) = (-1)^3 2^2 = -4$, $\quad u(4) = (-1)^4 2^3 = 8$

   The elements $-1, 2, -4, 8$ form an alternating sequence. The graph is shown in the figure.

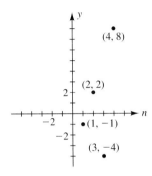

2. $a_5 = 5a_4 = 5(24) = 120$;
   $a_6 = 6a_5 = 6(120) = 720$
   Hence, 5 factorial is 120 and 6 factorial is 720.

3. (a) For $a_n = 2n - 1$, $a_1 = 2(1) - 1 = 1$,
   $a_2 = 2(2) - 1 = 3$, $a_3 = 2(3) - 1 = 5$,
   $a_4 = 2(4) - 1 = 7$
   (b) For $a_n = 3^{n-1}$, $a_1 = 3^0 = 1$,
   $a_2 = 3^1 = 3$, $a_3 = 3^2 = 9$, $a_4 = 3^3 = 27$
   (c) For $a_n = \dfrac{n+1}{n+2}$, $a_1 = \dfrac{1+1}{1+2} = \dfrac{2}{3}$,
   $a_2 = \dfrac{2+1}{2+2} = \dfrac{3}{4}$, $a_3 = \dfrac{3+1}{3+2} = \dfrac{4}{5}$,
   $a_4 = \dfrac{4+1}{4+2} = \dfrac{5}{6}$
   (d) For $a_n = (-1)^n x^{4n-3}$, $a_1 = (-1)^1 x^{4-3} = -x$,
   $a_2 = (-1)^2 x^{8-3} = x^5$, $a_3 = (-1)^3 x^{12-3} = -x^9$,
   $a_4 = (-1)^4 x^{16-3} = x^{13}$

4. $\displaystyle\sum_{k=1}^{27} \dfrac{k}{k+1} = \dfrac{1}{2} + \dfrac{2}{3} + \dfrac{3}{4} + \cdots + \dfrac{n}{n+1} + \cdots + \dfrac{27}{28}$

5. Replacing i with $k - 1$, we obtain
   $$\sum_{i=1}^{\infty} (i+1)x^{i+1} = \sum_{k-1=1}^{k=\infty} [(k-1)+1]x^{(k-1)+1}$$
   $$= \sum_{k=2}^{\infty} kx^k$$

Exercises

1. $3, 0, -3, -6, -9$ 3. $1, -2, 3, -4$

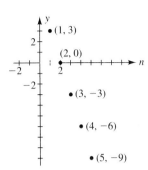
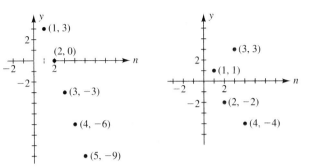

5. $2, 2, \dfrac{32}{9}, 8, \dfrac{512}{25}, \cdots$

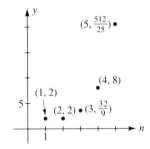

7. $1, 6, 11, 16, 21, 26$ 9. $-6, 12, -24, 48, -96, 192$

11. $3, \dfrac{5}{2}, \dfrac{7}{3}, \dfrac{9}{4}, \dfrac{11}{5}, \dfrac{13}{6}$ 13. $2, 0, 2, 0, 2, 0$

15. $1, 3, 3, 7, 5, 11$ 17. $3, 6, 12, 24, 48, 96$

19. $1, 2, 1, -1, -2, -1$ 21. $a_n = n^2$

23. $a_n = (-1)^{n+1}(2n)$ 25. $a_n = n + 2$

27. $a_n = \dfrac{2n-1}{2n}$ 29. $a_n = 2(0.1)^{n-1}$

31. $a_n = (-1)^n \dfrac{x^n}{2n-1}$ 33. $2 + 4 + 6 + 8 + 10$

35. $-1 + 8 - 27 + 64 - 125 + 216 - 343 + 512 - 729 + 1000$

37. $1 + 4 + 27 + 256 + 3125 + \cdots$

39. $-\dfrac{1}{2} + \dfrac{1}{2} + \dfrac{3}{2} + \dfrac{5}{2} + \cdots + \dfrac{2n-3}{2} + \cdots + \dfrac{97}{2}$

41. $-2 + x - \dfrac{2x^2}{3} + \dfrac{x^3}{2} - \dfrac{2x^4}{5} + \cdots$

43. $\displaystyle\sum_{i=1}^{7} 2i$ 45. $\displaystyle\sum_{i=1}^{30} (5i - 3)$ 47. $\displaystyle\sum_{i=1}^{\infty} (-1)^{i+1}$

49. $\sum_{i=1}^{10} \frac{(-1)^i}{2^{i-1}}$ 51. $\sum_{i=1}^{\infty} \frac{2^i}{3^{i-1}}$

53. $\sum_{i=1}^{5} (i+2)x^{2i-1}$ 55. $\sum_{k=0}^{3} (2k-3)$

57. $\sum_{k=0}^{\infty} \frac{(-1)^{k+1}(2x^k)}{k+1}$ 59. $\sum_{k=0}^{23} \frac{k+2}{(k+2)^2+1}$

61. One possibility is
$$a_n = (2n-1) + (n-1)(n-2)(n-3),$$
which generates the sequence $1, 3, 5, 13, 33, \ldots$.

63. (a) $u(n) = 4n - 6$ for domain $\{1, 2, 3, 4, 5\}$
    (b) $u(n) = -3n + 13$ for domain $\{1, 2, 3, 4\}$

65. (a) $1, 1, 2, 3, 5, 8, 13, 21$
    (b) Horizontal lines through the tree intersect the branches in the sequence $1, 1, 2, 3, 5, 8$, as shown in the figure.

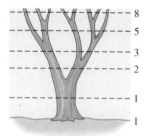

SECTION 11.2

◆ Problems

1. $\sum_{i=1}^{1000} 4 = 1000(4) = 4000$

2. $\sum_{i=1}^{50} (2i - 1) = (50)^2 = 2500$

3. (a) $\sum_{i=1}^{19} i = \frac{19(19+1)}{2} = 190$
   (b) $\sum_{i=1}^{19} i^2 = \frac{19(19+1)[2(19)+1]}{6} = 2470$
   (c) $\sum_{i=1}^{19} i^3 = \left[\frac{19(19+1)}{2}\right]^2 = 36{,}100$

4. $\sum_{i=1}^{20} (5i - 2) = 5\sum_{i=1}^{20} i - \sum_{i=1}^{20} 2$
   $= 5\left[\frac{20(20+1)}{2}\right] - 20(2) = 1010$

5. The sum of the interior angles of a 12-sided convex polygon is $(12 - 2)180° = 1800°$.

◆ Exercises

1. 150   3. 100π 5. -20 7. 247   9. $\frac{76}{15}$

29. 800   31. 16,206   33. 20,825   35. 5440

37. $\frac{499}{500}$ 39. 2046   41. 5460   43. 45,057   45. $\frac{255}{256}$

47. 3825   49. 9680   51. 35,490   53. 4065

55. -2822.4 57. 87,423

59. Approximately 5.3 square units

61. (a) log 120   (b) log 6

65. (a) $a_n = (2n-1)(2n) = 4n^2 - 2n$
    (b) 11,060

67. (a) $a_n = (3n-2)(3n-1)(3n)$
    $= 27n^3 - 27n^2 + 6n$
    (b) 1,114,470

SECTION 11.3

◆ Problems

1. For the sequence $3, 7, 11, 15, \ldots$, we have $a_1 = 3$ and $d = 4$. Hence, the general element is
$$a_n = a_1 + (n-1)d$$
$$= 3 + (n-1)4$$
$$= 4n - 1$$

2. Using $a_n = a_1 + (n-1)d$ with $n = 60$, we have
$$a_{60} = a_1 + 59d$$
Hence,
$$a_1 = a_{60} - 59d = -98 - (59)(-4) = 138$$

3. If $a_1 = -8$ and $d = 7$, then
$$a_n = a_1 + (n-1)d$$
$$= -8 + (n-1)7$$
$$= 7n - 15$$
Hence,
$$a_5 = 7(5) - 15 = 20 \quad \text{and} \quad a_{11} = 7(11) - 15 = 62.$$

4. $\sum_{i=0}^{40} (4i - 3) = 41\left(\frac{-3 + 157}{2}\right) = 3157$

5. Using $S_n = \frac{n}{2}[2a_1 + (n-1)d]$ with $n = 10$, we have
$$S_{10} = \frac{10}{2}[2a_1 + (10-1)d]$$
$$215 = 5[2(8) + 9d]$$
$$43 = 16 + 9d$$
$$27 = 9d$$
$$d = 3$$

6. Using $a_n = a_1 + (n - 1)d$, we find
$$a_{24} = \$68 + (24 - 1)(-\$0.75) = \$50.75.$$
Hence, $S_{24} = 24 \left(\dfrac{\$68.00 + \$50.75}{2} \right) = \$1425.$

Exercises

1. 3, 7, 11, 15, 19   3. 2, $\frac{4}{3}$, $\frac{2}{3}$, 0, $-\frac{2}{3}$
5. $-1, -1 + \pi, -1 + 2\pi, -1 + 3\pi, -1 + 4\pi$
7. $a_n = 6n - 3$ 9. $a_n = 18 - 7n$
11. $a_n = \dfrac{5n - 7}{3}$ 13. $a_n = \dfrac{(3 - 2n)\pi}{6}$
15. $a_n = x + (2n - 6)$ 17. $a_{28} = 193$
19. $a_1 = -307$ 21. $d = -\frac{1}{2}$ 23. $a_{80} = 31.85$
25. $d = -\frac{18}{7}$ 27. 2520   29. -8149 31. $\frac{273}{2}$
33. 4895   35. -273 37. $S_{31} = 2697$ 39. $a_{32} = 26$
41. $d = 8$ 43. $S_{62} = 7006$ 45. $a_{33} = 100$
47. 144.56   49. 1072.4   51. -16.2052 53. 2550
55. 166   57. $75,000
59. (a) 128 seats   (b) 2000 seats
61. (a) $a_1 = \$1050$, $a_2 = \$1042.50$, $a_3 = \$1035$
    (b) $a_n = 1057.50 - 7.50n$ (c) $49,725
63. Yes; common difference is $\ln a$ 65. $x = 0.00505$
67. $\frac{108}{5}, \frac{121}{5}, \frac{134}{5}, \frac{147}{5}$
69. (a) 24 payments   (b) $2600

SECTION 11.4

Problems

1. For the sequence 5, 15, 45, 135, . . . we have $a_1 = 5$ and $r = 3$. Thus, the general element is
$$a_n = a_1 r^{n-1} = 5(3)^{n-1}$$

2. Using $a_n = a_1 r^{n-1}$ with $n = 9$, we have $a_9 = a_1 r^8$. Hence,
$$a_1 = \dfrac{a_9}{r^8} = \dfrac{\frac{1}{81}}{\left(\frac{1}{3}\right)^8} = \dfrac{\left(\frac{1}{3}\right)^4}{\left(\frac{1}{3}\right)^8} = 81$$

3. If $a_1 = 96$ and $r = \frac{1}{2}$, then
$$a_n = a_1 r^{n-1} = 96 \left(\tfrac{1}{2}\right)^{n-1}$$
Hence,
$$a_3 = 96\left(\tfrac{1}{2}\right)^2 = 24 \quad \text{and} \quad a_6 = 96\left(\tfrac{1}{2}\right)^5 = 3$$

4. Using $S_n = \dfrac{a_1(1 - r^n)}{1 - r}$ with $n = 12$, we have
$$S_{12} = \dfrac{a_1(1 - r^{12})}{1 - r}$$
Hence,
$$a_1 = \dfrac{S_{12}(1 - r)}{1 - r^{12}} = \dfrac{8190(1 - 2)}{1 - 2^{12}} = 2$$

5. Using $a_n = a_1 r^{n-1}$ with $a_1 = 3$, $r = 4$, and $a_n = 201,326,592$, we find
$$201,326,592 = 3(4)^{n-1}$$
$$67,108,864 = 4^{n-1}$$
$$\ln 67,108,864 = (n - 1) \ln 4$$
$$n - 1 = \dfrac{\ln 67,108,864}{\ln 4}$$
$$n = \dfrac{\ln 67,108,864}{\ln 4} + 1 = 14$$
Hence, the series has 14 terms.

6. The amount of $5000 earns interest at the rate of 9% compounded monthly, for 10 years. Using the compound interest formula (Section 5.1), we find
$$A = P\left(1 + \dfrac{r}{n}\right)^{nt} = \$5000\left(1 + \dfrac{0.09}{12}\right)^{120}$$
$$= \$5000(1.0075)^{120}$$
$$= \$12,257$$

Adding this amount to the amount of the annuity ($29,027) yields a total amount of $41,284.

Exercises

1. 3, 12, 48, 192, 768   3. $\frac{1}{4}, -\frac{1}{6}, \frac{1}{9}, -\frac{2}{27}, \frac{4}{81}$
5. $-1, -\pi, -\pi^2, -\pi^3, -\pi^4$ 7. $a_n = 3^{n-1}$
9. $a_n = -64\left(-\tfrac{1}{4}\right)^{n-1}$ 11. $a_n = \tfrac{1}{4}\left(\tfrac{2}{3}\right)^{n-1}$
13. $a_n = -\sqrt{2}(-\sqrt{3})^{n-1}$ 15. $a_n = 3(2x^2)^{n-1}$
17. $a_9 = 26,244$ 19. $a_1 = -\frac{1}{216}$ 21. $r = \pm\frac{3}{2}$
23. $a_{13} = \frac{243}{32}$ 25. $r = \pm\frac{1}{2}$ 27. -364 29. $\frac{605}{729}$
31. 46.66666662   33. 1,398,101   35. $\frac{1023}{8}$
37. $S_{11} = \frac{683}{64}$ 39. $a_1 = 2$ 41. $S_6 = \frac{665}{144}$
43. $S_7 = -\frac{2653}{64}$ 45. $a_6 = \frac{729}{8}, -\frac{9375}{8}$ 47. -0.0000807
49. 3.35   51. 13   53. Approximately 182,976   55. $500
57. Choose the scholarship that doubles the amount each month for one year. It pays $40,950, whereas the other scholarship pays only $40,000.
59. $18,294.60   61. $81,397.92

63. (a) Geometric sequence with $r = e^{3x}$ (b) $a_n = e^{(3n-1)x}$

65. $x = 10^{-3}$ 67. 24, 36, 54 69. $r = -\frac{1}{5}, -5$

SECTION 11.5

◆ Problems

1. The infinite series $1 + 0.1 + 0.01 + 0.001 + \cdots$ has $a_1 = 1$ and $r = 0.1$. Since $|r| < 1$, this series converges and its sum S is

$$S = \frac{a_1}{1 - r} = \frac{1}{1 - 0.1} = \frac{1}{0.9} = \frac{10}{9}$$

2. The repeating decimal $0.88888\ldots$ can be written as an infinite series with $a_1 = 0.8$, and $r = 0.1$. Thus,

$$S = \frac{0.8}{1 - 0.1} = \frac{8}{9}$$

3.
$$\begin{array}{r} 1 + 2x + 4x^2 + 8x^3 + \cdots \\ 1 - 2x \overline{)\, 1 } \\ \underline{1 - 2x} \\ 2x \\ \underline{2x - 4x^2} \\ 4x^2 \\ \underline{4x^2 - 8x^3} \\ 8x^3 \\ \underline{8x^3 - 16x^4} \\ 16x^4 \end{array}$$

4. If the initial hammer impact drives the nail $\frac{7}{8}$ inch, then the total distance the nail can be driven is given by

$$\frac{7}{8} + \frac{7}{12} + \frac{7}{18} + \frac{7}{27} + \cdots$$

The sum of this infinite geometric series is

$$S = \frac{a_1}{1 - r} = \frac{\frac{7}{8}}{1 - \frac{2}{3}} = \frac{21}{8} = 2\frac{5}{8} \text{ inches.}$$

Since the nail is $2\frac{1}{2}$ inches long and $2\frac{5}{8} > 2\frac{1}{2}$, it is possible for the head of the nail to be flush with the board. In fact, this will occur during the eighth impact.

◆ Exercises

1. $1, \frac{5}{3}, \frac{19}{9}, \frac{65}{27}; 3\left[1 - \left(\frac{2}{3}\right)^n\right]$

3. $2, -4, 14, -40; \frac{1}{2}[1 - (-3)^n]$

5. $-243, -81, -189, -117; -\frac{729}{5}\left[1 - \left(-\frac{2}{3}\right)^n\right]$

7. $\frac{9}{8}, \frac{15}{8}, \frac{19}{8}, \frac{65}{24}; \frac{27}{8}\left[1 - \left(\frac{2}{3}\right)^n\right]$

9. Diverges 11. Converges to the sum $\frac{1}{30}$

13. Converges to the sum $\frac{5}{11}$ 15. Diverges

17. Converges to the sum -4

19. Converges to the sum $\frac{25}{4}(\sqrt{10} + \sqrt{2})$

21. $r = \frac{2}{3}$ 23. $S = \frac{81}{16}$ 25. $S = -\frac{256}{7}$ 27. $a_1 = 6, 12$

29. $\frac{1}{3}$ 31. $\frac{1}{11}$ 33. $\frac{78}{37}$ 35. $\frac{500627}{33330}$ 37. $\frac{2669}{3330}$

39. Interval of convergence $(-5, 5)$; $S = \dfrac{5x}{x + 5}$

41. Interval of convergence $\left(-\frac{1}{2}, \frac{1}{2}\right)$; $S = \dfrac{1}{1 - 4x^2}$

43. Interval of convergence $(1, 3)$; $S = \dfrac{1}{x - 1}$

45. Interval of convergence $\left(-\infty, -\frac{1}{2}\right) \cup \left(\frac{1}{2}, \infty\right)$; $S = \dfrac{4x^2}{2x - 1}$

47. (a) 54 m (b) 36 m (c) 90 m

49. No; he runs only 25 miles. 51. 140.875 cm

53. $x = -\frac{1}{6}$

55. (a) Does not exist (b) 0 (c) 1; The associative property of real numbers does not apply to infinite series.

57. $r = \frac{1}{3}$

59. (a) 1152 sq cm (b) $\left(192 + 92\sqrt{2}\right)$ cm

SECTION 11.6

◆ Problems

1. The number of possible orders of finish for the first four positions is

$$_8P_4 = \frac{8!}{(8 - 4)!} = \frac{8!}{4!} = 1680 \text{ orders}$$

2. A team that plays ten games can end the season with 3 wins, 5 losses, and 2 ties in $\dfrac{10!}{3!\,5!\,2!} = 2520$ permutations.

3. A five-card poker hand consisting of two queens and three kings can be dealt in $_4C_2 \cdot {_4C_3} = \dfrac{4!}{2!\,2!} \cdot \dfrac{4!}{3!\,1!} = 24$ combinations.

4. The fifth term in the expansion of $(x + y)^9$ is $\binom{9}{4} x^{9-4} y^4 = 126 x^5 y^4$.

◆ Exercises

1. (a) 10! (b) $k!$ 3. (a) 20! (b) $(k-1)!$
5. (a) $\dfrac{11}{10!}$ (b) $\dfrac{1+k}{k!}$ 7. (a) 120 (b) 24
9. 2520 11. 2520 13. 84
15. (a) 56 (b) 30 (c) 9
17. 1,816,214,400 19. 726,485,760 21. 42 23. 1
25. 1 27. 15 29. 84 31. 1260 33. 0
35. $x^5 + 5x^4y + 10x^3y^2 + 10x^2y^3 + 5xy^4 + y^5$
37. $64x^6 - 192x^5 + 240x^4 - 160x^3 + 60x^2 - 12x + 1$
39. $81a^4 + 216a^3b + 216a^2b^2 + 96ab^3 + 16b^4$
41. $128x^{14} - 448x^{12} + 672x^{10} - 560x^8 + 280x^6 - 84x^4 + 14x^2 - 1$
43. $x^4 + 16x^3\sqrt{x} + 112x^3 + 448x^2\sqrt{x} + 1120x^2 + 1792x\sqrt{x} + 1792x + 1024\sqrt{x} + 256$
45. $220x^9y^3$ 47. $20{,}412n^5$ 49. $5670x^4y^8$
51. $13{,}440t^2$ 53. $n = 14$ 55. $n = 9$ 57. $n = 8$
61. $(n-1)!$
65. (a) $x^{10} + 5x^9 + 15x^8 + 30x^7 + 45x^6 + 51x^5 + 45x^4 + 30x^3 + 15x^2 + 5x + 1$
    (b) $x^{10} - 5x^9 + 5x^8 + 10x^7 - 15x^6 - 11x^5 + 15x^4 + 10x^3 - 5x^2 - 5x - 1$
67. (a) 1, 7, 21, 35, 35, 21, 7, 1
    (b) 1, 8, 28, 56, 70, 56, 28, 8, 1
    (c) 1, 9, 36, 84, 126, 126, 84, 36, 9, 1
69. (a) $1 - x + x^2 - x^3 + \cdots$
    (b) $1 + 2y + 3y^2 + 4y^3 + \cdots$
    (c) $1 + \dfrac{z}{2} + \dfrac{3z^2}{8} + \dfrac{5z^3}{16} + \cdots$
    (d) $1 + \dfrac{x}{3} - \dfrac{x^2}{9} + \dfrac{5x^3}{81} - \cdots$
71. Power series yields $e \approx 2.718281526$; $\boxed{e^x}$ key yields $e \approx 2.718281828$

CHAPTER 11 REVIEW EXERCISES

1. $2, \frac{3}{2}, \frac{4}{3}, \frac{5}{4}, \frac{6}{5}, \frac{7}{6}$

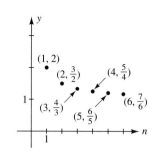

3. $-4, -1, 2, 5, 8, 11$ 5. $1, -2, 6, -24, 120, -720$
7. 0, 0, 2, 4, 4, 16
9. (a) Neither (b) $a_n = (-1)^n (2n - 1)$
11. (a) Arithmetic (b) $a_n = 6n - 3$
13. (a) Geometric (b) $a_n = 768 \left(\frac{1}{4}\right)^{n-1}$
15. (a) Neither (b) $a_n = \dfrac{(-1)^{n+1}(2n-1)}{2n}$
17. (a) Geometric (b) $a_n = -4\left(-\frac{1}{2}\right)^{n-1}$
19. (a) Arithmetic (b) $a_n = \dfrac{17 - 4n}{3}$
21. (a) Neither (b) $a_n = n^2 - 1$
23. (a) Arithmetic (b) 1260
25. (a) Geometric (b) $-\dfrac{12{,}117}{64}$
27. (a) Neither (b) 6327
29. (a) Geometric (b) 10
31. (a) Arithmetic (b) 93
33. (a) Neither (b) 36,664
41. 8120 43. 101,025
45. (a) $\sum\limits_{i=1}^{250} (4i + 1)$ (b) $\sum\limits_{i=0}^{249} (4i + 5)$
47. $\sum\limits_{i=1}^{\infty} 243 \left(-\frac{2}{3}\right)^{i-1}$
49. (a) 384, 288, 312, 306; $\dfrac{1536}{5}\left[1 - \left(-\frac{1}{4}\right)^n\right]$ (b) $\dfrac{1536}{5}$
51. -83 53. 27,090 55. 96
57. $x = 4, -\frac{3}{2}$ 59. $\frac{81}{16}$ or $\frac{81}{80}$
61. (a) $\frac{100}{33}$ (b) $\frac{3031}{1110}$

63. 42,340   65. (a) $12,160.67   (b) $19,782.09

67. Yes, during the 9th second   69. 16   71. 78   73. 0

75. $729x^6 - 2916x^5 + 4860x^4 - 4320x^3 + 2160x^2 - 576x + 64$

77. $120a^{-7}b^3$ 79. 60,060

81. (a) 635,013,559,600   (b) 1,677,106,640
    (c) 740,999,259

Cumulative Review: Chapters 10, 11

1. (a) 420   (b) 2,097,150

3. (b) 722,666   5. 4

7. (a) $(2, -1, 3)$ (b) $\left(1, 2, -3, -\frac{3}{2}\right)$

9. $A^{-1} = \dfrac{1}{17}\begin{bmatrix} -17 & -2 & 28 \\ -17 & 2 & 23 \\ 17 & 3 & -25 \end{bmatrix}$;

   $X = \begin{bmatrix} 3 \\ 2 \\ -2 \end{bmatrix}$

11. $x = 8, y = 12$ or $x = \frac{1}{2}, y = -3$

13. Arithmetic sequence with common difference $\log r$

15. (a) $\begin{bmatrix} 7 & 5 & 3 & -7 \\ 8 & 4 & -4 & 5 \\ 9 & 10 & 6 & -3 \\ 0 & 5 & 9 & -1 \end{bmatrix}$

(b) $\begin{bmatrix} 11 & -15 & 14 & -21 \\ -1 & -13 & 8 & 5 \\ 27 & -10 & 8 & 6 \\ -30 & -25 & 17 & 12 \end{bmatrix}$

(c) $\begin{bmatrix} 19 & -10 & 4 & 4 \\ -8 & -24 & 18 & -16 \\ 60 & 5 & -13 & 18 \\ 90 & -1 & 11 & -24 \end{bmatrix}$

(d) $\begin{bmatrix} -37 & 1 & -7 & 7 \\ 19 & 37 & 7 & -11 \\ 28 & 96 & -9 & -8 \\ -15 & 21 & 36 & -33 \end{bmatrix}$

17. 0.3

19. (a) 64   (b) 128   (c) 0   (d) 924

21. 18, 24, 30

23. (a) 66   (b) 220

25. $1 + \dfrac{1}{2} + \dfrac{1}{3} + \dfrac{1}{4} + \cdots + \dfrac{1}{n} + \cdots$

27. $\begin{bmatrix} 16 & -4 \\ -11 & 3 \end{bmatrix}$

29. 3 mi/h uphill, 8 mi/h on level ground, 12 mi/h downhill

31. About 29 years

INDEX

A

Abscissa, 76
Absolute inequality, 61
Absolute value:
    of a complex number, 586
    properties of, 7
    of a real number, 7
Absolute value function, 113
Acre, 570 (Exercises 41, 42)
Acute angle, 413
Addition:
    of algebraic fractions, 53–55
    of complex numbers, 31
    of matrices, 654
    of polynomials, 37
    of vectors, 576
Additive identity property, 3
Additive inverse property, 3
A-frame structure, 531
Algebra of calculus, 55–57
Algebraic definition:
    of a circle, 345
    of an ellipse, 367
    of a hyperbola, 380
    of a parabola, 172, 354
Algebraic expression:
    definition of, 35
    operations with, 37–41
Algebraic fraction:
    definition of, 50
    operations with, 51–55
Algebraic function, 294
Alternating current (AC) circuit, 591
Alternating sequence, 694
Ambiguous case, 553
Amount of an annuity, 732
Amplitude, 430
Analytic geometry, 78, 79
Analytic trigonometry, 486
Angle:
    acute, 413
    central, 350, 402
    complemetary, 415
    coterminal, 397
    definition of, 396
    degree measure of, 397–399
    of depression, 471
    direction, 572
    of elevation, 471
    initial side of, 396
    interior, of an n-sided convex polygon, 711
    measure of, 397
    negative, 396
    obtuse, 551
    positive, 396
    quadrantal, 419–421
    radian measure of, 399–401
    right, 397, 413
    special (30°, 45°, 60°), 421
    in standard position, 396
    straight, 397
    terminal side of, 396
    trigonometric ratios for, 407
    vertex of, 396
Angular speed, 403
Annuity, simple, 731–733
Applied problems (*see Contents: Applied Problems and Models*, xi)
Approximate number, 763
Arc, 399
Arc length formula, 402
Arccosine function (arccos), 456
Archimedean spiral, 614 (Exercises 59, 60)
Archimedes, 614 (Exercises 59, 60)
Arcsine function (arcsin), 454
Arctangent function (arctan), 458
Area:
    formulas for, 66
    of a rhombus, 535 (Exercise 85)
    of a sector, 406 (Exercise 71)
    of a triangle, 565–567
Argument of a complex number, 586
Arithmetic mean, 724 (Exercise 67)
Arithmetic sequence, 714–717
Arithmetic series, 718–721
Associative property:
    for matrices, 658
    for real numbers, 3
    of scalar multiplication, 578
    for vector addition, 578
Asymptote:
    horizontal, 116, 265–268
    of a hyperbola, 379
    oblique, 268, 269
    parabolic, 276 (Exercise 60)
    vertical, 161, 262–264, 441, 445, 448
Augmented matrix, 651
Autumnal equinox, 507
Average cost function, 272–274
Axis (axes):
    coordinate, 76
    of an ellipse, 366
    of a hyperbola, 379
    imaginary, 586
    polar, 604
    real, 586
    of symmetry of a parabola, 172, 353

B

Base:
    definition of, 13
    of an exponential function, 294, 296
    of a logarithmic function, 305, 306
Base e, 296, 297
Base e conversion formula, 307
Base 10 conversion formula, 321
Bearing, 581
Bending moment of a beam, 223–225
Binomial:
    cube of, 39
    definition of, 36
    like, 45
    square of, 39
Binomial coefficient symbol, 749
Binomial series, infinite, 754 (Exercise 69)
Binomial theorem, 752
Boundary number, 188, 254, 281
Boyle's law, 144 (Exercise 28)
Braces, 40
Brackets, 40
Branches of a hyperbola, 379
Brightness scale, 322, 324 (Exercises 49, 50)

C

Calculator:
  general discussion of, 84, 85
  usage of, (*see Contents: Graphing Technology Usage*, viii)
Capacitive reactance, 591
Cardioid, 613
Cartesian coordinates, 76
Cartesian plane, 76
Catenary, 337 (Exercise 91)
Cayley, Arthur, 654, 662 (Exercise 71)
Center:
  of a circle, 345
  of an ellipse, 366
  of a hyperbola, 379
Centi- (c), 18
Central angle, 350, 402
Change of base formula, 321, 322
Chord of a circle, 482 (Exercise 91), 510 (Exercise 90)
Circle:
  algebraic definition of, 345
  area of, 66
  center of, 345
  chord of, 482 (Exercise 91), 510 (Exercise 90)
  circumference of, 66
  concentric, 352 (Exercise 43)
  as a conic section, 344
  equation of,
    in general form, 347
    in parametric form, 618, 619, 626 (Exercise 57)
    in polar form, 608, 611, 613
    in standard form, 345
  geometric definition of, 345
  radius of, 345
  sector of, 406 (Exercise 71)
  tangent to, 172 (Exercise 60), 346
  unit, 418
Circular permutation, 754 (Exercise 61)
Clockwise rotation, 396
Closed interval, 8
Coefficient:
  binomial, 749
  leading, 218
  numerical, 36
Coefficient matrix, 670
Cofunction, 415
Cofunction identities, 415, 541
Collecting like terms, 36
Collinear, 78
Column of a matrix, 649
Combination, 747–749
Combining like terms, 36
Common difference of an arithmetic sequence, 714
Common formulas, 66

Common logarithm, 306
Common logarithmic function, 306
Common ratio of a geometric sequence, 725
Common term factoring, 44
Commutative property:
  for matrices, 658
  for real numbers, 3
  for vector addition, 578
Complementary angle, 415
Complete graph, 220
Completing the square, 175
Complex conjugate, 32, 594 (Exercise 64)
Complex fraction, 56
Complex number:
  absolute value of, 586
  argument of, 586
  conjugate of, 32, 594 (Exercise 64)
  exponential form of, 594 (Exercise 67), 604 (Exercise 67)
  modulus of, 586
  nth root of, 598–601
  operations with, 31–33, 588–591
  power of, 33, 34, 595, 596
  in standard form, 30, 585
  in trigonometric form, 587
Complex number system, 30
Complex plane, 586
Component form of a vector, 572
Components, vector, 572
Composite function, 125–127
Composition properties for trigonometric functions, 460
Compound interest formula:
  for continuous compounding, 301
  for n compoundings per year, 300
Compress and stretch rules, 118–120
Computing the functional value, 96
Concentric circles, 352 (Exercise 43)
Concurrent force, 580
Condensing logarithmic expressions, 320
Conditional equation, 60
Conditional inequality, 61
Cone, 67
Conic sections, 344
Conjugate:
  of a complex number, 32, 594 (Exercise 64)
  of a radical expression, 56
Conjugate axis of a hyperbola, 379
Conjugate hyperbola, 391 (Exercise 59)
Conjugate pair, 246
Conjugate pair theorem, 246, 247
Constant:
  definition of, 2
  of variation, 138, 139
Constant function, 113, 120, 121, 157
Constant matrix, 670
Constant term, 36
Constraint, 679

Constructions for SSA, 555
Continuous compounding, 301
Continuous function, 216
Convergence:
  of an infinite geometric series, 736–738
  interval of, 739, 740
Converse of the Pythagorean theorem, 79
Conversion factor:
  definition of, 17
  for degrees, minutes, seconds and decimal degress, 398
  for radians and degrees, 400
Conversion formula:
  base e, 307
  base 10, 321
Convex polygon, 711
Convex region, 679
Coordinate axes, 76
Coordinate plane, 76
Coordinate of a point:
  in the Cartesian plane, 76
  in the polar plane, 605
  on the real number line, 5
Coplanar force, 580
Cosecant function (csc):
  of an angle, 407
  graph of, 442
  inverse of, 465 (Exercise 69)
  properties of, 442
  as the ratio of the sides in a right triangle, 414
  of a real number, 419
  restricted, 465 (Exercise 69)
Cosine function (cos):
  of an angle, 407
  graph of, 429
  inverse of, 456
  properties of, 430
  as the ratio of the sides in a right triangle, 414
  of a real number, 419
  restricted, 455
Cosines, law of, 562
Cotangent function (cot):
  of an angle, 407
  graph of, 442
  inverse of, 466 (Exercise 71)
  properties of, 443
  as the ratio of the sides in a right triangle, 414
  of a real number, 419
  restricted, 466 (Exercise 71)
Coterminal angle, 397
Counterclockwise rotation, 396
Cube of a binomial, 39
Cube root function, 113
Cubic formula, 261 (Exercises 71–74)
Cubic function, 216
Cubing function, 113

Cycle, 429
Cylinder, 67

D

Decibel scale, 322, 325 (Exercises 51, 52)
Decimal number, 738
Decreasing function, 120
Degenerate circle, 349
Degenerate conic sections, 344
Degenerate ellipse, 370
Degenerate hyperbola, 383
Degree (°), 397
Degree:
    of a polynomial, 36
    of a polynomial function, 216
    of a term, 37
De Moivre, Abraham, 595
De Moivre's theorem:
    extension of, 597
    statement of, 595
Denominator:
    least common, 54
    rationalizing, 27, 28, 56
Dependent system, 638
Dependent variable, 82
Depreciation, linear, 169
Depression, angle of, 471
Descartes, René, 30, 76
Determinant:
    definition of, 662
    expansion of, 668
    of a matrix in echelon form, 663
    row operations on, 664
    of a 2 X 2 matrix, 663
Diagonal of a parallelogram, 81 (Exercises 37–40), 576
Difference:
    of complex numbers, 31
    of matrices, 655
    of polynomials, 37
    of real numbers, 4
    of two cubes, 47
    of two squares, 47
    of vectors, 577
Difference formula:
    for cosine, 512
    for sine, 514
    for tangent, 515
Difference quotient, 97
Direct variation, 138
Direction angle of a vector, 572
Directly proportional, 138
Directrix, 358
Discontinuous function, 216
Discriminant, 185
Displacement, 581
Distance:
    from a line to the origin, 209 (Exercise 61)
    between points, 7, 77, 615 (Exercise 70)
Distance formula:
    in the Cartesian plane, 77
    in the polar plane, 615 (Exercise 70)
Distinguishable permutation, 747
Distributive property:
    over addition, 3
    factoring by using, 44
    for matrices, 658
    of a scalar over vector addition, 578
    of a vector over scalar addition, 578
Divergence of an infinite geometric series, 736
Dividend in polynomial division, 228
Division:
    of algebraic fractions, 51, 52
    of complex numbers, 32, 590
    definition of, 4
    of polynomials, 228–232
    by zero, 4
Division algorithm for polynomials, 228
Divisor:
    in polynomial division, 228
    synthetic, 231
Domain, 94, 99–101, 103, 408
Dot product, 585 (Exercises 80–82)
Double-angle formulas, 524, 525
Double inequality, 6
Double subscript notation, 639, 649
Dummy variable, 698

E

e, the number, 296
Eccentricity:
    of an ellipse, 372
    of a hyperbola, 384
Echelon form of a matrix, 651
Elastic curve, 365 (Exercise 42)
Element:
    of a matrix, 649
    of a sequence, 692
    of a set, 2
Elementary row operations, 652
Elevation, angle of, 471
Elimination method, 198, 639
Ellipse:
    algebraic definition of, 367
    axes of, 366
    center of, 366
    as a conic section, 344
    eccentricity of, 372
    equation of,
        in general form, 368–370
        in parametric form, 618, 619, 626 (Exercise 57)
        in standard form, 367, 368
    foci of, 371
    geometric definition of, 371–374
    latus rectum of, 377 (Exercise 77)
    reflection property of, 374, 375
    vertices of, 366
Equality:
    of matrices, 650, 652
    of vectors, 572
Equation:
    conditional, 60
    definition of, 60
    equivalent, 61
    exponential, 325–328
    graph of, 82
    linear, in two variables, 157–159
    literal, 65
    logarithmic, 326, 328–330
    matrix, 670
    parametric, 615
    polar, 607
    polynomial, 249–253, 601
    quadratic,
        in one unknown, 183–188
        in two unknowns, 344
    radical, 107
    rational, 277–280
    root of, 60
    solution of, 60
    solving simultaneously, 197–204
    system of, 638
    trigonometric, 496
Equilateral triangle, 421
Equilibrium, 581
Equivalency of matrices, 652
Equivalent equation, 61
Equivalent fraction, 50
Equivalent inequality, 63
Equivalent matrix, 652
Equivalent system, 640
Euler, Leonhard, 30
Euler's formula, 594 (Exercise 68)
Even function, 104
Even-odd identities, 541
Expanded form of a series, 698
Expanding a logarithmic expression, 317–319
Expansion of a third-order determinant by its first row, 666
Explicitly defined sequence, 695
Exponent:
    definition of, 13
    laws of, 15
    negative integer, 14, 596
    properties of positive integer, 13
    rational, 22–25
    real, 297, 298
    zero, 14, 596
Exponential decay function, 332, 334
Exponential equation, 325–328
Exponential form:
    of a complex number, 594 (Exercise 67), 604 (Exercise 67)

Exponential form (continued):
   of an equation, 305
   of a radical expression, 22
   of repeated multiplication, 13
Exponential function:
   with base b, 294
   with base e, 296
Exponential growth function, 332–334
Extracting the square root, 63
Extraneous root, 107, 277, 502
Extremum, relative, 219

F

Factor, 13, 15, 51, 238
Factor theorem, 237–239
Factored completely:
   over the set of complex numbers, 49 (Exercises 83–86)
   over the set of integers, 44
   over the set of real numbers, 49 (Exercises 79–82)
Factorial, 696, 743
Factoring:
   common term, 44
   definition of, 44
   by grouping terms, 44, 45
   a polynomial, 44, 238, 244
   a trigonometric expression, 487
   a trinomial with binomial factors, 45–47
Factoring formulas, 47, 48
Fibonacci, Leonardo, 703 (Exercise 65)
Fibonacci sequence, 703 (Exercise 65)
Finite geometric series, 728–731
Finite sequence, 692, 693
Finite series, 692, 698
Fixed cost, 272
Focus (foci):
   of an ellipse, 371
   of a hyperbola, 383
   of a parabola, 358
Force:
   concurrent, 580
   coplanar, 580
   resultant, 580
   as a vector quantity, 579
Force system:
   definition of, 580
   in equilibrium, 581
Formula, 66
Four-leafed rose, 612
Fraction:
   algebraic, 50
   complex, 56
   equivalent, 50
   fundamental property of, 50
Fractional equation (see Rational equation)
Frequency, 438 (Exercise 49), 472
Function:
   absolute value, 113
   algebraic, 294
   arccosine (arccos), 456
   arcsine (arcsin), 454
   arctangent (arctan), 458
   average cost, 272–274
   common logarithmic, 306
   composite, 125–127
   constant, 113, 120, 157
   continuous, 216
   cosecant (csc), 442
   cosine (cos), 430
   cotangent (cot), 443
   cube root, 113
   cubic, 216
   cubing, 113
   decreasing, 120
   definition of, 94
   discontinuous, 216
   domain of, 94, 99–101, 103
   even, 104
   exponential, 294–297
   exponential growth and decay, 332–335
   graph of, 103
   greatest integer, 124 (Exercise 72)
   hyperbolic cosine, 304, (Exercise 56)
   hyperbolic sine, 304 (Exercise 56)
   identity, 113
   increasing, 120
   input value of, 94
   inverse, 128–132
   inverse cosecant (csc^{-1}), 465 (Exercise 69)
   inverse cosine (cos^{-1}), 456
   inverse cotangent (cot^{-1}), 466 (Exercise 71)
   inverse secant (sec^{-1}), 466 (Exercise 70)
   inverse sine (sin^{-1}), 454
   inverse tangent (tan^{-1}), 458
   linear, 152
   logarithmic, 305
   natural logarithmic, 306
   normal probability distribution, 304 (Exercise 60)
   objective, 679
   odd, 104
   one-to-one, 127, 128
   output value of, 94
   periodic, 429
   piecewise-defined, 98, 121
   piecewise linear, 159
   polynomial, 216
   power, 217
   quadratic, 173
   quartic, 216
   quintic, 216
   range of, 94, 99, 103
   rational, 262
   real zero of, 104
   reciprocal, 113, 262
   restricted cosecant, 465 (Exercise 69)
   restricted cosine, 455
   restricted cotangent, 466 (Exercise 71)
   restricted secant, 466 (Exercise 70)
   restricted sine, 453
   restricted tangent, 457
   retention, 312, 313
   secant (sec), 443
   sequence, 692
   sine (sin), 429
   square root, 113
   squaring, 113
   step, 124 (Exercise 72)
   tangent (tan), 443
   transcendental, 294
   trigonometric, 408
   value of, 96
   zeros of, 104, 234, 238, 240, 241
Function machine, 94
Functional notation, $f(x)$, 96–99
Functional value, 96
Fundamental principle of linear programming, 680
Fundamental properties of binomial coefficients, 750
Fundamental properties of real numbers, 2, 3
Fundamental properties of scalar multiplication and vector addition, 578
Fundamental property of fractions, 50
Fundamental theorem of algebra:
   extension of, 240
   statement of, 239
Fundamental trigonometric identities, 411–413

G

Galileo, 179, 622
Gauss, Carl Friedrich, 30, 652
Gaussian method, 652
General element:
   of an arithmetic sequence, 715–717
   of a geometric sequence, 725–728
   of a sequence, definition of, 694–697
General form:
   of the equation of a circle, 347
   of the equation of an ellipse, 368
   of the equation of a hyperbola, 381
   of the equation of a line, 157–159
   of the equation of a parabola, 356
   of a quadratic function, 175
General quadratic equation in two unknowns, 344
General solution of a trigonometric equation, 497
General term of a series, 697
Generating equivalent equations, 61–63
Generating equivalent inequalities, 63–65
Generating equivalent systems, 640
Generating row equivalent matrices, 652
Geometric definition:
   of a circle, 345

of an ellipse, 371–374
of a hyperbola, 383–386
of a parabola, 358–361
Geometric mean, 735 (Exercise 67)
Geometric sequence, 724–728
Geometric series:
    finite, 728–731
    infinite, 735
Geometry formulas, 66, 67
Graph:
    of an equation, 82
    of a function, 103
    horizontal shift of, 115, 116
    of an inequality, 676, 677
    of an inverse function, 129, 130
    of a parametric equation, 616
    of a polar equation, 607
    reflection of, 116–118
    of a sequence function, 694
    of a system of equations, 638, 639
    of a system of inequalities, 678, 679
    vertical shift of, 113, 114
    vertical stretch and compress of, 118–120
Graphical solution:
    of an equation, 106
    of an inequality, 107
Graphing calculator (see Calculator)
Greater than ($>$), 5
Greater than or equal to ($\geq$), 5
Greatest integer function, 124 (Exercise 72)
Grouping symbols, 40
Grouping terms, factoring by, 45

H

Half-angle formulas, 529–531
Half-life, 334
Half-open interval, 9
Halley's comet, 20 (Exercise 70), 376 (Exercise 40)
Harmonic motion, simple, 471–474
Hero of Alexandria (Heron), 566
Hero's formula:
    proof of, 571 (Exercise 52)
    statement of, 566
Homogeneous system, 649 (Exercises 51, 52)
Hooke's law, 143 (Exercise 23)
Horizontal asymptote, 116, 265–268
Horizontal line, equation of, 157, 610
Horizontal line test, 127
Horizontal shift rule, 115, 116
Hyperbola:
    algebraic definition of, 380
    asymptotes of, 379
    axes of, 379
    branch of, 379
    center of, 379
    as a conic section, 344
    eccentricity of, 384

equation of,
    in general form, 381–383
    in parametric form, 620, 621, 626 (Exercise 58)
    in standard form, 380, 381
foci of, 383
geometric definition of, 383–386
reflection property of, 386
vertices of, 379
Hyperbolic cosine function, 304 (Exercise 56)
Hyperbolic sine function, 304 (Exercise 56)
Hypotenuse, 414

I

i, the imaginary unit, 30
Identity:
    definition of, 60
    logarithmic, 307, 311
    trigonometric, 490
Identity element for multiplication, 14
Identity function, 113
Identity matrix, 666
Identity property:
    of matrix multiplication, 666
    of real numbers, 3
    of scalar multiplication, 578
    of vector addition, 578
Imaginary axis, 586
Imaginary number, 31
Imaginary part of a complex number, 30
Imaginary unit i, 30
Impedance of an AC circuit, 591
Inconsistent system, 638
Increasing function, 120
Independent variable, 82
Index:
    of a radical, 22
    shift of, 701
    of summation, 698
Induction, mathematical, 705–709
Inductive reactance, 591
Inequality:
    absolute, 61
    conditional, 61
    definition of, 60
    double, 6
    equivalent, 63
    graph of, 676, 677
    linear, in two variables, 676
    literal, 65
    polynomial, 249, 254–256
    quadratic, 188–191
    rational, 277, 280–283
    solution set of, 61
    system of, 678, 679
Inequality symbols, 5
Infinite binomial series, 754 (Exercise 69)
Infinite geometric series, 735
Infinite sequence, 692, 693

Infinite series, 692, 698
Infinity (∞), 9
Initial point of a vector, 572
Initial side of an angle, 396
Input value of a function, 94
Installment plan, 721, 722
Integer, 2
Integer exponent, 14–16
Intercept, 86
Intercept form of a line, 163 (Exercise 69)
Interest:
    compound, 299–302
    simple, 66
Interior angles of an n-sided convex polygon, 711
Intersection point:
    definition of, 197
    of a line and a parabola, 200, 201
    of other curves, 202, 203
    of two lines, 197–199
Intersection point method for solving an equation, 204, 205
Interval:
    of convergence, 739, 740
    definition of, 8
Interval notation, 8–11
Inverse:
    of a function, 130–132
    of a matrix, 666–669
Inverse cosecant function ($\csc^{-1}$), 465 (Exercise 69)
Inverse cosine function ($\cos^{-1}$), 456
Inverse cotangent function ($\cot^{-1}$), 466 (Exercise 71)
Inverse function, 129
Inverse method, 670
Inverse operation, 125
Inverse property:
    of matrix multiplication, 425
    of real numbers, 3
    of vector addition, 578
Inverse secant function ($\sec^{-1}$), 466 (Exercise 70)
Inverse sine function ($\sin^{-1}$), 454
Inverse tangent function ($\tan^{-1}$), 458
Inverse variation, 139
Inversely proportional, 139
Invertible matrix, 668
Irrational number, 2
Irrational zero of a polynomial function, 245
Isosceles right triangle, 421

J

Joint variation, 141

K

Kilo- (k), 18
kth term of $(A + B)^n$, 752

L

Latus rectum:
    of an ellipse, 377 (Exercise 77)
    of a parabola, 365 (Exercise 48)
Law of cosines, 562
Law of motion for freely falling objects, 179
Law of sines, 551
Laws of exponents, 15
Leading coefficient, 218
Least common denominator, LCD, 54
Left and right behavior, 218, 219
Legs of a right triangle, 414
Lemniscate, 613
Length of a line segment, 77
Less than ($<$), 5
Less than or equal to ($\leq$), 5
Like binomials, 45
Like terms, 36
Limaçon, 613
Line:
    equation of,
        in general form, 157–159
        in intercept form, 163 (Exercise 69)
        in parametric form, 616, 617, 626 (Exercise 56)
        in point-slope form, 163–165
        in polar form, 609, 610, 614 (Exercise 62)
        in slope-intercept form, 156
    horizontal, 157
    parallel, 165, 166
    perpendicular, 167, 168
    slope of, 153–155
    vertical, 157
Line segment:
    length of, 77
    midpoint of, 78
Linear depreciation, 169
Linear equation:
    solving simultaneously, 197–199
    system of, 396
    in two variables, 157–159
Linear function, 152
Linear inequality:
    system of, 678
    in two variables, 676
Linear interpolation, 91
Linear programming, 679–683
Linear speed, 402
Lissajous figure, 626 (Exercise 55)
Literal equation, 65
Literal inequality, 65
Lithotripter, 377 (Exercise 42)
Logarithm:
    common, 306
    condensing, 320
    evaluating, 306–308
    expanding, 317–319
    Napierian, 306
    natural, 306
    properties of, 316, 317
Logarithmic equation, 326, 328–330
Logarithmic form of an equation, 305
Logarithmic function:
    definition of, 305
    graph of, 308–311
Logarithmic identities, 307, 311
Logarithmic scales, 322
Logistic law, 338 (Exercise 94)
Long division of polynomials, 228–230
Lowest terms, 50, 51

M

Magnitude of a vector, 572
Main diagonal, 662
Major axis of an ellipse, 366
Malthus, Thomas, 333
Malthusian model, 333
Mathematical induction:
    extension of, 711
    proof by, 705–709
Matrix (matrices):
    addition of, 654
    augmented, 651
    coefficient, 670
    column of, 649
    constant, 670
    definition of, 649
    determinant of, 663
    echelon form of, 651
    element of, 649
    equality of, 650, 652
    equivalency of, 652
    identity, 666
    inverse of, 667
    invertible, 668
    main diagonal of, 662
    multiplication of, 657
    order of, 649
    properties of, 658
    row of, 649
    row equivalent, 652
    row operations for, 652
    scalar multiple of, 655
    second-order determinant of, 662
    singular, 668
    square, 662
    square of, 661 (Exercise 67)
    subtraction of, 655
    transpose of, 662 (Exercises, 69, 70), 688 (Exercise 56)
    variable, 670
    zero, 655
Matrix equation, 670
Maxima, relative, 219, 448
Maximum functional value, 136
Maximum value of a quadratic function, 177, 178

Mean:
    arithmetic, 724 (Exercise 67)
    geometric, 735 (Exercise 67)
Measure of an angle, 397
Median of a triangle, 81 (Exercise 29)
Mega- (M), 18
Method of linear interpolation, 91
Micro- (μ), 18
Midpoint, 78
Midpoint formula, 77, 78
Milli- (m), 18
Minima, relative, 219, 448
Minimum value of a quadratic function, 177, 178
Minor axis of an ellipse, 366
Minute ('), 398
Mixed algebraic expression, 59 (Exercise 63)
Models (*see Contents: Applied Problems and Models*, xi)
Modulus of a complex number, 586
Mollweide's formula, 560 (Exercises 47–50)
Monomial, 36
Motion:
    law of, 179
    projectile, 622–624
    uniform, 66, 283–285
Multiple-angle formulas, 524
Multiplication:
    of algebraic fractions, 51, 52
    of complex numbers, 32, 590
    of matrices, 656–658
    of polynomials, 37, 38
    scalar,
        with matrices, 655
        with vectors, 575
Multiplicative identity, 3, 14, 35 (Exercise 45)
Multiplicative inverse, 3, 14, 35 (Exercise 46), 594 (Exercise 63)
Multiplicative property of zero, 578
Multiplicity of a zero, 240

N

Napier, John, 306
Napierian logarithm, 306
Natural logarithm, 306
Natural logarithmic function, 306
Natural number, 2
Negation, properties of, 4
Negative angle, 396
Negative infinity ($-\infty$), 9
Negative integer exponent, 14–17, 596–598
Newton, Sir Isaac, 338 (Exercises 95, 96)
Newton's law of cooling, 338 (Exercises 95, 96)
Nonsquare system, 644, 645
Normal probability distribution function, 304 (Exercise 60)
n-sided convex polygon, 711

nth partial sum, 736
nth root:
   of a complex number, 598
   of a real number, 22
nth root formula, 599
Number:
   complex, 30
   e, 296
   i, 30
   imaginary, 31
   integer, 2
   irrational, 2
   natural, 2
   quadrantal, 419
   rational, 2
   real, 2
   whole, 2
Number line, real, 4
Numerator, rationalizing, 27, 28, 56
Numerical coefficient, 36

O

Objective function, 679
Oblique asymptote, 268, 269
Oblique triangle, 550
Obtuse angle, 551
Odd function, 104
One-to-one correspondence, 5, 76
One-to-one function, 127, 128
Open interval, 8
Order of a matrix, 649
Order of operations, 40
Ordered n-tuple, 639
Ordered pair, 76
Ordered quadruple, 644
Ordered triple, 641
Ordinate, 76
Origin:
   in a coordinate plane, 76
   symmetric with respect to, 85
Output value of a function, 94

P

Parabola:
   algebraic definition of, 354
   axis of symmetry of, 172, 353
   as a conic section, 344
   directrix of, 358
   equation of,
      in general form, 356–358
      in parametric form, 617, 618
      in standard form, 354–356
   focus of, 358
   geometric definition of, 358–361
   as the graph of a quadratic function, 173
   latus rectum of, 365 (Exercise 48)
   reflection property of, 362, 363
   tangent line to, 200

   vertex of, 172, 353
   vertex formula for, 175, 176
Parabolic asymptote, 276 (Exercise 60)
Parallel lines, 165, 166
Parallelogram, diagonals of, 81 (Exercises 37–40), 576
Parallelogram law for vectors, 576
Parameter, 616
Parametric equation:
   definition of, 615
   of an ellipse (or circle), 618, 619, 626 (Exercise 57)
   of a hyperbola, 620, 621, 626 (Exercise 58)
   of a line, 616, 617, 626 (Exercise 56)
   of a parabola, 617
   of projectile motion, 622–624
Parametric form of a polar equation, 622
Parentheses, 40
Partial sums, sequence of, 736
Particular solution of a trigonometric equation, 497
Pascal, Blaise, 754 (Exercise 67)
Pascal triangle, 754 (Exercise 67, 68)
Pattern recognition, 696
Perfect square trinomial, 47
Period:
   definition of, 428, 429
   of a pendulum, 29 (Exercise 93)
Period formula:
   for cosecant, 448
   for cosine, 431
   for cotangent, 444
   for secant, 448
   for sine, 431
   for tangent, 444
Periodic function, 429
Permutation, 744–747
Perpendicular bisector, 171 (Exercise 49)
Perpendicular lines, 167, 168
Petal of a rose, 612
pH of a liquid, 314 (Exercises 75, 76)
Phase shift, 432, 520
Piecewise-defined function:
   definition of, 98
   graph of, 121
Piecewise linear function, 159
Pixel, 85
Point of curvature (PC), 346
Point of intersection:
   definition of, 197
   of a line and a parabola, 200, 201
   of other curves, 202, 203
   of two lines, 197–200
Point-plotting method, 82–84
Point-slope form, 163–165
Point of tangency (PT), 346, 350
Polar axis, 604
Polar coordinate system, 604
Polar coordinates, 605

Polar equation:
   of a cardioid, 613
   of a circle, 608, 611, 613
   definition of, 607
   graph of, 607
   of a lemniscate, 613
   of a limaçon, 613
   of a line, 609, 610, 614 (Exercise 62)
   in parametric form, 622
   of a rose, 613
Pole, 604
Polygon:
   in the complex plane, 601
   interior angles of, 711
Polynomial:
   definition of, 36
   degree of, 36, 37
   operation with, 37–40
Polynomial division, 228–230
Polynomial equation, 249–253, 601
Polynomial function:
   definition of, 216
   factors and zeros of, 237–239
   graph of, 220–222
   imaginary zeros of, 246, 247
   irrational zeros of, 245
   multiplicity of zeros of, 240, 241
   rational zeros of, 242–244
   relative extrema of, 218–220
Polynomial inequality, 249, 254–256
Positive angle, 396
Positive integer:
   definition of, 2
   sums of powers of, 708, 709
Power, 13
Power function, 217, 218
Power reduction formulas, 528
Power series for e^x, 755 (Exercise 71)
Powers:
   of a complex number, 595–598
   of i, 33, 34
   sums of, 708, 709
Predator-prey relationship, 434–436
Prefix, symbol and meaning, 18
Prime polynomial, 44
Principal nth root of a, 22
Principal square root of $-a$, 30
Product:
   of complex numbers, 32, 589
   dot, 585 (Exercises 80–82)
   of matrices, 657
   of polynomials, 37, 38
   of a scalar and a vector, 574
   special, 38, 39
   of a sum and difference, 39
Product-to-sum formulas, 536
Projectile motion:
   parametric equations of, 622–624
   range of, 627 (Exercise 64)
Proof by mathematical induction, 705–709

Properties:
  of absolute value, 7
  of binomial coefficients, 750
  of cosecant function 442
  of cosine function, 430
  of cotangent function, 443
  of fractions, 50
  of logarithms, 316, 317
  of matrices, 658
  of negation, 4
  of positive integer exponents, 13
  of radicals, 25
  of real exponents, 297, 298
  of real numbers, 3
  of secant function, 443
  of sine function, 429
  summation, 267
  of tangent function, 441
  of vectors, 578
Pythagorean theorem:
  converse of, 79
  statement of, 66, 414

Q

Quadrant, 76
Quadrantal number (angle), 419
Quadratic equation:
  in standard form, 182, 183
  in two unknowns, 344
Quadratic formula, 185
Quadratic function:
  in general form, 175
  in standard form, 173
Quadratic inequality, 183, 188
Quartic function, 216
Quintic function, 216
Quotient:
  of complex numbers, 32, 589
  in polynomial division, 228

R

Radian, 399
Radical:
  properties of, 25
  simplified form of, 26
  symbol for, 22
Radical equation, 107
Radicand, 22
Radius of a circle, 345
Range:
  of a function, 94, 99, 103
  of projectile motion, 627 (Exercise 64)
Rate of interest, 299
Rate of work, 285, 286
Ratio, trigonometric, 407
Rational equation, 277–280
Rational exponent, 22, 23
Rational expression, 50, 277

Rational function, 262
Rational inequality, 277, 281–283
Rational number:
  definition of, 2
  as a repeating decimal, 738
Rational zero theorem, 242
Rational zeros of a polynomial function, 242–245
Rationalizing a denominator, 27, 28, 56
Rationalizing factor, 27
Rationalizing a numerator, 27, 28, 56
Ray, 396
Reactance in an AC circuit, 591
Real axis, 586
Real exponents, 297, 298
Real number line, 4
Real numbers:
  properties of, 3
  set of, 2
Real part of a complex number, 30
Real zeros, 104
Reciprocal:
  of a complex number, 35 (Exercise 46), 594 (Exercise 63)
  of a real number, 4
  of a trigonometric function, 409
Reciprocal function, 113, 262
Rectangle, 66
Rectangular coordinates, 76
Rectangular solid, 67
Recursively defined sequence, 695
Reduced to lowest terms, 50, 51
Reflection property:
  of an ellipse, 374, 375
  of a hyperbola, 386
  of a parabola, 362, 363
Reflection rule, x- axis and y- axis, 116–118
Regular n-sided polygon, 601
Relative extrema rule, 219
Relative extremum, 219
Relative maxima, 219, 448
Relative minima, 219, 448
Remainder in polynomial division, 228
Remainder theorem, 232, 233
Repeated root of multiplicity two, 186
Repeating decimal, 738
Resistance of an AC circuit, 591
Restricted cosecant function, 465 (Exercise 69)
Restricted cosine function, 455
Restricted cotangent function, 466 (Exercise 71)
Restricted secant function, 466 (Exercise 70)
Restricted sine function, 453
Restricted tangent function, 457
Resultant, 576
Resultant force, 580
Retention curve, 312
Retention function, 312, 313

Rhombus, 535 (Exercise 85)
Richter, Charles, 322
Richter scale, 322, 323
Right angle, 397, 413
Right triangle:
  definition of, 413
  isosceles, 421
  solving, 467
  trigonometric ratios for, 414
Root:
  of a complex number, 598
  of an equation, 60
  extraneous, 107, 277, 502
  nth, 21
  of a polynomial equation, 234, 238
  principal nth, 22
  repeated, 186
Rose, 612, 613
Rotation, angle formed by, 396
Rounding error, 552, 764
Row equivalent matrices, 652
Row of a matrix, 649
Row operations:
  on a determinant, 664
  for a matrix, 652

S

Scalar:
  with matrices, 655
  with vectors, 574
Scalar multiple, 655
Scalar multiplication, 574, 575
Scientific notation, 16, 763, 764
Secant function (sec):
  of an angle, 407
  graph of, 442
  inverse of, 466 (Exercise 70)
  properties of, 443
  as the ratio of the sides in a right triangle, 414
  of a real number, 419
  restricted, 466 (Exercise 70)
Secant line, 391 (Exercise 58)
Second ("), 398
Second-order determinant, 662
Sector, 406 (Exercise 71)
Semiaxis:
  of an ellipse, 366
  of a hyperbola, 379
Sequence:
  alternating, 694
  arithmetic, 714–717
  definition of, 692, 693
  explicitly defined, 695
  Fibonacci, 703 (Exercise 65)
  finite, 692
  general element of, 694–697
  geometric, 724–728
  graph of, 694

Series:
  infinite, 692, 693
  of partial sums, 736
  recursively defined, 695
Series:
  arithmetic, 718–721
  definition of, 692
  expanded form of, 698
  general term of, 697
  geometric, 728–731
  infinite binomial, 754 (Exercise 69)
  infinite geometric, 735
  power, 755 (Exercise 71)
  sigma form of, 698
  sum of, 703–710
Set:
  definition of, 2
  element of, 2
  of real numbers, 2
  subset of, 2
  union of, 10
Set-builder notation, 8
Shift of index, 701
Shift rule:
  horizontal, 115, 116
  vertical, 113, 114
Sigma form of a series, 698
Significant digits, 763, 764
Similar triangles, 137, 408, 575
Simple annuity, 731–733
Simple harmonic motion, 471–474
Simple interest, 66
Simultaneous equations, 197–204
Sine function (sin):
  of an angle, 407
  graph of, 429
  inverse of, 454
  properties of, 429
  as the ratio of the sides in a right triangle, 414
  of a real number, 419
  restricted, 453
Sines, law of, 551
Singular matrix, 668
Sketch the graph, 82
Slope:
  of a line, 153–155
  of parallel lines, 166
  of perpendicular lines, 168
  of a tangent line, 171, (Exercises 59, 60)
Slope formula, 153
Slope-intercept form, 156
Snell's law, 560 (Exercises 51, 52)
Solution:
  of an equation, 60
  of a system of equations, 638
Solution set:
  of an inequality, 61
  of a system of inequalities, 678
Solve:
  an equation, 61
  an inequality, 63
  a right triangle, 467
Solving equations simultaneously, 197–204
Solving a right triangle, 467
Special angles (30°, 45°, 60°), 421, 497
Special triangles (30°-60°-90° and 45°-45°-90°), 421
Special products, 38, 39
Speed:
  angular, 403
  linear, 402
Sphere, 67
Square of a binomial, 39
Square matrix, 662
Square root, 22, 30
Square root function, 113
Square set viewing rectangle, 89
Square system, 644
Squaring function, 113
SSA, constructions for, 535
Standard form:
  of a complex number, 30, 585
  of the equation of a circle, 345
  of the equation of an ellipse, 367
  of the equation of a hyperbola, 380
  of the equation of a parabola, 354
  of a polynomial equation, 249
  of a polynomial inequality, 249
  of a quadratic equation, 183
  of a quadratic function, 173
  of a quadratic inequality, 183
  of a rational inequality, 280
Standard position:
  of an angle, 396
  of a vector, 572
Step function, 124 (Exercise 72)
Stirling's formula, 755 (Exercise 72)
Straight angle, 397
Straight line (see Line)
Stretch and compress rule, vertical, 118–120
Stretch point, 445
Subset, 2
Substitution, 135, 494
Substitution method, 197
Subtraction:
  of algebraic fractions, 53–55
  of complex numbers, 31
  definition of, 4
  of matrices, 655
  of polynomials, 37
  of vectors, 577
Sum:
  of complex numbers, 31
  of a finite arithmetic series, 719, 720
  of a finite geometric series, 729, 730
  of an infinite geometric series, 737–739
  of matrices, 654
  nth partial, 736
  of polynomials, 37
  of powers of positive integers, 709
  of a series, 703, 704
  of two cubes, 47
  of vectors, 576
Sum and difference formulas:
  for cosine, 512
  for sine, 514
  for tangent, 515
Sum-to-product formulas, 538
Summation properties, 709, 710
Summation variable, 698
Summer solstice, 507
Surface area formulas, 67
Surveying, 471, 555
Symmetry:
  axis of, 172, 353
  tests for, 86
Synthetic division, 230–232
Synthetic divisor, 230
System:
  dependent, 638
  equivalent, 640
  homogeneous, 649 (Exercises 51, 52)
  inconsistent, 638
  of linear equations, 638
  of linear inequalities, 678
  nonsquare, 644, 645
  solving simultaneously, 197–204
  square, 644

T

Table of values, 82
Tangent:
  to a circle, 172 (Exercise 60), 346
  to a parabola, 200
  to the x-axis, 241
Tangent function (tan):
  of an angle, 407
  graph of, 441
  inverse of, 458
  properties of, 441
  as the ratio of the sides in a right triangle, 414
  of a real number, 419
  restricted, 457
Tax rate schedule, 159, 160
Temperature formula, 66
Term:
  of an algebraic expression, 12, 15, 36, 51
  degree of, 37
  like, 36
  of a series, 697
Terminal point of a vector, 572
Terminal side of an angle, 396
Test number method, 190, 256, 282
Tests for symmetry, 86
Total cost, 272
Transcendental function, 294
Transit, 470

Transpose of a matrix, 662 (Exercises 69, 70), 688 (Exercise 56)
Transverse axis of a hyperbola, 379
Trapezoid, 66
Triangle:
  area of, 66, 565–567
  equilateral, 421
  isosceles right, 421
  median of, 81, (Exercise 29)
  oblique, 550
  right, 413
  similar, 137, 408, 575
  special (30°-60°-90° and 45°-45°-90°), 421
Trigonometric composition properties, 460
Trigonometric equation:
  conditional, 496
  definition of, 490
  formulas for solving, 504–506
  general solution of, 497
  particular solution of, 497
  procedure for solving, 498
  techniques of solving, 501–503, 518, 526, 540
Trigonometric expression:
  factoring, 487
  operations with, 486
  simplifying, 488
Trigonometric form of a complex number, 587
Trigonometric formulas (identities):
  cofunction, 415, 541
  double-angle, 524
  even-odd, 541
  fundamental, 412
  half-angle, 529
  power reduction, 528
  product-to-sum, 536
  sum and difference, 512–515
  sum-to-product, 538
  summary of, 541, 542
  triple-angle, 527
Trigonometric function:
  algebraic signs of, 411
  of an angle, 407
  calculator usage with, 424
  composition properties of, 460
  domains of, 408
  evaluating, 419–424
  fundamental identities of, 412
  introductory comments of, 396
  inverses of, 453–459
  as the ratio of the sides of a right triangle, 414
  of a real number, 419
  reciprocals of, 409
Trigonometric identity:
  definition of, 490
  verification of, 490–493, 518, 525, 530, 539
  (see also Trigonometric formulas)

Trigonometric ratios:
  of an angle, 407
  for a right triangle, 414
Trigonometric substitution, 494
Trinomial:
  definition of, 36
  factoring, 45–47
  perfect square, 47
Triple-angle formula for sine, 527

U

Unbounded convex region, 681
Unbounded interval, 9
Uniform motion, 66, 283–285
Union of two sets, 10
Unit circle:
  definition of, 422
  special values on, 422, 497
Unit vector, 578

V

Value of a function, 96
Variable:
  definition of, 2
  dependent, 82
  dummy, 698
  independent, 82
  summation, 698
Variable cost, 272
Variable matrix, 670
Variation:
  direct, 138
  inverse, 139
  joint, 141
Variation constant, 138
Vector:
  component form of, 572
  components of, 572
  difference of, 577
  direction angle of, 572
  dot product for, 585 (Exercises 80–82)
  equality of, 572
  initial point of, 572
  magnitude of, 572
  negative of, 575, 577
  parallelogram law for, 576
  properties of, 578
  resultant, 576
  scalar multiplication of, 575
  standard position of, 572
  sum of, 576
  terminal point of, 572
  unit, 578
  zero, 572
Vector addition, 576
Vector diagram, 579
Vector quantities, 572
Vector subtraction, 577

Vernal equinox, 507
Vertex (vertices):
  of an angle, 396
  of a convex region, 678
  of an ellipse, 366
  of a hyperbola, 379
  of a parabola, 172, 353
Vertex formula, 175, 176
Vertical asymptote, 116, 262–264, 441, 445, 448
Vertical line, equation of, 157, 610
Vertical line test, 94–96
Vertical shift rule, 113, 114
Vertical stretch and compress rule, 118–120
Viewing rectangle:
  of a graphing calculator, 84
  square set, 89
Volume formulas, 67

W

Wheatstone bridge, 689 (Exercise 70)
Whole number, 2
Winter solstice, 507
Work, rate of, 285, 286

X

x-axis:
  in the coordinate plane, 76
  reflection rule, 117
  symmetric with respect to, 85, 86
x-coordinate, 76
x-intercept, 86
x-intercept method for solving an equation, 106, 204
x-intercept method for solving an inequality, 107

Y

y-axis:
  in the coordinate plane, 76
  reflection rule, 117
  symmetric with respect to, 85, 86
y-coordinate, 76
y-intercept, 86

Z

Zero:
  division by, 4
  as an exponent, 14, 596
  of a function, 104
  of multiplicity k, 240, 241
  of a polynomial function, 234, 238
Zero factorial, 744
Zero matrix, 655
Zero product property, 89
Zero vector, 572

Photo Credits

1 National Optical Astronomy Observatories;   18 National Optical Astronomy Observatories;   75 Amy C. Etra/PhotoEdit;   91 Robert Finken/Photo Researchers, Inc.;   109 E. R. Degginger;   138 Amy C. Etra/PhotoEdit;   151 Michael Newman/PhotoEdit;   169 Michael Newman/PhotoEdit;   178 E. R. Degginger;   179 Jim Zerschling/Photo Researchers, Inc.;   215 Don Spiro/Tony Stone Images;   256 Don Spiro/Tony Stone Images;   257 Alan Oddie/PhotoEdit;   273 David Oscher/Tony Stone Images;   284 Jose Carrillo/PhotoEdit;   285 Charles McNulty/Tony Stone Images;   293 Mark Richards/PhotoEdit;   300 Spencer Grant/Photo Researchers, Inc.;   323 Mark Richards/PhotoEdit;   334 Michael Rosenfeld/Tony Stone Images;   335 David Woodfall/Tony Stone Images;   343 Tony Freeman/PhotoEdit;   362 Tony Freeman/PhotoEdit;   395 Bill Bachmann/PhotoEdit;   403 Chad Slattery/Tony Stone Images;   435 R. Van Nostrand/Photo Researchers, Inc.;   463 Tony Freeman/PhotoEdit;   470 Bill Bachmann/PhotoEdit;   471 Mark Segal/Tony Stone Images;   485 Jeremy Walker/Tony Stone Images;   507 Jeremy Walker/Tony Stone Images;   520 Alfred Pasieka/Science Photo Library/Photo Researchers, Inc.;   532 E. R. Degginger;   549 Focus on Sports, Inc.;   567 Dave Cannon/Tony Stone Images;   579 Mario Colonel/Photo Researchers, Inc.;   623 Focus on Sports, Inc.;   637 Didier Dorval/Photo Researchers, Inc.;   645 Didier Dorval/Photo Researchers, Inc.;   682 John McDermott/Tony Stone Images;   691 Chip Henderson/Tony Stone Images;   721 Michael Newman/PhotoEdit;   732 Chip Henderson/Tony Stone Images;   740 Henley and Savage/Tony Stone Images.